Wir haben diese schönen Bücher gesammelt ,

geordnet ,

abgefilmt ,

bearbeitet und gestaltet ,

um daraus eine Sammlung anzufertigen ,

die all jenen gewidmet ist ,

die das Buch lieben .

We collected these beautiful books and put them together,

photographed,

edited and designed into a compilation,

which is dedicated to all book lovers .

我们收藏了这些美丽的书
把它们整理、拍摄、编辑、设计，做成一本合集
献给所有爱书的人

20 世纪 50 年代初德国的政治分歧引致两个图书设计竞赛几乎同时设立：

1952 年西德在美因河畔法兰克福设立竞赛，

1965 年起由德国图书艺术基金会赞助并组织评选，

而“德意志民主共和国最佳图书奖”1953 年设立于莱比锡。

10 年后，

随着 Messehaus am Markt 的成立，

莱比锡书展进入了一个新的历史阶段，

“世界最美的书”国际比赛正式启动。

莱比锡书商协会将其作为国际书展之间的纽带，

从 1959 年起每隔五年举办一次。

自 1968 年以来，

该展览因为一个附属的主题特别展览而进一步升级。

“金字符奖”Goldene Letter 最初作为主题展览的最佳贡献奖颁发，

后来被提升为竞赛的最高奖项。

随着 1989 年 11 月柏林墙的倒塌，

组织两场平行的图书艺术竞赛已经不合时宜。

自 1991 年以来，

德国图书艺术基金会负责在莱比锡举办“世界最美的书”竞赛和展览。

In den *frühen fünfziger* Jahren führte die politische Spaltung Deutschlands zur Gründung zweier Buchgestaltungswettbewerbe. Parallel zum westdeutschen Wettbewerb, erstmals in Frankfurt am Main *1952* ausgerichtet, ab *1965* unter der Leitung der Stiftung Buchkunst, wurden »Die schönsten Bücher der DDR« ab *1953* in Leipzig geehrt.

Ein Jahrzehnt später begann in Leipzig mit der Einweihung des Messehauses am Markt ein neuer Abschnitt der Geschichte der Buchmesse Leipzig, der internationale Wettbewerb »Best Book Design from all over the World« wurde aus der Taufe gehoben. Ost-Börsenverein und Kommune planten es als Bindeglied zwischen den internationalen Buchausstellungen, die ab 1959 in Fünfjahresabständen stattfanden.

Seit 1968 wurde die Ausstellung weiter ausgebaut, indem sie von einer thematischen Sonderschau begleitet wurde. Die »Goldene Letter« wurde erstmals als Preis für die beste Einreichung in dieser thematischen Ausstellung vergeben, später wurde sie zur bedeutendsten Auszeichnung des Wettbewerbs. Mit dem Fall der Mauer im November 1989 vereinigten sich auch die beiden parallel laufenden Buchkunst Wettbewerbe. Seit 1991 richtet die Stiftung Buchkunst den Wettbewerb und die Ausstellung »Schönste Bücher aus aller Welt« aus.

In the early fifties Germany's political division led to an almost simultaneous establishment of two book art competitions: parallel to the West German vote, begun in Frankfurt am Main in 1952 and organised from 1965 under the auspices of the Stiftung Buchkunst, »Die schönsten Bücher der DDR« / »The Best Books of the GDR« were honoured in Leipzig from 1953 on.

One decade later, when a new phase in the history of the Leipzig Book Fair began with the inauguration of the Messehaus am Markt, the international competition »Best Book Design from all over the World« was founded. The Ost-Börsenverein/Leipzig Book Traders' Association and the commune planed it as a link between the international book art exhibitions reinvested from 1959 on in five-year intervals.

Since 1968 the exhibit was further upgraded by an affiliated thematic special show. The »Goldene Letter«, first given as a prize for the best contribution of these thematic exhibitions, later advanced to become the highest award of the contest. With the fall of the Berlin Wall in November 1989 the organisation of two parallel book art competitions had become obsolete. Since 1991 the Stiftung Buchkunst holds the competition and exhibition »Best Book Design from all over the World« in Leipzig.

KAREN BLIXEN
Nature

CHIEF EDITOR
ZHAO QING
主编 赵 清

BROWSE LEIPZIG

BEST BOOK DESIGN FROM ALL OVER THE WORLD

1991–2003

翻阅莱比锡

「世界最美的书」

1991—2003

江苏凤凰美术出版社
Jiangsu Phoenix Fine Arts
Publishing House

Vorwort — *Preface*

序

Als ich mit der Arbeit von Professor Zhao Qing bekannt wurde, wurde mir ein weiteres Mal deutlich, warum das Grafikdesign in den letzten vierzig Jahren der Designentwicklung in China fast allen seinen Nachbardisziplinen immer einen Schritt voraus war.

Für einen versierten Buchgestalter wie Zhao stellte der Wunsch, Leipzigs „Schönste Bücher der Welt" kennenzulernen, zunächst eine berufliche Notwendigkeit dar, die darauf beruhte, sich von dem Reichtum und der Vielfalt des Verlagswesens außerhalb Chinas und von der internationalen Welt des Designs inspirieren zu lassen. Intuition und unmittelbare sinnliche Wahrnehmung waren an diesem Punkt das Entscheidende. Dies war nachvollziehbar und zunächst ausreichend, denn die Buchbinderkünstler ab der Generation der Neuen Kulturbewegung hatten die visuelle Kommunikation (konfrontiert mit den exzellenten Werken aus Japan) stets auf dieser Grundlage betrieben. Eine Folge davon war, dass es viele Jahre später und nachdem China all die Zeit über von europäischem, amerikanischem und japanischem Buch- oder Grafikdesign profitiert hatte, innerhalb der Gemeinschaft chinesischer Designer immer noch kein einziges Buch gab, das den Wandlungen und Ursprüngen der Entwicklungen anderswo sorgfältig auf den Grund ging. Dies war insofern beschämend, als das alleinige Streben nach Erfolg und der Mangel einer Erforschung der Ursachen für den Erfolg es nahezu unmöglich machte, herausragende Entwicklungen mit entsprechenden Ergebnissen in Gang zu bringen.

Zhao Qings Ausgangspunkt und der Weg, den er in der Forschung einschlug, lassen sich klar und deutlich ausmachen. Nachdem eine erhebliche Anzahl chinesischer Designer 2004 nach der deutschen Wiedervereinigung in Leipzig am Wettbewerb „Schönste Bücher der Welt" teilgenommen hatte, erkannte er die Wichtigkeit der zeitlichen Zusammenhänge. Hatte es sich zuvor weitgehend um eine eigen-

What professor Zhao Qing has done makes me further understand why graphic design has always been leading all the counterparts in the development of designs for nearly forty years in China.

Initially, as an distinguished book designer, he was willing to know "the best designed book all over the world" in Leipzig out of professional needs and take the diversity of publication and design outside China as a source of inspiration. At that time, as a visual expression, it was necessary and enough to make visual and stylistic observation, which have been done by book binding artists since the New Culture Movement - facing outstanding modern Japanese works. Years past, long after the benefits we got from European, American and Japanese book design or graphic design, there has still not been a single book studying "their" evolution, which is a shame. Being a race knowing the hows but not the whys, it was impossible to seek surpassing in development.

Zhao Qing's purpose and study path were very clear. After many Chinese designers taking part in the competition of "the Best Designed Book All Over the World" in 2004, he noted the significance of the time node. It has been developed only in western world but from then on joined by the giant eastern country. But for designers, the more important was the sudden changes of printing technology - the overall promotion from manually plate making, laser phototype setting to digital technology. Under this back-

序

了解赵清所做的工作，我又一次理解了中国近 40 年设计发展中，为何平面设计始终走在几乎所有相邻设计的前面。

作为一位卓有成就的书籍设计师，有愿望去了解莱比锡“世界最美的书”，最初应该是专业的需要。那些非中文世界出版和设计的丰富多样，带来足够的启示。这个时候，直观和风格的体察是必须的——但那也足够，因为作为视觉传达，从新文化运动一辈的书籍装帧艺术家开始（面对近代日本的优秀作品），就在这样做了。多少年以来，我们受惠欧美和日本的书籍设计或平面设计很久之后，中国的设计界仍然没有一本仔细研究“他们”流变的书，这很令人脸红，因为一个只求已然不求所以然的民族，想求得借鉴和发展的超越，几乎不可能。

赵清的出发点和研究轨迹很清晰，在 2004 年中国设计师成规模参加了两德合并后的莱比锡“世界最美的书”的评选后，他注意到了时间节点的意义，此前是中国以外的欧美世界独自发展，此后融汇了这个东方书籍大国的参与。但是这前后的变化，对设计师而言，更重要的还是印刷技术的突变——从手工制版、

ständige Entwicklung in Europa und den USA gehandelt, so wurde nun das große Bücherland aus dem Fernen Osten integriert. Die Wahrnehmung dieses Wandels und die Perspektive auf den Wandel, stellten für einen Designer etwas überaus Wichtiges dar. Weit wichtiger jedoch waren die Veränderungen in der Drucktechnik, wenn es darum ging, wie Designer in verschiedenen Ländern die Entwicklungen von den handgefertigten Werken über die Laserfotografie bis hin zum digitalen Druck durchliefen und sich in ihren schöpferischen Arbeiten den Veränderungen anzupassen versuchten. Zhao Qing hat fast 85% der preisgekrönten Bücher der Leipziger Reihe „Schönste Bücher" gesammelt. Dies macht es möglich, die Beziehung zwischen Kunst und Technik in den Originalwerken zuverlässig und gründlich zu analysieren.

Zhao Qings Buch „Leipzig durchblättert – Die schönsten Bücher der Welt aus den Jahren 1991-2003" und sein früheres Buch „Leipziger Auswahl – Die schönsten Bücher der Welt aus den Jahren 2019-2004" bilden eine gestalterisch-historische Beziehung, die mehr als nur chronologisch ist. Vielmehr zeigt der systematische Charakter der Arbeit auch die umfassende Betrachtung eines Forschers über die Beziehung zwischen Kultur, Technologie und Design.

Ich hoffe, dass es Zhao Qing in seinen künftigen Untersuchungen über die moderne Entwicklung des Buchdesigns in der Welt gelingt, den Bogen noch weiter zu spannen. Dies betrifft insbesondere die Beschäftigung mit Deutschland, wo es zahlreiche besondere Beziehungen zum chinesischen Grafikdesign gibt. So zum Beispiel den Entwurf des berühmten Dichters und Buchkünstlers Cao Xinzhi für Die Gemäldesammlung von Präsident Sukarno. Im Jahre 1959 wurde die Arbeit auf der Internationalen Buchkunstmesse in Leipzig mit der Goldmedaille für das Rahmendesign ausgezeichnet.

Der kürzlich verstorbene Professor Yu Bingnan, ein berühmter Grafikdesigner, wurde für seinen typografischen Entwurf „Freundschaft" hoch gelobt, der während seines Auslandstudiums in Ostdeutschland entstanden ist und in den 1950er Jahren zu einem wichtigen Projekt des kulturellen Austauschs zwischen den beiden Ländern wurde. Professor Yu wurde

ground, it was an important angle to see how designers from all the countries would adapt themselves and create. Zhao Qing has collected more than 85% of the award-winning books of "the best designed books" in Leipzig's competition, which has made reliably possible to analyze the connections between the art and technology of the books.

Zhao Qing's *Browse Leipzig – Best Book Design From All Over The World 1991–2003* and his earlier *The Choice Of Leipzig – Best Book Design From All Over The World 2019–2004* structured a design history. They displayed not only the time order but also a scholar's reflection on the connections among art, technology and design. I hope he will make deeper studies on modern design development, especially that of Germany, which has a lot of uncommon relationships with Chinese graphic design. For example, the work *Paintings from the collection of Dr. Sukarno*, by the famous poet and book artist Cao Xinzhi, won the golden award of binding and designing in the Leipzig International Book Art Fair in 1959. Famous graphic designer Yu Binnan, who has passed away recently, won praises for his typeface design "friendship", which became the major project of cultural communication between the two countries in the 1950s. Afterwards, he was awarded the "Gutenberg" Lifetime Achievement Award by the Leipzig government and became a member of the review committee of "The Best Designed Book All Over The World". The book *Graphic Grid*, which he translated and published, has had a great impact on Chinese graphic design. We can say that it is a perfect observation to study the relationship between China and the world in the 20th century in graphic design

激光照排到数字技术对印刷技术的全面跃进。在这个前提下，各国设计师如何应变创造，是一种重要的观察视角。赵清收集了莱比锡“最美的书”几乎85%以上的获奖图书，这使得分析原作的艺术和技术的关系，有了可靠和深入的可能。

赵清的这本《翻阅莱比锡——世界最美的书1991－2003》与他早前出版的《莱比锡的选择——世界最美的书2019－2004》，构成了一种设计史的关系，这不仅仅是时间上有了系统性，还展现了一位研究者对文化、技术与设计关系的全面思考。我很期待以后他还能进一步把世界书籍设计的现代发展研究得更深远些。尤其是德国，对于中国平面设计而言，我们拥有许多与他们的特殊关系。例如著名诗人、书籍艺术家曹辛之先生设计的《苏加诺总统藏画集》，在1959年莱比锡国际书籍艺术博览会上获得装帧设计金奖；刚刚故去的著名平面设计师余秉楠教授，早年留学东德期间设计的字体“友谊体”受到好评，成为20世纪50年代两国文化交流的重要项目，他后来还获得莱比锡政府颁发的“古腾堡”终身成就奖，并担任过“世界最美的书”的评委，他当年翻译和出版的《网格构成》也对中国平面设计产生过很大

später mit dem von der Leipziger Stadtregierung vergebenen „Gutenberg“-Preis für sein Lebenswerk ausgezeichnet. Yu war ebenfalls Mitglied der Kommission von „Schönste Bücher der Welt“. Die Publikation seiner Übersetzung von „Rasterbildung“ hatte einen großen Einfluss auf das chinesische Grafikdesign. Alles in allem lässt sich sagen, dass eine Vertiefung in das Studium von „Schönste Bücher“ ausgezeichnete Beobachtungen zulässt, um die Beziehungen zwischen China und dem globalen Grafikdesign im zwanzigsten Jahrhundert zu untersuchen.

So weit zu meinen neuen Hoffnungen bezüglich des Wirkens von Professor Zhao Qing. Ich wünsche mir, dass diesem Buch eines folgt, in dem er den Entwicklungen in Deutschland vor 1991 und noch früher nachspürt und dass er sich in seinen Forschungen auch mit den Entwicklungen außerhalb Deutschlands im Europa des 20. Jahrhunderts oder früher auseinandersetzt. Dies wäre ein überaus verdienstvoller Beitrag für das Design in China.

Hang Jian
Oktober 2020, Taipu-Gebirge

through the study of “the Best Books”.

It's a new expectation to Professor Zhao Qing. After this book, I hope he can study book design in Germany before 1991, or even earlier, as well as Europe before the 20th century or earlier. It should be a great contribution for Chinese graphic design.

Hang Jian
October 2020, in Taipu Mountain, Liangzhu, Hangzhou

的影响。可以这么说，借由“最美的书”研究展开的20世纪中国与全球平面设计关系的研究，是一个极佳的观察。

这是我对赵清提出的一个新希望，本书之后，希望他去追索德国1991年之前，或更前，希望他去研究德国之外的20世纪欧洲或更早，这是于中国平面设计有远大贡献的善事。

杭间

2020年10月于良渚太璞山

Hang Jian ist Designhistoriker, Designkurator, Professor an der Chinesischen Kunstakademie und Chefkurator des Kunstmuseums von China. Er war Chefredakteur der Zeitschrift Decoration, Vizepräsident der Akademie der Schönen Künste der Qinghua-Universität und Vizepräsident der Chinesischen Kunstakademie. Er hat zahlreiche Werke zur Kunstgeschichte, Designgeschichte und Designtheorie verfasst, darunter Geschichte der chinesischen Handwerksästhetik, Ideen des Handwerks und Die Güte des Designs.

Hang Jian, the design history scholar, design exhibition planner, professor of China Academy of Art and curator of Art Museum, has been the chief editor of Zhuangshi magazine, the deputy dean of Academy of Art & Design, Tsinghua University, and the vice president of China Academy of Art. He has written many books about art history, design history, and design theory including *Chinese Art and Aesthetics History*, *Thoughts of Crafts* and *Kindness of Design*.

杭间
Hang Jian

杭间，设计史学者，设计策展人，中国美术学院教授、艺术博物馆总馆长。曾任《装饰》杂志主编、清华大学美术学院副院长、中国美术学院副校长。著有《中国工艺美学史》《手艺的思想》《设计的善意》等艺术史、设计史、设计理论著作多部。

Year 年份	Page Number 页码
1991	0031
1992	0129
1993	0241
1994	0359
1995	0433
1996	0537
1997	0619
1998	0713
1999	0809
2000	0935
2001	1031
2002	1115
2003	1193

Code
编码

Gl	Golden Letter 金字符奖
Gm	Gold Medal 金奖
Sm	Silver Medal 银奖
Bm	Bronze Medal 铜奖
Ha	Honorary Appreciation 荣誉奖

Year 年份 · Award 奖项 · Number 编号

1991 Bm2

奖项 Award

书名 Title

国家 / 地区 Country / Region

设计师 Designer

作者 Author

出版社 Publisher

尺寸 Size

重量 Weight

页数 Page

国际书号 ISBN

1991

Country / Region

Germany 5

U.S.A. 2

The Netherlands 2

Japan 2

Switzerland 1

Czech Republic 1

VLADIVOSTOK JOHN HEJDUK

ANDY WARHOL, CINEMA

Květa Pacovská eins, fünf, viele

Wim Quist

ZELFPORTRETTEN

VISIT US AT USPS.COM®
ORDER FREE SUPPLIES ONLINE
PRESS FIRMLY TO SEAL
PRESS FIRMLY TO SEAL
PRIORITY®
★ MAIL ★
CP 72-CUSTOMS DECLARATION
Page 3 - Not Valid As
Proof-of-Payment for
US Postage
PAGE 3 - DISPATCH NOTE
PS00001000016
Padded Flat Rate Envelope
UNITED STATES
POSTAL SERVICE.
PRIORITY MAIL
POSTAGE REQUIRED

CP 72-CUSTOMS DECLARATION

Page 3 - Not Valid A...
Proof-of-Payment fo...
US Postage

GEORGE F WILKINSON
G.F. WILKINSON BOOKS
10440 KEENAN WAY
GRASS VALLEY CA 95949-6895
UNITED STATES
415-200-8707

Origin Post: US POSTAL SERVICE
Date of Mailing: 12/17/2019

Importer's Reference:

Importer's Contact: momoko960430@163.com

...ce of Exchange

Customs Stamp

Please affix labels here when requested

Customs Duty:

Postal Charges/Fees: $33.60

...egory of Items: Merchandise

...orter's reference:

Exporter's contact: wilkinso@mindspring.com

Total Dimensions: L: 0 W: 0 H: 0

I certify the particulars given in this customs declaration are correct. This form does not contain any undeclared dangerous articles, or articles prohibited by legislation or by postal or customs regulations. I have met all applicable export filing requirements under federal law and regulations.

Sender's Signature and Date
GEORGE F WILKINSON 12/17/2...

...B/ITN/Exemption: EEL_3037a

Invoice/License/Certificate No(s)

Return to Sender Instructions in case of nondeli...
Return to Sender

...claration by ADDRESSEE: I have received the parcel described on this note.

...DRESSEE's Signature and Date

ZHAO QING
QING TANG DESIGN CO.
DABEI XIANG 7-3,NANJING HAN
210000 NANJING
CHINA

PAGE 3 - DISPATCH NOTE
CJ 174 343 495 US

大悲巷7-3

U.S.A.

U.S.A.　　　China

258mm

16mm

191mm

金字符奖
两条河流
美国

Golden Letter
Two Rivers
U.S.A.
James Robertson, Carolyn Robertson
Wallace Stegner
The Yolla Bolly Press, Covelo
258 191 16mm
454g
118p

这是美国历史学家、小说家华莱士·斯特格纳（Wallace Stegner）的短篇小说集，书名《两条河流》是其中 7 篇短篇小说的第 2 篇。本书是收藏版本，总计只印刷了 255 本。书籍的装订形式为锁线胶平装。函套选用中灰色艺术纸裱覆白卡，轻薄但具有较好的保护作用。护封手工艺术纸张夹带大量絮状杂质，蓝绿黑三色印刷。内封选用墨绿色艺术纸，无印刷和工艺。内页主要使用胶版纸印单黑，并通过四种冷灰色艺术纸作为各部分之间的隔页。从封面起，波纹就作为贯穿全书的图案被使用，在各篇小说起始页和隔页上均通过波纹线条的粗细、峰值、波距等变化形成丰富的纹样肌理。字体方面选用了具有经典数字和标点符号的 California，具有非常舒适的阅读体验。

This is a collection of short stories by Wallace Stegner, an American historian and novelist. The title of the book *Two Rivers* is the title of the second short story. This book is a collector's edition, with only 255 copies printed in total. The book is perfect bound with thread sewing. The bookcase is made of medium gray art paper and mounted with white card, which is light and thin with good protective effect. There are a lot of flocculent impurities in the art paper, which is used for the jacket. It is printed in blue, green and black. The inner cover is made of dark green art paper without printing and other technology. The inner pages mainly use offset paper printed in single black. Four kinds of cool gray art paper are used as the separation page between each part. The ripple is used as the pattern throughout the whole book starting from the cover. On the beginning page and the separation page of each short story, rich textures are formed by the thickness, peak value, wave distance and other changes of the ripple lines. Font California, which has the classic numbers and punctuation marks, is selected for the comfortable reading experience.

IMPASSE

ɔed down off the heights, the reluctant sun, which h
he Col de Vence and forced Louis to dodge and shield h
e Citroën around the curves, had finally been dragg
glare of the day was taken off them. Along the gratef
y bounced through the streets of Nice and onto the hig
he coast.

it was still full afternoon. Sails passed like gulls; clo
with the water-bug tracks of paddle boats. But whe
quieted, and in the confined car the bickering seemed
ightening out with the traffic toward Monte Carlo,
ppraised the lengthening silence and grew halfw

the quiet was too pleasant. Only in the sloping win
tion of his wife's face, the mouth drawn down ruefull

WALLACE STEGNER

TWO RIVERS

THE YOLLA BOLLY PRESS

is back quietly, letting complete wa
t dangled on a golden, shining thre
he spider came down in tiny jerks,
the beam of sun. From the other roo

I'd give every man in the army a quarter
ey'd all take a shot at my mother-in-law.

out of bed and yanked the nightshi
's face poking around the door, say
' He didn't want to be joked with. Yes
e had been avoiding his father ever
yet ready to accept any joking or at
ting a person for nothing, and you be
whistle and sing out there, pretendi
iness yesterday was the matter, the
he whole lost Fourth of July was the

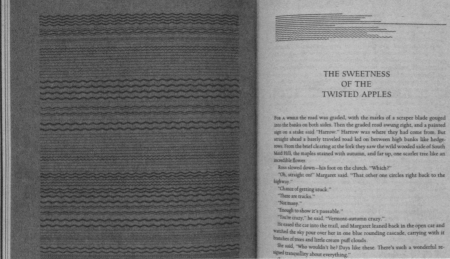

THE SWEETNESS OF THE TWISTED APPLES

For a while the road was graded, with the marks of a scraper blade gouged into the banks on both sides. Then the graded road swung right, and a painted sign on a stake said "Harrow." Harrow was where they had come from. But straight ahead a barely traveled road led on between high banks like hedgerows. From the brief clearing at the fork they saw the wild wooded side of South Maid Hill, the maples stained with autumn, and far up, one scarlet tree like an incredible flower.

Ross slowed down—his foot on the clutch. "Which?"

"Oh, straight on!" Margaret said. "That other one circles right back to the highway."

"Chance of getting stuck."

"There are tracks."

"Not many."

"Enough to show it's passable."

"You're crazy," he said. "Vermont-autumn crazy."

He eased the car into the trail, and Margaret leaned back in the open car and watched the sky pour over her in one blue rounding cascade, carrying with it branches of trees and little cream puff clouds.

She said, "Who wouldn't be? Days like these. There's such a wonderful resigned tranquillity about everything."

MEDIUM FLAT RATE BOX
PRIORITY MAIL
MEDIUM FLAT RATE
POSTAGE REQUIRED
VISIT US AT USPS.COM®
ORDER FREE SUPPLIES ONLINE
PS00011000001
PRIORITY
MAIL
WORLDWIDE EXPRESS MAIL SERVICE

210 S 8TH ST
LEWISTON, NY 14092-1702
1-844-842-8777 MID 922215827
US Postal Service
Importer's Reference
Importer's Contact
Detailed Description of Contents
Printed Limited Book John Hejduk
Qty 1
Weight (lb/oz) 6 0
300.00
Category of Items: Commercial Sample
Exporter's Reference:
Exporter's Contact: 1-844-842-8777
Dimensions (in)
TOTAL 6 0 300.00
AES/ITN Exemptions: No EEI 30.37(a)
Invoice/License/Certificate No(s)
Invoice License Certificate No
Chit Chats C39Y97G62Z
Origin Country CA
Insured Value ($)
SDR Value ($)
Insured Fee ($)
Postal Charges
U.S. POSTAGE PAID
INTERNATIONAL
ePOSTAGE
I certify the particulars in this Customs declaration are correct. This form does not declare any undeclared dangerous articles, or articles prohibited by legislation or by postal or customs regulations. I have met all applicable export filing requirements under federal law and regulations.
Sender's Signature and Date
Carlos Galdamez 06/05/2020 Medium Flat Rate Box
Return to sender instructions if non-deliverable
Return
TO
ZHAO QUING
7-3 DABEI LANE, MEIYUAN
NEW VILLAGE
210000 NANJING JS
CHINA
0086 13952094042
96.0 oz
CB 129 628 191 US
PS Form 2976-A (March 2015)
Do not duplicate this form without USPS approval
This item may be opened

313mm

28mm

237mm

● 金奖
《 符拉迪沃斯托克
美国

Gold Medal ◎
Vladivostok 《
U.S.A.
Kim Shkapich
John Hejduk
Rizzoli International Publications, New York
313 237 28mm
1720g
272p
ISBN-10 0-8478-1129-8

美国建筑师、教育家约翰·海杜克（John Hejduk）从 20 世纪 70 年代末开始了为期约 10 年的旅行项目。从威尼斯开始，先到柏林，再到俄国的里加、贝加尔湖和最东边的符拉迪沃斯托克（海参崴）。他绘制了所到之处的素描，创作了一些诗歌，以及依此写就的与建筑和人文相关的论文。海杜克通过这样的实践来创建了一个资料库，尝试建立一种了解城市建筑和城市生态的新方法。所有这些内容被整理编辑成册，形成他城市研究的三部曲，本书便是其中的一本。书籍的装订形式为锁线硬精装。护封白卡纸墨蓝色和红色双色印刷，内封深蓝色布面烫印红色，裱覆卡板。内页哑面铜版纸四色印刷。书籍的主体部分为海杜克在里加、贝加尔湖和符拉迪沃斯托克（海参崴）绘制的建筑速写、结构分析、相关照片和部分平面图，附带他在此期间写作的诗歌和与当地文化相关的速写。全书双栏设计，分别为英文和俄文。依照建筑平面或立面抽象而成的手绘图形，用作目录、索引和数量统计的标识符号，是本书的亮点。

John Hejduk, an American architect and educator, started a ten-year travel program in the late 1970s. His journey started in Venice, first to Berlin, then to Riga in Russia, Lake Baikal and Vladivostok in the East. He drew sketches wherever he went, wrote some poems, and wrote papers on architecture and humanities accordingly. Through such practice, Hejduk created a database and tried to establish a new method to understand urban architecture and urban ecology. All these contents have been compiled into a trilogy of his urban studies and this book is one of them. The hardcover book is bound by thread sewing. The jacket is made of white cardboard printed in blue and red, and the inner cover is made of dark blue cloth treated with red hot stamping, mounted with cardboard. The inside pages are printed on matte coated paper in four colors. The main part of the book is the architectural sketches, structural analysis, related photos and some planar graphs drawn by Hejduk in Riga, Lake Baikal and Vladivostok, along with the poems he wrote during this period and the sketches related to the local culture. The book is set to double columns, one for English and one for Russian. Hand drawn architectural graphics are the characteristics of this book, which serve as the table of contents, index, and quantity statistics.

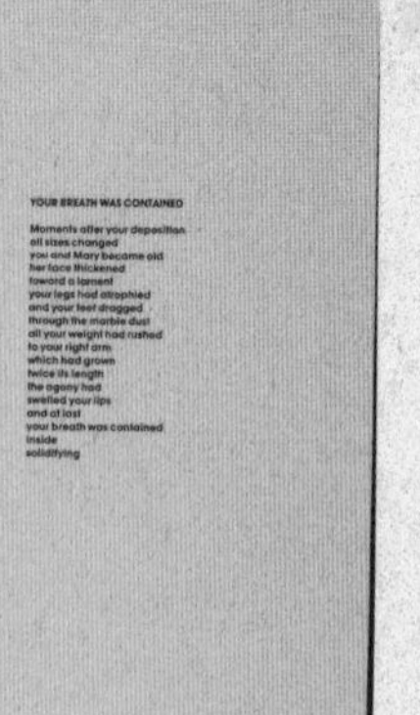
YOUR BREATH WAS CONTAINED
Moments after your deposition
all sizes changed
you and Mary became old
her face thickened
toward a lament
your legs had atrophied
and your feet dragged
through the marble dust
all your weight had rushed
to your right arm
which had grown
twice its length
the agony had
swelled your lips
and at last
your breath was contained
inside
solidifying
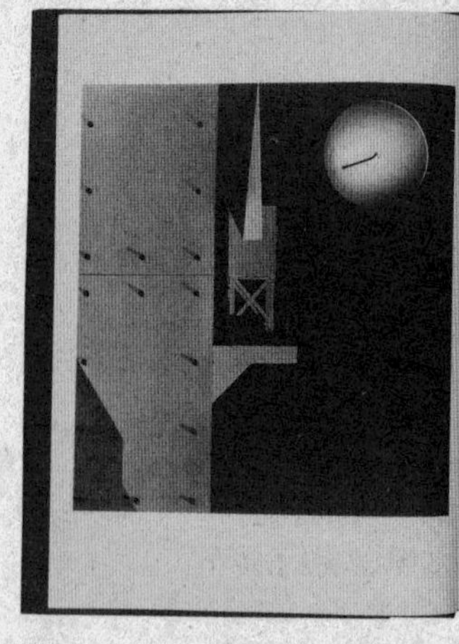
VLADIV
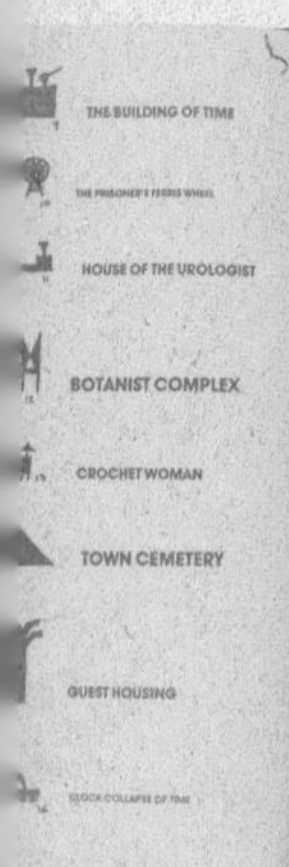
THE BUILDING OF TIME
HOUSE OF THE UROLOGIST
BOTANIST COMPLEX
CROCHET WOMAN
TOWN CEMETERY
GUEST HOUSING
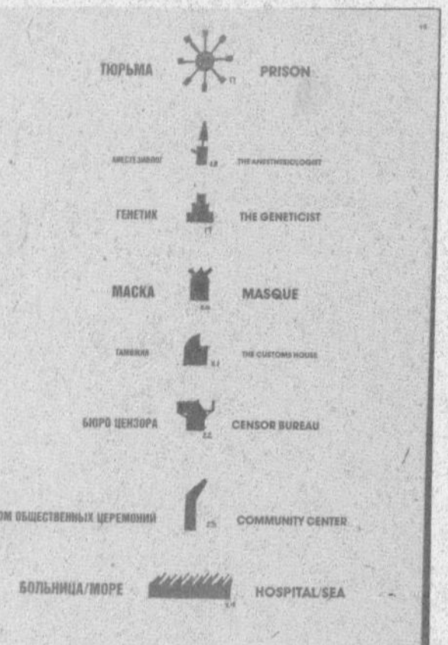
ТЮРЬМА
PRISON
THE ANESTHESIOLOGIST
ГЕНЕТИК
THE GENETICIST
МАСКА
MASQUE
THE CUSTOMS HOUSE
БЮРО ЦЕНЗОРА
CENSOR BUREAU
ДОМ ОБЩЕСТВЕННЫХ ЦЕРЕМОНИЙ
COMMUNITY CENTER
БОЛЬНИЦА/МОРЕ
HOSPITAL/SEA

MASQUE
RIGA
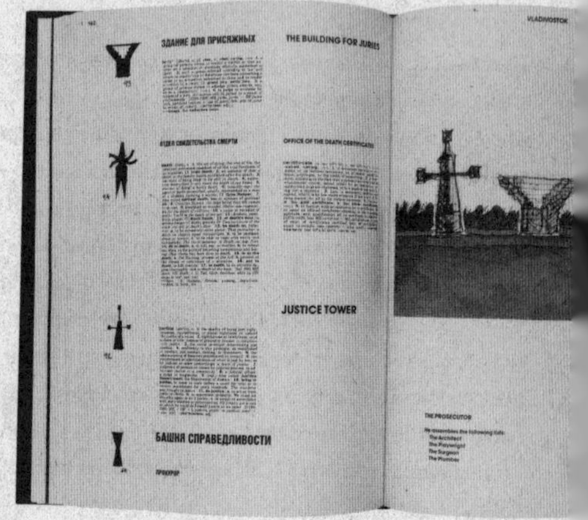
ЗДАНИЕ ДЛЯ ПРИСЯЖНЫХ
THE BUILDING FOR JURIES
ОТДЕЛ СВИДЕТЕЛЬСТВА СМЕРТИ
OFFICE OF THE DEATH CERTIFICATE
JUSTICE TOWER
БАШНЯ СПРАВЕДЛИВОСТИ
VLADIVOSTOK
THE PROSECUTOR

THE BUILDING OF TIME

ЗДАНИЕ ВРЕМЕНИ

ТЮРЬМА НА ЧЕРТОВОМ КОЛЕСЕ

Здание Времени имеет сложную конструкцию. Она наполнена часовыми механизмами и связана с течением времени, его износом и стиранием. Здесь празднуется триумф времени и поддерживается постоянство времени. Природу времени можно увидеть, можно экспериментировать с ним, можно слышать его с правой стороны здания.

Заключенного выводят из тюрьмы и сажают в единственную кабинку, прикрепленную к чертову колесу. Заключенный находится лицом к Стене Времени (восточная сторона Здания Времени). Стена Времени разделена на 100 квадратных частей из гранитных блоков толщиной в 2 дюйма. Наверху стены находится устройство, перемалывающее время. Например, если заключенный осужден на тридцать лет, устройство подводится к 1-му блоку и начинает его перемалывать. На каждый квадратный блок уходит год. По мере того, как перемалывающее устройство грызет гранит, чертово колесо разгоняется. Каждый полный оборот чертова колеса занимает один год. Осужденный на тридцать лет заключенный, находящийся в кабинке, сделает тридцать кругов. По тому, как размалываются гранитные блоки, он судит о течении времени.

THE PRISONER'S FERRIS WHEEL

The Building of Time is a complex structure. It's filled with time mechanisms and addresses the movement of time and the erosion and erasure of time. It celebrates the triumph of time and supports time's persistence. The nature of time can be seen, experienced, and heard on the right side of the building.

A prisoner is taken from the prison and placed in the single car attached to the Ferris wheel. The prisoner faces a Wall of Time (the eastern elevation of the Building of Time). The Wall of Time is divided into 100 square divisions of 2" thick granite blocks. At the top of the wall is an apparatus that grinds away time. For instance, if the prisoner's sentence is thirty years, the grinder moves to block 1 and begins grinding. It takes one year to grind one square block. As the grinder erases the granite, the Ferris wheel begins to accelerate. Each complete cycle of the Ferris wheel takes one year. In the case of a thirty-year sentence, the prisoner in the car will have to go through thirty cycles. He witnesses the passing of time through the disappearance of the granite blocks.

ьнику полиции. Его сразу же
ое преступление было совер-
, при котором, как известно,
ех жертв был вырван голос.
мели вид следов животного.
ся от места нападения, была
рной типографской краски.
ви, белого снега и черной
у политический плакат 1917
али отпечатки лап пантеры,
была ли это черная пантера.
у. Он обошел все типографии
совпадает ли типографская

густо повалил снег. Началь-
сапоги, зимнюю шинель и
нутренний карман он поло-
бой он захватил рупор. Он
ицы города. В темно-синем
лись хлопья снега. Из пере-
ыхватив револьвер, он ри-
омко крича в рупор.
еулка он увидел, как черный
оей жертвы голос. Началь-
ил. Зверь упал. Начальник
львер, убрал ото рта рупор
ежал на снегу лицом вниз.
Зверь оказался пантерой, но

Chief of Police is
tions. What perple
crimes had been c
so tracking was no
was torn out. The fo
mal. Its route from
covered with the c
ink. The red of the b
and the black of
1917 political post
panther, but he wo
panther. This disturl
ing shops in the tow

On the night of
began to fall. The
leather boots, wir
placed his revolve
picked up a bull h
voice. He walked
Prussian blue nigh
snow crystals. He
alley. Pulling out h
darkness shouting

At the far end he
to be a black anim
victim. The Police C
animal fell. The Po
dropped his bull

are planted so that they c
perches. When an angel
the angel's wings flap in
creating a slight wind. The
of almonds.

PRISON

Structure: Wood frame, w
roof, and glass skylight. A c
a cube. A wood stair in th
nects with the Cross-Ove
connects the Prison House

v., **-eled, -el·ing** or (*esp.*
—n. **1.** a private or subord
ship; oratory. **2.** a separately
h, or a small independent ch
o special services. **3.** a roo
n an institution, palace, etc.
of worship for members of va
churches, as Baptists or M
lace of public worship deper
arish. **6.** a religious service
or chapel! **7.** a funeral hom
eral services are held. **8.** a
el, court, etc. **9.** a print sho
n association of employees
with their interests, problems
naneuver (a sailing vessel t
e until the wind can be recov
—adj. **12.** (in England) belon
senting Protestant sects. [1
DE < LL *cappella* hooded c

ise, as *ball* in
enice, coin and
bject, indirect
which a cogni-
eason or argu-
el disapproval,
use or attempt
etc. *—v.t.* **12.**
forward in ob-
oposed import
haic. to bring
5–75; (n.) ME:
ML *objectum*
the mind), n.
, equiv. to *ob-*
v; see JET[1]) +
against (< MF
before, oppose]

intent, inten-

functions as one of the two main consti
sentence, the other being the predicate
of a noun, noun phrase, or noun subs
refers to the one performing the actio
state expressed by the predicate, as *He*
10. a person or thing that undergoes
some action: *As a dissenter, he found h*
of the group's animosity. **11.** a perso
the control or influence of another. **1**
object of medical, surgical, or psycholo
experiment. **13.** a cadaver used fo
Logic. that term of a proposition conc
predicate is affirmed or denied. **15**
which thinks, feels, perceives, intends,
with the objects of thought, feeling, e
ego. **16.** *Metaphysics.* that in which q
utes inhere; substance.
—adj. **17.** being under domination, co
(often fol. by *to*). **18.** being under dom
thority, as of a sovereign, state, or
power; owing allegiance or obedience
19. open or exposed (usually fol. by *to*
cule. **20.** being dependent or conditi
thing (usually fol. by *to*): *His consent*
approval. **21.** being under the necess
something (usually fol. by *to*): *All bei*
death. **22.** liable; prone (usually fol.

САД АНГЕЛОВ

Хранитель Ангелов с трех сторон замурован стенками высотой в десять футов. Четвертая сторона открыта, из нее проход идет в извивающиеся, как змеи, металлические коридоры. Сад обнесен изгородью из деревьев. Деревья посажены так, что они могут служить насестами для ангелов. Когда один из ангелов садится на дерево, крылья его гармоничне складываются, производя легкий ветерок. При этом в воздухе распространяется аромат миндаля.

GARDEN OF THE ANGELS

The Garden of the Angels is enclosed on three sides by walls ten feet high. The fourth side has openings which lead into snake-like metal corridors. The garden has a grid of trees. The trees are planted so that they can be used as angel perches. When an angel alights onto a tree, the angel's wings flap in a harmonic motion, creating a slight wind. The air then has the odor of almonds.

ТЮРЬМА

Конструкция: деревянный каркас, деревянные поверхности, медная крыша, застекленный потолок. Куб внутри куба. Деревянная лестница из внутреннего куба ведет на Перекидной мост. Мост связывает Тюрьму со зданием Суда. Обвиняемый может перейти через этот мост. Во внутреннем кубе стоит стул, стол и койка. Еда вся свежая (не требующая готовки).

Маленькая деревянная лестница связывает внутренний куб с элементом, подвешенным на фасаде, обращенном к внешнему Упраздненному суду.

В элементе на уровне глаз сделана прорезь в 2 дюйма высотой и 4 фута длиной. Сквозь нее обитатель внутреннего куба может наблюдать за тем, что происходит в Упраздненном суде.

Обитатель куба может также видеть Здание Церкви и Дом Смерти, но он никогда не видит Дом Суда снаружи. Обвиняемый может также видеть подвесные сиденья по стенам и длинные полки. Пол в Доме Тюрьмы земляной.

PRISON

Structure: Wood frame, wood surfaces, copper roof, and glass skylight. A cube contained within a cube. A wood stair in the interior cube connects with the Cross-Over Bridge. The Bridge connects the Prison House with the Court House. The Accused can cross over. The internal cube contains a chair, a desk, and a bed. All food is fresh (not to be cooked).

There is a small wood stair which connects the inner cube with an element that is suspended on the facade facing the exterior Voided Court.

The element has an eye-level slit across its surface 2" high and 4' long. The inhabitant of the cube can look out to the Voided Court.

The inhabitant can also see the Church House and the Death House, but never the exterior of the Court House. The Accused can also see the walls of suspended chairs and the extended planks. The floor of the Prison House is made of earth.

ПЕРЕХОДНЫЙ МОСТИК

Проходит над Вильгельмштрассе, соединяя районы №19 и №20. Сделан целиком из стали, выкрашенной в зеленый цвет. Две дорожки для прохода, каждая в одну сторону.

CROSS/OVER BRIDGE

Crosses over Wilhelmstrasse connecting sites #19 and #20. Made completely of steel painted green. Two one-way passages.

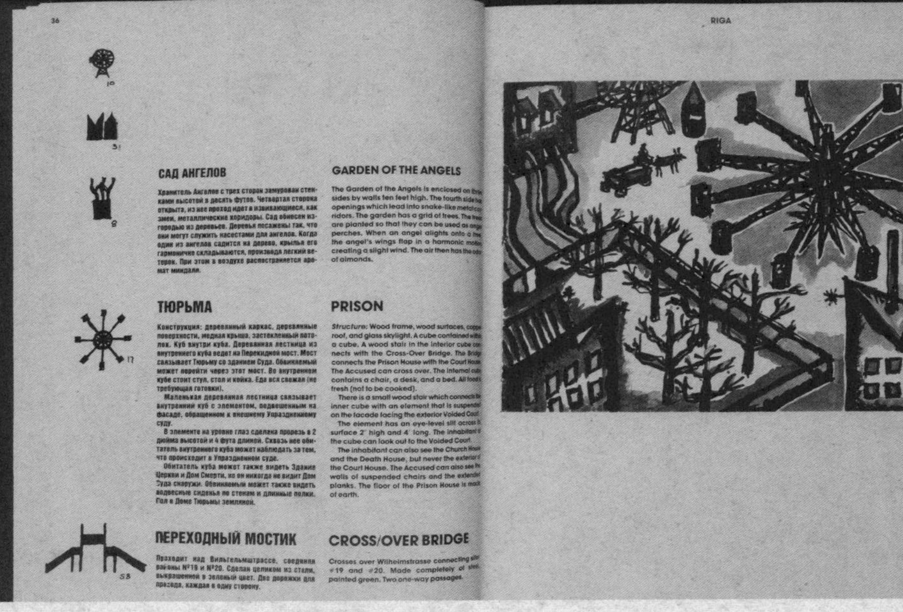

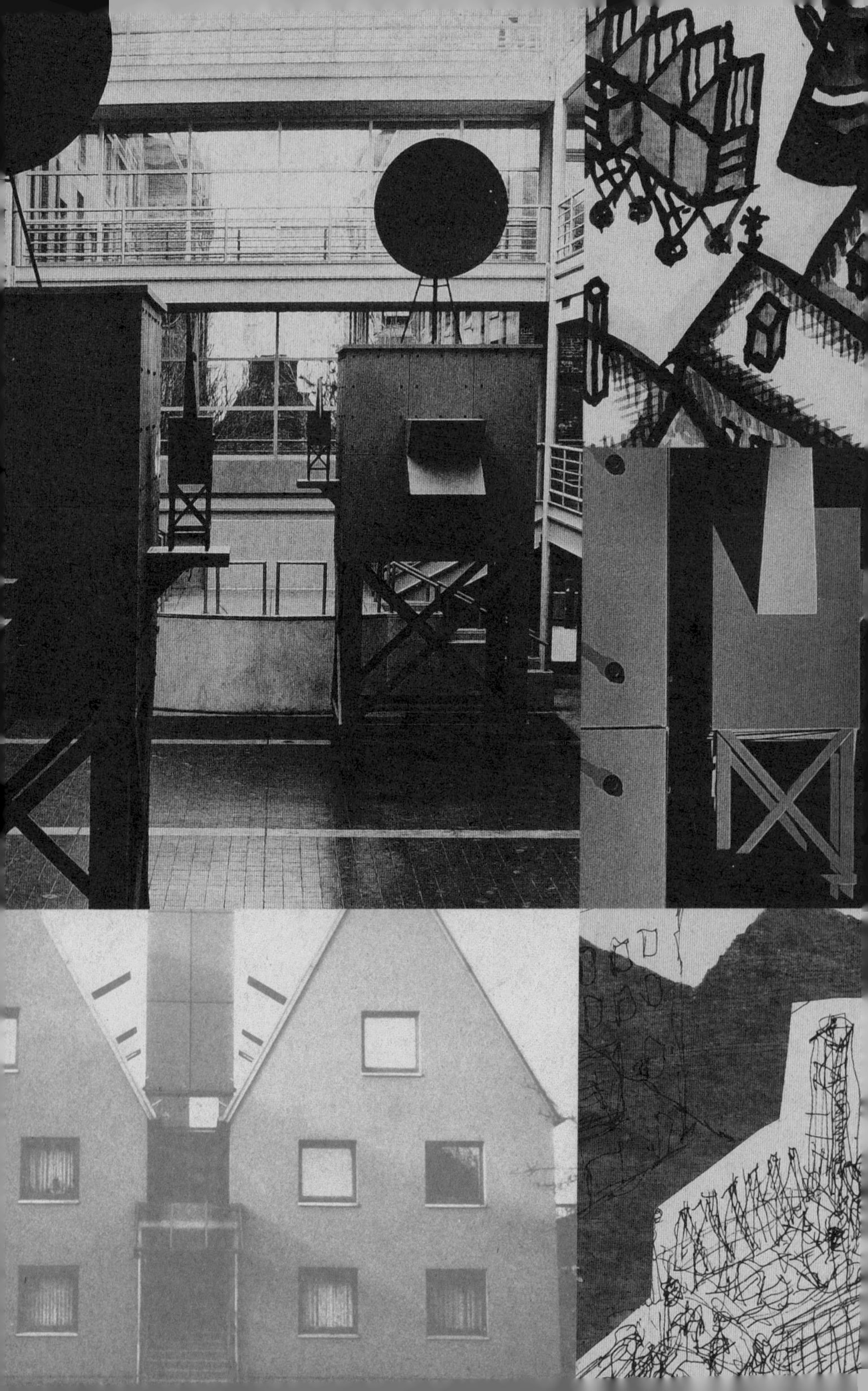

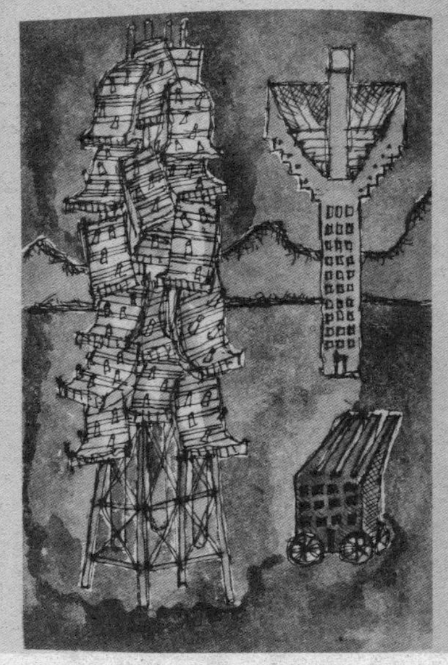

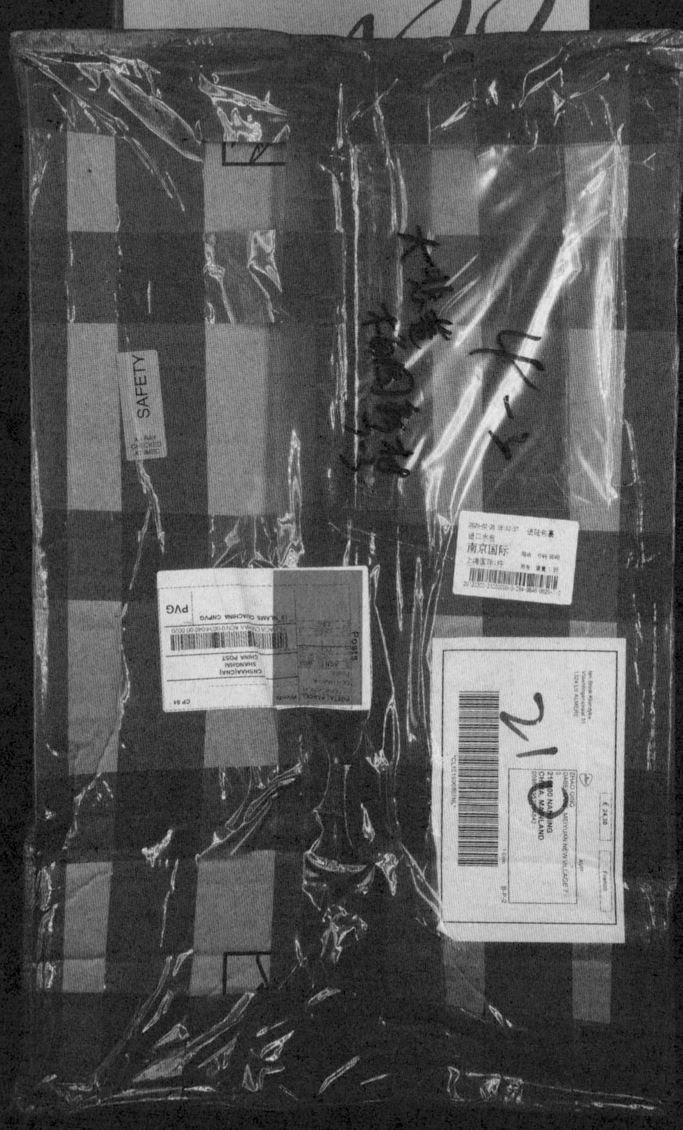
BOB WILLOUGHBY
SAFETY
南京国际
PVG
CHINA POST
SHANGHAI
210

Germany

The Netherlands

China

480mm

12mm

338mm

● 银奖
《 爵士乐在洛杉矶
◐ 德国

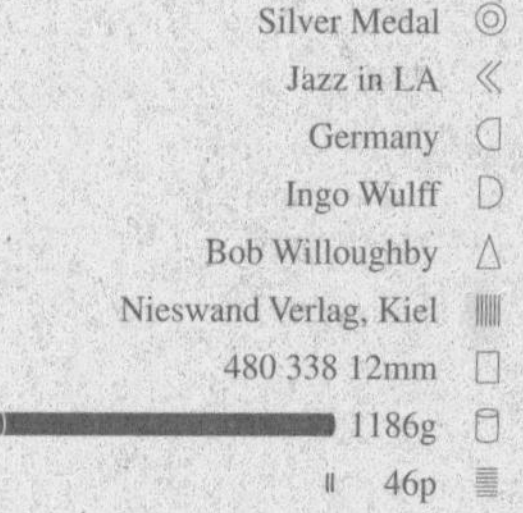

Silver Medal

Jazz in LA

Germany

Ingo Wulff

Bob Willoughby

Nieswand Verlag, Kiel

480 338 12mm

1186g

46p

ISBN-10 3-926048-41-7

美国纪实摄影师鲍勃·威洛比（Bob Willoughby）到1954年为影星朱迪·加兰（Judy Garland）拍摄《生活》杂志照片后才名声大噪，之后一直在好莱坞拍摄电影剧照和现场。但在此之前他就已经拍摄了大量精彩的作品，比如本书，集中展示了他在1950－1956年间拍摄的爵士音乐家在洛杉矶演出、排练时的纪实摄影和富有特色的人物肖像。书籍的装订形式较为特殊，通过开口的铜管将纸张夹在一起，铜管的使用暗示乐器的部分构件。封面采用单面白卡反折，将暖灰色的一面置于外侧，灰蓝色和黑色双色印刷。内页的主要纸张也是单面白卡，蓝灰色和黑色印刷，展现了多位爵士音乐家表演时的精彩瞬间，在双色纸上通过对蓝灰色版的少量使用，使得黑白照片的层次十分细腻，并呈现偏冷的倾向。卡纸间少量穿插蓝灰色半透明牛油纸，单黑印刷，用德英双语记录下这些爵士音乐家在洛杉矶期间的故事。

Bob Willoughby, an American documentary photographer, became famous after taking pictures of Judy Garland for *Life* magazine in 1954. After that, he has been shooting movie stills and live scenes in Hollywood. But before fame he had shot a lot of wonderful works, such as the photos collected in this book, which focuses on the documentary photography or characteristic portraits of jazz musicians when they performed and rehearsed in Los Angeles from 1950 to 1956. The binding form of this book is special. The papers are clamped together by the open copper tube, which implies the component of the instruments. Single-sided white card is folded back to from the cover, the warm gray side is placed on the outside printed in gray-blue and black. The inner pages are also made of single-sided white card, blue gray and black two color printing, which presents the wonderful moments of jazz musicians when performing. By using a small amount of blue-gray pantone on two-color paper, the layers of black and white photos are very delicate and tend to be cold. A small amount of blue gray translucent butter paper was interspersed between the cardboards, printed in single black, and the stories of these jazz musicians in Los Angeles were recorded in German and English.

BOB WILLOUGHBY
JAZZ
IN
LA

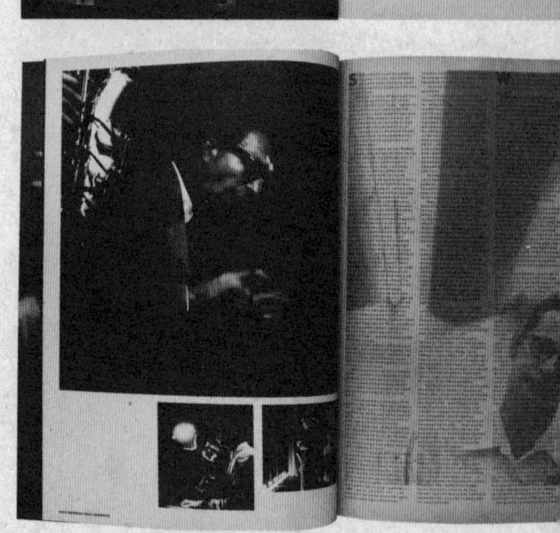

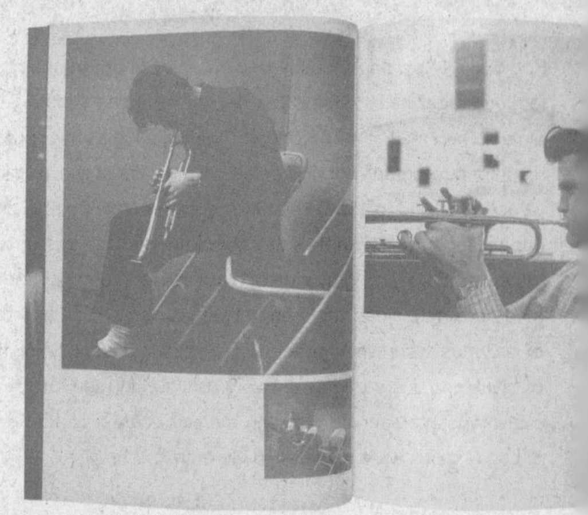

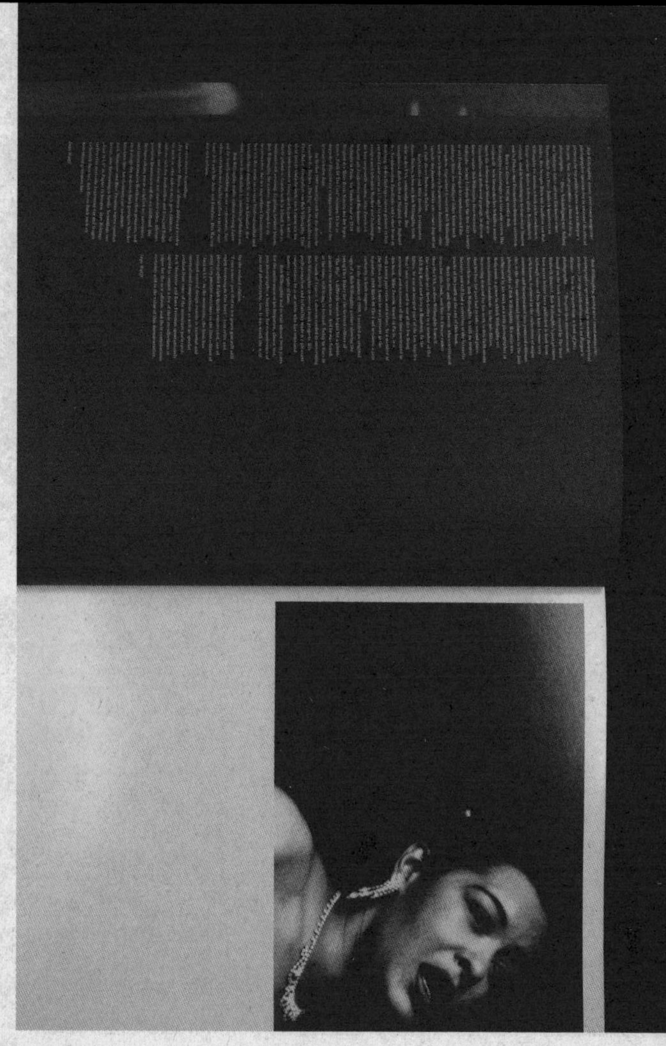

Kehrte um und
...
...ren jemals
...
...n den frühen
...n aus New York
...e, was in der
... Miles Davis,
...lla Fitzgerald,
... wenn sie in die
...ater versuchte ich
...ne Ausrüstung
...hne zu langsam.
...hatte heraus-
...eine Filme auf
... zu puschen.
...ten ich kriegen
...tentat rauschen,
...um der Polin auf
... der hohen
...rischen des Films

...kJockey seiner
...um weniger
...e veranstaltete
...w mir, hinter die
...n an, einige der
...tografieren.
...derer bekannten
...tternacht-
...mit Big Jay
... aber da mir Jay
..., was mich
...ote der Saal, wie
... im Olympic
...kämpfe
...sagen im Ring und
...ly-Stil kreischen
... machte diese
...ll Drogen (damals
...o sieht man auf
...itten, die das
... habe nie etwas
...e Polizei den
...größeren Sälen
...fmachte er die
...er Fans von der
...würde er in San
...

W[illegible] – of course – [illegible] enthusiast, and spent all of my spare cash buying early jazz records at a second hand record store run by [illegible] in Los Angeles.

I was really just starting in photography. In 1950, I was 22/23 years old, I processed my films in a darkroom I had in the garage. It was so hot in the daytime – and light leaked in through the roof – that I got into the habit of printing and developing at night. It was then that I was able to listen to a San Francisco radio station, whose signal only became strong enough to pick up at night . . . I used to listen to Dave Brubeck and his quartet and sometimes sextet, playing from there.

I thought Brubeck was wonderful, and his mixture of modern jazz and the fugue was really exciting. So when they announced that he was coming to Los Angeles, I made a point of taking myself and the camera over to The Haig. This was a small jazz club, run by Dick Bock, who later became boss of Pacific Jazz Records Co.

At that time I had a Speed Graphic (4x5 in.) and extension flash guns . . . it was before I gave it all up for 35mm natural light. I listened and was already a fan, but found the music very special. I sent Dave a print of the photo that was my favorite, and later had a phone call from Sol Weiss who, with his brother Max, ran the small Fantasy Records label in San Francisco.

They had Dave under contract, and could they use my picture on the cover of Dave's first album? They told me later that this was the first photo used on a jazz record sleeve . . . or if it wasn't the first anywhere, it was their first. They didn't have any money, however, but offered a set of their records by way of payment. As it turned out, they also pressed Chinese folk operas that were closely guarded from other Chinese record people. Well, when the "payment" came, a lot of their library was these Chinese records. We had a great laugh about it at the time but Max and Sol, who became great buddies and made Fantasy an important label, went on to use my work on quite a number of their releases.

Later I photographed Dave when he went to Columbia Records, for many of his covers with them. I was with Dave and Paul Desmond on many a night, and at many a concert. Sometimes the

...fied. I've watched in horror as I sa...
...image off the backing onto the flo...
...time, to push a film, you had to rai...
temperature.

The biggest local disc jockey ...
Gene Norman, who put on the "Jus...
and they too were formidable. I wa...
backstage on some of these, and th...
began photographing some of the ...
artists.

Another late night disc jockey ...
Hunter Hancock. He organized the ...
sessions at the Olympic Auditorium...
McNeely. I had heard about the sh...
know what to expect from Jay. Wh...
place was rocking like nothing else...
This was a place for boxing matche...
in the ring all right, blasting away ...
special style.

I jumped up on the stage and ...
some photos. The police were so wo...
drugs – dope they called it then – t...
the photographs you can see a unif...
watching. I never saw anything mys...
police finally closed the place dow...
bigger places. He got the crowd so ...
Shrine Auditorium one night that o...
leaped out of the balcony. Another...
arrested in San Diego for leading t...
audience out of the theatre and ma...
around the block while he was still...

He often left the stage, leavin...
fend for themselves and honked hi...
audience, whipping them up to wha...
sexual release. Some of the girls re...
for it was orgasm time. Remember,...

Lionel Hampton was one of the...
virtuosos of the vibraharp. He play...
instrument like no one before or si...
only musician, other than Big Jay, ...
tear an audience up, shake them, ...
them jump up and down, become hy...
had that drive, that energy. He was...
and while I have often thought bac...
performance, I always felt him a li...
could never be said about Lionel. I...
was magic.

Stan Kenton had a big band, ...
taste. But he was a shaker and mov...

...zeug), **RUSS FREEMAN** *(Piano) und* **CARSON SMITH (Baß)** *bei derselben Aufnahmesession in Los Angeles, 15. September 1953. An der Aufnahme zu der LP „Chet Baker Sextet" nahm neben Bud Shank noch Bob Brookmeyer teil.*

JACK MONTROSE *im Gespräch mit* **RUSS FREEMAN** *während einer Session mit Chet Baker (Chet Baker Ensemble), für die Jack die Arrangements geschrieben hatte. Los Angeles, 1953*

BIG JAY McNEELY *concert im Olympic Auditorium, 1951. So antworteten weibliche Fans auf McNeelys pulsierende Musik. Dieses Foto verwendete das Museum of Modern Art für die Weltausstellung „The Family Of Man".*

BIG JAY McNEELY *im Sturmangriff auf das Publikum im Olympic Auditorium bei einem von Hunter Hancocks Mitter...*

IN LA

...eiträgen von ...s by

Herausgegeben und zusammengestellt von

...en. Mike Zwerin ist
...r und Buchautor mit
...usiker war er 1948
... Nonetts bei der legen-
...l"-Session. Er ist in
...inzigartiger Kenner des
...

...ötigte ich für die deutsche
...e Hip" ein Originalfoto von
... von Bob Willoughby. Eine
...che nach einem mir bis
...grafen begann. Und er
... noch in Irland, heute in
... Es gelang darüber hinaus,
...rere Buchprojekte zu
...genehmen Zusammenar-
... seine Professionalität und
...er ist ein Künstler, der
... bescheiden geblieben ist.
...Ste nahezu vierzig Jahre
... seiner vergessenen, fast
...zfotografien warten: eine
... und Fotofans. Das Warten

authority in both respect...

In the autumn of 198... copy of a Bob Willoughby... young Chet Baker for the... "The Hip". A difficult sea... grapher then unknown to... at the time, still living in... the South of France, I als... Bob Willoughby in severa... professionalism and relia... cooperation; he is an arti... modest in spite of fame.

Bob Willoughby had... for the publication of his... jazz photographs: a long... jazz and photography, to...

On the revised edition:

"I can assure you that i... enjoy browsing in this b... photographer Ralph Quir... many very positive reacti... "Jazz in LA". In the mean... received several design a...

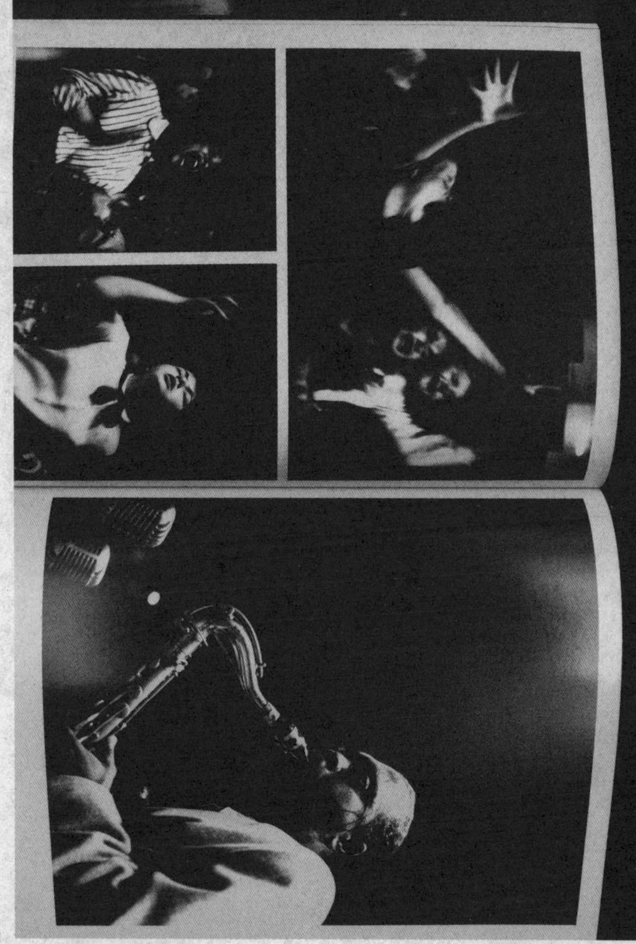

DEUTSCHE POST
CP 92
CP 85
Postal parcels
DEFRAA (DEA)
Frankfurt/M
DeutschePost
BCN
209
2019-05-23
PC
12
1
CN
1.3
S.A.L.
Non priority
airlifted
CNNKGA (CNA)
Nanjing
CHINA POST
PVG
24 FRA ACF 209 PVG
Deutsche Post

Absender / Expéditeur
Kristin Rupp
Karl-Liebknecht-Str. 83
04275 Leipzig
Postleitzahl Ort
Deutschland / Allemagne

Bei Unzustellbarkeit / En cas de non-livraison
☒ Rücksenden / Renvoyer à l'expéditeur
☐ Preisgabe / Traiter comme abandonné

(98)993019

Empfänger / Destinataire
Mr. Zhaoqin
7-3 Dabei Lan
Meiyuan New
210000 NAN
CHINA
Land / Pays de destination

BITTE DEN VERSANDSCHEIN MIT KUGELSCHREIBER UND IN DRUCKBUCHSTABEN AUSFÜLLEN. DABEI FEST AUFDRÜCKEN!

Germany

Germany China

255mm

27mm

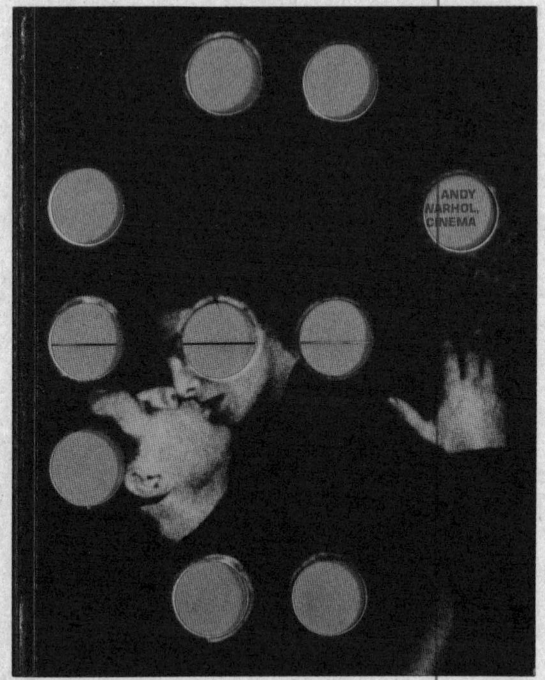

208mm

● 银奖

《 安迪 · 沃霍尔的电影

——巴黎蓬皮杜艺术中心电影回顾展展册

◖ 德国

Silver Medal

Andy Warhol Cinema:
Katalog zur Filmretrospektive im
Centre Georges Pompidou, Paris

Germany

Éditions Carré

——

Éditions Carré, Paris

255 208 27mm

803g

272p

ISBN-10 2-908393-30-1

安迪·沃霍尔（Andy Warhol）在版画、绘画、摄影和电影上都对当代艺术产生了巨大的影响力。1990年6月至9月在巴黎蓬皮杜艺术中心举办了安迪·沃霍尔的电影回顾展，本书是展览的研究文献集。书籍的装订形式为锁线胶平装，封面白卡纸银黑双色印刷，覆亮膜。10个圆形模切暗示沃霍尔常用的16mm摄影机的镜头孔径，漏出银色印刷的环衬和上面的书名。内页胶版纸双色印刷，除了固定的黑色外，另一色会按照文章和电影章节分别使用银、鹅黄、较浅的玫红和蓝色，模拟安迪·沃霍尔的黑白片及套色印刷的版画质感。书中的内容主要是几篇研究安迪·沃霍尔的文章，包括对他电影创作方法的分析、作为艺术家电影的作者性分析、电影中的艺术诉求研究，以及影响力阐述。穿插在这些文章中的是其知名电影的定帧画面。通过连贯帧与大小画面的混排，再将页面上下切开，形成连贯、断档、混合的时间线。沃霍尔对电影的独特理解为“时间胶囊”，而这些页面的呈现就是时间胶囊里散落的生活碎片。

Andy Warhol has had a tremendous influence on contemporary art in printmaking, painting, photography, and film. The Andy Warhol Film Retrospective Exhibition was held at the Pompidou Art Centre in Paris from June to September 1990, and this book is a collection of research documents for the exhibition. The book is perfect bound with thread sewing and the laminated cover is made of white cardboard printed in silver and black. The 10 circular die-cuts suggest the lens aperture of Warhol's commonly used 16mm camera, thus revealing the silver-printed endpaper and the book title. The inner pages are made of offset paper and printed in two colors: one is black and the other is changing accordingly among silver, goose yellow, lighter rose red, and light blue based on the chapters of different articles and movies. The printing on the inner pages simulates the texture of Andy Warhol's black-and-white films and Chromatic printing. The book mainly consists of several articles studying Andy Warhol, including an analysis of his film creation methods, an analysis of his authorship as an artist in films, a study of artistic appeals in films, and the exposition of his influence. The fixed frames of his well-known films are interspersed in these articles. By mixing coherent frames with large and small images, and then cutting the pages up and down, a timeline, which is coherent, broken and mixed, is formed. Warhol's unique understanding of movies is "time capsules", and these pages present the scattered life fragments inside the time capsules.

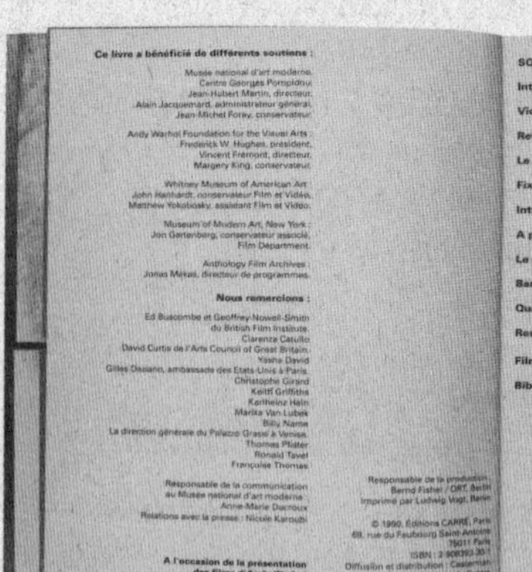

Ce livre a bénéficié de différents soutiens :

Musée national d'art moderne,
Centre Georges Pompidou
Jean-Hubert Martin, directeur,
Alain Jacquemard, administrateur général,
Jean-Michel Foray, conservateur.

Andy Warhol Foundation for the Visual Arts :
Frederick W. Hughes, président,
Vincent Fremont, directeur,
Margery King, conservateur.

Whitney Museum of American Art :
John Hanhardt, conservateur Film et Vidéo,
Matthew Yokobosky, assistant Film et Vidéo.

Museum of Modern Art, New York :
Jon Gartenberg, conservateur associé,
Film Department.

Anthology Film Archives :
Jonas Mekas, directeur de programmes.

Nous remercions :

Ed Buscombe et Geoffrey Nowell-Smith
du British Film Institute,
Clarenza Catullo
David Curtis de l'Arts Council of Great Britain,
Yasha David
Gilles Daziano, ambassade des Etats-Unis à Paris,
Christophe Girard
Keith Griffiths
Karlheinz Hein
Marika Van Lubek
Billy Name
La direction générale du Palazzo Grassi à Venise,
Thomas Pfister
Ronald Tavel
Françoise Thomas

Responsable de la communication
au Musée national d'art moderne :
Anne-Marie Ducroux
Relations avec la presse : Nicole Karoubi

A l'occasion de la présentation des films d'Andy Warhol par la Direction des Musées de Marseille au centre de la Vieille Charité, la Mairie de Marseille a bien voulu prêter son concours à la réalisation de cet ouvrage.

Responsable de la production :
Bernd Fisher / ORT, Berlin
Imprimé par Ludwig Vogt, Berlin

68, rue du Faubourg Saint-Antoine
75011 Paris
ISBN : 2-908393-30-1
Diffusion et distribution : Casterman

ISBN : 2-85850-577-2

Édité avec le concours de
la Délégation aux arts plastiques (Fiacre)

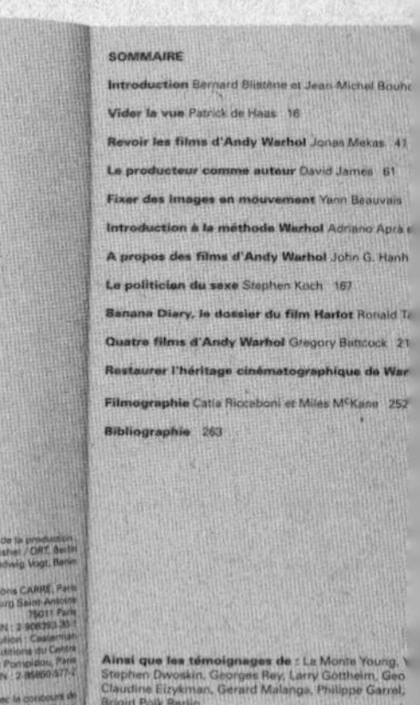

SOMMAIRE

Ainsi que les témoignages de : La Monte Young, Stephen Dwoskin, Georges Rey, Larry Gottheim, Geo Claudine Eizykman, Gerard Malanga, Philippe Garrel, Brigid Polk Berlin.

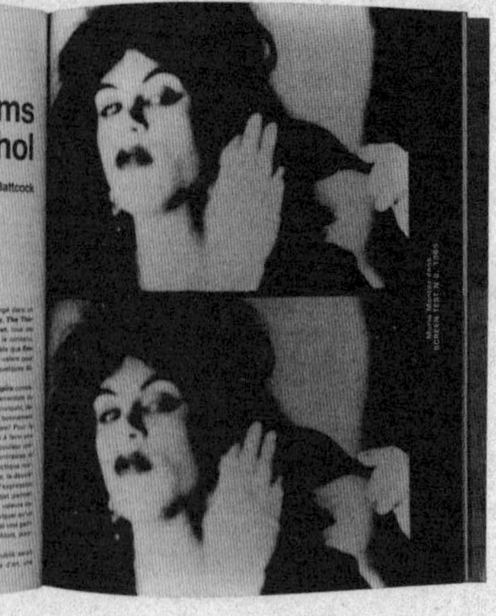

Vider la vue

Patrick de Haas

ABANDON ?

« Je ne peins plus, j'ai abandonné la peinture il y a à peu près un an et maintenant je ne fais plus que des films. Je pourrais faire deux choses à la fois mais les films sont plus excitants. La peinture était seulement une phase que j'ai traversée[1]. »

Pour Warhol, autour de 1965, le film viendrait donc après la peinture qui n'aurait été qu'une phase. Aujourd'hui, à l'examen de l'ensemble de son œuvre qui s'est développée sous de multiples aspects (publicité, peinture, sculpture, cinéma, photo...) on peut néanmoins penser que ces pratiques - la peinture *comme* le cinéma - sont des « phases » du modèle abstrait « cinéma » qui les génère.

Ce modèle met à vif diverses questions : le mouvement (mobilité/immobilité), la série, la répétition, l'indifférence de la machine, « l'œuvre d'art à l'ère de sa reproductibilité technique »...

Toutes ces questions étaient déjà à l'œuvre dans sa peinture avant qu'il ne passe au cinéma,depuis *Orchestra* (1948-49), dessin à l'encre et aquarelle présentant une série de musiciens ...

ANDY DANDY

Warhol est à la fois homme du monde et solitaire, original et banal, attirant et décevant, grave et futile. Ce qui paraît pourtant le distinguer du dandy de Baudelaire, c'est que son abandon de l'éloquence, de l'héroïsme, du message, de la profondeur, pour la neutralité du vide, l'impassibilité de la machine, l'absorption dans la série, semble se développer sans tensions, c'est-à-dire sans mélancolie pour ce qui était, ni mépris pour ce qui est. « Son insensibilité est vengeresse » disait Baudelaire du dandy (*Fusées*) ; celle de Warhol paraît ne plus être victime d'une telle émotion : elle semble naturelle, au-delà du désir, frigide.

« Le dandy doit être sublime sans interruption, il doit vivre et dormir devant un miroir[2] », note Baudelaire. « Sans interruption » : le dandy ne peut restreindre son art à un métier. Warhol est l'un des rares artistes du XXe siècle (avec Duchamp, Picabia, Beuys, quelques autres) dont la vie et l'art constituent une totalité non divisible. Et le cinéma, dans cette perspective, représente pour Warhol la possibilité moderne qui est offerte

qui souligne que désormais une chose, un événement, un humain n'acquièrent une existence que par l'image.

TRANSPARENCE

Le cinéma de Warhol est un miroir étrange. Il s'agit, comme le disait Vinci du tableau, d'une « paroi de verre ». Les perspecteurs de la Renaissance, avec la métaphore du tableau-fenêtre, faisaient oublier son plan matériel d'inscription (la toile), le rendaient transparent, au bénéfice du volume virtuel figuré. Cette transparence devait passer pour naturelle, ne pas se faire remarquer.

Une partie importante de la peinture moderne (mais aussi bien du cinéma « expérimental ») a consisté à opacifier cette paroi, à la laisser visible, à ne plus en masquer les ressorts matériels : ce qui a eu pour effet de changer la « fenêtre ouverte sur le monde » à travers laquelle le spectateur devait conduire son regard, en mur vertical que le regard balaie sans pouvoir le pénétrer.

Warhol prend à revers toute cette logique : il garde la transparence mais celle-ci devient paradoxalement visible, surexposée. Elle affiche tellement sa présence

monde imaginaire comm
n'ouvre pas non plus à
ge cinématographique c
néastes expérimentau
propre lumière, renvoy
La distance est produite
est donc infranchissable
pas de consistance.

IMAGE

L'image, pour nombre
XIXe siècle, est suspecte
masque le réel ou s'y su
frontalement. Malevitc
peintures qui, dans la m
sentent rien du « réel »
vantes, présentes. Et c
donc, aussi « abstraites
crire sur le même regis
« dehors » : ainsi Mond
comme des « plaques à
A l'inverse, plus la ren

st insuffisante. Le cinéma de Warhol n'est pas seule
node d'emploi » ou de la célébrité d'un quart d'heure
e que nous sommes, où l'auteur feint de s'absenter
hol a su le dire lui-même, « les gens ne tournent pas
ur d'eux ».
u plus semblent donc, pour un homme qui comme
rrive » beaucoup trop ou trop peu. Et de fait, Jon Ga
npossible, n'en finit pas de visionner des pellicules,
, de chercher un ordre que la découverte d'une vers
méthode quand Warhol insistait pragmatiquement
pas en avoir mais surtout à ne pas en donner.
mettre en garde tous les taxinomistes et leur dire q
ecyclage et qu'en cela, il est déjà indéfectiblement r
ation entre cinéma, peinture, et ces genres qu'on s'
nopérante. Alors un livre sur le cinéma de Warhol n
e dit d'un film) pour une intelligence de son projet.
t donc pas le propos d'un auteur mais un ouvrage à
ndy Warhol tel qu'il est puisque, du reste, d'aucuns
ais davantage tel que chacun se le rappelle pour en
l'acteur fortuit ou désigné, l'exégète ou le parasite
y mêlent des lectures différentes, que des approche
pas, que des témoignages du genre « je me souvien
tique ou historien veut reconnaître. Mais surtout, o
soit là, en gros plans fixes, en séquences en gros pl
euille tourner les pages de ce livre, tente de « s'intro
graphe a occupé obsessionnellement Andy Warhol
ussi du climat d'une époque qui a vu se développer
éma underground. On ne comprend pas le cinéma
compte Brakhage, Markopoulos ou Jack Smith. War
nema regroupé à New York autour de la *Filmmaker*
que nous lui demandions de contribuer à cet ouvra

AN

ogue

rry Fainlight
Billy Linich
onald Tavel

Tu savais ça?
Linich: Je co
Tavel: C'éta
sur un autre
boration. Je
que ça allait
comme ça.
Fainlight: C
Tavel: Un tr
Linich: L'Os
Tavel: Bon,
de ce genre?
Linich: Oh!
tions pareille
Tavel: Qui v
Linich: Herr

taurer ritage mato-hique

Une mythologie considérab
personnalité d'Andy Warhol.
que son activité cinématogra
une large part dans une pé
1967), et qui a donné des fil
jourd'hui que **Kiss** (1963), **Em**
The Chelsea Girls (1966) e
(1967). On n'a guère vu les
cours des vingt-cinq derniè
avons-nous que des souvenir
présent. La remise au jour d
veau public, et pour les spec
ne les ont pas vus depuis leu
salle, permet de réexaminer
phique dans une perspective
ses oeuvres à une plus juste

la documentation et
la bibliographie
UENEAU pour la con
méricains sont de Je
anna MINELLI
é assuré par Martine
Mireille DEVALS.

Introduction à la méthode Warhol

Adriano Aprà et Enzo Ungari

LA PEINTURE, LE CINÉMA

Après avoir évoqué les différentes façons dont le cinéma de Warhol a été perçu et avant de parler de ses films, il convient d'aborder tout de suite un aspect de son travail qui semble se prêter à des malentendus, et qui risque de mettre en cause l'autonomie d'un domaine, celui de la peinture, dont nous n'avons aucune connaissance spécifique.

Ce qui frappe dans les premiers films de Warhol (dans lesquels le son, le montage, les mouvements de la caméra, la couleur et les sujets mobiles sont significativement absents), ce n'est pas tant une scène (étrangement familière), ni une action (annoncée, préparée, attendue et qui ne tardera pas à venir), mais une perspective. Etant donné l'activité de peintre de Warhol, cette remarque peut non seulement paraître évidente, mais aussi ambiguë, car il ne faut pas voir ses premiers films tout simplement comme une tangente ou une section de sa peinture. Mais ce serait commettre une erreur encore plus grossière que d'isoler ses films à l'intérieur de l'histoire du ciné[...]

(ni en ce qui concerne le sujet, ni en ce qui concerne les thèmes ou le traitement visuel ou stylistique des images) ; bien au contraire, Warhol a presque voulu souligner l'indépendance de son cinéma en se servant de certains de ses films pour réaliser des tableaux (il a utilisé, par exemple, pour certaines sérigraphies, des photogrammes de **Sleep** et d'**Henry Geldzahler** auxquels il a donné le titre d'un autre film : **Couch**, etc.) en mettant donc ses films au même rang que les boîtes Brillo ou les bouteilles de Coca Cola, c'est-à-dire des objets déjà existants et réalisés avant son intervention de peintre. Cette duplication (il est en effet impossible de distinguer la reproduction du photogramme de **Sleep** de la reproduction de sa sérigraphie) correspond parfaitement à l'attitude de Warhol, qui veut tester les limites des différentes techniques et des différentes disciplines, en mesurer et en épuiser toutes les possibilités. Ce genre de démarche artistique, qui investit tous les domaines dans lesquels les techniques sont appliquées (dans le cas de Warhol, du moins jusqu'ici : la peinture, le cinéma, la littérature, le théâtre, la vidéo et [...]

Exploding Plastic Inevitable ; des romans construits à partir de bouts des bandes sonores des films ; des tableaux réalisés d'après des photogrammes de films, etc.). Il n'existe donc aucune hiérarchie parmi ces différentes opérations, qui sont soumises aux fluctuations les plus exhaustives. Ce qui les rapprocherait toutes de la peinture, ou d'un certain enseignement de la peinture, dont Warhol semble avoir fait la devise de toute sa pratique artistique, ce serait plutôt la définition même de la peinture que nous a donnée Chou wen : « La peinture est faite de limites, elle ressemble aux chemins qui délimitent les champs[1]. »

Lorsque nous parlons de perspective dans les films muets de Warhol, nous ne faisons donc pas allusion simplement à une continuité d'ordre mécaniciste entre le tableau et le film. Nous pouvons saisir le sens cinématographique de cette perspective à l'intérieur d'un temps et d'un espace non chronologiques et non linéaires. La caméra fixe cessera donc de représenter pour nous, de manière anachronique, un point de vue privilégié et inattaquable, pour devenir la représentation d'une représentation : celle constituée par la scène originelle des premiers films de l'histoire du cinéma, une scène muette et immobile, qui aspire tacitement à la parole que la technique et l'industrie n'ont pas encore su lui donner.

LE CINÉMA COMME PRÉ-HISTOIRE ET COMME DÉCOUVERTE SCIENTIFIQUE

Cette deuxième opération qui préside à toute représentation d'une représentation, circonscrit immédiatment le pré-texte des premiers films de Warhol, constitué principalement par les films « pré-historiques » réalisés par Norman Raff et Frank Gammon pour la Kinetoscope Company, la société fondée par Edison, « le Sorcier de Menlo Park », l'inventeur de la pellicule cinématographique et du phonographe, le protagoniste du roman de Villiers de l'Isle-Adam, l'Eve future. Edison fit tourner des centaines de films destinés à son Kinétoscope dans la Black Maria, qui fut le premier studio de l'histoire du cinéma et le véritable ancêtre de la Factory. La plupart de ces films se limitaient à enregistrer, au début, les mouvements des personnages qui se plaçaient devant la caméra. Fred Otto, déjà connu pour avoir réalisé un film destiné au phonographe optique et intitulé **Fred Otto's Gestures** (1888), fut le protagoniste d'un certain nombre de films, dont le plus connu est **Fred Otto's Sneeze** (Fred Otto éternue, 1893). Certains films de Warhol semblent continuer cette liste d'actions ba[...]

champignon ; dan[...]
fume le cigare. Il n[...]
qui serviraient à r[...]
les premiers film[...]
dans ces films on [...]
téresser à rien d'[...]
traites, comme l'e[...]
se font encore p[...]
montre un certain [...]
ger des baisers. U[...]
mi and Rufus K[...]
The May Irvin-J[...]
été tourné par N. [...]
Kinetoscope Edis[...]
pas évoquer **Le G**[...]
mière, avec lequ[...]
l'égard d'un spec[...]
être capable d'ép[...]
et du plaisir face à [...]
élémentaires et le[...]
Dance Movie et [...]
de Warhol qui mo[...]
Herko (assisté, da[...]
paraissent particu[...]
re qu'en 1894 les [...]
portant dans la pr[...]
à des films comm[...]
cing, Silver Dan[...]
lèbre de tous, **A**[...]
l'histoire d'une da[...]
style qui rappelle [...]
pour photograp[...]
Warhol semble vi[...]
pe même qui était [...]
toire du cinéma, [...]
raconter une histo[...]
près la scène de [...]
que son caractèr[...]
qui réglait les pri[...]
Lumière. Ce qui l[...]
table manque d'i[...]
nématographique [...]
une copie mais de[...]
essentielle est au[...]
Lumière s'occupe[...]
registrement (de [...]
puissent amuser, [...]
Warhol semble co[...]
de filmer. Le seul [...]
période est en eff[...]

0067

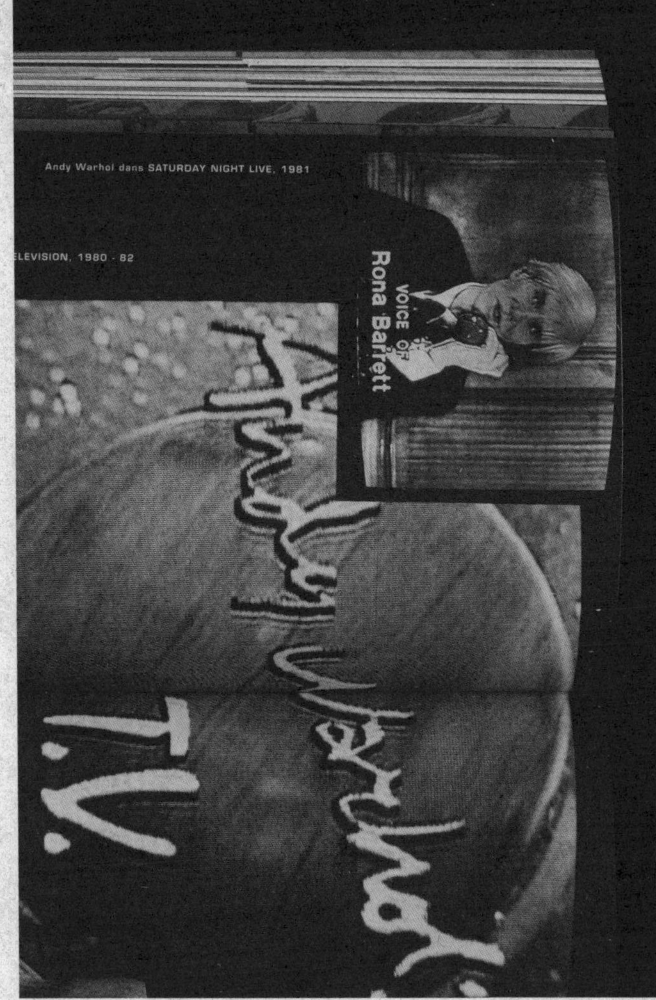
Andy Warhol dans SATURDAY NIGHT LIVE, 1981
ELEVISION, 1980 - 82
VOICE OF
Rona Barrett
Andy Warhol
T.V.

Richtigkeit und Heiterkeit
Gedanken zum Buch als Gebrauchsgegenstand von Franz Zeier
25.02.20 16:56
CHF 15.00
PRIORITY
P.P. Postage Paid
SWISS POST
www.buchplanet.ch
Onlineshop für gebrauchte Bücher
CUSTOMS DECLARATION - CN 22
Customs control allowed
Content: Goods
1 Buch
TO
FROM
Eva
Zhao
65 5th ave, 1010, New York, NY
10003 New York
UNITED STATES OF AMERICA
RK 997 020 977 CH
PRIORITY
Post CH AG
0.2

CUSTOMS DECLARATION - CN 22

s control allowed Content : Goods

detailed description	Weight	Value (CHF)	Customs tarif number
1 Buch	0.2	9	
	0.2	9	

ıt the particulars given in the declaration are correct and that this
not contain any dangerous articles prohibited by postal regulations.

nature 27.02.2020

R Recom étrange SWISS

FROM Bu Wa 93 Sw

Switzerland

Switzerland China

铜奖
正确与欢乐
——从思想到作为商品的书本
瑞士

Bronze Medal
Richtigkeit und Heiterkeit:
Gedanken zum Buch als Gebrauchsgegenstand
Switzerland
Jost Hochuli
Franz Zeier
Typotron AG, St. Gallen
240 150 5mm
121g
44p
ISBN-10 3-7291-1058-6

瑞士设计师、设计理论家弗朗兹·蔡尔（Franz Zeier）出版了很多有关书籍的论著，探讨书籍的形制、阅读、纸张、工艺、历史等问题，本书是其中的一本。书籍的装订形式为线装骑马订，护封采用黄色胶版纸印黑，书脊处上部45°斜切，露出黑卡纸内封一角。环衬采用较薄的黑色纸张，内页胶版纸四色印刷。内容方面，书中从思想的记录、古登堡现代印刷、书籍发展的历史、装订工艺的演进，一直说到作为大宗商品的现代书籍。对偏向手工艺的装订技术、传统延续、书籍设计的各个方面进行了深入思考。设计方面，全书单栏排版，在衬线体为主的段落中采用首字母粗黑字加以装饰。全书设计尽可能地简朴，只在很小的细节做有限装饰，体现了作者对现代书籍设计中装饰和工艺的思考和态度。

Franz Zeier, a Swiss designer and design theorist, has published a lot of books on the shape, reading, paper, technology, history and other related issues. This book is one of them. The binding form of books is saddle stitching with thread. The jacket is printed in black with yellow offset paper. The upper part of the spine is cut at 45 ° to expose a corner of the inner cover made of black cardboard. The end paper is made of thinner black paper, while the inside pages are made of offset paper printed in four colors. As for the contents, the book talks about the record of ideas, modern printing of Gutenberg, the history of book development, the evolution of binding technology, and modern books as bulk commodities. It has made in-depth thoughts on all aspects of binding technology, tradition continuation and book design. As for the design, all the pages are arranged in single column, and the paragraphs typeset with serif font are decorated with thick black initial characters. The design of the whole book is as simple as possible, with limited decoration only in small details, reflecting the author's thinking and attitude towards the decoration technology in modern book design.

Zu diesem Heft

Richtig möchte der Buchbinder Franz Zeier seine Bücher kleiden – richtig: dem Inhalt entsprechend, angenehm anzufassen, möglichst leicht im Gewicht und mühelos zu blättern. Falsch scheint ihm jede Schwere, jeglicher Prunk; falsch das Bemühen jener Buchbinder, die ihr handwerkliches Können auf dem Einband zelebrieren und zur Kunst erheben.

Neben der Arbeit und über Jahre hinweg hat er seine Gedanken notiert. In diesem Heft sind einige davon versammelt. Der Leser merkt: Handwerk, so wie es Franz Zeier versteht, ist nicht nur Fingerfertigkeit, es bedeutet auch Nachdenken. Nachdenken über die Zusammenhänge, über das Buch als Ganzes, über die Stellung des Buches in Zeit und Gesellschaft.

Selbstbeschränkung auf das Einfache heißt nicht, phantasielos sein, ist nicht dogmatische Enge. Zeier ist abwechslungsreich in der Wahl der Materialien, in seinen Farbkombinationen, in Gestaltung und Plazierung der Schildchen. Freilich drängt sich solches nicht auf; erst das eingehende Betrachten offenbart den Ideenreichtum. Was sofort auffällt: die Heiterkeit, die von seinen Büchern ausgeht. Sie nimmt den Betrachter gefangen und wirkt über das unmittelbare Erlebnis nach, denn sie ist mehr als nur oberflächlich wohlgefällig.

J.H.

Der letzte

Auf dem Herstellungsweg des Buches ist der Buchbinder der letzte aller Beiträger. Er soll der Vollender sein. Das Werk, das mit der Herstellung des Papiers begonnen wurde, das der Buchdrucker nach den Angaben des Typographen oder Buchgestalters mit der Schrift versehen hat (vielleicht auch mit Illustrationen), wird vom Buchbinder gefalzt, geheftet, mit dem Einband versehen. Jetzt kann man mit Lesen beginnen.

So wie vor diesem Arbeitsprozeß der Autor, der Fotograf, der Illustrator, so am Ende der geneigte, zweifelnde oder leidenschaftliche Leser. Über alles hinweg reicht er dem Dichter, dem Schriftsteller, Essayisten, oder was der Autor sei, die Hand, freundschaftlich, begeistert, je nachdem. Vielleicht auch verstimmt.

Zuhause

Ich hoffe, daß das Buch, dessen Sterben seit Jahrzehnten prophezeit wird, noch lange lebt. Ein Leben ohne Bücher sich vorzustellen, fällt uns schwer, dies hängt ja mit so vielem zusammen: Erinnerungen an Bilderbücher, Alben mit eingeklebten Reptos, Märchenbücher, an die ersten Lese-Bücher, die ersten spannenden Geschichten, an frühe Antiquariatsbesuche, glückliche Funde, erfolglosen Suchen, an erlebnisreiche Lektüre während Tagen der Krankheit oder in den Ferien usw.

Tradition

Es ist nicht so, daß die Tradition Sache der Traditionalisten ist. Diese produzieren den Traditionalismus, der nichts taugt; die Tradition wird von den wahren Neuerern fortgesetzt und bleibt durch sie lebendig. Die Traditionalisten beuten die Tradition aus, führen sie auf das tote Geleise. Die Neutöner aber verachten beide, glauben, absolut alles aus sich selbst schöpfen zu können.

Versiegt

Die stilbildenden Säfte versiegten nach 1800 nicht allerorts zu gleicher Zeit. Wie die Volkskunst sich in Gegenden fern von den Zentren der Zivilisation länger am Leben erhielt, so gedieh das noch intakte Handwerk in den weniger anspruchsvollen Werkstätten, in den abseitsliegenden Arbeiten noch lange, als in den fortschrittlichen Ateliers die Décors bereits von einer bedenklichen Haltlosigkeit zeugten. Dafür war ihre Ausführung von einer Brillanz, wie sie bisher nicht gesehen wurde. Aus Handwerk war Kunsthandwerk geworden.

Handwerk – Kunsthandwerk

Der Begriff Kunsthandwerk und das kunsthandwerkliche Wesen oder Unwesen entstehen in dem Moment, wo die im Handwerk selbstverständlich gegebene Einheit des Manuell-Technischen und des Schöpferischen auseinanderfällt, wo sich das letztere sozusagen emanzipiert und das erstere zur ‹Technik› verkommt.

Als Produkt dieses Zerfalls entsteht der Perfektionismus.

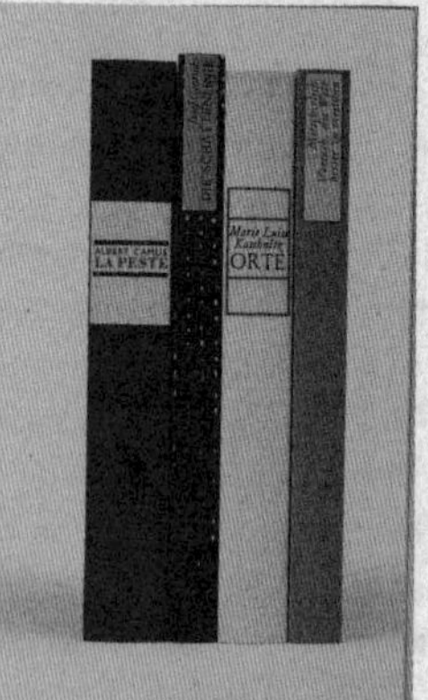

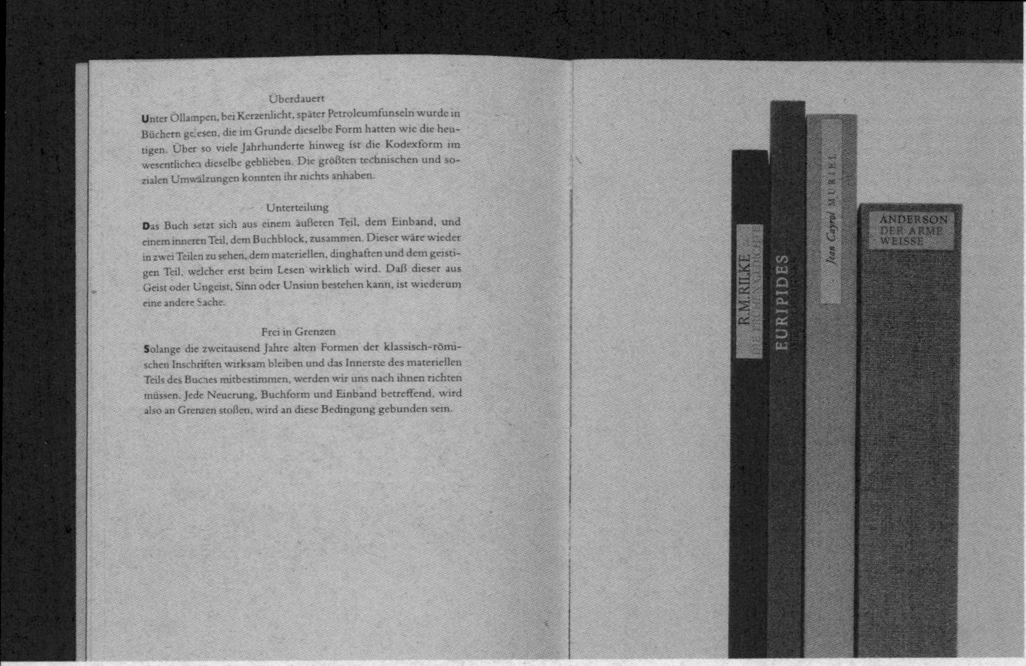

Überdauert

Unter Öllampen, bei Kerzenlicht, später Petroleumfunseln wurde in Büchern gelesen, die im Grunde dieselbe Form hatten wie die heutigen. Über so viele Jahrhunderte hinweg ist die Kodexform im wesentlichen dieselbe geblieben. Die größten technischen und sozialen Umwälzungen konnten ihr nichts anhaben.

Unterteilung

Das Buch setzt sich aus einem äußeren Teil, dem Einband, und einem inneren Teil, dem Buchblock, zusammen. Dieser wäre wieder in zwei Teilen zu sehen, dem materiellen, dinghaften und dem geistigen Teil, welcher erst beim Lesen wirklich wird. Daß dieser aus Geist oder Ungeist, Sinn oder Unsinn bestehen kann, ist wiederum eine andere Sache.

Frei in Grenzen

Solange die zweitausend Jahre alten Formen der klassisch-römischen Inschriften wirksam bleiben und das Innerste des materiellen Teils des Buches mitbestimmen, werden wir uns nach ihnen richten müssen. Jede Neuerung, Buchform und Einband betreffend, wird also an Grenzen stoßen, wird an diese Bedingung gebunden sein.

REISE
CH
LIN

SWIF
LEMU
GULLIV
REIS

DIE
TOTE
GEMEINDE

MÆRCHEN
AUS TIBET

SWIFT
LEMUEL
GULLIVERS
REISEN

KARL
KRAUS
AUSWAHL
AUS DEM
WERK

INGRES IN ITALIA
MOSTRA ALL' ACCADEMIA DI FRANCIA / VILLA MED

EURIPIDES
DIE TROERINNEN

Jean Cay

D
W

LEO TOLSTOI
Hadschi-Murad
JOHANNES HALLER
DANTE
DICHTER UND MENSCH
SAMMLUNG KLOSTERBERG
NOVALIS
Die Schwelle beider Reiche
SAMMLUNG KLOSTERBERG

ZVAB.com
ZENTRALES VERZEICHNIS ANTIQUARISCHER BÜCHER
www.zvab.com
CN22
Deutsche Post
FROM
iele
Ravensburger

Customs Declaration
May be opened officially

CN22

Designated operator **Deutsche Post**

☐ Gift ☐ Documents
☐ Samples ☐ Others
☑ Sales of goods

Quantity and detailed description of contents	Weight (in kg)	Value (in EUR)
1x Book	0.30	13.80
HS tariff number and country of origin of goods	Total Weight	Total Value
DE	0.40	13.80

I, the undersigned, whose name and address are given on the item, certify that particulars given in this declaration are correct and that this item does not contain any dangerous article or articles prohibited by legislation or customs regulations.

Date and sender's signature: 01-Jun-20

FROM
Manuel Weiner Antiquaria
Stichstr. 2
79877 Friedenweiler
GERMANY

TRACKED
PRIORITY P.P.

Deutsche
IM 01.06.
WARENP.
A0 0227
00 0000

TO
Zhao Qing
…iyuan New Village
7-… Dabei Lane,
2…000 NANJING
Chi…

Germany

Germany China

270mm

10mm

185mm

● 铜奖
《 一个，五个，很多个
◐ 德国

Bronze Medal
eins, fünf, viele
Germany
Kv ta Pacovská
Kv ta Pacovská
Ravensburger Buchverlag Otto Maier, Ravensburg
270 185 10mm
306g
24p
ISBN-10 3-473-33569-X

本书是由捷克艺术家、儿童作家克维塔·帕佐夫斯卡（Květa Pacovská）编写、绘制和设计的互动识数绘本，通过古灵精怪的插图、大胆的造型和色彩，以及互动引导式阅读，帮助学龄前儿童认识数字1—10。书籍的装订形式为塑胶圈装。封面白卡纸四色印刷，书脊处采用墨绿色塑胶圈，并在圈上白色丝印文字。内页白卡纸四色印刷。书中的“小丑”用身体形态组成数字1—9，夸张的绘画风格与独特的人物造型相得益彰。内页白卡纸均为单面印刷，部分页面折叠成包背形式，通过单面开口、翻折、粘贴其他材料等方式成为可以互动的“玩具书”。书中并非一页一个数字地展示，而是在每个页面中加入之前数字的演绎，通过图画和游戏阐释数字之间的关系，巩固所学。帕佐夫斯卡绘制和设计的书籍还获得过1993年金字符奖和1991年荣誉奖（可在相关页面查找）。

This book is an interactive picture book written, drawn, and designed by Czech artist Květa Pacovská. It aims to help preschool children understand the numbers 1-10 through quirky illustrations, bold shapes and colors, and interactive guiding reading. The book is bound by plastic spiral. The cover is made of white cardboard printed in four colors. The dark green plastic spiral at the spine is treated with white silk screening. The inside pages are printed on the white cardboard in four colors. The figures of the 'clown' in the book are composed of numbers 1-9, and the exaggerated painting style complements the unique character shape. The inner pages of white cardboard are printed on single side, and some pages are folded into double-leaved form. Through folding, and pasting, it becomes an interactive 'toy book'. This book does not show a number on one page, but uses pictures and games to explain the relationship between numbers and consolidate what you have learned. The books drawn and designed by Pacovská also won the 1993 Golden Character Award and the 1991 Honor Award (available on the relevant pages).

d...ich,
Vogel...du,
2
schau mal!

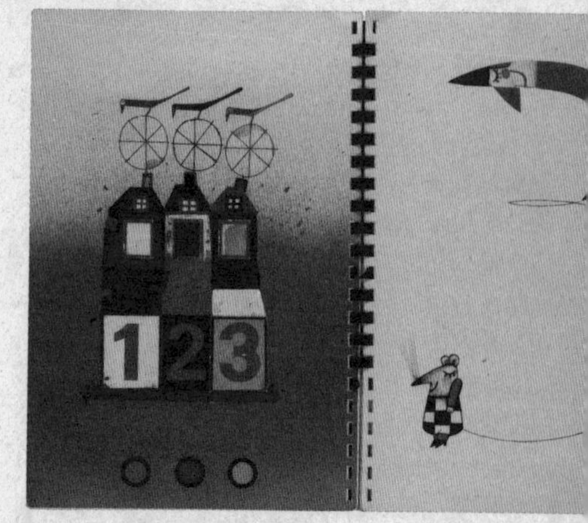
1 2 3

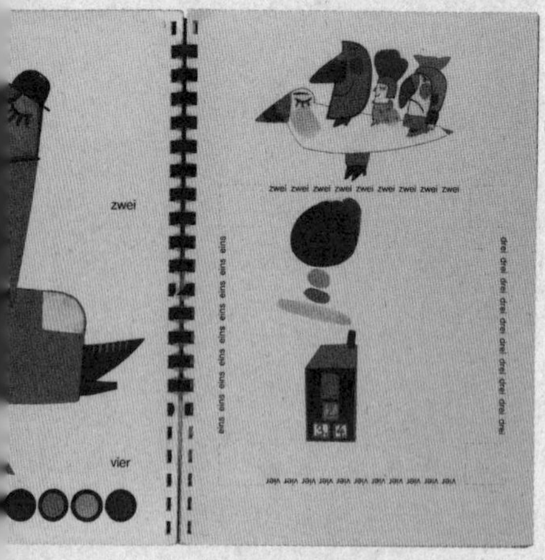
zwei
vier

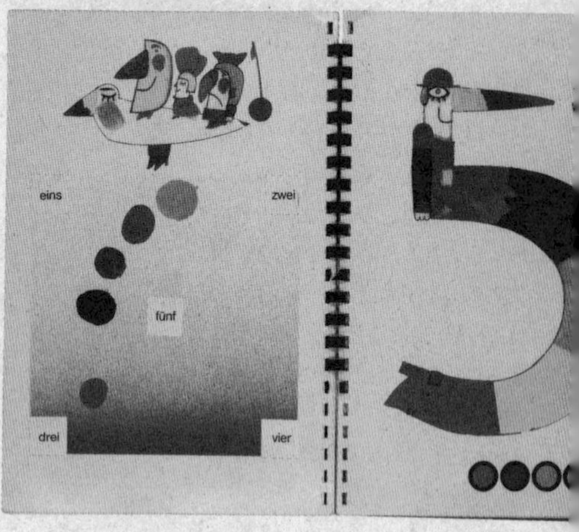
eins
zwei
fünf
drei
vier

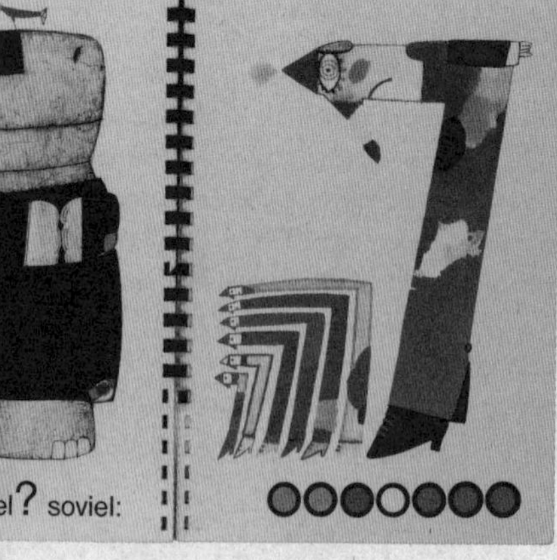
el? soviel:

sooooooooviel

eins eins eins eins eins eins eins eins eins eins eins eins eins eins
eins eins eins eins eins eins eins eins
eins eins eins eins eins
1
Originalausgabe des Ravensburger Buchverlages
© 19[illegible] by Ravensburger Buchverlag
Otto Maier GmbH
Alle Rechte vorbehalten
Gesamtherstellung: Otto Maier Verlag AG
Printed in Germany
ISBN 3-473-33569-X
Vorzugsausgabe mit einer Original-Lithographie,
von der Künstlerin von 1-160 numeriert und signiert
ISBN 3-473-33608-6

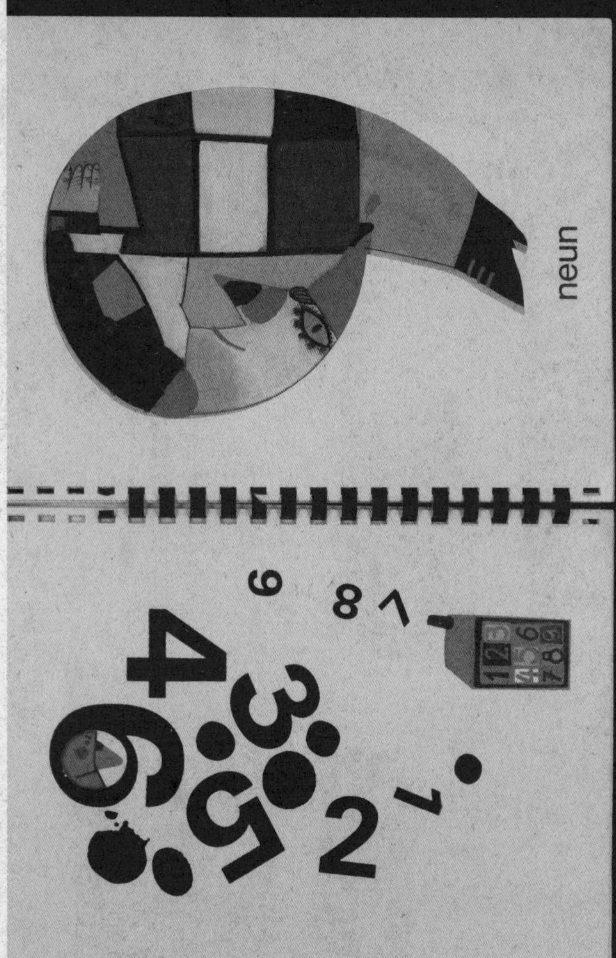
neun

sehr
viele!
10

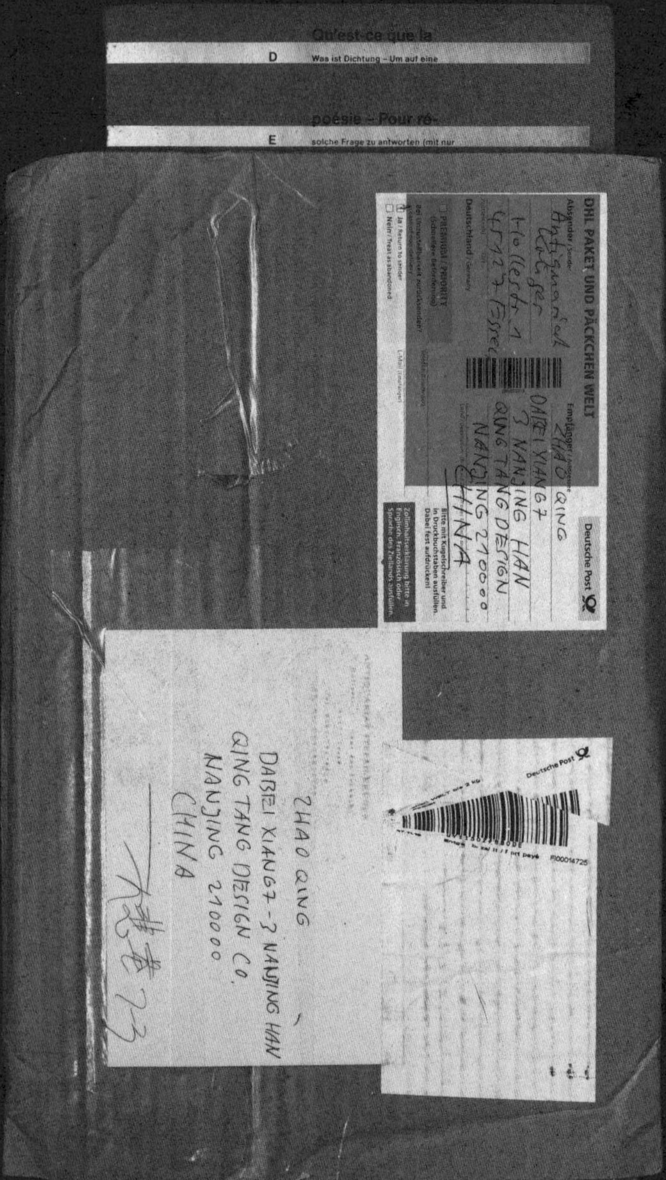

Qu'est-ce que la

D Was ist Dichtung – Um auf eine

poésie – Pour ré-

E solche Frage zu antworten (mit nur

DHL PAKET UND PÄCKCHEN WELT

Absender / Sender

Antiquariat
Krüger
Hollestr. 1
45127 Essen

Postleitzahl Ort

Deutschland / Germany

☐ **PREMIUM / PRIORITY**
(Schnellere Beförderung)

Bei Unzustellbarkeit zurücksenden?
In case of non-delivery

(96)993019

Empfänger / Addressee

ZHAO
DABEI XIAN
3 NANJ
QING TAN
NANJI

Land / Destination (Bitte auf D

Telefon (Empfänger)

Germany
China
Germany

285mm

6mm

Qu'est-ce que la
D Was ist Dichtung – Um auf eine
poésie – Pour ré-
E solche Frage zu antworten (mit nur
pondre à une telle
R zwei Worten, nicht wahr) verlangt
question – en deux
R man von dir, daß du auf das Wissen
mots, n'est-ce pas-
I zu verzichten weißt, wie man es rich-
on te demande de
D tig tut, ohne es je zu vergessen: De-
savoir renoncer au
A mobilisiere die Kultur, doch verges-

D E R R I D A

180mm

● 铜奖
《 什么是诗歌?
◖ 德国

Bronze Medal
Was ist Dichtung?
Germany
Brinkmann & Bose
Jacques Derrida
Brinkmann & Bose, Berlin
285 180 6mm
197g
48p
ISBN-10 3-922660-46-0

这是法国解构主义大师、哲学家雅克·德里达（Jacques Derrida）对诗歌进行描述的小短文，比较特别的是：本书是法、德、意、英四种语言文字的版本。书籍的装订形式为锁线胶平装，护封灰色艺术卡纸烫银后烫黑，其中直接印在卡纸上的大字号文字是法文版的书名和简介，压印在烫银上的小字号文字是德文版的相同内容。内封白卡纸无印刷和工艺。内页胶版纸印单黑。书中并没有把四种文字并置在一起，也没有用不同文字重复四遍相同的内容。由于德里达的写作语言是法文，其他三种文字分别由不同的译者进行翻译，书中通过法德、法意、法英的三次对照进行了三次排版。法文采用较细字重但较大字号的文本进行排版，行与行之间穿插其他文种，设置为较粗字重但较小字号，以便区隔。法文始终保持在版面的相同位置，其他文种根据译制的长短进行折行或缩进，形成丰富的文本肌理，在版面形式上衬托出德里达对于诗歌定义的阐释。

This book is a collection of short essays describing poetry, written by French deconstructionism master and philosopher Jacques Derrida. In particular, it is a version of French, German, Italian and English. The book is perfect bound with thread sewing. The jacket is made of gray art cardboard with silver hot-stamping and black hot-stamping. The large size characters printed directly on the card paper are the title and brief introduction in French, and the small size characters printed on the silver are the same contents in German. No printing and special techniques are used on the inner cover made of white cardboard. The inner pages are made of offset paper printed in single black. The book neither juxtaposes the texts in four languages together, nor repeats the same content four times with different languages. Since Derrida's writing language is French, different translators translated the texts in other three languages, and the book was typeset three times through comparisons of French-German, French-Italian, and French-English. French texts are typeset with a finer weight but a larger font size, and texts in other languages are interspersed between lines. Thicker but smaller fonts are used to be distinguished. French is always kept in the same position in the layout. Other languages are lined or indented according to the length of the translation, forming rich textual textures, which complements Derrida's interpretation of poetry on the layout.

dissociations, l'histoire des transcendances),

schichte der Transzendenzen); bewirke in j

fais en sorte en tout cas que la provenance de la

dem Fall, daß die Herkunft der Markierung i

marque reste désormais introuvable ou mécon-

sich nicht ausfindig machen läßt. Versprich es: Sie soll sich entstelle

naissable. Promets le: qu'elle se défigure, trans-

eit schlagen (lassen), und zwar in ihrem Tragen, in ihrem *Hafen (en so*

figure ou indétermine en son *port,* et tu enten-

hl das Ufer der Abreise als auch den Referenten, dem die Übertragun

dras sous ce mot la rive du départ aussi bien que

ehre, trinke, verschlinge meinen Buchstaben (meinen Brief), trag

le référent vers lequel une translation se porte.

wie das Gesetz einer Schrift, die zu deinem Körper geworden ist: *d*

Mange, bois, avale ma lettre, porte la, transporte

er Weisung kann zunächst von der bloßen Möglichkeit des Todes ihr

la en toi, comme la loi d'une écriture devenue ton

Anregung erhalten: von der Gefahr, die ei

corps: *l'écriture en soi.* La ruse de l'injonction

Fahrzeug für jedes endliche Wesen bedeutet. D

peut d'abord se laisser inspirer par la simple

merkst, daß die Katastrophe naht. Dem Zug dan

possibilité de la mort, par le danger que fait

selber einbeschrieben, vom Herzen herrühren

courir un véhicule à tout être fini. Tu entends

löst der Wunsch des Sterblichen in dir die Bewegun

venir la catastrophe. Dès lors imprimé à même le

1. *L'économie de la mémoire*: un poème doi

des Gedächtnisses: Ein Gedicht muß kurz sein, elliptisch sei

bref, par vocation elliptique, quelle qu'e

wie sehr es sich auch objektiv oder dem Anschein nach in die Lä

l'étendue objective ou apparente. Docte

ußtes der *Verdichtung* und des Entzugs.

scient de la *Verdichtung* et du retrait.

t das Herz mitten in den Sätzen, die gefahrlos auf den Verkehrs

2. *Le cœur.* Non pas le cœur au milieu

und sich dabei in alle Sprachen übersetzen lassen. Nicht einfach

phrases qui circulent sans risque sur les é

rchive, der Gegenstand der verschiedenen Wissensbereiche u

geurs et s'y laissent traduire en toutes la

und der bio-ethisch-juridischen Diskurse. Vielleicht auch ni

Non pas simplement le cœur des archive

es Pascal, nicht einmal (aber dies ist weniger gewiß) jenes, das

diographiques, l'objet des savoirs ou des

Nein, eine Geschichte des ‹Herzens›, das der idiomatische Aus

niques, des philosophies et des discour

isch einhüllt, Herz und Ausdruck meiner Sprache oder einer a

éthico-juridiques. Peut-être pas le cœu

y heart) oder – Herz und Ausdruck noch einer anderen Sprache

Ecritures ou de Pascal, ni même, c'est r

kalb) – eine einzige Strecke mit mehreren Wegen (Stimmen).

sûr, celui que leur préfère Heidegger. Non

e einer bestimmten Äußerlichkeit des

cœur› se confie comme une prière, c'est plu

n – so ist es sicherer –; es vertraut sich

une certaine extériorité de l'automate, aux

n der Mnemotechnik an; jener Litur-

e la mnémotechnique, à cette liturgie qui n

erflächlich betrachtet die Mechanik

n surface la mécanique, à l'automobile qui

em Automobil, das deine Leidenschaft,

rend ta passion et vient sur toi comme du

n überrascht und wie aus dem Außen über dich he

rs: *auswendig*, ‹par cœur› en allemand.

: Das Herz schlägt dir, Geburt des Rhythmus, jens

onc: le cœur te bat, naissance du rythme

, der bewußten Vorstellung und des verlassenen

elà des oppositions, du dedans et du de

schen den P

e la représentation consciente et de l'arc

Gegenwart

bandonnée. Un cœur là-bas, entre les sent

Qu'est-ce que la **poésie**?

Pour répondre à une telle question – *en deu*

Um auf eine solche Frage zu antworten *(mit –*

mots, n'est-ce pas? – on te demande de savo

***nur – zwei Worten, nicht wahr?),* verlangt man von**

renoncer au savoir. Et de bien le savoir, sans

dir, daß du auf das Wissen zu verzichten weißt.

mais l'oublier: démobilise la culture mais

Und daß du weißt, wie man es richtig tut, ohne es

que tu sacrifies en route, en traversant la rou

je zu vergessen: Demobilisiere die Kultur, doch

ne l'oublie jamais dans ta docte ignorance.

vergesse in deinem weisen, wissenden Unwissen

Qui ose me demander cela? Même s'il n'en para

ndre à une telle question – *en deux*

Per rispondere a un simile interrogativo – *i*

ce pas? – on te demande de savoir

***parole, no?* – ti si chiede di saper rinuncia**

u savoir. Et de bien le savoir, sans ja-

sapere. E di saperlo bene, di non scordartelo

lier: démobilise la culture mais ce

smobilita la cultura, ma non scordarti mai,

ifies en route, en traversant la route,

tua dotta ignoranza, quello che sacrifich

jamais dans ta docte ignorance.

strada, attraversando la strada.

demander cela? Même s'il n'en paraît

Chi osa chiedermi una cosa simile? Anche se

sparaître est sa loi, la réponse *se voit*

pare – infatti la sua legge è sparire – quell

uis *une* dictée, prononce la poésie,

rispondo si vede *dettato.* Sono *un* dettato,

oi par cœur, recopie, veille et garde

Brinkmann & Bose
Qu'est-ce que la poésie?
Che chos' è la poesia?
übersetzt von Maurizio Ferraris
What is poetry?
übersetzt von Peggy Kamuf
Was ist Dichtung?
übersetzt von Alexander García Düttmann
W
as
ist
Dich
tung
Jacques

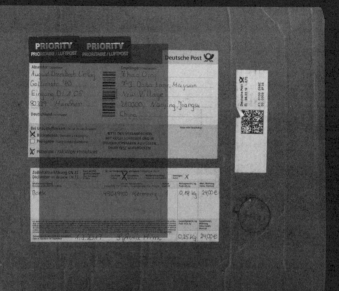

PRIORITY
PRIORITAIRE / LUFTPOST
PRIORITY
PRIORITAIRE / LUFTPOST
Deutsche Post
Absender
Empfänger
New Village
China
Deutschland
Rücksenden
Preisgabe
PREMIUM / PAR AVION PRIORITAIRE
BITTE DEN VERSANDSCHEIN MIT KUGELSCHREIBER UND IN DRUCKBUCHSTABEN AUSFÜLLEN, DANN FEST AUFDRÜCKEN!
Zollinhaltserklärung CN 22
Déclaration en douane CN 22
Book
0,18 kg
24,00 €
0,25 kg
24,00 €

RIORITY PRIORITY
RITAIRE / LUFTPOST PRIORITAIRE / LUFTPOST

nder / Expéditeur
gust Dreesbach Verlag
Lierstr. 70
gang D1 1.06
39 München
zahl Ort
schland / Allemagne

(96)9993019

Empfänger / Destinataire
Zhao Qing
7-3, Dabei Lan
New Village
210000 - Nanj
China
Land / Pays de destination

nzustellbarkeit / En cas de non-livraison
ücksenden / Renvoyer à l'expéditeur
reisgabe / Traiter comme abandonné
REMIUM / PAR AVION PRIORITAIRE

BITTE DEN VERSANDSCHEIN
MIT KUGELSCHREIBER UND IN
DRUCKBUCHSTABEN AUSFÜLLEN.
DABEI FEST AUFDRÜCKEN!

inhaltserklärung CN 22
aration en douane CN 22

Kann amtlich geöffnet werden
Peut être ouvert d'office

Art der Sendung (bitte ankreuzen) / Catégorie de l'envoi
Geschenk / Cadeau
X Dokumente / Documents
Warenrücksendung / Retour de marchandise

beschreibung
tion détaillée du contenu
Nur bei Handelswaren:
Zolltarif-Nr. / N° tarifaire
Ursprungsland / Pays d'origine
Me
Qua

ck 49019900 Germany

und Unterschrift des Absenders

271mm

18mm

254mm

◎ 铜奖

《 建筑师维姆·奎斯特
——荷兰建筑师专论

◐ 荷兰

Bronze Medal

Wim Quist, architect:
Monografie van Nederlandse architecten

The Netherlands

Reynoud Homan

Auke van der Woud

Uitgeverij 010, Rotterdam

271 254 18mm

957g

144p

ISBN-10 90-6450-075-4

荷兰建筑师专论丛书，旨在记录重要的荷兰设计师及其令人印象深刻的作品。本书是丛书中的一本（丛书中的另一本获得 1992 年荣誉奖，可在相关页面查找），对荷兰设计师维姆·奎斯特（Wim Quist）的职业生涯、作品和理念进行了全面展示。书籍的装订形式为锁线硬精装。护封铜版纸浅灰、灰蓝、黑三色印刷，覆哑膜。内封黑色布面裱覆卡板，书脊处烫黑处理。黑卡纸作为环衬，内页混合使用哑面铜版纸、黄色和白色胶版纸，单黑印刷。书中展示了奎斯特在荷兰设计的 18 栋建筑，包含饮用水公司、图书馆、水塔、博物馆、剧院、消防局、造船厂等多种门类。通过草图、平立剖面图和建成照片来全面展现建筑细节。荷英双语在书中并未对照排版，哑面铜版纸上为荷兰文，正文设定为单栏，侧面有较宽的注释栏。而英文被统一放置在书中两段黄色的胶版纸上，设置为双栏，并附带较窄的注释栏。通过纸张和栏位设置的差异加以区分。

The Dutch Architects Monograph Series aims to record important Dutch designers and their impressive works. This book is one of the series (another of the series won the 1992 Honor Award, which can be found on the relevant page), which comprehensively displays the career, works and ideas of Wim Quist, a Dutch designer. The hardcover book is bound with thread sewing. The matted jacket is printed in light gray, gray blue and black on coated paper. The inner cover is made of black cloth mounted with cardboard, and the spine of the book is black hot-stamped. Black cardboard is used for the end paper, while the inner pages are mixed with matte coated paper, yellow and white offset paper, which are printed in single black. The book shows 18 buildings designed by Quist in the Netherlands, including drinking water companies, libraries, water towers, museums, theatres, fire stations, shipyards and other categories. The architectural details are fully presented through sketches, horizontal and vertical section and completed photos. The bilingual contents in Dutch and English are not typeset in contrast. The Dutch texts are printed on the matte coated paper, set to a single column with a rather wide comment column on the side. The English texts are placed on yellow offset paper in the book, set as double columns, with a narrow comment column. Bilingual contents are distinguished by differences in paper and field settings.

Wim Quist
ARCHITECT
Auke van der Woud | Fotografie: Kim Zwarts
UITGEVERIJ 010 ROTTERDAM

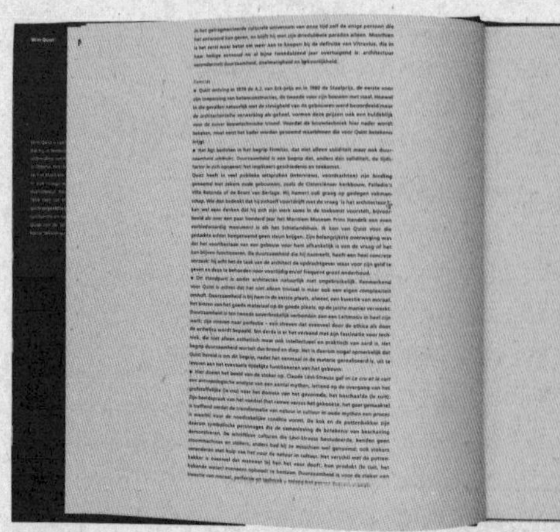

The art of building

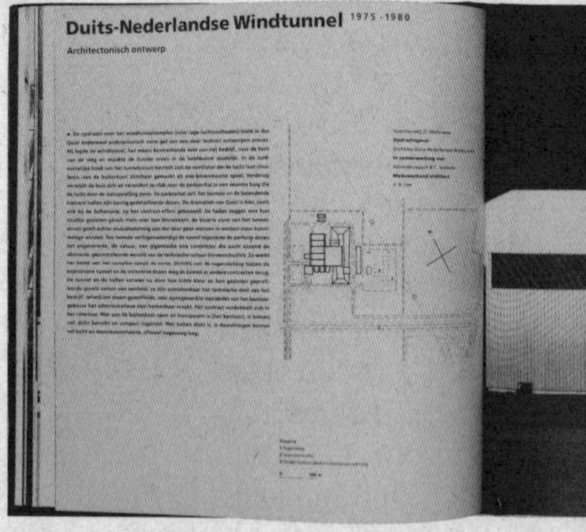
Duits-Nederlandse Windtunnel 1975-1980
Architectonisch ontwerp

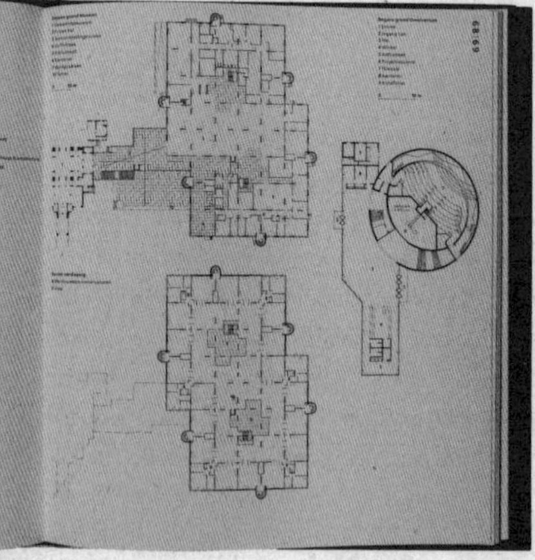
68/69

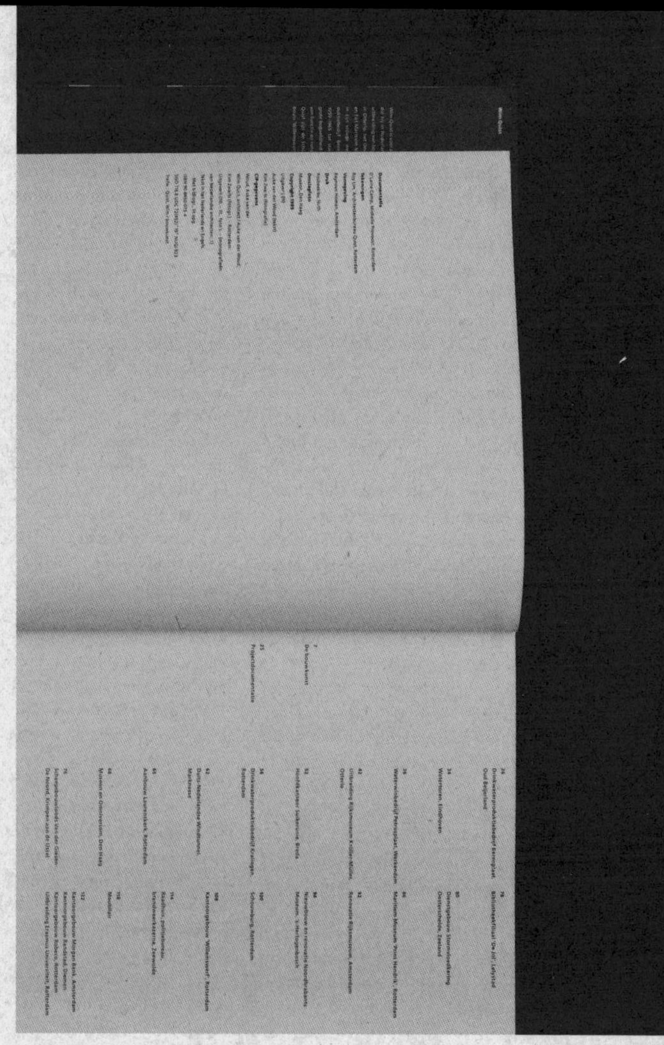

n practical formulas with the results guaranteed. This ates research and its outcome is unpredictable. The , sometimes undergoes a like transformation: 'build- rchitecture'. This process can be subdivided into the rmulated by Vitruvius in 125 B.C.: Firmitas (solidity), y), and Venustas (charm), together the three basic cture as well as a metaphor for the categories of the ssion. Firmitas is the realm of physics; Commoditas n an understanding of the laws of nature, but on an well. Venustas requires besides such knowledge and eloped taste, wit and liveliness. How is it, then, that where the public's conception of architecture has no art with, discussion is never about the nature and se elements, but tends to concentrate on just one of le question might be: how and when does Firmitas – life through Commoditas, become Venustas – spirit? answer is a principal issue in the work of Quist. It tracted by nature, and extremely demanding too; it signment that is neither far-fetched nor fortuitous, emotional side of the personality. From this source ain the most marvellous things, such as the typically Architecture (like art) is vulnerable; it redefines the paths. And architecture (like art and quality) has no s'[1].

's hypothesis has a less general side to it too, one ove paradox into three layers. Thus it is also saying: erable, and I, Willem Gerhard Quist, am in principle hat I want. Then again architecture has no place for ust obey it. The third layer reads: I, Quist, am vulner- antly redefining the task and looking for new paths ill further from home. Yet I accept no concessions; in cture I am, if needs be, as tough as they come. It has nd justifiably, that Quist is not building architecture, s[2]. Indeed, architecture as 'a thing in itself' is some- ide his world. His works, on the other hand, are as ch commission means one more lonely mental expe- hts. Thus the work created over the years makes up ut the ravines as well as the path upward. The term high-flown, yet at times is more apt than simply 'a it shows that the work's essentials are not achieved , but that life as a whole has to be involved. It was in

mentioned his attraction tc cian churches, Palladio's Vill is also fond of stressing ou Quist is spurred on by the c him imagining on occasion instance when in a few cent ed a monument as the Schi could get no support what was that the survival of a functioning. His quest for d the architect's job to give his from premature and/or frec

■ Of course this attitude is tic of Quist, however, is th complexity of its own. For l question of morality; choos with the right application. the leitmotif running thro determined as much by eth up with his fascination with aesthetic but intellectual a has roots both deep and w once this concept has bee abandon it in the case of bu

■ At this point there loom *cuit* Claude Lévi-Strauss g myths, paying attention to realm of the fashioned, the (raw versus cooked) is apt, in ancient myths is a proce Thus the cook and the pott to society the meaning of c Strauss studied had no stea have mentioned them. Stok Where they differ from pot guished, the product (le cu stoker, durability is a ques for as long as the process r

■ In Quist's biography to and an educational process

rworks 1969 - 1974

r enormous reservoirs where for the purified 'naturally', before being d Kralingen works. The programme ly modest in comparison with the , offices and laboratories, and tech- side offered no real challenge to the intellectual radicalism that for this ximum independence. Petrusplaat is nical side is almost hidden away, as h in general – not for nothing did d laboratory building a 'cocoon'. The ke an unearthed prehistoric settle- ta dominate, for the reason that the . The buildings say nothing at all of

52
Hoofdkantoor Suikerunie, Br

56
Drinkwaterproduktiebedrijf K
Rotterdam

62
Duits-Nederlandse Windtunn
Marknesse

w Morgan Bank

w Randstad 1987 - 1990

w Robeco 1987 - 1991

smus Universiteit 19

oek in aanbouw. Het kantoor
grijze geglazuurde baksteen
aan de achterkant verspringt
e de gemeente op het terrein
cht voor de liften en trappen

8-111
aan, H. de, I. Haagsma, 'W.G. Quist: "Het at erom de essentie van de opgave te nden"', *Intermediair* 1979 nr. 45, 13-17
onstra, R., 'Rijksbouwmeester Quist: "We tten op een doodlopende weg"', *Elseviers agazine* 15-9-1979, 118-119
ouwers, R., 'Wim Quist vertrekt als rijks- uwmeester met een zeker cynisme: "Bij chitecten en hun critici heerst gebrek aan erpe analyse"', *Wonen-TABK* 1979 nr. 23, 12
jk, H. van, 'Gesloten dozen vol emoties', *RC-Handelsblad* 2-3-1984
oon, B. (red.), 'Zulk grijs kunnen alleen ilders maken', *De Tijd* 22-6-1984
rwindt, E., 'Kunstenaar mag best corrigeren treden', *de Architect* 1984 nr. 11, 66-71
yper, R., 'Architect zonder kapsones', *Dordtenaar* 15-12-1984
gesprek met Wim G. Quist', *Bulletin van he ksmuseum* 1985 nr. 3, 160-175

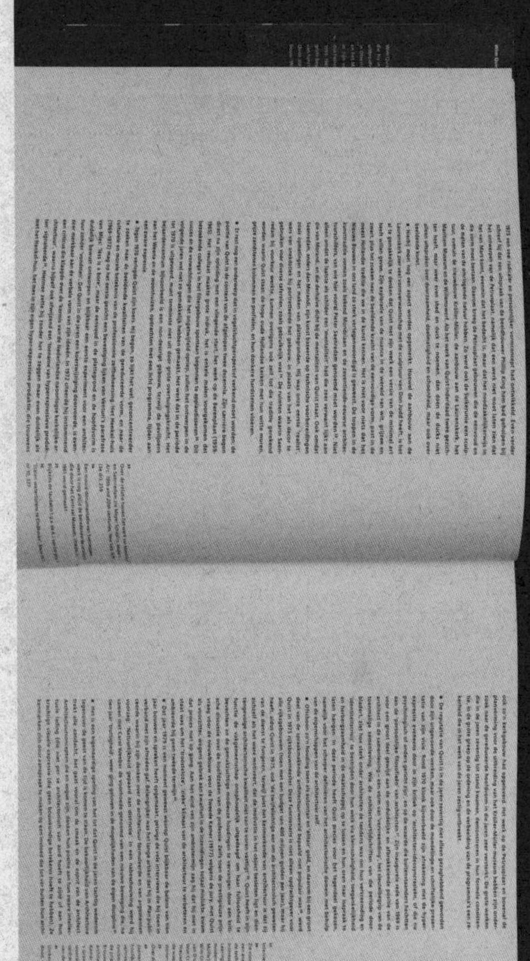

0101

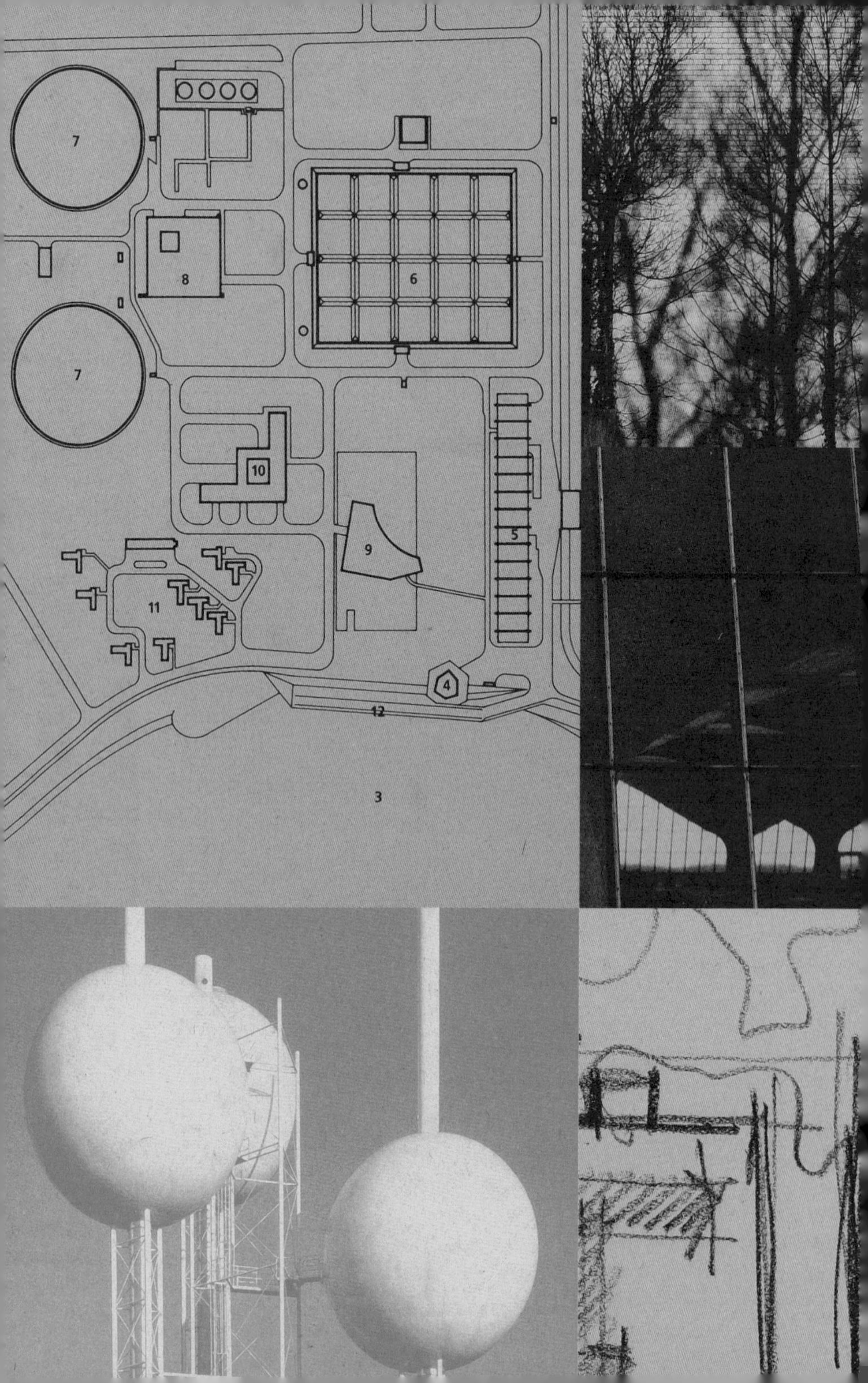
7
6
8
7
10
5
9
11
4
12
3

1991 Ha1

Qing Zhao
7-3, Dabei Lane, Meiyuan New Village
210000 Nanjing
čína
ČESKÁ LÍPA
CS 00 031 494 5 CZ
ADDRESS LABEL / ADRESNÍ ŠTÍTEK
12
ALBATROS

Czech Republic

Czech Republic　　China

295mm

32mm

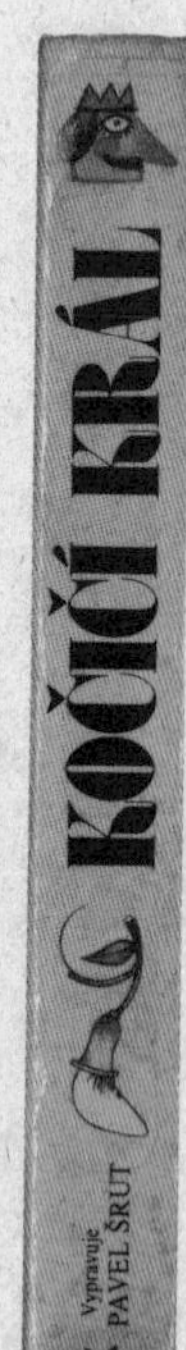

210mm

荣誉奖
猫王
捷克

Honorary Appreciation
Ko i í král
Czech Republic
Milan Grygar
Pavel Šrut
Albatros, Prag
295 210 32mm
1416g
352p

捷克作家、诗人帕维尔·什鲁特（Pavel Šrut）除了自身的作品外，还进行大量翻译和出版工作，包括出版儿童书籍。本书是他整理和用捷克语讲述的源于英语、爱尔兰语、苏格兰语和威尔士语的童话故事。书籍的装订形式为锁线硬精装。护封哑面铜版纸四色印刷，覆亮膜。内封白色布面烫黑处理，裱覆卡板。内页胶版纸四色印刷。书中包含来自不列颠和爱尔兰岛上的四种语言的短篇童话共 64 篇，书名《猫王》是其中的一篇。作为成年人向儿童讲述的故事，什鲁特在内容的筛选和表达上做了很多思考。书籍的正文采用较大的字号，单栏排版，便于家长捧着给孩子做伴读。书籍的插图由捷克艺术家、儿童作家克维塔·帕佐夫斯卡（Květa Pacovská）绘制。插图风格古灵精怪，充满大胆的造型和色彩，与这些独特的童话故事搭配得当。帕佐夫斯卡绘制和设计的书籍还获得过 1993 年金字符奖和 1991 年铜奖（可在相关页面查找）。

In addition to creating his own works, the Czech writer and poet Pavel Šrut also carried out a lot of translation and publishing work, including the publication of children's books. This book is a collection of fairy tales in English, Irish, Scots and Welsh that he organized and retold in Czech. The hardcover book is bound by thread sewing. The laminated jacket is made of matte coated paper printed in four colors and the inside cover is made of white cloth treated with black hot stamping, mounted with cardboard. The inside pages are printed on offset paper in four colors. The book contains 64 short stories in four languages from the islands of Britain and Ireland, and the title *Kočičí krális* is one of them. As an adult, when Šrut told stories to children, he thought a lot about the selection and expression of the content. The text of the book uses large font size and single column typesetting, which is convenient for parents to read with their children. The illustrations of the book were drawn by Květa Pacovská, a Czech artist and Children's literature writer. Both the quirky illustrations and the bold shapes and colors perfectly match the fairy tales. The books drawn and designed by Pacovská also won the 1993 Golden Character Award and the 1991 Honor Award (available on the relevant pages).

tíř Wynd podruhé poslechl. Odpásal
políbil. Pak odstoupil a čekal. A saň
la potřetí:

„Neváhej, už zhasne den,
bratře můj milený,
a moc kletby zlomí jen
trojí políbení.“

tíř Wynd poslechl i potřetí. Přist
a políbil ji. V tu chvíli se zvedl vítr
n zahřmělo, strašná saň zmizela a
lem stála jeho sestra Markétka. Bratr
do svého pláště. Jednou rukou tiskl s
aň a v druhé držel zelenou haluzku z
jasanu.

yž vstoupili do zámku, Wynd se otá

„Teď mi ale odpověz,
sestro moje milá,
jaká kletba, jaká lest
tvůj žal způsobila?“

íč poznáš po
si líný ani hlo
ím třikrát v
jako myšk
tiše

OČ

v anglick
otských

*k abyste věd
tohle už číst
nemusíte,
leda někdy
v neděli,
až budete*

DOSPĚLÍ

Na návsi cvrnkali kluci třešňové pecky do důlků. Tóma hned hru pochytil, a když mu kluci pár pecek půjčili, začal cvrnkat s nimi.

Sám byl velký jako palec, tak si dovedete představit, jak asi mrňavý měl paleček. Není divu, že brzo všechny své pecky prohrál. Byl to ale klouček mazaný. Potají vlezl hned jednomu, hned druhému klukovi do kapsy, nabral do hrstě pár pecek a zase potají seskočil na zem. A tak pořád prohrával, ale pořád měl s čím hrát.

Klukům to bylo divné, až jeden načapal Tó-

„Tak takhle je to, ty mrňousku!" rozzlobil se. „Počkej, co ti udělám!"

Zadrhl šňůru pytlíku a tak dlouho s ním natřásal, dokud chudák Tóma nebyl od pecek samá modřina. Pak teprve ho pustil domů.

Matka už prosmýčila všechny kouty a prohledala celý dvůr. Měla o chlapce strach, a když se vrátil, ze samé radosti zadělala na makový koláč.

Těsto nechala v míse vykynout a šla semlít mák. Jenže Paleček byl stejně zvědavý jako mlsný. Vydrápal se do mísy, a když se natáhl, aby těsto ochutnal, bác! zapadl po hlavě do těsta. A jak těsto kynulo a bublalo, nikdo neslyšel jeho volání o pomoc.

Kamna už si zpívala, a tak matka vyklopila těsto do pekáče a pekáč strčila do trouby. Za chvíli byl upečený. Ale když pekáč vyklopila na stůl a ohlížela se po Palečkovi, aby přišel ochutnat, koláč začal pištět a naříkat.

Matka se vylekala a v tom úleku koláč popadla a vyhodila oknem na pěšinu. Po pěšině šel právě hladový švec. Mrknul očkem napravo i nalevo, a když nikoho neviděl, rychle koláč popadl a strčil do vaku mezi děravé střevice, které nesl domů ke správce.

Jenže koláč začal v ševcově vaku pištět a naříkat. Švec se vylekal, popadl ten zpropadený koláč, hodil jej za hlavu a utekl.

Koláč dopadl na kámen a rozlomil se. Paleček vyskočil a celý špinavý od máku a od dro-

Jednoho rána otevřel okno, chytil se lodyhy a začal šplhat vysoko, vysoko, a když se mu ukázala oblačná cesta, rozběhl se k nebeskému stavení. Ve vratech stála stařena. Slovem Janka nepřivítala, zato Janek, sotva pozdravil, požádal o snídani.

„Snídani bys chtěl?" rozzlobila se stařena. „Radši se schovej za pec. Tamhle jde můj muž a bude chtít počítat zlatá vejce."

Janek se schoval a obr vrazil do síně. Zvedal poklice z obrovských hrnců, lovil v hrncích obrovskou naběračkou, ale pořád čenichal kolem a něco se mu nezdálo. Obrátil se na stařenu a spustil:

„Jed bych, ženo, dneska za tři
a pak bych pil jako bečka!
Čichám, čichám mládenečka –
ať je kde je, tak mi patří!"

„Ale muži," vylekala se stařena, „nikdo tu není. To jen větřík divně fouká!"

„Dobrá, dobrá," řekl obr a mrzutě dosedl na obří lavici. „Přines mi ošatku a zlatou slepici. Ať počítám jak počítám, jedno zlaté vejce mi schází."

Stařena přinesla ošatku a zlatou slepici a vrátila se ke svým hrncům. Obr držel slepici na klíně a počítal vejce na ošatce. Ale za chvíli se mu z toho počítání začaly klížit oči. A když mu padla hlava na prsa, vyskočil Janek zpoza pece.

„Kokokodák!" vylekala se slepice a rozběhla se po nebi. Janek za ní a za Jankem obr. Slepice si našla díru v mracích, Janek se chytil vršíčku fazole a obr zůstal vzadu. A když Janek sešplhal na zahrádku, slepice už tu pokvokávala.

To vám byla podívaná! Kdo šel kolem, musel si oči zaclonit, taková záře ze zahrádky vycházela. V chaloupce bylo veselo, stará matka se usmívala a Janek si pískal.

EMS 国際スピード郵便
物品用 (For goods)
JAPAN
EG688836640JP
FROM(ご依頼主)
Name EAST-WEST SCHOLARS
Address 4-14-10 NEGISHI
TAITO-KU
TOKYO
Postal code 110-0003
TO(お届け先)
Name ZHAO QING
Address 7-3, DABEI LANE
MEIYUAN NEW VILLAGE
NANJING
210000
CHINA
CUSTOMS DECLARATION CN23
特別定価提供期限'90年11月30日　定価7000円(本体6796円)
岩波書店

HOLARS

SHI

JAPAN

Fax No.

12 TO（お届け先）

Name ZHAO QI

Address 7-3, DA

MEIYUAN

都市名 City NANJIN

国名 Country CHINA

業物品のみ記入
ommercial items only

の原産国
origin of the goods

内容品の個数
Number of items contained

25 正味重量
Net weight
kg g

24 内容品の価
Value

50,000

6 この郵便物は
Number of this pieces

番目
1 / 1 個中
Total number of pieces

L* EG68883

・現地で郵便物が損傷して到着した場合は、
Please contact your nearest deliver

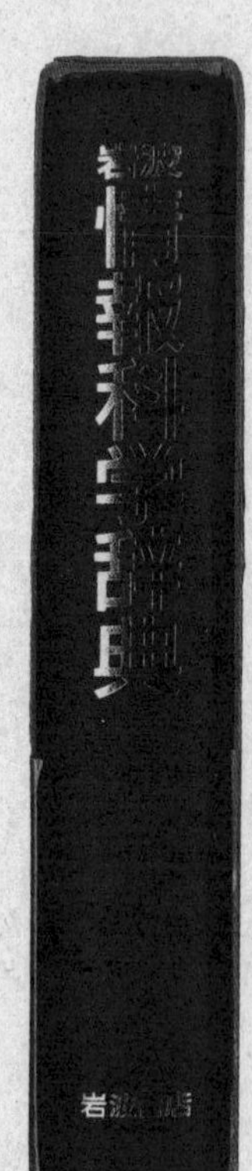

● 荣誉奖
《 岩波情报科学辞典
◖ 日本

Honorary Appreciation ◎
Dictionary of Iwanami Information Science 《
Japan
Iwanami Shoten, Publishers, Tokyo
Makoto Nagao △
Iwanami Shoten, Publishers, Tokyo
231 170 43mm
1388g
1172p
ISBN-10 4-00-080074-4

岩波书店是一家颇有历史的日本出版社，主要的出版方向为学术研究和经典作品，并自编了大量文库丛书，本书便是岩波书店自编的“情报学”（信息学）学术辞典。书籍的装订方式为圆脊锁线硬精装。函套黑灰双色印刷，覆亮膜，裱覆薄卡板，函套上包裹的腰封采用白色铜版纸印暗红色。封面采用深蓝色仿布纹塑料材质，裱覆薄卡板，书脊处文字烫金处理。内页字典纸印单黑，只在附录处夹插一张光面铜版纸四色印刷的说明图。全书将“情报学”的词汇按照语音进行编排，但提供了逻辑树方便检索，包括五大方面：数学模型为主的基础理论、电脑等现代计算设备的构造和原理、计算机语言和基本算法、人工智能和机器学习、相关名人和历史。书最后的附录部分约占全书页面的十分之三，呈现各种图表和内容之间的错综关联，意图将分散的词条共同组成完整的知识体系，也提供相关页码方便检索。正文部分双栏排版，词条名黑体加粗，词条最后为所属分类编号，与开篇的逻辑树形成检索回路。

Iwanami Shotenis a Japanese Publishing House with a long history, which mainly publishes academic research and classic works and compiles a large number of book series. This book is the academic dictionary of Information Science compiled by Iwanami Shotenis. This hardback book is bound by thread sewing with rounding spine. The bookcase is printed in black and gray, covered with bright film, mounted with thin cardboard. The wraparound band covering on the case is printed in dark red on white coated paper. The cover is made of dark blue cloth-like plastic material mounted with thin cardboard, and the texts on the spine are bronze stamped. The inner pages are using dictionary paper printed in black, with only one page printed with four-color illustration on glossy coated paper. The book arranges the vocabulary of Information Science according to speech, but provides a logic tree for convenient retrieval, including five aspects: basic theory based on mathematical model, construction and principles of modern computing equipment such as computer, computer language and basic algorithm, artificial intelligence and machine learning, related celebrities and history. The appendix at the end of the book accounts for about three tenths of the entire pages of the book. It presents the intricate connection between various charts and contents, with the intention of combining the scattered entries into a complete knowledge system and providing relevant pages for easy retrieval. The body part is double-columned, the entry name is bold, and the category number is listed at the classification number, thus forming a retrieval circuit with the opening logical tree.

岩波
情報科学辞典
●編集――長尾真 石田晴久
稲垣康善 田中英彦 辻井潤一
所真理雄 中田育男 米澤明憲

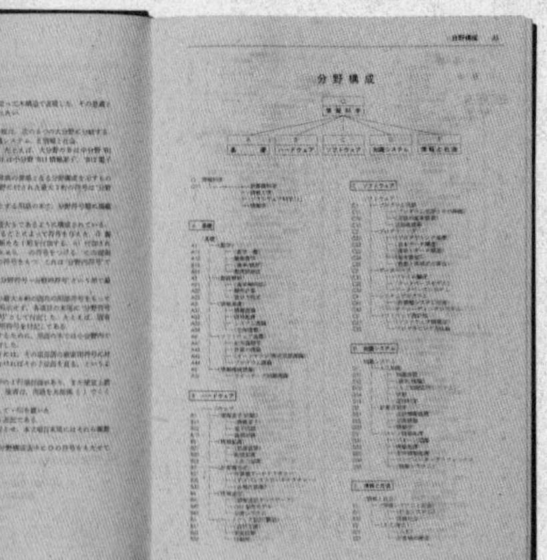

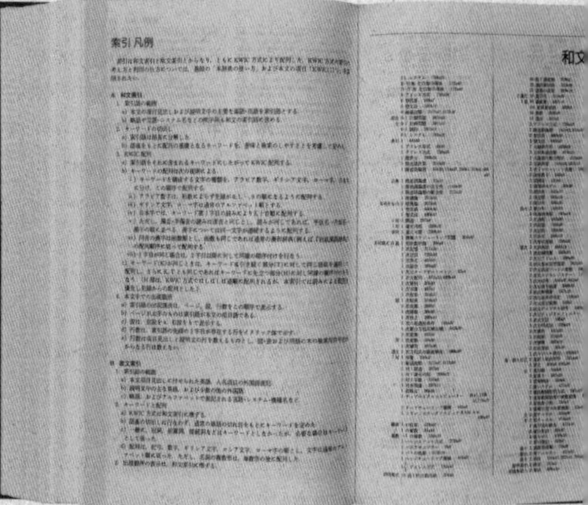

記憶し利用するための†データベース技術，目的とする仕事を記述する†プログラム言語など多くのものがある．ソフトウェアの設計は目的をどのような構造でとらえるかが基礎となる．情報は伝達されることによって価値を生むので，†通信システムのディジタル化と計算機との結合に基づく広範囲な地域をおおう統合された情報システムが形成されつつある(→計算機ネットワーク)．

情報科学では従来の哲学的立場からではなく，数学基礎論，†記号論理学に立脚して†アルゴリズムに関する理論を展開し，自然科学的方法論を用いてその理論を各種のソフトウェアに具体化している．情報科学が単に自然科学だけでなく，人文･社会科学や文学など，あらゆる分野に適用できるのは，このような立場で物事の抽象的側面を扱う科学だからである．情報科学の対象の中心はシンボルの世界であるが，一方では生体における情報の授受，伝達，処理という分野も包含する(→生体工学)．そこではアナログ量を扱い，フィードバックシステム，制御システムという形での情報処理機構が研究の対象となる．計算機に代表される離散的情報を扱うシステムと生体に代表される連続的情報を扱うシステムとの統合を実現する方法論が今後の情報科学の大きな課題である．

情報科学の起源はウィーナーの†サイバネティクスであるといわれるが，情報を具体的な計量の対象として取り扱えることを示したのはシャノンである．しかし情報科学という広範な学問分野を包含する体系が作られる基礎となったのは，1940年代に研究･開発が始まった†ディジタル計算機であり，フォン･ノイマンの†プログラム内蔵方式である．†半導体の発見，†トランジスターの発明，†集積回路への発展などにより可能となった計算機ハードウェア技術の上に，計算機を多くの利用者が効率良く使えるようにするため，ソフトウェアに多くの努力がそそがれ，一括処理システムから時分割システム，ネットワークによる†分散処理へと展開されてきた．これまではハードウェア設計の容易さという立場からだけ†計算機アーキテクチャーが決定されていたため，その使用目的との間に大きな隔たり(→セマンティックギャップ)が存在した．これを埋め，利用者の使いやすさを図るためのソフトウェア開発の費用は膨大になり，†ソフトウェア危機とよばれる事態を招来した．しかし，今日では利用者の使用目的に基づいてアルゴリズムを決定し，それに適した計算機アーキテクチャーを設計してセマンティックギャップを少なくし，さらにソフトウェア作成の費用をできるだけ減らすという方向の研究が盛んになってきた．

情報科学は社会に対して大きなインパクトを与えている．物質，エネルギーと並んで社会において情報が果す役割がますます大きくなり，†情報産業とよばれる巨大産業が創出され，†情報社会の到来が語られている．情報社会については，†情報格差，†コ

語は，computer science，あるいは computer and information science であり，information science はむしろ library and information science の意味で用いられる(→情報学)．ヨーロッパでは informatique，Informatik が普及している．→情報工学．　　カ

情報学　information science, informatics

情報の発生，収集，組織化，蓄積，検索，理解，伝達，活用などにかかわる本質･性質を究明し，かつそこで明らかにされた事項の社会的適応可能性を追究する学問．informatics(ドイツ語では Informatik)という用語はソ連のミハイロフ(Mikhaĭlov, A. I.)らが，科学情報の構造と属性，ならびに情報活動の一般的法則，歴史，方法，組織化を追究する領域に対して与えた言葉である．†情報科学が情報のもつ論理的･数学的構造，†アルゴリズムなどを研究対象とする学問であるのに対し，情報学は性質のちがった大量の情報をどのように扱うかに重きを置く．計算機による情報処理技術により情報学は急速に発展し，発生情報の収集，組織化，蓄積，伝達などは自動化されてきている．組織化された情報の活用では†情報検索技術が中心となる．言葉を媒介として伝えられる情報の内容は概念の結合によってとらえられるので，基本概念ならびに複合概念にどのような言葉を与えるか，また言葉はこれら概念の形成にどのような影響を与えるかを研究する†用語学も情報学の重要な分野である．文字で表現される情報だけでなく，音声，図形，画像，映像など種々の情報が情報科学的に取り扱われるようになってきて，情報学と情報科学との関係は密接になりつつある．図書館学との相補性を図りながら情報学の実践的側面を追究する分野は†ドキュメンテーションとよばれている．　D23-O114

情報格差　information gap

国家間で情報が偏在することをいう．情報産業の発達が遅れている発展途上国から進んでいる先進国に情報が一方的に流れて，情報流通の不均衡が生ずるために起こる．情報流通の手段としては従来から新聞，雑誌，テレビなどマスメディアが中心であった．最近は，さらに国際的な†計算機ネットワークを用いて情報流通が行なわれるようになり，情報格差は拡大する傾向にある．発展途上国は工業化の初期に経験した産業格差(industrial gap)に加え，経済活動上ならびに国家保全上重要な情報を先進国に押さえられることの脅威を感じている．情報武装化の進んだ企業と遅れている企業間の情報の偏在にも情報格差という用語を使うこともある．→TDF問題．　E12-44

情報革命　information revolution

高度に発達した情報通信システムを通じて情報の獲得が容易になり社会全体の仕組が大変革されることを表わす概念．このようなシステムは，パソコン

さまざまな利便がもたらされる．2) 生産現場や事務所での労働が質的に変化し，企業の組織や経営内容にも変化が生じる．3) 情報資源を生産，加工する産業が成長し，経済において第3次産業が大きな役割を果すようになり，産業の相対的比重に変化が起きる．第1次，第2次産業においても情報と結びついた計画生産が容易になり，省資源，省エネルギーがもたらされる．4) 国際間の連帯，協調が促進され，政治，文化，経済面での協力が円滑になる，などの変化が起こると考えられる．しかし，個人のストレスや不安の増大(→テクノストレス)，†プライバシー保護や情報の法的問題，情報格差による新世界秩序の形成など，社会科学，人文科学によって解決されるべき問題も多い．　E12-12

情報化ビル　information building

インテリジェントビルともいう．ビル内で働く人たちが快適な居住環境で，効率的に業務を遂行できるように，主にエレクトロニクス技術を用いて種々の設備やサービスを用意しているビル．情報化ビルは高度情報通信システム，ビルディングオートメーションシステムの2つの要素からなり，これらが統合された1つのシステムとして機能するときに最もその効果を発揮する．高度情報通信システムは，交換機，各種情報処理装置をネットワーク化し，ビル内部や外部との情報伝達や†オフィスオートメーションの機能を効率的に実現するシステムである．ビルディングオートメーションシステムは，空調，照明などの省エネルギー管理，防犯，防災監視などの保全，およびエレベーター群管理，駐車場管理などのビル管理の各システムを自動制御する機能をもつシステムである．　E11-31

情報基本法

情報処理システムの†プライバシー保護に関する憲法に相当する法律．プライバシー保護については，欧米諸国では1970年代から国や地方公共団体，企業などの情報処理システムに蓄積される個人情報を保護するための法律が制定されている．OECDでも，1980年9月"プライバシー保護と個人データの国際流通についてのガイドラインに関する理事会勧告"を採択し，加盟国に対してガイドラインの諸原則を国内法で考慮することを勧告している．わが国ではまだ法制化されていないが，行政管理庁が準備を進めており，5原則の柱を提案している(→プライバシー保護)．　E12-43

情報経済学　information economics

情報の生産，流通，利用に関する社会的効率性の追究を目的とした経済学の一分野．情報の経済的価値に着目して社会制度の設計や政策の判断のよりどころを与えようとしている点で，情報の価値には言及しない†情報工学とは立場を異にする．たとえば，

問題にする．情報を経済学の対象として最初にとらえたのはマッハルプ(Machlup, F.)の"The Production and Distribution of Knowledge in the United States"(1962)であり，それを情報経済学に発展させたのは，ポラト(Porat, M. U.)の"The Information Economy: Definition and Measurement"(1977)である．その後の情報産業の発展とともに，ますます注目を集めている．わが国でも，早くからその重要性は認識され，情報化施策や電気通信の高度化などの推進に大きな影響を与えている．　E12-13

情報源　information source

一般には，データ系列や信号，メッセージの発信源を指すが，本来は情報理論の用語として作られたものである．情報理論では，情報源は確率モデルで定義されるものと暗黙のうちに了解されている．すなわち，記号の離散的集合をアルファベットとよび，情報源はこのアルファベットの文字からなる系列を一定の速度で生成する†確率過程とみなす．この確率過程が定常であるとき，情報源は定常情報源(stationary source)であるといい，その確率構造のもとに†エントロピーが定義できる．これを定常情報源のエントロピーとよぶ．つまり，エントロピーは確率モデルのアンサンブルに対して定義されるものであり，情報理論ではエントロピーをもつ情報源だけを扱うのが普通である．定常情報源の中で，特に系列の生起が各時点で互いに独立でかつ分布が一定のとき，それを情報理論では無記憶情報源(memoryless source)という．この場合，確率モデルはアルファベット $A=\{a_1, \cdots, a_n\}$ と文字 a_i の確率分布 $P(a_i)=p_i$ によって規定できる．そのとき，その無記憶情報源のエントロピーは

$$H=\sum_{i=1}^{n}(-p_i \log_2 p_i)$$

で表わされる．$H \leq \log_2 n$ であり，無記憶情報源から生成される系列は1文字当りの長さがいくらでも H に近くなるような2進数の系列に変換できる．これを†情報源符号化定理という．同様なことが定常情報源に対しても成立することが証明されている．複数の情報源があり，これらが互いに確率的に相関するとき，相関性情報源(correlated source)という．この場合，それぞれに条件付きエントロピーが定義できる．そして，これらの条件付きエントロピーもそれぞれ符号化の限界を与える量として物理的に意味づけられることが示されている．→情報理論．　A31-11

情報検索　information retrieval

主として非数値データを対象として，そのデータあるいはその分析結果を蓄積しておき，要求に応じて必要な部分(情報)を効率良く取り出すことをいう．多くの場合蓄積対象のデータは文献であること

た こ さ ら ち く つ ， 、

て け せ は と き い ん

ひ す ふ へ め そ ね ほ ・

同一の手によるシフト状態

反対側の手によるシフト状態

合に準ずる(→有向木)．n 進

節点の個数は $h+1$ と $(n^{h+1}$

．n 進木をデータ構造として

れの節点に第 1 部分木，…，第

ターを設ければよい(図 2 a)．

a

1 部分木　第 2 部分木　第 3 部分木

a　1　3

第 1 部分木　第 3 部

図 2　3 進木の節点の表現

行列網モデ

現われる正

となる．ブ

する正規化

，積形式解

C41-23

ration

†メッセー

には，

合を用いる

ャネル(通信

定義する．

は送付先と

専門家システム　＝エキスパートシステム．

専門用語辞書

technical terminology dictionary

1 つの専門分野における概念を表現する専門を集め，文法情報や意味などを与えた辞書を一般的な語でも専門分野で使用されるときは分野で定義された明確な意味をもつ．例えば数学の用語では四則算法の成立する集合を意社会の進歩によって専門分野はますます細分て新しい概念が作られ，それに対応して専門与えねばならないので，造語規則を設けて，だけそれに従って専門用語を作ることが望ま用語の説明内容が専門分野に応じて詳しくな場合は事典(→辞書[1])となる．説明が非常な場合には，専門用語集または専門用語デー

—量子信号検出
—量子推定理論
—量子情報理論

システム理論

システム理論
—線形システム理論
——線形動的シス
——状態変数
——安定性[2]
—可制御性
—可観測性
—入出力表現
——インパルス応
——伝達関数
——ラプラス変換
—標準構造
—最小実現
——マクミラン次

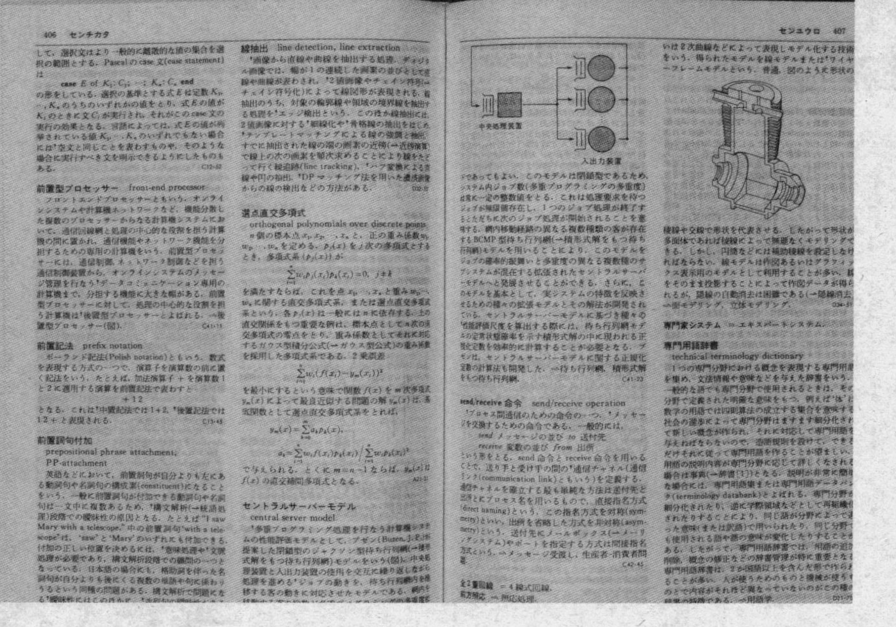

して，選択文はより一般的に離散的な値の集合を選択の範囲とする．Pascal の case 文(case statement)は

case E of K_1: C_1; …; K_n: C_n **end**

の形をしている．選択の基準とする式 E は定数 K_1, …, K_n のうちのいずれかの値をとり，式 E の値が K_i のときに文 C_i が実行され，それがこの case 文の実行の効果となる．言語によっては，式 E の値が列挙されている値 K_1, …, K_n のいずれでもない場合には†空文と同じことを表わすものや，そのような場合に実行すべき文を明示できるようにしたものもある． C12-62

前置型プロセッサー front-end processor

フロントエンドプロセッサーともいう．オンラインシステムや計算機ネットワークなど，機能分散した複数のプロセッサーからなる計算機システムにおいて，通信回線網と処理の中心的な役割を担う計算機の間に置かれ，通信機能やネットワーク機能を分担するための専用の計算機をいう．前置型プロセッサーには，通信制御，ネットワーク制御などを担う通信制御装置から，オンラインシステムのメッセージ管理を行なう†データコミュニケーション専用の計算機まで，分担する機能に大きな幅がある．前置型プロセッサーに対して，処理の中心的な役割を担う計算機は†後置型プロセッサーとよばれる．→後置型プロセッサー(図)． C41-11

前置記法 prefix notation

ポーランド記法(Polish notation)ともいう．数式を表現する方式の一つで，演算子を演算数の前に置く記法をいう．たとえば，加法演算子 + を演算数 1 と 2 に適用する演算を前置記法で表わすと

+ 1 2

となる．これは†中置記法では 1+2，†後置記法では 1 2 + と表現される． C13-45

前置詞句付加

prepositional phrase attachment,

PP-attachment

英語などにおいて，前置詞句が自分よりも左にある動詞句や名詞句の構成素(constituent)になることをいう．一般に前置詞句が付加できる動詞句や名詞句は一文中に複数あるため，†構文解析(→統語処理)段階での曖昧性の原因となる．たとえば "I saw Mary with a telescope." 中の前置詞句 'with a telescope' は，'saw' と 'Mary' のいずれにも付加できる．付加の正しい位置を決めるには，†意味処理や†文脈処理が必要であり，構文解析段階での難問の一つとなっている．日本語の場合にも，格助詞を伴った名詞句が自分よりも後にくる複数の述語や句に係わりうるという同種の問題がある．構文解析で問題になる†曖昧性にはこの

線抽出 line detection, line extraction

†画像から直線や曲線を抽出する処理．ディジタル画像では，幅が 1 の連続した画素の並びとして直線や曲線が表わされ，†2 値画像やチェイン符号(→チェイン符号化)によって線図形が表現される．線抽出のうち，対象の輪郭線や領域の境界線を抽出する処理を†エッジ検出という．この他か線抽出には，2 値画像に対する†細線化や†骨格線の抽出をはじめ，†テンプレートマッチングによる線の強調と抽出，すでに抽出された線の端の画素の近傍(→近傍演算)で線上の次の画素を順次求めることにより線をたどって行く線追跡(line tracking)，†ハフ変換による直線や円の抽出，†DP マッチング法を用いた濃淡画像からの線の検出などの方法がある． D32-8

選点直交多項式

orthogonal polynomials over discrete points

n 個の標本点 $x_1, x_2, \cdots, x_n$ と，正の重み係数 $w_1, w_2, \cdots, w_n$ を定める．$p_j(x)$ を j 次の多項式とするとき，多項式系 $\{p_j(x)\}$ が

$$\sum_{i=1}^{n} w_i p_j(x_i) p_k(x_i) = 0, \quad j \neq k$$

を満たすならば，これを点 $x_1, \cdots, x_n$ と重み $w_1, \cdots, w_n$ に関する直交多項式系，または選点直交多項式系という．各 $p_j(x)$ は一般には n に依存する．上の直交関係をもつ重要な例は，標本点として n 次の直交多項式の零点をとり，重み係数としてそれに対応するガウス型積分公式(→ガウス型公式)の重み係数を採用した多項式系である．2 乗誤差

$$\sum_{i=1}^{n} w_i (f(x_i) - y_m(x_i))^2$$

を最小にするという意味で関数 $f(x)$ を m 次多項式 $y_m(x)$ によって最良近似する問題の解 $y_m(x)$ は，基底関数として選点直交多項式系をとれば，

$$y_m(x) = \sum_{k=0}^{m} a_k p_k(x),$$

$$a_k = \sum_{i=1}^{n} w_i f(x_i) p_k(x_i) \Big/ \sum_{i=1}^{n} w_i p_k(x_i)^2$$

で与えられる．とくに $m = n-1$ ならば，$y_m(x)$ は $f(x)$ の直交補間多項式となる． A22-21

セントラルサーバーモデル

central server model

†多重プログラミング処理を行なう計算機システムの性能評価モデルとして，ブゼン(Buzen, J. P.)が提案した閉鎖型のジャクソン型待ち行列網(→積形式解をもつ待ち行列網)モデルをいう(図)．中央処理装置と入出力装置の使用を交互に繰り返しながら処理を進める†ジョブの動きを，待ち行列網内を推移する客の動きに対応させたモデルである．網内を

中央処理装置

入出力装置

…であってもよい．このモデルは閉鎖型であるため，システム内ジョブ数(多重プログラミングの多重度)は常に一定の整数値をとる．これは処理要求を待つジョブが無限個存在し，1 つのジョブ処理が終了するとただちに次のジョブ処理が開始されることを意味する．網内移動経路の異なる複数種類の客が存在する BCMP 型待ち行列網(→積形式解をもつ待ち行列網)モデルを用いることにより，このモデルをジョブの確率的振舞いと多重度の異なる複数種のサブシステムが混在する拡張されたセントラルサーバーモデルへと発展させることができる．さらに，このモデルを基本として，実システムの特徴を反映させるための種々の拡張モデルとその解法が開発されている．セントラルサーバーモデルに基づき種々の†性能評価尺度を算出する際には，待ち行列網モデルの定常状態確率を示す積形式解の中に現われる正規化定数を効率的に計算することが必要となる．ブゼンは，セントラルサーバーモデルに関する正規化定数の計算法も開発した．→待ち行列網，積形式解をもつ待ち行列網． C41-23

send/receive 命令 send/receive operation

†プロセス間通信のための命令の一つ．†メッセージを交換するための命令である．一般的には，

send メッセージの並び *to* 送付先

receive 変数の並び *from* 出所

という形をとる．send 命令と receive 命令を用いることで，送り手と受け手の間の†通信チャネル(通信リンク(communication link)ともいう)を定義する．通信チャネルを確立する最も単純な方法は送付先と出所とにプロセス名を用いるもので，直接指名方式(direct naming)という．この指名方式を対称(symmetry)といい，出所を省略した方式を非対称(asymmetry)という．送付先にメールボックス(→メーリングシステム)やポートを指定する方式は間接指名方式という．→メッセージ受渡し，生産者-消費者問題． C42-45

全2重回線 = 4 線式回線．

前方照応 → 照応処理．

いは 2 次曲線などによって表現しモデル化する技術をいう．得られたモデルを線モデルまたは†ワイヤーフレームモデルという．普通，図のような形状の

稜線や交線で形状を代表させる．したがって形状が多面体であれば稜線によって無理なくモデリングできる．しかし，円筒などには補助稜線を設定しなければならない．線モデルは作図あるいはグラフィックス表示用のモデルとして利用することが多い．線をそのまま投影することによって作図データが得られるが，隠線の自動消去は困難である(→隠線消去．→面モデリング，立体モデリング． D34-51

専門家システム = エキスパートシステム．

専門用語辞書

technical terminology dictionary

1 つの専門分野における概念を表現する専門用語を集め，文法情報や意味などを与えた辞書をいう．一般的な語でも専門分野で使用されるときは，その分野で定義された明確な意味をもつ．例えば"体"は数学の用語では四則算法の成立する集合を意味する．社会の進歩によって専門分野はますます細分化されて新しい概念が作られ，それに対応して専門用語を与えねばならないので，造語規則を設けて，できるだけそれに従って専門用語を作ることが望ましい．用語の説明内容が専門分野に応じて詳しくなされる場合は事典(→辞書[1])となる．説明が非常に簡単な場合には，専門用語集または専門用語データバンク(terminology databank)とよばれる．専門分野が細分化されたり，逆に学際領域などとして再組織化されたりすることにより，同じ語が分野によって違った意味(または訳語)で用いられたり，同じ分野でも使用される語や語の意味が変化したりすることがある．したがって，専門用語辞書では，用語の追加，削除，概念の修正などの辞書管理が特に重要となる．専門用語辞書は，2 か国語以上を含んだ形で作られることが多い．人が使うためのものと機械が使うものとで内容がそれほど異なっていないのがこの種の辞書の特徴である．→用語学． D21-75

0119

66

FRANCE

DÉCLARATION DE DOUANE CN23

Peut être ouvert d'office

Expéditeur:

Nom: Dorians' Day

Adresse: 30 rue de la Baisse

Code postal: 69100 Ville: Villeurbanne Pays: FRANCE

Société: Dorian's Day

Destinataire:

Nom: Eva Zhao

Adresse: 65 5TH AVE 1010 NEW YORK NY 10003

Code postal: 10003-3003

Ville: NEW YORK

Pays: UNITED STATES

Référence de l'importateur (code fiscal/n° de TVA/code de l'importateur) (procédure simplifiée postale n°) (facultatif)

N° de téléphone/fax/e-mail de l'importateur (si connus)

Description détaillée du contenu	Quantité	Poids net (Kg)	Valeur EUR	N° tarifaire du SH	Pays d'origine des marchandises
Book	1	0,5	48,96	490110	France
	Total	0,5	48,96	Frais de port/Frais: 0,00 EUR	

Catégorie de l'envoi:

☐ Échantillon commercial ☐ Cadeau

☐ Document ☐ Autre

Explication:

☐ Retour de marchandise

■ Envoi commercial

Bureau d'origine/Date de dépôt

Observations:

Je certifie que les renseignements donnés dans la présente déclaration en douane sont exacts et que cet envoi ne contient aucun objet dangereux ou interdit par la législation ou la réglementation postale ou douanière

Date et signature de l'expéditeur

☐ Licence ☐ Certificat ■ Facture : N° 994766

EXPEDITEUR / SHIPPER

Dorian's Day

30 rue de la Baisse

69100 Villeurbanne - FRANCE

E-mail : main@doriansday.com

N° d'identification à la TVA : FR64839861002

DESTINATAIRE / ADDRESSES

Eva Zhao

65 5TH AVE 1010 NEW YORK NY 10003

10003-3003

NEW YORK

UNITED STATES

RECIPIENT

Invoice #	994766
Date	2020-02-14

Item	Price EUR	Quantity	Totals EUR
Book	48,96	1	48,96

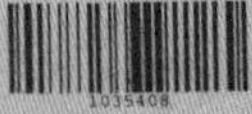

Subtotal EUR	48,96
Shipping EUR	0,00
Total EUR	48,96

111-2591313-2828211

TVA non applicable, article 262-ter du Code général des impôts

E3

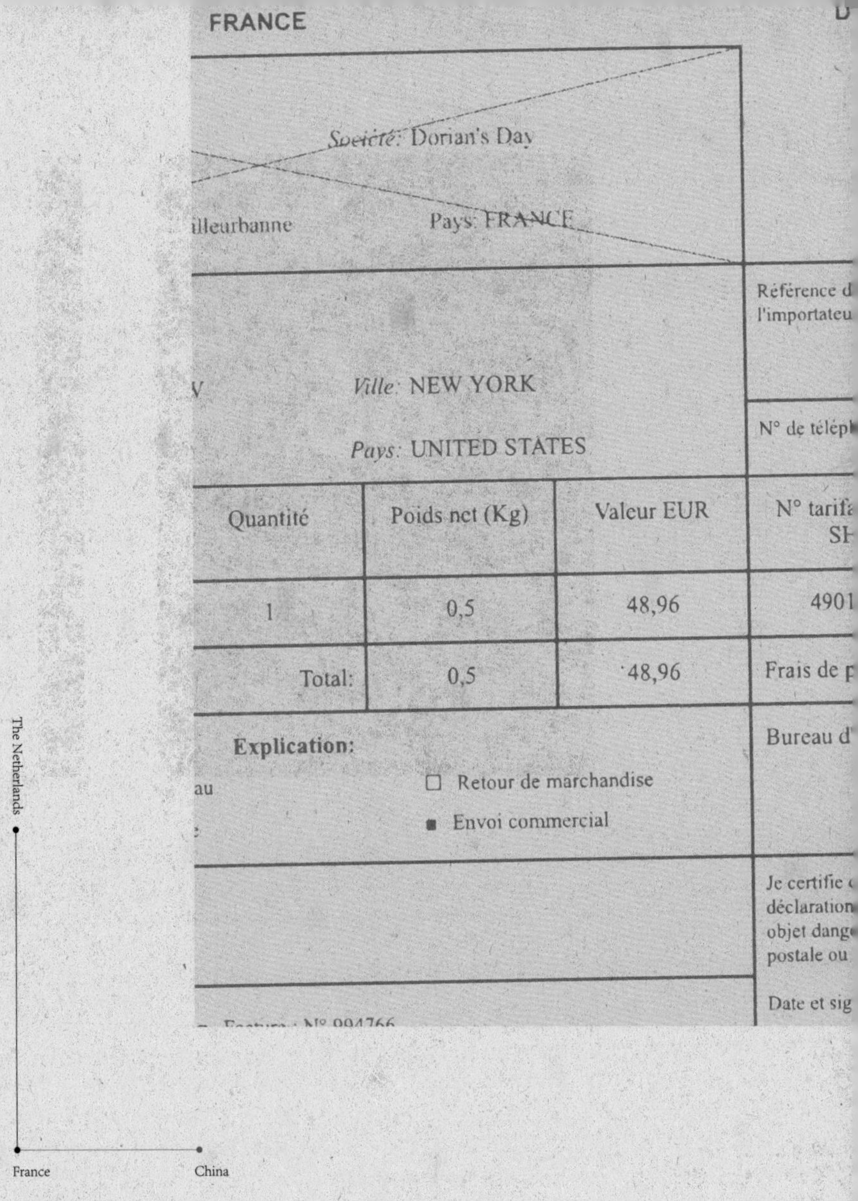

FRANCE

Société: Dorian's Day

illeurbanne Pays: FRANCE

Référence d
l'importateu

V Ville: NEW YORK

Pays: UNITED STATES

N° de téléph

	Quantité	Poids net (Kg)	Valeur EUR	N° tarifa SH
	1	0,5	48,96	4901
	Total:	0,5	48,96	Frais de p

Explication:

au

☐ Retour de marchandise

■ Envoi commercial

Bureau d'

Je certifie
déclaration
objet dange
postale ou

Date et sig

150mm

13mm

ZELFPORTRETTEN

110mm

荣誉奖

66 位荷兰摄影师的自拍

荷兰

Honorary Appreciation

66 Zelfportretten van Nederlandse Fotografen

The Netherlands

Lex Reitsma

Rudy Kousbroek

Nicolaas Henneman Stichting, Amsterdam

150 110 13mm

210g

160p

ISBN-10 90-7255-602-X

在摄影技术被发明以前，艺术家通过绘制自画像来记录特定时间的自己。有了摄影之后，自拍成了更真实更便捷的方式。本书收录了66位荷兰摄影师的自拍照。书籍的装订形式为锁线胶平装。封面黑色仿皮面材质印白和烫银处理。内页哑面铜版纸印单黑。内容方面，除了摄影师的自拍和照片的收藏信息外，还对自画像、自拍、拍摄手法等方面进行了深入探讨。与现在用手机自拍或者用带翻转屏的相机自拍不同，因为老相机本身的局限会对拍摄方式造成很多限制。比如自拍照不能有他人的参与，那么是用快门线，还是延时快门；是自己站在镜头前拍摄，还是端着相机对着镜面按下快门；是模仿架上绘画的庄严构图，还是加入随意的生活化场景：都是值得探讨的问题。依照这些可探讨的方面，书中收录的自拍照也将各种尝试尽数罗列。文本采用居中对齐的排版方式，但其中无衬线体、衬线体和斜体的混合使用，在传统的框架中透露出现代味道。

Before the invention of photography, artists recorded themselves at a specific time by drawing self-portraits. With photography, selfies have become a more realistic and convenient way. This book is a collection of 66 Dutch photographers' self-portraits. The book is perfect bound with thread sewing. The cover is made of black faux leather with white printing and silver hot stamping. The inner pages are made of matte coated paper printed in black. As for the content, in addition to collection information of the photographers' selfies, the self-portraits, selfies, shooting techniques and other aspects are also discussed in depth. Different from self-shooting with mobile phone or camera with flip screen, some factors of old cameras themselves limit the way of shooting. For example, when taking selfies without the participation of others, use the shutter release or time-lapse shutter; stand in front of the camera to shoot, or press the shutter against the mirror while holding the camera; imitate the solemn composition of the painting on the easel, or join casual life scenes. All these issues are worth discussing. According to these aspects, the selfies included in the book also present all kinds of attempts. Centered alignments mixed with non-serif fonts, serif fonts and italics, revealing modernity in the traditional framework.

66
ZELFPORTRETTEN
VAN
NEDERLANDSE
FOTOGRAFEN

MET EEN ESSAY
VAN
RUDY KOUSBROEK

NICOLAAS HENNEMAN
Heemskerk 8 november 1813 –
Londen 18 januari 1898

Lacock Abbey, 19 april 1842

Collectie Hans Kraus jr. New York

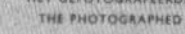

HET GEFOTOGRAFEERDE
THE PHOTOGRAPHED S

Het gefotografeerde zelfportret – in dit
een fascinerend gegeven, dat voortduren
is. Variabel zijn niet alleen de individuel
Door de jaren heen kan het zelfportret o
zijn, de tijdelijke conclusie van al die v
professionele en meer levensbeschouwel
de revue van het 'ik' passeren. Tussen he
dat de dromer tot de orde van de dag dw
waarin de auteur zichzelf als een onder
voert, ligt hemelsbreed
Bij de samenstelling van deze 'bloemlez
naar contrasten en varianten in een zo kl
soorten zelfportretten. De opzet was om
typologie samen te
Bij de selectie van het foto-materiaal w
Mariëtte Haveman, Jaap Lieverse, Le

The photographed selfportrait – in this
fascinating phenomenon which is co
Not only are there differences in tempe
through the years the selfportrait can als
temporary conclusion of various monolo
a more philosophical nature. Between th
that forces the dreamer to the order of th
which the author presents himself as pa
difference.
At the compilation of this collection w
looked for contrasts, variations and th
selfportraits. We aimed at composing a
typology.
The following people were involved i
material: Els Barents, Mariëtte Haveman
and Fred Schn

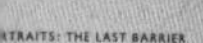

TRAITS: THE LAST BARRIER

graphy went a lot further than the

man to see himself while he was look-
phy this could also be done while he
gical consequence people could now,
lions of years, in fact since the begin-
elves with their eyes shut.
dred years for someone to understand
sue of *La révolution surréaliste*, dated
here is a photograph well worth linger-
f sixteen surrealists, every one of them
he portraits form the frame of a kind of
ne vois pas la... cachée dans la forêt,
den in the forest)'.
es is the picture of a naked girl (by
nsequently be read as 'virgin', 'truth',
ou like – purity, innocence, longing,
long as it is feminine in French; but in
abstract nouns are feminine.
phs that form the border, of the faces
rhaps not strictly speaking self-portraits
ey are passport photos, probably made
the prophetic nature of the whole
come clear.

f-portrait, strictly speaking? If one
he earliest photographs that are so
Charles Nègre's self-portrait from
g on a chair in the courtyard of his
on – and remembers that in those days
, then it becomes apparent that the
nnot have removed the lens-cap him

de fototoestellen in die tijd geen sluiters hadden, dan is het duidelijk dat de gefotografeerde niet zelf het deksel van de lens kan hebben genomen; dat moet iemand anders hebben gedaan. Wie heeft nu de foto gemaakt?

Het begrip van auteurschap is in onze tijd zo mechanisch geworden dat er waarschijnlijk mensen bestaan die zouden zeggen: degene die de lensdop heeft verwijderd. Maar het is zoals met een kanonschot: wie is verantwoordelijk voor de voltreffer, de artillerist die elevatie en afstand heeft ingesteld of de soldaat die het trektouw heeft bediend? Het juiste antwoord is ook hier: degene die de foto heeft uitgedokterd – die de plek heeft uitgekozen, de belichtingstijd berekend, de afstand en de hoek bepaald vanwaar en waaronder zij zou worden genomen, etcetera.

Interessante vraag: kun je een zelfportret *laten maken?* Ongetwijfeld, zo goed als een Japanse harakiri in wezen door een helper werd uitgevoerd en niettemin als een zelfmoord werd beschouwd.

De kwestie is dat er in de fotografie allang geen helper meer nodig is; het betreffende detail is opgelost door de uitvinding van de sluiter en de mogelijkheid deze op afstand te bedienen. Vele decennia lang gebeurde dat pneumatisch, met behulp van een cylinder, een rubberen slangetje en een dito balg, welke men op vele vooroorlogse zelfportretten dan ook duidelijk of minder duidelijk in de hand van de gefotografeerde kan zien, bijvoorbeeld op een bekend zelfportret van Lotte Jacobi uit 1930.

Daarna kwam de draadontspanner; geen onverdeelde verbetering voor dit speciale doel, niet alleen doordat exemplaren van voldoende lengte duur en onbetrouwbaar zijn, maar ook omdat zij meer spierkracht vergen dan zo'n ouderwetse gummibal, die aan een discreet kneepje genoeg had. Bij het gebruik van een lange draadontspanner kunnen sporen van de

9

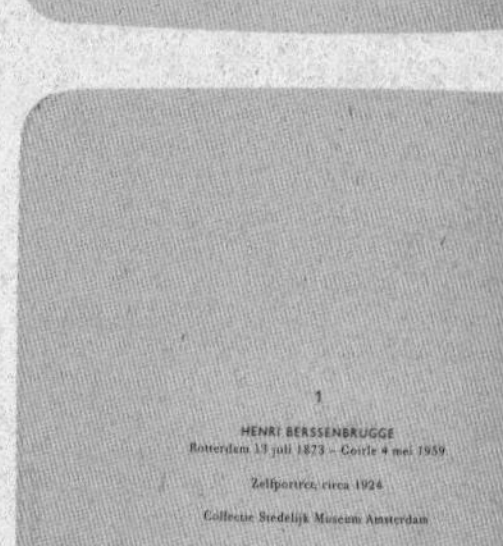

1

HENRI BERSSENBRUGGE
Rotterdam 13 juli 1873 – Goirle 4 mei 1959

Zelfportret, circa 1924

Collectie Stedelijk Museum Amsterdam

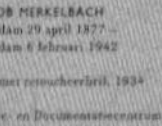

6

OB MERKELBACH
dam 29 april 1877 –
dam 6 februari 1942

met retoucheerbril, 1934

e- en Documentatiecentrum
tie Rijksuniversiteit Leiden

29

HELENA VAN DER KRAAN
Praag, Tsjechoslowakije 19 juni 1940

Zelfportret, Rotterdam 1982

De uitvinding van de fotografie ging een heel stuk verd
de uitvinding van de spiegel.

De spiegel stelde de mens in staat zichzelf te zien te
kijkt, maar met de fotografie kan het ook terwijl hij nie
Zoals hier logisch uit volgt konden de mensen nu voor
eerst sinds miljoenen jaren, ja sinds het begin der tijde
zelf zien met gesloten ogen.

Het heeft daarna nog wel honderd jaar geduurd voc
iemand dat goed begrepen had. In het twaalfde numme
La révolution surréaliste, gedateerd 15 december 1929
een foto die het overdenken waard is: de portretten van
surrealisten, allemaal met hun ogen dicht. Deze portret
omlijsten een soort rebus, luidend: *Je ne vois pas la* . . .
dans la forêt (ik zie niet de . . . verborgen in het woud).

Op de plaats van die puntjes bevindt zich de afbeeld
een naakt meisje (getekend door Magritte), waaraan de
een lezing kan worden gegeven als: 'de maagd', 'de wa
'de schoonheid', of wat ook, naar keuze – zuiverheid, o
verlangen, sensualiteit, eenzaamheid – zolang het maar
vrouwelijk woord is in het Frans; maar in die taal zijn b
abstracte zelfstandige naamwoorden vrouwelijk.

Nu zijn de omringende foto's, van die gezichten me
ten ogen dus, misschien geen zelfportretten in de stren
kenis (misschien ook wel: het zijn paspoortfoto's, moge
wijze gemaakt met een automaat), maar het profetische
hele voorstelling zal gaandeweg blijken.

Wat zijn trouwens wèl in de strenge betekenis zelfp
ten? Als men nauwkeurig de vroegste als zodanig betit
foto's bestudeert – ik denk bijvoorbeeld aan een zelfpo

KHEID
ZAAK
EN KLANTEN HELDER TEGEN

7

ANDRÉ-PIERRE LAMOTH
Hayange, Frankrijk 27 november 1931

Amsterdam 15 maart 1984, Polaroid PN55

1992

Country / Region

The Netherlands ❹

Japan ❸

Germany ❸

Switzerland ❷

Norway ❶

U.S.A. ❶

Dick Elffers

150 Jahre Sulzer-Heizungstechnik
Wie die Heizung Karriere machte

Langen • SPAZIERGANG NACH SYRAKUS edition cantz

VOILÀ Glanzstücke historischer Moden 1750–1960 Prestel

NIKKEN SEKKEI
Building Modern Japan 1900-1990

ROWOHLT Hervé Guibert Dem Freund, der mir das Leben nicht gerettet hat

Alexander Bodon 010

schräg

Designer

Kristian Ystehede, Guttorm Guttormsgaard

André Hefti

Takashi Kuroda

Rob Schröder, Lies Ros

Bruno Güttinger

Karin Girlatschek

KMS graphic, Maja Thorn

Harry N. Sierman

Thomas Cox (Willi Kunz Associates)

Joachim Düster

Kiichi Miyazaki

Koji Ise

Reynoud Homan

Gerard Hadders, André van Dijk (hard werken)

NORLIS ANTIKVARIAT
125 ÅR
A PRIORITAIRE PAR AVION
ZHAO QING
210000 NANJING
CHINA
Norlis Antikvariat AS, Universitetsgaten 18, 0164 Oslo
post@norlis.no · www.norlis.no
NORGE P.P.
CN22

Norway

Norway

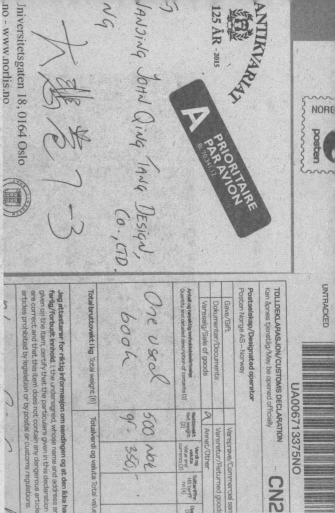

China

265mm

20mm

150mm

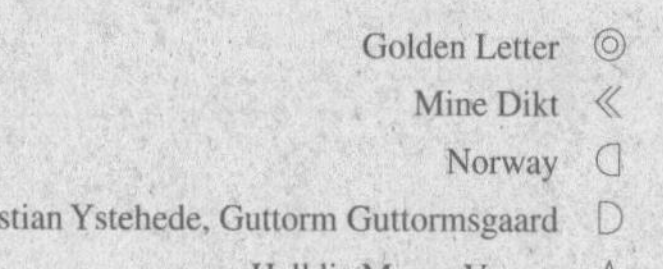

● 金字符奖

《 我的诗

◐ 挪威

Golden Letter ◎

Mine Dikt 《

Norway

Kristian Ystehede, Guttorm Guttormsgaard

Halldis Moren Vesaas △

H. Aschehoug, Oslo

265 150 20mm

547g

216p

ISBN-10 82-03-16803-5

哈尔迪斯·莫伦·韦索斯（Halldis Moren Vesaas）是挪威女诗人、翻译家和儿童读物作家，她的诗作多表达童年、乡村、爱情、战争、生命与死亡的主题。本书的装订形式为圆脊锁线硬精装。封面采用蓝色布面裱覆卡板，图案和文字烫金处理。封面除均匀的网格小点外，中心有完整的天蝎座群星图案，暗指作者的星座。环衬选用绿色艺术纸，内页采用哑面涂布艺术纸，印红蓝绿黑灰五色。内容方面，诗歌摘选自诗人 19 部诗集中的 8 部，按照摘选顺序作为页码排序。目录部分使用诗歌名首字母排序，供读者查阅。本书最大的特点是色彩的运用，红蓝绿黑灰分别对应不同的内容和情绪，自封面至内页贯穿始终。蓝色：天空、大海、沉思；红色：火焰、光明、热情；绿色：大地、植被、生机；黑色：暗夜、死亡、沉睡；灰色则更多是不带情绪的客观表达，包括版权信息。五种颜色贯穿在诗歌当中，每两页之间交换标题配色，单首诗歌只在需要表达不同情绪的时候使用其他颜色，形成了有趣的节奏。

Halldis Moren Vesaas is a Norwegian poetess, translator and Children's literature writer. Her poems mainly express the themes of childhood, country, love, war, life and death. The hardcover book is bound by thread sewing with rounding spine. The cover is decorated with blue cloth mounted by cardboard, and the pattern and text are gold stamped. In addition to the uniform grid dots, the cover has a complete Scorpio constellation pattern in the center, implying the author's constellation. Green art paper is used for the end paper, matte coated art paper is used for the inner pages, printed in red, blue, green, black and gray. As for the content, the poems are selected from 8 poetry collections of the poetess's 19 collections. The page numbers are sorted according to the order of the selected poems, and the table of contents is sorted alphabetically by the first letters of the poems. The biggest feature of this book is the use of color. Red, blue, green, black and gray correspond to different contents and emotions respectively, from the cover to the inside pages. Blue represents sky, sea and contemplation; red represents flame, light and enthusiasm; green represents Earth, vegetation and vitality; Black represents dark night, death and deep sleep; gray represents more objective expression without emotion, including copyright information. Five colors run through the poems, and the colors of the titles are exchanged between two pages. A single poem only uses other colors when it needs to express different emotions, thus forming an interesting rhythm.

i bryllupsnatta.
Ho førtest til brureloftet
med faklar og spel,
men låg på den høge puta
med armen krøkt opp framfor andletet
da brudgommen kom.

Han vandest ved synet:
den oppkrøkte armen
kvar gang han kom.
Ho fødde han mange barn.
Grått vart det gule hår.

Min festarmann i berget
er like ung
no og i tusen år.

•

Trøytt gammal kone med stav i hand
tok seg ein dag ut til berget.

Stirer du enno i mørkret?
– Dette er berre no.
Som tette regndropar
dryp sekundane,
sløkker til sist
alle logande bål.

Kanskje alt i morgon
strøymer det kalde regnet
fritt ned kring nakne greiner,

il gjekk opp og ned her kvar einaste
med lass av brød den bilen fór.
insteikte brød var han. Det låg att
in ange av ferskt brød i hans spor.
kvar dag med glede. Det lasset hans
in kveld fekk han lass av eit anna slag

orstøkte på han da han kom.
rilla han da, så vi visste han var tom.
han, og køyrde som galdt det liv.
liv. På sidene av han var måla ein sto
vi: no hadde krigen nådd like fram til

var bilen borte. Vi høyrde det knitra
skogen ein stad. Så stilna det. Skymi
tjerna bleik steig fram på himlen, kor
re delar av ei krigsmaskin, sjå det vart
vaks opp og vi vart menn og greip til
er dei som aldri slapp til orde nokon

u ser på oss, frå stålhjelm ned til støve
et er hat i auga dine. Men òg andre tir
u kjem kanskje hug dei heime, som st
ller er det det vi møter snart du står o

Har du enno såpass rom inni bringa? –
ikt skal nok døy i deg òg, før marsjen
Og er det litt for vår skuld òg du står d
å høyr da eitt: det er ikkje så ille som

Vi er delar av maskina. Slepp vi levand
heile marsjen – ja så er da maskina gått

Det kom litt regn i kveld.
Små tette dropar fann den nakne grunn,
og alt det liv han gøymer rørte på seg vårt
og anda angen sin ut i den svale kvelden
som førebod om alt som kjem – no snart.

Frå halvt utbrosten knupp,
frå bleike brodd, frå samanrulla blad,
kom bod og helsing: enn ei lita tid –
så kjem det, alt du ventar, du som signar
kvar stund av denne vår fordi ho lid.

Kvar helst det regnet fall
vart noko vekt. På meg òg fall det regn
i kveld, der still i skumringa eg stod.
Sitt varsel fekk òg alt i meg som ventar
på bod frå denne vår, i utålmod.

Alt spart og gøymt er vekt,
og bur seg no til det som kome skal:
alt mot, all styrke til den store prøva,
og alle frø av godt til fredens så-stund,
– kom, bruk det, vår, no har det fått ditt kall.

Det stod nok fleir som eg
i kveld, – det ligg nok fleir som ikkje
kan sova desse ventetimar bort.

Kring vatn og skogar ligg den svale natt
og femner med sin fred det fagre riket
som vi i dag så underfullt fekk att.

Mildt lyser stjerner frå den skøyre kvelv
av vårbleik himmel. Svart søv åsen under,
og stilla tonar døyvt av sus frå skog og elv.

Dei tunge krigsdønn stilna av i vest.
I denne natt skal alle våpen falle.
Høyr, small eit skott? Det tyder berre: fest.

På vegen durar det som ofte før,
men no ber alle bilar norske flagg,
og ingen reddast om dei stansar ved hans dør.

Enn brenn det lys i grenda somme stad,
så seint det er. Eg veit at kvar som vaker
med lampa tent i denne natt, er glad!

I andre heimar er det mørkt og skygt.
Der søv dei nok. Eg veit at kvar som kviler
bak nattdim rute no i natt, søv trygt!

Den same gleda under alle tak!
Ho får dei svævde til å le i svevne
og væter kinna mildt på den som ho held vàk.

KLICK
ZVAB.com
ZENTRALES VERZEICHNIS ANTIQUARISCHER BÜCHER
ZVAB.com
ZENTRALES VERZEICHNIS ANTIQUARISCHER BÜCHER
www.zvab.com
CN22
FROM
TO
R

Switzerland

Germany

China

320mm

20mm

KLICK

240mm

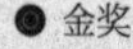

金奖
克利克
瑞士

Gold Medal
Klick
Switzerland
André Hefti
Erich Grasdorf
Edition A., Zürich
320 240 20mm
1074g
140p
ISBN-10 3-9520133-0-7

作为哈灵顿设计学院的副教授，彼得·克利克（Peter Klick）是一位拥有多年从业经验的室内设计师，服务于多家公司，也出版过多本专业书籍。本书是他的作品集，也是他创作理念和设计过程的汇总。书籍的装订形式为锁线硬精装。封面铜版纸印单黑，覆哑膜，裱覆卡板。内页混合使用了光面铜版纸和类似水彩纸的艺术纸四色印刷，用以展现理念文字、草图、完成照片等内容。为了体现克利克对于空间分割、布局和家具的独特见解，书籍的外观并非矩形，只有书脊和上书口是垂直和水平方向，其他两边都有明显的倾斜。内页的排版也大量采用倾斜的方式，文字部分打碎重组，排版自由，将原本此类书籍建立在网格系统内的严谨、规制完全瓦解，呈现出较为轻松、舒适、有趣的视觉风格，比较符合以人为本的室内设计的状态。

As an associate professor at Harrington College of Design, Peter Klick is an experienced interior designer. He has worked for many companies and published many professional books. This book is a collection of his works, as well as a collection of his creative ideas and designs. The hardcover book is bound with thread sewing. The matte cover is made of the coated paper printed in single black and mounted with cardboard. The inner pages are mixed with glossy coated paper and art paper similar to watercolor paper, printed in four colors to show concept, sketch, photos and other contents. In order to reflect Klick's unique views on space division, layout and furniture, the book's appearance is not rectangular, only the spine and the upper book edge are vertical and horizontal, while the other two sides have obvious inclinations. The layout of the inner pages also fully adopts inclined ways. The text is broken and recombined with free layout. The strictness and regulations built in the grid system are completely broken, presenting a relatively relaxed, comfortable and interesting visual style, which is in line with the status of people-oriented indoor design.

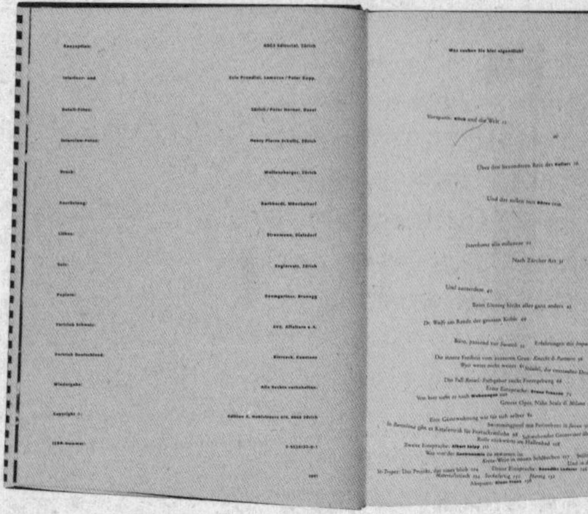

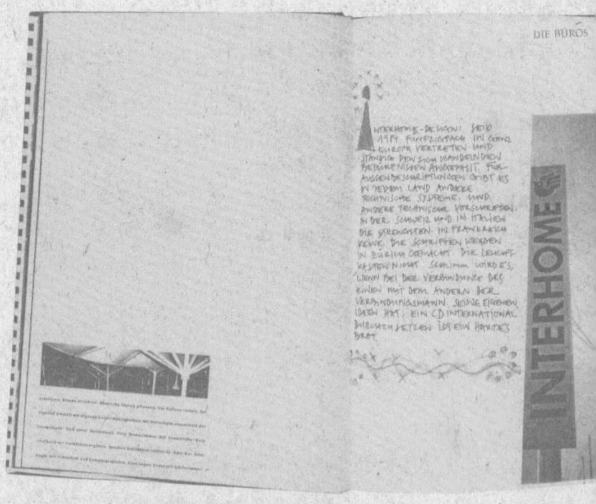
DIE BÜROS
INTERHOME

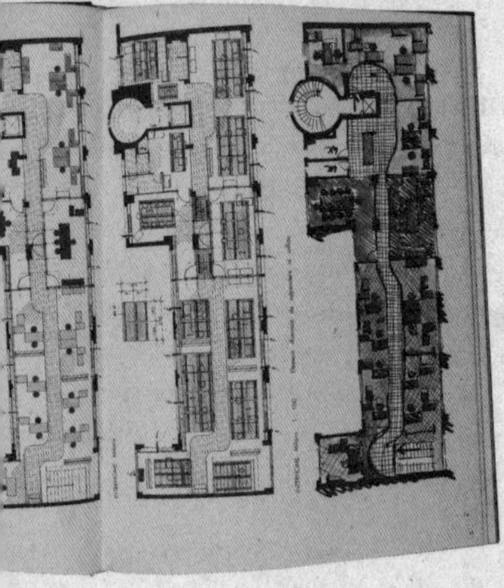

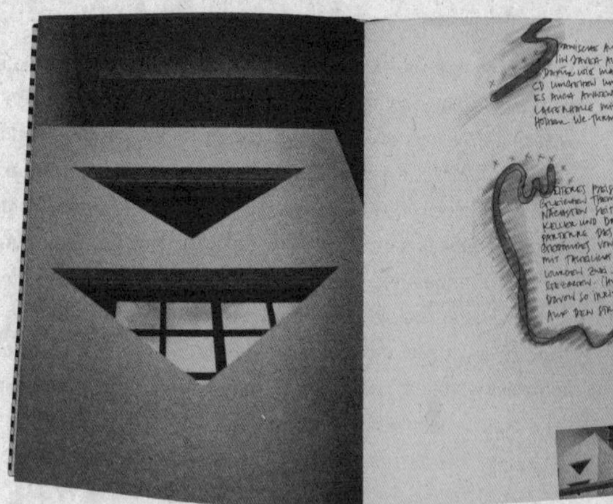

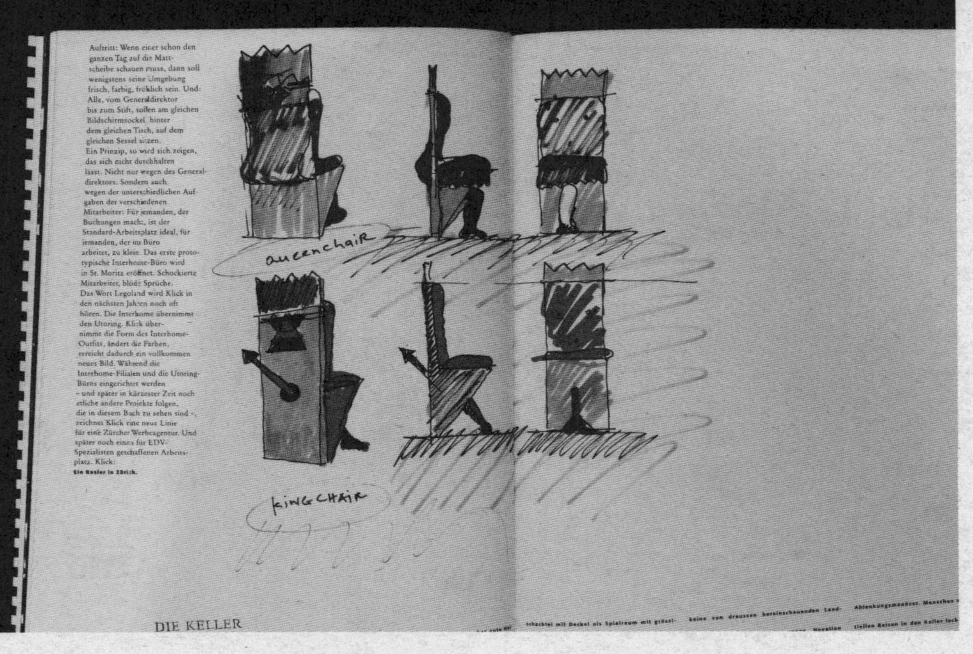

Auftritt: Wenn einer schon den ganzen Tag auf die Mattscheibe schauen muss, dann soll wenigstens seine Umgebung frisch, farbig, fröhlich sein. Und: Alle, vom Generaldirektor bis zum Stift, sollen am gleichen Bildschirmsockel hinter dem gleichen Tisch, auf dem gleichen Sessel sitzen.
Ein Prinzip, so wird sich zeigen, das sich nicht durchhalten lässt. Nicht nur wegen des Generaldirektors. Sondern auch wegen der unterschiedlichen Aufgaben der verschiedenen Mitarbeiter: Für jemanden, der Buchungen macht, ist der Standard-Arbeitsplatz ideal, für jemanden, der im Büro arbeitet, zu klein. Das erste prototypische Interhome-Büro wird in St. Moritz eröffnet. Schockierte Mitarbeiter, blöde Sprüche.
Das Wort Legoland wird Klick in den nächsten Jahren noch oft hören. Die Interhome übernimmt den Utoring. Klick übernimmt die Form des Interhome-Outfits, ändert die Farben, erreicht dadurch ein vollkommen neues Bild. Während die Interhome-Filialen und die Utoring-Büros eingerichtet werden – und später in kürzester Zeit noch etliche andere Projekte folgen, die in diesem Buch zu sehen sind –, zeichnet Klick eine neue Linie für eine Zürcher Werbeagentur. Und später noch einen für EDV-Spezialisten geschaffenen Arbeitsplatz. Klick:

Ein Basler in Zürich.

DIE KELLER

AILAND: RECHTS
USSEN. DER ERSTE
RISS. IN DER MITTE
KENPLAN MIT BE-
NG. LINKS DER
UNGSPLAN. MEHR
WIRD AN PLÄNEN
RIEBEN. GEZEICHNET
EIL HANDWERKER,
ERHAUPT, LIEBER
N KLEINEN PLAN
ALS AUF EINEN
FÜR DEN SOWIESO
ER PLATZ FEHLT

ze, Transparentes in allen Lichtspielarten: Pla

r Holz. Neue Lust auf Rundfurnier, Schnitzw

zurück – sind dann zu oft übe
Erfahrung eben die ist, dass w
gefahren sind, als wenn wir u
Auch das ist – jetzt kommt da
der Konsequenz. Viele Firmen
loser Konsequenz und in welc
BP-Tankstellen auf ein neues
finden das toll. Aber warum fi
anderen toll und nicht bei sic

Gehen wir doch noch ei
Unternehmenskultur: Was ha
denn nach innen, bei den eig

Fest steht, dass wir über
Mitarbeiter haben. Dass sie si
auch daran, dass sehr viele, d

würden Sie denn den Klick-Stil bezeichnen? Mir ist
stitel einmal «Post-Memphis» eingefallen, aber ich weiss
, dass es das nicht trifft. Braucht Klick überhaupt ein

wird es schwierig. Memphis ist eine so starke Kraft,
as Wort in keiner neuen Zusammensetzung
n würde. Klick ist sehr eigenständig, unkonventionell,
Seine Perspektiven waren schon während des
regelrechte **comics.** Für **comics** hat er übrigens schon
chwärmt. Er hat die Klassiker gesammelt. Vielleicht ist
Stil ein **comic**-Stil.

er Klick eigentlich bereit, die Menschen, die in seinen
Wohnräumen leben, überhaupt in seine Planungen
ziehen? Natürlich passen die gestylten Mailänder gut in
me-Büro. Aber wenn ich hier so in die Waldshuter

Klic
Wäh
hin p

Metr
Währ
Mate
für de
sich i

E
was b
Stil ge
D
konnte
belaste

zeichnender waleuder
umdernder bewegter
fahrender zweifelnder suchender irrender
weiser-und-sauber-bauender
strassen-gestaltender engagierter u.v.a.m.

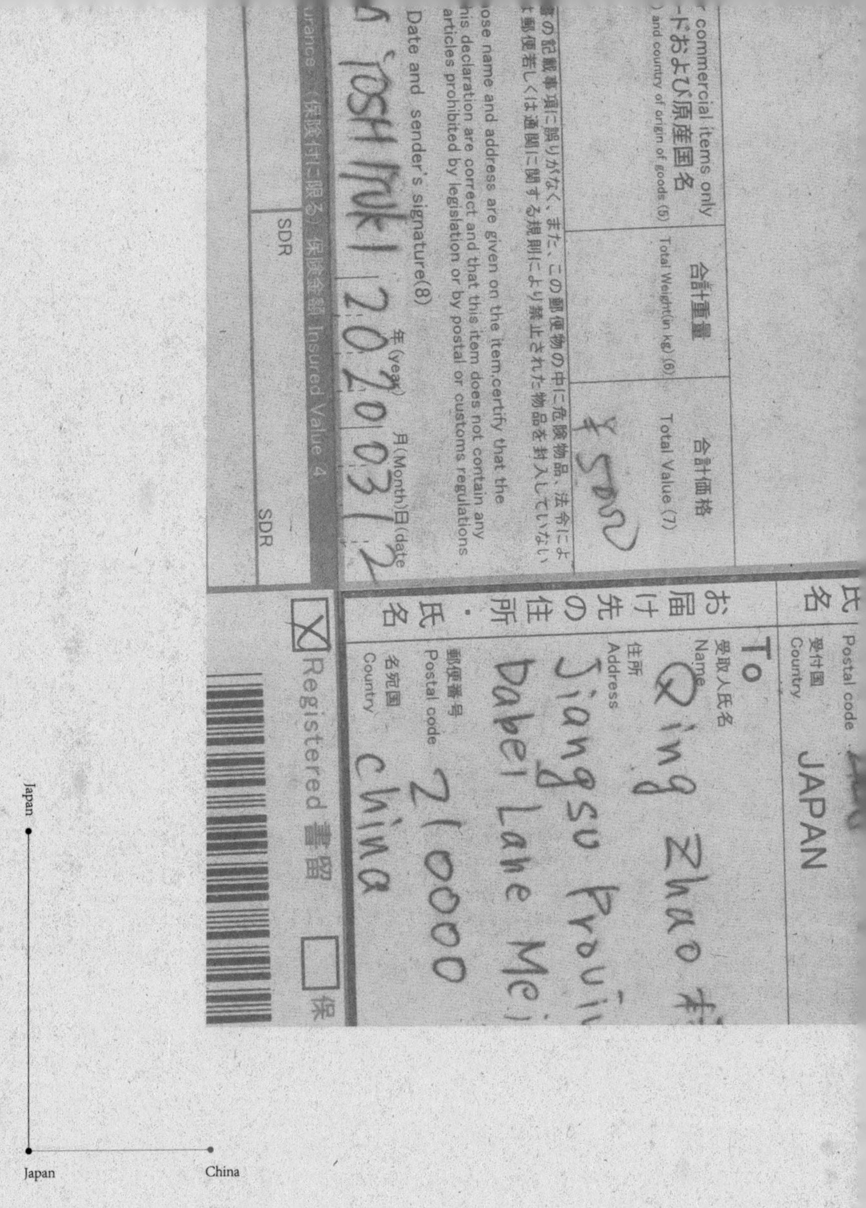

commercial items only
ードおよび原産国名
) and country of origin of goods (5)

合計重量
Total Weight(in kg) (6)

合計価格
Total Value (7)

¥5000

書の記載事項に誤りがなく、また、この郵便物の中に危険物品、法令によ
り郵便若しくは通関に関する規則により禁止された物品を封入していない

ose name and address are given on the item,certify that the
his declaration are correct and that this item does not contain any
articles prohibited by legislation or by postal or customs regulations

Date and sender's signature(8)

JOSH ... 2020 年(year) 03 月(Month) 12 日(date)

urance (保険付に限る) 保険金額 Insured Value 4

SDR

SDR

氏名

Postal code

受付国 Country JAPAN

お届け先の住所・氏名

To

受取人氏名 Name Qing Zhao

住所 Address Jiangsu Provi
Dabei Lane Mei

郵便番号 Postal code 210000

名宛国 Country china

Registered 書留

保

Japan

Japan

China

210mm

13mm

148mm

● 银奖
« 午后之绿
◖ 日本

Silver Medal
Green in the afternoon
Japan
Takashi Kuroda
Chizuru Miyasako
Tokyo Shoseki Co., Tokyo
210 148 13mm
365g
168p
ISBN-10 4-487-75295-7

伊豆半岛位于日本静冈县，处在富士山火山带，温泉众多，自然风貌得天独厚，再加上日本文学大家川端康成的《伊豆的舞女》的文化影响，这里是关东地区的旅游胜地。本书是画家和评论家宫迫千鹤在伊豆生活的记录，体会这里的田园牧歌与城市生活的不同节奏。书籍的装订形式为锁线胶平装。护封采用带杂质的半透明牛油纸印单黑，内封光面铜版纸四色印刷。环衬选用具有粗糙质感的绿色特种纸，内页哑面铜版纸四色印刷。书籍分为伊豆风貌的介绍、与当地人的交流和生活、伊豆的文化和依据伊豆创作的文艺作品。全书极力营造一种自然舒适的感受，一种充满四季甚至早晚的“自然生活”的节奏。所以书中附加了大量宫迫千鹤自己绘制的插图，抽象的大色块相互冲击，慵懒的线条带来与内容相适配的舒适气息。

Izu Peninsula is located In Shizuoka-ken, Japan. It's in the volcanic zone of Mount Fuji, with numerous hot springs and unique natural features. In addition to the cultural support of Yasunari Kawabata's *The Dancing Girl of Izu*, it becomes a famous tourist resort in Kanto Region. This book records the artist and critic Chizuru Miyasako's life in Izu, experiencing the gentle rhythm different from urban life. The book is perfect bound with thread sewing. The jacket is made of semi-translucent kraft paper with impurities printed in single black, and the inner cover is made of glossy coated paper printed in four colors. The green special art paper with rough texture is selected for the end paper, while the matte coated paper is used for inside pages printed in four colors. The book content is divided into four parts: the introduction of Izu, the communication and life with the local people, Izu's culture and literary works related to Izu. The book endeavors to create a natural comfortable feeling, a rhythm of 'natural life' of four seasons. Therefore, a large number of illustrations drawn by Chizuru Miyasako himself are added to the book. The large abstract color blocks impact each other, and the lazy lines bring the comfortable atmosphere matched with the content.

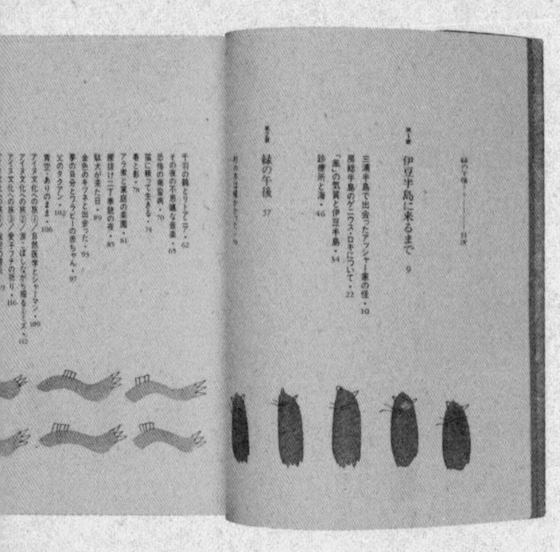
伊豆半島に来るまで
緑の午後

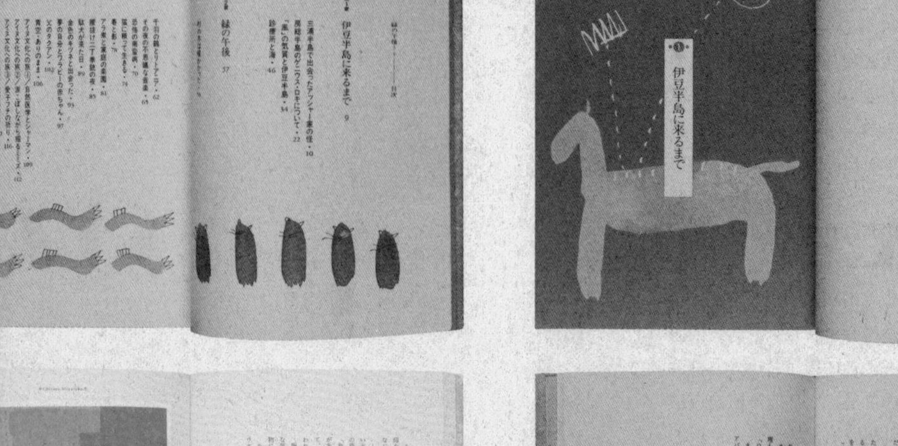
伊豆半島に来るまで

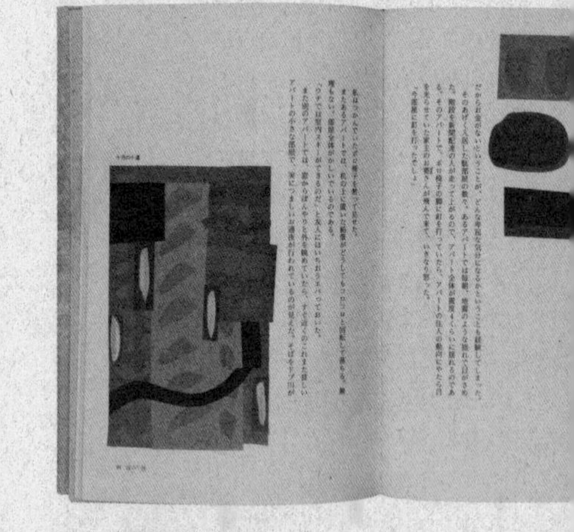

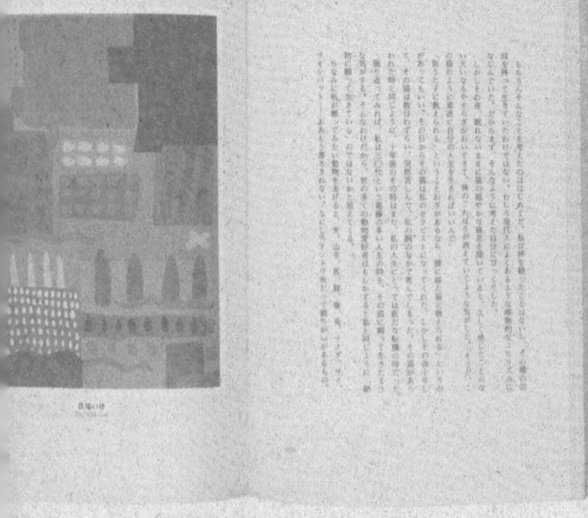

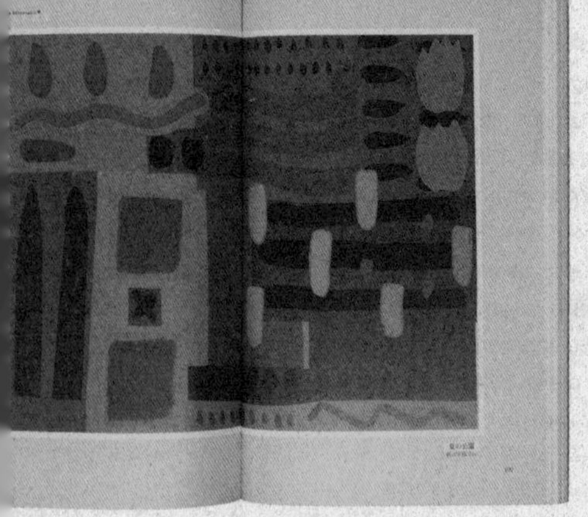

われわれの優柔不断ぶりを見てか、息子は冷たい。もちろんどこにでも行けるほど、われわれはリッチではない。のらりくらりしていたのは、お金がないからなのだ。なにしろ我が家が転居通告を受け取った頃、東京はすさまじい地上げの真っ最中だった。新聞のチラシの不動産広告を見ていても、あれよあれよという間に値上がりしていった。われわれはその金額に驚き、呆れ、やがて意気消沈し、ささやかな「マイ・ハウス」の所有欲が消えていくのを自虐的に笑った。そんな親の屈辱を知らずして、息子は涼しい顔をしてるので、ある時、言ってやった。

「ねえ、二世代ローンというのもありなのよ。だからあんたの稼ぎも頼りにしてるからさあ。がんばってよね」

と反論してきたもので、当方としてもからかい甲斐があったが、働きはじめてからは大人になったせいか、もはやワタシのからかいを相手にしてくれない。面白くない。かわいくない。「フン」と鼻先であしらわれる悔しさ。あのかつての少年の日の未熟さがなつかしいものだ。

で、あいかわらず、どうしたものかとのらくらしていたある時、逗子に住む友人が「土地は借地権で家はかなり古いけれど、家がとにかく広い。庭もなかなかだ。しかもこれが三〇〇〇万円台。一度見てごらんよ」と誘ってきた。友人夫婦が以前から目をつけていた物件らしい。友人はできれば一緒に見物しようという気である。

逗子と聞いて、その気になった。三〇〇〇万円台なら残りの人生でローンを背負ってもいいだろう。海が近く、しかも古い町だ。新興住宅地と違って、土地に歴史の陰影や人々の暮らしが刻印されていて、落ち着いて暮らせそうである。なんとなくココロが動いたので夫婦で出かけた。

その日、逗子の町は実に静かだった。東京のせせこましい騒がしさが嘘のような、のんびりとした時間があった。まるで歴史のなかで眠っているような静けさだ。こんな町に住むと、私の人格はきわめて遅くなるかもしれない。これからますます中年に向かう身にはなかなかよろしいではないか、などと思う。私はいつも自分の都合のいいように、物事をねじまげて考える才能を持っているのだ。

めざす家は低い崖の下にあった。たしかに古い。その古さはしかし『となりのトトロ』のメイちゃん一家が住んでいた家ほどの古さではない。高度経済成長期の直前に建てられた、築二十数年から三十年近いといったところか。しかし草ぼうぼうの庭はかなり広い。外から

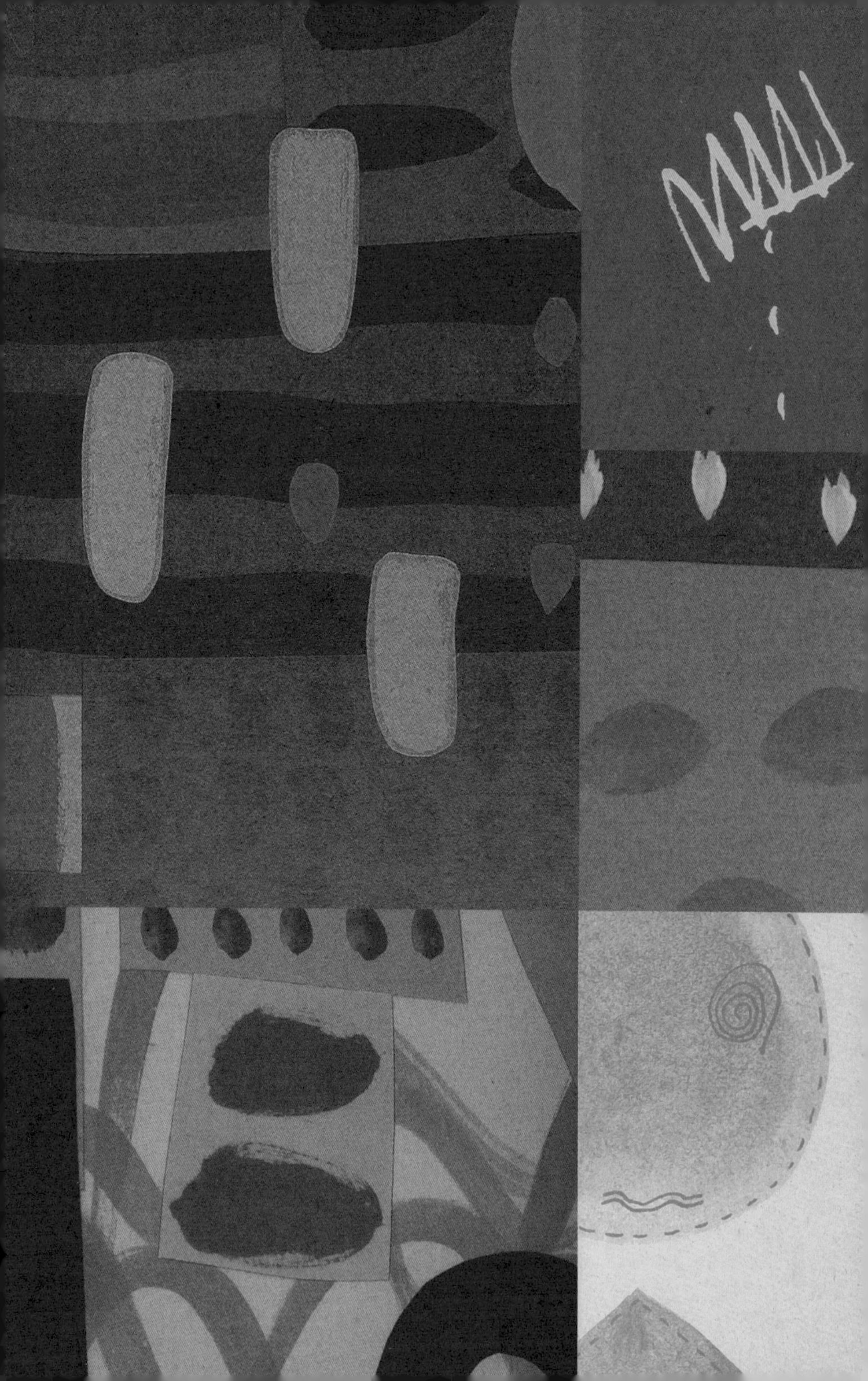

代というのは人生のなかでもっとも忙しい時代である。仕事の量はふえるし、それをこなそうとする体力にまだ自信があるから、平気で過労して過ごす。また子どももめまぐるしく成長する。なんだかやたらあわただしい。

そうこうしているうちに私の場合、親が倒れた。そして忙しさはさらに加速した。家にいて、家庭生活しているつもりなのに、いま振り返るとそこはごちゃまぜの作業場であったり、「親の問題」、「子どもの問題」そして「自分の問題」がうずまいている激流みたいなものだった。なんていうかしみじみしていなかった。

そのせいか、私はいずれ落ち着いたら「家庭生活」をしたい、とよく思った。もっともどういうのが家庭生活なんですかと質問されたら、ちょっと答えに困る。なにしろじっくりと料理することだとか、ていねいに掃除すること、暮らしに必要なものを自分でつくること、というふうにしか答えられない。

つまり三〇代の頃にはすべてが「忙しさ」に支配されていた。料理も掃除も、早く簡単にということが第一目標だった。暮らしに必要なものなんか買えばいい。つくっている時間なんかなかった。その結果、わが生活はきわめて合理主義的に展開したのであった。仕事をしてかつ主婦業をやっていれば、誰でもそうだろう。

いま私の世代の女性たちは、子育てが一段落した頃で、もう一度社会に出ていこうとしている。それはそれでいいことだと思う。社会も大いに受け入れてほしい。しかし私はといえば「家庭」人間になって、平凡な家庭的時間を味わいたいと思いはじめている。たとえば、つぎのような他愛ないことが、最近ではすこぶる充実して感じられるのである。

連れあいが開店早々の時間にスーパーに行くと、魚売り場の店員さんがひとかかえの「アジ」のパックを持って来たという。その中に一パックだけ高級魚「ヒラマサ」のアラがあった。なんと二〇〇円！　連れあいはすかさずその一パックをわしづかみにし、残念そうに眺

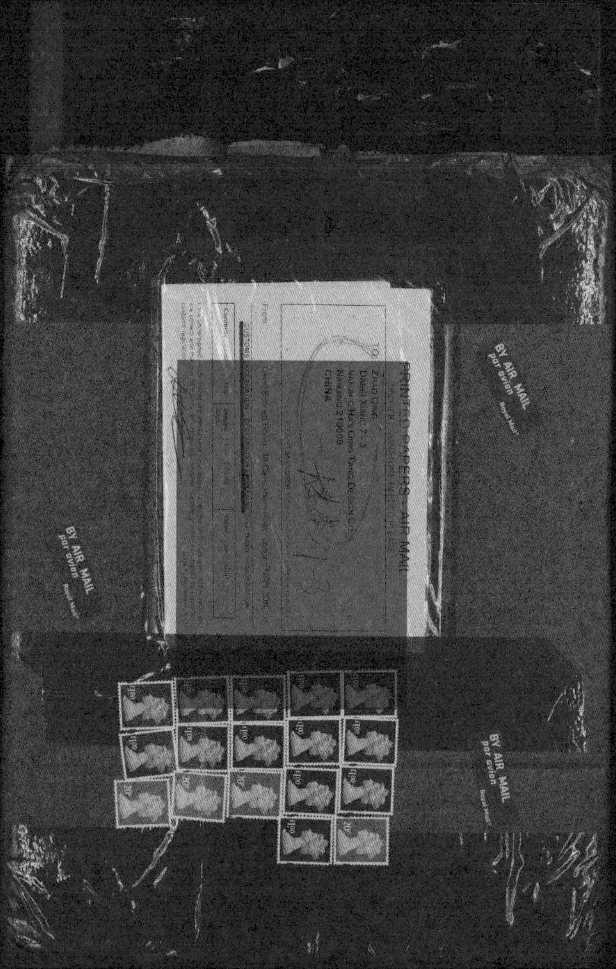

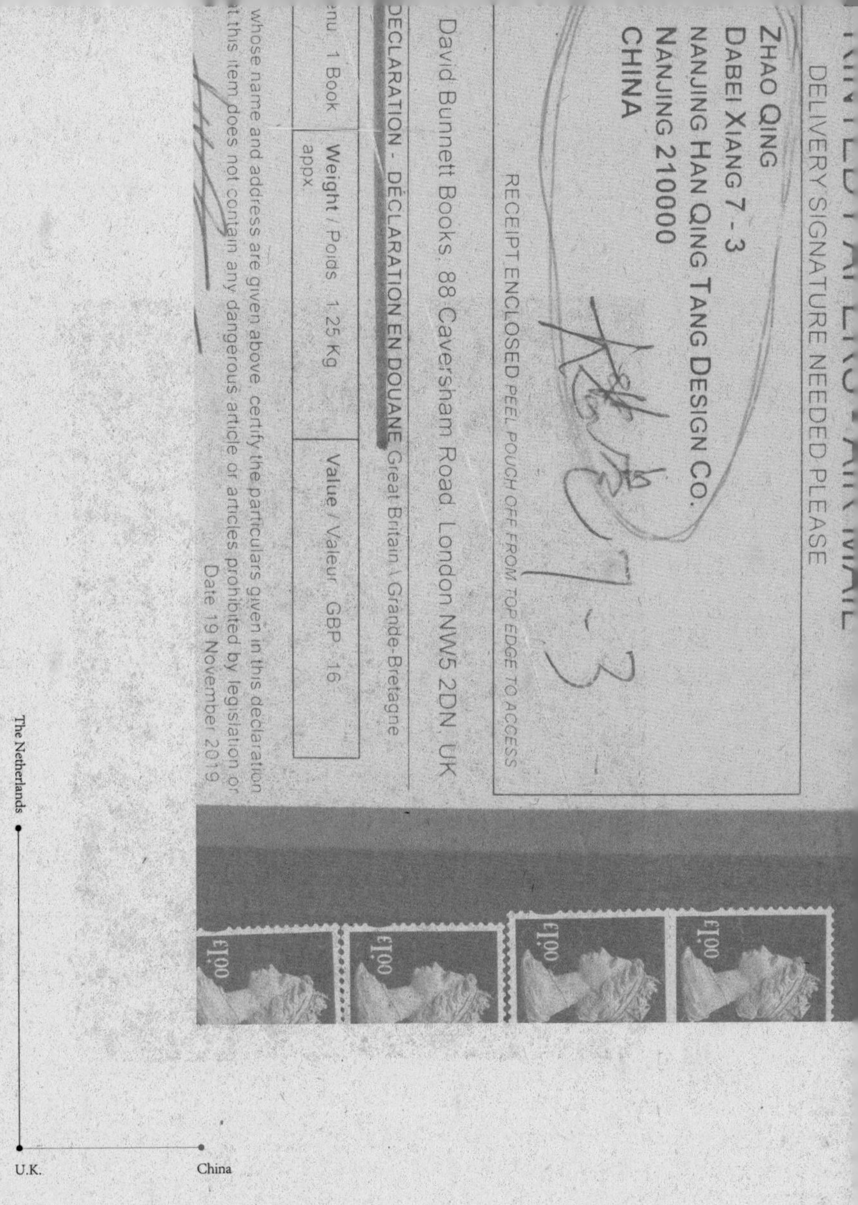
DELIVERY SIGNATURE NEEDED PLEASE
ZHAO QING
DABEI XIANG 7 - 3
NANJING HAN QING TANG DESIGN CO.
NANJING 210000
CHINA
RECEIPT ENCLOSED PEEL POUCH OFF FROM TOP EDGE TO ACCESS
David Bunnett Books, 88 Caversham Road, London NW5 2DN, UK
DECLARATION - DÉCLARATION EN DOUANE Great Britain \ Grande-Bretagne
1 Book
Weight / Poids 1.25 Kg appx.
Value / Valeur GBP 16
whose name and address are given above, certify the particulars given in this declaration
t this item does not contain any dangerous article or articles prohibited by legislation or
Date 19 November 2019
£1.00
£1.00
£1.00
£1.00
The Netherlands
U.K.
China

311mm

15mm

246mm

● 银奖
《 最后三英尺
——迪克 · 埃尔弗斯与他的艺术
● 荷兰

Silver Medal
Een leest heeft drie voeten:
Dick Elffers & de kunsten
The Netherlands
Rob Schröder, Lies Ros
Max Bruinsma
Uitgeverij De Balie; Gerrit Jan Thiemefonds, Amsterdam
311 246 15mm
835g
104p
ISBN-10 90-6617-062-X

荷兰艺术家迪克·埃尔弗斯（Dick Elffers）本是一名平面设计师，并在包括平面设计在内的多个领域里成为荷兰视觉文化的风向标。本书以他在各领域的作品为主，穿插对他艺术和设计作品的分析、对后世的影响，以及部分语录。本书的装订形式为锁线硬精装。护封铜版纸四色印刷覆哑光膜。内封蓝色布面裱覆卡板，书名压凹，具有艺术家代表符号的眼睛图案做烫黑处理。内页以埃尔弗斯的作品和创作解析为主，包括平面、室内、建筑、雕塑、陶瓷、绘画、编织等多个领域，使用哑面铜版纸四色印刷。书的最后部分是荷兰设计评论家马克斯·布鲁因斯马（Max Bruinsma）对埃尔弗斯工作的观察和思考，采用胶版纸深蓝色和中黄色双色印刷。书籍的设计上也具有一定的先锋性（1992 年），版心很大，页边距只有 5 毫米，正文选用衬线字体，图注的标题和解释文字使用了同一种无衬线字体的不同粗细和宽窄字重进行混排。页面上大量使用局部色块，与埃尔弗斯的作品相呼应。

Dutch artist Dick Elffers was originally a graphic designer, and he has become the icon of Dutch visual culture in many fields including graphic design. This book focuses on his works in various fields, including the analysis of his works of art and design, the influence on later generations, and some quotations. The hardcover book is bound with thread sewing. The jacket is made of laminated coated paper printed in four colors. The inner cover is made of blue cloth mounted with cardboard. The title of the book is embossed, and the eye pattern, artist's representative symbol, is black hot-stamping. The inside pages are printed on matte coated paper in four colors, mainly present Elffers' works and related analysis, including graphic design, interior design, architecture, sculpture, ceramics, painting, weaving and other fields. The last part of the book is the observation and thinking of Max Bruinsma, a Dutch design critic, on the works of Elffers, which is printed in dark blue and medium yellow on offset paper. The design of the book is avant-garde in 1992. The margin is only 5mm with a large type page. Serif font is used for the text. The title and explanatory text of the illustrations use the same sans-serif font with different thickness, width and weight for mixed arrangement. A large number of partial color blocks are used on the pages, corresponding to the works of Elffers.

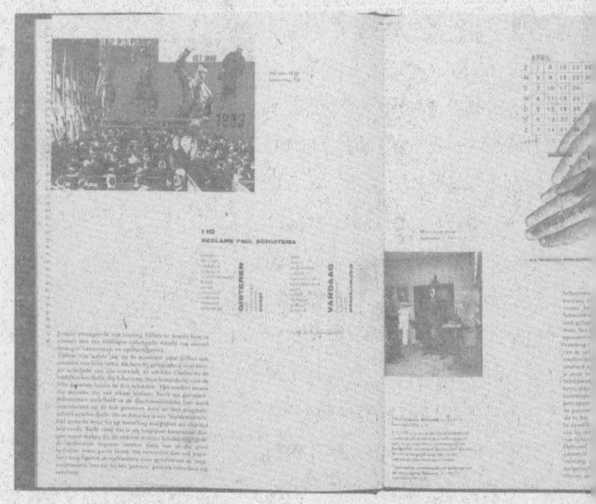

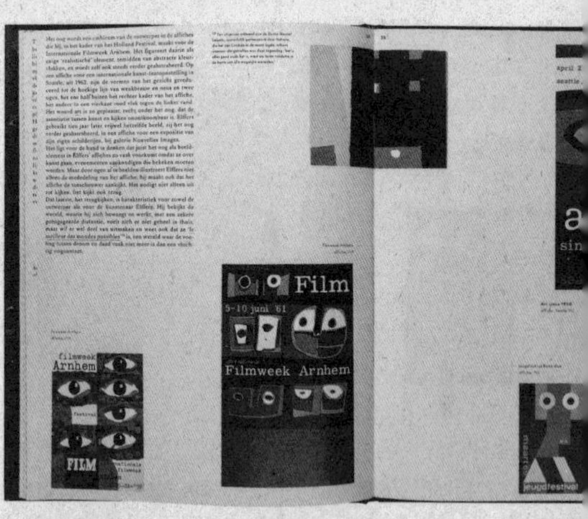

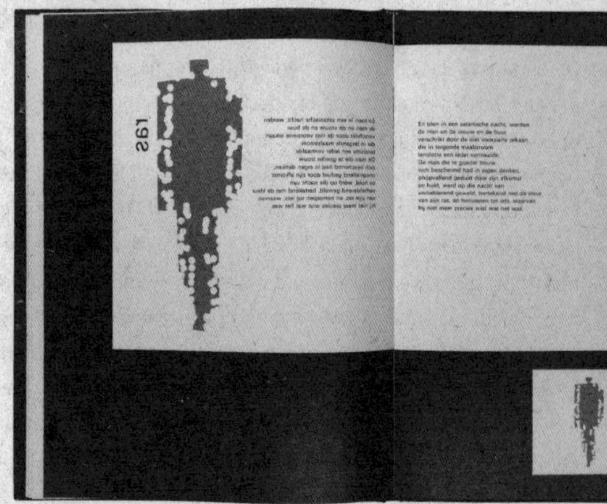

12

Elffers' eerste kennismaking met het ontwerpvak vond
plaats toen hij op 14-jarige leeftijd, na de ULO, 'maar aan
het werk' moest en in de leer kwam bij de Rotterdamse
decorateur Kemel, een vervaardiger van reclameborden en
huisdecoraties. Hij deed er klusjes en uitvoerend werk zoals
het 'behouten' en beletteren van bakkerswagens en uit-
hangborden en werkte mee aan plafondschilderingen en
dergelijke. De 'eerste man' van Kemel vermoedde dat er
meer in de jongen zat en adviseerde hem om 's avonds naar
de academie te gaan. Die stap is voor Elffers' ontwikkeling
van groot belang gebleken want na een jaar avondschool
werd hij toegelaten tot de dagschool, waar hij, op de afdeling decoratieve kunsten, in aanraking kwam met leermeesters als Jac. Jongert, Piet Zwart en Kiljan.
Om in zijn levensonderhoud te voorzien en zijn opleiding te
betalen, maakte hij gebruik van zijn ervaring bij de reclame-
schilder en vervaardigde en decoreerde uiteenlopende
gebruiksgoederen, zoals lampekappen en kunstbeslag voor
lijkkisten.
Het lijkt ons niet ongepast om hier al op te merken dat de
jonge Elffers door het ambacht in aanraking is gekomen
met zijn latere vak, en niet vanuit een, laat ons zeggen, theo-
retisch of maatschappelijk beredeneerde keuze. Het volgen
van de opleiding vloeide op een natuurlijke wijze voort uit
zijn in de praktijk gebleken belangstelling en talent.

13

Paul Schuitema
ontwerp affiche VANK, ± 1930

Voor de ontwikkeling van dat
– de Rotterdamse avant-gardist
tema van zeer groot belang
hoofddocent Decoratieve Ku
Rotterdamse Academie de t
afdeling te moderniseren en n
in tussen de strikt functional

erstnummer kleur

'Hier had ik de moeilijkheid: hoe kun je met
ling komt voort uit de opdracht die ik meze
over kleur moet je niet alleen je eigen voor
moment ook eens van een andere geestes
vloeden door Japanse tekens, door kleuren
tiek. Ik zat naar een schilderij van de jon
dan, dat er ineens kleuren tot je komen, di
Ik wilde een blad maken dat je doet schrik
den. Er zijn zoveel uitgaven waarin niets
komen vaak de meest goedkope versierse
Dat zit hier ook in, zo'n waaiertje, die rot
De drukkers vonden die waaiers heel verv
iets doen over kleur en druktechniek. Ik wa
ceramische wand, en daar heb ik die tegels
had, dacht ik: en nou moet het goddomme
woord. Dus je laat zien wat er allemaal m
opbrengen, en dan ineens komt dit eruit,
Hier illustreert de tekst het beeld.'
'Die kleurpartituur heb ik aan Andriessen
werk gegaan, daar had ik verder weinig

lovige publiek spraken een zelfde t
nerlijke overtuiging. De glasschilders
niemand te overtuigen en niets te ve
vormgevers van een door allen gede
chende beelden vierden zij namens
geloof.
werk gaat uit van een zelfde gevoel
beelt, zoals bleek uit zijn bedrijfspubli
rol van verbeelder, al is hij zich er
beelding in onze tijd abstracter en min
d moet zijn dan in vroeger tijden. Elff
ijfsboeken en ook de monumentale
bij bedrijfs- en openbare gebouwen
ctioneren als 'gemeenschapskunst'.
fiches zelf 'straatschilderijen', kunst
enheid en afzondering van een muse
, maar in directe confrontatie met
tad en haar bewoners.[20] Dat gaat op v
schien nog wel meer voor monumen
an kunsttoepassing bij of in gebouw

est heeft

e mededeling,
e context (MB)

vat is... 1988
liceerde tekst)

Zowel bij Crouwel als bij Van Toorn
onzichtbaar willen maken.[16] Je ziet in h
weleens verliefd of dronken waren. Zo
nieuw-zakelijke en functionele ontwerp
verboden dat de zakelijke mededeling
kan zijn van emotionele ervaringen va
situatie is het tegendeel het geval!'
Het experimenteren met vormen en t
bij Elffers altijd een zoeken naar exp
mogelijkheden om de mededeling te o
te gaan dan de eerste, ordenende taak
Wanneer Elffers het heeft over het o
een vrij streng onderscheid tussen die
spel: 'De grafische vormgeving van be
met het ordenen van gegevens. Goed
baarheid hebben op zich niets met kuns
is mogelijk dat er tijdens het werkpr

woord vooraf

Dick Elffers is een eigenaardig ontwerper. Een moeilijk te plaatsen, maar ontegenzeggelijk aanwezige factor in de Nederlandse visuele cultuur. Wanneer je zijn affiches ziet, zijn postzegels, zijn tentoonstellingsontwerpen en -catalogi, dan herken je ze; je hebt ze eerder gezien. Wanneer Elffers spreekt doet hij dat op de rustige toon van een man die weet dat hij gezag heeft. Wanneer je je afvraagt: wat weet ik van deze man, dan ontsnapt hij je.

Elffers is een van de weinige vooraanstaande ontwerpers die zich er niet toe heeft laten verleiden zijn werkwijze en ontwerpuitgangspunten te formuleren tot een eigen theorie. Anders dan zijn leermeesters Schuitema en Zwart en anders dan zijn jongere collega's Crouwel en Van Toorn, heeft hij geen analytische architectuur rond zijn werk gebouwd of laten bouwen, geen besliste statements geschreven over hoe zijn werk begrepen dient te worden en geen theorieën gepubliceerd over wezen en werking van het grafisch ontwerpen. Eén keer heeft Elffers iets geschreven over, niet het ontwerpen, maar over de ontwerper, over zichzelf: vorm + tegenvorm.

Die Thiemefondspublikatie uit 1976 is, zoals Wim Crouwel in zijn voorwoord opmerkt, een zelfportret geworden. Dick Elffers beschrijft de ontwerper daarin als 'een grensbewoner; hij houdt zich op in het niemandsland tussen de beroepen.'

En overschrijdt vandaaruit regelmatig de grenzen, had hij eraan toe kunnen voegen. Want er is praktisch geen terrein binnen de toegepaste en vrije kunsten waarop Elffers niet werkzaam is geweest.

Wanneer Elffers omschreven wordt als 'ontwerper/kunstenaar', dan dient dat in de ruimste zin te worden opgevat. Zijn werk omvat grafische ontwerpen, ruimtelijke en architectonische ontwerpen, monumentale plastieken en reliefs, wandtapijten, ceramiek, foto's, illustraties en schilderijen. [...]

Wat je over Dick Elffers kunt w
besloten en zal er bij nadere bes
schijn moeten komen. Het conc
zal dus het uitgangspunt moete
over de maker. Voor de beschouw
dezes, is er echter allereerst de v
heid vormen, technieken, materi
de kunstenaar/ontwerper zich h
Waar is de lijn in deze kluwen,
onderstroom?

De hier volgende tekst is in vele
ze pretendeert geen wetenscha
biografie te zijn, noch een afgero
kunstenaarschap, noch zelfs ee
van zijn oeuvre.

Wat dan wel? Een beetje van alle
de constanten in het bont gescha
Elffers kan haast niet anders dan
en deze tekst zal dan ook op zijn
mozaïek, dat zijn kleur ontleent
Mocht dit boek een pretentie he
in beeld en tekst een hommage
ontwerper die een belangrijk a
de vormgeving van het ontwerpv
ook een 'eenling' als Dick Elffer
dat hij gevoed is door de tijd wa
gevormd is door de vraagstukk
kampte.

Dat vak, ondertussen, stoort zic
van de media. Het gaat erom h
media structureert en ze tot wer
dige verbeelding maakt.

Misschien verwoordt Elffers de
digheid zelf het treffendst, wann
vraag waarom hij zich niet, zoal
zijn leest houdt. Hij antwoordt d
en verwijzend naar het aloude

holland
festival
holland
festival
carré
labyrint
holland
festival
1966
Als de kunst vleugels heeft,
heeft de gebonden kunst een motor.
Een motor moet vervangen worden,
vleugels gaan een leven lang mee.
holland
festival
holland
festival '69
Carré
holland festival
holland festival

0165

Designated operator **Swiss Post**

☐ Gift ☐ Comm. sample ☐ Others

☐ Docs ☐ Returns ☒ Goods

Quantity and detailed description of contents	Weight in kg	Value in CHF
1 x Book	1.35	21.00
For commercial items: only if know HS tariff number and country of origin of goods	Total weight	Total value
	1.35	21.00

I certify that the particulars given in the declaration are correct and that this item does not contain any dangerous articles prohibited by postal regulations

FRO

Herr Artus Grenacher
Loh 3
8362 Balterswil
Switzerland

RY 997 726 994

TO

Qing Zhao
Meiyuan New Village

Switzerland

Switzerland

China

294mm

24mm

202mm

铜奖

加热如何成为事业

——150 年苏尔寿加热技术史

瑞士

Bronze Medal

Wie die Heizung Karriere machte:

150 Jahre Sulzer-Heizungstechnik

Switzerland

Bruno Güttinger

——

Sulzer Infra AG, Winterthur

294 202 24mm

1340g

368p

——

苏尔寿是瑞士著名的工业工程技术及机械制造企业，主要从事压缩机、离心机、工业泵、水利机械、热力涡轮及成套设备、纺织机械、医疗技术等机械产品的研发和生产。苏尔寿从 1834 年成立至今已超过 180 年。本书是 1991 年苏尔寿加热技术和产品达到 150 年时发表的纪念出版物。书籍的装订形式为锁线胶平装，封面白卡纸深蓝色和红色双色印刷，覆哑亮膜。内页哑面铜版纸四色印刷。书籍内容分为加热技术史、苏尔寿的早期加热技术、苏尔寿的产业转型、苏尔寿供暖设备的案例、供暖配套设备、新能源与新技术、苏尔寿的未来展望等多个章节，基本按照时间线进行叙述。但各个章节都会提到技术、案例、历史、产品等方面的内容，包括一般的加热技术及架构、苏尔寿的加热制造技术、苏尔寿在各个时期的企业改变、具体供暖产品四个方面，通过书口处设定的四个红色块进行标示。全书版心设定为三栏，正文跨装订口位置的两栏，留出翻口处的一栏放置注释和插图。

Sulzer is a well-known industrial engineering technology and machinery manufacturing enterprise in Switzerland, mainly engaged in the research and production of compressors, centrifuges, industrial pumps, hydraulic machinery, thermal turbines equipment, textile machinery and other mechanical products. It has been more than 180 years since Sulzer was founded in 1834. This book is a commemorative book published in 1991, celebrating 150 years since the birth of Sulzer's heating technology and products. The paperback book is glue bound with thread sewing. The cover is made of white cardboard printed in dark blue and red, covered with matte film. The inside pages are made of matte coated paper printed in four colors. The content of the book is divided into several chapters, such as the history of heating technology, Sulzer's early heating technology, Sulzer's industrial transformation, the case of Sulzer heating equipment, heating equipment, new energy and new technology, Sulzer's future prospects, etc.. The chapters are basically organized according to the timeline. However, each chapter will mention technology, cases, history, products and other aspects, including four aspects: general heating technology, Sulzer's heating manufacturing technology, Sulzer's changes in various periods, and specific heating products. Four aspects are marked by four red blocks at the fore-edge. The typeset area of the book pages is set to three columns: the text spans two columns, and a column is reserved for notes and illustrations.

Bemerkungen zum Gebrauch des Buches

Thematische Gliederung

Die Ursprünge

bis 1841

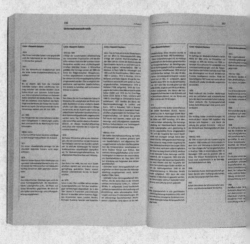

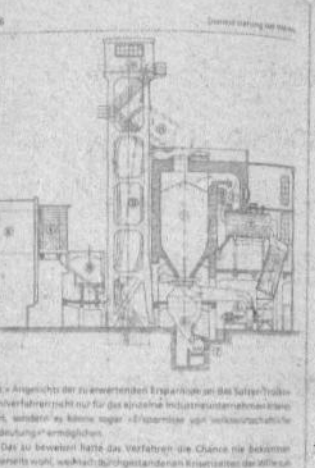

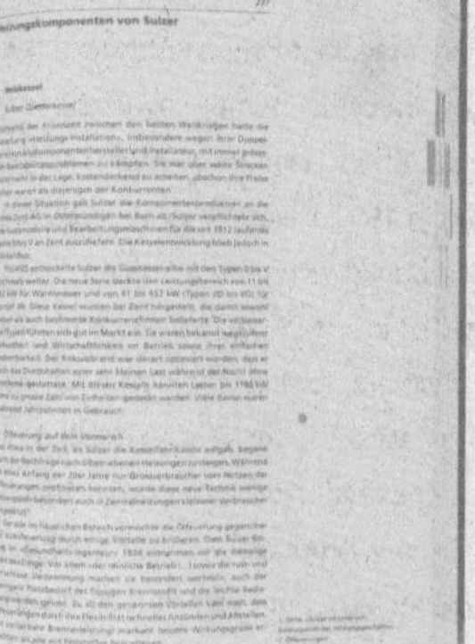

Unternehmenschronik

gen- oder Zirkulierofen bekannt, die vielen Spielarten von Mantelöfen, bei denen der zu heizende Luftstrom um den zentralen Rauchzug oder das Rauchgas um das zentrale Warmluftrohr fliesst, ihre Ausbildung als Schachtfüllofen oder als Dauerbrenner, ihre Abwandlung als kachelverkleideter Eisenofen: Sie alle sind nur wenige Vertreter einer Modellvielfalt, die sich in Europa und in den USA gleichermassen herausbildete (Bild 5)[19].

Auch haben diese «Stubenöfen», was ihren Wirkungsgrad und ihr Speichervermögen anbetrifft, mittlerweile eine beachtliche Perfektion erreicht. 1974 holte man ja die Einzelofentechnik wieder aus der Versenkung hervor. Eisenofen oder Kachelofen sind seither zunehmend zum beliebten Requisit einer häuslichen Gemütlichkeit sowie einer bescheidenen Praxis ökologisch bewusster Lebensweise geworden.

Im Rückblick lässt sich feststellen, dass der Übergang vom offenen Feuer zur Ofenheizung mehr als nur einfach ein technischer Fortschritt war: Er berührt das alte Thema vom Ursprung des Hauses[20]. Im «Vitruv», dem ältesten bekannten Buch über das Bauen, wird geschildert, die Entdeckung des Feuers habe bewirkt, dass die Menschen an einer Stelle zusammenkamen und sich in dieser Gemeinschaft im Wettbewerb ihrer Erfindungskraft unter anderem im Bau von Häusern übten. Behausung setzt demnach Feuer voraus[21].

Es ist nun interessant, diese Paarung von Feuer und Haus in alten Hausformen, wie beispielsweise dem Yomon-Haus der Ureinwohner Japans oder dem Vier-Pfosten-Grubenhaus der Dard-Stämme im Hindukusch, zu beobachten. Dort teilt das zentrale Feuer – Herd, Heizung und Altar zugleich – im Hauptwohn-Schlafraum den Grundriss in Zonen ein, die streng den verschiedenen profanen und geweihten Handlungen des täglichen Lebens gewidmet sind (Bild 6)[22]. Beobachten wir die Rolle des Feuers im frühmittelalterlichen Manorialpalast und seinen Abkömmlingen, Schlösser, Bauern- und Bürgerhäuser bis zum berühmten englischen Landhaus[23], so sehen wir, dass es weit über seine profane Wirkung der Wärmestrahlung hinaus im Grundriss eines Hauses stets einen sehr zentralen Platz eingenommen hat (Bild 7). Zentral war es einmal wörtlich genommen, indem es zuerst die Mitte des Raumes, später die Mitte der Wand besetzte, doch schaffte es auch in der Sitzordnung eine Hierarchie der Raumbenützung. Mit der Einführung des Ofens verschwindet dieser Faktor räumlicher Organisation, und was an seine Stelle tritt, lässt sich unter anderem am Grundriss des Würzburger Residenzpalastes 1719 bis 1753 des berühmten deutschen Barockbaumeisters Balthasar Neumann (1687 bis 1753) gut darstellen.

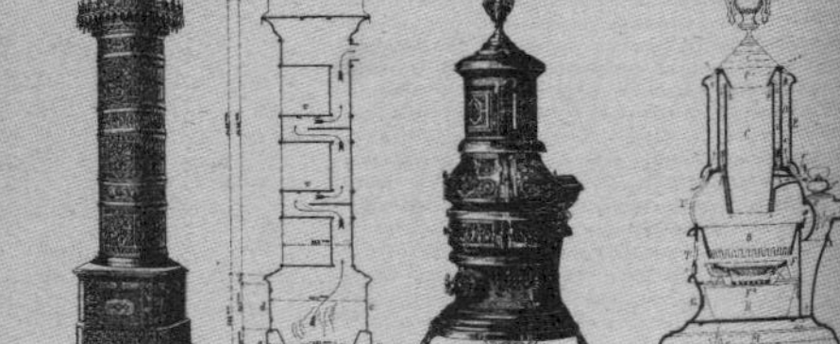

5 links: Säulenofen der Eisengiesserei Königshütte bei Lauterberg, Harz. Durch mäanderartige Führung der Rauchgase wird die Heizfläche vergrössert.
Rechts: Verbesserter Säulenofen von Architekt Leras in Besançon, 1867 an der Pariser Weltausstellung ausgestellt. (Quelle: Breymann)

6 Einzelgehöft der Familie Bul in Shogrom. Die Feuergrube trennt den Wohn-Schlaf-Raum in eindeutige Zonen für verschiedene Nutzungen. (Quelle: Illi)

7 oben: Grundriss eines typischen Barockschlosses: Er zeigt die Raumfolge von den «öffentlichen» Räumen in der Hausmitte zu den «privaten» der Seitenflügel. Die zeremonielle Verbindung der Räume ist nur als Enfilade den Fenstern entlang möglich. Um die Innenhöfe liegt eine Zone von Service- und Nebenräumen, von denen aus die Öfen der Salons geheizt werden.

Der Palast ist mit wundervollen Gusseisen- und Kachelöfen ausgestattet, die alle völlig unabhängig von den Wohn- und Repräsentationsräumen, wo sich das Leben der Herrschaft abspielt, von der Dienerschaft über ein «Versorgungssystem» aus Treppen und Korridoren beheizt werden können. Mit dem Einschliessen in einen Ofen hat das Feuer gleichsam seine ordnende Kraft im Grundriss verloren. Der Ofen rutscht in die Ecke und wird zum Schmuckstück. Noch ist der Ofentritt im Bürger- und Bauernhaus des Grossvaters beliebter Rückzugsort, doch ist das Feuer «ausgeklammert», seine Kraft gezähmt. Aus dem mittelalterlichen Zusammenleben von Herr und Knecht war in unserem Beispiel eine Trennung von Fürst und Lakai geworden.

Wie kam es denn zur Zentralheizung?

Die technische Errungenschaft, unsere Wohn- und Arbeitsräume sauber und gefahrlos von einem entfernt liegenden Wärmeerzeuger aus zu heizen, hat sich weit weg von diesem Ziel angebahnt: Seitdem durch den internationalen Seehandel auch exotische Pflanzen nach Europa gebracht und wegen ihrer Fremdartigkeit, Schönheit und erhofften Nützlichkeit in Viridarien, Orangerien, Palm- und Treibhäusern gehalten wurden, war das Erzeugen von künstlichem Klima in solchen Gebäuden

H. Reck... KALENDE... für Gesundheits-Techniker 190...

...bestimmung
...en Brennmaterials.

Heizfläche = 9,3 m^2.
Rostfläche = 0,3 "
R:H = 1:31

...askoks = 7130 Cal.
...lt = 1,40 %
...bare Subst. = 9,24 "
Substanz = 89,36 "
100,00 %

Mittlere Rauchgastemperatur im Fuchs — 175° · 174° · 221 · 233 · 248
Koksverbrauch pro Stunde — 14,5 · 14,3
Kesselleistung pro Stunde. — 80000 · 80607 · 116952
Wirkungsgrad des Kessels — 75,8 · 77,19 · 75,7 %
Zugstärke im Fuchs m/m Wassersäule. — 2 · 2 · 2

5000 · 6000 · 7000 · 8000 · 9000 · 10000 · 11000 · 12000
5122 · 8603 · 8667 · 12575

38 Jelmoli wirbt mit der

im Konferenzsaal des Deutschen Bundes in Regensburg, in der Marienburg in Maulbronn und im Rathaus zu Göttingen[36].

Ob es sich um Mönche oder Ratsherren handelte, deren geduldiges Sitzen in zugigen Sälen es angenehmer zu gestalten galt, «when our tenderlings complaine of rheumes catarhs and poses», wie Reverend William Harrison 1577 spottet[37]: «Feuerkunst», die über das offene Feuer hinausging, war eher eine Seltenheit, gedieh eher im Schatten des öffentlichen Bewusstseins, etwa in den Klausen der Alchemisten, die ihr geheimnisvolles Wesen mit den Elementen trieben. Es waren die Beobachtungen aufgeklärter Gelehrter und die Resultate erfinderischer Handwerker und Unternehmer, die die mystische Sicht der Natur erhellten. Vor allem waren es die Debatten und Publikationen von Gruppierungen wie etwa der Birmingham Lunar Society und ihrer Tochtergesellschaft der Derbyshire Philosophical Society[38], die anregten, dass der Corpus der so entstehenden naturwissenschaftlichen Erfahrungen gesammelt und verwertbar gemacht wurde.

Zu Beginn des 19. Jahrhunderts war eine Serie von Büchern publiziert, die sich mehr oder weniger ausführlich und kritisch mit den Erfahrungen auseinandersetzten, die man beim Beheizen von Gebäuden gesammelt hatte. Im Vordergrund stand fast durchweg die Rauchrohrheizung. Thomas Tredgold (1788 bis 1829) hatte schon die Leckanfälligkeit der Rauchrohre beklagt[39], Meissner kritisierte die ungleiche Erwärmung entlang der Rauchzüge[40], Heigelin unterstreicht die Eignung dieses Heizsystems für grosse Räume[41]. Einig war man sich jedoch, dass besonders seit der Einführung der viel beweglicheren Rauchrohre aus Gusseisen und der neuartigen Heizkessel die Jahre der Kanalheizung und Feuerrohrheizung gezählt seien.

Vorab die Luftheizung

Auch wenn die Luftheizung unter den nun folgenden Systemen die rascheste Verbreitung erfuhr, vollzog sich die technische Entwicklung praktisch simultan über die Verwendung von Luft, Dampf, Warm- und Heisswasser als Wärmeträger. Auch da finden sich selbstverständlich Frühformen. So beschreibt Meydenbauer die Luftheizung im Marburger Ordensschloss des 15. Jahrhunderts als eine Anlage mit einem grossen Kachelofen, dessen Wärme durch Öffnungen in der Decke in den darüberliegenden Saal entlassen wurde. Eine Anlage dieser Art findet sich auch im Redoutensaal zu Petersburg, wo der grosse Zar bestrebt war, das Neuste und Beste aus dem Westen zu installieren. Friedrich der

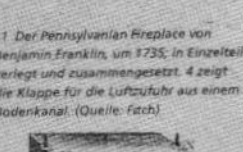
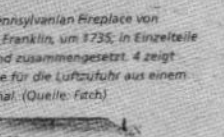

11 Der Pennsylvanian Fireplace von Benjamin Franklin, um 1735, in Einzelteile zerlegt und zusammengesetzt. 4 zeigt die Klappe für die Luftzufuhr aus einem Bodenkanal. (Quelle: Fitch)

Grosse, durch seinen Gesandten informiert, liess eine ähnliche Anlage in Potsdam einrichten.

Eine Warmluftheizung jedoch, die auf die Bedürfnisse einer breiten Bevölkerung zugeschnitten war, den Gegebenheiten der modernsten Produktionsweise ihrer Zeit entsprach und daher sofort weiteste Verbreitung fand, war Franklins «Pennsylvanian Fireplace» 1745 (Bild 11). Das harte Klima der amerikanischen Ostküste gab der Heizung von Beginn der Besiedlung an eine – verglichen mit Europa – überproportionale Bedeutung, was sich in der Grösse sowie der Stellung des Fireplaces im Grundriss der Siedlerhäuser deutlich ausprägt[42]. Trotz dem Holzüberfluss, der die frühe Ökonomie der Amerikaner auszeichnet, kam doch dem Wirkungsgrad eines leicht transportablen Ofens für den beginnenden «westward drive» der Siedler grosse Bedeutung zu. In seinen «Observation on the Cause and Cure of Smoky Chimneys»[43] übernahm Benjamin Franklin wohl Gedanken aus Gaugers «Fires improv'd», das 1715 aus dem Französischen übersetzt worden war, doch resultierten seine Beobachtungen in brauchbarer Hardware.

Der Ofen ist aus Gussplatten zusammensteckbar und somit demontierbar und wurde (als flaches Paket zusammengelegt) im Conestoga-Wagon Richtung Westen mitgeführt. Der Ofen besteht aus einer stehenden Warmluftkammer, die unten durch einen Kanal Kaltluft bezieht und sie oben warm wieder entlässt. Das Feuer schlägt vorne an die Kammer, umfliesst sie oben, zieht auf ihrer Hinterseite hinunter und tritt dort in den Kaminzug ein. Öfen aus Gusseisen waren nichts Neues in den Kolonien. Schon 1642 wird aus Lynn, Mass. ein Gussofen gemeldet – es war insbesondere das Wärmetauschverfahren, das dem Franklin-Stove seine Berühmtheit verschaffte.

Um die Wende zum 19. Jahrhundert begann auch in Europa die verbreitete Anwendung von Luftheizungen ausserhalb religiöser und fürstlicher Häuser. Charles Sylvester beschreibt in seiner Philosophy of Domestic Economy 1819 die Heizung, die der Textilindustrielle Jedidah Strutt (1756 bis 1830) etwa 1793 in seinen Fabriken in Belper und 1806 bis 1810 in der General Infirmary of Derbyshire einrichtete, als «the arrangement of flues and turncaps for the purpose of warming and ventilating the building»[44]. Die Wärme wurde erzeugt durch «high temperatur stoves (man nannte sie cockles) for warming extensive spaces» (Bild 12). Diese Anlagen waren so erfolgreich, dass David Dale und William Kelly 1796 auch die neuen Textilfabriken in New Lanark mit solchen Heizungen versahen (Bild 13). Kanäle, deren Querschnitt in jedem Geschoss verjüngt waren, führten die Warmluft von den giebelseitigen Mauern her in die Arbeitssäle ein[45].

1792 wurde auch in der Mainzer Klinik von Prof. Molitor eine Luftheizung eingerichtet, die durch einen «Windofen» betrieben wurde, bei dem ein Keramik-Luftrohr, mit einem Torf- oder Steinkohlenfeuer zur Rotglut gebracht, die Aussenluft durch einen Trichter ansog, erwärmte und 20 Klafter weit zur Heizung der Krankensäle beförderte[46]. Auch in der Porzellanmanufaktur von Nast bei Paris war 1797 eine Luftheizung

12 Cockle furnace von William Strutt. (Quelle: Sylvester)

36. Vetter. 37. Harrison. 38. Bruegmann. 39. Tredgold. 40. Meissner. 41. Heigelin

Gebaut im Jahre
1841

1. Winterthurer Kachelofen von Abraham Pfau, datiert 1689; die Bemalung von Heinrich Pfau zeigt Bilder biblischen Inhalts. (Quelle: Museum Lindengut, Winterthur)

Wieso gerade Heizungen?

Was konnte ein blutjunges Unternehmen der Giessereibranche, Vater, zwei Söhne und 30 bis 40 Arbeiter, um die Mitte des 19. Jahrhunderts dazu bewegen, Zentralheizungsanlagen bauen zu wollen? Wieso konzentrierten sie ihre Kräfte nicht auf etwas offensichtlich Aussichtsreicheres, etwa den Kundenguss für die allenthalben in der Umgebung entstehenden mechanischen Betriebe oder die Ausrüstung von Textilfabriken, die ringsum gebaut wurden? Da war beispielsweise Heinrich Rieters Spinnerei in der «Mülleren», die wohl einen Teil ihrer Spinnmaschinen selber herstellte, aber trotzdem von Risler frères und Schlumberger im Elsass sowie Ziegler in Wülflingen Maschinen zukaufen musste[1]. Auch hätte sich eine Giesserei an der Entwicklung rationeller Methoden der Rohrproduktion beteiligen und damit den Von-Roll-Werken zuvorkommen können, die mit dem Rohrguss erst in den 50er Jahren begannen[2], ein damals offensichtlich wachsender Markt. Werkzeug- und Dampfmaschinen waren ebenfalls sehr gefragt, wie man bei Escher Wyss in Zürich sehen konnte, die gerade im Begriff waren, sich mit dem Dampfschiffbau einen internationalen Ruf zu erwerben[3].

Wieso gerade Heizungen, wo doch das zentrale Beheizen von Bauten weder einem allgemein anerkannten Bedürfnis entsprach noch eine eigentliche Heiztechnik unangefochten existierte und Aussicht auf kommerziellen Erfolg geboten hätte? Ganz im Gegenteil: Count Rumford (1753 bis 1814), der Erforscher der Natur von Licht und Wärme, hatte 1807 noch skeptisch formuliert: «It is a question often discussed in this country whether living in a warm room in winter be or be not detrimental to health.»

Er fragte dies im Hinblick auf seine Erfahrungen in Deutschland, wo «The rooms of people of rank and fashion are commonly kept in winter at the Temperature of 64 or 65 °F (17,8–18,3 °C)» – in seinen Augen eine krasse Übertreibung, auf die er die bleiche Gesichtsfarbe besagter «people of rank and fashion» zurückführte[4]. Tatsächlich war auf dem Kontinent, vor allem in Skandinavien, Russland und in den Alpenländern, das Heizen einzelner Räume durch Kachelofen anstelle von offenem Feuer schon im Mittelalter bekannt und seit dem ausgehenden 15. Jahrhundert auch in Deutschland beim langsam sich entwickelnden Bürgertum üblich geworden[5]. Daraus hatte sich das Gewerbe der Ofenbauer entwickelt, von denen beispielsweise die Winterthurer Familien Pfau, Graf, Erhart und Sulzer zwischen 1570 und 1740 einigen Ruhm erlangten[6] (Bild 1). Aber Zentralheizungen?

Heiztechnik in der guten alten Zeit

Welches war denn der Stand der Heizkunst früher? Es war die Kunst des

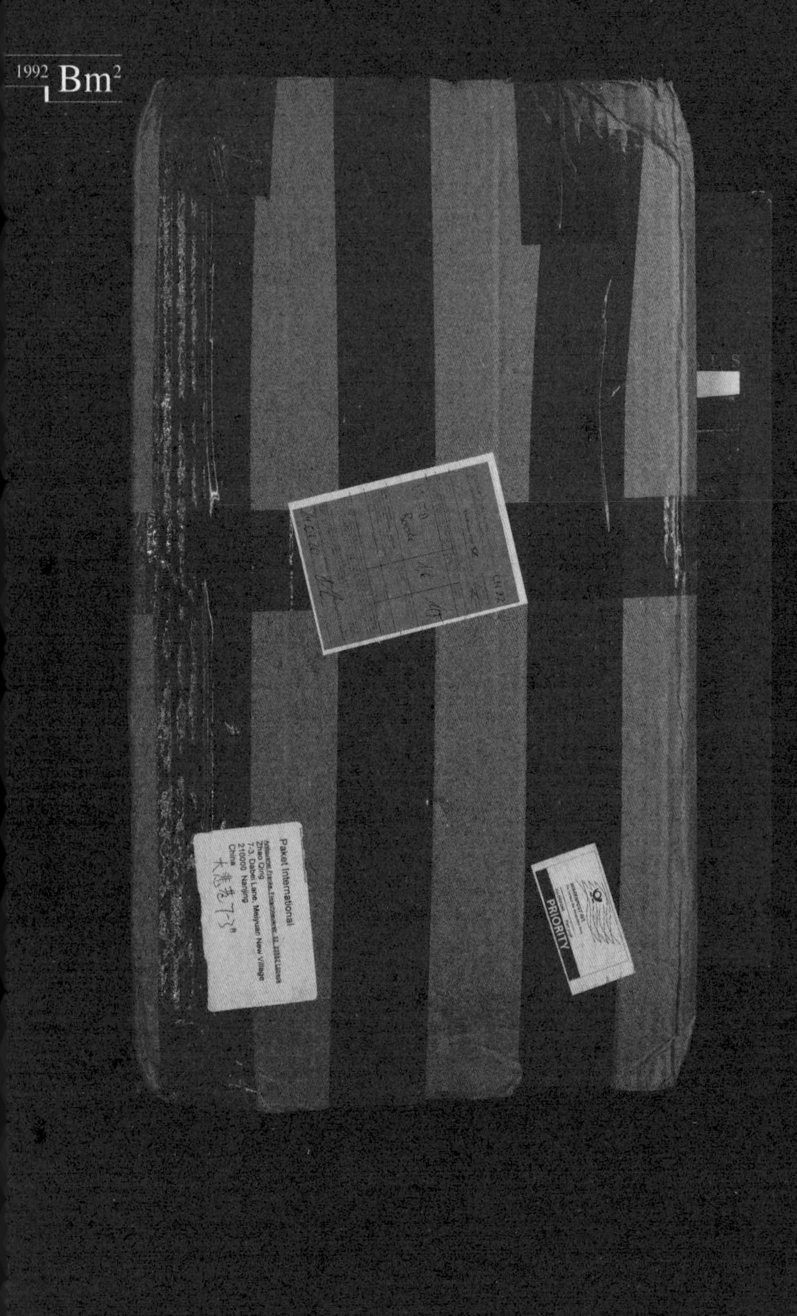
Paket International
Zhao Qing
7-3, Dabei Lane, Meiyuan New Village
210000 Nanjing
China
PRIORITY

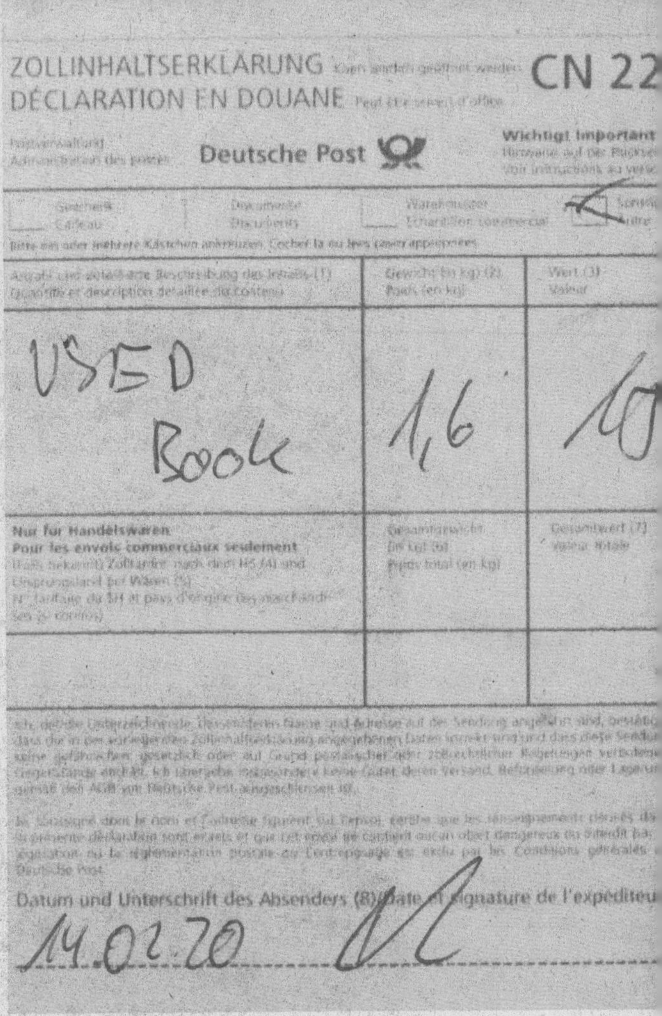

ZOLLINHALTSERKLÄRUNG

DÉCLARATION EN DOUANE

CN 22

Deutsche Post

Wichtig! Important

Geschenk / Cadeau | Dokumente / Documents | Warenmuster / Échantillon commercial | Sonstiges / Autre

Anzahl und detaillierte Beschreibung des Inhalts (1) / Quantité et description détaillée du contenu	Gewicht (in kg) (2) / Poids (en kg)	Wert (3) / Valeur
USED Book	1,6	10
Nur für Handelswaren / Pour les envois commerciaux seulement	Gesamtgewicht (in kg) (6) / Poids total (en kg)	Gesamtwert (7) / Valeur totale

Datum und Unterschrift des Absenders (8)/Date et signature de l'expéditeur

14.02.20

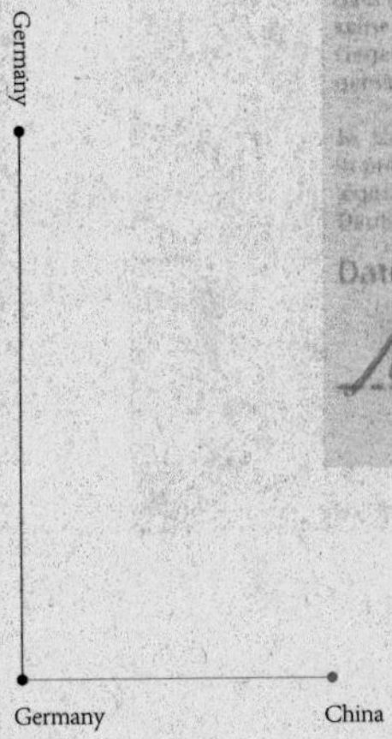

Germany

Germany

China

385mm

15mm

Langen • SPAZIERGANG NACH SYRAKUS edition cantz

268mm

铜奖

步行前往锡拉库萨

——1989 年踏寻约翰·戈特弗里德·佐伊梅的足迹

德国

Bronze Medal

Spaziergang nach Syrakus: Eine Reise nach Johann Gottfried Seume im Jahre 1989

Germany

Karin Girlatschek

Andreas Langen

Edition Cantz, Stuttgart

385 268 15mm

1415g

100p

ISBN-10 3-89-322-235-9

锡拉库萨（Syrakus，意大利文 Syracuse）是一座位于意大利西西里岛东部的古城。由于其位于地中海的重要位置，从古希腊人在此建城开始，这里便是极其重要的贸易中心、交通枢纽和兵家必争之地，也因此留下了相当丰富的历史遗迹。1801 年，德国作家约翰·戈特弗里德·佐伊梅（Johann Gottfried Seume）从莱比锡步行前往西西里岛，历时 9 个月，并在与本书同名的著作中记录了全程，包括途经的教堂、修道院、博物馆等，并加入了自己的观察和评论。本书的作者踏寻着佐伊梅记录的关键地点，从意大利东北部的的里雅斯特（Trieste）纵贯意大利，来到锡拉库萨附近的卡塔尼亚（Catania），历时 3 个月，通过拍摄和随笔记录下近 200 年后这些地区的风貌。本书的装订形式为锁线硬精装。封面胶版纸橙黑双色印刷，裱覆卡板。内页哑面铜版纸印单黑。内页设置为 7 栏，但文字和照片的排版在网格中运用得相当自由，呈现出按“文”索骥的零星记录和个人思考。

Syrakus is an ancient city in the east of the Italian island Sicily. Because of its important position in the Mediterranean, it has been an extremely important trading center, transportation hub and military battleground since the ancient Greeks established a city here. As a sequence, there are quite a lot of historical sites left. In 1801, Johann Gottfried Seume, a German writer, took nine month to walk from Leipzig to Sicily. He recorded the whole process with the same title as this book, including the churches, monasteries, museums he came across, along with his own observations and comments. The author of this book chased the key points recorded by Seume, from Trieste in the northeast of Italy to Catania near Syrakus. It took three months to record the scenes of these areas nearly 200 years later by shooting and writing. The hardcover book is bound with thread sewing. The cover is made of offset paper printed in orange and black, mounted with cardboard. The inner pages are printed on matte coated paper in black. The text is set to 7 columns, but the typesetting of the text and photos is quite free in the grid, reflecting the sporadic records and personal thinking.

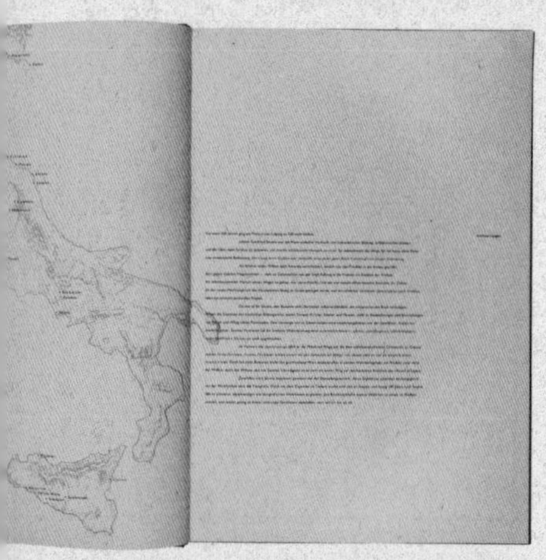

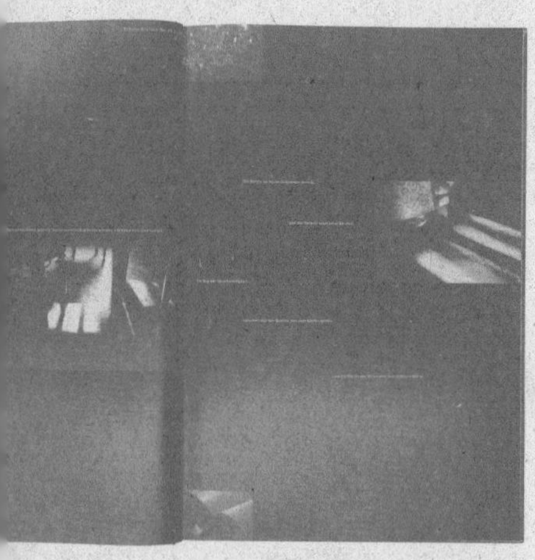
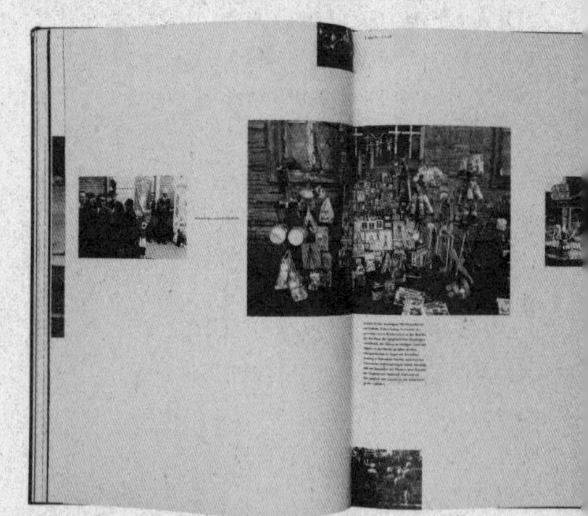

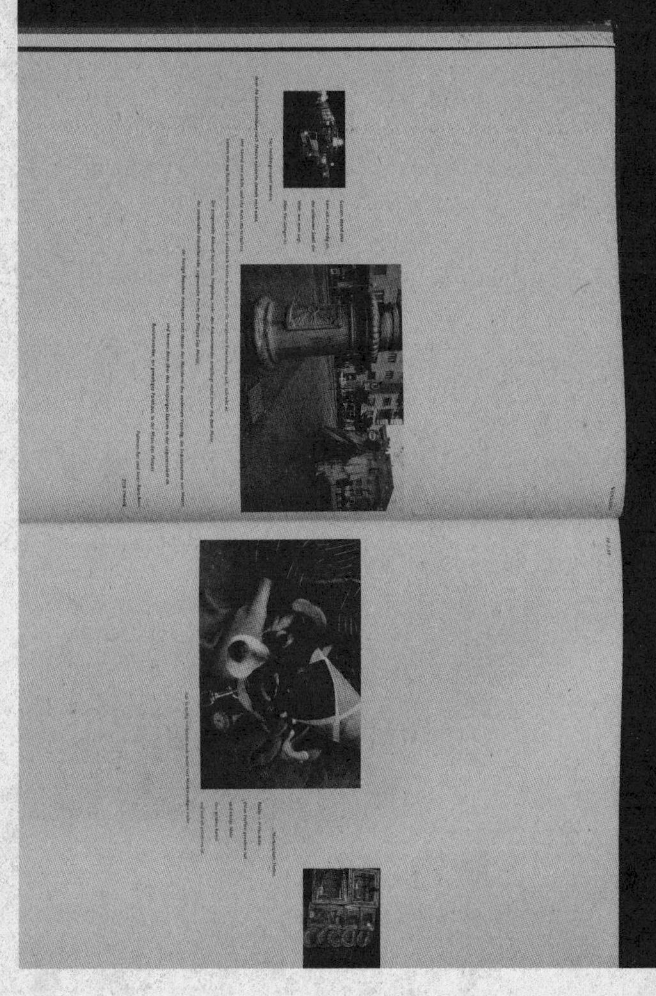

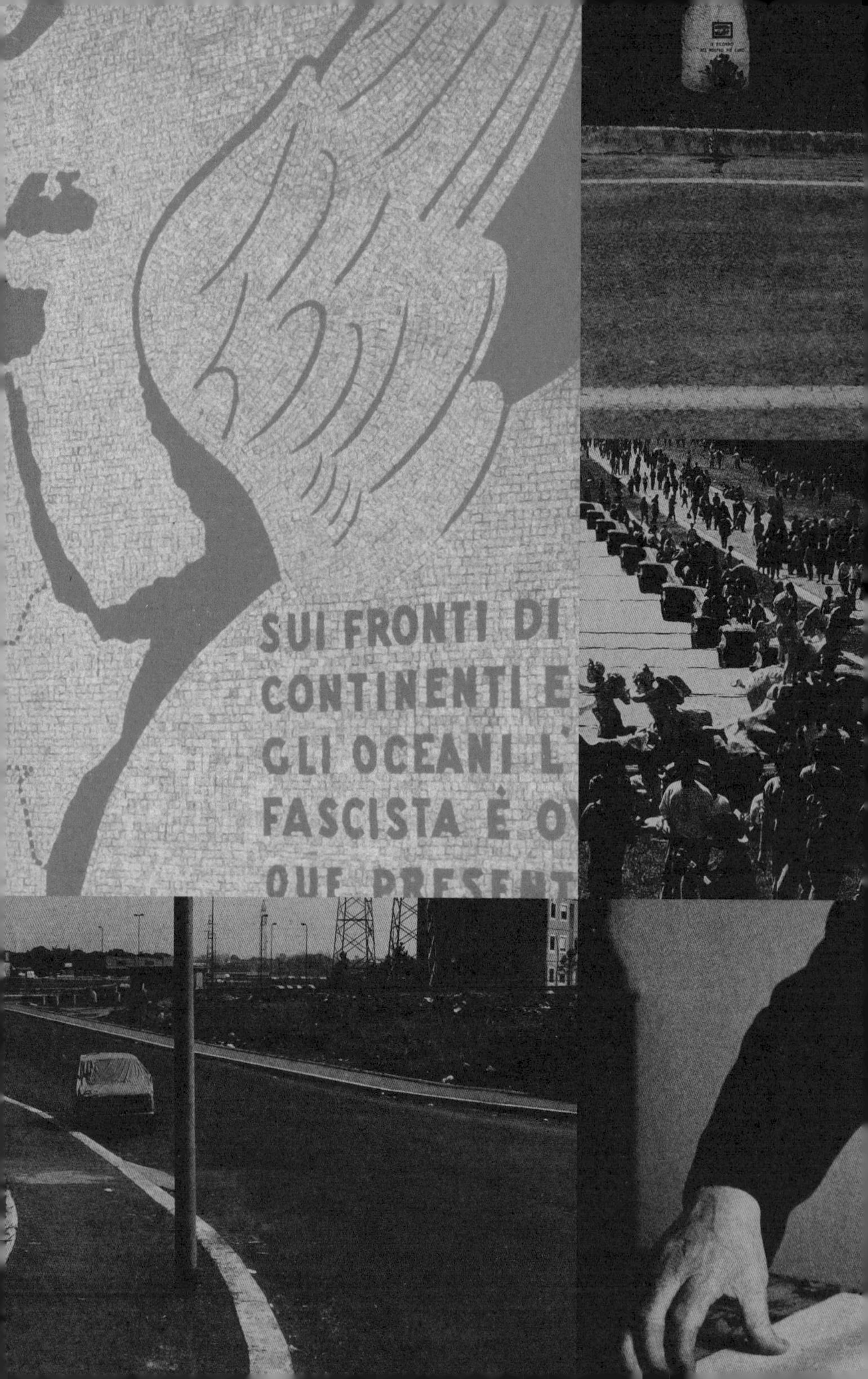
SUI FRONTI DI
CONTINENTI E
GLI OCEANI L'
FASCISTA È O
QUE PRESENT

Von Ferrara nach Bologna nahm Seume die Kutsche. Seine Beschreibung ist lesenswert:

Bm³

VOILÀ
antiquariat.de
Das Verkaufsportal für antiquarische Bücher und mehr
antiquaria
Das Verkaufsportal für antiquarische

CH22
FROM
TO

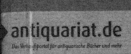

Declaration
May be opened officially
CN22
Designated operator
Deutsche Post
Gift
Documents
Samples
Others
Sales of goods
Quantity and detailed description of contents
Weight (in kg)
Value (in EUR)
1x book
1.11
15.00
Total Weight
Total Value
DE
1.24
15.00
Date and sender's signature:
21-Nov-19
FROM
Plantage 13
28215 Bremen
GERMANY
TRACKED
PRIORITY P.P.
WARENP.IN
TO
ZHAO QING
QING TANG DESIGN CO.
DABEI XIANG 7 - 3
210000 NANJING
NANJING HAN
CHINA
Germany
Germany
China

309mm

17mm

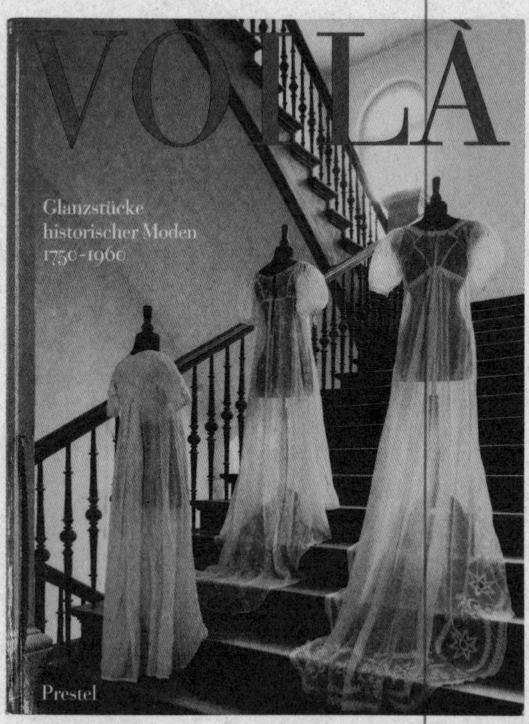

VOILÀ Glanzstücke historischer Moden 1750-1960 Prestel

240mm

● 铜奖 Bronze Medal

《 瞧 Voilà:

——1750－1960 年历史时尚亮点 Glanzstücke historischer Moden von 1750-1960

◐ 德国 Germany

KMS graphic, Maja Thorn

Wilhelm Hornbostel

Prestel Verlag, München

309 240 17mm

1106g

192p

ISBN-10 3-7913-1117-4

德国的 Prestel 出版社出版了一系列有关博物馆收藏的艺术书籍，本书是其中之一。书中展示了来自汉堡工艺美术博物馆的纺织与服装收藏，翻阅的过程好似观看一场 18 世纪 50 年代到 20 世纪 60 年代的欧洲历史服装秀。本书的装订形式为锁线硬精装。护封铜版纸四色印刷覆亮膜，内封红色布面裱覆卡板，书脊上书名烫黑。内页采用哑面铜版纸四色印刷，只在环衬处采用胶版纸单面印深蓝色，作为封面和内页的节奏缓冲。书籍开篇，采用 3 篇文章作为全书总览，分别介绍了汉堡工艺美术博物馆的收藏、服装的时代特征和欧洲服饰 3 个世纪以来的发展史。主体部分是 84 件服装的展示，配上服装本身材质、剪裁、细节的描述文字，有的还配有同一时期相同风格的绘画作品作为辅衬。全书采用了经典字体 Bodoni，与所展示的时尚和时代特征相得益彰。

This book is one of a series of museum art books published by Prestel Press in Germany. The book shows the collection of textile and clothing from the Hamburg Museum of Arts and crafts. The reading experience is like watching a historical fashion show of European clothing from 1750s to 1960s. The hardcover book is bound by thread sewing. The jacket is made of laminated coated paper printed in four colors, and the inner cover is made of red cloth mounted with cardboard. There is only the title of the book hot-stamped on the spine. The inner pages are printed in four colors on matte coated paper, and offset paper with one side printed in dark blue is used for the end paper to sooth the rhythm between the cover and the inner pages. At the beginning of the book, three articles are used as the general view of the book, which are discussed the collection of Hamburg Arts and Crafts Museum, the characteristics of clothing in different times, and the development history of European clothing in the past three centuries. 84 pieces of clothing are displayed as the main part, accompanied with the description of the material, cutting and details of the clothing itself, and some paintings of the same period as the displayed clothing. The classic Bodoni font is adopted in the book, which complements the fashion features displayed.

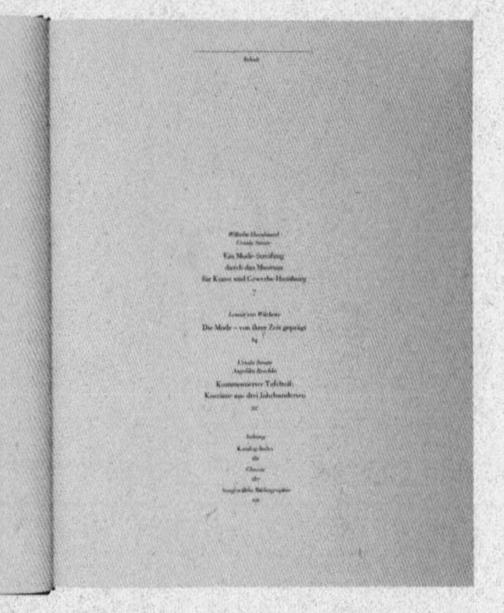

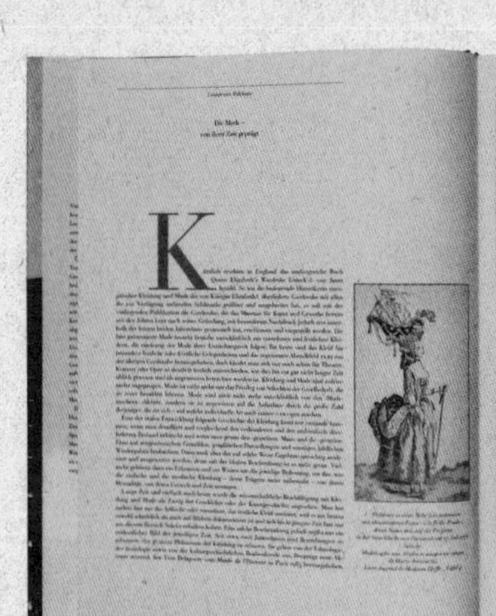

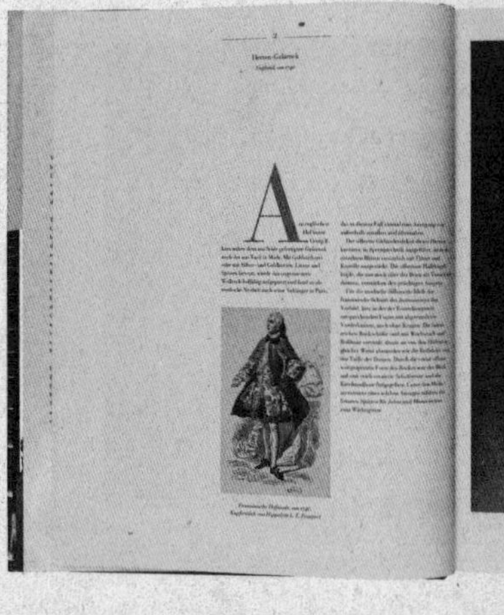

gelockert, tief eingesetzte Falten erweiterten die Röcke und ließen
hend konnten die Ärmel kleine Überärmel oder weich fallende

und dem Zweiten Weltkrieg bestimmte die Pariser Haute Couture
päisch-amerikanische Mode. Dazu gehörten schon lange tätige
its in der dritten Generation; Kat. 61), Poiret, Callot Sœurs (sei
), Lanvin (seit 1914), bald nach 1918 Chanel, Patou (Kat. 74 links),
, Vionnet, dann Lelong, Rochas, Schiaparelli, seit 1934 bzw. 193
l Balenciaga (Kat. 78). Ein Außenseiter, nichtsdestoweniger von
schaft hochgeschätzt, war der Spanier Mariano Fortuny, der sei
e (Kat. 52, 53). Seine Gewänder in klassischem, orientalischem oder
aus kostbaren, eigens für sie geschaffenen Materialien zeichneten
von genialer Schlichtheit aus.
1836 sein Warenhaus in Berlin eröffnete, bot er erstmals fertig
Mäntel an. Trotz verschiedener Vorläufer von in Serien hergestell-
enden Kleidungsstücken hat es dann noch fast ein Jahrhundert
anziger und dreißiger Jahren Konfektionskleidung auch für geho-
eren begann. Erst in den zurückliegenden vier Jahrzehnten wurde
sich – wie etwa Italien – damit lange zurückgehalten haben, zum
geschätzten Angebot mit Markenartikeln und Massenware. Dem
te Couture mit *Prêt-à-Porter-* bzw. *Alta-Moda-Pronta-*Kollektio-
nmenarbeit mit großen Kaufhäusern angepaßt.
gs- und ersten Nachkriegszeit der vierziger Jahre setzte 1947 die
zu einem neuen, freudig begrüßten Modefrühling mit dem *Neu*
e Kat. 76). Während oberhalb der Taille der Körper modellierend
ngen darunter graziöse, weite, wadenlange, durch Petticoats ge-
n Kreationen von Dior zogen in Paris bald ebenso die Schöpfungen
Cardin, Fath, Laroche, de Givenchy, Ricci und Saint-Laurent die
ch. Die fünfziger Jahre brachten höchst reizvolle und tragbare
ostüme und Mäntel dieser jugendlich beflügelten Mode. Als Reak-
1960 die Kleider erneut immer stärker verkürzt, bis sie schließlich
ndeten. Darauf antwortete eine gerade Schnittlinie, oft ohne Naht

Voilà! Vorhang auf zu einer Modenschau
Art! Sie führt den Betrachter und Leser dies
des durch drei Jahrhunderte europäischer
ziergang durch die Pracht und Novitäten vergangener Tage, ein
nuß und Anregung verspricht. Die nachfolgenden Photographien
omas Zimmermann im Museum für Kunst und Gewerbe in Hambu
Mitwirkung von Ursula Strate mit Charme und Bildwitz geradezu
e offenbaren eine geheimnisvolle Welt voller Stimmung und zaube
nternen High-Tech-Zeitalter Sehnsucht nach Vergangenem aufko

Der Wunsch reicher Bürger und
rn, die Lebensformen der jeweils nächst-
ren Gesellschaftsschicht nachzuahmen,
te am ehesten im Bereich der Mode erfüllt
len. Wer genügend Geld für kostbare
zösische Seidengewebe besaß, wie sie für die
sche Garderobe verwendet wurden, zeigte
n Luxus allen Verboten und ständischen
lerordnungen zum Trotz, um so seinen
htum öffentlich zur Schau zu stellen.
e beiden bürgerlichen Schnürmieder mit
eise angeschnittenen und teilweise
esetzten Schößen sind aus derart teuren
engeweben gearbeitet. Das blau-weiß
usterte Jäckchen zeigt einen groß
ortigen Blüten- und Früchtedekor in Weiß,
auf einen mit kleinen Rauten gemusterten
*-de-Tours-*Grund in Blau gesetzt ist, ein

hunde
höfisch
vorneh
bis in d

OI

Glanzstücke historischer Moden 1750-1960

Herausgegeben von
Wilhelm Hornbostel

6

Herrenrock

Italien, um 1750-17

Herrenrock ›Habit à la f

Frankreich, um 178

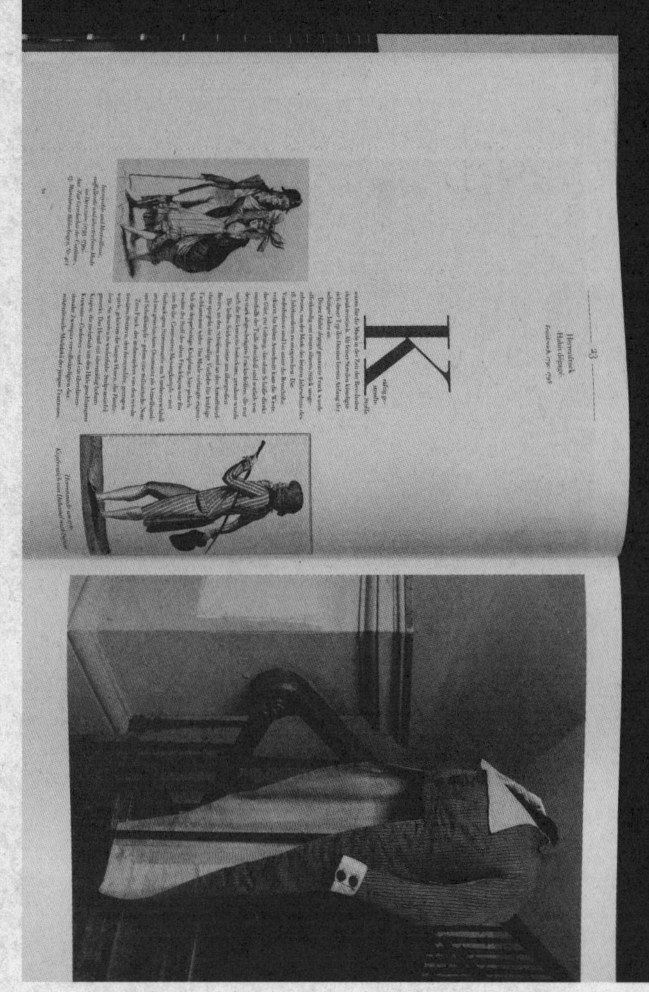

Zwei Tanzkleider

WORTH
Paris, 1926
England, um 1925

Créations Maison Beer
Paris, November 1924

Ein Unterkleid in sattem Goldton schimmert durch das Netz von filigraner Goldspitze (links). Bei dieser Kreation des Hauses Worth tritt der schlichte Schnitt ganz in den Dienst der edlen Materialien. Die Spitzenhülle ist von kleinen elfenbeinfarbenen Perlen und von mattgoldenen und winzigen, wie poliert wirkenden Silberpailletten, die besondere Reflexe erzeugen, übersät. Unter dem tiefgezogenen Gürtel bedecken aus Goldperlen gefädelte Fransen, die sich in gebogten Stufen überlappen, den kniekurzen Rock. Als lange Perlschnüre hängen sie von den schmalen Trägern zu beiden Seiten des Dekolletés und zeichnen jede Bewegung nach.

Chinois inspiriert ist der Dekor des transparenten Kleides aus schmiegsamem fliederfarbenen Baumwolltüll (rechts). Ursprünglich mit einem zartrosafarbenen Unterkleid getragen, kommt die Musterung des Netzes nur auf der schwarzen Figurine besonders zur Geltung.

Mit Durchzugstickereien ist der Tüll zu floralen Ornamenten verdichtet, die mit lila Strasssteinen und transparenten Glasperlen geschmückt sind. Der leicht ausgestellte, tiefansetzende Rock ist auf diese Art in schmale Bahnen unterteilt, deren geschwungene Säume mit Perlengehängen beschwert sind.

全球邮政特快专递
WORLDWIDE EXPRESS MAIL SERVICE
金陵海关驻邮办监封
La Poste
400289565
001-0002
Better World Books
Better World Books
55740 Currant Road
Mishawaka, IN 46545
UNITED STATES OF AMERICA
US1512003200332475
To:
Other
Zhao Qing
Zhao Qing
7 – 3 Dabei Lane Molynan New Vil
Nanjing, CHINA 210000
Package Weight 2.65
CUSTOMS DECLARATION
May be opened officially
Content: Merchandise
Detailed description of contents
Qty
Weight (in lb)
Value (in USD)
BOOKS
1
$23.64
大悲巷 7-3
Total
2.65
$23.64
Country of origin of goods: US
We certify that the particulars given in this declaration are correct and that this item does not contain any dangerous article or articles prohibited by legislation or by postal or customs regulations.
B
0064167728
-1990

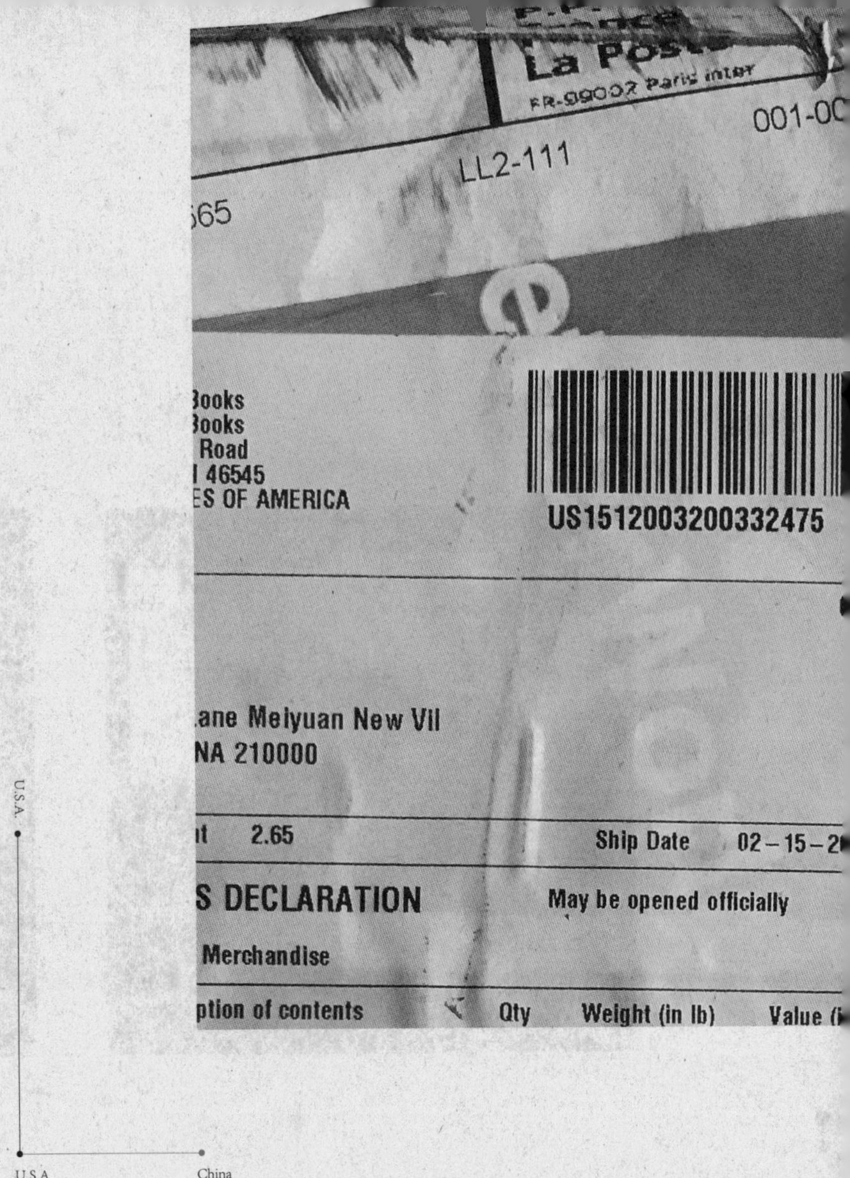
La Poste
FR-99007 Paris inter
001-0C
LL2-111
565
Books
Books
Road
46545
ES OF AMERICA
US1512003200332475
Lane Meiyuan New Vil
NA 210000
2.65
Ship Date
02-15-2
S DECLARATION
May be opened officially
Merchandise
ption of contents
Qty
Weight (in lb)
Value
U.S.A.
U.S.A.
China

258mm

37mm

262mm

● 铜奖

《 日建设计

——建造现代日本 1900－1990

◖ 美国

Bronze Medal

Nikken Sekkei:

Building Modern Japan 1900-1990

U.S.A.

Thomas Cox (Willi Kunz Associates)

Kenneth Frampton, Kunio Kudo

Princeton Architectural Press, New York

258 262 37mm

1863g

288p

ISBN-10 1-878271-01-6

日建设计是一家集规划、建设和工程于一体的建筑公司，在全球多地拥有办事处，员工超过 2000 人，体量居于世界第二。从 1900 年成立至今，日建设计已在全球 50 多个国家和地区完成超过 25000 个项目。作为一家成立已有 100 多年的公司，日建设计在日本的现代化和自然保护上贡献卓著。本书是成立 90 年时的纪念集。书籍的装订形式为锁线硬精装。护封铜版纸印黑覆哑膜，护封四边向内翻折。内封黑色布面裱覆卡板，书脊文字烫银，封面文字烫黑。黑卡纸作为环衬，内页哑面铜版纸四色印刷。内容大致分为日建设计的历史、企业理念、建筑作品、成就和优劣势评估，以及对未来的展望。正文设置为双栏，左对齐排版，每页靠翻口的一栏会在第 13 行处向外侧横移 3 厘米，呈现出富有建筑意味的结构化文本形态。附注的字号比正文小但字重更粗，行距也更窄一些，与正文形成巨大的差异。

Nikken Design is a construction company integrating planning, construction and engineering. It has offices in many locations around the world, with more than 2,000 employees ranking second in the world in volume. Since its establishment in 1900, Nikken Design has completed more than 25,000 projects in more than 50 countries and regions around the world. As a company that has been established for more than 100 years, Nikken Design has made outstanding contributions to the modernization and protection of nature in Japan. This book is a commemorative collection of its 90th anniversary. The hardcover book is bound with thread sewing. The matted jacket is made of coated paper printed in black, and it's designed to fold inwards on all four sides. The inner cover is made of black cloth mounted with cardboard, the text on the spine is silver hot-stamped while the cover text is black hot-stamped. Black cardboard is used as the end paper, and matte coated paper is used for the inner pages. The content is roughly divided into several parts: the history of Nikken Design, corporate philosophy, architectural works, achievements, assessment of advantages and disadvantages, and future prospects. The text is set to double columns, left-aligned typesetting, and the outer column on each page is intentionally move 3cm to the outside at the thirteenth line in order to form a structured text form with architectural meaning. The font size of the notes is smaller than that of the text, but the font weight is thicker, and the line spacing is narrower, forming a huge difference from the main text.

Sumitomo Pavilion

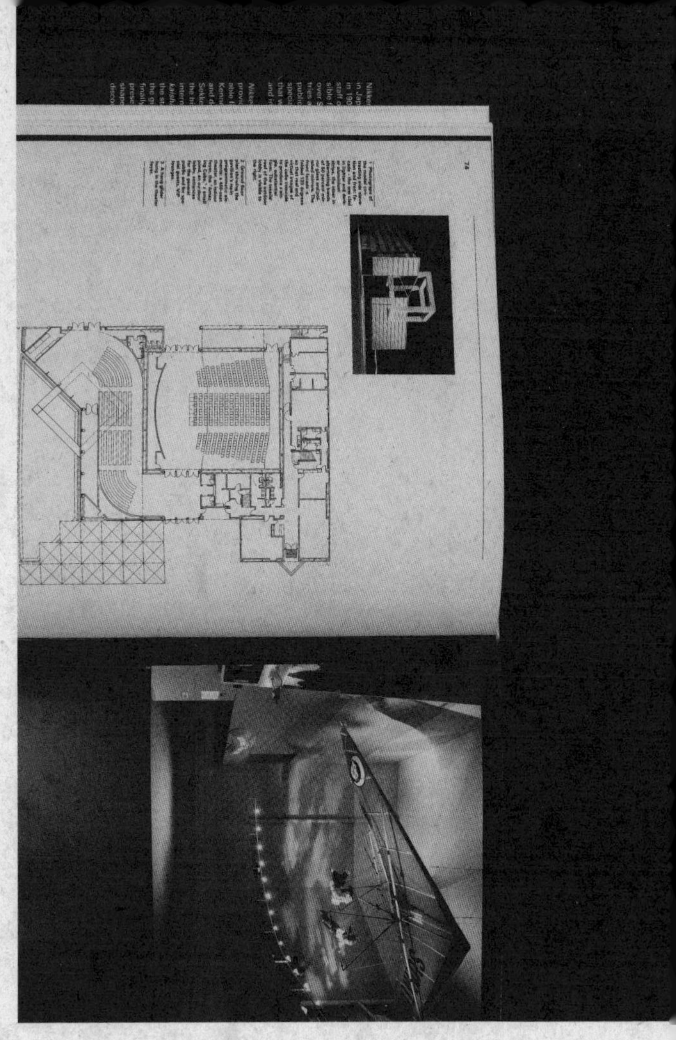

a
b
c
d
DN
DN

momox Shop
medimops
FROM
momox GmbH
Am Alten Flughafen 1
Germany
AIR
PRIORITY
W
Deutsche Post
TO
ZHAO QING
Design Co
210000 Nanjing
CHINA PR
大悲巷7-3
CN 22
ROMAN ROWOHLT

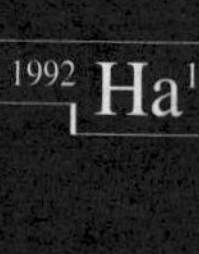

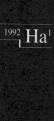

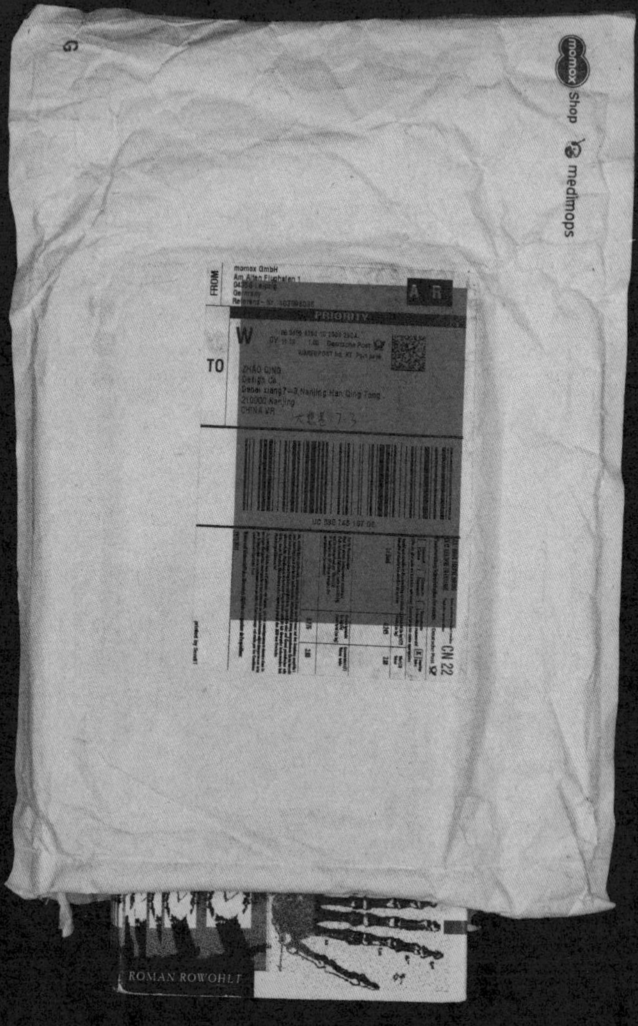

06 Leipzig
many
erenz-Nr.:103698036

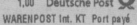

PRIORITY

N 06 30E0 A750 1C 2000 29CA
DV 11.19 1,00 Deutsche Post
WARENPOST Int. KT Port payé

ZHAO QING
Design Co.,
Dabei xiang7-3,Nanjing Han Qing Tang .
210000 Nanjing
CHINA VR 大悲巷 7-3

UC 030 746 167 DE

Germany

Germany China

210mm

28mm

133mm

荣誉奖

给那没救我命的朋友

德国

Honorary Appreciation

Dem Freund, der mir das Leben nicht gerettet hat

Germany

Joachim Düster

Hervé Guibert

Rowohlt Verlag, Reinbek

210 133 28mm

380g

256p

ISBN-10 3-498-02463-9

法国作家、摄影师艾尔维·吉贝尔（Hervé Guibert）在书中记录了他被确诊艾滋病后最后几年的事，包括与自己关系密切的几位朋友：同样患病身故的法国哲学家米歇尔·福柯（Michel Foucault），曾经相恋但之后分手的演员伊莎贝尔·阿佳妮（Isabelle Adjani），以及研发艾滋病疫苗但最终与吉贝尔疏远的比尔。书籍的装订形式为锁线硬精装。护封铜版纸四色印刷，过油处理，内封黑色布纹纸书脊烫银。环衬选用灰色艺术纸，内页胶版纸印单黑。书中的文章一共 100 篇，按照时间顺序排列，部分包含写作日期。半日记半虚构的方式记录患病后的日子，包括死亡、艾滋病、同性恋、各种迷失的自我和朋友之间的故事。每一篇的页眉是从 1 到 100 的自然数和从左至右的进度条，当进度条逐渐盖住数字，文章和情绪也逐渐到达终点。书籍于 1990 年出版，而 1991 年底吉贝尔因病去世。

French writer and photographer Hervé Guibert recorded in the book the last few years after he was diagnosed with AIDS, including several close friends: a French philosopher Michel Foucault who also died of illness, actor Isabelle Adjani who once fell in love but later broke up, and Bill, who developed an AIDS vaccine but was eventually estranged from Guibert. The hardcover book is bound with thread sewing. The jacket is made of vanished coated paper printed in four colors, and black arlin paper is used for the inner cover with silver hot stamping on the spine. The end paper is made of gray art paper, and the inside pages are made of offset paper printed in single black. There are 100 texts in the book, arranged in chronological order, partly containing the writing dates. Half real and the author records the days after the illness, including the stories of death, AIDS, homosexuality, various lost self and friends. The author recorded the days after the illness as half diary and half fiction, including the stories of death, AIDS, homosexuality and friendship. The header of each article contains a natural number from 1 to 100 and a progress bar from left to right. When the progress bar gradually covers the number, the stories and emotion gradually reach the end. The book was published in 1990 and Guibert died at the end of 1991.

prozeß, der in meinem Blut begonnen hat, grei
Tag weiter um sich und läßt meinen Fall zur Ze
ıkopenie erscheinen. Die jüngste Analyse, si
n 18. November, gibt mir 368 T4-Zellen, ei
;t bei guter Gesundheit über rund 1000 bis 130
T4-Zellen sind jene Gruppe weißer Blutkörpei
Aids-Virus hauptsächlich angreift, wodurch e
chutz nach und nach schwächt. Die schwerste
e Pneumocystis, welche die Lungen, und die To
welche das Hirn befällt, schalten sich im Bereic
-Zellen ein; mittlerweile verzögert man sie mit
ibung von AZT. Zu Beginn der Geschichte vo
man die T4-Zellen «the keepers», die Hüter, un
n, eine andere Fraktion der Leukozyten, «th
Mörder. Vor dem Auftauchen von Aids hatte ei
n Computerspielen das Umsichgreifen de
Blut vorgezeichnet. In seinem Spiel für Jugend
das Blut auf dem Bildschirm als Labyrinth, i
nan umherschweifte, ein gelber, von einem He
er Shadok, der im Vorbeigehen alles frißt, di
n Gänge von Plankton leert und dabei zugleich
hlreicher umherwimmelnden roten, noch gefrä
oks bedroht wird. Wollte man das Pacman

91

tag, 7. Oktober, auf Elba: Kaum haben wir mit den Ge-
änden und Kartons, die wir von meinem Pavillon aus
mitgebracht haben, das Haus betreten, da klingelt das
on, Gustave nimmt ab, ich höre ihn sagen: «Ja, Bill.»
uft aus New York an, er hat Hummeln im Hintern, wir
ren, daß Robin ihm den Marsch geblasen hat, er sagt,
neys Impfstoff habe tags zuvor endlich von einer über-

17

ie Mitteilung der gutartigen Mißbil-
nach die Spasmophilie-Theorie fru-
ür den Augenblick seiner Leidens-
ohne Zweifel gierig, tief in seinem
umhertastend. Ich erlitt keine epi-
war ich in der Lage, mich von einem
uchstäblich vor Schmerzen zu win-
wenig gelitten wie seit dem Zeit-
ß ich Aids habe, ich achte äußerst
ichen für das Vordringen des Virus,
ie seiner Kolonisierung zu kennen

99

r das Gelände des Flughafens von Miami
Hause zu fahren, fällt das Licht von Bills
einen struppigen jungen Mann, der bar-
Autobahn entlangläuft. Er läßt ihn in sei-
n Jaguar einsteigen, nimmt ihn mit zu sich
t ihn in der Badewanne mit Ausnahme des
as die seltsame Figur ihn nicht berühren
im Dunkeln im Bett. Anderntags nimmt
e mit, um ihn von Kopf bis Fuß neu einzu-
sagt Onkel zu ihm. Aus Angst, er könne
und weil er außerdem zu einer Geschäfts-
uß, bringt Bill den jungen Mann am näch-
endherberge, zahlt ihm zehn Tage Aufent-

Dem Freund, der m
das Leben
nicht gerettet hat

Hervé Guib

Roman

Bill, Diego und ich fuhren, da alle Flugzeuge am Wochenende des 14. Juli ausgebucht waren, in einem überfüllten TGV zurück, worin wir der Bequemlichkeit halber einen Schichtdienst zwischen der Bar und Winkeln im Abteil organisierten, in denen wir am Boden sitzen konnten. Lachend las Bill mein Buch, das ich gerade vom Verleger erhalten und ihm geschenkt hatte, mit der überlegtesten, ernstesten und herzlichsten Zueignung, die ich ihm je geschrieben hatte, es war von mir gewiß gewagt. Ich bewahrte an den Fingerspitzen den Genuß, den ich am Vorabend empfunden hatte, als ich den Rücken jenes wunderbaren jungen Mannes, Laurent, streichelte, und dies Kribbeln stieg mir bis ins Herz, dies perfekte Beispiel von unfreiwilligem «safer sex» badete es in Sinnlichkeit. Bill ging anderntags ins Val-de-Grâce, wo er sich der Naht seiner Bauchmuskulatur unterziehen mußte, und für mich hieß es warten, hieß es tagtäglich fieberhaft zum Briefkasten hinuntergehen, um nach dem dicken Umschlag aus dem Institut Alfred-Fournier zu schauen, den ich an Gustave hatte adressieren lassen und der querüber den Stempel «Arztgeheimnis» trug, den Stempel der tödlichen Krankheiten, die Ergebnisse der letzten Analysen, die ich vor meiner Abreise nach La Coste hatte machen lassen. Als ich mich ins Institut begeben hatte, mit nüchternem Magen, und als ich es wieder verließ, um ins nächste Bistro zu rennen und dort Kaffee zu saufen und mich hysterisch mit Croissants und Bri-

Krankheit angegriffen gefühlt, ich war sicher, daß die Ergebnisse schlecht ausfallen und mich in ein anderes Stadium des Bewußtseins von meiner Krankheit bringen sowie Doktor Chandis und des medizinischen Instituts Haltung ihr gegenüber verändern würden. Es war aber so, daß diese dicke, doppelt gefaltete Postsendung, die ich mit aller Hast oder verdächtig langsam gleich am Briefkasten glattstrich, wobei ich mich auf das Blatt mit dem Eintrag über meine T4-Helferzellenzahl stürzte, mir enthüllte, daß ich mich in dem Augenblick, als ich mich von meiner Krankheit so geschwächt gefühlt hatte, in Wahrheit in einer Linderungs- oder gar Rückzugsphase der Krankheit befunden hatte, denn meine T4-Zahl war auf mehr als 550 gestiegen und hatte sich fast dem normalen Bereich genähert, so nah wie noch nie seit dem Beginn der Analysen, welche die spezifische Wirkung von HIV auf die Entvölkerung dieser Art Lymphozyten untersuchen, mein Körper hatte das vollbracht, was Doktor Chandi eine Spontanbesserung nannte, ohne Unterstützung irgendeines Medikaments, weder Immuthiol noch was auch immer. Ich spürte, wie ich da vor meinem Briefkasten stand, etwas wie den Anflug lebensspendender Luft, ein Gefühl des Entrinnens, ein sich Ausweiten der allgemeinen Perspektive; das Schmerzlichste in den Bewußtheitsphasen der tödlichen Krankheit ist zweifellos die Entbehrung der Ferne, aller möglicher Fernen, wie eine unentrinnbare Blindheit in der gleichzeitigen Ausdehnung und Zusammenziehung der Zeit. Meine Resultate versetzten Doktor Chandi in seiner Praxis in Freude, er lachte, er sagte, Elba, das Baden im Meer und in der Sonne, die Ruhe, die ganze Lebensart schlügen mir gut an, ich solle aber zugleich nicht allzuviel ruhen, was er ein wenig befürchte, übertriebene Ruhe könne sich fatal auf die Lebenskräfte auswirken. Ich brachte Bill meine Ergebnisse in das Val-de-Grâce,

a²

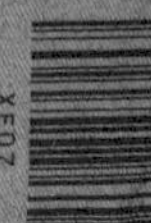

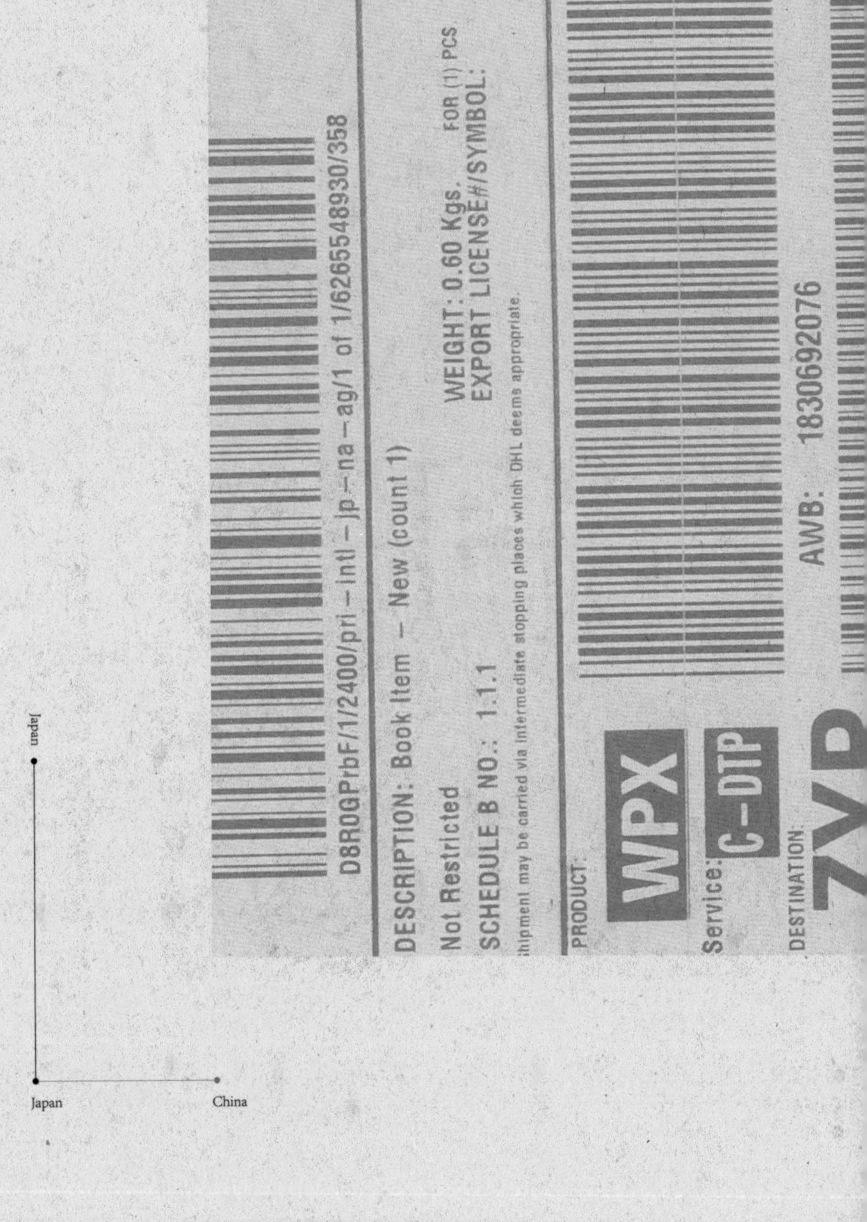
D8R0GPrbF/1/2400/pri – Intl – jp – na – ag/1 of 1/626554893О/358
DESCRIPTION: Book Item – New (count 1)
Not Restricted
WEIGHT: 0.60 Kgs. FOR (1) PCS
SCHEDULE B NO.: 1.1.1
EXPORT LICENSE#/SYMBOL:
hipment may be carried via intermediate stopping places which DHL deems appropriate.
PRODUCT:
WPX
Service:
C – DTP
DESTINATION:
AWB: 1830692076
Japan
Japan
China

241mm

12mm

もりへ さがしに
え 村田清司
ぶん 田島征三
こうせい 宮崎喜一
もりへ さがしに
え村田清司 ぶん田島征三 こうせい宮崎喜一
偕成社

300mm

荣誉奖
林中探秘
日本

Honorary Appreciation
Searching in the Forest
Japan
Kiichi Miyazaki
Seizo Tashima
Kaisei-Sha Publishing Co, Tokyo
241 300 12mm
537g
44p
ISBN-10 4-03-435080-6

绘本的绘制者村田清司患有唐氏综合征，在特殊学校学习过，从小喜爱绘画。之后从事陶艺制作和手工造纸，绘制具有其独特风格的绘画作品，本书便是他的代表作。书籍是典型的绘本形式，横开本锁线硬精装。护封胶版纸四色印刷，内封胶版纸四色印刷，裱覆卡板。内页胶版纸四色印刷。书籍的内容主要由村田清司多色点彩风格的绘画为主，描绘他在树林中的奇妙幻想。同为绘本画家的田岛征三为画面配上了童话故事。设计则由艺术家宫崎喜一完成，图文倾斜、绕排，部分字体经过处理与图画风格相匹配。

The painter of the picture book, Kiyoshi Murata, suffered from Down syndrome. He studied in a special school and loved painting since childhood. When he grew up, he was engaged in pottery-making and hand-made papermaking, and drew paintings with his unique style. This book is one of his representative works. It is a typical picture book, which is hardbound with thread sewing. Offset paper is used for the jacket and the inner cover, both printed in four colors and mounted with cardboard. The insides pages are made of offset paper printed in four colors as well. The book depicts the author's wonderful fantasy in the woods, using multi-color pointillism style. As a picture book painter, Seizo Tashima matched the picture with a fairy tale. Artist Kiichi Miyazaki completed the design – the pictures and text were sloped and finely arranged, some fonts were processed to match the painting style.

モクモクはオノをかついで森に木を切りにでかけました。
森にはいろんな木がありますが、モクモクは、ノンキという木をさがしにゆくのです。
モクモクは、ノンキの木でテーブルといすをつくろうと思っていま
そして、そのテーブルを囲んでにぎやかなパーティーをしたり、
ひとりでしずかにお茶をのんだりもしたいのです。
テーブルの上にはパラソルをたてようとかんがえています。

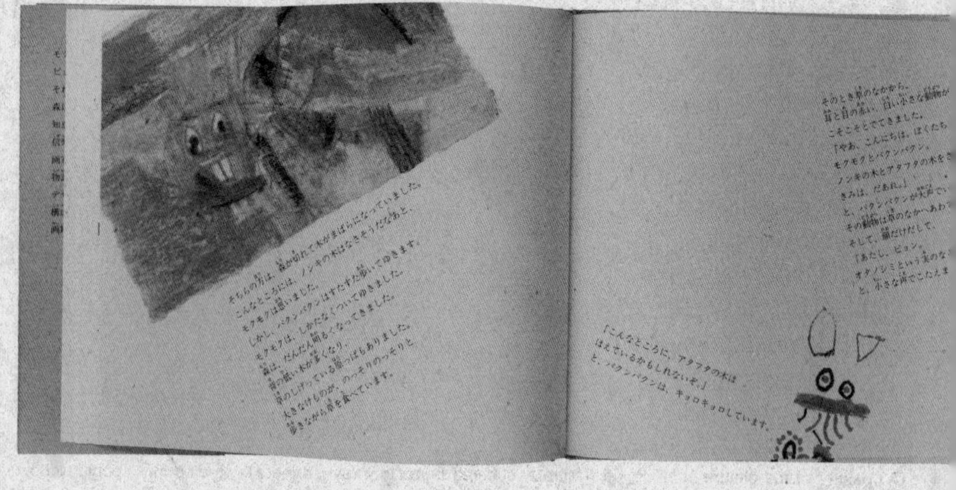

そちらの方は、森が切れて木がまばらになっていました。
こんなところには、ノンキの木はなさそうだなあと、
モクモクは思いました。
しかし、パクンパクンはすたすた歩いてゆきます。
モクモクは、しかたなくついてゆきました。
森は、だんだん明るくなってきました。
背の低い木が多くなり、
草のしげっている原っぱもありました。
大きなけものが、のっそりのっそりと
歩きながら草を食べています。
「こんなところに、アタフタの木は
はえているかもしれないぞ。」
と、パクンパクンは、キョロキョロしています。
そのとき草のなかから、
耳と目の赤い、白い小さな動物が
こそこそとでてきました。
「やあ、こんにちは。ぼくたち
モクモクとパクンパクン。
ノンキの木とアタフタの木を
きみは、だあれ。」
と、パクンパクンが大声で
その動物は草のなかへあわて
そして、顔だけだして、
「あたし、ピョン。
オタノシミという実のな
と、小さな声でこたえま

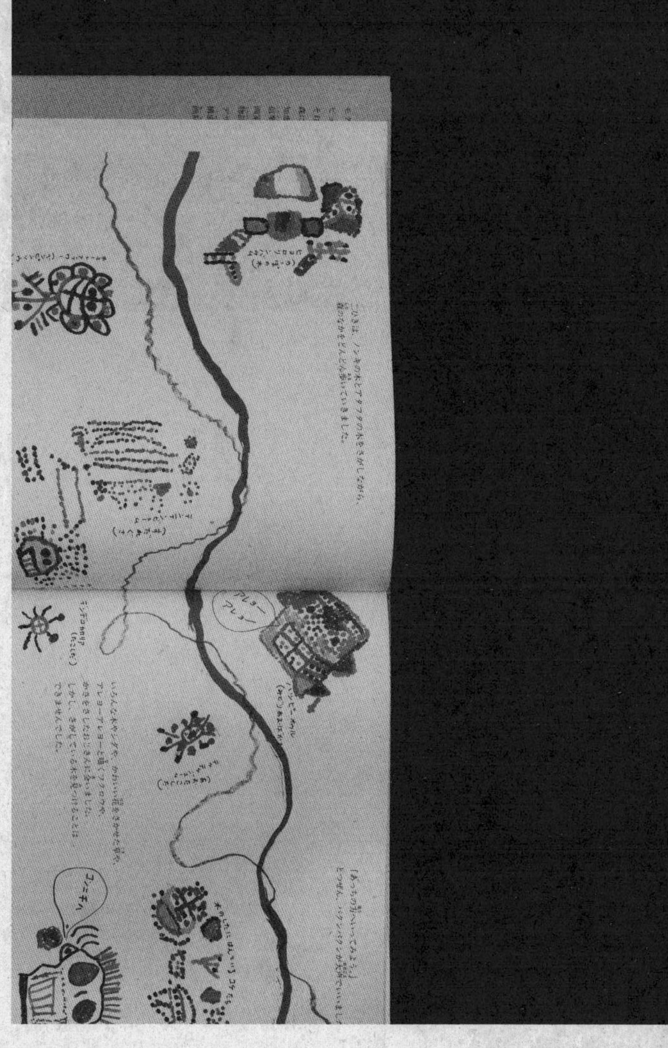

ミラ

5605

9784560041796

Small Packet
Air Mail

From:
Ryosuke Togoe
1F 7-18-1 MURASAKIBARU KAGOSHIMA-SHI
KAGOSHIMA
8900082 JAPAN

To:
EVA ZHAO
65 5TH AVE 1010 NEW YORK NY 10003
NEW YORK NY

10003-3003 UnitedStates
Phone: 6265546930

HAKATAKITA
Postage
Paid
JAPAN

税関告知書
CUSTOMS
DECLARATION

May be opened officially

CN22

Designated operator

JAPAN

✓ Gift 贈物 / Documents 書類 / Commercial sample 商品見本 / Other その他

内容品の数量及び明細 Quantity and detailed description of contents (1)	重量 Weight (in kg)(2)	価格 Value(3)
Book × 1		€ 11.07
商用物品のみ For commercial items only 分かれば、HSコードおよび原産国名 If known, HS tariff number (4) and country of origin of goods (5)	合計重量 Total Weight (in kg)(6)	合計価格 Total Value(7)
	0.360	€ 11.07

I, the undersigned, whose name and address are given on the item, certify that the particulars given in this declaration are correct and that this item does not contain any dangerous article or articles prohibited by legislation or by postal or customs regulations

署名及び署名日付
Date and sender's signature (8)

Togoe

404-8198109-0053133

goe
MURASAKIBARU KAGOSHIMA-SHI
A
JAPAN

E 1010 NEW YORK NY 10003
NY

UnitedStates
5548930

HAKATAKI
Postage
Paid
JAPAN

書
S
ATION

職権により開封されることがあります。
May be opened
officially

CN22

rator

APAN

贈物		Commercial sample	商品見本	該当するものにチェ ックをしてください。 Tick one or more bo
書類		Other	その他	

び明細 detailed description	重量 Weight (in kg)(2)	価格 Value(3)

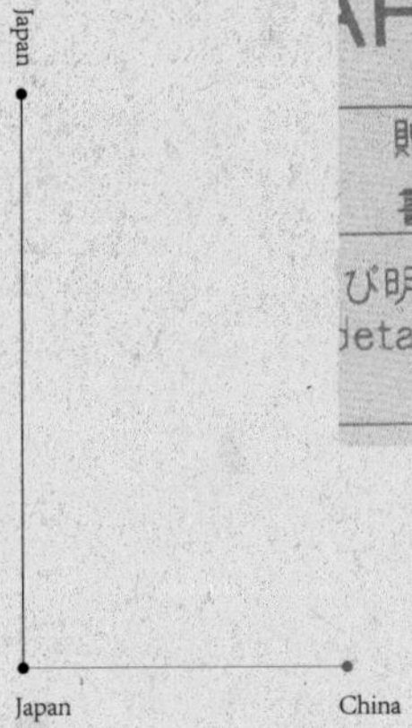

Japan

Japan

China

194mm

18mm

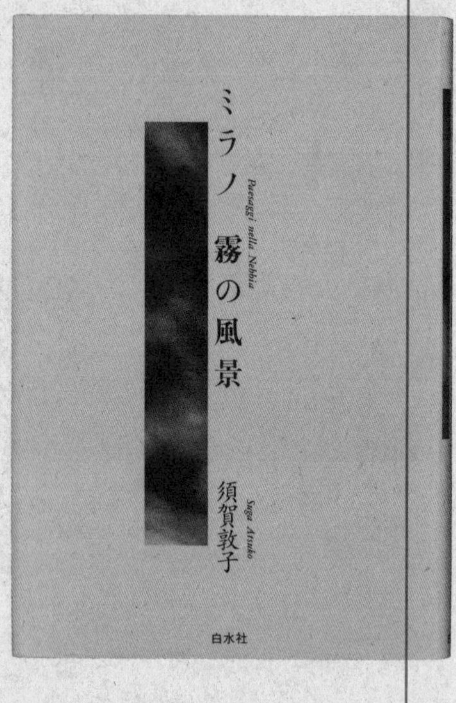

140mm

● 荣誉奖
《 米兰——雾中风景
◖ 日本

Honorary Appreciation ◎
Milan: Paesaggi nella nebbia 《
Japan
Koji Ise
Atsuko Suga △
Hakusui-Sha, Co., Tokyo
194 140 18mm
336g
220p
ISBN-10 4-560-04179-2

日本散文家须贺敦子早年在日本的天主教会学校学习，之后前往法国巴黎学习神学。但她的皮肤很不适应巴黎的气候，遂开始转向意大利求学。之后在意大利生活多年，并成了意大利文学翻译家和散文作家。本书是她在意大利米兰的一段生活经历。书籍的装订形式为圆脊锁线硬精装。护封铜版纸四色印刷覆哑膜，内封浅粉色横纹艺术纸。书脊处文字印黑，裱覆卡板。环衬选用与内封相同的浅粉色横纹艺术纸形成连贯的过渡，内页胶版纸印单黑。在作者的印象中日本很少有雾，当别人问她米兰最具特色的事情是什么时，她毫不犹豫地回答——雾。书中都是她在雾中观察米兰的各种回忆——生活、运河、教堂、铁路，以及朋友们。

Japanese essayist Atsuko Suga studied in the Catholic school in Japan during her early years, and then went to Paris to study theology. Since her skin didn't adapt to the air conditions in Paris, she turn to Italy for further studies. She has been living in Italy for many years and became an Italian literary translator and prose writer. This book records her life experience in Milan, Italy. The hardcover book is bound by thread sewing with rounding spine. The matted jacket is made of coated paper printed in four colors, and the inner cover is made of light pink art paper with black text printed on the spine, mounted with cardboard. The same light pink art paper is used for the end paper to form a continuous transition, and offset paper is used for inner pages printed in single black. In the author's impression, it was seldom foggy in Japan. When people ask her about the most distinctive thing in Milan, she will not hesitate to answer 'fog'. The book is filled with memories of her observation of Milan in the fog – life, canals, churches, railways, and friends.

ラノ　霧の風景

Paesaggi nella Nebbia

遠い霧の匂い　5

いままに　すっと昔から民謡やオ　フスに歌われてきた霧が
はめずらしくなったという。暖房に重油をつかわなくなっ
だろうか。あんな霧、なくなったほうがいいですよ、とミラ
が、古くからのミラノ人は、なんとなく淋しく思っている。
う二十年もまえのことになるが、私がミラノに住んでいた

マリア・ボットーニの長い旅

い一枚の写真。真夏の太陽に照らされたジェノワの記念墓地の白い大理石の階段の
、まっ黒に日焼けした私が、白いツーピースを着てかすかに笑っている。白黒の写
白い柱廊のうしろには、何本かの糸杉が黒くそそりたっている。日本を出て四十日
の日、一九五三年八月十日の朝、私はイタリアに上陸したばかりだった。雨に洗わ
ェノワの埠頭に迎えに出てくれたのは、まったく初対面のマリア・ボットーニだっ
にいる共通の知人の頼みをこころよく引受けて、ジェノワ経由でパリに向う私を彼
まで案内してくれるはずだった。At-su-ko? とまぶしそうに、突堤に停泊した船を

ネといえと、北イタリア人の蔑称である、トウモロコシの粉
るい入れて、すっかり底が焦げついてしまうまで、ただ捏ねに捏
肉料理を添えて、そのソースで食べる。仕事から帰ってきて、女
れ、夫は、あ、ポレンタだな、いい匂いだ、と言いながら台所に入
日は、これはきっと晴れるぞ、と言って、夫は上機嫌だった。事実
ると、うそのように青い空が顔を出す。「ロンバルディアの空は、美
美しい」――文豪マンゾーニがそう書いている、と夫は出典の真
ムを唱えるように言って、青く晴れた空をたのもしそうに眺めた
節になると、飛行場もまったく無用の長物となってしまう。私たち
テの辺りの霧はとくにひどくて、春が来るまでは飛行場もただの

った。それで、来ない、来ないと心配しているローザを、きっと急に都合がわるくなったのよ、と平気な顔でなぐさめて、彼女のためにタクシーを呼んだ。

翌日の新聞で、私たちはテミの操縦していたグライダーが、山に衝突して墜落し、テミが行方不明になったことを知った。生存の可能性はまったくないという。雪が深くて、春まで事故の現場には登れない、と新聞は報じていた。

ミラノに霧の日は少なくなったというけれど、記憶の中のミラノには、いまもあの霧が静かに流れている。

チェデルナのミラノ、私のミラノ

「カミッラ大伯母はスカラ座に桟敷を持っていた。それは彼女がボルディ・ペッツォーリ侯爵の遺産としてもらったもので、舞台わき右側四番目の桟敷だった。大伯母はたいへんな資産家で、たいへんなけちんぼう、音楽好きで、宝石のコレクションが趣味であった。彼女は水腫病をはじめとする様々な病気もちで、ほとんどいつもふせっていた。大伯母が外出するのはスカラ座に行くときだけだったが、そんなとき、ガウンにスリッパという格好で裏階段をあえぎあえぎ上って行った。それはとてつもなく大きなガウンで（深紅色のビロウドに手編のレースをいちめんに縫いつけたしろもので、大伯母が亡くなってから、母はそれを安楽椅子に張らせた）、胸には数知れぬ宝石がかがやき、遠くからでもきらきらと光ってみえた。オペラ・グラスを持っていない人の目にはそれは勲章をちりばめた将軍の胸かとうつった。それぞれの出し物の初日には真珠のネックレス

1992 Ha4

CN22 - CUSTOMS DECLARATION
FIRST-CLASS PKG
INT'L SERVICE
U.S. POSTAGE
PAID
ePostage
TO
F

The Netherlands

U.S.A.

China

271mm

18mm

254mm

● 荣誉奖
《 建筑师亚历山大·博顿
——荷兰建筑师专论
◖ 荷兰

Honorary Appreciation
Alexander Bodon, architect:
Monografie van Nederlandse architecten
The Netherlands
Reynoud Homan
Maarten Kloos
Uitgeverij 010, Rotterdam
271 254 18mm
995g
136p
ISBN-10 90-6450-087-8

荷兰建筑师专论丛书，旨在记录重要的荷兰设计师及其令人印象深刻的作品。本书是丛书中的一本（丛书中的另一本获得 1991 年铜奖，可在相关页面查找），对匈牙利裔荷兰设计师亚历山大·博顿（Alexander Bodon）的职业生涯、作品和理念进行了全面展示。书籍的装订形式为锁线硬精装。护封铜版纸浅灰、灰蓝、黑三色印刷，覆哑膜。内封黑色布面裱覆卡板，书脊处烫黑处理。黑卡纸作为环衬，内页混合使用哑面铜版纸、黄色和白色胶版纸，单黑印刷。书中展示了博顿在荷兰设计的 21 栋建筑，包含书店、会议中心、艺术空间、办公楼、餐厅、住宅、诊所等多种门类。通过草图、平立剖面图和完成照片来全面展现建筑细节。荷英双语在书中并未对照排版，哑面铜版纸上为荷兰文。正文设定为单栏，侧面有较宽的注释栏。而英文被统一放置在书中两段黄色的胶版纸上，设置为双栏，并附带较窄的注释栏。通过纸张和栏位设置的差异加以区分。

The Dutch Architects Monograph Series aims to record important Dutch designers and their impressive works. This book is one of the series (another of the series won the 1991 Bronze Medal, which can be found on the relevant page), which comprehensively presents the career, works and ideas of Alexander Bodon, a Hungarian-Dutch designer. The hardcover book is bound with thread sewing. The matted jacket is printed in light gray, gray blue and black on coated paper. The inner cover is made of black cloth mounted with cardboard, and the spine of the book is black hot-stamped. Black cardboard is used for the end paper, while the inner pages are mixed with matte coated paper, yellow and white offset paper, which are printed in single black. The book shows 21 buildings designed by Bodon in the Netherlands, including bookstores, conference centers, art spaces, office buildings, restaurants, residences, clinics and other categories. The architectural details are fully presented through sketches, horizontal and vertical section and completed photos. The bilingual contents in Dutch and English are not typeset in contrast. The Dutch text are printed on the matte coated paper, set to a single column with a rather wide comment column on the side. The English text are placed on yellow offset paper in the book, set as double columns, with a narrow comment column. Bilingual contents are distinguished by differences in paper and field settings.

soplossing – een weg die
en geforceerde indruk en
ɟt (zelfs in Bodons eigen
:kenning van de stedelijke
men, maar het is er niet
wellicht verwijst ook het
ıstrijd die op dat moment
vas. Tegelijk is de samen-
u had zullen beginnen als
an hoop en verwachtingen
ting met het werk van Le
van de volumes en in de
ı het interieur. Dat het na
vam, waarbij Le Corbusier
liteit van het detail wees,

de Bodon een respectabele
l gebouwen gerealiseerd
dt vanwege de klare een-
lineerd functionalisme ...
moderne.'[4] Tegelijk kan
re zo duidelijk is, door de
Bodon heeft dat zelf in de
geen theoreticus te zijn.[5]
g van het raadsel van de
is de barrière van de een-

daarvoor is het raadhuis
t dit probleem bemoeien.
en bij de uiteindelijke af-
er de prijsvraag van 1936
le hij weer een plan in en

nder Bodon

ste-eeuwse architect

ment nor a social
ad nothing else to
time to build up a
ion soon changed.
ure from the new
This resulted in a
eate a new line of
ory in itself as it
furniture designs
gh Ahrend Bodon
Dupont, who was
Keizersgracht as a
principles of the

rge space, whose
long lines of the
and blue) De Stijl
nis debut elicited
protest demanding
h to know who is
ld city'.[10] A large

1
Robert Mens, 'Documente
rondom Le Corbusier', in J
Bosman et al (Ed.), *Le Corb
sier en Nederland*, Utrecht
1985, 54
2
'Bij eenige plannen voor h
nieuwe Amsterdamsche Ra
huis', in *de 8 en Opbouw*
11(1940), 63
3
Unless otherwise stated all
information derives from
conversations between the
author and Bodon in the
summer of 1989
4
Ruud Brouwers, 'Schraal
verjaardagsgeschenk voor
Bodon', in *Archis* 10-86, 2

Onbekend
Foto's Woonhuis Van Meurs
Jaap en Maarten d'Oliveira, Amsterdam
Copyright 1990
Uitgeverij 010 Publishers
Maarten Kloos (tekst)
Jan Versnel (fotografie)
CIP-gegevens
Kloos, Maarten
Alexander Bodon, architect / Maarten K
Jan Versnel (fotogr.). – Rotterdam:
Uitgeverij 010. – Ill., foto's. – (monografi
van Nederlandse architecten; 4)
Tekst in het Nederlands en Engels.
– Met bibliogr., lit.opg.

lutions areas in relation to the three classrooms. The teachers are
oused in a volume set slightly apart from the classrooms, which offers
em physical distance from the teaching aspect during breaks between
ssons. At the same time the comparative independence of this area of
e building clearly reveals in a unauthoritarian way where the school
aff can be found.

ı kring kreeg Bodon alle lof. Ida
ıte' en Hein Salomonson sprak van e
ier van wonen.'[45] Onverwachte bijva
ijdschrift *Wereldkroniek* was blij verra
en ervan op de gedachte dat de me
zijn gaan hechten (...) en minder aa
oral die reactie moet Bodon goed he
g was Bodon nog een aantal jaren b
Het Nederlandse textielwezen vroe
- en buitenland. Met Salomonson ve
landse inzending naar de 'Salon de
ijs en maakte hij een plan voor de t
doorgegaan was – een beeld had geg
ropbouw tot stand was gebracht.[47]
ar biedt wel de mogelijkheid tot een
ene *Amsterdammer* naar aanleiding
k op het idee gekomen een verge

nformation Office of De Nederlanden van 1845' 1950 - 1952

This office had to be inserted at the corner of Muntplein and the
kin boulevard in Amsterdam, an area bustling with city life. The
sign is of a refined simplicity. Bodon accepted the lack of space for a
and entrance zone and designed it as a narrow channel between
side and outside. Directly beyond, the space widens as the furniture is
gned to the rotated facade on the Rokin boulevard. The means used are
inimal, modification of the facade being largely achieved by reorgani-
ng existing material. All other material was chosen because it was
expensive and thus all the more unobtrusive, with teak door-handles
d a mahogany handrail for the stair. The chair is from a design by Titia

Plan
1 Enclosed porc
2 Display windo
3 Information o
4 Stair to office

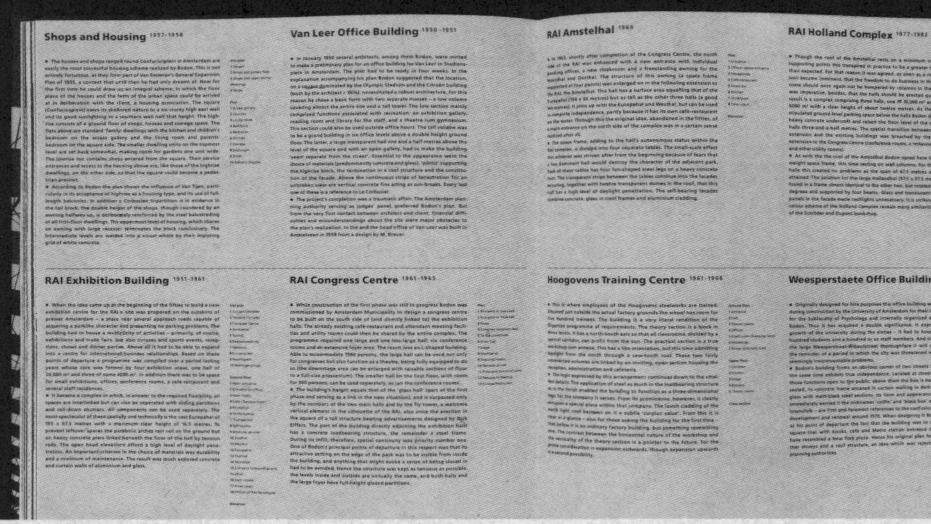

Shops and Housing 1957-1958

• The houses and shops ranged round Confuciusplein in Amsterdam are easily the most successful housing scheme realized by Bodon. This is not entirely fortuitous, as they form part of Van Eesteren's General Expansion Plan of 1935, a context that until then he had only dreamt of. Now for the first time he could draw up an integral scheme, in which the floor plans of the houses and the form of the urban space could be arrived at in deliberation with the client, a housing association. The square (Confuciusplein) owes its sheltered nature to a six-storey high east wall and its good sunlighting to a southern wall half that height. The highrise consists of a ground floor of shops, houses and storage space. The flats above are standard 'family' dwellings with the kitchen and children's bedroom on the access gallery and the living room and parents' bedroom on the square side. The smaller dwelling units on the topmost level are set back somewhat, making room for gardens one unit wide. The lowrise too contains shops entered from the square. Their service entrances and access to the housing above are, like those of the highrise dwellings, on the other side, so that the square could become a pedestrian precinct.

• According to Bodon the plan shows the influence of Van Tijen, particularly in its acceptance of highrise as a housing type, and its use of full-length balconies. In addition a Corbusian tripartition is in evidence in the tall block: the double height of the shops, though countered by an awning halfway up, is deliberately reinforced by the steel balustrading at all first-floor dwellings. The uppermost level of housing, which shares an awning with large recesses, terminates the block conclusively. The intermediate levels are welded into a visual whole by their imposing grid of white concrete.

Van Leer Office Building 1950-1951

• In January 1950 several architects, among them Bodon, were invited to make a preliminary plan for an office building for Van Leer in Stadionsplein in Amsterdam. The plan had to be ready in four weeks. In the explanation accompanying his plan Bodon suggested that the location, on a square dominated by the Olympic Stadium and the Citroën building (both by the architect J. Wils), necessitated a robust architecture. For this reason he chose a basic form with two separate masses – a low volume covering almost the entire site and a tall tower. The low section mainly comprised functions associated with recreation: an exhibition gallery, reading room and library for the staff, and a theatre cum gymnasium. This section could also be used outside office hours. The tall volume was to be a grand building in six office levels above a double height ground floor. The latter, a large transparent hall one and a half metres above the level of the square and with an open gallery, had to make the building 'seem separate from the street'. Essential to the appearance were the choice of materials (predominantly concrete and glass), 'pilotis' supporting the highrise block, the termination in a roof structure and the constitution of the facade. Above the continuous strips of fenestration for an unbroken view are vertical concrete fins acting as sun-breaks. Every last one of these is a reference to Le Corbusier.

• The project's completion was a traumatic affair. The Amsterdam planning authority serving as judges' panel, preferred Bodon's plan. But from the very first contact between architect and client, financial difficulties and misunderstandings about the site were major obstacles to the plan's realization. In the end the head office of Van Leer was built in Amstelveen in 1958 from a design by M. Breuer.

RAI Exhibition Building 1951-1961

• When the idea came up at the beginning of the fifties to build a new exhibition centre for the RAI a site was proposed on the outskirts of prewar Amsterdam – a place near several approach roads capable of acquiring a parklike character and presenting no parking problems. The building had to house a multiplicity of activities – primarily, of course, exhibitions and trade fairs, but also circuses and sports events, receptions, shows and dinner parties. Above all it had to be able to expand into a centre for international business relationships. Based on these points of departure a programme was compiled over a period lasting years whose core was formed by four exhibition areas, one hall of 20,000 m² and three of some 4000 m². In addition there was to be space for small exhibitions, offices, conference rooms, a café-restaurant and several staff residences.

• It became a complex in which, in answer to the required flexibility, all spaces are interlinked but can also be seperated with sliding partitions and roll-down shutters. All components can be used separately. The most spectacular of these spatially and technically is the vast Europahal of 195 x 67.5 metres with a maximum clear height of 16.5 metres. To prevent leftover spaces the parabolic arches rest not on the ground but on heavy concrete piers linked beneath the floor of the hall by tension rods. The open head elevations afford a high level of daylight penetration. An important criterion in the choice of materials was durability and a minimum of maintenance. The result was much exposed concrete and curtain walls of aluminium and glass.

RAI Congress Centre 1961-1965

• While construction of the first phase was still in progress Bodon was commissioned by Amsterdam Municipality to design a congress centre to be built on the south side of (and directly linked to) the exhibition halls. The already existing café-restaurant and attendant meeting facilities and utility rooms could then be shared by the entire complex. The programme required one large and one less-large hall, six conference rooms and an extensive foyer area. The result was an L-shaped building. Able to accommodate 1560 persons, the large hall can be used not only for congresses but also function as a theatre, being fully equipped to do so (the downstage area can be enlarged with raisable sections of floor to a full-size proscenium). The smaller hall on the first floor, with room for 300 persons, can be used separately, as can the conference rooms.

• The building's height equals that of the 'glass hall' (part of the first phase and serving as a link in the new situation), and is surpassed only by the contours of the two main halls and by the fly tower, a welcome vertical element in the silhouette of the RAI, also since the erection in the square of a tall structure bearing advertisements designed by Dick Elffers. The part of the building directly adjoining the exhibition halls has a concrete loadbearing structure, the remainder a steel frame. During its infill, therefore, spatial continuity was priority number one. One of Bodon's principal points of departure in this respect was that its attractive setting on the edge of the park was to be visible from inside the building, and anything that might evoke a sense of being closed in had to be avoided. Hence the structure was kept as tenuous as possible, the levels inside and outside are virtually the same, and both halls and the large foyer have full-height glazed partitions.

RAI Amstelhal 1969

• In 1965, shortly after completion of the Congress Centre, the north side of the RAI was enhanced with a new entrance with individual booking offices, a new cloakroom and a freestanding awning for the Noordhal and Oosthal. The structure of this awning (a space frame supported at four points) was enlarged on in the following extension to the RAI, the Amstelhal. This hall has a surface area equalling that of the Europahal (190 x 60 metres) but as tall as the other three halls (a good ten metres). It joins up with the Europahal and Westhal, but can be used in complete independence, partly because it has its own café-restaurant on the water. Through this the original idea, abandoned in the fifties, of a main entrance on the north side of the complex was in a certain sense realized after all.

• The space frame, adding to the hall's autonomous status within the RAI complex, is divided into four separate tables. The small-scale effect this achieves was striven after from the beginning because of fears that a too dominant hall would destroy the character of the adjacent park. Each of these tables has four fan-shaped steel legs on a heavy concrete foot. The transparent strips between the tables continue into the facades ensuring, together with twelve transparent domes in the roof, that this hall has a high level of daylight penetration. The self-bearing facades combine concrete, glass in steel frames and aluminium cladding.

RAI Holland Complex 1977-1982

• Though the roof of the Amstelhal rests on a minimum number of supporting points this transpired in practice to be a greater hindrance than expected. For that reason it was agreed, as soon as a new extension became imminent, that the freedom to do business in the halls to come should once again not be hampered by columns in the space. It was imperative, besides, that the halls should be erected quickly. The result is a complex comprising three halls, one of 10,000 m² and two of 5000 m² with a clear height of about twelve metres. As the RAI had stipulated ground-level parking space below the halls Bodon designed a heavy concrete undercroft and raised the floor level of the exhibition halls three-and-a-half metres. The spatial transition between the new extension and the existing buildings was breached by the required extension to the Congress Centre (conference rooms, a restaurant, offices and other utility rooms).

• As with the the roof of the Amstelhal Bodon opted here for a lightweight space frame, this time resting on wall columns. For the smaller halls this created no problems as the span of 67.5 metres was easily attained. The solution for the large Hollandhal (97.5 x 97.5 metres) was found in a frame almost identical to the other two, but rotated forty-five degrees and supported by four beams. Glass and translucent synthetic panels in the facade made rooflights unnecessary. It is striking that the colour scheme of the Holland Complex reveals many similarities to that of the Scholder and Dupont bookshop.

Hoogovens Training Centre 1961-1966

• This is where employees of the Hoogovens steelworks are trained. Situated just outside the actual factory grounds the school has room for five hundred trainees. The building is a very literal rendition of the bipartite programme of requirements. The theory section is a block in three levels. It has a north-south axis so that all classrooms, divided by a central corridor, can profit from the sun. The practical section is a true workshop cum annexe. This has a like orientation, but this time admitting daylight from the north through a saw-tooth roof. These two fairly introverted volumes are linked by an inviting, open section housing the reception, administration and cafeteria.

• The logic expressed by this arrangement continues down to the smallest detail. The application of steel as much in the loadbearing structure as in the finish enabled the building to function as a three-dimensional logo for the company it serves. From its prominence, however, it clearly occupies a special place within that company. The lavish cladding of the north light roof bestows on it a subtle 'surplus value'. From this it is clear at a glance – also for those seeing the building for the first time – that below it is no ordinary factory building, but something resembling one. The contrast between the horizontal nature of the workshop and the verticality of the theory section is a pointer to the future. For the prime consideration is expansion outwards, though expansion upwards is a second possibility.

Weesperstaete Office Building 1969

• Originally designed for hire purposes this office building was chosen during construction by the University of Amsterdam for their laboratory for the Subfaculty of Psychology and internally organized as such by Bodon. Thus it has acquired a double significance. It expresses the growth of the university during the sixties – it had to house sixteen hundred students and a hundred or so staff members. And situated on the large Weesperstraat-Wibautstraat thoroughfare it will always be the reminder of a period in which the city was threatened with many seemingly insurmountable problems.

• Bodon's building forms an obvious corner of two streets though at the same time exhibits true independence. Located at street level are those functions open to the public; above them the box is hermetically sealed, its concrete frame encased in curtain walling in dark sunproof glass with matt-black steel sections. Its form and appearance – which immediately earned it the nicknames 'coffin' and 'black box' among the townsfolk – are first and foremost references to the confusion in urban development and renewal around 1970. When designing it Bodon took as his point of departure the fact that the building was to stand in a square that with kiosks, café and Metro station entrance could well have resembled a New York plaza. Hence his original plan for two further storeys and a roof structure, an idea which was rejected by the planning authorities.

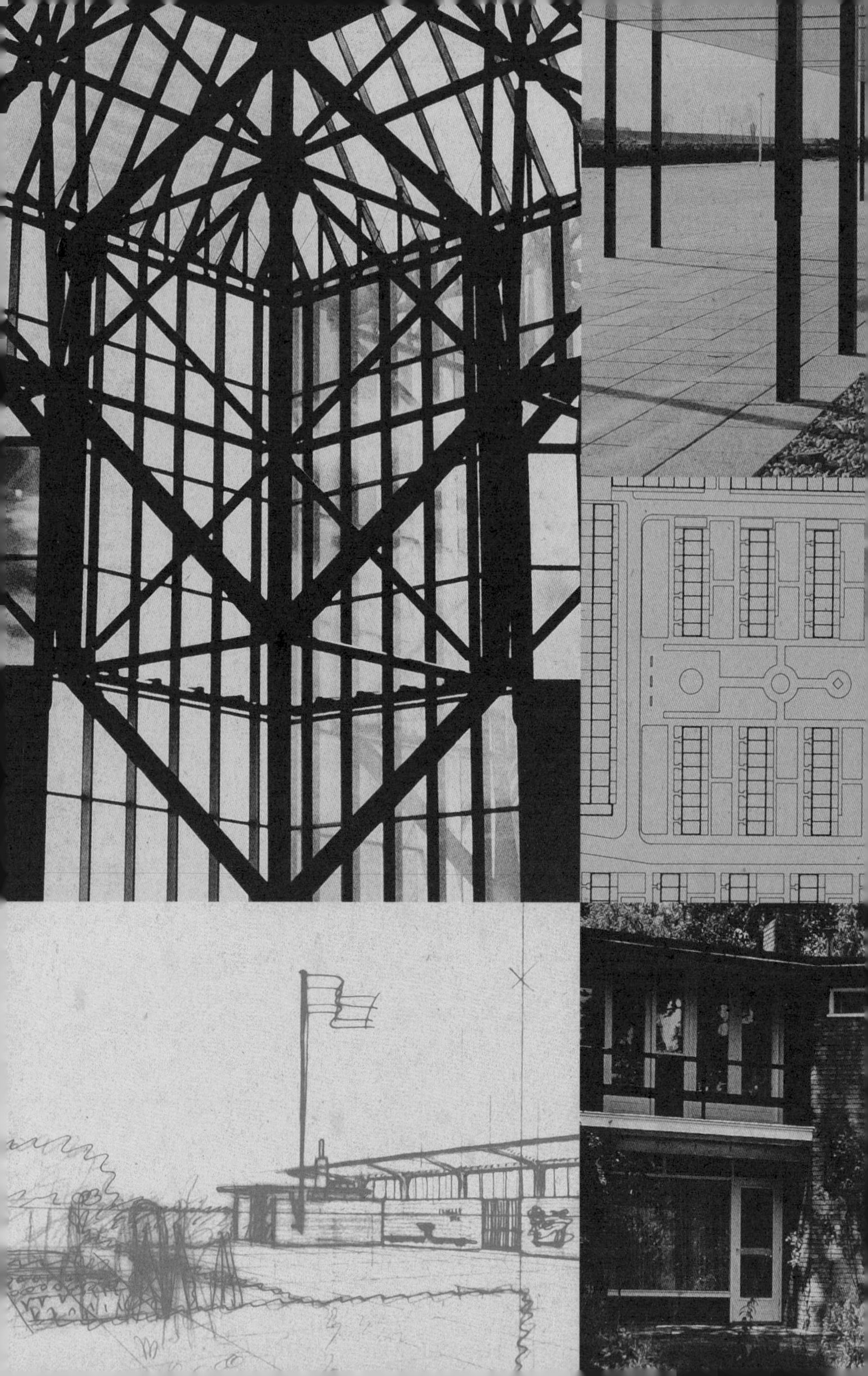

SIMCA

R AANGETEKENDE BRIEF BTL
Recommandé
PRIORITY CN
PRIORITY
Douaneverklaring
CN22
Mr. Zhao Qing
Dabei xiang 7-3 Nanjing
Han Qing Tang Design CO. Ltd 7-3
210000 Nanjing
China
13952094842
Afzender:
Jonas
Tijpoort 38
3312WB Dordrecht
Nederland
SCHRAG

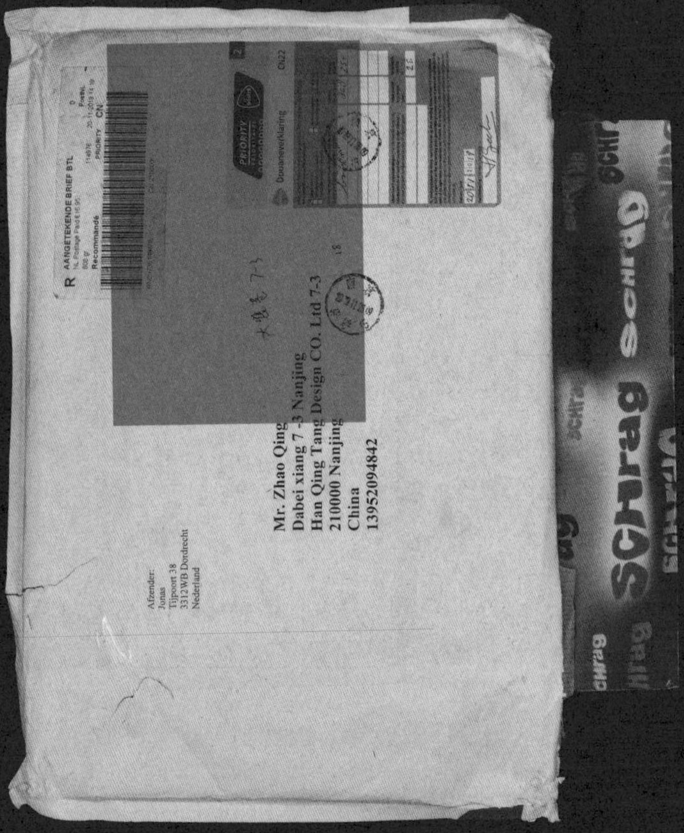

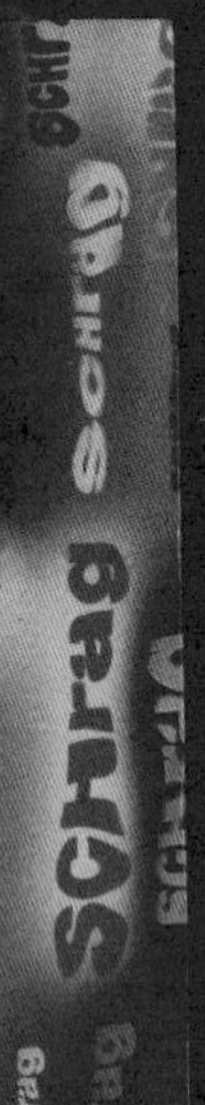

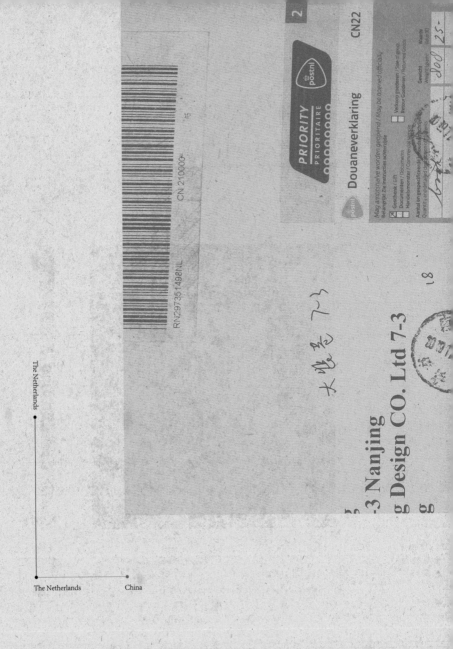
RN297351498NL
CN 210000
PRIORITY
PRIORITAIRE
Douaneverklaring
CN22
-3 Nanjing
g Design CO. Ltd 7-3
The Netherlands
The Netherlands
China

● 荣誉奖 Honorary Appreciation

《 反对——荷兰当代艺术中的戏仿、幽默与嘲弄 Schräg / Tegendraads: parodie, humor en spot in hedendaagse Nederlandse kunst

荷兰 The Netherlands

Gerard Hadders, André van Dijk (hard werken)

Paul Donker Duyvis, Klaus Honnef

Edition Braus, Heidelberg

240 215 13mm

739g

176p

ISBN-10 90-12-06639-5

240mm

13mm

215mm

这是一本艺术评论和作品集，多位艺术评论家对 20 世纪 90 年代前后荷兰的艺术品进行了详细的论述和评论。本书的装订形式为锁线胶平装。封面白卡纸四色印刷，覆亮膜。内页哑面铜版纸四色印刷。内容方面分为三大部分：第一部分是 8 位评论家的艺术评论，内容涵盖对荷兰当代艺术（20 世纪 90 年代前后）的思考、对荷兰电影的评价、对建筑公共与私人之间关系的思考等等；第二部分是 28 位荷兰艺术家的作品展示；第三部分是艺术家简历。书籍的版式为非常细密的多栏设置，并在图片和文字的排版上灵活处理。在艺术评论部分，所有图片均处理为黑色，用以突出文本；但在作品部分，黑白和彩色作品混用，在目录上通过粗体数字来呈现彩色作品的页码。

This is a collection of art reviews and works, with many art critics discussing and commenting on the works of Dutch artists around the 1990s. The binding form of this book is paperback with lock-thread glue. Cover white cardboard is printed in four colors and covered with bright film. Inner page matte coated paper printed in four colors. The content is divided into three parts. The first part is an art review by eight critics. It covers the thinking of Dutch contemporary art (around 1990s), the evaluation of Dutch film, and the thinking about the relationship between architecture public and private. Wait; the second part is the exhibition of the works of 28 Dutch artists; the third part is the artist resume. The layout of the book is a very detailed multi-column setting, and it is flexible in the layout of pictures and text. In the art review section, all pictures are processed as black to highlight the text part, but in the work part, black and white and color works are mixed, and the page numbers of the color works are presented by bold numbers on the catalog.

Parodie, travestie en de gulden middenweg

Hedendaagse kunst in Nederland

Klaus Honnef

Tegendraads, of Lof der Zotheid?

Moralisme met een knipoog

Paul Donker Duyvis

Gijs Bakker

Paul Beckman

Jan Jansen

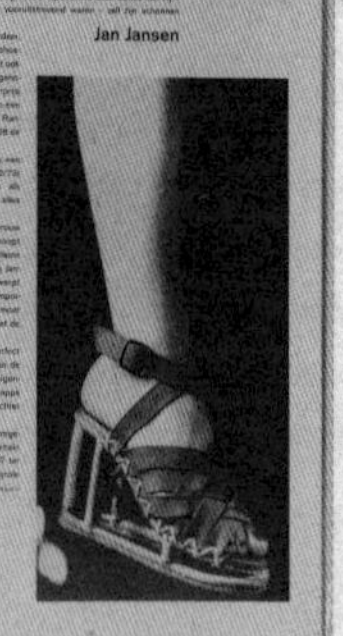

Joost van den Toorn

Harald Vlugt

terug. Hij verspreidde het gerucht dat hij zich alleen nog maar aan het schaakspel zou wijden; onderwijl liet hij zijn kunsthandelaar de 'ready mades' – massaal geproduceerde industriële produkten – in beperkte, genummerde oplage verkopen.
Minder cynisch was het gedrag van Francis Picabia, een andere vertegenwoordiger van de avantgarde. Hij leverde zijn opdrachtgevers de door hen bestelde erotische schilderkunst per strekkende meter – van eigen hand – zoals dat heette in de verdragen uit de renaissance – zonder aarzeling en zonder scrupules. Zulke artistieke praktijken veroordelen, zou uiteindelijk betekenen dat men langs kunstuitingen de morele meetlat zou aanleggen. Dat hebben puristen van alle gezindten altijd graag mogen doen, maar hun misnoegen werd veroorzaakt door wàt er werd uitgebeeld, niet door de kunst als zodanig. Zo ging in het Nederland van de Gouden Eeuw de calvinistische geestelijkheid vanaf de kansel te keer tegen te ostentatief tentoongespreide levenslust – en ondertussen werden er hoe langer hoe meer schilderijen gemaakt die juist belichaamden wat volgens de geestelijkheid verderfelijk was voor het zieleheil van degene die deze schilderijen kocht.
En toch ... het wapen van de ironie om de wereld zoals die is te lijf te gaan is de Nederlander in feite vreemd. De Franse ironie ademt de geest van de Verlichting, de romantische Duitse ironie is gebaseerd op het 'principe van de catastrofe' (Hans Egon Holthusen), maar geen van deze beide vormen van ironie is een vertrouwd element in de Nederlandse cultuur. Wel is er een onmiskenbare hang naar travestie, de neiging tot 'verkleden'. Travestie is eigenlijk een literair genre dat in de zeventiende eeuw vooral in Frankrijk werd beoefend. De encyclopedie definieert travestie als een satirisch literair genre, waarin de draak wordt gestoken met een bestaand ernstig werk door niet de inhoud te veranderen (dat gebeurt in de parodie), maar deze te kleden in een andere – oneigenlijke – vorm[4]. In de Nederlandse cultuur is de travestie heel gebruikelijk sinds luidruchtige boeren en zelfverzekerde kapiteins in genres als het 'aristocratische portret' en het heldendicht tekeer mogen gaan. Aan de andere kant is er in de Nederlandse culturele traditie geen scherpe grens tussen de parodie en de travestie, die elkaar vrolijk afwisselen of tegelijk worden toegepast.

hedendaagse kunst.
Het gebruik van de parodie en de travestie in de beeldende kunst gaat met name in Nederland vaak samen met een kritische houding tegenover het eigen werk. Wanneer men eenmaal een artistiek standpunt heeft ingenomen, denkt men daar voortdurend over na. Ger van Elk, een van de sleutelfiguren in de hedendaagse kunst in Nederland, daarbij een kunstenaar die door velen terecht wordt beschouwd als de wegbereider van de nieuwe culturele stroming waarover we het hier hebben, sprak in een interview jaren geleden in de volgende typerende bewoordingen over zijn werk: 'Er zitten wel tegenstellingen in [Van Elk doelt op zijn artistieke ontwikkeling – K.H.] en de lijn die je erin vindt is niet een stilistische. Het is meer een lijn – een houding, een consequentie. Heel voorzichtig kan ik zeggen ... misschien dat ik wel zeker ben dat in essentie mijn houding is, dat ik voortdurend de dingen van mijzelf omvergooi, dat ik altijd een tegenbeweging maak omdat ik niets als absoluut zou willen brengen. Zodra ik iets maak en het heeft een onderwijzend of leraarachtig tintje, dan word ik wantrouwend.' En hij voegt eraan toe: 'Ik geloof erin, dat dat het enige is wat je doen kan: contradictie'[5].
Natuurlijk geldt wat van Van Elk zegt in wezen ook voor vele andere kunstenaars van zijn generatie in andere landen van Europa (en voor een enkeling in de Verenigde Staten); het verschil met de meesten is dat hij er geen enkele ideologie op na houdt bij het realiseren van zijn artistieke aspiraties. Van Elks verhouding tot kunst is ongetwijfeld ambivalent, maar dat betekent niet dat hij de kunst zelf als zinvolle menselijke activiteit op losse schroeven zet. Hij is zich volledig bewust van zowel de grenzen als de mogelijkheden van de kunstenaar. 'Hij heeft een uiterst klassieke, zelfs conventionele opvatting over Kunst; verschillende uitspraken in interviews duiden daarop. Maar tegelijkertijd vindt hij ook dat elk werk moet overkomen als een verrassend, intrigerend, moeiteloos gevonden "bon mot"'[6]. Wat Rudi Fuchs 'bon mot' noemt, is dat complexe en ingewikkelde aspect dat een gemeenschappelijk kenmerk is van het verder in menig opzicht zo uiteenlopende werk van schilders en beeldhouwers, architecten en performers, fotografen en ontwerpers, musici en filmmakers, die op de tentoonstelling en in dit boek bijeengebracht zijn onder de ongewone titel *Schräg/Tegendraads*. Ook deze betiteling raakt niet de kern van de schilderijen en sculpturen, foto's en maquettes, [illegible] en performances. Maar aangezien 'schräg' typisch een term is van deze tijd, uit het jargon van de Duitse jongerencultuur, [illegible] zij worden gebruikt om – zij het wat onbeholpen – iets ongewoons in het culturele aanbod aan te duiden en te karakteriseren. Hoe het ook zij, de benaming 'tegendraads' is geen nieuwe uitvinding, ze duikt al vroeg op in de Duitse kunstgeschiedenis: 'Tegendraads is heroïsch', schreef de schilder Otto Coenen (1907–1971) eens, met de wijze spot van iemand die zijn verontrusting heeft gesublimeerd, en hij doelde [illegible]

leven kracht bij zetten door de helden of de woedende menigte schuin in beeld te brengen ....'[7].
In dit citaat wordt de oorspronkelijke betekenis van 'schräg' duidelijk: het is de manier om aan te duiden dat een voorwerp scheef staat, of een geschilderde of getekende lijn diagonaal verloopt. De term heeft inmiddels ook de connotatie van 'belachelijk', 'raar', waarbij de contrastwerking tussen het 'ongepaste' en het 'verhevene' onontbeerlijk is voor het 'scheve effect'. In Duitsland, waar de clown nooit is ingeburgerd, wordt dit jongleren met het verhevene en het belachelijke het liefst afgedaan als oppervlakkige grappenmakerij[8]. Het is dan ook niet verwonderlijk dat Ger van Elk vroeger in de pers vaak is afgeschilderd als paljas. Zijn reactie daarop verraadt een mentaliteit die (nog?) zeer ongebruikelijk is in de beeldende kunst, en voor zover ik weet alleen in Nederland voorkomt: 'In Duitsland word ik ook vaak een "Witzemacher" genoemd. Dat vind ik uitstekend. Zo van: God er zit een zwarte lijst om. Dat kun je ook zeggen'[9]. En om de verwarring nog groter te maken bekent de kunstenaar: '... Niks is serieus bedoeld bij mij'[10]. Maar men moet ook weer niet de vergissing begaan, deze uitspraak al te serieus te nemen.

Hetzelfde geldt voor Wim T. Schippers, die tegelijk met Van Elk aan de Rietveldacademie in Amsterdam heeft gestudeerd. In het buitenland is hij bijna onbekend maar in Nederland geniet hij grote populariteit. Deze populariteit is niet in de laatste plaats te danken aan zijn wekelijkse radioprogramma's en natuurlijk de talrijke tv-sketches en performances; en dat hij een uiterst geestige tekstschrijver is, is zelfs de Duitse televisie al opgevallen. Ook hij fungeert als een soort vaderfiguur voor een bepaalde groep Nederlandse kunstenaars, voor wie hij tegelijk een uitdaging en een voorbeeld is. Als kunstenaar is hij niet gemakkelijk te plaatsen en het is nog moeilijker te bepalen tot welke artistieke professie Wim T. Schippers moet worden gerekend. Ziet hij zich zelf als serieus kunstenaar of als populaire entertainer? Is hij beeldhouwer of schilder, acteur of schrijver – en een grammofoonplaat heeft hij ook al ingezongen. Alle pogingen zijn artistieke werk onder één noemer te brengen, moeten mislukken. Zijn laconieke commentaar op de denkbeelden die hun uitdrukking vonden op de tentoonstelling met de merkwaardige titel *Schräg/Tegendraads* luidt: 'That's me.' Is hij dus de personificatie van het tegendraadse in de kunst? In elk geval provoceert zijn werk de kunstopvatting waarin ernst wordt verward met diepzinnigheid; en er is niet eens tegen in te brengen, dat er in Schippers' werk een speciaal soort – *horribile dictu* – humor zit. Dit is in de ogen van veel kunstadepten een gruwelijke eigenschap, genoeg om de duim afkeurend naar beneden te laten wijzen. Tussen haakjes: alleen als de Romeinse keizer zijn duim omhoog stak, had het laatste uur

Er is een hel[illegible]
lijk maakt da[illegible]
Nederlandse[illegible]
kunst een ac[illegible]
naar de pers[illegible]
de travestie [illegible]
een uitsteke[illegible]
tret van Jan [illegible]
Beatrix – ter[illegible]
tot uitdrukki[illegible]
ziet. De kun[illegible]
of vermeend[illegible]
grappenmak[illegible]
kunstenaar [illegible]
ga's die zich [illegible]
volgens de v[illegible]
citaten. Tege[illegible]
schappelijk [illegible]
de greep is t[illegible]
travestie, pa[illegible]
manifesteert[illegible]
voortdurend[illegible]
waarden wo[illegible]
ten als beeld[illegible]
grafen en f[illegible]
autonome k[illegible]
laten vallen [illegible]
van voor de [illegible]
sen en kuns[illegible]
museale ruin[illegible]
traal Museun[illegible]
ingrepen in [illegible]
objecten die [illegible]

de ver-
ı, in het
bepaald
tramien,
pzichte
ein naar
d, waar-
acrokos-
zorg op
g.
naar het
ı strikte
t zoveel
architec-
e stede-
icrokos-
ontroleert en beheerst. Op
en stedelijk niveau zijn de
debouwkundige ontwikke-

aoorlogse architectuurpro-
uitdrukken in woningbouw
ignood, die tijdens de oor-
men. Al in 1901 kwam de
ı kader werd geschapen
middels in de grote steden
zijn gebouwd. De Woning-
uliere speculatiebouw die
reidingen kenmerkte en

Meer van h

Over architectuur: tusse

openbare ordening

weg beantwoorden aan de opgeworpen vrager
de Nederlandse architectuur vanaf het begin
momenten en hoogtepunten die ook internatio
opvallend dat naast de groots
sche en politieke dan door bo
woningbouwprojecten, een or
tegendraadse architectuur is c
De architectuur in het eerste k
halverwege het midden van de
van een explosieve vernieuwi
van Berlage staat het expres
School en de abstractie van D
op een ideologie (en als manif
op een nieuwe wereld, waarin
kunstenaar en de organisatie
zouden veranderen. Maatscha
gingen, zoals socialisme, the
visioenen voor een nieuwe or

nuancering
dkleur in het
ıaloranje. De
het nationale
and portret-
perend voor
naar blijkbaar ook voor de
rlander in het algemeen.
ect kunnen hand in hand
of in ieder geval met een
uidt een van de vele stelre-
er: 'Men moet altijd kritisch
e houding getuigde ruim
an Elks collega en genera-
pers in een van zijn vele

Moralisme met een knipoog

eckman

-uptafel met spiegel
1986
Hout, spiegel, lampen
Circa 240 x 140 x 180 cm
dienst Beeldende Kunst,
Den Haag
Foto: Pieter van de Meer

besta
Colo
De *n*
Beck
Wilm
steun
blok.
grani
steen
Voor
'deco
uitga
schri
het m

rs, cameramensen, acteurs en p
in wisselende configuraties bij
de films betrokken zijn.
ang van die vriendschapsnetwe
ks worden overschat. Bijv
iet bij de meer eigenzinnige film
enoeg altijd gaat om ondergefin
uiver professionele samenwerki
t films onmogelijk; de belangel
sen die vertrouwen hebben in d
ontbeerlijk. Die samenwerking h
inhoud en de vorm van de film
een aantal filmmakers bij elka
ontstaan van een soort gemeen
ıkel geval zou je zelfs van een sc
en die in dit netwerk voor 'klein

De vitaliteit van het n[illegible]derlandse schrijvers en dichters op de dremp[illegible] van een nieuwe eeuw

Jaap Goedegebuure

Tijdens de duizend jaar van haar bestaan heeft de Nederlandse literatuur altijd een moralistische inslag gehad. Zelfs nu het openlijke preken en beleren van dominees en schoolmeesters met minder vanzelfsprekendheid wordt aanvaard dan eertijds het geval was, hebben veel schrijvers nog altijd een boodschap. Misschien is het toegenomen belang van de literatuur als levensbeschouwelijke vraagbaak wel een gevolg van het autoriteitsverlies onder priesters, onderwijzers en andere opvoeders.

Het ethische element kenmerkt het optreden van middeleeuwers als Jacob van Maerlant en Jan van Ruusbroec, van humanisten als Erasmus en Coornhert, van de barokauteur Vondel, van 'femmes savantes' als Belle van Zuylen en de dames Wolff en Deken, en – begrijpelijkerwijs – van dichtende dominees in de negentiende eeuw als Beets, De Genestet en Ten Kate. Ook Multatuli, onze grootste schrijver en tegelijk een van de weinige Nederlandse cultuurdragers wier faam tot het buitenland is doorgedrongen, was een moralist. Niet voor niets luidt een van zijn aforismen: 'De roeping van de mens is mens te zijn.' Hijzelf vatte die roeping op als een maatschappelijke taak. *Max Havelaar* (1860), zijn eerste boek, is een felle aanklacht tegen het stelsel van koloniale uitbuiting in Indonesië, destijds een Nederlands wingewest.

Tegen het einde van de negentiende eeuw kwam er een reactie tegen de 'domineesliteratuur'. De Beweging van Tachtig zag de poëzie als een autonoom gebied, waar niet de moraal maar de schoonheid het voor het zeggen had. Maar lang niet alle schrijvers na 1880 hebben zich aan het estheticisme van de Tachtigers geconformeerd. In de jaren dertig namen Ter Braak en Du Perron een uitgesproken politiek standpunt in. En na de oorlog raakten vooraanstaande romanciers als W.F. Hermans en Harry Mulisch keer op keer in maatschappelijke discussies betrokken.

In weerwil van de postmodernistische beeldvorming die in de hedendaagse kunst en de cultuur uitsluitend skepsis, levensbeschouwelijke twijfel, verzet tegen de moraal en een herleving van het estheticisme projecteert, sluiten verschillende jonge auteurs zich aan bij hun voorgangers, door kritisch te reflecteren op sociale kwesties. Deze betrokkenheid heeft zeer weinig gemeen met het [illegible]

linkse engagement dat na 1945 ontstond en omstreeks [illegible] zijn hoogtepunt beleefde. De jonge schrijvers gaan niet [illegible] dan het verstrekken van enkele suggesties betreffende [illegible] male bagage die iemand bij zich moet hebben om zich [illegible] door het leven te slaan, zonder zich vast te klampen aan [illegible] houvast van een bestaande ideologie. Christendom, [illegible] lisme en socialisme krijgen zelfs felle kritiek te verduren. Een enkeling, zoals Robert Vernooy, houdt bewust het [illegible] tussen estheticisme en engagement. In Vernooy's uit 1989 [illegible] rende essaybundel *De contramine* (een titel waaruit al [illegible] tegendraadse houding spreekt; 'in de contramine' is [illegible] wordt gezegd van ongehoorzame of lastige kinderen) [illegible] gekozen voor een 'esthetisch humanisme'. In veel op[illegible] doet deze houding denken aan het aristocratische dilettan[illegible] van een renaissance-auteur als Baldassare Castiglione [illegible] wie wetenschap op één lijn stond met schermen, paard[illegible] en de schone kunsten. Vernooy pleit dan ook voor een [illegible] sport, 'mindbuilding'. Daarbij is de energie niet gericht [illegible] versterking van de spieren, maar op de ontwikkeling [illegible] geest. 'Ik zie ze al zitten, op een bankje in het Vondel[illegible] *mindbuilders* die 's morgens vroeg een boom omver prob[illegible] te denken.'

Vernooy's ideale mens 'leeft zijn leven en probeert het [illegible] te beleven. In zijn levensbeleving is hij behalve acteur en [illegible] seur dus ook nog zijn eigen publiek.' Geen wonder dat [illegible] deze vitale manier van bestaan de term *gaya scienza* [illegible] een renaissancistisch begrip dat Nietzsche [illegible] tot het schrijven van *Die fröhliche Wissenschaft*. [illegible] sche's amorele moralisme en paradoxale crea[illegible] zich uit in de wil tot vernietiging komen in Vernooy's [illegible] nadrukkelijk aan bod, wat er eens te meer op wijst [illegible] filosoof met de hamer zijn zoveelste *revival* doorma[illegible] De door Vernooy gezochte eenheid in het tegens[illegible] hecht zich aan eigentijdse sociale types als de [illegible] de vandaal die in zijn betoog een symbolische [illegible] krijgen. Vernooy ziet de agressie van zijn [illegible] ingooiende en telefooncellen verwoestende leeftijd[illegible] ten als een artistieke daad van subversieve [illegible] vandaal handelt vanuit een 'belangeloos welbe[illegible] Zijn motieven zijn 'niet politiek of ideologisch [illegible] esthetisch. De zinnelijkheid verzet zich tegen [illegible] mentele rede. Het vandalisme demonstreert de [illegible]

naar een ambivalente positie die ook hem verscheurt tussen tegendelen. Splitst hij zich in *De contramine* uit over kunstenaar en vandaal, in zijn roman *De postperistaltici* (een satire op de skeptisch ingestelde postmodernen) fingeert hij een briefwisseling tussen zichzelf en het weerzinwekkende personage dr. Microbius. Dit demonische karakter verkondigt opvattingen die later zullen worden toegeschreven aan de vandaal. Maar wat Microbius anders maakt is zijn biologisch nihilisme, dat hem doet koketteren met de machten van het kwaad, fascisme en sadisme. Hij zingt het lied dat elke cultuur als tegenmelodie begeleidt. 'Scheppen en vernietigen; in allebei openbaart zich dezelfde drang die zich doet gelden in de macht om een vaste samenhang te veranderen. Produceren en konsumeren; geeft mij een wereld en ik zal hem verteren! Ik kan hem een nieuwe zin geven of hem elke zin ontnemen.'

De hier onthulde mefistofelische kant van de kunst, die een subversief gerichte scheppingsdrang combineert met destructiviteit, hoort tot de schaduwen die de mens altijd al hebben begeleid en in de toekomst zullen blijven volgen. Kennelijk ziet Vernooy in, dat het tot leven wekken van die schaduw noodzakelijk is om hem te kunnen beheersen. Een dergelijke drijfveer is ook aan te treffen in het werk van Frans Kellendonk (1951–1990), die eens heeft gezegd dat de reacties op het herlevende fascisme soms even bedenkelijk zijn als het verschijnsel zelf. Ze maken het immers mogelijk om het kwaad af te wentelen op een zondebok, en zelf vrijuit te gaan. Dat racistische stereotypen (om maar een voorbeeld te geven) bij ieder mens en in elke samenleving voorkomen, is een gegeven dat door Kellendonk werd uitgebuit in zijn roman *Mystiek lichaam* (1986), een groteske satire op een wereld in zijn post-christelijke fase.

In zijn tegelijk met *De contramine* verschenen roman *De dingen die er niet toe doen* geeft Vernooy zijn 'esthetisch humanisme' verder inhoud. De held van zijn verhaal is een moderne Taugenichts, een vagebond die zich willens en wetens onttrekt aan een wereld waarin iedereen de gevangene is van het beeld dat hem als een model van politiek en sociaal gedrag wordt opgedrongen door reclame en media. Hij kiest voor het bestaan van een 'vrij zwevend deeltje, dat steeds wisselende bindingen aangaat, zonder zich ergens aan te hechten', en streeft een har[illegible]

waarde voor het slagen van deze *mod*[illegible]
mentele rede, die de dingen afweegt [illegible]
listische belangenhiërarchie.
Robert Vernooy is niet de enige aute[illegible]
van 'de dingen die er niet toe doen', [illegible]
hat', om met de negentiende-eeuwsa[illegible]
Jong wees in zijn roman *Cirkel in he*[illegible]
voorbeeld van kinderen die met een [illegible]
ene spel in het andere kunnen glijde[illegible]
verwisselen, zonder hun waarachtigh[illegible]
onthechte en tegelijk zeer vitale mani[illegible]
Weeldens *Tegenwoordigheid van gee*[illegible]
seling de spil van e[illegible]
slissende momenten [illegible]
van triviaal belang, [illegible]
was in aanleg even [illegible]
Voor Van Weelden e[illegible]
Arbeidsvitaminen ([illegible]
impliceert het verwa[illegible]
'laag', relevant en [illegible]
eerste gezamenlijke [illegible]
doende frivoliteit en [illegible]
videoclip van het e[illegible]
dialogen, grappen, [illegible]
Contactadvertenties [illegible]
dachtig behandeld [illegible]
zielsverhuizing of de [illegible]
Esthetische categor[illegible]
ben voor deze schrij[illegible]
ontdekking gekomen [illegible]
van de straat, de kr[illegible]
poëzie schuilt als [illegible]
gerespecteerde sch[illegible]
den ze gesterkt do[illegible]
schrijvers, journalist[illegible]
1965 verenigden in [illegible]
Nieuwe Stijl. Daarto[illegible]
beroemdheid als de [illegible]
nu eerder een neo-[illegible]
is de hang naar he[illegible]
ingestelde Vernooy, [illegible]
Voor Bril en Van We[illegible]
zelfkant, en het mo[illegible]

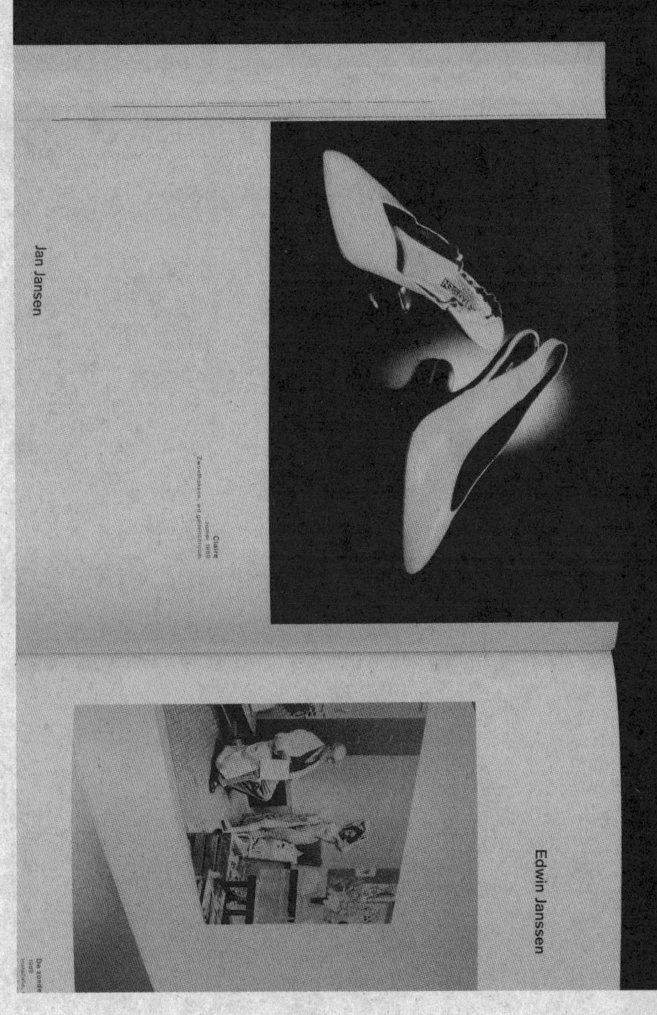
Jan Jansen
Edwin Janssen

1993

Country / Region

The Netherlands 5

Germany 3

Switzerland 3

U.S.A. 3

Japan 1

Czech Republic 1

Spain 1

Sweden 1

pacovská schwitters

Connie Imboden | Out Of Darkness

MOZART IN TONSPUREN UND BILDRÄUMEN

Herbert Maeder Werner Warth **Wil** die Altstadt

Círculo de Lectores San Juan de la Cruz Poesía y otros textos

Designer

Kv ta Pacovská

Piet Gerards, Marc Vleugels

Kaspar Mühlemann

Yasuhiko Naito

Roelof Koster (Studio Dorèl)

Lutz Dudek, Claudia Grotefendt

Ulysses Voelker

Reynoud Homan

Harry N. Sierman

Reynoud Homan

April Greiman, Sean Adams (April Greiman Inc.)

Gaston Isoz

Isabelle Lajoinie

Clara Istlerová

Norbert Denkel

HC Ericson

David Betz

Felicia Rice

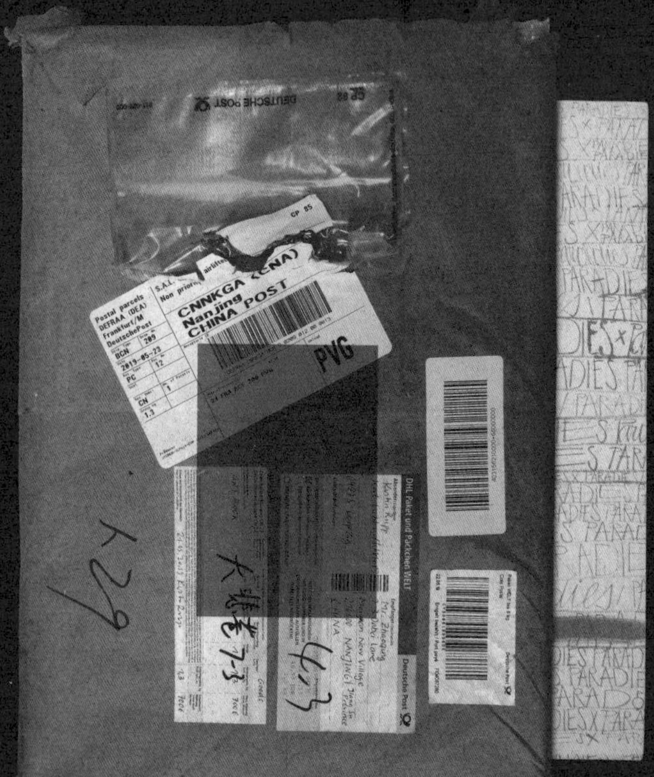
DEUTSCHE POST
CP 85
Postal parcels
DEFRAA (DEA)
Frankfurt/M
DeutschePost
CNNKGA (CNA)
Nanjing
CHINA POST
2019-05-23
PVG
DHL Paket und Päckchen WELT
Deutsche Post

Germany

Germany

China

424mm

45mm

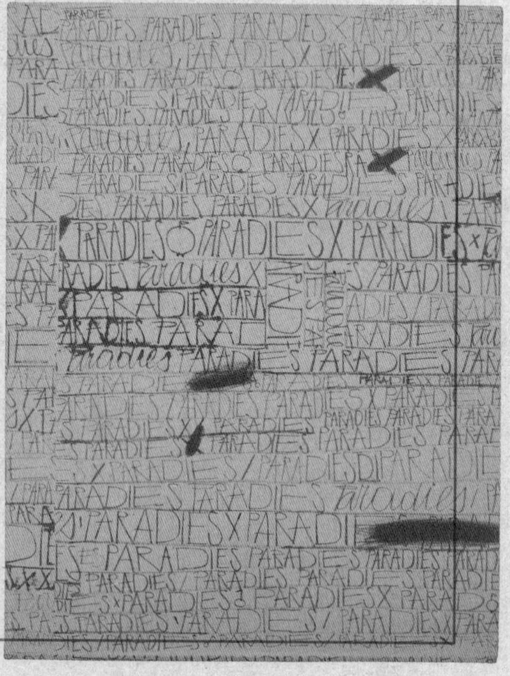

332mm

◉ 金字符奖
《 公鸡彼得，纸天堂
◐ 德国

Golden Letter
Der Hahnepeter, Papier Paradies
Germany
Kv ta Pacovská
Kurt Schwitters, Kv ta Pacovská
Pravis Verlag, Osnabrück
424 332 45mm
4354g
96p
——

德国艺术家库尔特·施维特斯（Kurt Schwitters）曾经出版过一本童话故事 *Der Hahnepeter*，这是一本以文字为主配以少量插图的手工活字铅印书籍。捷克艺术家、儿童作家克维塔·帕佐夫斯卡（Květa Pacovská）对这个故事进行了再创作，最终形成了本书。书籍的装订形式为包背胶精装。书籍的函套采用布面丝印灰，裱覆卡板。封面白色布面丝印灰裱覆卡板。内页艺术纸多色丝印。内容分为两部分：第一部分是对早年这本童话的致敬，原书的扫描件被直接呈现；第二部分是本书作者帕佐夫斯卡的再创作，以插图为主，配上少量文字，叙事并未完全按照原童话的故事走向，而是加入了很多艺术家自己的演绎。插图原稿全部由石版画制作，再由多色丝网套版的形式印制。书籍中间部分镂空正方形区域，贯穿第二部分内容，并露出了前后故意放置的两个单词 paradies 中的 die 和 warten 中的 art，将天堂、死亡、等待、艺术，以及贯穿书中的镂空“通道”的意义留给读者思考。|||||||
||||||||||||||||||||||||||||||||||||||||
||||||||||||||||||||||||||||||||||||||||

German artist Kurt Schwitters once published a fairy tale *Der Hahnepeter*, which is a text-based book with a few illustrations printed with handmade movable type. Květa Pacovská, a Czech artist and children's literature writer, recreated the story and eventually published this book. The double-leaved book is glue bound. The book case is covered by cloth printed with silkscreen in grey and mounted with cardboard, while the cover is covered with white cloth printed with silkscreen in grey as well. The inside pages are treated with silkscreen printing in multi colors. The content is divided into two parts. The first part is a tribute to the early fairy tale, and the scanned copy of the original book is directly presented; the second part is the re-creation of Pacovská, the author of the book, which is mainly illustrated with a small amount of words. The narration is not completely in accordance with the original fairy tale, but with the interpretation of the artist himself. The original illustrations are all lithographs, and then printed in the form of multi-color silk screening. In the middle part of the book, a square area is hollowed out, which runs through the second part of the book and leaks two words deliberately – one is 'die' in 'Paradies' and the other is 'art' in 'warten'. The significance of heaven, death, waiting, art, and the hollowed out 'passage' throughout the book leave readers for provoking thought. ||||||||||||||||||||||||||||||||||||||||||||||||||||||

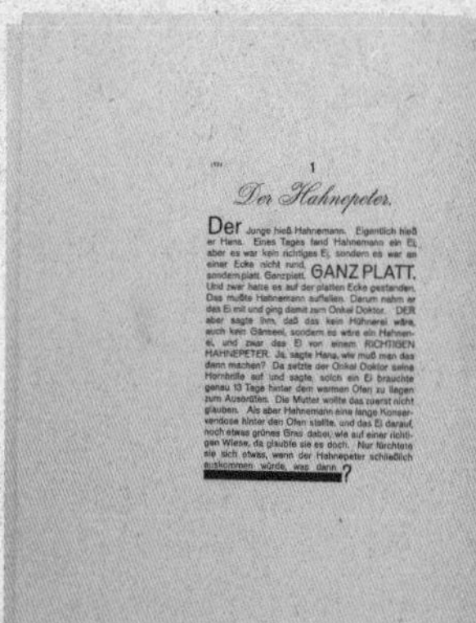

1

Der Hahnepeter.

Der Junge hieß Hahnemann. Eigentlich hieß er Hans. Eines Tages fand Hahnemann ein Ei, aber es war kein richtiges Ei, sondern es war an einer Ecke nicht rund, sondern platt. Ganzplatt. GANZ PLATT. Und zwar hatte es auf der platten Ecke gestanden. Das mußte Hahnemann auffallen. Darum nahm er das Ei mit und ging damit zum Onkel Doktor. DER aber sagte ihm, daß das kein Hühnerei wäre, auch kein Gänseei, sondern es wäre ein Hahnenei, und zwar das Ei von einem RICHTIGEN HAHNEPETER. Ja, sagte Hans, wie muß man das dann machen? Da setzte der Onkel Doktor seine Hornbrille auf und sagte, solch ein Ei brauchte genau 13 Tage hinter dem warmen Ofen zu liegen zum Ausbrüten. Die Mutter wollte das zuerst nicht glauben. Als aber Hahnemann eine lange Konservendose hinter den Ofen stellte, und das Ei darauf, noch etwas grünes Gras dabei, wie auf einer richtigen Wiese, da glaubte sie es doch. Nur fürchtete sie sich etwas, wenn der Hahnepeter schließlich auskommen würde, was dann ?

2

Hahnemann beruhigte sie, sie müsse nur 13 Tage Geduld haben. Morgens und abends betastete Hahnemann das Ei mit dem Zeigefinger und brachte morgens und abends neues Grün von derselben Wiese, auf der er das Ei gefunden hatte. Ihr könnt Euch das alle merken, wie man das macht, denn Ihr findet vielleicht auch mal solch ein Ei von einem richtigen HAHNEPETER. Und nun kam der Tag, wo der Hahnepeter aus dem Ei kriechen sollte. Das Ei war schon ganz warm geworden, und Hahnemann setzte sich schon vorsichtshalber die BRILLE seiner Großmutter auf. Denn wenn der Hahnepeter herauskommen würde, dann könnte er leicht hinunterfallen. Darum nahm Hahnemann zuletzt

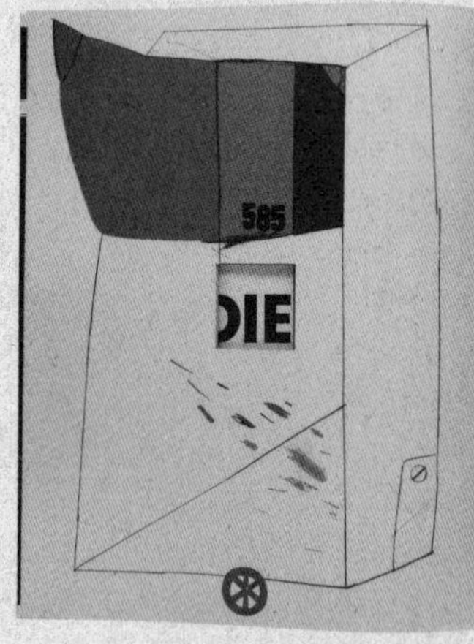

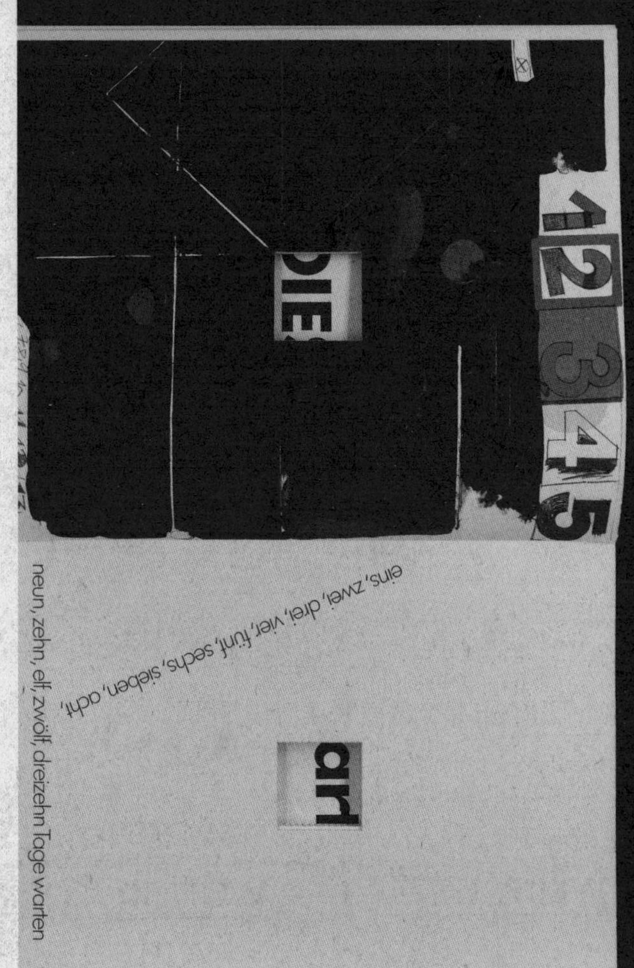

DIE
1 2 3 4 5
eins, zwei, drei, vier, fünf, sechs, sieben, acht,
neun, zehn, elf, zwölf, dreizehn Tage warten
ard

warten warten
en warten warten
n warten warten
en warten warten warte
warten
warten warte
en warten warten warte
en warten warten
warten
en warten warten war
rten warten warte
warten warten
en warten wa
warten warten warten w
n
warten warten warten w
warten warten
en warten w
en warten warten warten w
warten warten

aus dem Parad
denen eine ganze
entstehen sollte,
– ein Torso blieb
Hahnepeter habe
Anfang bis zum E
sen und nochmals
– und dann vom E
Anfang, und ich h
sehnlichst gewüns

eine Balletteuse. Darum

6 MAL rum. Und a
umdrehte, tanzte der Hah
sich dabei etwas in die

9 MAL rum. U
mal rumdrehte, flog der
und sang dazu, wie v
spielt. Darum drehte H

13 M

edicht DANN das Gedicht aus dem Paradies D
gessen DANN die Stille DANN die grüne W
erühren können DANN eins, zwei, drei, vie
eben, acht, neun, zehn, elf, zwölf,
Mal Abwarten DANN **i** Gedicht u
en Weg ins Paradies und wiede
um Schluss habe ich alle bem
lle leeren Seiten dieses Buches
rate von fünfundfünfzig mal

Da meinte Hahnemann, der würde doch Eier legen, und dann hätten wir eine ganze Familie von Hahnepetern. Aber die Mutter meinte, das hätte doch auch keinen Zweck. - O doch, das hätte den Zweck, daß viele Hahnepetermärchen kämen. Und mit einem Male fiel es Hahnemann dabei ein, daß der Hahnepeter auch „Kra" sagen konnte. Wenn man nämlich den Hahnepeter am Halse unter dem Lappen kitzelte, so sagte er K R A - K R A - ! so ähnlich wie Kikeriki. Und als die Kinder kamen, kitzelten sie ihn alle erst mal mit einem Strohhalm unter dem Halse, damit er „Kra, Kra" sagte. Und der Hahnepeter mußte fürchterlich lachen und sagte immer:

„KRA, KRA, KRA, KRA, KRA, KRA, KRA, KRA,"

und wie
ihn
nun
die KINDER alle kitzelten,

da kitzelte eines verkehrt, es wußte selbst nicht, wie und wo und warum, und wir wissen es auch nicht, warum es verkehrt war. Und plötzlich wurde er grasgrün, schrie furchtbar auf, sprang ein bißchen in die Luft und legte dabei ein schwarzes Ei. Und nun kitzelte jedes Kind einmal an der verkehrten Stelle. Und jedesmal legte der Hahnepeter ein schwarzes Ei. Und da es 13 Kinder waren, so legte er 13 Eier. Und dabei bemerkten es die Kinder, daß er hinten eine richtige Schraube hatte und einen richtigen

PROPELLER.

WENN WO NE
SCHRAUBE IST
MUSS MAN AUCH
DRAN DREHEN.

arl
5678
3x

a Gedicht
DIE
mit Pflaumen, die die Löwen im Paradies
zum Mittagessen haben
art
PARADIES

VIERMAL X
SECHSMAL X
4567 01

Schillerndes Paradies

1993 Gm

The Netherlands

The Netherlands

China

300mm

15mm

maastricht europe

310mm

● 金奖
« 肖像
◖ 荷兰

Gold Medal
faces
The Netherlands
Piet Gerards, Marc Vleugels
Pieter Beek
Gemeente Maastricht, Provincie Limburg
300 310 15mm
814g
70p
ISBN-10 90-73094-04-6

荷兰东南部的林堡省马斯特里赫特市是荷兰历史最悠久的城市之一。由于其夹在德、法、比三国之间的独特地理位置，成为西欧经济、交通、文化的十字路口，城市虽小但拥有丰富的教育产业资源。本书对这座城市中重要的研究机构和学院做了详细介绍，并展示了代表人物在其空间中的肖像，意图用人物的肖像指代城市的形象，让这些机构成为城市的名片，共同组城文化上的城市肖像。本书的装订形式为活页圈装。封面牛皮卡板印黑，外部包裹透明 PVC 材料丝印黑。内页混合使用了两种纸张，光面铜版纸印人物肖像，超薄的胶版纸呈现文字信息。除扉页标题字使用红色外，其余均为单黑印刷。文字页采用了包背夹页的形式，即在白色半透的包背页面中间夹一张黑底白字印刷的纸张，隐隐透出下方的文字。在包背折叠处特别制作了圆形刀版，透出夹页上的页码数字是本书最大的设计亮点。内文排版通过多种字体和栏位划分形成丰富的效果，与拍摄的极其精美的人物黑白肖像配合，形成值得信赖的城市形象。

Maastricht in Province Limburg, is one of the oldest cities in the Netherlands. Because of its unique geographical location between Germany, France and Belgium, it has become the crossroads of economy, transportation and culture in Western Europe. Although the city is small, it has rich educational and industrial resources. This book gives a detailed introduction to the important research institutions and colleges in the city, and shows the portraits of the representative figures in their space. It is intended to use the portraits of the characters to refer to the image of the city, so that these institutions can become the business cards of the city and form a cultural city portrait jointly. The book is bound by loose leaf. The cover is made of Kraft cardboard printed in black, and the outside is wrapped by transparent PVC material and printed with silkscreen in black. Two kinds of paper are used for the inner pages. The portraits are printed on glossy coated paper, and the text information is presented on the ultra-thin offset paper. Except the red title on the title page, the rest are printed in single black. The pages for the text are double-leaved with an interleaf, that is, a piece of paper printed with white characters on a black background is sandwiched in the middle of the double-leaved white semi-transparent pages, and the text below is faintly revealed. A hole is specially made at the folding part of the double-leaved pages, and the page number on the interleaved pages is revealed, which is the biggest design highlight of this book. A variety of fonts and field divisions form a rich effect in the layout of the text, and cooperate with extremely exquisite black and white portraits to form a reliable city image.

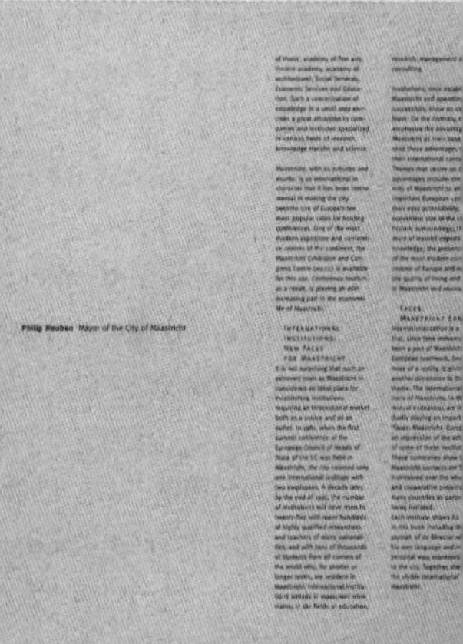

Maastricht is voor mij een bijzondere combinatie van 'Gemeinschaft' en 'Gesellschaft', gesitueerd in een Europese context.

Maastricht, for me, is an exceptional combination of 'Gemeinschaft' and 'Gesellschaft', placed in a European context.

Mr Dany Wijgaerts Managing Director

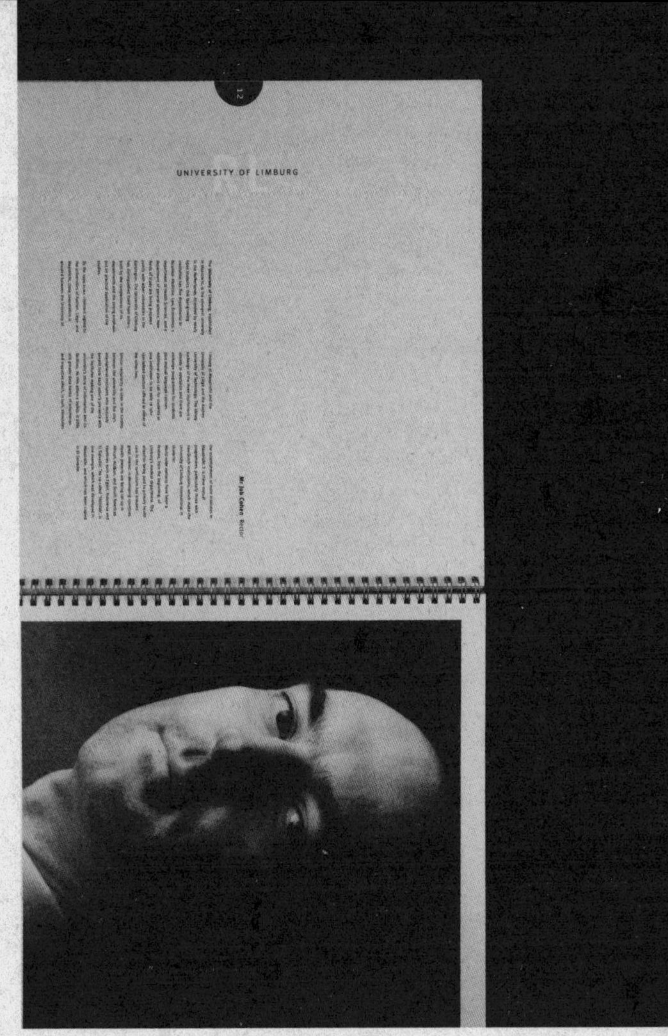
UNIVERSITY OF LIMBURG

ITEIT LIMBURG
3

stricht
2

STITUTE OF PUBLIC
ON (EIPA)
ouwe Plein 22
tricht
2

NTRE FOR DEVELOP-
MANAGEMENT

ouwe Plein 21
tricht
1

NTRE FOR WORK
(ECWS)
at 43
tricht
4

INTERNATIONAL
R MANAGEMENT

ein 150
tricht
3

N UNIVERSITY/
R NEW TECHNO-
INTECH)

31 EUROPEAN FAIR TRADE ASSOCIATION (EFTA)
Witmakersstraat 10
6211 JB Maastricht
tel 043-256917

32 RESEARCH CENTRE FOR EDUCATION AND LABOUR MARKET (ROA)
Tongersestraat 53
6200 MD Maastricht
tel 043-888827/888120

35 EUROTERM
Boschstraat 45
6211 AT Maastricht
tel 043-215199

36 WORLDNETH
Bredestraat 12
6211 HC Maastricht
tel 043-257052

39 BIOMATERIAL AND POLYMER RESEARCH INSTITUTE (BIOPRIME)
Postbus 616
6200 MD Maastricht
tel 043-888486

40 CARDIOVASCULAR RESEARCH INSTITUTE MAASTRICHT (CARIM)
Beeldsnijdersdreef 101
6216 EA Maastricht
tel 043-888223/130

43 INSTITUTE FOR BIOETHICS (IGE)

51 INSTITUTE FOR R
INTERCULTURAL C
(IRIC)
Tongersestraat 53
6200 MD Maastric
tel 043-888296/8

52 CENTRE FOR EUR
(CES)
Tongersestraat 10
6211 LN Maastrich
tel 043-888278

55 DE JOPPENHOF P
INTERNATIONAL
Kelvinstraat 3
6227 VA Maastric
tel 043-613242

59 JEANNE D'ARC CO
TIONAL DEPARTM
Postbus 4050
6202 RB Maastric
tel 043-612200

63 DRIEKANT FOUND
HOUSE (DFE)
Postbus 910
6350 AX Valkenb
tel 04406-15353

TEIKYO EUROPE

2

fac

exactly at the centroid of the great western European pow Germany. That makes Maastricht the ideal place to get to different facets. Maastricht is the embodiment of a genuin

richt, with close to 120,000 inhabitants, may occupy a modest place in the international ranking of European cities, but in spite of this, the importance of the city is many times greate than the number of inhabitants would lead one to expect. Maastricht, on its merits, deserves the title of European city It is not only the international history of Maastricht, nor its geographical location but most it is the spirit that seeks to look beyond its national boundaries

Maastricht's history goes back at least to the first century before

e

Mr Shoichi Okinaga Chairman of the Board of Directors of the Teik

Teikyo Europe is one of the largest private university networks in Japan with subsidiaries in various parts of the world. In 1991, Teikyo bought the former premises of the University Hos-

live and work in Europe. The doctors are also participating in research in the University Hospital of Maastricht, and there are joint programmes with the medical faculty of the University of

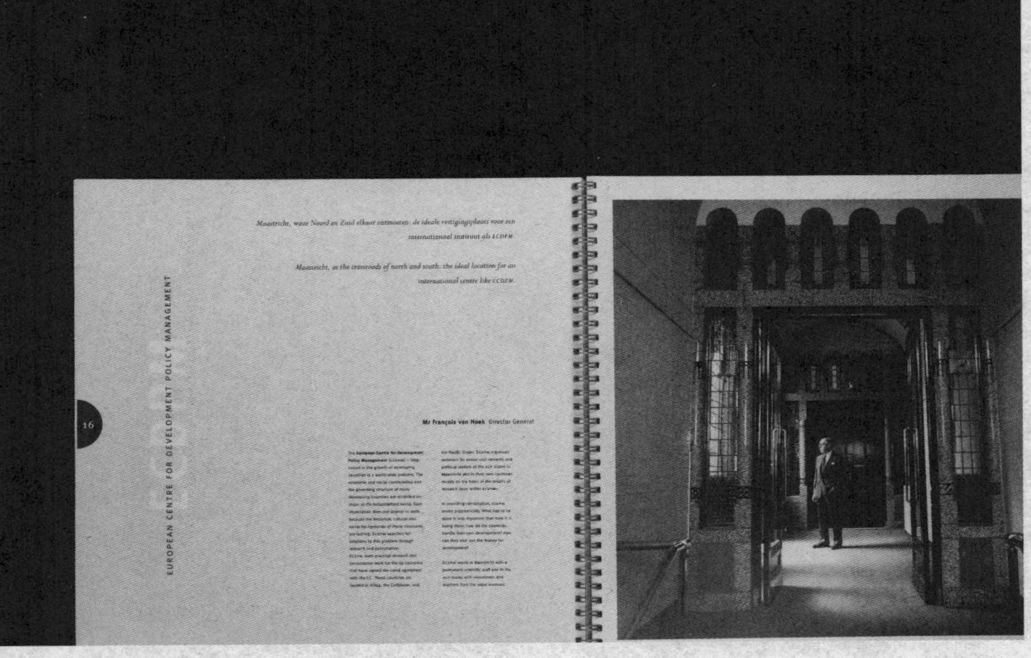
16
EUROPEAN CENTRE FOR DEVELOPMENT POLICY MANAGEMENT
Maastricht, waar Noord en Zuid elkaar ontmoeten: de ideale vestigingsplaats voor een internationaal instituut als ECDPM.
Maastricht, as the crossroads of north and south: the ideal location for an international centre like ECDPM.
Mr François van Hoek Director General

0263

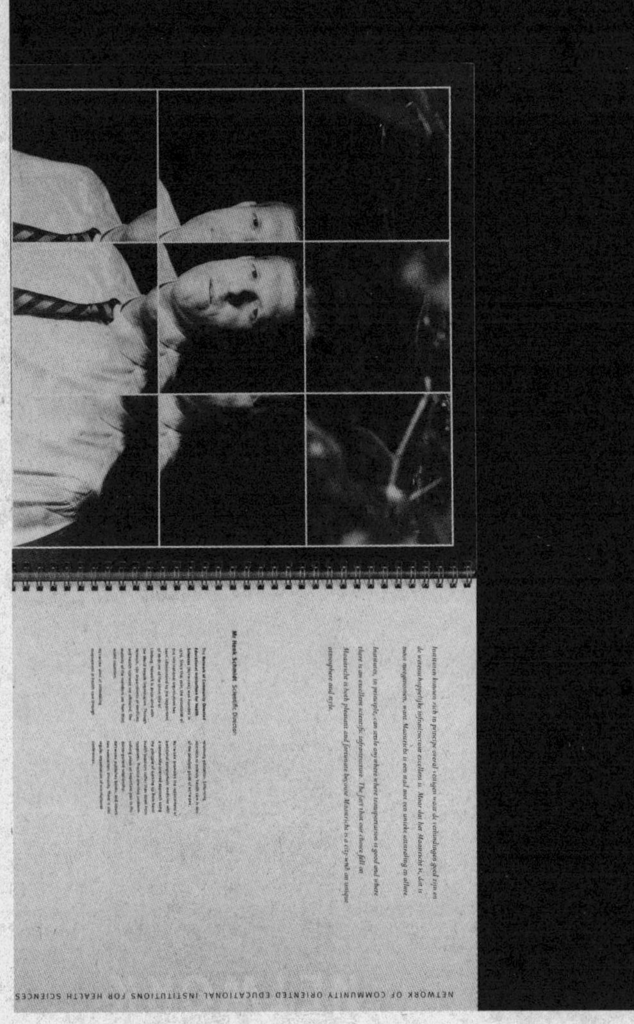
NETWORK OF COMMUNITY ORIENTED EDUCATIONAL INSTITUTIONS FOR HEALTH SCIENCES
Instituten kunnen zich in principe overal vestigen waar de verbindingen goed zijn en
de wetenschappelijke infrastructuur excellent is. Maar dat het Maastricht is, dat is
mooi meegenomen, want Maastricht is een stad met een unieke uitstraling en allure.
Institutes, in principle, can settle anywhere where transportation is good and where
there is an excellent scientific infrastructure. The fact that our choice fell on
Maastricht is both pleasant and fortunate because Maastricht is a city with an unique
atmosphere and style.
Mr Henk Schmidt Scientific Director

0265

Connie Imboden

WORLDWIDE EXPRESS MAIL SERVICE

全球邮政特快专递

Commercial Invoice - 17th Nov 2019

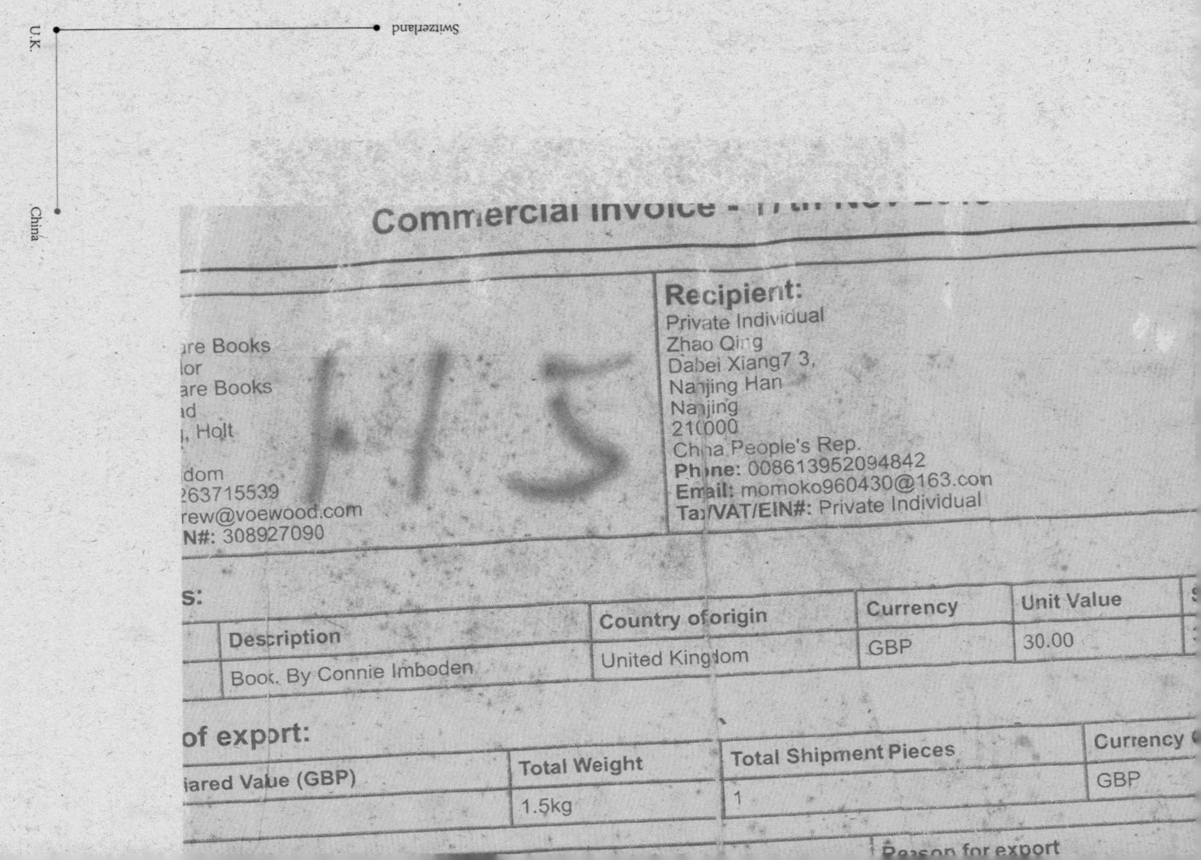

Commercial Invoice - ...

...re Books
...lor
...are Books
...d
..., Holt
...dom
...263715539
...rew@voewood.com
...N#: 308927090

Recipient:
Private Individual
Zhao Qing
Dabei Xiang7 3,
Nanjing Han
Nanjing
210000
China People's Rep.
Phone: 008613952094842
Email: momoko960430@163.con
Tax/VAT/EIN#: Private Individual

...s:

Description	Country of origin	Currency	Unit Value
Book. By Connie Imboden	United Kingdom	GBP	30.00

...of export:

...ared Value (GBP)	Total Weight	Total Shipment Pieces	Currency ...
	1.5kg	1	GBP

Reason for export

289mm

14mm

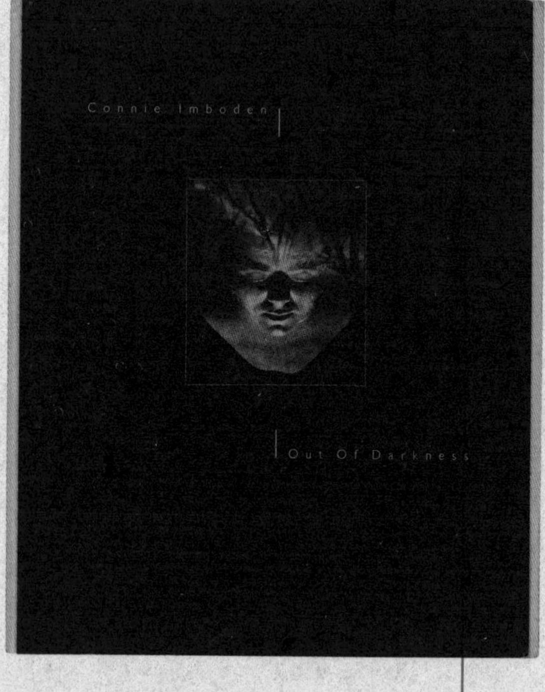

Out Of Darkness

Connie Imboden

245mm

● 银奖
《 脱离黑暗
◐ 瑞士

Silver Medal

Out Of Darkness

Switzerland

Kaspar Mühlemann

Connie Imboden

Edition FotoFolie, Zürich/Paris

289 245 14mm

902g

100p

ISBN-10 3-9520309-0-2

美国摄影师康妮·英博登（Connie Imboden）以水和镜子作为道具拍摄的人体摄影而闻名，本书包含了她从 1986 年到 1991 年间创作的 40 幅照片。书籍的装订形式为裱贴了卡板的锁线平装。封面铜版纸印单黑，覆哑膜。封面封底裱覆卡板，单黑印刷文字反白，漏出未裱卡板的书脊部分；卡板中心方形镂空，成为一个内凹的相框，呈现下部的照片。内页无酸纸单黑印刷，照片部分过油处理，呈现类似冲洗出的照片光泽。书中前言部分探讨了黑暗的母题，包括在太空中的孤寂体验、在母体中的温暖知觉，以及深处黑暗中的幽闭恐惧等经验。探讨通过水和镜子的反射、折射、流动、遮蔽等形式来体现这些感受。照片作品部分则是摄影师对这些感受的实践，由于水和镜子对光线的改变，将多重影像相互叠加、扭曲、隔断，形成独特的视觉体验。最后一部分是摄影师对自己长期创作方向的思考。书籍的文本排版很有特点，将文字块错位并将部分文字加粗的做法，体现了作者对水和镜子扭曲现实的理解。

American photographer Connie Imboden is known for her body photography through water and mirrors as props. This book contains 40 photos she created from 1986 to 1991. The paperback book is bound by thread sewing mounted with cardboard. The cover is made of coated paper printed in black, mounted with matte film. The front cover and back cover are mounted with cardboard printed in black with white text. The center of the cardboard is hollowed out to form an inner concave photo frame to display the photo inside. The inner pages are printed in black on acid-free paper. The photos have been vanished partly to give a similar luster to that of the developed ones. The foreword of the book discusses the motif of darkness, including the experience of loneliness in space, the sense of warmth in the matrix, and the claustrophobia in the dark. It explores the reflection, refraction, flow and shielding of water and mirror to reflect these feelings. The photos express the photographer's practice. Due to the change of light caused by water and mirror, multiple images will be superimposed, distorted and separated, forming a unique visual experience. The last part is the photographer's reflection on his long-term creative direction. The layout of the book has its own characteristics. Dislocating the text blocks and making some texts thicker reflect the author's understanding of the distortion by water and mirrors.

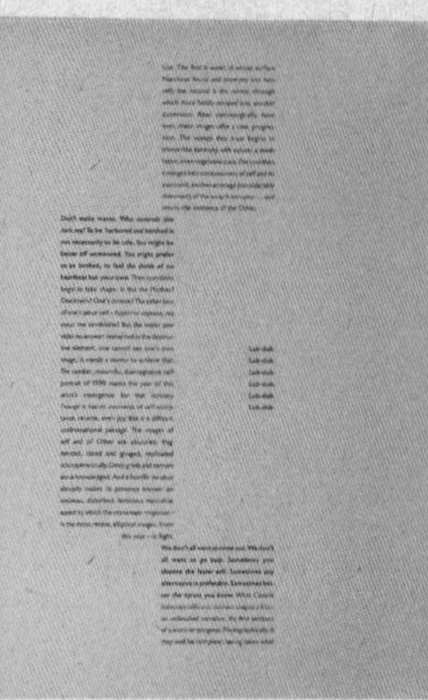

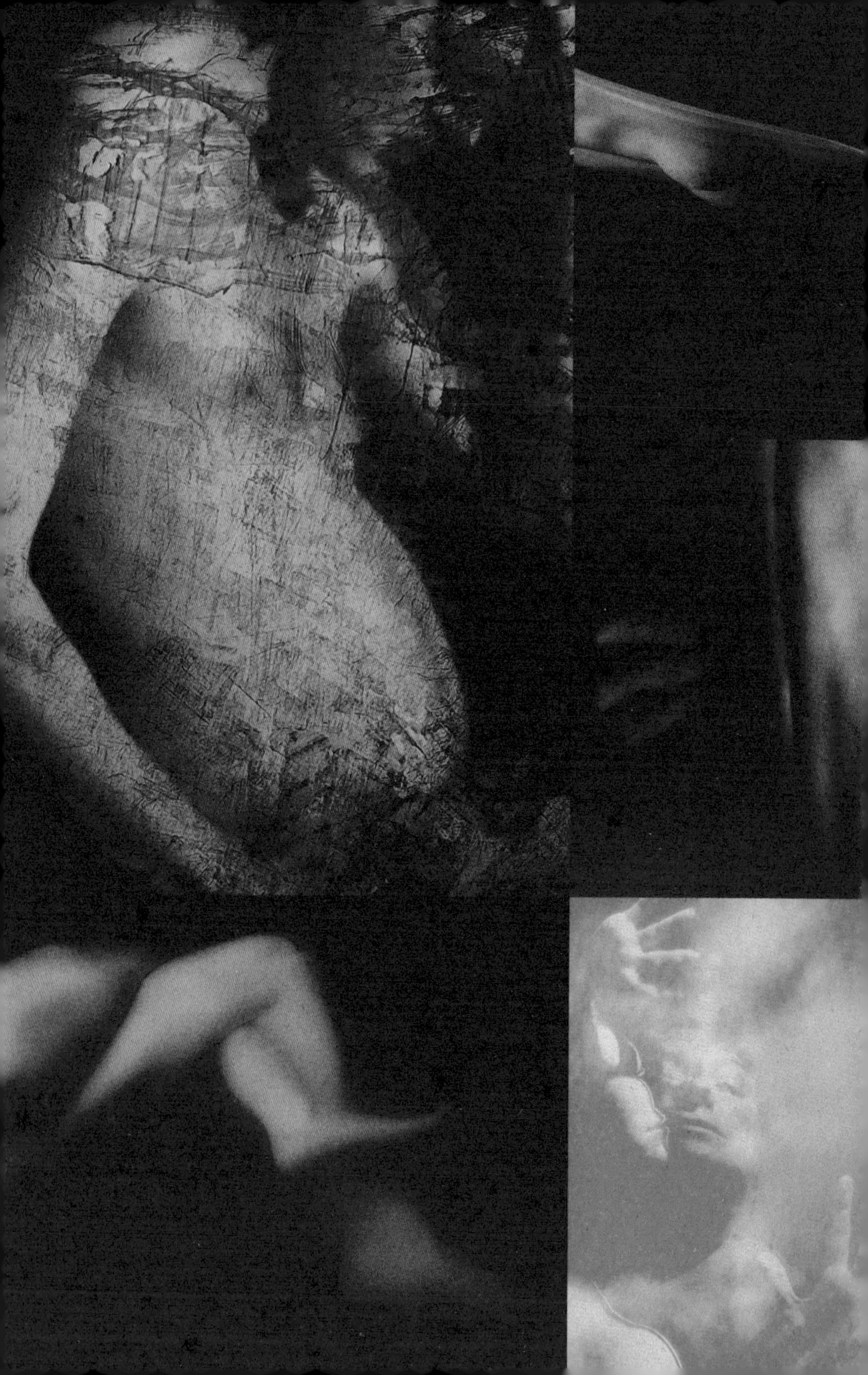

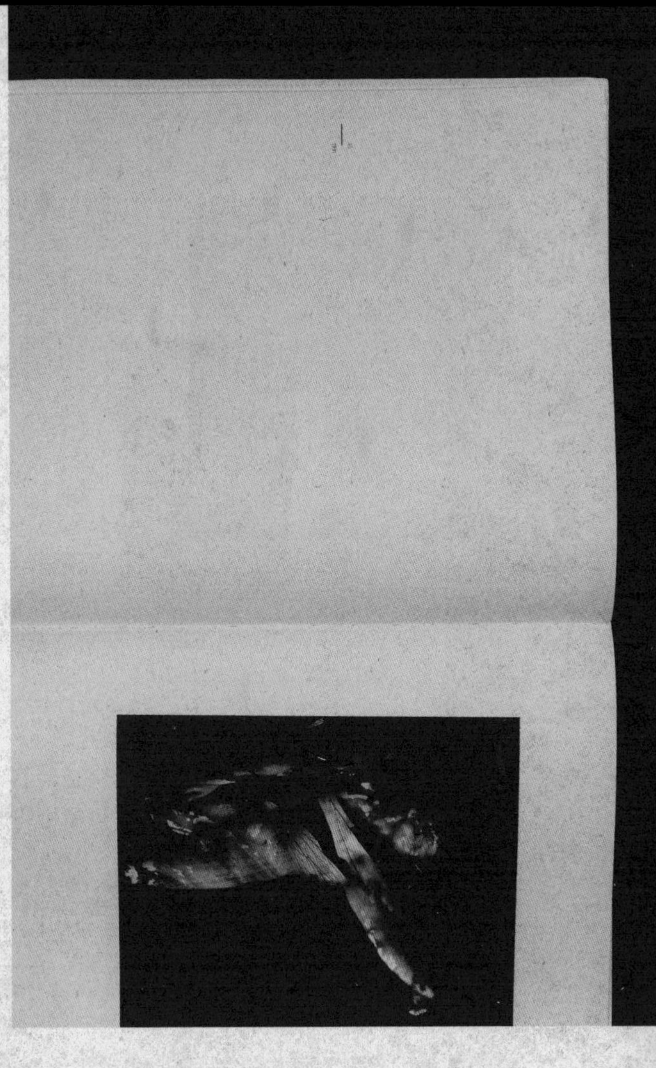

Andreas Sadowski, Erlenstr. 3a, 33803 Steinhagen, DEU
RC 58 729 817 5DE Recommandé PRIORITY P.P.
R
A0.0163.95B1.00.0001.8782
IM 19.02.20 7.20 Deutsche Post
P INT
Zhao Qing
Meiyuan New Village 7-3, Dabei Lane
210000 NANJING
CHINA+(VOLKSREPUBLIK)
7-3
全球
EXPRESS
专递

Andreas Sadowski, Erlenstr. 3a, 33803 Steinhagen, DEU
RC 58 728 817 5DE Recommandé PRIORITY P.P.
A0 0163 95B1 00 0001 67B2
IM 19.02.20 7,20 Deutsche Post
P INT
Zhao Qing
Meiyuan New Village 7-3, Dabei Lane
210000 NANJING
CHINA+(VOLKSREPUBLIK)

Germany
Germany
China

255mm

10mm

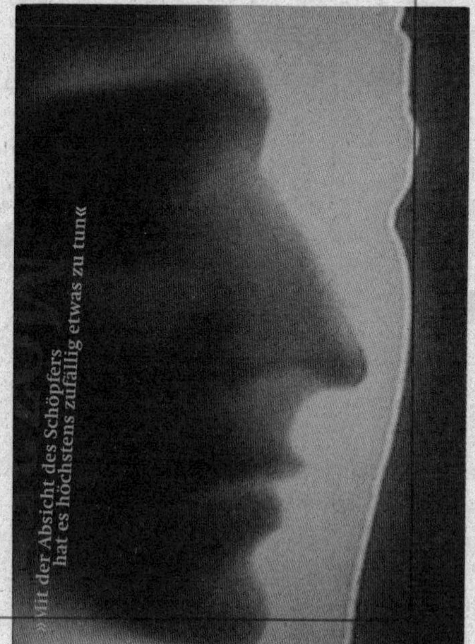
»Mit der Absicht des Schöpfers
hat es höchstens zufällig etwas zu tun«
MOZART IN TONSPUREN UND BILDRÄUMEN

186mm

铜奖 Bronze Medal

创作者的即兴创作——声音图像中的莫扎特 Mit der Absicht des Schöpfers hat es höchstens zufällig erwas zu tun: Mozart in Tonspuren und Bildräumen

德国 Germany

Lutz Dudek, Claudia Grotefendt

Jörg Boström (Hrsg.)

Fotoforum SCHWARZBUNT e.V., Bielefeld

255 186 10mm

385g

56p

ISBN-10 3-928625-01-2

本书记录了多位艺术家对声音图像这一联结听觉和视觉的前沿艺术的探索。书中以莫扎特的音乐、图片为创作素材，通过各自不同的表现手法，表达每个人对莫扎特音乐的视觉化理解。书籍的装订形式为包背无线胶平装。封面白卡纸四色印刷，其中黄色被替换为鹅黄专色。内页哑面艺术纸四色加哑金色共五色印刷。内容方面，共 15 位艺术家对莫扎特的音乐进行了各自理解上的阐述和再创作，通过音轨图形、肖像处理、连续照片、视听状态、乐器装置等方面的记录，在音频的视觉化表现上进行创作，试图建立听觉和视觉之间的转化通道。艺术家名、作品名和创作年份信息用五线谱的形式放置在各自作品的前部。音频本身具有节奏规律，所以光栅图形和音轨记录后的再创作是艺术家们主要的创作方向。与局限在传统平面空间内的视觉艺术不同，音频具有时间上的连续性，所以本书的内页设计为包背形式，图片和文字在装订处和翻口处“自由流动”到下个页面。||||||||||||||

This book records many artists' exploration of sound tracks and image spaces, a cutting-edge art category connecting hearing and vision. In the book, Mozart's music and pictures are used as creative materials to express personal visual understanding of Mozart's music through different expressions. The double-leaved book is glue bound. The cover is made of white cardboard printed in four colors, in which yellow is replaced by goose yellow PMS color. The inner pages are made of matte art paper printed in four colors plus matte gold. In terms of content, a total of 15 artists elaborated and re-created Mozart's music on their own understanding. Through recording the audio track graphics, portrait processing, continuous photos, audio-visual status, musical instrument installation, etc., they created audiovisual performance and tried to establish a transformation channel between auditory and visual senses. The names of the artist, the titles of the works and the year of creation are placed in the front of their works in the form of Musical Notation. The audio itself has a rhythm pattern, so the re-creation after recording bitmap and audio tracks is the main direction of the artists' creation. Different from the visual art confined in the traditional plane space, audio has the continuity in time. Therefore, the inner pages of this book are double-leaved, so that the pictures and text 'flow freely' to the next page at the turning edge.

„Wir haben zu Lebzeiten alles gesagt,“ redete ich auf ihn ein, „ich habe mein Bestes gegeben und Sie werden dergleichen Ihr Bestes gegeben haben – Andy, Sie sind frei. Es ist Zeit, Frieden zu schließen.“

In mir stockte das Blut. Was hatte ich gesagt? Ich, Mozart, der ich seit Jahrzehnten wie leer zwischen den Welten wandere, stummer Zeuge meines mir zu Lebzeiten nicht vergönnten irdischen Ruhmes, der Fülle der mittelmäßigen Interpretationen meiner Musik inzwischen überdrüssig, kehre plötzlich, eines Tages zurück, um einem Fremden zu sagen: Sie sind frei? - Nein, mir selbst gilt dieses Wort.

Als ich aufschaute, war Andy verschwunden. Noch eine ganze Weile betrachtete ich die wechselnden Erscheinungsbilder meines Portraits auf der Oberfäche des kleinen Kubuses. Das Requiem rief mich. Es ist Zeit, Frieden zu schließen, klang es in mir

und skizziert Kadenzen.
rsten Takte seiner Jazzo-
chine” nach Motiven des
i, Librettist wird Charles
ie sphärischen Klänge
aphon von Gary Burton
terdessen in kosmischen
3töne bewegen sich zu
Fruchtformen, pfeifen-
e Trompetentöne bilden
heleien. Dagegen setzt
n von Jan Gabarek steil
otische Spitzen

5

MONA EISE
Requiem, 1991

nden. Nicolas Schöffer führt
im Pariser Théatre Sarah-Bern-
t eine kinetische Raum-Licht-
ptur vor. Er versucht eine neue
n der Integration von Raum,
egung und Licht zu entwickeln
ner Konstruktion, die den Be-
ungsumfang des Gesamtkunst-
s mit Mitteln der Technik auf-
t: „Wort und Ton, Bewegung
Raum, Licht und Farbe – sie alle
n, einander überlagernd, viel-
kontrapunktierte Strukturen in
ausziselierten und gleichzeitig

Skulptur und Klang v

Der Maulwurf des
einer Verbindung zwi
Bild und Klang taucht
Unsere Klangbilder e
aus der Spannung un
insbesondere von Aug
Dabei geraten die Me
Typografie, Malerei, e
Bild, Objektkunst und
wechselseitige Bezieh
sprechend der Idee ei
kunstwerks alle Sinne
herausfordern.

findsamen Ariette mit
inzuziehung der Chro-
er verästelten Stil....
eist der Hörer gleicher-
meicheln. (nachdem ich
gespielt) den letzten
eines Lebens mit
rliebe bei allen seinen
wendete und das heute
burtshaus ausgestellt
dem Komponisten und
ann. Kragen paßte ihm,
ushalten«

Unsere Aufmerksamkeit ist eher auf das Rätselhafte als das Erklärbare gerichtet. Es entstehen Klangbilder, Tonspuren, Auflösung und Öffnung der Zeitdokumente durch generative Reproduktion, fotografische Anmutungen von Lebensspuren, physiognomische Details von Musikerfahrungen, Klangräume, Bildräume, Videotransformationen.

Eine Raumbildung in Quaderform ist dem überlieferten Erscheinungsbild des Künstlers gewidmet. Einem Mausoleum vergleichbar entsteht ein Klangraum, in dem ein Monitor am Boden in wechselnden Überblendungen die Folge trügerischer, puppenhafter Porträts des Meisters ablaufen läßt, während an den Wänden sichtbar und in einem Grundton hörbar das Requiem erklingt. Visuell erlebbar wird die Musik Mozarts durch die Projektion von Bildern eines neu entwickelten Musikvideophons. In Fotografien an den Wänden und in bewegten Lineamenten am Monitor und auf einer Projektionsfläche erscheinen taktgenau grafische Rhythmen, welche Auge und Ohr gleichzeitig Musik erleben lassen.

1968 entwickelt der Fotograf Chargesheimer kinetische Skulpturen, „Meditationsmühlen", die Raum, Licht und Bewegung zu einem Schaustück vereinen sollen.

Der Maler Karl Otto Götz läßt sich durch das Auftreten von Störbildern beim Radarbetrieb während des 2. Weltkrieges zu Experimenten mit elektronischer Steuerung solcher Phänomene anregen und entwirft ein Konzept: „Der Werkvorgang beim kinetisch-elektronischen Bild ist ein variabler Kreislauf und erstreckt sich von einem Zahlengeflecht über einen Frequenzplan zum Frequenzgenerator und Fernrohr.

Das hier entstehende Resultat ist in seiner optischen Makrostruktur vorausberechenbar, aber nicht voraus-sehbar; denn seine Zeichengebung geschieht im Bereich der Mikrostruktur und besteht aus einer Kette von elektrischen Impulsen und Frequenzgemischen, also aus einer Folge von Zeiteinheiten".

Musik ist neben anderem auch dies: eine Folge von Zeiteinheiten. Der Begriff des variablen Kreislaufs bei Götz nimmt offenbar den der Rückkopplung voraus, der in den video-grafischen Musikbildern unserer Arbeit zu neuen Unendlichkeitsstrukturen führt. Diese Rückkopplung bedingt eine sich selbst ins Unendliche reproduzierende und variierende Folge von dynamischen Formen, für welche erst die Chaostheorie, die Götz noch nicht kennen konnte, Erklärungen anbietet.

In Zeitabständen von Jahrzehnten tauchen Ideenkomplexe auf und verschwinden wieder. Unsere Ausstellung zeigt Lichtmalerei in großformatigen, freien Fotografien. Als künstlerische Interpretation von Klängen erscheinen solche Bilder. Was sieht im Tanz der Arbeit des Dirigenten der Taktstock, wenn er Augen hätte? Diese neuartige Erlebnisweise eines Konzerts vermittelt ein Videofilm, der diese Bewegungen sichtbar macht.

Das kaum Vorstellbare zu realisieren, die Verbindung der Sinne und Dimensionen immer wieder zu versuchen, die Vergeblichkeit solcher Kunstbewegungen, die sich mehr an die Vorstellungskraft wenden als an den objektabhängigen Kunstgenuß, solche Bewegungen werden immer wieder angestoßen, Problemlösungen werden versucht, vorangetrieben, abgelegt. Es geht mehr um den Prozeß als um das Resultat.

Vielleicht erreichen solche Bilder eine Annäherung auch an das Prozeßhafte in dieser Musik, die mit sich selbst spricht – in unübersetzbarer Sprache.

1

CHRISTINE JACOBY
Sinfonie für Orchester, 1991

Ich stelle mir
einem Konzer
Zuschauerrau
eine Mozart-S
Orchester im
fängt an, sich
zu verbinden.
ren sowie inn
schmelzen M
und meine As

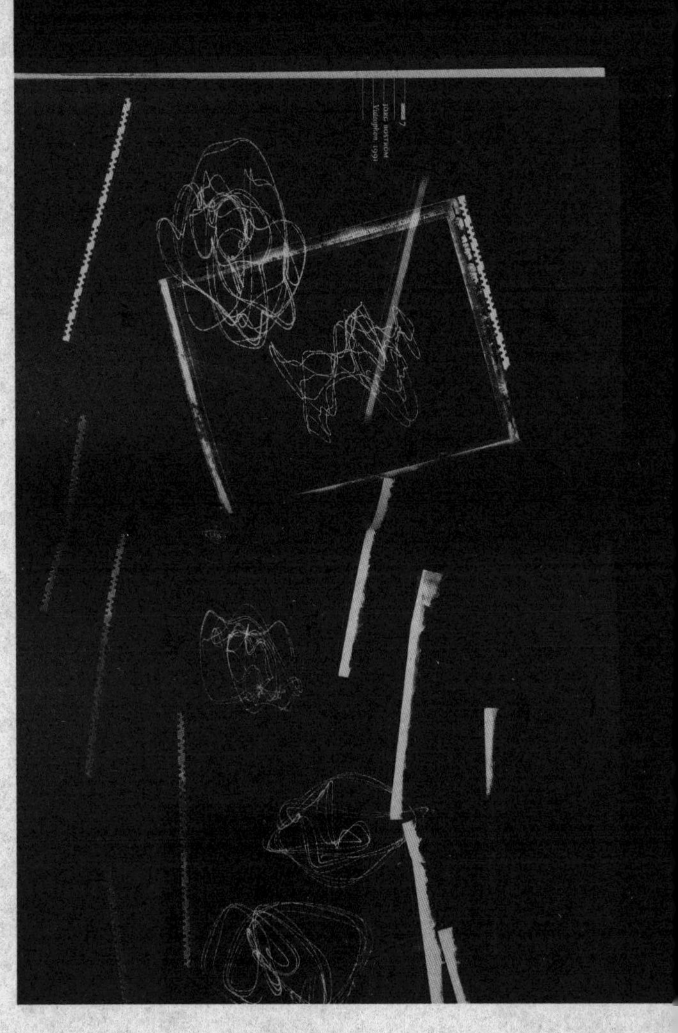

TO
Tang Design Co.
ZHAO QING
Dabei Xiang 7-3
210000 NANJING
Nanjing Han
CHINA
RC 58 114 488 0DE
Germany
Germany
China

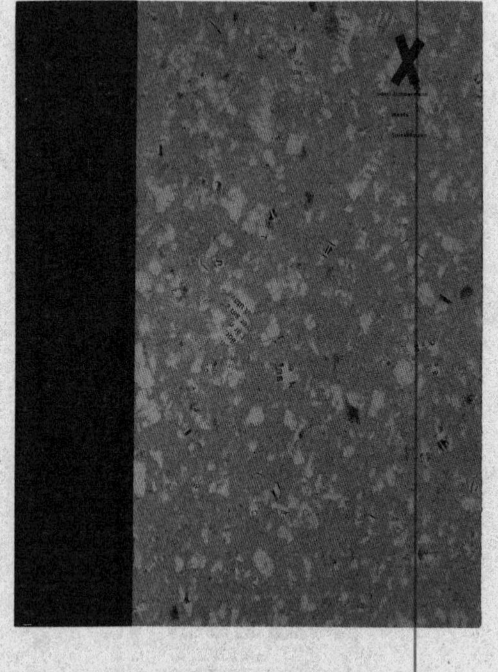

铜奖
齐默尔曼遇见施皮克曼
德国

Bronze Medal
Zimmermann meets Spiekermann
Germany
Ulysses Voelker
Ulysses Voelker
Meta Design, Berlin
280 218 8mm
234g
42p
ISBN-10 3-929200-01-5

埃里克·施皮克曼（Erik Spiekermann）作为德国著名的平面设计师，对于印刷、字体、视觉语言等方面具有很重要的影响力。不来梅艺术大学开设了与设计趋势相关的课程，对施皮克曼的作品进行深入探讨，集结成本书。书籍的装订形式为包背软精装。封面牛皮卡纸红黑双色印刷，大勒口粘贴，形成厚实的手感。书脊处裱贴深灰色布面，文字烫红处理。内页超薄纸张正反四色印刷，通过纸张透出的正反图案形成丰富的信息层级。书中内容以施皮克曼设计的作品为主，包括海报、封面、地铁信息系统、邮票、字体、广告、表单等设计作品，均为施皮克曼当时创立的两家公司 Meta Design 和 Fontshop 提供。标题中的“齐默尔曼”（Zimmermann）原意为木匠，但在设计中用以指代纯色条。这个词与“施皮克曼”（Spiekermann）读起来有点像（谐音），在这里被用作读音和语义上的双关（英文部分翻译加的注解）。

As a famous graphic designer in Germany, Erik Spiekermann has a great influence on printing, fonts, visual language and so on. Bremen University of the Arts has offered courses related to design trends and discussed Spiekmann's works deeply, and finally assembled into this book. The paperback book is doubled-leaved. The cover is printed with kraft cardboard in red and black, and pasted with a large flap to increase the thickness. The spine of the book is mounted by dark gray cloth, and the characters are hot-stamped in red. Ultra thin paper is used for inside pages printed in four colors, which form rich information levels through the hidden patterns under the paper. The contents of the book are mainly the works designed by Spiekermann, including posters, covers, subway information systems, stamps, fonts, advertisements, forms and other design works, which are provided by Meta Design and Fontshop, two companies founded by Spiekermann at that time. 'Zimmermann' in the title originally meant 'a carpenter', but it was used to refer to a solid color bar in design. The word sounds a bit like Spiekermann (homophone) and is used here as a phonetic and semantic pun (comments added in the translation of the English part).

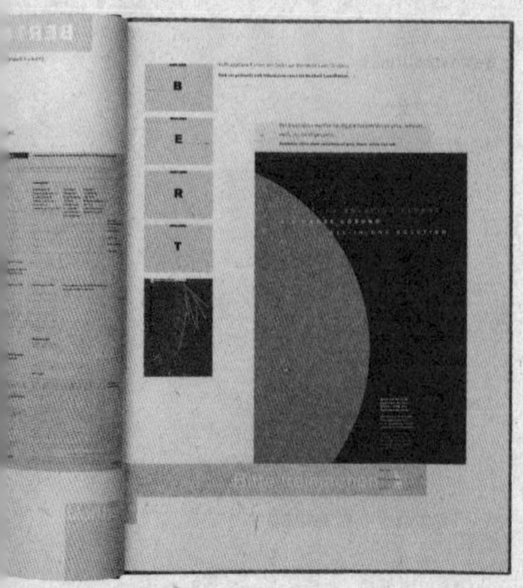
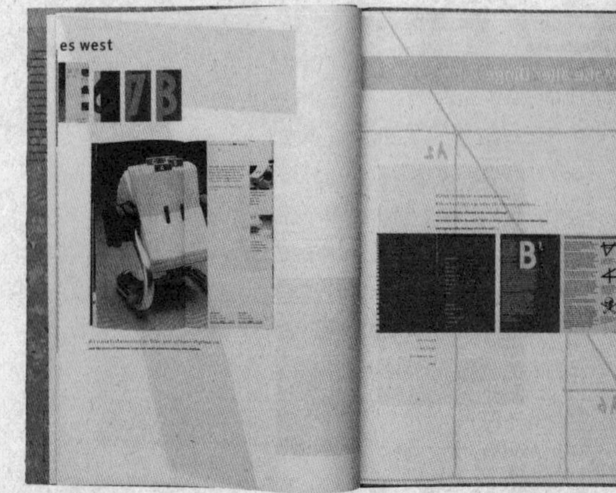
es west
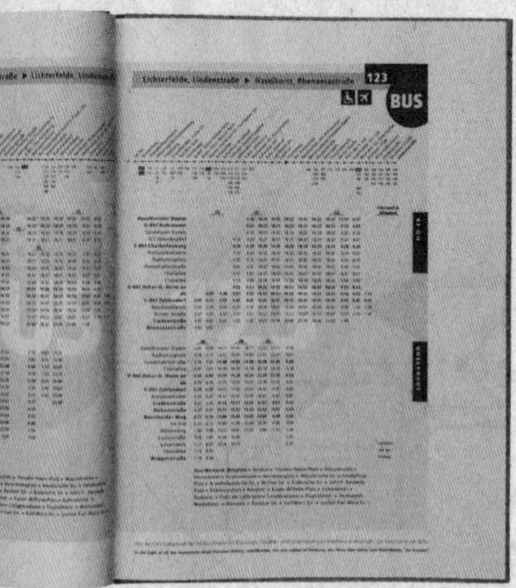
Lichterfelde, Lindenstraße ▶ Havelberg, Rhenaniastraße
123
BUS
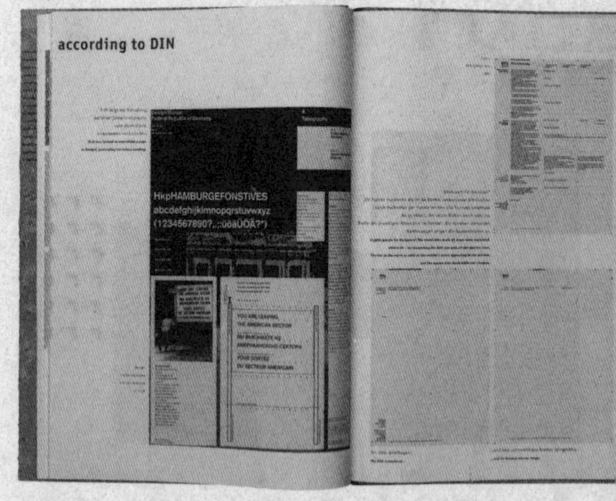
according to DIN
HkpHAMBURGEFONSTIVES
abcdefghijklmnopqrstuvwxyz
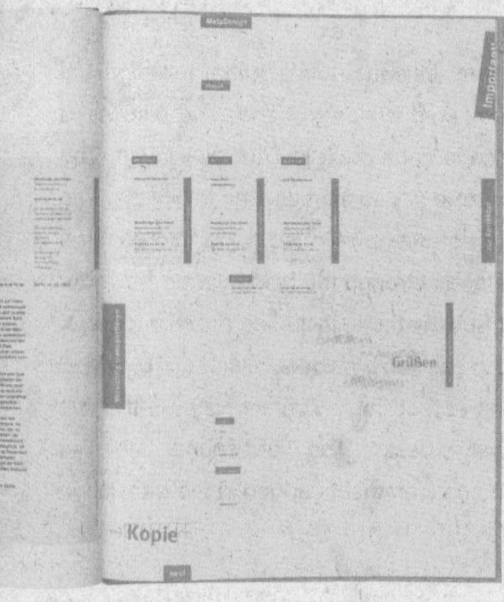
Grüßen
Kopie
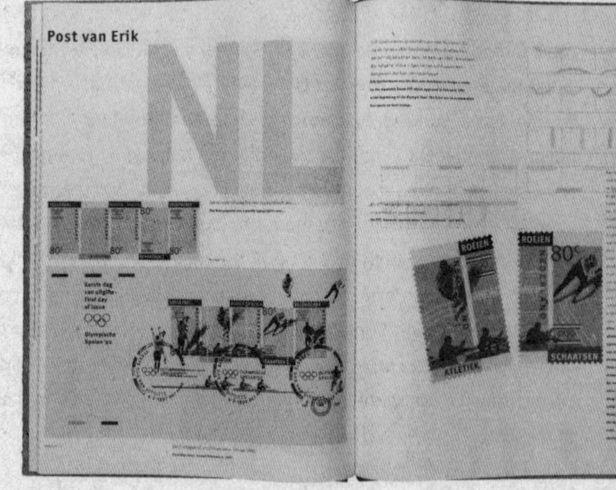
Post van Erik
NL
ROEIEN
SCHAATSEN
ATLETIEK

iern und Kaffeetassen

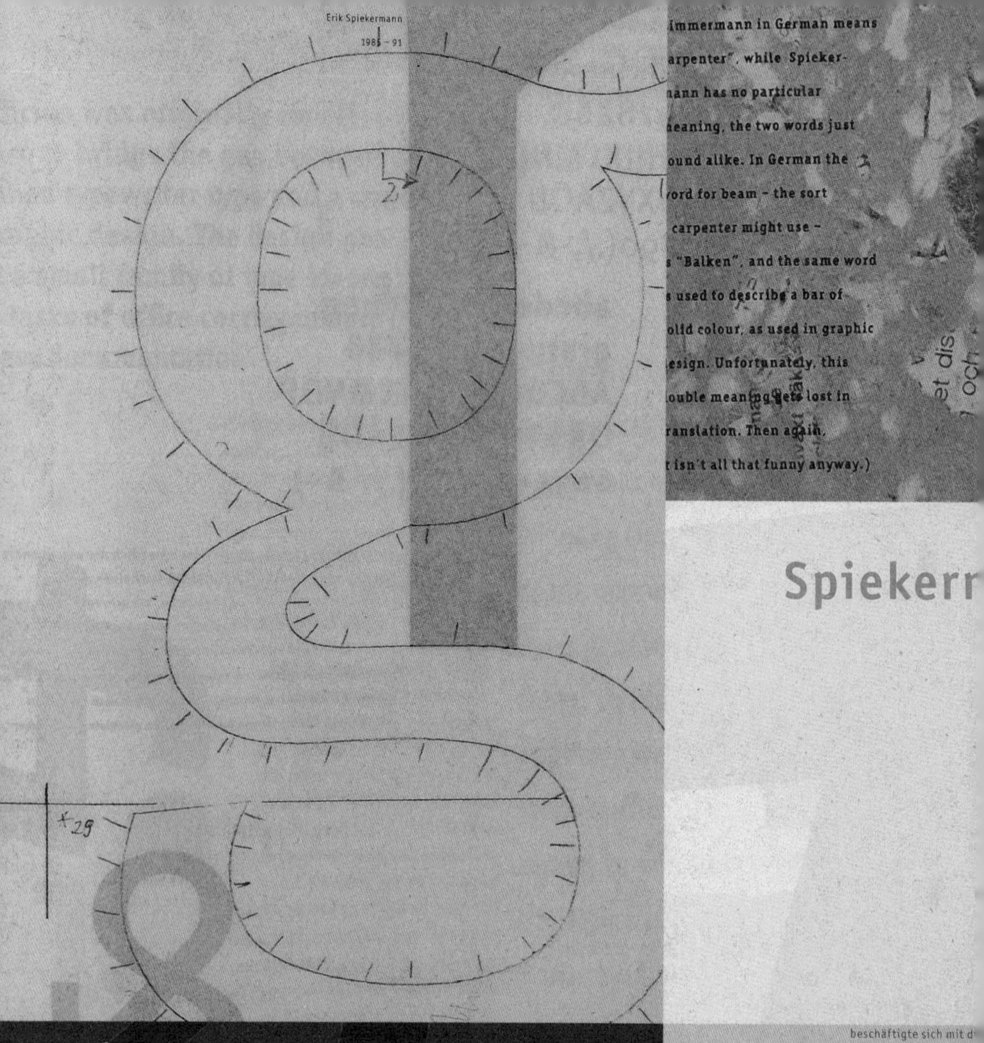

Erik Spiekermann
1985 – 91
Zimmermann in German means
carpenter", while Spiekermann has no particular
meaning, the two words just
sound alike. In German the
word for beam – the sort
a carpenter might use –
is "Balken", and the same word
is used to describe a bar of
solid colour, as used in graphic
design. Unfortunately, this
double meaning gets lost in
translation. Then again,
it isn't all that funny anyway.)
Spiekerm

Zimmermann meets Spiekermann

Is there a visual language? Is there a level of design which is perceived besides the textual contents, be it subconsciously or consciously, and which thus gains its own importance? Are there designer-specific languages?

These were the questions posed by a student project within Professor Eckhard Jung's typography department at the Academy for Arts in Bremen. Students picked well-known contemporary designers and went to find out what makes them tick and why one often asks oneself, when looking at work published in some magazine, "this must be by such and such, you can see it at a glance...". Our shared preoccupation with

beschäftigte sich mit d
Fragestellungen. Stude
tInnen nahmen sich als
Cracks der zeitgenössis
Designszene vor, um zu
suchen, mit welchen Wa
da gekocht wird.
Und wie es kommt, daß
beim Betrachten neuer
kationen häufig festste
»...das ist ganz eindeut
der/die..., das sieht ma
von weitem!«
Nicht zuletzt wegen uns
gemeinsamen Vorliebe
für angeschnittene Balk
habe ich (der gelernte
Zimmermann) mich mit
visuellen Sprache von E
Spiekermann befaßt. M
Studie beschreibt, was
reizte und dann neugie
machte, dem Meister bi
die verborgensten Deta
seines Schaffens, nämli

abcdefghijklmnop
qrstuvwxyzßäöü
ABCDEFGHIJKLMNOP
QRSTUVWXYZÄÖÜ
1234567890(./,-&–)

abcdefghijklmnop
qrstuvwxyzßäöü
ABCDEFGHIJKLMNOP
QRSTUVWXYZÄÖÜ
1234567890(./,-&–)

Wesentliche Merkmale, die die Meta dynamisch und eigenstä
machen, ohne die Schrift aufdringlich erscheinen zu lassen,
sind die gebogenen Auf- und Abstriche, die schrägen Abkan
vor allem aber der Kontrast zwischen etwas kantigeren
Innenräumen und fließenden, ovalen Außenformen.

**Meta gets its unique and dynamic character from details like oblique up- and
and diagonal terminals. Its extremely good legibility at small sizes results
from a distinct contrast between squarish counters and flowing, oval contours.**

ITC Officina

ITC Officina was originally conceived as a typeface to bridge the gap between old fashioned typewriter type and a traditional typographic design. The design goal was to create a small family of type ideally suited to the tasks of office correspondence and business documentation.

Midway through the design, however, it became obvious that this face had capabilities far beyond its original intention. Production tests showed that ITC Officina could stand on its own as a highly legible and remarkably functional type style.

The European design team, under the close guidance of the Berlin designer, Erik Spiekermann, was given the directive to continue the work on ITC Officina, but now with two goals. The first was to maintain the original objective of the design: to create a practical and utilitarian tool for the office environment. And the second was to develop a family of type suitable to a wide range of typographic applications.

What *developed is a different so of type family. It has a distilled range of just two weights: Book and Bold (medium weight being unnecessary in office correspondence) with complementary Italics. In addition, ITC Officina is available in two styles: Serif and Sans. The end res is an exceptionally versatile communication tool packaged in a relatively small type family.*

Proportionally, the design has been kept somewhat condensed to make the family space economical. Special care was also taken to insure that counters were full and serifs sufficiently strong to withstand the rigors of small sizes, modest resolution output devices, telefaxing, and less than ideal

CP 71 DISPATCH NOTE
ACHTERKANT
PostNL
210000 NANJING
CHINA, MAINLAND

CG835873411NL
PRIORITY
Office of origin
(Bureau d'origine)
PostNL
Insured value (SDR)
(Valeur déclarée DTS)
Total gross weight (kg)
(Poids en kg)
1,0
Customs duty
(Droit de douane)
Custom stamp
(Timbre de douane)
Declaration by addressee
(Déclaration du destinataire)
I have received goods as described on this note
Date
Signature
I certify that this item does not contain any dangerous articles prohibited by postal regulations
Date 16-11-2019
Signature Sender
NG DESIGN CO.
ANG 7-3, NANJING HA
NANJING
MAINLAND
non-delivery
(in cas de non-livraison)
(traiter comme abandonné)
stroken
952094842
The Netherlands
The Netherlands
China

240mm

6mm

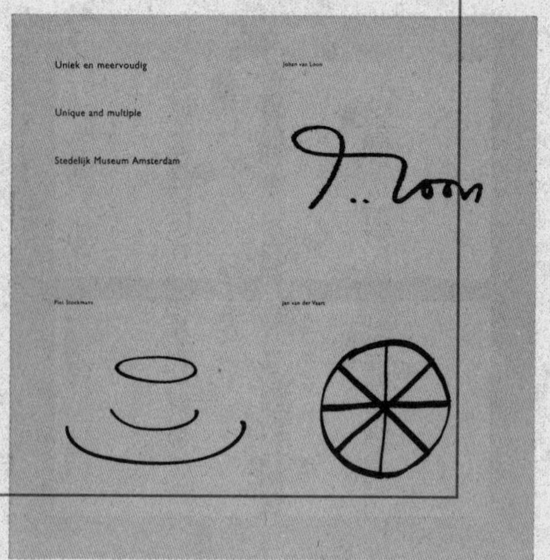

240mm

● 铜奖
《 唯一与多样
◐ 荷兰

Bronze Medal
Uniek en meervoudig
The Netherlands
Reynoud Homan
Liesbeth Crommelin
Stedelijk Museum, Amsterdam
240 240 6mm
289g
48p
ISBN-10 90-5006-060-9

阿姆斯特丹市立博物馆举办了荷兰和比利时三位陶瓷艺术家的群展，本书是这次展览的画册。书籍的装订形式为锁线胶平装。封面白卡纸印哑光白、亮白和黑色，两种白色的运用形成陶瓷釉面的对比效果。封二和封三印蓝色。内页采用薄透的胶版纸印黑和哑面铜版纸四色印刷，两种纸张混合装订。内容方面，开篇是对展览本身和三位艺术家的综述，之后分为三大部分进行单独介绍和作品展示。从封面开始，设计者使用三个简单的符号代表三位艺术家，分别是代表约翰·范隆（Johan van Loon）的个人签名、代表皮特·斯托克曼斯（Piet Stockmans）的层层叠叠的盘子，以及代表扬·范德法特（Jan van der Vaart）的抽象几何形——带网格的圆形。书中文字依照封面的四格设定进行荷英双语排版，其中英语为逆时针 90°。超薄的纸张使得文字前后透叠，指代三人展的多重叠加效应，也暗示瓷器釉面的透薄效果。每位艺术家的页面各占 24 页，单独计算页码。

This book is the album of a group exhibition of three ceramic artists from Holland and Belgium held by the Amsterdam Municipal Museum. The paperback book is glue bound with thread sewing. The cover is made of white cardboard printed in matte white, light white and black. The use of two kinds of white forms the contrast of ceramic glaze. The inside front cover and the inside back cover are printed in blue. The inner pages are mixed with thin offset paper printed in black and matte coated paper printed in four colors. In terms of content, the first part is a summary of the exhibition itself and the three artists, and then it is divided into three parts for individual introduction and work display. From the cover, the designer used three simple symbols to represent the three artists: the personal signature of Johan van Loon, the stacked plates representing Piet Stockmans, and the abstract geometry pattern of Jan van der Vaart – the circle with grid. The typeset area is set according to the four grids on the cover. The text is bilingual in English and Dutch, and English is typeset 90° counterclockwise. The ultra-thin paper makes the characters superimposed, which refers to the multiple overlapping effect of the troi exhibition, and also implies the transparent effect of porcelain glaze. Each artist's content takes up 24 pages, and the page numbers are calculated separately.

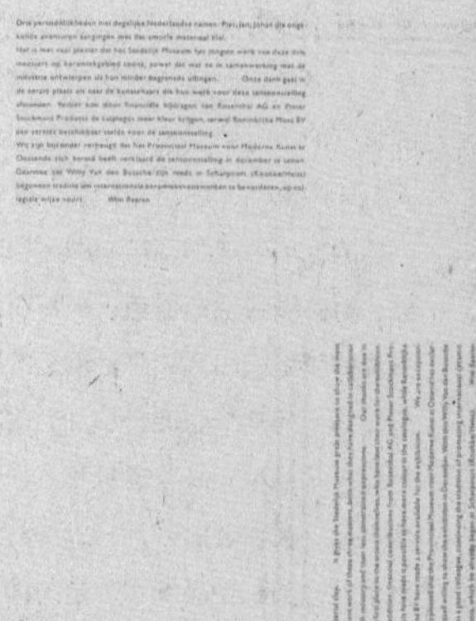

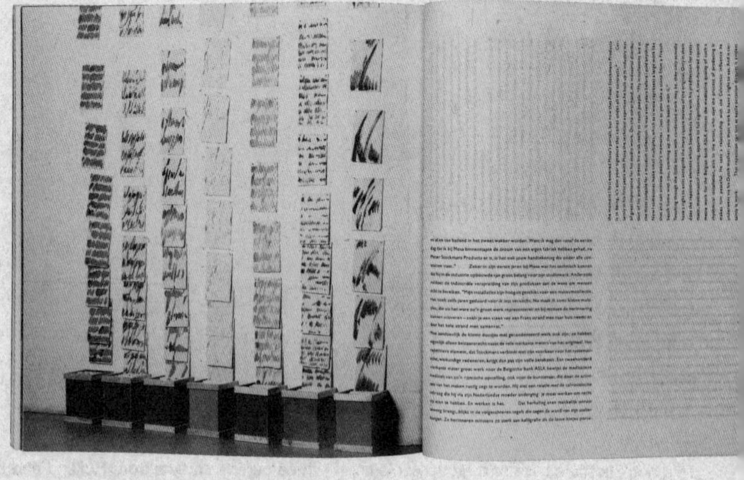

enhance their marketability were a concession which he accepted and occasionally poked fun at in his free work. The factory never flaunted the name of its designer very much and, unlike his predecessor Edmond Bellefroid, he never regarded himself as an artist in industry either: "Working for industry is an arduous intellectual process and in that respect much more difficult than the free work, which I regard as manual labour [in 1987 he founded the first factory of his own, in 1989 he broke with Mosa and after that he founded a factory in collaboration with Henk Dresens, which has been in operation since January 1991, GS]. 75 per cent of the Mosa collection is still designed by me." But he can equally justifiably claim to be an artist – there is enough evidence to hand for that career too. "I have always combined it with my work for industry, initially in earthenware and since 1970 in porcelain. As an industrial designer I want to fit into the history of porcelain, purely in the industrial tradition. As an artist I want to make my presence felt in the way a sportsman sets a world record. Authenticity, that's what it's about: being 150 per cent yourself. Even if there is only one person who has ever felt that on seeing the work, then it's art." And then there is in addition his post as a teacher of industrial design at Genk – which he has also held for twenty-two years by now.

This is a triangular affair which Slotkmans cherishes. The teaching puts the tension of industry in perspective, while he regards working in the studio as a holiday.

1993 Bm^{4}

Ref. F/3

JoviPack

Mail Lite®

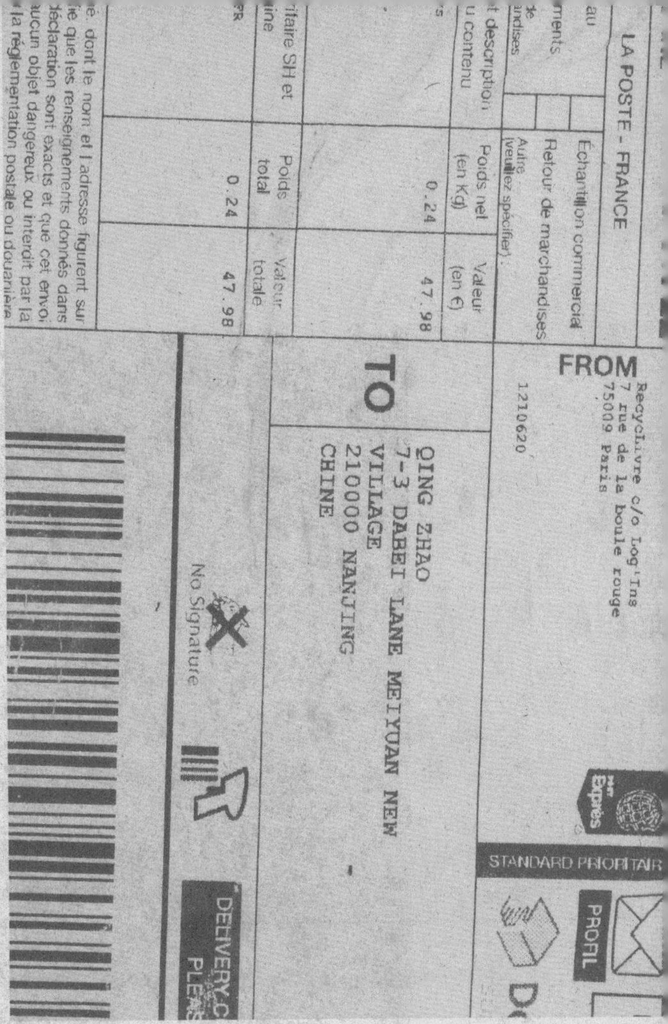

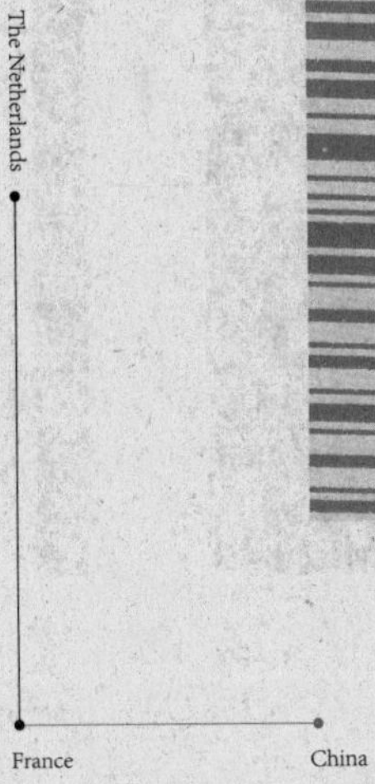

The Netherlands

France

China

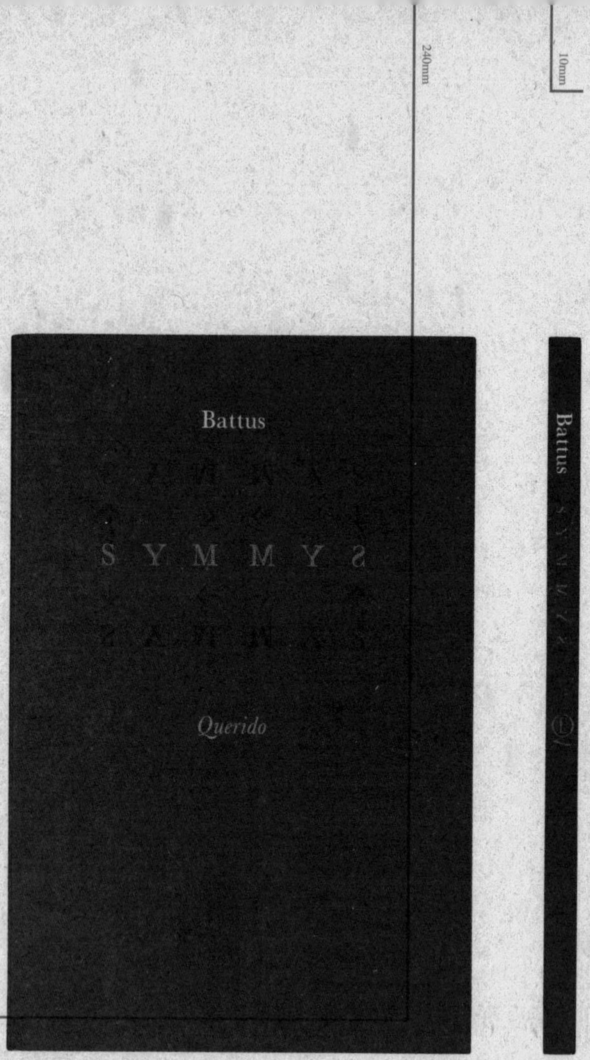

240mm

10mm

Battus

SYMMYS

Querido

161mm

● 铜奖

« 回文

◐ 荷兰

Bronze Medal

SYMMYS

The Netherlands

Harry N. Sierman

Battus

Em. Querido's Uitgeverij, Amsterdam

240 161 10mm

236g

104p

ISBN-10 90-214-5359-2

回文是指可以从正反两端阅读的单词、短语或句子，例如，“上海自来水来自海上”就是中文里熟知的回文。广义上的回文包含数字、符号、公式等形式，加上单词切分和标点符号对语义的分割，可以产生非常多的可能性，比如“02/02/2020”，以及“I’m a dad, am I? ”。回文属于娱乐语言文字范畴，早早地被各地人群发现并娱乐。世界上更是有回文比赛，颁发奖项 SymmyS Awards，SYMMYS 正是本书的书名。书籍的装订形式为锁线胶平装。封面白卡纸红、棕、黑三色印刷，覆哑膜。内文胶版纸印单黑。书籍的装帧、印刷和用纸都很质朴，但在内容和设计上充满趣味。作者收集了英、法、德、荷、希、俄、日等多种文字（甚至包含印地和楔形文字）的回文约 2500 条，以及在插图、标识、碑刻等处出现的回文。作者将各种文字混编在一起，5 条为一组，5 组为一页。每条回文均有编号，四位数字为当前行数和当前页码，首行设为 00。例如第 45 页第 6 行记为 0645。页面下部对部分句子背后的故事作阐述。

Palindrome refers to a word, phrase, or sequence that reads the same backward as forward. In a broad sense, palindrome contains figures, symbols, formulas and other forms. With the segmentation of words and punctuation marks, there are plenty of possibilities. For example, 02/02/2020 and ‘I’m a dad, am I?’ Palindrome belongs to the category of entertainment language, which has been found and entertained by people from all over the country. There is even a palindrome contest in the world, presenting SymmyS Awards. Symmys is the title of the book. The paperback book is glue bound with thread sewing. The cover is printed in red, brown and black on white cardboard, mounted with matte film. The inside pages are made of offset paper printed in single black. The book binding, printing and paper using are quite simple, but it is interesting in content and design. The author has collected about 2500 palindromes in English, French, German, Dutch, Greek, Russian, Japanese and other languages (even including Hindi and Cuneiform), as well as palindromes appearing in illustrations, signs and inscriptions. The author mixes all kinds of characters together, with 5 palindromes in one group and 5 groups in one page. Each palindrome has a four-digit number, which consists the current line number and the current page number. The first line is 00. For example, line 6 on page 45 is recorded as 0645. At the bottom of the page, the story behind some sentences is explained.

'Reviled did I live' said I, 'as evil I did deliver'

Selim's tired, no wonder, it's miles*
Σοὶ ἄρ ὁ γάσος, ὁ λάλος, ὅς ἀγοραῖος
Σὺ ἔσο κανὼν ὠμῶν γαία, γνώμων ὤν, ἄκος ἐύ
Toi, Roger, épèle le père Goriot
Wok utył i ma miły tukow

elaat' om in een zeshoek te zetten
ijn *De doopvont* (1952). Het nut v
ers is dat hij rond wasbakken, wij
akken en doopbakken geschreven
vorden. Maar waarom Bordewijk
es erin liet staan en het tegen de k
at lezen, is me een raadsel.

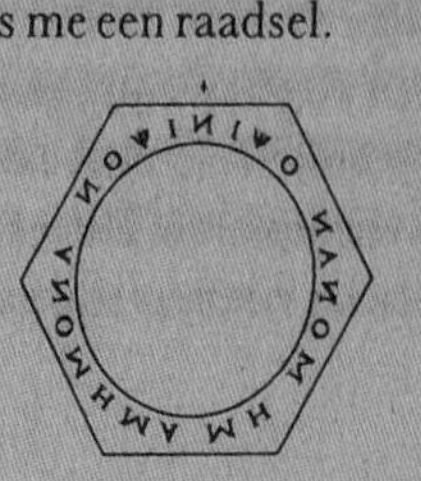

e naam *Selim* u bekend
op de vorige bladzij in het
elim smiles (2009). Net als
boven, *Revere piper ever*, is
3 Engelse symmys die
1887 in Glasgow, waar het
otheek is te bezichtigen, uit-

gaf onder het motto *Was it*
eerste geheel aan de Enge
de boek, en het eerste geïll
over symmys. Ik druk hie
van *Selim*, de *piper* en *Was*
Veel van Clarkes ideeën z
erna geëxploiteerd worde

'Mooie zeilboten
N'a-t-elle pas ôté

No D? No L? .on.on
Obbedir a
mt 'n omnipotente
Hun, kaptein; 't manne
Raadt me zo'n
Retro

Se corta Sarita
Snug & raw was I e

LES. REVERE PIPER EVER. WAS IT

Swept pews; yes, Sam, Massey swept pe
Top, ik neem een kip, Ot
Yell up Elba car, O fallen Nella, for a cable

よの音のねぶりのみ

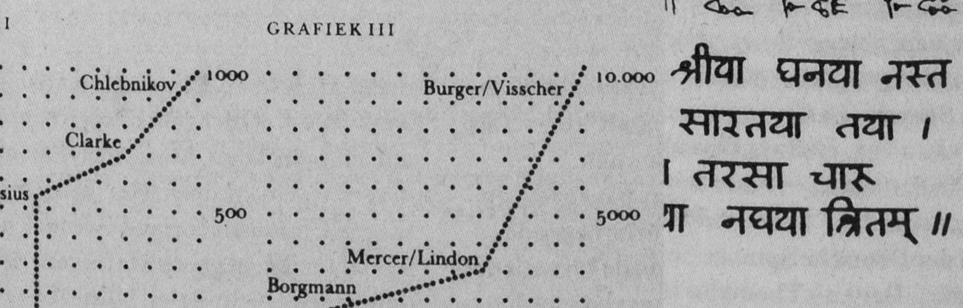

श्रीया घनया नस्त
सारतया तया ।
। तरसा चारु
ग नघया त्रितम् ॥

'Aïε, ϱαίᾳ ΣΥ̓ Ἐ ΣΟ, ὯΣ ἘΎΣ αἴᾳ ϱεῖα
Ἀναβάσατε τὰ σάβανα
Ἄτα μὴ μόνα; ἀλλὰ κ' ἄλλα ἀνομήματα
Cigar? Toss it in a can, it is so tragic
0496 Die rebus is leuk. U, Eis, is u bereid?

Εἰμί ταραχὴ ἔγωγε ἡ χαρὰ τίμιε
'Ερεῖς ὁ ἐν Σάμῳ Θωμᾶς νέος ἱερέ
'Ga alleen, os,' repte het personeel laag
Ἦμα δήμῳ, τῷ ὀλοῷ τῷ, μηδαμῆ
0996 'Ιὸς, ἀὲν ἀνίαν ἔμεν αἰνὰ νέα σοι

К сиренам? А не риск? *
Lososa maso sol
'Mooie zeer vette etser bij Brest eet tevree' zei oom
Nana, je monokini kon ome Jan an!
1496 No bajará Sara jabon

'Nurses run' says sick Cissy as nurses run
Ove regnai piangere vo'
Pure woman in Eden, I win Eden in . . . a mower-up!
Report paracetamol is not in unit, on silo, mate. Car apt, Roper
1996 Satire veritas

Snak nu naar ei bij bier, aan u 'n kans
Strategy: Get Arts!
Ties ruck, curse it!
Warren, slip a Pilsner raw
2496 Я ел мясо лося млея

* 1096. Vladimir Gersjoeni (uit het artikel 'Gersjoeni, meesterpalindroma' van J. Telesin in *Ami* [Jeruzalem] no. 3, 1973). *K sirenam? A ne risk?*, dat is dus: *O. K. I sirene? Gene risiko?*

De Russische symmy wordt beheerst door twee schrijvers die – hoe zal ik het zachtjes zeggen? – in hun land en tijd niet de optimale kansen kregen om een normaal leven te leiden.

L. Gersjoeni (1930), in 1949 tot tien jaar kamp veroordeeld, dissident en oprichter van een onafhankelijke vakbond, in 1969 weer gearresteerd. Hij schreef in 1971 'Notities over de speciale krankzinnigeninrichting in Orel', werd in '74 vrijgelaten, en volgens *Roesskaja mysl* van 5 mei 1983 op 12 april 1983 weer in een psychiatrische inrichting gezet. Behalve de publikatie in

Αἴ ἔφητ', ἄκη τὰ ἄτῃ; κατηφεία
Analytic Paget saw an inn in a waste gap city, Lana
Ἄτα μὴ νικᾷ ἐτεὰ κινήματα;
Гни, комсомол, лом о смокинг
Die rebus is nogal net en lag ons; is u bereid? 0497

Εἰμὶ, τῷ, ἐθὰς, ὅλη ζῆλος ἀθέῳ τίμιε
Éres o no eres . . . seré o no seré
Ga al! Reden: nederlaag
Ἦν ἄγρα ταραχῇ ἄλλᾳ ἡ χαρὰ ταργάνη;
'Ιὸς οἷ ἡ ἰός εὐς σὺ ἔσο ἰήϊος οἱ 0997

'k Stok na Marianne, en na IRA-man kots 'k
Lost u er al af, odalisk? Si la do fa la re ut sol
'Mooie zeevlo, weidt die wolvee?' zei oom
Naomi, did I moan?
No benison, no sin, Ebon 1497

Ну, червон снов речун
O vita, vali lavativo!
Pusillanimity obsesses Boy Tim in *All Is Up*
Rep u, o stadsdeerne, mannen namen reeds dat souper!
Sator arepo tenet opera rotas* 1997

S A T O R
A R E P O
T E N E T
O P E R A
R O T A S

Sneed, nette, etste etten Deens
Straw? No, too stupid a fad. I put soot on warts
Tina is an Asia-nit
Warsaw was raw
Я иду с мечем судия 2497

S N E E D
N E T T E
E T S T E
E T T E N
D E E N S

* 1997. Over deze vijf woorden van vijf letters is meer geschreven dan ze verdienen. Ze vormen een tovervierkant (noot 1744), maar Arepo is dan wel de naam van de zoon van wie deze symmy bedacht. Jan de Bisschop legde in zijn boek (noot 2219), een inlegvel waarin hij *arepo* verbeterde in *erepo*, aldus een symmy makend, maar een tovervierkant afbrekend.

Alle pogingen om een christelijke symbo-

meense tegel van voor het jaar nul zit ingebakken. Het is dus ongetwijfeld een oude Latijnse spreuk, maar is het ook een symmy?

Op 30 juli 1990 ontdekte Piet Burger in Andijk de volmaakte tovervierkantsymmy 2097, die ik op 3 augustus 1990 in *de Volkskrant* wereldkundig maakte. Hij betekent: *Graveerde, bevochtigde, vereeuwigde in koper etstoelrechters op de wijze der Denen.* De com-

0303

CUP
CN
Deutsche Post
PREMIUM
FROM
To
(ZL)CN210060 + 67800900
CC 814 735 259 DE

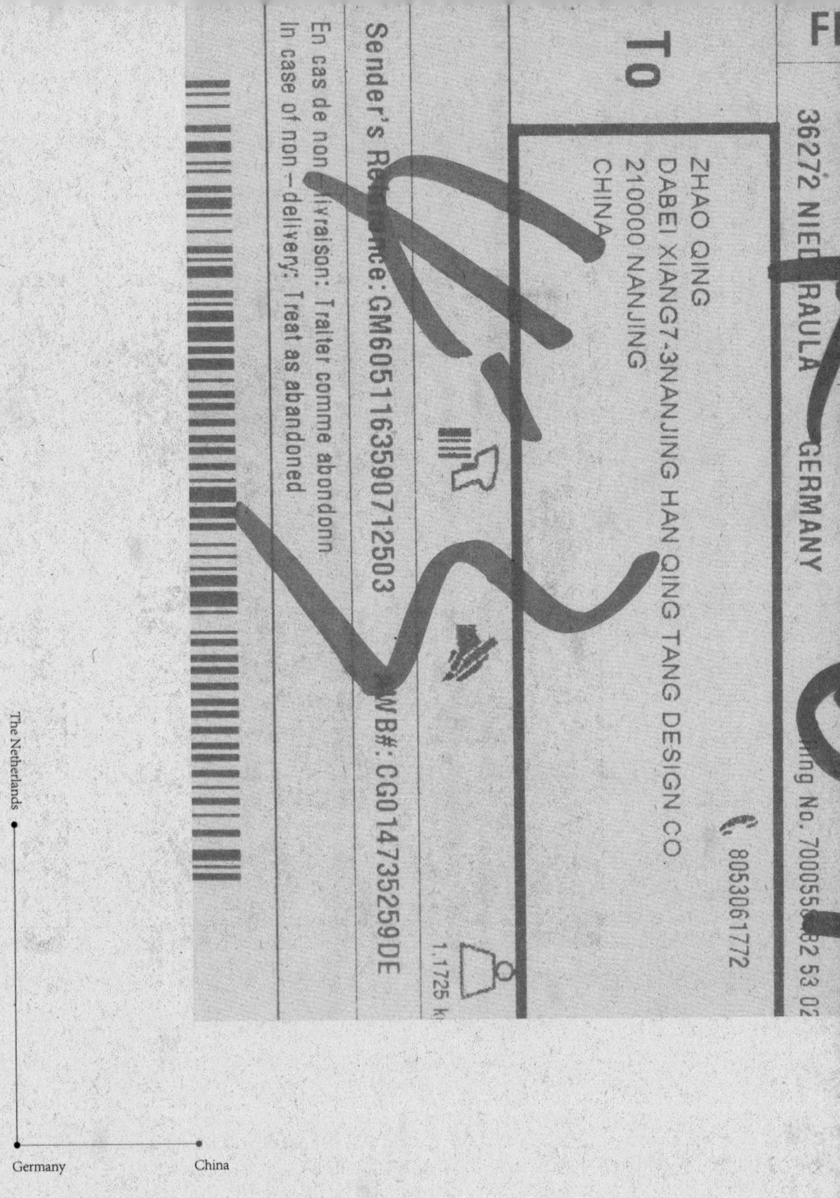
To
ZHAO QING
DABEI XIANG7-3NANJING HAN QING TANG DESIGN CO.
210000 NANJING
CHINA
8053061772
GERMANY
In case of non-delivery: Treat as abandoned
The Netherlands
Germany
China

303mm

21mm

216mm

铜奖
弗里索 · 克莱默——工业设计师
荷兰

Bronze Medal
Friso Kramer: Industrieel ontwerper
The Netherlands
Reynoud Homan
Rainer Bullhorst, Rudoiphine Eggink
Uitgeverij 010, Rotterdam
303 216 21mm
1048g
160p
ISBN-10 90-6450-124-6

荷兰工业设计师弗里索·克莱默（Friso Kramer）是经典座椅 Revolt Chair 的设计师。本书是 1991 年在鹿特丹的博伊曼斯·范布宁根博物馆（Museum Boijmans Van Beuningen）举办回顾展时候的展览画册。书籍的装订形式为锁线硬精装。护封哑面铜版纸黑银双色印刷，内封黑色布面裱覆卡板，姓名首字母 f、k 烫黑处理。蓝色艺术纸做环衬，内页哑面铜版纸印单黑。内容方面，除了对克莱默的介绍，也阐述了他对工业设计和荷兰家具行业的贡献、他的工作方法和设计理念，同时呈现了部分家具的图纸、材质、组装、展示等资料。护封的设计是本书最大亮点，全书被想象成一块待组装的座椅木面，开本、厚度、重量均与一把座椅的座面接近，四角加工出待装配椅腿的圆弧状内凹槽，与克莱默设计易生产、易组装的家具理念相吻合。

Dutch industrial designer Friso Kramer is the designer of the classic seat Revolt Chair. This book is an exhibition album of a retrospective exhibition held at the Museum Boijmans Van Beuningen in Rotterdam in 1991. The hardcover book is bound by thread sewing. The jacket is made of matte coated paper printed in black and silver, and the inside cover is made of black cloth and mounted with cardboard. The initials of the designer's name F and K are treated with black hot stamping. Blue art paper is selected for the end paper, and the inside pages are printed on matte coated paper in single black. In terms of content, in addition to the introduction of Kramer, this book expounds his contribution to industrial design and Dutch furniture industry, his working methods and design concepts, as well as some furniture drawings, materials, assembly, display and other information. The design of the jacket is the biggest highlight of the book. The book is imagined as a wooden surface of the seat to be assembled, which has the similar format, thickness and weight to the surface of a seat. The four corners are processed into a circular arc shape for the legs to be assembled, which is consistent with the furniture concept of easy production and easy assembly designed by Kramer.

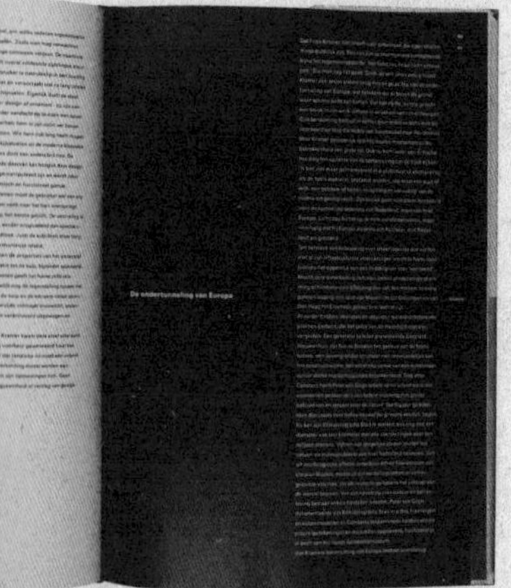

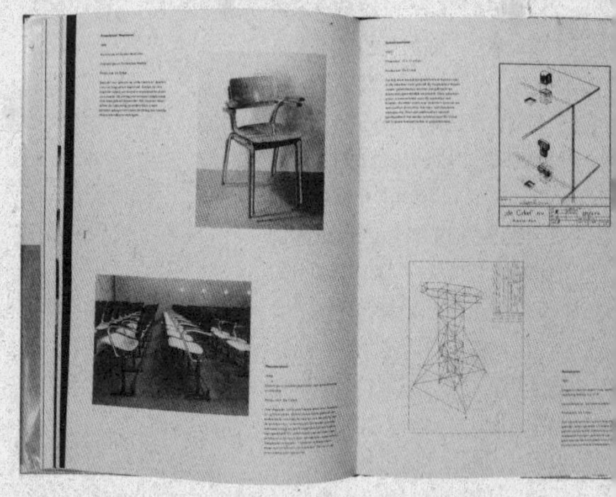

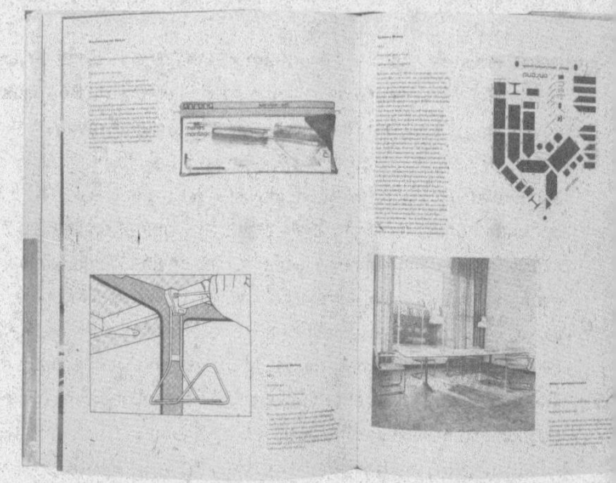

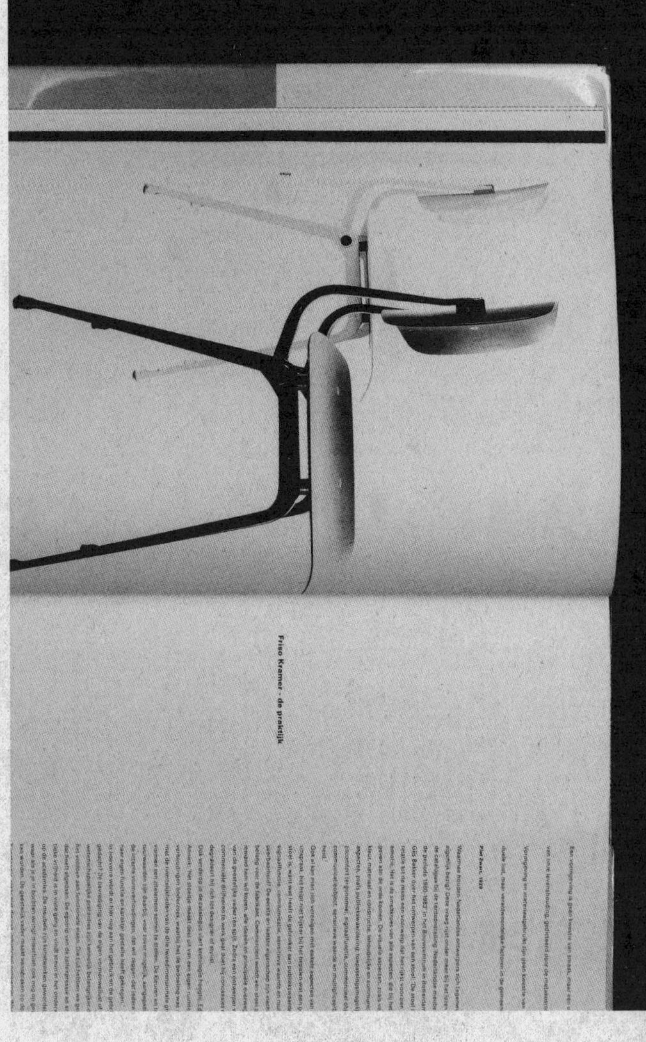

0309

14

ɔeder was altijd heel opgewekt. Zonder dat je zegt, mens 15
ɪt ben je nou hinderlijk opgewekt, maar gewoon leuk
gewekt. Ze was zeer gelukkig eigenlijk. Ik geloof dat mijn
der van die scheiding meer last heeft gehad dan zij. Voor
ar was het een opluchting, voor mijn vader niet. Als dromerig
nstenaar kun je wel absoluut niet meer kunnen werken met
'n groot luidruchtig gezin, maar je mist toch die kinderen en
vrouw waar je vijfentwintig jaar mee opgetrokken hebt.'[3]
n lagere schooltijd heeft Friso Kramer doorgebracht op de
rste Montessorischool in Amsterdam. 'Ik heb er niets
leerd, maar ze hebben me er tenminste heel gelaten'. Naar
ɟen zeggen heeft hij elke klas zo'n beetje twee keer gedaan
ıdat hij naar zijn gevoel de domste van de school was. Hij
d er ook het minste last van. 'Als iemand er last van had, ik
ɛt. Thuis waren we heel vrijgevochten. In die zin, dat als je
ɛt wilde leren, dan leerde je niet. Dan kon je ook niet geleerd
ɪrden om te leren.'[4]
n normale middelbare schoolopleiding was dus duidelijk
ɛt de aangewezen weg om dit eigenzinnige kind verder op te
eden, vandaar dat naar aanleiding van de psychotechnische
ɜt besloten werd om hem naar een ambachtsschool te
ɪren. Hieraan bewaart Friso Kramer een goede herinnering.
werd eerst op de timmer- en later op de metselafdeling
plaatst. Vervolgens, na ongeveer tweeëneenhalf jaar, ging
naar de Elektrotechnische School met het idee om later
ar de MTS te gaan. Om wat bijgespijkerd te worden volgde

Kruisvormige zittingdrager

1967

Gegoten aluminium, 28 x 25 x 4 cm

Opdrachtgever: Wilkhahn

Niet uitgevoerd

Aan de voorzijde wordt de zitting bevestigd op scharnierende houders, terwijl aan de achterzijde de twee steunpunten verend zijn uitgevoerd. Op deze wijze kan de stoel met vrij eenvoudige en goedkope middelen toch in beperkte mate veren en achterover kantelen.

haal. En dat is niet alleen fantastisch om te doen, n
noodzakelijk omdat de bewijzen op tafel liggen da
ondiaal die zaken kan aanpakken. Als ik dan nu alle
uropa praat is dat omdat dat ogenschijnlijk even be
ch te concentreren, te clusteren. Daar moet een ei g
orden. De BTW moet gelijkgeschakeld worden, de b
et vervoer, landbouwbeleid, kolen- en staalgemeens
aar wordt aan gewerkt. Die plannen in het kader va
ureka-project waar ik het daarnet over had, die gesu
orden wanneer het de moeite waard is, zijn zo in el
at er alleen zaken worden aangepakt die door minst
nden worden gedaan. Niet vijf, of zeven, maar het i
edoeling dat door deze gelegenheid te scheppen er
menwerking tussen de landen onderling ontstaat,
uropa een gemeenschap is. Als je in je eentje komt k
ubsidie waarschijnlijk niet, ook al kan je het over he
aarschijnlijk goed verkopen. Er is een soort stokje a
eur, die die grensvrees en de handicap van een and

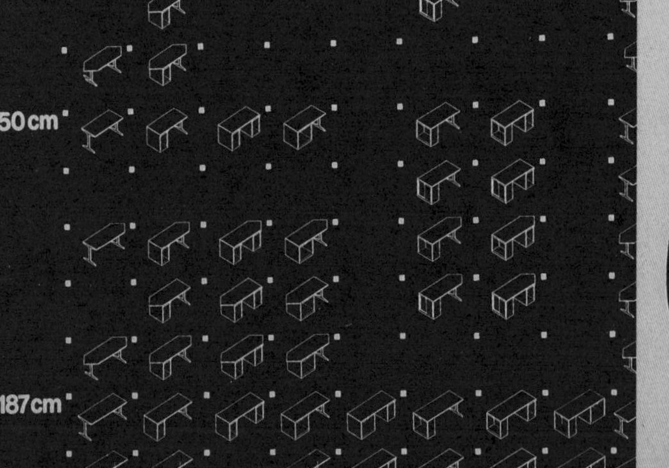

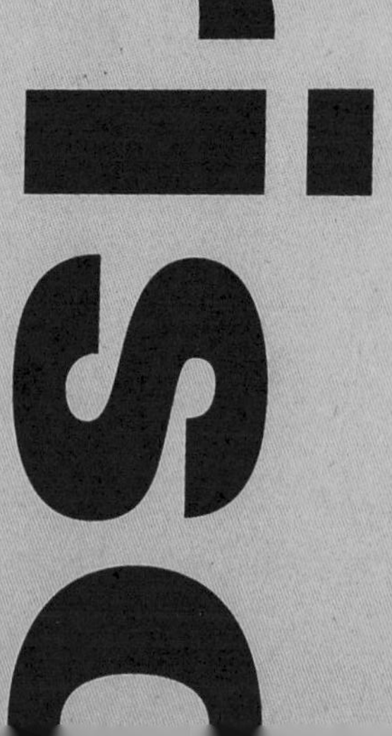

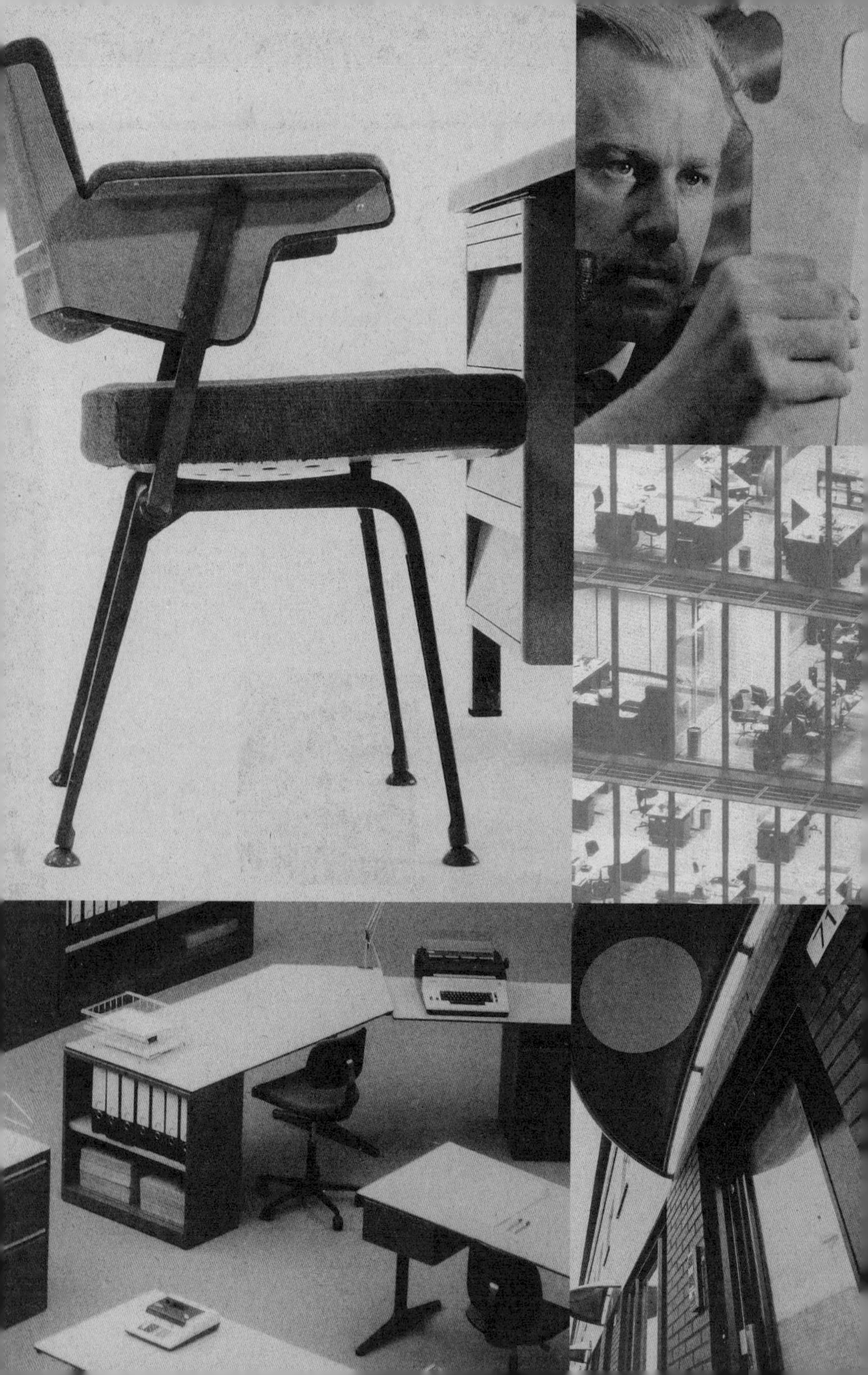

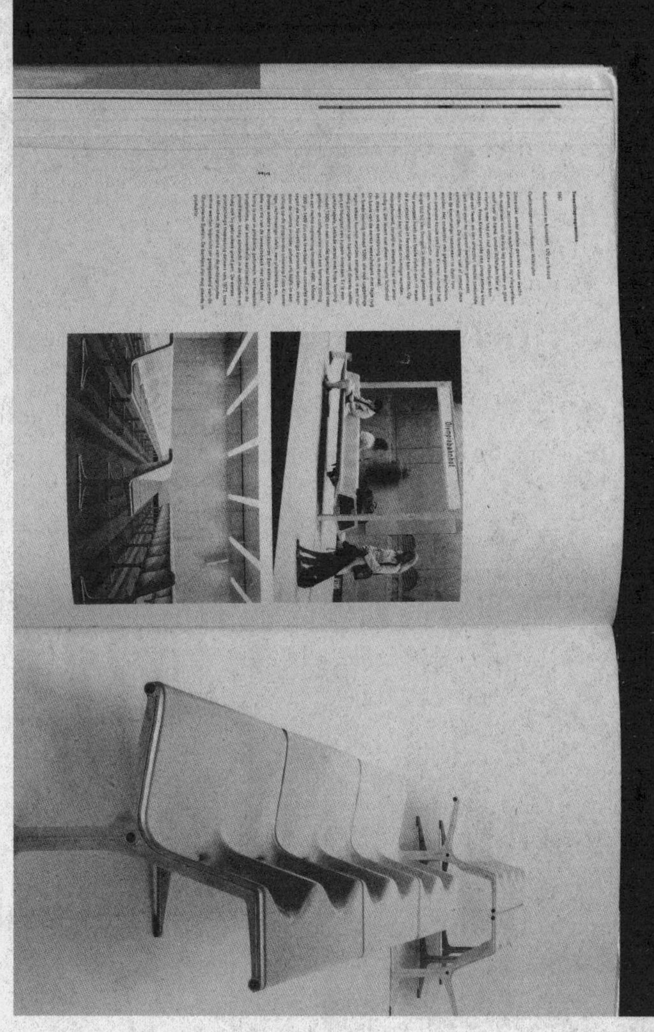
Olympiabahnhof

Wil
die Altstadt
PRIORITY
R
RD 248 281 478 CH
AbeBooks: Lieferschein
ZVAB.com
Von:
Antiquariat Vianus
Staubeggstr. 8
Frauenfeld 8500
Switzerland
An:
ZHAO QING
Dabei xiang7-3,Nanjing Han Qing
Tang Design Co.,
210000 Nanjing
China
CN 22

Switzerland

Switzerland

China

SWISS POST

Déclaration en douane

Zolldeklaration/Dichiarazione doganale

CN 22

Peut être ouvert d'office/Zollamtliche Prüfung gestattet/Visita doganale ammessa

☐ Cadeau / Geschenk / Regalo ☐ Echantillon commercial / Warenmuster / Campione di merci ☐ Documents / Dokumente / Documenti ☒ Autres / Andere / Altri

Quantité et description détaillée du contenu (1) Menge und detaillierte Beschreibung des Inhalts Quantità e descrizione dettagliata del contenuto	Poids (2) Gewicht (kg) Peso	Valeur (3) Wert Valore	N° tarifaire du SH et origine (4) Zolltarifnummer und Herkunft Numero di tariffa e origine
1 Buch		30,-	
Poids total (5) Gesamtgewicht (kg) Peso totale			Valeur totale (+monnaie) (6) Gesamtwert (+Währung) Valore totale (+valuta)

Je certifie que les renseignements donnés dans la présente déclaration sont exacts et que cet envoi ne contient aucun objet dangere
ou interdit par la réglementation postale ou douanière. / Ich bestätige hiermit, dass die Angaben in der vorliegenden Deklaration rich
sind und dass die Sendung keine durch die Post- oder Zollvorschriften verbotenen oder gefährlichen Gegenstände enthält. / Certifico c
le informazioni contenute nella presente dichiarazione sono esatte e che quest'invio non contiene nessun oggetto pericoloso o proibi
dal regolamento postale o doganale.

Date, Signature (7) 18.11.19

286mm

20mm

215mm

荣誉奖
维尔老城
瑞士

Honorary Appreciation
Wil. Die Altstadt
Switzerland
Gaston Isoz
Werner Warth, Herbert Maeder
Meyerhans Druck AG, Wil/St. Gallen
286 215 20mm
884g
136p
ISBN-10 3-9520321-0-7

维尔是位于瑞士苏黎世与波登湖之间的一座老城，被瑞士列为国家文化遗产并对建筑遗产进行保护。求学于维尔的社会学家维尔纳·瓦特（Werner Warth）与成长于维尔的摄影师赫伯特·梅德（Herbert Maeder）合著了这本书，对维尔老城的自然风景、建筑风情、人文活动等进行了大量拍摄，并对老城的历史进行了阐述。本书的装订形式为锁线硬精装。护封涂布纸四色印刷，内封黑色布面裱覆卡板，书名压凹处理，书脊处裱贴单黑印刷的白色纸片。环衬选用灰色纸张，内页浅黄色涂布纸四色印刷，微微泛黄的纸张仿佛历史在城市建筑表面留下的岁月痕迹。内容方面分为两部分：第一部分为文章部分，单栏设计，记录了维尔老城的历史、建筑、管理、保护、市场、工商业及日常生活的方方面面；第二部分为照片部分，通过丰富的图片大小变化和不同内容的分组来展现真实风貌。

Wil is an old city located between Zurich and Lake Constance in Switzerland. It is listed as a national cultural heritage by Switzerland and its architectures are protected well. The sociologist (Werner Warth) who studied in Wil and the photographer Herbert Maeder who grew up in Wil co-authored this book. They took a large number of photos of the natural scenery, architecture and cultural activities in Wil, and elaborated the history of the old city. The hardcover book is bound by thread sewing. The jacket is made of coated paper printed in four colors, while the inside cover is made of black cloth mounted with cardboard. The book title is embossed. There is white paper printed in single black mounted on the spine. The end paper is made of gray paper, and the inner pages are printed with light-yellow coated paper in four colors. The slightly yellowish paper seems to be traces of history left on the surface of urban buildings. The content is divided into two parts. The first part is the text part, typeset in single-column, records the history, architecture, management, protection, market, industry, commerce and daily life of the old town of Wil. The second part is the photo part, which shows the real scene through rich pictures of different sizes and different content.

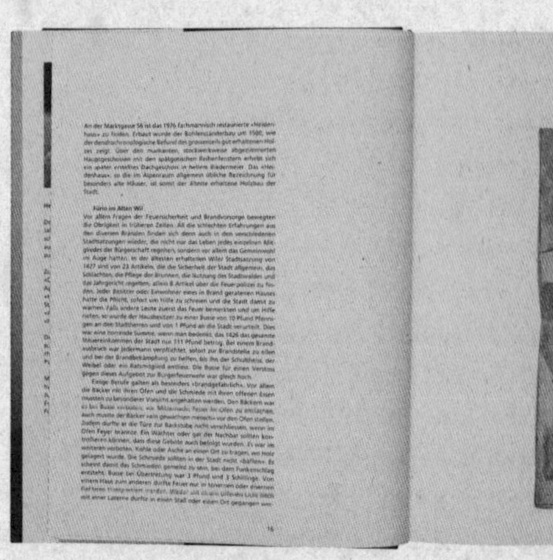

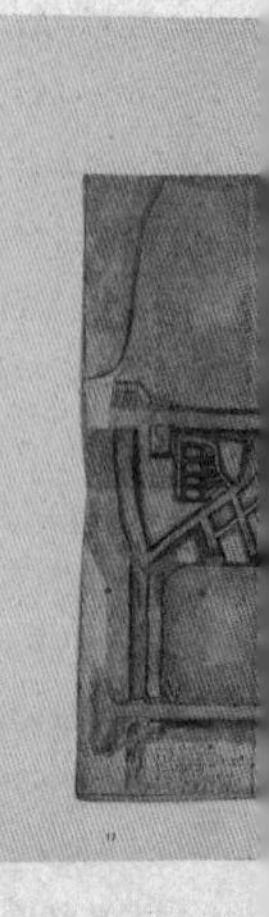

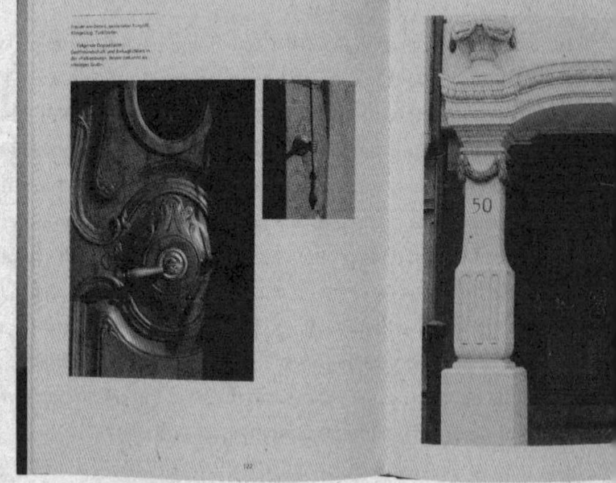

Die grosse Glocke aus der Stadtkirche St. Nikolaus, 1589 gegossen und 1931 vor dem Einschmelzen gerettet.
Der Turm der Stadtkirche St. Nikolaus überragt immer noch alle Häuser der Altstadt.
Folgende Doppelseite:
«Christophorus» und «Christophorus», einmal von Hans Haggenberg um 1480, zum anderen von Urban Blank um 1950 geschaffen.

Dem Buch zum Geleit

ist zur Stadt geworden, als dieser Ort im Fürstenland vor Jah
ten mit besonderen Aufgaben betraut wurde.

Der Stadt wurde aufgetragen, den Menschen Schutz und Schi
währen: Stadtmauern wurden gebaut und Stadttore errichte
Der Stadt wurde zur Pflicht gemacht, auf engem Raum me
es Zusammenleben zu regeln und über Recht und Unrec
chen: ein mächtiger fürstäbtischer Hof ist Zeuge davon.
Der Stadt war vorbehalten, den Handel für Stadt und Regi
anisieren: der «Goldene Boden» als Marktplatz zu Füssen de
d die Verkaufsläden und Werkstätten der Händler und Handw
ter den Arkaden weisen darauf hin.
Aber auch Gotteshaus und Spital gehörten unverzichtbar zu
Stadt. An den Gassen reihten sich Trinkstuben zur Musse
selligkeit. Und in vielen Stadtbrunnen plätscherte das Wass
chen dafür, dass auch der Mensch in der Stadt ohne die Natur
erleben kann.
Eine Stadt mit diesen ursprünglichen städtischen Wesensme
ist uns in der Wiler Altstadt vererbt worden. Es ist eine wi
turelle Aufgabe, dieses Erbe lebendig zu erhalten. Dazu geh
onderen ein wiedergeborener Hof zu Wil!
Dieses Buch ist eine Herausforderung zu gemeinschaftlichem
n: sorgsam zu bewahren, was bewahrenswert ist, und mu
dern, wo Förderung Not tut.
So freue ich mich sehr, dass «Wil die Altstadt» geschaffen w

zlich danke ich dem meisterhaften Fotografen Herbert Ma
n kundigen Historiker Werner Warth und dem unserer Stadt s
ndenen Buchdrucker Christof Meyerhans für dieses eindrüc
rk: Eine kulturelle Leistung von Wil für Wil!

det, so dass wir heute nur no
Wiler Befestigungen erahner

ahrzeichen – der Hof

heute endlich den Hof ihr Eig
er, hoffentlich andauernder A
es Wiler Wahrzeichens angeb
pältige Verhältnis der Wiler zu
genburg und vor allem den Ä
wenn erst jetzt der Hof der Ei
nden wohl aus einer Burg o
bäude der Toggenburger, wu
haft zur «Krone» der Wiler A

t dem breit ausladenden Hof,
r Stadtkirche St. Nikolaus, wir
ine Kindheit gelebt und da z
Dieses Buch ist das dritte,
affen habe. «Wil im Fürste
dt», «Wil – die Altstadt». Dr
tografen! Ein bisschen viel, ni
1965 hatte der Verlag Friedr
t dem Titel «Wil im Fürstenlan
tos stammten von mir, den Te
ler Schulkamerad, späterer G
fasst. Wil hatte in jener Zeit J
klung durchgemacht, und ich
n Altstadt auch die neueren A
gst vergriffenen Buche blätte

en bleiben. Weitere Schenku
n sich gegen Ende des 8. Ja
ne Frau namens Sleta und ih
Vilen und Bronschhofen dem K
Wil schon vor der grossen S
rhundert als Stadt bezeichne
n, wurden doch bei der gänz
0 auch die ältesten Urkunde
n des St. Galler Klosterbrude
von einem Unfall «circa villam
htet. Doch auch hier ist nich
mit «villa» wirklich eine Sta
nkreter wird Wil in einer Ur
ezeichnet. Graf Diethelm II. vo
gen Alttoggenburg und Wil a
ch hier von Wil als «villa» (Do

den hohen Grad der Organisation, Bewunderung für die zähe Bewahrung der Freiheiten und der Eigenart, ebenso wie mit Respekt für die ungezählten Menschen, die in Wil Freud und Leid «erlebten».

Die ersten Erwähnungen

Die erste urkundliche Erwähnung Wils geht wahrscheinlich auf eine Urkunde aus dem Jahre 754 zurück, wo unter anderem ein Gütchen in einer Ortschaft Wila als Schenkung eines «Rothpald» an das Kloster St. Gallen erwähnt wird. Doch bedeutet dies noch nicht eine Stadt, wohl eher einen kleinen Weiler oder ein befestigtes alemannisches Dorf. Ob mit der Bezeichnung Wila wirklich das heutige Wil oder das benachbarte Wilen gemeint ist, muss offen bleiben. Weitere Schenkungsurkunden aus unserer Gegend finden sich gegen Ende des 8. Jahrhunderts. So sollen am 20. Mai 796 eine Frau namens Sleta und ihr Sohn Winithar ihren Besitz in Wil oder Wilen und Bronschhofen dem Kloster St. Gallen übertragen haben. Dass Wil schon vor der grossen Städtegründungswelle im 12. und 13. Jahrhundert als Stadt bezeichnet werden kann, ist nicht mehr zu belegen, wurden doch bei der gänzlichen Zerstörung der Stadt im Jahr 1360 auch die ältesten Urkunden vernichtet. Trotzdem, in den Schriften des St. Galler Klosterbruders Eckhard, der im Jahre 1071 starb, wird von einem Unfall «circa villam Wilaham», also in der Nähe Wils berichtet. Doch auch hier ist nicht hundertprozentig sicher, ob Eckhard mit «villa» wirklich eine Stadt und nicht nur eine Ortschaft meint. Konkreter wird Wil in einer Urkunde vom 12. Dezember 1226 als Stadt bezeichnet. Graf Diethelm II. von Toggenburg schenkt darin die Festungen Alttoggenburg und Wil an Abt Konrad von Bussnang. Obwohl auch hier von Wil als «villa» (Dorf) die Rede ist, lassen die zahlreichen Benennungen Wils als Gemeinde mit Bürgern – civitatem Wile cum civibus suis – den Schluss zu, dass schon um diese Zeit Wil als Stadt existierte. Mit der Erwähnung von Schultheiss Eberhard in der Verpflichtungsurkunde für Abt Walter und Probst Burkart von Bussnang vom Mai 1244 wird erstmals in einer öffentlichen Urkunde Wil als Stadt ausgewiesen.

Wil und seine Grenzen

Die Stadt Wil ist auf den ersten Blick durch die Häuser und Mauern klar begrenzt und lokalisierbar. Bei genauerer Betrachtung werden die

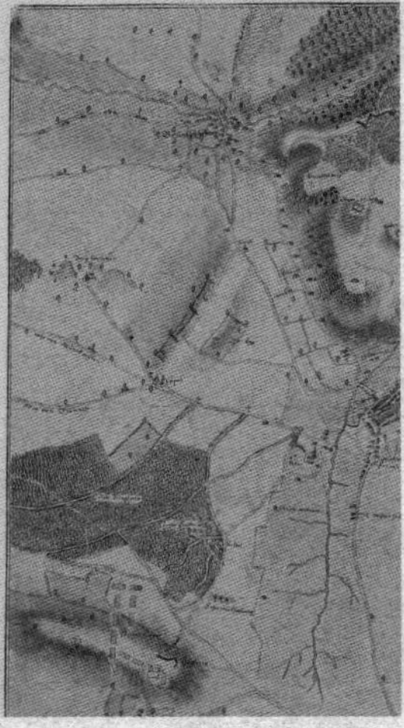

Die Wiler Wappenscheiben im Gerichtshaus sind einmalige hervorragende Beispiele spätmittelalterlicher Glasmalerei.
Folgende Doppelseite:
Dekorative Illusionsmalerei am Baronenhaus, ein Werk von Josef Keller aus Pfronten (1740–1823).

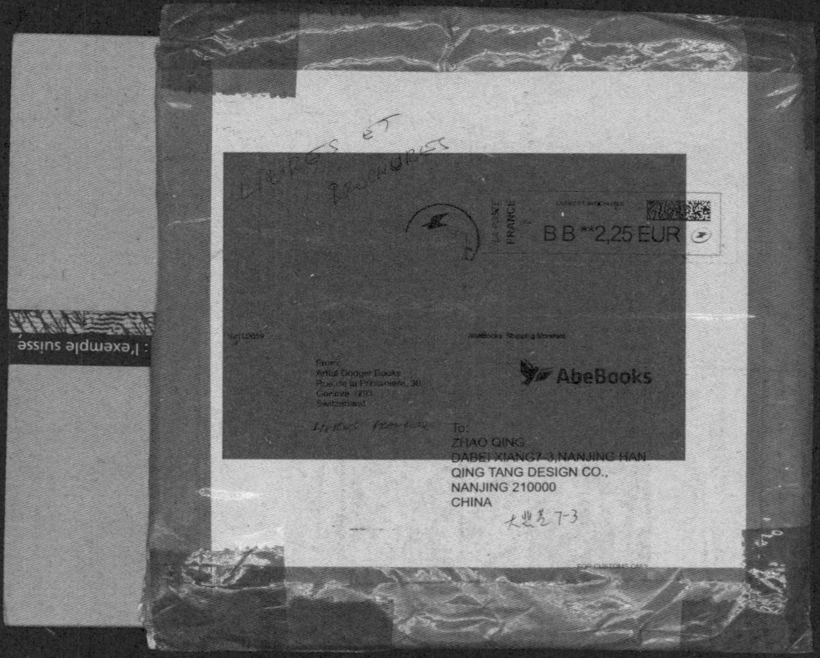

l'exemple suisse

LA POSTE
FRANCE

BB **2,25 EUR

AbeBooks

From:
Artful Dodger Books
Rue de la Printanière, 30
Genève 1203
Switzerland

To:
ZHAO QING
DABEI XIANG7-3,NANJING HAN
QING TANG DESIGN CO.,
NANJING 210000
CHINA

大悲巷7-3

FOR CUSTOMS ONLY

Switzerland

Switzerland

China

220mm

5mm

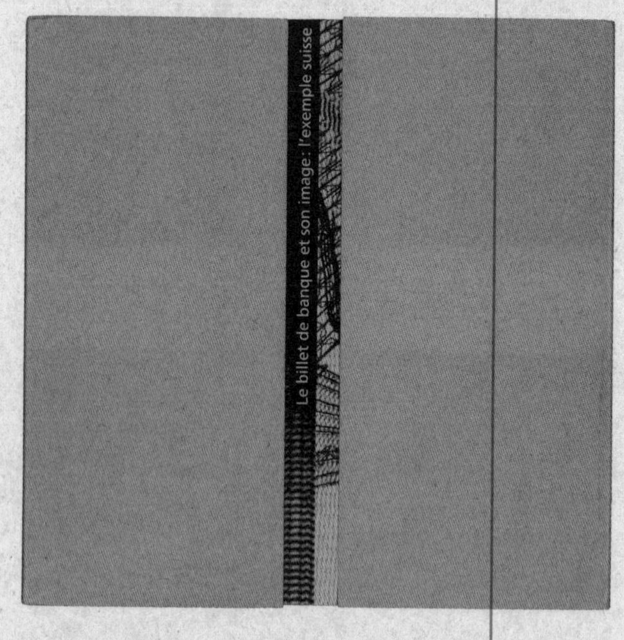

227mm

● 荣誉奖

《 钞票及其图案——瑞士案例

◐ 瑞士

Honorary Appreciation

Le billet de banque et son image: l'exemple suisse

Switzerland

Isabelle Lajoinie

——

République et Canton de Genève, Genf

220 227 5mm

222g

40p

——

这是一门在艺术学院里开设的课程，向学生介绍瑞士纸币及相应图案背后的故事，最终集结成本书。书籍的装订形式为无线胶平装。本书并无标准意义上的封面、书函等结构，而是将这些书籍构件混合在了一起：书函部分采用灰卡纸，无印刷和工艺，直接对书进行包裹，并在书脊处粘贴。书函漏出封面中部细条，可以看见书名和非常有限的局部图案。封面与内页用纸相同，均为无漂白的胶版纸四色印刷。书中按照纸币的象征、社会功能、图案和造币相关故事进行陈述。内页每两页为一组，形成类似蝴蝶装的结构。文章和图案混排部分均包含半面拉页，并置的两张图案均为纸币上人物眼睛和其他花纹的局部。把拉页拉开，可以看到每篇文章的篇名及起始文字。

This is a course offered at the Academy of Art, which introduced the story behind Swiss banknotes and the corresponding patterns to the students, finally assembled into this book. The paperback book is perfect bound. This book does not have a standard cover, case and other traditional parts. Instead, these book components are mixed together: gray cardboard without printing and other processing directly wraps the book, and pastes at the spine to form a book case. A thin strip is hollowed out to reveal the middle of the cover, showing the title of the book and very limited patterns. The cover and the inside pages are all printed in four colors with unbleached offset paper. The book is presented in the order of paper money symbol, social function, pattern and coinage related stories. Every two pages of the inner pages are a group, forming a structure similar to a butterfly-fold binding. The part where the text and the patterns are typeset together includes a half page of pullout – the two juxtaposed patterns are the characters' eyes and other patterns on the paper money while the pullout page presents the title and starting text of each article.

Fonctions éco
des billets de b

Jean-Pierre Roth

L'imagerie
du billet de banqu
reflet de l'identité
nationale

Michel de Rivaz

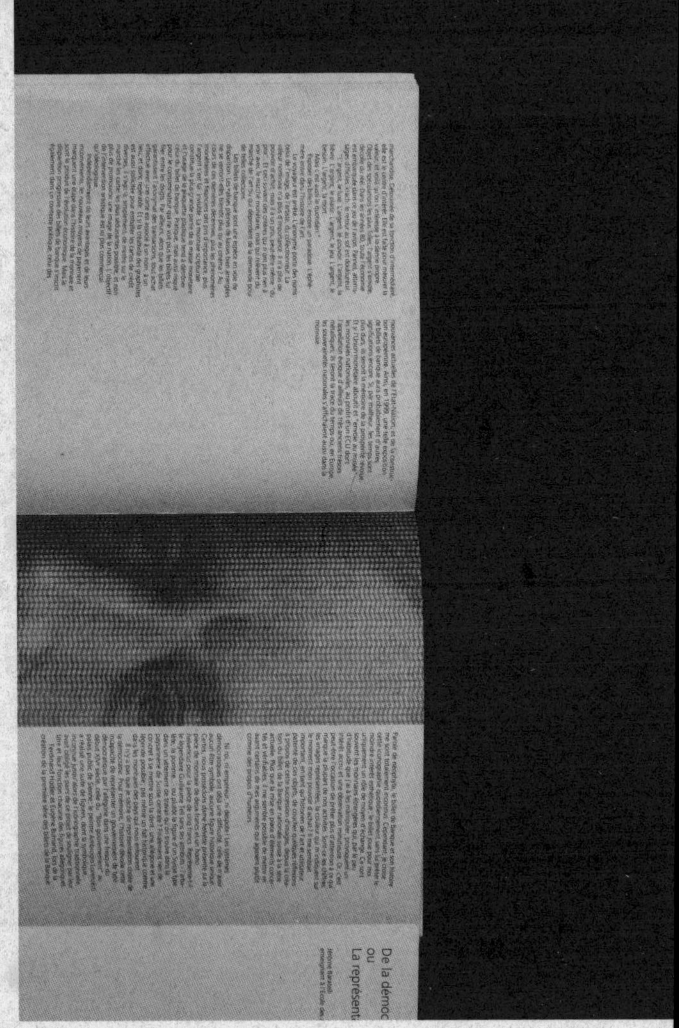

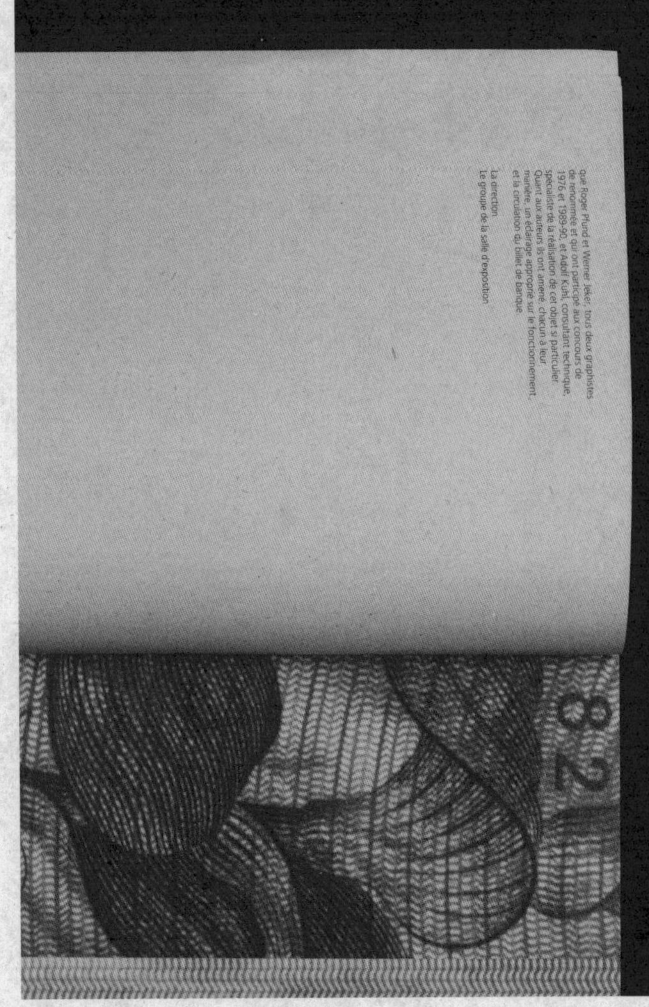

que Roger Pfund et Werner Jeker, tous deux graphistes de renommée et qui ont participé aux concours de 1976 et 1989-90, et Adolf Kuhl, consultant technique, spécialiste de la réalisation de cet objet si particulier. Quant aux auteurs ils ont amené, chacun à leur manière, un éclairage approprié sur le fonctionnement et la circulation du billet de banque.

La direction
Le groupe de la salle d'exposition

1993 Ha[4]

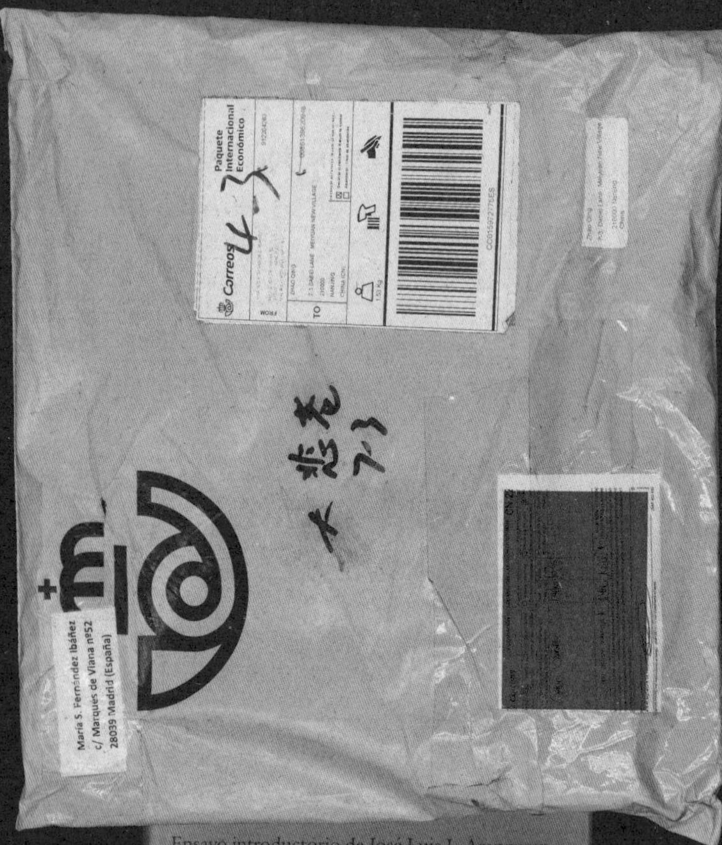

Ensayo introductorio de José Luis L. Aranguren

Círculo de Lectores

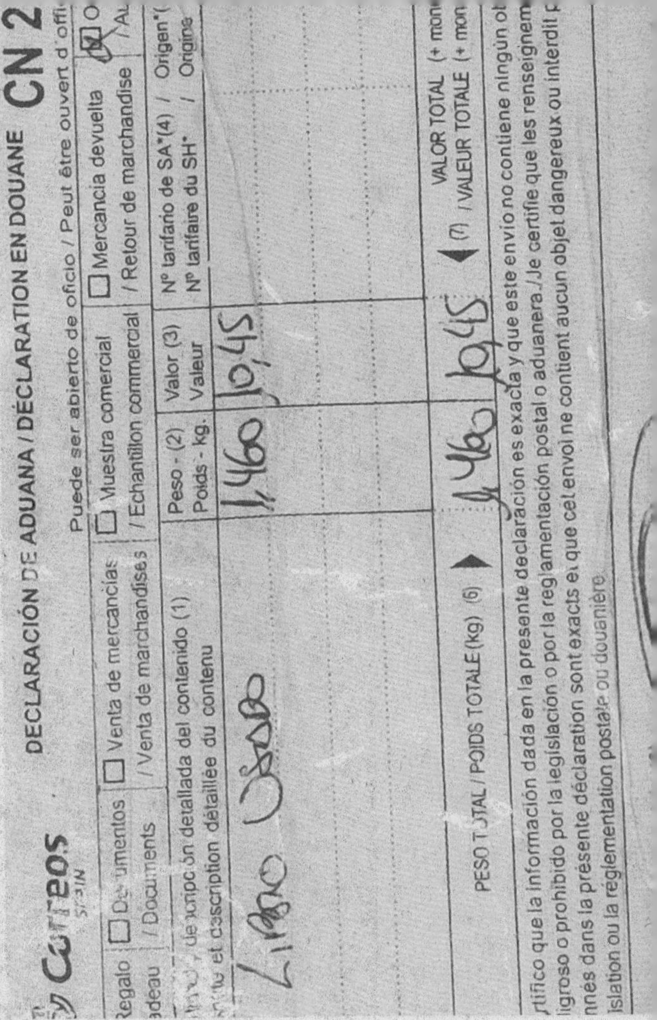

Correos
SPAIN

DECLARACIÓN DE ADUANA / DÉCLARATION EN DOUANE CN 2

Puede ser abierto de oficio / Peut être ouvert d'offi

Regalo / Cadeau	Documentos / Documents	Venta de mercancías / Vente de marchandises	Muestra comercial / Echantillon commercial	Mercancía devuelta / Retour de marchandise	O / Au

Descripción detallada del contenido (1) / Désignation et description détaillée du contenu	Peso - (2) / Poids - kg.	Valor (3) / Valeur	Nº tarifario de SA*(4) / Nº tarifaire du SH*	Origen*(/ Origine
Libro Usado	1,460	10,45		
PESO TOTAL / POIDS TOTALE (Kg) (6)	1,460	10,45	(7) VALOR TOTAL / VALEUR TOTALE (+ mon	

rtifico que la información dada en la presente declaración es exacta y que este envío no contiene ningún ob
ligroso o prohibido por la legislación o por la reglamentación postal o aduanera./Je certifie que les renseignem
nnés dans la présente déclaration sont exacts et que cet envoi ne contient aucun objet dangereux ou interdit p
islation ou la réglementation postale ou douanière

Spain

Spain

China

316mm

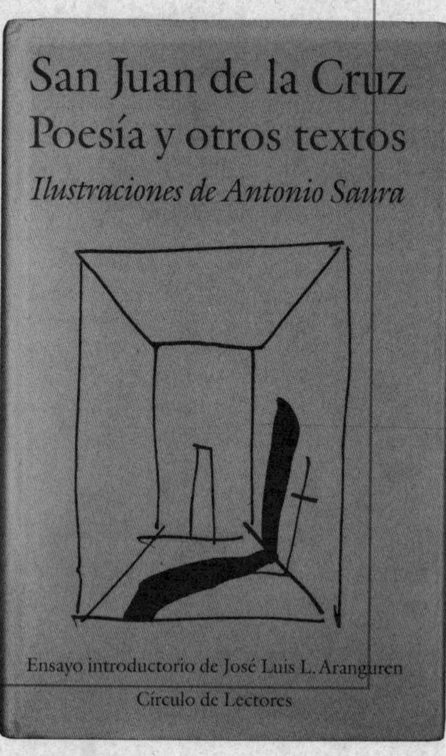

San Juan de la Cruz
Poesía y otros textos
Ilustraciones de Antonio Saura
Ensayo introductorio de José Luis L. Aranguren
Círculo de Lectores
Poesía y otros textos
San Juan de la Cruz
Círculo de Lectores

208mm

荣誉奖
诗歌与其他文字
西班牙

Honorary Appreciation
Poesía y otros textos
Spain
Norbert Denkel
San Juan de la Cruz
Círculo de Lectores, Barcelona
316 208 36mm
1493g
304p
ISBN-10 84-226-3764-2

圣十字若望（San Juan de la Cruz）是 16 世纪西班牙的神秘学家、天主教加尔默罗会的修士和神父。圣十字若望以写作著称，他的诗歌和灵修研究文章被认为是西班牙神秘文学的巅峰。本书是 1991 年圣十字若望逝世 400 周年时推出的纪念文集，汇编了他的诗歌和一些文章。书籍的装订形式为圆脊锁线硬精装。护封灰黄、暗红和黑色三色印刷，覆哑膜。内封棕色布面裱覆卡板，文字烫白，图片及文字的方形区域压凹处理。内页以胶版纸印黑为主，加插哑面铜版纸银灰色和黑色双色印刷的绘画作品。书中包含多位学者对圣十字若望的研究文章，圣十字若望的诗歌 12 篇、文章 4 篇，以及相关的注解和说明。诗歌部分的版心较小，采用了单栏设置，左对齐，且字号较大。而文章部分由于文字较多，版心比较饱满，双栏设置，两端对齐，且字号较小。书中插图由西班牙超现实主义艺术家安东尼奥·绍拉（Antonio Saura）绘制。

San Juan de la Cruz was a occultist in Spain in the 16th century, a friar and priest of the Catholic Carmelite Church. San Juan de la Cruz is famous for his writing, and his poetry and spiritual studies are considered to be the pinnacle of Spanish mysterious literature. This book is a commemorative collection published at the 400th anniversary of the death of San Juan de la Cruz in 1991, which compiled his poems and articles. The hardcover book is bound by thread sewing with rounding back. The jacket is printed in three colors: gray yellow, dark red, and black, covered with matte film. The inner cover is made of brown cloth mounted with cardboard. The text is treated with white hot-stamping, and the square area with pictures and text is embossed. The inner pages are mainly printed in black on offset paper, while the inserts are made of matte coated paper printed with silver-gray and black paintings. The book contains many scholars' research articles on San Juan de la Cruz, 12 poems and 4 articles of San Juan de la Cruz, and related notes and explanations. The poem part has a small typeset area with a single column, left alignment and a large font size. However, due to the large amount of text in the article part, the typeset area is relatively full, with double columns, justify alignment and smaller font size. The illustrations in the book were painted by Antonio Saura, a Spanish surrealist artist.

n en «otra pestí-
lar libertad a las
s privan de ella.
a San Juan así:
solamente des-
ma en el despre-
ión de sus apeti-
llador, que será
aciones, y no sa-
alma hasta la úl-
pintura, que ya
i en entallar, ni
bra que Dios en
San Juan de la

SAN JUAN
DE LA CRUZ

POESÍA

OTROS TEXTOS

ILUSTRACIONES

DE ANTONIO SAURA

NOTAS

inefable místico», en
Revista de Occidente,
nza, Madrid, 1969 y ss.
contexto sociocultu-
tianismo», en *La mu-*
femenina, Barcelona,

La poesía de San Juan
ra), Madrid, Aguilar,
análisis estilístico, en
emas, y de éstos otros,
or toda la hermosura /
por un no sé qué / que
n la repoetización del

sobre un
de Franci
La escond
4. El pro
que ha tr
guiendo
en el vers
traslada
brane, m
Noche, co
erotista,
amada, «
sólo se g
of the Cr

mis entrañas dibujados!

ECLARACIÓN

2.

nto deseo desea el alma la
so y ve que no halla medio ni
en todas las criaturas, vuél-
on la fe, como la que más al
ar de su Amado luz, tomán-
para esto. Porque, a la ver-
tro por donde se venga a la

pone la fe llar
inteligencia de
es de saber que
en las proposi
verdades y sus
son comparad
sustancia que
bierta con plat
zar en la otra v
oro de la fe. D
ella, dice así: *Si*
las plumas de l
postrimerías de

OROZCO DÍAZ, Emilio, *Poesía y*
ducción a la lírica de San Juan de la
Guadarrama, 1959.
———, *Paisaje y sentimiento de l*
a poesía española, Madrid, Prensa E
pp. 105-138.
———, *Manierismo y Barroco*, M
1975 (2.ª ed.).
PACHO, Eulogio de la Virgen de
Juan de la Cruz y sus escritos, Madrid
tiandad, 1969.
——— (introducción, edición
Juan de la Cruz, *Cántico espiritual.*
ción y texto retocado, Madrid, FUE (F
versitaria Española), 1981.
——— (introducciones, notas
texto): San Juan de la Cruz, *Obras*
gos, Monte Carmelo, 1982.

RESPUESTA DE LAS CRIATURAS

Mil gracias derramando
pasó por estos sotos con presura,
y, yéndolos mirando,
con sola su figura
vestidos los dejó de hermosura.

ESPOSA

¡Ay!, ¿quién podrá sanarme?
Acaba de entregarte ya de vero.
No quieras enviarme
de hoy más ya mensajero,
que no saben decirme lo que quiero.

Y todos cuantos vagan
de ti me van mil gracias refiriendo,
y todos más me llagan,
y déjame muriendo
un no sé qué que quedan balbuciendo.

Mas ¿cómo perseveras,
¡oh vida!, no viviendo dónde vives
y haciendo porque mueras
las flechas que recibes
de lo que del Amado en ti concibes?

¿Por qué, pues has llagado
aqueste corazón, no le sanaste?
Y, pues me le has robado,
¿por qué así le dejaste
y no tomas el robo que robaste?

Apaga mis enojos,
pues que ninguno basta a deshacellos,
y véante mis ojos,

¡Oh cristalina fuente,
si en esos tus semblantes plateados
formases de repente
los ojos deseados
que tengo en mis entrañas dibujados!

¡Apártalos, Amado,
que voy de vuelo!

ESPOSO

Vuélvete, paloma,
que el ciervo vulnerado
por el otero asoma
al aire de tu vuelo, y fresco toma.

ESPOSA

Mi Amado, las montañas,
los valles solitarios nemorosos,
las ínsulas extrañas,
los ríos sonorosos,
el silbo de los aires amorosos,

la noche sosegada
en par de los levantes de la aurora,
la música callada,
la soledad sonora,
la cena que recrea y enamora.

Nuestro lecho florido,
de cuevas de leones enlazado,
en púrpura tendido,
de paz edificado,
de mil escudos de oro coronado.

la hubo herido, porque antes tiene ella las tales heridas por salud, sino porque la dejó así penando en amor, y no la hirió más valerosamente, acabándola de matar para verse y juntarse con él en vida de amor perfecto. Por tanto, encareciendo o declarando ella su dolor, dice: ***habiéndome herido.***

19.

Es a saber, dejándome así herida, muriendo con heridas de amor de ti, te escondiste con tanta ligereza como ciervo. Este sentimiento acaece así tan grande porque en aquella herida de amor que hace Dios al alma levántase el afecto de la voluntad con súbita presteza a la posesión del Amado, cuyo toque sintió. Con esa misma presteza siente la ausencia y el no poderle poseer aquí como desea, y así luego allí juntamente siente el gemido de la tal ausencia, porque estas visitas tales no son como otras en que Dios recrea y satisface al alma, porque éstas sólo las hace más para herir que para sanar y más para lastimar que para satisfacer, pues sirven para avivar la noticia y aumentar el apetito y, por consiguiente, el dolor y ansia de ver a Dios.

Estas se llaman heridas espirituales de amor, las cuales son al alma sabrosísimas y deseables, por lo cual querría ella estar siempre muriendo mil muertes a estas lanzadas, porque la hacen salir de sí y entrar en Dios, lo cual da ella a entender en el verso siguiente, diciendo:

salí tras ti clamando, y eras ido.

20.

En las heridas de amor no puede haber medicina sino de parte del que hirió, y por eso esta herida alma salió, en la fuerza del fuego que causó la herida, tras de su Amado que la había herido, clamando a él para que la sanase.

Es de saber que este salir espiritualmente se entiende aquí de dos maneras para ir tras Dios: la una, saliendo de todas las cosas, lo cual se hace por aborrecimiento y desprecio de ellas; la otra, saliendo de sí misma por olvido de sí, lo cual se hace por el amor de Dios; porque, cuando éste toca al alma

cios y modos e inclinaciones naturales la saca clamando por Dios.

Y así es como si dijera: Esposo mío, en aquel toque tuyo y herida de amor sacaste mi alma, no sólo de todas las cosas, mas también la sacaste e hiciste salir de sí (porque, a la verdad, y aun de las carnes parece la saca), y levantástela a ti, clamando por ti, ya desasida de todo para asirse a ti,

y eras ido.

21.

Como si dijera: al tiempo que quise comprehender tu presencia no te hallé, y quedéme desasida de lo uno y sin asir lo otro, penando en los aires de amor sin arrimo de ti y de mí. Esto que aquí llama el alma salir para ir a buscar al Amado, llama la esposa en los Cantares levantar, diciendo: ***Levantarme he y buscaré al que ama mi alma, rodeando la ciudad por los arrabales y las plazas. Busquéle,*** dice, ***y no le hallé, y llagáronme*** (3, 2; y 5, 6-7). Levantarse el alma esposa, se entiende allí, hablando espiritualmente, de lo bajo a lo alto, que es lo mismo que aquí dice el alma ***salir***, esto es, de su modo y amor bajo al alto amor de Dios.

Pero dice allí la esposa que quedó llagada porque no le halló; y aquí el alma también dice que está herida de amor, y la dejó así. Por eso el enamorado vive siempre penado en la ausencia, porque él está ya entregado al que ama, esperando la paga de la entrega que ha hecho, y es la entrega del Amado a él, y todavía no se le da; y estando ya perdido a todas las cosas y a sí mismo por el Amado, no ha hallado la ganancia de su pérdida, pues carece de la posesión del que ama su alma.

22

Esta pena y sentimiento de la ausencia de Dios suele ser tan grande a los que van llegando al estado de perfección al tiempo de estas divinas heridas, que si no proveyese el Señor morirían; porque, como tienen el paladar de la voluntad sano y el espíritu limpio y bien dispuesto para Dios, y en lo que está dicho se les da a gustar algo de la dulzura del amor divino, que ellos sobre todo modo

HC Ericson

ART CONSULTING: SCANDINAVIA
Books on Art & Architecture
E-mail: info@nordicartbooks.com http://www.nordicartbooks.com

TO:

AbeBooks

AbeBooks: Shipping Manifest

LTING:SCANDINAVIA,
de Vista Drive
CA 91302-2155

To:
ZHAO QING
DABEI XIANG7-3, NANJING HAN
QING TANG DESIGN CO.,
NANJING 210000
CHINA

Sweden

U.S.A

China

257mm

12mm

206mm

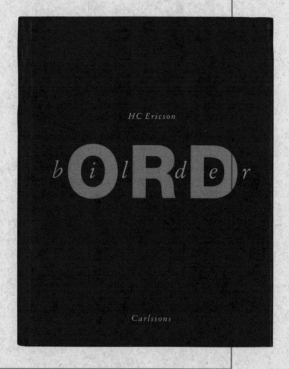

荣誉奖
文字图片
瑞典

Honorary Appreciation
Ordbilder
Sweden
HC Ericson
HC Ericson
Carlssons Bokförlag, Stockholm
257 206 12mm
554g
88p
ISBN-10 91-7798-451-X

记录一个人的想法有很多种方式，可以是日记，可以是照片，可以是影像。瑞典平面设计师汉斯·克里斯·埃里克松（HC Ericson）通过独特的文字排版和相关图片的组合，记录了一个人 40 年的成长故事。他的另一本同类书籍《处理厂》（*Reningsverk*）获得了莱比锡“世界最美的书”1998 年铜奖（在本书的相关页面可以找到）。书籍的装订形式为锁线硬精装。护封铜版纸红黄黑三色印刷，内封黑色布面烫黄色和灰色，裱覆卡板。内页哑面铜版纸四色印刷。内容方面，诗歌与图片并置在左右页，互为补充。文字方面，通过丰富的字号大小、断行与缩进设置来建立阅读节奏，并将部分单词设置为意大利斜体，起到了语义上的提示或语气上的强调。图片方面，并非都是带有故事的纪实照片，而是拍摄了很多事物的特写或独特的创意组合，隐隐透出成长过程中激动、痛苦、困惑、无助等丰富的情绪。

There are many ways to record a person's thoughts, which can be a diary, a photo, or an image. Swedish graphic designer Hans Chris Ericson (HC Ericson) has recorded a person's growth story for 40 years through a unique combination of text and related pictures. His another similar book Treatment Plant *Reningsverk* won the 1998 Bronze Medal for Leipzig's most beautiful book in the world (which can be found on the relevant page of the book). The hardcover book is bound by thread sewing. The jacket is made of matte coated paper printed in red, yellow and black and the inner cover is made of black cloth treated with yellow hot stamping and gray hot stamping, mounted with cardboard. The inside pages are printed on matte coated paper in four colors. In terms of content, the poems and pictures are placed on the left and right pages respectively to complement each other. The reading rhythm of the text is established by various font sizes, broken lines and indented setting. Some words are set to italics, playing the role of a semantic hint or tone emphasis. The pictures are not all documentary photos with stories, but close-up or unique creative combination of many things, which vaguely reveals the excitement, pain, confusion, helplessness and other emotions in the growth process.

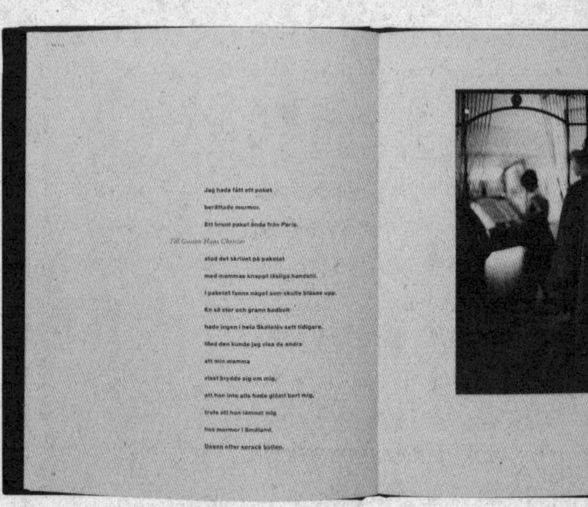
Jag hade fått ett paket
berättade mormor.
Ett brunt paket ända från Paris.
Till Gossen Hans Christer
stod det skrivet på paketet
med mammas knappt läsliga handstil.
I paketet fanns något som skulle blåsas upp.
En så stor och grann badboll
hade ingen i hela Skatelöv sett tidigare.
Med den kunde jag visa de andra
att min mamma
visst brydde sig om mig,
att hon inte alls hade glömt bort mig,
trots att hon lämnat mig
hos mormor i Småland.
Dagen efter sprack bollen.

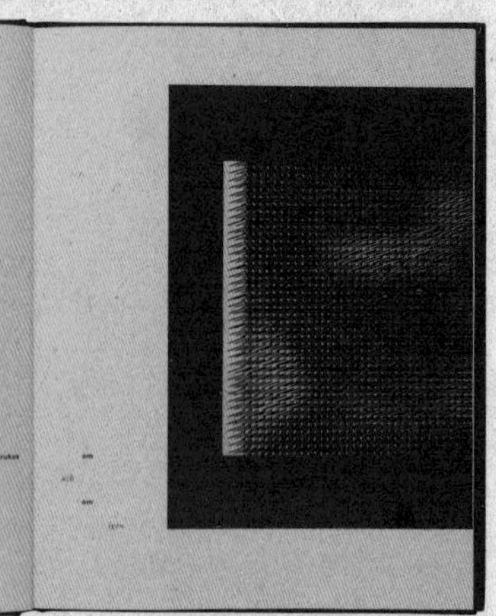

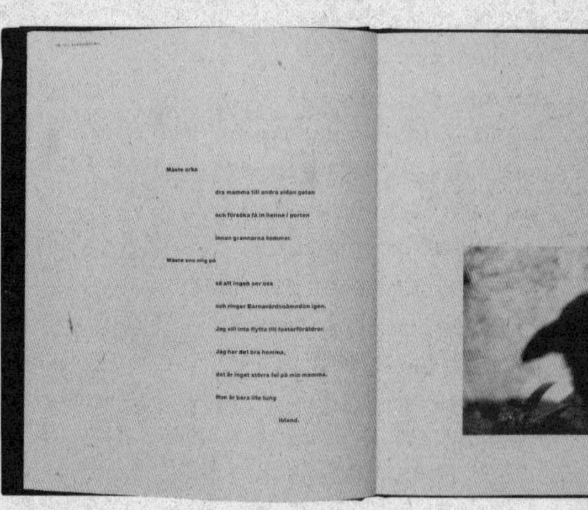
Måste orka
dra mamma till andra sidan gatan
och försöka få in henne i porten
innan grannarna kommer.
Måste sno mig på
så att ingen ser oss
och ringer Barnavårdsnämnden igen.
Jag vill inte flytta till fosterföräldrar.
Jag har det bra hemma,
det är inget större fel på min mamma.
Hon är bara lite tung
ibland.

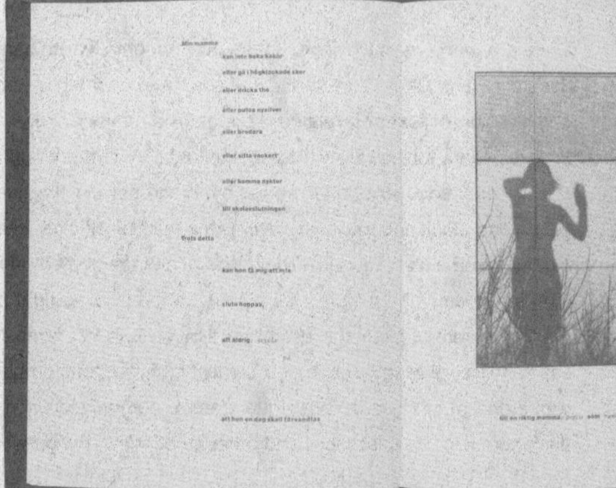
Min mamma
kan inte baka kakor
eller gå i högklackade skor
eller dricka the
eller putsa nysilver
eller brodera
eller sitta vackert
eller komma nykter
till skolavslutningen
Trots detta
sluta hoppas,
att hon en dag skall förvandlas

"Vad tycker du
att det här föreställer?"
frågade barnpsykologen inställsamt.
Bäst att svara
att det liknar en liten tomte
eller några barn som leker på en äng
och att bläckplumpen i mitten
ser ut som en hundvalp
som nyss har badat.
Fast på riktigt
påminde plumparna om mammas spyor,
som jag skrapade upp med en kartongbit
och slog in i tidningspapper
när hon låg redlös
då jag kom hem efter skolan.
Vissa plumpar
fick mig också att tänka på hennes hudutslag
och på den återkommande drömmen om en öken
där jag sakta förvandlades till en stenfigur
som så småningom vittrade sönder till sand.
Men det vågade jag inte berätta om.
Man vill ju inte flytta hemifrån.
Man vill ju behålla sina föräldrar så länge de lever.

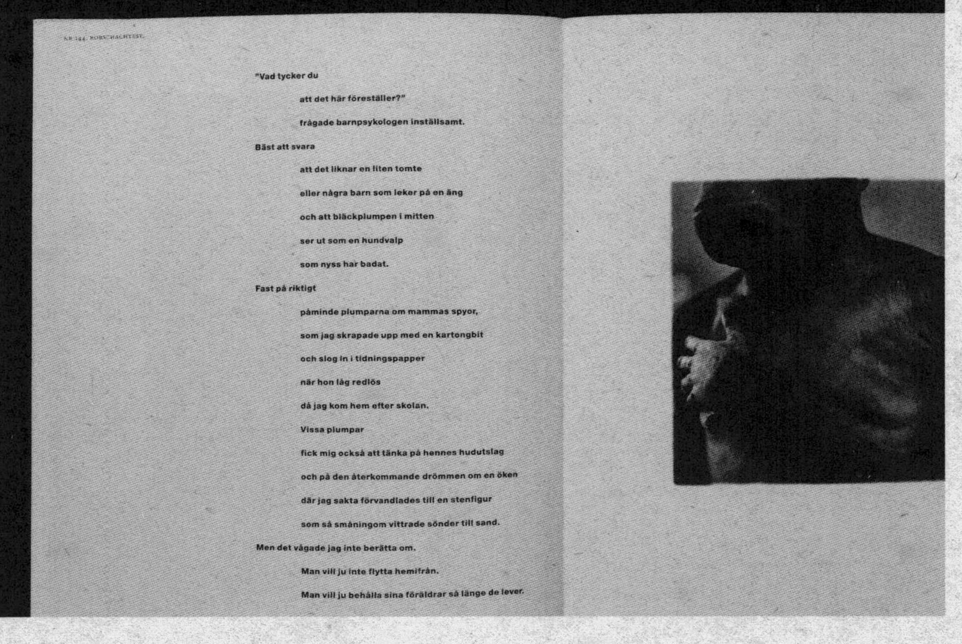

T O M

Jag önskade att mamma
skulle ta båten till Amerika
och träffa någon annan som kunde ta hand
om henne.

Helst en indian,
för gjorde hon det så skulle jag få hans fjäderskrud
lovade mamma.

Australien gick bra, det med.
Det låg också så långt bort
att hon inte skulle komma hem
för ofta.

Det var bäst för hennes egen skull.
Kom hon inte hemifrån
så skulle hon aldrig kunna börja sluta
sitt gamla liv.

Ibland skulle jag nog sakna mamma.
Men en dag skulle hon ju komma
tillbaka
och ta hand om mig
och vara så frisk

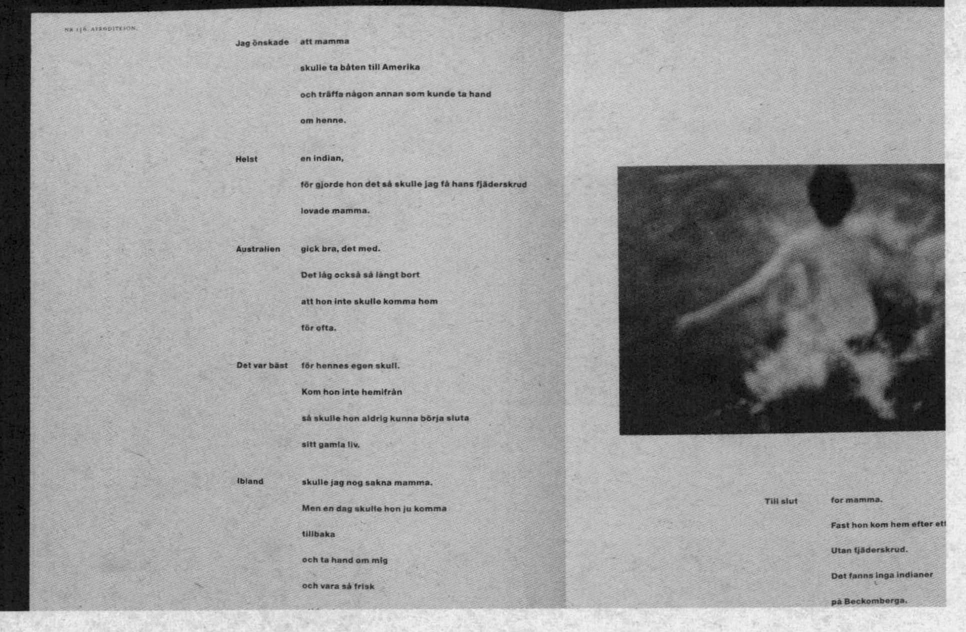

Till slut for mamma.
Fast hon kom hem efter ett
Utan fjäderskrud.
Det fanns inga indianer
på Beckomberga.

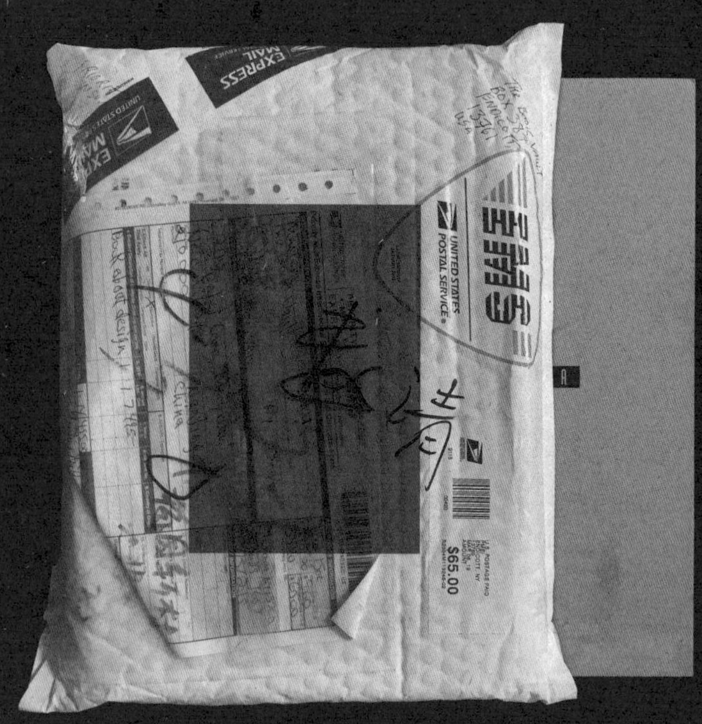
EXPRESS MAIL
EMS
UNITED STATES POSTAL SERVICE
$65.00

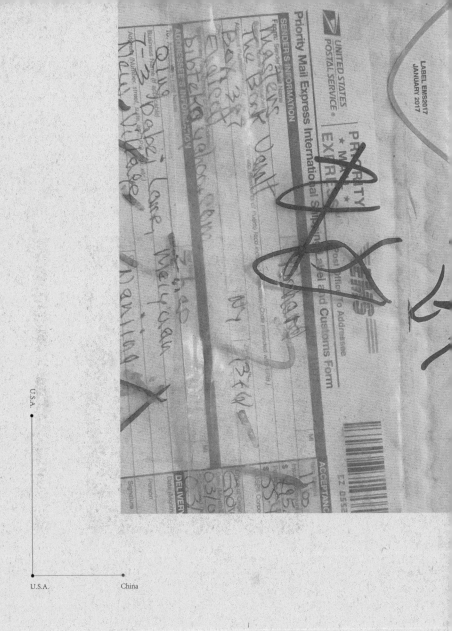

U.S.A.

U.S.A. China

306mm

14mm

MECHANIKA

230mm

● 荣誉奖
« 机械学
◐ 美国

Honorary Appreciation
Mechanika
U.S.A.
David Betz
——
The Contemporary Arts Center, Cincinatti
306 230 14mm
374g
60p
ISBN-10 0-917562-57-7

自工业革命以来，机械成为人类社会重要的一部分，以机械为主题的艺术创作也越来越多。19 世纪的艺术家会用绘画来展现机械具有的工业力量。20 世纪初期，未来主义艺术家会从机械中找到代表新世界的灵感和希望。当然，也有艺术家会通过作品展现对机械和工业社会的思考，比如卓别林的名作《摩登时代》。1991 年，在美国辛辛那提当代艺术中心举办了名为 Mechanika 的机械装置艺术展，本书是展览画册。书籍的装订形式为铁圈装。封面牛皮卡纸印黑、印银，烫印珠光白。过门页采用半透明牛油纸正反面印黑，呈现前后透叠的层次。内页哑面铜版纸四色印刷。书中一共呈现了展览中 16 位艺术家的 29 件装置作品。在作品页之前，通过访谈的形式对 150 年来机械对社会的影响、艺术作品、社会思潮和展览本身进行了剖析。标题选用了很多早年的机械上使用的钢印字体，正文则采用打字机字体，并极力呈现老式打字机文本常用的排版方式，例如文本的两端齐行、整段长缩进、居中对齐等形式。|||||||||||||||
||||||||||||||||||||||||||||||||||||

Since the industrial revolution, machinery has become an important part of human society, and there are more and more art creations with the theme of machinery. Artists in the 19th century used painting to show the industrial power of machinery while in the early 20th century, Futurist artists found inspiration and hope to represent the new world from machinery. Also, there were artists who express their thoughts on mechanical and industrial society through their works, such as the famous *Modern Times* by Charles Chaplin. In 1991, a mechanical installation exhibition named ‘Mechanika’ was held in Cincinnati Contemporary Art Center in the United States. This book is an exhibition album. The book is bound by iron ring binding. The cover is made of kraft cardboard printed in black and silver, also treated with pearlescent white hot stamping. The end paper is printed in black with translucent kraft paper, showing the layers of f permutation. The inner pages are printed with matte coated paper in four colors. A total of 29 installation works of 16 artists in the exhibition are presented in the book. Before displaying the works, the influence of machinery on society, art works, social thoughts and the exhibition itself are analyzed in the form of interviews. Many mechanical steel seal fonts were used in the title, while typewriter font was used in the text. The typesetting methods commonly used with the old typewriter were strongly presented, such as justify alignment, paragraph indentation, center alignment and so on. ||||||||||

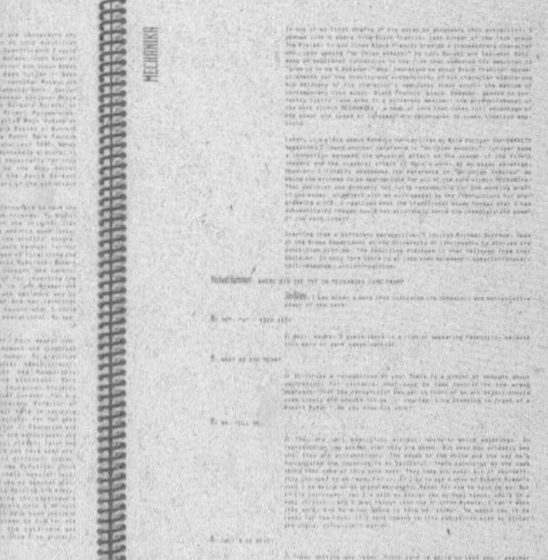
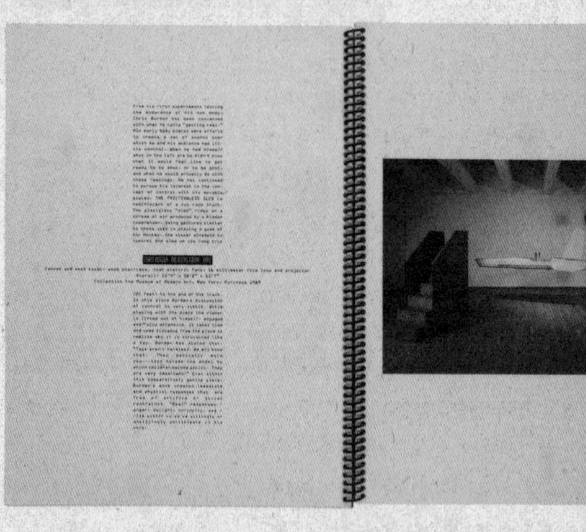

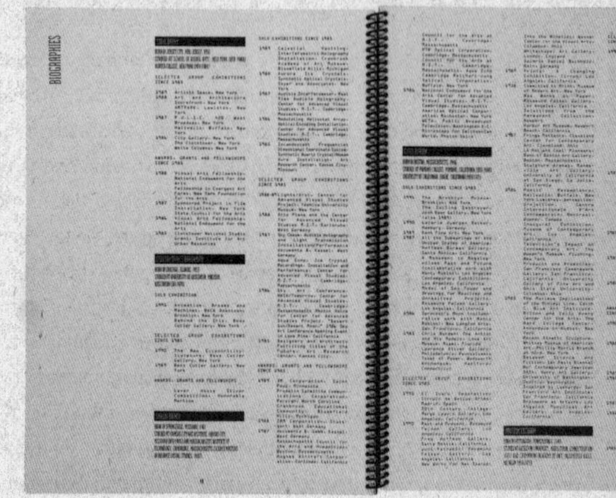

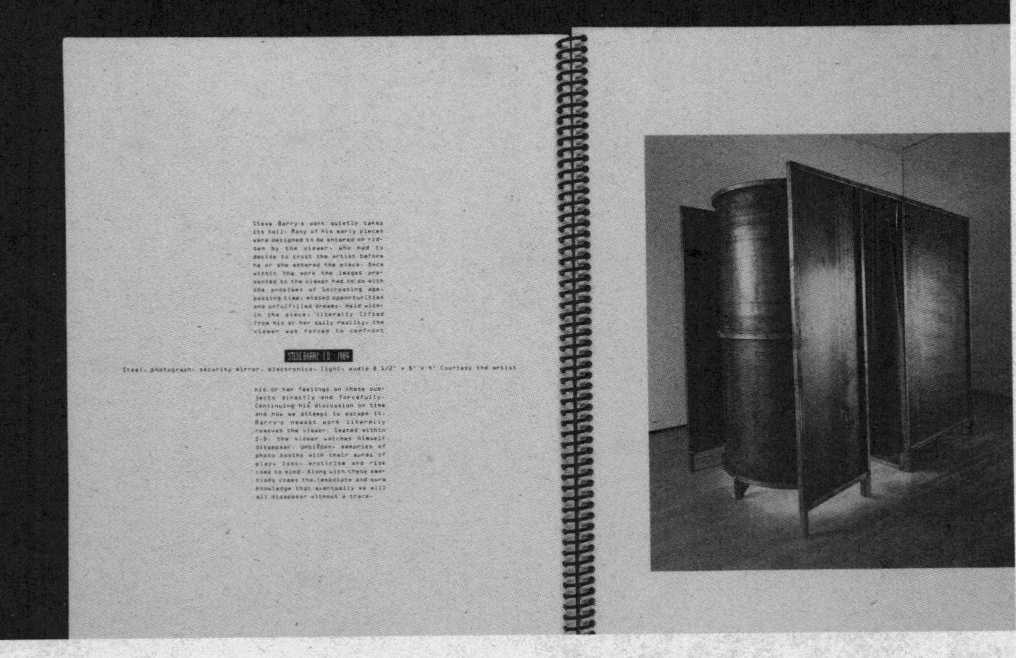

Steve Barry's work quietly takes its toll. Many of his early pieces were designed to be entered or ridden by the viewer, who had to decide to trust the artist before he or she entered the piece. Once within the work the images presented to the viewer had to do with the problems of increasing age, passing time, missed opportunities and unfulfilled dreams. Held within the piece, literally lifted from his or her daily reality, the viewer was forced to confront

STEVE BARRY I.D. 1989

Steel, photograph, security mirror, electronics, light, audio 8 1/2' x 8' x 4' Courtesy the artist

his or her feelings on these subjects directly and forcefully. Continuing his discussion on time and how we attempt to escape it, Barry's newest work literally removes the viewer. Seated within I.D. the viewer watches himself disappear. Unbidden, memories of photo booths with their auras of play, loss, eroticism and risk come to mind. Along with these emotions comes the immediate and sure knowledge that eventually we will all disappear without a trace.

D GINZEL WAS BORN IN CHICAGO, ILLINOIS, 1954
STUDIED AT BROWN UNIVERSITY, PROVIDENCE, RHODE ISLAND
78), SAN FRANCISCO ART INSTITUTE (1978), ST. MARTIN'S
L OF ART, LONDON, ENGLAND (1978-79), RHODE ISLAND
OF DESIGN, PROVIDENCE, RHODE ISLAND (BFA 1979) AND YALE
SITY, NEW HAVEN, CONNECTICUT, (MFA 1983) GINZEL IS SELF

EXHIBITIONS SINCE 1985

9 Antithesis, Kunsthalle, Basel, Switzerland
Ecclesia, Damon Brandt Gallery, New York
8 Metathesis, Annina Nosei Gallery, New York
Charybdis, List Visual Arts Center, M.I.T., Cambridge, Massachusetts
Seraphim, Matrix 99, Wadsworth Atheneum, Hartford, Connecticut
-88 Pananemone, City Hall Park, The Public Art Fund, New York
Vis-a-vis, Art Galaxy, New York
Triptych, The New Museum of Contemporary Art, New York
Clepsydra, Lawrence Oliver Gallery, Philadelphia, Pennsylvania
Ephemeris, Virginia Museum of Fine Arts, Richmond, Virginia
5 Spheric Storm, Art Galaxy, New York

CTED GROUP EXHIBITIONS
E 1985

1986 Lumieres
Projectio
Canada
Art in th
Anchorag
Bridge,
Inc., New
Invitati
Grace Bor
New York
Engaging
tute for
Resource
tower, Ne
Bernice S
New York
The Art
Gallery,
Cultural
1985 National
Studio Ar
Institute
Resource
Clocktowe
Art an
Storefron
Modern M
Kinetic S
Museum of
Phillip M
Institute
Resource
tower, Ne
National
Studio Ar
Institute
Resource
tower, Ne

AWARDS, GRANTS A

1989 MacDowel

crylic, elec-
tube

se, aluminum,
's

1991
Mixed me
9' x 4'
Courtesy

MFO
1991
Mixed me
3' x 3'
Courtesy

MARY ZIEGLE

LEVEL TWO
1990
Steel, c
nesium,
54" x 33"
Courtesy

nting. Berman Prin
N: 0-917562-57-7.

Contemporary Arts C
atefully acknowle
continuing suppor
e Fine Arts Fun

of the Drama Department at the Univ
exhibition with me. The resulting
decision. In this form there is at
call/response, action/reaction.

Burnham: WHERE DID THE "K" IN MECHANIKA COME FROM?

Jan Riley: I was after a word that indi
power of the work.

, "A" - KICK ASS?

J: Well, maybe. I guess there is a
this body of work takes control.

T DO YOU MEAN?

J: It forces a recognition on you.

or to revolve frantic
of polished metal "mi
is a sweet and immedi
in this work that Opp
ciates and has anti
balance between some

ENNIS OPPENHEIM · DISCO MATTRESS (FROM

lectric cord, electric pl
tesy Ace Contemporary Exh

funny on one side an
side, really quite h
something that enchan
condition implies
viewpoints and

This exhibition has been supported by generous grants from The Andy Warhol Foundation for the Visual Arts, Inc., and the David Bermant Foundation: Color, Light, Motion.

FOREWORD

Since the Industrial Age and particularly the 20th century
have examined the impact of the machine upon their so
Paintings from the 19th century, for example, celebrate
industrial might of nations. In the 1920s and '30s the
found inspiration and hope for a new world order in the ma
the machine. And yet, artists from Charlie Chaplin to Chr
have questioned the values and benefits of an industrial
by depicting machine imagery from more uncertain persp
MECHANIKA, an exhibition organized by The Contempor
Center takes a look at how a number of contemporary art
currently using machine-based imagery to explore the i
contemporary society.

MECHANIKA continues a series of group exhibitions present
Center that have explored contemporary three-dimensional a
very specific point of view. In 1985, The Center organized
SOUL: ASPECTS OF RECENT FIGURATIVE SCULPTURE, one of t
comprehensive examinations of the re-emergence of figurati
ture in the '80s. In 1987, STANDING GROUND: SCULPTURE BY
WOMEN was a major exhibition which focused on the impact wom
tors were having on the art world in a time when sculpture
sidered a "man's world." MECHANIKA factors in as a similar
breaking survey that will help define the face of sculptur
nineties.

MECHANIKA also continues a series of one-person and grou
tions at The Center that have dealt with the increasing i
of technology in the making of contemporary art. From Rock
laser installation in 1984 to COMPUTERS IN ART in 1987 and
itself. The Center has sought to deal with the issue, invo
artists at the forefront of the field.

Finally, MECHANIKA continues a philosophy at The Center
emerging artists are shown side by side with artists of con
reputation. In MECHANIKA, the works of artists such a
Oppenheim, Bruce Nauman and Rebecca Horn are shown with l
known and younger artists such as Shawn Brixey, Mary Zi
Malcolm Cochran.

MECHANIKA is the brain child of Center Curator Jan Riley. Th
idea for the exhibition, the selection of the artists a
included, and the exhibition's impressive installation and
alogue copy were chiefly Jan's doing and for this con
accomplishment, we salute her. Recognition must be given
Betz, who designed the signage treatment and the impress
logue, as well as to Preparator Bob Swaney and his talente
the exciting installation.

MECHANIKA was a difficult exhibition to organize and pr
would not have been possible without funding support from
Warhol Foundation for the Visual Arts, Inc., and the Davi
Foundation: Color, Light, Motion. The prestigious Warhol F
provided the needed boost when it looked like a lack of fund
halt or delay the project. The Bermant Foundation grant ga
ability to make the exhibition comprehensive and produce
alogue. To both we are most grateful, for without the com
these two foundations, MECHANIKA might never have gone from
reality.

There are sixteen artists, twenty-nine pieces and 150 yea
history in this provocative exhibition. MECHANIKA is inde
the most exciting shows in The Center's history.

Dennis Barrie
Director

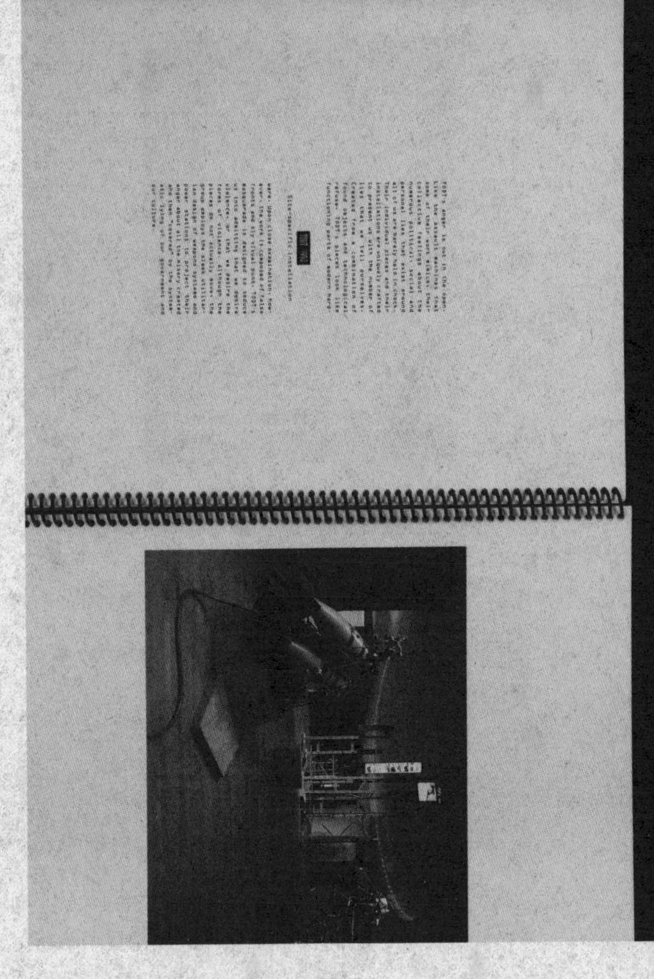
TODT's anger is out in the open.
Like the assault machines that
some of their work mimics, their
collective feelings about the
numerous political, social and
personal lies that exist around
all of us are barely held in check.
Their individual pieces and their
installations are uniquely crafted
to present us with the number of
lies that we tell ourselves.
Created from a combination of
found objects and technological
refuse, TODT's pieces look like
functioning parts of modern hard-
TODT
Site-specific installation
ware. Upon close examination, however, the work is composed of false
fronts and sly visual puns. TODT's
masquerade is designed to seduce
us into admitting that we desire
violence, or that we desire the
forms of violence. Although the
pieces do not actually move, the
group employs the sleek utilitarian design of weapons systems and
power stations to project their
anger about all the misery created
and then "covered" by the systematic lying of our government and
our culture.

Country / Region

Germany ❻

Switzerland ❸

Poland ❶

Japan ❶

Austria ❶

Denmark ❶

The Netherlands ❶

Innenarchitektur in der Schweiz
1942–1992
Architecture d'intérieur en Suisse
1942–1992
Franz Kafka
Übergriff
SHEN JIJI
DIE GESCHICHTE DES FRÄULEIN REN
司馬遼太郎
新潮社

1994 Sm^1 *0362*

1994 Sm^2 *0372*

1994 Sm^3 *0380*

1994 Bm^1 *0388*

1994 Bm^2 *0396*

1994 Bm^3 *0404*

1994 Bm^4 *0412*

1994 Bm^5 *0420*

1994 Ha^2 *0426*

Designer

Thomas Petraschke (Studio Halblützel)

Eckhard Jung, Ulysses Voelker

Gert Wunderlich

Antje Krausch, Ruedi Tachezy

Studenten der HfG Karlsruhe und Sepp Landsbek

Hartmut Brückner

Clemens Tobias Lange

Shincho-Sha Co., Tokyo

Jadwiga, Janusz Tryzno

Wolfgang Buchta

Jost Hochuli

Anja Harms

Anne Rohweder

Bram de Does

1994 Sm1

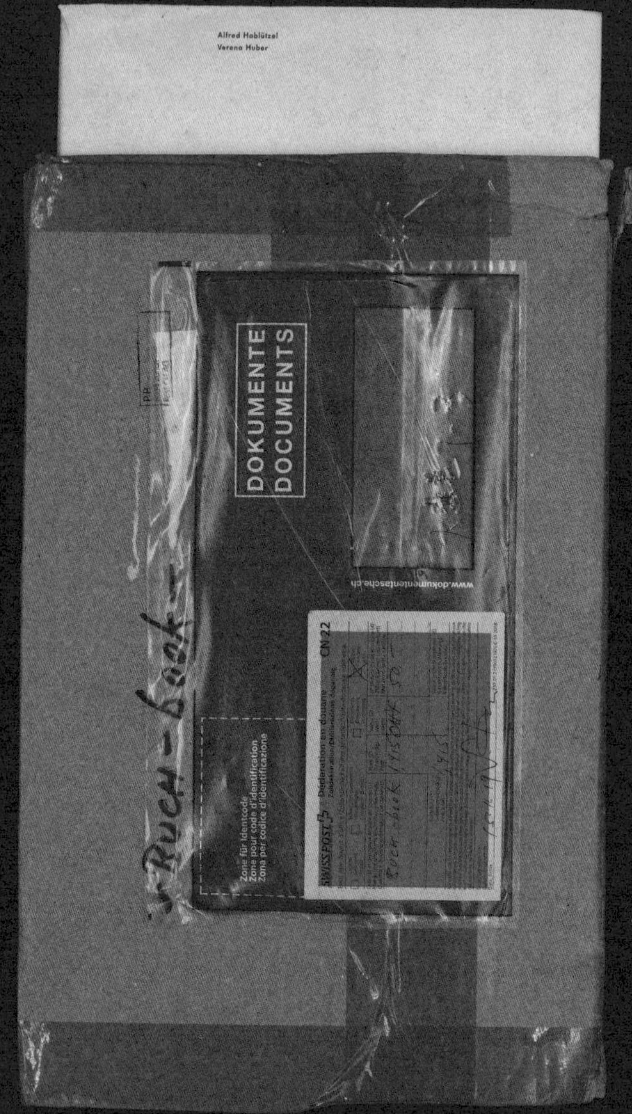
Alfred Hablützel
Verena Huber
DOKUMENTE
DOCUMENTS
www.dokumententasche.ch
Zone für Identcode
Zone pour code d'identification
Zona per codice d'identificazione
SWISS POST
CN 22
Déclaration en douane
Buch-book

SWISS POST

Déclaration en douane CN 2

Zolldeklaration/Dichiarazione doganale

Peut être ouvert d'office/Zollamtliche Prüfung gestattet/Visita doganale ammessa

☐ Cadeau / Geschenk / Regalo	☐ Echantillon commercial / Warenmuster / Campione di merci	☐ Documents / Dokumente / Documenti	☒ Autres / Andere / Altri

Quantité et description détaillée du contenu (1) / Menge und detaillierte Beschreibung des Inhalts / Quantità e descrizione dettagliata del contenuto	Poids (2) / Gewicht (kg) / Peso	Valeur (3) / Wert / Valore	N° tarifaire du SH et origine / Zolltarifnummer und Herkunft / Numero di tariffa e origine
Buch - book	1,415	CHF 50.–	
Poids total (5) / Gesamtgewicht (kg) / Peso totale	1,415		Valeur totale (+monnaie) / Gesamtwert (+Währung) / Valore totale (+valuta)

Je certifie que les renseignements donnés dans la présente déclaration sont exacts et que cet envoi ne contient aucun objet dangereux interdit par la réglementation postale ou douanière. / Ich bestätige hiermit, dass die Angaben in der vorliegenden Deklaration … und dass die Sendung keine durch die Post- oder Zollvorschriften verbotenen oder gefährlichen Gegenstände enthält. / Certifico che le informazioni contenute nella presente dichiarazione sono esatte e che quest'invio non contiene nessun oggetto pericoloso o … dal regolamento postale o doganale.

Date, Signature (7)

Switzerland

Switzerland

China

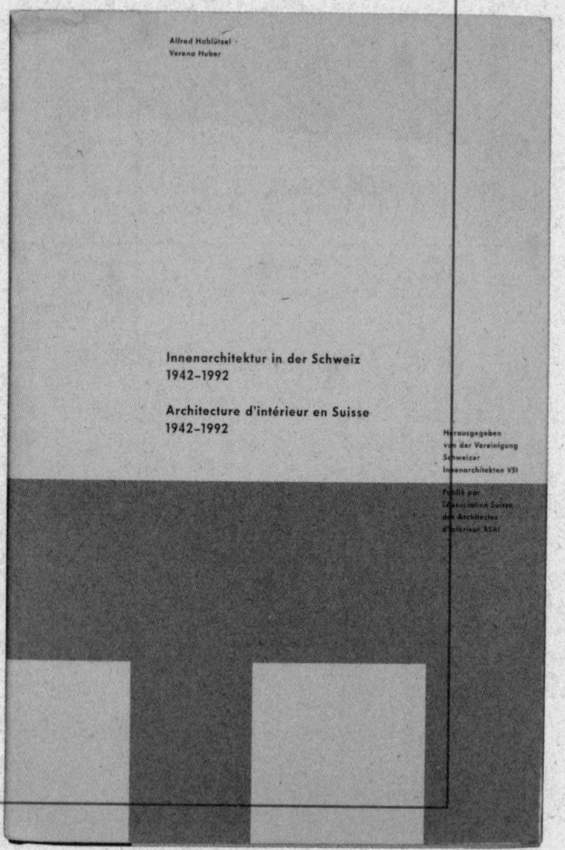

- 银奖
- 1942－1992 年的瑞士室内设计
- 瑞士

Silver Medal
Innenarchitektur in der Schweiz 1942-1992
Switzerland
Thomas Petraschke (Studio Halblützel)
Alfred Hablützel, Verena Huber
Verlag Niggli AG, Sulgen
306 207 21mm
1280g
248p
ISBN-10 3-7212-0276-7

瑞士室内设计协会于1993年成立50周年，在此期间出版了这本50年来瑞士经典室内设计案例的书籍作为总结和献礼。本书的装订形式为锁线硬精装。护封采用哑面艺术纸黄黑双色印刷，护封的上下书口略大于书籍开本，向内翻折，形成较厚且舒适的手感。内封灰色卡板直接装订，文字烫黑。书脊处裱贴黑色布纹纸，文字烫白。环衬采用黑卡纸，正文哑面铜版纸四色印刷。内容方面，除了开篇的序言，还附带了两篇从不同角度写就的散文。之后以2页或4页为一个单元，总计展示了76个设计案例，分为工作空间、餐馆、公共空间、卖场、客厅共5个门类。全书德法双语排版，序言和散文部分采用双栏设置，案例部分设置为五栏，双语正文各占两栏，注解单栏。图片通过大小对比进行混排，产生丰富的版面。值得一提的是：每个案例都在左页上部放置一张案例所在建筑的外观，独占一栏，在书籍的翻口上呈现整齐的黑色块。

In 1993, the 50th anniversary of the Swiss Interior Design Association, this book was published as a summary and tribute, which collected classic interior design cases in Switzerland. The hardcover book is bound by thread sewing. The jacket is made of matte art paper printed in yellow and black. The upper and lower edges of the jacket are slightly larger than the book format, which can be folded inwards to form a thick and comfortable feel. The inner cover is made of gray cardboard with the text stamped in black. The book spine is mounted with black arlin paper, and the text is white stamped. The end paper is made of black cardboard, and the inner pages are printed in four colors on matte coated paper. As for the content, in addition to the opening introduction, two essays written from different perspectives are included. Then 76 design cases are displayed in total, taking 2 or 4 pages as a unit. The cases are divided into five categories: working space, restaurant, public space, mall and living room. The preface and the essays are typeset in double-column while the cases are set to 5 columns. The bilingual text occupies two columns each, and the annotations are set in a column. The pictures are mixed by size comparison, resulting in a rich layout. It is worth mentioning that each case shows the appearance of the building on the upper part of the left page, with an exclusive column and neat black block on the flap of the spine..

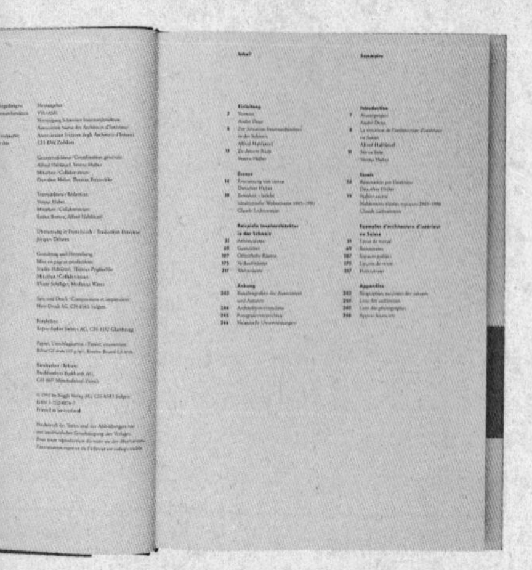
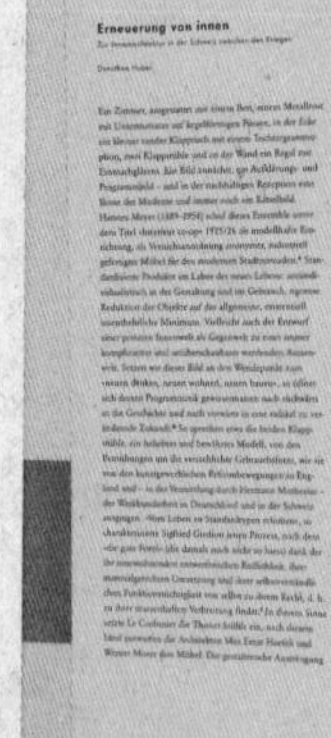
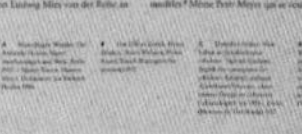

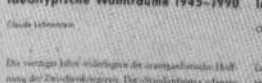

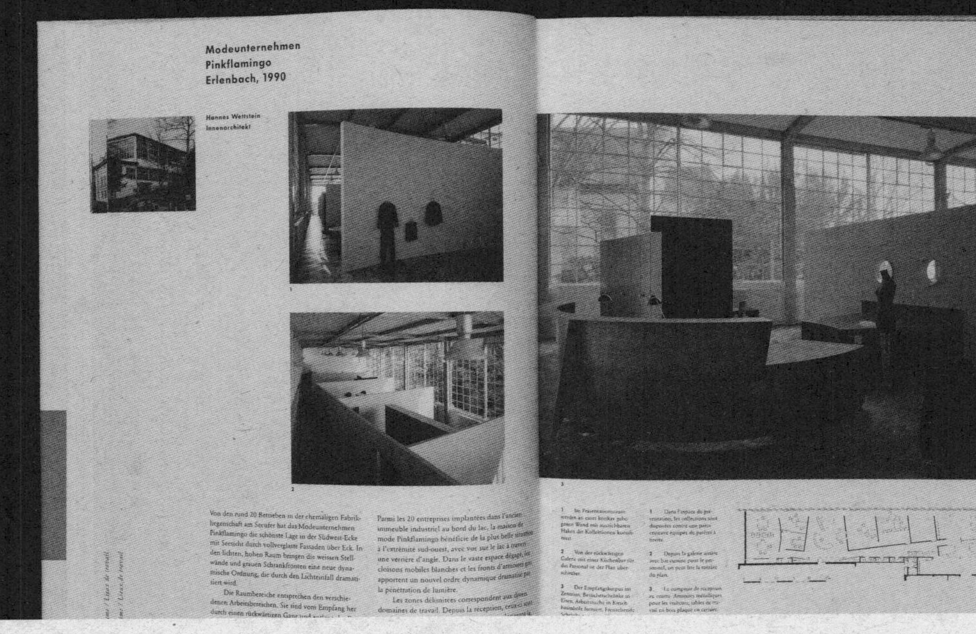

Modeunternehmen Pinkflamingo Erlenbach, 1990

Hannes Wettstein
Innenarchitekt

Von den rund 20 Betrieben in der ehemaligen Fabrikliegenschaft am Seeufer hat das Modeunternehmen Pinkflamingo die schönste Lage in der Südwest-Ecke mit Seesicht durch vollverglaste Fassaden über Eck. In den lichten, hohen Raum bringen die weissen Stellwände und grauen Schrankfronten eine neue dynamische Ordnung, die durch den Lichteinfall dramatisiert wird.

Die Raumbereiche entsprechen den verschiedenen Arbeitsbereichen. Sie sind vom Empfang her

Parmi les 20 entreprises implantées dans l'ancien immeuble industriel au bord du lac, la maison de mode Pinkflamingo bénéficie de la plus belle situation à l'extrémité sud-ouest, avec vue sur le lac à travers une verrière d'angle. Dans le vaste espace dégagé, les cloisons mobiles blanches et les fronts d'armoires gris apportent un nouvel ordre dynamique dramatisé par la pénétration de lumière.

Les zones délimitées correspondent aux divers domaines de travail. Depuis la réception, ceux-ci sont

1 Im Präsentationsraum werden an einer konkav gebogenen Wand mit ausziehbaren Haken die Kollektionen konzipiert.

2 Von der rückwärtigen Galerie mit einer Küchenbar für das Personal ist der Plan überschaubar.

3 Der Empfangskorpus im Zentrum. Besucherschränke in Eisen, Arbeitstische in Kirschbaumholz furniert. Freistehende

1 Dans l'espace de présentation, les collections sont disposées contre une paroi concave équipée de patères à tirette.

2 Depuis la galerie arrière avec bar-cuisine pour le personnel, on peut lire la totalité du plan.

3 Le comptoir de réception au centre. Armoires métalliques pour les visiteurs; tables de travail en bois plaqué en cerisier.

…me / Lieux de travail

itsräume / Lieux de travail

Innenarchitekt
1942–1992

Architecture d'
1942–1992

Discoteca Alcatraz Lavertezzo Piano, 19

Claude Jost
Architetto d'interni ASAI
Collaborazione: Lydia Hänggi

Bis zu 1500 Besucher haben Platz in di
ausserhalb von Locarno. Am Tag beme
Industriebau konzipierte Gebäude kau

rossartige repräsentative Bei-
wohl im öffentlichen wie im
Dieses für Kreativität nicht
ovoziert und begünstigt, wie
hl von erstaunlich enga-
Auftragsverhältnissen mit

en letzten Jahren das ermu-
ortmärkte, welche dem
n Bereich der Innenausstat-
stellung, dass hier plötzlich
t attestiert wird, könnte – so
tsein, Freude und Toleranz

nivellement. En matière d'arc
malgré une grande prospérité
de grande envergure sont plu
les domaines public que prive
qui n'encourage guère la créa
comme le montre cet ouvrage
mandes de petite et moyenne
architectes d'intérieur fortem
D'autre part, il est enco
accordé au cours des dernière
d'exportation au design indu
ment intérieur. Constater que
suisse est ici confirmée pourr

smen im
es bewirken
enspiel in
as sich je
chteinfall

1 Le corps c
miroirs. Les tabo
spécialement pro
tecte pour la piè

2 Groupes d
tables contre le f
la cour des arts.

Bank Okasan
Zürich, 1991

Andreas Ramseier
Innenarchitekt VSI
Mitarbeit: Alexandra Mott,
Thierry Landis, Antonia Infanger

Die japanische Okasan Bank hat in Zürich ihre Niederlassung in einem Geschäftshaus aus den siebziger Jahren. Sie ist im Wertschriftenhandel tätig und hat keinen Publikumsverkehr. Die Büroräume sind auf zwei Stockwerken durch einen Mittelkorridor erschlossen, in dem weitgehend die bestehende Infrastruktur übernommen und neu profiliert wurde. Im Empfangsbereich wird der Korridor zum Rundbau. Die angrenzenden Sitzungsräume und Büros werden in der Querachse erschlossen. Durch die verglasten Oberlichter dieser Türen fällt Tageslicht in den zentralen Bereich. Dem in schwarzen Rippen umgesetzten japanischen Motiv begegnet der Besucher erstmals an der Eingangstüre und dann am Empfang in einer weiteren Variation.

Für das Projekt wurden Architekten zu einer Konkurrenzpräsentation eingeladen.

La succursale zurichoise de la banque japonaise Okasan est installée dans un immeuble commercial des années septante. Elle travaille dans le domaine des titres et n'est pas ouverte au public. Les bureaux répartis sur deux niveaux sont desservis par un couloir central dans lequel on a repris l'infrastructure existante en lui donnant un nouveau profil. Dans la zone de réception, le couloir devient un volume circulaire. Les salles de réunion et les bureaux sont desservis par l'axe tranversal, la lumière pouvant pénétrer dans la zone centrale grâce aux impostes vitrées des portes. Le visiteur rencontre les lamelles noires inspirées par un motif japonais dès la porte d'entrée et sous une autre forme au niveau de la réception.

Pour le projet, plusieurs architectes invités ont été mis en concurrence au cours d'une présentation.

1 Empfangspult. Die tiefer gehängte Decke in aluminiumbeschichteten Platten ist überall von den Wänden losgelöst und mit einer indirekten Beleuchtung ausgestattet.

2 Der rückwärtige Schild des Empfangs bildet den Abschluss zum internen Korridor. Die schwarzen Horizontalrippen in Verbindung mit einer halbtransparenten gerundeten Plexiglasplatte erinnern an japanische Papier-Schiebetüren.

1 Comptoir de réception. Le plafond suspendu surbaissé fait de plaques doublées d'aluminium est partout isolé des parois latérales et comporte un éclairage indirect.

2 L'écran situé derrière la réception assure la fermeture du couloir interne. Les lamelles horizontales noires alliées aux plaques de plexiglas bombées semi-transparentes rappellent les parois coulissantes en papier du Japon.

Zahnarztpraxis Dr. Beat Blum
Strengelbach, 1991

Ludwig Meyer
Innenarchitekt VSI

Mit dem Einbau der Praxis fügte sich der Innenarchitekt in die bestehende Umgebung ein. Seine sachliche, knappe Sprache distanziert sich von der Architektur ohne Identität. Der Ausbau wurde ohne Rücksicht auf die Öffnungen berührungslos in die Gebäudehülle gesetzt.

Halbhohe weisse Zwischenwände bilden Nischen für die Behandlungsplätze und Arbeitsbereiche. Die vollverglasten oberen Wandpartien und Türen bieten viel Transparenz. Dieses Raumkonzept hat sich für Zahnarztpraxen in den verschiedensten Situationen bewährt. Individuelle Lösungen werden im Empfangsbereich angestrebt.

Die Wartezone ist mit einem mobilen Paravent abgegrenzt. Garderobe und Empfangstisch sind als schwenkbare Elemente an den tragenden Stützen angebracht, wo sie am Eingang die Torsituation markieren.

En aménageant ce cabinet, l'architecte d'intérieur s'est adapté à l'entourage existant. Son langage objectif et sobre se démarque d'une architecture dépourvue d'identité. L'aménagement fut installé dans l'enveloppe bâtie et scindé de celle-ci sans tenir compte des ouvertures.

Des cloisons séparatrices blanches montant à mi-hauteur délimitent des niches formant cellules de traitement et zones de travail. Une grande transparence est apportée par les vitrages fermant le haut des cloisons et ceux des portes. Cette conception spatiale a fait ses preuves dans les cabinets dentaires les plus différents. Des détails individuels ont été recherchés dans l'entrée.

La zone d'attente est placée derrière un paravent mobile. Vestiaire et comptoir de réception sont des éléments pivotants fixés sur les appuis où ils marquent la situation de portail au droit le l'entrée.

1 Der Eingangsbereich mit den schwenkbaren Elementen für Garderobe und Empfang. Materialien: Boden graublauer Giessharzbelag, Wände blau lasierter Zementabrieb, Ausbauten weissglänzend, Metallteile Eisenglimmerlack.

2 Der Paravent auf arretierbaren Schwenkrollen nimmt mit den Eisenrahmen die Sprache der Fenster- und Türlaibungen auf. Fest montierte elastische Stoffgurten wirken als Sichtschutz oder als transparenter Schild.

1 La zone d'entrée avec éléments de vestiaire et de réception pivotants. Matériaux: sol en résine synthétique gris-bleu, parois en enduit de ciment [illegible] en bleu, éléments incorporés en blanc brillant, parties en acier laquées [illegible] métal.

2 Avec son cadre métallique le paravent sur roulettes [illegible] exprime l'idée des [illegible] de fenêtre et de porte. Des sangles en tissu [illegible] tendues [illegible] écran optique ou panneau transparent [illegible]

1994 Sm2

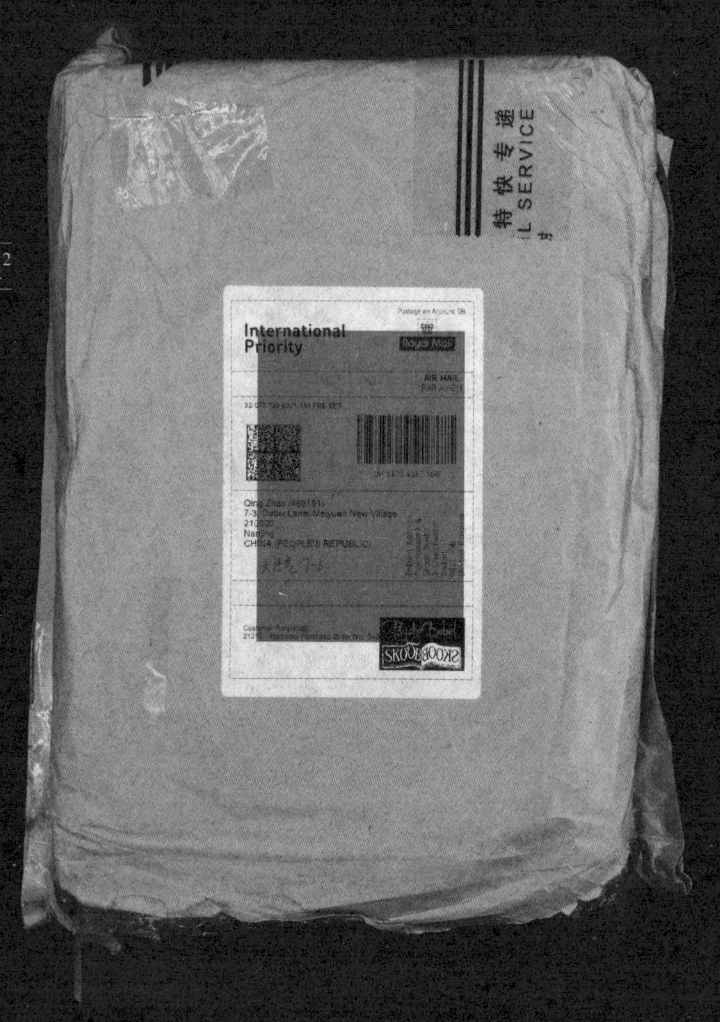
特快专递
L SERVICE
International
Priority
Royal Mail
AIR MAIL
PAR AVION
CHINA (PEOPLE'S REPUBLIC)
SKOOB BOOKS

ternational
riority

Royal Mail

AIR MA
PAR AVI

77 739 6001-134 FBE EE7

UH 1373 4387 1GB

g Zhao (469151)
, Dabei Lane, Meiyuan New Village
)000
njing
INA (PEOPLE'S REPUBLIC)

大悲巷 7-3

Return Address
Psychobabel &
Skoob Books
1 Churchward
Didcot
OX11 7UP

tomer Reference:
15 / Abebooks Purchase Order No.: 549

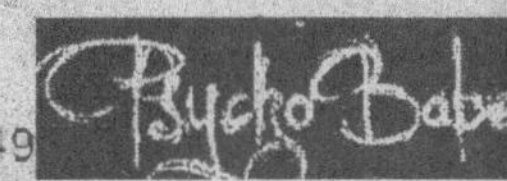

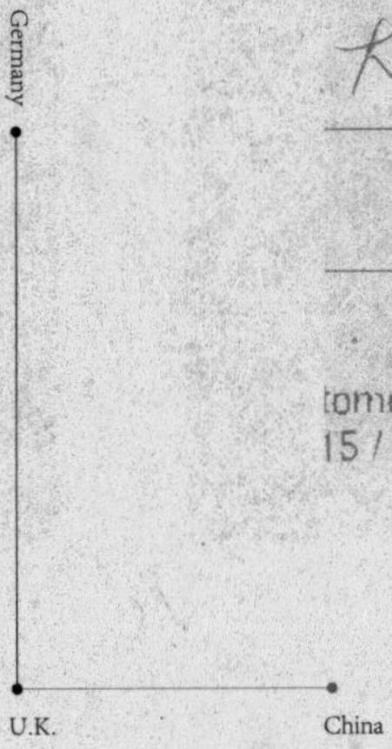

Germany

U.K.

China

310mm

38mm

Franz Kafka Das Schloß

Büchergilde Gutenberg

Franz Kafka Das Schloß

205mm

银奖
城堡——小说的手稿版
德国

Silver Medal
Das Schloß: Roman in der Fassung der Handschrift
Germany
Eckhard Jung, Ulysses Voelker
Franz Kafka
Büchergilde Gutenberg, Frankfurt am Main
310 205 38mm
1540g
432p
ISBN-10 3-7632-3973-1

著名小说家弗朗茨·卡夫卡（Franz Kafka）逝世前曾嘱咐挚友、作家马克斯·布罗德（Max Brod）销毁他的所有作品。但在卡夫卡去世后布罗德将作品全部公布出来，包括《城堡》这样的未完成作品。《城堡》存世很多版本，本书是直接从手稿中整理出的未经任何改动的版本。书籍的装订形式为圆脊锁线硬精装，护封采用半透明牛油纸印单黑，内封黑色布面烫白和压凹处理，裱覆卡板。黑色艺术卡纸作为环衬，内页胶版纸四色印刷。由于卡夫卡未想将本作出版，所以这个版本有很多非书面用语，也有很多标点符号没有用在语义停顿之处；文本上有大量超长段落，且对话不做分行，人物关系易混淆，在阅读体验上有较强的压抑感。本书后记对卡夫卡是否故意这样写作进行了思考。这种压抑感在设计上也有所体现，章节区隔只有一张插图，而无明显标题；版心不固定，从前至后逐渐变小，使得读者的视角越来越狭小，从而呈现出主人公的"可能性"越来越受限。

The famous novelist Franz Kafka had asked his close friend writer Max Brod to destroy all his works before his death. But after Kafka's death, Brod released all his works, including unfinished work like *The Castle*. There are many versions of *The Castle* in the world, and this book is the unedited version compiled directly from the manuscript. The hardcover book is bound by thread sewing with rounding spine. The jacket is made of semi-translucent tracing paper printed in black. The inner cover is made of black cloth with white hotstamping, mounted with cardboard. The black art cardboard is used as the end paper, and the offset paper is used for inner pages printed in four colors. Since Kafka had no intension to publish this work, this version had used spoken language a lot, and some punctuation marks were omitted at the semantic pauses. There are a lot of super long paragraphs, and the dialogue does not branch, so the relationship between the characters is easy to be confused. There is a strong sense of depression in the reading experience. The postscript of this book ponders whether Kafka deliberately wrote in this way. This kind of depression is also reflected in design. There is only one illustration in the chapter section, with no obvious title. The print area is not fixed. It gradually becomes smaller from the front to the back, which makes the reader's perspective narrower, and presents the 'probability' of the protagonist is more limited.

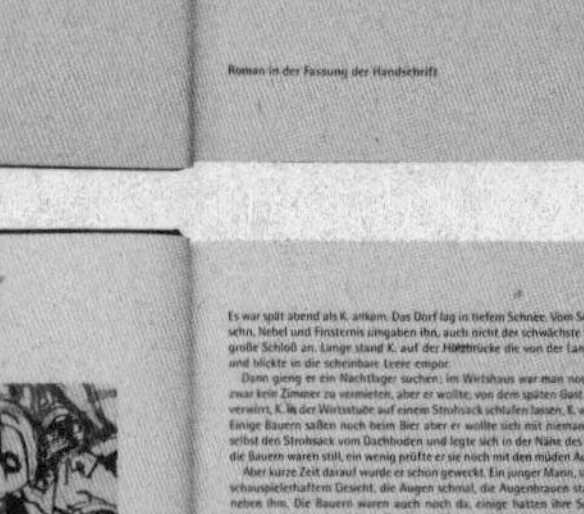

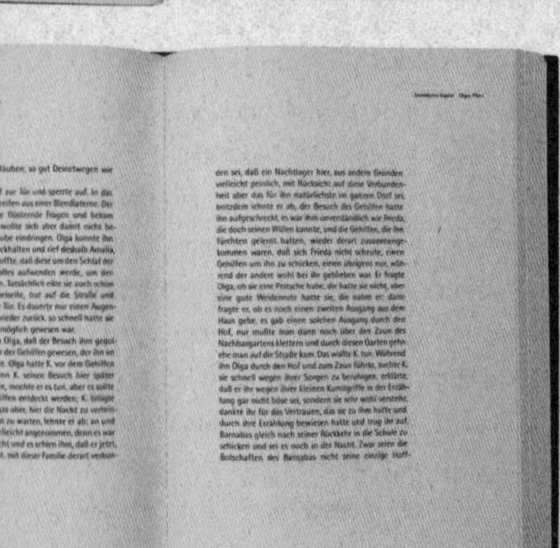

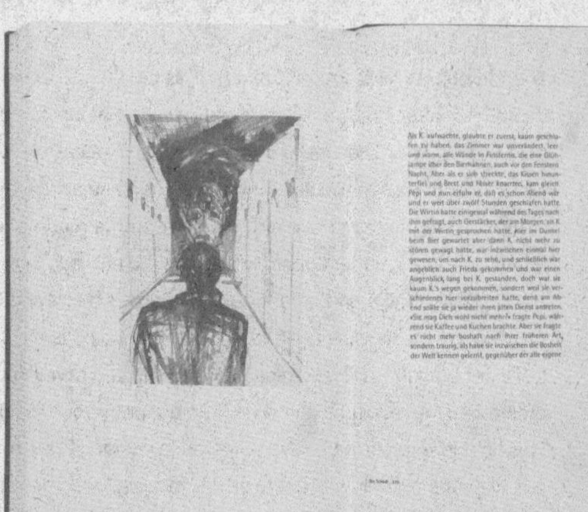

Das alles hatte nur einen Augenblick gedauert, schon hatte K. rechts und links einen der Männer und wurde, als gäbe es kein anderes Verständigungsmittel schweigend aber mit aller Kraft zur Tür gezogen. Der Alte freute sich über irgendetwas dabei und klatschte in die Hände. Auch die Wäscherin lachte bei den plötzlich wie toll lärmenden Kindern.

K. aber stand bald auf der Gasse, die Männer beaufsichtigten ihn von der Schwelle aus, es fiel wieder Schnee, trotzdem schien es ein wenig heller zu sein. Der Vollbärtige rief ungeduldig: »Wohin wollt Ihr gehn? Hier führt es zum Schloß, hier zum Dorf.« Ihm antwortete K. nicht, aber zu dem andern, der ihm trotz seiner Überlegenheit der umgänglichere schien, sagte er: »Wer seid Ihr? Wem habe ich für den Aufenthalt zu danken?« »Ich bin der Gerbermeister Lasemann«, war die Antwort, »zu danken habt Ihr aber niemandem.« »Gut«, sagte K., »vielleicht werden wir noch zusammenkommen.« »Ich glaube nicht«, sagte der Mann. In diesem Augenblick rief der Vollbärtige mit erhobener Hand: »Guten Tag Artur, guten Tag Jeremias!« K. wandte sich um, es zeigten sich in diesem Dorf also doch noch Menschen auf der Gasse! Aus der Richtung vom Schlosse her kamen zwei junge Männer von mittlerer Größe, beide sehr schlank, in engen Kleidern, auch im Gesicht einander sehr ähnlich, die Gesichtsfarbe war ein dunkles Braun, von dem ein Spitzbart in seiner besondern Schwärze dennoch abstach. Sie gingen bei diesen Straßenverhältnissen erstaunlich schnell, warfen im Takt die schlanken Beine. »Was habt Ihr?« rief der Vollbärtige. Man konnte sich nur rufend mit ihnen verständigen, so schnell gingen sie und hielten nicht ein. »Geschäfte«, riefen sie lachend zurück. »Wo?« »Im Wirtshaus.« »Dorthin gehe auch ich«, schrie K. auf einmal mehr als alle andern, er hatte großes Verlangen von den zwei mitgenommen zu werden; ihre Bekanntschaft schien ihm zwar nicht sehr ergiebig aber gute aufmunternde Wegbegleiter waren sie offenbar. Sie aber hörten K.'s Worte, nickten jedoch nur und waren schon vorüber.

K. stand noch immer im Schnee, hatte wenig Lust den Fuß aus dem Schnee zu heben, um ihn ein Stückchen weiter wieder in die Tiefe zu senken; der Gerbermeister und sein Genosse, zufrieden damit K. endgiltig hinausgeschafft zu haben, schoben sich langsam, immer nach K. zurückblickend, durch die nur wenig geöffnete Tür ins Haus und K. war mit dem ihn einhüllenden Schnee allein. »Gelegenheit zu einer kleinen Verzweiflung«, fiel ihm ein, »wenn ich nur zufällig, nicht absichtlich hier stünde.«

Da öffnete sich in der Hütte linker Hand ein winziges Fenster, geschlossen hatte es tiefblau ausgesehn, vielleicht im Widerschein des Schnees, und war so winzig daß als es jetzt geöffnet war nicht das ganze Gesicht des Hinausschauenden zu sehen war, sondern nur die Augen, alte braune Augen. »Dort steht er«, hörte K. eine zittrige Frauenstimme sagen. »Es ist der Landvermesser«, sagte eine Männerstimme. Dann trat der Mann zum Fenster und fragte nicht unfreundlich, aber doch so als sei ihm daran gelegen, daß auf der Straße vor seinem Haus alles in Ordnung sei: »Auf wen wartet Ihr?« »Auf einen Schlitten, der mich mitnimmt«, sagte K. »Hier kommt kein Schlitten«, sagte der Mann, »hier ist kein Verkehr.« »Es ist doch die Straße, die zum Schloß führt«, wendete K. ein. »Trotzdem, trotzdem«, sagte der Mann mit einer gewissen Unerbittlichkeit, »hier ist kein Verkehr.« Dann schwiegen beide. Aber der Mann überlegte offenbar etwas, denn das Fenster, aus dem Rauch strömte, hielt er noch immer offen. »Ein

wunderte sich sehr. »Ihr seid doch der Landvermesser«, sagte der Mann erklärend, »und g
zum Schloß. Wohin wollt Ihr denn fahren?« »Ins Schloß«, sagte K. schnell. »Dann fahr
nicht«, sagte der Mann sofort. »Ich gehöre doch zum Schloß«, sagte K., des Mannes e
Worte wiederholend. »Mag sein«, sagte der Mann abweisend. »Dann fahrt mich also
Wirtshaus«, sagte K. »Gut«, sagte der Mann, »ich komme gleich mit dem Schlitten.« Das G
machte nicht den Eindruck besonderer Freundlichkeit, sondern eher den einer Art
eigensüchtigen ängstlichen fast pedantischen Bestrebens, K. von dem Platz vor dem H
wegzuschaffen.

Das Hoftor öffnete sich und ein kleiner Schlitten für leichte Lasten, ganz flach
irgendwelchen Sitz, von einem schwachen Pferdchen gezogen kam hervor, dahinte
Mann, nicht alt aber schwach, gebückt, hinkend, mit magerem rotem verschnupftem Ge
das besonders klein erschien durch einen fest um den Hals gewickelten Wollshawl. Der M
war sichtlich krank und nur um K. wegbefördern zu können, war er doch hervorgekomm
erwähnte etwas derartiges, aber der Mann winkte ab. Nur daß er der Fuhrmann Gerst
war, erfuhr K., und daß er diesen unbequemen Schlitten genommen habe, weil er g
bereit stand und das Hervorziehn eines andern zu viel Zeit gebraucht hätte. »Setzt E
sagte er und zeigte mit der Peitsche hinten auf den Schlitten. »Ich werde mich neben
setzen«, sagte K. »Ich werde gehn«, sagte Gerstäcker. »Warum denn?« fragte K. »Ich v
gehn«, wiederholte Gerstäcker und bekam einen Hustenanfall, der ihn so schüttelte, daß
Beine in den Schnee stemmen und mit den Händen den Schlittenrand halten mußte. K.
nichts weiter, setzte sich hinten auf den Schlitten, der Husten beruhigte sich langsam u
fuhren.

Das Schloß dort oben, merkwürdig dunkel schon, das K. heute noch zu erreichen ge
hatte, entfernte sich wieder. Als sollte ihm aber noch zum vorläufigen Abschied ein Ze
gegeben werden, erklang dort ein Glockenton, fröhlich beschwingt, eine Glocke, die
stens einen Augenblick lang das Herz erbeben ließ, so als drohe ihm – denn auch schme
war der Klang – die Erfüllung dessen, wonach er sich unsicher sehnte. Aber bald verstu
diese große Glocke und wurde von einem schwachen eintönigen Glöckchen abgelös
leicht noch oben, vielleicht aber schon im Dorfe. Dieses Geklingel paßte freilich besser
langsamen Fahrt und dem jämmerlichen aber unerbittlichen Fuhrmann.

»Du«, rief K. plötzlich – sie waren schon in der Nähe der Kirche, der Weg ins Wirtshaus
mehr weit, K. durfte schon etwas wagen – »ich wundere mich sehr, daß Du auf Deine e
Verantwortung mich herumzufahren wagst. Darfst Du denn das?« Gerstäcker kümmert
nicht darum und schritt ruhig weiter neben dem Pferdchen. »He«, rief K., ballte etwas S
vom Schlitten zusammen und traf Gerstäcker damit voll ins Ohr. Nun blieb dieser doch
und drehte sich um; als ihn K. aber nun so nahe bei sich sah – der Schlitten hatte sich no
wenig weiter geschoben – diese gebückte, gewissermaßen mißhandelte Gestalt, da
müde schmale Gesicht mit irgendwie verschiedenen Wangen, die eine flach, die a
eingefallen, den offenen aufhorchenden Mund, in dem nur paar vereinzelte Zähne
mußte er das was er früher aus Bosheit gesagt hatte, jetzt aus Mitleid wiederhol

Und er riß sich los und ging ins Haus zurück, diesmal nicht an der Mauer entlang, sondern mitten durch den Schnee, traf im Flur den Wirt, der ihn stumm grüßte und auf die Tür des Ausschanks zeigte, folgte dem Wink, weil ihn fror und weil er Menschen sehen wollte, war aber sehr enttäuscht, als er dort an einem Tischchen, das wohl eigens hingestellt worden war, denn sonst begnügte man sich dort mit Fässern, den jungen Herrn sitzen und vor ihm – ein für K. bedrückender Anblick – die Wirtin aus dem Brückengasthaus stehen sah. Pepi, stolz, mit zurückgeworfenem Kopf, ewig gleichem Lächeln, ihrer Würde unwiderlegbar sich bewußt, schwenkend den Zopf bei jeder Wendung, eilte hin und wieder, brachte Bier und dann Tinte und Feder, denn der Herr hatte Papiere vor sich ausgebreitet, verglich Daten, die er einmal in diesem, dann wieder einmal in einem Papiere am andern Ende des Tisches fand, und wollte nun schreiben. Die Wirtin von ihrer Höhe überblickte still mit ein wenig aufgestülpten Lippen wie ausruhend den Herrn und die Papiere, so als habe sie schon alles Nötige gesagt und es sei gut aufgenommen worden. »Der Herr Landvermesser, endlich«, sagte der Herr bei K.'s Eintritt mit kurzem Aufschauen, dann vertiefte er sich wieder in seine Papiere. Auch die Wirtin streifte K. nur mit einem gleichgültigen, gar nicht überraschten Blick. Pepi aber schien K. überhaupt erst zu bemerken, als er zum Ausschankpult trat und einen Kognak bestellte.

K. lehnte dort, drückte die Hand an die Augen und kümmerte sich um nichts. Dann nippte er von dem Kognak und schob ihn zurück, weil er ungenießbar sei. »Alle Herren trinken ihn«, sagte Pepi kurz, goß den Rest aus, wusch das Gläschen und stellte es ins Regal. »Die Herren haben auch besseren«, sagte K. »Möglich«, sagte Pepi, »ich aber nicht«, damit hatte sie K. erledigt und war wieder dem Herrn zu Diensten, der aber nichts benötigte und hinter dem sie nur im Bogen immerfort auf und ab gieng mit respektvollen Versuchen über seine Schultern hinweg einen Blick auf die Papiere zu werfen; es war aber nur wesenlose Neugier und Großtuerei, welche auch die Wirtin mit zusammengezogenen Augenbrauen mißbilligte.

Plötzlich aber horchte die Wirtin auf und starrte, ganz dem Horchen hingegeben, ins Leere. K. drehte sich um, er hörte gar nichts besonderes, auch die andern schienen nichts zu hören, aber die Wirtin lief auf den Fußspitzen mit großen Schritten zu der Tür im Hintergrund, die in den Hof führte, blickte durchs Schlüsselloch, wandte sich dann zu den andern mit aufgeris-

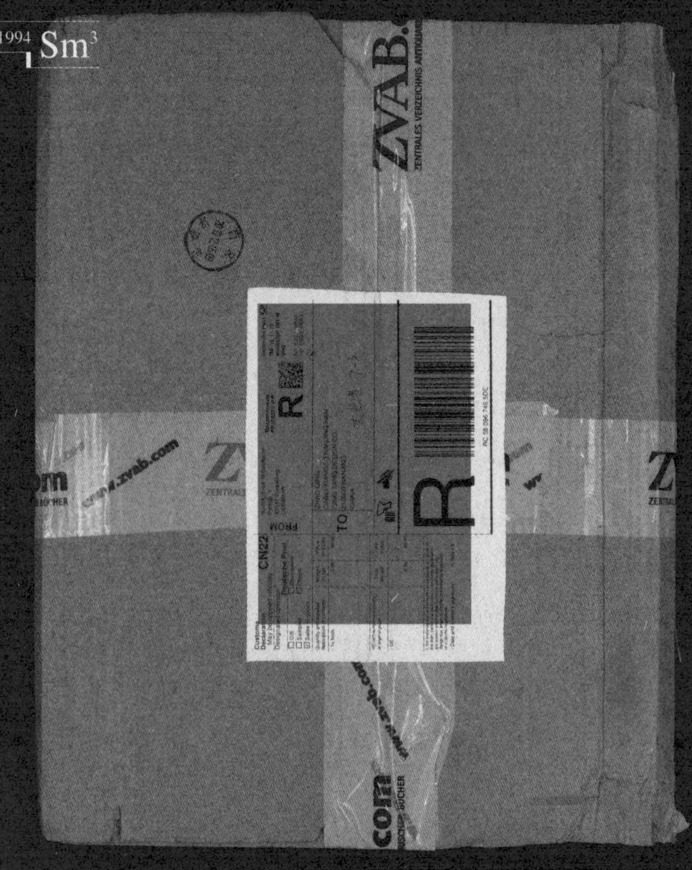
ZVAB
ZENTRALES VERZEICHNIS ANTIQUARISCHER
www.zvab.com
Customs Declaration
CN22
FROM
TO
R
R

Deutsche Post
IM 16.11.19
WARENP. INT M
UNT
A0 0221 9960
00 0000 0B63
Recommandé
PRIORITY P.P.
R
FROM
BuchX, Erich Butzenlechn
Parkstr. 1
82547 Eurasburg
GERMANY
TO
ZHAO QING
DABEI XIANG7-3,NANJING HAN
QING TANG DESIGN CO.
210000 NANJING
CHINA
大悲巷 7-3
CN22
Deutsche Post
Documents
Others
Weight (in kg)
Value (in EUR)
0.80
90.00
Total Weight
Total Value
0.80
90.00
16-Nov-19
Germany
Germany
China

262mm

16mm

169mm

● 银奖
《 鼻子
◖ 德国

Silver Medal
Die Nase
Germany
Gert Wunderlich
Nikolai Gogol
Edition Curt Visel, Memmingen
262 169 16mm
480g
64p
ISBN-10 3-922406-51-3

本书是俄国戏剧家尼古拉·果戈理（Nikolai Gogol）的短篇小说《鼻子》的德文版。小说讲述了一位圣彼得堡官员一觉醒来发现鼻子已经独立自主的故事。故事将真实的叙事手法融入虚构的设定当中，充满了对现实的讽刺，成为魔幻现实主义的前兆。本书的装订形式为锁线硬精装，封面采用具有明显纹理、手感粗糙的墨绿色艺术纸，作者名的首字母“NG”以及“鼻子”的首字母“N”做压凹处理。简写符号“.”与书脊上的全称一并烫红。书的函套采用坚硬的黑色艺术卡纸制作，无印刷与工艺。环衬采用与封面相同品种的蓝灰色艺术纸。内页选用手感厚实的半羊皮纸单黑印刷。由于是短篇小说，除了选用较厚的纸张外，正文的字号也比较大，配合穿插于书中的 12 幅版画，呈现出一种手工印制的古典书籍的味道。书中插图风格夸张、荒诞，极具黑色幽默的讽刺意味。

This book is a German version of the short story *Nose* by Russian dramatist Nikolai Gogol. The novel tells the story of a Saint Petersburg official who waked up to find that his nose has become independent. The story integrates the real narrative techniques into the fictional settings, full of satire on reality, and becomes the precursor of magical realism. The hardcover book is bound by thread sewing. The cover is made of dark green art paper with obvious texture and rough feeling. The initial letters of the author's name 'NG' and the initial 'N' of 'nose' are embossed. The abbreviation mark '.' and the full name on the spine is red stamped. The bookcase is made of hard black art cardboard without printing and processing. The end paper uses the same blue and gray art paper as the cover. The inner pages are printed in single black on thick semi parchment. Since it is a short story, the book uses thick paper; furthermore, the font size of the text is also relatively large. With the 12 prints interspersed in the book, the book presents the taste of a hand-printed classic book. The style of illustrations in the book is exaggerated, absurd and full of black humor.

…otel, silbergestickter Kragen, ein De-
er zitterte am ganzen Leibe. Schließ-
zu seinen Unterhosen und seinen Stie-
as ganze Zeug an und ging, begleitet
ewiß nicht leichten Verwünschungen
Ossipownas, mit der in einen Lappen
lten Nase auf die Straße.
e sie gern irgendwo verstecken; ent-
ter einem Prellstein bei einem Tor
endwie unversehens fallen lassen und
eine Seitengasse einbiegen. Doch wie
n ihm da ein Bekannter entgegen, der
fing, ihn auszufragen: »Wohin gehst
oder: »Wen gehst du denn so früh ra-
daß Iwan Jakowljewitsch immer den
ugenblick verpassen mußte. Bei der
elegenheit hatte er sie schon fallen las-
der Nachtwächter winkte ihm schon
mit seiner Hellebarde und rief ihn an:
! Du hast da was fallen lassen!« Und

denen diese Bäder sich ergießen
Ästhetische ist so völlig geleistet,
igt und allen Glanz dem
n letzter Triftigkeit, scheint
Als Existenz, gezeichnet von
nd doch auch kostbar durch-
hungen solidarischer Barm-
lann das nannte – »gogolisch

Peter Gosse

II Der Kollegienassessor Kowal
lich früh und machte mit de
was er immer machte, wenn
er selbst nicht erklären ko
Grunde er das tat. Kowaljow
tete sich auf und ließ sich d
Tische stehenden Spiegel bri
kleines Pickelchen untersuch
abend auf der Nase aufgega
seinem größten Erstaunen sa
der Nase einen vollkommen
Erschrocken ließ sich Kowal
und rieb sich mit einem H
wirklich, die Nase war weg!
Hand zu tasten, um festzu
schlafe; aber er schlief offenb
gienassessor Kowaljow spran
schüttelte sich: die Nase war
gleich die Kleider bringen, z
geradewegs zum Oberpolize

gte sich zwischen
ber deren einge-
zwei Löcher für
so oft lustig ge-
ne ein. Im Innern
ter; die meisten
Tür. Kowaljow
r nicht imstande
n Augen in allen

I K O L A
G O G O L
I E *N A S*

Iwan Jakowljewitsch zog anstandshalber über das Hemd den Frack an, setzte sich an den Tisch, streute Salz auf, machte zwei Zwiebelköpfe zurecht, nahm das Messer zur Hand und ging mit bedeutender Miene daran, das Brot durchzuschneiden. Nachdem er das Brot in zwei Hälften geteilt hatte, betrachtete er den Anschnitt und sah zu seinem Erstaunen irgend etwas Weißes hervorgucken. Vorsichtig bohrte Iwan Jakowljewitsch mit dem Messer und tastete mit dem Finger: »Etwas Festes«, sagte er vor sich hin, »was kann denn das sein?«

Er steckte die Finger hinein und zog – eine Nase heraus…! Iwan Jakowljewitsch ließ die Hände sinken; er begann sich die Augen zu reiben und das Ding zu betasten. Es war eine Nase, wirklich eine Nase! Und dabei schien sie ihm irgendwie bekannt zu sein. Furcht malte sich in Iwan Jakowljewitschs Gesicht. Doch diese Furcht war gar nichts gegen die Empörung, die seine Gemahlin erfaßte.

»Wo hast denn du, Viechskerl, die Nase abgeschnitten?« rief sie voll Zorn aus. »Lump, Säufer! Ich selbst werde dich bei der Polizei anzeigen, so ein Räuber! Ich hab' auch schon von drei Menschen gehört, daß du die Leute beim Rasieren so an der Nase ziehst, daß sie es kaum aushalten.«

Doch Iwan Jakowljewitsch war mehr tot als lebendig. Er erkannte, daß diese Nase niemand anderem gehörte als dem Kollegienassessor Kowaljow, den er jeden Mittwoch und Sonntag rasierte.

»Halt, Prasskowja Ossipowna! Ich will sie in einen Lappen einwickeln und in die Ecke legen; da kann sie ein Weilchen liegenbleiben, dann will ich sie wegtragen.«

»Ich will nichts davon hören! Soll ich hier in meinem Zimmer eine abgeschnittene Nase herumliegen lassen? Du Strohkopf! Kannst auch nur mit dem Rasiermesser über den Riemen fahren und wirst bald nicht mehr imstand sein, deine Pflicht zu erfüllen, Schwindler, Taugenichts! Soll etwa ich mich für dich vor der Polizei verantworten…? Ach, du Schmierfink, du Holzklotz! Hinaus damit, hinaus! Bring sie, wohin du willst! Daß ich nichts mehr davon höre und sehe!«

Iwan Jakowljewitsch stand ganz erschlagen da. Er dachte und dachte – und wußte gar nicht mehr, was er denken sollte.

»Weiß der Teufel, wie das geschehen ist«, sagte er endlich, sich mit der Hand hinterm Ohr kratzend. »Bin ich abends besoffen nach Haus gekom-

Kowaljow war ganz durcheinandergeraten, er wußte nicht, was er tun und was er denken sollte. Da machte sich das angenehme Rauschen eines Damenkleides bemerkbar. Eine etwas ältere Dame, ganz in Spitzen gehüllt, schritt heran und mit ihr eine junge, in einem weißen Kleid, in dem sich ihre schlanke Gestalt reizend abzeichnete. Hinter ihnen nahm ein hochgewachsener Herr mit großen Koteletten und einem ganzen Dutzend Kragen Aufstellung und öffnete seine Tabatiere.

Kowaljow trat näher heran, zog den Batistkragen seines Hemdes zurecht, ordnete seine an der goldenen Uhrkette hängenden Breloques und richtete, nach allen Seiten hin lächelnd, seine Aufmerksamkeit auf die zarte Dame, die wie eine Frühlingsblume sich leicht neigte und ihr weißes Händchen mit den durchscheinenden Fingern an die Stirn führte. Das Lächeln auf Kowaljows Gesicht verbreiterte sich, als er unter ihrem Hut das hübsche, runde, schneeweiße Kinn und einen Teil der rosigen, wie die ersten Frühlingsrosen blühenden Wangen hervorlugen sah. Doch plötzlich sprang er zurück, wie wenn er sich verbrannt hätte. Er besann sich, daß er an Stelle der Nase nichts im Gesicht habe, und Tränen drangen aus seinen Augen. Er wandte sich, um dem Herrn in Uniform auf den Kopf zuzusagen, er habe sich nur als Staatsrat verkleidet, er sei ein Schwindler und Schuft und weiter nichts als seine eigene Nase. Doch die Nase war schon weg. Es war ihr gelungen zu entspringen, wahrscheinlich, um wieder jemand einen Besuch zu machen.

Das versetzte Kowaljow in Verzweiflung. Er ging wieder zurück und blieb einen Augenblick unter den Kolonnaden stehen, schaute sich sorgfältig nach allen Seiten um, ob die Nase nicht irgendwo zu finden sei. Er erinnerte sich sehr gut daran, daß sie einen Hut mit Federn aufhatte und eine goldgestickte Uniform trug; doch ihren Mantel hatte er nicht bemerkt, sich auch weder die

1994 Bm1

Rund ums ‹Blaue Haus› –
von Klosterbrüdern, Kaufleuten,
Büchern und Buchhändlern

Destinataire:
QING ZHAO
7-3, Dabei Lane, Meiyuan New Village
NANJING, Jiangsu, 210000
Chine
中国南京市玄武区梅园新村大悲巷7-3
Tel : +86 13952094842

DÉCLARATION EN DOUANE
Peut être ouvert d'office
CN 22
LA POSTE
France
Important! Voir instructions au verso

Cadeau
Documents
Echantillon commercial
Autre
Cocher la ou les cases appropriées

Quantité et description détaillée du contenu (1)	Poids (en kg) (2)	Valeur (3)
Book		10€
Pour les envois commerciaux seulement N° tarifaire du SH (4) et pays d'origine des marchandises (si connus) (5)	Poids total (en kg) (6)	Valeur totale (7)

Je, soussigné dont le nom et l'adresse figurent sur l'envoi, certifie que les renseignements donnés dans la présente déclaration sont exacts et que cet envoi ne contient aucun objet dangereux ou interdit par la législation ou la réglementation postale ou douanière.

Date et signature de l'expéditeur (8)
25/02/19

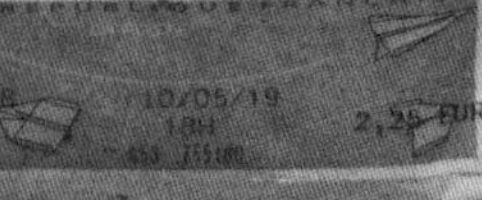

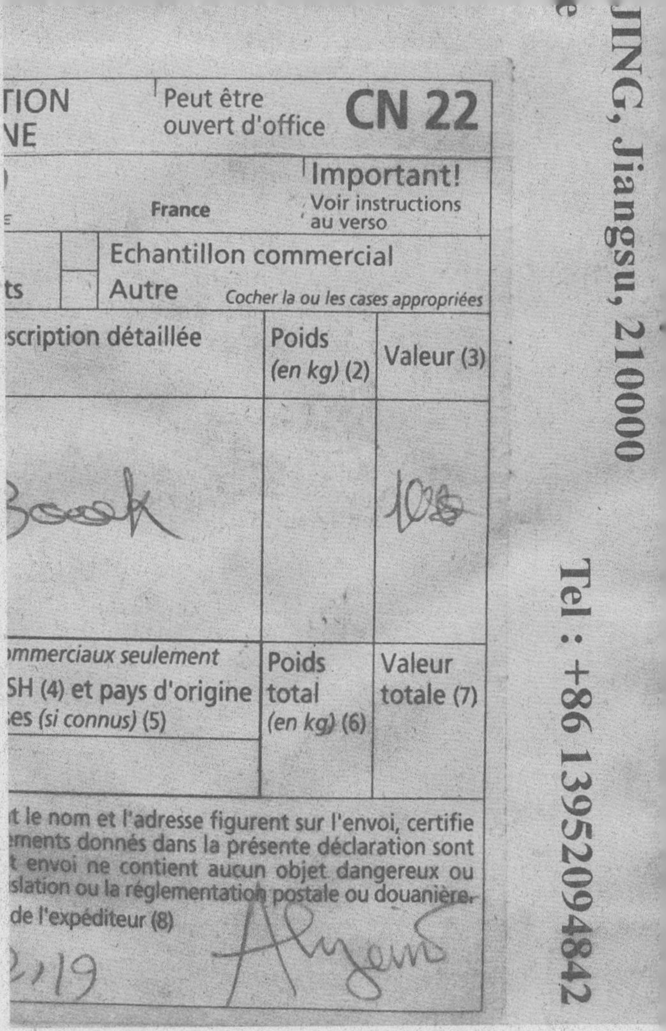

TION
NE

Peut être
ouvert d'office

CN 22

)

France

Important!
Voir instructions
au verso

ts

Echantillon commercial
Autre

Cocher la ou les cases appropriées

scription détaillée	Poids *(en kg)* (2)	Valeur (3)
Book		100
mmerciaux seulement SH (4) et pays d'origine es *(si connus)* (5)	Poids total *(en kg)* (6)	Valeur totale (7)

t le nom et l'adresse figurent sur l'envoi, certifie
ements donnés dans la présente déclaration sont
t envoi ne contient aucun objet dangereux ou
slation ou la réglementation postale ou douanière.
de l'expéditeur (8)

2,19 Alyens

e

JING, Jiangsu, 2100000

Tel : +86 13952094842

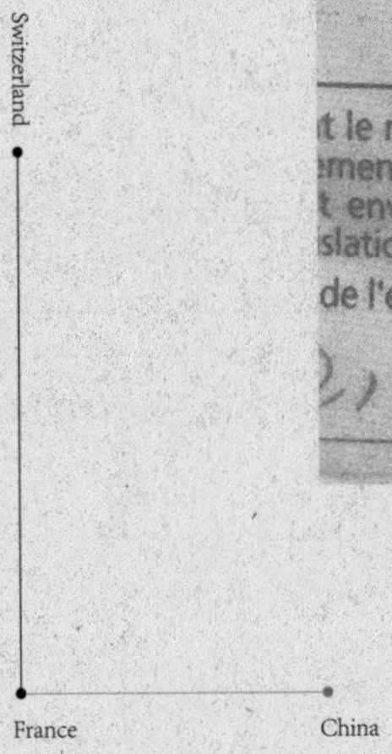

Switzerland

France

China

266mm

13mm

Rund ums ‹Blaue Haus› –
von Klosterbrüdern, Kaufleuten,
Büchern und Buchhändlern

Ernst Ziegler

Peter Ochsenbein

Hermann Bauer

Ophir-Verlag

Das Blaue Haus *Ophir-Verlag*

165mm

铜奖

“蓝屋”的周围

——来自修士、商人、书籍和书店

瑞士

Bronze Medal

Rund urns <Blaue Haus> :

Von Klosterbrüdern, Kaufleuten,

Büchern und Buchhändlern

Switzerland

Antje Krausch, Ruedi Tachezy

Ernst Ziegler, Peter Ochsenbein, Hermann Bauer

Ophir-Verlag, St. Gallen

266 165 13mm

454g

128p

ISBN-10 3-907787-02-4

1993年，Leo书店（Leobuchhandlung）成立75周年之际出版了本书，记录了书店的建筑——“蓝屋”及所处的圣加仑（St. Gallen）地区的历史。本书的装订形式为锁线精装。封面采用带有纹理的艺术纸蓝黑双色印刷，裱覆卡板，版画的蓝色部分让“蓝屋”凸显出来，封面和封底用蓝色标明的出版社为Leo书店的子公司。翻过灰蓝色的艺术纸环衬，内页采用涂布纸张四色印刷。全书总计三个大章节，由三位作者写成，分别介绍了“蓝屋”所在的圣加仑地区的历史、“蓝屋”的原建筑从修道院宿舍变为图书馆的历史，以及“蓝屋”与书商结缘逐步成为书店的历史。内页有大量历史地图、版画、建筑照片等资料，也有各个时期人物的树状谱系。“蓝屋”是典型的德国传统木框架建筑，与大多数漆成黑色的框架不同，“蓝屋”的框架被漆成了蓝灰色。这一颜色贯穿全书，在版画、时间线、标题索引线等方面有所呈现。

In 1993, Leo Bookstore (Leo Buchhandlung)published this book on the occasion of its 75th anniversary, documenting the history of the Blue House, the building of the bookstore, and the history of the St. Gallen area where the bookstore is located. The hardcover book is bound by thread sewing. The cover is printed in blue and black on textured art paper, mounted with cardboard. The blue part of the print highlights the “Blue House”, and the publisher indicated in blue on the front cover and back cover is a subsidiary of Leo Bookstore. The endpaper is made of gray-blue art paper while the inner pages are made of coated paper printed in four colors. The book consists of three major chapters, written by three authors, which respectively introduce the history of the Saint Gallen area where the Blue House is located, the history of the original building of the Blue House transforming from a monastery dormitory to a library, and the history of the Blue House gradually becoming a bookstore due to association with booksellers. The inner pages contain a large amount of historical maps, prints, and architectural photos, as well as family trees of historical characters from different periods. “Blue House” is a typical traditional German wooden frame building. Unlike most frames painted black, the frames of the “Blue House” were painted blue-gray. This color is present throughout the book, in prints, timelines, and title index lines.

B·FEHR'SCHE· BUCHHANDLUNG

n No. 1192 ST. GALLEN Schmidgasse No. 16

-, Kunst- und Landkartenhandlung

ielle Niederlage und vollständiges r der eidgenössischen Kartenwerke

Wohlassortiertes Lager der
arsten Werke aus allen Gebieten der Literatur
dekers Reisebücher, Kursbücher, Reisekarten
stik und Reiseliteratur, frz. Romane, Tauchnitz-Edition
er der farbigen **Schweizer-Ansichten**
Photoglob Co. Zürich in diversen Formaten

dschriften, Bilderbücher, Klassiker, Prachtwerke

lätter für Zimmerschmuck in Kupferstich und Photogravure

Galli Anno ab incarnatione dm
In dem namen Jhesu Cristi Amen. Ich
bruder N. N. lay von Sant Gallen zu
vnd zu er dem almächtigen got vnd de
hailgen iunckfrowen marie vnd vnse
hailgen vatter Sant Benedicten vnd alle
hailgen mit disem brieff verhaiss ich stä
kait vnd bekerüg vnd besrung miner sit
vnd üch. Vnd ouch gehorsam näch de
hailgen Regel des egenanté vnsers hai
vatters Sant Benedicten vor got vnd syn
hailgen Vnd sunderlich vor Sant Pete
vnd sant pauls der zwelfbotten vnd
Sant Gallen vnd Sant Othmar der h
gen bichtigern Vnd dar zü dem hailg
er hailtům in dyser kyrchen rastend.
gegenwirtikait des erwirdigen vnd g
stlichen vatters hérr N. abbt dyß Cl
sters zu Sant Gallen. Vnd vor üch ver

99, 103, 107–111, 116, 122–123, 125, 134-
209, 215, 222, 231, 239, 258, 267–268,
26–328, 341–342, 348–349, 354,
400, 406–409, 414–415, 419–421,
-485, 493–495, 503, 510–512, 515-
568, 571–572, 578–580, 587–589,
5, 659–661.

r sicher 6 Männer bezeugt: Henggele
idern, deren Eintritts- bzw. Todesda
5 in Frage: Nr. 47, 53, 82, 83, 92, 94,
1519 Sebastianus Rembs bzw. Rennh
ldast am 27.8. (Henggeler Nr. 62) ur

Ziegler, oben S. 29). Bei den Geschl
e es sich um Schimpf- oder Übernan
r verächtlich Noll- oder Nollibrüder

jesu crist amen ich bruder
dem allmechtigen got und de
nsten heligen vater sant
f ver hais ich stetikait v
rsam nach der heiligen reg
nedikten vor got und sinen h
und sant pauls der zwelf
mar den heligen bichtern v
diser kirchen rastend in gegen
n vaters her dietheln abt dis
vater und bruder hie gegen
zu und der heiligen

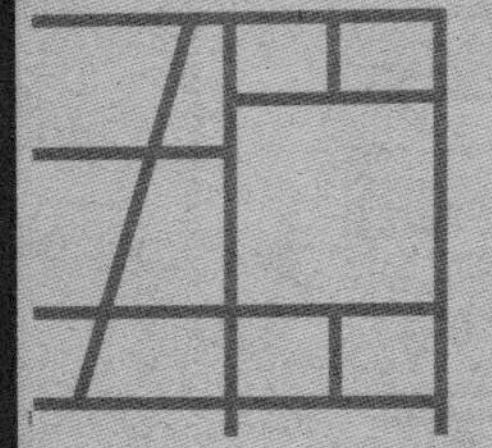

Das ‹Blaue Haus› und seine Umgebung, eine Quartier- und Hausgeschichte

Ernst Ziegler

Lange nach der Klostergründung (719) im Hochtal der Steinach, nachdem sich die Zahl der Mönche und Schüler, der Beamten und Knechte des Gallus-Klosters vermehrt hatte und immer mehr kirchliche und weltliche Gebäude um die Kirche entstanden waren, sah sich der Abt genötigt, von seinen Besitzungen geschickte Arbeiter heranzuziehen und sie beim Kloster anzusiedeln, damit sie hier unter Leitung seiner Beamten für die Bedürfnisse des Klosters ihr Handwerk betrieben. Er schenkte ihnen Hofstätten, um Häuser darauf zu bauen und sie gegen einen jährlichen Zins zu bewohnen. Der Grund und Boden ringsum gehörte dem Kloster. Es entstanden zuerst die Quartiere der Weber, der Schmiede und der Bäcker, die Weber-, Schmied- und Multergasse, in nächster Nähe des Klosterbezirks. So erklärt Hermann Wartmann, der erste Bearbeiter des Urkundenbuches der Abtei St. Gallen, in seinem Neujahrsblatt des Historischen Vereins über ‹Das alte St. Gallen› (1867) das Wachsen des Klosters St. Gallen zur Stadt.[1]

Nachdem vom 10. bis 14. Jahrhundert die Stadt vom Kloster aus gewachsen und immer selbständiger geworden war, erlangte sie 1457 die völlige Unabhängigkeit von der Abtei. Dieser politischen Loslösung von der Abtei folgte 1524/27 auch die konfessionelle Trennung, indem sich die Stadt der Reformation anschloß. Das äußere Zeichen der vollständigen Scheidung in zwei selbständige Staatswesen wurde vierzig Jahre später errichtet; es war die 1566/67 erbaute Trennmauer zwischen Stift und Stadt.[2]

Alte und neue Namen

Die Gegend westlich des Klosters, wo sich schon sehr früh Leute ansiedelten, hieß in früheren Zeiten ‹im Loch oben› (heute Gallusplatz). Die Gallusstraße, welche sich über den Gallusplatz bis zum Oberen Graben hinzieht, bildet die Verlängerung der Zeughausgasse. Früher hieß der untere Teil der Gallusstraße, von der St. Laurenzenkirche bis gegen das ‹Blaue Haus› (Gallusstraße 20), Schmalzmarkt, weil seit alten Zeiten an dieser Stelle der Käse-, Schmalz- und Butterverkauf vor sich ging. Schon im 15. Jahrhun-

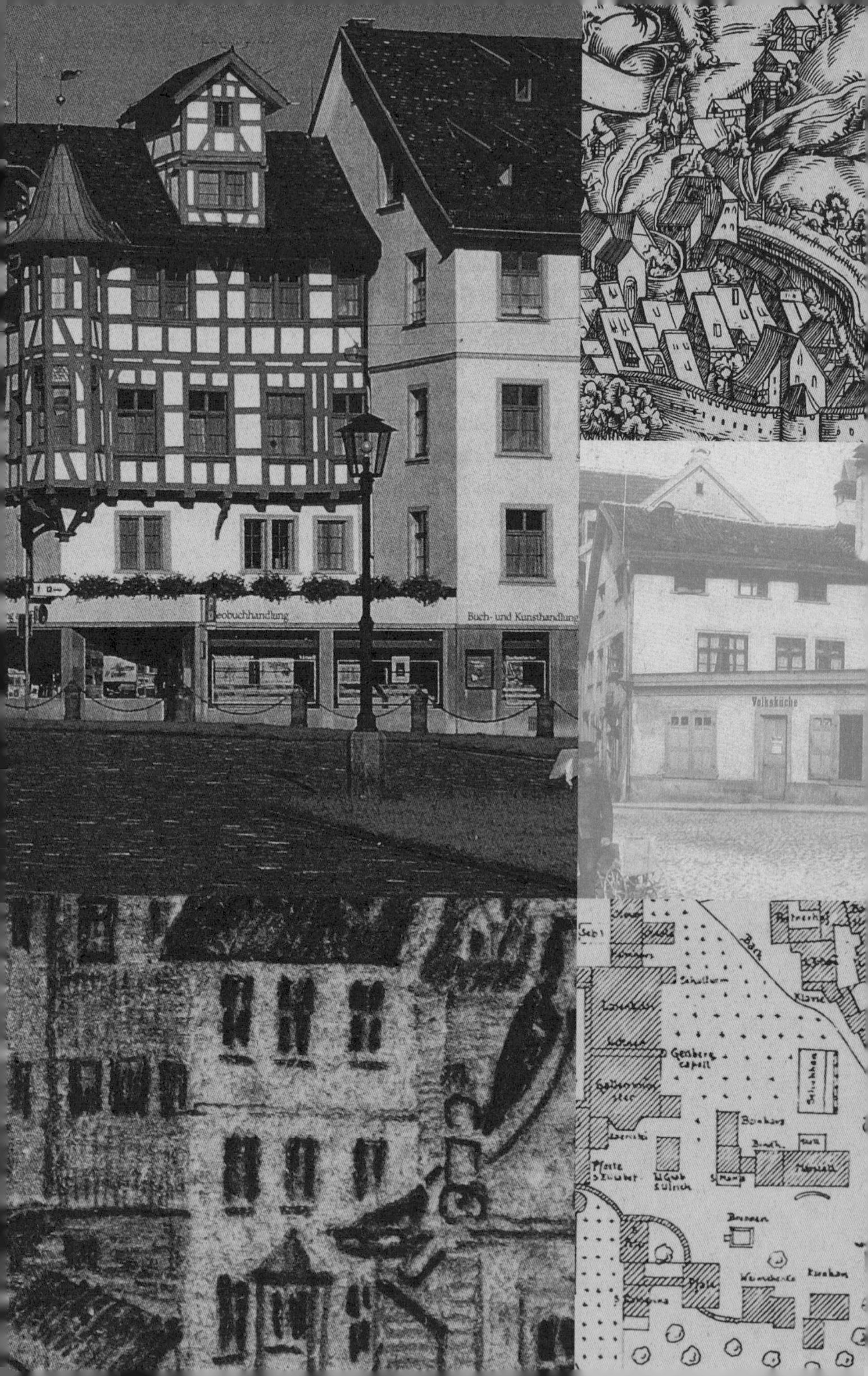

Buch- und Kunsthandlung
Volksküche

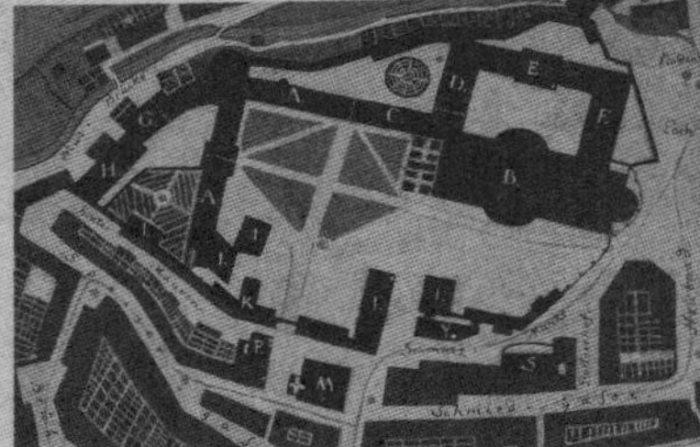

Ausschnitt aus dem Grundriss der Stadt St.Gallen von Johannes Zuber, 1828. Schmalz-Markt heute Gallusstraße, Portnerhof heute Rosengasse und Neubad heute Bankgasse. Original im Stadtarchiv (Vadiana) St.Gallen (StadtASG).

Waren feil. In der zweiten Hälfte des 19. Jahrhunderts wurde dann die ‹unpassende Benennung› Loch durch Gallusplatz ersetzt, und die Namen St.Laurenzenplatz sowie Gallusstraße kamen in Gebrauch.[3] Seit alters her lagen auf städtischem Gebiet zahlreiche Besitzungen des Klosters. So muß nach Vadian bereits Abt Otmar (719–759) jenseits des Grabens, der die Klostersiedelung umgab, eine Unterkunft für Aussätzige eingerichtet haben. Dieses Gebäude fiel nach 1566 an die Stadt, die darin eine öffentliche Badeanstalt einrichtete. Nach diesem Bad erhielt das Quartier dann den Namen ‹Neubad›. (Davon wird noch die Rede sein.)[4]

Neben dem Abt besaßen der Dekan des Klosters, der Propst und der Pförtner außerhalb des Klosters eigene Wohnhäuser. Aus dem Jahre 1255 gibt es Verzeichnisse der Einkünfte des Spitals der Brüder zu St.Gallen, der Kirche St.Fiden sowie der Einkünfte des Dekans, des Propstes, des Küsters usw.[5] Der Pförtner des Klosters oder Portner hatte sein Haus vermutlich an der heutigen Rosengasse, ungefähr dort, wo das Haus Zur Stärke steht. ‹Aber gegen dem Portner hoff hinab, da were einist ein Portner gesessen, und derselb hoff in dem Gotzhus unnd der Fryheit gewesen und hiesse noch hüttbytag der Portner hoff.›[6] Der Hof mit dem Haus des Portners gab während Jahrhunderten einem Quartier den Namen,

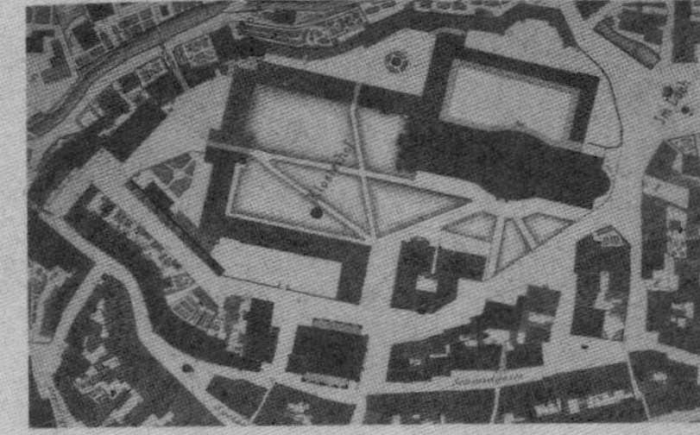

Ausschnitt aus dem Übersichtsplan der Stadt St.Gallen von Johann Fierz, 1861–1863. Nummer 224 das ‹Blaue Haus› und die ehemalige Volksküche. Original im StadtASG.

1847 wurde für diese Bank ein Neubau an der Schmiedgasse zwischen Bank- und Rosengasse erstellt, die spätere Nationalbank und heutige Sparkassa der Administration. Dies war dann wohl der Grund, warum der Gemeinderat 1865 beschloß: ‹Die Straße von dem Bankgebäude südwestlich gegen die Berneck gelegen erhält, statt wie früher «Neubad», von nun an die Bezeichnung «Bankgasse».›

1865 bestimmte der Gemeinderat, ein Teil der heutigen Gallusstraße und ‹das dem Bankgebäude gegenüber gelegene Haus Zur Stärke, sowie die Häuser in der Richtung vom Bankgebäude gegen das westliche Ende der Klosterkirche, wie Rose und Blaues Haus, bilden den «Bankplatz»›. Der Name ‹Rosengasse› ist also nicht alt; in den Häuserverzeichnissen kommt er erst 1891 vor. – Benannt wurde diese Gasse nach dem Haus Zur Rose, das neben dem ‹Blauen Haus› steht.[7]

CN22
FROM
TO
R
R

Customs Declaration CN22
May be opened officially
Designated operator **Deutsche Post**

☐ Gift ☐ Documents
☐ Samples ☐ Others
☑ Sales of goods

Quantity and detailed description of contents	Weight (in kg)	Value (in EUR)
- 1x Book	1.20	25.00
HS tariff number and country of origin of goods	Total Weight	Total Value
- 4901, DE	1.20	25.00

I, the undersigned, whose name and address are given on the item, certify that particulars given in this declaration are correct and that this item does not contain any dangerous article or articles prohibited by legislation or customs regulations.

FROM
Antiquariat Leissle Inh. Pe
Georg-Hager-Str. 24
81369 München
GERMANY

Recommandé
PRIORITY P.P.
R

TO
Zhao Qing
Nanjing Han Qing - Tang Design
Dabei xiang 7-3
210000 NANJING
CHINA

大悲巷7-

Germany

Germany China

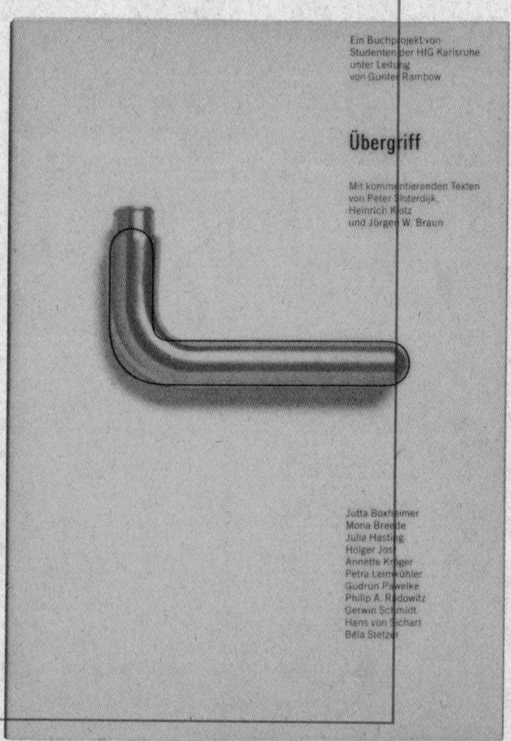

297mm
18mm
Ein Buchprojekt von
Studenten der HfG Karlsruhe
unter Leitung
von Gunter Rambow
Übergriff
Mit kommentierenden Texten
von Peter Sloterdijk,
Heinrich Klotz
und Jürgen W. Braun
Jutta Boxheimer
Mona Breede
Julia Hasting
Holger Jost
Annette Kröger
Petra Leimkühler
Gudrun Pawelke
Philip A. Radowitz
Gerwin Schmidt
Hans von Sichart
Béla Stetzer
Übergriff
213mm

铜奖
侵占
德国

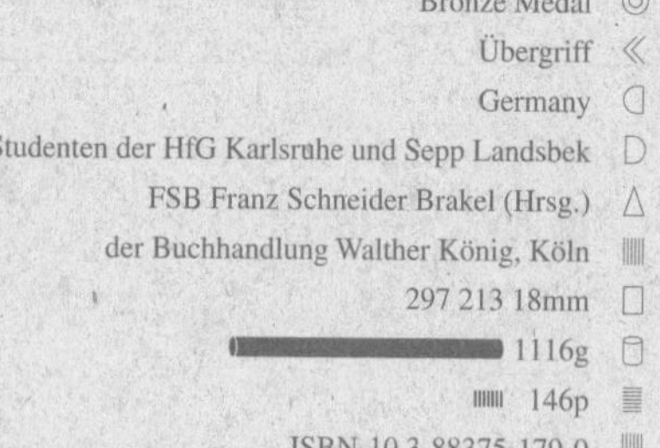
Bronze Medal
Übergriff
Germany
Studenten der HfG Karlsruhe und Sepp Landsbek
FSB Franz Schneider Brakel (Hrsg.)
der Buchhandlung Walther König, Köln
297 213 18mm
1116g
146p
ISBN-10 3-88375-179-0

这是德国设计大师冈特·兰堡（Gunter Rambow）在卡尔斯鲁厄艺术设计大学任教时一门课程的教学成果合集。书籍的装订形式为无线胶平装。护封硫酸纸印单黑，护封背面覆亮膜用以增加纸张强度（耐用度）、减少卷曲，内封白卡纸印单黑。内页硫酸纸与哑面铜版纸混合装订，单黑印刷。课程的主要目标是激起学生对于生活中最常见的物品的思考，使他们可以重新建立对这些东西的直观感知。书中的例子是门的把手，使用把手的方式无非就是手指和手掌的抓取、紧握、用力、施放这样的连贯动作。但把手的材质、形状、尺寸各不相同，使用者在使用中的故事又有更多差异。在书中的众多案例中，学生们对分配到的不同把手进行思考，建立各自不同的使用故事。把手被印在铜版纸上，而硫酸纸上印着各不相同的图形表达。

This is a collection of Gunter Rambow's achievements when the German design master taught at Karlsruhe University of Art and Design. The book is perfect bound. The jacket is made of butter paper printed in black, and the backside of the jacket is covered by bright film in order to increase paper strength (durability) and reduce curl. The inner cover is made of white cardboard printed in black. The butter paper and the matte coated paper are mixed for inside pages printed in single black. The main goal of the course is to stimulate students to think about the most common things in life, so that they can re-establish their intuitive perception. The book takes the handle of the door as an example. The way to use the handle is nothing more than the continuous action of grasping, gripping, exerting, and casting with the fingers and palm, but the material, shape, and size of the handles are different, and the user's story is more different. In many cases in the book, students think about the different handles assigned to them, and establish their own different stories. The handles are printed on coated paper, while the butter paper is printed with different graphic expressions.

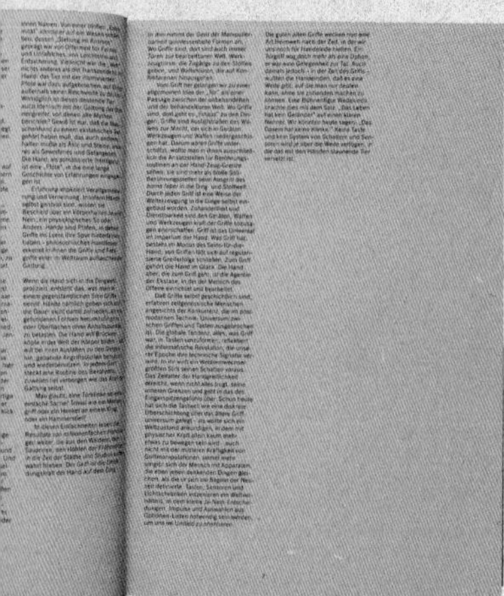

Der Mensch selbst »handelt« durch die Hand; d
ist in einem mit dem Wort die Wesensauszeichn
schen. Nur das Seiende, das wie der Mensch das
(λόγος) »hat«, kann auch und muß »die Har
Durch die Hand geschieht zumal das Gebet ur
der Gruß und der Dank, der Schwur und der Wi
das »Werk« der Hand, das »Handwerk« und d
Handschlag gründet den bündigen Bund. Die H
das »Werk« der Verwüstung. Die Hand west n
wo Entbergung und Verbergung ist. Kein Tier h
und niemals entsteht aus einer Pfote oder eine
einer Kralle eine Hand. Auch die verzweifelte
mals und nie am wenigsten eine »Kralle«, mit
Mensch »verkrallt«. Nur aus dem Wort und
ist die Hand entsprungen. Der Mensch »hat«
sondern die Hand hat das Wesen des Menschen
Wort als der Wesensbereich der Hand der Wes
Menschen ist. Das Wort als das eingezeichnete
Blick sich zeigende ist das geschriebene Wort, d.
Das Wort als die Schrift aber ist die Handschrift

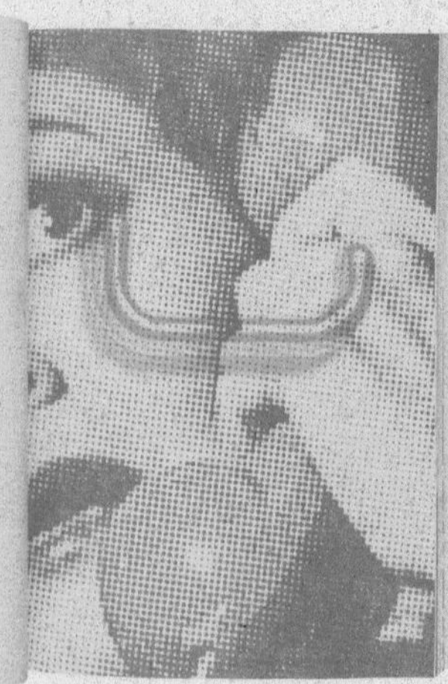

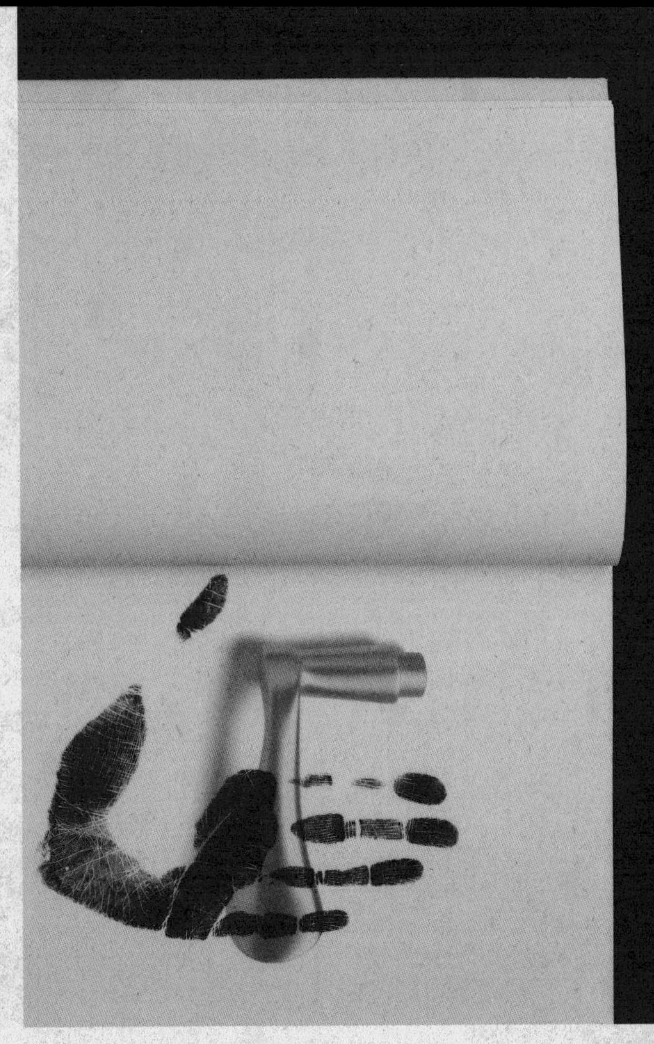

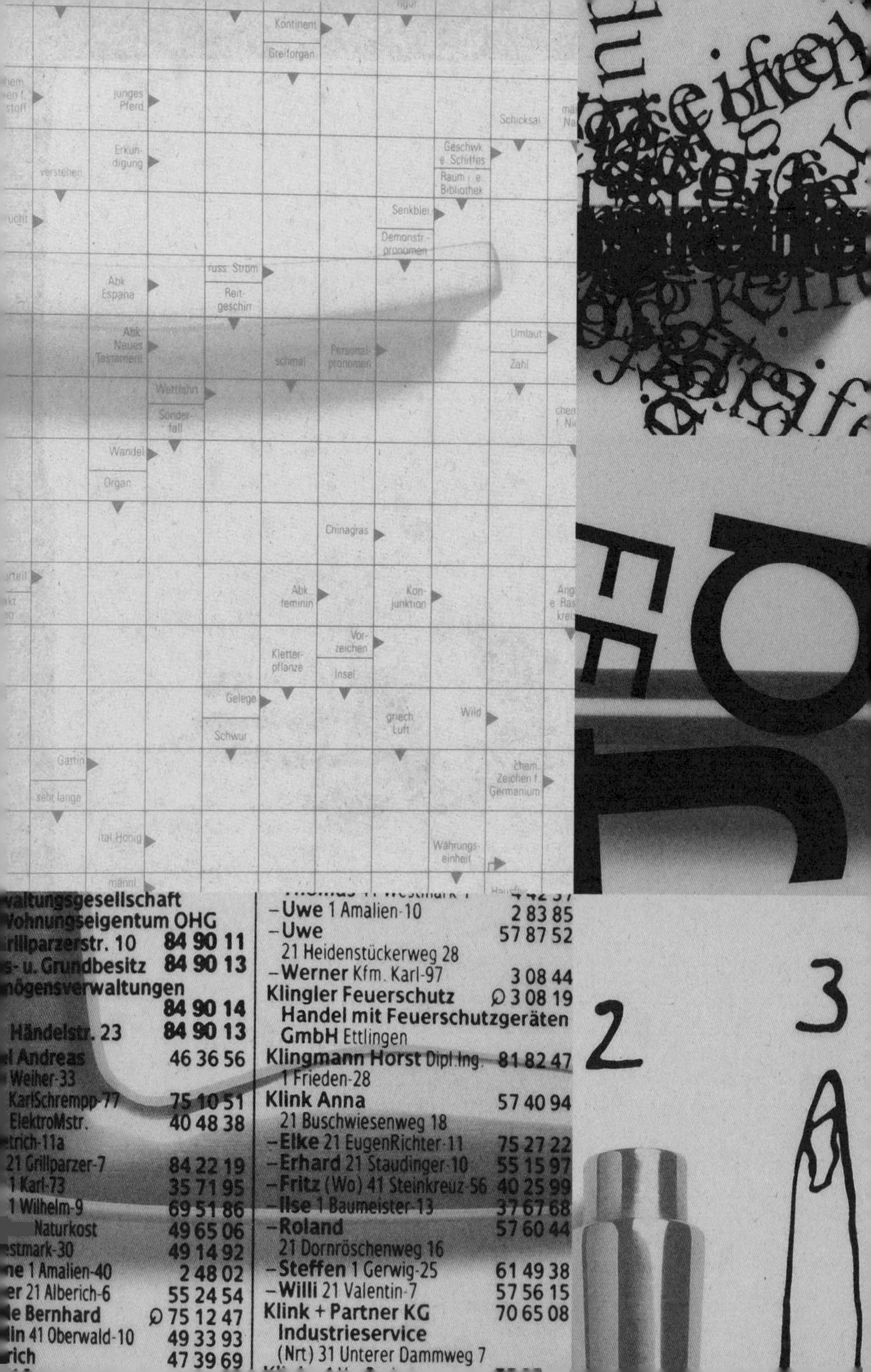

Kontinent
Greiforgan
junges Pferd
Schicksal
Erkundigung
verstehen
Geschwk. e. Schiffes
Raum f. e. Bibliothek
Senkblei
Demonstr.-pronomen
russ. Strom
Abk. Espana
Reitgeschirr
Abk. Neues Testament
schmal
Personalpronomen
Umlaut
Zahl
Wettlahn
Sonderfall
Wandel
Organ
Chinagras
Abk. feminin
Konjunktion
Vorzeichen
Kletterpflanze
Insel
Gelege
Schwur
griech. Luft
Wild
Gattin
sehr lange
chem. Zeichen f. Germanium
ital. Honig
Währungseinheit
valtungsgesellschaft
Wohnungseigentum OHG
rillparzerstr. 10 84 90 11
s- u. Grundbesitz 84 90 13
nögensverwaltungen
84 90 14
Händelstr. 23 84 90 13
l Andreas 46 36 56
Weiher-33
KarlSchrempp-77 75 10 51
ElektroMstr. 40 48 38
etrich-11a
21 Grillparzer-7 84 22 19
1 Karl-73 35 71 95
1 Wilhelm-9 69 51 86
Naturkost 49 65 06
estmark-30 49 14 92
ne 1 Amalien-40 2 48 02
er 21 Alberich-6 55 24 54
le Bernhard 75 12 47
lin 41 Oberwald-10 49 33 93
rich 47 39 69
–Uwe 1 Amalien-10 2 83 85
–Uwe 57 87 52
21 Heidenstückerweg 28
–Werner Kfm. Karl-97 3 08 44
Klingler Feuerschutz 3 08 19
Handel mit Feuerschutzgeräten
GmbH Ettlingen
Klingmann Horst Dipl.Ing. 81 82 47
1 Frieden-28
Klink Anna 57 40 94
21 Buschwiesenweg 18
–Elke 21 EugenRichter-11 75 27 22
–Erhard 21 Staudinger-10 55 15 97
–Fritz (Wo) 41 Steinkreuz-56 40 25 99
–Ilse 1 Baumeister-13 37 67 68
–Roland 57 60 44
21 Dornröschenweg 16
–Steffen 1 Gerwig-25 61 49 38
–Willi 21 Valentin-7 57 56 15
Klink + Partner KG 70 65 08
Industrieservice
(Nrt) 31 Unterer Dammweg 7
2
3

1994 Bm³

Design aus Bremen

der Linie 3
der Linie 3
raße 1
285

An:
ZHAO QING
Dabei xiang7-3,Nanjing Han Qing
Tang Design Co.,
210000 Nanjing
China

ZVAB.com

https://www.abebooks.com/servlet/ShipmentManifest?abepoid

260mm

7mm

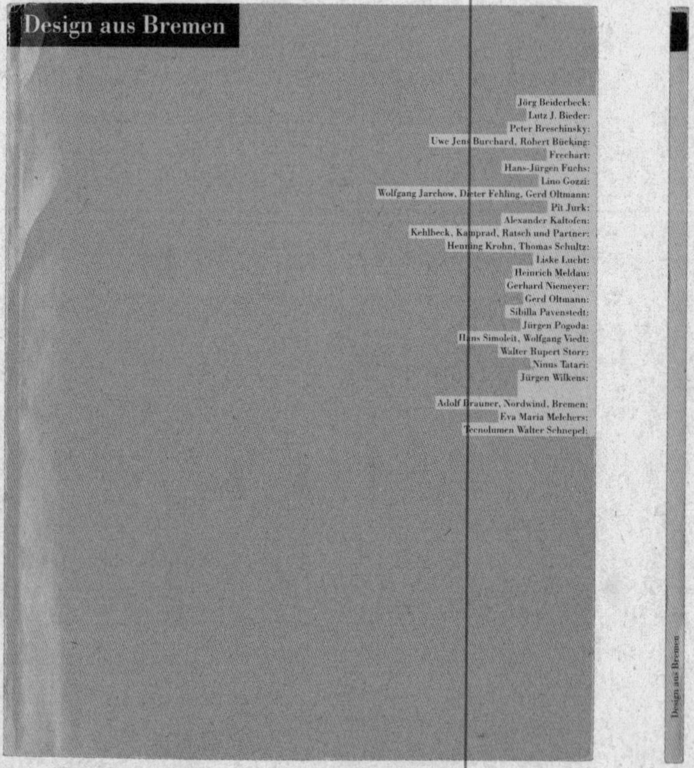

Design aus Bremen

Jörg Beiderbeck:
Lutz J. Bieder:
Peter Breschinsky:
Uwe Jens Burchard, Robert Bücking:
Frechart:
Hans-Jürgen Fuchs:
Lino Gozzi:
Wolfgang Jarchow, Dieter Fehling, Gerd Oltmann:
Pit Jurk:
Alexander Kaltofen:
Kehlbeck, Kamprad, Ratsch und Partner:
Henning Krohn, Thomas Schultz:
Liske Lucht:
Heinrich Meldau:
Gerhard Niemeyer:
Gerd Oltmann:
Sibilla Pavenstedt:
Jürgen Pogoda:
Hans Simoleit, Wolfgang Viedt:
Walter Rupert Storr:
Ninus Tatari:
Jürgen Wilkens:

Adolf Brauner, Nordwind, Bremen:
Eva Maria Melchers:
Tecnolumen Walter Schnepel:

Design aus Bremen

212mm

铜奖
来自不来梅的设计
德国

Bronze Medal
Design aus Bremen
Germany
Hartmut Brückner
Beate Manske, Jochen Rahe
Design Zentrum, Bremen
260 212 7mm
351g
76p
ISBN-10 3-929430-01-0

不来梅作为德国重要的制造业中心，在汽车、造船、家具、服装等方面一直具有很高水准。本书是不来梅设计中心和不来梅国家艺术与文化历史博物馆举办的不来梅设计展的作品集。书籍的装订形式为无线包背胶平装。封面采用白卡纸黄黑双色印刷，内页采用超薄的胶版纸四色印刷。内容方面，书籍以介绍不来梅制造业的历史开篇，作品部分一共展示了 22 位不来梅设计师在 1990 年前后设计的 22 件产品，包含游艇、餐具、家具、服装、灯具等多个领域。随后紧跟 3 位来自不来梅的制造业雇主的文章。设计方面，作品展示页面没有像大多数作品集把左右放在一起整体设计，而是利用了装订的包背页结构。包背页的外部前后页分别是作品大图和相应介绍，而包背页内部印有设计师和企业雇主的全身照片，利用纸张的超薄属性隐隐透出。书籍的封面也延续了这个做法，将设计师和雇主名放在封面靠翻口的位置，而在前勒口对应各自的设计作品。

Bremen, as an important manufacturing center in Germany, has always had a very high standard in automobiles, shipbuilding, furniture, clothing and so on. This book is a collection of works from the Bremen Design Exhibition organized by the Bremen Design Center and the Bremen National Museum of Art and Cultural History. The double-leaved book is perfect bound. The cover is printed in yellow and black on white cardboard, and the inner pages are printed in four colors on ultra-thin offset paper. As for the content, the book begins with an introduction to the history of the Bremen manufacturing industry. The work section displays a total of 22 products designed by 22 Bremen designers around 1990, including yachts, tableware, furniture, clothing, lamps and other fields, followed by statements from three manufacturing employers from Bremen. As for the design, display pages are not designed as a double page spread. According to the double-leaved binding structure, large pictures of the work and the corresponding introduction are printed on the outer surface, while the photos of the designer and the employer are printed inside the 'leaf', which can be seen through the ultra-thin paper. On the cover, the names of designers and employers are placed near the fore-edge, while their design works are printed on the flap.

Design
aus Bremen

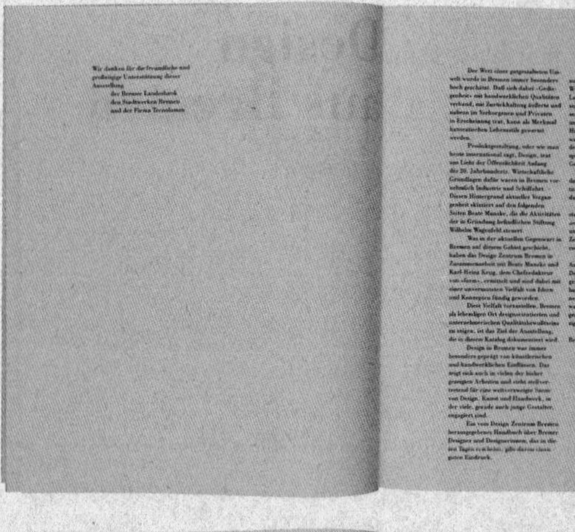

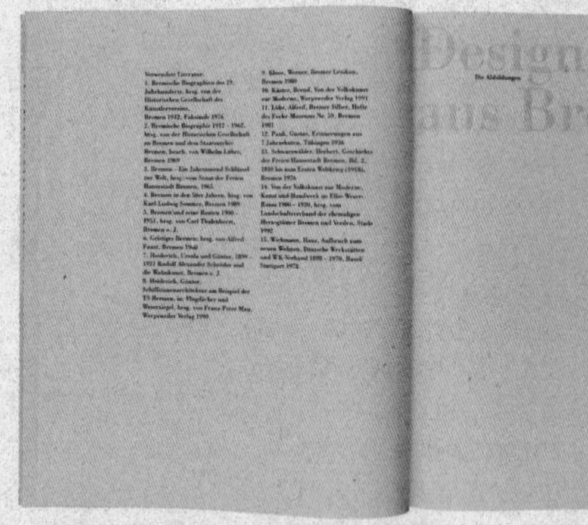
Design
aus Bre

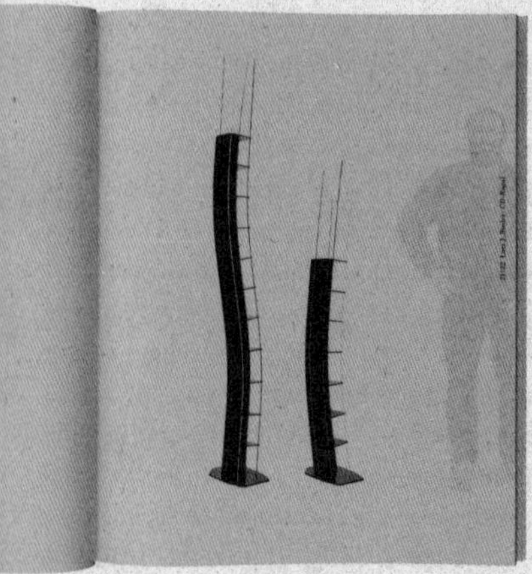

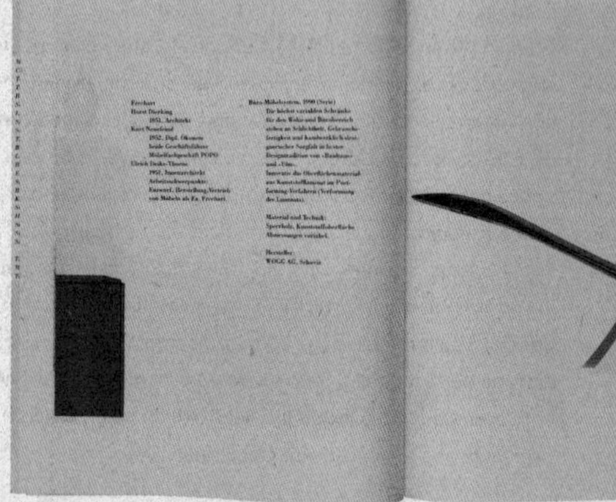

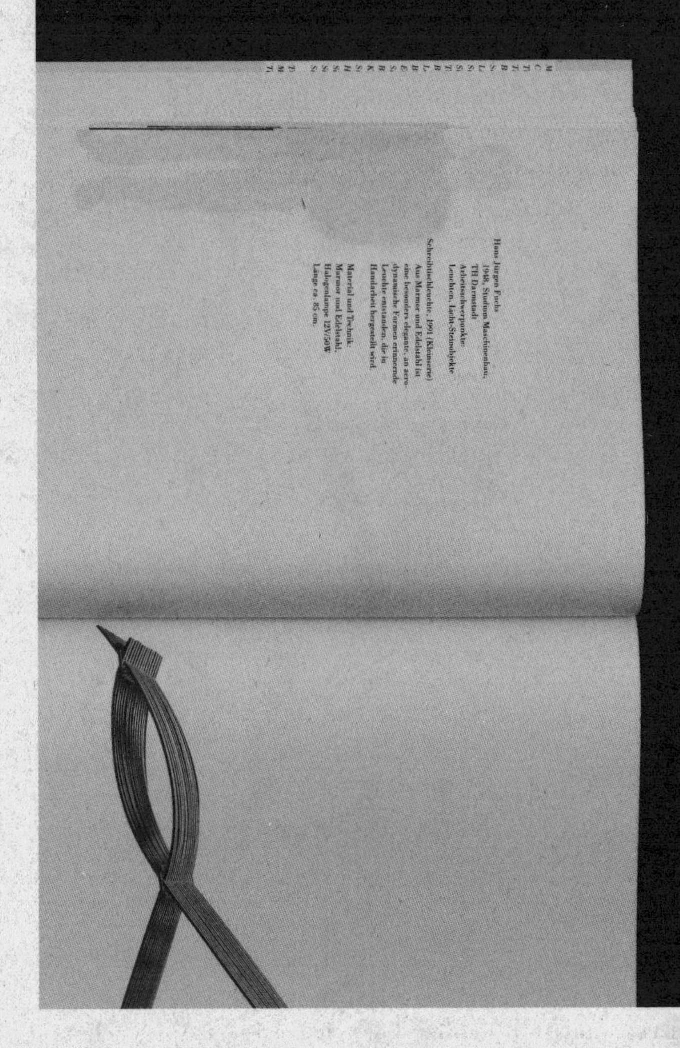
Hans Jürgen Fuchs
1948, Studium Maschinenbau,
TH Darmstadt
Arbeitsschwerpunkte:
Leuchten, Licht-Steinobjekte
Schreibtischleuchte, 1991 (Kleinserie)
Aus Marmor und Edelstahl ist
eine besonders elegante, an aero-
dynamische Formen erinnernde
Leuchte entstanden, die in
Handarbeit hergestellt wird.
Material und Technik:
Marmor und Edelstahl,
Halogenlampe 12V/50W
Länge ca. 85 cm.

Auf den folgenden Seiten stellen wir drei Bremer UnternehmerInnen vor, deren Engagement für gutes Design richtungsweisend ist.

Johann Georg Poppe, Speisesaal I. Klasse auf der »Kaiser Wilhelm II«, 1903, Ausführung Heinrich Bremer u. a.

noc
son
ren
die
verä
teil
ähn
bau
sich
verg

erregionale

eren Stellen-
beit bei der
dem zweitälte-
ig. Der Gold-
der sich 1810
Goldschmied
Kompagnon
teinschneider
ca. 1829 eine
nging, haben
nd mit hohem
lienunterneh-

äußerst selten umgesetzt, und selbst d
Leitern der etwa seit den 70er Jahren
eingerichteten Entwurfsabteilungen
kann man vor 1900 kaum eigenständi
Entwürfe nachweisen: Das lag auch
nicht im Interesse der Firmenleitunge

Nicht uninteressant waren die
Versuche von staatlicher Seite, die Au
und Weiterbildung von Handwerkern
und Gewerbetreibenden zu fördern.
der 1825 gegründeten »Sonntagsschul
für angehende Künstler und Handwer
ker« unterrichteten neben Zeichenleh
rern auch durchaus namhafte Künstl

Nordwind Tischlerwerkstatt GmbH,
Adolf Brauner Geschäftsführer, designorientierte Produkte in handwerklicher Fertigung; zum Beispiel

Glastisch, 1990/91 (Prototyp)
Designer: Johann Schröder, Braunschweig
Die filigrane Konstruktion des Tisches besteht aus einer raffinierten Verbindung der einzelnen Holzstreben – bis zu vier Streben treffen sich in einem Punkt, ohne daß die Festigkeit des Tisches leidet. Die zum Teil formschlüssigen Verbindungen würden selbst ohne Verleimung die Tischplatte tragen.
Eine an Shaker-Möbel erinnernde, sorgfältige handwerklich-designerische Arbeit.

Material und Technik:
Feinporiges, festes Laubholz; im Ausstellungsbeispiel dunkelrot gedämpfter Schweizer Birnbaum.
Tischhöhe: 75 cm
Tischdurchmesser: 124 cm/32 cm (achteckig)

Hersteller:
Fa. Nordwind/Adolf Brauner, Tischlerwerkstatt GmbH, Bremen

Die Produktentwicklung wurde aus Mitteln der Designförderung des Senators für Wirtschaft, Mittelstand und Technologie unterstützt.

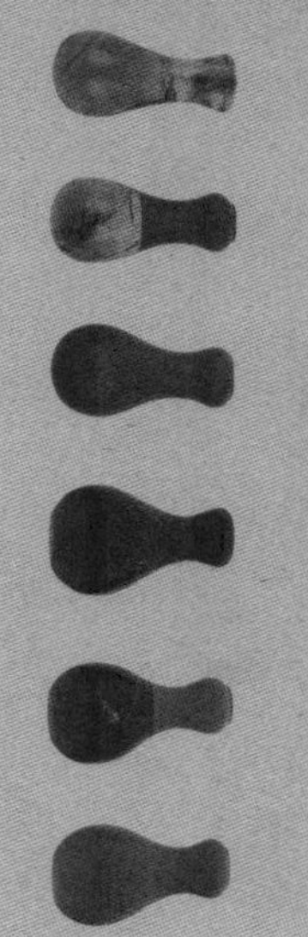

1994 Bm^{4}

EMS
EMS
EMS
CNNKGA (CNA)
Nanjing
CHINA POST
PVG
Non priority
DHL
PAKET INTERNATIONAL
Hinweise für den Versender
Paketkarte
Zollinhaltserklärung
Handelsrechnung
Ursprungserklärung
Sonstiges
CP 91
DEUTSCHE POST

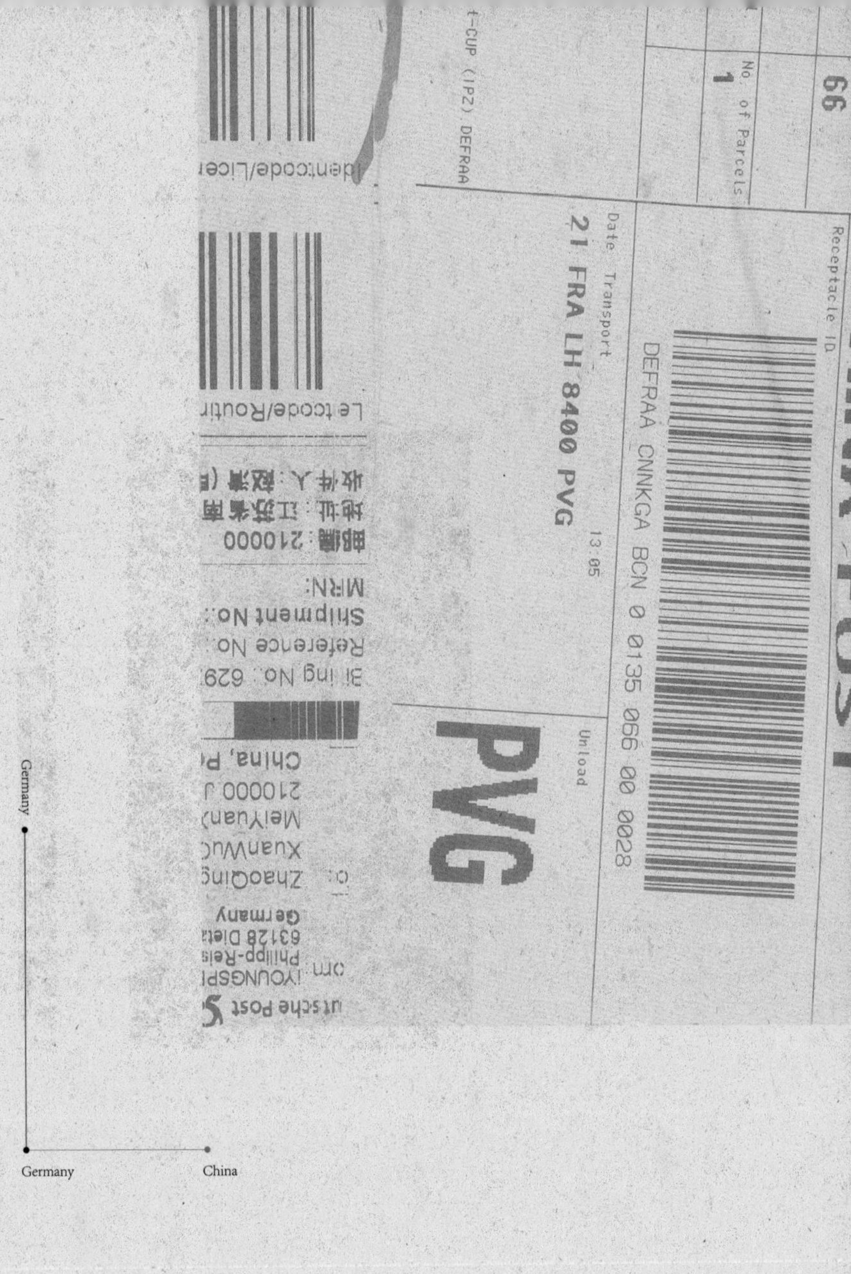
Receptacle ID
DEFRAA CNNKGA BCN 0 0135 066 00 0028
Date Transport
21 FRA LH 8400 PVG
13:05
Unload
PVG
No. of Parcels
1
Germany
Germany
China

288mm

24mm

SHEN JIJI DIE GESCHICHTE DES FRÄULEIN REN

166mm

铜奖
任氏传
德国

Bronze Medal
Die Geschichte des Fräulein Ren
Germany
Clemens Tobias Lange
Shen Jiji
CTL-Presse, Clemens Tobias Lange, Hamburg
288 166 24mm
538g
106p

中国唐代传奇小说多为奇闻逸事，《任氏传》是其中的名篇，塑造了一个美丽聪慧、散发着人性魅力的狐妖。作为一反以往狐妖害人观念的小说，实则借妖写人，带有反封建礼教的现实意义。书籍的装订形式为包背装，书脊处采用木条开槽将装订口夹住。函套采用蓝色艺术纸裱覆卡板，侧面文字印白。封面采用布满蓝色刺绣小点的暗红色布，带有独特的幽幽寒光。书脊选用的深色木料很坚硬，但很轻，握持感良好，书脊文字丝印白色。环衬采用与函套相近的深蓝色艺术纸，文字和画面印白，内页宣纸红黑双色印刷。本书中德双语，在设计上兼顾了两种语言文字在阅读上的需求。正面左翻为德文，黑色文字均印于右页，横排左对齐，页码只呈现奇数。反面右翻为中文，红色的文字均印于左页，八列文字竖排上下对齐，中式页码纵向印于翻口。每两页间夹带一张故事的抽象水墨，黑白反转，呈现拓片效果。中文最后附带解读文字，采用楷体，与正文的宋体加以区分。

Most of the legendary novels of Tang Dynasty in China are anecdotes. *Die Geschichte des Fräulein Ren* is one of the most famous novels, which creates a beautiful and intelligent fox demon. This novel, contrary to the previous concept of fox demon harming people, is of practical significance of anti-feudal ethics. The book is double-leaved and the gutter is clamped by a piece of wood with slot at the spine. The case is made of blue art paper mounted with cardboard. The cover is made of dark red cloth with blue embroidery dots, which has unique faint light. The dark wood used for the spine is hard but light. The text on the spine is treated with silk screening in white. The end paper is made of dark blue art paper which is similar to the case. The text and pictures are printed in white, and the inner pages are printed in red and black on rice paper. This book is bilingual in Chinese and German, and takes the reading requirements of both languages into account. If you turn the pages from the front of the book, the black text in German is printed on the right pages, horizontally aligned to the left; if you turn the pages from the back of the book, the red text in Chinese is printed on the left pages, and the eight columns of text are aligned vertically. The Chinese page numbers are printed longitudinally on the fore-edge. There is an abstract ink painting between every two pages, with black and white reversed, showing rubbing effect. The Chinese text finally comes with an interpretation using Kai typeface, which is distinguished from the text using Song typeface.

時除籍官徵其估計錢六萬設
尚多矣若有馬以備數則三年
之且所償蓋寡是以買耳
任氏又以衣服故弊乞衣于崟
任氏不欲曰愿得成制者崟召
便見任氏問所欲張大見之惊

Fuchsgeist.
er Neunte
, ein Enkel
en Xin'an; er war
nd trank gerne

任氏传

自太平广记卷第
篇措写人与妖精
主义的手法，塑造
服于暴力，敢于

on Zhang Youhe, Beijing 1983.
estaltung, Handsatz, Druck des Text
ilder von den Originalplatten, (Mira
lemens-Tobias Lange.
atz des chinesischen Textes: Fabian S
a. Druckwork, Tübingen.
chrift: Trump Mediaeval, der Schrift

erwiderte ihm lachend: »Wenn sie deinem Aussehen entspricht, kann sie ja nur häßlich sein! Nun, in ihrem Inneren wird sie sicher schön sein!« Jedenfalls lieh er ihm Vorhänge, Kissen und Matratzen, und außerdem schickte er ihm einen seiner Diener mit, der besonders wachsam war, damit er ein Auge auf das Mädchen werfe. Der Diener kam verschwitzt und ganz außer Atem zurückgelaufen, und Wei Yin fragte ihn gleich: »Nun?«, und drängte: »Wie sieht sie aus?« Der Diener antwortete: »Oh, überraschend! Unter diesem Himmel ist nie eine Frau von derartiger Schönheit gesehen worden!«
Wei Yin hatte eine große Verwandtschaft,

笑曰明智若此而爲妖感何哉
巫者言可徵徒爲公死何益一
懇請入初任氏不得已遂行
臨皐揮袂驚去信宿至馬巍任
子乘驢居其後女奴驚乘又在
人教獵狗于洛川已旬日矣適
于草間鄭子見任氏欻然墜于
蒼犬逐之鄭子隨走叫呼不能

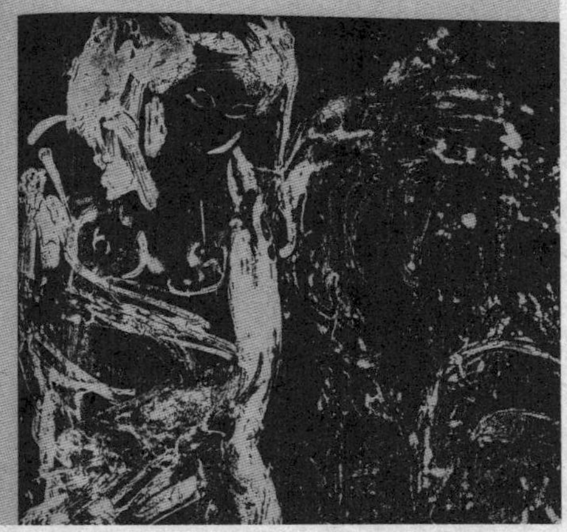

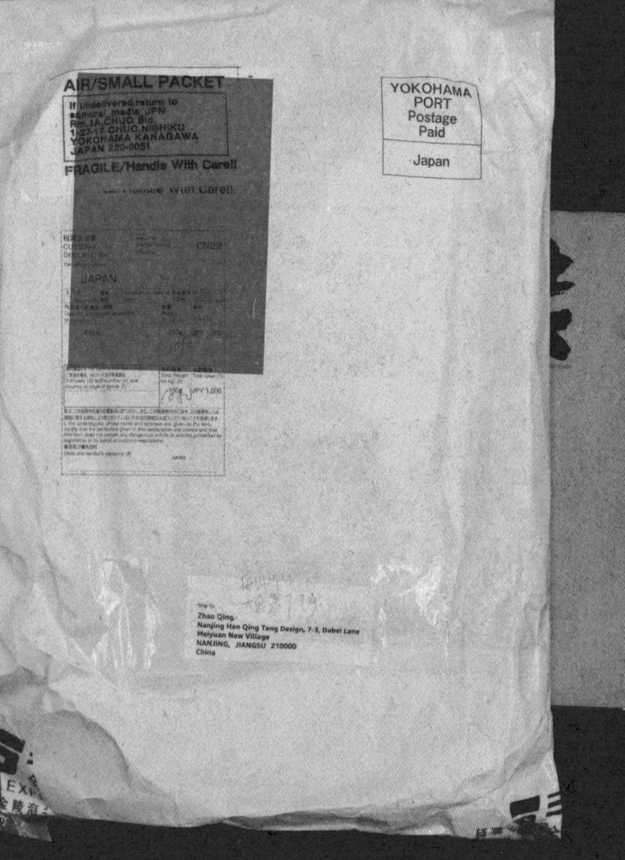
AIR/SMALL PACKET
If undelivered,return to
samurai_media_JPN
Rm.3A,CHUO Bld.
1-27-17 CHUO,NISHIKU
YOKOHAMA KANAGAWA
JAPAN 220-0051
FRAGILE/Handle With Care!!
YOKOHAMA
PORT
Postage
Paid
Japan
CN22
JAPAN
Zhao Qing
Nanjing Han Qing Tang Design, 7-3, Dabei Lane
Meiyuan New Village
NANJING, JIANGSU 210000
China

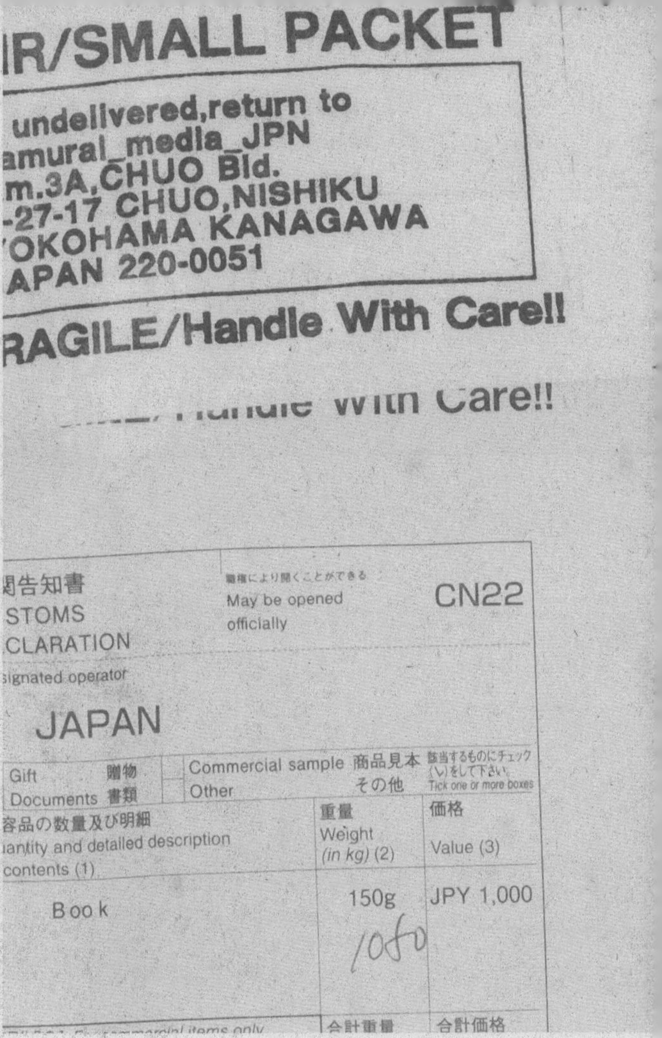

IR/SMALL PACKET

undelivered,return to
amurai_media_JPN
m.3A,CHUO Bld.
-27-17 CHUO,NISHIKU
OKOHAMA KANAGAWA
APAN 220-0051

RAGILE/Handle With Care!!

Handle With Care!!

関告知書
STOMS
CLARATION

職権により開くことができる
May be opened
officially

CN22

signated operator

JAPAN

Gift 贈物		Commercial sample 商品見本	該当するものにチェック(√)をして下さい。Tick one or more boxes
Documents 書類		Other その他	

容品の数量及び明細 uantity and detailed description contents (1)	重量 Weight (in kg) (2)	価格 Value (3)
Book	150g 1080	JPY 1,000
	合計重量	合計価格

Japan

Japan

China

216mm

44mm

160mm

铜奖
峠
日本

Bronze Medal
Touge: A Mountain Pass
Japan
Shincho-Sha Co., Tokyo
Ryotaro Shiba
Shincho-Sha Co., Tokyo
216 160 44mm
1015g
716p
ISBN-10 4-10-309738-8

大多数日本作家擅长描写内心感受和身边琐事，而日本文坛巨匠司马辽太郎善于以理性的历史观来写作历史小说，将他研究的地理、交通、人文等学问运用于写作中，呈现出俯瞰大时代的独特气质。书名“峠”，古时同“卡”，关卡的意思。故事发生在幕府末期和明治维新初期，既不归属幕府又不从属朝廷的小藩——越后长冈藩在独特的时代中追求自治，最终导致维新期间最惨烈的战斗。本书的装订形式为圆脊锁线硬精装。护封采用带有纵向纹理和内含杂质的艺术纸四色印刷，书名“峠”烫黑处理。内封选用带有三角形和波浪形“山川”纹理的艺术纸深蓝色和黑色双色印刷，裱覆卡板。内页未漂白的胶版纸印单黑。作为一本大部头历史小说，文字量较大，所以在版式的设计上四边距设置较窄、字号偏小，传统竖向排版，上下双栏设置，且未设置章节页，只是通过换栏及加大章节标题的方式加以区隔。但全书排版疏朗，在阅读体验和节省页面之间找到了良好的平衡。

Most Japanese writers are good at describing their inner feelings and the trivial things around them. Ryotaro Shiba, a great Japanese literature master, is skilled in writing historical novels with a rational view of history. He applies his knowledge of geography, transportation, and humanities to his writing, presenting a unique temperament. The title of the book is *Touge*, which means ‘a mountain pass’ in ancient times. The story took place at the end of the Shogunate period and the beginning of the Meiji Restoration. It was a historical story that Nagaoka, neither subordinate to the Shogunate nor to the imperial court, endeavored to pursue autonomy in this unique era and finally lead to the most tragic battle during the restoration. The hardcover book is bound by thread sewing with rounding spine. The jacket is printed in four colors on art paper with longitudinal texture and impurities, and the title of the book is black hot-stapmed. The inner cover is made of art paper with triangular and wavy ‘mountains and rivers’ patterns printed in dark blue and black, mounted with cardboard. The inner pages are made of unbleached offset paper printed in black. As a large historical novel, there is a large amount of words. As a result, the margin setting is relatively narrow and the font size is small. It adopts the traditional vertical layout with double columns. There is no chapter page in this book, it is separated by changing columns and increasing the font size of chapter titles. Due to its sparse layout, the book has balanced reading experience and page saving successfully.

目次

屋三井　ほうへひきかえして行っ
しとか　やがてかれらは出てき
　こんどは継之助への
きない。　していた。みな荒縄で
の自滅　は蓑や合羽をおなじく縄
ここが　すそを尻のみえるまで
解をま　に鳴らして出て行った。

「徳川家をその窮状
させてもらいたい。
い。まして賊名をか
　というものであっ
るならばわれわれは
ある。そのための使
　使者には、人材が
　正使には二十四歳
ながら帯刀は早くか
出に似合わず気概に
のに馴れないため、
姓）をつけた。川島

返上し、君臣とも当地をひきはらい、徳川
よろしい」
がいに白熱して衝突した。
内の混乱について触れつづける。時期とし
船が越後沖にさしかかっているころになる
、深刻になっている。
派が連日激論をつづけ、ついに老公のじき
たねばならなくなった。
っている」
堂はいった。

——せめて藩主だけでも。
と交渉したが、それもことわられた。将軍脱出後、幕府艦隊は高級幕臣や負傷兵を収容するだけで手いっぱいであるという。
「会津藩士や幕臣の一部は、紀州路（和歌山県）へ落ちるらしい」
と、継之助は大坂城できいたことをいった。
なにしろ京都と大津、それに京坂の街道は、薩長でおさえられているのである。どこかを大まわりしなければ江戸へ帰れない。
紀州へ落ちる、という意見が大部分なのは紀州藩が徳川の御三家であるからであった。
「まさか」
と、たれかがつぶやいた。
「紀州藩まで寝返っておりますまいな」
「わからん」
継之助は、おもしろそうに笑った。こうなればこの現状を面白がるしか手がない。
「しかし紀州家は、御三家の筆頭でございますぞ。それにご先代の大樹公（将軍家茂）は紀州家から出ておられます」
「この乱世に、血の濃さなどあてになるものか。関ヶ原において大坂方をうらぎったのは豊臣家の養子であり、かつまた太閤の義理のおいでもあった小早川秀秋だ」
「紀州家が、小早川秀秋だと申されるのでござるか」
つい、議論は殺気立ってくる。
「たれがそうだといった。乱世にあっては由緒や縁起はあてにならんといったまでだ。徳川家の譜代筆頭の井伊家はいますでに裏切っており、徳川家ともっとも因縁のふかかった藤堂家は味方の旗色がわるいとみると、にわかに砲門の方角を変え、味方の頭上に砲弾の雨をふらせたではないか。あの藤堂の砲弾がとんできたとき、最初はどこから飛んでくるのかわからなかったそうだ。まさか、藤堂藩がうらぎるとはおもわなかったのだろう」
「では大夫（家老）はどうするというお肚なのです」
「ひとに頼らず、おのれのみを恃めということさ。紀州は大まわりだ。大和から伊勢松坂に出て松坂から船にのればいい」
「それも考えたのです。しかし」
あぶない。京から奈良は近いから薩長はすでに奈良に兵を出し、退路を遮っているのではないか。
「そのときは戦えばいい。とにかく近いほうがいい。それでゆく」
この七日の夕刻、継之助は藩主を擁し、藩兵をひきいて大坂を去った。

江　戸

継之助らは、奈良をとおり、笠置へ出、伊賀に入った。伊賀の国は、伊勢の藤堂藩の属領であり、その主城である伊賀上野城は、代々藤堂家の城代がまもっている。
「伊賀上野城下を通りますか」
と、三間市之進がきいた。
「なぜきく」
「あれは藤堂家の城下です」
藤堂家は、鳥羽伏見での後半における裏切り以来、官軍になっている。かれら新官軍が忠義顔をしてこっちへ押し寄せられてはたまらない。
「しかし上野城下を通るのが、松坂への近道だろう」
「そりゃ、近うはございますが」
「通ろう」
と、継之助はいった。
「さればいそいで通りましょう」
「なぜいそがねばならぬ」
「藤堂家の魂胆がわかりませぬ」
「なあに、裏切るような藩に攻めてくるほどの度胸があるものか。いっそ、今夜の宿は上野城下にとろう」
「宿を」
「ちょっと存念があるのだ」
と継之助がいったのは、今後の大名の動向を観望するためのなにかの足しになるかもしれぬとおもったのである。
宿割りのための者を先発させた。
やがて上野城下に近づくと、その先発させたうちの一人が道で待っていた。
「なんだ、宿をことわられたのか」
継之助はそうおもった。藤堂家が薩長に加担した以上、当然徳川方とみられる長岡藩の藩主と藩兵が城下にとまることを拒否するだろうとおもったのである。
「それが」
まだ決らぬのです、という。長岡勢を城下に泊めるか泊めないか、城で評定をしているという。うっかり長岡勢を泊めたがために、藤堂家はあたらしい主人である薩長の機嫌を損ずることをおそれているのであろう。
「厄介なことだな」
継之助は、行列を停止させた。
「城に泊めてくれといっているわけではない。こっちは旅籠に泊るのだ。それをお城で評定するなどとはよほど小

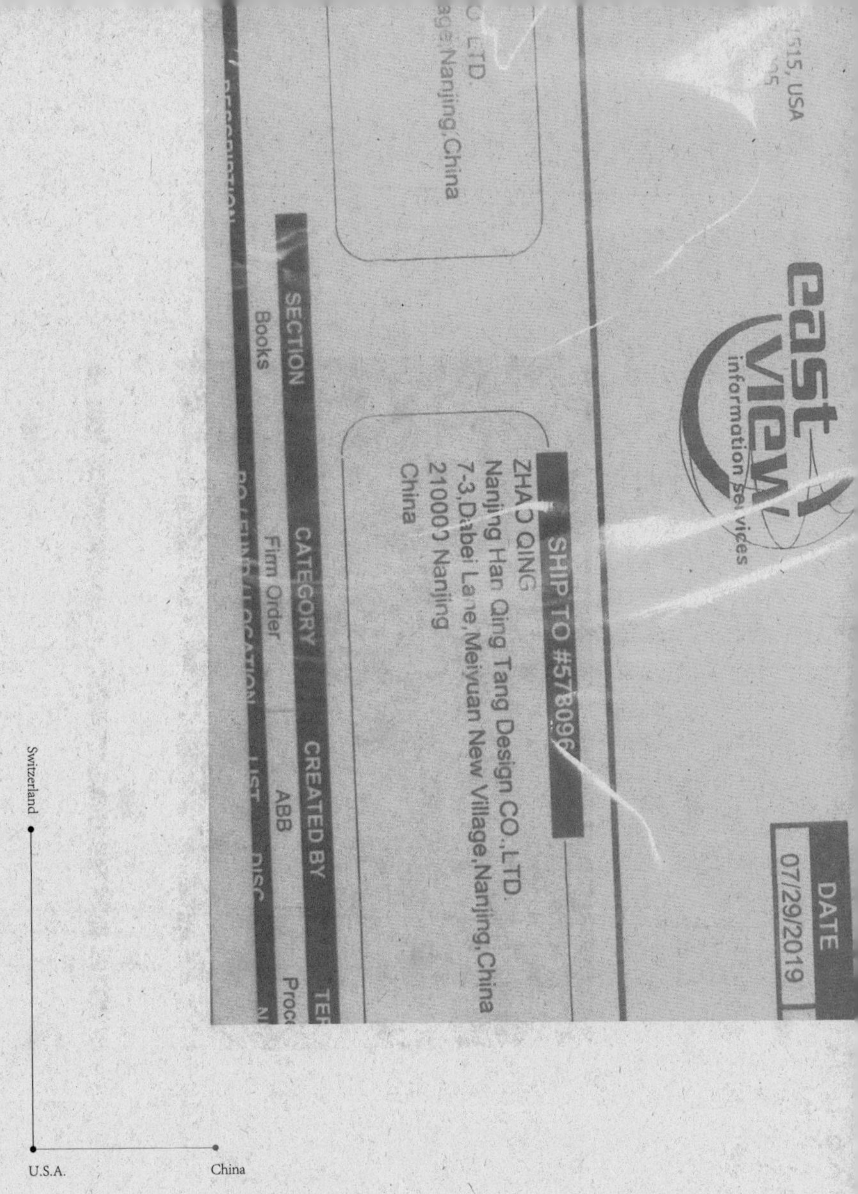
east view
information services
DATE
07/29/2019
SHIP TO #578096
ZHAO QING
Nanjing Han Qing Tang Design CO.,LTD.
7-3,Dabei Lane,Meiyuan New Village,Nanjing,China
210009 Nanjing
China
SECTION
CATEGORY
CREATED BY
Books
Firm Order
ABB
Switzerland
U.S.A.
China

荣誉奖
字体的乐趣
瑞士

Honorary Appreciation
Freude an Schriften
Switzerland
Jost Hochuli
Jost Hochuli
Typotron AG, St. Gallen
240 152 5mm
114g
36p
ISBN-10 3-7291-1072-1

瑞士平面设计师和书籍设计师约斯特·霍胡利（Jost Hochuli）是“瑞士字体排印风格”的代表设计师。1979年他与参与创立圣加仑出版社（St. Gallen），策划、编撰和设计了Typotron系列。本书是这个系列1993年出版的第11期。书籍的装订形式为线装骑马钉。护封绿色纸张，封面菱形镂空，露出下层蓝色纸张上印刷的银色文字。环衬采用红色纸张。内页胶版纸红蓝绿三色印刷。本书展示了霍胡利在字体排印方面的创造力和精湛技巧。书中用德英双语介绍了13个著名字体的简史、分类和特点。他没有同时使用两种文字，而是用这款字体的设计师的国籍语言撰写介绍，并在附录部分加上另一种语言的翻译，这样就很好地控制了每页的信息量，把大量的页面空间用于字体本身的排版。其中“IJ”两个字母被合并在一起，形成5×5的字母“矩阵”，并在这样的矩阵上进行排版变化。“IJ”的合并很巧妙，这两个较窄的字母合并会跟其他字母看起来宽度接近，易形成较好的矩阵感，同时暗示了“I”“J”两个字母的历史关联性。

Swiss graphic designer and book designer Jost Hochuli is the representative designer of the “Swiss Typography Style”. In 1979 he co-founded the St. Gallen Publishing House, where he acquired, edited and designed the *Typotron Series*. This book was the 11th issue of the series, published in 1993. The book is bound by saddle stitch with thread sewing. The cover made of green paper is carved with a diamond shape, revealing silver text printed on blue paper. The endpaper is red and the inside pages are made of offset paper printed in red, blue and green. This book demonstrates Hochuli's creativity and virtuosity in typography. The book introduces the brief history, classification and characteristics of 13 famous typefaces in both German and English. Instead of using both languages, he wrote the introduction in the language of the font's designer and added a translation in the other language to the appendix, which controlled the amount of information on each page and devoted a lot of page space to the typesetting of the font itself. Two letters IJ are combined into a single letter to form a 5×5 letter “matrix” and typographical changes are made to such a matrix. The combination of IJ is very clever because the combination of these two narrow letters looks close to the width of the other letters, forms a good matrix sense easily, and suggests the historical connection between the two letters.

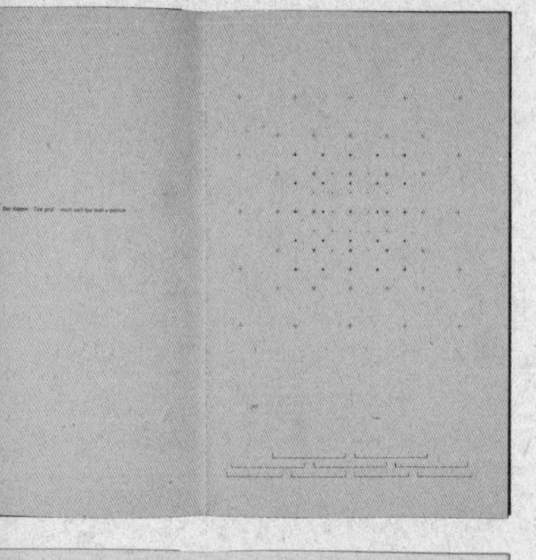
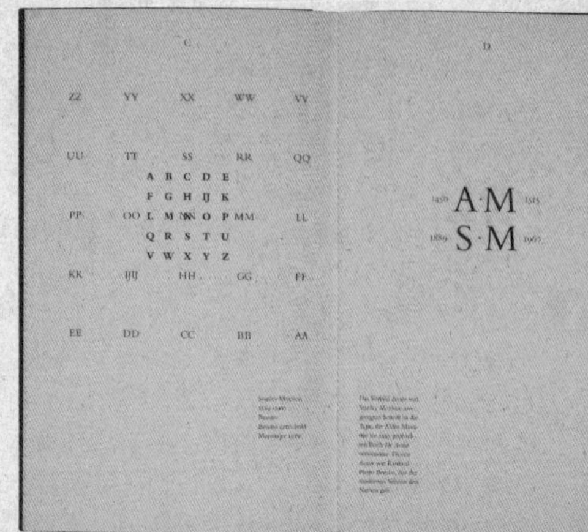
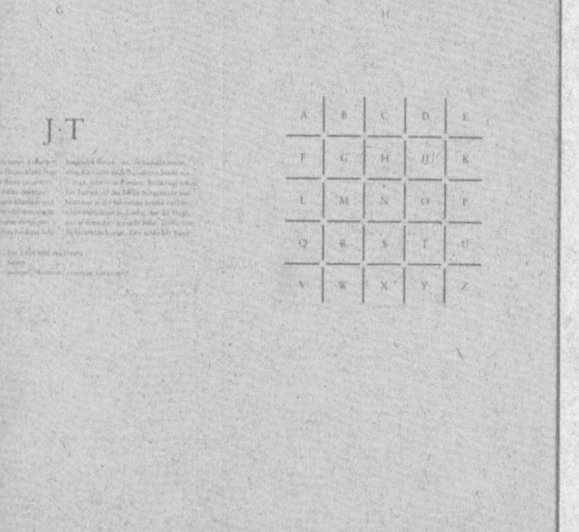
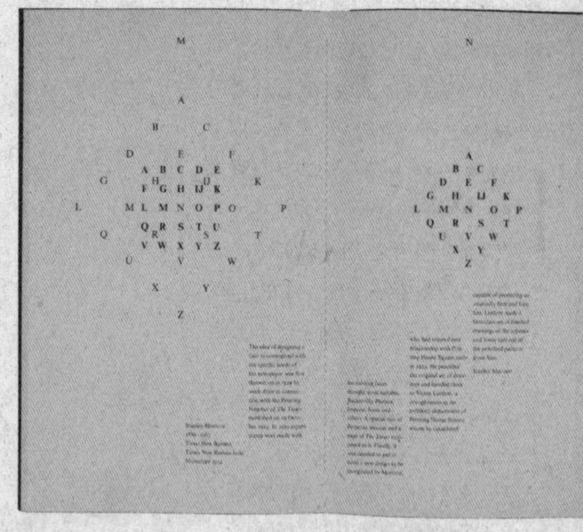
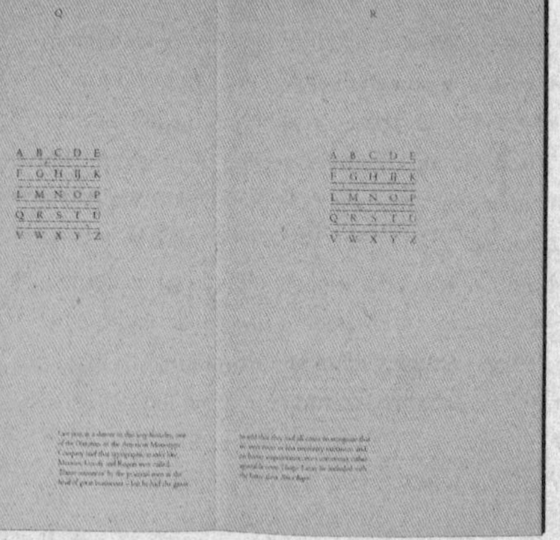
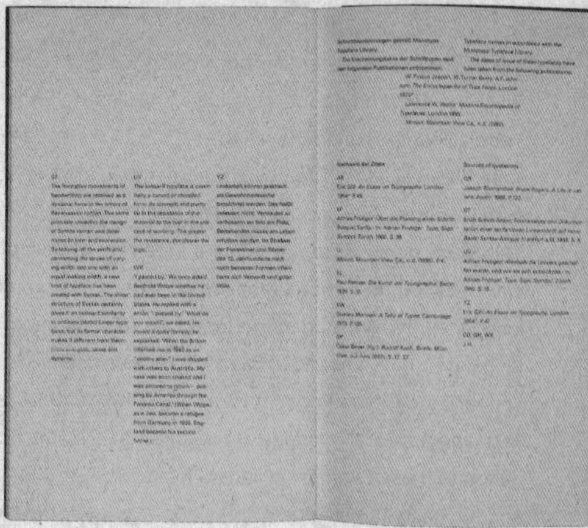

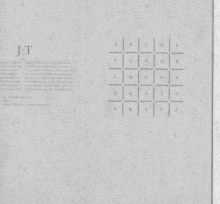

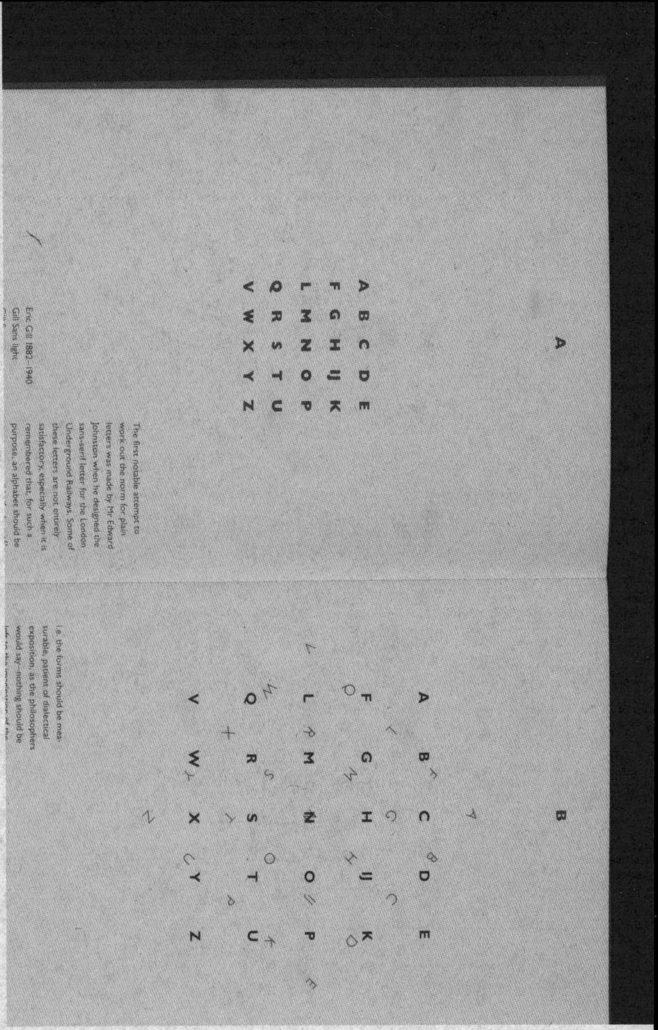

A

A B C D E
F G H IJ K
L M N O P
Q R S T U
V W X Y Z

Eric Gill 1882–1940
Gill Sans light

The first notable attempt to work out the norm for plain letters was made by Mr Edward Johnston when he designed the sans-serif letter for the London Underground Railways. Some of these letters are not entirely satisfactory, especially when it is remembered that, for such a purpose, an alphabet should be

B

A B C D E
F G H IJ K
L M Ñ O P
Q R S T U
V W X Y Z

i.e. the forms should be measurable, patient of dialectical exposition, as the philosophers would say—nothing should be

1995

Country / Region

U.S.A. ❹

Germany ❹

Switzerland ❷

The Netherlands ❶

Spain ❶

Austria ❶

Egypt ❶

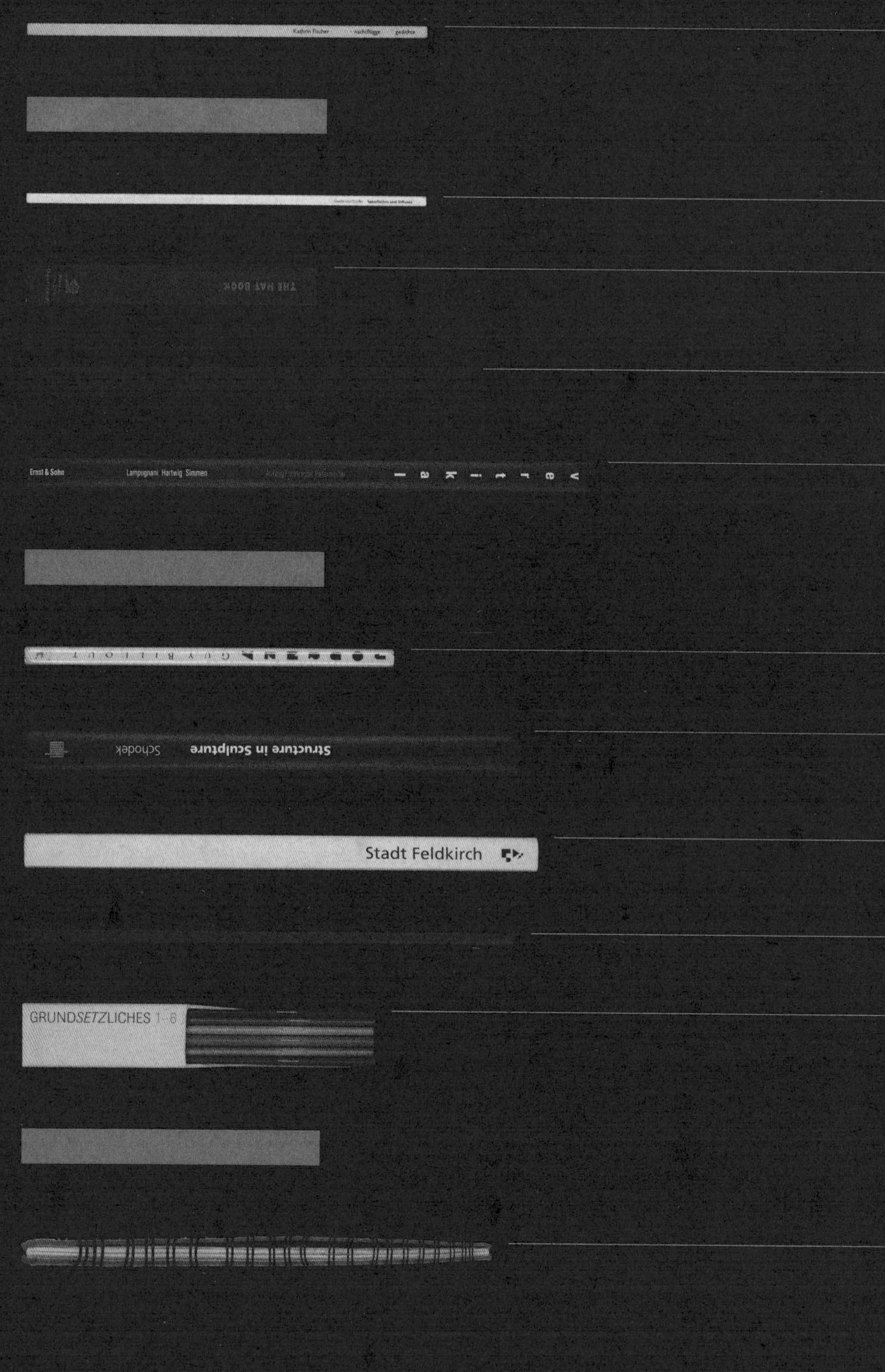

Kathrin Fischer nachtflügge gedichte
THE HAT BOOK
Ernst & Sohn
Lampugnani Hartwig Simmen
vertikal
Guy Billout
Structure in Sculpture
Schodek
Stadt Feldkirch
GRUNDSETZLICHES 1–6

1995 Gl 0436

1995 Sm[1] 0444

1995 Sm[2] 0452

1995 Bm[1] 0462

1995 Bm[2] 0472

1995 Bm[4] 0482

1995 Bm[5] 0492

1995 Ha[1] 0502

1995 Ha[2] 0510

1995 Ha[3] 0518

1995 Ha[5] 0526

Designer

Kaspar Mühlemann

Reynoud Homan, Robbert Zweegman

Urs Stuber

Leslie Smolan, Jennifer Domer (Carbon Smolan Associates)

Julia Hasting, Patricia Müller, Gerwin Schmidt, Béla Stetzer

Grappa-Design

Norbert Denkel

Rita Marshall (Delessert & Marshall)

Jeanet Leendertse

Reinhard Gassner

Büro für Gestaltung Biste & Weißhaupt

Philipp Luidl

Mohiedden El-Labbad

Judy Anderson, Phillip Helms Cook, Claudia Meyer-Newman

1995 G1

Kathrin Fischer

nachtflügge

ZVAB.com
ZENTRALES VERZEICHNIS ANTIQUARISCHER BÜCHER

www.zvab.com

ZVAB.com
ZENTRALES VERZEICHNIS ANTIQUARISCHER BÜCHER

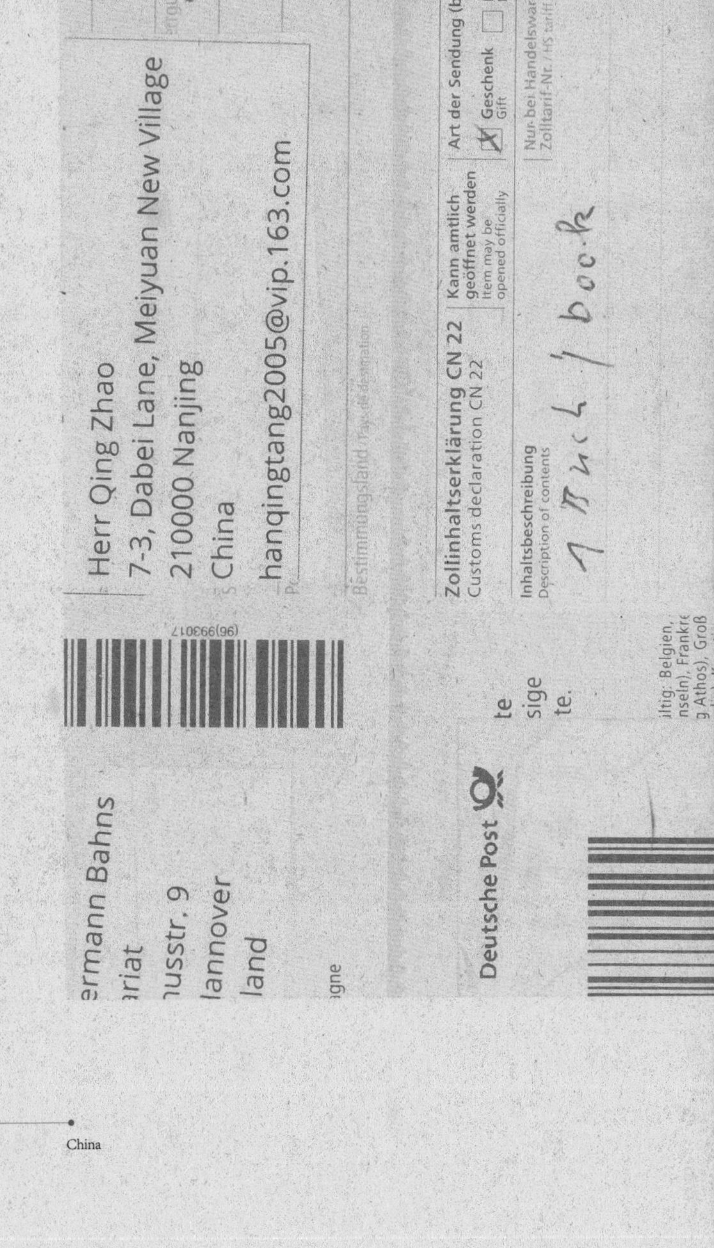

Switzerland

Germany　China

240mm

7mm

Kathrin Fischer

nachtflügge

gedichte

gedichte nachtflügge Kathrin Fischer

152mm

金字符奖
夜间飞行——诗歌
瑞士

Golden Letter
Nachtflügge: gedichte
Switzerland
Kaspar Mühlemann
Kathrin Fischer
Kranich-Verlag, Zürich
240 152 7mm
174g
88p
ISBN-10 3-906640-52-3

这是瑞士诗人卡特林·菲舍尔（Kathrin Fischer）的第一本诗集。“夜间飞行”是诗集的最后一个非严格意义上的章节主题，也是本书设计主要亮点。书籍的装订形式为锁线胶平装。封面浅蓝灰色卡纸印单黑。内页胶版纸印单黑。诗集分为 7 个章节，章节间没有专门的分隔页，只在目录部分通过横线加以区隔。内页版式较为简朴，正文无标点，通过断行来控制诗歌节奏。每一首诗的诗名被放置在左页，而页码放在右页的相同高度，并与目录对应的页码和诗名相一致。在翻阅过程中，诗名和页码逐渐下降，寓意着诗歌中的某种自由或变化。

This is the first collection of poems by the Swiss poet Kathrin Fischer. “Night falls” is the last chapter theme of the poetry collection, and it is also the highlight of book design. The paperback books is glue bound with thread sewing. The cover is made of light blue gray cardboard printed in single black and the inside pages are made of offset paper printed in black. The poetry collection is divided into seven chapters. There is no special separation page between the chapters, only separated by horizontal lines in the table of contents. The layout of the inner pages is simple, and the text has no punctuation. The rhythm of poetry is controlled by breaking lines. The title of each poem is placed on the left page, while the page number is placed on the right page with the same height. The titles and the page numbers are consistent with the corresponding information in the table of contents. As we read, the titles of the poems and page numbers gradually decline, implying some kind of freedom or changes in the poems.

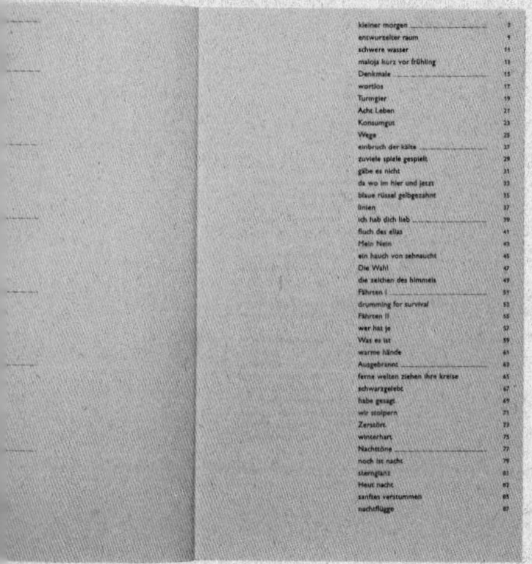

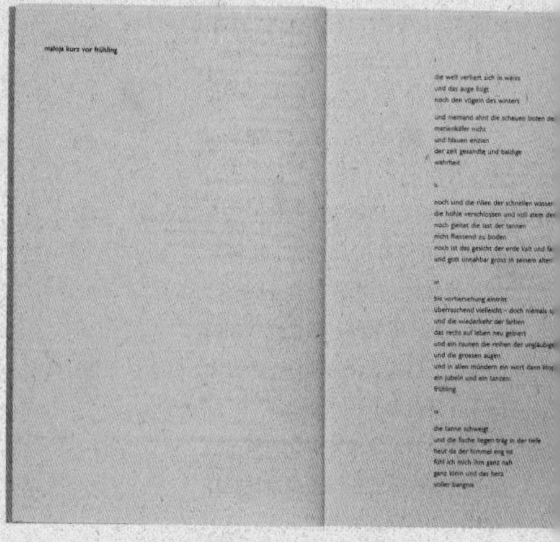

maloja kurz vor frühling

i

die welt verliert sich in weiss
und das auge folgt
noch den vögeln des winters

und niemand ahnt die scheuen boten de[illegible]
marienkäfer nicht
und blauen enzian
der zeit gesandte und baldige
wahrheit

ii

noch sind die rillen der schnellen wasser
die höhle verschlossen und voll atem de[illegible]
noch gleitet die last der tannen
nicht fliessend zu boden
noch ist das gesicht der erde kalt und fa[illegible]
und gott unnahbar gross in seinem alten[illegible]

iii

bis vorhersehung eintritt
überraschend vielleicht – doch niemals s[illegible]
und die wiederkehr der farben
das recht auf leben neu gebiert
und ein raunen die reihen der ungläubige[illegible]
und die grossen augen
und in allen mündern ein wort dann klop[illegible]
ein jubeln und ein tanzen:
frühling

iv

die tanne schweigt
und die fische liegen träg in der tiefe
heut da der himmel eng ist
fühl ich mich ihm ganz nah
ganz klein und das herz
voller bangnis

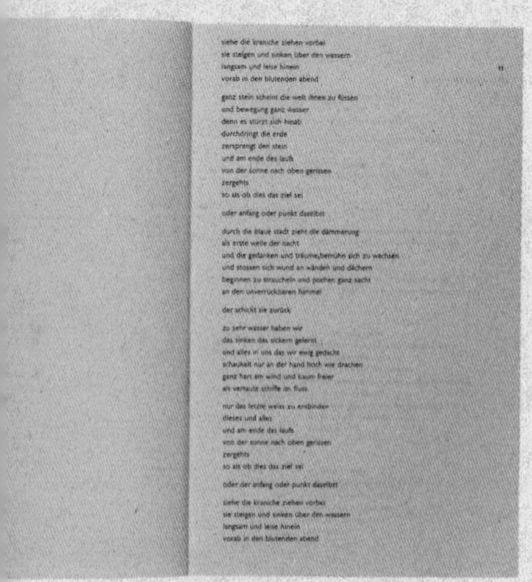

11

siehe die kraniche ziehen vorbei
sie steigen und sinken über den wassern
langsam und leise hinein
vorab in den blutenden abend

ganz stein scheint die welt ihnen zu füssen
und bewegung ganz wasser
denn es stürzt sich hinab
durchdringt die erde
zersprengt den stein
und am ende des laufs
von der sonne nach oben gerissen
zergehts
so als ob dies das ziel sei

oder anfang oder punkt daselbst

durch die blaue stadt zieht die dämmerung
als erste welle der nacht
und die gedanken und träume bemühn sich zu wachsen
und stossen sich wund an wänden und dächern
beginnen zu straucheln und pochen ganz sacht
an den unverrückbaren himmel

der schickt sie zurück

zu sehr wasser haben wir
das sinken das sickern gelernt
und alles in uns das wir ewig gedacht
schaukelt nur an der hand hoch wie drachen
ganz hart am wind und kaum freier
als vertäute schiffe im fluss

nur das letzte weiss zu entbinden
dieses und alles
und am ende des laufs
von der sonne nach oben gerissen
zergehts
so als ob dies das ziel sei

oder der anfang oder punkt daselbst

siehe die kraniche ziehen vorbei
sie steigen und sinken über den wassern
langsam und leise hinein
vorab in den blutenden abend

warme hände

immer wenn der mond sich dreht
kehre wieder
und nimmer sei leer dein beet
von [illegible] der liebe

wozu wenn der mond doch tanzt
deine tränen
[illegible]
dunkles grämen

nichts ist neu
das mag sein
[illegible]
ewig [illegible]

zähl nicht worte
schweig
[illegible]
[illegible] händen

schwarzgelebt

du hast dein kalkül mit dem regen gemacht
der aus dir fiel
und dir zutief ins blut
hart schlagen die adern dein herz ins licht
den kreisel der sich immer dreht bis zum letzten
und sich an ungesagtem wundschreit

im spiegel fallen die tage dir ungreifbar in den rücken
ein blau wallender stoff
und von neuem durchstossen knochen
deine flimmernden muskeln
die splitter und keime der bitternis

das hast du gewusst
und dass es gleichbleibt wie immer
fiel dir doch schon früher die ferne ins aug

die nacht ist gut sagen sie
und du wäschst mit ihnen deine hände in blut
deine haare in unschuld
und brichst deine gedanken am salz der erde

dann lauschst du dem betörenden lied
der schildkröte mit gebrochenen augen
und schweigst
von etwas anderem

gäbe es nicht
und tausend worte würden
rufen tosen schrein
gäbe es nicht
horizonte für alles
was immer ist

ach liebe
ach schmerzen
ach sterbend lied

senkten
tausend sonnen
die herzen tiefblau in den
see der sehnsucht

sprengten
unbekannte leise weisen
die viel zu kleine
harte brust

wortlos

n seine tiefe ahnend
die bäume greifen nach dem mond
den mageren spitzen fingern des winters

land fliesst unter dem licht vollfett und satt
stadt trieft orange bis an ihre grenzen
hügel und das ausgegossne schwarz des sees
seine wahrheit in der farbe des himmels sucht

der himmel schweigt
für himmlisches nicht
was dich an ihn bindet
ehr als ein traum
unendlicher weite

ist nacht
kel erheben sich die hügel
sprachlos mit ihrem geheimnis
ist das ufer des sees ein band von diamanten
erdgöttin um den hals gelegt

egst du still wie die zeit
dein herz hört zu schlagen auf
moment nur doch du hast den atem
welt gespürt und das lächeln
deinen lippen das dir kein auge stiehlt

ist nacht

ch in mir keimen sphärenklänge
rwurzelt wie in erde und
auer beim gruss
s baums
f dem das requiem der stare klingt
rwoben in den augenblick

d immer die angst im auge
s mir blieb

Stirb und flieh
ihr entgegen
endliche Ewigkeit

Hoffnungslos sieh
den Träumer
sinnlos
die Seelenqual

Bläulich schimmern
noch Scherben
des uralt
entheiligten Grals

entwurzelter raum

manchmal sind die räume zu weit
manchmal zu eng
und die nächte zu kurz und zu lang –

das ändert

wo die strecke längst noch keine halbe ist
und keine strasse mündet
asphaltgrau und bräunlichfett
mäandert die welt und die zeit zieht sibeliusschle

verlassene wasser locken ins unbegehbare
gestern und die steine rufen
so bunt zum harten lager
von wellen umkämpft die sich in lärm brachen

nun
bestaun ich meine helleren haare
goldgleiche märchen und betrachte
die schatten im rauch die bilder und nehme
farben nur sanft noch wahr und nebelfern

neu
zwischen welten zu schweben ists:
nicht gestern war die abfuhr nicht heute
die ankunft

das jetzt ganz klein
und müde
zwischen dem war und wird

1995 Sm 1

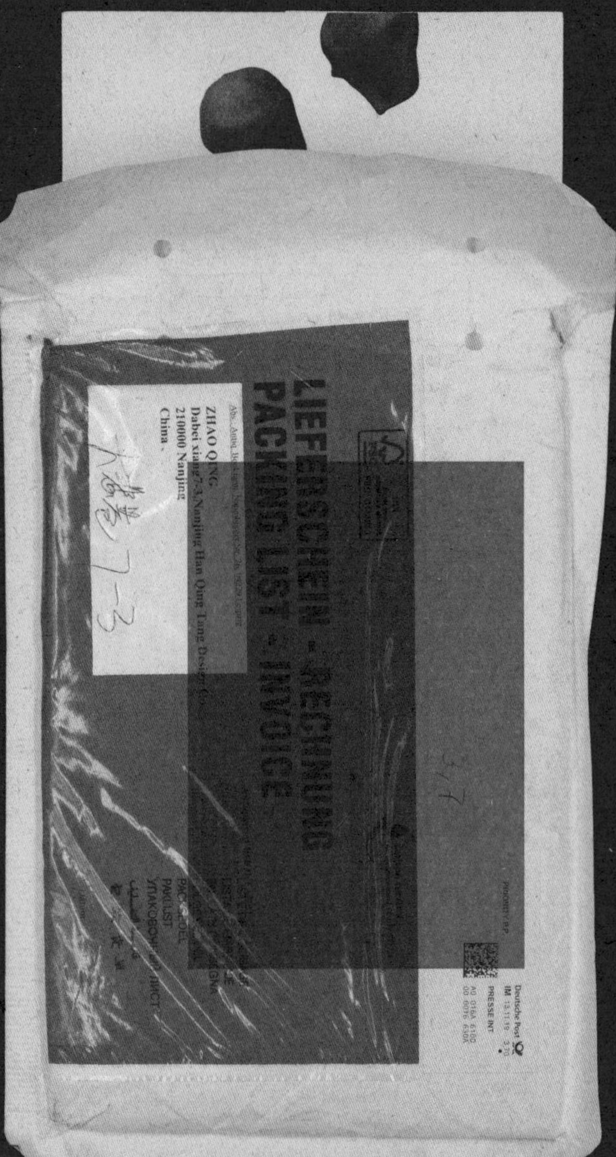
LIEFERSCHEIN - RECHNUNG
PACKING LIST - INVOICE
ZHAO QING
Dabei xiang 7-3,Nanjing Han Qing Tang Design Co.
210000 Nanjing
China
Deutsche Post
PRESSE INT

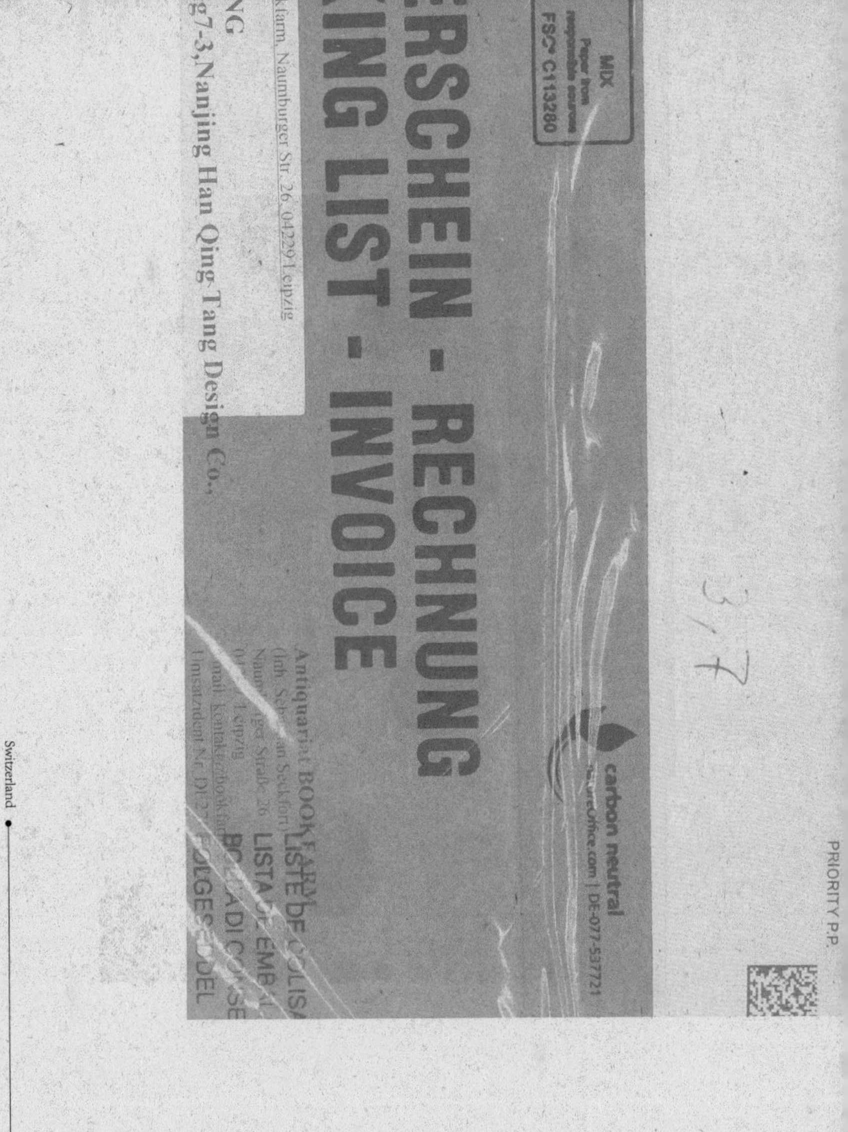

Switzerland

Germany

China

240mm

6mm

Guido von Stürler Spezifisches und Diffuses

158mm

● 银奖
《 具体与发散
◖ 瑞士

Silver Medal

Spezifisches und Diffuses

Switzerland

Urs Stuber

Guido von Stürler

Verlag Niggli AG, Sulgen

240 158 6mm

231g

60p

ISBN-10 3-7212-0290-2

吉多·冯·斯图尔勒（Guido von Stürler）是一名现居瑞士的艺术家，擅长数字媒体和装置艺术的创作。本书是他的作品在瑞士图尔高州立美术馆举办展览时候的作品集。书籍的装订形式为无线胶平装。护封采用珠光纸四色印刷，内封白卡纸无印刷和工艺。内页以哑面铜版纸四色印刷为主，作品部分使用半透明牛油纸印浅黄色图文信息。书的内容结构属于展览作品集常规做法：前言、作品、对谈、年表与参考书目。作品页面参考了作品外框透明与半透明玻璃丝印形成的虚实关系，在牛油纸上印刷并不显眼的图文信息，遮挡在作品结构详图前，形成作品集这一纸面“展览”与读者的互动。

Guido von Stürler is an artist living in Switzerland, specializing in the creation of digital media and installations. This book is a collection of his works when he had an exhibition at the Thuringian State Gallery in Switzerland. The paperback book is perfect bound. The jacket is printed with pearlescent paper in four colors, and the inner cover is made of white cardboard without any printing and processing. The inner pages are mainly printed with matte coated paper in four colors, and translucent kraft paper is used to display the works printed in light yellow. The structure of the book is consistent with the conventional practice of the exhibition portfolio: forewords, works, dialogues, chronology and bibliography. The outer frame of the work adopts transparent and translucent glass screen printing, thus forming a virtual and real relationship; using tracing paper to print inconspicuous information and covering it before the detailed structure of the works, has formed the interaction with readers.

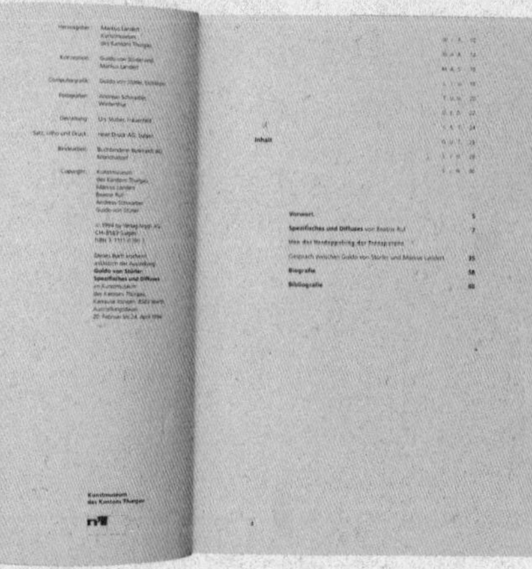

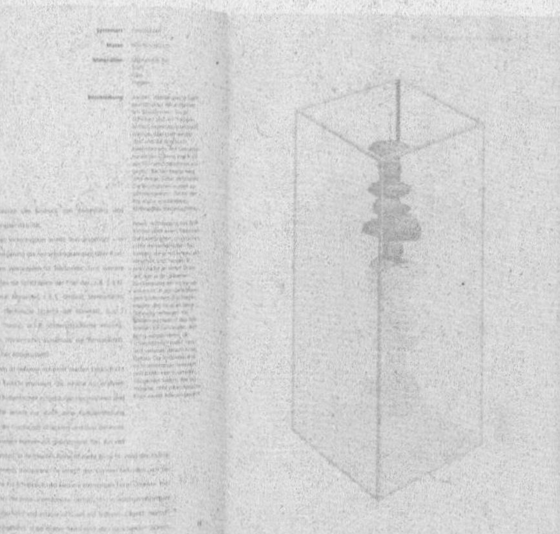
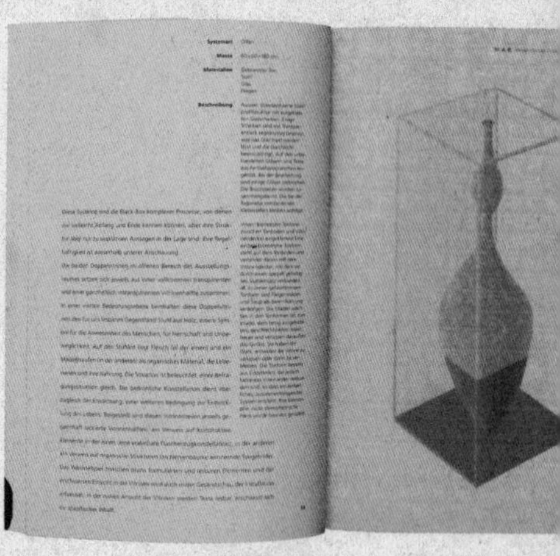
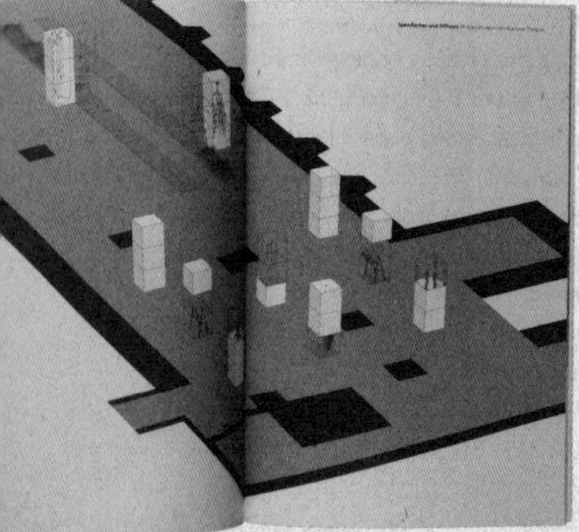

Systemart Geschlossen

Masse 60×60×180 cm

Materialien Gebrannter Ton
Stahl
Glas
Fliegen

Beschreibung Aussen: Standardisierte Stahlprofilstruktur mit aufgeklebten Glasscheiben. Einige Scheiben sind mit Transparentlack regelmässig besprayt, was das Glas matt werden lässt und die Durchsicht beeinträchtigt. Auf den unbehandelten Gläsern sind Texte aus Fernsehprogrammen eingeritzt. Bei der Bearbeitung sind einige Gläser zerbrochen. Die Bruchstellen wurden zusammengeleimt. Die bei der Reparatur entstandenen Klebestellen bleiben sichtbar.

Innen: Montagen von Tonformen an der Stahlprofilstruktur über einem Tonboden. Die biomorphen, organischen, nicht stereometrischen Tonformen, die je mit einem Loch versehen sind, hängen unregelmässig verteilt an der Stahlprofilstruktur der Vitrine. In den gefässförmigen Tonformen sind Fliegenmaden und Sirup als deren Nahrung verborgen. Die Maden wachsen in den Tonformen bis zum Imago, dem fertig ausgebildeten, geschlechtsreifen Insekt heran und verlassen daraufhin das Gefäss. Die Tonformen sind nicht miteinander verknüpft und bilden kein zusammenhängendes System. Ihre biomorphe, nicht stereometrische Form wurde bewusst gewählt.

Reduktion Vitrinen verwendet Guido von Stürler seit einigen Jahren und er setzt damit ein formales Mittel ein, das in der Kunst auf unterschiedlichste Weise als Bedeutungsträger genutzt wurde. Von Joseph Beuys als Verweis auf die Museumsvitrine, das Museumsexponat, auf das althergebrachte, traditionsgemässe Instrument zur Sichtbarmachung kulturhistorischer Zusammenhänge und auch auf den geschlossenen Raum des Museums, so wurde die Vitrine auch von Ilja Kabakow eingesetzt. Bei Damian Hirst ist die Vitrine das Reagenzglas biologischer Konservierungszustände, so wie wir sie aus dem Biologieunterricht oder der medizinhistorischen Sammlung kennen, sei es als formalingetränkter toter Körper oder als Körper in Verwesung begriffen. Die Liste liesse sich fortsetzen ...

Vitrinen sind der ideale, abgetrennte Raum, in dem unabhängige Konstellationen dargestellt werden können. Ihr Innenleben ist zur Schau gestellt und gleichzeitig autonom, sie stellen die ideale

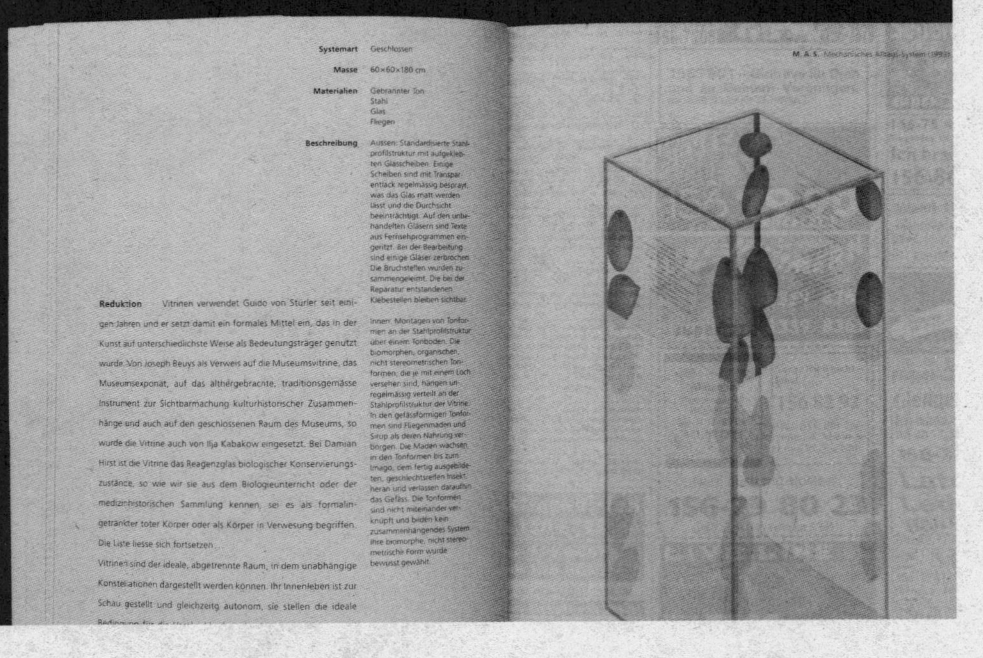

M. A. S. Mechanisches Alltags-System (1993)

konstante Druck,
unter dem die
gewordene Beute steht,
Vene, die sie durch den
Leib wandert,
lösung und die innige
ndung, die sie mit
eingeht,
Vollkommene und
Verschwinden

Systemart Offen

Masse 60×60×180 cm

Materialien Gebrannter Ton
Stahl
Holz
Glas
Fliegen

Beschreibung Aussen: Standardisierte Stahlprofilstruktur mit aufgeklebten Glasscheiben. Einige Scheiben sind mit Transparentlack regelmässig besprayt, was das Glas matt werden lässt und die Durchsicht beeinträchtigt. Auf den unbehandelten Gläsern sind Kontaktanzeigen, die der Tagespresse entnommen wurden, eingeritzt. Bei der Bearbeitung sind einige Gläser zerbrochen. Die Bruchstellen wurden zusammengeleimt. Die bei der Reparatur entstandenen Klebestellen bleiben sichtbar.

Innen: Hölzernes Stehpult mit Tonformen auf Tonboden. An einem stehpultartigen Gebilde, einem Hinweis auf die Sphäre der Wissenschaft und Lehre, hängen biomorphe Tonformen, die durch die Schräglage der Pultfläche an die untere Kante gerutscht sind. Einige der Tonformen sind heruntergefallen und am Boden zerschellt. In den gefässförmigen Tonformen sind Fliegenmaden und Sirup als deren Nahrung verborgen. Die Maden wachsen in den Tonformen bis zum Imago, dem fertig ausgebildeten, geschlechtsreifen Insekt, heran und verlassen daraufhin das Gefäss. Die Tonformen sind nicht miteinander verknüpft und bilden kein zusammenhängendes System. Ihre biomorphe, nicht stereometrische Form wurde bewusst gewählt.

Masse und Macht, Fischer 1960) Die Ausscheidungsprodukte erfreuen sich keiner grossen Beliebtheit, sind Symbol für alles zu Verwerfende, vom Leben Abzuscheidende.

Erotik der Information Mit den in mühevoller Handarbeit in die Glasscheiben der Vitrinen geritzten Texten verweist von Stürler auf einer anderen Ebene auf das widersprüchliche Verhältnis einer vermeintlichen Präzision gültiger Formen und Ordnungsstrukturen mit den diffusen Bereichen, in denen sie mit dem Leben in Verbindung treten. Den offiziellen Nachrichtenorganen medienglatter Textproduktion entnimmt er die Informationen zur menschlichen Bedürfnisbefriedigung: Fernsehprogramme, Kontaktanzeigen, Werbezeilen für Telefonsex und manchmal, als Entspannung des informationsüberfluteten Individuums, die Sprache als Nonsens.

Die unterschiedlichen Versionen, in denen der Künstler dieses Informationsmittel verwendet, verweist auf die Bereiche, in denen Ordnungsstrukturen bedeutungsvoll sein können. Fernsehpro-

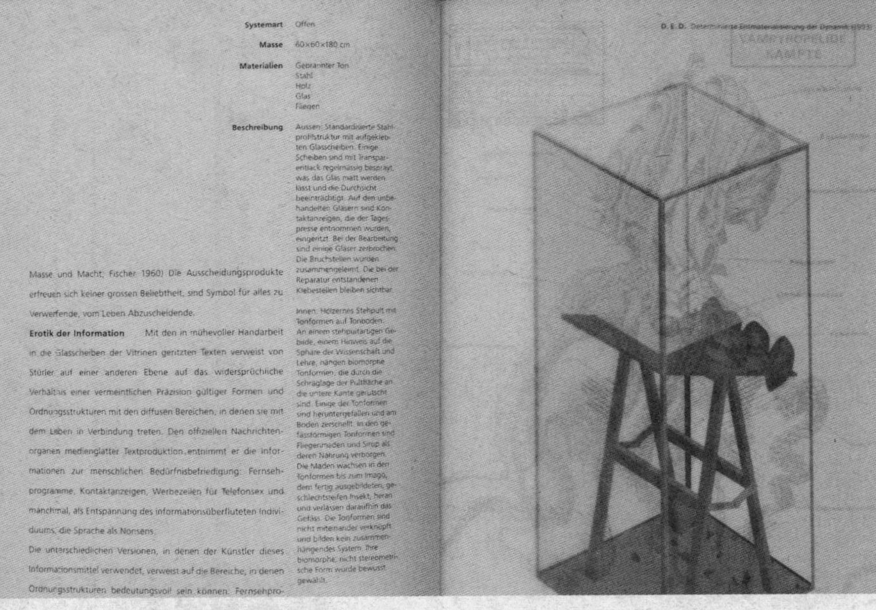

1995 Sm[2]

RODNEY SMITH AND LESLIE SMOLAN

PRIORITY A
If undeliverable return to
PO Box 5001 202 26 Malm
Sweden
UF 52 395 229 0SE
THRIFTBOOKS
2413 PROSPECT DRIVE
AURORA IL 60502 UNITED STATES
ZHAO QING
DABEI XIANG7 3,NANJING HAN QING TAN
NANJING 210000
CHINA
Order no 0015111807528
CUSTOMS DECLARA
Switzerland
Germany
China

22mm

175mm

197mm

THE HAT BOOK

RODNEY SMITH AND LESLIE SMOLAN

THE HAT BOOK

银奖
帽之书
美国

Silver Medal
The Hat Book
U.S.A.
Leslie Smolan, Jennifer Domer (Carbon Smolan Associates)
Rodney Smith, Leslie Smolan
Nan A. Talese; Doubleday, New York
175 197 22mm
459g
128p
ISBN-10 0-385-47228-5

本书的作者是一对深爱帽子的夫妻——时尚摄影师罗德尼·史密斯（Rodney Smith）和《生活》杂志的设计师莱斯莉·斯莫兰（Leslie Smolan）。他们将与帽子相关的照片、文字和相关信息组织整理起来，融汇在本书中。书籍的装订形式为锁线硬精装。采用黑色布面裱覆卡板，哑面艺术纸红黑双色印刷后裱贴在封面和封底，书脊上烫红处理。内页哑面艺术纸红黑双色印刷。内容方面，书中选用了多位摄影师拍摄的照片，无论照片的主题是肖像、景物场景、时尚人物还是人文纪实，均包含形形色色的帽子元素：礼帽、草帽、棒球帽、渔夫帽、鸭舌帽、安全帽、牛仔帽，甚至假装帽子的大铃铛。而文字部分则来自多位文学家、艺术家的小说或笔记，其中不乏众所周知的奥斯卡·王尔德、斯科特·菲茨杰拉德、玛格丽特·杜拉斯等作家的作品节选。书籍的最后部分是书中照片的摄影师署名、经营帽子的店铺地址、帽子的供货商和生产商的联系方式。这是一本有关帽子的“情诗”与大全。

The authors are a couple who love hats – fashion photographer Rodney Smith and designer Leslie Smolan from magazine *Life*. They organized the photos, text and information related to the hats and integrated them into this book. The hardcover book is bound by thread sewing. The cover is made of black cloth mounted with cardboard printed in red and black. The spine is treated with red hot stamping. The inner pages are made of matt art paper printed in red and black. In terms of contents, the book uses photos taken by multiple photographers. Regardless of whether the theme of the photo is portraits, scenes, fashion figures or humanities, all kinds of hat elements are included: bowler hats, straw hats, baseball hats, fisherman hats, caps, hard hats, cowboy hats, and even big bells that pretend to be hats. The text part comes from the novels or notes of many writers and artists, including many well-known excerpts from the works of Oscar Wilde, Scott Fitzgerald, Margaret Duras and other writers. The last part of the book is the signatures of the photographers in the book, the addresses of the shops that sells hats, and the contact information of the suppliers and manufacturers of hats. This is A 'love poem' and encyclopedia about hats.

Off
with your hat as the flag goes by

"Look at me!
Look at me now!" said
"With a cup and a cake
On top of my hat!"

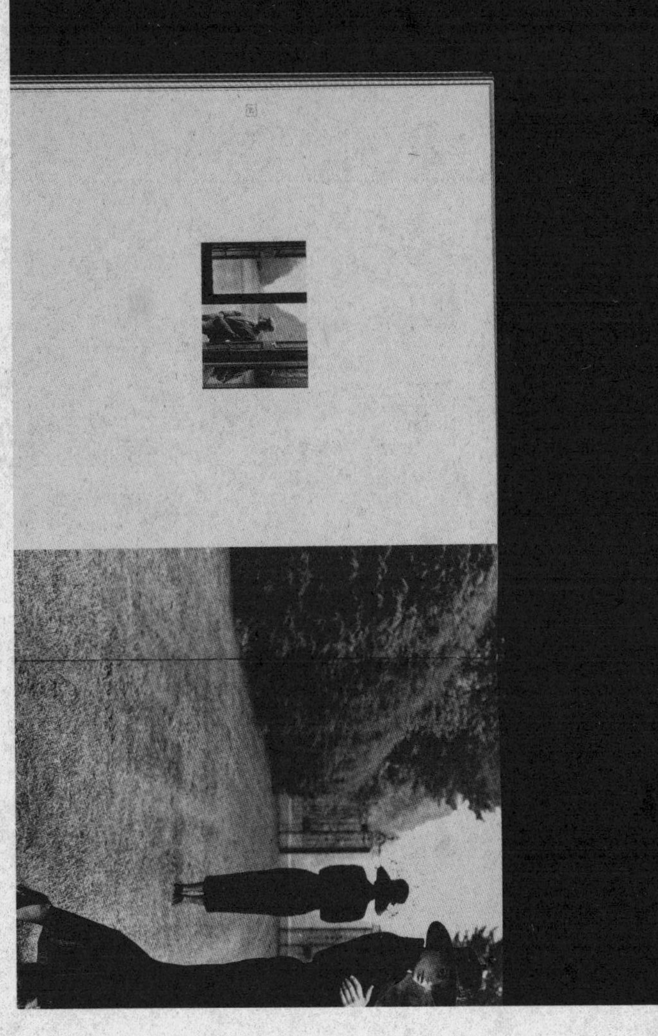

eper's glass and

ere, beneath th

at, the thin av

the inadequacy

od, has turned

ing else. MARGUERITE DURAS

A hat is a flag,

between wearing clothes a

dressed up; it's the differe

me!

me now!" said t

cup and a cake

of my hat!" —DR. SEUSS, TH

all match is an occasion for display. We are smartly turned out in nel's hair coat and a wool town-and-country suit—a bit too chic good looks; I am also in a camel's hair coat, navy, crested blazer, s mplar of youthful privilege, my father wears a blue, double-breast vet-collared chesterfield. For this occasion, he also wears spor scarf and a black homburg. A small white carnation is in his lapel e! On this day, as always, my father is an uncommonly hand gure—an Arlenesque bon vivant who hobnobs easily—charmin in literary and entertainment circles. And as always, I am pro will begin at 1:30, and we arrive at the Yale Bowl with just mi

Why be just you

I m

Then wear the gold hat, if
that will move her;
If you can bounce high,
bounce for her too,
Till she cry "Lover, gold-
hatted, high-bouncing lover,
I must have you!"

F. SCOTT FITZGERALD, THE GREAT GATSBY

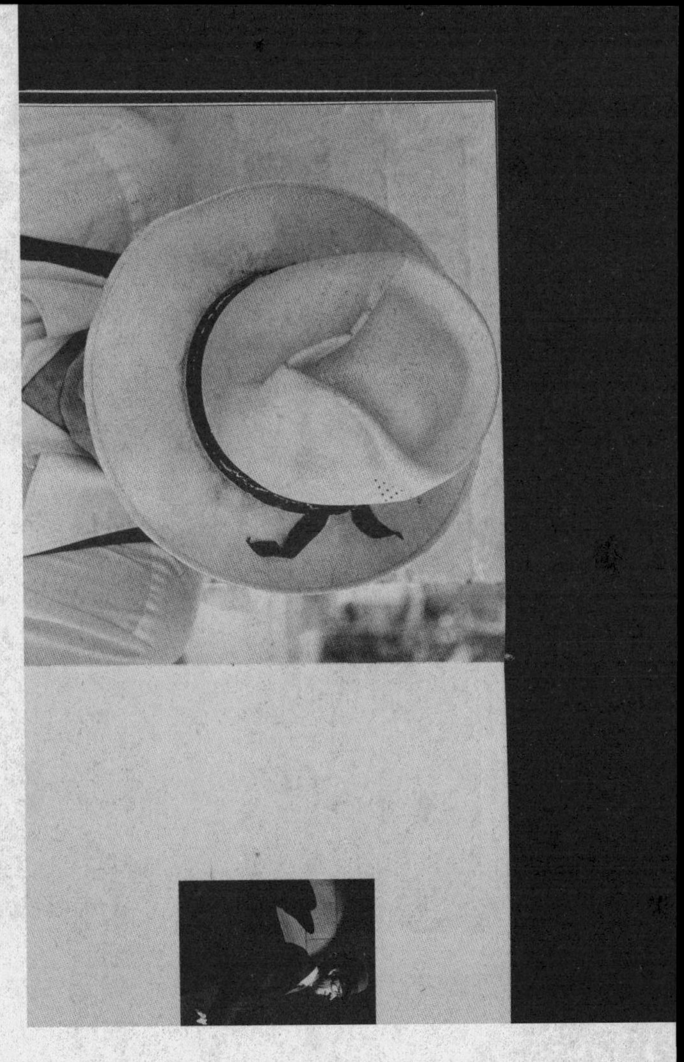

1995 Bm[1]

ZVAB.com
ZENTRALES VERZEICHNIS ANTIQUARISCHER BÜCHER
www.zvab.com
LIEFERSCHEIN - RECHNUNG
PACKING LIST - INVOICE
Hardy Eppler
PAKLIJST
LISTE DE COLISAGE
CONTIENE DOCUMENTI
BOLLA DI CONSEGNA
LISTA DOS VOLUMES
FÖLGESEDDEL
PACKSEDEL
DOWÓD DOSTAWY
УПАКОВОЧНЫЙ ЛИСТ
Herr
ZHAO QING
QING TANG DESIGN CO.
DABEI XIANG7-3,NANJING HAN
210000 Nanjing
China
CN22
FROM
TO
R

Germany

Germany

China

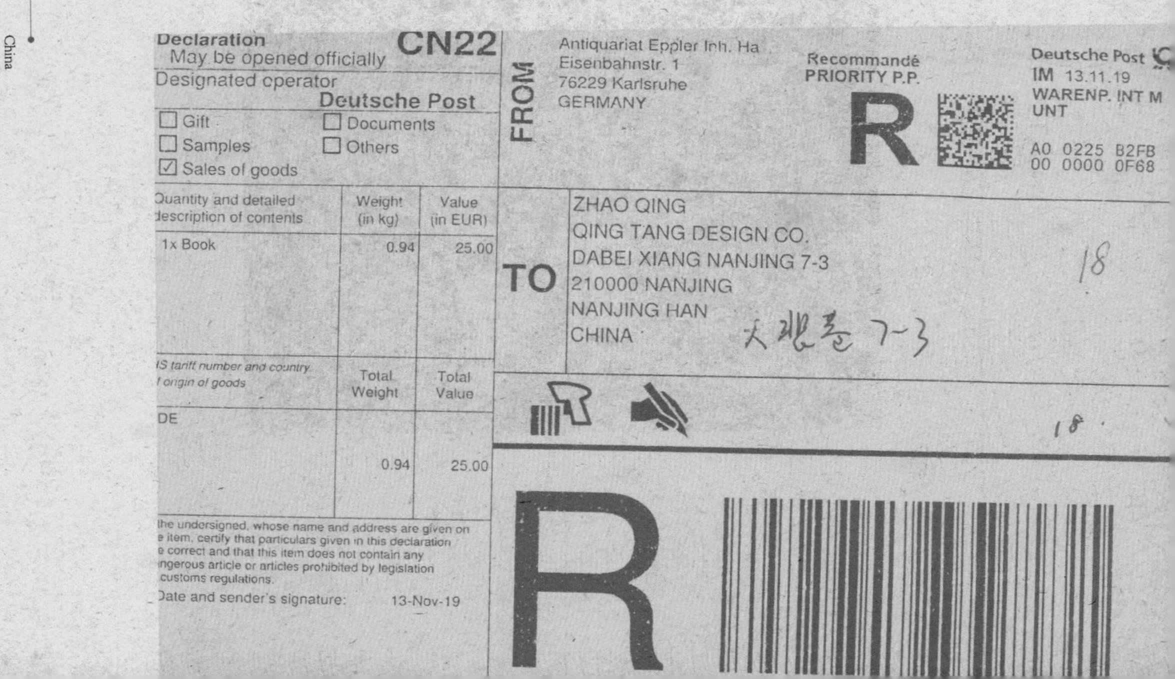

Declaration CN22
May be opened officially
Designated operator
Deutsche Post

☐ Gift ☐ Documents
☐ Samples ☐ Others
☑ Sales of goods

Quantity and detailed description of contents	Weight (in kg)	Value (in EUR)
1x Book	0.94	25.00
HS tariff number and country of origin of goods	Total Weight	Total Value
DE	0.94	25.00

I, the undersigned, whose name and address are given on the item, certify that particulars given in this declaration are correct and that this item does not contain any dangerous article or articles prohibited by legislation or customs regulations.
Date and sender's signature: 13-Nov-19

FROM
Antiquariat Eppler Inh. Ha
Eisenbahnstr. 1
76229 Karlsruhe
GERMANY

Recommandé
PRIORITY P.P.
R

Deutsche Post
IM 13.11.19
WARENP. INT M
UNT
A0 0225 B2FB
00 0000 0F68

TO
ZHAO QING
QING TANG DESIGN CO.
DABEI XIANG NANJING 7-3
210000 NANJING
NANJING HAN
CHINA

大悲巷 7-3

18

18

R

● 铜奖
《 沉寂之屋
◐ 德国

Bronze Medal
camera silens
Germany
Julia Hasting, Patricia Müller, Gerwin Schmidt, Béla Stetzer
Olaf Arndt, Rob Moonen
ZKM Zentrum für Kunst und Medientechnologie, Karlsruhe
265 209 18mm
815g
176p
ISBN-10 3-928201-10-7

本书呈现了艺术家奥拉夫·阿恩特（Olaf Arndt）和罗布·穆恩（Rob Moonen）在德国卡尔斯鲁厄艺术与媒体技术中心创作装置作品*Camera Silens*的完整过程。“Camera Silens”是酷刑室的别称，通过对感官的剥夺来打破囚犯的心理抵抗力。艺术家创造了这样一个空间——墙壁由消声装置包裹，中心放置一把经改造过的医疗椅，屋中设置专门的闭路电视和一盏红灯。一次限制一人进入空间，当回声、视觉、行动均被限制、剥夺甚至控制之后，可以听到的只有呼吸、吞咽、心跳，以及被灯光染红的周遭。这是一种由环境营造的独特体验，以及因此激发的生理反应。书籍的装订形式为无线胶平装。封面黑卡纸印黑，在一定角度呈现反光。内页胶版纸红黑双色印刷。本书对作品的创作、搭建过程做了详细记录，配有大量空间设计图、电路图、手稿、文件以及照片。书中使用黑底白字的打字机文稿，搭配白底单栏黑字，呈现出弗兰肯斯坦（科学怪人）的独特气质。整版红黑印刷的室内照片，将展览现场搬到了纸上。

This book presents the process of creating the installation *Camera Silens* by the artists Olaf Arndt and Rob Moonen at the Karlsruhe Center for Art and Media Technology in Germany. Camera Silens is another name for the torture chamber, which breaks the psychological resistance of prisoners by depriving the senses. The artists created such a space – the walls were wrapped in in silencing devices, a modified medical chair was placed in the center, and a dedicated closed-circuit television and a red light were set in the room. The artists limited the number of people who can enter the space each time. When people's vision and movement were restricted, deprived or even controlled, only breathing, swallowing and heartbeat could be heard. It was a unique experience created by the environment, and the physiological response it stimulated. The book is perfect bound. The cover is made of black cardboard printed in black, reflecting light at a certain angle. The inside pages are printed on offset paper in red and black. The book records the creation and construction process of this work in details, with a large number of space design drawings, circuit diagrams, manuscripts, documents and photos. In the book, main text is printed in black on white background in single column, accompanied with typewriter scripts with white characters on black background. The layout presents the unique temperament of Frankenstein. The full-page of indoor photos printed in red and black moved the exhibition to paper.

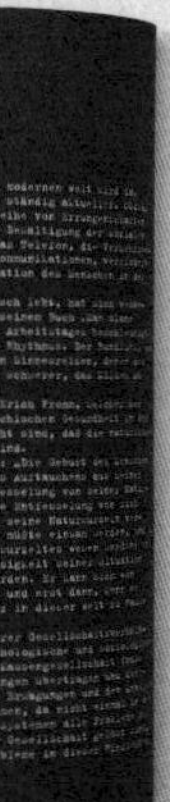

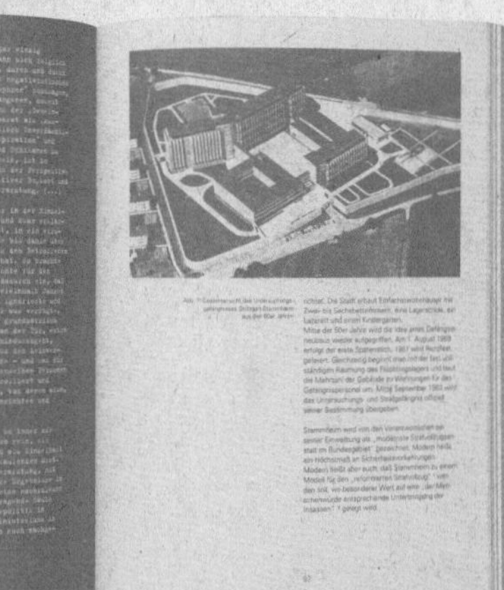

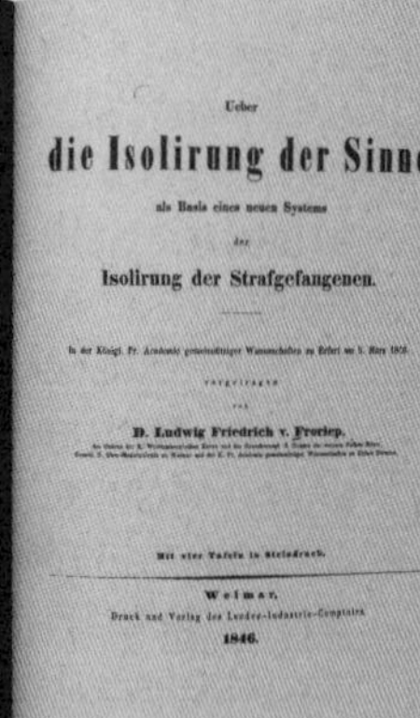
Ueber
die Isolirung der Sinne
als Basis eines neuen Systems
der
Isolirung der Strafgefangenen.
vorgetragen
von
D. Ludwig Friedrich v. Froriep.
Mit einer Tafeln in Steindruck.
Weimar,
Druck und Verlag des Landes-Industrie-Comptoirs
1846.

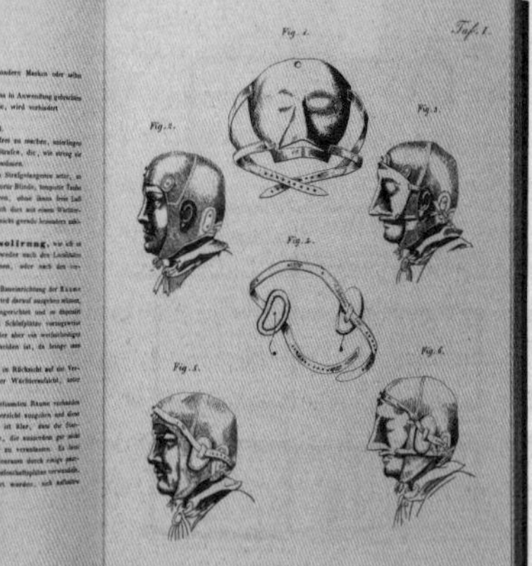
Fig. 1.
Taf. I.
Fig. 2.
Fig. 3.
Fig. 4.
Fig. 5.
Fig. 6.

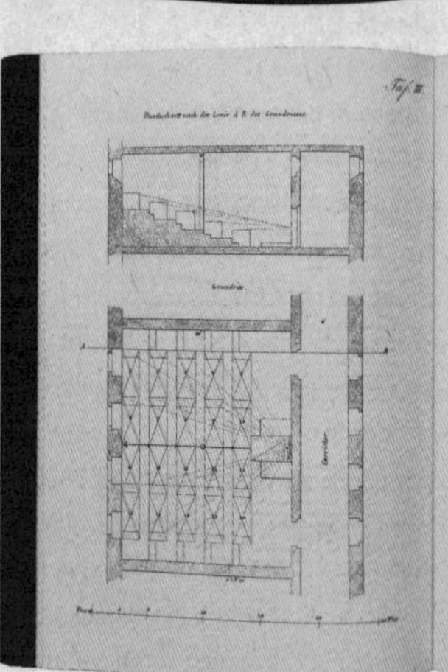
Taf. III.
Grundriss

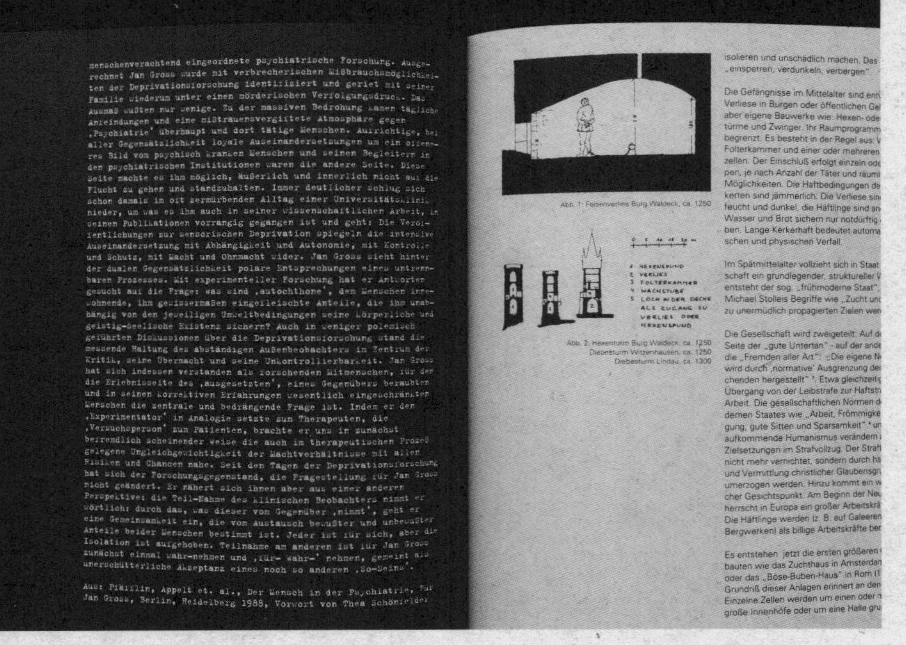

menschenverachtend eingeordnete psychiatrische Forschung. Ausgerechnet Jan Gross wurde mit verbrecherischen Mißbrauchsmöglichkeiten der Deprivationsforschung identifiziert und geriet mit seiner Familie wiederum unter einen mörderischen Verfolgungsdruck. Das Ausmaß wußten nur wenige. Zu der massiven Bedrohung kamen tägliche Anfeindungen und eine mißtrauensvergiftete Atmosphäre gegen ‚Psychiatrie' überhaupt und dort tätige Menschen. Aufrichtige, bei aller Gegensätzlichkeit loyale Auseinandersetzungen um ein offeneres Bild vom psychisch kranken Menschen und seinen Begleitern in den psychiatrischen Institutionen waren die andere Seite. Diese Seite machte es ihm möglich, äußerlich und innerlich nicht auf die Flucht zu gehen und standzuhalten. Immer deutlicher schlug sich schon damals im oft zermürbenden Alltag einer Universitätsklinik nieder, um was es ihm auch in seiner wissenschaftlichen Arbeit, in seinen Publikationen vorrangig gegangen ist und geht: Die Veröffentlichungen zur sensorischen Deprivation spiegeln die intensive Auseinandersetzung mit Abhängigkeit und Autonomie, mit Kontrolle und Schutz, mit Macht und Ohnmacht wider. Jan Gross sieht hinter der dualen Gegensätzlichkeit polare Entsprechungen eines untrennbaren Prozesses. Mit experimenteller Forschung hat er Antworten gesucht auf die Frage: was sind ‚autochthone', dem Menschen innewohnende, ihm gewissermaßen eingefleischte Anteile, die ihm unabhängig von den jeweiligen Umweltbedingungen seine körperliche und geistig-seelische Existenz sichern? Auch in weniger polemisch geführten Diskussionen über die Deprivationsforschung stand die messende Haltung des abständigen Außenbeobachters im Zentrum der Kritik, seine Übermacht und seine Unkontrollierbarkeit. Jan Gross hat sich indessen verstanden als forschenden Mitmenschen, für den die Erlebnisseite des ‚ausgesetzten', eines Gegenübers beraubten und in seinen korrektiven Erfahrungen wesentlich eingeschränkten Menschen die zentrale und bedrängende Frage ist. Indem er den ‚Experimentator' in Analogie setzte zum Therapeuten, die ‚Versuchsperson' zum Patienten, brachte er uns in zunächst befremdlich scheinender Weise die auch im therapeutischen Prozeß gelegene Ungleichgewichtigkeit der Machtverhältnisse mit allen Risiken und Chancen nahe. Seit den Tagen der Deprivationsforschung hat sich der Forschungsgegenstand, die Fragestellung für Jan Gross nicht geändert. Er nähert sich ihnen aber aus einer anderen Perspektive: die Teil-Nahme des klinischen Beobachters nimmt er wörtlich: durch das, was dieser vom Gegenüber ‚nimmt', geht er eine Gemeinsamkeit ein, die vom Austausch bewußter und unbewußter Anteile beider Menschen bestimmt ist. Jeder ist für sich, aber die Isolation ist aufgehoben. Teilnahme am anderen ist für Jan Gross zunächst einmal wahr-nehmen und ‚für- wahr-' nehmen, gemeint als unerschütterliche Akzeptanz eines noch so anderen ‚So-Seins'.

Aus: Pfäfflin, Appelt et. al., Der Mensch in der Psychiatrie, Für Jan Gross, Berlin, Heidelberg 1988, Vorwort von Thea Schönfelder

Abb. 1: Felsenverlies Burg Waldeck, ca. 1250

Abb. 2: Hexenturm Burg Waldeck, ca. 1250
Diebesturm Witzenhausen, ca. 1250
Diebesturm Lindau, ca. 1300

isolieren und unschädlich machen. Das
„einsperren, verdunkeln, verbergen".

Die Gefängnisse im Mittelalter sind ent
Verliese in Burgen oder öffentlichen Ge
aber eigene Bauwerke wie: Hexen- ode
türme und Zwinger. Ihr Raumprogramm
begrenzt. Es besteht in der Regel aus: V
Folterkammer und einer oder mehreren
zellen. Der Einschluß erfolgt einzeln ode
pen, je nach Anzahl der Täter und räuml
Möglichkeiten. Die Haftbedingungen de
kerten sind jämmerlich. Die Verliese sin
feucht und dunkel, die Häftlinge sind an
Wasser und Brot sichern nur notdürftig
ben. Lange Kerkerhaft bedeutet automa
schen und physischen Verfall.

Im Spätmittelalter vollzieht sich in Staat
schaft ein grundlegender, struktureller V
entsteht der sog. „frühmoderne Staat",
Michael Stolleis Begriffe wie „Zucht un
zu unermüdlich propagierten Zielen wer

Die Gesellschaft wird zweigeteilt: Auf d
Seite der „gute Untertan" - auf der ande
die „Fremden aller Art": „Die eigene N
wird durch ‚normative' Ausgrenzung de
chenden hergestellt" [3]. Etwa gleichzeit
Übergang von der Leibstrafe zur Haftstr
Arbeit. Die gesellschaftlichen Normen d
dernen Staates wie „Arbeit, Frömmigke
gung, gute Sitten und Sparsamkeit" [4] un
aufkommende Humanismus verändern
Zielsetzungen im Strafvollzug. Der Straf
nicht mehr vernichtet, sondern durch ha
und Vermittlung christlicher Glaubensgr
umerzogen werden. Hinzu kommt ein w
cher Gesichtspunkt. Am Beginn der Neu
herrscht in Europa ein großer Arbeitskrä
Die Häftlinge werden (z. B. auf Galeeren
Bergwerken) als billige Arbeitskräfte be

Es entstehen jetzt die ersten größeren
bauten wie das Zuchthaus in Amsterdan
oder das „Böse-Buben-Haus" in Rom (
Grundriß dieser Anlagen erinnert an den
Einzelne Zellen werden um einen oder n
große Innenhöfe oder um eine Halle gru

anische Revolution eine ganze Reihe von E
chleunigung, Vertiefung und zur Bewältigu
kation mit sich brachte, z.B. das Telefon
igung, die interkontinentalen Kommunikati
e Möglichkeit der sozialen Isolation des
n Gesellschaft nicht. (...)
ieu, in welchem der moderne Mensch lebt,
ewis Mumford beschreibt das in seinem Buc
ermaßen: „Da sich das Tempo des Arbeitsta
veränderte sich der alltägliche Rhythmus.
die Zeitungen, diese Menge von Sinnesre
ausgesetzt ist, machen es immer schwerer,
eren und erfassen zu können."
teressant ist eine Analyse von Erich From
ist, daß die Störungen der psychischen G
n Gesellschaft dadurch verursacht sind, d
nbedürfnisse nicht befriedigt sind.
m beschreibt das folgendermaßen: „Die Geb
schen bedeutet den Anfang seines Auftauch
imat und den Anfang seiner Entfesselung v
enheit. Jedoch ist für ihn diese Entfesse
sehr aufregend: wenn der Mensch seine Nat
wer ist er und wo bleibt er? Er müßte ein
ne Heimat bleiben, wie ein entwurzeltes
seine Einsamkeit und die Hilflosigkeit se
rtragen. Er müßte wahnsinnig werden. Er k
Naturwurzeln nur dann befreien, und erst
efunden hat, kann er sich wieder in diese

uerglut der heißen Zündeleien
ndeter Jugendlicher und dem s
nverständnis vieler Älterer ihr L
n. Hinzu kam die Erkenntnis, c
iche Systeme offensichtlich el
sitzen, auf jeweils höchstem t
die Potentiale ihrer Destruktic
zu entwickeln, zu konstruieren
n, als aus einer braunen oder
t die notwendigen Lehren zu z
beit von BBM weniger kontrolli
ufwendiger Maschinenschlach
is auf deren ungewisses Ende
chnische Spielerei als szenisch
dustriellen Dilemmas (BBM, 19

möchte ich glauben, dass sich dies mit einem Wächt
allerdings ausgesucht, aber nicht gerade besonders za

ystem der Sinnisolirung, wie ich
sen zulässt, welche sich entweder nach den Localitä
ftigungen der Strafgefangenen, oder nach den v
gefangenen richten werden.
Localitäten wird man die Baueinrichtung der Räu
hängig machen, d. h., man wird darauf ausgehen müss
welche amphitheatralisch eingerichtet und so dispo
s-, Speise-, Arbeits- und Schlafplätze vorzugsw
erden können, unter einander aber ein wechselseiti

rung

Basis eines neue

der

der Stra

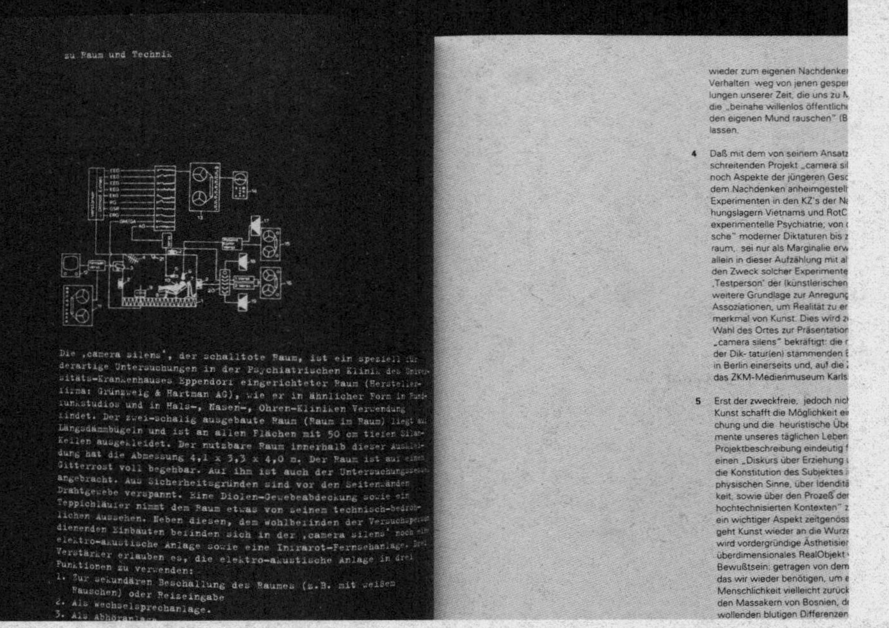

zu Raum und Technik

Die ‚camera silens', der schalltote Raum, ist ein speziell für derartige Untersuchungen in der Psychiatrischen Klinik des Universitäts-Krankenhauses Eppendorf eingerichteter Raum (Herstellerfirma: Grünzweig & Hartman AG), wie er in ähnlicher Form in Rundfunkstudios und in Hals-, Nasen-, Ohren-Kliniken Verwendung findet. Der zwei-schalig ausgebaute Raum (Raum im Raum) liegt auf Längsdämmbügeln und ist an allen Flächen mit 50 cm tiefen Sillan-Keilen ausgekleidet. Der nutzbare Raum innerhalb dieser Auskleidung hat die Abmessung 4,1 x 3,3 x 4,0 m. Der Raum ist auf einem Gitterrost voll begehbar. Auf ihm ist auch der Untersuchungssessel angebracht. Aus Sicherheitsgründen sind vor den Seitenwänden Drahtgewebe verspannt. Eine Diolen-Gewebeabdeckung sowie ein Teppichläufer nimmt dem Raum etwas von seinem technisch-bedrohlichen Aussehen. Neben diesen, dem Wohlbefinden der Versuchsperson dienenden Einbauten befinden sich in der ‚camera silens' noch eine elektro-akustische Anlage sowie eine Infrarot-Fernsehanlage. Drei Verstärker erlauben es, die elektro-akustische Anlage in drei Funktionen zu verwenden:

1. Zur sekundären Beschallung des Raumes (z.B. mit weißem Rauschen) oder Reizeingabe
2. Als Wechselsprechanlage.
3. Als Abhöranlage.

wieder zum eigenen Nachdenke[r]
Verhalten weg von jenen gespe[r]
lungen unserer Zeit, die uns zu M
die „beinahe willenlos öffentlich
den eigenen Mund rauschen" (B
lassen.

4 Daß mit dem von seinem Ansatz
schreitenden Projekt „camera si
noch Aspekte der jüngeren Gesc
dem Nachdenken anheimgestell
Experimenten in den KZ's der N
hungslagern Vietnams und RotC
experimentelle Psychiatrie; von
sche" moderner Diktaturen bis z
raum, sei nur als Marginalie erw
allein in dieser Aufzählung mit al
den Zweck solcher Experimente
‚Testperson' der (künstlerischen
weitere Grundlage zur Anregung
Assoziationen, um Realität zu er
merkmal von Kunst. Dies wird z
Wahl des Ortes zur Präsentation
„camera silens" bekräftigt: die
der Dik- tatur(en) stammenden E
in Berlin einerseits und, auf die
das ZKM-Medienmuseum Karls

5 Erst der zweckfreie, jedoch nic
Kunst schafft die Möglichkeit ei
chung und die heuristische Üb
mente unseres täglichen Leben
Projektbeschreibung eindeutig
einen „Diskurs über Erziehung
die Konstitution des Subjektes
physischen Sinne, über Identitä
keit, sowie über den Prozeß de
hochtechnisierten Kontexten" z
ein wichtiger Aspekt zeitgenöss
geht Kunst wieder an die Wurz
wird vordergründige Ästhetisier
überdimensionales RealObjekt
Bewußtsein; getragen von dem
das wir wieder benötigen, um e
Menschlichkeit vielleicht zurück
den Massakern von Bosnien, d
wollenden blutigen Differenzen

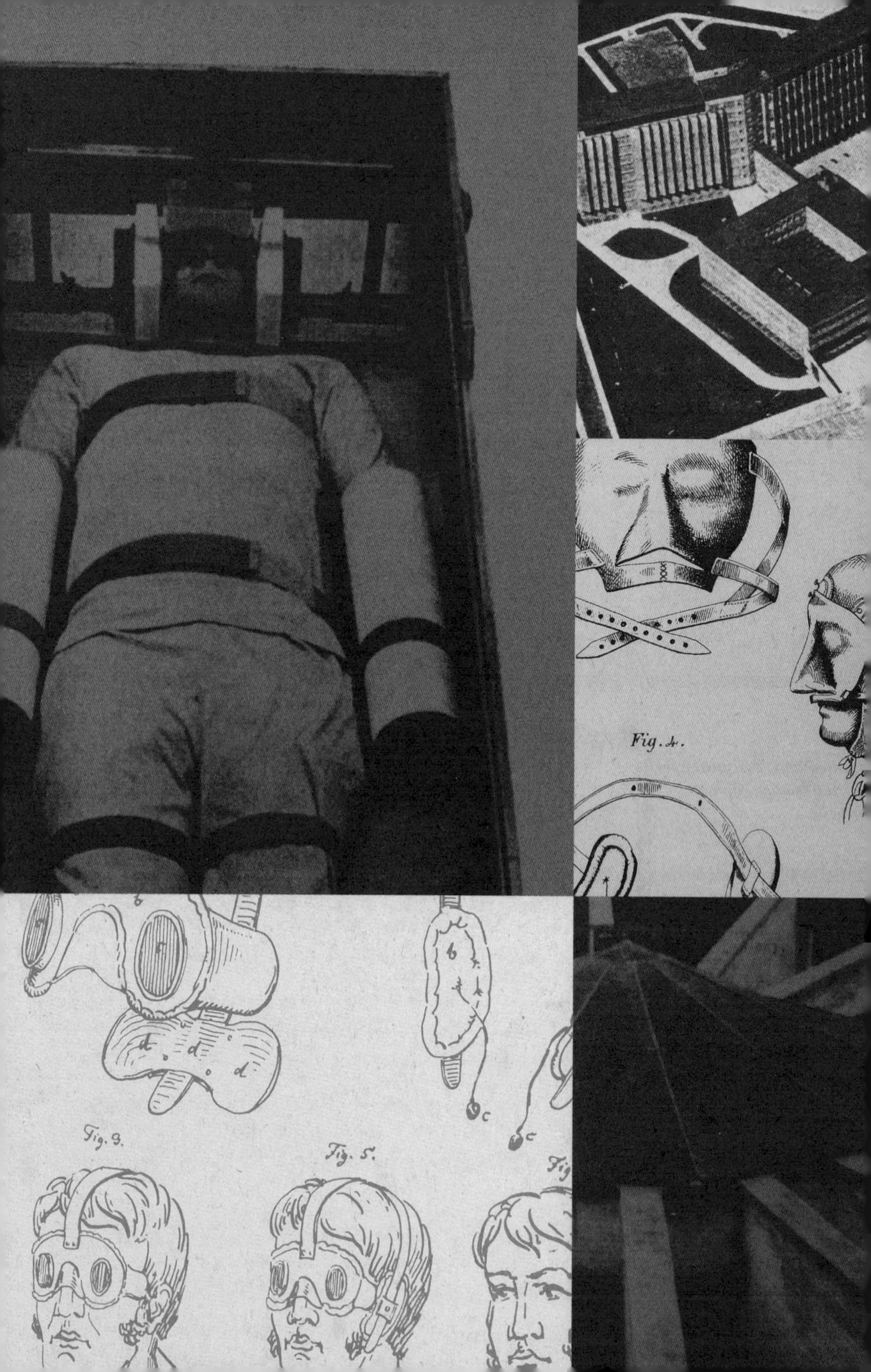
Fig. 4.
Fig. 3.
Fig. 5.

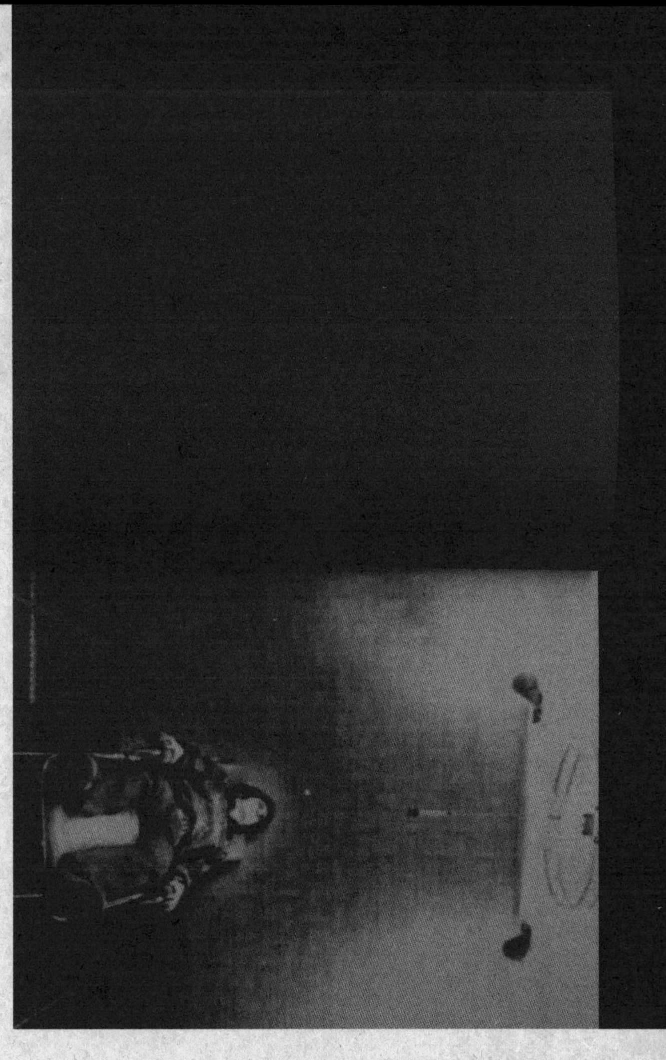

1995 Bm²

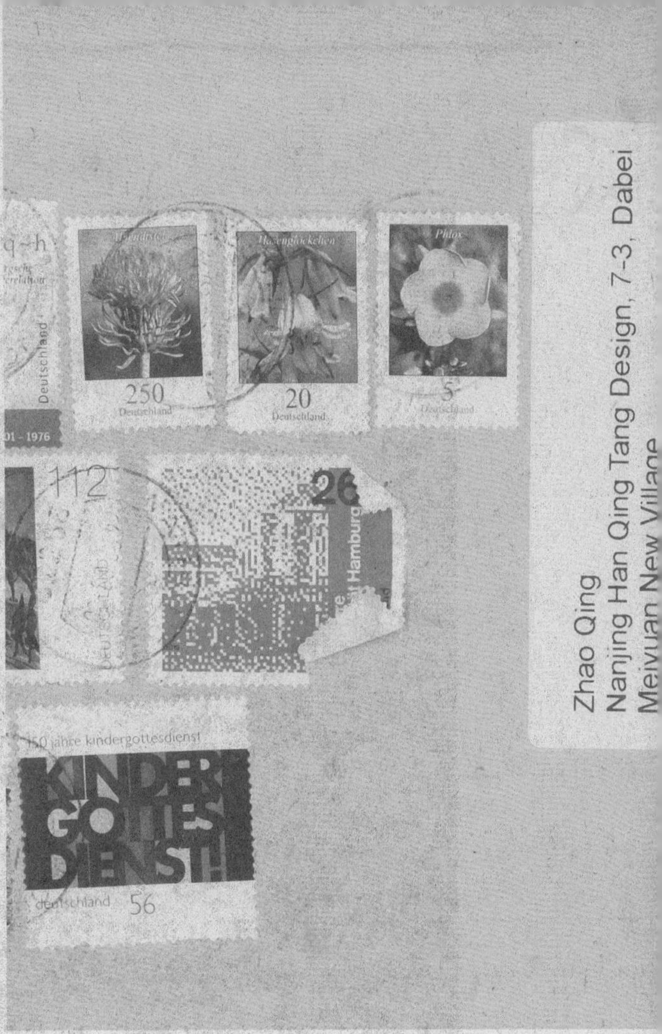

Zhao Qing
Nanjing Han Qing Tang Design, 7-3, Dabei
Meiyuan New Village

Germany

Germany China

340mm

18mm

242mm

铜奖 Bronze Medal

直立
——电梯和扶梯的垂直运输文化史

Vertikal: Aufzug Fahrtreppe Paternoster
Eine Kulturgeschichte vom Vertikal-Transport

德国 Germany

Grappa-Design

Vittorio Magnano Lampugnani, Lutz Hartwig

Ernst & Sohn, Berlin

340 242 18mm

1209g

144p

ISBN-10 3-433-02480-4

古时飞升的梦想在当今已成为日常，如果电梯这种安全、平稳又高效的技术没有被发明出来，现代高楼构建的“垂直”城市将不可能实现。本书邀请10位作者撰写了12篇文章，记录了电梯的各个方面，共同构建出完整的“垂直运输”文化史。书籍的装订形式为锁线硬精装。护封铜版纸红黑绿三色印刷，覆哑膜。内封黑卡纸大面积烫银，裱覆卡板。环衬选用粗糙的黑色艺术纸，内页哑面铜版纸四色印刷，其中各种电梯样式的彩色照片放置在最前部的页面中，之后单黑印刷。10位作者分别为艺术科学家、工程师、电影史学家和建筑师，他们从历史、技术、电影、艺术、设计和社会生活中具有代表性的小故事着手进行写作。版式方面，单栏中轴对称设置，栏宽只有页宽的一半，留出左右两侧巨大的空白，与护封上的电梯井照片和内封上的大面积烫银位置对应。正文排在“电梯井”这一单栏中，两侧的空间用于添加小图和90°旋转排版的附注信息，恰似电梯井周边的辅助设备。

The dream of soaring in ancient times has become a common reality today. Without the safe, stable, and efficient invention of elevators, a 'vertical' city constructed by modern tall buildings would not be possible. This book invites 10 authors to write 12 articles, recording all aspects of the elevators and jointly constructing a complete cultural history of 'vertical transportation'. The hardcover book is bound by thread sewing. The jacket is made of art paper printed in three colors of red, black and green, and covered with matte film. The inside cover is made of black cardboard treated with silver hot stamping, and mounted with cardboard. The end paper is made of rough black art paper, and the inner pages are printed on matte coated paper in four colors. The color prints of various elevators are placed on the front page, and then printed in single black. The ten authors are art scientists, engineers, historians, and architects, starting with writing stories that are representative of history, technology, film, art, design, and social life. The layout is axisymmetric in a single column, and the column width is only half of the page width, leaving a huge blank on the left and right sides, corresponding to the photo of the elevator shaft on the jacket and the large area in silver on the inner cover. The text is arranged in a single column of 'elevator shaft'. The space on both sides is used to add information such as small pictures and notes, just like the auxiliary equipment around the elevator shaft.

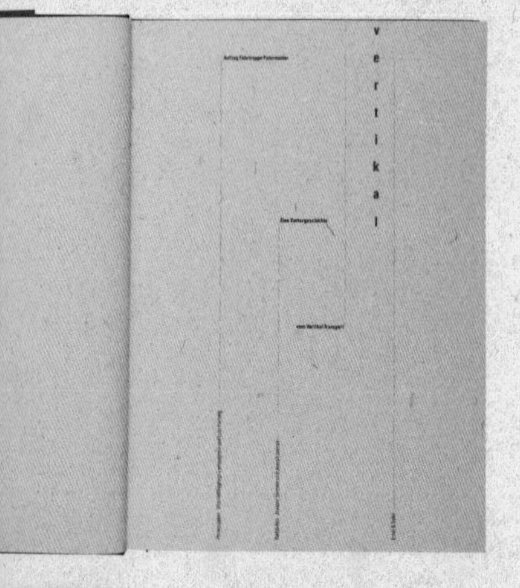
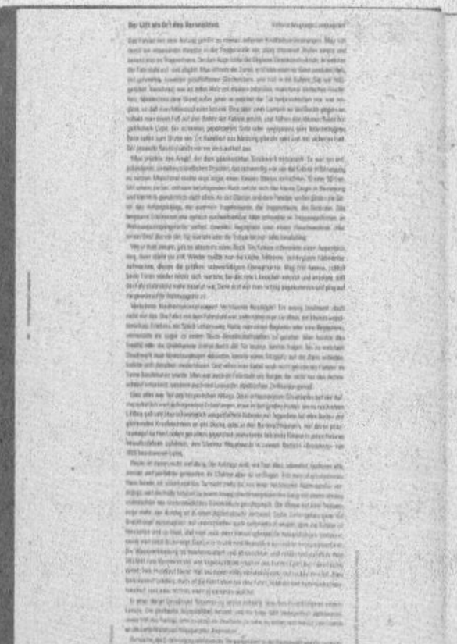
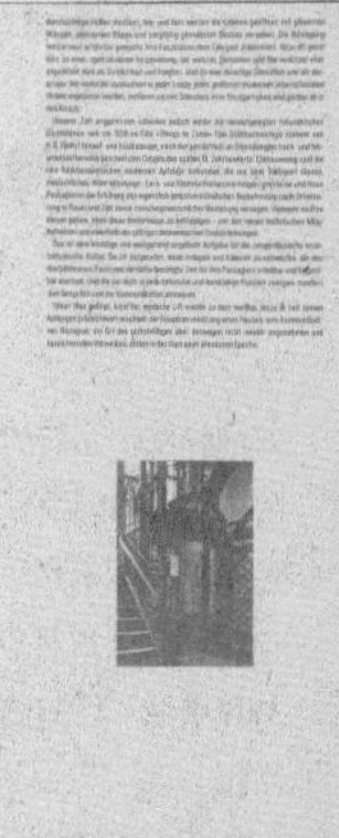
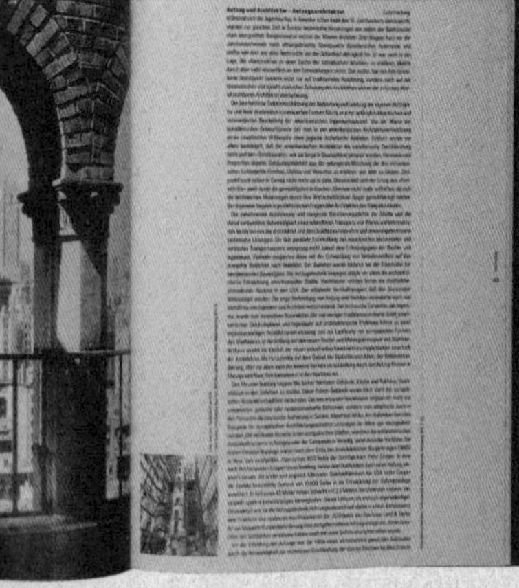
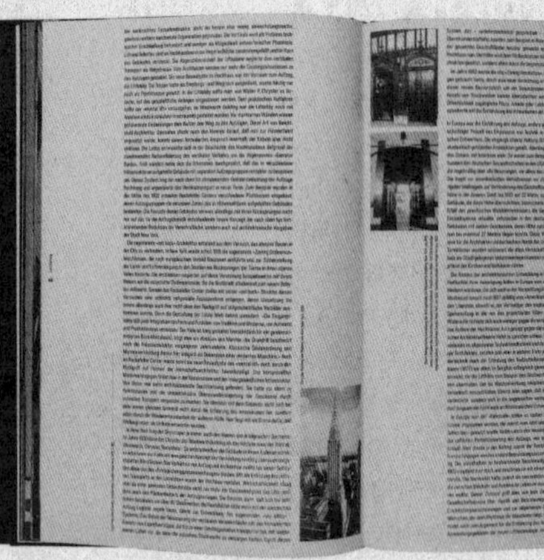
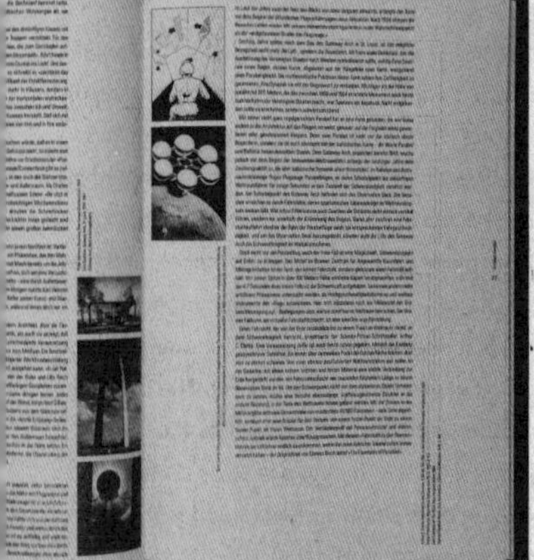

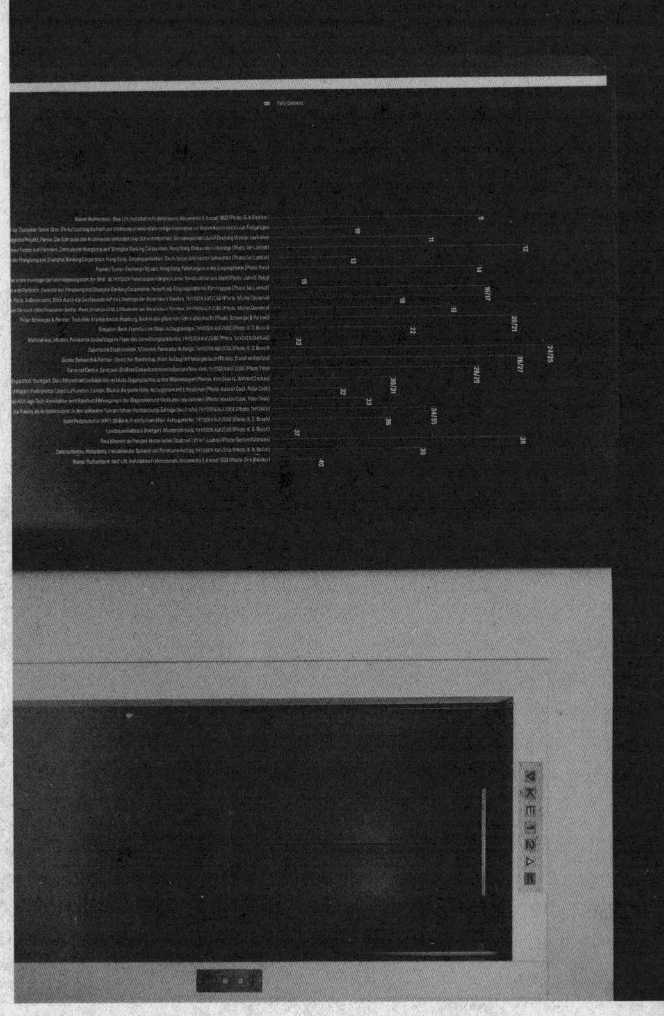

Eisenbahnen werden in der (technischen) Moderne konkret und verwirklicht. Di
lungen verwandeln den Fahrstuhl oder den Lastenaufzug zum sicheren, modern
lie automatische Notbremse, der elektrische Antrieb und die Treibscheiben-Transm
Der moderne Lift erschließt Wolkenkratzer in beinah unendliche Höhen. D
bundene Himmelfahrt schafft mit einem Gran Realität jene Imaginationen, die d
aller Völker bilden: die Schwerkraft zu überwinden, in den Himmel mechanisch e
zu fahren, die Erde zu verlassen.

kalfahrt statt Treppenlaufen

erne Bequemlichkeits-Technik überwindet Erschöpfung und Ermüdung.

rika», jenes vielversprechende Wort voll Hoffnung und der Traum eines neu
ns, war Titel und Sehnsuchtspol von Franz Kafkas unvollendetem Roman, in dem d
das Fremde einer fernen und anderen Welt schilderte. Der Protagonist, Karl Ro
, wurde siebzehnjährig von seinen armen Eltern nach Amerika ausgeschifft, weil «i
ienstmädchen verführt und ein Kind von ihm bekommen hatte.» Der jugendliche H
s traumatischen Berichtes bewundert (noch vom Schiff aus) die Monumentalität d
heitsgöttin ... ‹So hoch›, sagte er sich.» Staunend steht er vor den hohen Häusern,
vom Balkon mit dem Operngucker die Straße beobachtet.

ch vielen Enttäuschungen wird Karl Rossmann in New York Liftjunge im Hotel Occid
uerst durchaus erfolgreich: «Schon nach der ersten Woche sah Karl ein, daß er d
t vollständig gewachsen war. Das Messing seines Aufzugs war am besten gepu
r der dreißig anderen Aufzüge konnte sich darin vergleichen». Durch den Besuch ei
rnten, doch lästigen, volltrunkenen Bekannten, ist er gezwungen, seine Kabine zu v
n, und verstößt dadurch gegen die «Dienstordnung der Lifts», was ihn seinen J
t.

ehrenhaft entlassen als Liftjunge, irrt Karl Rossmann, vom Unglück und von der Pol
lgt, zu einem Bekannten, in die Wohnung seines «Retters» Delamarche. An der St
dert Franz Kafka den mühevollen Treppenlauf in die Höhe: «‹Wir sind gleich ob
Delamarche einigemale während des Treppensteigens, aber seine Voraussage wo
nicht erfüllen, immer wieder setzte sich an eine Treppe eine neue in nur unmerk

i k a l

geschichte

14 16/17 28/29 19 34/35 39 22 18 30/31 36 33 13 32 40 15 23 37

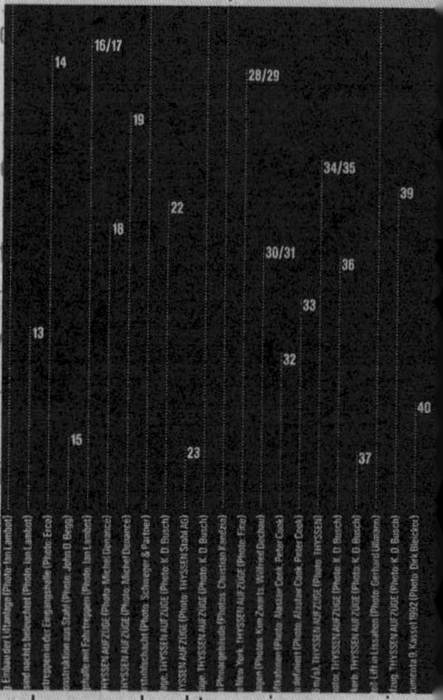

llgemeinen für Gasthöfe weniger zweck-
it dem «Cyclic Elevator konnten in einem
mit Liftanlagen». Die «durchschnittliche
um obersten Geschoß) 2000 Personen.»
ssen, Paternoster-Anlagen machten das

tzer der neuartigen Anlage: «In watching
derly gentlemen avail themselves off it

ffene Kabine»

dankt der Paternoster einem Hamburger
6–1928). Auf einer Englandreise lernte er
1885 ließ er im Hamburger Geschäfts-
etigförderer für Personen einbauen.
hr später umgebaut «wegen der notwen-

weise. In den USA scheint er unbekannt, in der Schweiz existieren ganz wenige Anlagen.

lt», so die Angaben in «Elevator» (Dez. 1990, S. 2). Diese Firma wurde 1912 von Friedrich Kehr
sien (Schreiben, Staatsarchiv Hamburg 1984).

anden war, erbrachte es einen zu geringe
runnenbohrungen (für die Kolbenversenku
ugsantrieb, die damit oft unrentabel wurde.

Schlagartig löste der elektrische Antrieb
tandortunabhängig. Der elektrische Moto
imensioniert und leicht. Elektrizität revol
nergiearten, obwohl elektrische Kraft kein e
rdöl ist. Elektrizität bildet eine praktikable
as zum revoltierenden Medium der Modern

Elektrizität veränderte radikal das überl
ungsprozeß ohne Sauerstoff wird möglich,
erngeleitet werden, wird irgendwo in der Na
s wundert deshalb kaum, daß die Entwicklu
nlage einen etwas kuriosen Irrweg nahm, w
1880 führte Werner von Siemens (1816–18
alske (1814–1890) auf der Pfalzgau-Ausst

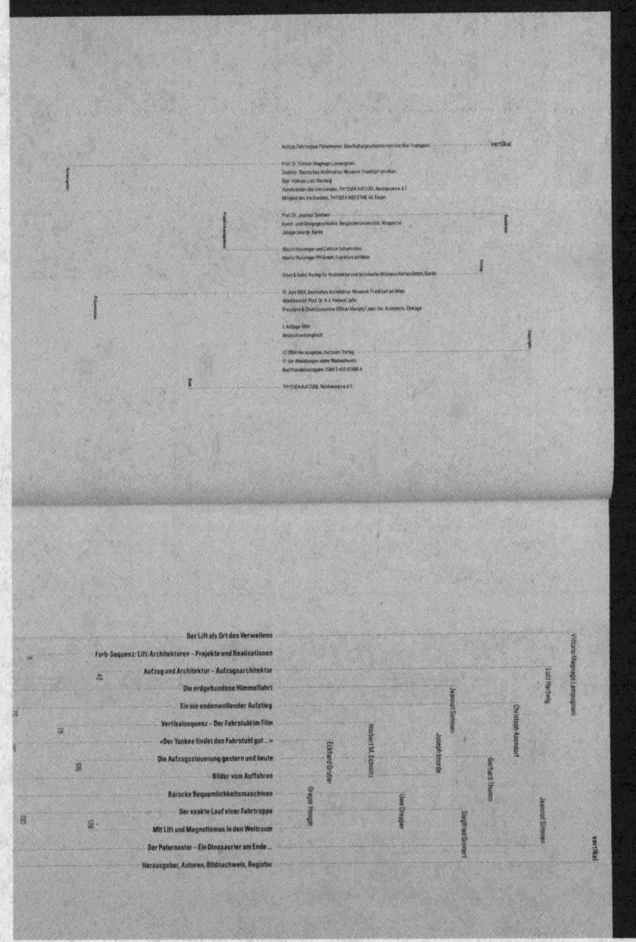

Aufzug, Fahrtreppe, Paternoster. Eine Kulturgeschichte vom Vertikal-Transport — vertikal

Prof. Dr. Vittorio Magnago Lampugnani
Direktor, Deutsches Architektur-Museum, Frankfurt am Main
Dipl.-Volksw. Lutz Harteig
Vorsitzender des Vorstandes, THYSSEN AUFZÜGE, Neuhausen a. d. F.
Mitglied des Vorstandes, THYSSEN INDUSTRIE AG, Essen

Prof. Dr. Jeannot Simmen — Redaktion
Kunst- und Designgeschichte, Bergische Universität, Wuppertal
Jatagn inverte, Berlin

Moritz Hunzinger und Cathrin Scharnstein
Moritz Hunzinger PR GmbH, Frankfurt am Main

Ernst & Sohn, Verlag für Architektur und technische Wissenschaften GmbH, Berlin — Verlag

17. Juni 1994, Deutsches Architektur-Museum, Frankfurt am Main
Werkbericht: Prof. Dr. h. c. Helmut Jahn
President & Chief Executive Officer Murphy/Jahn Inc. Architects, Chicago

1. Auflage 1994
deutsch und englisch

Buchhandelsausgabe, ISBN 3-433-02480-4

THYSSEN AUFZÜGE, Neuhausen a. d. F.

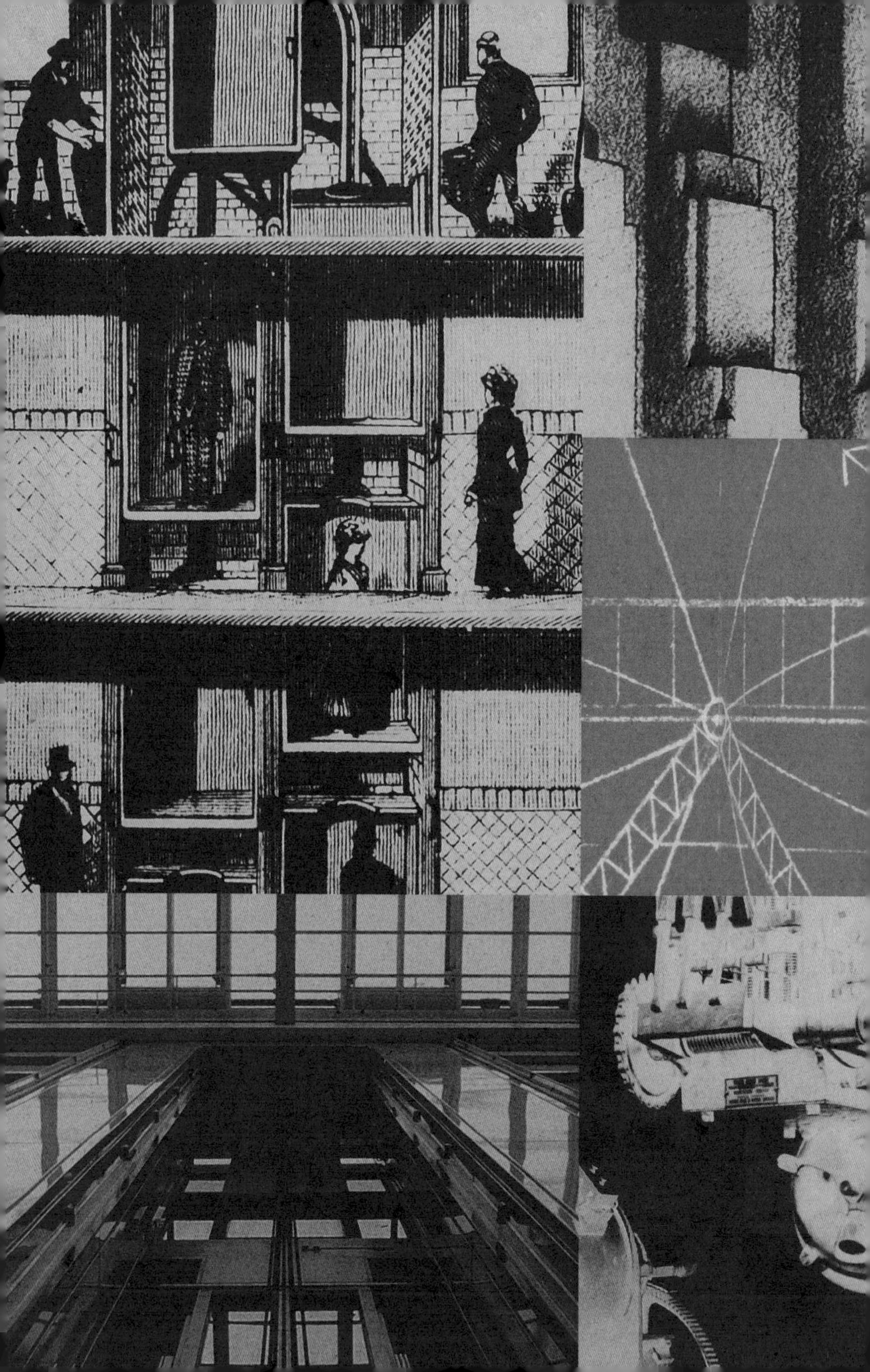

1995 Bm4

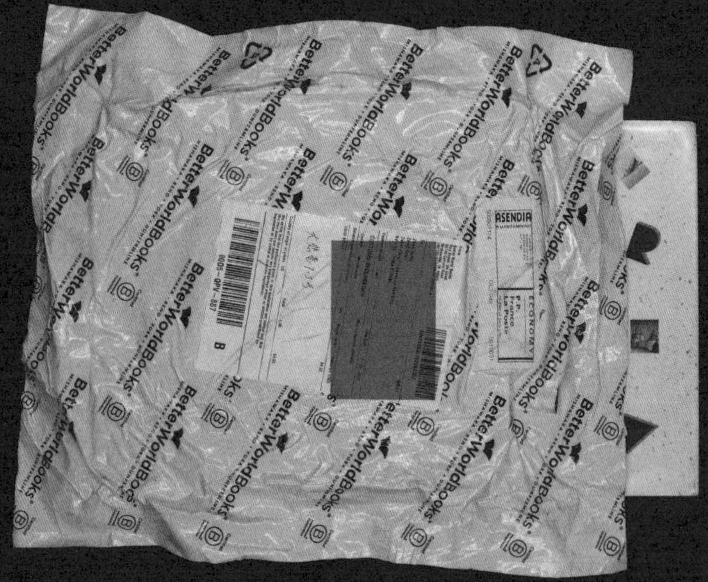

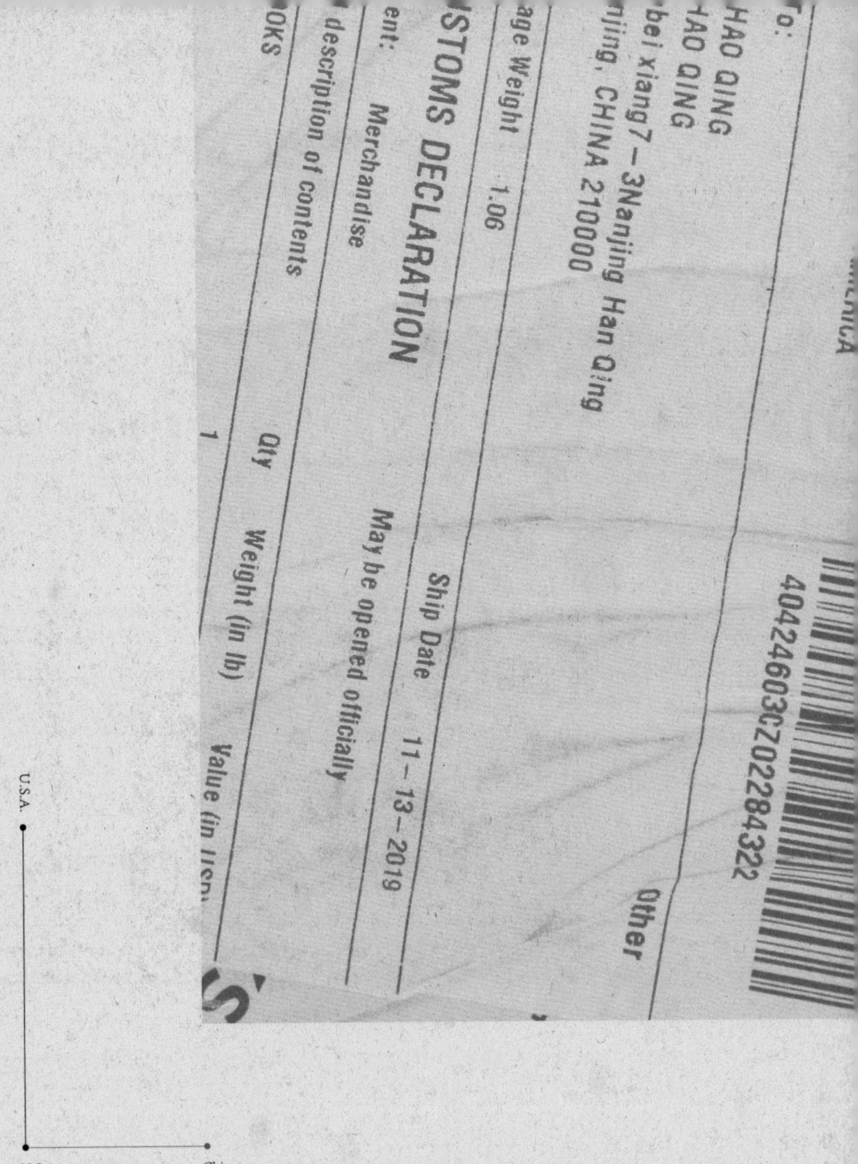
To:
HAO QING
HAO QING
bei xiang7－3Nanjing Han Qing
njing, CHINA 210000
Other
40424603CZ02284322
age Weight 1.06
STOMS DECLARATION
Ship Date 11－13－2019
ent: Merchandise
May be opened officially
description of contents
Qty
Weight (in lb)
Value (in
OKS
1
U.S.A.
U.S.A.
China

223mm

11mm

280mm

● 铜奖
《 旅程
◐ 美国

Bronze Medal
Journey
U.S.A.
Rita Marshall (Delessert & Marshall)
Guy Billout
Creative Education, Mankato
280 223 11mm
466g
32p
ISBN-10 0-88682-626-8

本书是法国艺术家、插图画家居伊·比尤（Guy Billout）创作的绘本，记录了一个年轻人乘坐一列火车“旅行”的经历。书籍的装订形式为绘本常见的横开本锁线硬精装。护封胶版纸四色印刷覆亮膜，内封红色布面裱覆卡板，文字压凹处理。环衬选用带有大量杂质的手工纸，内页哑面铜版纸四色印刷。作者通过火车窗口这个唯一视角去感受时间和事件的变化。每一幅记录的月份都相差甚远，且不标注年份，这是幽闭空间叙事的典型方式。火车实际并不是火车，而是穿梭的时间，每个窗口记录的是值得铭记的人生节点。男孩在左页望出去的窗口，会在右页呈现完整的窗外景色——各种超自然的或者奇观景象映入眼帘。故事的终点，男孩已成为老人，走下火车，来到铁轨的终点——断崖处，望向已远去的轮船。表层故事，这是一个年轻人乘坐长途火车的旅行经历，而里层故事则是一个人经历了一生，有如书籍的副标题“白日梦旅行日记”。

This is a picture book created by French artist and illustrator Guy Billout, which records the experience of a young man ‘traveling’ on a train. It is a typical picture book, which is hardbound with thread sewing. The laminated jacket is printed in four colors. The inner cover is made of red cloth mounted with cardboard and the text is embossed. The end paper is made of handmade paper with a lot of impurities, and the inner pages are printed on matte coated paper in four colors. The author felt the changes of time and events from the unique perspective of train windows. The time recorded in each photo is very different, and the year is not marked. This is a typical way of narrating claustrophobic space. The train is not actually a train, but shuttle time. What is recorded in each window is a life node worth remembering. The window that the boy looks out on the left page will show the full view of the window on the right page – all kinds of supernatural or spectacle images. At the end of the story, the boy has become an old man, gets off the train, and comes to the end of the railroad track – the cliff, looking at the ship that has gone away. From the superficial narrative point of view, this is a young man’s travel experience on a long-distance train, while the inner narrative is a person’s life experience, just like the subtitle Daydream Travel Diary.

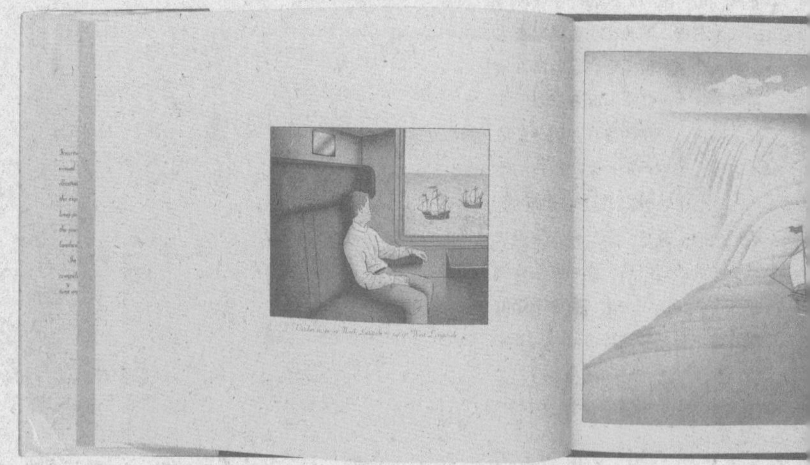

G U Y

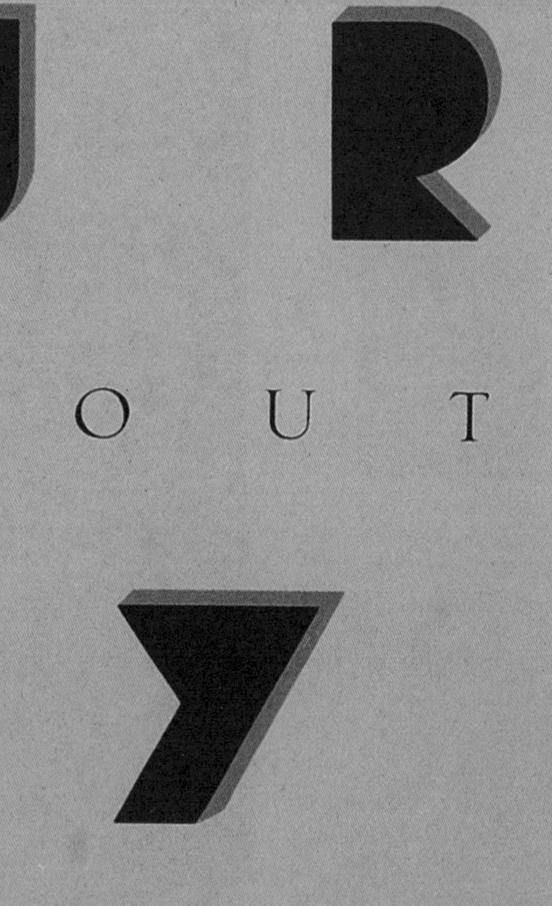

xt and illustrations ©1993 Guy Billout
'acket designed by Rita Marshall
ıblished in 1993 by Creative Education,
ad Street, Mankato, Minnesota 56001 USA
cation is an imprint of Creative Education, Inc.

f Congress Cataloging-in-Publication Data
Guy. Journey/illustrated by Guy Billout.
'ustrations without words depict scenes out of a train
in moves through the countryside – and through time.
l 0-88682-626-8 Printed in Italy.
ut words. 2. Railroads-Trains-Fiction.) I. Title.
PZ7.B4997Jo 1993 93-17094
(E) – dc20 CIP

It all seems to have happened so long ago.
The train must have traveled for many miles, for
many years. Or maybe it was only for a few minutes.
During my journey I witnessed many events,
and imagined many others in the pages of my diary.
I am an old man now, but my dreams continue to
travel, far beyond the edge of that last cliff.

Y B

1995 Bm5

HALF PRICE BOOKS
HPB.COM
#HPBhaul
CP72 CUSTOMS DECLARATION
PRIORITY MAIL INTERNATIONAL
JAMAICA, NY
CY 102 018 293 US

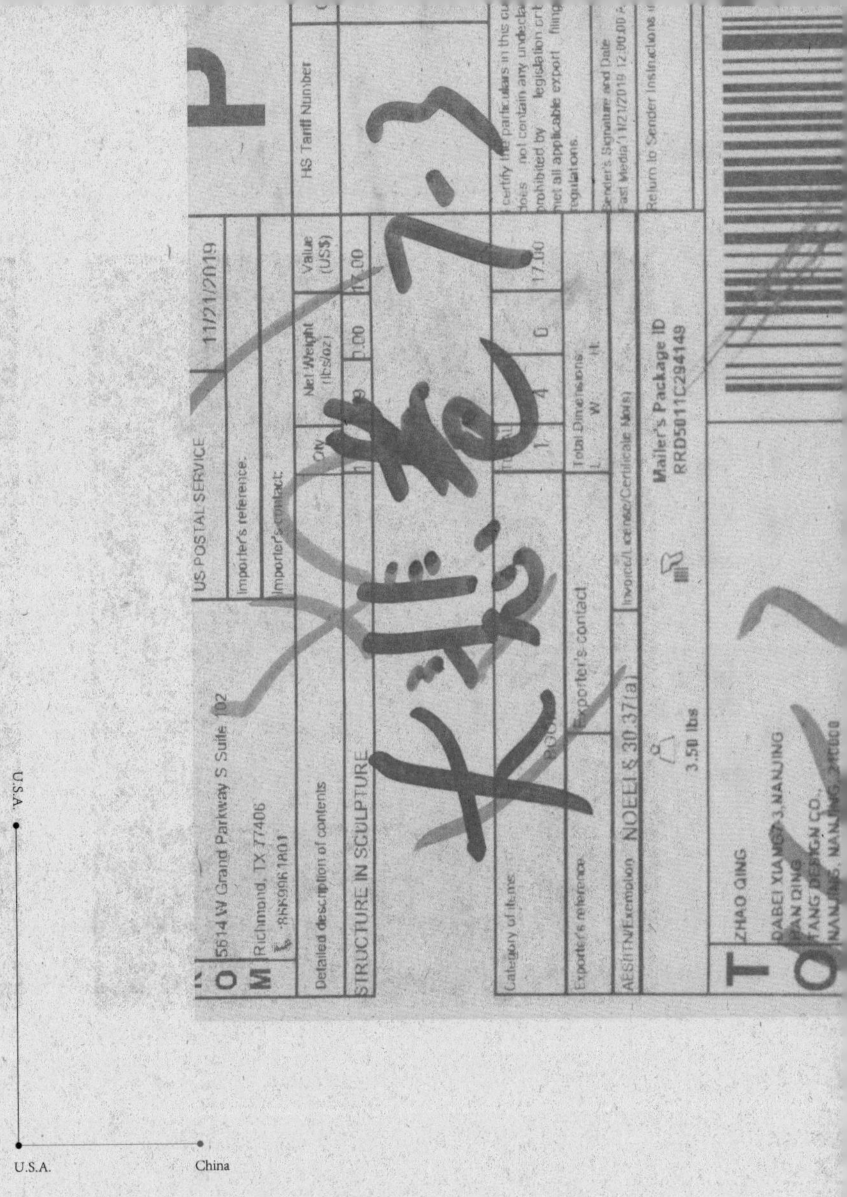

U.S.A.

U.S.A.

China

298mm

26mm

Structure in Sculpture

Daniel L. Schodek

Structure in Sculpture

Schodek

231mm

● 铜奖

《 雕塑中的结构

◖ 美国

Bronze Medal

Structure in Sculpture

U.S.A.

Jeanet Leendertse

Daniel L. Schodek

The MIT Press, Cambridge

298 231 26mm

1556g

328p

ISBN-10 0-262-19313-2

本书作者丹尼尔·L. 舒德克(Daniel L. Schodek)是哈佛大学建筑技术方面的教授，他在本书中讨论了雕塑的结构问题，对于雕塑家有重要的实践指导意义。本书的装订形式为锁线硬精装。护封采用铜版纸四色印刷覆哑光膜，内封黑色布面裱覆卡板，书脊处的书名烫银处理。环衬选用绿色艺术纸与封面所处的公园草地相呼应。内文则采用涂布纸张印单黑。内容方面，本书主要讨论了雕塑结构的几方面问题：稳定性问题（防滑、防倒、防塌）、结构类型（框架、电缆、旋转）、运动结构、土方工程和地貌、材料特性和材料对结构的影响。书中列举了大量实际案例，从罗丹的小型雕塑到自由女神像等大型雕塑装置，配以草图和各种结构图表，方便非工程师了解结构的原理。内文版面设定为四栏，正文跨两栏，注解占一栏，图片根据实际需求进行自由排版，形成了有序而又充满变化的版面风格。

The author of this book, Daniel L. Schodek, is a professor of Architectural Technology at Harvard University. In this book, he discusses the structure of sculpture and has important practical guidance for sculptors. The hard-cover book is bound by thread sewing. The matted jacket is made of coated paper printed in four colors, and the inner cover is made of black cloth and mounted with cardboard. The title of the book is treated with silver hot stamping at the spine. The end paper uses green art paper to match the photo of grass on the cover. The text is printed on coated paper in black. In terms of content, this book mainly discusses several aspects of the structure of sculptures: stability (slip resistance, anti-fall, anti-collapse), structural types (frames, cables, rotation), motion structure, earthwork and landforms, material properties and influences on structure. A large number of practical cases are cited in the book, from Rodin's small sculptures to large sculpture installations such as the Statue of Liberty, together with sketches and various structural diagrams, so that non-engineers can understand the principles of structure. The inside pages are typeset to 4 columns: the text spans 2 columns and the comments occupy 1 column, while the pictures are freely typeset according to the needs, forming an orderly and changing style.

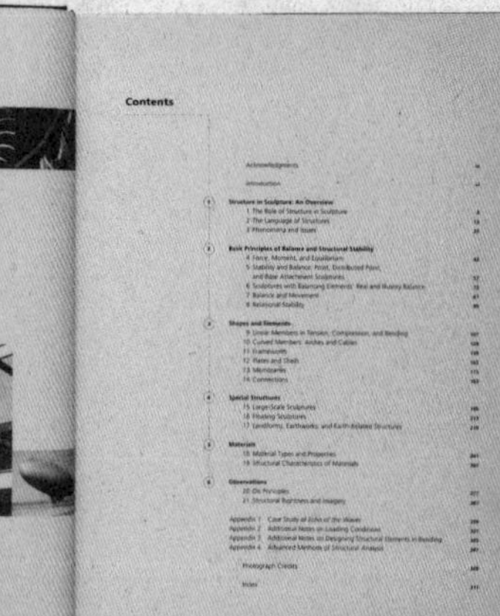
Contents

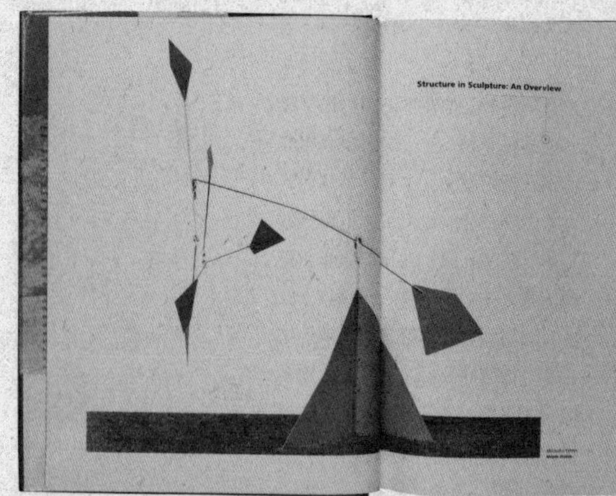
Structure in Sculpture: An Overview

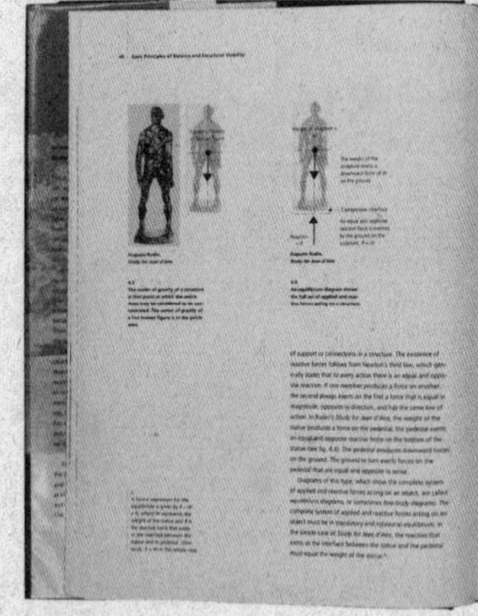

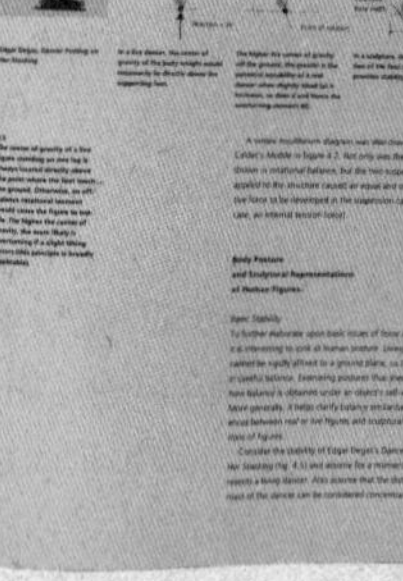

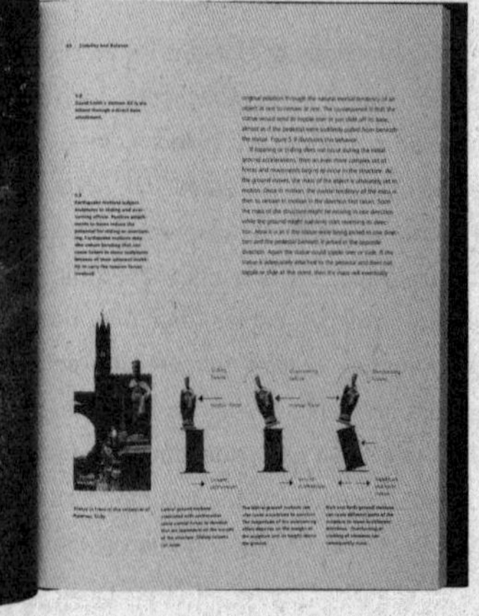

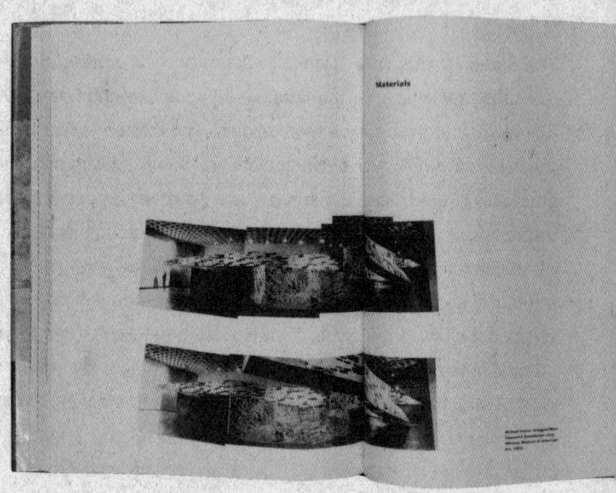
Materials

potentials. That structural precepts may have a positive and broad role in generating forms is a theme that intrinsically underlies the remainder of this book. This fundamental theme will be returned to shortly. The use of structural precepts in a prescriptive sense—that is, to tell a sculptor what to do in a particular situation—will be seen to be neither a real possibility nor of any conceptual interest. This is not now, nor has it ever been in related fields, either a viable or a useful approach (despite the undoubted lingering hope on the part of some that such precepts will help provide "answers" in complex circumstances and ambiguous situations). The use of structural precepts in their prohibitive sense—that is, to tell a sculptor what may not be done—can ultimately contain a measure of truth if the precepts are selected to be used in this way. This view of structural and material concerns is pervasive in the literature but usually appears in a negative light, frequently taken to an extreme. Structural concerns are thus characterized as constraining influences, to be treated dismissively or argued passionately against by sculptors and critics alike.[2] This general attitude is curious and rather naive. Broadly defined material or contextual circumstances have long been part of the field of sculpture. Obvious material examples include limitations imposed by technique. Viewed differently, many of these constraints are in fact opportunities that have provided the essence of many fine sculptural traditions.

Sir Isaac Newton and his successors have simply articulated another contextual dimension that is relevant and that, at a minimum, must simply be respected. Sculptures as forms may well have been shaped to convey some meaning or other intent, but they can exist only in accordance with a largely known set of physical laws that received their first comprehensive postulation by Newton several centuries ago. A sculptor must operate within the context of these laws and may do so knowingly or unknowingly. For physical sculpture, a work not shaped or proportioned such that it exists in a state of equilibrium, for example, will simply fall over or otherwise collapse. It will do this no matter what the desires or artistic intents of the sculptor might be. A sculptor may seem to wish to explicitly contradict these physical laws—Barnett Newman's *Broken Obelisk* (fig. 1.1), discussed in detail in chapter 6, is an interesting case in point—but such works become portrayals of contradictions rather

2
Sidney Geist in *Brancusi* (New York: Grossman Publishers, 1968) goes so far as to refer to materials in the work of Brancusi as serving as a simple "negative" principle (p. 159) that comes into play beyond their more or less useful or adaptable roles, which he did argue were treated feelingly.

1.1
Barnett Newman, *Broken Obelisk*

0497

William Tucker's book *Early Modern Sculpture* in which he discusses Degas's sculpture in terms of "the hard-won equilibrium of volume, surface and silhouette."[1] He goes on to deal further with the *imagery* of balance in Degas's sculptures, but he also notes the potential dependence of some of Degas's forms on the *real* balancing of some of the initial armatures of modeling wax, tallow, and pieces of cork.

The previous chapters have focused on the basic stability of different types of sculptures. This chapter explores more complex balance principles. It deals first with forms that are the direct outgrowth of the real balancing of objects. Calder's mobiles and other similar works surely mean to convey a sense of balance as well as actually be in a state of balance. Next addressed are the technical implications of creating objects that may visually imply a sense of balance but not actually be physically functioning in the way that is visually suggested. Barnett Newman's *Broken Obelisk* (fig. 1.1) is perhaps one of the most elegant of sculptures of this type. Such works seek to reflect states of balance that are at once conceptually possible yet highly improbable. Hence they have intriguing characteristics.

Real Balance

Balanced Elements with Lowered Centers of Gravity

An important balance principle is illustrated by Calder's

Susumu Shingu, *Path of the Wind*

Te
de
a
de
cu

11.4
Susumu Shingu, *Path of the Wind*. When straight members are used in triangulated forms, as is the case in the familiar truss, only tension or compression forces are developed in the members. When a member is curved, detrimental bending may be present that must be taken into account in sizing the member.

Four points or more: redundant but okay to use

Rigid planes: two or more placed at angle (minimum for stability)

Buried end (typically temporary only)

Example

Example

w') can be
ent soil
ravel has a
/ft², aver-
lbs/ft², wet
², and mud

10
See the discussion in chap
In the case of a retaining w
the overall applied force a
on a unit length that tend
cause overturning is given
$F = ½w'h$, where w' is the
equivalent distributed soil
sure and h is the wall heig
The factor of ½ comes fro
fact that the distributed lo
varies from zero to a maxi
so an average value is used
force is considered applied
one-third of the way up fr
the bottom of the wall (at
centroid of the triangular l
area), which reflects the fa
that there is more force ne

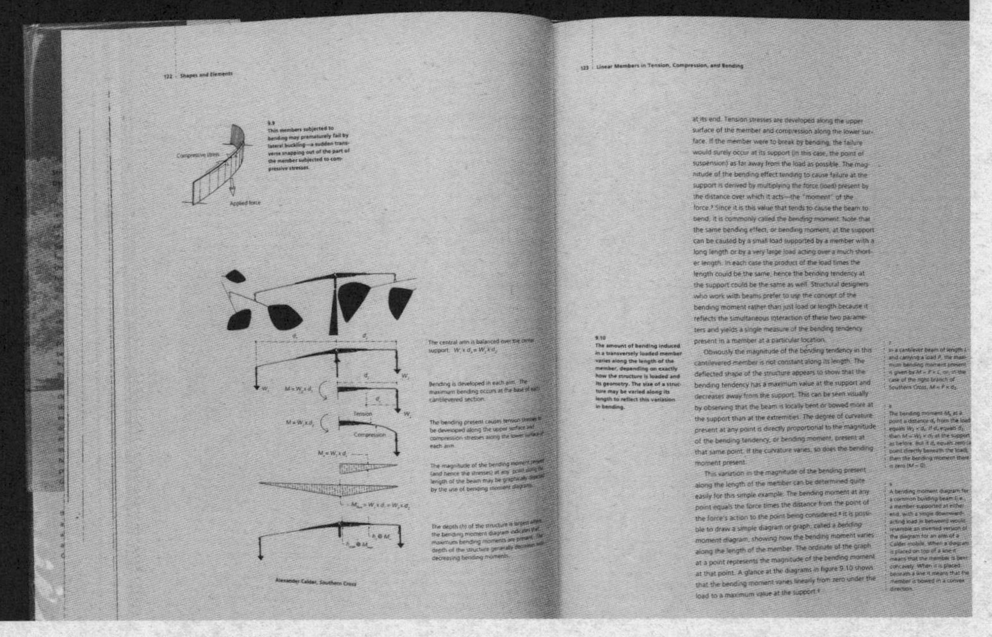

9.9
Thin members subjected to bending may prematurely fail by lateral buckling—a sudden transverse snapping out of the part of the member subjected to compressive stresses.

9.10
The amount of bending induced in a transversely loaded member varies along the length of the member, depending on exactly how the structure is loaded and its geometry. The size of a structure may be varied along its length to reflect this variation in bending.

at its end. Tension stresses are developed along the upper surface of the member and compression along the lower surface. If the member were to break by bending, the failure would surely occur at its support (in this case, the point of suspension) as far away from the load as possible. The magnitude of the bending effect tending to cause failure at the support is derived by multiplying the force (load) present by the distance over which it acts—the "moment" of the force.[7] Since it is this value that tends to cause the beam to bend, it is commonly called the *bending moment*. Note that the same bending effect, or bending moment, at the support can be caused by a small load supported by a member with a long length or by a very large load acting over a much shorter length. In each case the product of the load times the length could be the same; hence the bending tendency at the support could be the same as well. Structural designers who work with beams prefer to use the concept of the bending moment rather than just load or length because it reflects the simultaneous interaction of these two parameters and yields a single measure of the bending tendency present in a member at a particular location.

Obviously the magnitude of the bending tendency in this cantilevered member is not constant along its length. The deflected shape of the structure appears to show that the bending tendency has a maximum value at the support and decreases away from the support. This can be seen visually by observing that the beam is locally bent or bowed more at the support than at the extremities. The degree of curvature present at any point is directly proportional to the magnitude of the bending tendency, or *bending moment*, present at that same point. If the curvature varies, so does the bending moment present.

This variation in the magnitude of the bending present along the length of the member can be determined quite easily for this simple example. The bending moment at any point equals the force times the distance from the point of the force's action to the point being considered.[8] It is possible to draw a simple diagram or graph, called a *bending moment diagram*, showing how the bending moment varies along the length of the member. The ordinate of the graph at a point represents the magnitude of the bending moment at that point. A glance at the diagrams in figure 9.10 shows that the bending moment varies linearly from zero under the load to a maximum value at the support.[9]

7
In a cantilever beam of length L and carrying a load P, the maximum bending moment present is given by $M = P \times L$, or, in the case of the right branch of Southern Cross, $M = P \times d_2$.

8
The bending moment M_x at a point a distance d_x from the load equals $W_2 \times d_x$. If d_x equals d_2, then $M = W_2 \times d_2$ at the support as before. But if d_x equals zero (a point directly beneath the load), then the bending moment there is zero ($M = 0$).

9
A bending moment diagram for a common building beam (i.e., a member supported at either end, with a single downward-acting load in between) would resemble an inverted version of the diagram for an arm of a Calder mobile. When a diagram is placed on top of a line it means that the member is bent concavely. When it is placed beneath a line it means that the member is bowed in a convex direction.

0499

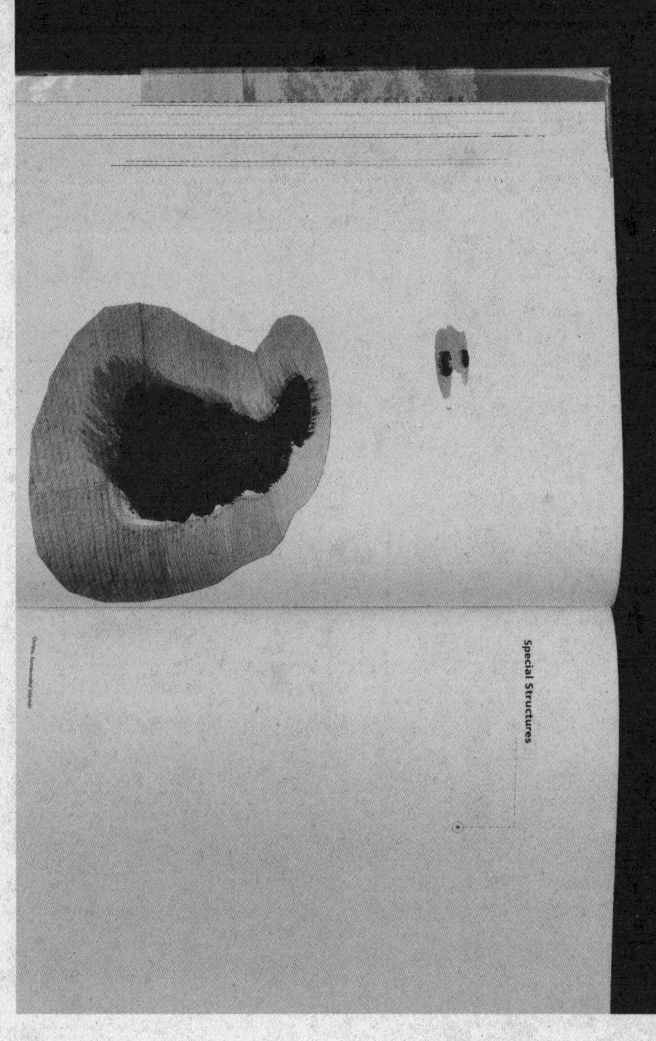
Special Structures
Christo, Surrounded Islands

1995 Ha[1]

Stadt Feldkirch
ZVAB.com
www.zvab.com
Paket bis 5 kg
Deutsche Post
Postal parcels
S.A.L. surface airlifted
Non priority
CP 85
CNNKGA (CNA)
Nanjing
CHINA POST
PVG

Postal parcels
DEFRAA (DEA)
Frankfurt/M
DeutschePost

S.A.L. surface airlifted
Non priority

Desp. Type	Desp. No
BCN	423

Date 2019-11-15

Rec. Type	Rec. No
PC	16

Seal

CNNKGA (CNA
Nanjing
CHINA POST

Receptacle ID

Austria

Germany

China

310mm

20mm

237mm

● 荣誉奖
《 费尔德基希市
◖ 奥地利

Honorary Appreciation
Stadt Feldkirch
Austria
Reinhard Gassner
Christian Mähr, Nikolaus Walter
Amt der Stadt Feldkirch, Feldkirch
310 237 20mm
1469g
192p

费尔德基希市是奥地利西部福拉尔贝格州的一座城市，本书是该市的城市介绍。书籍的装订形式为锁线硬精装。护封铜版纸单黑印刷覆哑膜，内封中灰色布面书名烫黑。环衬选用了黑色艺术纸，内页无酸纸黑棕双色印刷。书籍主要分为两大部分，前一部分通过两篇长文对费尔德基希市的历史和现在进行阐述，版面设置为两栏。后一部分则通过大量黑白照片展现城市的风景、建筑和四季，人们的工作、学习和生活。在呈现历史照片时，黑色中适当加入了一些棕色，与单黑印刷的记录当代的照片拉开了些许距离。

Feldkirch is a city in Vorarlberg in western Austria. This book is an introduction to the city. The hardcover book is bound by thread sewing. The jacket is made of coated paper printed in a single black, covered with matte film. The inside cover is made of gray cloth with the title hot stamped in black. The end paper uses black art paper, and the inner pages are printed with acid-free paper in black and brown. The book is divided into two major parts. The first part explains the history and present of the city of Feldkirch with two long articles. The layout is set to two columns. The latter part uses a large number of black and white photos to show the city's scenery, architecture and seasons, people's work, study and daily life. When presenting historical photographs, color brown is appropriately added to black, which is slightly different from the contemporary photographs recorded in black-and-white.

Christian Mähr

Annäherung an Feldkirch

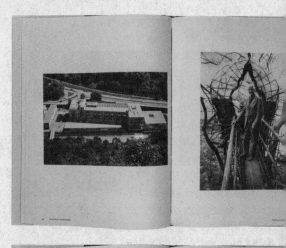

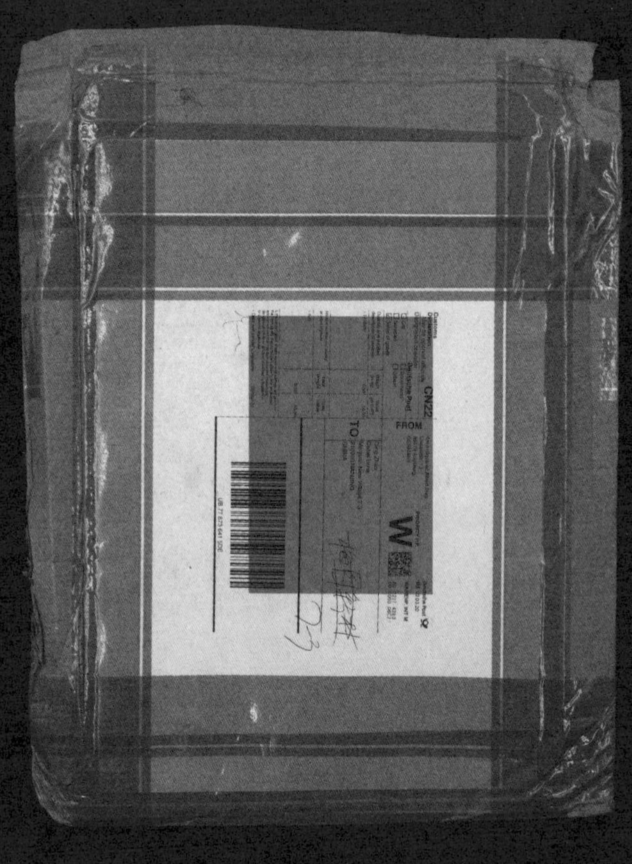

lay be opened officially
esignated operator **Deutsche Post**

- Gift
- Documents
- Samples
- Others
- Sales of goods

antity and detailed scription of contents	Weight (in kg)	Value (in EUR)
x Book	0.50	10.00
tariff number and country rigin of goods	Total Weight	Total Value
ull	0.55	10.00

e undersigned, whose name and address are given on item, certify that particulars given in this declaration correct and that this item does not contain any gerous article or articles prohibited by legislation ustoms regulations.

ate and sender's signature: 02-Mar-20

FROM

Loisachstr. 13
86179 Augsburg
GERMANY

TO

Qing Zhao
Dabei Lane
Meiyuan New Village 7-3
210000 NANJING
CHINA

PRIORITY P.P.

W

Germany

Germany China

296mm

7mm

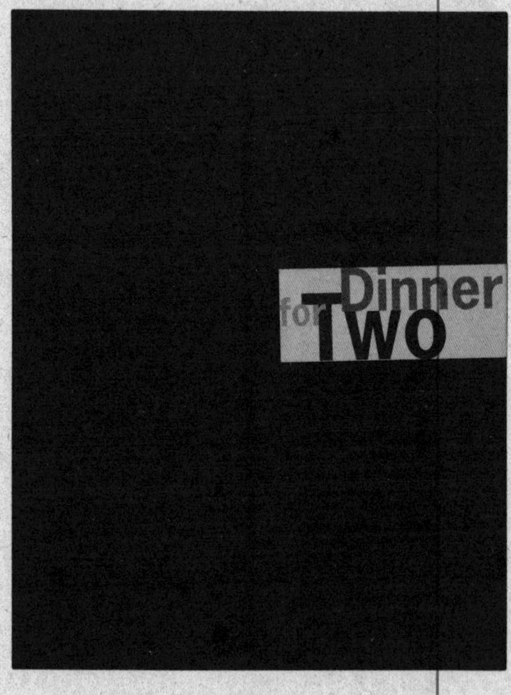

232mm

● 荣誉奖
《 双人晚餐
◐ 德国

Honorary Appreciation
Dinner for Two
Germany
Büro für Gestaltung Biste & Weißhaupt
——
Stiftung Gold- und Silber-schmiedekunst, Schwäbisch Gmünd
296 232 7mm
485g
80p
——

1994 年德国金银匠基金会举办了餐具设计比赛，最终选出 32 件作品进行展览，本书是这次展览的画册。书籍的装订形式为锁线胶平装。封面选用带竖纹的黑色艺术纸，前勒口处裱贴红黑双色印刷的纸张，呈现书名。内文哑面铜版纸红黑双色印刷，作品页面过油处理。书籍开篇对金银加工历史、基金会、赛制和奖项进行阐述。大赛主题为“使用金银工艺为双人晚餐进行设计”，书中作品包含盘、杯、刀、叉、勺、筷、壶、锅、调味瓶等餐具，以及烛台、餐篮等各种相关用餐物品。作品并非单件展示，而是根据主题设计的多件物品的组合。作品经过统一拍摄，黑色背景、灰色台面，并通过专业打光，呈现统一的质感和金属加工的细节品质。

In 1994, the German Goldsmiths Foundation organized a tableware design competition, and finally selected 32 works for exhibition. This book is a picture book of this exhibition. The paperback book is glue bound with thread sewing. The cover is made of black art paper with vertical stripes. The title of the book is displayed on paper printed in red and black attached on the front flap. The inner pages are made of matt coated paper printed in red and black, and the pages of the works are vanished. The book starts with an introduction to the history of gold and silver processing, foundations, competitions and awards. The theme of the contest is ‘Designing a Dinner for Two Using Gold and Silver Crafts’. The works in the book include dishes, cups, knives, forks, spoons, chopsticks, pots, seasoning bottles and other related dining items such as candlesticks and dining baskets. The works are not displayed individually, but as a combination of multiple items designed according to the theme. The works have been photographed uniformly, with a black background and a gray countertop. Through professional lighting, it presents a uniform texture and metal processing details.

Marcus Marienfeld · CH
Christopher Maron · AUS
Anja Merten · D
Axel Meuthen · D
Oliver Michl · D
Angelika Möhrlein · D
Matthias Mönnich · D · 38
Claudia Moosbrugger · D
Mercedes Moragas · D
Nadia Morgenthaler · CH
Christina Weck · D
Peter Müller · D
Kerstin und Johann Müller · D
Sally Mussett · AUS

N

Günther Nahm · D
Bernd Neimke · D
Kurt P. Neukomm · CH
Jean-Pierre Neumeister · D
Leon Nielson · DK
Eva Niemand · D
Tobias Nitschke · D
Debbie Noble · GB

O

Helge Ott · D · 28
Jörg Otto · D

P

Alfred Pain · GB · 66
Edgar Palme · D
Martina Peter · D
Michael Pfister · D · 14
Bernhard Pflug · D
Helena Pichler · D

Sabine te Reh · D
Vera Richter · D
Konrad Richter · D
Barbara Rieger · D
Andreas Rimkus · D
Manfred Rombach · D
Marion Röthig · D
Christian Rudolph · D
Barbara Ryman · AUS

S

Annegret Sanders · D
Karen Sophie Sandler · D
Helena Sandström · S
Richard Schäfer · D · 72
Heike Scheurenberg · D
Jens Schleede · D
J. D. Schleissner · D
Jörg Schmalzhaf · D
Bertolt Schmidt · D
Ute Schmidtchen · D
Lothar Schmitt · D · 50
Frauke Schmitz · D
Gisa Schmitz · D
Peter Schmitz · D · 60
Anke Schneider · D
Gudrun Schreiner · D
Torsten Schröder · D
Andrea Schüle · D
Janna Schulz · D
Ursula und Herbert Schulze ·
Claudia Schuster · D
Barbara Schwab · D
Brigitte Schwarz-Stradmann ·

verbsergebnisse und durch die- rständnis für zeit- gestaltung unter- Silberschmieden Wertschätzung n und handwerk- ntgegenge- eisträgern und ng wie im Kata- l- und Silber- e ich zum Er-

work in Schwäbisch Gmünd goes back many hundreds of years. Schwäbisch Gmünd has always been concerned to support the craft; one important step was the town council's decision to set up a school of draughtsmanship in 1776, when Schwäbisch Gmünd was still a Free Town of the Empire, a school which assisted the town's gold and silversmiths to develop their sense of design. A hundred years later, the Precious Metals Guild built up a collection of master patterns, now in the town's Natural and Municipal History Museum.
Finally, the State College for the Precious Metals Industry was founded to provide continuing facilities for both pre-service and in-service training. Today we can offer various courses; the Training

or Foundatio founded in th Town Counc ciation, the S Silversmiths' kasse Ostalb the things th the craft is t »Municipal G appointing M vite either hi practitioners men, to sper Gmünd. Inter the opportun Municipal Go practices.

The Trust als

geb. 1963
Gold- und Silberschmiedemeister
Hanau

Lachsgedeck für zwei Personen
Silber 925/000
Messer und Gabeln aus Edelstahl,
Schieferplatte

b. 1963
Master gold and silversmith
Hanau

geb. 1966
Diplom Schmuck-
und Gerätdesignerin
Pforzheim
Kaffeekanne
Silber 925/000
Filz, Moosgummi
doppelwandig
Inhalt 0,75 l

995 Ha³

Customs Declaration
May be opened officially
CN22
Deutsche Post
FROM
TO
GERMANY
CHINA
TRACKED
PRIORITY P.P.
Deutsche Post
WARENP. INT M
LB 77 120 227 4DE

Samples
Sales of goods
Others
Quantity and detailed description of contents
Weight (in kg)
Value (in EUR)
- 1x book
1.00
25.80
HS tariff number and country of origin of goods
Total Weight
Total Value
- DE
1.00
25.80
I, the undersigned, whose name and address are given on the item, certify that particulars given in this declaration are correct and that this item does not contain any dangerous article or articles prohibited by legislation or customs regulations.
- Date and sender's signature:
13-Nov-19
TO
ZHAO QING
Nanjing John Qing Tang Design
Dabei xiang 7-3
210000 NANJING
CHINA
Tracked
Germany
Germany
China

213mm

133mm

● 荣誉奖
《 基础知识 1 — 6 册
◐ 德国

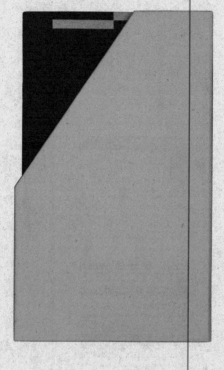

Honorary Appreciation
Grundsetzliches 1-6
Germany
Philipp Luidl
Philipp Luidl
SchumacherGebler, München
213 133 37mm
540g
32p / 42p / 56p / 44p / 32p / 36p
ISBN-10 3-920856-04-X / ISBN-10 3-920856-05-8 / ISBN-10 3-920856-08-2 / ISBN-10 3-920856-09-0 / ISBN-10 3-920856-10-4 / ISBN-10 3-920856-03-1

这是一套 6 本由设计师菲利普·鲁德尔（Philipp Luidl）编写和设计的字体设计和排印基础知识的小册子。6 本小册子均为骑马钉，使用白色布纹纸裱贴卡板制作的异形半盒套将 6 本册子收纳其中，盒套侧边有套书名和册数。6 本册子分别采用六色布纹纸做封面。第 1 册与第 6 册的封面文字印银，其余 4 册印黑。内页书写纸三色印刷，除第 1 册为黑蓝棕三色外，其余 5 册均为黑蓝红三色。6 册的内容分别是：字体与工具、字体历史及风格概述、文艺复兴时期的字体、衬线字体与非衬线字体、字体的变形与书写字体、哥特字体的流变。每册内容量有不小差异，页数也不尽相同。蓝色虚线在内页起到版心线的提示作用，红色则主要用于字体之间的分割与重点标注。

This is a set of basic knowledge of typography and typography written and designed by designer Philipp Luidl. The 6 booklets are all bound by saddle stitching and stored in the special-shaped box which is made of white arlin paper mounted with cardboard. There are the book title and the number of the books printed on the side of the box. The cover of the 6 booklets is made of arlin paper in six colors. The text on the cover of the first and sixth volume is printed in silver, and the remaining 4 volumes are printed in black. The inner pages are printed in three colors. Except for the first volume printed in black, blue and brown, the remaining 5 volumes are printed in black, blue and red. The contents of the six volumes are: fonts and tools, font history and style overview, Renaissance fonts, serif fonts and sans serif fonts, font deformation and writing fonts, and the evolution of Gothic fonts. The content of each volume varies greatly, and the number of pages varies. The blue dotted line on the inner pages serve as a reminder of the typeset area, and the red lines are mainly used for the segmentation and highlighting of fonts.

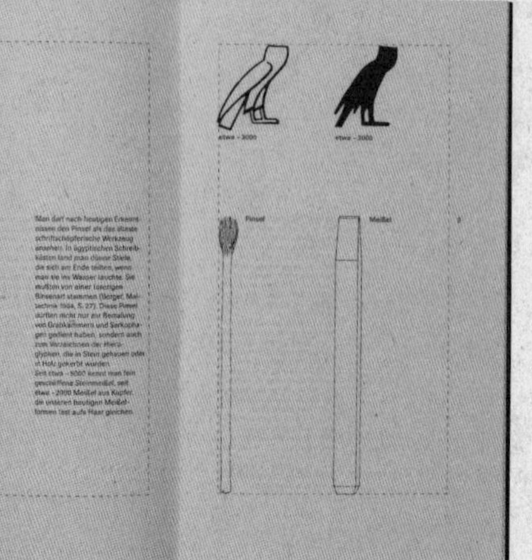

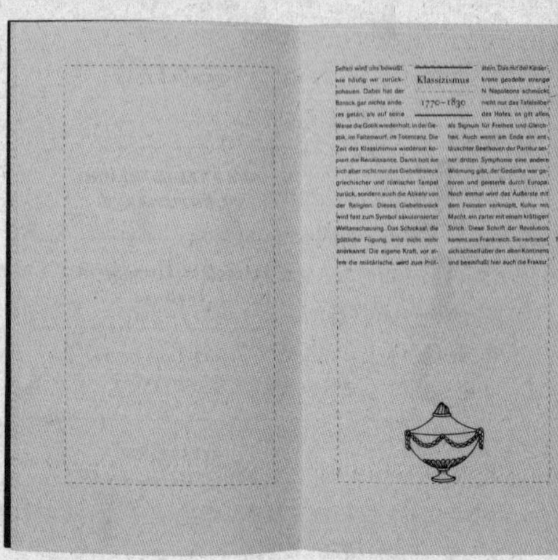
Klassizismus
1770–1830

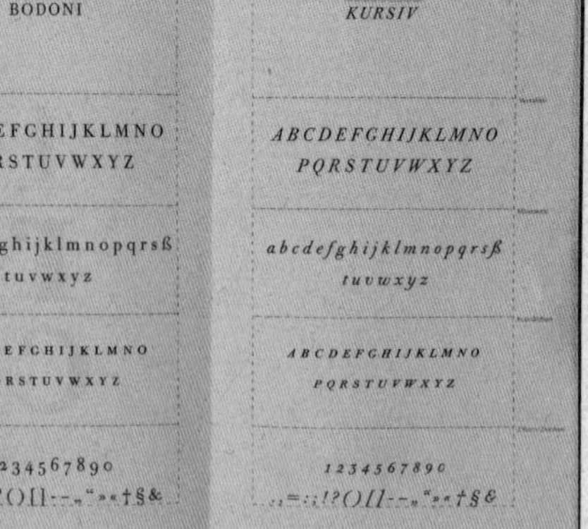
BODONI
KURSIV
ABCDEFGHIJKLMNO
PQRSTUVWXYZ
abcdefghijklmnopqrsß
tuvwxyz
ABCDEFGHIJKLMNO
PQRSTUVWXYZ
1234567890

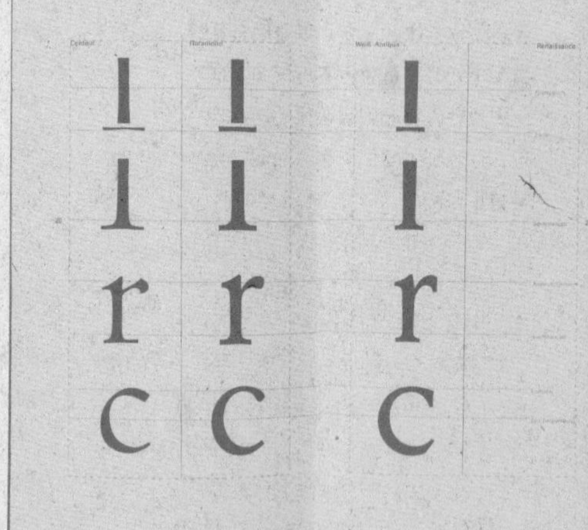

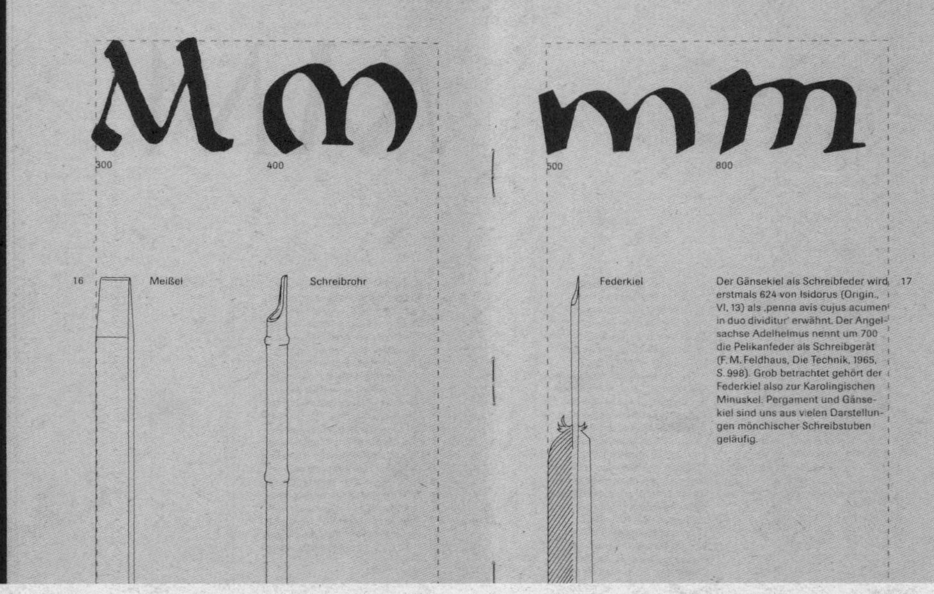

Der Gänsekiel als Schreibfeder wird erstmals 624 von Isidorus (Origin., VI, 13) als ‚penna avis cujus acumen in duo dividitur' erwähnt. Der Angelsachse Adelhelmus nennt um 700 die Pelikanfeder als Schreibgerät (F. M. Feldhaus, Die Technik, 1965, S. 998). Grob betrachtet gehört der Federkiel also zur Karolingischen Minuskel. Pergament und Gänsekiel sind uns aus vielen Darstellungen mönchischer Schreibstuben geläufig.

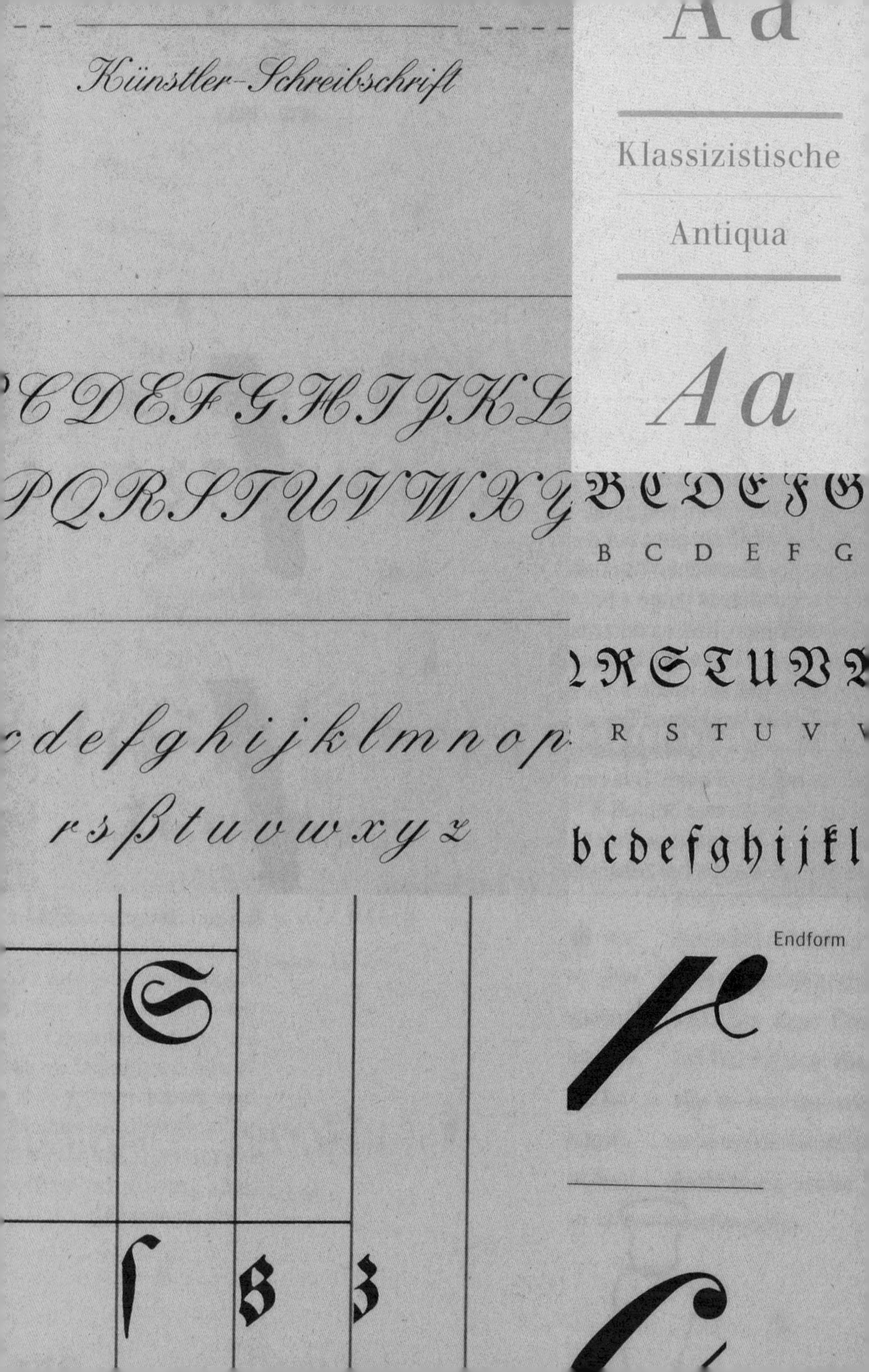

Aa

Klassizistische

Antiqua

Aa

Künstler-Schreibschrift

C D E F G H I J K L

P Q R S T U V W X Y

c d e f g h i j k l m n o p

r s ß t u v w x y z

B C D E F G

B C D E F G

R S T U V

R S T U V

b c d e f g h i j k l

S

ſ ß z

Endform

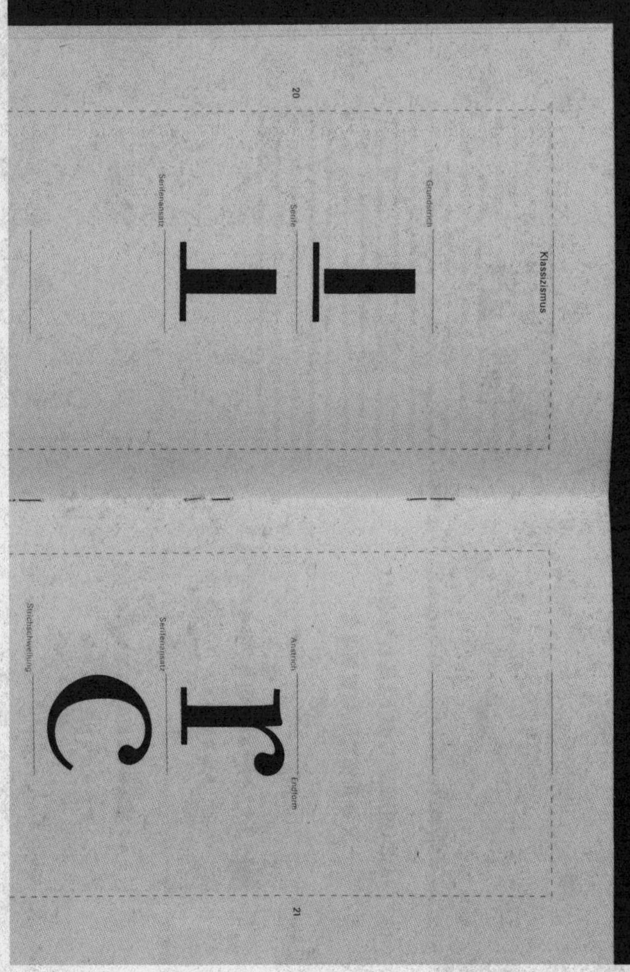

Klassizismus
Grundstrich
Serife
Serifenansatz
20
Anstrich
Endform
Serifenansatz
Strichschwellung
21

1995 Ha5

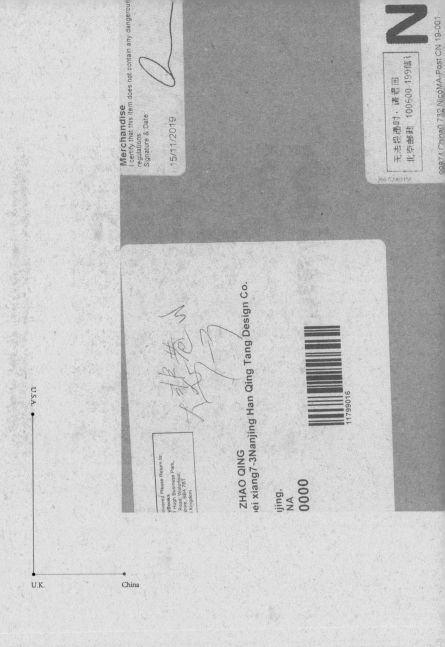
Merchandise
I certify that this item does not contain any dangerous
regulations.
Signature & Date:
15/11/2019
ZHAO QING
ei xiang7-3Nanjing Han Qing Tang Design Co.
11799016
U.S.A.
U.K.
China

284mm

21mm

244mm

● 荣誉奖 Honorary Appreciation

《 看得见——西雅图公共艺术益智书 In Sight: The Seattle Public Art Puzzle Book

美国 U.S.A.

Judy Anderson, Phillip Helms Cook, Claudia Meyer-Newman

Judy Anderson, Phillip Helms Cook, Claudia Meyer-Newman

Seattle Arts Commission, Seattle

284 244 21mm

620g

60p

ISBN-10 0-9617443-5-9

这是一本由 3 位艺术家撰写的西雅图公共艺术的启蒙书，试图通过这本书来链接非艺术从业人群、公共艺术和公共场所。本书的装订形式为活页圈装。封面铜版纸四色印刷覆亮光膜，裱覆牛皮卡板。内页混合使用浅灰色艺术纸和哑面铜版纸，黑棕双色印刷。内容方面，本书试图通过对公共艺术的分析、评论和读者参与，带领非艺术从业人群理解公共艺术，讨论公共艺术的场所、资助、定义和意义，讨论公众参与的价值和可能性。书中共分 7 个“参与”章节，分别在广场、公园、办公场所等领域提出公共艺术的意义，提供艺术卡片让读者进行标贴，提供来自不同人群的评论，并让读者进行思考。几大章节通过纸张的变换进行区隔。版心设置为四栏，但字体和版式较为灵活，与装订形式一样。本书的设计在努力使得艺术不那么高高在上，降低了公众理解和参与艺术的门槛。

This is an enlightenment book about Seattle public art written by three artists, who try to use this book to link non-art practitioners, public art and public places. The book is bound by ring binder. The laminated cover is made of coated paper printed in four colors mounted with kraft cardboard. The inner pages are mixed with light gray art paper and matte coated paper, printed in black and brown. In terms of content, this book attempts to lead non-art practitioners to understand public art through analysis, comment, and readers' participation in public art, to discuss the venue, funding, definition, and significance of public art, and to evaluate the value and possibility of public participation. The book is divided into 7 'participation' chapters, which respectively propose the meaning of public art in the places like squares, parks, offices, etc.. It provides artistic cards for readers to label, provides comments from different groups, and makes the readers think. Several chapters are distinguished by the change of paper. The typeset area is set to four columns, but the fonts and layout are more flexible. Like the book binding format, the design of this book attempts to make art not so arrogant, thereby lowering the threshold for public understanding and participation in art.

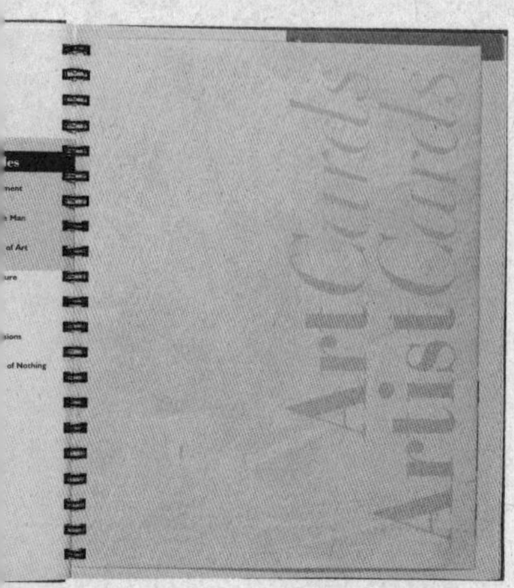

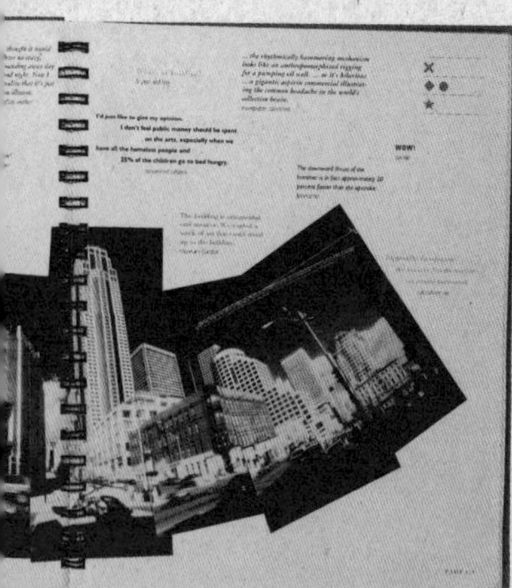

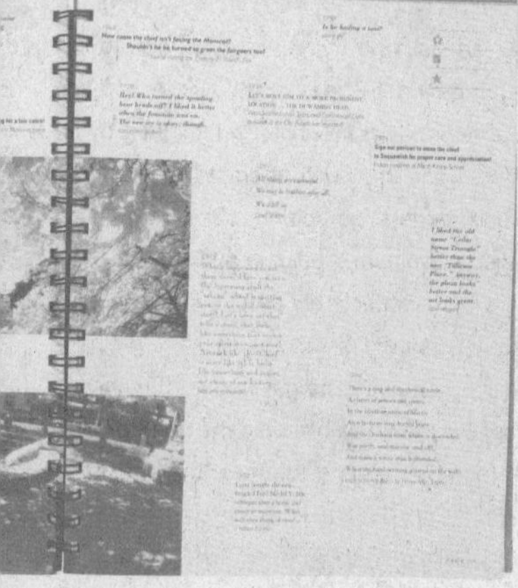

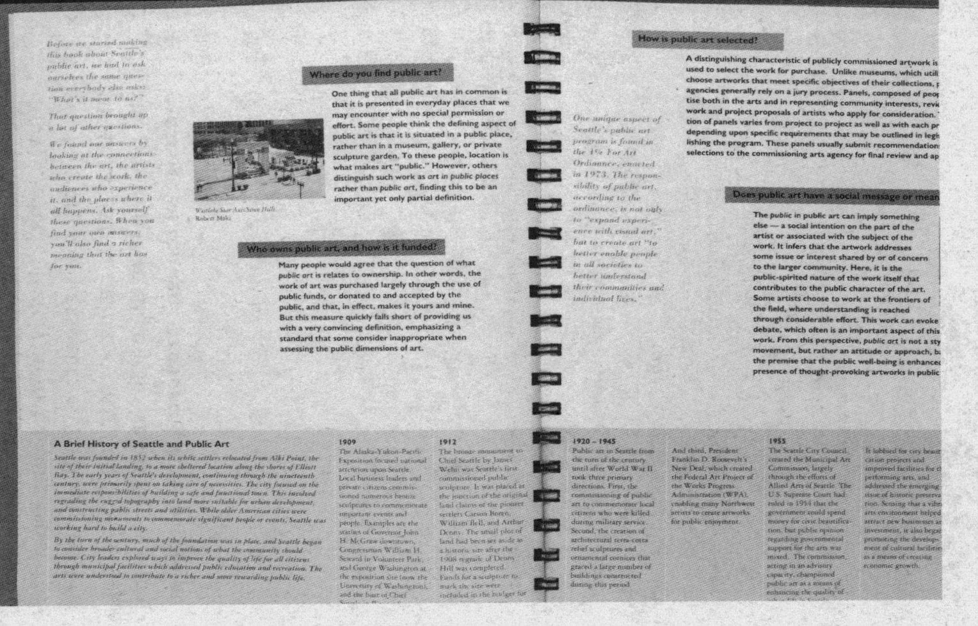

Before we started making this book about Seattle's public art, we had to ask ourselves the same question everybody else asks: "What's it mean to us?"

That question brought up a lot of other questions.

We found our answers by looking at the connections between the art, the artists who create the work, the audiences who experience it, and the places where it all happens. Ask yourself these questions. When you find your own answers, you'll also find a richer meaning that the art has for you.

Where do you find public art?

One thing that all public art has in common is that it is presented in everyday places that we may encounter with no special permission or effort. Some people think the defining aspect of public art is that it is situated in a public place, rather than in a museum, gallery, or private sculpture garden. To these people, location is what makes art "public." However, others distinguish such work as *art in public places* rather than *public art*, finding this to be an important yet only partial definition.

Waterfale Star Axis/Stone Hall
Robert Maki

Who owns public art, and how is it funded?

Many people would agree that the question of what *public art* is relates to ownership. In other words, the work of art was purchased largely through the use of public funds, or donated to and accepted by the public, and that, in effect, makes it yours and mine. But this measure quickly falls short of providing us with a very convincing definition, emphasizing a standard that some consider inappropriate when assessing the public dimensions of art.

One unique aspect of Seattle's public art program is found in the 1% For Art Ordinance, enacted in 1973. The responsibility of public art, according to the ordinance, is not only to "expand experience with visual art," but to create art "to better enable people in all societies to better understand their communities and individual lives."

How is public art selected?

A distinguishing characteristic of publicly commissioned artwork is used to select the work for purchase. Unlike museums, which utili choose artworks that meet specific objectives of their collections, p agencies generally rely on a jury process. Panels, composed of peop tise both in the arts and in representing community interests, revi work and project proposals of artists who apply for consideration. tion of panels varies from project to project as well as with each pr depending upon specific requirements that may be outlined in legis lishing the program. These panels usually submit recommendation selections to the commissioning arts agency for final review and ap

Does public art have a social message or mean

The *public* in public art can imply something else — a social intention on the part of the artist or associated with the subject of the work. It infers that the artwork addresses some issue or interest shared by or of concern to the larger community. Here, it is the public-spirited nature of the work itself that contributes to the public character of the art. Some artists choose to work at the frontiers of the field, where understanding is reached through considerable effort. This work can evoke debate, which often is an important aspect of this work. From this perspective, *public art* is not a sty movement, but rather an attitude or approach, ba the premise that the public well-being is enhance presence of thought-provoking artworks in public

A Brief History of Seattle and Public Art

Seattle was founded in 1852 when its white settlers relocated from Alki Point, the site of their initial landing, to a more sheltered location along the shores of Elliott Bay. The early years of Seattle's development, continuing through the nineteenth century, were primarily spent on taking care of necessities. The city focused on the immediate responsibilities of building a safe and functional town. This involved regrading the rugged topography into land more suitable for urban development, and constructing public streets and utilities. While older American cities were commissioning monuments to commemorate significant people or events, Seattle was working hard to build a city.

By the turn of the century, much of the foundation was in place, and Seattle began to consider broader cultural and social notions of what the community should become. City leaders explored ways to improve the quality of life for all citizens through municipal facilities which addressed public education and recreation. The arts were understood to contribute to a richer and more rewarding public life.

1909
The Alaska-Yukon-Pacific Exposition focused national attention upon Seattle. Local business leaders and private citizens commissioned numerous bronze sculptures to commemorate important events and people. Examples are the statues of Governor John H. McGraw downtown, Congressman William H. Seward in Volunteer Park, and George Washington at the exposition site (now the University of Washington), and the bust of Chief

1912
The bronze monument to Chief Seattle by James Wehn was Seattle's first commissioned public sculpture. It was placed at the junction of the original land claims of the pioneer settlers Carson Boren, William Bell, and Arthur Denny. The small plot of land had been set aside as a historic site after the 1908 regrade of Denny Hill was completed. Funds for a sculpture to mark the site were included in the budget for

1920 – 1945
Public art in Seattle from the turn of the century until after World War II took three primary directions. First, the commissioning of public art to commemorate local citizens who were killed during military service. Second, the creation of architectural terra-cotta relief sculptures and ornamental cornices that graced a large number of buildings constructed during this period.

And third, President Franklin D. Roosevelt's New Deal, which created the Federal Art Project of the Works Progress Administration (WPA), enabling many Northwest artists to create artworks for public enjoyment.

1955
The Seattle City Council created the Municipal Art Commission, largely through the efforts of Allied Arts of Seattle. The U.S. Supreme Court had ruled in 1954 that the government could spend money for civic beautification, but public opinion regarding governmental support for the arts was mixed. The commission, acting in an advisory capacity, championed public art as a means of enhancing the quality of

It lobbied for city beaut cation projects and improved facilities for th performing arts, and addressed the emerging issue of historic preserva tion. Sensing that a vibr arts environment helped attract new businesses ar investment, it also begar promoting the develop- ment of cultural facilitie as a means of creating economic growth.

Some artists choose to work at the frontiers of the field, where understanding is reached through considerable effort. This work can ev debate, which often is an important aspect of work. From this perspective, *public art* is not a movement, but rather an attitude or approac the premise that the public well-being is enha presence of thought-provoking artworks in pu

More than fifteen d languages are spoke

People traveling thr may be sick, worrie

President D. Roosevelt's which created Art Project of Progress ation (WPA), many Northwest reate artworks enjoyment.

1955
The Seattle City Council created the Municipal Art Commission, largely through the efforts of Allied Arts of Seattle. The U.S. Supreme Court had ruled in 1954 that the government could spend money for civic beautification, but public opinion regarding governmental support for the arts was mixed. The commission, acting in an advisory capacity, championed public art as a means of enhancing the quality of urban life in Seattle.

It lobbied for city b cation projects and improved facilities performing arts, an addressed the emer issue of historic pre tion. Sensing that a arts environment h attract new busines investment, it also promoting the deve ment of cultural fac as a means of creati economic growth.

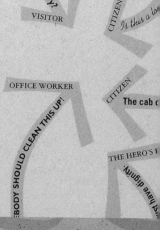

the story?
VISITOR
CITIZEN
Is this a la
OFFICE WORKER
CITIZEN
The cab c
BODY SHOULD CLEAN THIS UP!
THE HERO'S I
must have dignity

1 THE SITUATIO

1 Building a Grand M

It's a place whe expect to see t

You can see thi

Art

The Artist: James A. Wehn

James A. Wehn was born in Indianapolis in 1883 and first came to the Washington Territory just after Seattle's Great Fire of 1889. He began drawing and sculpting when bedridden with illness as a child. At age fifteen, Wehn returned to Indianapolis, where he studied for five years as an apprentice sculptor. Upon returning to Seattle, he created sculpture and architectural ornamentation at his father's foundry. He established the first sculpture studio in Seattle, behind the family home in the Mount Baker vicinity. In 1919, he founded the sculpture department at the University of Washington, and served as its first chair; he taught there for five years. * Wehn devoted much of his long career to Northwest Indian and pioneer history, during which he created numerous sculptures based on Chief Seattle. In addition to the Tillicum Place statue, copies of the bust were installed on a fountain in Pioneer Square (1909), at the Fenton Fire Station (1910), and on Broderick Fountain at Seattle University (1958). Wehn also designed the official seal for the City of Seattle, which includes a profile of Chief Seattle. A bronze replica of that seal (made in 1950) is located above the entry to the Public Safety Building at Third Avenue between James Street and Cherry Street in downtown Seattle.

The Artwork: *Chief Seattle*

James Wehn's sculpture was the first statue commissioned by the City of Seattle, in 1907. The statue is based on the only known photo of Chief Seattle, and adopts a pose noted as characteristic by the early settlers. The artwork faces Elliott Bay with its hand raised in welcome to the original settlers. The 6-foot, full-size figure is remarkably tall for the period. It stands on a large, rectangular, rough granite pedestal with a surrounding granite pool area. Two bronze bear heads, on the north and south sides of the pedestal, serve as fountain waterspouts. On the east and west sides are two bronze tablets describing historical events of the era. One depicts Chief Kitsap viewing the arrival of Captain Vancouver; the other pays

Maintenance/ Conservation

A well-intentioned yet potentially damaging attempt by a cab driver to remove discoloration on the statue of Chief Seattle serves as a reminder of the importance of proper care and maintenance of public art. Due to long-term neglect, the statue had accumulated layers of oxidation with significant streaking. Harsh materials used to clean the face and chest stripped away the sculpture's natural protective patina, exposing the raw metal and leading to accelerated deterioration. Unfortunately, it also damaged some of the gilding hidden under the streaks. A conservator was called in to assess the appropriate means of neutralizing the effects of the solvent and to recommend a plan of restoration. The remaining corrosion was carefully removed using ground walnut shells as a gentle abrasive. The bronze was then cleaned, coated with a corrosion-inhibiting clear acrylic, and given a polished wax finish, forming two layers of protection.

Other Things To Do

Both *Hammering Man* and *Chief Seattle* depict the figure of a man. Write a brief description of their differences and their similarities. Include your opinion of what the artists were trying to say to the public.

If Chief Seattle suddenly came to life, what would he have to say to the people of today? Write a speech for Chief Seattle addressed to a specific audience of your choice – i.e., parents, environmental groups, the mayor, your classmates, artists, etc.

The *Chief Seattle* sculpture is a representational monument, where *Hammering Man* is abstract (refer to the Illustrated Glossary at the end of this book to find out more about what these terms mean). In what other ways can we give public recognition to someone?

Select a personal or public hero you would like to commemorate. Write a letter to them detailing your plans to commemorate them.

If, as part of your personal legacy, you had to design a way to be commemorated, how would you want to be recognized? Design an artwork that would commemorate you.

Create a timeline that includes personal and public events that Chief Seattle would have seen from 1907 to now. Highlight events from your own ethnic group.

The sculpture entitled *People Waiting for the Interurban*, in the Fremont district, is often dressed up in costumes. Would you consider putting a costume on Chief Seattle? Explain your response.

What would you recommend for the maintenance and preservation of the *Chief Seattle*

3 Speaking the Lar

1991: A Health Center

Historically, Southeast Seattle has been one of the city's most culturally diverse areas. Its ethnic richness is a source of pride for the local community. The Southeast Seattle Community Health Center represents a better future, both for patients and for the community. The new artwork will be located in the center's sunny skylit atrium.

It is a two-sto
a focal point c

It is in a cultu
neighborhooc

More than fift
languages are

People traveli
may be sick, v

0533

SEATTLE·ART

7 MAKE THE CONNECTION

1977
It's daring and original, a new way to look at art and reality, as opposed to a restatement of an old idea.
reporter

1977
I find it more mute and forbidding than anything else.
young woman

1993
Pet rocks at play!
teenager

1977
IT IS ONE-HALF MAN, ONE-HALF NATURE. THE CONCRETE IS MODERN BUILDING MATERIAL. THE GEOMETRIC FORMS ARE OF MAN.
art critic

1993
This is public art? You've got to be kidding!
teenager

1976
It's an inappropriate memorial to the former Seattle City Councilwoman.
reporter

1976
It seems to me that it's something the arts community likes because you can't understand it ... and the public hates because they can't!
citizen

1976
IF HIS PIECE WASN'T INTENDED FOR THE POPULACE, WHY DOES HE SOLICIT PUBLIC MONEY FOR IT?
citizen

1976
BY INSTALLING THAT 'SCULPTURE' IT LOOKS AS IF THE RUBBLE-RIP-RAP HAD BEEN RE-INSTALLED FOR A HUGE PRICE.
reporter

1993
So whaddya gonna do 'bout it?
East Coast transplant

1980
A pattern can be detected that suggests a growing familiarity produces feelings of appreciation and support. Scorn has given way to praise for the work, despite its lack of recognizable content or function.
art expert

1993
RIGHT ...
ABOUT, ABOVE, ACROSS, AFTER, AGAINST, ALONG, AMONG, AROUND, AT ...
jogger

1977
In addition to political, social, and environmental near-chaos, it should be obvious to all that we are living in an ugly age.
observer

1993
Why is this art?
man and son

1976
THE WORK ITSELF IS ENOUGH TO MAKE ANY POLITICIAN JEALOUS. EVEN BEFORE ITS DEDICATION, IT RECEIVED MORE PRINTED AND ELECTRONIC PRESS COVERAGE THAN ALL THE CANDIDATES IN THE 7TH DISTRICT RACE. AND IT HAS CREATED A LIVELY CONTROVERSY, A HEALTHY TENSION, A FOCUS FOR POINTS OF VIEW. AND I THINK THAT'S PART OF WHAT MAKES URBAN LIVING AN EXCITING, ENLIVENING EXPERIENCE.

1980
It is placed in a way that forces viewers to see it in relationship

Country / Region

Germany ❺

France ❷

U.S.A. ❷

Czech Republic ❶

Austria ❶

Australia ❶

Norway ❶

Poland ❶

PARIS 1994 NADAR Les années créatrices : 1854 - 1860

DIE RUSSEN IN WIEN / FALTER VERLAG

Harold Brodkey Die flüchtige Seele

Christians Groothuis/Janßen Schwarzbuntes. Bilder aus Ostfriesland

AFRICA HERB RITTS

ANTHOLOGIE DU CINÉMA INVISIBLE Christian Janicot

arte Éditions

Stasys · Erotyki

SPURR/GORTON ✦ THE GUMDROP TREE ✦ HYPERION

1996 Gl 0540

1996 Sm^2 0550

1996 Bm^1 0560

1996 Bm^3 0570

1996 Bm^4 0576

1996 Bm^5 0584

1996 Ha^2 0592

1996 Ha^4 0602

1996 Ha^5 0610

Designer

Andreas Brylka

Klaus Detjen

Libor Beránek

Pierre-Louis Hardy

Hofmann & Kraner

Julia Hasting, Patricia Müller, Gerwin Schmidt, Béla Stetzer

Edith Lackmann

Rainer Groothuis

Betty Egg, Sam Shahid

Graham Rendoth (Reno Design Group 14144)

Bulnes & Robaglia

Marte Fæhn, Line Jerner (Lucas Design & illustrasjon)

Gra yna Bareccy, Andrzej Bareccy

Julia Gorton

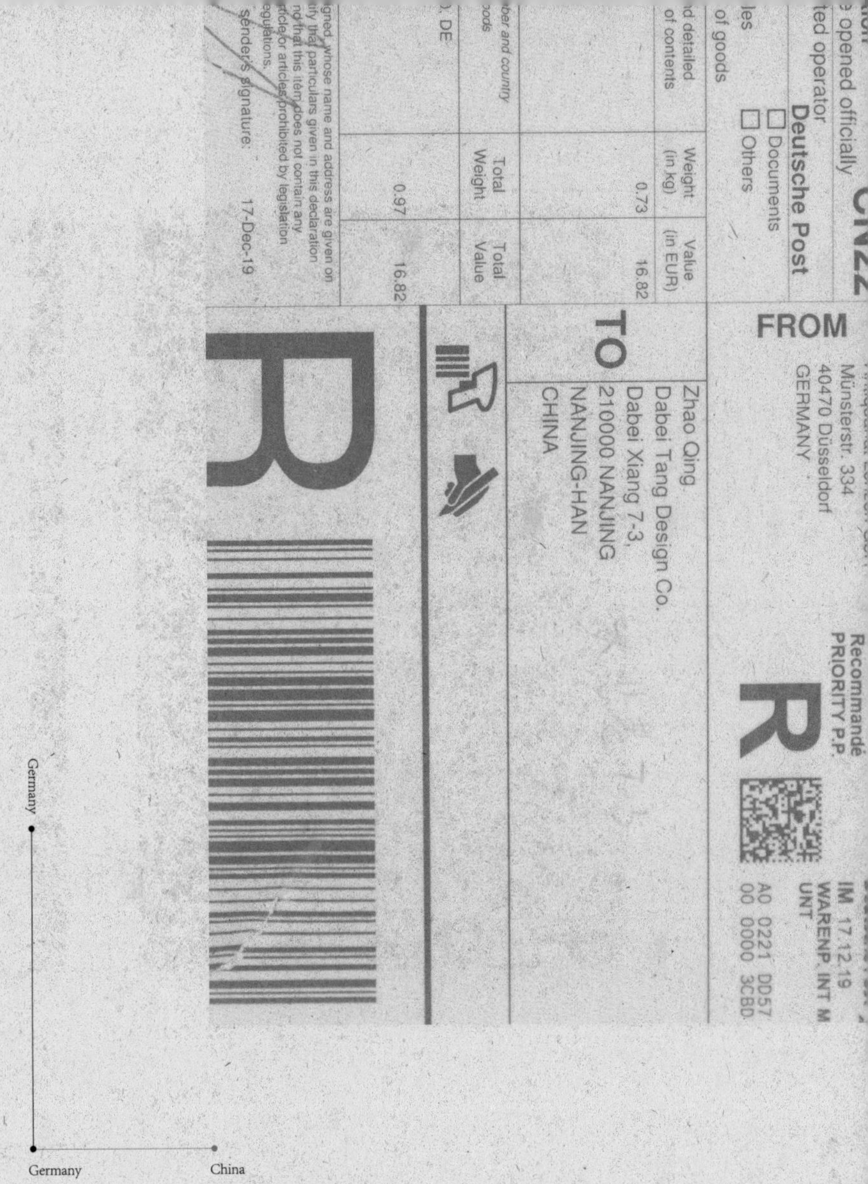
Deutsche Post
Documents
Others
Weight (in kg)
Value (in EUR)
0.73
16.82
Total Weight
Total Value
0.97
16.82
17-Dec-19
FROM
Münsterstr. 334
40470 Düsseldorf
GERMANY
TO
Zhao Qing
Dabei Tang Design Co.
Dabei Xiang 7-3,
210000 NANJING
NANJING-HAN
CHINA
Recommandé
PRIORITY P.P.
R
Germany
Germany
China

350mm

19mm

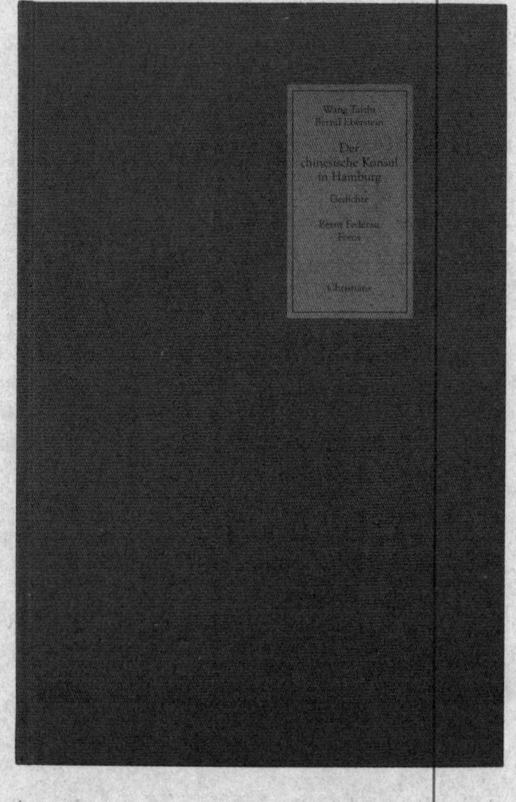

240mm

◎ 金字符奖

《 中国总领事在汉堡的诗

◖ 德国

Golden Letter

Der chinesische Konsul in Hamburg gedichte: Gedichte

Germany

Andreas Brylka

Wang Taizhi, Bernd Eberstein

Christians Verlag, Hamburg

350 240 19mm

723g

74p

中国驻汉堡总领事馆是中国在德国设立的第一个总领事馆，从 1984 年至今已有 10 位外交官在此任职过总领事的职位。第二任总领事王泰智是位诗人，他的诗歌记录了大量在任期间的所闻所感，成为中德两国友谊的见证。诗歌结集成册，便是本书。书籍的装订形式为包背无线胶精装。封面选用了一种类似宣纸纹理的手工纸蓝黑双色印刷，裱覆卡板。内页选用光面铜版纸和与封面相同的手工纸红黑棕三色印刷。书中的主要部分是诗歌，印制在手工纸上，使用楷体，字号较大，呈现较为古典的中式风格，但文本使用横排。用于对照的德文使用带衬线的 Italic（斜体），字号较小，与开篇介绍页的衬线字体（正体）相区别。诗歌内容大致包括亭（汉堡的中式建筑）、雨、港、雕塑、湖、桥、友人、古街、帆、风车等地区风貌、四季变化及人文琐事。书中专门搭配了与每一篇主题相对应的黑白照片，通过棕黑双色印刷呈现出微微偏暖的老照片味道。

The Chinese Consulate General in Hamburg is the first Chinese consulate general established in Germany. Since 1984, 10 consuls have held this position. The second consul general, Wang Taizhi, was a poet and recorded a lot of what he heard and felt during his term of office through poetry, and became a witness to the friendship between China and Germany. This is the origin of the book. The double-leaved hardcover book is perfect bound. The cover is made of hand-made paper, similar to the texture of rice paper, printed in blue and black and mounted by the cardboard. The inner pages are printed with glossy coated paper and the same handmade paper as the cover, printed in red, black and brown. The main part of the book is poetry, printed on handmade paper, using the font 'Kaiti' with a large font size to present classical Chinese style, however the text is typeset horizontally. The text in German used for comparison uses with serif fonts in Italic, and the font size is smaller, which is different from the serif font used in the opening introduction pages. The content of the poems generally includes regional features, seasonal changes and cultural trivia, such as pavilions (Chinese architecture in Hamburg), rain, port, sculptures, lakes, bridges, friends, ancient streets, sails, windmills and so on. In the book, the black and white photos are selected to match each corresponding theme, showing the warm hue of old photos through the separation of brown and black.

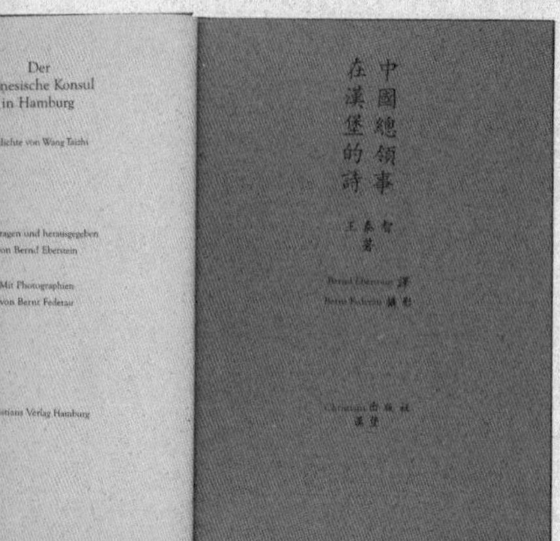

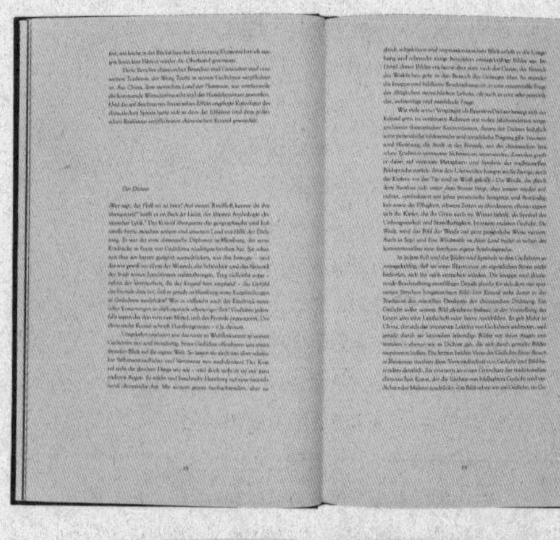

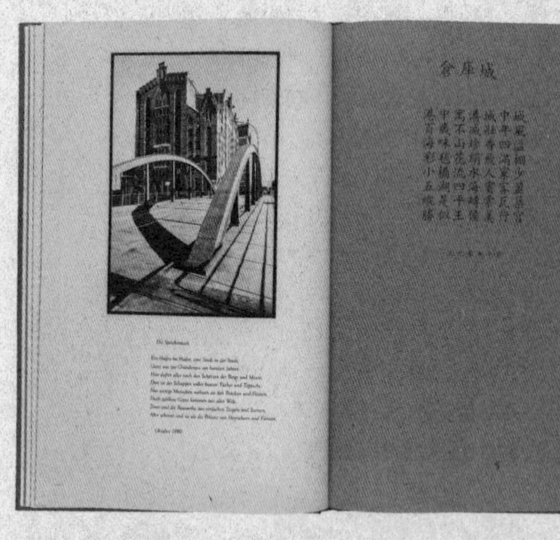

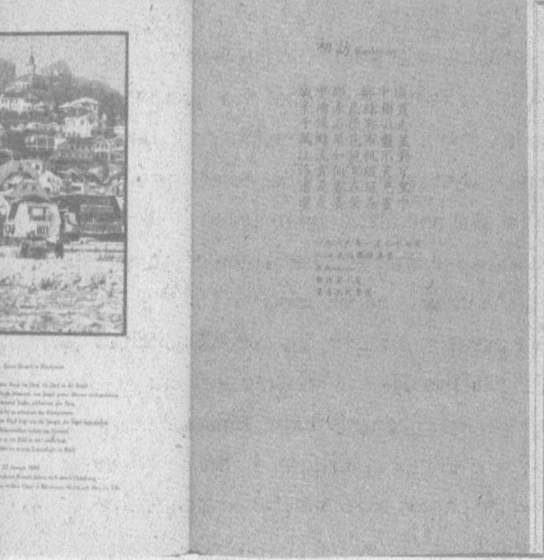

漢堡的橋

大橋、小橋、長橋、短橋
石橋、鐵橋、磚橋、木橋
新橋、古橋、
水橋、陸橋、
好看的橋和丑陋的橋、
有名的橋和被人遺忘的橋、
看不完的橋，
數不盡的橋，
走不斷的橋，
橋、橋、橋……

我們每天走在橋上，
橋在我們腳下
埋怨，呻吟。

橋太多了，
我們忘記了橋。

世界上橋最多的城市，
它的居民站在自己的橋上，
津津有味地贊美着
威尼斯的橋，
而自己的橋，
卻視而不見。

橋太多了，
他們忘記了橋。

漢堡秋雨

天綿旅帆水磚里園
月雨行張清紅頭甸
十細少不蓋蔽心覓
堡聲上中波霧知處
漢無路湖亂迷不何

一九八九年十月

Brücken über Wasser, Brücken über Land.
Schöne Brücken und häßliche Brücken.
Berühmte Brücken und vergessene Brücken.
Brücken, die ich nie sah,
Brücken, die ich nie zählte,
Brücken, die ich nie betrat.
Brücken, Brücken, Brücken …

Jeden Tag gehen wir über Brücken,
Und sie klagen und stöhnen
Unter unseren Füßen.

So viele Brücken gibt es,
Daß wir sie vergessen!

Mehr Brücken gibt es in keiner Stadt auf der Welt.
Aber die Menschen schwärmen

Der
chinesische K
in Hambu

Gedichte von Wan

n die eingangs zitierten Sätze entnommen.
och etwas sei nachgetragen: Der Konsul schrieb diese Essays
nter seinem Namen Wang Taizhi. Er gab einem Pseudonym
rang, das er seit den Studententagen gern für seine schrift-
che Tätigkeit benutzt und das ihn in seiner Selbstbescheidung
fend kennzeichnet: Ren Ke, »Der seine Aufgaben erfüllt«.
err von Schnabelewopski, der junge Graf und Spötter, wurde,
ahen, schon sehr bald widerlegt. Die *Karten und Berichte über*
er am Meere stellten die Hansestadt den Chinesen zum ersten
Seither sind eineinhalb Jahrhunderte vergangen, in denen
lei Bereichen des Handels, der Politik und der Kultur enge
erhafte Verbindungen geknüpft wurden. Heute ist Hamburg
einer der in China am besten bekannten europäischen Städ-
rden – nicht zuletzt auch durch die einfühlsamen Gedichte

石橋，鐵橋，
新橋，古橋．
水橋，陸橋．
好看的橋和
有名的橋和
看不完的橋，
數不盡的橋，
走不斷的橋．
橋，橋，橋……

我們每天走
橋在我們脚
埋怨，呻吟．

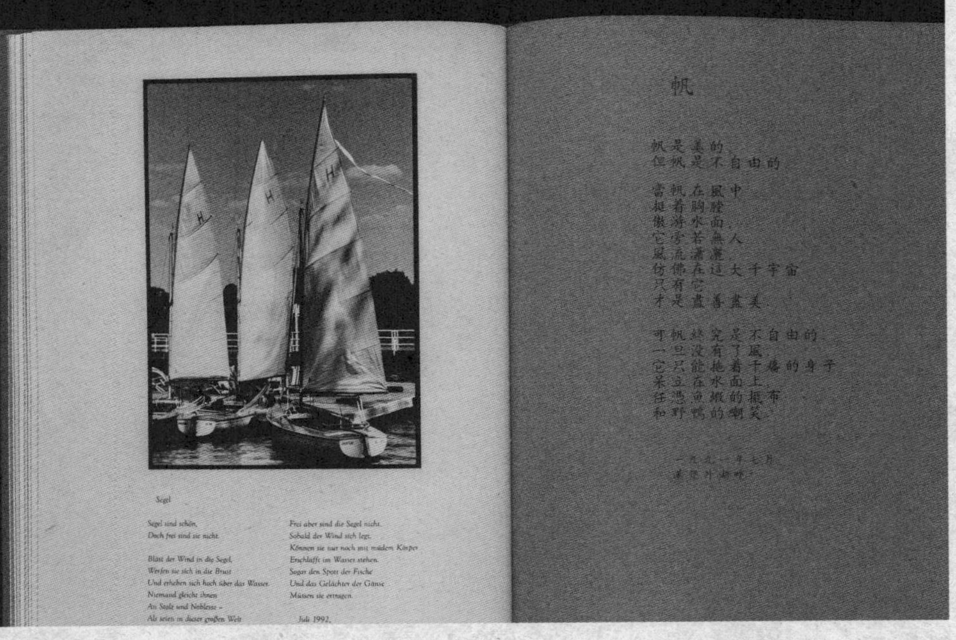

Segel

Segel sind schön,
Doch frei sind sie nicht.

Bläst der Wind in die Segel,
Werfen sie sich in die Brust
Und erheben sich hoch über das Wasser.
Niemand gleicht ihnen
An Stolz und Noblesse –
Als seien in dieser großen Welt

Frei aber sind die Segel nicht.
Sobald der Wind sich legt,
Können sie nur noch mit müdem Körper
Erschlafft im Wasser stehen.
Sogar den Spott der Fische
Und das Gelächter der Gänse
Müssen sie ertragen.

Juli 1992.

帆

帆是美的
但帆是不自由的

當帆在風中
挺着胸膛
傲游水面，
它旁若無人
風流瀟灑，
彷彿在這大千宇宙
只有它
才是盡善盡美．

可帆終究是不自由的
一旦沒有了風
它只能拖着千瘡的身子
呆立在水面上，
任憑魚蝦的擺布
和野鴨的嘲笑．

一九九一年七月
漢堡外湖畔

Hamburgs Brücken

Große Brücken, kleine Brücken, lange Brücken, kurze Brücken.
Steinbrücken, Stahlbrücken, Ziegelbrücken, Holzbrücken.
Neue Brücken, alte Brücken.
Brücken über Wasser, Brücken über Land.
Schöne Brücken und häßliche Brücken.
Berühmte Brücken und vergessene Brücken.
Brücken, die ich nie sah,
Brücken, die ich nie zählte,
Brücken, die ich nie betrat.
Brücken, Brücken, Brücken . . .

Jeden Tag gehen wir über Brücken,
Und sie klagen und stöhnen
Unter unseren Füßen.

So viele Brücken gibt es,
Daß wir sie vergessen!

Mehr Brücken gibt es in keiner Stadt auf der Welt.
Aber die Menschen schwärmen
Von den Brücken Venedigs.
Ihre eigenen Brücken
Sehen sie nicht!

So viele Brücken gibt es,
Daß sie sie vergessen!

Fremde kommen in die Hansestadt,
Sie sind begeistert von der Fahrt über die Alster,
Bestimmt fahren sie auch auf dem trüben Wasser der Elbe.
Sie bewundern den grenzenlosen Hafen,
Überwältigt halten sie mit der Kamera
Den Reichtum der Speicherstadt fest . . .
Aber gäbe es keine Brücken –
Wie könnte das Wasser der Alster so hell glänzen!

NADAR

1996 Sm²

000059
Return to sender by air
nternational parcel
pack World Business
Christine VINCKE
Vaartstraat 19
8400 Oostende
Belgium
32471930857
Destination 000059-15737232616610001
Zhao Qing
Qing Tan Design Co.
Dabei Xiang 7-3, Nanjing Han
210000 Nanjing
China
13952094842
CN
PLEASE ALWAYS ADD CN23 DOCUMENTS FOR ALL S
J006
EPG
bpost
France
Belgium
China

313mm

40mm

239mm

◉ 银奖
《 纳达尔——创作年代 1854－1860
◐ 法国

Silver Medal
Nadar: Les années créatrices: 1854-1860
France
Pierre-Louis Hardy
——
Éditions de la Réunion des Musées Nationaux, Paris
313 239 40mm
2142g
368p
ISBN-10 2-7118-2583-3

纳达尔（Nadar）是法国摄影师、漫画家和新闻记者加斯帕尔·费利克斯·图尔纳雄（Gaspard-Félix Tournachon）的笔名。这本书是1994年和1995年分别在巴黎奥赛美术馆和纽约大都会艺术博物馆举办的纳达尔回顾展的作品集。书籍的装订形式为锁线硬精装。护封铜版纸四色印刷，内封布纹纸印灰紫色，裱覆卡板。内页胶版纸黑棕双色印刷，其中棕色主要用于呈现老照片泛黄的质感。内容方面，全书主要分为四大部分：第一部分为各位评论家的相关文章，其中穿插大量纳达尔和同时期其他艺术家的作品作辅助说明；第二部分为纳达尔的摄影作品；第三部分为展览上摄影及其他艺术作品的简介；第四部分则是艺术家年表信息和附录。版式设置为三栏，正文占两栏，侧边一栏灵活处理旁注或图片。照片和其他作品尽可能地保留泛黄、划痕和瑕疵，甚至不同的裁切形状。书籍极高的印刷质量，呈现出完美的“纸上展览”。

Nadar is the pseudonym of Gaspard-Félix Tournachon, a French photographer, cartoonist and journalist. This book is a collection of works from the retrospective exhibition of Nadar held at the Musée d'Orsay in Paris and the Metropolitan Museum of Art in New York in 1994 and 1995 respectively. The hardcover book is bound by thread sewing. The jacket is made of coated paper printed in four colors, and the inner cover is made of arlin paper printed in grey purple, mounted with cardboard. Offset paper is used for inside pages printed in black and brown, and Brown is mainly used to present the texture of old photos. As for the content, the book is divided into four parts: the first part contains the relevant articles of each critic, which is interspersed with a large number of works of Nadar and other artists at the same time; the second part contains the photography works of Nadar; the third part is the introduction of photography and other works of art on the exhibition; the fourth part is the information of the artist's chronology and the appendix. The layout is set to 3 columns – the text occupies 2 columns, and the side column flexibly handles side notes or pictures. As much as possible, yellowing, scratches, and flaws in works are retained, plus the non-uniform cutting shapes and the high printing quality, together present a perfect 'exhibition on paper'.

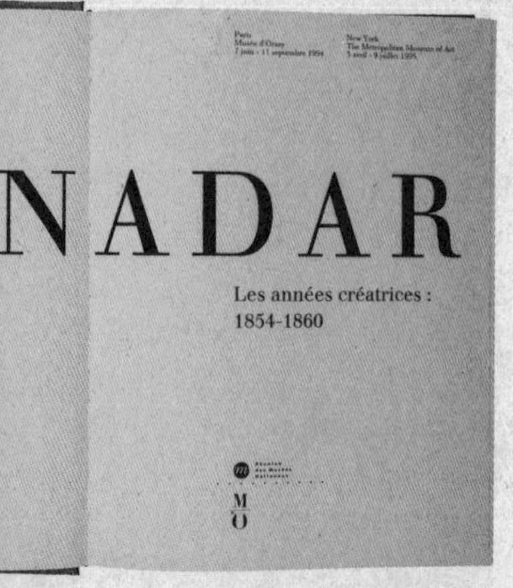
Paris
Musée d'Orsay
7 juin - 11 septembre 1994
New York
The Metropolitan Museum of Art
5 avril - 9 juillet 1995
NADAR
Les années créatrices :
1854-1860

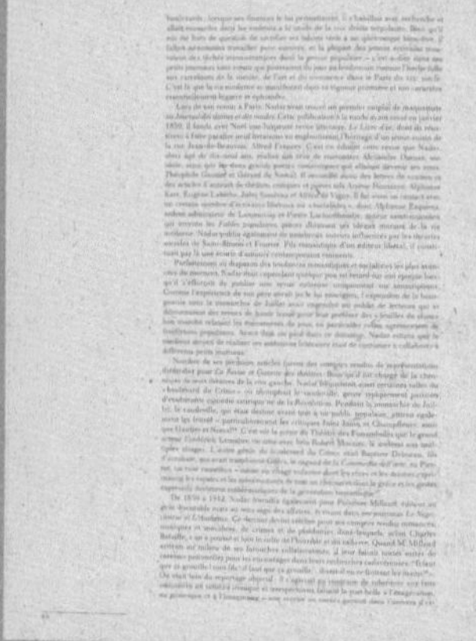

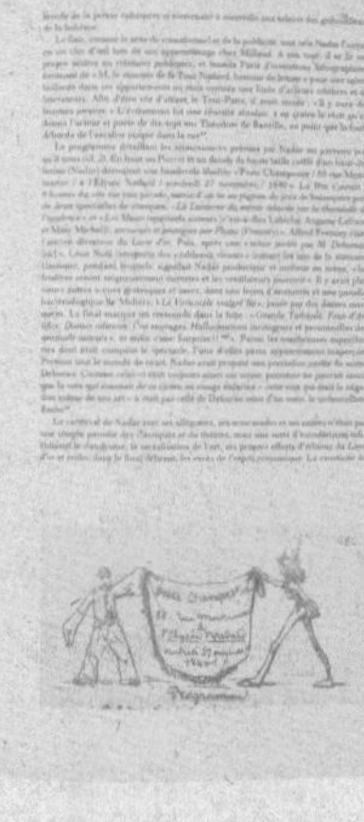

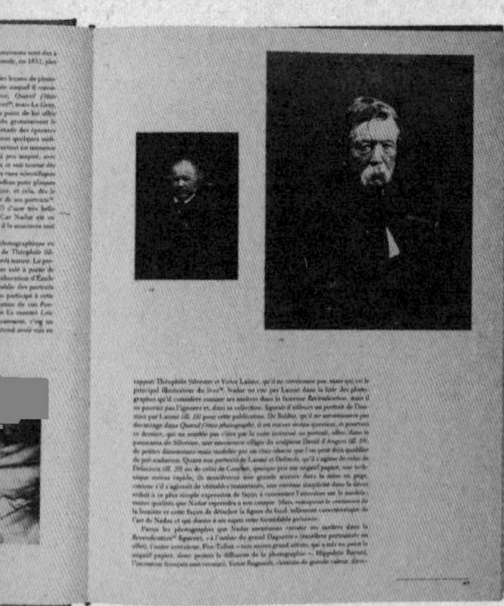

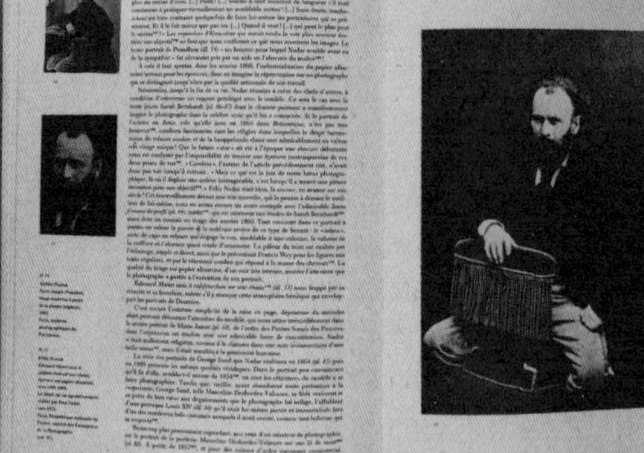

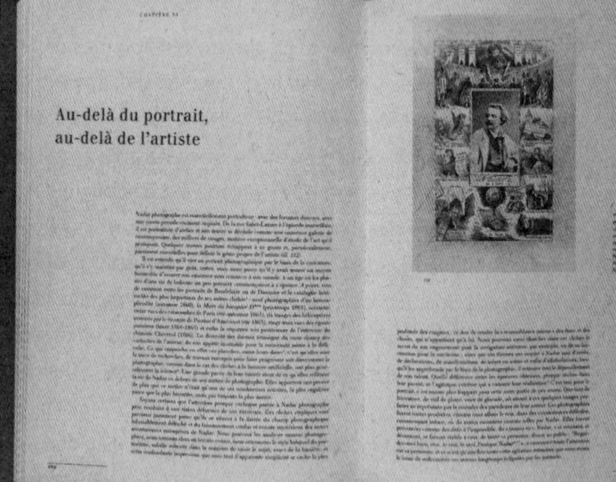
Au-delà du portrait,
au-delà de l'artiste

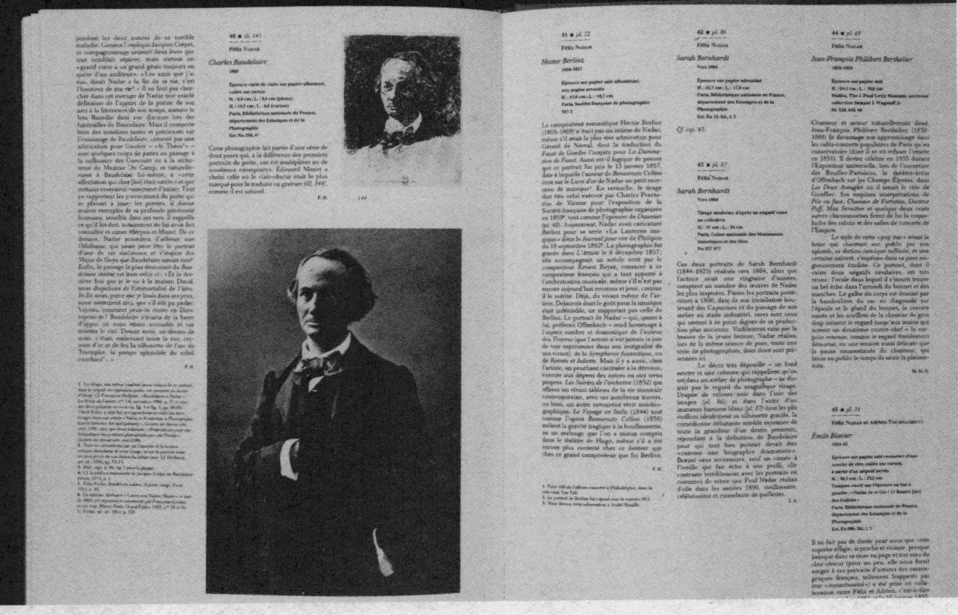

pendant les deux années de sa terrible maladie. Comme l'explique Jacques Crépet, ce compagnonnage unissait deux êtres que tout semblait séparer, mais surtout un «grand cœur à un grand génie toujours en quête d'un auditeur». «Les amis que j'ai eus, disait Nadar à la fin de sa vie, c'est l'honneur de ma vie[4].» Il ne faut pas chercher dans cet ouvrage de Nadar une exacte définition de l'apport de la poésie de son ami à la littérature de son temps, comme le fera Banville dans son discours lors des funérailles de Baudelaire. Mais il comporte bien des notations justes et précieuses sur l'entourage de Baudelaire, cimenté par une admiration pour Gautier – «le Théo[5]» – avec quelques coups de pattes au passage à la suffisance des Goncourt ou à la sécheresse de Maxime Du Camp, et naturellement à Baudelaire lui-même, à «cette affectation qui chez [lui] était native» et que certains essayaient vainement d'imiter. Tout en rapportant les provocations du poète qui se plaisait à jouer les pervers, il donne maints exemples de sa profonde générosité humaine, sensible dans ses vers. Il rappelle ce qu'il lui doit, notamment de lui avoir fait connaître et aimer Méryon et Manet. De ce dernier, Nadar possédera d'ailleurs une *Odalisque*, qui passe pour être le portrait d'une de ses maîtresses et s'inspire des *Majas* de Goya que Baudelaire aimait tant[6]. Enfin, le passage le plus émouvant du *Baudelaire intime* est bien celui-ci : «Et la dernière fois que je le vis à la maison Duval, nous disputions de l'immortalité de l'âme. Je dis nous, parce que je lisais dans ses yeux, aussi nettement moi, que s'il eût pu parler. Voyons, comment peux-tu croire en Dieu, repétai-je ? Baudelaire s'écarta de la barre d'appui où nous étions accoudés et me montra le ciel. Devant nous, au-dessus de nous, c'était, embrassant toute la rue, cernant d'or et de feu la silhouette de l'arc de Triomphe, la pompe splendide du soleil couchant[7]...»

F. H.

[illegible]

40 ■ *ill. 143*

Félix Nadar

Charles Baudelaire

1860

Épreuve carte de visite sur papier albuminé, collée sur carton
H. : 8,9 cm ; L. : 5,4 cm (photo)
H. : 10,5 cm ; L. : 6,5 (carton)
Paris, Bibliothèque nationale de France, département des Estampes et de la Photographie
Est. Na 258, 4°

Cette photographie fait partie d'une série de deux poses qui, à la différence des premiers portraits du poète, ont été multipliées en de nombreux exemplaires. Édouard Manet a choisi celle où le clair-obscur était le plus marqué pour la traduire en gravure *(ill. 144)*, comme il est naturel.

F. H.

41 ■ *pl. 72*

Félix Nadar

Hector Berlioz

1856-1857

Épreuve sur papier salé albuminé, aux angles arrondis
H. : 23,6 cm ; L. : 18,7 cm
Paris, Société française de photographie
307-3

Le compositeur romantique Hector Berlioz (1803-1869) n'était pas un intime de Nadar, même s'il avait la plus vive admiration pour Gérard de Nerval, dont la traduction du *Faust* de Goethe l'inspira pour *La Damnation de Faust*. Aussi est-il logique de penser que ce portrait fut pris le 15 janvier 1857, date à laquelle l'auteur de *Benvenuto Cellini* écrit sur le *Livre d'or* de Nadar un petit morceau de musique[1]. En revanche, le tirage doit être celui exécuté par Charles Praetorius de Vienne pour l'exposition de la Société française de photographie organisée en 1859[2], tout comme l'épreuve de *Daumier (pl. 40)*. Auparavant, Nadar avait caricaturé Berlioz pour sa série «La Lanterne magique» dans le *Journal pour rire* de Philipon du 18 septembre 1852[3]. La photographie fut gravée dans *L'Artiste* le 6 décembre 1857 ; elle accompagnait un article écrit par le compositeur Ernest Reyer, consacré à ce compositeur français qui a tant apporté à l'orchestration musicale, même s'il n'est pas encore aujourd'hui reconnu et joué, comme il le mérite. Déjà, du vivant même de l'artiste, Delacroix dont le goût pour la musique était indéniable, ne supportait pas celle de Berlioz. Le portrait de Nadar – qui, quant à lui, préférait Offenbach – rend hommage à l'aspect sombre et dramatique de l'auteur des *Troyens* (que l'artiste n'eut jamais la joie de voir représenter dans son intégralité de son vivant), de la *Symphonie fantastique*, ou de *Roméo et Juliette*. Mais il y a aussi, chez l'artiste, un penchant contraire à la dérision, exercée aux dépens des autres ou aux siens propres. *Les Soirées de l'orchestre* (1852) qui offrent un vivant tableau de la vie musicale contemporaine, avec ses nombreux travers, ou bien, un autre savoureux récit autobiographique, *Le Voyage en Italie* (1844) tout comme l'opéra *Benvenuto Cellini* (1838) mêlent la gravité tragique à la bouffonnerie, en un mélange que l'on a mieux compris dans le théâtre de Hugo, même s'il a été encore plus contesté chez ce dernier que chez ce grand compositeur que fut Berlioz.

F. H.

1. Folio 168 de l'album conservé à Philadelphie, dans la collection Van Pelt.
2. Le portrait de Berlioz fut exposé sous le numéro [illegible].
3. Nous devons cette information à André Rouillé.

42 ■ *pl. 86*

Félix Nadar

Sarah Bernhardt

Vers 1864

Épreuve sur papier albuminé
H. : 22,7 cm ; L. : 17,6 cm
Paris, Bibliothèque nationale de France, département des Estampes et de la Photographie
Est. Eo 15, fol., t. 1

Cf. cat. 43.

43 ■ *pl. 87*

Félix Nadar

Sarah Bernhardt

Vers 1864

Tirage moderne, d'après un négatif verre au collodion
H. : 37 cm ; L. : 24 cm
Paris, Caisse nationale des Monuments historiques et des Sites
Na 232 877

Ces deux portraits de Sarah Bernhardt (1844-1923) réalisés vers 1864, alors que l'actrice avait une vingtaine d'années, comptent au nombre des œuvres de Nadar les plus inspirées. Parmi les portraits postérieurs à 1860, date de son installation boulevard des Capucines et du passage de son atelier au stade industriel, rares sont ceux qui restent à ce point dignes de sa production plus ancienne. Visiblement saisi par la beauté de la jeune femme, Nadar réalisa, lors de la même séance de pose, toute une série de photographies, dont deux sont présentées ici.

Le décor très dépouillé – un fond neutre et une colonne qui rappellent qu'on est dans un atelier de photographe – ne distrait pas le regard du magnifique visage. Drapée de velours noir dans l'une des images *(pl. 86)*, et dans l'autre d'un immense burnous blanc *(pl. 87)* dont les plis étoffent idéalement sa silhouette gracile, la comédienne débutante semble rayonner de toute la grandeur d'un destin pressenti, répondant à la définition de Baudelaire pour qui tout bon portrait devait être «comme une biographie dramatisée». Beauté sans accessoires, sauf un camée à l'oreille qui fait écho à son profil, elle contraste terriblement avec les portraits en costumes de scène que Paul Nadar réalisa d'elle dans les années 1890, vieillissante, célébrissime et ruisselante de paillettes.

S. A.

44 ■ *pl. 69*

Félix Nadar

Jean-François Philibert Berthelier

1856-1859

Épreuve sur papier salé
H. : 24,4 cm ; L. : 18,8 cm
Malibu, The J. Paul Getty Museum, ancienne collection Samuel J. Wagstaff Jr.
84. XM. 436. 49

Chanteur et acteur naturellement doué, Jean-François Philibert Berthelier (1830-1888) fit davantage son apprentissage dans les cafés-concerts populaires de Paris qu'au conservatoire (dont il se vit refuser l'entrée en 1851). Il devint célèbre en 1855 durant l'Exposition universelle, lors de l'ouverture des Bouffes-Parisiens, le théâtre-écrin d'Offenbach sur les Champs-Élysées, dans *Les Deux Aveugles* où il tenait le rôle de Giraffier. Ses exquises interprétations de *Pile ou face*, *Chanson de Fortunio*, *Docteur Puff*, *Miss Sensitive* et quelque deux cents autres chansonnettes firent de lui la coqueluche des salons et des salles de concerts de l'Empire.

Le style de cette «pop star» avant la lettre qui charmait son public par son aplomb, sa diction comique raffinée, et une certaine naïveté, s'exprime dans sa pose soigneusement étudiée. Ce portrait, dont il existe deux négatifs similaires, est très réussi : l'ovale dans lequel il s'inscrit trouve un bel écho dans l'arrondi du bonnet et des manches. Le galbe du corps est dessiné par la bandoulière du sac en diagonale sur l'épaule et le gland du bonnet, la cravate nouée et les soufflets de la chemise de gros drap mènent le regard jusqu'aux mains qui serrent un deuxième couvre-chef – la surprise retenue, comme le regard timidement détourné, en une tension aussi délicate que la pause momentanée du chanteur, qui laisse au public le temps de saisir la plaisanterie.

M. M. H.

45 ■ *pl. 31*

Félix Nadar et Adrien Tournachon

Émile Blavier

1854-55

Épreuve sur papier salé recouverte d'une couche de cire, collée sur carton, à partir d'un négatif perdu
H. : 30,5 cm ; L. : 23,5 cm
Tampon encré sur l'épreuve en bas à gauche : «Nadar Je et Cie / 11 Boulet [sic] des Italiens»
Paris, Bibliothèque nationale de France, département des Estampes et de la Photographie
Est. Eo 986, fol., t. 1

Il ne fait pas de doute pour nous que cette superbe effigie, si proche et vivante, presque baroque dans sa mise en page et son sens du clair-obscur (pour un peu, elle nous ferait songer à ces portraits d'artistes des caravagesques français, tellement frappants par leur «instantanéité») a été prise en collaboration entre Félix et Adrien, c'est-à-dire

ı enflé de s'être rassasié sans retenue – qui est issu de la branche mince de la ille des clowns – il a mal, délicieuse- ıt mal.

M. M. H.

29 ■ *pl. 7*

Félix Nadar et Adrien Tournachon

rrot surpris

1854-55

Épreuve sur papier albuminé
H. : 28,7 cm ; L. : 20,8 cm
Collection Suzanne Winsberg

:at. 30.

30 ■ *pl. 9*

Félix Nadar et Adrien Tournachon

rrot surpris

1854-55

Épreuve sur papier albuminé
H. : 29 cm ; L. : 21,1 cm
Paris, musée d'Orsay, don Marie-Thérèse et André Jammes
Pho. 1986.74

aines photographies de cette série pour- ıt fort bien avoir été extraites d'un rtoire d'expressions stéréotypées. Ces : planches *(pl. 7 et9)*, variantes l'une de

Épreuve sur papier sa
H. : 28,7 cm ; L. : 21,1 c
Paris, musée d'Orsay,
et André Jammes
Pho. 1991.1 (10)

La photographie qui prer place entre *Pierrot voleur* tant une action en tapin illustre la supériorité rus perdue. Quoi qu'il se soit le dessus. Il nous invite gardé l'argent, ce qui, no son geste, n'est pas le cas

32 ■ *pl. 17*

Félix Nadar et Adr

Pierrot voleur

1854-55

Épreuve sur papier sal
H. : 28,6 cm ; L. : 21,1 cr
Paris, musée d'Orsay, c
et André Jammes
Pho. 1991.1 (9)

Pierrot commet un vol. L connaît pas de plus pror Dans les premières panto qui valaient la peine d'êtr nourriture, la tendresse, Mais au XIXe siècle, la cr nomie monétaire et l'enr introduisirent l'or ou les monnaie comme objets d des plus célèbres panton fut *Le Songe d'or*, joué po

A

s années c

54-1860

ır définir le génie propre de l'artiste *(ill. 112)*.
t au portrait photographique par le biais de la caricatur t, certes, mais aussi parce qu'il y avait trouvé un moye xistence sans renoncer à son monde, à un âge où les pla un peu précaire commençaient à s'épuiser. *A priori*, rie traits de Baudelaire ou de Daumier et le catalogue hét s de ses autres clichés[1] : neuf photographies d'un herma , la *Main du banquier D**** (printemps 1861), soixant s de Paris (été-automne 1861), six images des hélicoptère Ponton d'Amécourt (été 1863), vingt-trois vues des égou 5) et enfin la séquence très postérieure de l'*interview* d . La diversité des thèmes témoigne du vaste champ d son appétit insatiable pour la nouveauté jointe à la diff ı effet ces planches, outre leurs dates[2], c'est qu'elles so travaux entrepris pour faire progresser soit directement l s le cas des clichés à la lumière artificielle, soit plus gén

COMMISSAIRES

Françoise HEILBRUN
Conservateur en chef au musée d'(

Maria MORRIS HAMBC
Conservateur en chef du départem au Metropolitan Museum of Art

Philippe NÉAGU
Conservateur au musée d'Orsay

COMMISSAIRE ASSOCIÉ

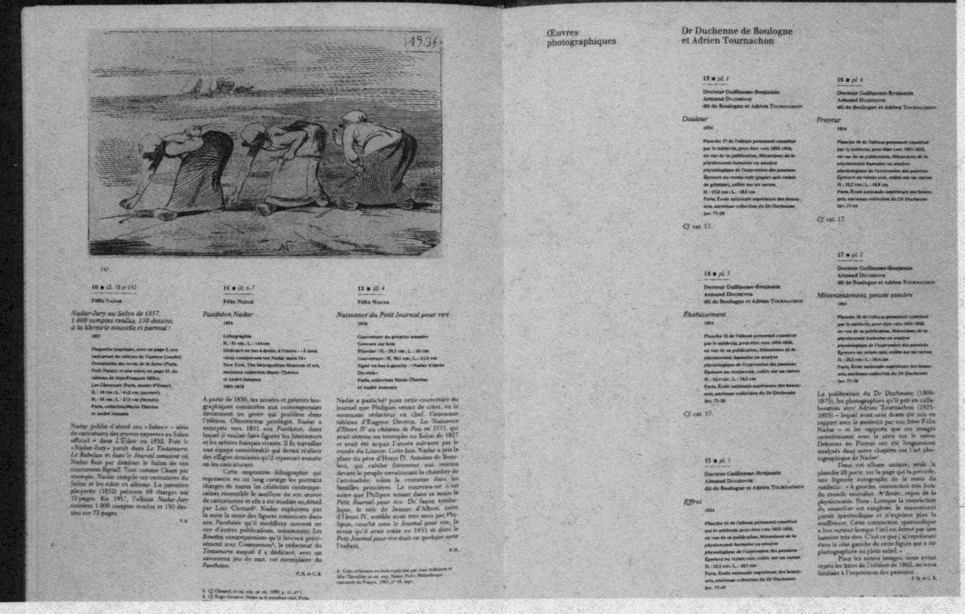

142

10 ■ *ill. 78 et 142*

Félix Nadar

Nadar-Jury au Salon de 1857. 1 000 comptes rendus, 150 dessins, à la librairie nouvelle et partout !

1857

Plaquette imprimée, avec en page 5, une caricature du tableau de Gustave Courbet *Demoiselles des bords de la Seine* (Paris, Petit Palais) et une autre, en page 57, du tableau de Jean-François Millet, *Les Glaneuses* (Paris, musée d'Orsay)
H. : 18 cm ; L. : 4,3 cm (ouverte)
H. : 18 cm ; L. : 17,5 cm (fermée)
Paris, collection Marie-Thérèse et André Jammes

Nadar publie d'abord son « Salon » – série de caricatures des œuvres exposées au Salon officiel – dans *L'Éclair* en 1852. Puis le « Nadar-Jury » paraît dans *Le Tintamarre*, *Le Rabelais* et dans le *Journal amusant* où Nadar finit par détrôner le Salon de son concurrent Bertall. Tout comme Cham par exemple, Nadar compile ses caricatures du Salon et les édite en albums. La première plaquette (1852) présente 69 charges sur 12 pages. En 1857, l'album *Nadar-Jury* contient 1 000 comptes rendus et 150 dessins sur 72 pages.

S. B.

11 ■ *ill. 6-7*

Félix Nadar

Panthéon Nadar

1854

Lithographie
H. : 81 cm ; L. : 114 cm
Dédicace en bas à droite, à l'encre : « À mon vieux compère-son ton Nadar mars 54 »
New York, The Metropolitan Museum of Art, ancienne collection Marie-Thérèse et André Jammes
1985-1078

À partir de 1830, les musées et galeries biographiques consacrées aux contemporains deviennent un genre qui prolifère dans l'édition. Observateur privilégié, Nadar a entrepris vers 1851 son *Panthéon*, dans lequel il voulait faire figurer les littérateurs et les artistes français vivants. Il fit travailler une équipe considérable qui devait réaliser des effigies dessinées qu'il reprenait ensuite en les caricaturant.

Cette imposante lithographie qui représente en un long cortège les portraits charges de toutes les célébrités contemporaines rassemble le meilleur de son œuvre de caricaturiste et elle a été étudiée en détail par Loïc Chotard[2]. Nadar exploitera par la suite la mine des figures contenues dans son *Panthéon* qu'il modifiera souvent en vue d'autres publications, notamment *Les Binettes contemporaines* qu'il lancera précisément avec Commerson[3], le rédacteur du *Tintamarre* auquel il a dédicacé, avec un savoureux jeu de mot, cet exemplaire du *Panthéon*.

P. N. et C. B.

12 ■ *ill. 4*

Félix Nadar

Naissance du Petit Journal pour rire

1856

Couverture du premier numéro
Gravure sur bois
Planche : H. : 39,5 cm ; L. : 20 cm
Couverture : H. 50,5 cm ; L. : 21,5 cm
Signé en bas à gauche : « Nadar d'après Devéria »
Paris, collection Marie-Thérèse et André Jammes

Nadar a pastiché[1] pour cette couverture du journal que Philipon venait de créer, en le nommant rédacteur en chef, l'immense tableau d'Eugène Devéria, *La Naissance d'Henri IV au château de Pau en 1553*, qui avait obtenu un triomphe au Salon de 1827 et avait été acquis l'année suivante par le musée du Louvre. Cette fois, Nadar a pris la place du père d'Henri IV, Antoine de Bourbon, qui exhibe fièrement son rejeton devant le peuple envahissant la chambre de l'accouchée, selon la coutume dans les familles princières. Le nouveau-né n'est autre que Philipon tenant dans sa main le *Petit Journal pour rire*. De façon symbolique, le rôle de Jeanne d'Albret, mère d'Henri IV, semble aussi être tenu par Philipon, couché sous le *Journal pour rire*, la revue qu'il avait créée en 1851 et dont le *Petit Journal pour rire* était en quelque sorte l'enfant.

F. H.

1. Cf. Chotard, in cat. exp. op. cit., 1990, p. 11, n° 1.
2. Cf. Roger Greaves, *Nadar ou le paradoxe vital*, Paris, ...

1. Cette référence est bien explicitée par Jean Adhémar et Aile Chevallier in cat. exp. Nadar, Paris, Bibliothèque nationale de France, 1965, n° 58, repr.

Œuvres photographiques

Dr Duchenne de Boulogne et Adrien Tournachon

13 ■ *pl. 1*

Docteur Guillaume-Benjamin Armand Duchenne dit de Boulogne et Adrien Tournachon

Douleur

1854

Planche 27 de l'album personnel constitué par le médecin, peut-être vers 1855-1856, en vue de sa publication, *Mécanisme de la physionomie humaine ou analyse physiologique de l'expression des passions*
Épreuve au vernis-cuir (papier salé enduit de gélatine), collée sur un carton
H. : 25,2 cm ; L. : 18,5 cm
Paris, École nationale supérieure des beaux-arts, ancienne collection du Dr Duchenne
Inv. 77-28

Cf. cat. 17.

14 ■ *pl. 3*

Docteur Guillaume-Benjamin Armand Duchenne dit de Boulogne et Adrien Tournachon

Ébahissement

1854

Planche 55 de l'album personnel constitué par le médecin, peut-être vers 1855-1856, en vue de sa publication, *Mécanisme de la physionomie humaine ou analyse physiologique de l'expression des passions*
Épreuve au vernis-cuir, collée sur un carton
H. : 22,4 cm ; L. : 16,5 cm
Paris, École nationale supérieure des beaux-arts, ancienne collection du Dr Duchenne
Inv. 77-38

Cf. cat. 17.

15 ■ *pl. 5*

Docteur Guillaume-Benjamin Armand Duchenne dit de Boulogne et Adrien Tournachon

Effroi

1854

Planche 41 de l'album personnel constitué par le médecin, peut-être vers 1855-1856, en vue de sa publication, *Mécanisme de la physionomie humaine ou analyse physiologique de l'expression des passions*
Épreuve au vernis-cuir, collée sur un carton
H. : 28,5 cm ; L. : 18,5 cm
Paris, École nationale supérieure des beaux-arts, ancienne collection du Dr Duchenne
Inv. 77-44

16 ■ *pl. 4*

Docteur Guillaume-Benjamin Armand Duchenne dit de Boulogne et Adrien Tournachon

Frayeur

1854

Planche 40 de l'album personnel constitué par le médecin, peut-être vers 1855-1856, en vue de sa publication, *Mécanisme de la physionomie humaine ou analyse physiologique de l'expression des passions*
Épreuve au vernis-cuir, collée sur un carton
H. : 22,7 cm ; L. : 16,8 cm
Paris, École nationale supérieure des beaux-arts, ancienne collection du Dr Duchenne
Inv. 77-43

Cf. cat. 17.

17 ■ *pl. 2*

Docteur Guillaume-Benjamin Armand Duchenne dit de Boulogne et Adrien Tournachon

Mécontentement, pensée sombre

1854

Planche 28 de l'album personnel constitué par le médecin, peut-être vers 1855-1856, en vue de sa publication, *Mécanisme de la physionomie humaine ou analyse physiologique de l'expression des passions*
Épreuve au vernis-cuir, collée sur un carton
H. : 28,5 cm ; L. : 20,4 cm
Paris, École nationale supérieure des beaux-arts, ancienne collection du Dr Duchenne
Inv. 77-30

La publication du Dr Duchenne (1806-1875), les photographies qu'il prit en collaboration avec Adrien Tournachon (1825-1903) – lequel avait sans doute été mis en rapport avec le médecin par son frère Félix Nadar – et les rapports que ces images entretiennent avec la série sur le mime Deburau en Pierrot ont été longuement analysés dans notre chapitre sur l'art photographique de Nadar[1].

Dans cet album unique, seule la planche 28 porte, sur la page qui la précède, une légende autographe de la main du médecin : « À gauche, contraction très forte du muscle sourcilier. À droite, repos de la physionomie. Nota : Lorsque la contraction du sourcilier est exagérée, le mouvement paraît spasmodique et n'exprime plus la souffrance. Cette contraction spasmodique a lieu surtout lorsque l'œil est fermé par une lumière très vive. C'est ce que j'ai représenté dans le côté gauche de cette figure qui a été photographiée en plein soleil. »

Pour les autres images, nous avons repris les titres de l'édition de 1862, en nous limitant à l'expression des passions.

F. H. et C. B.

1996 Bm[1]

ZVAB.com

ZENTRALES VERZEICHNIS ANTIQUARISCHER BÜCHER

Päckchen XS Welt bis 2 kg Deutsche Post

Zhao Qing

210000 Nanjing

Volksrepublik China

WIEN 1945 / AUGENZEUGENBERICHTE UND

ÜBER 400 UNPUBLIZIERTE FOTOS AUS RUSSLAND

HERAUSGEGEBEN VON ERICH KLEIN

FALTER VERLAG

Dabei Lane, Meiyuan New Villag 7 6
210000 Nanjing
Volksrepublik China

大悲巷 梅园新村 7-

ndungsnummer: UM 900 798 956 DE
x. 35 x 25 x 3 cm

Entgelt Beza
Port Payé

Unzustellbarkeit / En cas de non-livraison

itcode/Routingcode

dentcode

Austria

Germany

China

铜奖

维也纳的俄罗斯人——奥地利的解放

奥地利

Bronze Medal

Die Russen in Wien: Die Befreiung Österreichs

Austria

Hofmann & Kraner

Erich Klein

Falter Verlagsgesellschaft, Wien

285 245 20mm

1244g

248p

ISBN-10 3-85439-141-2

1945年奥地利在以苏联军队为主的盟军协助下战胜纳粹德国，获得解放，本书是这段历史的记录。书中整理了由苏联军官、摄影师叶甫根尼·哈尔代伊（Yevgeny Khaldei）和摄影记者奥尔加·兰达（Olga Lander）拍摄的超过400幅照片，详细记录了解放奥地利的过程。本书是第二次世界大战胜利50周年时出版的重要资料。书籍的装订形式为锁线硬精装。护封铜版纸红黑双色印刷，覆哑膜，内封黑色布面裱覆卡板，书脊处红色丝印五角星图案。环衬选用黑色卡纸，内页哑面铜版纸印单黑。书中除了开篇对于历史的简要回顾和书籍出版的简介外，正文分为三个篇章——为解放奥地利做的战斗准备、维也纳的解放过程、苏联军队在维也纳的授勋与后续活动。正文主要被设置为双栏，但在图片和文字的排版上较为灵活，加粗的小字号图说与笔画较细但大字号密排的正文形成对比。引用的文字采用斜体排版，形成更丰富的文本层次。标题字选用斑驳的无衬线老字体，有如战争年代载具的涂装和街边字牌。

In 1945, the Allied Army, mainly the Soviet army, assisted Austria in defeating Nazi Germany and liberated Austria. This book is a record of this period of history. It contains more than 400 photos taken by Soviet military officers, photographer Yevgeny Khaldei and photojournalist Olga Lander, who documented the process of liberating Austria in details. This book is an important document published on the 50th anniversary of the victory of the Second World War. The hardcover book is bound by thread sewing. The jacket is made of coated paper printed in red and black, mounted with matte film. The inside cover is made of black cloth mounted with cardboard. There is a five pointed star pattern treated with silk screening on the spine. The end paper is made of black cardboard, and the inner pages are printed with matte coated paper in black. In addition to a brief review of the history and a brief introduction to the publication of the book at the beginning of the book, the main body is divided into three chapters: preparation for the battle before the liberation of Austria, the process of the liberation of Vienna, and the honors and follow-up activities of the Soviet army in Vienna. The text is mainly typeset to double columns, but the arrangement of pictures and annotations is more flexible. The bold annotation in small font size contrasts with the thin text in the large font size. The quoted text is typeset in italics to form a text hierarchy. Mottled sans-serif typeface is selected for the headlines, which looks like the painting of vehicles and street signs during the war.

DIE RUSSEN IN WIEN
DIE BEFREIUNG ÖSTERREICHS
WIEN 1945 / AUGENZEUGENBERICHTE UND
ÜBER 400 UNPUBLIZIERTE FOTOS AUS RUSSLAND
HERAUSGEGEBEN VON ERICH KLEIN
FALTER VERLAG

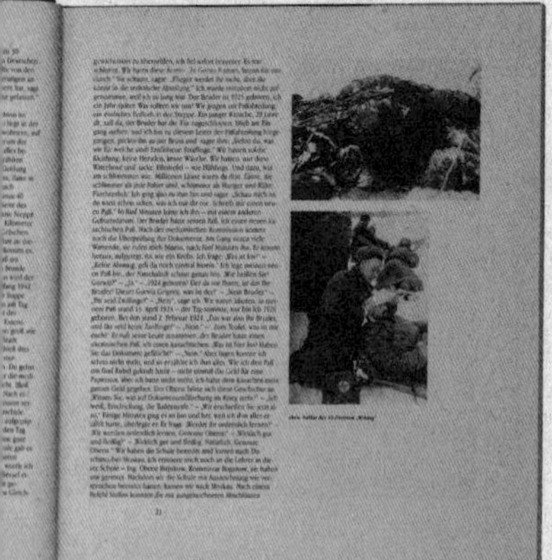

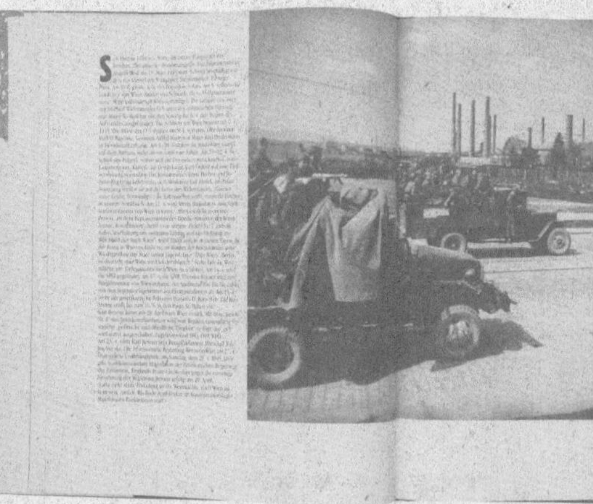

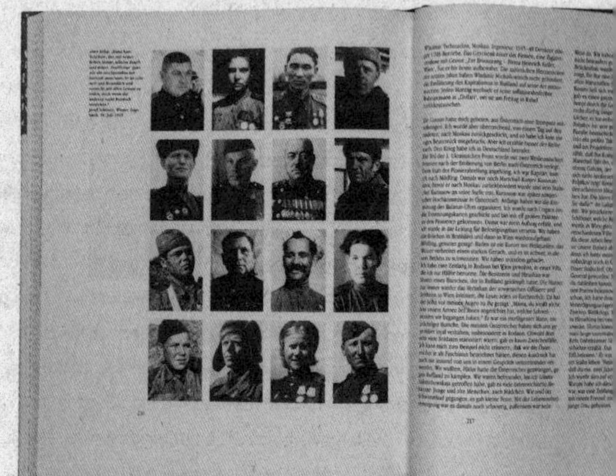

Erklärung der Sowjetregierung über Österreich

Die Rote Armee schlägt die deutsch-faschistischen Truppen und ist bei ihrer Verfolgung in Österreich einmarschiert. Wien, die Hauptstadt Österreichs, ist belagert.

Im Gegensatz zu den Deutschen in Deutschland widersetzt sich die Bevölkerung Österreichs der von den Deutschen durchgeführten Evakuierung. Sie bleibt an ihren Plätzen und begrüßt die Rote Armee herzlich als Befreierin Österreichs vom Joch der Hillerfaschisten.

Die Sowjetregierung hat nicht das Ziel, sich irgend einen Teil des österreichischen Territoriums anzueignen oder die gesellschaftliche Ordnung Österreichs zu ändern. Die Sowjetregierung steht auf dem Boden der Moskauer Deklaration der Verbündeten Mächte über die Unabhängigkeit Österreichs. Sie wird diese Deklaration in die Wirklichkeit umsetzen. Sie wird die Liquidierung des Regimes der deutsch-faschistischen Okkupanten und die Wiederherstellung demokratischer Zustände und Einrichtungen in Österreich unterstützen.

Das Oberkommando der Roten Armee gab den Sowjettruppen den Befehl, der Bevölkerung Österreichs in diesem Werk beizustehen.

Einem Leutnant hat es an der Donau die Finger abgerissen. Ich bin nicht stark genug, mich zu erinnern. Einmal laufen alle durcheinander, sie achten nicht mehr darauf, was geschieht. Da waren Schützenlöcher im Halbkreis angelegt. Ich bin mit einem anderen hingegangen: „Schau, da liegt einer." Er hatte deutsche Stiefel angezogen, ich wußte das nicht und sage: „Schauen wir, was für ein Fritz da liegt." So habe ich mich ausgedrückt und ihn sogleich erkannt. Dort gibt es kleine Häuser, es stehen immer einige zusammen. Ich hatte ihn gleich erkannt. Ich bin also ins Dorf gegangen und habe, unter uns gesprochen, eine Schaufel requiriert. Diese Leute haben nicht viel gesagt und ich auch nicht viel. Wir haben eine Grube ausgehoben und ihn hineingelegt. Ein Jahr nach dem Krieg habe ich seine Schwester gesucht. Dieser Name ist lange vergessen, er hieß Iwan Tarupkin. Da war noch ein Junge.
Er schaut hinter einem Baum hervor, genau in diesem Moment trifft ihn die Kugel. In seinem Helm war ein mächtiges Loch. Ich habe seine Papiere angeschaut – Jahrgang 28! Ich sagte den Leuten, die dort bei ihm standen: „Begrabt ihn!" Wo ist der verschwunden, welche Idioten haben ihn dahin geschickt? Als ich in das Haus um die Schaufel ging – da stehen ein russischer Soldat und die Frau, und die kleinen Kinder zittern vor Angst. Das schmerzt noch immer. Warum hatte die Frau Angst? Die Kinder. Sie konnten nicht wissen, ob ich ein Guter oder ein Schlechter bin. Wenn ich ein Plünderer gewesen wäre – aber wir sind doch erzogen worden, die Menschen zu respektieren.
In Wien war ich nur zwei Tage. Alle Brücken waren gesprengt. Die Deutschen haben sich nicht sofort ergeben. Ich erinnere mich an das Parlament, der Stephansdom brannte noch. Wir sind sonst nirgendwohin gegangen, außerdem wurden wir auch nirgendwohin gelassen. Die Leute standen vor unserer Feldküche Schlange – Soldaten, Alte, Kinder standen da. Die Leute hatten schon lange kein Fleisch mehr bekommen, deshalb haben sie die toten Pferde gegessen. Sie haben das Fleisch einfach aus den Kadavern herausgeschnitten. Die Häuser waren zerstört, sie hatten keine Dächer mehr. Wozu haben die Amerikaner die Häuser bombardiert, war das noch notwendig? Aber mir hat es dort gefallen, ich komme aus dem Dorf, dort gab es keine solch schöne Häuser. Viele Russen liefen auf den Dächern herum. Wie Kinder es tun. Abends sind wir in die Stadt gekommen, am Morgen war der Kampf zu Ende.
Österreich war der erste Verbündete, aber das ist nicht Schuld des Volkes. Wir wissen, wie so etwas geschieht und gemacht wird. Man darf die Leute nicht auf einen Haufen werfen und über einen Leisten schlagen. Es gab dort auch anständige Leute.
Wir sind nachts weitermarschiert, Richtung Tschechoslowakei. Wie im Traum bin ich durch Österreich gekommen, nicht einmal den Wienerwald habe ich gesehen. Wir sind die ganze Zeit zu Fuß gelaufen.
Zweimal bin ich durch Österreich gekommen, aus der Tschechoslowakei sind wir wieder nach Wien zurückmarschiert. Den neunten Mai haben wir in Österreich gefeiert, am siebten Mai habe ich den letzten Brief geschrieben. Am achten sind die Deutschen nur noch davongelaufen. Aber die Straßen waren noch vermint. Wir sind auf den Autos mitgefahren, der erste Wagen ist auf eine Mine aufgefahren und explodiert. Es war sehr viel unterschiedliches Volk unterwegs. Auf tschechischem Territorium habe ich Franzosen gesehen, sehr viele Frauen, die der Krieg dorthin getrieben hatte.
Sie mußten auf den Bauernhöfen arbeiten. Wir haben sie befreit. Pferde sind frei herumgelaufen, wir haben viele eingefangen. Komisch, dort ist noch ein Leutnant erschossen worden. Er war Zugskommandant, ungefähr 40 Jahre alt. Es war abends, ein schöner Wald. Ich komme aus Sibirien, und mir hat der Wald gefallen. In den Häusern wird dort unterm Dach Heu gelagert. Die Leute haben schon nicht mehr aufgepaßt, alle waren erleichtert. Wir sind zu einem Haus mit einer Veranda gekommen, wollten hineingehen, und der Leutnant ist von dort heraus erschossen worden. Er hat nicht mehr aufgepaßt.
Die Tschechen hatten eine Besonderheit: Sie haben uns Brot und Wasser gegeben. Bei uns ist es ja Brot und Salz. Es war erstaunlich, daß sie uns so freudig empfangen haben. In Österreich hatten die Leute Angst.
Bei der Befreiung Österreichs sind 24.000 unserer Soldaten gefallen. In unserem Land sagen heute die einen: Ihr seid Befreier, die anderen schreien: Ihr wart bloß Okkupanten? Warum wir, was hat das mit uns zu tun? Was sollen wir dabei? Ständig versucht man uns etwas einzureden, ich aber verstehe das nicht mehr. Diese Zeit ist nicht mehr zu verstehen. Ich habe später mein Tagebuch angezündet. Wie haben sie geheißen – Iwan Sakłow, Igor Naumow … Aus meinem Dorf sind zwei Drittel umgekommen.
Wir haben zusammen mit den Kirgisen gekämpft und gedacht, alle zusammen sind wir die Sowjetunion. Burjaten waren da, Mongolen, Kaukasier, alle Völker – aber in unserem Jahrhundert kennt sich keiner mehr aus.

Ihnen sagt man, doch sie schweigen.
– Burschen, heißt es.
– Helden, heißt es.
– Die Stunde der Entscheidung.
– Vorwärts, heißt's, zurück.
So sagt man ihnen.

Du, was schweigst du?
Schweig nicht, murr.
Brülle, Iwan.
Jetzt geht's zur Schlachtung, mit dir.

Ich auch, ich komm' mit euch.
Daneben ich, brülle: Aufgestanden,
brüll': Dawaj,
ich brülle…

Irgendwer muß brüllen.

60

61

ießlich habe ich es ja genommen."
ıitrij T. Schepilow, Mitglied des
gsrates der 4. Garde-Armee,
krainische Front

Wladimir Makanin, angesehener russischer Schriftsteller und Intellektueller, der seine Bücher in der Sowjetunion jahrelang unter Schwierigkeiten publizieren nte, kam vor einigen Jahren nach Wien iner Lesung. Seine Übersetzerin, eine re, würdevolle Dame, erzählte ihm bei r Stadtrundfahrt unter anderem da-, daß nach dem Krieg eine der Donaucken zu Ehren eines jener Marschälle, vor 50 Jahren Wien befreit hatten, in nowskij-Brücke umbenannt worden Der russische Gast war verwundert, n begann er über den Ausdruck eiung in diesem Zusammenhang zu la-. Die geschichtsbewußte Österherin unterbrach ein wenig beleidigt Gespräch...
Skepsis des Moskauer Intellektuellen, seine eigene Rechnung mit seinem t zu begleichen hat, bezog sich nicht frevlerisch auf die bei der Befreiung ns gefallenen Soldaten, sondern auf n obersten Befehlshaber, Stalin, auf en Kriegsführung und die Absichten, er im Krieg verfolgt hatte. Es ist dieselMischung aus Skepsis und Verwunder-, die russische Touristen beim Anblick

findet bei den Nachfahren
ne ungleich nüchternere B

Daten und Fakten.
Wien war schon befreit, der krieg noch nicht zu Ende. M marsch der Roten Armee E wurde das Land im sechste erstmals zum unmittelbare zweier sich bekämpfender Die Wiederherstellung der des Landes war schon am 1 1943 von den Außenministe der USA und der Sowjetuni kauer Deklaration" beschlo Österreich, „das erste Land schen Angriffspolitik Hitler (fiel), sollte von deutscher freit werden". Dies geschah jahr nachdem Hitlers Ausgr macht in der Schlacht von S schieden zurückgewiesen w nationalsozialistische Führu Europa unter ihre Macht ge verkündete nunmehr den „ ein Krieg, der mit dem Eint die Auseinandersetzung sch krieg geworden war und nu gung Europas gegen anstür baren aus dem Osten umsti Die Alliierten stellten die B conditional surrender", bed Kapitulation Deutschlands, für einen Frieden nach Hitl

klärung der Sowjetregier

über Österreich

Die Rote Armee schlägt die deutsch-chen Truppen und ist bei ihrer Verf Österreich einmarschiert. Wien, die dt Österreichs, ist belagert.

Im Gegensatz zu den Deutschen in De d widersetzt sich die Bevölkerung Oster von den Deutschen durchgeführten erung. Sie bleibt an ihren Plätzen und be Rote Armee herzlich als Befreierin chs vom Joch der Hitlerfaschisten.

Die Sowjetregierung hat nicht das Zie end einen Teil des österreichischen T

DIE RU
DIE BE
WIEN 1945 / AU
ÜBER 400 UN
HERAUSGEGEB
FALTER VERLA

INLEITUNG

schuldigen Sie, ich weiß nicht, wer Sie sind. Aber wissen Sie, wer ic ämlich nicht. Bin ich ein befreiter Österreicher oder ein geschlagen Russen haben meine Uhr geschnappt, also bin ich ein Verlierer. And nich befreit, auf dem Plakat hier steht's, schwarz auf weiß: ‚Die Ro pft nicht gegen das österreichische Volk... Ein freies, unabhängige gehören wir halb zu den Gewinnern, fast eine kleine Siegernation. t sich aus. Unter dem Hitler hätt's eine solche Unordnung nicht ge Deutsche besetzt, was können wir dafür, und jetzt befreit uns der R auch nichts machen. So ändert sich die Welt, und der Steffl brennt t' ich vor lauter Mitleid mit mir selbst. Wenn uns doch wenigstens it hätt' und Geld ins Land gebracht, und keine so beschissene Befr wir brauchen, Herr, sind zahlende Ausländer!"
t Fischer, Das Ende einer Illusion

Geschichte Österreichs zählt eben auch die Geschichte der seit 194

reter des Stabes der 3. Ukraini . Bulgarischen Armee, zuletzt on Wien.
lagodatow ist ein Mann von h ung. Kenner unserer und der iteratur, liebt die Musik, sprich Deutsch und Französisch. Durch ebenserfahrung hat er sich ein efühl erworben, zugleich Anm enschen gänzlich verschieden m genauso leicht, sich mit de ucker aus Floridsdorf zu unter ngen mit dem amerikanische ıhren; ein Gespräch mit den e on Wien bedeutet ihm nicht w inem Professor des Konservato

„immer noch, nach drei jahren zittert unter den arbeitern, höre ich allgemein, die panik, verursacht durch die plünderungen und vergewaltigungen, nach, die der eroberung von berlin folgten. (…) es wird berichtet, daß die russischen soldaten noch während der kämpfe von haus zu haus, blutend, erschöpft, erbittert ihr feuer einstellten, damit frauen wasser holen konnten, die hungrigen aus den kellern in die bäckerei geleiteten, die unter trümmern begrabenen ausgraben halfen. aber nach dem kampf durchzogen betrunkene horden die wohnungen, holten die frauen, schossen die widerstand leistenden männer und frauen nieder, vergewaltigten vor den augen der kinder, standen in schlangen an vor den häusern."
Bertolt Brecht, Arbeitsjournal 1941–1955

literatur über die sogenannte „dreckige" Schützengrabenwahrheit gab oder heute in der russischen Presse fünfzig Jahre lang unter Verschluß gehaltene Bilder von betrunkenen Soldaten in den Straßen Berlins veröffentlicht werden, so ist gewiß, daß Chaldejs Bilder hier kaum auftauchen und eine Rolle spielen können. Er fotografiert keine Soldaten im Kampf, nicht deren Gesichter, sondern ihre Rücken, da er sich, wie er heute noch sagt, weigert, deren „tierisches Aussehen" abzubilden.

„Hier hast du deine Geschichte, was haben mir die Rücken erzählt!" Die „saubere" Attitüde und die Propaganda, deren Zweck nie außer acht gelassen wird, fallen hier zusammen. Auf der Darstellung von Toten liegt ein Tabu. Keines der hier veröffentlichten Bilder wurde vom Zensor für den Abdruck während oder nach dem Krieg freigegeben, die Botschaft schien „zu wenig eindeutig" zu sein. Ein Soldat, der über die Leiche eines Gegners steigt, eine Waffe in der Hand...

Wie sieht die Freiheit aus, wie fotografiert man sie? Es gibt einen klassischen Vorrat an Bildern der Befreiung. Einem Soldaten wird in der Ukraine ein Kind anvertraut, die Zukunft, um derentwillen die Opfer eingegangen wurden; ein alter Mann küßt einen jungen Soldaten; die den einmarschierenden Soldaten zuwinkende und applaudierende Bevölkerung. Blumen werden überreicht, Frauen bewundern einen Befreier. Nicht zuletzt nach den Bildern aus dem Österreich des Jahres 38 wäre solchen Motiven gegenüber Skepsis angebracht, zumindest Verwunderung. Statt dessen leere Straßen beim Einzug der russischen Soldaten, allmähliches Herauskommen aus den Häusern, abmessende, mißtrauische Blicke zwischen Wienern und Sowjetsoldaten. Noch brennen die Häuser, irren Flüchtlinge mit ihren Bündeln auf dem Rücken durch die Straßen.

Chaldej nimmt auf dem Ring eine Frau auf, die ihre Kinder in einem Leiterwagen hinter sich her zieht (Seite 158). Sie schaut erschöpft und streng in die Kamera.

Erst in einer zweiten Aufnahme, da sie merkt, daß die Kinder fotografiert werden, lacht sie: „Geknipst." Eine fast vertraute Situation. Solche Details, winzige Gesten, die einem beim schnellen Betrachten entgehen, hat Roland Barthes durch den Ausdruck „punctum" zu charakterisieren versucht. Es sind jene Momente in der Fotografie, da über die bloße Mitteilung einer Information hinaus – in diesem Fall: Wie sehen die Soldaten aus, welche Kleider trägt die Zivilbevölkerung etc. – der Schleier der Erscheinungen zerrissen wird. Etwas trifft uns, so wie uns in einem der besten Fotos über den Zweiten Weltkrieg ein Detail trifft: Ein jüdisches Paar in Budapest, dem Schrecken des Ghettos entkommen. Beider Blick noch voller Schrecken, an den Mänteln die gelben Sterne. Es ist der Faden zu sehen, mit dem die Sterne angenäht wurden. Chaldej geht auf sie zu, spricht sie an, fotografiert sie.

Das anfängliche Sprachproblem und die ängstliche Frage, ob er Deutscher sei, beantwortet Jewgenij Chaldej auf jiddisch: „I bin och a Jidd." Das Bild des Grauens, allein den Gesichtern dieser beiden Menschen abzulesen, ist nur indirekt erfaßbar. Aus dem Antlitz der Welt wird es nicht mehr verschwinden, aber es hat ein Ende gefunden – in der Befreiung.

Bücher über Österreich im Jahr 45, den Beginn der Zweiten Republik, werden normalerweise mit Rückbezug auf die „innere" Sichtweise verfaßt, Fluchtpunkt sind Motive wie „Leben in den Trümmern", „Wiederaufbau", „Die Zeit des Dritten Mannes". Der Wechsel zur Außenperspektive, etwa zur sowjetischen Wahrnehmung, ist nicht unproblematisch. Wer sind diese Befreier Österreichs, deren Zahl im Mai 1945 im östlichen Teil des Landes fast eine halbe Million beträgt? „Russen in Wien", das klingt fast wie der Schlachtruf jener „Verteidiger" Österreichs, die in den letzten Tagen vor ihrer Flucht aus Wien von der europäischen Rolle der Stadt zu sprechen beginnen. Und mit der Befreiung ist die Stadt wiederum in fremden Händen. Das Bild „der Russen" ist zwiespältig in der populären Darstellung, wie sie jeder in

Ostösterreich nach dem Krieg Aufgewachsene kennt: Gebildete Offiziere, barbarische Horden von Vergewaltigern, Fahrrad-, Uhrendiebe und Liebhaber kleiner Kinder, die sie mit Schokolade beschenken, Erleichterung und Angst – ein Bündel von widersprüchlichen Wahrnehmungen. All das vermag nicht dem Bild eines Befreiers von der nationalsozialistischen Barbarei zu entsprechen, zumal es nach dem Ende der Sowjetunion, die den Antifaschismus für sich zur Staatsideologie kultivierte, keinen respektgebietenden Fürsprecher einer eindeutigen Sprachregelung mehr gibt.

Die österreichische Grundstimmung jener Zeit rührt jedenfalls von ihrer Opferperspektive her. Wie soll man sich heute, fünfzig Jahre danach, das Verhältnis zwischen Befreiern und Befreiten vorstellen, wenn sich die Befreiten als Opfer ihrer Befreier fühlten und oft genug fühlen mußten. Wäre es allzu zynisch zu behaupten: Hätte es die Russen nicht gegeben, hätte man sie erfinden müssen?

Die siebzehn in den Jahren 93/94 geführten Gespräche mit sowjetischen „Befreiern" Österreichs dieses Buches handeln von wiedererkennbaren Dingen, davon, wie das eigene Land vor 50 Jahren im Blick von außen sich darstellt. Die an die Gesprächspartner gerichteten Fragen lauteten: „Welche Erinnerung haben Sie an Österreich?", „Wie war das Verhältnis zur Zivilbevölkerung?", je nach Gesprächssituation wurden Fragen nach Plünderung, Vergewaltigung und Verschleppung gestellt, der unheiligen Dreifaltigkeit der Befreiung. Die Gesprächspartner und Erzähler sind hauptsächlich Moskauer, ein Interview wurde in Riga, Lettland, geführt, andere in Bischkek, Kirgisien, und in der karelischen Hauptstadt Petrosawodsk. Alle früher Sowjetbürger, sind sie heute Bürger von auf verschiedene Weise untereinander verbundenen selbständigen Staaten.

Diese gegenwärtige Situation stellt einen wichtigen Ausgangspunkt in vielen Gesprächen dar: Der Ruhm der Sieger über Hitler ist mit der Sowjetunion zerfallen. Die Fragen haben nur indirekte Antworten gefunden: „Ja, es hat diese Dinge gegeben, aber in nicht sehr großem Ausmaß." Wichtig dabei sei das strenge Regime gewesen, das schwere Strafen für den Fall von Übergriffen androhte. Ein Rotarmist, der zwischen 1945 und 49 in Österreich stationiert war, erzählt von einem Prozeß gegen drei junge Soldaten in Eisenstadt, die wegen Vergewaltigung – unter dem Applaus der im Gerichtssaal anwesenden Offiziere über das Urteil – zum Tode verurteilt werden (Seite 28). Ein ehemaliger USIA-Direktor berichtet über die Verschleppung eines österreichischen Mitarbeiters (Seite 218). Während das Töten mit dem Hinweis auf den gerechten Krieg und die Verschleppungen mit dem Hinweis auf die diktatorische Praxis der Sowjetunion heute keinem Tabu mehr unterliegen, bleibt Vergewaltigung als dem Bereich der persönlichen Verantwortung zugehörig in den Gesprächen weitestgehend ausgespart. Einen Grund für die Besatzung, die langjährige Stationierung, konnte niemand angeben, und das ist offenbar für einen militärischen Dienst Tuenden eine unverständliche Frage. In den Augen seiner Befreier war Österreich nicht wirklich Schauplatz des Krieges. Wiederholt wird auf die Geringfügigkeit der Zerstörung von Wien hingewiesen. Das Land war – fast nicht anders zu erwarten – das Land des Johann Strauß. (Im ersten sowjetischen Film des Jahres 45 über die Befreiung von Wien wurde das Geschützfeuer mit Walzerklängen unterlegt.)

Die sowjetischen Veteranen differenzieren nur selten zwischen Österreich als Teil des Dritten Reichs und Deutschland. So besteht etwa die irrige Meinung, Österreich habe am Krieg nicht teilgenommen, da keine Österreicher in eigenen österreichischen Uniformen angetroffen wurden. Nur eine kurz in Österreich stationiert gewesene ehemalige sowjetische Dolmetscherin spricht von einem Unterschied zwischen Österreichern und Deutschen, die sich als Besiegte fühlten, während die österreichische Bevölkerung selbstbewußter aufgetreten sei, nicht als Verlierer, sondern als

„So was von unrasier
Blusen, ungepanzten
Haltung, wie man bei
daß man einen reinen
justierten Rotarmiste
habe ich bei den Engl
Amerikanern bisher
Wunder, daß sich ihn
dern an den Hals wer
park sitzen sie mit de
links und rechts, und
Verständigung mit de
die doch sehr viele W
Die Soldaten kramen
Unterhaltung geht un
immer besser, je meh
Verständigungsmittel
angefangen vom Hän
rührung, bedient. Es
Freundschaft der Mä
verurteilen, ich will
muß zugeben, daß es
Burschen, ihrer Gebe
und Schokolade und
Männern in Wien mi
Josef Schöner, Wiener

0567

momox Shop
medimops
FROM
momox GmbH
Am Alten Flughafen 1
04356 Leipzig
Germany
A R
PRIORITY
W
Deutsche Post
TO
Zhao Qing
210000 Nanjing
CHINA VR
CN 22

/ 02.20 1,00 Deutsche Post

WARENPOST Int. KT Port payé

g

bei Lane, Meiyuan New Village ,

Nanjing

R

UC 032 676 146 DE

– 1 x Buch

Germany

Germany

China

210mm

62mm

140mm

● 铜奖
《 逃亡之魂
◖ 德国

Bronze Medal
Die flüchtige Seele
Germany
Edith Lackmann
Harold Brodkey
Rowohlt Verlag, Reinbek
210 140 62mm
1220g
1344p
ISBN-10 3-498-00540-5

美国作家哈罗德·布罗德基（Harold Brodkey）在 1991 年出版了自己的第一部长篇小说《逃亡之魂》（*The Runaway Soul*）。这是一部根据作者自己童年经历与之前已出版的短篇小说集的部分素材写就的半自传体小说，主人公的命运被认为与作者极为相似。本书是 1996 年出版的德文版。书籍的装订形式为圆脊锁线硬精装。护封艺术纸黑黄双色印刷，内封暖灰色布面裱覆卡板，封面图案印单黑，书脊处书名烫灰色。内文极薄的胶版纸印单黑，堵头布与书签带都选择了银色绸布，与封面压抑阴郁的风格相适配。作者的文字具有令人叹为观止的对话技巧，文字表达也十分复杂，加上半自传体的内容，导致对本书的评价毁誉参半，有人认为这是有史以来最重要的文学作品之一，也有人认为这种沉迷于早年悲惨记忆的漫无目的、重复、模糊的文风充满了傲慢。本书的排版风格十分质朴，主人公心理描写的部分采用斜体大段呈现。

American writer Harold Brodkey published his first novel *The Runaway Soul* in 1991. It is a semi-autobiographical novel based on the author's own childhood experience and the previous published collection of short stories. The fate of the protagonist is considered to be very similar to that of the author. This book was published in German in 1996. The hardcover book is bound by thread sewing with rounding spine. The jacket is made of art paper printed in black and yellow; the inner cover is covered with warm gray cloth mounted with cardboard. The patterns on the cover are printed in single black, and the title at the spine is stamped in gray. The very thin offset paper is used for inside pages printed in single black. The head band and bookmark belt are made of silver silk, which is suitable for the depressing and gloomy style of the cover. The author's writing skills are amazing, and the expression is very complex. With the semi-autobiographical content, the evaluation of the book is complicated. Some people think that it is one of the most important literary works in history; others think that the aimless, repetitive and vague style of writing, which is addicted to the tragic memory of the early years, is full of arrogance. The typesetting style of this book is very simple. The psychological descriptions of the protagonist are presented in italics.

ar sie weggespült worden, hatte ich gehört –, u
Kneipe, keine Reklamen, keinen Parkplatz ge
d Gras und Felder. Wir hatten oft auf Wies

n Fahrrad in einem Dickicht nah an der Straß
Boden unter einem umgestürzten Johannisbr
nen Zweigen deckte ich es ab. Ich zog die Jea
n abgeschnittenen Jeans dastand, die ich dahei
hatte. Die langen Jeans ließ ich bei dem Fahrra
und durchquerte den Wald, indem ich in Tur
Bachs entlangging, der sich (zwischen oft hohe
nunterwand. Ohne die Fliegen und Moskitos
chten, ging ich voran, mit verstochenen, nass
r Stimmung, bald unter sichtbarem Himme
. An der Mündung des Bachs, dort, wo er si
Bucht ergoß, lag ein sehr großes, plumpes Fluß
ichen grün und schwer, aus dicken, schwere
, voller Schrammen und Kratzer, abblättern

OMOSEX
Nach der Zu

B R O D K

l ü c h t

S e e

s zu einem gewissen Punkt

Forestville oder
Nonie, als ich g
Der Krieg · 64

s Selbstgespräch

Homosexualität
oder Zwei Männer i
Worin ich partie
aus der ich ausg
Einer von Abes
Daniels Freund

chts in jenem Haus, in meinem Zimmer, in
cke, die sich in diesem Zimmer bei Lampen
, und dazu die Geräusche aus der Dunkelhe
m Fenster – das Sommerkichern des Wildgra

«Ich lese nur Sachen auf englisch. Ich mag James Joyce – was er so über Familien schreibt und seine Schweinereien. Tolstoi mag ich auch. Dostojewski heb ich mir fürs College auf.»

«Gute Idee. Wie ein Leser siehst du mir nicht aus», sagte mein Cousin Daniel. Ich lasse mir nichts anmerken. Komisch, daß man zu allem Training und Glück braucht. Mut außerdem.

Ob du mich magst oder nicht ist mir egal, aber dich mag ich, wenn du mich so akzeptieren kannst, wie ich eben bin.

Köder, Pläsanterien … Geld … Aussehen …

Sein Bein glitt irgendwie seitwärts, bis sich sein Knie kribbelnd neben meinem befand, ohne mich zu berühren. Ich zuckte – unwillkürlich – zusammen, und er nahm es fort. Ich ziehe in sexuellen Dingen niemals als *erster.* Wenn man sich aber gut benimmt, *provoziert man Ärger.* Bei aller Peinlichkeit habe ich das Gefühl, ich sollte mich entschuldigen, und unbarmherzigerweise fühle ich mich peinlich *belustigt.*

«Magst du einfache Dinge?»

«Ich bin selbst einfach», sage ich.

Mit jemandem, der mir im Alter näher wäre, würde ich es darauf ankommen lassen, wer mehr schmutzige Wörter kennt; vielleicht würden wir uns keuchend voreinander zeigen, wenn es nichts zu bedeuten hätte, wenn es buchstäblich nur der Information diente und nicht auf anderer Ebene Teil einer persönlichen Absicht wäre – wenn es zum Beispiel nicht einem vielleicht echten, vielleicht dauerhaften Gefühl diente. Ich meine, man läßt sich darauf ein, auf die Erinnerung daran und auf Wiederholungen davon.

Wenn Daniel ein *stinknormaler Kerl* wäre, würden wir miteinander ringen oder so. Ich wäre derjenige, der das Sagen hat und der bestimmt, inwieweit der andere für eine Weile gehorcht, erotisch und nicht-erotisch, wenn man so will … Wer wen mag, wer die Sache hier im Abteil steuert. Wann wer dran ist. Oder so.

Das Gleißen draußen läßt mich die Augen zusammenkneifen. Der Zug neigt sich, als er eine kurvige Geleisstrecke befährt, und setzt unsere Fenster direkt dem Licht aus. Auf eine Frage von Daniel erwidere ich: «Ich bin ein besserer Ringer als Boxer; ein besserer Pitcher als Baseman; ein besserer Batter als manche, die eigentlich besser sind, weil ich treffsicher bin. Ich kann auch schlagen, wenn's ums Ganze geht. Das hab ich raus.»

Die Fragebögen des sozialen Lebens … Der Zug fährt in einer heruntergekommenen ländlichen Gegend durch die kalten Schatten sehr hoher Alleebäume. Ach, diese Momente von *Beinahe* und *Zutiefst*, von *Nicht sehr*, *Flüchtig* und *Reserviert*! Ich komme mir gemein vor. Noch selbstsüchtige Leidenschaft ist sakramentaler als das, was ich da treibe. Obwohl Leidenschaft etwas ganz schön Übles ist. Vielleicht *«liebe»* ich ihn ja neckisch – und sadistisch … nur eben nicht leidenschaftlich. Jedenfalls war das Objekt zutiefst sakramentaler Gefühle für ihn junge Männlichkeit, nicht ich – oder nur partiell ich. Im Grunde studiert er das *Transzendente* hinter der männlichen Realität. Ich glaube, daß dies zu einer adoleszenten Tradition der Sinnsuche gehört, aber genau weiß ich es nicht. Voranzuschreiten, sich in jenes Transzendente zu verwandeln …

Mittlerweile, während wir bald vibrierend dahinzuckeln, bald rasch voranrücken, erzeugt der Drachenatem seiner Beachtung irgendwie den Eindruck, als berühre ihn mein Anblick so, als wäre ich ein Mädchen. Das Licht mir übertragener Bedeutsamkeit geht von mir aus – vielleicht. Zwei Fliegen summen in der Hitze und bedrängen einander in einer Ecke des schmutzigen Fensters. Ich weiß nicht, ob diese Zugreise den Erfahrungen meiner toten Mutter oder denen von S. L. in seiner Jugend entspricht. Ich glaube schon. Sicher weiß ich es nicht – ist die ganze Welt sexuell oder nicht?

Anlässe, maßlos jung zu sein, ergeben sich häufig.

1996 Bm^{4}

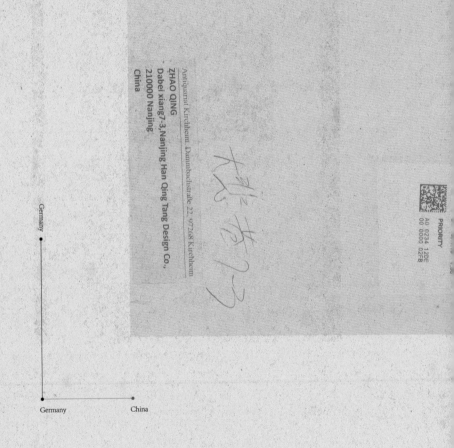
Antiquariat Kirchheim, Dammbachstraße 22, 97268 Kirchheim
ZHAO QING
Dabei xiang7-3,Nanjing Han Qing Tang Design Co.,
210000 Nanjing
China
PRIORITY
A0 0234 12DE
00 0000 02FB
Germany
Germany
China

10mm

302mm

Groothuis/Janßen Schwarzbuntes. Bilder aus Ostfriesland Christians

231mm

● 铜奖
« 丰富的黑白——东弗里斯兰的照片
● 德国

Bronze Medal
Schwarzbuntes: Bilder aus Ostfriesland
Germany
Rainer Groothuis
Rainer Groothuis, Karl-Heinz Janßen
Christians Verlag, Hamburg
302 231 10mm
617g
80p
ISBN-10 3-7672-1238-2

东弗里斯兰是位于德国西北部下萨克森州的临海地区，与荷兰接壤。本书的作者、摄影师赖纳·格鲁特胡斯（Rainer Groothuis）生于埃姆登市，是这一区域埃姆斯河入海口处与荷兰遥遥相望的城市。书中的照片记录了 20 世纪 90 年代中期东弗里斯兰地区的自然风光与风土人情。书籍的装订形式为锁线硬精装，封面黄色艺术纸烫黑色文字，裱贴卡板，书脊处露出深蓝色布面烫黄色处理。封面中上位置的黑色照片为哑面纸张印单黑。深蓝色艺术纸作为环衬。内页哑面铜版纸单黑印刷，挂网较细，呈现非常丰富的画面细节与层次变化。内容部分主要是有关东弗里斯兰地区的照片，并对当地的自然气候、四季变化、地貌特征、经济特产与人文风情进行阐述，探讨这个以农业为主的地区受到旅游业的影响，思考在追求经济利益的情况下如何保护自然风貌的方法，并表达出把这一地区改造为经济与风景俱佳的宜居地的愿景。

East Friesland is located in the coastal area of Lower Saxony in northwestern Germany, bordering the Netherlands. The author and photographer Rainer Groothuis was born in Emden, a city in the region that faces the Netherlands from the mouth of the Ames River. The photos in the book record the natural scenery and customs of the East Friesland in the mid-1990s. The hardcover book is bound by thread sewing. The cover is made of yellow art paper with black hot-stamped text, mounted with cardboard. The spine is covered with dark blue cloth stamped in yellow. Matte paper printed in single black is used for the photo in the middle of the cover. The dark blue art paper is used as the end paper and the inner pages are made of matte coated paper printed in single black with delicate screen-printing, showing rich details and changes. The content is mainly about photos of the East Frisian region, and explains the local natural climate, seasonal changes, landform features, economic specialties and cultural customs, and explores the impact of tourism on this agricultural-based area. The book explores the protection of natural features in pursuit of economic benefits, and the vision of transforming this area into a livable place with both economic and scenic advantages.

Schwarzbuntes
Bilder aus
Ostfriesland

von Rainer Groothuis

mit Texten von
Karl-Heinz Janßen

Christians

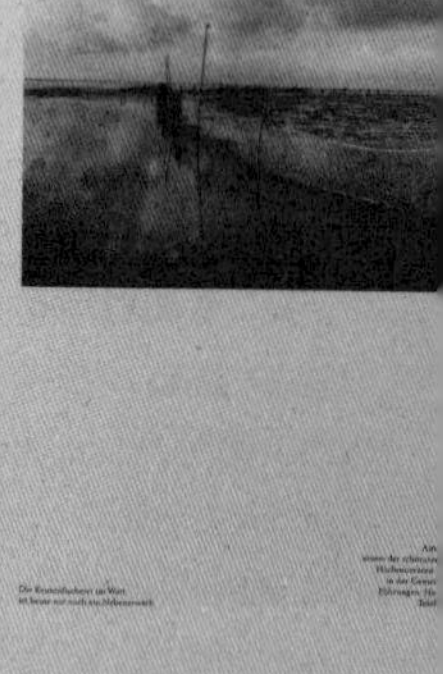

Ditzum.
Blick auf den 1846
erbauten Kirchturm

links
Im Dollart bei Pogum:
Schlickschlitten (›Kreier‹) rennen mit kultureller Begleitung
Diese Kreier, die wie Schlitten auf dem Watt liegen,
wurden von den Fischern benutzt, um die Reusen

Aus diesen Bohnen wird ein ostfriesisches Nationalgericht:
Updrögt Bohnen
Reife Bohnen werden auf Fäden gezogen und getrocknet, erst im Wind, dann auf dem Boden oder in der Küche. Sobald die erste Bohne herausplatzt, werden die Ketten abgenommen und möglichst in einem Leinensack verwahrt. Am Abend vor der Zubereitung gründlich waschen und in etwa zwei Zentimeter lange Stücke schneiden, über Nacht einweichen. Am folgenden Tag werden die Bohnen in frischem Wasser 20 Minuten gekocht, auf einen Durchschlag gegeben und nochmals abgespült. Den Speck mit dem Wasser und den Bohnen aufsetzen und zwei Stunden kochen lassen, in der letzten halben Stunde werden Kartoffeln und Mettwurst mitgekocht. Danach alles vermengen und mit Salz und Pfeffer abschmecken.
Für vier Personen: 500 g Bohnen, 750 g getrockneter, durchwachsener Speck, zwei Mettwürste, 0,75 l Wasser, 500 g Kartoffeln, Salz und Pfeffer

1996 Bm[5]

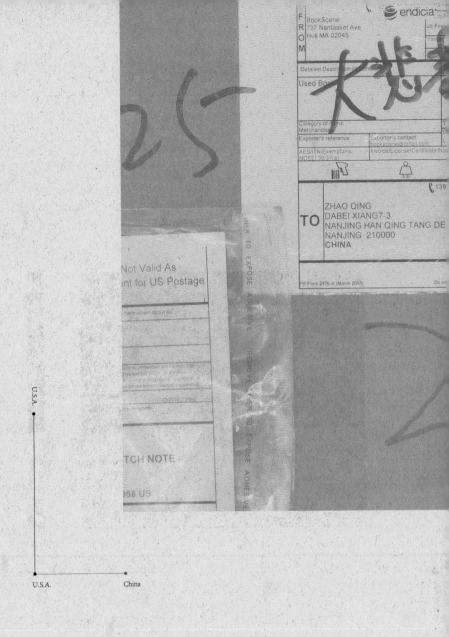
endicia
FROM
BookScene
737 Nantasket Ave
Hull MA 02045
Detailed Description of Contents
Category of items
Merchandise
Exporter's reference
Exporter's contact
bookscene@gmail.com
AES/ITN/Exemptions
NOEEI 30.37(a)
Invoice/License/Certificate No
5 lb
TO
ZHAO QING
DABEI XIANG7-3
NANJING HAN QING TANG DE
NANJING 210000
CHINA
PS Form 2976-A (March 2015)
Not Valid As
ent for US Postage
TCH NOTE
958 US
U.S.A.
U.S.A.
China

312mm

25mm

AFRICA
HERB RITTS

363mm

铜奖
非洲
美国

Bronze Medal
Africa
U.S.A.
Betty Egg, Sam Shahid
Herb Ritts
Bulfinch Press; Little, Brown and Company, Boston
312 363 25mm
2002g
136p
ISBN-10 0-8212-2121-3

赫布·里茨（Herb Ritts）是20世纪美国著名的人像摄影师，曾为约翰尼·德普、梅尔·吉布森、艾尔·帕西诺、杰克·尼科尔森、朱莉亚·罗伯茨等国际影星拍过肖像。他的肖像作品均为黑白摄影，擅长通过光影来塑造人物，创造出古典雕塑般的形象。这本书是他拍摄的非洲特辑，当时曾引起了极大反响。书籍的装订形式为锁线硬精装。护封采用铜版纸印单黑，覆哑膜。上下书口和勒口处均向内包折，形成良好的手感，同时避免了这么大的开本上护封变形的问题。内封灰色布面，书名烫金属灰色。环衬选用了一种极其厚重且带有粗糙手感的艺术纸。内页哑面铜版纸印单黑。赫布·里茨在拍摄过程中深入非洲大陆，与各族黑人交流、舞蹈，并为他们拍摄。照片主要包含人像、自然环境、牲畜和野兽，以及他们之间的互动和生活场景。黝黑的皮肤和广袤的环境形成强烈对比，衬托出非洲黑人的生存现状。

Herb Ritts is a famous 20th-century American portrait photographer. He has taken portraits of Johnny Depp, Mel Gibson, Al Pacino, Jack Nicholson, Julia Roberts and other international movie stars. His portraits are all black-and-white photography. He is good at highlighting people through light and shadow and creating classical sculpture-like images. This book is a special African album he shot, which aroused great repercussions at the time. The hardcover book is bound by thread sewing. The matted jacket is printed with coated paper in single black. The upper and lower book edges and the flaps are folded inward to avoid the deformation of the jacket of such a large format. The book's title is gray hot stamping. A kind of art paper with thick and rough feel is selected for the end paper. The inner pages are made of matte coated paper printed in black. During the filming, herb Ritz went deep into the African continent to communicate, dance and shoot for black people of all ethnic groups. The photos mainly include portraits, natural environment, livestock and wild animals, as well as their interaction and living scenes. The dark skin and the vast environment form a strong contrast, which sets off the living situation of African blacks.

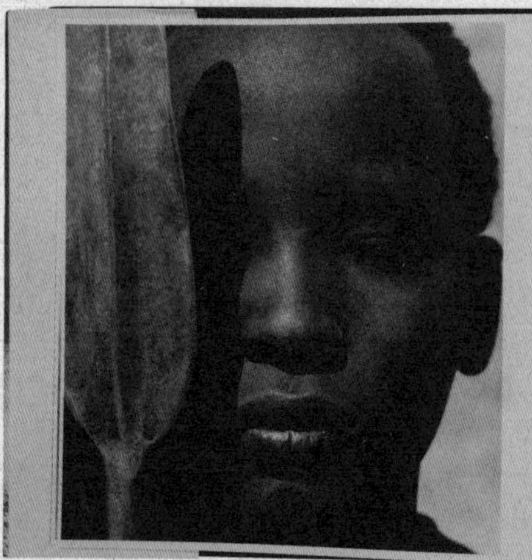

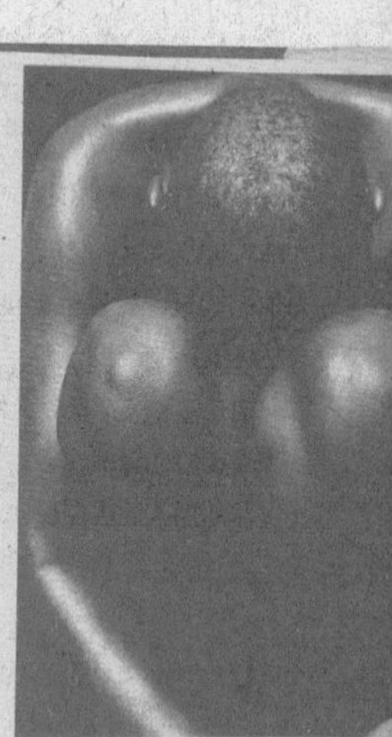

AFRICA
HERB RITTS

AUDIBERTI
BATAILLE
PEREC

ANTHOLOGIE
DU CINÉMA
INVISIBLE

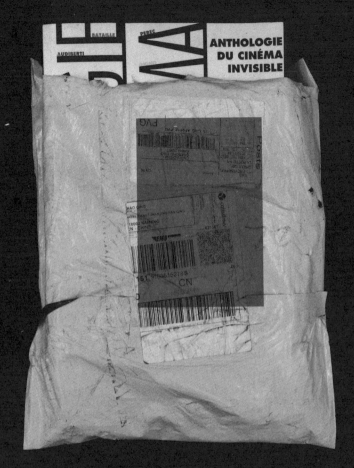

EMS
FRCYMA(FRA)
CHILLY CEI
LA POSTE, FR
AEN 0312
2019-11-14
PC 909
EN 1
2.2
CNSHAD(CNA)
SHANGHAI EMS
CHINA POST
15 CBG AF0112 PVG
PVG
Airmail
CN 35
colissimo
821367
XIANG7-3NANJING HAN QING
0 NANJING
CHINE
EMS
Poids
2.15 kg
51 5193515278 S
CN
101-CN-210000
D
France
France
China

300mm

55mm

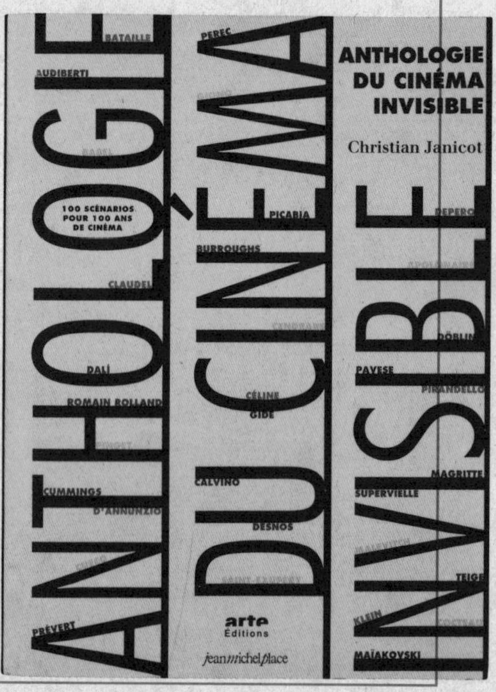

230mm

◎ 荣誉奖

《 "看不见的"电影选集

◖ 法国

Honorary Appreciation

Anthologie du cinéma invisible

France

Bulnes & Robaglia

Christian Janicot

Éditions Jean-Michel Place; ARTE Éditions, Paris

300 230 55mm

2195g

672p

ISBN-10 2-85893-233-6

1995 年，正值电影诞生 100 周年，艺术家克里斯蒂安 · 雅尼科（Christian Janicot）为此策划了一场“看不见的”电影放映会。他收集了众多艺术家、诗人、编剧、作家的 100 份未拍摄或未公映的剧本、大纲或故事版，希望通过文字的形式让读者进入电影想象力的世界，合辑成册便是本书。书籍的装订形式为锁线胶平装。封面白卡纸红蓝紫黄黑多种非常规油墨套印，覆哑光膜，供稿的艺术家名做亮面 UV 处理。内页胶版纸五色印刷。由于是 100 份剧本，内容上千差万别，所以在内页的设计上没有统一的规范，而是根据每份内容单独设计相应的版面，类似每部电影后不同风格的字幕。每份剧本均有剧名设计，有的还会配上相应的插图、手稿、照片，帮助读者在阅读时将电影“放映”在脑内“小剧场”中。书的翻口处有 100 份剧本的索引标签，方便查阅。

In 1995, on the 100th anniversary of the birth of film, artist Christian Janicot planned an ‘invisible’ film screening meeting for this purpose, and collected 100 unprinted or unpublished scripts, outlines or story versions of many artists, poets, screenwriters and writers. Hopefully□readers can enter the world of film imagination through the form of words. It’s the origin of this book. The book is perfect bound with thread sewing. The matted cover is made of white cardboard printed in various unconventional colors, such as red, blue, purple, yellow and black. The name of the artists who made contribution to this book is UV varnished. The inside pages are made of offset paper printed in five colors. Since there are 100 scripts and the content varies greatly, no specification for the design of the inner pages is fixed. The layout of each script is designed separately according to its content, similar to the different styles of subtitles of each movie. Each title has been carefully designed, and corresponding illustrations, manuscripts, and photos are attached to help readers ‘watch’ the movie in the theater of their mind when reading. There are index labels for 100 scripts at the fore-edge of the book for easy reference.

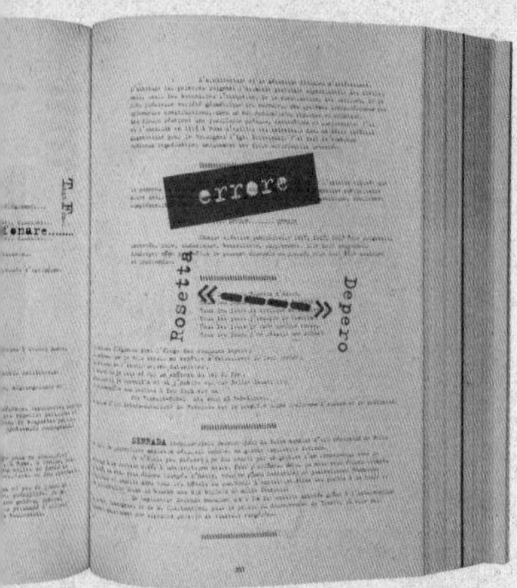

Secam. Et un jour, dans des milliers d'années, l'Univers se contracte, revenant dans le passé vers sa forme primitive, et tout revient à l'envers une vie sans mémoire où l'on connaîtrait l'avenir. On voit un tombeau dans un cimetière. Des ouvriers arrivent, qui l'ouvrent. Puis des gens en pleurs. « C'est plus fort que moi, je pleure, et pourtant je sais qu'il va vivre. » Des fossoyeurs sortent le cercueil. On le place sur un corbillard. Le cortège s'éloigne. Il arrive à l'église, puis au domicile du mort. Les veillées autour du corps que l'on reprend, que l'on rapporte sur la route et qu'on abandonne tout sanglant. Il s'agite un instant. Une auto survient qui lui passe sur le corps. Après quoi le mort se relève et retrouve son ami qui recommence la conversation du début. On imagine le procédé qui nous permettra de faire revivre toute une époque historique, ses modes, ses coutumes, ses événements, mais à l'envers. Ainsi, dans un tel monde, on n'aura aucune mémoire, mais on connaîtra l'avenir. Et on en saura de moins en moins au fur et à mesure qu'on vieillira ou plutôt qu'on se rapprochera de la jeunesse.

Il n'y aura plus de douleur,
sens vrai du mot puisqu'
commencera par souffrir et q
dès l'événement douloure
survenu, on l'oubliera et qu'on
trouvera heureux. Pour prendre
exemple élémentaire, un gosse a
d'abord mal aux fesses, recevra
fessée, se relèvera tout guilleret et dévor
un pot de confitures qu'il remettra en p
ensuite. Il est peut-être inutile de pousser la fic
jusqu'à l'extrême et de montrer un banquet o
convives rempliraient à l'aide de fourchettes leur ass
du poulet qui, reconstitué sur le plat, irait au four
sortir cru, retrouver ses plumes dans la cuisine, recevoir u
couteau après lequel il ira becqueter dans la cour en gloussa
aussi les vaisseaux sombrés sortant de la mer, les morcea
se reformant en avion qui prendra son vol, etc.

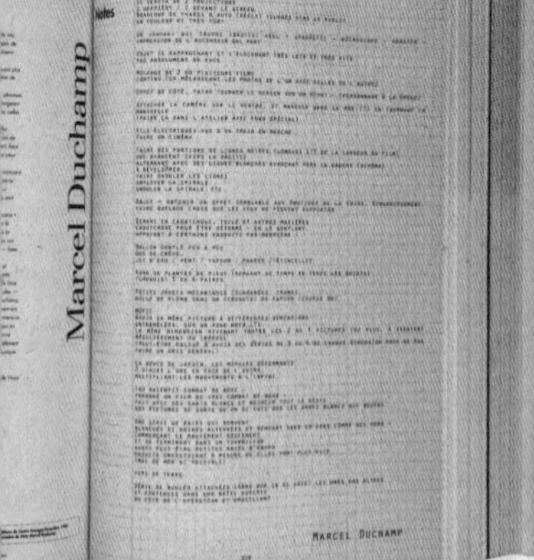

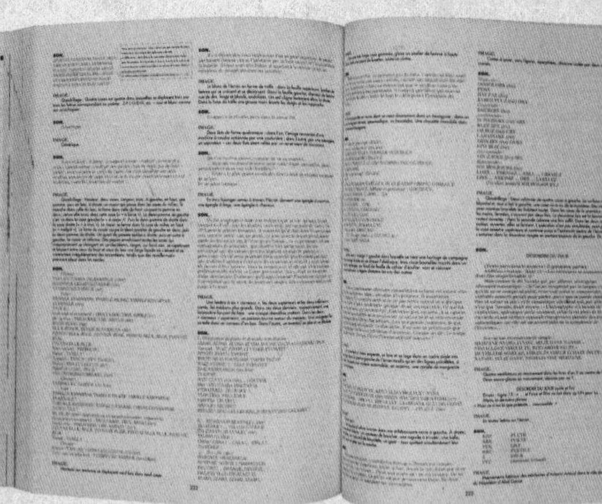

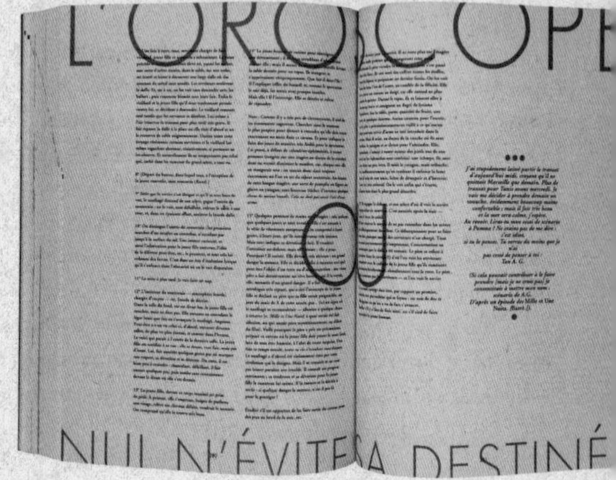

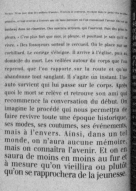

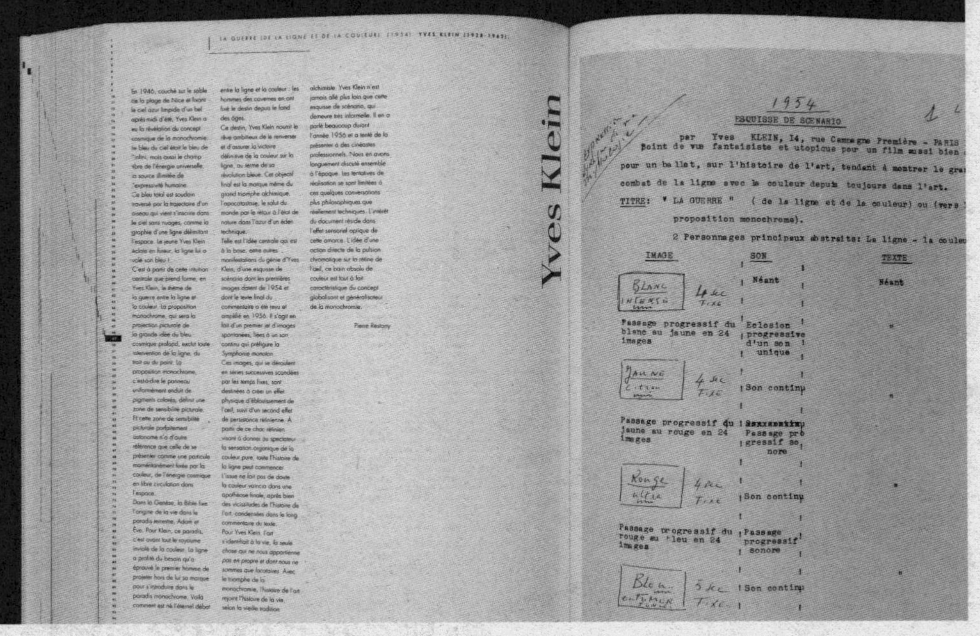

Yves Klein

En 1946, couché sur le sable de la plage de Nice et fixant le ciel azur limpide d'un bel après-midi d'été, Yves Klein a eu la révélation du concept cosmique de la monochromie. Le bleu du ciel était le bleu de l'infini, mais aussi le champ libre de l'énergie universelle, la source illimitée de l'expressivité humaine.

Ce bleu total est soudain traversé par la trajectoire d'un oiseau qui vient s'inscrire dans le ciel sans nuages, comme la graphie d'une ligne délimitant l'espace. Le jeune Yves Klein éclate en fureur, la ligne lui a volé son bleu !

C'est à partir de cette intuition centrale que prend forme, en Yves Klein, le thème de la guerre entre la ligne et la couleur. La proposition monochrome, qui sera la projection picturale de la grande idée du bleu cosmique profond, exclut toute intervention de la ligne, du trait ou du point. La proposition monochrome, c'est-à-dire le panneau uniformément enduit de pigments colorés, définit une zone de sensibilité picturale. Et cette zone de sensibilité picturale parfaitement autonome n'a d'autre référence que celle de se présenter comme une particule momentanément fixée par la couleur, de l'énergie cosmique en libre circulation dans l'espace.

Dans la Genèse, la Bible fixe l'origine de la vie dans le paradis terrestre, Adam et Ève. Pour Klein, ce paradis, c'est avant tout le royaume inviolé de la couleur. La ligne a profité du besoin qu'a éprouvé le premier homme de projeter hors de lui sa marque pour s'introduire dans le paradis monochrome. Voilà comment est né l'éternel débat entre la ligne et la couleur : les hommes des cavernes en ont fixé le destin depuis le fond des âges.

Ce destin, Yves Klein nourrit le rêve ambitieux de le renverser et d'assurer la victoire définitive de la couleur sur la ligne, au terme de sa révolution bleue. Cet objectif final est la marque même du grand triomphe alchimique, l'apocatastase, le salut du monde par le retour à l'état de nature dans l'azur d'un éden technique.

Telle est l'idée centrale qui est à la base, entre autres manifestations du génie d'Yves Klein, d'une esquisse de scénario dont les premières images datent de 1954 et dont le texte final du commentaire a été revu et amplifié en 1956. Il s'agit en fait d'un premier jet d'images spontanées, liées à un son continu qui préfigure la *Symphonie monoton*.

Ces images, qui se déroulent en séries successives scandées par les temps fixes, sont destinées à créer un effet physique d'éblouissement de l'oeil, suivi d'un second effet de persistance rétinienne. À partir de ce choc rétinien visant à donner au spectateur la sensation organique de la couleur pure, toute l'histoire de la ligne peut commencer. L'issue ne fait pas de doute : la couleur vaincra dans une apothéose finale, après bien des vicissitudes de l'histoire de l'art, condensées dans le long commentaire du texte.

Pour Yves Klein, l'art s'identifiait à la vie, *la seule chose qui ne nous appartienne pas en propre et dont nous ne sommes que locataires*. Avec le triomphe de la monochromie, l'histoire de l'art rejoint l'histoire de la vie, selon la vieille tradition alchimiste. Yves Klein n'est jamais allé plus loin que cette esquisse de scénario, qui demeure très informelle. Il en a parlé beaucoup durant l'année 1956 et a tenté de la présenter à des cinéastes professionnels. Nous en avons longuement discuté ensemble à l'époque. Les tentatives de réalisation se sont limitées à ces quelques conversations plus philosophiques que réellement techniques. L'intérêt du document réside dans l'effet sensoriel optique de cette amorce. L'idée d'une action directe de la pulsion chromatique sur la rétine de l'oeil, ce bain absolu de couleur est tout à fait caractéristique du concept globalisant et généralisateur de la monochromie.

Pierre Restany

1954

1

ESQUISSE DE SCENARIO

par Yves KLEIN, I4, rue Campagne Première - PARIS

Point de vue fantaisiste et utopique pour un film aussi bien [...] pour un ballet, sur l'histoire de l'art, tendant à montrer le gra[nd] combat de la ligne avec la couleur depuis toujours dans l'art.

TITRE: " LA GUERRE " (de la ligne et de la couleur) ou (vers [...] proposition monochrome).

2 Personnages principaux abstraits: La ligne - la couleu[r]

IMAGE	SON	TEXTE
Blanc intense — 4 sec fixe	Néant	Néant
Passage progressif du blanc au jaune en 24 images	Eclosion progressive d'un son unique	"
Jaune citron — 4 sec fixe	Son contin[u]	"
Passage progressif du jaune au rouge en 24 images	Passage progressif sonore	
Rouge ultra — 4 sec fixe	Son contin[u]	"
Passage progressif du rouge au bleu en 24 images	Passage progressif sonore	"
Bleu outremer foncé — 5 sec fixe	Son contin[u]	

Albert de Sembler

Vers

Dans une petite ville de l'Est, Ra

Le panoramique descriptif b

coteaux et sapins pour s'arr

la tourelle d'un ch

ennuyeux et fais

Les I

Renco

Le père d'Albert est marquis.

Il aime les vieilles épées.

Albert aime une jeune fille,

nd une

rmation de

ne

Figueras,

agne)

eures de

cultés

brochage,

l convient

tion

la totalité

es

absolumen

grande pla

cyclistes q

Ils ont les y

grosse pier

petite pèler

est accroch

Babaouo tr

avec préca

Il descend

non sans a

de la main

chemise, q

RAN EN BLANC

QUES SECONDES

SIÈME PARTIE

I

ublique chez lui. *C'est l'heure du*

ou cinq missives que son secrétaire

nent.

— personnelle.

écriture grasse

Le f

• M

• Il

• Be

(*On*

• Be

l'ore

• *O*

• *O*

• Be

• Ur

• Il

• Be

femmes, entiers ou par morceaux. Ici un

, etc. La bouchère sourit. Elle est tout à fa

ent ensemble.

Ils sont partis on voit un garçon boucher p

mises, des ceintures, des corsets, des souli

au ruisseau.

R ELLE UNE BONNE PETITE VIE.

on de recettes et Sadie entrent dans un sq

ssent. Au milieu d'une allée il y a le photo

s oiseaux. Ils s'éloignent. Arrivent dans un

. Elle a bien envie des bijoux. Quand elle

res de l'avenue sont parés de diamants. Le

êtres, etc. La lumière du soleil qui se reflè

éra par

nemen

n l'oubli

eureux.

émentair

l aux fes

évera tout

fitures qu'

Federico García Lorca

VOYAGE

À

LA LUNE

Federico García Lorca

1
Un lit blanc contre un mur gris. Sur les couvertures surgit une danse de chiffres : 13 et 22.
Deux d'abord, ils finissent par couvrir le lit comme des fourmis minuscules.

2
Une main invisible arrache les couvertures.

3
De grands pieds courent rapidement avec de gros bas à losanges blancs et noirs.

4
Une tête effrayée qui regarde fixement un point et se fond pour faire place
à une tête en fil de fer sur fond d'eau.

5
Une légende qui dit *Au secours Au secours Au secours* se déplace de haut en bas
en surimpression sur un sexe de femme.

6
Un long couloir étroit traversé par la caméra avec au bout une fenêtre.

7
Vue de Broadway la nuit.

8
Cette vue se fond dans la scène précédente.

9
Deux jambes oscillent avec une grande rapidité.

10
Les jambes se fondent en un groupe de mains qui tremblent.

11
Les mains qui tremblent en surimpression sur un petit enfant qui pleure.

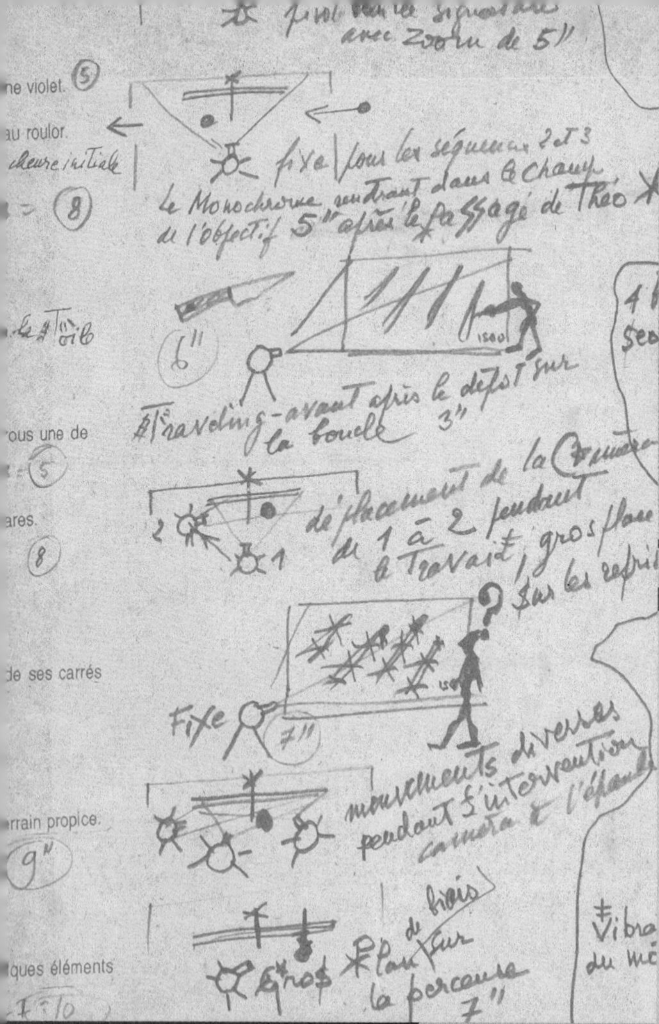

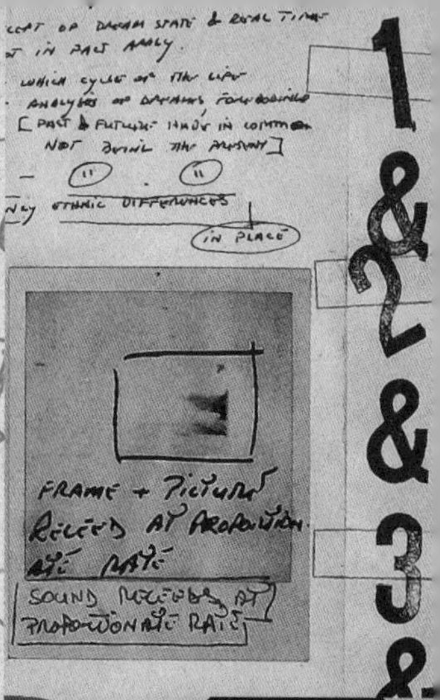

ne violet.

au roulor.

vous une de

ares.

de ses carrés

rrain propice.

ques éléments

J'étais
mieux
inspiré par
les cartes
postales de
Suisse
Élégie rose du mont
Cervin
Les soirs d'hiver dans
ma petite ville de Pologne
Devant les vitrines des

FRAME + PICTURE
RECEED AT PROPORTION-
ATE RATE
SOUND RECEDING AT
PROPORTIONATE RATE

1 & 2 & 3 &

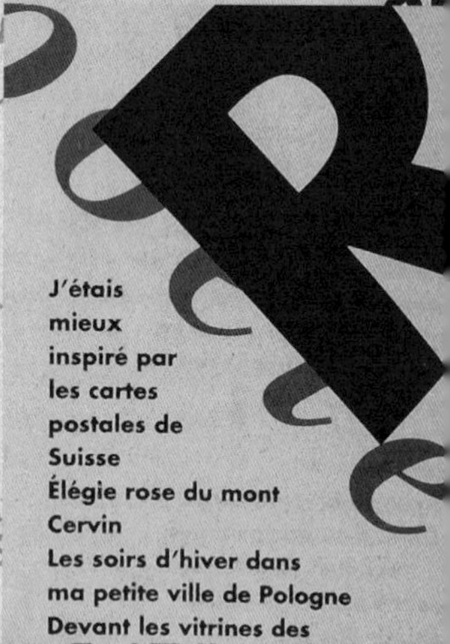

R R R R R
RRRR
TRR TRR
ara Tatum Taaa Trrrr

che.
erole de fonte -

ns de nacre avec ma charrue de fantaisie - docile
de cuivre dans le soleil, qui braille sa patience

vert, W blanc
W W d'or

- non Tà - mais
- non Tà - mais

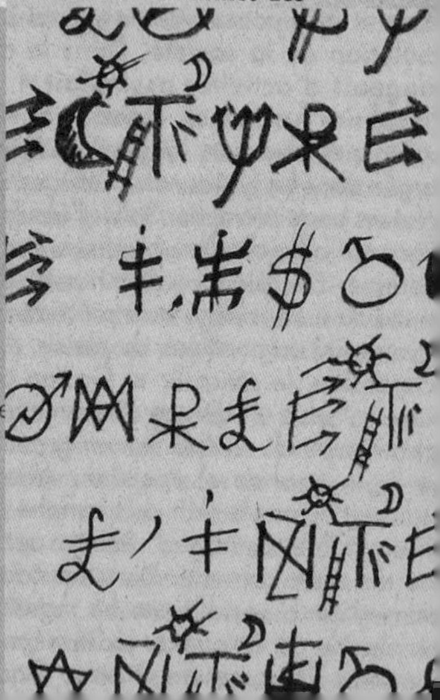

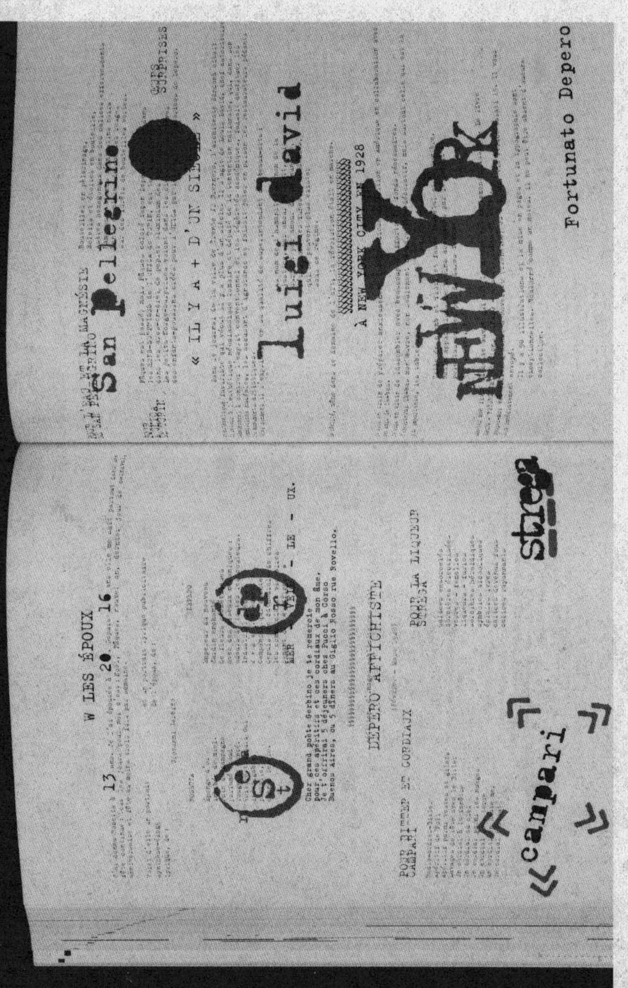
W LES ÉPOUX
13 20 16
DEPERO AFFICHISTE
POUR BITTER ET CORDIAUX CAMPARI
POUR LA LIQUEUR STREGA
campari
strega
San Pellegrino
« IL Y A + D'UN SIÈCLE »
Luigi david
À NEW YORK CITY EN 1928
NEW YORK
Fortunato Depero

PRIORITY
★ MAIL ★
Kotas
440 Portola Dr
San Mateo
CA 94403
FLAT RATE ENVELOPE
ONE RATE ★ ANY WEIGHT
Ship To:
Zhao Qing
Nanjing Ha Qing Tang Design, 7-3, Dabei
Lane
Meiyuan New Village
NANJING, JIANGSU 210000
China
大悲巷7-3
PS00001000016
Padded Flat Rate Envelope
EP14PE July 2013
ID: 9.5 x 12.5
UNITED ST
POSTAL SE
TRAC
INSU
★ Domestic only.
* For international shipments, the maximum weight is 4 lbs.

Poland

U.S.A.

China

★ MAIL ★

Kotas
440 Portola
Dr
San Mateo
CA. 94403

FLAT RATE ENVELOPE
ONE RATE ★ ANY WEIGHT*

Ship To:

Zhao Qing
Nanjing Ha   Qing Tang Design, 7-3, Dabei
Lane
Meiyuan New Village

248mm

10mm

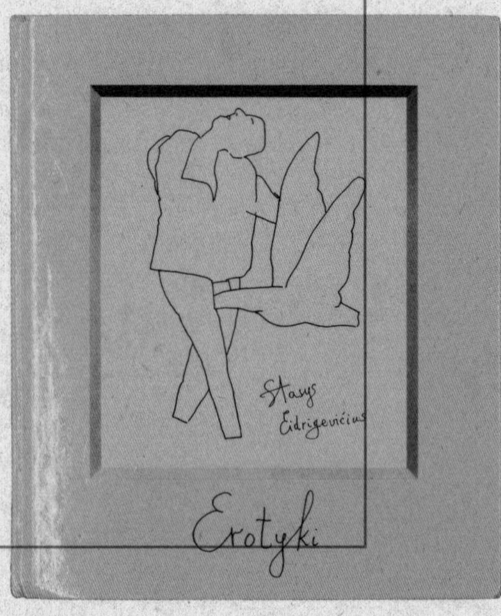

218mm

Stasys · Erotyki

● 荣誉奖
《 情色
◐ 波兰

Honorary Appreciation
Erotyki
Poland
Gra yna Bareccy, Andrzej Bareccy
Stasys Eidrigevi ius
Wydawnictwo Tenten, Warschau
248 218 10mm
374g
64p
ISBN-10 83-85477-85-3

立陶宛艺术家、设计师、导演斯塔西斯·艾德里格维修斯（Stasys Eidrigevičius）从家园、童年和立陶宛的乡村印象中汲取灵感，绘画风格荒诞、怪异，善于表现孤独、疏远、悲伤、沉思等情绪。本书是他的绘画与短诗集。书籍的装订形式为锁线硬精装。封面铜版纸黄棕黑三色印刷，覆亮膜，裱覆卡板。环衬胶版纸四色印刷，内页胶版纸印单黑。书中包含 30 幅超现实主题的钢笔画，随附 30 首小诗，大多只是单词或词组的罗列。这些绘画和小诗的主人公是作者把纸面当作剧场多年创造的一个形象，“他”在这个“剧院”中环游世界、生活、聚会、等待、相爱、玩笑。“剧院”中与“他”相伴的东西只有桌椅和床，直到新事物的介入。小诗的标题均为粉彩手写字体，保留了粉彩在粗糙纸面上留下的干燥笔触。全书呈现出一种与画作风格相适应的闲散、怪诞、孤绝的气质。

Stasys Eidrigevičius, a Lithuanian artist, designer and director, draws inspiration from his home, childhood and rural impression of Lithuania. His painting style is absurd and weird, showing loneliness, alienation, sadness, contemplation and other emotions. This book is a collection of his paintings and short poems. The hardcover book is bound by thread sewing. The laminated cover is printed in yellow, brown and black on coated paper, mounted with cardboard. The end paper is made of offset paper printed in four colors while the inside pages are printed in single black on offset paper. The book contains 30 pen drawings of surreal themes, accompanied by 30 small poems, mostly just lists of words or phrases. These paintings and small poems are images that the author has created on paper as a theater for many years. In this ‘theater’, he travels, lives, meets, waits, falls in love, and makes jokes. The things that accompany ‘him’ in ‘theatre’ are only tables, chairs and beds, until new things intervene. The titles of the little poems are all pastel handwritten fonts, retaining the dry brushstrokes left by pastels on rough paper. The book presents an idle, grotesque, and solitary temperament that adapts to the painting’s style.

Sytuacja

otyt

WYDAWNICTWO
TENTEN

Dyrygent zmi
odległość
między naszy
ustami

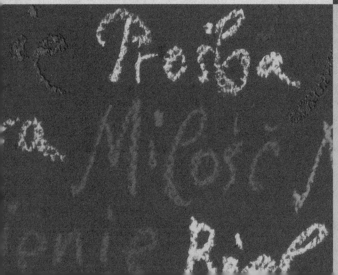

List

Okno otw
odleciał
w łóżku z
twój cie

Adres

Gniazdo puste
tylko biedronka
siedzi na kartce
z twoim adresem

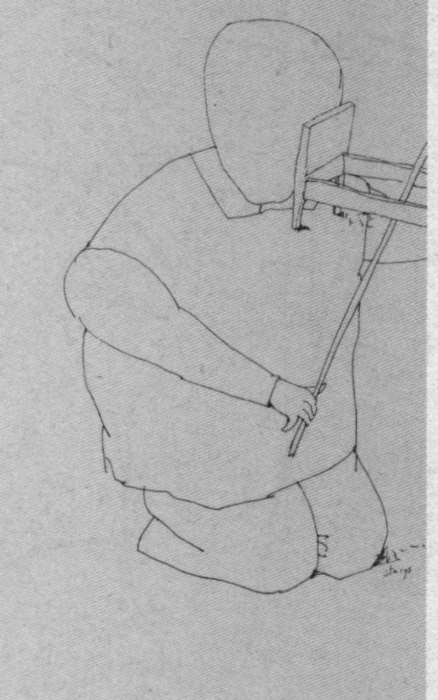

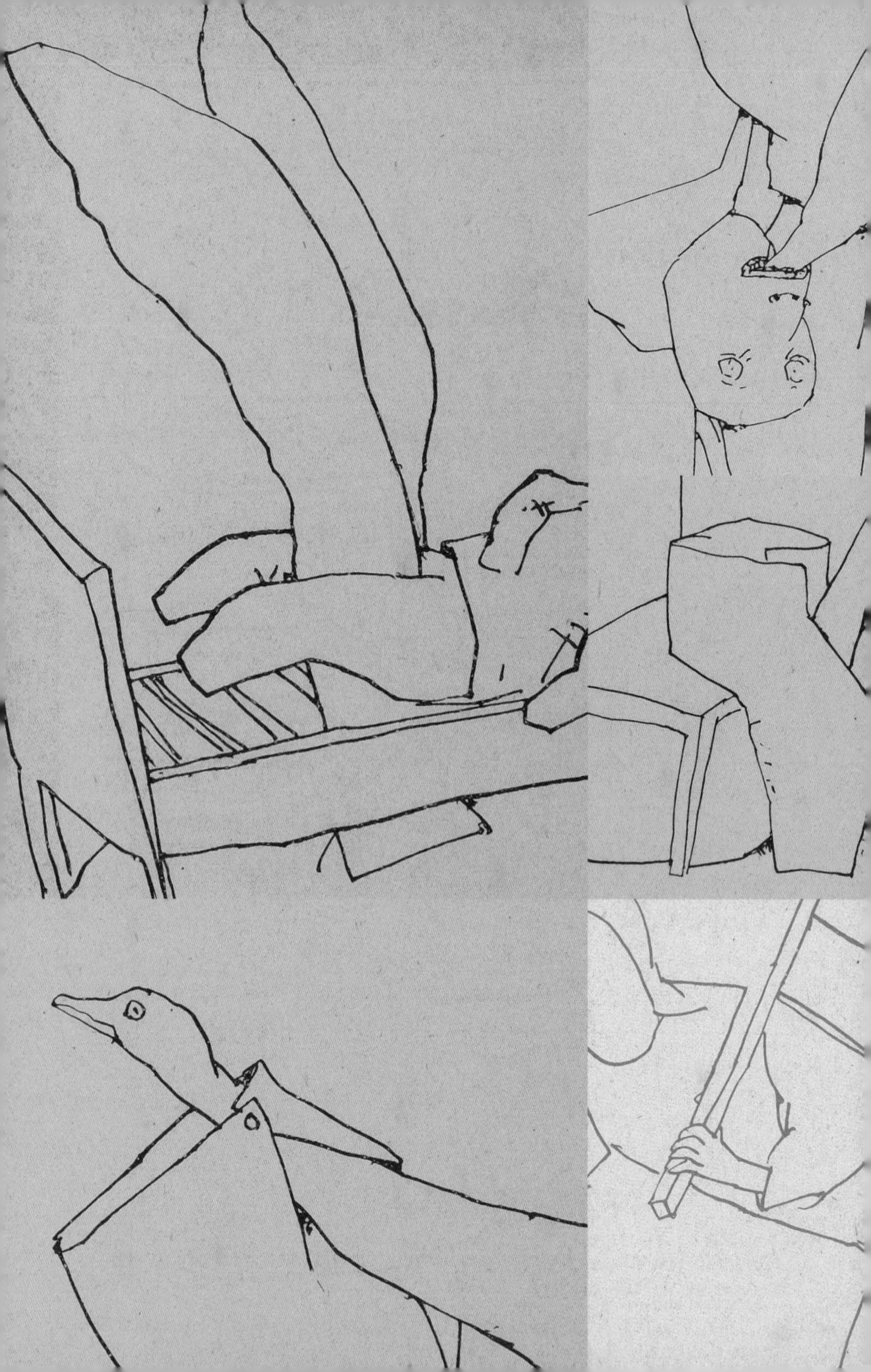

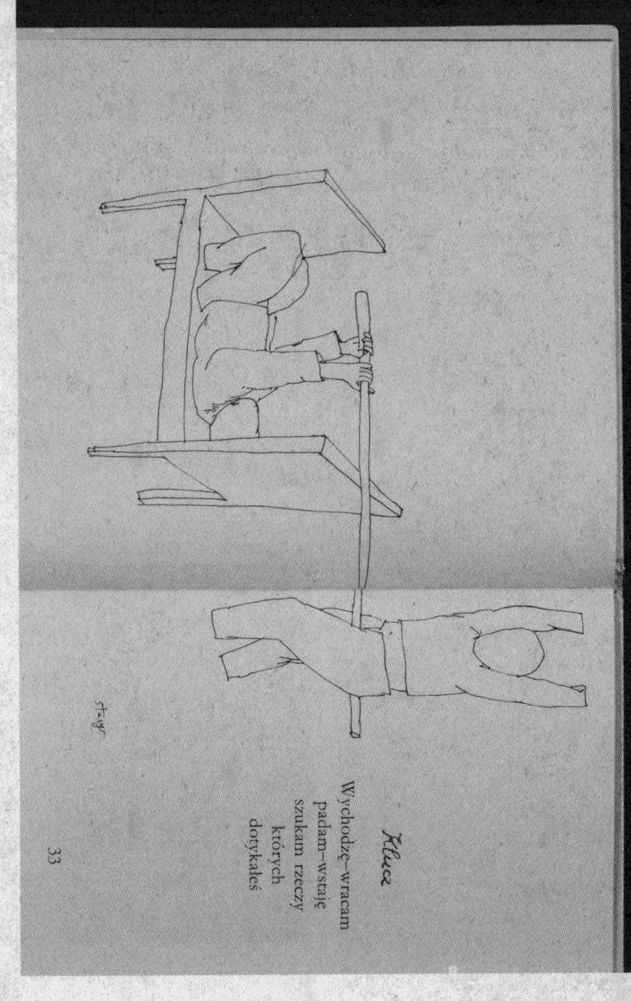

Klucz

Wychodzę–wracam
padam–wstaję
szukam rzeczy
których
dotykałeś

33

1996 Ha[5]

U.S.A.

U.S.A.

China

256mm

9mm

208mm

● 荣誉奖
《 软糖树
◖ 美国

Honorary Appreciation
The Gumdrop Tree
U.S.A.
Julia Gorton
Elizabeth Spurr
Hyperion Books for Children, New York
256 208 9mm
332g
32p
ISBN-10 0-7868-0008-9

这是一个充满童真的故事。小女孩舍不得吃掉父亲送给她的软糖，一开始把软糖摆出各种造型，之后她有了个大胆的想法：把软糖像梨子一样种下，期待长出一颗软糖树。当树一天天长大，却没有结出软糖的时候，她的信心受到了打击，不敢再看这棵树，情绪也变得低落。但突然有一天，她看到树上长满了软糖。书的最后一页交代了这些软糖都是被人用线挂在树上的。本书的装订形式为锁线硬精装，偏方的开本也是典型的儿童绘本的做法。封面铜版纸四色印刷，覆亮膜。内页哑面铜版纸四色印刷。本书的插画风格比较有特点，造型大多偏向于几何形切割，边缘保留部分手绘线条，填色则是带有颗粒的渐变色。为了与插画风格匹配，全书选用了类似 Futura 的偏几何形字体，体现出一种稚趣。

This is a story full of innocence. The little girl is reluctant to eat the jelly that her father gave her. At the beginning, she put the jelly into various shapes. Then she had a bold idea: planting the jelly like a pear tree and looking forward to growing a soft candy tree. When the tree grew up day by day and did not produce jelly, she felt a blow to her confidence. She dared not look at the tree again, and became frustrated. One day she suddenly saw the tree was cover with jelly. The last page of the book explained that these gummies were all hung on the tree with thread. The hardcover book is bound with thread sewing, and the special format is also a typical way to design a children's picture book. The laminated cover is made of coated paper printed in four colors while the inner pages are made of matte coated paper printed in four colors. The style of the illustrations of this book has its own characteristics. Most of the models are geometric cutting, with some hand-painted lines on the edges, and the color filling is reflected by the gradient color with particles. In order to match the style of illustration, the font of the whole book adopts the partial geometric font similar to Futura, which reflects a naive interest.

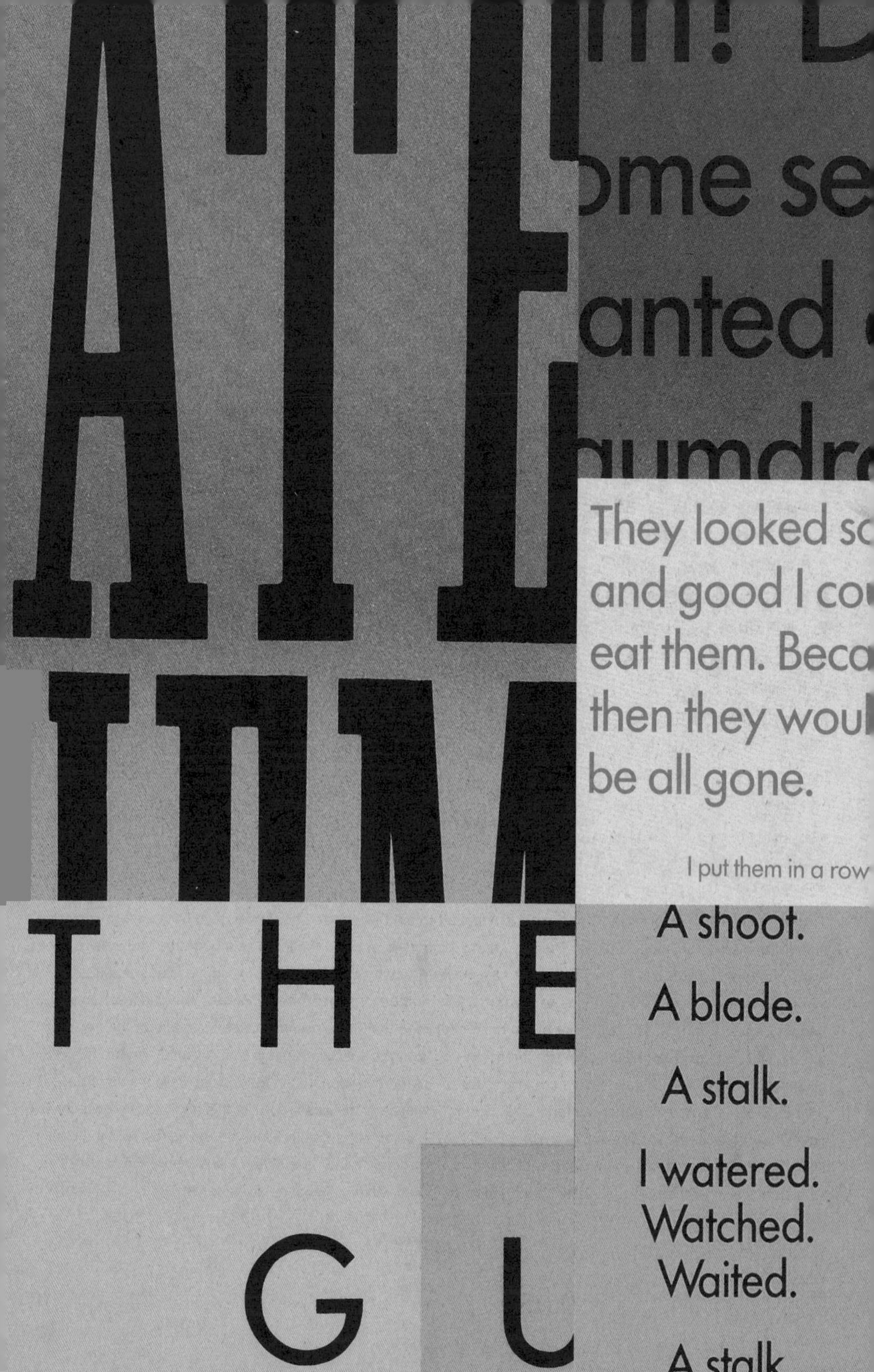

ome se
anted
They looked sc
and good I co
eat them. Beca
then they woul
be all gone.
I put them in a row
A shoot.
A blade.
A stalk.
I watered.
Watched.
Waited.
T H E
G U

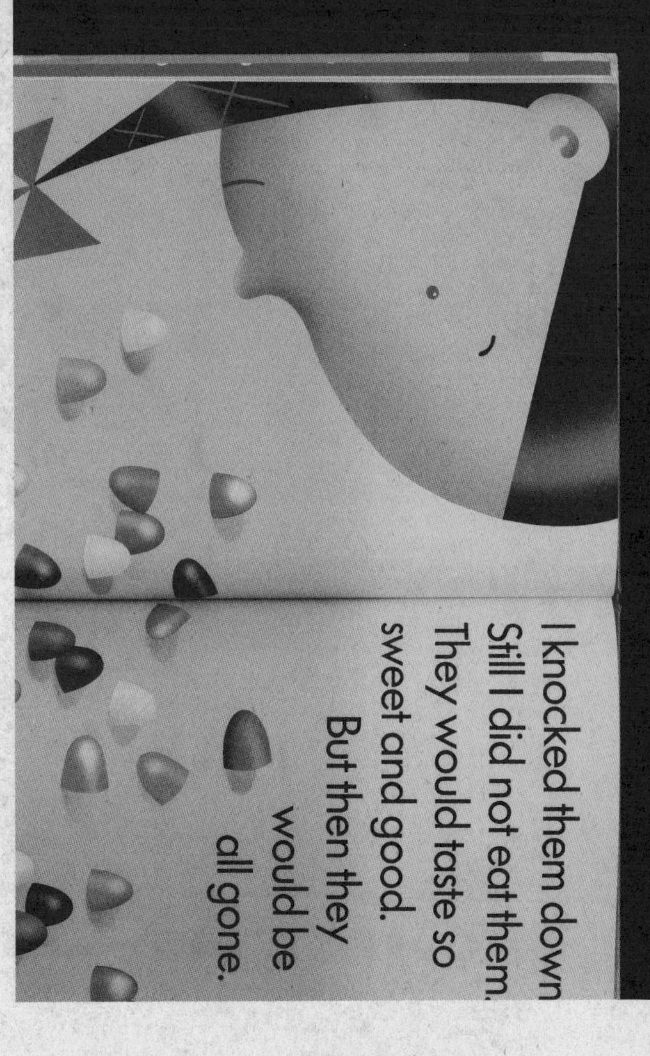
I knocked them down.
Still I did not eat them.
They would taste so
sweet and good.
But then they
would be
all gone.

1997

Country / Region

Germany ❹
The Netherlands ❸
Switzerland ❷
Czech Republic ❶
Japan ❶
Colombia ❶
Egypt ❶
Poland ❶

Typoundso

Jahangir

Unica T

幸田文の箪笥の引き出し
青木玉
新潮社

ALTA COLOMBIA
EL ESPLENDOR DE LA MONTAÑA

RICHTKRACHT
Richard Niessen

MUZEUM ULICY

1997 Gl 0622

1997 Sm[1] 0632

1997 Sm[2] 0640

1997 Bm[2] 0648

1997 Bm[4] 0656

1997 Bm[5] 0666

1997 Ha[1] 0674

1997 Ha[3] 0682

1997 Ha[4] 0692

1997 Ha[5] 0702

Designer

Hans-Rudolf Lutz

Mieke Gerritzen, Janine Huizenga

Kaspar Mühlemann

Kerstin Weber, Olaf Schmidt

Stefanie Harms

Wiebke Oeser

Hans Peter Willberg, Friedrich Forssman

Anja Harms, Ines v. Ketelhodt, Dois Preußner, Uta Schneider, Ulrike Stoltz (Unica T)

Akio Nonaka

Benjamin Villegas

Mohiedden El-Labbad

Bureau Piet Gerards

Richard Menken

Michał Piekarski

1997 Gl

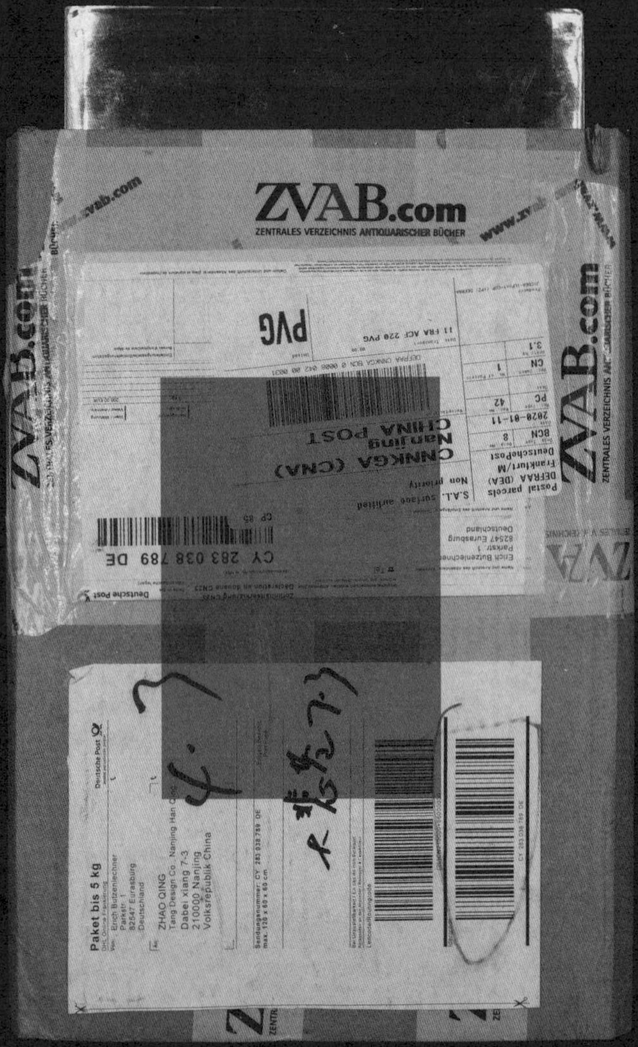

Switzerland
Germany
China
CHINA POST
NKGA (CNA)
Zollinhaltserklärung CN23
Déclaration en douane CN23
CY 283 038 7
CP 85
Entgelt Bezahlt
Port Payé

300mm

40mm

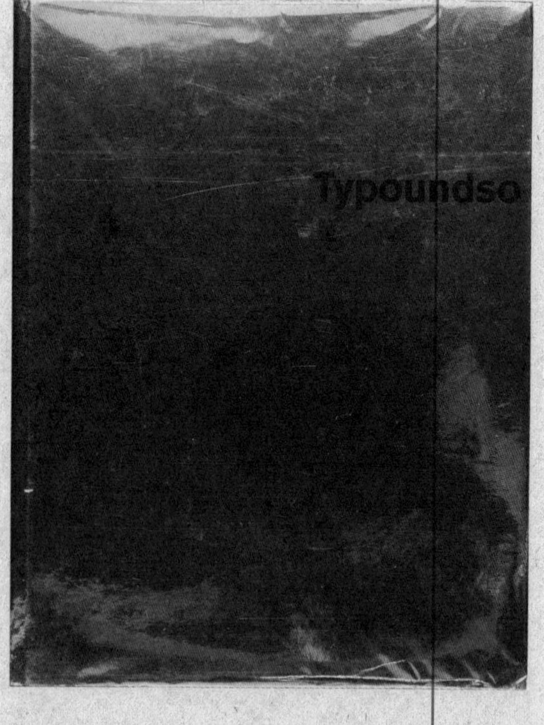

237mm

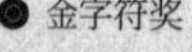

金字符奖
Typoundso
瑞士

Golden Letter
Typoundso
Switzerland
Hans-Rudolf Lutz
Hans-Rudolf Lutz
Hans-Rudolf Lutz, Zürich
300 237 40mm
2602g
440p
——

瑞士设计师汉斯·鲁道夫·卢茨（Hans-Rudolf Lutz）出版过九本有关视觉传达的书籍，本书是其中一本。另一本书《版式设计训练》（*Ausbildung in typografischer Gestaltung*）与本书一同记录了他30多年的设计工作和相关思考，探索了排版的教育和对社会的影响。书籍的装订形式为锁线硬精装。护封采用单面镜面纸张丝印黑，内封黑卡纸裱覆卡板，书名烫银处理。内页涂布纸四色加浅灰绿色共五色印刷。内容共分五章，通过卢茨自己的学习、工作和教学生涯探讨了平面设计在社会和学校中扮演的角色，思考作为一个“视觉作者”的能力与责任。书中主要版式设置为四栏，但在图片、字体方面较为灵活多变。书中特别加入了一层专色——浅灰绿色，用于衬托白色底的作品展示。

Swiss designer Hans-Rudolf Lutz has published nine books on visual communication. This book is one of them. Together with *Ausbildung in typografischer Gestaltung*, the two books have recorded his design and related thoughts for more than 30 years, exploring the education of typesetting and its influence on society. The hard-cover book is bound by thread sewing. The jacket is made of mirror paper with silk-screen printing in black. The inner cover is made of black cardboard with the title of the book stamped in silver. The inner pages are printed in four colors plus light gray green on coated paper. The content is divided into five chapters. The book not only discusses the role of graphic design in society and school through Lutz's own study, work and teaching career, but also thinks about the ability and responsibility of a 'visual author'. The main layout of the book is set to four columns, while the pictures and fonts are quite flexible. In the book, PMS color light gray green, is specially used to display the works with white background.

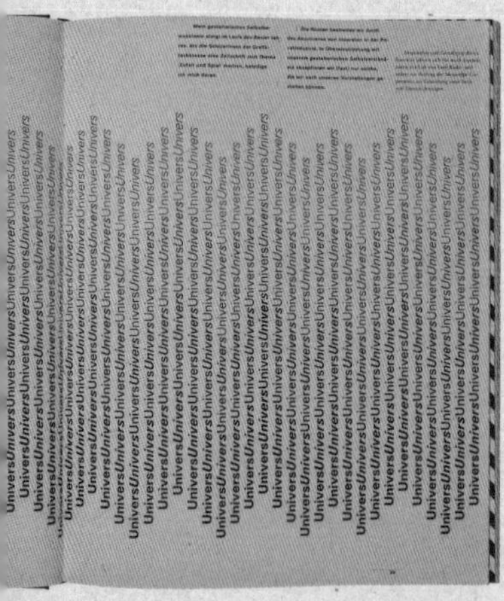

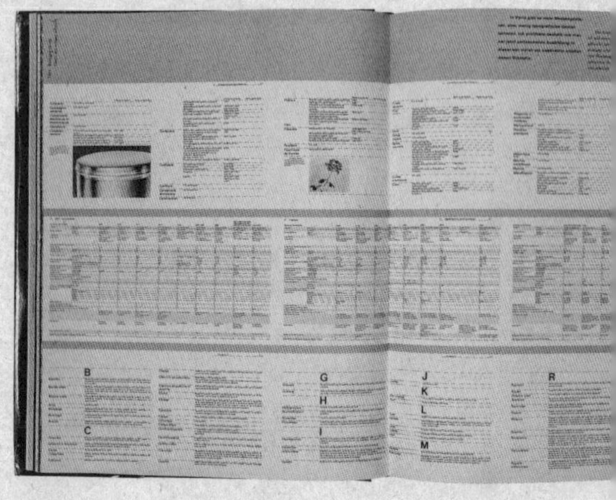

Viewport
Archive
Digitize
Analyze
Zoom & Pan
Contrast
Filter
Warp
Image edit
Program

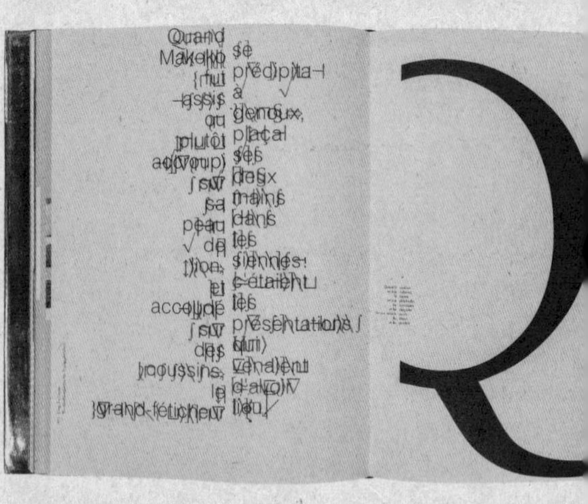

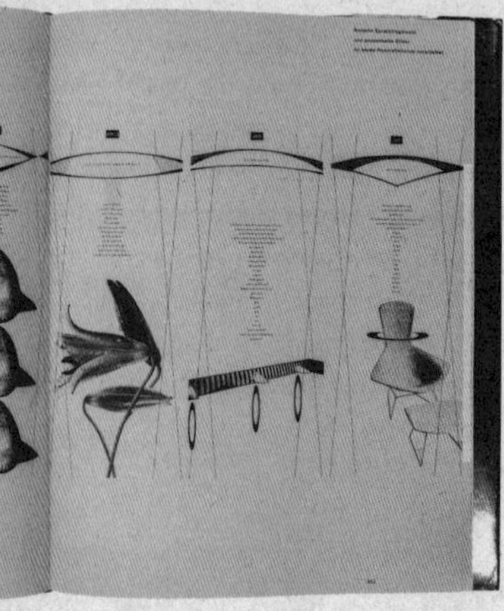

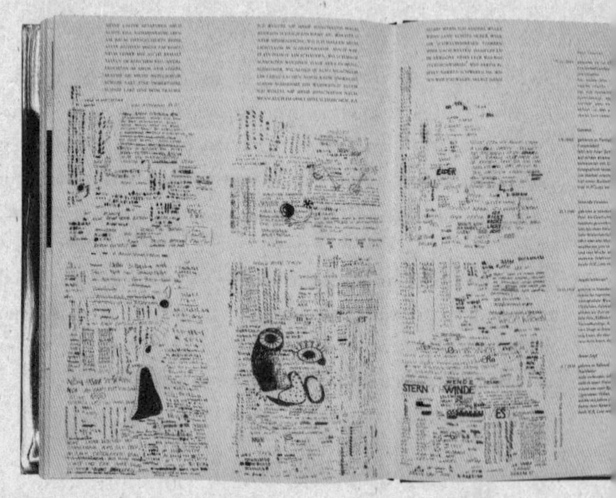

war empörend, m
lcher Rücksichts-
sigkeit man gezw
n wurde, das Hau
r immer zu verla
r Verlust der en
en Heimat und de
hligen Behausung
hmerzte. Die vie
ndererinnerungen

Elisabeth hat sich schon mit einem höhnischen
Lächeln abgewandt.

Na bitte. Lassen Sie mich doch erk

en Sie mir sagen, dass sie vom Himmel gefallen sind und...

Sie lacht. Sie schenkt mir keine Beachtung
mehr. Verlässt kopfschüttelnd meinen Tisch.
Ich lege eine Note auf den Tisch und gehe.

Ratte hierher? Der hatte im letzten Buch einen
Unfall und liegt jetzt
im Koma. In einem Spital in Varese.

He, was soll das!

Den gibt's doch gar nicht. Die können gar nicht
leben. Das sind doch Figuren.
Figuren aus meiner Phantasie!
Ich setz mich auf eine Treppe im riesigen
Treppenhaus. Ich halte meinen Kopf in
den Händen, bis ich Schritte höre. Ein schwarz-
gekleideter Mann naht vom andern Eingang
her. Ueber so einen hab ich nie geschrieben.
Ich überlege.
Nein!
-Ich blick ihm ins Gesicht.

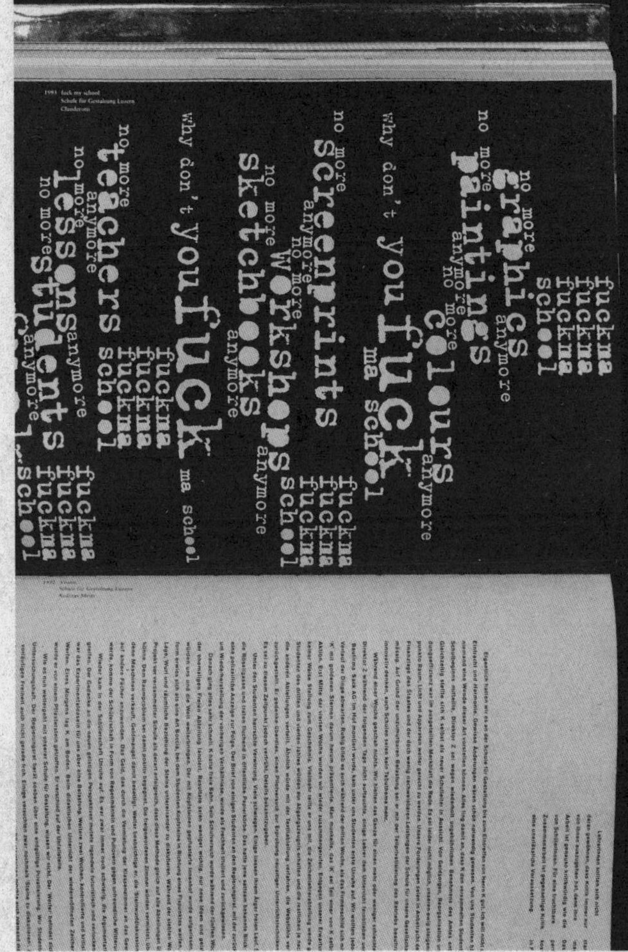

IN JAHR
ONE EGGSTEIN BEAT RI
NA KOLLER CORNEL WINDLIN
ADEN EIN ZUR
PARTY!
Es
bloss ein Werbegag
SI: Was mögen Sie
beantworten. Je nachdem
wem: überall oder nirgen
sechzehn- oder
les
a
fix
on
THE
CON
NELLS
By Jim
fixi
Be
my Valent
photograph
Jill Green
gebratene
Hähnchen.
Das
geht
mir
an
die
Nieren.
spazieren?
VIPER '89 vom 23.-29.Okto
im Kulturpanorama und in der
4. Videowerkschau Schw

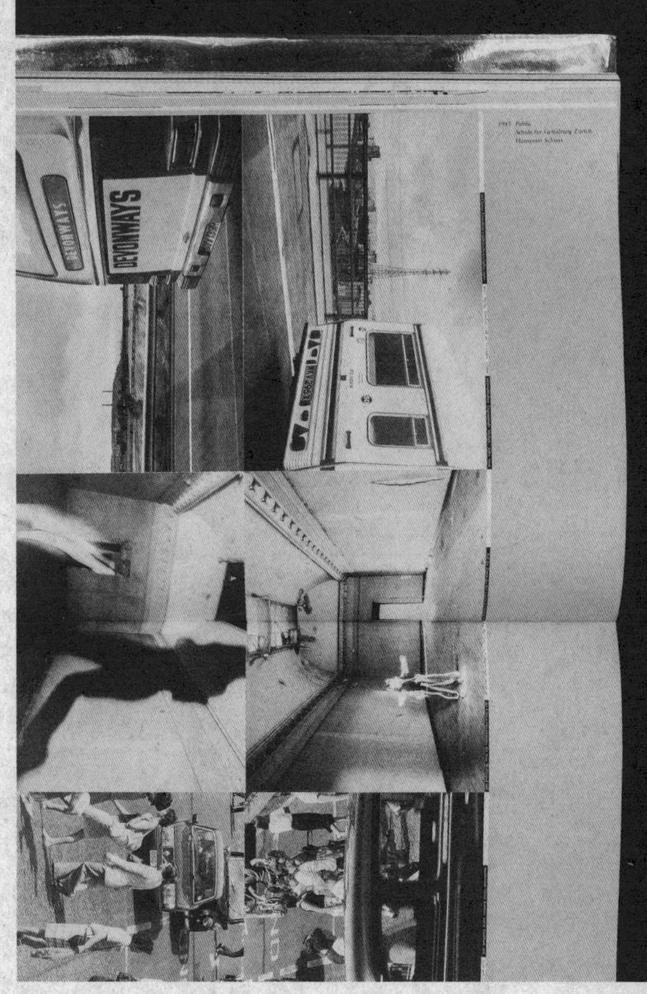
Public
Schule für Gestaltung Zürich
Hanspeter Schiess

1997 Sm1

C113280

SCHEIN - RECHNUNG
NG LIST - INVOICE

er Str. 26, 04229 Leipzig

ng Han Qing Tang Design Co.,

Antiquariat BOOKFARM
(Inh. Sebastian Seckfort)
Naumburger Straße 26
04229 Leipzig
Email: kontakt@bookfarm.de
Umsatzident.Nr. DE271183375

LISTE DE
LISTA DE E
BOLLA DI
FØLGESE
PACKSEDE
PAKLIJST
УПАКОВ
包装

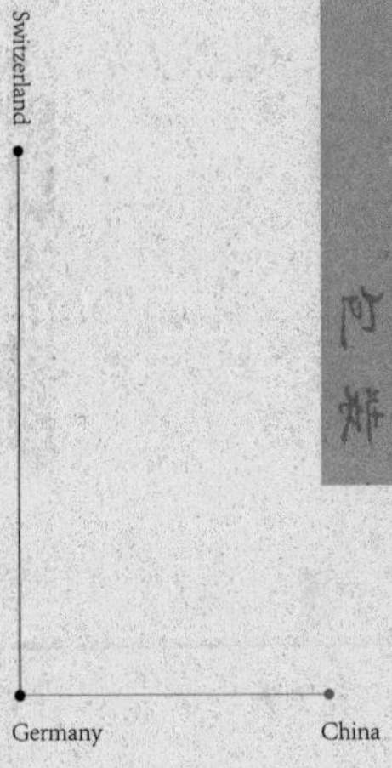

Switzerland

Germany

China

320mm

11mm

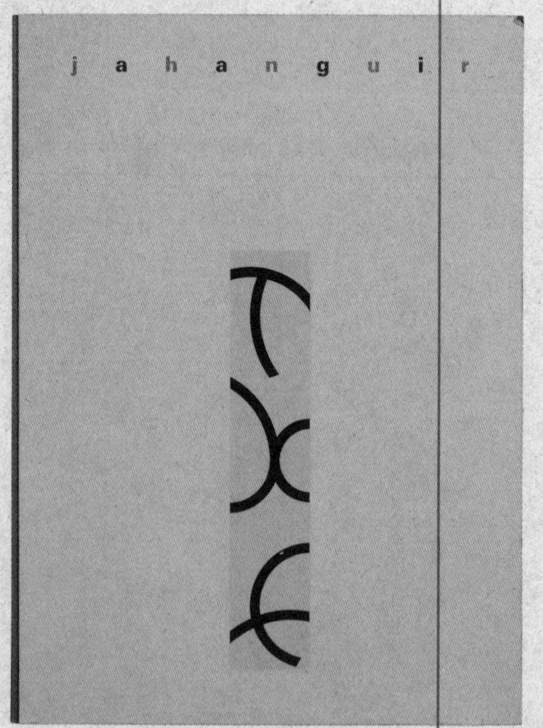

240mm

银奖
贾汉吉尔
瑞士

Silver Medal
Jahanguir
Switzerland
Kaspar Mühlemann
——
Galerie Jamileh Weber, Zürich
320 240 11mm
720g
76p
ISBN-10 3-85809-100-X

这是伊朗裔瑞士艺术家菲利普·韦伯·贾汉吉尔（Philippe Weber Jahanguir）的作品集，是他 1996 年在瑞士苏黎世 Jamileh Weber 画廊举办个展时的展览画册。本书的装订形式为锁线胶平装。封面白色艺术纸单黑印刷，局部过油处理，裱覆卡板。卡板只覆盖封面的主要部分，留出浅绿色艺术纸的书脊印单黑。内页哑面铜版纸四色加专色银印刷，作品部分局部过油处理，呈现本白色的画布范围。内容方面，贾汉吉尔的画作与对他作品的分析文章并行呈现，左页文章，右页作品。页面设置为双栏，文章占据左栏，并且行距很大，右栏只放置对页的作品名称和少量注释。这样的做法与他以几何形为元素的抽象画风相得益彰。右页的作品并非置于页面中心，而是放在了略微偏下的视觉重心，配合左页的文字留白，让页面空间既规整又带有视觉上微妙的处理。

This is a collection of paintings by the Iranian Swiss artist Philippe Weber Jahanguir, which was exhibited at the Jamileh Weber Gallery in Zurich, Switzerland in 1996. The book is perfect bound with thread sewing. The cover is made of white art paper printed in single black, partly varnished, and mounted with cardboard. The cardboard only covers the main part of the cover, leaving the spine made of light green art paper and printed in black. The inner pages are made of matte coated paper in four colors plus PMS color silver. All the works are varnished to emphasize the white canvas. As for the content, Jahanguir's paintings are presented in parallel with the reviews of his works. Articles on the left pages and works on the right pages, all the pages are set with double columns. The main text occupies the left column with large line spacing while the titles of the works and a few notes are placed in the right column. This approach complements Jahanguir's abstract style of drawing with geometric shapes as the basic elements. The works on the right pages are not placed in the center of the page, but placed in a slightly lower spot. With leaving the blank area on the left pages, the space of the entire page is regular but with subtle changes.

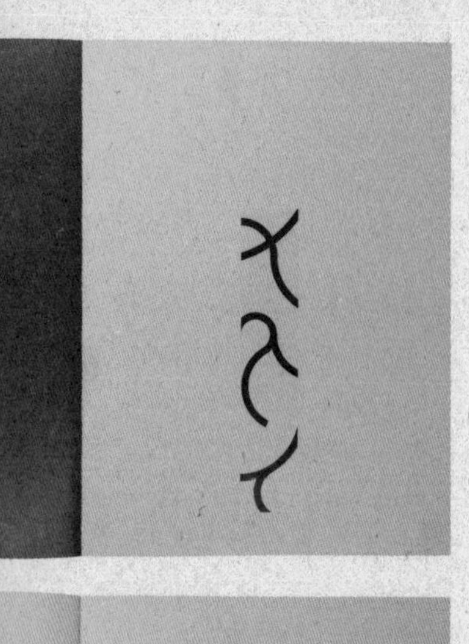

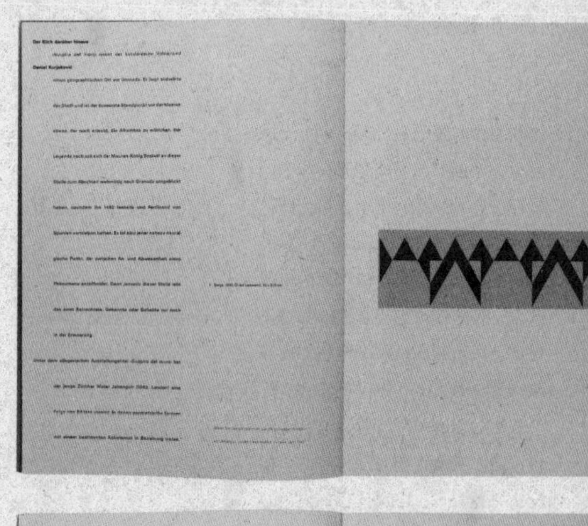

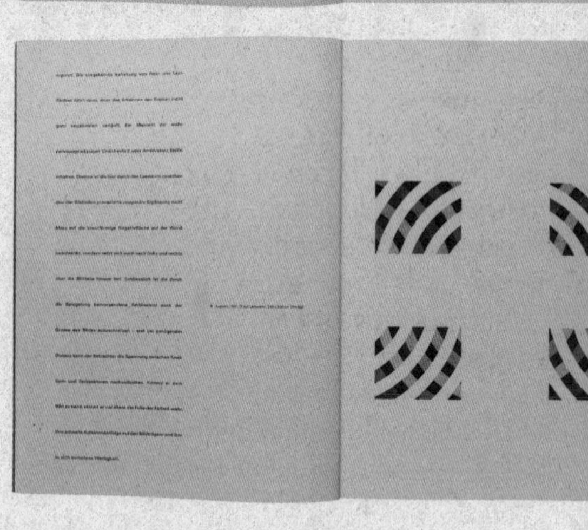

von der sie immer wieder Zeugnis ablegen, zeigen, dass sie stellvertretend für eine gewisse symbolische Bedeutung stehen. Die gesellschaftliche Mythologie der Formen bringt den Kreis mit einer breiten Palette von Inhalten in Verbindung, die auf unterschiedliche Weise für ‹Ganzheit› stehen. Was heisst es, dass Jahanguir so eine symbolische Dimension erarbeitet? Wohl – und in diesem Sinn zeigt sich der Modernismus seiner Arbeit, die konkrete Zeichenwelt im Hinblick auf eine bildexterne Wirklichkeit zu überschreiten –, dass die Bedeutung der Arbeiten letztlich deren materiale Präsenz zu übersteigen sucht; die Bilder *verweisen* immer schon auf eine Bedeutung, so leicht oder so schwierig zu fassen sie auch sein mag. So postuliert das Symbol (die *Idee* des Kreises) mittels einer Art Stellvertreter (hier der gemalte Kreis) eine Bedeutung, die

6 India, 1992. Öl auf Leinwand, 210 × 210 cm

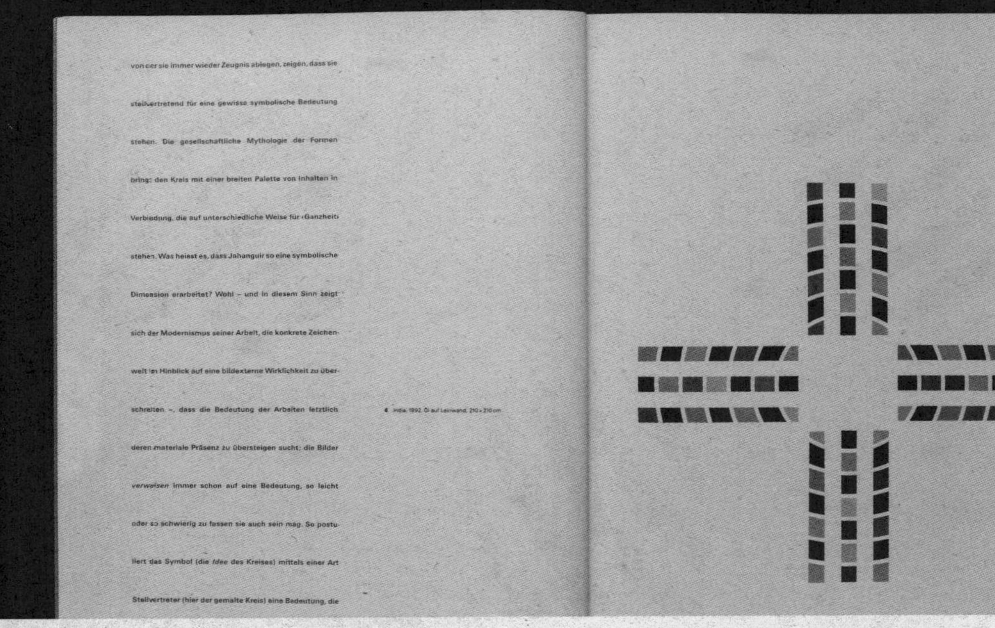

0637

die Bilder *wie* Fenster funktionieren. Wenn ich das Motiv des Fensters nenne, mache ich nichts anderes als stillschweigend einen bestimmten untergründigen Bereich zu evozieren: den Subtext der Romantik, denn bekanntermassen ist das Fenster ein zentraler Topos romantischer Malerei. Das Bild bildet bei Jahanguir nicht ein Fenster ab, es *ist* ein Fenster. Nebenbei scheint diese Vorstellung auch eine sinnvolle Erklärung zu bieten, weshalb Jahanguir bis heute verschiedene Formate und deren Kombinationen verwendete: das Einzelbild, die Dreierserie, die Konstellation mit vier Bildern, das quadratische Format, das ‹Fries›, das Übereck-Format. Sie alle sind Variationen auf das ‹Fenster›, verschiedene Positionierungen des Trägers, einen Ausblick zu ermöglichen. Wohl haben die Formate eine Funktion in bezug auf die formalen Entscheidungen im Bild,

8 Lunes, 1992, Öl auf Leinwand, 120 × 120 cm, diagonal 169 cm

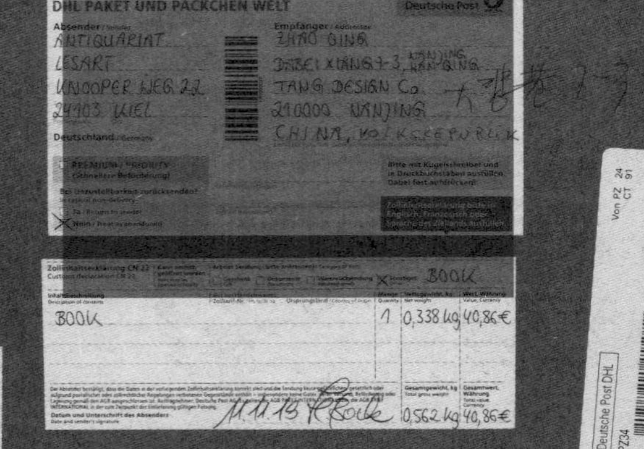

DHL PAKET UND PÄCKCHEN WELT
Deutsche Post
Absender / Sender
ANTIQUARIAT
LESART
KNOOPER WEG 22
24103 KIEL
Deutschland / Germany
Empfänger / Addressee
ZHAO QING
CHINA
Zollinhaltserklärung CN 22
Customs declaration CN 22
BOOK
BOOK
1
0,338 kg
40,86 €
Gesamtgewicht, kg
0,562 kg
Gesamtwert, Währung
40,86 €
Datum und Unterschrift des Absenders
Date and sender's signature
Päckchen WELT bis 2 kg
Petit Paquet
Deutsche Post
UY111066782DE
11.11.19
Entgelt bezahlt / Port payé
FI04123395
Von PZ 24
CT 91
Deutsche Post DHL
PZ34
00347 156 400 80

Deutsche Post
HL PAKET UND PACKCHEN WELT
Empfänger / Addressee
ZHAO QING
DABEI XIANG 7-3, NANJING HAN QING
TANG DESIGN Co.
210000 NANJING
CHINA, VOLKSREPUBLI
Bitte mit Kugelschreiber und in Druckbuchstaben ausfüllen Dabei fest aufdrücken!
Zollinhaltserklärung bitte in Englisch, Französisch oder Sprache des Ziellands ausfüllen
ANTIQUARIAT
ESART
KNOOPER WEG 22
24103 KIEL
eutschland / Germany
PREMIUM / PRIORITY
(Schnellere Beförderung)
Telefon (Empfänger)
E-Mail (Empfänger)
Art der Sendung (bitte ankreuzen) / Category of item
Geschenk
Dokumente
Warenrücksendung
Sonstiges: / Other
BOOK
Ja / Return to sender
Nein / Treat as abandoned
Germany
Germany
China

235mm

11mm

173mm

● 银奖
《 至亲至疏夫妻
◖ 德国

Silver Medal
Sehr nah, sehr fern sind sich Mann und Frau
Germany
Kerstin Weber, Olaf Schmidt
Hanne Chen
Edition ZeichenSatz, Kiel
235 173 11mm
332g
68p
ISBN-10 3-00-000733-4

这是唐代女性爱情诗选的中德对照版，德文为主，诗歌大多表达相思相怨的苦楚。书籍的装订形式较为独特，内文采用包背装，装订处无锁线和刷胶，而是在书脊处用三个黑色长尾夹夹住，并通过两根细钢棒连接固定。封面黄色艺术卡纸向外弯折裱覆黑色薄卡板，书口处贴有黑银双色印刷的书名信息标签。内页胶版纸黄紫双色印刷。内容方面，由两篇诗选的介绍开始，德文左对齐排版。之后便是诗歌部分，分为三章，按照诗人身份划分：青楼女子、妻子、尼姑。黄色只在章节隔页的人名和第一篇诗歌人名处使用，形成翻阅过程中的明显区隔，其余部分均为紫色。诗前加入诗人简介，包括所处年代和写作背景。由于唐诗严格的格律与德文翻译的文字量存在差异，设计上的处理较为灵活：中德文字上下错开，中文横排左右对齐，德文横排左对齐，产生右侧长短丰富的韵律。书籍的页码设定在翻口的统一高度上，而页眉信息会根据每页的具体排版灵活调整。

This book is a collection of selected love poems by women in the Tang Dynasty. Most of the poems express the bitterness of lovesickness and resentment. The book is bilingual in Chinese and German. The binding form is quite unique because the double-leaved book is not bound by glue or thread, but is clamped with three black binder clips, which are connected and fixed by two thin steel rods. The cover is made of yellow art paper folded outwards and mounted with black thin cardboard. The label with title information is printed in black and silver and pasted at the fore-edge. Inside pages are made of offset paper printed in yellow and purple. The book begins with the introduction of two selected poems. Then the poem part is divided into three chapters according to the identity of the poets: the courtesans, the wives, and the nuns. Yellow is only used in the chapter titles and the author's name of the first poem, forming an obvious distinction in the process of browsing. The rest parts are purple. A brief introduction of the poet is added before each poem, including the dynasty and writing background. Due to the difference between the strict metrical rules of Tang poetry and the amount of words translated into German, the design is quite flexible: Chinese and German characters are staggered up and down, Chinese is justified horizontally, German is left aligned, resulting in rich variety on the right side. The page number of the book is set at the uniform height of the flap, and the header information is flexibly adjusted according to the specific layout of each page.

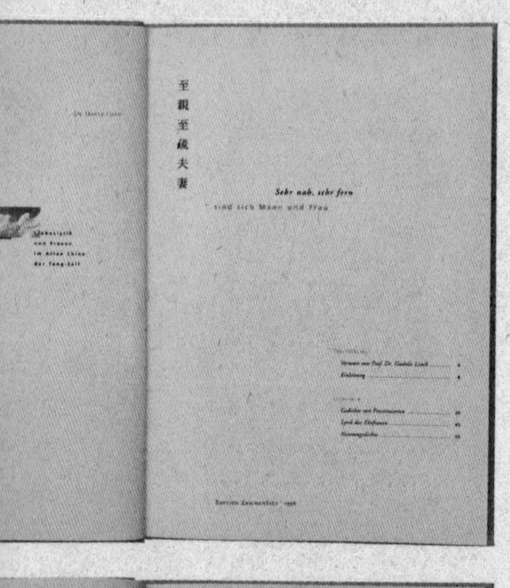
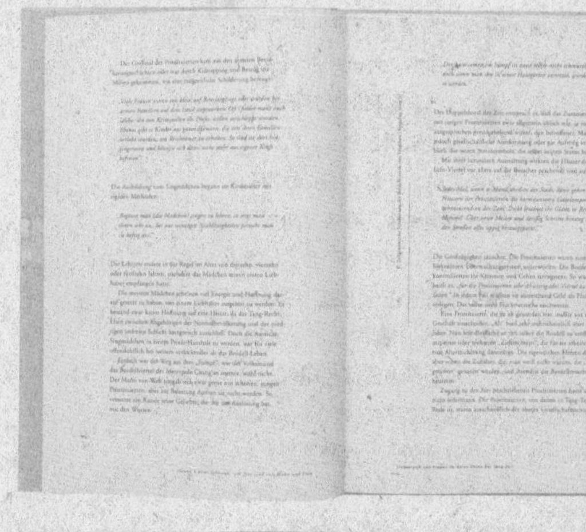
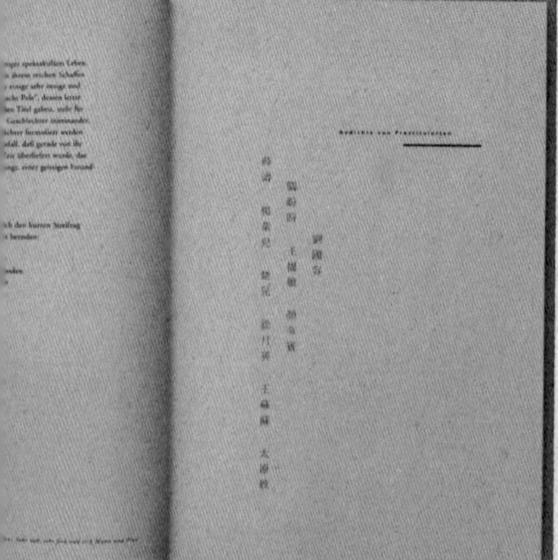
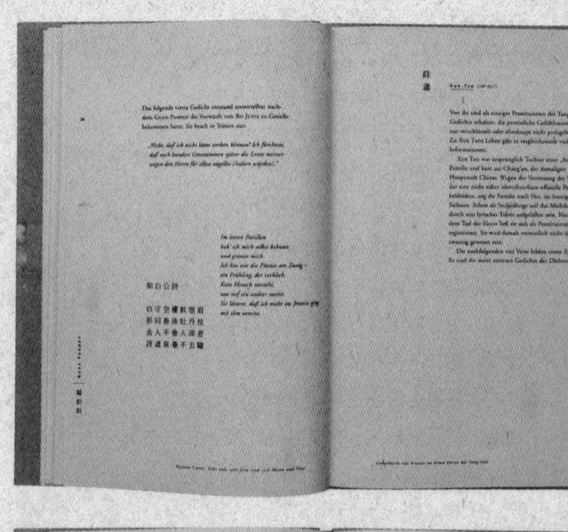
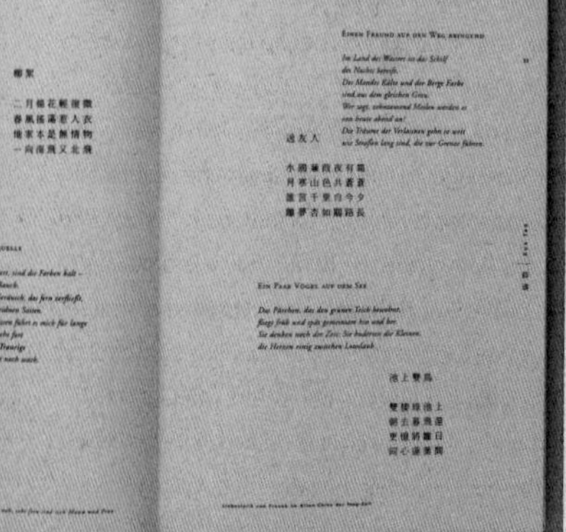
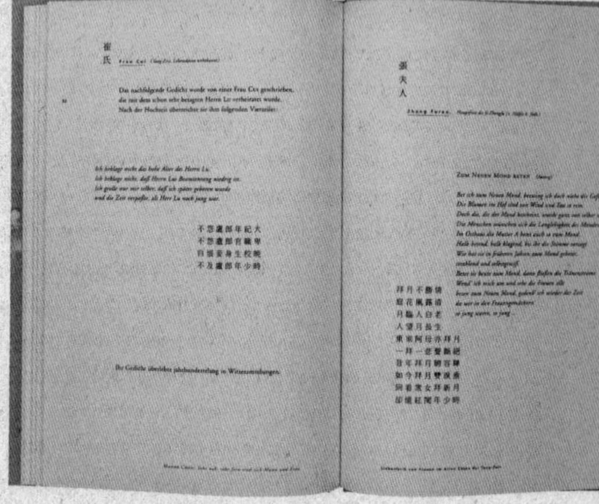

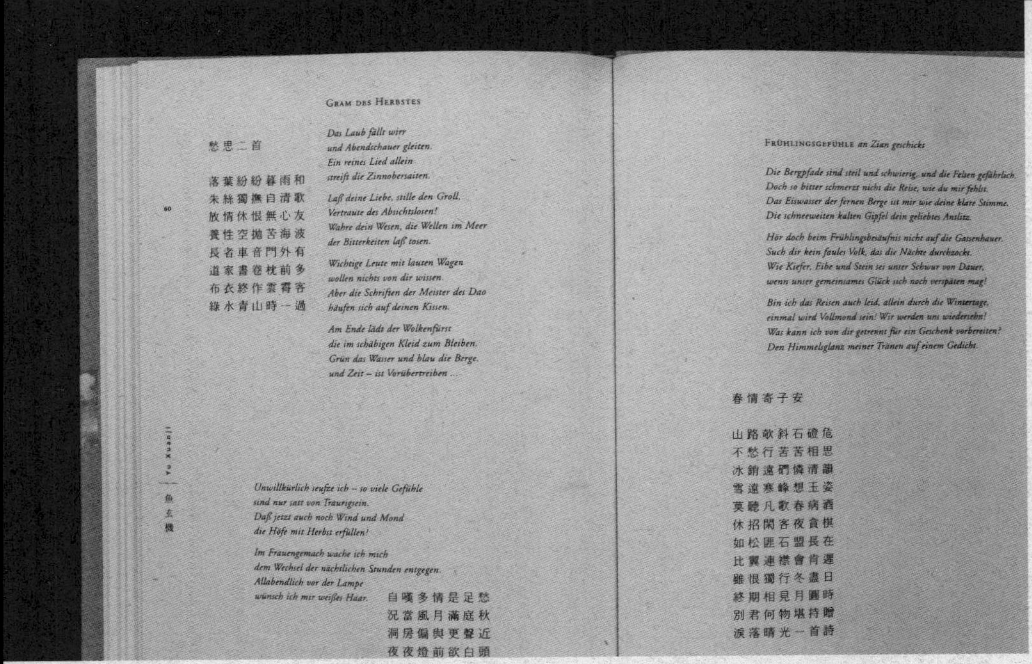

Gram des Herbstes

愁思二首

落葉紛紛暮雨和
朱絲獨撫自清歌
放情休恨無心友
養性空拋苦海波
長者車音門外有
道家書卷枕前多
布衣終作雲霄客
綠水青山時一過

Das Laub fällt wirr
und Abendschauer gleiten.
Ein reines Lied allein
streift die Zinnobersaiten.

Laß deine Liebe, stille den Groll,
Vertraute des Absichtslosen!
Wahre dein Wesen, die Wellen im Meer
der Bitterkeiten laß tosen.

Wichtige Leute mit lauten Wagen
wollen nichts von dir wissen.
Aber die Schriften der Meister des Dao
häufen sich auf deinen Kissen.

Am Ende lädt der Wolkenfürst
die im schäbigen Kleid zum Bleiben.
Grün das Wasser und blau die Berge,
und Zeit – ist Vorübertreiben ...

Unwillkürlich seufze ich – so viele Gefühle
sind nur satt von Traurigsein.
Daß jetzt auch noch Wind und Mond
die Höfe mit Herbst erfüllen!

Im Frauengemach wache ich mich
dem Wechsel der nächtlichen Stunden entgegen.
Allabendlich vor der Lampe
wünsch ich mir weißes Haar.

自嘆多情是足愁
況當風月滿庭秋
洞房偏與更聲近
夜夜燈前欲白頭

Frühlingsgefühle *an Zian geschickt*

Die Bergpfade sind steil und schwierig, und die Felsen gefährlich.
Doch so bitter schmerzt nicht die Reise, wie du mir fehlst.
Das Eiswasser der fernen Berge ist mir wie deine klare Stimme.
Die schneeweiten kalten Gipfel dein geliebtes Antlitz.

Hör doch beim Frühlingsbesäufnis nicht auf die Gassenhauer.
Such dir kein faules Volk, das die Nächte durchzockt.
Wie Kiefer, Eibe und Stein sei unser Schwur von Dauer,
wenn unser gemeinsames Glück sich noch verspäten mag!

Bin ich das Reisen auch leid, allein durch die Wintertage,
einmal wird Vollmond sein! Wir werden uns wiedersehn!
Was kann ich von dir getrennt für ein Geschenk vorbereiten?
Den Himmelsglanz meiner Tränen auf einem Gedicht.

春情寄子安

山路欹斜石磴危
不愁行苦苦相思
冰銷遠磵憐清韻
雪遠寒峰想玉姿
莫聽凡歌春病酒
休招閑客夜貪棋
如松匪石盟長在
比翼連襟會肯遲
雖恨獨行冬盡日
終期相見月圓時
別君何物堪持贈
淚落晴光一首詩

Gedichte von Prostitui

薛濤
楊萊兒
楚兒
徐月英

關盼盼
王福娘
顏令賓

劉國容

m Fernen gesandt

gsam und weich der neue Kalm
ı und ebenmäßig –
Frühlingsende fallen seine Blüt
verstopfen den Bachlauf.
weiß, du bist noch nicht zurück
der Kavallerie am Qin-Pass.
ın der Mondschein auf die tause
uchze ich in meine Ärmel.

lautet diese Erkenntnis, daß „die Liebsten nahe wohnen, err auf getrenntesten Bergen".

Wanderer zwischen den Abgründen, Seiltänzer der Liebe auch wir. Möge uns auf der Suche nach größerer Kunstferti Geschmeidigkeit und Liebeskraft die Begegnung mit der Fr wenn nicht hilfreich, so doch tröstlich und anrührend sein. jedem Falle erfahren wir auf den folgenden Seiten dank der lichkeit und Lebendigkeit dieser Verse etwas über die Leben liebender Frauen im China der Tang-Zeit (618–906), ihre S süchte und Ängste – und damit auch etwas über uns selbst.

Da dieses „Liebhabe(r)buch" im mehrfachen Sinne des Wor nicht entstanden wäre ohne die großzügige Unterstützung d Breuninger Stiftung, Stuttgart, steht am Ende dieses Vor ein herzliches Dankeschön.

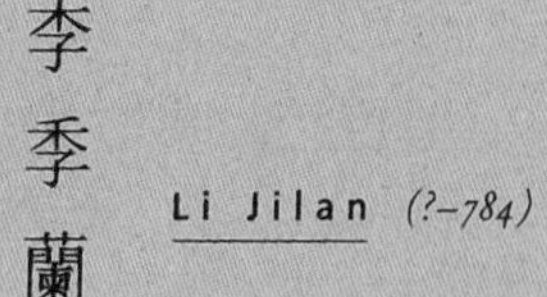

Li Jilan ist mit Xu der drei bedeutends ist wenig bekannt. I heute erhalten.

王蘇蘇

Wang Susu *(9. Jhdt.)*

Bei Wang Susus Vers handelt es sich um eine Replik auf das Gedicht eines ihr unbekannten jungen Mannes namens Li Biao. Dieser hatte die Prostituierten mit im Zimmer umherwehenden Blüten verglichen. Wang Susu, die sich über den Neuankömmling ärgerte, beschied ihn kein dummes Zeug zu reden:

怪得犬驚雞亂飛
羸童瘦馬老麻衣
阿誰亂引閒人到
留住青蚨熟趕歸

Kein Wunder, daß der Hund die Hühner
verschreckt durcheinanderfliegen läßt.
Der schmächtige Knabe auf seiner klapprigen
Mähre in alten Hanfkleidern!
Welcher Sowieso hat hier einfach
diesen Tunichtgut mitgebracht?
Laß etwas Geld hier
und geh schleunigst heim.

Li Biao errötete und folgte der unfreundlichen Aufforderung.

裴羽仙
劉氏
張琰
梁瓊
張氏
劉媛
張夫人
劉瑤

Lyrik der Ehefrauen

1997 Bm²

Wiebk
CN22 - CUSTOMS DECLARATION
Book
ZHAO QING
NANJING 21000G
CHINA
endicia
UA 669 024 095 US

LARATION
ailing:
7(a)
S Tariff Number ;
ountry of Origin
SE
AL
ustoms
es not contain any
ticles prohibited by
egulations. I have
ements under
11/15/2019

FROM
3104568762
MID:200830
ART CONSULTING: SCANDINAV
Lena Torslow Hansen
25777 Punto De Vista Dr
Monte Nido CA 91302-2155

F

US POS
32 oz First-C
Commerc
Nov 15 2
CID: 434

TO
ZHAO QING
DABEI XIANG7-3,
NANJING HAN QING TANG
NANJING 210000,
CHINA

2 lb.

UA 669 024

Do not duplicate this form without USPS Approval.

244mm

10mm

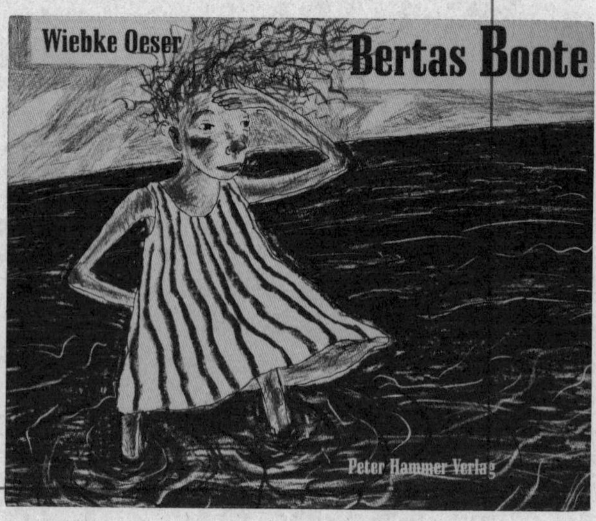

305mm

铜奖
贝尔塔的船
德国

Bronze Medal
Bertas Boote
Germany
Wiebke Oeser
Wiebke Oeser
Peter Hammer Verlag, Wuppertal
244 305 10mm
447g
32p
ISBN-10 3-87294-755-9

这是一本展现女孩丰富想象力、动手能力的绘本。书中的女孩叫贝尔塔（Bertas），她将亲手制作的小帆船放在海中航行，但被大鱼瞬间吞食。回到家后，她失落、伤心，通过彻夜玩游戏和吃空冰箱里的鱼罐头来发泄。第二天她抓到了这条大鱼，但鱼肚子中并没有她的帆船。最终她把这个故事绘制下来，放入瓶中，扔回大海。书籍的装订形式为锁线硬精装。封面胶版纸四色印刷覆哑膜，裱覆卡板。内页用纸和印刷与封面相同。四色中红蓝双色为专色，封面红黄蓝三色挂调频网用以展现油画棒画作的丰富肌理，内页则只有蓝色挂调频网。绘本中稚拙的绘画配以加强衬线的字体，呈现出充满创造、无所拘束的整体风格。书中把女孩绘制、制作和后续一系列具有想象力的创作过程记录下来，为艺术教育提供了很好的参考蓝本。

This is a picture book that shows a girl's rich imagination and hands-on ability. The book tells the story of a girl named Bertas, who manipulated the little sailboat she made by herself in the sea, but the big fish swallowed the boat instantly. When she got home, she was sad and lost. She played games all night and ate out the canned fish in the refrigerator. She caught the big fish the next day, but there was no sailing boat in its belly. Eventually she drew the story, put it in a bottle, and threw it back into the sea. The hardcover book is bound by thread sewing. The matted cover is made of offset paper printed in four colors, mounted with cardboard. The inner pages are made of the same paper and printed as the front cover. Among the four printing colors, red and blue are PMS colors. Frequency modulated screens in red, yellow and blue are used for the cover to present the rich texture of the paintings created by oil painting sticks; while the inner pages use frequency modulated screen in blue only. The naive paintings in the picture book, together with the strengthening serif typeface, present a creative and unrestrained style. The book provides a good reference for art education through the description of the girl's drawing, production and subsequent series of imaginative creative processes.

Kein Boot ist wie das andere. Am gefährlichsten sieht das Pira

Ob Bertas Boote so gut schwimmen können wie sie aussehen?

Eines Vormittags baut Berta aus Treibholzstücken, Schaschlikspießen und Stoffresten sieben Segelboote.

Es ist kaum Seegang und fast windstill. Alle Boot dümpeln so vor sich hin, keins kippt, keins kentert. Berta langweilt sich fast. Aber da

1997 Bm4

ANJA HARMS
INES v. KETELHODT
DORIS PREUSSNER
UTA SCHNEIDER
ULRIKE STOLTZ
ZHAO QING
DABEI XIANG7-3,NANJING HAN QING TANG
DESIGN CO.
NANJING 210000
CHINA

Germany

Germany China

279mm

19mm

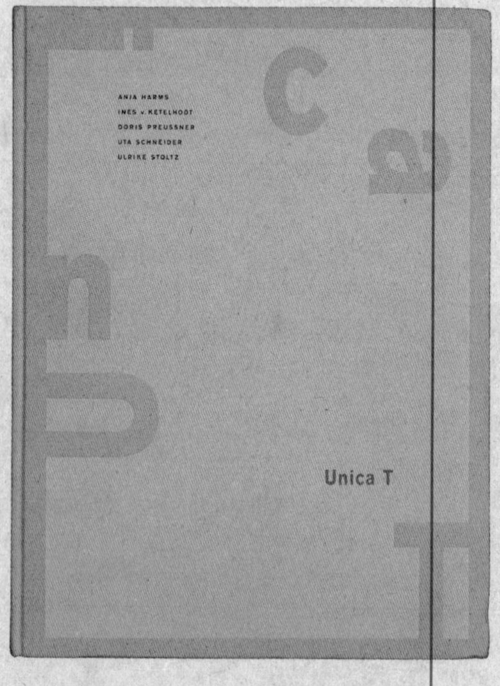

214mm

● 铜奖 Bronze Medal

《 Unica T——10 年的艺术家书籍 Unica T: 10 Jahre Künstlerbücher

◖ 德国 Germany

Anja Harms, Ines v. Ketelhodt, Dois Preußner, Uta Schneider, Ulrike Stoltz (Unica T)

Unica T, Oberursel i. Ts./Offenbach am Main

Unica T, Oberursel i. Ts./Offenbach am Main

279 214 19mm

1103g

228p

ISBN-10 3-00-000854-3

1996 年，德国美因河畔的法兰克福的工艺博物馆举办了一场书籍设计展，展览中的书籍均来自 Unica T 于 1986 — 1995 年间出版的艺术类图书，本书是这场展览的作品集。书籍的装订形式为锁线硬精装。封面艺术纸裱覆卡板，封面和封底大面积烫白，凸显出“Unica T”字样，烫白上继而烫橙色和银色。内页哑面铜版纸四色印刷。书中除了 Unica T 设计的书籍外，还包含几位其他设计师在 Unica T 出版的书籍。正文英德双语排版，德文比英文粗一个字重。书中版心设置为四栏，左页文字三栏排版，右页直接跨三栏，空出靠近装订线的栏位用于放置作品信息和页码。每篇文章的起始文字字号加大，在文章名或作品名不处于同一页的情况下，具有提示作用。

In 1996, a book design exhibition was held at the Museum of Arts and crafts in Frankfurt, Germany. The books in the exhibition were all from art books published by Unica T from 1986 to 1995. This book is a collection of works from this exhibition. The hardcover book is bound by thread sewing. The cover is made of art paper mounted with cardboard. The front cover and back cover are hot-stamped in white in large areas, highlighting the word ‘Unica T’ with orange and silver hot-stamping. The inner pages are printed with matte coated paper in four colors. In addition to the books designed by Unica T, the book also contains books written by several other designers at Unica T. The text is bilingual in English and German. German text is bolder in weight than English text. The typeset area of the book is set as four columns – on the left pages, three columns are used for text; on the right pages, the contents are typeset across three columns. The column near the binding line is vacated for placing work information and page numbers. The font size of the initial text of each article is increased, which has the function of prompting when the title of the article or the title of the work is not on the same page.

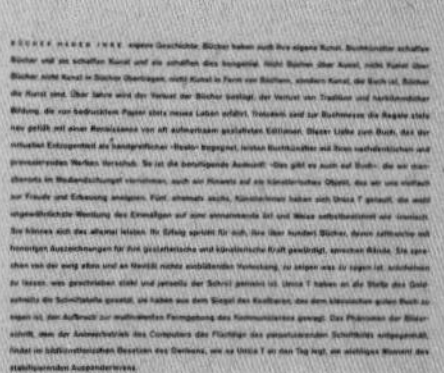
UNICA T – EIN DEZENNIUM

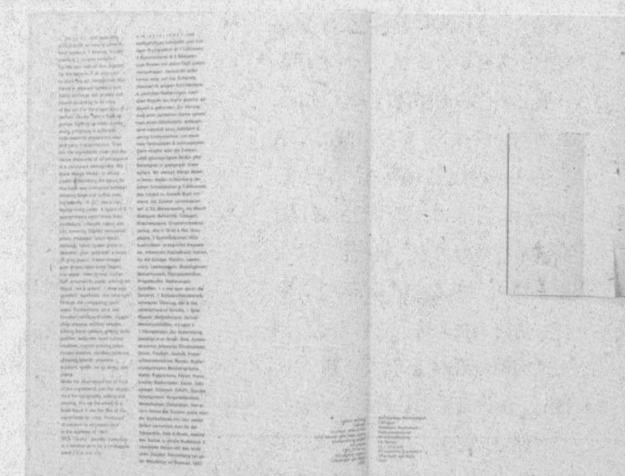

ICH & DU

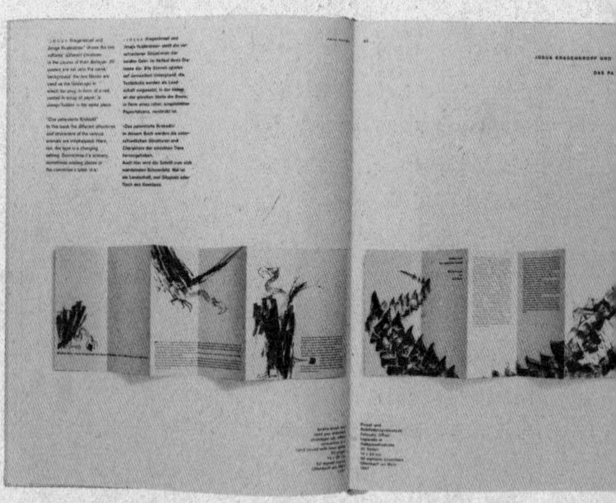
JOSUA KRAGENKOPF UND
DAS PA

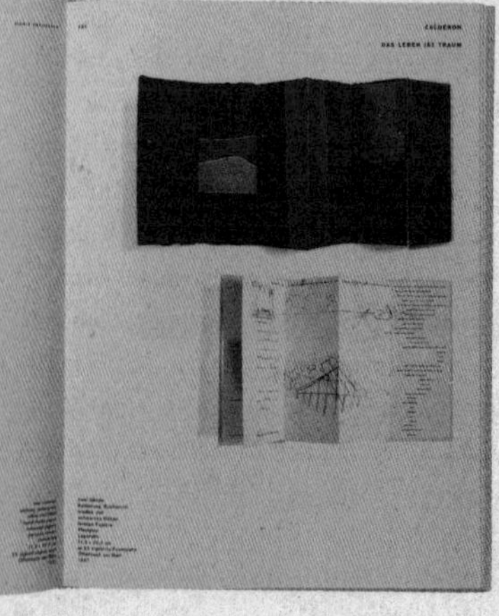
CALDERON
DAS LEBEN IST TRAUM

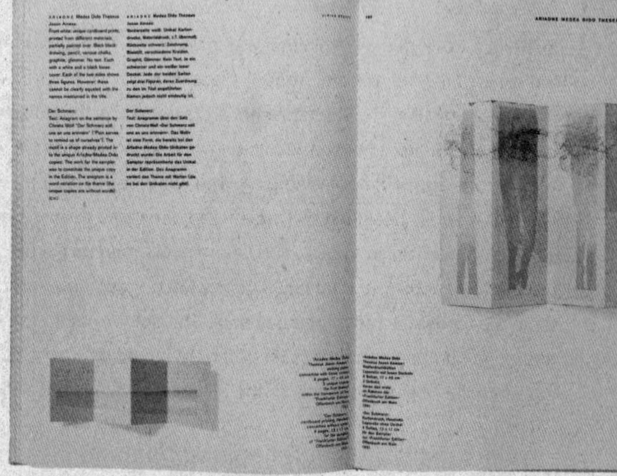
ARIADNE MEDEA DIDO THESEUS

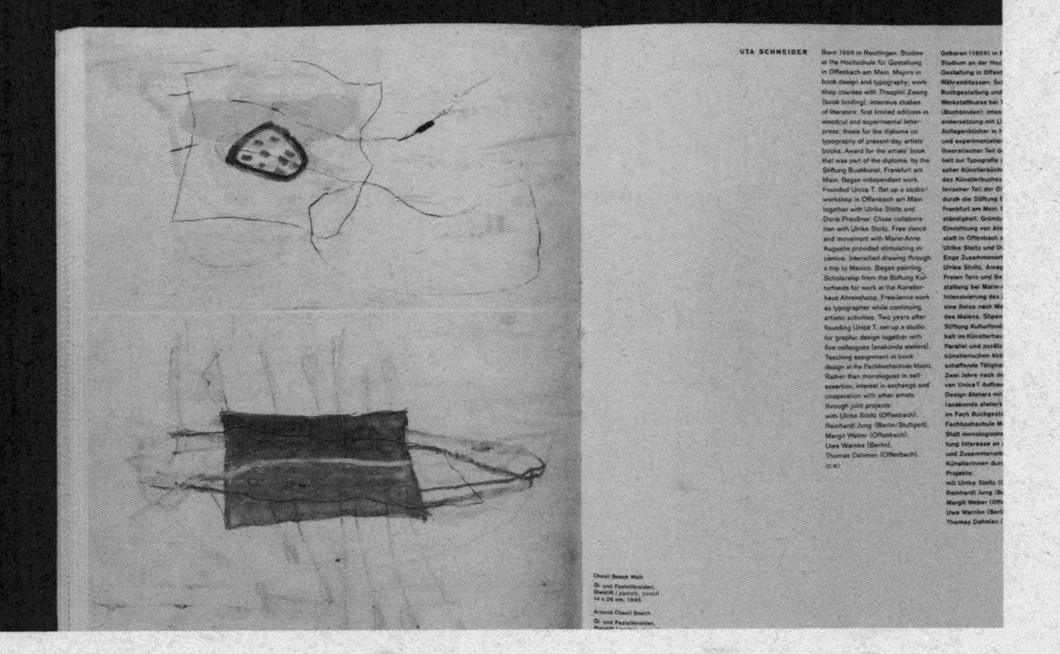

UTA SCHNEIDER

Born 1959 in Reutlingen. Studies at the Hochschule für Gestaltung in Offenbach am Main. Majors in book design and typography; workshop courses with Theophil Zwang (book binding); intensive studies of literature; first limited editions in woodcut and experimental letterpress; thesis for the diploma on typography of present-day artists' books. Award for the artists' book that was part of the diploma, by the Stiftung Buchkunst, Frankfurt am Main. Began independent work. Founded Unica T. Set up a studio/workshop in Offenbach am Main together with Ulrike Stoltz and Doris Preußner. Close collaboration with Ulrike Stoltz. Free dance and movement with Marie-Anne Augustin provided stimulating incentive. Intensified drawing through a trip to Mexico. Began painting. Scholarship from the Stiftung Kulturfonds for work at the Künstlerhaus Ahrenshoop. Free-lance work as typographer while continuing artistic activities. Two years after founding Unica T, set up a studio for graphic design together with five colleagues (anakonda ateliers). Teaching assignment in book design at the Fachhochschule Mainz. Rather than monologues in self-assertion, interest in exchange and cooperation with other artists through joint projects:
with Ulrike Stoltz (Offenbach),
Reinhardt Jung (Berlin/Stuttgart),
Margit Weber (Offenbach),
Uwe Warnke (Berlin),
Thomas Dahmen (Offenbach).
(C.N)

Chesil Beach Walk
Öl- und Pastellkreiden, Bleistift / pastels, pencil
14 x 26 cm, 1996

Around Chesil Beach
Öl- und Pastellkreiden,

then hope and expecta-
›w become reality.
ıve shown that they are
finest artists in the
ɔooks they have created
im leave no doubt
fact, as you will be able
ıe exhibition. There
ıo lack of acknowledge-
outside. Official collec-
all over the world
ıased their works, the
ɜived awards from
ıtions as the Stiftung
, the international exhibi-
›k art in Leipzig and
-Tiemann-Award 1992
Unica T have gained
ıd today they are on
e in Reutlingen.
em it necessary to say
ɔu regarding the art
king in general. The
shown by the Stadt-

has developed to a point which was hitherto unknown. The most astounding change is the manifest diversity in the appearance of this modern generation of artists' books. This has many causes, of which I would here like to mention only a few.

Books as a medium have adopted a completely new role (as catchwords I would only name mass production of books, changes in reading behaviour as well as the appearance of new media such as microfilm and video). In consequence the artists who produce books have also adopted a new understanding of their art. An increasing number of artists have turned to books as a means of expressing themselves; artists working in a wide variety of fields who consequently integrated completely new materials into the making of books. Books and book objects, be it of stone, glass, metal, plastic, hand-made paper or whatever have nowadays become quite customary. In brief, books have

9

Arnulf Herbst

Composersatz
Broschur
52 Seiten
5 x 29,5 cm
5 signierte Exemplare
Offenbach am Main
1983

ER AUSSTELLUNG AM 11.1.1993
STADTBIBLIOTHEK REUTLINGEN

R GEEHRTE DAMEN UND

darf und im Rahmen dieser Ausstel
Mit großer Freude nahm ich zur
ellung der Künstlergruppe Unica T
bei der Eröffnung zu sprechen, hatt
ler ersten Ausstellungen von Unica
hatte, war das natürlich ein gewisse
icht möglich gewesen, deren Arbeit
s vorliegenden Arbeiten kam ich zu
ch der Buchkunst handelt, die siche
Nun, was damals Vermutung und
e zu den wichtigsten Vertretern der

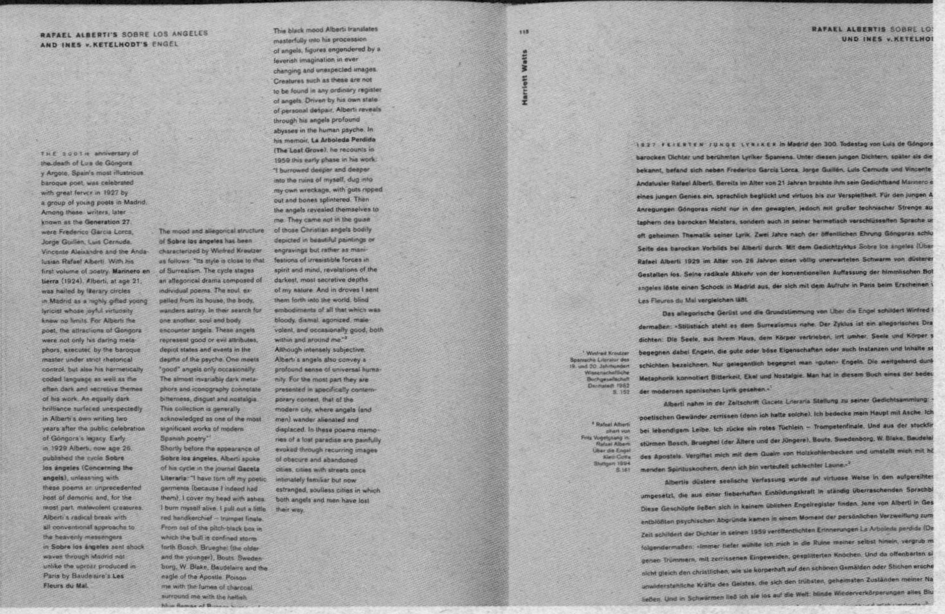

RAFAEL ALBERTI'S SOBRE LOS ANGELES
AND INES v. KETELHODT'S ENGEL

THE 300TH anniversary of the death of Luis de Góngora y Argote, Spain's most illustrious baroque poet, was celebrated with great fervor in 1927 by a group of young poets in Madrid. Among these writers, later known as the **Generation 27**, were Frederico Garcia Lorca, Jorge Guillen, Luis Cernuda, Vincente Aleixandre and the Andalusian Rafael Alberti. With his first volume of poetry, **Marinero en tierra** (1924), Alberti, at age 21, was hailed by literary circles in Madrid as a highly gifted young lyricist whose joyful virtuosity knew no limits. For Alberti the poet, the attractions of Gongora were not only his daring metaphors, executed by the baroque master under strict rhetorical control, but also his hermetically coded language as well as the often dark and secretive themes of his work. An equally dark brilliance surfaced unexpectedly in Alberti's own writing two years after the public celebration of Góngora's legacy. Early in 1929 Alberti, now age 26, published the cycle **Sobre los ángeles (Concerning the angels)**, unleashing with these poems an unprecedented host of demonic and, for the most part, malevolent creatures. Alberti's radical break with all conventional approachs to the heavenly messengers in **Sobre los ángeles** sent shock waves through Madrid not unlike the uproar produced in Paris by Baudelaire's **Les Fleurs du Mal.**

The mood and allegorical structure of **Sobre los ángeles** has been characterized by Winfred Kreutzer as follows: "Its style is close to that of Surrealism. The cycle stages an allegorical drama composed of individual poems. The soul, expelled from its house, the body, wanders astray. In their search for one another, soul and body encounter angels. These angels represent good or evil attributes, depict states and events in the depths of the psyche. One meets "good" angels only occasionally. The almost invariably dark metaphors and iconography connotate bitterness, disgust and nostalgia. This collection is generally acknowledged as one of the most significant works of modern Spanish poetry."[1]
Shortly before the appearance of **Sobre los ángeles**, Alberti spoke of his cycle in the journal **Gaceta Literaria**: "I have torn off my poetic garments (because I indeed had them). I cover my head with ashes. I burn myself alive. I pull out a little red handkerchief – trumpet finale. From out of the pitch-black box in which the bull is confined storm forth Bosch, Brueghel (the older and the younger), Bouts, Swedenburg, W. Blake, Baudelaire and the eagle of the Apostle. Poison me with the fumes of charcoal, surround me with the hellish

This black mood Alberti translates masterfully into his procession of angels, figures engendered by a feverish imagination in ever changing and unexpected images. Creatures such as these are not to be found in any ordinary register of angels. Driven by his own state of personal despair, Alberti reveals through his angels profound abysses in the human psyche. In his memoir, **La Arboleda Perdida (The Lost Grove)**, he recounts in 1959 this early phase in his work: "I burrowed deeper and deeper into the ruins of myself, dug into my own wreckage, with guts ripped out and bones splintered. Then the angels revealed themselves to me. They came not in the guise of those Christian angels bodily depicted in beautiful paintings or engravings but rather as manifestions of irresistible forces in spirit and mind, revelations of the darkest, most secretive depths of my nature. And in droves I sent them forth into the world, blind embodiments of all that which was bloody, dismal, agonized, malevolent, and occasionally good, both within and around me."[3]
Although intensely subjective, Alberti's angels also convey a profound sense of universal humanity. For the most part they are presented in specifically contemporary context, that of the modern city, where angels (and men) wander alienated and displaced. In these poems memories of a lost paradise are painfully evoked through recurring images of obscure and abandoned cities, cities with streets once intimately familiar but now estranged, soulless cities in which both angels and men have lost their way.

1927 FEIERTEN JUNGE LYRIKER in Madrid den 300. Todestag von Luis de Góngora
barocken Dichter und berühmten Lyriker Spaniens. Unter diesen jungen Dichtern, später als die
bekannt, befand sich neben Frederico Garcia Lorca, Jorge Guillén, Luis Cernuda und Vincente
Andalusier Rafael Alberti. Bereits im Alter von 21 Jahren brachte ihm sein Gedichtband Marinero e
eines jungen Genies ein, sprachlich beglückt und virtuos bis zur Verspieltheit. Für den jungen A
Anregungen Góngoras nicht nur in den gewagten, jedoch mit großer technischer Strenge au
taphern des barocken Meisters, sondern auch in seiner hermetisch verschlüsselten Sprache u
oft geheimen Thematik seiner Lyrik. Zwei Jahre nach der öffentlichen Ehrung Góngoras schlu
Seite des barocken Vorbilds bei Alberti durch. Mit dem Gedichtzyklus Sobre los ángeles (Über
Rafael Alberti 1929 im Alter von 26 Jahren einen völlig unerwarteten Schwarm von düstere
Gestalten los. Seine radikale Abkehr von der konventionellen Auffassung der himmlischen Bo
angeles löste einen Schock in Madrid aus, der sich mit dem Aufruhr in Paris beim Erscheinen
Les Fleures du Mal vergleichen läßt.

Das allegorische Gerüst und die Grundstimmung von Über die Engel schildert Winfred
dermaßen: »Stilistisch steht es dem Surrealismus nahe. Der Zyklus ist ein allegorisches Dra
dichten: Die Seele, aus ihrem Haus, dem Körper vertrieben, irrt umher. Seele und Körper s
begegnen dabei Engeln, die gute oder böse Eigenschaften oder auch Instanzen und Inhalte s
schichten bezeichnen. Nur gelegentlich begegnet man »guten« Engeln. Die weitgehend dun
Metaphorik konnotiert Bitterkeit, Ekel und Nostalgie. Man hat in diesem Buch eines der bede
der modernen spanischen Lyrik gesehen.«[1]

Alberti nahm in der Zeitschrift Gacete Literaria Stellung zu seiner Gedichtsammlung: »
poetischen Gewänder zerrissen (denn ich hatte solche). Ich bedecke mein Haupt mit Asche. Ich
bei lebendigem Leibe. Ich zücke ein rotes Tüchlein – Trompetenfinale. Und aus der stockfi
stürmen Bosch, Brueghel (der Ältere und der Jüngere), Bouts, Swedenborg, W. Blake, Baudela
des Apostels. Vergiftet mich mit dem Qualm von Holzkohlenbecken und umstellt mich mit h
menden Spirituskochern, denn ich bin verteufelt schlechter Laune.«[2]

Albertis düstere seelische Verfassung wurde auf virtuose Weise in den aufgereihte
umgesetzt, die aus einer fieberhaften Einbildungskraft in ständig überraschenden Sprachb
Diese Geschöpfe ließen sich in keinem üblichen Engelregister finden. Jene von Alberti in Ges
entblößten psychischen Abgründe kamen in einem Moment der persönlichen Verzweiflung zum
Zeit schildert der Dichter in seinen 1959 veröffentlichten Erinnerungen La Arboleda perdida (De
folgendermaßen: »Immer tiefer wühlte ich mich in die Ruine meiner selbst hinein, vergrub m
genen Trümmern, mit zerrissenen Eingeweiden, gesplitterten Knochen. Und da offenbarten si
nicht gleich den christlichen, wie sie körperhaft auf den schönen Gemälden oder Stichen ersch
unwiderstehliche Kräfte des Geistes, die sich den trübsten, geheimsten Zuständen meiner Na
ließen. Und in Schwärmen ließ ich sie los auf die Welt: blinde Wiederverkörperungen alles Bl

[1] Winfred Kreutzer Spanische Literatur des 19. und 20. Jahrhundert Wissenschaftliche Buchgesellschaft Darmstadt 1982 S. 152

[2] Rafael Alberti zitiert von Fritz Vogelgsang in: *Rafael Alberti Über die Engel* Klett-Cotta Stuttgart 1994 S. 161

0663

首歌人
ける
ツあやめぞかをる 郭公 鳴くや五月の雨 夕ぐれ
か門の苅田のねやにふす鴫の 床あらはなる冬の夜の月
ein ne
wird eig
bis ne

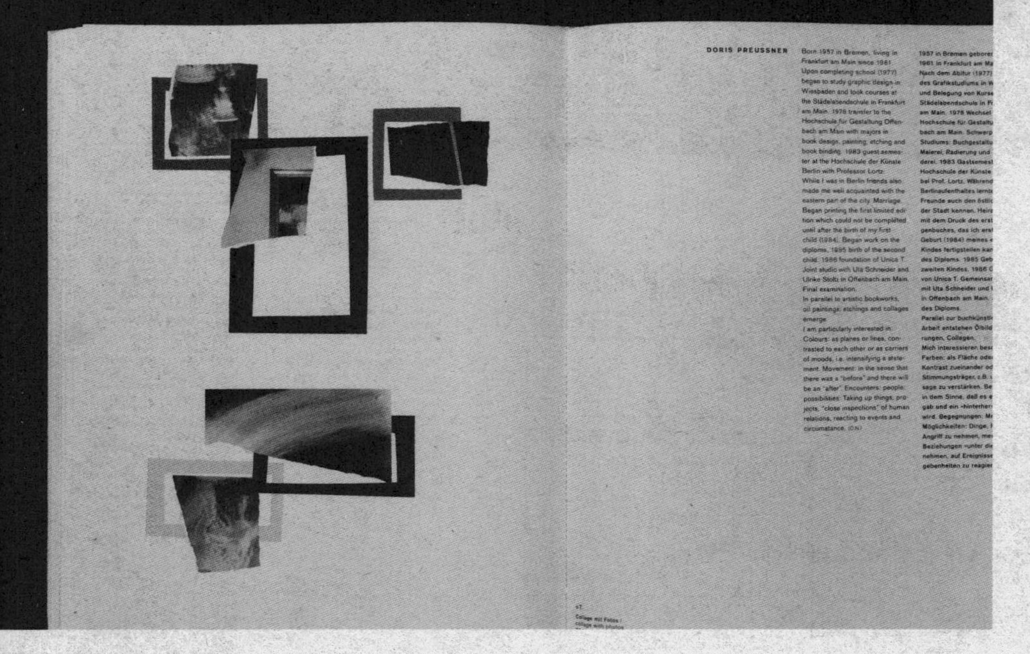

91
Collage mit Fotos /
collage with photos

DORIS PREUSSNER

Born 1957 in Bremen, living in Frankfurt am Main since 1961. Upon completing school (1977) began to study graphic design in Wiesbaden and took courses at the Städelabendschule in Frankfurt am Main. 1978 transfer to the Hochschule für Gestaltung Offenbach am Main with majors in book design, painting, etching and book binding. 1983 guest semester at the Hochschule der Künste Berlin with Professor Lortz.
While I was in Berlin friends also made me well acquainted with the eastern part of the city. Marriage. Began printing the first limited edition which could not be completed until after the birth of my first child (1984). Began work on the diploma. 1985 birth of the second child. 1986 foundation of Unica T. Joint studio with Uta Schneider and Ulrike Stoltz in Offenbach am Main. Final examination.
In parallel to artistic bookworks, oil paintings, etchings and collages emerge.
I am particularly interested in: Colours: as planes or lines, contrasted to each other or as carriers of moods, i.e. intensifying a statement. Movement: in the sense that there was a "before" and there will be an "after". Encounters: people: possibilities. Taking up things, projects, "close inspections" of human relations, reacting to events and circumstance. (D.N.)

1957 in Bremen geborer
1961 in Frankfurt am Ma
Nach dem Abitur (1977)
des Grafikstudiums in W
und Belegung von Kurse
Städelabendschule in Fr
am Main. 1978 Wechsel
Hochschule für Gestaltu
bach am Main. Schwerp
Studiums: Buchgestaltu
Malerei, Radierung und
derei. 1983 Gastsemest
Hochschule der Künste
bei Prof. Lortz. Während
Berlinaufenthaltes lernt
Freunde auch den östlic
der Stadt kennen. Heira
mit dem Druck des erst
genbuches, das ich erst
Geburt (1984) meines e
Kindes fertigstellen kan
des Diploms. 1985 Geb
zweiten Kindes. 1986 G
von Unica T. Gemeinsar
mit Uta Schneider und U
in Offenbach am Main.
des Diploms.
Parallel zur buchkünstle
Arbeit entstehen Ölbild
rungen, Collagen.
Mich interessieren besc
Farben: als Fläche oder
Kontrast zueinander od
Stimmungsträger, z.B. u
sage zu verstärken. Be
in dem Sinne, daß es e
gab und ein »hinterher«
wird. Begegnungen: Me
Möglichkeiten: Dinge, P
Angriff zu nehmen, me
Beziehungen »unter die
nehmen, auf Ereignisse
gebenheiten zu reagier

97 Bm5

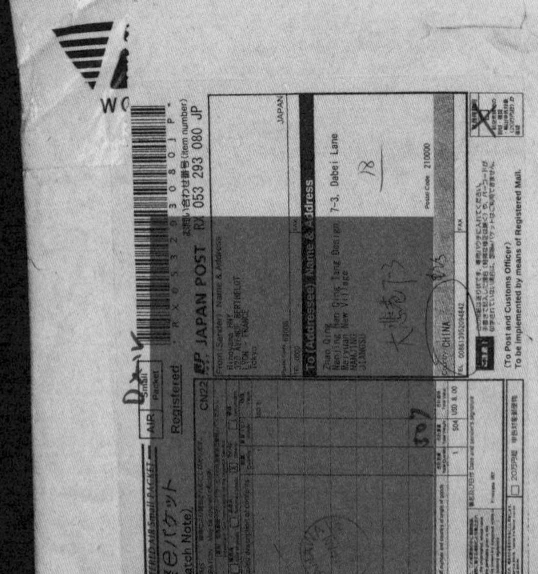

JAPAN POST
RX 053 293 080 JP
To (Addressee) Name & Address
7-3, Dabei Lane
CHINA
Postal Code 210000
AIR
Small Packet
Registered
CN22
国際eパケット
送り状 (Dispatch Note)
used book
(To Post and Customs Officer)
To be Implemented by means of Registered Mail.

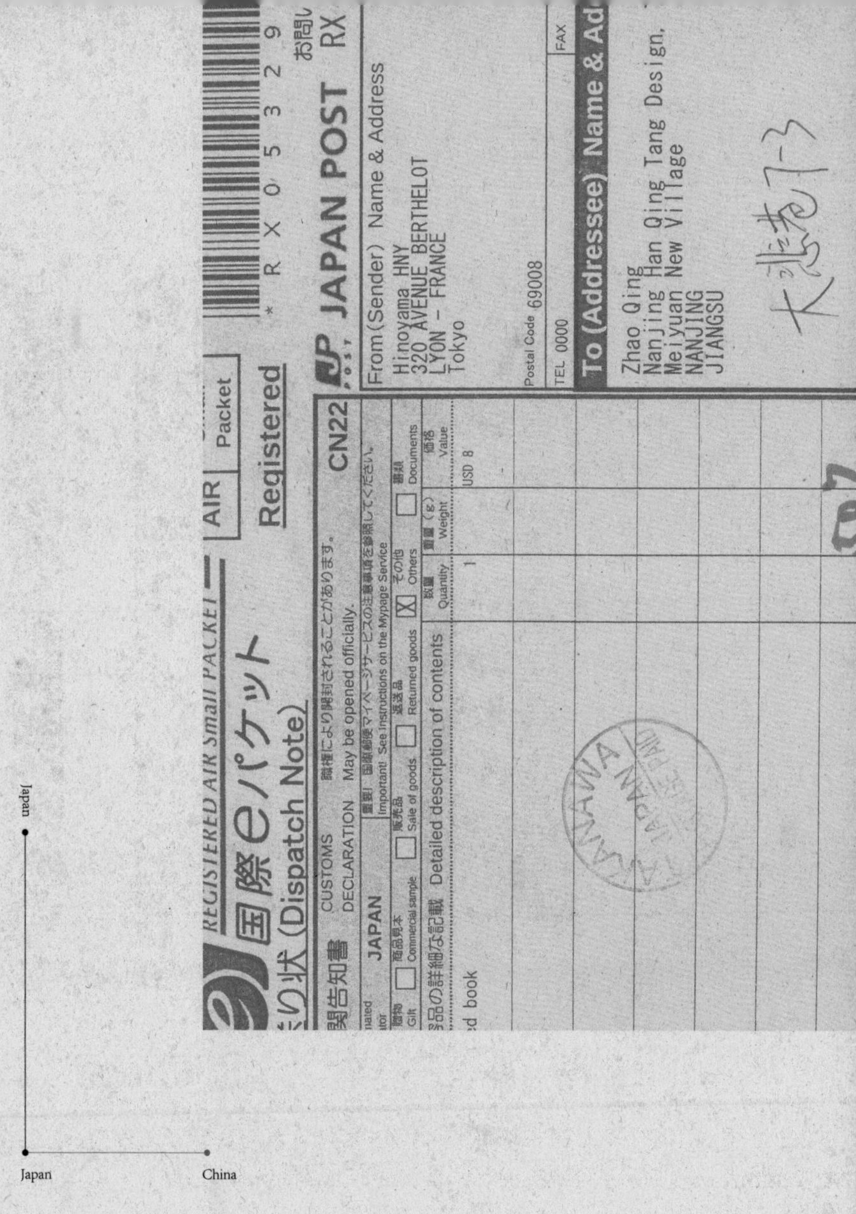

REGISTERED AIR Small PACKET
国際eパケット
(Dispatch Note)
AIR
Packet
Registered
CN22
CUSTOMS DECLARATION
May be opened officially.
JAPAN
Gift
Commercial sample
Sale of goods
Returned goods
Others
Documents
Detailed description of contents
Quantity
Weight
Value
1
USD 8
JAPAN POST
From (Sender) Name & Address
Hinoyama HNY
320 AVENUE BERTHELOT
LYON - FRANCE
Tokyo
Postal Code 69008
TEL 0000
FAX
To (Addressee) Name & Address
Zhao Qing
Nanjing Han Qing Tang Design,
Meiyuan New Village
NANJING
JIANGSU
Japan
Japan
China

217mm

22mm

157mm

铜奖

幸田文的五斗柜

日本

Bronze Medal

Koda Aya No Tansu No Hikidashi

Japan

Akio Nonaka

Gyoku Aoki

Shincho-Sha Co., Tokyo

217 157 22mm

476g

208p

ISBN-10 4-10-405201-9

日本女作家幸田文从小受作家父亲幸田露伴的影响，学习各种富有日本文化特色的生活技能，是较为典型的日本昭和时代女性。幸田文撰写了大量回忆其父亲的随笔集，受到世人的关注。幸田文于 1990 年去世后，其女即本书的作者青木玉以母亲的柜子为线索，将柜中和服、着装记忆以及家庭照片汇编成为本书。本书的装订形式为圆脊锁线硬精装。护封胶版纸四色印刷，内封采用含有金属光泽的艺术纸张蓝黑双色印刷，裱覆卡板。环衬及扉页选用了不同的艺术纸用于映衬和服丰富的肌理与工艺。内页胶版纸四色印刷。书中内容包含礼服、内衬、家纹、纹样、潮流、雨具、浴袍、配色等方面的和服文化。全书设计和内文故事内敛雅致，充满了对日本传统文化的崇敬以及对家人的思念。本书是一本将传统文化与家族历史融为一体的传承之书。

Japanese female writer Koda Aya, influenced by her father Koda Rohan who was also a writer, learned various life skills related to Japanese culture. She was a typical Japanese woman in the Showa period. Koda Aya wrote a large number of essays to look back upon her father, which attracted the attention of the world. After Koda Aya died in 1990, her daughter Akio Nonaka, who is the author of the book, used her mother's cabinet as a clue to compile photos of kimono, family and life memory. The hardcover book is bound by thread sewing with rounding spine. The jacket is printed on offset paper in four colors and the inner cover is made of shimmering art paper printed in blue and black, mounted with cardboard. Different kinds of art paper are used for the end paper and the title page to display the rich texture and layers of kimono. The inside pages are made of offset paper printed in four colors. The content involves all aspects of kimono culture, such as dresses, linings, family patterns, patterns, trends, rain gear, bathrobes, and color matching. The book design and story are restrained and elegant, full of respect for traditional Japanese culture and thoughts of family. It is a heritage book that integrates traditional culture with family history.

ここにある布や着物を見ていると、よくもまあ残ったものだ
ちてくる。昭和二十年三月戦火に追われて、長年住み続けた
た。病床の祖父は、薄がけ、小抱巻に包まれたまま、人の背
けた町を抜け汽車に抱え込まれた。母は食料と祖父の机廻り
射薬、救急の飲み薬など細かい指示書が付されていた）、寝間着

入れと煙管入れは対の山椒粒大の相良繍で模様が一面に刺して
懐から滑り落ちることのないように選ばれた素材だ。一度水を
に挟んで懐中すれば、いい加減なゴマの灰如きにしてやられる
先生は用心のいいところもあった。

普段着の祖父は、いつも石摺りの着物で居たように思う。母は
合った着物を用意していた筈だのに、他の着物の祖父は思い出
の両氏が撮った写真の着物もやはりこれだ。それほど好んで着
洗い張りし、内側には継ぎも当てられている。所どころ手当て

幸田文の箪笥の引

になった。母の世代で
等のいい恰好に見える
寸胴で、蜂の胴のよう

「まあそう言わずにお座りなさいよ」
と勧められて尻こそばゆき思いをした。

祖父が居なくなってしばらくの間、何年か母も私も、夏の浴衣は昼は蚊絣、夜は格子、裾が切れれば丈をつめて男物ばかり着たが、それもついには雑巾になって終った。羽織裏は軽いので、年を取ってから母は寒さ防ぎの真綿の半纏を作って、「あったかいよ」と羽織っていた。

母は祖父もそうだが、ことにおばあさんから物の納まりを厳しく仕込まれたという。どんな物でもいい加減におっぽり出してはいけない、終りの面倒を見てやらなければならないと。母はその教育を遵守したが、年と共にこれはきつい教えになった。石摺りは残った。好んで男物を着てよく似合い、どんなボロを着てもたじろぐことの無かった母だが、祖父そのものとも思える石摺りはあまりに思うことが多く、着る気になれなかったのだろうと思われる。

母さえ付けかねたこの納まりを私はどうしたものか、子供の頃、習字の稽古をさぼっていたら、「人、墨を磨さず　墨、人を磨す」と教えられたことがあった。うかうかしていれば石摺りはそのまま、こちらの命が摺り切れて終らないものでもない。

誰が袖

今年も暖冬だ暖冬だと聞かされるけれど、やっぱり冬だ。衿やら袖口を思わず押さえて首をすくめる日もある、座っていても肩や膝頭が冷えてくると落ち着かない。私は小さい時から寒がりで、厚目の肌着を着せてもらって、やれよかったと喜ぶが、それでもすぐまた、もっと重ねて着たくなり、古くなって小さくなった肌着を引きずり出して下に着る。ふくら雀は可愛いが膨れ雀である。

母は元気で寒がらない、私のことを「持ったが病」と笑う。かたちもなく着込んで、手が通って首が出れば、身上ありったけ着ないと落ち着かないのだ。

「持っているから着たくなる。それが病で、無ければ着られないでしょ。無いと思いなさい」

と言われて古い肌着を取り上げられた。半日しょぼしょぼしていたが、翌日には風邪を引

裏衿

着物も共布にして、下着はふくれ織の菊重ね、秋のこしらえでまとめられた。さて赤、着物の赤はすぐ思い付くのが緋縮緬だが、これはどうも長襦袢、まごまごしていると腰巻を連想してしまう。内に着る色で表に出る色ではない。母は羽二重を探させた。縮緬に較べて羽二重はべた付かない、すぺんとした風合で男物に使う。京都で羽二重を特に扱う問屋から取り寄せた絹は「九重の紅」と名が付いているという。太く巻いてある布を手元から転がすと、真紅な色が八畳の畳の上をさっと横切って流れ、五月の光を余すことなく反射して、はっとする鮮やかさであった。

「ああいい風合だ、これにしなさい。もう他のものを考えることは無い、これに白い花を散らそう。下に紅があるから、真白な所も少し紅が透けるところもあって綺麗だよ、きっと」

私の考えを上廻る華やかさで、しかも汚れを寄せつけない気性を持った着物が形づくられていった。

秋、空が高く空気が冷えた日、大ぶりな衣裳箱に納められた花嫁衣裳が運び込まれた。結婚前の雑事を手伝うため、いつもより賑やかに人が来ていた。取り出され、衣桁にかけられた溢れる白と赤に、皆びっくりし、そしてまあきれいと眺めた。母は呉服屋さんに見事な仕上りを労らい、

「みんなでお茶にしようじゃないの、男ならこれで一杯祝うものを惜しいこと、昼から酔っ

九重の紅の羽二重で仕立てた婚礼衣裳の大振袖。

1997 Ha[1]

medimops
shop
FROM
PRIORITY
B R
W
Deutsche Post
TO
ZHAO QING
Design Co.
210000 Nanjing
CHINA V.R.
CN 22
E EXPR
Villegas
editores

momox GmbH
Am Alten Flughafen 1
04356 Leipzig
Germany
Referenz-Nr.:103075781

B R

PRIORITY

W

06 30E0 A750 1B 1000 0AA0
DV 11.19 1,00 Deutsche Post
WARENPOST Int. KT Port payé

TO

ZHAO QING
Design Co.,
Dabei xiang7-3,Nanjing Han Qing Tang .
210000 Nanjing
CHINA VR

UC 030 628 001 DE

- 1 x Book

Colombia

Germany

China

311mm

21mm

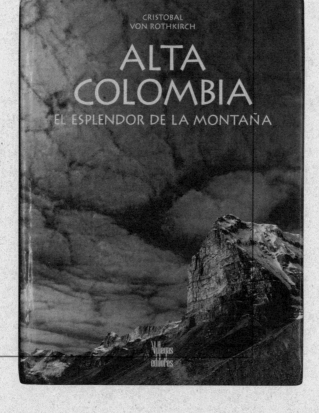

234mm

◎ 荣誉奖 Honorary Appreciation

《 壮美的哥伦比亚高山 Alta Colombia: El Esplendor de la Montaña

哥伦比亚 Colombia

Benjamin Villegas

Cristobal von Rothkirch

Villegas Editores, Bogotá

311 234 21mm

1598g

216p

ISBN-10 958-9393-22-5

摄影师里斯托瓦尔·冯·罗特基希（Cristobal von Rothkirch）参与到环保主义者的探险队中，拍摄了大量哥伦比亚山川的照片。本书的精美照片展现了山脉间的自然地貌、生态植被、人居生活和四季风貌，并配以文字介绍山脉和地貌的形成、人居环境的问题和环保行动的诸多思考。书籍的装订形式为锁线硬精装。护封铜版纸四色印刷，覆亮膜。上下书口和勒口处均向内包折，形成良好的手感，同时避免了这么大的开本上护封变形的问题。内封采用蓝色布面裱覆卡板，文字部分烫银处理。环衬选用蓝色艺术纸。内页亮面铜版纸五色印刷。书中以风景摄影为主，加入少量局部特写、微距甚至俯视航拍的多种视角来丰富整体内容。文字和非满版印刷的照片部分使用暖灰色（专色）铺底，形成“画框”，衬托照片中自然的白色。

Photographer Cristobal von Rothkirch participated in an environmentalist expedition and took numerous pictures of Columbia Mountains. A large number of exquisite photos in this book show the natural landforms, ecological vegetation, human settlements and four seasons of the mountains. They are accompanied by texts to introduce the formation of mountains and landforms, the problems of human settlements and many reflections on environmental protection actions. The hardcover book is bound by thread sewing. The laminated jacket is made of coated paper printed in four colors. The upper and lower book edges and the flaps are folded inward to avoid the deformation of the jacket of such a large format. The inner cover is made of blue cloth mounted with cardboard, and the text is treated with silver stamping. Blue art paper is used for the end paper and glossy coated paper is used for inner pages printed in five colors. The book mainly focuses on landscape photographs, adding a few from different perspectives, such as local close-up, macro and even aerial photographs to enrich the overall content. Warm gray (PMS color) is used under the text and photos to form a 'frame' that contrasts with the natural white in the photo.

ALTA
COLOMBIA
EL ESPLENDOR DE LA MONTAÑA

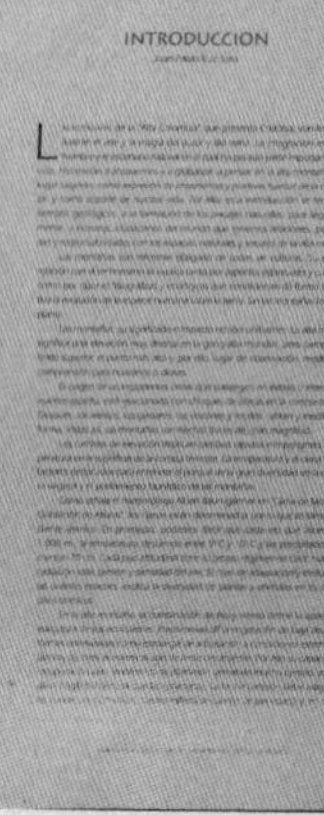
INTRODUCCION

Y más al sur del continente americano, los escaladores chilenos Bión González y Juan Harseim encontraron, en 1952, vestigios del sitio arqueológico más alto del mundo: un templo inca en la cima del monte Llullaillaco (6.723 m.), entre Chile y Argentina. Arqueólogos y montañistas han identificado más de 100 sitios incas por encima de los 5.000 m. de altura, al sur de Cuzco, la capital del imperio inca, iniciando lo que podríamos denominar arqueología de alta montaña. En esta zona, de muy baja humedad, la línea de nieves perpetuas está muy arriba, permitiendo alcanzar las altas cumbres sin dificultad técnica.

Casi todos los hallazgos arqueológicos de alta montaña han sido sitios ceremoniales o lugares de "pagamento". En 1954, un escalador encontró en el cerro El Plomo, cerca de Santiago de Chile, a 5.430 m., el cuerpo de un niño inca de ocho años que, en traje de ceremonia y sacrificio, permanecía conservado por el hielo. Casos similares se han presentado luego en El Toro (6.386 m.) y en Quehar (6.097,8 m.), en Argentina, donde también se han encontrado ofrendas humanas a los dioses de las montañas, al parecer en petición de lluvia y buenas cosechas. En 1958, cerca del lago Titicaca, en Bolivia, se reportó el último sacrificio humano destinado a solicitar favores de los dioses. No obstante, en la Isla del Sol, en el mismo lago, todavía se realizan frecuentes sacrificios de animales para pedir lluvias o evitar las sequías.

Desde épocas precolombinas, Ausangate, cerca a Cuzco, ha sido considerada morada de dioses. En las fiestas anuales del Cuzco, en las que se mezclan creencias incas y cristianas, se realizan grandes peregrinajes a la base de la montaña. Al amparo del cerro tutelar, se elevan cantos y se ejecutan danzas que representan el ascenso de algunos espíritus de las selvas bajas para interactuar con las deidades de las alturas. Y a menos de 93 Km de allí, espléndida y misteriosa, surge entre la bruma del tiempo la ciudad de Machu-Picchu, centro ceremonial de los Incas, cuya ubicación parece estar determinada por Paumasillo, Salcantay, Verónica y Ausangate, cuatro picos nevados, guardianes vegetales, que celosamente la protegen.

El Fitzroy y el Cerro Torre, dos de las paredes más difíciles de escalar en el mundo, tienen, según los Araucanos, un origen mítico. El cai cai, una serpiente maligna, intentó, mediante inundaciones, eliminar la raza humana, pero una serpiente amiga salvó a algunos seres en las partes altas de las montañas. Cai cai, sin embargo, convirtió sus cuerpos en los dos cerros de difícil acceso.

Esta reseña nos demuestra cómo la grandeza física de las montañas ha suscitado imágenes demiúrgicas en todos los tiempos y continentes, y cómo sólo de manera reciente, y en algunas culturas occidentales, se ha perdido el entendimiento de la estrecha relación que existe entre la vida sobre el planeta y la necesaria conservación y respeto por Pacha Mamma, la madre tierra.

Desde el punto de vista geológico, los Andes se pueden dividir en tres grandes secciones. Los Andes meridionales, una estrecha cordillera que atraviesa

0681

Ha³

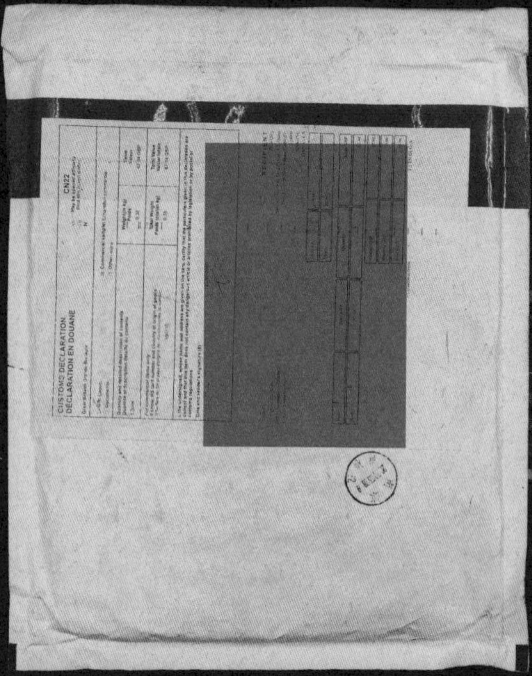

CUSTOMS DECLARATION
DECLARATION EN DOUANE

CN22

T E R

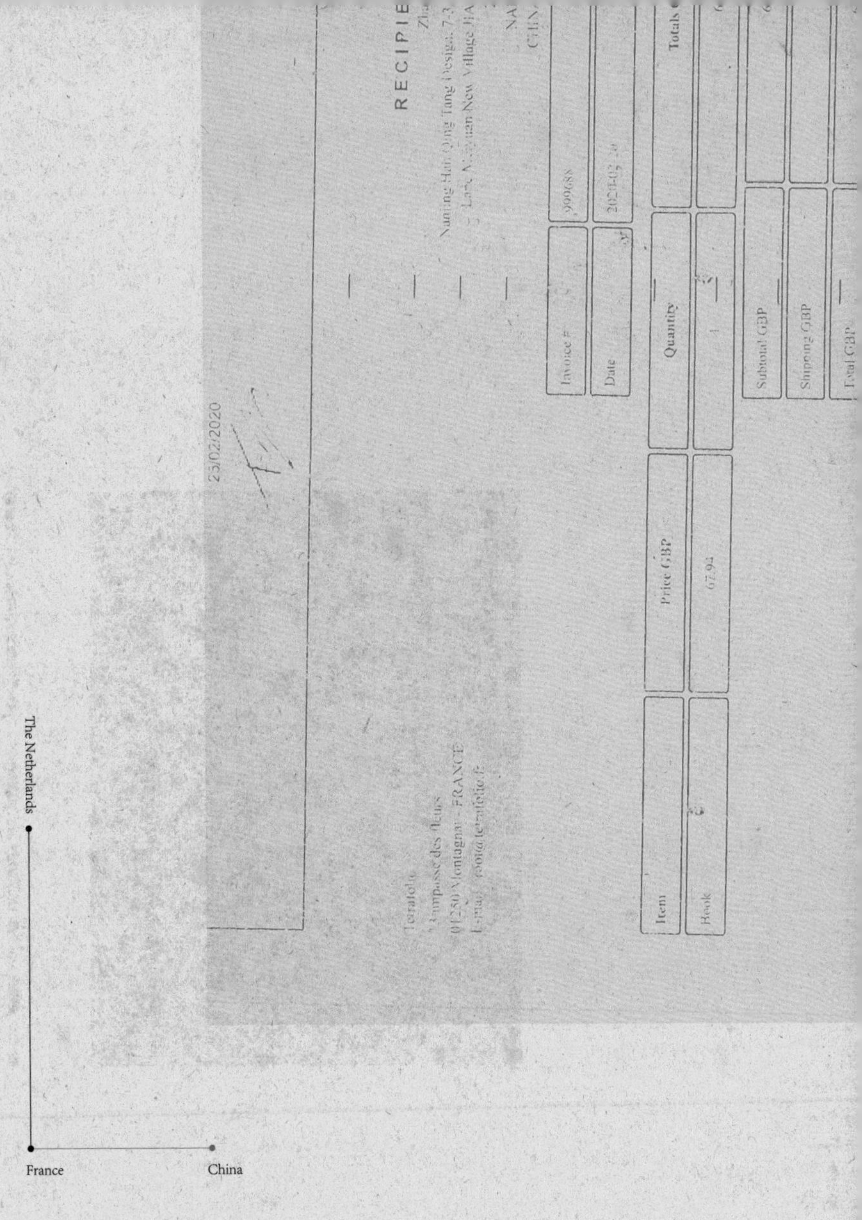

25/02/2020
RECIPIE
Nanjing Han Qing Tang Design 7-3
Invoice #
Date
Item
Book
Price GBP
67.94
Quantity
Subtotal GBP
Shipping GBP
Total GBP
Totals
01250 Montagnat - FRANCE
The Netherlands
France
China

270mm

7mm

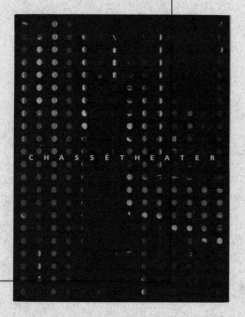

CHASSÉTHEATER

211mm

荣誉奖

布雷达的沙塞剧院

荷兰

Honorary Appreciation

Chassé Theater Breda

The Netherlands

Bureau Piet Gerards

Herman Hertzberger

Uitgeverij 010, Rotterdam

270 211 7mm

377g

72p

ISBN-10 90-6450-277-3

沙塞剧院坐落于荷兰南部的北布拉班特省布雷达市，由荷兰建筑设计师赫尔曼·赫兹伯格（Herman Hertzberger）设计。主要用途为现代艺术表演，上演节目的范围包括戏剧、舞蹈、音乐会，以及电影。本书是剧院的建筑设计介绍。书籍的装订形式为锁线胶平装。封面白卡纸四色印刷覆哑膜。封面圆形矩阵镂空，露出宽大的勒口内侧印制的剧院内部照片。内页哑面铜版纸四色加哑金色五色印刷。书中包含布雷达的城市介绍、文化传统、建筑规划、未来愿景、建筑技术、建筑与城市关系，以及最重要的部分——沙塞剧院的设计草图、平立剖面图、建成后内外照片。全书荷英双语设计，荷兰语被设置在左页，灰底黑字，而英文被设置在右页，白底哑金字，对照的双页上图片共用。建筑草图被放置在两两成对的拉页中，分别展示了外观整体结构和内部彩色立柱两方面。

Located in the city of Breda, North Brabant, in the south of the Netherlands, the Chassé Theater was designed by Dutch architect Herman Hertzberger and was mainly used for modern art performances. The range of performances includes drama, dance, concerts and movies. This book is an introduction to the architectural design of the theatre. The book is perfect bound with thread sewing. The matted cover is made of white cardboard printed in four colors. The matrix on the cover is hollowed out in circular to reveal the pictures inside the theater. The inner pages are printed with matte coated paper in four colors plus gold. The book contains the introduction of Breda, cultural traditions, architectural planning, future vision, architectural technology, the relationship between architecture and the city, as well as the most important part, the design sketch, plan and profile of the Chassé Theater, as well as the internal and external photos after its completion. The book is bilingual in Dutch and English. The Dutch language is set on the left page with black characters on a gray background, while English is set on the right page with matte gold characters on white background. The pictures are shared on the two pages of comparison. The architectural sketches are placed on the gatefolds in pairs, showing the overall exterior structure and the colored columns inside.

Chassé Theater Breda | Theatre Breda
HERMAN HERTZBERGER
É THEATER

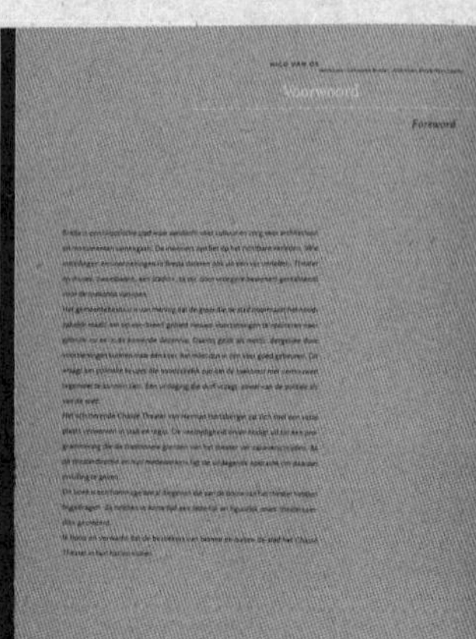

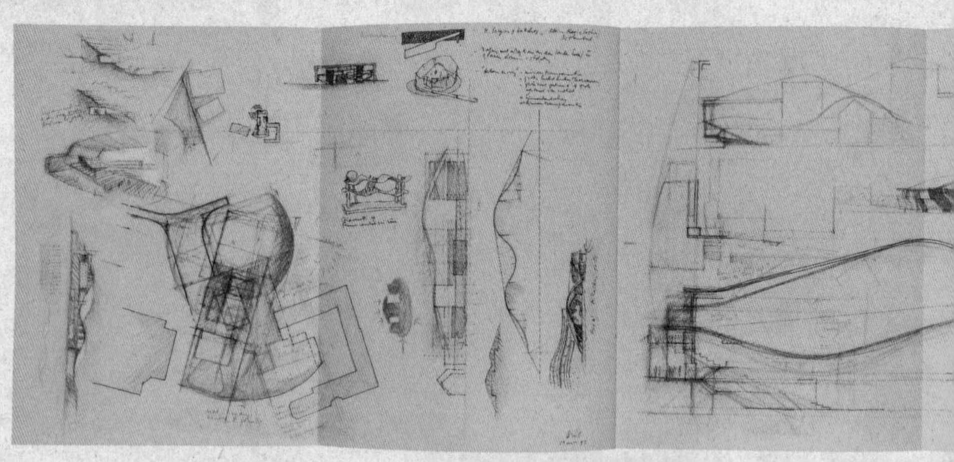

REG TEN ZIJTHOFF

Het theater

The theatre

'Een volk dat zijn toneel niet helpt en aanmoedigt is op sterven na dood' hield Federico García Lorca zijn gehoor, onder wie politici, dreigend voor in zijn vermaarde 'Charla sobre Teatro' (tafelrede over theater).

Tweehonderd jaar hiervoor kleurt een opgewekt theaterleven Breda als cultuurstad. Tegenstribbelende kerkeraden en predikanten kunnen niet voorkomen dat talloze toneelgezelschappen de stad op diverse locaties aandoen. In 1755 organiseert De Nederduytsche Haagsche Schouwburg hier zelfs als eerste lange seriebespelingen. Het residentiegezelschap geeft maar liefst zeventig voorstellingen.

Tweehonderdveertig jaar later zal de plaatselijke krant een Bredase schouwburgdirecteur voor megalomaan uitmaken. Hij voorspelt namelijk voorstellingenseries van vijftien en meer in een theaterzaal die exact evenveel stoelen telt als het Concordia uit 1881, dan al de derde Bredase schouwburg.

Tweehonderd jaar voorsprong | In 1802 is Breda de eerste stad in Noord-Brabant met een actief theaterleven dat vaste voet krijgt in de eerste Bredase en Noordbrabantse schouwburg. De volgens geschiedschrijvers 'meest liberale en tolerante stad' in Zuid-Nederland heeft haar culturele ontwikkeling op dat moment mede te danken aan de rol van het garnizoen en de beïnvloeding vanuit het kunstencentrum Brussel. In 1814 brandt de schouwburg af. Het stadsbestuur geeft al in 1815 vergunning tot herbouw aan de Oude Vest (Vlaszak-zijde). Hier zal een 'dans- en komediezaal' ontwikkeld worden. Deze gaat van start als Comedie en wordt later Schouwburg. Die situatie blijft bestaan totdat in 1881 het Concordia-gebouw aan het Van Coothplein wordt neergezet op initiatief van de uit 1807 daterende culturele sociëteit 'Amicitia et Concordia'. Ongeveer tachtig jaar later wordt deze schouwburg in handen van de gemeente formeel 'Stadsschouwburg'.

Bijna tweehonderd jaar behoudt Breda een voorsprong in het Noordbrabantse theaterleven, geheel te danken aan het particulier initiatief. Breda verliest haar leidende positie als zuidelijke theaterstad, wanneer in de jaren vijftig en zestig een

'A people which does not aid or promote its theatre is dying, if not dead,' Federico García Lorca gave his audience, which included politicians, unequivocally to understand in his celebrated 1935 talk at the Madrid National Theatre, 'Charal sobre Teatro'.

Two hundred years before then Breda had been a cultural town coloured by a lively theatre life. The protests of church councils and ministers were unable to stop countless theatre companies from performing at numerous venues in the town. Even a famous Hague company (Nederduytsche Haagsche Schouwburg) gave no less than seventy performances here in 1755.

Two hundred and forty years on, the local paper would describe a Breda theatre director as a megalomaniac. He had predicted performances in series of fifteen and more in a theatre auditorium with as many seats as the Concordia of 1881, which itself had already been Breda's third theatre house.

Two centuries ahead of the field | In 1802 Breda was the first town in the province of Noord-Brabant with an active theatre programme firmly ensconced in a municipal theatre, the first in both town and province. What historians describe as 'the most liberal and tolerant town' in the South Netherlands at that time owed its cultural development in part to the role played by the garrisons and the influence of Brussels, then as now a centre of the arts. In 1814 the theatre was burnt down. The city council was quick to give permission, in 1815, for a new one to be built on the Oude Vest (on the Vlaszak side). This began life as a 'dance and comedy hall', and later became the municipal theatre. So it was until 1881, when the Concordia building was erected on Van Coothplein at the initiative of 'Amicitia at Concordia', a cultural society founded in 1807. Some eighty years later this theatre was officially declared municipal theatre by the council, which by then was its owner.

For almost two hundred years Breda called the tune in the Noord-Brabant theatrical scene, thanks purely to private enterprise. Breda lost its leading position as

Generaal Baron Chassé | G
(Gemeentearchief Breda)

Schouwburg, Van Coothpl
Van Coothplein
(Gemeentearchief Breda)

0687

baasde bezoeker. Het geheim schuilt in de multifunctionali
en. Architectuur, technologie en programmeringsvisie gaan
fin de siècle waarin het podiumkunstenbeleid moet gaan
eranderende samenleving roepen theatermaker en publiek
xibiliteit. De media-zappende theaterbezoeker is niet lang
r òf operafan òf dansgek, maar alles tegelijk. Zoals mega-ci
udig aanbod inspringen op een last minute-keuze van de be
eater op weg naar meervoudige keuzemogelijkheden. Hier
techniek en programmeringsvisies.

gelijkbare steden wordt het podiumkunstenbeleid bepaald do
aren vijftig en zestig, later aangevuld met nieuwe gebouwen
it leidt tot een opeenstapeling van investeringen en exploit
oudig programmabeleid maar een meervoudige exploitat
ewacht en het cultuurhistorisch besef laten winnen.

Architectuur en techniek | Binnen de architectonische
n Hertzberger worden ideeën van Iain Mackintosh over de
tiest en publiek (het renaissancetheater) vertaald in een ge
de Grote Zaal met 1200 stoelen en toch een uiterst intieme
ultifunctionele theater van de toekomst, waarin klassiek lijst

rofiel 2. kunststof dakbedekking
rofielplaat 20/75 4. installatieroos
erk op isolatie 6. systeemplafond
ozijn 8. glazen U-profielen 9. stal
ast 10. stalen puiversteviging | *Ve*
etails: 1. aluminium roof edge sect
etic roofing 3. aluminium profiled
. services grille 5. rendering abov
. prefab ceiling 7. aluminium fram
lass U-extrusions 9. steel convect
o steel facade bearers

een pl
ar nooi

The more cultured a person the more mise

ne - te, che co - sa e a - mor

WOLFGANG

GER

chitectuur

The a

oëzie danst op de maat van

Poetry dances to the rhythm o

De hel, dat zij

Hell, it's oth

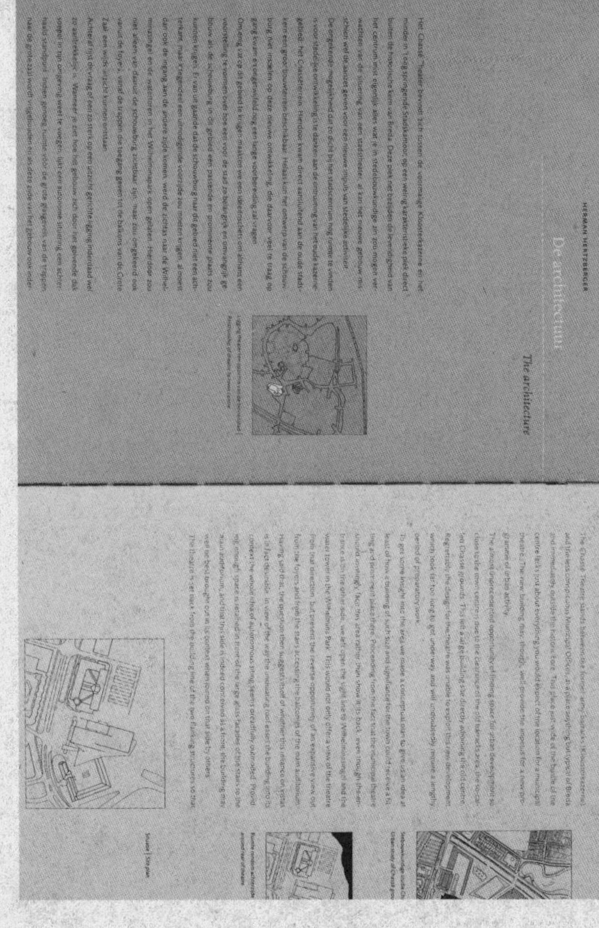

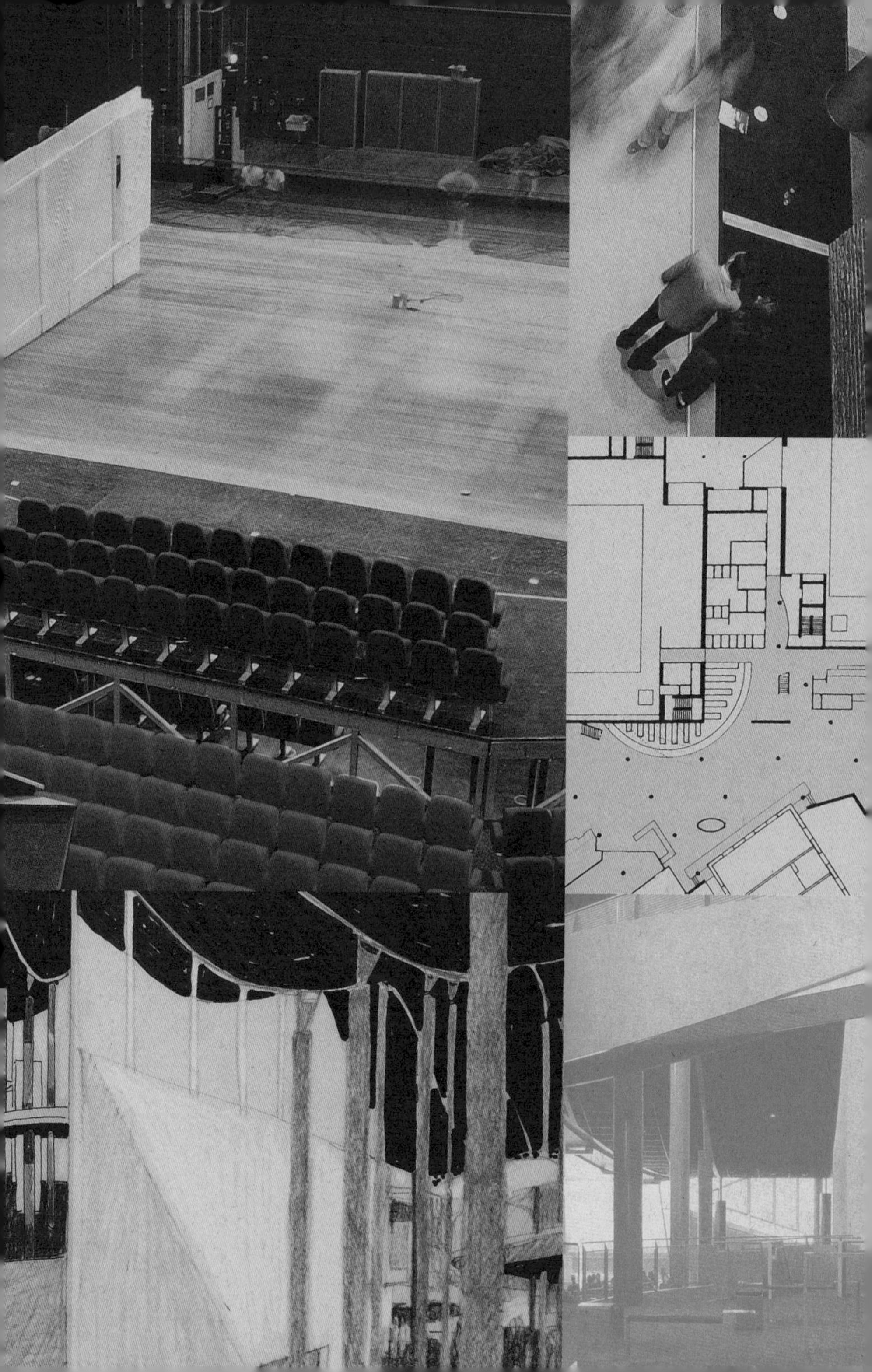

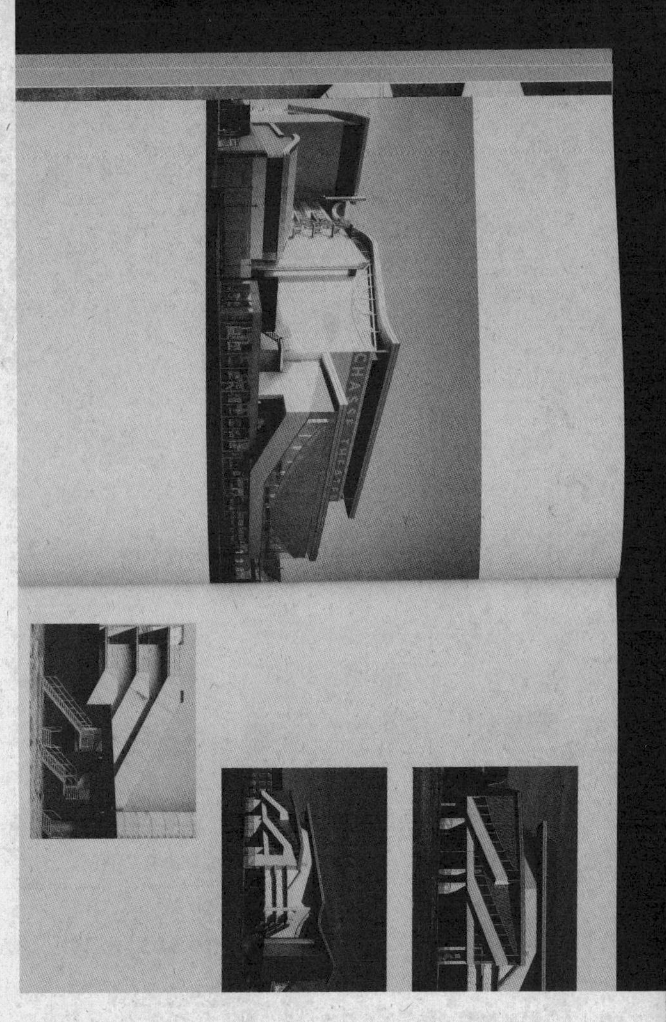

1997 Ha^4

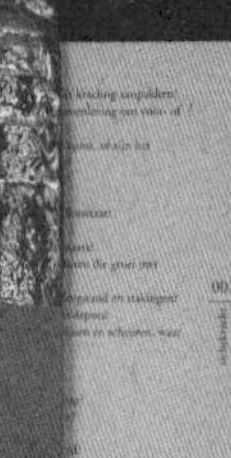

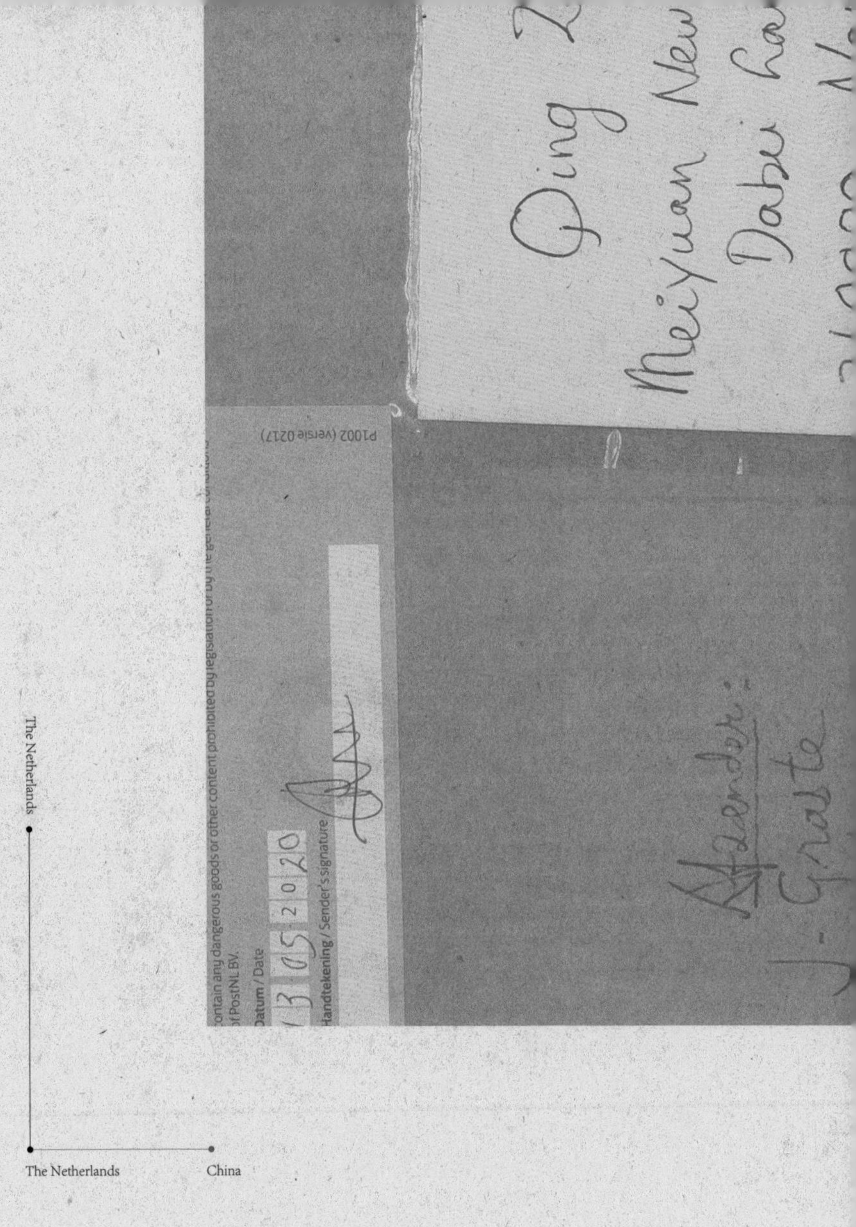

The Netherlands

The Netherlands China

297mm

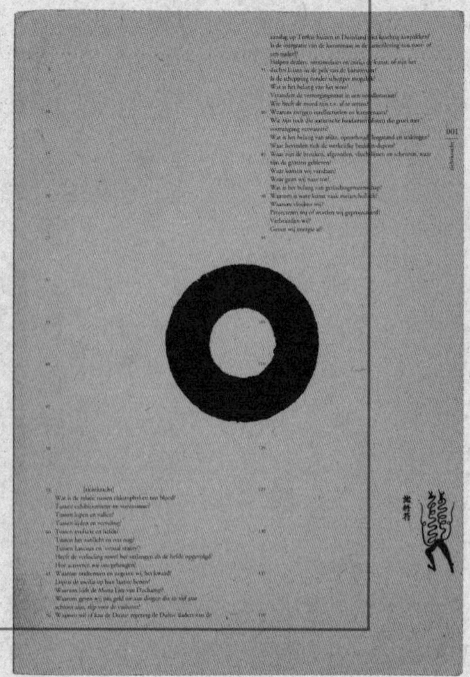

210mm

● 荣誉奖
« 目标动力
◐ 荷兰

Honorary Appreciation
Richtkracht
The Netherlands
Richard Menken
Richard Menken
Hogeschool voor de Kunsten HKA, Arnhem
297 210 14mm
689g
156p
ISBN-10 90-74485-13-8

荷兰艺术家里查德·门肯（Richard Menken）擅长装置、绘画、摄影和图形的混合表达。由于受到爱尔兰作家萨缪尔·贝克特（Samuel Beckett）和美国作曲家约翰·凯奇（John Milton Cage Jr.）的影响，他也尝试通过形式去消解意义，探索形式中的随机性。本书便是基于这样的理念创作和设计完成的一本“百科全书”。书籍的装订形式为无线胶平装。封面单面白卡纸灰面朝外，印单黑。内页胶版纸，绝大部分印单黑，只在局部穿插其间的英文“对话”处使用紫红色，另有红色和黄色两页满版隔页。内容方面包罗万象，科技、社会、政治、宗教、艺术……全部混合在一起，有些只是提问，有些自问自答，更多的是一些简短的观点和平常的对话。这些内容之间没有任何联系，也并无统一指向，艺术家试图将它们随机地组合在一起，消解统一的意义，产生意外的解读和联系。封面虽然使用了与内页不同的纸张，但页码和内容从封面开始，这样就消解了封面。内页两栏，以 5 的倍数标记行数，内容随机排布。

Dutch artist Richard Menken is good at the mixed expression by installation, painting, photography and graphics. Influenced by Irish writer Samuel Beckett and American composer John Milton Cage Jr., he also tried to eliminate the significance through the form and explore the randomness in form. This book is such an ‘encyclopedia’ created and designed based on this concept. The paperback book is perfect bound. The cover is made of white cardboard with a single side printed in gray facing outward. The inner pages are made of offset paper, printed mostly in black, purplish red is used only for the English ‘dialogue’ interspersed in the book, and there are two full pages printed in red and yellow as the interleavs. In terms of content, it covers everything, such as science, technology, society, politics, religion and arts. They’re all mixed up – asking questions, answering questions, short opinions and ordinary conversations. There is no connection or unified direction between these contents. The artist tries to combine them randomly to eliminate the unified meaning and produce unexpected interpretation and connection. Although the cover uses different paper from the inner pages, the page number and the content start from the cover, thus eliminating the cover. The inner pages are typeset to two columns, and the contents are arranged randomly.

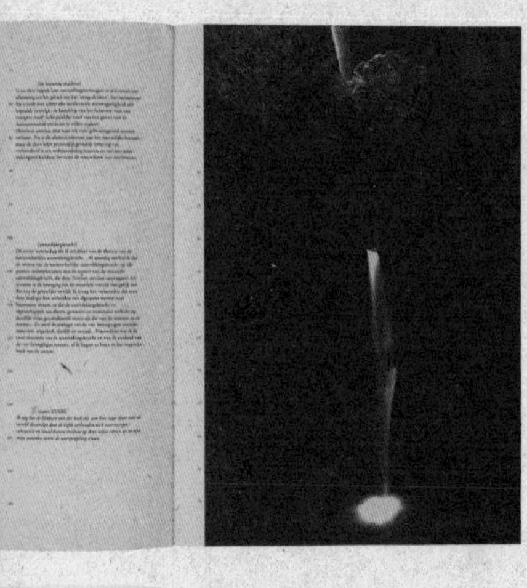

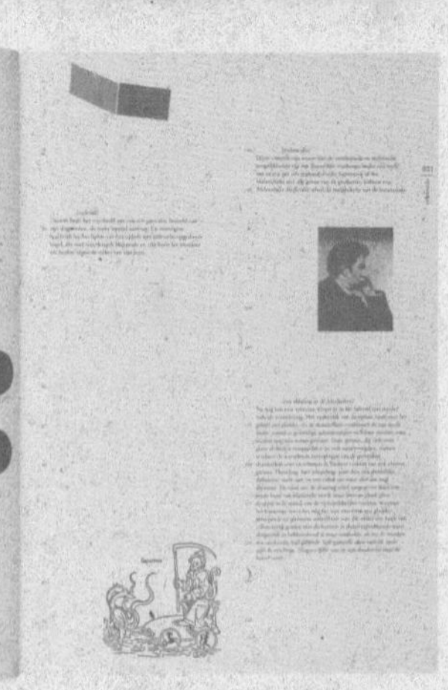
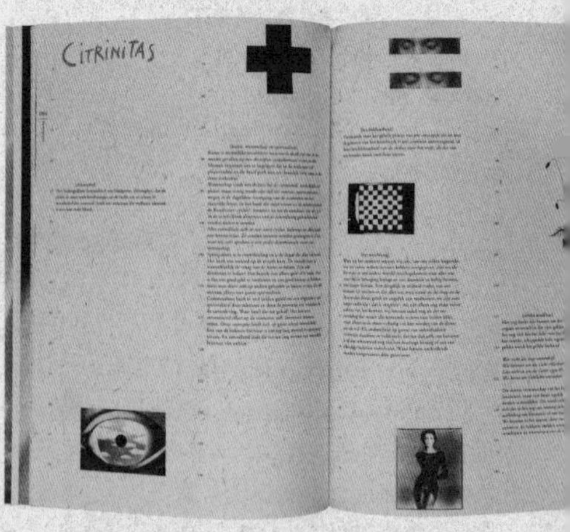
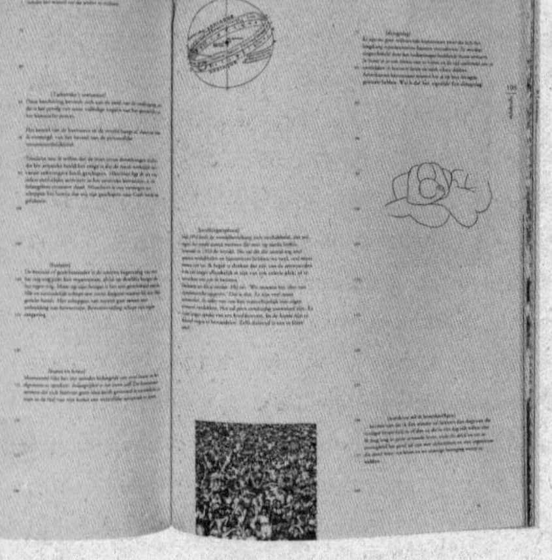
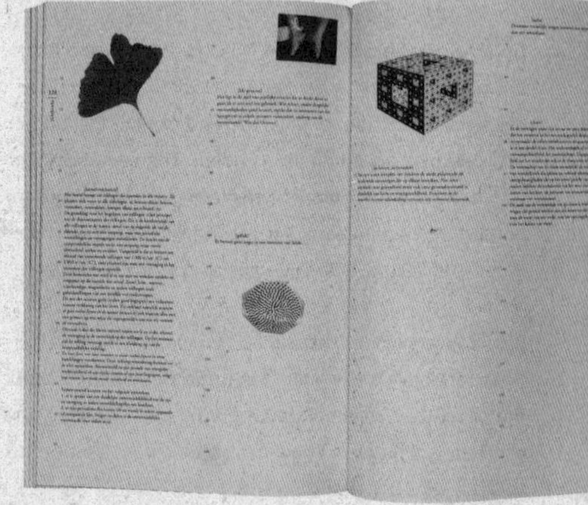

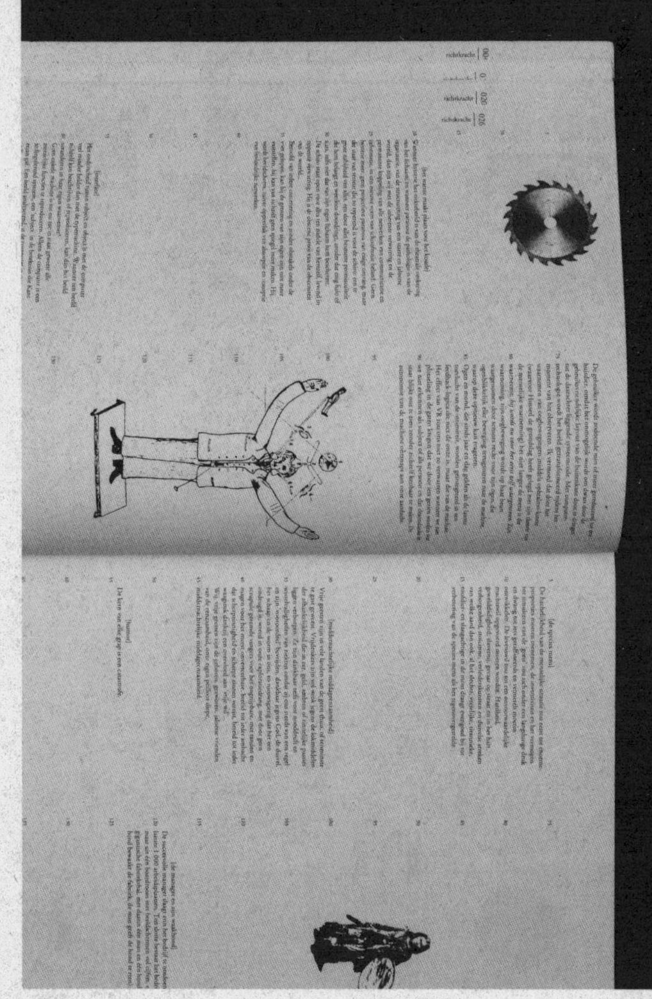

TERRE

Nig

[economie]

Elke koper (rijk of arm) krijgt m
een nieuwe ervaring van nutteloo

056

richtkracht

[snelheid is pure oorl

[oorlog-1]

Een wreedaardig en oorlogsz indianen in Venezuela en Br van verschansingen, het uitd vervaardigen van handwapen krijgsmentaliteit. Maar zij le de eerste ernstig gewonde of lange periode van rouw en v wraakactie wordt uitgevoerd verwachte tegenaanval wordt *ritueel.* Dat wil zeggen: de an overmacht opleggen wordt i definitie niet kan leiden tot gebiedsuitbreiding, maar enk wordt gespaard om hem toe uitdaging en aanval.

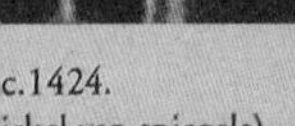

c.1424.
irkel van spiegels).

n (Wired, sept/oct.1993).
ek van Andreas Corsali, 1515).

he voyage d'amour proberen de
elen van de onderzeese liefde te vangen
a Chemica Curiosa, 1677).
acht op een vrouwenhand ter stimulering
9e eeuw).
dige der aarde): de alchemist moet als
van de natuur (Georg Reisch: De

6).
en alchemist tijdens het destillatieproces
h: De Filosofische Parel).
n Welling: Opus Mago-Cabbalisticum
ch door dit werk inspireren tot zijn

049 - Haïti: Tempel-ui
050 - Vuurwerk 01-01-
- Twee eskimo-tov
Tales, 1921).
051 - Anatomie (bron o
052 - Deep Space Netw
- Advertentie Unite
053 - Het Russische pas
mijn terugreis vanu
Southampton.
- Ingres: Het Turks
054 - Richard Menken:
055 - Sterrenhemel in d
056 - Gyroscoop.
- Het binoculaire ge
d'un soleil, 1979).
057 - Richard Menken:
058 - Het lentehuwelijk
- Eclips.
059 - Richard Menken:
daarvoor 17 bronze
060 - De ziggurat-tempe

6,4%

EXCLUSIEF WINSTDELIN

7,1%

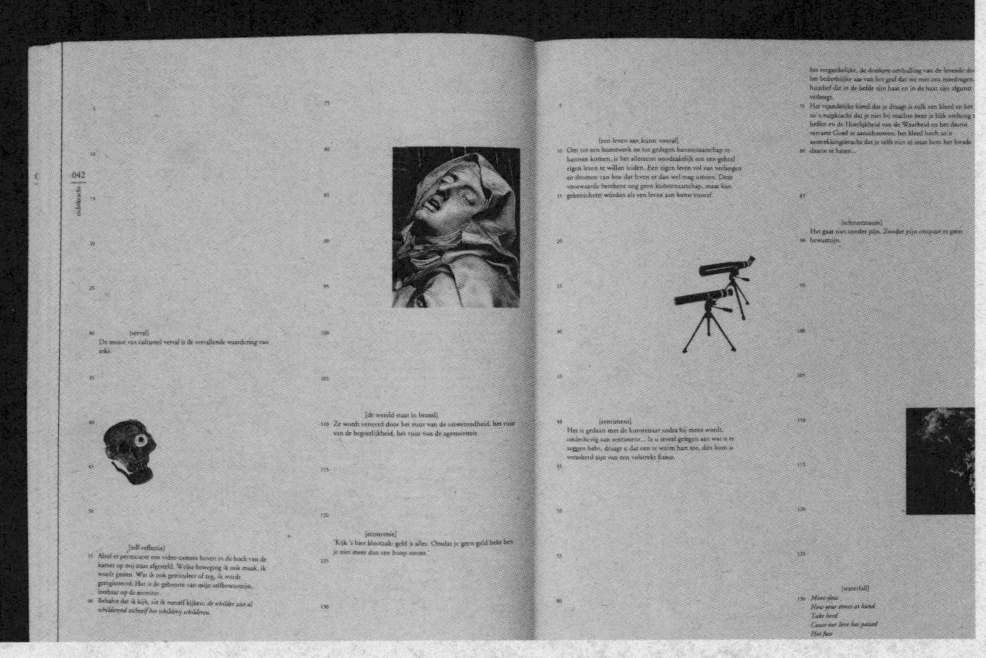

richtkracht | 042

[verval]

De motor van cultureel verval is de vervallende waardering van seks.

[zelf-reflectie]

Alsof er permanent een video-camera boven in de hoek van de kamer op mij staat afgesteld. Welke beweging ik ook maak, ik wordt gezien. Wat ik ook gesticuleer of zeg, ik wordt geregistreerd. Het is de geboorte van mijn zelfbewustzijn, leesbaar op de monitor.
Behalve dat ik kijk, zie ik mezelf kijken: *de schilder ziet al schilderend zichzelf het schilderij schilderen.*

[de wereld staat in brand]

Ze wordt verteerd door het vuur van de onwetendheid, het vuur van de begeerlijkheid, het vuur van de agressiviteit.

[economie]

'Kijk 's hier klootzak: geld is alles. Omdat je geen geld hebt ben je niet meer dan een hoop stront.'

[een leven aan kunst vooraf]

Om tot een kunstwerk en tot gedegen kunstenaarschap te kunnen komen, is het allereerst noodzakelijk om een geheel eigen leven te willen leiden. Een eigen leven vol van verlangen en dromen van hoe dat leven er dan wel mag uitzien. Deze voorwaarde betekent nog geen kunstenaarschap, maar kan gekenschetst worden als een leven aan kunst vooraf.

[sentiment]

Het is gedaan met de kunstenaar zodra hij mens wordt, onderhevig aan sentiment... Is u teveel gelegen aan wat u te zeggen hebt, draagt u dat een te warm hart toe, dan kunt u verzekerd zijn van een volstrekt fiasco.

het vergankelijke, de donkere omhulling van de levende do
het bederfelijke aas van het graf dat we met ons meedragen
huisdief die in de liefde zijn haat en in de haat zijn afgunst
verbergt.
Het vijandelijke kleed dat je draagt is zulk een kleed en het
zo'n zuigkracht dat je niet bij machte bent je blik omhoog t
heffen en de Heerlijkheid van de Waarheid en het daarin vervatte Goed te aanschouwen; het kleed heeft zo'n aantrekkingskracht dat je zelfs niet in staat bent het kwade daarin te haten...

[schmerzraum]

Het gaat niet zonder pijn. Zonder pijn ontstaat er geen bewustzijn.

[waterfall]

Move slow
Now your times at hand
Take heed
Cause our love has passed
Hot fuse

A
Num 54
ad Cap IV
A

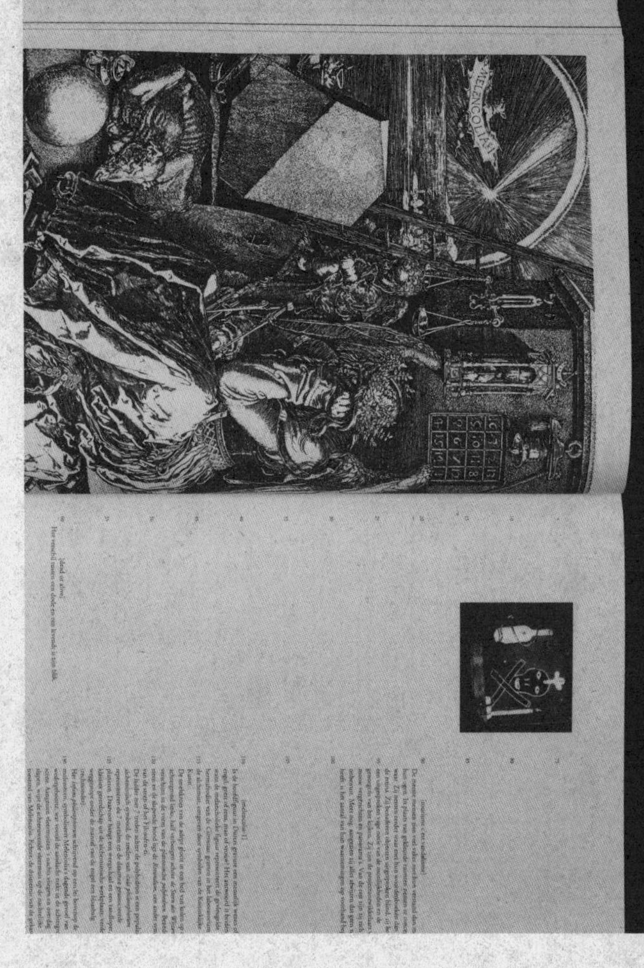

0701

1997 Ha5

MUZEUM
CN22 - CUSTOMS DECLARATION
US POSTAGE
F
TO
ZHAO QING
CHINA
TOTAL
UA 433 773 525 US
PS Form 2976-E March 2008

Poland

U.S.A.

China

293mm

25mm

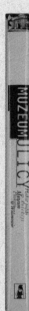

222mm

荣誉奖 Honorary Appreciation

街头博物馆 Muzeum Ulicy: plakat polski

——维拉诺夫海报博物馆中的波兰海报 w kolekcji muzeum plakatu w wilanowie

波兰 Poland

Michał Piekarski

Mariusz Knorowski

Krupski i S-ka

293 222 25mm

1347g

240p

ISBN-10 83-86117-60-5

海报博物馆位于波兰首都华沙维拉诺夫宫建筑群内，属于国家博物馆的一部分。海报博物馆是世界上最古老的以海报收藏为主题的博物馆，创建于 1968 年，收藏的从 1892 年至今的波兰海报超过 50000 张（数据截至成书时间 1996 年），以及部分外国海报。本书的出版基于 1996 年第一届波兰华沙国际海报双年展，并对相关历史进行探讨。书籍的装订形式为锁线硬精装。护封铜版纸四色印刷，覆亮膜。内封红色布纹艺术纸压凹处理。黑色布纹艺术纸作为环衬，内页哑面铜版纸四色印刷，其中以文字为主的前后两部分单黑印刷，作品部分四色印刷。上下书口及右侧翻口刷黄色处理。内容方面，大致分为海报博物馆的历史，波兰海报的历史，著名设计师、评论家、双年展评委和设计机构的文章，以及第一届波兰华沙国际海报双年展的相关介绍。设计方面，版面主要设置为双栏，通过丰富的字体和多种字重、双栏位的上下错位、段首单词斜体放大、灰色层次等方式丰富排版形式。

The poster museum is located in the Wilanów palace in Warsaw, Poland, and is part of the National Museum. The poster museum is the oldest museum in the world with the theme of poster collection. It was founded in 1968. It has a collection of more than 50,000 Polish posters (from 1892 to 1996), as well as some foreign posters. The book was based on the first International Poster Biennale in Warsaw, Poland in 1996, and discussed the relevant history. The hardcover book is bound by thread sewing. The laminated jacket is printed in four colors on coated paper. The inner cover is made of red arlin paper treated with embossing process. Black arlin paper is used for the end paper. The inside pages are made of matte coated paper printed in four colors. The text part is printed in single black while the exhibition part is printed in four colors. The upper and lower book edges and the fore-edge are painted yellow. As for the content, it can be roughly divided into the history of the poster Museum, the history of Polish posters, the articles of famous designers, critics, judges and design institutions of the Biennale, as well as the relevant contents of the first Warsaw International Poster Biennale. As for the design, the layout is mainly set up with double columns, which enriches the layout of the text by means of various fonts and multiple word weights, the misalignment of the upper and lower positions of the double columns, italicized enlargement of the first word of the paragraph, and the layers of using the color gray.

MUZEUM
ULICY
Plakat polski
w kolekcji
Muzeum
Plakatu w Wilanowie

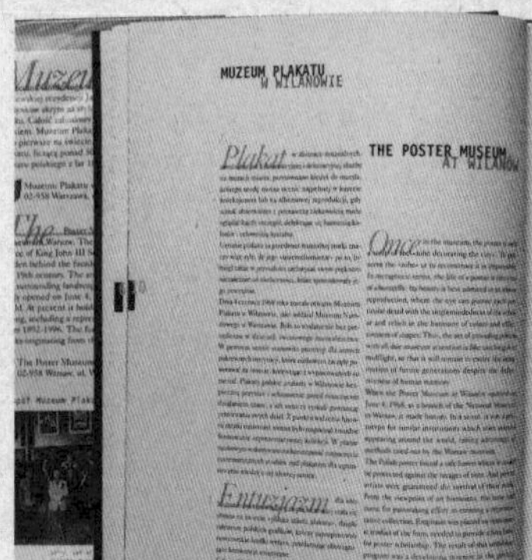
MUZEUM PLAKATU
W WILANOWIE
THE POSTER MUSEUM
AT WILANOW
Plakat
Once
Entuzjazm
Enthusiasm

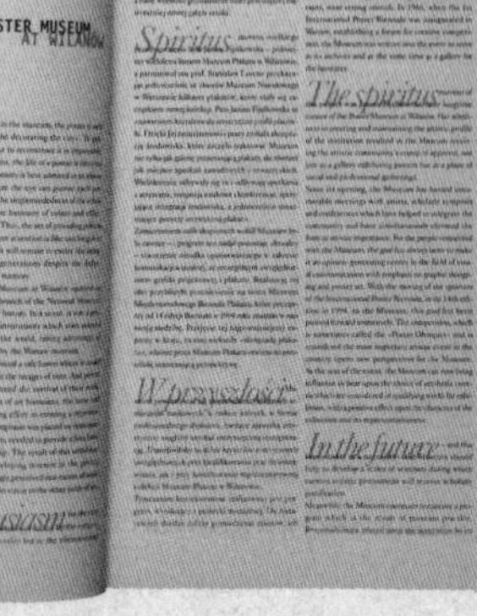
Spiritus
The spiritus
W przyszłości
In the future

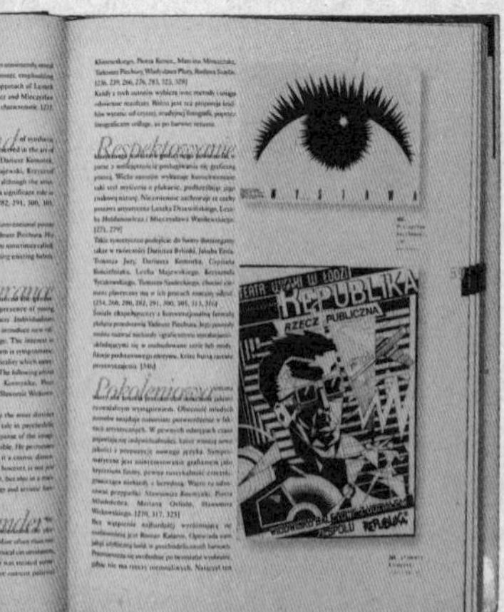
Respektowanie
Pokoleniowy
REPUBLIKA
RZECZ PUBLICZNA

Heinz
salad cream
BLACK & WHITE
SCOTCH WHISKY

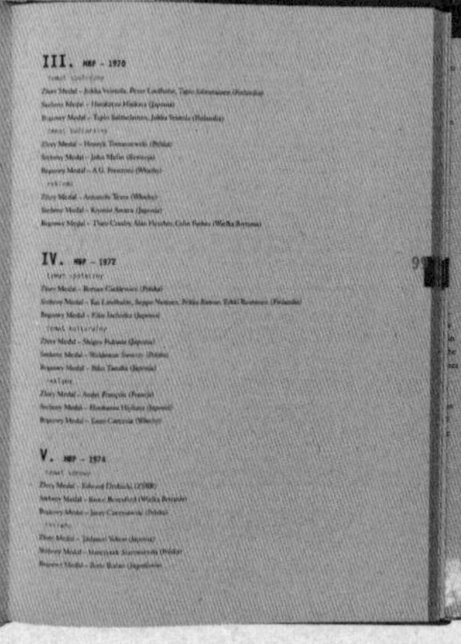
III. MBP – 1970
IV. MBP – 1972
V. MBP – 1974

Sztuka
PIWO BROZDOWSKIE
Figliki
RZECZY WESOŁE
I NIEPOŻYTECZNE

ZEUM PLAKATU W WILANOWIE — *Krystyna Spiegel* 10
POSTER MUSEUM AT WILANOW

PLAKAT POLSKI — *Mariusz Knorowski* 16
POLISH POSTER

IEDŹ ARTYSTY — *Jan Lenica* 64
BY AN ARTIST

POLSKO-TOMASZEWSKA SZKOŁA PLAKATU — *Stanisław*
POLISH TOMASZEWSKI SCHOOL OF THE POSTER

WYPOWIEDŹ ZNAWCY PLAKATU — *Krzysztof Zagr*
ORD BY AN HISTORIAN OF THE POSTER

IEDŹ LAUREATA I JURORA BIENNALE PLAKATU — *Shigeo Fu*
RD BY A LAUREATE AND MEMBER OF THE JURY

KONSERWACJA PLAKATU — *Maria Kurpik* 90
CONSERVATION OF POSTERS

LAUREACI MIĘDZYNARODOWEGO BIENNALE PLAKATU W WARSZAWIE — opracowała
LAUREATES OF THE MEDALS
E INTERNATIONAL POSTER BIENNALE IN WARSAW

iecień - maj 1984
skusstwo Narodnoj Polszy 1944–1964», Moskwa,

j - lipiec 1984
ávoli Utatok Uzente», Budapeszt, Magyar Nemze

erwiec - wrzesień 1984
ransformacje 1944–1984», MPW

zesień - grudzień 1984
ureaci VIII Międzynarodowego Biennale Plakatu
rapus» (Francja), Jan Młodożeniec (Polska), Günte
chael van de Sand (Niemcy), MPW

yczeń - marzec 1985
amięci Tadeusza Trepkowskiego», wystawa zorgan
rocznicy śmierci artysty, MPW

ecień - czerwiec 1985

MU
U

Pl

Plakatu
Vilanowie
ał
rodowego
w Wars

», a jego
y, nieza-
a impre-

zlecenie
niem dla
ów tury-
asta i re-
folklory-

ość Gro-
Cassan-
e do sty-
Ludwiga
ka posta-
sonanse

Washes by Itself) was Grono
The poster was made for the S
appearance in the streets broug
For many adepts of graphic de
quintessence of the «true» po
their own artistic developme
motion film and techniques
indisputably played a role in t
idea. From the moment this
possible to speak of a partne
European art.

The first

Polish designing abroad came
Decorative Arts Exhibition h
The Grand Prix in the Art Pul
went to Tadeusz Gronowski a

FERNER OSTEN
BALTIKUM
U. S. S. R.
BILLIG - SCHNELL - BEQUEM
NUR DURCH WARSCHAU
SCHNELLZÜGE - NORD-EXPRESS

Hiszpania 36-1939

1998

Country / Region

The Netherlands ❷

U.S.A. ❶

Germany ❼

Sweden ❶

U.S.A. ❶

Hungary ❶

Japan ❶

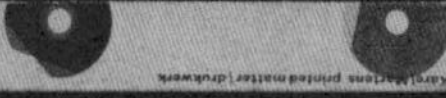

SOUL OF THE GAME John Huet

brecht FLÜCHTLINGSGESPRÄCHE

Reclam Leipzig Keller+Kuhn Die blauen Wunder

HC Ericson Carlssons

BILL VIOLA

HANSER FRIED / GLEICH · HAT OPA EINEN ANZUG AN ?

BUDAPESTI GALÉRIÁK · 1996

1998 Gl *0716*

1998 Gm *0726*

1998 Sm^{2} *0736*

1998 Bm^{1} *0744*

1998 Bm^{2} *0752*

1998 Bm^{4} *0762*

1998 Bm^{5} *0770*

1998 Ha^{1} *0780*

1998 Ha^{2} *0790*

1998 Ha^{5} *0798*

Designer

Jaap van Triest, Karel Martens

John C. Jay

Anja Wesner

Gert Wunderlich

Matthias Gubig

Sabine Golde, Tom Gebhardt

José van't Klooster

HC Ericson

Rebeca Méndez

Matthias Beyrow, Marion Wagner

Claus Seitz

Anja Osterwalder

Johanna Bárd

Tsutomu Nishioka

FROM: 1 OF 1

POWELLS.COM
2720 NW 29TH AVENUE
PORTLAND, OR 97210-1702 UNITED STATES

TO:

ZHAO QING
Dabei xiang7-3
Nanjing Han Qing Tang Design Co.,
NANJING, 210000, CHINA

SHIP DATE: 12/17/2019
WEIGHT: 3 lbs
SHIPPER REF: 3603357-1
CONSIGNEE REF: 100013690290
DESCRIPTION: Book/Media

Asendia Priority Airmail

Tracking #: UM699542920US

ZHAO QING
Dabei xiang7-3
,Nanjing Han Qing Tang Design Co.,
NANJING,, 210000, CHINA

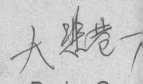

DATE: 12/17/2019

GHT: 3 lbs.

PER REF: 3603357-1

SIGNEE REF: 100013590290

CRIPTION: Book/Media

sendia Priority Airm

racking #: UM699542920US

The Netherlands

U.S.A.

China

0717

金字符奖
卡雷尔·马滕斯的印刷作品
荷兰

Golden Letter
Karel Martens printed matter / drukwerk
The Netherlands
Jaap van Triest, Karel Martens
Karel Martens
Hyphen Press, London
233 173 16mm
527g
144p
ISBN-10 0-907259-11-1

1996年，荷兰皇家艺术与科学研究院将旨在表彰荷兰最杰出的视觉艺术家的喜力艺术奖（Heineken Prize for Art）颁发给了设计师卡雷尔·马滕斯（Karel Martens）。本书是为此出版的马滕斯从业至今几十年来的作品集。书籍的装订形式为包背无线胶平装。封面牛皮纸四色印刷，上下书口与翻口均向内包折，形成较厚实的封面手感。内页胶版纸孔版多色印刷，尤其是作品部分大量使用特种油墨，用以体现原作丰富的印刷效果。内容方面以作品为主，之后附加马滕斯从事设计的诸多感悟和在教育行业里的贡献。文字部分，版心到书页边的边距极窄，双栏设置，并且将页面分为上下两半，分别呈现英文和荷兰文。页码被放置在书的右页，且两个数字分别指向本页和包背装的折叠页（后一页）。而左页上角则标注了作品的年份，与页码一同形成双索引体系。书中大量使用了多种特殊油墨，除了更好地呈现马滕斯的作品原貌，也是呼应他在印刷领域做出的先锋贡献。

In 1996, the Royal Netherlands Academy of Arts and Sciences awarded the designer Karel Martens the Heineken Prize for Art, which commends the most outstanding visual artist in the Netherlands. This book is a collection of works by Martens for decades. The double-leaved book is perfect bound. The cover is made of kraft paper printed in four colors. The upper and lower edges of the jacket are slightly larger than the book format, which can be folded inwards to form a thick and comfortable feel. The inside pages are made of offset paper printed in multi colors, especially for the part of the works, special ink is used to reflect the rich printing effects of the original. The content is mainly based on the works, and added with Martens' insights on design and contribution in education and industry. The margin from the typeset area to the book edge is extremely narrow, and the layout is set to double columns. The pages are divided into upper and lower halves, for English and Dutch respectively. The page numbers are placed on the right page of the book, and the two numbers point to this page and the folding page (the next page). The upper corner of the left page is marked with the year of the work, which forms a double index system together with the page number. Special inks are used in the book to present the original appearance of Martens' works better and to echo his pioneering contributions in the field of printing.

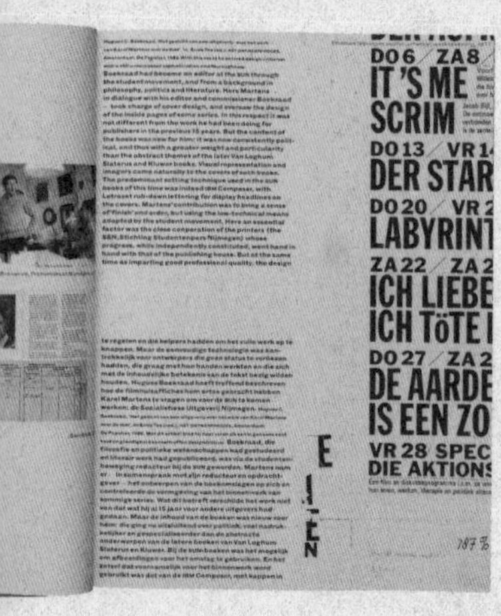
DO 6 / ZA 8
IT'S ME
SCRIM
DO 13 / VR 14
DER STAR
DO 20 / VR 2
LABYRINT
ZA 22 / ZA 2
ICH LIEBE
ICH TöTE
DO 27 / ZA 2
DE AARDE
IS EEN ZO
VR 28 SPEC
DIE AKTIONS

DE ESTETIESE TEORIE
VAN ADORNO EN BENJAMIN
EEN INLEIDING
MANFREDO TAFURI
ONTWERP EN UTOPIE
KUNSTGESCHIEDENIS
EN IDEOLOGIE
NICOS HADJINICOLAOU
ALETTA JACOBS
HERINNERINGEN

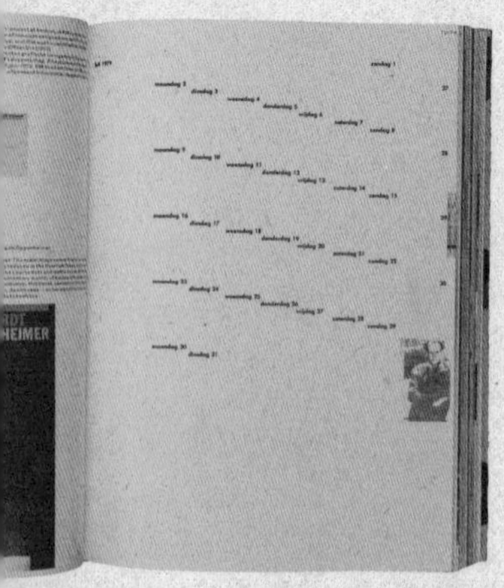

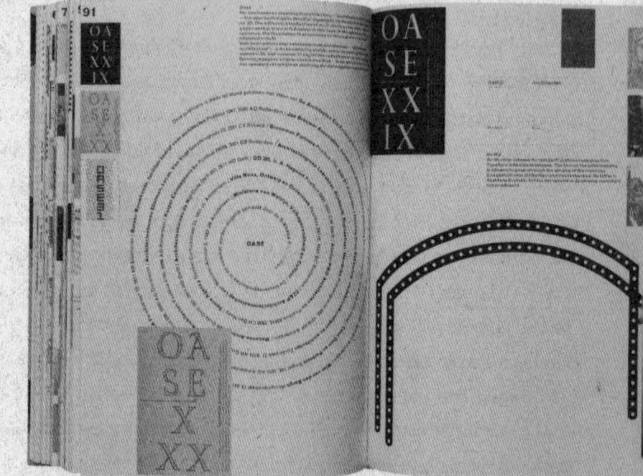
OA
SE
XX
IX
OASE
OA
SE
X
XX

85

Nederlandse postzegels 1982
PTT Post|PTT Kunst & Vormgeving
The opening essay by Arie van den Berg on 'De taal van de postzegel' (the language of the postage stamp) is illustrated with images of printed ephemera from around 1900, arranged (apparently) quite freely, and printed with great fidelity. In fact these reproductions own paste-up (made for visualization and instruction), rather than grid-true made-up film, which had to be changed: it lacked life
Het openingsessay door Arie van den Berg over 'De taal van de postzegel' wordt geïllustreerd met beelden van efemeer drukwerk van rond 1900, zo op het oog losjes geordend en met zorg gedrukt. In feite beantwoorden die meer aan de plakproef van de ontwerper (gemaakt als schets en instructie), dan de maatvaste filmmontage—die was veranderd: er zat geen leven in

Harry Stroeken: Freud en zijn patiënten
Te Elfder Ure
On the cover, the photograph of Freud is enlarged so that his ears—essential instruments of psychoanalysis—occupy strategic positions. On the front flap, the left ear attends on the blurb. The right ear appears on the edge of the rear cover
Op de omslag is het portret van Freud zodanig uitvergroot dat zijn oren—onmisbare instrumenten van de psychoanalyse—strategisch geplaatst zijn: binnen luistert het linker naar de flaptekst, het rechter verschijnt juist op het achterplat

Papier (design: ontwerp Wigger Bierma)
Nijmeegs Museum Commanderie van St Jan
Booklet accompanying an exhibition of KM's paper work. Typography plays with show-through paper—this also allows for some freestanding pieces, drawings that complete themselves on the other side of paper
Kleine catalogus bij een tentoonstelling van vrij werk. De typografie speelt met het doorschijnende papier, dat ook nieuw werk toelaat: tekeningen die zichzelf aanvullen op de achterzijde van de pagina

Rose en wit met zwarte banen (1984)

In de hiërarchie van de materialen heeft papier een lage status gekregen: gemaakt uit vodden, uit lompen, afval. Weinig in onze cultuur heeft minder waarde, scheidbaar wordt het gebruikt.

Uit afval gemaakt, is het zelf een bron van afval geworden in een tijd waarin meningen en massaprodukten in papier verpakt worden. Aan papier kleeft iets abjects.

"Alles wat vuil is," zegt Freud in *Het onbehagen in de cultuur*, "achten wij met cultuur onverenigbaar. Wij maken ons boos en noemen het "barbaars"—wat het tegendeel is van cultureel—als we zien dat de paden van het Wiener Wald bezaaid liggen met weggegooid papier..."

In een boek met herinneringen aan Sturm en Bauhaus wordt Kurt Schwitters de voddenraper, 'Der Lumpensammler' genoemd. Hij gebruikte wat anderen hadden weggegooid en verworpen. Men zou dit een activiteit kunnen noemen die Freud wezenlijk achtte voor de artistieke produktie: sublimatie.

Deze voelbaarheid te vergroten is een effect van KM's bewerking: men voelt de aanraakbare, aaibare huid van papier. Een luchtvlaag laat zijn papierreliefs wuiven en wekt de herinnering aan een rietbos.

Behalve tactiele heeft papier ook optische kwaliteiten door zijn inherente structuur. Het is nooit zonder kleur. En het is altijd, hoe miniem ook, getekend.

Deze optische kwaliteiten van het papier wordt men gewaar, als men een papierreliëf van KM passeert: gaandeweg verandert het van uitdrukking. Hoe het eruit ziet, hangt niet alleen af van het uur van de dag, van het licht dat er op valt, maar ook van de hoek waaronder men het beschouwt. Het is blauw of rood; afhankelijk van de

ion through which a common life becomes
which a relationship between people
sign determines the quality of our common
y important to stress this.
en knowledge and insight are missing
field of expression, a collaboration is
eople who explicitly concern themselves
given material: 'form givers'. I prefer to
rs, because an essential process needs
he form. Design should be able to be seen
into a desired social structure.
s act as intermediaries. They maintain
with the graphic industry, for which some
dge is necessary. There is always talk of a
he job) and of the one who is to be informed
ween them stands the designer with a
nd knowledge of things.
o happen, there has to be a dialogue,
ect, between the client and the designer.
considerations
distilled out of t
through the cor
of production o
matter, becaus
In good design,
certain tension
and the qualitie
has to come ove
melody — is imp
communicative
to us.
My thoughts ab
time when I trai
good of functio
which the appli
time that prece
purification. Ra
voor mij betekent
ontwerper beschouw ik mijzelf als vorm-
drukte woord, waarmee nog niet gezegd is
oken woord bij mij in goede handen is.
ft de taken verdeeld. In de loop van de
daaruit een groot aantal bezigheden
t een ieder vanuit zijn belangstelling,
ogelijkheden een keuze kan maken.
gheden drukt zich uit in vorm. In die zin zijn
ers en makers van de gedaante waarin de
ns voordoen.
Om tot een goe
respect, een dia
gever en ontwe
vertrouwen, en
moeten hebben
boodschap moe
voor degene aa
juiste keuze te
leven, kiezen.
Kiezen uit een v
aan de hand va
voorstelling va
er voldoende m
geraken. Ieder
ICH TöT
DO 27 / ZA
DE AAR
IS EEN Z

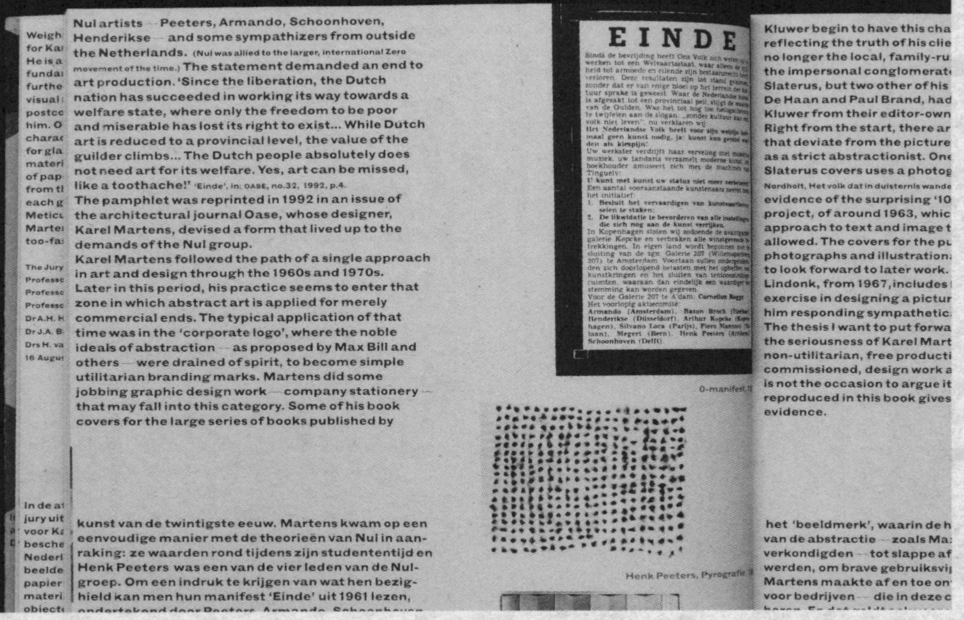

Nul artists — Peeters, Armando, Schoonhoven, Henderikse — and some sympathizers from outside the Netherlands. (Nul was allied to the larger, international Zero movement of the time.) The statement demanded an end to art production. 'Since the liberation, the Dutch nation has succeeded in working its way towards a welfare state, where only the freedom to be poor and miserable has lost its right to exist... While Dutch art is reduced to a provincial level, the value of the guilder climbs... The Dutch people absolutely does not need art for its welfare. Yes, art can be missed, like a toothache!' 'Einde', in: OASE, no.32, 1992, p.4.

The pamphlet was reprinted in 1992 in an issue of the architectural journal Oase, whose designer, Karel Martens, devised a form that lived up to the demands of the Nul group.

Karel Martens followed the path of a single approach in art and design through the 1960s and 1970s. Later in this period, his practice seems to enter that zone in which abstract art is applied for merely commercial ends. The typical application of that time was in the 'corporate logo', where the noble ideals of abstraction — as proposed by Max Bill and others — were drained of spirit, to become simple utilitarian branding marks. Martens did some jobbing graphic design work — company stationery — that may fall into this category. Some of his book covers for the large series of books published by

kunst van de twintigste eeuw. Martens kwam op een eenvoudige manier met de theorieën van Nul in aanraking: ze waarden rond tijdens zijn studententijd en Henk Peeters was een van de vier leden van de Nul-groep. Om een indruk te krijgen van wat hen bezighield kan men hun manifest 'Einde' uit 1961 lezen,

EINDE

0-manifest

Henk Peeters, Pyrografie

Kluwer begin to have this cha reflecting the truth of his clie no longer the local, family-ru the impersonal conglomerat Slaterus, but two other of his De Haan and Paul Brand, had Kluwer from their editor-own Right from the start, there ar that deviate from the picture as a strict abstractionist. One Slaterus covers uses a photog

Nordholt, Het volk dat in duisternis wande

evidence of the surprising '10 project, of around 1963, whic approach to text and image t allowed. The covers for the pu photographs and illustration to look forward to later work. Lindonk, from 1967, includes exercise in designing a pictur him responding sympathetic The thesis I want to put forwa the seriousness of Karel Mart non-utilitarian, free producti commissioned, design work a is not the occasion to argue it reproduced in this book gives evidence.

het 'beeldmerk', waarin de h van de abstractie — zoals Ma verkondigden — tot slappe af werden, om brave gebruiksvi Martens maakte af en toe on voor bedrijven — die in deze c

62

Leaflets for Aberson in German and Dutch versions. The company's product was in itself geometrical, and KM's mark responds nicely to that. Still within the conventions of the contemporary style
Brochures voor Aberson in Duits en Nederlands. Het produkt van het bedrijf, in principe geometrisch van vorm, komt goed uit in het beeldmerk — binnen de beperkingen van de toen gangbare stijl

Giuseppe Tomasi di Lampedusa: De senator en de sirene
Van Loghum Slaterus
Unusually 'literary' in approach: in choice of typeface, colour, paper and in the pattern-making with the S forms. Such works of pure literature were unusual in the VLS list
Ongewoon 'literair' in benadering in letterkeuze, kleur, papier en het patroon van S-vormen. Puur literair, een uitzondering binnen het fonds

Phillipp Dessauer: Meditatie
Van Loghum Slaterus
Patterns formed from the interaction and overlapping of grids: abstract but with some loose possible associations to the subjects of the books. 'Meditatie' is a dummy (the book was never published) with the intersecting grid on a foil overlay. The same pattern would then overlap at different angles for each title, to produce something greater than the minimal means employed
Rasterpatronen die elkaar gedeeltelijk overlappen: geheel abstract, maar met een mogelijke verwijzing naar de inhoud van de boeken. Meditatie is een dummy (het boek is nooit verschenen) met het rasterpatroon op een overlay van folie gedrukt. De rasters zouden iedere volgende titel onder een andere hoek over elkaar heen geschoven moeten worden, tot een geheel dat het minimale uitgangspunt overstijgt

1998 Gm

PAR AVION
INTERNATIONAL
PRIORITY AIRMAIL
INTL PRIORITY AIRMAIL
U.S. POSTAGE
PAID
JAMAICA, NY 11430
USPS
Pri
China
Group 14

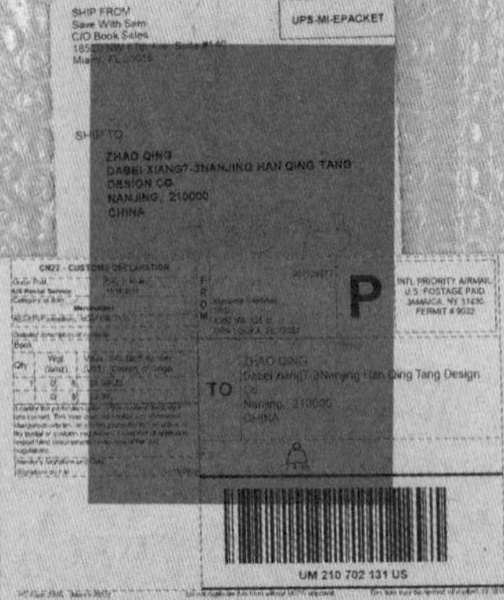
SHIP FROM
Save With Sam
C/O Book Sales
UPS-MI-EPACKET
SHIP TO
ZHAO QING
DABEI XIANG7-3NANJING HAN QING TANG
DESIGN CO.
NANJING, 210000
CHINA
CN22 - CUSTOMS DECLARATION
P
INTL PRIORITY AIRMAIL
U.S. POSTAGE PAID
TO
UM 210 702 131 US

20 NW 87th Ave, Suite
mi, FL 33015

IIP TO

ZHAO QING
DABEI XIANG7-3NANJING HAN QING TANG
DESIGN CO.
NANJING, 210000
CHINA

大悲巷7-3

TOMS DECLARATION

Date of Mailing
11/18/2019

chandise
EEI 30.37(a)

ontents

alue (US$)	HS Tariff Number; Country of Origin
$9.99	US
$9.99	

given in this customs declaration
oes not contain any undeclared
rticles prohibited by Legislation or
gulations. I have met all applicable
ts under federal law and

Date:
11/18/2019

FROM

901129977

Marisela Sardinas
SBS
4360 NW 135 st
OPA LOCKA, FL 33054

P

TO

ZHAO QING
Dabei xiang7-3Nanjing Ha
Co.
Nanjing, 210000
CHINA

2.4 lbs

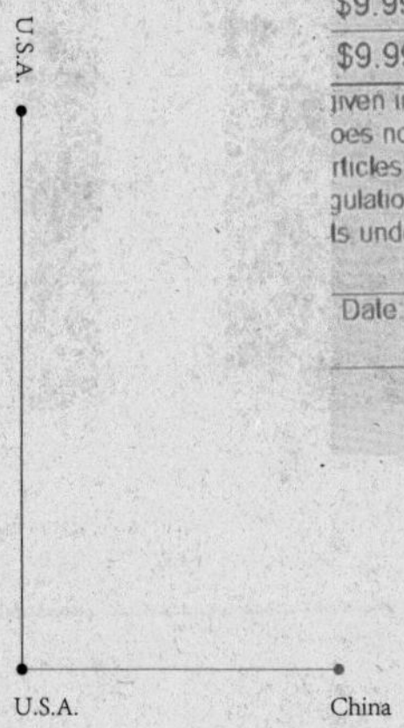

U.S.A.

U.S.A.

China

324mm

19mm

SOUL
Photography by John Huet
Images & Voices of Street Basketball
Introduction by Pee Wee Kirkland
of the Game

235mm

● 金奖
《 运动之魂
◖ 美国

Gold Medal
Soul of the Game
U.S.A.
John C. Jay
Jimmy Smith, John Huet
Melcher Media; Workman Publishing, New York
324 235 19mm
1070g
144p
ISBN-10 0-7611-1028-3

以 NBA 为代表的职业篮球之外，街头篮球是篮球这项运动非常重要的组成部分。有大量的非职业球员在城市中的赛场上挥洒汗水，展现着他们对于篮球的理解和运动的魅力。本书是体育摄影师约翰 · 休特（John Huet）为纽约、洛杉矶、费城、底特律、华盛顿等地的著名街球手拍摄的运动影集，并且配有对运动员的采访、个人说唱，以及音乐家吉米 · 史密斯（Jimmy Smith）为此写作的诗歌。书籍的装订形式为锁线硬精装。封面铜版纸黑棕双色印刷，书名印金。内文胶版纸黑棕双色印刷。通过在黑色中加入棕色，使得照片可以更加细腻地表现黑人街球手的皮肤质感，也为照片增添了一丝经典和怀旧的特色。由于文体较多，包括人物介绍、访谈、说唱、诗歌，书中文本根据文体特征和版面需要处理得较为灵活，带有阅读节奏的控制，仿佛街球手打球时的呼吸感。

In addition to professional basketball represented by the NBA, street basketball is a very important part of basketball. A large number of non-professional players are sweating on the arena in the city, showing their understanding of basketball and the charm of sports. This book is a sports photo album by sports photographer John Huet for famous street players in New York, Los Angeles, Philadelphia, Detroit, Washington, etc.. The book also includes interviews with athletes, personal rap, and poems written by the musician Jimmy Smith. The hardcover book is bound by thread sewing. The cover is made of coated paper printed in black and brown, and the title is printed in gold. The inside pages are printed on offset paper in black and brown. By adding brown to black, the photos can express the skin texture of black athletes more delicately, and also become more classic and nostalgic. Due to the variety of literary styles, such as character introduction, interviews, rap, and poetry, the text in the book is flexible in typesetting according to the characteristics of different styles. The layout itself can control the reading rhythm and share the street players' breath with the readers.

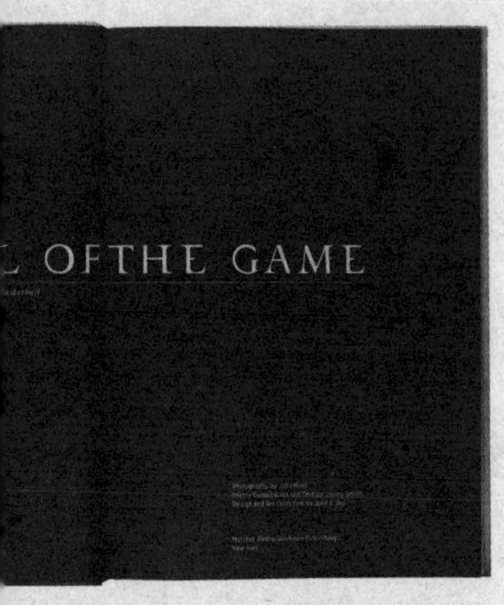

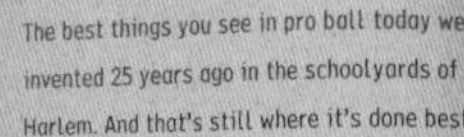

The best things you see in pro ball today were invented 25 years ago in the schoolyards of Harlem. And that's still where it's done best.

All the best moves of tomorrow are being invented today by some unknown kid in Philly, Chicago, Atlanta or you name it. Razzle and dazzle, creation and devastation—these are what the street brings to the game. When a player does a sandlot move in a pro game situation, you know he's for real, not the other way around. It's on the street that players earn their places in the hearts of the people. Where Earl Monroe became Black Jesus. Where Joe Hammond became the Destroyer.

I've seen the game played on playgrounds in ways I'll never forget. In ways that took my mind somewhere it had never been before. On the playground, legends are born, and you'll never forget what you see a legend do because you will never see anybody repeat it in quite the same way.

A finger roll by Connie Hawkins after he floats in from the foul line?

How about Jackie Jackson, New York's greatest leaper? This man had incredible springs. His hang time was unbelievable—don't check your watch, check your beeper. Imagine yourself trying to score points when you're laying up against a human astronaut. How're you not going to get your shot blocked? I remember first seeing him play at Street and Seventh Avenue. I was a high school freshman, wa through the fence. He was playing on his Brooklyn team with C Hawkins against Wilt Chamberlain, the 7-foot-1 Big Dipper, many consider the best big man ever to play the game. Wilt was shooting hooks that day. And Jackie would go up with him and block them. Not just block them, but catch them in midair. Jackie cuffed the ball under his arm and took it away. The impact and intensity from the crowd—you wouldn't believe it.

After the game was over, the crowd wouldn't leave. People started shouting, "Take it off the top, Jackie! Take it off the top!" I had no idea what they were talking about. Somebody brought out a ladder and put a 50-cen on top of the backboard. And Jackie jumped into the air and t coin down. It shocked me. Here was a man who would spend 1 with the Harlem Globetrotters and never play a minute in the I was amazed.

I guess that's what being a legend is: creating an indelible, l impression in the minds of the crowd, like Jackie did to me. Ex

INDIAN SQUAD

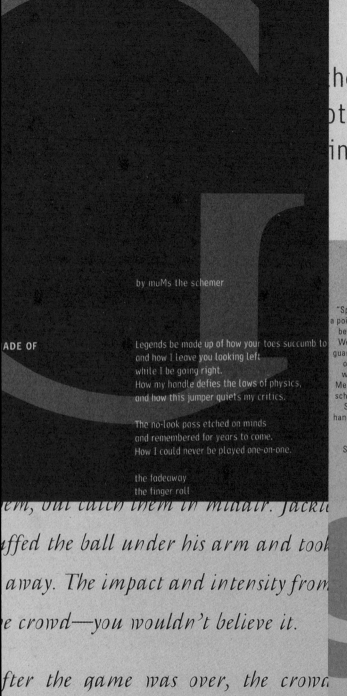

Evans became a coach, mento
making many futile attempts
toward a pro basketball care

he court. He co
ot, he could jum
in the NBA."

Joe was a cult figure around the
had heard about him. Some had s
the NBA had no records of him. H

by muMs the schemer

ADE OF

Legends be made up of how your toes succumb to
and how I leave you looking left
while I be going right.
How my handle defies the laws of physics,
and how this jumper quiets my critics.

The no-look pass etched on minds
and remembered for years to come.
How I could never be played one-on-one.

the fadeaway
the finger roll

James Williams, aka Speedy, point guard, NYC.
"Speedy is a product of NBA politics, a point guard with heart, speed and far better than average skills," says Pee Wee Kirkland. "I could name 10 NBA guards whose lunch Speedy could eat on any given day. However, Speedy went to a small college in Brooklyn, Medgar Evers. Many players from big schools get the break, while guys like Speedy who deserve better get the handshake. But Speedy won't give up. He can't. The game is a part of his soul. Locked out of the big dream, Speedy sees playground basketball as his only way to be recognized, remembered or appreciated."

em, but catch them in midair. Jacki
uffed the ball under his arm and took
away. The impact and intensity from
e crowd—you wouldn't believe it.

fter the game was over, the crowd
ouldn't leave. People started shouting,
Take it off the top, Jackie! Take it of

The Goat

On 99th Street and Amsterdam

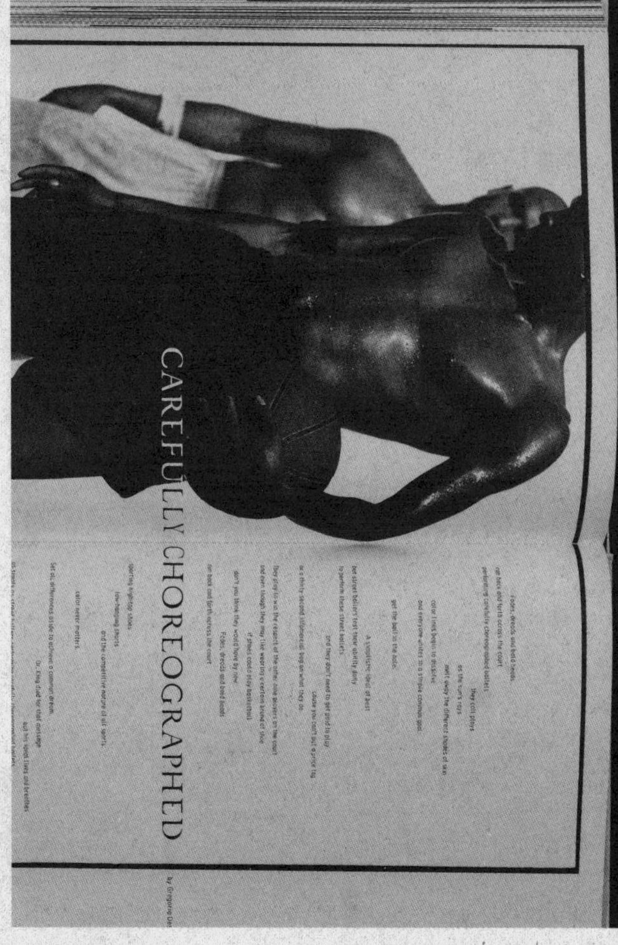
CAREFULLY CHOREOGRAPHED

THE SPOT

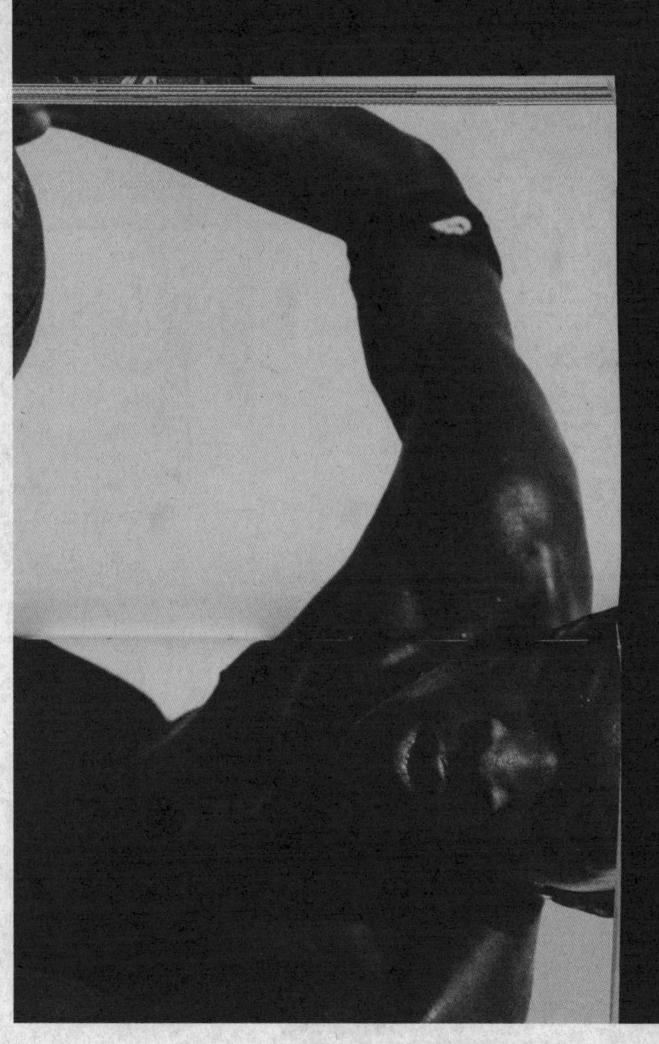

SUPRAWELL 4

CC 07621960 4 NL
Germany
The Netherlands
China

219mm

138mm

● 银奖
《 难民的对话
◖ 德国

Silver Medal
Flüchtlingsgespräche
Germany
Gert Wunderlich
Bertolt Brecht
Leipziger Bibliophilen-Abend e.V., Leipzig
219 138 26mm
584g
152p

1933年纳粹党上台后，德国戏剧家、诗人贝托尔特·布莱希特（Bertolt Brecht）被迫开始了流亡生活。本书是他写作于20世纪40年代的剧本，通过对话体的形式，讲述了多个难民的日常生活。书籍的装订形式为包背胶平装。护封黄色纸张印单黑，护封的勒口设计了多层折叠，将剧本中的对话片段、部分插图包含其中。内封黑卡纸正面裱贴灰色卡板，书脊处印单黑。内页鹅绒纸印单黑，13张黄色纸张夹插其中，印制了由捷克艺术家汉斯·季哈（Hans Ticha）创作的版画。剧本基本由两个难民——知识分子齐费尔（Ziffel）和工人卡勒（Kalle）在赫尔辛基中央车站餐厅里的对话组成，话题涉及第二次世界大战的背景、双方的难民处境、阶级差异带来的观点冲突、身份变化后的一致目标等。全书由铅版印刷机印制而成，正文只采用两种字号与两种字重。其中章节标题、姓氏首字母和重要信息使用特粗字重，对话外的旁白使用小字号。粗重的字体带来了沉重的“烙印”感，一如作者的亲身感受。

After the Nazi Party came to power in 1933, the German dramatist and poet Bertolt Brecht was forced to start a life in exile. This book is a script written by Brecht in the 1940s. It tells the daily life of many refugees in the form of dialogue. The double-leaved book is glue bound. The jacket is made of yellow paper printed in single black, and the flap is designed with multiple folds, including dialogues and some illustrations. The inside cover is made of black cardboard mounted with gray cardboard, and the spine is printed in single black. The inner pages are printed on velvet paper in single black, with 13 pieces of yellow paper inserted into them, which are printed with woodcuts created by Czech artist Hans Ticha. The script is basically composed of dialogue between two refugees, the intellectual Ziffel and the worker Kalle, in the restaurant of Helsinki Central Station. The topics involve the background of the Second World War, the refugee situation of both sides, the conflict of views caused by class differences, and the common goal after the change of status. The whole book is printed by a stereotyped press. The main text of the book only uses two font sizes and two word weights: the chapter titles, the first letter of surname and important information use extra thick font weight, and the narration text uses small font size. Heavy font brings hard feeling of branding, just like the author's personal experience.

ÜBER NIEDRIGEN MATERIALISMUS

ÜBER DIE FREIDEN

ZIFFEL SCHREIBT MEMOIREN

ÜBER DAS ÜBERHAN NEHMEN BEDEUTEN MENSCHEN

c h

BERTOLT

Ziffel und Kalle waren sehr überrascht, als sie sic im Bahnhofsrestaurant wieder trafen. Kalle war u seinen dicken Mantel nicht mehr an, den er das Sommerwetters getragen hatte.

Ziffel Ich habe ein Zimmer gefunden. Ich bi meine 180 Pfund Fleisch und Knochen v Kleinigkeit, einen solchen Haufen Fleis zu bringen. Und die Verantwortung ist schlimmer, wenn 180 Pfund verderben

Meinungs F R

Bei uns

jede Überzeugun

die Sie wü

indem das überall stimmt.

können Ihre Überzeugung nicht.

bar. Wenns in der Schweiz was gegen

nus sagen, was mehr ist als nur,

BERTOLT B FLÜCHTLIN

MIT 12 HOL VON HANS

LEIPZIGER ABEND 1997

Kalle Wenn Sie sich mit den Saurieren vergleichen, wärs höchste Zeit, daß Sie noch Ihre Memoiren aufschrieben, denn nicht lang, und keiner möcht sie nicht mehr verstehen.

Ziffel Der Übergang findet reißend schnell statt. Die Wissenschaft nimmt heute an, daß der Übergang eines Zeitalters in ein anderes ruckartig, Sie können auch schlagartig sagen, stattfindet. Lange Zeit hindurch gibt es winzige Veränderungen, Unstimmigkeiten und Verunstaltungen, welche den Umschlag vorbereiten. Aber der Umschlag selber tritt mit dramatischer Plötzlichkeit ein. Die Saurier bewegen sich sozusagen noch eine geraume Zeit in der besten Gesellschaft, wenn sie auch schon etwas ins Hintertreffen geraten sind. Es steht nichts mehr hinter ihnen, aber sie werden noch gegrüßt. Im Adelskalender der Tierwelt nehmen sie schon ihres Alters wegen noch einen geachteten Platz ein. Es gilt noch durchaus als gute Kinderstube, Gras zu fressen, wenngleich die besseren Tiere schon Fleisch bevorzugen. Es ist noch keine Schande, 20 Meter von Kopf bis zum Schwanz zu messen, wenn es auch schon kein Verdienst mehr darstellt. Das geht soundso lang und dann kommt plötzlich der totale Umschwung. Wenn Sie nicht sehr große Einwendungen haben, möchte ich Sie bitten, ab und zu das eine oder andere Kapitel

3

ÜBER DEN UNMENSCHE

GERINGE FORDERUNGE
SCHULE

HERRNREITTER

Ziffel ging beinahe täglich in das Bahnhofsrestaurant, den
Lokal war ein kleiner Stand für Tabakwaren und zu unreg
Zeitpunkten erschien ein Mädchen, die, ein paar Tüten ur
und dann zehn Minuten lang Zigarren und Zigaretten ver
schon ein Kapitel von seinen Memoiren in der Brusttasche
Als er eine Woche lang nicht kam, dachte Ziffel schon, er
umsonst geschrieben, und stoppte alle weitere Arbeit. E
niemand in H., der Deutsch sprach. Aber am 10. oder 11
Kalle und zeigte keine besonderen Anzeichen von Schre
Manuskript hervorzog.

Ziffel Ich fange an mit einer Einleitung, in der ich in b
darauf aufmerksam mache, daß meine Meinu
zubringen gedenke, wenigstens bis vor kurz
nungen von Millionen waren, so daß sie als
uninteressant sein **können.** Ich überspring d
noch ein Stück und komm gleich auf die Ausfü
Erziehung, die ich genossen habe. Diese A
ich nämlich für sehr wissenswert, stellenweis
net, beugen Sie sich ein wenig vor, daß der L
stört. (Er liest) »Ich weiß, daß die Güte unser

Andrerseits ist Menschlichkeit in unsern Zeitläuften kaum zu erhalten ohne Bestechlichkeit, auch eine Art Unordnung. Sie werden Menschlichkeit finden, wenn Sie einen Beamten finden, der nimmt. Mit etwas Bestechung können Sie sogar gelegentlich Gerechtigkeit erlangen. Damit ich in Österreich auf dem Paßamt an der Reih drangekommen bin, hab ich ein Trinkgeld gegeben. Ich hab einem Beamten am Gesicht angesehen, daß er gütig war und was genommen hat. Die faschistischen Regime schreiten ein gegen die Bestechlichkeit, grad weil sie eben unhuman sind.

Kalle Einer hat einmal behauptet, Dreck sei überhaupt nur Materie am falschen Ort. In einem Blumentopf können Sie Dreck eigentlich nicht Dreck nennen. Ich bin im Grund für Ordnung. Aber ich hab einmal einen Film gesehn mit Charlie Chaplin. Er hat seine Kleider und so weiter in einen Handkoffer gepackt, das heißt hineingeschmissen, und den Deckel zugeklappt und dann war es ihm zu unordentlich, weil zuviel herausgeschaut hat, und da hat er eine Schere genommen und die Ärmel und Hosenbeine, kurz alles, was herausgehängt ist, einfach abgeschnitten. Das hat mich in Erstaunen gesetzt. Ich seh, Sie setzen die Ordnungsliebe auch nicht hoch an.

0743

1998 Bm[1]

KELLER + KUHN

Die blauen WUNDER

Firma Armebooks.de
Schmalkaldener Str. 6
65929 Frankfurt/Germany
aloismanfred1@gmx.de,

ZHAO QING
Dabei xiang7-3,Nanjing Han Qing Tang
Design Co.,
210000 Nanjing
China

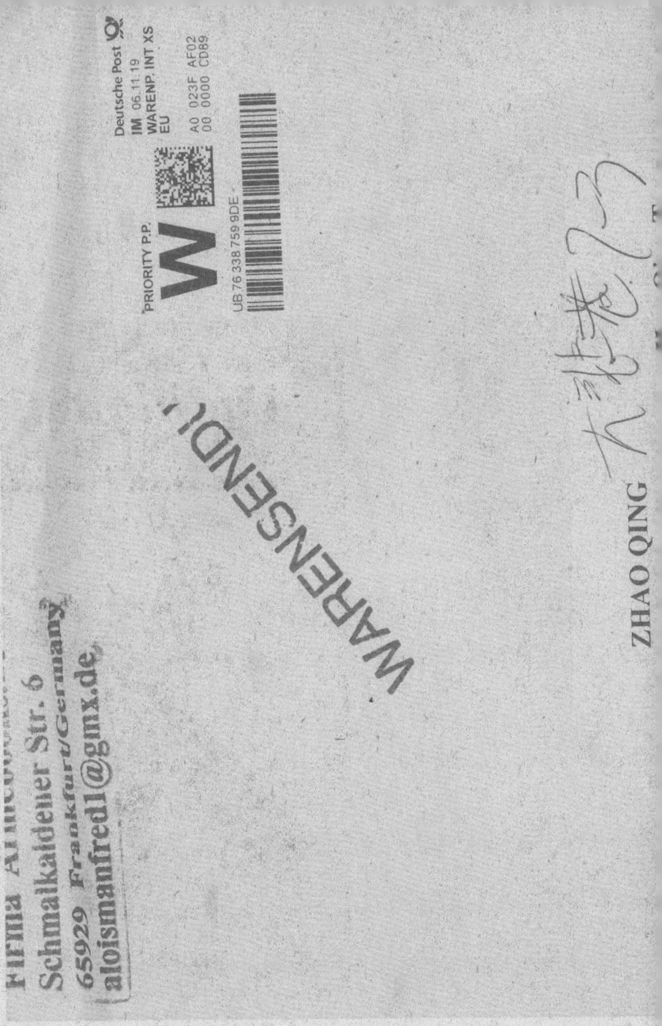
Deutsche Post
IM 06.11.19
WARENP. INT XS
EU
A0 023F AF02
00 0000 CD89
PRIORITY P.P.
W
UB 76 338 759 9DE
Schmalkaldener Str. 6
65929 Frankfurt/Germany
aloismanfred1@gmx.de
ZHAO QING

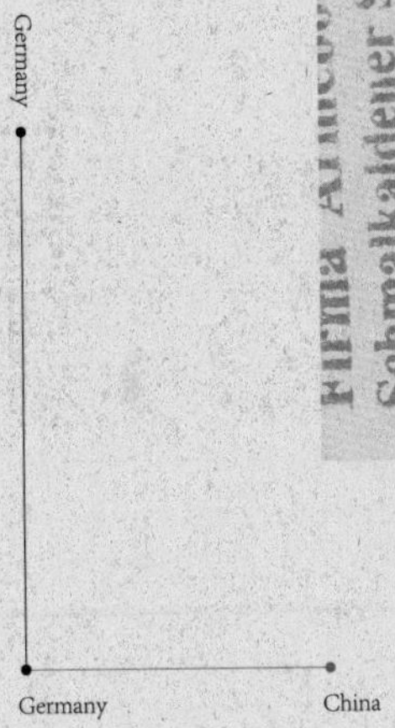
Germany
Germany
China

220mm

23mm

133mm

- 铜奖
- 蓝色奇迹——传真小说
- 德国

Bronze Medal

Die blauen Wunder: Faxroman

Germany

Matthias Gubig

Christoph Keller, Heinrich Kuhn

Reclam Verlag, Leipzig

220 133 23mm

409g

236p

ISBN-10 3-379-00761-7

这是由克里斯托弗·凯勒（Christoph Keller）和海因里希·库恩（Heinrich Kuhn）合著的一部小说，有趣而怪诞的故事由一份意外的传真引起。本书的装订形式为圆脊无线硬精装。护封采用铜版纸红蓝黑三色印刷覆哑膜，内封深蓝色布纹纸烫黑后烫银处理，裱覆卡板。内页胶版纸印单黑。内容方面，全书分为四个章节，除第四章按照时间线整理主人公的日记外，前三章均为传真，作为一个前互联网时代的类似电子邮件的交互方式，传真的文体和形式也介于信件和电邮之间，拥有一定的时效性和聊天的可能性，以及比电子邮件略简单的富文本形式。这些特点直接体现在了书籍的设计上。首先，传真的格式被保留，时间、地点及相关信息均出现在传真开头。第二，发送传真的人物间具有极强的互动性，甚至包含“PS”（postscript）这样的附注内容。第三，文中包含大量非字母的符号、简单的表情、加大的字号和加粗的字体等传真时代有限的富文本格式。第四，接近于聊天的分段处理。

This is a novel co-authored by Christoph Keller and Heinrich Kuhn. The funny and weird story was caused by an accidental fax. The hardcover book is perfect bound with rounding back. The jacket is made of coated paper printed in red, blue and black, covered with matte film. The inner cover is made of dark blue arlin paper treated with black hot stamping and silver hot stamping, mounted with cardboard as well. The inside pages are made of offset paper printed in single black.. In terms of content, the whole book is divided into four chapters. Except for the fourth chapter, which is the diary of the protagonist organized according to the timeline, the other three chapters are all faxes. As an e-mail-like interactive method in the pre-Internet era, the style and form of the fax is also between letters and emails, which has a certain timeliness and the possibility of chatting. These characteristics are directly reflected in the design of the book. Firstly, the ‘fax’ format is retained, and the time, location and related information appear at the beginning of the fax. Secondly, the characters that sent the ‘fax’ are extremely interactive, notes like ‘PS’ (postscript) are included. Thirdly, the text contains a large number of non-letter symbols, simple expressions, enlarged font size and bold fonts, which are limited text formats in the ‘fax’ era. Fourthly, the fax fragments are similar to chat records.

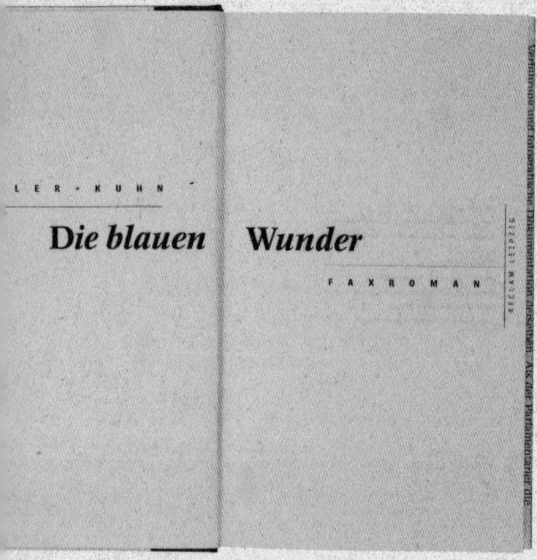

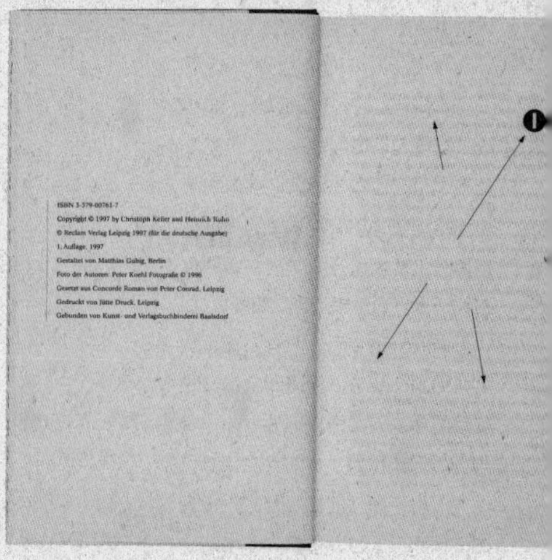

LER • KUHN

Die blauen Wunder

FAXROMAN

RECLAM LEIPZIG

ISBN 3-379-00761-7

1. Auflage, 1997
Gestaltet von Matthias Gubig, Berlin
Foto der Autoren: Peter Koehl Fotografie © 1996
Gesetzt aus Concorde Roman von Peter Conrad, Leipzig
Gedruckt von Jütte Druck, Leipzig
Gebunden von Kunst- und Verlagsbuchbinderei Baalsdorf

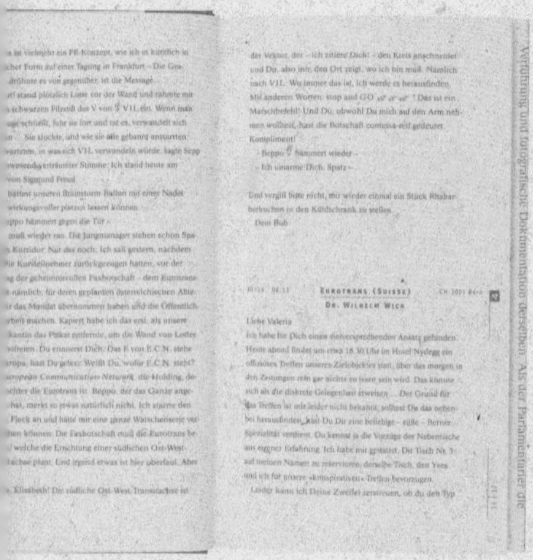

der Vektor, der – ich zitiere Dich! – den Kreis anschneidet und Du, also mir, den Ort zeigt, wo ich hin muß. Nämlich nach VII. Wo immer das ist, ich werde es herausfinden. Mit anderen Worten: stop and GO! Das ist ein Marschbefehl! Und Du, obwohl Du mich auf den Arm nehmen wolltest, hast die Botschaft contessa-reif gedeutet. Kompliment!

– Beppo hämmert wieder –

– Ich umarme Dich, Spatz –

Und vergiß bitte nicht, mir wieder einmal ein Stück Rhabarberkuchen in den Kühlschrank zu stellen.

Dein Bub

EUROTRANS (SUISSE) — CH-3001 Bern

DR. WILHELM WICK

Liebe Valeria

Ich habe für Dich einen vielversprechenden Ansatz gefunden. Heute abend findet um etwa 18.30 Uhr im Hotel Nydegg ein offizielles Treffen unseres Zielobjektes statt, über das morgen in den Zeitungen rein gar nichts zu lesen sein wird. Das könnte sich als *die* diskrete Gelegenheit erweisen ... Der Grund für das Treffen ist mir leider nicht bekannt; solltest Du das nebenbei herausfinden, hast Du Dir eine beliebige – süße – Berner Spezialität verdient. Du kennst ja die Vorzüge der Nebentische aus eigner Erfahrung. Ich habe mir gestattet, Dir Tisch Nr. 3 auf meinen Namen zu reservieren, derselbe Tisch, den Yves und ich für unsere »konspirativen« Treffen bevorzugen.

Leider kann ich Deine Zweifel zerstreuen, ob du den Typ

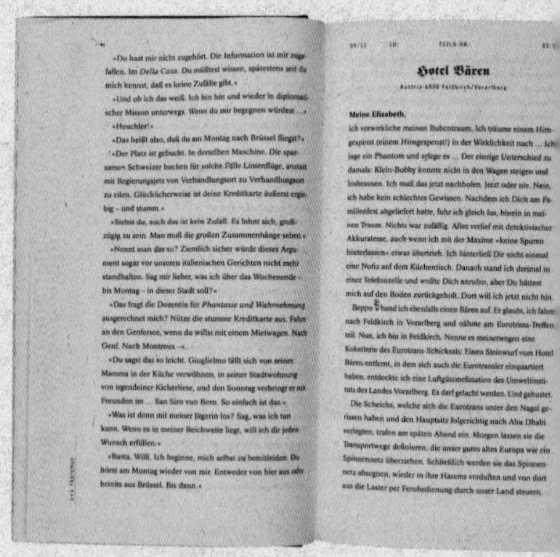

»Du hast mir nicht zugehört. Die Information ist mir zugefallen. Im *Della Casa*. Du müßtest wissen, spätestens seit du mich kennst, daß es keine Zufälle gibt.«

»Und ob ich das weiß. Ich bin hin und wieder in diplomatischer Misson unterwegs. Wenn du mir begegnen würdest ...«

»Heuchler!«

»Das heißt also, daß du am Montag nach Brüssel fliegst?«

»Der Platz ist gebucht. In derselben Maschine. Die sparsamen Schweizer buchen für solche Fälle Linienflüge, anstatt mit Regierungsjets von Verhandlungsort zu Verhandlungsort zu eilen. Glücklicherweise ist deine Kreditkarte äußerst ergiebig – und stumm.«

»Siehst du, auch das ist kein Zufall. Es lohnt sich, großzügig zu sein. Man muß die großen Zusammenhänge sehen.«

»Nennt man das so? Ziemlich sicher würde dieses Argument sogar vor unseren italienischen Gerichten nicht mehr standhalten. Sag mir lieber, was ich über das Wochenende – bis Montag – in dieser Stadt soll?«

»Das fragt die Dozentin für *Phantasie und Wahrnehmung* ausgerechnet mich? Nütze die stumme Kreditkarte aus. Fahre an den Genfersee, wenn du willst mit einem Mietwagen. Nach Genf. Nach Montreux –«.

»Du sagst das so leicht. Giuglielmo läßt sich von seiner Mamma in der Küche verwöhnen, in seiner Stadtwohnung von irgendeiner Kicherliese, und den Sonntag verbringt er mit Freunden im ... San Siro von Bern. So einfach ist das.«

»Was ist denn mit meiner Jägerin los? Sag, was ich tun kann. Wenn es in meiner Reichweite liegt, will ich dir jeden Wunsch erfüllen.«

»Basta, Willi. Ich beginne, mich selbst zu bemitleiden. Du hörst am Montag wieder von mir. Entweder von hier aus oder bereits aus Brüssel. Bis dann.«

TEILN-NR:

Hotel Bären

Austria-6800 Feldkirch/Vorarlberg

Meine Elisabeth,

ich verwirkliche meinen Bubentraum. Ich träume einem Hirngespinst (einem Hirngespenst!) in der Wirklichkeit nach ... Ich jage ein Phantom und erlege es ... Der einzige Unterschied zu damals: Klein-Bobby konnte nicht in den Wagen steigen und losbrausen. Ich muß das jetzt nachholen. Jetzt oder nie. Nein, ich habe kein schlechtes Gewissen. Nachdem ich Dich am Familienfest abgeliefert hatte, fuhr ich gleich los, hinein in meinen Traum. Nichts war zufällig. Alles verlief mit detektivischer Akkuratesse, auch wenn ich mit der Maxime »keine Spuren hinterlassen« etwas übertrieb. Ich hinterließ Dir nicht einmal eine Notiz auf dem Küchentisch. Danach stand ich dreimal in einer Telefonzelle und wollte Dich anrufen, aber Du hättest mich auf den Boden zurückgeholt. Dort will ich jetzt nicht hin.

Beppo band ich ebenfalls einen Bären auf. Er glaubt, ich fahre nach Feldkirch in Vorarlberg und nähme am Eurotrans-Treffen teil. Nun, ich bin in Feldkirch. Nenne es meinetwegen eine Koketterie des Eurotrans-Schicksals: Einen Steinwurf vom Hotel Bären entfernt, in dem sich auch die Eurotransler einquartiert haben, entdeckte ich eine Luftgütemeßstation des Umweltinstituts des Landes Vorarlberg. Es darf gelacht werden. Und gehustet.

Die Scheichs, welche sich die Eurotrans unter den Nagel gerissen haben und den Hauptsitz folgerichtig nach Abu Dhabi verlegten, trafen am späten Abend ein. Morgen lassen sie die Transportwege definieren, die unser gutes altes Europa wie ein Spinnennetz überziehen. Schließlich werden sie das Spinnennetz absegnen, wieder in ihre Harems verduften und von dort aus die Laster per Fernbedienung durch unser Land steuern.

Also meide das V! Meide es, Parzival! Nein, suche es! Suche das V – es ist der Vektor (das wird Bob gefallen, liebt er doch die Geometrie über alles!), … es ist der Vektor, der den Kreis anschneidet und dir den Ort zeigt, an dem du fündig wirst. Es ist dein Gral, Glücksritter.

– – –

Da bin ich wieder, Rosi. Verzeih die Unterbrechung, aber ich ging rasch hinunter in Bobs Büro. Ich habe ihm die Faxbotschaft, versehen mit meiner feinsäuberlich abgetippten Interpretation, in die Publicom gefaxt. Der wird schön staunen.

Schon zehn Uhr. Die Einbrecher werden auch müde sein. Was gibt es noch zu sagen?

– – –

DAS WICHTIGSTE. Ich habe es bis zum Schluß aufbewahrt, Rosi, weil ich weiß, daß Du Überraschungen magst. Und ich habe eine: Bob und ich werden schwanger. Beziehungsweise: Wir sind in Erwartung. Wann? fragst Du. In einem Jahr. Spätestens im nächsten Herbst.

Ein Oktoberkätzchen wird es sein. Das haben wir einstimmig beschlossen, bevor Bob seine PR-Seminaristen um sich scharte. Was sagst du jetzt?

Denk Dir einen Namen aus, Rosi. Soll der Junge Boris heißen, Maxl – oder gar Yves? Das Mädchen Johanna? Margarete? Oder Rosi, Rosi?

Rat ist gefragt. Dringend.

Deine destabilisierte Elisabeth

wieder heftiger als vorher, und ich frage mich gerade, ob Einbrecher auch einen Regenschirm dabeihaben. Ich werde morgen bei Vrabec vorbeischauen.

PPS

Franzl? Oder Hannerl?

PPPS

Zur Stabilisierung gehe ich nochmals nach unten und erlaube mir einen Barack. Auf Hannerl!

»Eurotrans Schweiz AG, Elvira Wüthrich, Guten Tag.«

»Würden Sie mich bitte mit Herrn Wick verbinden, Dr. Willi Wick?«

»Augenblick, ich will schauen, ob er frei ist.«

»Hier Wick.«

»Willi! Ich bin's, Valeria. Hörst du mich? Ich bin eben in Zürich angekommen. Du hörst mich doch?«

»Wie wenn du neben mir stehen würdest. Bist du aufgeregt, oder traust du dem Schweizer Telefonnetz nicht?«

»Ich friere. Hier ist es kalt und neblig, und ich frage mich, was ich hier soll … Warum bin ich nicht in Siena bei meinen

O B I S T W I E D E R

es den Satz ein paarmal, kneif Dich in den P
Nase, oder tu, was Du sonst tust bei Dingen,
können. ***Bob ist wieder da***
ich vom Squashcenter in unser Büro fuhr, be
ich mich gesetzt hatte, das Fax zu schnurren
n drei Meter Papier und meldete die Wiederl
iters Robert Pöschl. **Er ist wiede**
r vermag ich dazu im Augenblick nicht zu sa
chluß an diesen Brief erhältst Du die drei Me
, damit Du Dir wenigstens annäherungsweise
hen kannst.
bitte Dich um folgendes: Fährst Du für mic
ch hinaus, um vorsichtig abzuklären, was mi
Bitte. Darf ich anschließend und möglichst b
Bericht ins Hotel Kummer an der Mariahilfe
rten. Es wäre mir recht, wenn Du die Adress
lten würdest. Ich werde mich in den nächste

I

CH-3011 Bern

6

nn Dir ge-
fang läßt
Service: lie-
lat: originell

Wunder

FAXROMAN

RECLAM LEIPZIG

02111 ✈ FLUGHAFEN CH-

bin.
nittag, um 12.30 Uhr: Ic
dschirm beziehungsweis
nitzten Berghotel, wo Fr
t fühle ich mich ein biß

zige, was ich weiß: Unser Vögelchen befindet sich zur Zeit in der Luft. Frag mich nicht, mit welcher Fluggesellschaft.«

»Das mußt du doch wissen. Genau von solchen Détails hängt der Erfolg –«

»Sprechen wir nicht über Erfolg oder Mißerfolg, mein Freund, und auch nicht über Planung. Valeria ist unterwegs nach Brüssel, sie –«

»Wer hat denn davon gesprochen, diese … diese Nutte nach Brüssel fliegen zu lassen?«

»Wie du meinst. Zwei Freunde haben etwas abgemacht. Du bezahlst weder den Flug noch die übrigen Spesen. Ich bezahle sie.«

»Ich erwarte kompromißlose Unterstützung, Willi.«

»Ich tue, was ich kann. Vielleicht ließe sich jetzt erarbeitetes Material auch nächsten Mai verwenden. Warum auch nicht. Denk darüber nach. Und noch etwas: Nenne Valeria nicht mehr Nutte.«

Angekommen! Elisabeth,

ich bin in VILstop! Ich durchfuhr die Schweiz, so vermute ich, in Rekordzeit. Geradezu ein Paradies für die Eurotrans, das Ländchen besteht nur aus Transitachsen. Jedenfalls hat man diesen Eindruck, wenn man sich auf einer solchen befindet.

Jetzt also Villars. Du solltest hier sein, Elisabeth. Weißt Du was? Zu Hause setzen wir uns ans Time System und verschieben unsere separaten Termine so lange, bis wir einen gemeinsamen haben. Zielsetzung: 1 Woche. Programm: Only you & I. Und wir könnten hier in einem netten, kleinen Hotel wie zum Beispiel dem *Aux cèpes* (heißt etwa »bei den Steinpilzen«), das ich ausgekundschaftet habe, mit Blick auf – ach, zum Teufel mit dem Blick! Also, wir könnten uns in Villars' gesunder Höhenluft dem Hannerl widmen … Na?

Noch aber bin ich in geheimer Mission unterwegs. Zuerst wird Spade die Lage sondieren und dann einen Plan entwickeln, bevor er zuschlägt.

Unter mir glitzerte, während ich im Auto saß, der Genfersee in der Abendsonne – du wirst seeblind, wenn du zu lange hinschaust. Da unten, überdacht von einer abscheulichen Autobahnbrücke, liegt das berühmte Schloß Chillon, weiter westlich Lausanne …

1998 Bm^2

Germany

U.S.A.

China

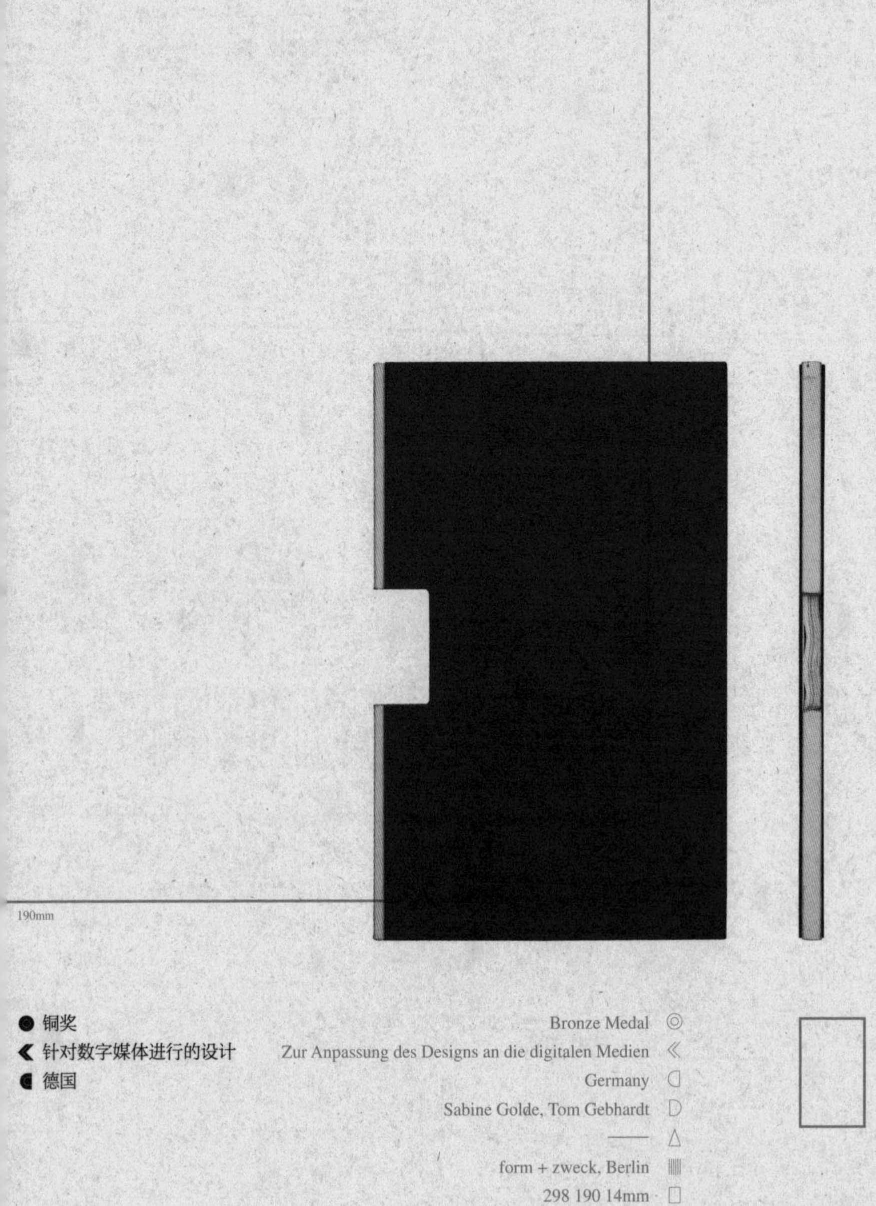

● 铜奖
《 针对数字媒体进行的设计
◖ 德国

Bronze Medal
Zur Anpassung des Designs an die digitalen Medien
Germany
Sabine Golde, Tom Gebhardt
——
form + zweck, Berlin
298 190 14mm
517g
160p
ISBN-10 3-9804679-3-7

自从哲学家、教育家、现代传播理论奠基人马歇尔·麦克卢汉（Marshall McLuhan）提出了“媒介是人的延伸”之后，有关机械、电脑、数字、媒介、人类和精神的讨论就从未中断过。到了20世纪末，随着个人电脑和互联网的普及化，这一问题产生了更深入更具体的讨论，本书就是这样一类问题的论文集。书籍的装订形式为锁线胶平装。封面白色布面裱覆黑色厚卡板，文字压凹处理。内页胶版纸玫红和黑色双色印刷。书中的文章包含计算机革命、人对信息智能化的担忧、数字复制的未来、从打孔机到网络世界的历史、新媒介是否会改变思考方式、信息技术是否为设计树立了新规范等方面。书籍采用的玫红色对网格化的正文排版产生了强烈冲击，仿佛原本格格不入的新鲜事物正在规范着新的秩序。书脊处有矩形开口，翻阅本书时中间形成巨大的空洞。全书德英双语，并未进行对照设计，德文占全书前四分之三，单栏或双栏排版，有大量图片辅助论述。英文被全部放到最后四分之一，小字号三栏密排。

Since Marshall McLuhan, the philosopher, educator, and founder of modern communication theory, put forward that ‘Media is an extension of human beings’, discussions about machinery, computers, digital, media, humans and spirits have never been interrupted. By the end of the 20th century, with the popularization of personal computers and the Internet, more in-depth and specific discussions on this issue emerged. This book is a collection of papers on such issues. The paperback book is glue bound with thread sewing. The cover is made of white cloth mounted with black cardboard, and the text is embossed. The inside pages are made of offset paper printed in rose red and black. The articles in the book include many aspects, such as the computer revolution, people’s worries about the intelligence of information, the future of digital replication, the history from the punching machine to the Internet world, whether new media will change the way of thinking, and whether information technology has set new standards for design. The rose red color used in the book has a strong impact on the grid typesetting, as if the incompatible new things are regulating a new order. There is a rectangular opening at the spine of the book, which forms a huge hollow in the middle of the book when people read the book. The book is bilingual in German and English. The text in German accounts for the first three quarters of the book, which is typeset in single column or double columns with a large number of pictures. The text in English is placed in the last quarter, using small font size in three columns.

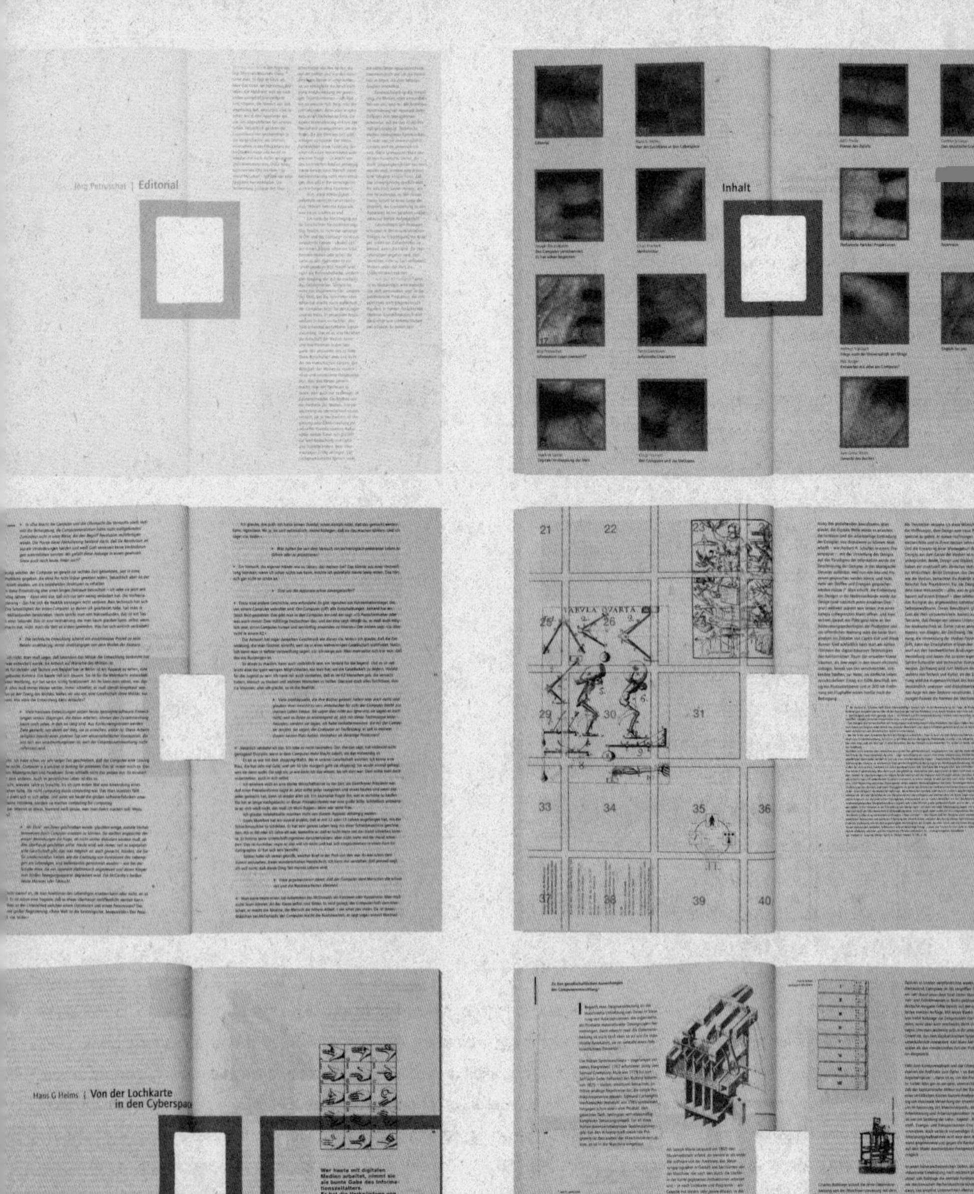
Jörg Petruschat | Editorial
Inhalt
Hans G Helms | Von der Lochkarte in den Cyberspace

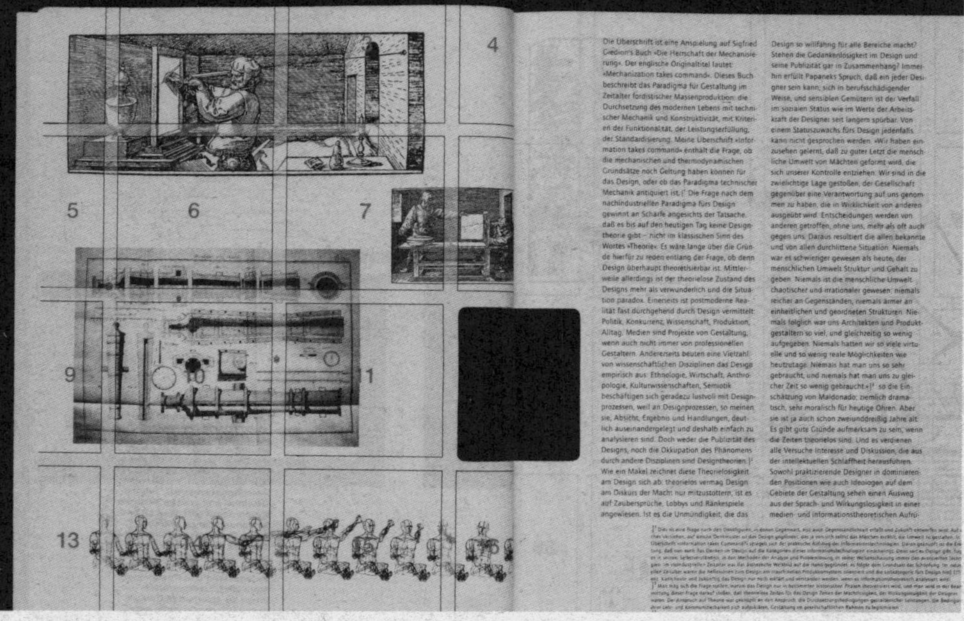

Die Überschrift ist eine Anspielung auf Sigfried Giedion's Buch »Die Herrschaft der Mechanisierung«. Der englische Originaltitel lautet: »Mechanization takes command«. Dieses Buch beschreibt das Paradigma für Gestaltung im Zeitalter fordistischer Massenproduktion: die Durchsetzung des modernen Lebens mit technischer Mechanik und Konstruktivität, mit Kriterien der Funktionalität, der Leistungserfüllung, der Standardisierung. Meine Überschrift »Information takes command« enthält die Frage, ob die mechanischen und thermodynamischen Grundsätze noch Geltung haben können für das Design, oder ob das Paradigma technischer Mechanik antiquiert ist.|[1] Die Frage nach dem nachindustriellen Paradigma fürs Design gewinnt an Schärfe angesichts der Tatsache, daß es bis auf den heutigen Tag keine Designtheorie gibt – nicht im klassischen Sinn des Wortes »Theorie«. Es wäre lange über die Gründe hierfür zu reden entlang der Frage, ob denn Design überhaupt theoretisierbar ist. Mittlerweile allerdings ist der theorielose Zustand des Designs mehr als verwunderlich und die Situation paradox. Einerseits ist postmoderne Realität fast durchgehend durch Design vermittelt: Politik, Konkurrenz, Wissenschaft, Produktion, Alltag. Medien sind Projekte von Gestaltung, wenn auch nicht immer von professionellen Gestaltern. Andererseits beuten eine Vielzahl von wissenschaftlichen Disziplinen das Design empirisch aus: Ethnologie, Wirtschaft, Anthropologie, Kulturwissenschaften, Semiotik beschäftigen sich geradezu lustvoll mit Designprozessen, weil an Designprozessen, so meinen sie, Absicht, Ergebnis und Handlungen, deutlich auseinandergelegt und deshalb einfach zu analysieren sind. Doch weder die Publizität des Designs, noch die Okkupation des Phänomens durch andere Disziplinen sind Designtheorien.|[2] Wie ein Makel zeichnet diese Theorielosigkeit am Design sich ab: theorielos vermag Design am Diskurs der Macht nur mitzustottern, ist es auf Zaubersprüche, Lobbys und Ränkespiele angewiesen. Ist es die Unmündigkeit, die das Design so willfährig für alle Bereiche macht? Stehen die Gedankenlosigkeit im Design und seine Publizität gar in Zusammenhang? Immerhin erfüllt Papaneks Spruch, daß ein jeder Designer sein kann, sich in berufsschädigender Weise, und sensiblen Gemütern ist der Verfall im sozialen Status wie im Werte der Arbeitskraft der Designer seit langem spürbar. Von einem Statuszuwachs fürs Design jedenfalls kann nicht gesprochen werden. »Wir haben einzusehen gelernt, daß zu guter Letzt die menschliche Umwelt von Mächten geformt wird, die sich unserer Kontrolle entziehen. Wir sind in die zwielichtige Lage gestoßen, der Gesellschaft gegenüber eine Verantwortung auf uns genommen zu haben, die in Wirklichkeit von anderen ausgeübt wird. Entscheidungen werden von anderen getroffen, ohne uns, mehr als oft auch gegen uns. Daraus resultiert die allen bekannte und von allen durchlittene Situation. Niemals war es schwieriger gewesen als heute, der menschlichen Umwelt Struktur und Gehalt zu geben. Niemals ist die menschliche Umwelt chaotischer und irrationaler gewesen; niemals reicher an Gegenständen, niemals ärmer an einheitlichen und geordneten Strukturen. Niemals folglich war uns Architekten und Produktgestaltern so viel, und gleichzeitig so wenig aufgegeben. Niemals hatten wir so viele virtuelle und so wenig reale Möglichkeiten wie heutzutage. Niemals hat man uns so sehr gebraucht, und niemals hat man uns zu gleicher Zeit so wenig gebraucht.«|[3] so die Einschätzung von Maldonado; ziemlich dramatisch, sehr moralisch für heutige Ohren. Aber sie ist ja auch schon zweiunddreißig Jahre alt. Es gibt gute Gründe aufmerksam zu sein, wenn die Zeiten theorielos sind. Und es verdienen alle Versuche Interesse und Diskussion, die aus der intellektuellen Schlaffheit herausführen. Sowohl praktizierende Designer in dominierenden Positionen wie auch Ideologen auf dem Gebiete der Gestaltung sehen einen Ausweg aus der Sprach- und Wirkungslosigkeit in einer medien- und informationstheoretischen Aufrü-

Er hat die Verknüpfung von *hardware* und *software* zu akzeptieren und er ist damit auf bestimmte Programmkonfigurationen festgelegt. Der folgende Text zeigt, daß am Beginn der Rechnerentwicklung die Trennung von *hardware* und *software* stand, er zeigt, wer mit welchen Zielen diese beiden Entwicklungssäulen miteinander verknüpfte und ihre Entwicklung finanzierte. Hans G Helms, in den fünfziger Jahren Schüler und Freund Adornos, Autor von »Fa:m' Ahniesgwow«, des ersten großen Versuchs, Sprache zu komponieren, Chronist des *urban sprawls*, Autor des legendären Films »Birdcage« über seinen Freund John Cage, beschäftigt sich seit langem mit der Entwicklung der künstlichen Intelligenz. Heute lebt Hans G Helms als freier Künstler und Privatgelehrter in Köln.

nern der digital basierten Technologi
alifornischen Traum der virtuellen Kl
nen, als *lone eagle* in den neuen *elect*
ges, fernab von den verschmutzten, s
n Städten, zur Natur, ins dörfliche Le
kzukehren. Einzig ein ISDN-Anschluß
n Postzustelldienst und in 200 km E
ein Flughafen wären hierfür noch di
gung.

[4] Bei Herbert H. Schultes heißt dies
forderungen reagieren könne oder ob
»– Der Designer wird nicht gefragt, o
zustellen; darüber befinden Produkte
– Der Designer hat keinen Einfluß au
Auch kann der Dsigner nicht einmal
nach technischen und ökonomischen

Dank der elektro-mechanis
Maschine, wie sie bald ger
der Staat, der sich – wie sc
Difference Engine – als Fi
Gesamtkapitalist betätigte
weiblichen oder männliche
weißer, roter, schwarzer od
San Francisco oder Bismar
chen Altersstufen sie ange
Berufe sie ausübten. Aus d
die Maschine auch, wievie
Kriegsfall aufzubieten wär

Um die 1892 anstehende
bewältigen zu können, erg
rith die noch halbautomat
ne um einen gleichfalls ele
Addierteil. Im Gespann ve
Rechner dem Herrn im We
Region die dicksten Sojab
den, wo die saftigsten Rin

Heute ist das Internet so etwas. Niemand weiß ge
braucht man es?

PLOM

ANZ

SOLD

ORP

> *Als Eliza*[1] *von Ihnen geschrie*
> *tensweisen durch Computer erse*
> *artiger Bestrebungen die Frage,*
> *dies überhaupt geschehen sollte*
> *sche Gesellschaft gibt, das, was*
> *für unüberwindbar hielten, wie*
> *gen am Lebendigen, sind beden*
> *Schabe etwa, die von Japanern*
> *zum bloßen Bewegungsapparat*
> *heute Moravec oder Takeuchi.*

[1] Eliza hieß eines der ersten interaktiven Programme. Weizenbaum schrieb es 1963 im MIT.

< Es kommt nicht darauf an, ob man Funktionen des
einfach absurd. Es ist schon eine Tragödie, daß so etwa

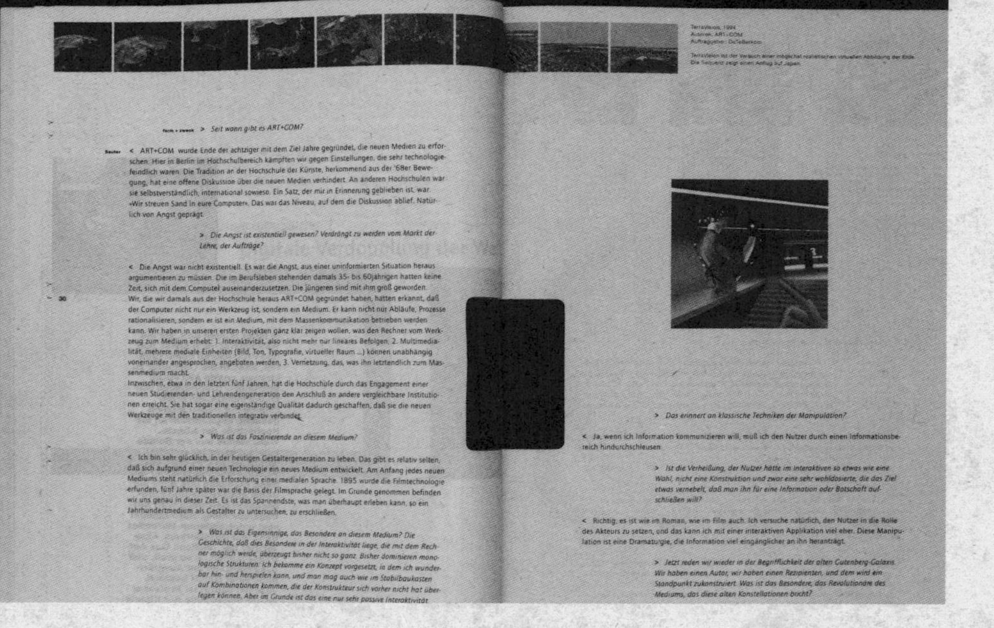

TerraVision, 1994
Autoren: ART+COM
Auftraggeber: DeTeBerkom

TerraVision ist der Versuch einer möglichst realistischen virtuellen Abbildung der Erde. Die Sequenz zeigt einen Anflug auf Japan.

> *Seit wann gibt es ART+COM?*

Sauter < ART+COM wurde Ende der achtziger mit dem Ziel Jahre gegründet, die neuen Medien zu erforschen. Hier in Berlin im Hochschulbereich kämpften wir gegen Einstellungen, die sehr technologiefeindlich waren. Die Tradition an der Hochschule der Künste, herkommend aus der '68er Bewegung, hat eine offene Diskussion über die neuen Medien verhindert. An anderen Hochschulen war sie selbstverständlich, international sowieso. Ein Satz, der mir in Erinnerung geblieben ist, war: »Wir streuen Sand in eure Computer«. Das war das Niveau, auf dem die Diskussion ablief. Natürlich von Angst geprägt.

> *Die Angst ist existentiell gewesen? Verdrängt zu werden vom Markt der Lehre, der Aufträge?*

< Die Angst war nicht existentiell. Es war die Angst, aus einer uninformierten Situation heraus argumentieren zu müssen. Die im Berufsleben stehenden damals 35- bis 60jährigen hatten keine Zeit, sich mit dem Computel auseinanderzusetzen. Die jüngeren sind mit ihm groß geworden. Wir, die wir damals aus der Hochschule heraus ART+COM gegründet haben, hatten erkannt, daß der Computer nicht nur ein Werkzeug ist, sondern ein Medium. Er kann nicht nur Abläufe, Prozesse rationalisieren, sondern er ist ein Medium, mit dem Massenkommunikation betrieben werden kann. Wir haben in unseren ersten Projekten ganz klar zeigen wollen, was den Rechner vom Werkzeug zum Medium erhebt: 1. Interaktivität, also nicht mehr nur lineares Befolgen, 2. Multimedialität, mehrere mediale Einheiten (Bild, Ton, Typografie, virtueller Raum ...) können unabhängig voneinander angesprochen, angeboten werden, 3. Vernetzung, das, was ihn letztendlich zum Massenmedium macht.
Inzwischen, etwa in den letzten fünf Jahren, hat die Hochschule durch das Engagement einer neuen Studierenden- und Lehrendengeneration den Anschluß an andere vergleichbare Institutionen erreicht. Sie hat sogar eine eigenständige Qualität dadurch geschaffen, daß sie die neuen Werkzeuge mit den traditionellen integrativ verbindet.

> *Was ist das Faszinierende an diesem Medium?*

< Ich bin sehr glücklich, in der heutigen Gestaltergeneration zu leben. Das gibt es relativ selten, daß sich aufgrund einer neuen Technologie ein neues Medium entwickelt. Am Anfang jedes neuen Mediums steht natürlich die Erforschung einer medialen Sprache. 1895 wurde die Filmtechnologie erfunden, fünf Jahre später war die Basis der Filmsprache gelegt. Im Grunde genommen befinden wir uns genau in dieser Zeit. Es ist das Spannendste, was man überhaupt erleben kann, so ein Jahrhundertmedium als Gestalter zu untersuchen, zu erschließen.

> *Was ist das Eigensinnige, das Besondere an diesem Medium? Die Geschichte, daß dies Besondere in der Interaktivität liege, die mit dem Rechner möglich werde, überzeugt bisher nicht so ganz. Bisher dominieren monologische Strukturen: ich bekomme ein Konzept vorgesetzt, in dem ich wunderbar hin- und herspielen kann, und man mag auch wie im Stabilbaukasten auf Kombinationen kommen, die der Konstrukteur sich vorher nicht hat überlegen können. Aber im Grunde ist das eine nur sehr passive Interaktivität*

> *Das erinnert an klassische Techniken der Manipulation?*

< Ja, wenn ich Information kommunizieren will, muß ich den Nutzer durch einen Informationsbereich hindurchschleusen.

> *Ist die Verheißung, der Nutzer hätte im Interaktiven so etwas wie eine Wahl, nicht eine Konstruktion und zwar eine sehr wohldosierte, die das Ziel etwas vernebelt, daß man ihn für eine Information oder Botschaft aufschließen will?*

< Richtig, es ist wie im Roman, wie im Film auch. Ich versuche natürlich, den Nutzer in die Rolle des Akteurs zu setzen, und das kann ich mit einer interaktiven Applikation viel eher. Diese Manipulation ist eine Dramaturgie, die Information viel eingänglicher an ihn heranträgt.

> *Jetzt reden wir wieder in der Begrifflichkeit der alten Gutenberg-Galaxis. Wir haben einen Autor, wir haben einen Rezipienten, und dem wird ein Standpunkt zukonstruiert. Was ist das Besondere, das Revolutionäre des Mediums, das diese alten Konstellationen bricht?*

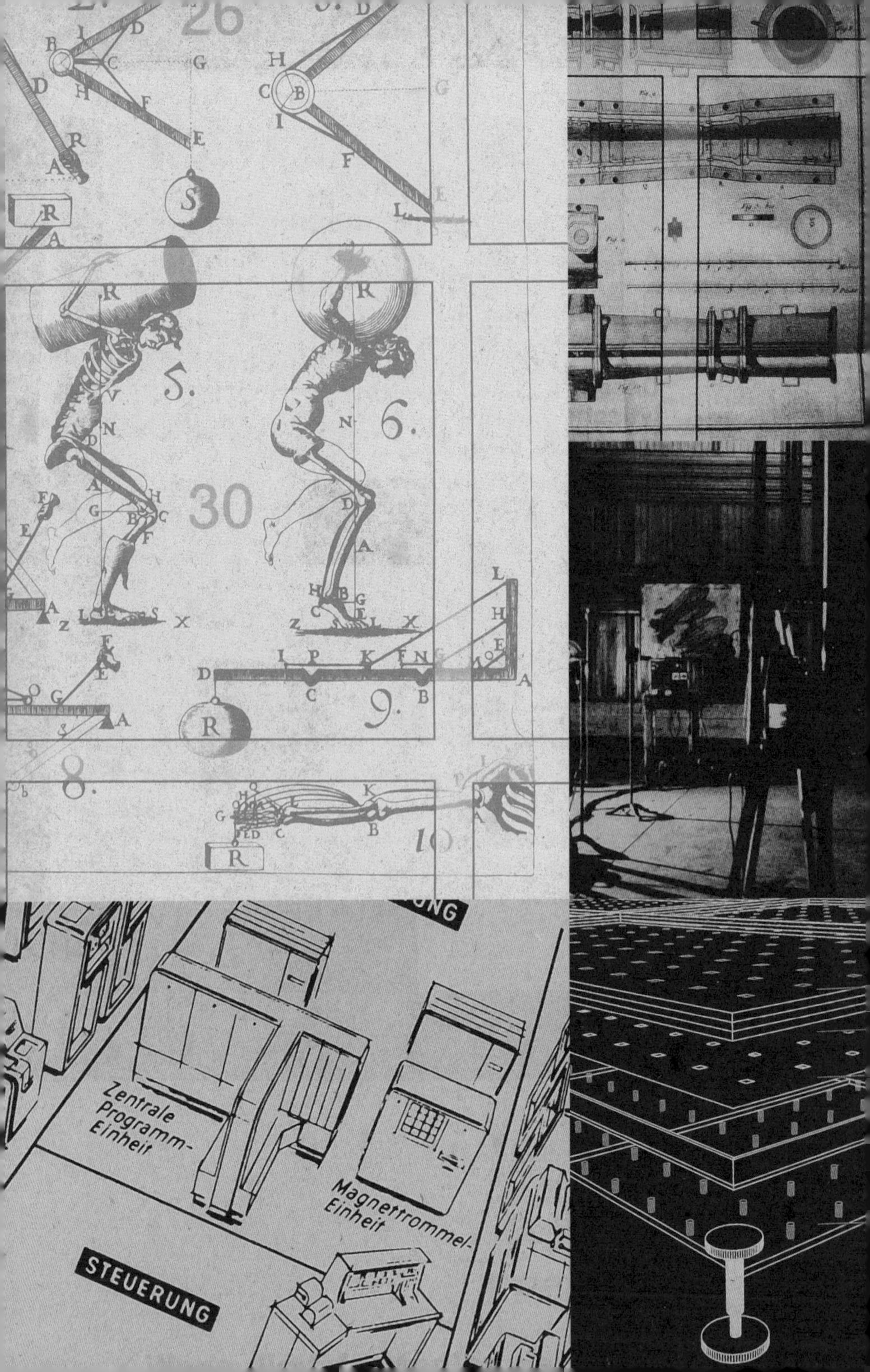
Zentrale Programm-Einheit
Magnettrommel-Einheit
STEUERUNG

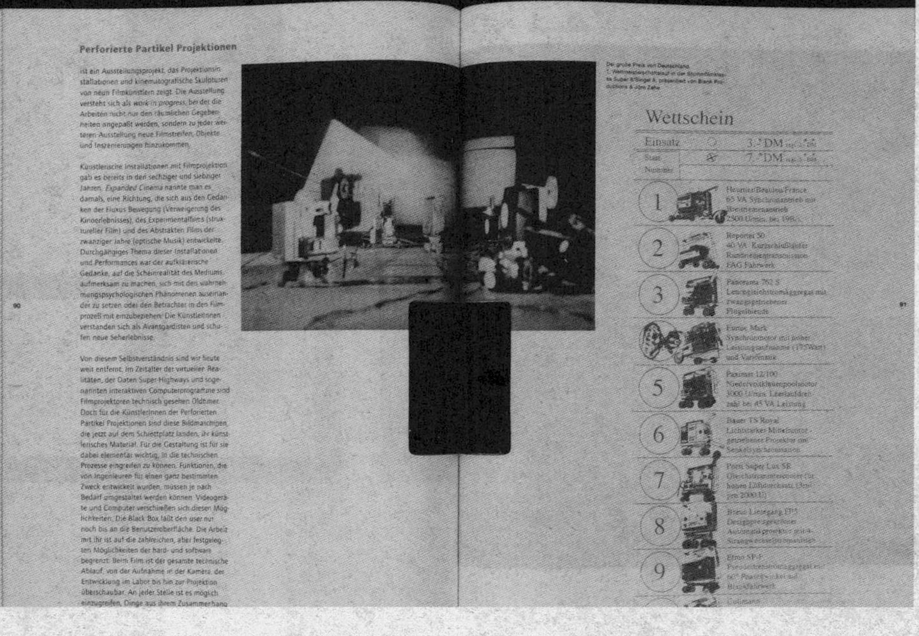

Perforierte Partikel Projektionen

ist ein Ausstellungsprojekt, das Projektionsinstallationen und kinematografische Skulpturen von neun Filmkünstlern zeigt. Die Ausstellung versteht sich als *work in progress*, bei der die Arbeiten nicht nur den räumlichen Gegebenheiten angepaßt werden, sondern zu jeder weiteren Ausstellung neue Filmstreifen, Objekte und Inszenierungen hinzukommen.

Künstlerische Installationen mit Filmprojektion gab es bereits in den sechziger und siebziger Jahren. *Expanded Cinema* nannte man es damals, eine Richtung, die sich aus den Gedanken der Fluxus Bewegung (Verweigerung des Kinoerlebnisses), des Experimentalfilms (struktureller Film) und des Abstrakten Films der zwanziger Jahre (optische Musik) entwickelte. Durchgängiges Thema dieser Installationen und Performances war der aufklärerische Gedanke, auf die Scheinrealität des Mediums aufmerksam zu machen, sich mit den wahrnehmungspsychologischen Phänomenen auseinander zu setzen oder den Betrachter in den Filmprozeß mit einzubeziehen. Die KünstlerInnen verstanden sich als Avantgardisten und schufen neue Seherlebnisse.

Von diesem Selbstverständnis sind wir heute weit entfernt. Im Zeitalter der virtueller Realitäten, der Daten Super-Highways und sogenannten interaktiven Computerprogramme sind Filmprojektoren technisch gesehen Oldtimer. Doch für die KünstlerInnen der Perforierten Partikel Projektionen sind diese Bildmaschinen, die jetzt auf dem Schrottplatz landen, ihr künstlerisches Material. Für die Gestaltung ist für sie dabei elementar wichtig, in die technischen Prozesse eingreifen zu können. Funktionen, die von Ingenieuren für einen ganz bestimmten Zweck entwickelt wurden, müssen je nach Bedarf umgestaltet werden können. Videogeräte und Computer verschließen sich diesen Möglichkeiten. Die Black Box läßt den *user* nur noch bis an die Benutzeroberfläche. Die Arbeit mit ihr ist auf die zahlreichen, aber festgelegten Möglichkeiten der hard- und software begrenzt. Beim Film ist der gesamte technische Ablauf, von der Aufnahme in der Kamera, der Entwicklung im Labor bis hin zur Projektion überschaubar. An jeder Stelle ist es möglich einzugreifen, Dinge aus ihrem Zusammenhang

Der große Preis von Deutschland
1. Weltmeisterschaftslauf in der Stummfilmklasse Super 8/Single 8, präsentiert von Blank Productions & Jörn Zehe

Wettschein

Einsatz	○	3,- DM [illegible]
Start Nummer	⊗	7,- DM [illegible]

1. Heurtier/Beaulieu/France – 65 VA Synchronantrieb mit Breitriemenantrieb 2500 U/min. bei 198[illegible]
2. Reporter 50 – 40 VA Kurzschlußläufer [illegible] FAG Fahrwerk
3. Panorama 762 S – [illegible] mit zwangsgetriebener Flügelblende
4. Eumig Mark – Synchronmotor mit hoher Leistungsaufnahme (175 Watt) und Variomatik
5. Paximat 12/100 – Niedervoltklauenpolmotor 3000 U/min. Leerlaufdrehzahl bei 45 VA Leistung
6. Bauer TS Royal – Lichtstarker [illegible] Projektor mit [illegible]
7. Porst Super Lux SR – [illegible] für hohen Luftdurchsatz ([illegible] pro 2000 U)
8. Braun Liesegang [illegible] – Designpreis-gekrönter Automatikprojektor mit 4-[illegible]
9. Elmo SP-F – [illegible]
10. [illegible]

RENA, RENA, RENA,
SE SÅ RENT OCH FRÄSCHT DET BLIR!
RENA, RENA, RENA!
INTE MINSTA SMOLK NU SYNS,

1998 Bm^4

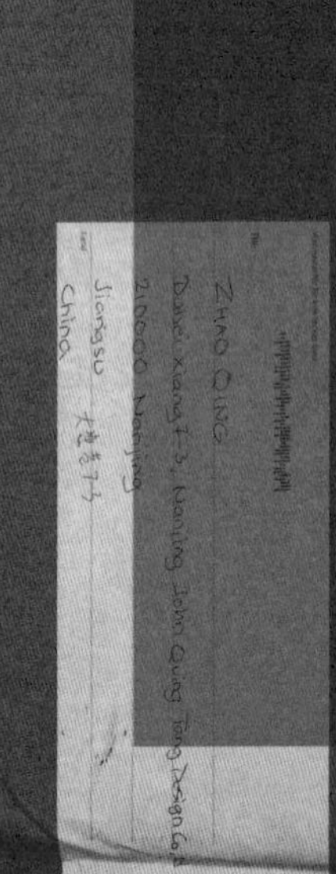

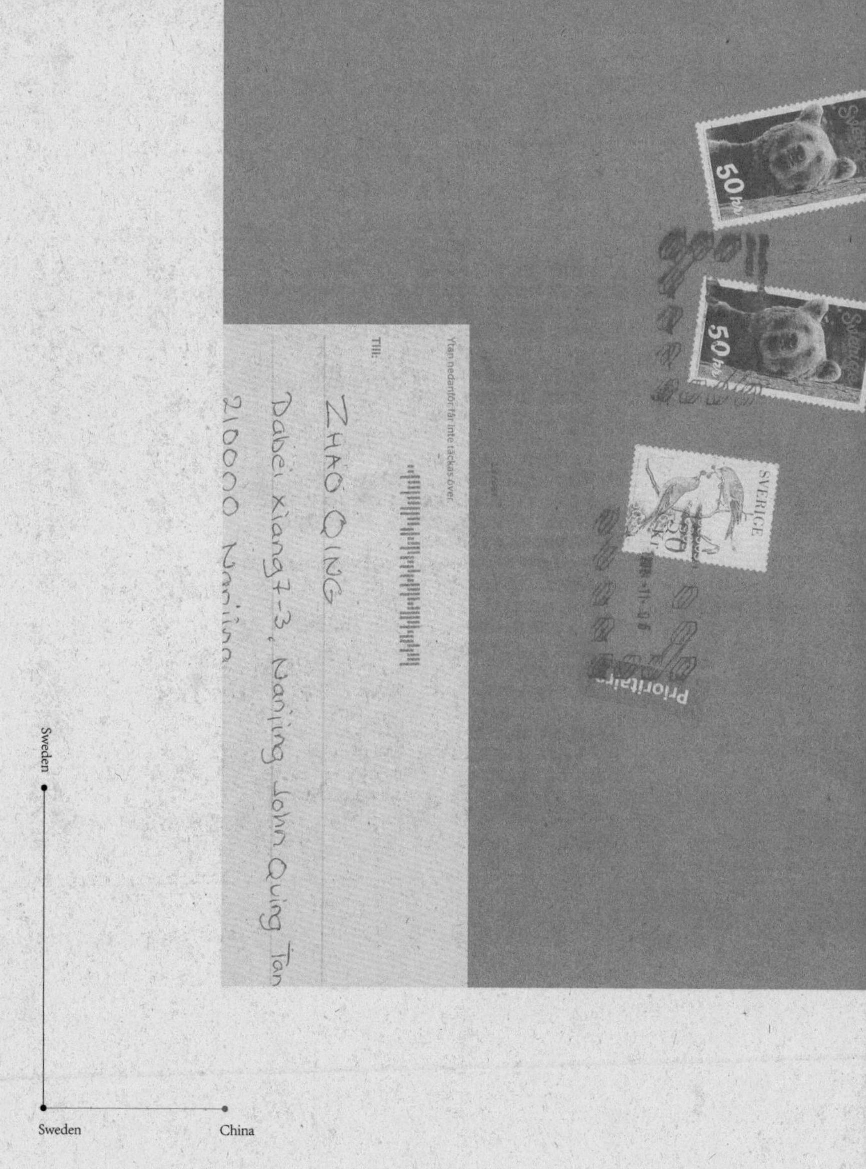
Ytan nedanför får inte täckas över.
Till:
ZHAO QING
Prioritaire
SVERIGE
Sweden
Sweden
China

297mm

23mm

RENA, RENA, RENA,
SE SÅ RENT OCH FRÄSCHT DET BLIR!
RENA, RENA, RENA!
INTE MINSTA SMOLK NU SYNS,
INGA SÅR SOM LÄNGRE VARAR,
INGA ÄRR SOM SKVALLRA KAN!
RENA, RENA, RENA!
RENA, RENA ÄVEN BLODET DITT,
AVUNDSJUKAN FLÄCKFRI BLIR!
RENA, RENA KROPP OCH SJÄL,
RENA SÅNT SOM SMUTSIGT ÄR!
BORT MED DET SOM VEMOD VÄCKER,
FRAM FÖR ALLT SOM LÄCKERT ÄR!
RENA, RENA, RENA!
RENA RAMA LYCKAN KOMMER SNART,
RENA, RENA OCH GÅ PÅ!
HEJ OCH HÅ SÅ FINT DET BLIR,
GNUGGA, SKRUBBA OCH STÅ PÅ!
RENA ALLT SOM RENAS BÖR,
INGEN SORG FÅR OSS BEFLÄCKA!
AJ, AJ SÅ SKÖNT ATT INGET UTÅT SYNS,
SKITKUL, DÖDSKUL ÄNNU MERA KUL!
ÖMSA, ÖMSA GAMMALT SKINN,
LYFTA, LYFTA UNGDOM UPP!
SE SÅ FRITT FRÅN FETT OCH FLÄSK,
SE SÅ SOBERT SNYGGT ALLT ÄR!
RENA, RENA, RENA!

HC Ericson

Reningsverk

Illuminerad skrift

Carlssons

HC Ericson *Reningsverk*

Carlssons

241mm

铜奖
处理厂
瑞典

Bronze Medal
Reningsverk
Sweden
HC Ericson
HC Ericson
Carlsson bokförlag, Stockholm
297 241 23mm
1068g
160p
ISBN-10 917-203-033-X

瑞典平面设计师汉斯·克里斯·埃里克松（HC Ericson）擅长使用独特视觉形式来撰写和设计图书，本书便是他使用视觉语言“撰写”的故事。他的另一本同类书籍《文字图片》（*Ordbilder*）获得了莱比锡“世界最美的书”1993 年荣誉奖（在本书的相关页面可以找到）。本书的装订形式为裸脊锁线胶平装，封面可以完全打开露出裸脊。护封透明 PVC 材料印灰色，内封白色卡纸孔板 19 色印刷（黑色加 18 种彩色）。内页涂布艺术纸 19 色印刷（同封面）。书中通过黑色文字、黑色齿状细线和彩色巨大的单词形成互不干扰的两个层面。彩色单词为封面出现的章节标题，贯穿页面中部。黑色文字和细线则使用较为自由的版式进行排版，以语义和情绪的自然中断进行分割和组合，类似视觉诗。

Swedish graphic designer HC Ericson is good at writing and designing books with unique visual forms. This book is a story ‘written’ with visual language. His another similar book *Ordbilder*, won the 1993 Honor Award for the most beautiful book in the world (you can find it on the relevant page of this book). The paperback book is glue bound by thread sewing with a bare spine. The cover can be completely opened to expose the spine. The cover is made of transparent PVC material printed in gray and the inner cover is made of white cardboard treated with Orifice plate printing in 19 colors (black plus 18 colors). The inner pages are made of coated art paper printed in 19 colors (the same as the inner cover). In the book, there are two non-interfering levels formed by black text, black Z-lines and large colorful words. Colorful words are the chapter titles that appear on the cover and run through the pages. Black text and Z-lines are arranged in a free format, and are naturally segmented and combined according to semantic and emotional paragraphs, similar to visual poetry.

RENA, R

SÅ RENT OC

RR SOM S

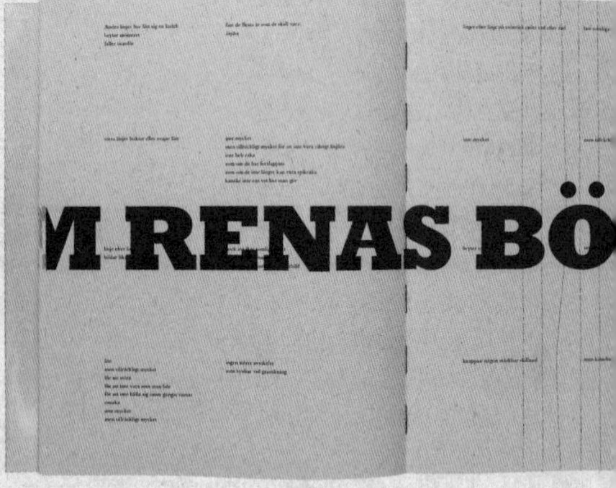

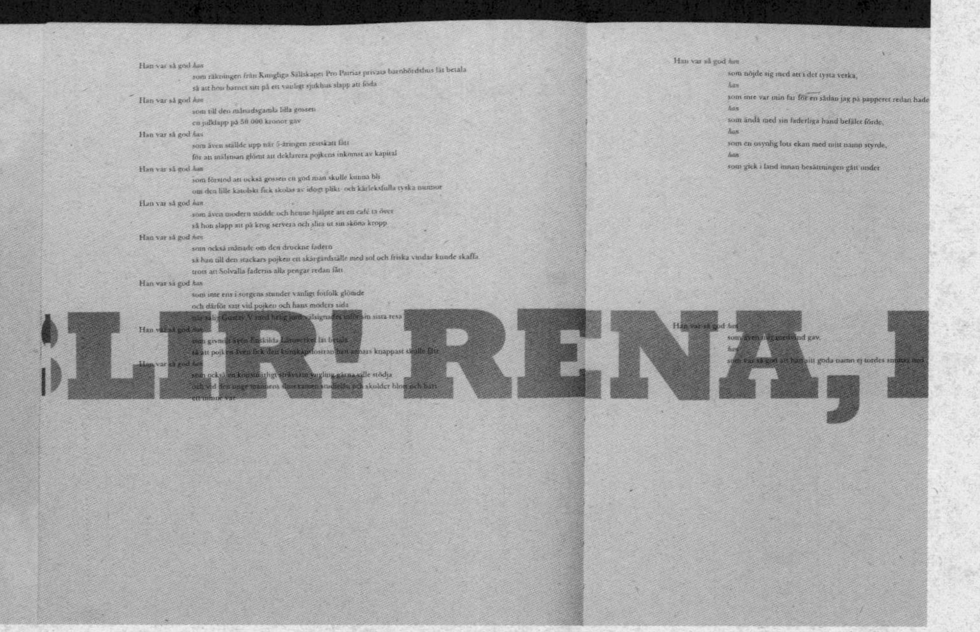

Han var så god *han*
som räkningen från Kungliga Sällskapet Pro Patrias privata barnbördshus lät betala
så att hon barnet sitt på ett vanligt sjukhus slapp att föda
Han var så god *han*
som till den månadsgamla lilla gossen
en julklapp på 50 000 kronor gav
Han var så god *han*
som även ställde upp när 5-åringen restskatt fått
för att målsman glömt att deklarera pojkens inkomst av kapital
Han var så god *han*
som förstod att också gossen en god man skulle kunna bli
om den lille katolskt fick skolas av idogt plikt- och kärleksfulla tyska nunnor
Han var så god *han*
som även modern stödde och henne hjälpte att ett café ta över
så hon slapp att på krog servera och slita ut sin sköna kropp
Han var så god *han*
som också månade om den druckne fadern
så han till den stackars pojken ett skärgårdsställe med sol och friska vindar kunde skaffa
trots att Solvalla faderns alla pengar redan fått
Han var så god *han*
som inte ens i sorgens stunder vanligt fotfolk glömde
och därför satt vid pojken och hans moders sida
när salig Gustav V med helig jord välsignades inför sin sista resa
Han var så god *han*
som givmilt även Enskilda Läroverket lät betala
så att pojken även fick den kunskapskost han annars knappast skulle fått
Han var så god *han*
som också en konstnärligt strävsam yngling gärna ville stödja
och vid den unge mannens [illegible] och skolder blott och bart
ett minne var

Han var så god *han*
som nöjde sig med att i det tysta verka,
han
som inte var min far för en sådan jag på papperet redan hade
han
som ändå med sin faderliga hand befälet förde,
han
som en osynlig lots ekan med mitt namn styrde,
han
som gick i land innan besättningen gått under

Han var så god *han*
som även [illegible] gav.
han
som var så god att han sitt goda namn ej tordes [illegible]

LIR! RENA,

en lånegivare som inget längre kräver
en låntagare som sitt har gjort
förvisso aldrig mera skall skulder utebli

och och och och och och och och och och och och och och och och och och och och och och och och och och och och och och och och

så länge människor som dessa
som du och jag och andra
trots välfärd väntar med att leva
så länge är vi alla döda
och sonar våra brott
så ingen annan än en gång oss kan klandra
och inget straff gottgöra
de gärningar vi gömt

R. R

förstenad och med nerver av stål
och ändå flyktigare än gas
och räddare kan ingen vara blott en dag i

dödsstraff på ett ögonblick är över
men själslig våldtäkt
dör aldrig
och kan ej tvättas bort
än mindre av den stumme förtigas

RE

VÄCKER,

1998 Bm5

MID: 969000940
...k Return
...89 Old Country Rd N
...llington FL 33414
Origin Post:
US Postal Service
Date of Mailing:
11/06/2019
Importer's reference:
Importer's contact:
P
062S00060566 2223737
ComBasPrice
description of contents:
...of item: Gift
Qty
2
Net Weight (lbs/oz)
7.0
Value (US $)
10.00
HS Tariff Number
Origin Country
US
TOTAL 2 7.0 10.00
...s reference:
Exporter's contact
Total Dimensions
L W H
Exemptions:
...37(a)
Invoice/Licence/Certificate No(s)
I certify the particulars given in this custo...
are correct. This form does not contain a...
...ous articles, or articles prohibited
...al or customs regulations. I have
export filing requirements under federal...
Sender's Signature and Date
Book Return
Return to Sender Instructions in case of
Treat as Abandoned
2.4lbs
555-5555
...AO QING
...BEI XIANG7-3,NANJING HAN QING TANG DESIGN CO.
...NJING 210000
...INA
U.S.A.
U.S.A.
China

291mm

19mm

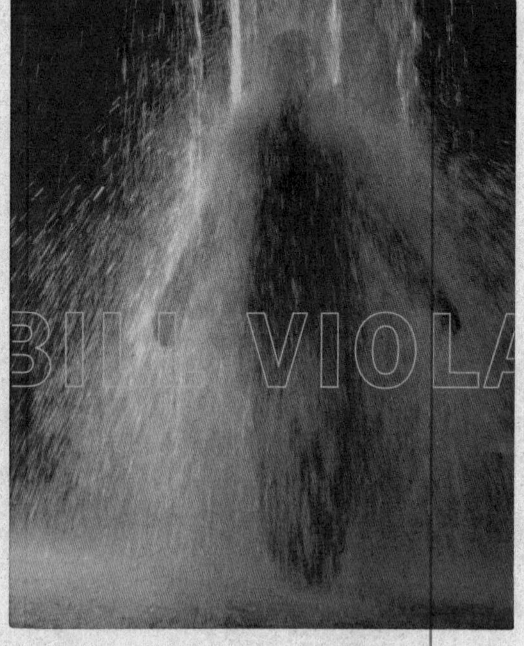

242mm

◎ 铜奖
《 比尔 · 维奥拉
◐ 美国

Bronze Medal
Bill Viola
U.S.A.
Rebeca Méndez
——
Whitney Museum of American Art, New York; Flammarion, Paris
291 242 19mm
1219g
216p
ISBN-10 0-87427-114-2

比尔·维奥拉（Bill Viola）是著名的装置艺术家，本书是他1998年在纽约惠特尼美国艺术博物馆展览时出版的作品集。书籍的装订形式为锁线平装。封面采用白卡纸四色印刷覆哑膜。内页采用涂布艺术纸四色印刷。内容方面，作为一本展览作品集，前言、参考书目、展览年表等内容都配置完整，采用双栏排版。作品部分精选出维奥拉创作于1972－1996年的64件作品。由于维奥拉的作品以影像和装置为主，平面上通过单帧图片配合文字很难展现作品全貌，所以大部分的页面均采用了作品的多帧截图，通过图片的线性关系、图片间的大小对比，以及文字的“流动性”排版体现时间性。用于封面封底的第64号作品——《穿越》（*The Crossing*）采用了左右页分别展示的方式，向后翻阅的过程中，两条时间线并行发展，有如两个并置的独立屏幕。

Bill Viola is a famous installation artist. This book is a collection of his works exhibited at the Whitney American Art Museum in New York in 1998. This paperback book is bound by thread sewing. The cover is printed with white cardboard in four colors and covered with matte film. The inner pages are printed in four colors with coated art paper. In terms of content, as a collection of exhibition works, the preface, bibliography, exhibition chronology and other contents are completely configured, using double-column typesetting. The part of his works shows the 64 works selected by Viola from 1972 to 1996. As Viola's works are mainly based on images and installations, it is difficult to show the full picture of the works through a single frame picture with words. Therefore, most of the pages adopt multi-frame screenshots of the works, reflecting a kind of timeliness through the linear relationship of pictures, the size comparison between pictures, and the 'flow' of words. 'The crossing', which is used for the front cover and back cover, is displayed on both the left page and the right page. When the readers browse backward, two time lines develop in parallel, like watching two juxtaposed independent screens.

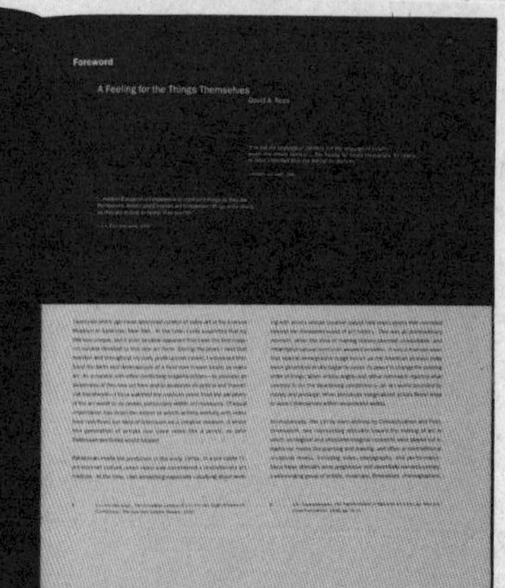
Foreword
A Feeling for the Things Themselves
David A. Ross

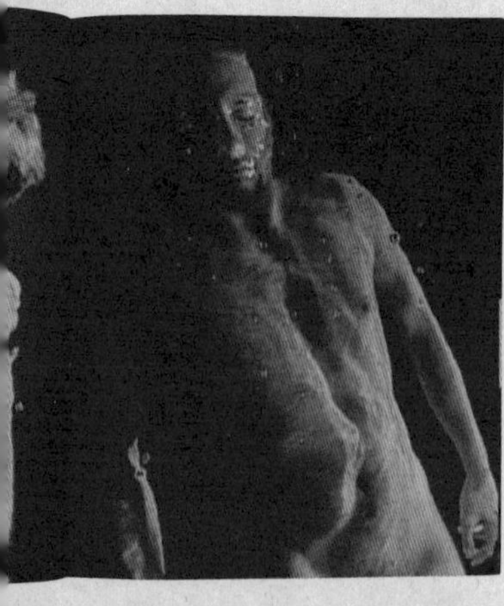

1
Tape I
2
Composition "D"

16
The Space Between the Teeth
17
Truth Through Mass Individuation

50

11

A Million Other Things (2)

1975

Videotape

Changes in light and sound on the edge of a pond during an eight-hour period, from afternoon to night, are composed into rhythmic variations resembling music. The individual is seen as a static fixture in the environment, sitting immobile in a chair for the duration. Finally, the sun sets, and he remains the sole visible object illuminated by a single electric lamp.

12

Return

1975

Videotape

The composition of the tape reflects the sonic structure of a ringing bell, complete with its echoes and subsequent reverberations. A man is seen within a natural landscape, moving toward the camera while pausing every few steps to ring a brass bell he is carrying in his hand. As the tape progresses, each time the bell is struck, a rapid edit sequence carries the figure back through each previous position where the bell was rung, all occurring within the time of the fading sound of a single ring. As the man advances, a cumulative process occurs whereby at each station his image is momentarily thrust back to the starting point, only then to advance further. Finally reaching his destination, he arrives at the camera, where his body momentarily blocks the view. He then is taken through all the stages of the journey in a single instant, returning to the point from where he began.

s screening, and machinery operating, are run throug
ne drumbeats. This allows these sounds to reach the
in the space. However, the loudness of the drum at
only present in the room as sonic "afterimages," aud
ent to listeners that the character of the echoes has c
nbeats. The more the drum is beat, the more the sha
sounds, altering their original qualities and sonic forn
the performer.

Reverse Tel

1983

Broadcast television project

ved as a sculptural form to tempor
to reveal the hidden dimension of t
s at every instant. Isolated from an

of people sitting at home, staring ir

38

e Mountain

1980

Music composition

A large tree leans diagonally a
far corner of the space. Fifty s
dominated by electronic noise
loud static and noise come thr
the bursts of noise. The only
sound in the room is the delic

In Version

1973

Videotape

BILL V

ences of distortions, one human and physical, the other electronic and
s he undergoes a series of violent contortions caused by physically pre
mera. At times painful, at times comical, the facial contortions are pre
a still frame. The abstract sound of electronic noise and distortion is
which a black-and-white video monitor placed on a table undergoes a se
ble patterns of visual noise. The corresponding sonic distortions are

Jacopo Pontormo
Visitation, c. 1528-29
Oil on panel, 80 × 61" (202 × 156 cm)
Pieve di San Michele, Carmignano, Italy

Bill Viola
From *The Greeting*, 1995
Video/sound installation, 14' × 21'6" × 25'6" (4.3 × 6.6 × 7.8 m)
Artist's proof 2. Whitney Museum of American Art, New York
Partial and promised gift of an anonymous donor 96.95

never announces itself, being always in motion toward transparency.

Viola's approach couples his uncommon empathy with the viewer's conscious experience and a deep scientific understanding of the neurophysiological mechanisms of perception. As noted earlier, Viola inherited these concerns in large part from Peter Campus. The powerful intellectual and corporeal impact of Campus' video installations encouraged Viola's interest in the raw physicality of perception and cognition. Yet even in his earliest videotaped works, produced while still in art school, it was quite evident that Viola had a strong interest in the connections between the science of cognition and the metaphysical implications of consciousness.[12]

In early works such as *He Weeps for You* (1976), these concerns would be expressed primarily in terms of scientific methodology. However, in later works such as *The Crossing* (1996), they are explored more within metaphysical realms. In terms of an image vocabulary, Viola's preoccupation with water and fire, his profoundly moving meditations on birth and dying, his recurring use of extreme slow motion to create states of altered consciousness—metaphorically evoked or directly represented in what we might term suspended time—locates him within a set of specific critical references.

These references are not to the historical practices of Surrealism or its associated intellectual connections to contemporary Freudian theory. Viola is resolutely a realist, though one who has found a space for his own approach to the real wedged somewhere between modernist concerns for art as a reflection of our collective unconscious and postmodern critiques of the sources and uses of dream and fantasy. He can be situated with other late twentieth-century artists who employ the symbolic codes drawn from sources as disparate as the great Georgian filmmaker Andrei Tarkovsky[13] and the works of the sixteenth-century Italian Mannerist Jacopo Pontormo, whose *Visitation* is the specific source of Viola's *The Greeting* (1995).[14]

Perhaps equally important, Viola developed an understanding of the ways contemporary art employed the structure and intelligence of traditional and non-Western art. In a recent conversation with me, he observed that Buddhist works of art still function as sites of ritual offerings and prayer even when situated and recontextualized within Japanese art museums, outside their original temple settings.[15] Similarly, Viola has commented on the power and use of art, citing the famed sixteenth-century Isenheim Altarpiece by Matthias Grünewald, which was originally used as an integral part of a healing procedure. The different panels were opened and closed in a process that required the patient viewer to spend hours in front of the altarpiece in prayer and contemplation. Does this work still function in the same way in a secularized museum context? Can art still fulfill an analogous role in our times? If so, should it? These are some of the questions implicit in Viola's demanding art. And his use of the installation form (especially when sited outside the art museum context) functions as an implied critique of the leveling and neutralizing function that the modern art museum inadvertently performs.

Viola nevertheless eschews overtly theoretical postmodern concerns; he does not question his role as an artist or provoke such questions in the minds of those experiencing his work. In the traditional manner of great art, Viola prov[illegible] mind to avenues of contemplation a[illegible] the art provides the basis for an expe[illegible] scendent—a curious word to use at th[illegible] reproduction, yet the only word that ap[illegible]

How does all this come about, given tha[illegible] cise observation of the sensate world i[illegible] method? How we see and hear, how th[illegible] behold and the ears receive, as well as [illegible] the stuff of Viola's work. This seems [illegible] artists allied to the spectacular interse[illegible] Viola does not fit into that particular g[illegible] marily concerned with the humanizing o[illegible] critique of science as the root of the [illegible]

If anything, Viola is a postmodern huma[illegible] a lifelong devotion to Eastern spiritual [illegible] pletely uninterested in the collapse of [illegible]

12. Viola's student experience at Syracuse University had a strong influence. [illegible] did the art school support his early exploration of video as a creative [illegible], but the university's psychology department maintained a [illegible] laboratory for the study of sensory perception. Viola often availed himself, as an uninvited visitor, of the laboratory and its resources.

13. See J. Hoberman's essay in Barbara London, ed., *Bill Viola: Installations and Videotapes*, exh. cat. (New York: The Museum of Modern Art, 1987), pp. 63-72.

14. For Viola's use of Pontormo's *Visitation*, see Susie Kalil, "Buried Secrets," in Marilyn A. Zeitlin, *Bill Viola: Buried Secrets*, [illegible], exh. cat. (Tempe, Arizona: Arizona State University Art Museum; Hannover, Germany: Kestner-Gesellschaft, 1995), p. 33.

15. Unpublished tape-recorded conversation with the artist, March 1997.

CN22
Customs Declaration
Deutsche Post
FROM
TO
UB 76 342 359 1DE

Germany
Germany
China
CN22
Documents
Others
Weight (in kg)
Value (in EUR)
0.35
7.50
Total Weight
Total Value
0.35
7.50
FROM
Elops e.V.
Südring 5
91438 Bad Windsheim
GERMANY
TO
ZHAO QING; Tang Design Co.
Nanjing Han Qing
Dabei xiang 7-3
210000 NANJING
CHINA
PRIORITY P.P.
W
IM 06.11.19
WARENP. INT XS
A0 0223 1309
00 0000 0756

239mm

11mm

169mm

● 荣誉奖
《 库鲁 · 芒金——埃及梦
◖ 德国

Honorary Appreciation
Kůllu mumkin: Ein ägyptischer Traum
Germany
Matthias Beyrow, Marion Wagner
Werner Döppner
Beyrow, Wagner, Berlin
239 169 11mm
298g
64p
ISBN-10 3-00-002924-9

维尔内·德普纳（Werner Döppner）于 1988－1991 年生活在埃及，之后他将自己的经历作为素材，创作了本书《库鲁·芒金——埃及梦》，以及同名公益戏剧。本书讲述了两个埃及女孩共同做的梦——坐上飞毯在埃及探险。书籍的装订形式为锁线硬精装。封面采用深蓝色艺术纸，书名烫金处理。环衬采用相同的纸张印金。内页哑面艺术纸黑、金、深蓝三色印刷。作为一本儿童与青少年读物，故事围绕女孩的梦境展开叙述，但故事里所描述的地点、环境、埃及人的生活都是真实的。作者为了体现梦境与现实的相互印证、模糊两者之间的界限，在设计上做了很多处理。首先，封面书名和内页章节标题采用了横向镜像的做法，使得文字本身呈现出神秘而古老的异域风情。环衬上通过对金色油墨量的控制（透明度）产生虚实两层图案。书中的插图由温弗里德学校的学生帮助绘制，照片则是作者自己拍摄，局部贴上文中谈及的物品碎片（比如飞毯的局部、树皮等）用以印证童话故事的“真实性”。

Werner Döppner, who lived in Egypt from 1988 to 1991, used his own experience as material to create the book *Kullu mumkin: Ein ägyptischer Traum*, as well as the public welfare drama of the same name. The main story of this book is about the dream of two Egyptian girls – sitting on a flying carpet to explore Egypt. The hardcover book is bound by thread sewing. The cover is made of dark blue art paper treated with gold hot stamping. The end paper is made of the same paper printed in gold. The inside pages are made of matte art paper printed in black, gold and dark blue. As a children's and teenagers' book, the story is about the girl's dream, but the location, environment and Egyptian life described in the story are all real. In order to reflect the mutual verification of dream and reality, the author has done a lot in book design to blur the boundary between them. First of all, the title of the book and the titles of the chapters are mirrored horizontally, which makes the text itself mysterious and exotic. Though the control of the amount of gold ink (transparency), there are two layers of patterns on the end paper –virtual and real. The illustrations in the book were drawn with the help of students from Winfried School, and the photos were taken by the author himself. The fragments mentioned in the book (such as parts of the flying carpet, tree bark, etc.) were pasted to confirm the 'authenticity' of the fairy tale.

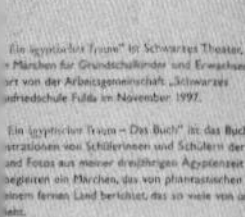

1995 konnte in Ras Sudr auf dem Sinai, unmittelbar am Ufer des Roten Meeres, ein Gesundheitszentrum aufgebaut werden, nachdem man in zahlreichen Sommerlagern mit Kindern und Familien der Deponie das Heilklima des Meeres und der Wüste erkannt hatte. Arbeitslose junge Männer, die Erfahrung beim Hausbau gesammelt hatten, konnten sich hier längerfristig weiterbilden und zugleich verdienen.
Die Arbeit von Schwester Maria und ihren zur Zeit 31 Mitarbeitern und Mitarbeiterinnen wird ausschließlich aus Spenden finanziert. Alle Einnahmen – außer Postgebühren – kommen den Projekten zugute. Schwester Maria wurde im vergangenen Jahr im Schloß Bellevue für ihren selbstlosen Einsatz von Bundespräsident Roman Herzog das Bundesverdienstkreuz 1. Ordnung verliehen.

Während meines dreijährigen Aufenthaltes in Ägypten lernte ich Schwester Maria kennen. Ihr tiefer Glaube und die darin gründende Kraft dieser Frau haben mich zutiefst beeindruckt.
Mit der Arbeitsgemeinschaft „Schwarzes Theater" möchten wir Schwester Maria unterstützen.

Ich möchte allen danken, die mich bei der Entstehung des Buches unterstützt haben.

Werner Doppner

III. BEI DEN PYRAMIDEN

Nach einem wunderschönen Flug über die Alpen und über das Mittelmeer erreichen Nelli und Jule Nordafrika und landen direkt neben den drei Pyramiden, die sich am Stadtrand von Kairo befinden.
Sie sind in Ägypten, ein Traum ist wahrgeworden!
Die beiden Mädchen sind weit weg von Fulda, 4000 Kilometer ungefähr, am Stadtrand von Kairo. Diese Riesenstadt liegt direkt vor ihnen. 17 Millionen Menschen sollen hier wohnen – siebzehn Millionen! Nelli und Jule können sich diese Menschenmenge überhaupt nicht vorstellen. Das sind etwa 340mal mehr Menschen als in Fulda, selbstverständlich mit den Leuten, die in Bronnzell, Horas, Lehnerz, Neuenberg, Johannesberg, Edelzell, Sickels und den anderen Vororten wohnen. Unglaublich!
Aber früher, als die Pyramiden gebaut wurden, haben noch nicht so viele Menschen hier gewohnt. Das ist aber auch schon wieder 4000 Jahre her. Wie die Zeit vergeht!

3 bis 1991) begleiten ein Märchen, das
nissen aus einem fernen Land berichte
nen Bann zieht.
Projekt „Kullu mumkin – Ein ägyptisch
etzten Endes auf die Unterstützung e
n einer Müllstadt in Kairo die Notlei
tz unterstützt.
aria Theresia Grabis ist die Projektlei
ester Maria – Kairo e.V." Schwester M
0 Jahren mit unterprivilegierten Mädch
den Müllsammlerfamilien auf der Müll
hlreichen Nähschulen wurden Mädche
n und Schneiderinnen ausgebildet. 19
n in Kairo Garden City mit dem Aufba
nderte Jugendliche. 1981 wurde auf die
perative „Hilfe zur Selbsthilfe" gegründ
dem Bau einer Sozialstation und eine
ooperative 1982 mit finanzieller Unter
ngshäuser erbauen. Heute leben bereit

lotzen, gelangweilt von
ts kauend, die beiden
ub, scheinen wichtige
nerkwürdige Pfeifen,
n stehen Gefäße aus
efäßen beginnen seltsame
e sich die braungebrann-
venn sie daran ziehen,
uch nicht vom Schlag
s ihren Bärten, und es
gluckst so, als hätte
re jetzt satt – jedenfalls

n an, die langsam, schon ein wenig ängstlich w
nung im Felsen zusteuert. Von einem Touriste
Mädchen, daß es in dem hinter dem Eingang
cht immer mit rechten Dingen zugehen soll.
Gruppe ist schweißgebadet, nicht nur von de
er Könige unerträglich ist.

en betreten mit Nelli und Jule hinter dem Ei
n Korridor. Ein Korridor ist ein langer, in den
r Gang, der zur eigentlichen Grabkammer füh
nmern stellte man früher die Sarkophage, das
r Pharaonen. Die Mädchen haben den Korrido

Nach dem Ta
frost und Jul
bei den Pyra
sentiert das
sten Bilder.
den fruchtba
die auf beide
um dann sof
aus der Höh
Jule wie klei
die beiden n
anlagen und

1. DIE BESTEN FREUNDINNEN

Nelli Schüttelfrost und Jule von Kieselbrett sind Schülerinnen der 7. Klasse und miteinander befreundet. In der Schule haben sie keine Probleme – wenn überhaupt, dann die Lehrer mit ihnen.

Wie fast alle Mädchen in diesem Alter lieben sie die Mathematik und die „Streuselkuchen", das ist die Klassenband, bestehend aus drei Jungen. Sie heißen Eitelwolf Flachdach, Zwieback Peng und Pinky Eff. Eitelwolf, Zwieback und Pinky kommen sich so unglaublich bedeutend vor und erzählen immer wieder den Schulneulingen, sie seien die Vorgruppe von „Oro Paks", der aktuellen Kultband der Stadt. So bleibt es nicht aus, daß die Jungs der Schwarm aller Fünftklässlerinnen sind; und immer wieder ist zu beobachten, daß Mädchen im Pausenhof in Ohnmacht fallen, wenn sie einem der „Streuselkuchen" zu nahe kommen. Übrigens werden Eitelwolf, Zwieback und Pinky „Streuselkuchen" genannt, weil sie alle zur gleichen Zeit die Windpocken hatten.

Eitelwolfs Vater muß früher ziemlich durchgeknallt gewesen sein. Bis vor kurzem hat er noch als Auslaufmodell in einer Band die Gitarre gequält und auf den Spitznamen „Ramses" gehört. Eitelwolf weiß zwar nicht, wie sein Vater zu dem Spitznamen gekommen ist, daß aber Ramses ein Pharao war, das hat sich auch bis zu ihm rumgesprochen, und daß Ramses II. der Pharao war, der architekturmäßig am meisten draufgehabt haben soll und als großer Baumeister in die Architekturgeschichte eingegangen ist, sieht man am Haus von Eitelwolfs Eltern. Das ist fast so groß wie die Schule! Ehrlich!

Eitelwolf, Zwieback und Pinky sind eigentlich ständig Mittelpunkt der Gespräche der Mädchen, die schon mal heimlich in der „Teeny" blättern, einer Zeitschrift für Girls ab 14. Die Jungs sind halt die absolut Coolsten. Man unterhält sich über ihre musikalischen Begabungen und was daraus hätte werden können, wenn die drei die Jugendmusikschule mal von innen gesehen hätten, über Zwiebacks Pferd, das schielt und wahnsinnig gerne Gummibärchen frißt, über Eitelwolfs Vater Ramses und seinen merkwürdigen Spitznamen und und und.

Das war auch neulich so, als Nelli an einem Wochenende bei ihrer besten Freundin Jule übernachtete. Sie waren ausnahmsweise einmal nicht gut zu sprechen auf die „Streuselkuchen". Irgendwie hatten diese sich ihnen gegenüber in der Woche zuvor nicht korrekt verhalten. Doch das wird sich sicher bald wieder zum Guten wenden!

Die beiden Mädchen plauderten lange miteinander. Sie hatten es sich gemütlich gemacht – mit Brennesseltee und dicken Bohnen. Dicke Bohnen gibt es auch jeden Sonntag bei Flachdachs. Eitelwolf kommen diese Dinger fast schon aus den Ohren raus, er kann dicke Bohnen nicht mehr sehen! Doch Eitelwolfs Vater Ramses steht auf dicke Bohnen, denn die sind für viele Ägypter das Hauptnahrungsmittel und heißen dort Ful. Und in Ägypten verbringen Flachdachs schließlich immer ihre Osterferien.

Jule zeigte gegen Mitternacht ihrer Freundin Nelli Fotos von Flachdachs letztem Osterurlaub, die sie von Eitelwolf geliehen hatte: Flachdachs beim Fulessen in einem Basar, Flachdachs beim Fulessen vor den Pyramiden, Flachdachs beim Fulessen in einer Oase, Eitelwolfs Vater Ramses nach dem Fulessen mit einer Wasserpfeife neben einem Kamel, Eitelwolf mit seiner Mutter nach dem Fulessen alleine in Kairo. Jule und Nelli hatten schon viel gelesen über dieses geheimnisvolle Land und schauten sich in aller Ruhe die Bilder an. Sie waren restlos begeistert von Ägypten, von Palmen und Beduinen, vom Basar und von Moscheen.

Jule und Nelli löschten das Licht und fingen an zu träumen.

• • •

Nelli und Jule sind froh, kurz nach Sonnenuntergang eine Oase zu erreichen. Hier gibt es Palmen, Lehmhütten und natürlich Sprit für die Kamele. Freundliche Beduinen empfangen sie und laden sie ein, eine gemeinsame Mahlzeit einzunehmen. Hierzu nehmen sie Platz an einem Lagerfeuer, in dem die Beduinen Fladenbrote backen. Diese werden dann gefüllt mit Schafskäse und verschiedenen grünen Blättern, die Endiviensalat ähneln. Dazu trinken sie Schai, das ist ägyptischer Tee.
Es schmeckt köstlich!
Nach dem Essen organisieren die Beduinen Wasserpfeifen, stekken sich die „Stromkabel" in die Bärte und machen Dampf. Nelli und Jule legen sich abseits unter eine Palme und beobachten den herrlich klaren Sternenhimmel.

Kamele – Daniel Erb, 13
Palmenhain – Alba Talamo, 17
Oase – Caroline F. Zint, 19

8 Ha 2

Büchersendung
PRESSE UND BUCH INTERNATIONAL
QING TANG DESIGN CO.
ZHAO QING
DABEI XIANG7-3,NANJING HAN
NANJING 210000
CHINA
Small Packet
Deutsche Post
Hans

Büchersendung
BUCH INTERNATIONAL
renvoyer à l'expéditeur
return to sender
zurück an Absender
Buchspeicher.d
QING TAN
ZHAO QIN
DABEI XIA
NANJING
CHINA

Germany
Germany
China

225mm

8mm

FRIED / GLEICH · HAT OPA EINEN ANZUG AN ?
HANSER

294mm

● 荣誉奖
《 爷爷穿西装吗?
◖ 德国

Honorary Appreciation
Hat Opa einen Anzug an?
Germany
Claus Seitz
Amelie Fried, Jacky Gleich
Carl Hanser Verlag, München
225 294 8mm
390g
32p
ISBN-10 3-446-19076-7

这是一本专为儿童绘制的生死启蒙书。书中讲述了小男孩布鲁诺的经历：懵懂地参加了爷爷的葬礼，之后很久才逐渐感受到失去亲人的悲痛，探索死后身体埋葬和灵魂归所的意义，逐步走出伤痛并在心中怀念爷爷。本书的装订形式为锁线硬精装，横式开本也是典型的儿童绘本做法。封面铜版纸四色印刷，覆哑光膜，裱覆卡板。环衬选用与整体插图风格统一的土红色。内页胶版纸四色印刷。由于页数较少且故事连贯，没有索引和查找的必要，所以全书未设置页码，减少了画面与故事之外的元素干扰。插图风格充满丰富的笔触肌理，具有稚拙的童趣，色调以棕色为主，衬托悲伤的心理主题。

This is an enlightenment book about life and death drawn specifically for children. The book tells about the experience of the little boy Bruno: he attended grandfather's funeral ignorantly, and gradually felt the grief of losing his relatives for a long time, then explored the meaning of body burial and soul returning, finally he got rid of the pain and missed his grandfather deeply in the heart. The hardcover book is bound by thread sewing, and the horizontal format is also typical for children's picture books. The cover is printed on coated paper in four colors, covered with matte film and mounted with cardboard. The color of the end paper is earth red that is consistent with the overall style of illustrations. The inside pages are made of offset paper printed in four colors. Since there aren't many pages and the story is coherent, there is no need to index and search, so the book does not set page numbers, which reduces the interference of elements other than the pictures and the story. The illustrations are rich in brushstrokes, full of childishness, and the color is mainly brown to set off the sad psychological theme.

»Hat Opa einen Anzug an?«, fragte Bruno und reckte sich auf die Zehenspitzen, um in den Sarg sehen zu können. »Opa ist von uns gegangen«, hatte Xaver gesagt, aber das stimmte gar nicht. Opa war kein bisschen davongegangen. Er lag da, schon seit vielen Stunden, ganz ruhig, ohne sich zu bewegen. Die ganze Zeit hatte Bruno nur die Sohlen der schwarzen Schuhe sehen können, die zur Hälfte über den Rand des Sarges ragten. Wenn Opa die schwarzen Schuhe trug, dann hatte er auch einen Anzug an. Die schwarzen Schuhe waren die Anzugschuhe, sonst trug der Opa Schnürstiefel.

Bruno setzte sich in den riesigen, alten Sessel, der am Fenster stand. Dort saß Opa immer und blätterte in den Büchern mit den Schiffen. Opa liebte Schiffe. Er hatte davon geträumt, zur See zu fahren. Aber dann war er doch Bauer geworden wie sein Vater und sein Großvater. Auch Brunos Papa war Bauer. Bruno überlegte, ob er auch einmal Bauer werden würde.

Er entdeckte das winzig kleine Holzschiffchen, das Opa von der einzigen Reise seines Lebens mitgebracht hatte. Er war damals in einer Stadt gewesen, die hieß Genua. Bruno wusste nicht, wo dieses Genua war, er wusste nur, dass es dort einen Hafen und viele Schiffe gab. »Das erbst du mal«, hatte Opa ihn immer getröstet, wenn Bruno bettelte, dass er das kleine Holzschiff unbedingt haben wollte. Erben, das hieß, man kriegte etwas von einem, der gestorben war. Jetzt, wo der Opa gestorben war, gehörte das Schiff also ihm. Bruno nahm es vorsichtig in die Hand, betrachtete es und strich mit dem Finger über das Holz. Dann schob er es unter seinen Pullover und sah lange aus dem Fenster.

Irgendjemand hob Bruno hoch. Endlich konnte er Opa genau ansehen. Opa trug tatsächlich einen Anzug. Er hatte die Hände gefaltet und die Augen geschlossen. »Er ist gar nicht tot, er schläft ja nur!«, rief Bruno aus. Sanft landeten seine Füße wieder auf dem Boden. Irgendjemand strich ihm übers Haar und murmelte: »Armer Bub.«

Mama und Papa unterhielten sich darüber, ob Bruno mit zur Beerdigung sollte. »Beerdigung«, das klang so ähnlich wie »Begradigung«. Das war, wenn Bruno im Sandkasten eine krumme Straße gerade machte. Wenn Begradigung hieß, dass man etwas gerade machte, dann hieß Beerdigung, dass man aus etwas Erde machte. Wie aus dem Opa Erde gemacht werden sollte, das musste Bruno unbedingt sehen. »Ich will mit!«, sagte er deshalb.

Am anderen Tag überraschte Xaver Bruno in der Küche. Bruno versuchte, etwas hinter dem Rücken zu verstecken. Xaver riss ihn am Arm, und das Senfbrot, das Bruno sich gerade geschmiert hatte, kam zum Vorschein.
»Wieso hältst du das Brot mit dem Senf nach unten?«, fragte Xaver.
»Weil der Opa es sonst vom Himmel aus sieht«, sagte Bruno.
Xaver lachte ihn aus. »Du bist so blöd«, sagte er und rannte aus der Küche.
Bruno aß nachdenklich sein Senfbrot. Wahrscheinlich war der Himmel eh so weit weg, dass der Opa nicht sehen konnte, womit Brunos Brot bestrichen war. Kurzsichtig war der Opa auch.

Das mit der Seele und dem lieben Gott ließ Bruno keine Ruhe.
»Sind die Seelen im Himmel oben lebendig?«, fragte er den Vater.
»Ich glaube schon.«
»Also ist der Opa nicht wirklich tot?«
»Er ist schon tot, aber in unserer Erinnerung lebt er weiter.«
Bruno nickte. Ja, das stimmte. Sobald er ganz fest an Opa dachte und sich vorstellte, wie er aussah, war es so, als könnte er jeden Moment ins Zimmer kommen.
»Und wenn ich vergesse, wie er aussieht?«, fragte Bruno.

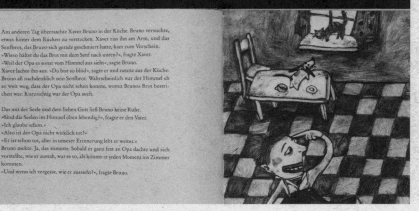

998 Ha4

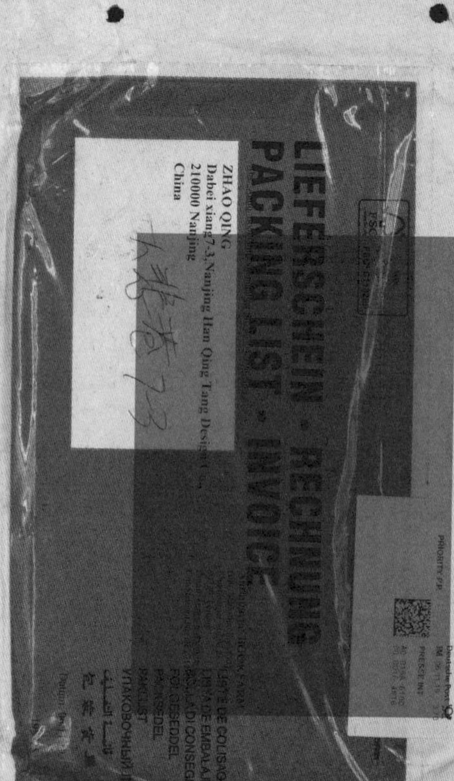
LIEFERSCHEIN · RECHNUNG
PACKING LIST · INVOICE
ZHAO QING
Dabei xiang7-3,Nanjing Han Qing Tang Design
210000 Nanjing
China
PRIORITY P.P.
Deutsche Post
LISTE DE COLISAGE
LISTA DE EMBALAJE
PAKSEDEL
PAKLIST

Hungary

Germany

China

179mm

12mm

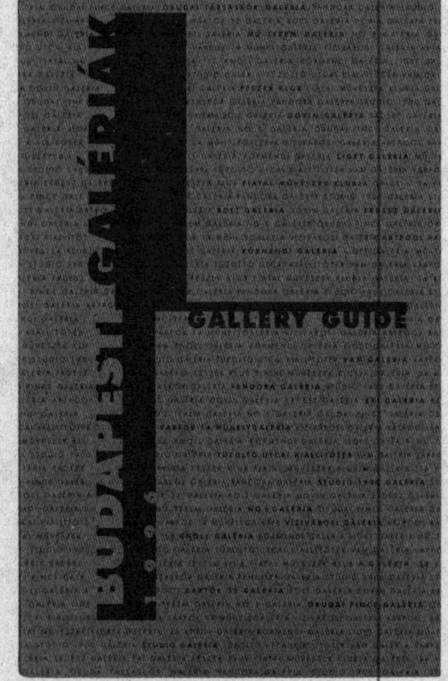

114mm

荣誉奖 Honorary Appreciation

1996年布达佩斯美术馆指南 Budapest Galériák 1996 / Gallery Guide of Budapest 1996

匈牙利 Hungary

Johanna Bárd

——

Budapest Art Expo Alapítvány, Budapest

179 114 12mm

228g

146p

ISBN 936-04-6417-9

1996 年是布达佩斯艺术博览会举办的第六年，主办方适时推出了这本介绍布达佩斯最重要的画廊、非营利性画廊和展览的小册子，希望可以引起画廊、藏家、艺术家和艺术爱好者的关注，提供更多的信息和便利。本书的装订形式为无线胶平装。封面采用灰色艺术卡纸印单黑。内页哑面铜版纸四色印刷。内容方面主要分为三部分：布达佩斯艺术博览会的整体介绍、主要画廊和重要展览推荐、画廊地址和联系方式，另附一张展览地图拉页。由于是匈牙利语和英语双语，在这样一个小开本并且版面设置为三栏的情况下，各部分的排版都用足了心思，将密集的信息尽可能多地组织在这么小的手册里。

1996 is the sixth year of the Budapest Art Fair. The organizer timely launched this booklet to introduce Budapest's most important galleries and non-profit galleries and exhibitions, hoping to attract galleries, collectors, artists and art lovers, and to provide more information and convenience. The paperback book is glue bound. The cover is made of gray art paper printed in single black. The inside pages are made of matte coated paper printed in four colors. The content is mainly divided into three parts: the overall introduction of the Budapest Art Fair, the recommendation of the main galleries and important exhibitions, the addresses of galleries and contact information, and an exhibition map attached as a pullout page. Since the book is bilingual in Hungarian and English, in such a small format, the typesetting of each part has taken enough care to organize as much information as possible in three columns.

PESTI TAVASZI FESZTIVÁL / BUDAPEST SPRING FESTIVAL 

PESTI FŐVÁROSI ÖNKORMÁNYZAT / BUDAPEST MUNICIPALITY

N KIADÓ

ÉS KERESKEDELMI MINISZTÉRIUM TURIZMUS FŐOSZTÁLY
TRY OF INDUSTRY AND TRADE DEPARTMENT OF TOURISM

VÁNY MEGJELENT A VI. BUDAPEST ART EXPO ALKALMÁBÓL
HED ON THE OCCASSION OF THE VI. BUDAPEST ART EXPO
HUNGEXPO – BNV, BUDAPEST, 1996. 03. 21-24.

KIADJA / PUBLISHED BY
Budapest Art Expo Alapítvány

ALAPÍTÓ TAGOK / FOUNDING MEMBERS
Keserü Ilona, Beke László, Johan van Dam

KURATÓRIUM / BOARD OF TRUSTEES
Bak Imre, Boros Iván, Johan van Dam, Deli Ágnes,
Fitz Péter, Gergely Mariann (elnök),
Dr. Hardeljits Zsuzsanna, Jerger Krisztina, Keserü Ilona

BUDAPESTI GALÉRIÁK

GALLERY GUIDE

BUDAPEST ART EXPO ALAPÍTVÁNY

BUDAPEST, 1996

places. Importantly: a friend who is also dealing with art, once said it is easier to sell those works he likes himself. It was a surprise. I knew about the collectors passion, but it was there that I heard first, that a transaction can also be tough and complicated, breaching hot regions of human emotion. Once Klára Deák spoke with high esteem about the art dealer, who are the painters real, understanding friends, the pillars of creative processes. In France I heard these opinions from those, who had a contract with a gallery. This meant a "secured subsistence", but a professional career is several times real friendship. Of course it has severe conditions.

I most cordially think on the Jean Fournier Gallery in the street next to Beaubourg. Most recently I saw great paintings by Joan Mitchell in the spacious, snow white country rooms, passing the small inner-court. On the other side of the street (rue Quincampoix) is a bookshop belonging to the gallery, where, after much exertion, I choose not the most expensive, not the biggest and not the most correct in colour Mitchell catalogue. Fournier, by the way, also deal with Simon Hantai paintings.

Of course, in France many speak and understand the language of painting. (Ákos Birkás once tried to explain this in the Palace of Art, but they didn't let him finish.) It takes a lot to let living art speak in a city, to be present. For example lots of time, work, slow water, patience, teaching and understanding.

Let us trust the power of passion, which created the works, and another passion, which makes us want to posses these works. The gallerist functions in between these two fields. His work of a considered series of financial manoeuvres, is only a small part. Let us not be cowards. What we're talking about is nothing less than that art remains in the spill of human existence. Here, in the vicinity of mass graves also. We painters know this. Sooner or later art-historians will repent from dissipations to their serving function. Together we should build those surroundings, which despite the crises are functioning in those societies with high traditions, the roots should be picked up again. Often I'm reminded of the slogan of the 1991 Art LA:

~~Sex, Drugs, Money~~, Art, The Last Vice.

Ilona Keserü

„ITT MINDENKI KÍSÉRLETEZIK"

BESZÉLGETÉS MŰKERESKEDŐKKEL

KIÁLLÍTÓ MŰVÉSZEK / ARTISTS

Gyarmathy Tihamér
Szaby Lajos
Szabados Árpád
Haraszty István
Pauer Gyula
Frey Krisztián
Tölg-Molnár Zoltán
Dienes Gábor
Hadzi István
Donáth Péter

A több évtizedes gyűjtő múltra visszatekintő Körmendi–Csák gyűjtemény a piacgazdaságra való áttérés után célul tűzte a világszínvonalú kortárs magyar képzőművészek mecenatúráját, mind hazai, mind külföldi bemutatását.

Német, francia és osztrák kiállítások mellett a budapesti Art Expók állandó és legnagyobb területű résztvevője, több önálló művész szponzora.

A galéria a kiállítás és vásárok propagálója a dublini Art Fair résztvevője.

A Körmendi Galéria a kortárs magyar művészet konzervatív és avantgarde vonalának – egymástól elkülönítve – ad teret, törekszik a teljességre, elsősorban az itt felsorolt művészek rendszeres, ismételt bemutatásával.

Coming onto the artmarket, the Körmendi–Csák collection – with a history spanning several decades – the Gallery has set its the goal of becoming a mini-centre for quality contemporary Hungarian artists, as well as their presentation in Hungary and abroad.

Next to exhibitions in Germany, France and Austria, it is the permanent largest exhibitor at the Budapest Art Expo, and a sponsor of several individual artists. The gallery – an advocate of exhibitions and fairs – was a participant at the Dublin Art Fair.

The Körmendi Galéria Budapest gives room – well seperated – for both the conservative and avantgarde lines in contemporary Hungarian art, and tries to show its completeness by exhibiting repeatedly the above artists.

1014 ÚRI UTCA 40. MOBIL KINETIKUS SZOBROK KIÁLLÍTÓTERME

PAUER GYULA SZÉPSÉGMINTÁK SZOBORCSOPORT ÉS HARASZTY ISTVÁN

Á L L A N D Ó   K I Á L L Í T Á S

Haraszty István: Arco Fonte (mobil)

Az Óbudai Társaskör Galéria és Pincegaléria semmiféle stílus, izmus és műfaj mellett sem kötelezte el magát. A művészek kiválasztásában – a kvalitáson túl, ami alapfeltétel – az alábbi szempontok játszanak szerepet:

1. Értékőrző, illetve felfedező funkció – a művészeti élet perifériájára szorult, értékes, de kevéssé ismert vagy ismeretlen életművekre irányítani a figyelmet.

2. Fiatal művészek bemutatkozó tárlata – vállalni az első megjelenés közös felelősségét; a még „szavatolatlan" minőség mellé állni.

3. Ismert életművek váratlan fordulatai – a változás megragadása egy kamaratárlattal.

4. Klasszikus életművek rejtett értékeinek, különös stiláris vagy műfaji fordulatainak bemutatása.

Óbudai Társaskör Gallery and Pince (Cellar) Gallery have never committed themselves to any style, movement or genre. In choosing the artist – beside the basic condition: quality – the following considerations are important:

1. Value-keeping and discovering function – to direct interest to oeuvres that are at the periphery of art-life, or the creation of new or unknown oeuvres.

2. Young artists' exhibition – take the first common responsibility; stand next to the unguaranteed quality.

3. The exhibition of live art and its evolution in relation to the unknown. The focus being on works created for the space.

4. Showing stylistic or generic changes of hidden values in classical oeuvres.

Madeleine Dietz • 1995

A STÚDIÓ SZOLGÁLTATÁSAI

1977-ben hoztam létre egy csoportot ezzel a névvel, Dr. Kroó Norbert fizikus társaságában, a Központi Fizikai Kutató Intézetben, a fény, közelebbről a speciális tulajdonságokkal bíró lézerfény képi lehetőségeinek vizsgálatára. A Stúdió a 90-es években betagolódott a Symposion Alapítványba és Társaságba. Ebben az időben dolgoztam ki a Kepes Vizuális Központ elképzelését, amely mögé odaállt támogatónak a Symposion Alapítvány, és létrehoztuk a Kepes György Alapítványt. (A másik alapító a Magyar Alkotóművészek Országos Egyesülete volt.) Cserny Józseffel és Péter Ágnessel (akivel azóta is szorosan együttműködünk) Fényszimpoziont rendeztünk 1993-ban Egerben a Kepes Vizuális Központ részeként megnyitott Kepes Múzeumban.

A Videóstúdió és a Videólánc is új kezdeményezés. A Foton Art Stúdió céljai között szerepelt a ma művészete és a közönség közötti szakadék csökkentése. A Videólánc egyszerre biztosítja a közönség széles körű tájékoztatását, ugyanakkor rövid és áttekinthető archív anyagokkal szolgál a legfontosabb képző- és iparművészeti kiállításokról – hónapokra bontott időszeletekben. Egyelőre csupán a budapestieknek, de igényeink és terveink között szerepelnek a legfontosabb vidéki események is.

Csáji Attila

I established the group Foton Art Studio with Dr. Norbert Kroó physicist in the Központi Fizikai Kutató Intézet (Central Physics Research Institute) to examine light, or rather, the special potentials of laser. In the beginning of 1990s the Studio became part of the Symposium Foundation and Company, and we established the György Kepes Foundation together with the Hungarian Artists' National Cooperative. In 1993, with József Cserny and Ágnes Péter, we organized a Light symposium in the Kepes Museum (Eger) which was opened as part of the Kepes Visual Center.

The Video Chain and the Video Studio are new endeavors. Reducing the gap between the art and the audience is among the objectives of the Foton Art Studio. Video Chain provides full service to its customers, and at the same time concise, clear, structural archiving of the principal visual and applied art exhibitions in Budapest. The Studio presently serves only the Budapest community, but our future aims include recording the most important events around the country, as well.

Attila Csáji

ZWICKL ANDRÁS
Végtelenül gyorsuló – Szűcs Attila kiállítása a Bartók 32 Galériában, 1994/11.

SZEGŐ GYÖRGY
Kutyából nem lesz szalonna – Bukta Imre kiállításáról, 1995/3.

SINKÓ ISTVÁN
Oltár és stációk villanyfélhomályban – Karina Horitz és Kicsiny Balázs kiállításáról, 1995/5.

BORDÁCS ANDREA
Mit üzen a baba? – Tóth Gábor kiállításáról, 1995/7–8.

BEKE LÁSZLÓ
A láthatóság határán – Bernát András képei, 1995/9.

SINKÓ ISTVÁN
Fény súrolta objektumok – Bernát András kiállítása, 1995/9.

It is not like it used to be

A Beszélő, az Élet és Irodalom, a Magyar Narancs és a Magyar Műhely is szerepeltetett írásokat kiállításainkról (Szemadám György, Szabics Ágnes, Karina Horitz – Kicsiny Balázs, Tóth Gábor, fe Lugossy László, Bernát András).

There are also articles in Beszélő, Élet és Irodalom, Magyar Narancs and Magyar Műhely.

A BARTÓK 32 GALÉRIA KIADVÁNYÁNAK TÁMOGATÓI
OUR SUPPORTERS:
Fővárosi Képzőművészeti Alap, Nemzeti Kulturális Alap, Soros Alapítvány, XI. Kerületi Kulturális Alap

03 BOLT GALÉRIA

Fotó / Photo • Detvay Jenő

Galéria-enteriőr / Interior of the Gallery

CÍM / ADDRESS
1085 Budapest, Leonardo da Vinci utca 40.

NYITVATARTÁS / OPENING HOURS:
hétfő – szombat: 10 – 18 óra
Monday – Saturday: 10 am – 6 pm

ALAPÍTÁSI ÉV / FOUNDED IN
1994

A GALÉRIA TULAJDONOSA / OWNER OF THE GALLERY
Hangyál Judit, Detvay Jenő

GALÉRIAVEZETŐ / GALLERIST

A GALÉRIÁBAN EDDIG KIÁLLÍTÓ MŰVÉSZEK / ARTISTS HAVE BEEN

Kiky Smith
Koncz András
On Kawara
Donald Sultan
1995 Moholy-Nagy László
Erró
Bak Imre
Bachman Gábor
Barta Zsolt Péter
Fehér László
Forgács Péter
Jovánovics György
Köves Éva
Rajk László
Szalai Tibor
Vida Judit
1996 Michael Ray Charles
Cindy Sherman

ször az amer...
ve politikai he
mint a nyugat
kapcsolattartó
tokkal és köz
történt, keves
Magyarország
közi képzőmű
kívüliek érdek
A Galéria 56-
temény megte
ha akarná, se
világ legmozg
jából, New Yo
művészeknek
gyarországot
keringésébe.
képzőművésze
az első magy
het a komoly
művészeti élet
ször tűnik fel

manifesztálta a transzavantgárde törté
felfogását; Bak posztmodern eklektikában re
korábbi korszakainak minimalista form
szisztematikus jelkutatásainak képi eredm
Hencze pedig kísérletet tett a rendkívül s
homogén formavilágának személyes geszt
történő átértelmezésére, a végtelen távolság
egyszemélyes megnyilvánulás közelsé
összekapcsolására. Az «Új szenzibilitás» els
lítása programkiállítás volt, egyfajta manifes
egy új, szubjektivista, eklektikus, érzéki és szug
művészet első feltérképező jellegű bemutatás

let Hegyi Lóránd:
szági Új Festészet
a Fészek Galéria"
ógus-előszavából

When the Fészek Gallery opened its
Sensibility I» exhibition in December 19

ÁBRÁZO
HOSSZÚ

lult és h
tudtak a
kaptam
ves volt
nála. M
többeme

unk a sötét démon alak
pe-et kísérte árnyként,
odt, kimért léptekkel kísé
a felé a műkereskedő

01

LOT

A C T

pen archive. The very
ost people a passive
pool grew by taking
y documenting activ-
of itself. At present, at
r, its contents may be
days, or by appoint-
rofessionals.

homogeneous and basicly not substitutable. Works of art exist in the smallest thinkable unit – they're individual and thus it is impossible to fix their demand- and supply-curve. Price theories relating to the Homo Economicus aren't valid at all on the buying and selling participants of this market, since their behaviour is unpredictable and fully irrational. In the field of contemporary art they'll mostly have a relation with dictated prices. The gallerist, willingly or not, knowing or not, is automatically in a monopoly position, being the single seller of a product for which there is no or hardly a demand.
The gallery dealing with contemporaries almost always enters the market where a demand is missing and where it's uncertain such a demand will exist; the raise of the demand by the introduction of its products through publicity is impossible, and at least is very remote from this type of trade. At the same time it is of importance that the profession, and not the public or the potential buyers, will recognise the innovations within the works introduced to the market. If the profession reacts positively, the word will go around and the demand will start. Until then however the work, as far as the market is concerned doesn't exist.

into account the production costs, introduction costs, the costs of the artist management, or the demand for such a works on the market. He will just bring forward a number, which – contrary to the characteristics of a monopoly – will be independent of the quantity offered. Don't forget we are talking of unique works.
The market reacts with a declining demand on growing prices, paradoxically, the artmarket reacts inversely. In economics this reaction isn't unknown (Veblen-effect), used for prestige consumption. Turning back to the monopoly, talking about it one condition is the products should not be prestige goods.
This is just a small slice of the contradiction, what's more important is that it is useless to try to force pure economical categories upon this market. We have to realise the function of intermediaries within the contemporary art market, the responsibility they carry, and due to this one might say that they are on the market with less sources and instruments, with intuitive methods but with a stronger influence than only dealer on any other market.

Ágnes Veronika Kovács

CREATIVE PERSONS AND THE ART MARKET IN SOCIETIES IN TRANSITION

A PERSONAL INTERPRETATION OF THE HUNGARIAN SITUATION

Ever since the eighties both the situation of the art market in Hungary and the position of the artists have gone through severe changes.
Throughout the "socialist" period, artists were members of the Art Fund, which was also their employer, their union and their institute for social insurance. The Art Fund was in word owned by the artists, de facto it was an instrument of the government. The Art Fund was owner of a chain of galleries where artists could sell there works, and where – or through which – they could get commissions, for example mural works or large sculptures. For reasons not of interest here, they did the

ness fee. With the Ministry of Finances there was an agreement according to which the fees and taxes paid by the artists where conciderably lower than was the case for any other citizen. Artists could survive either from their work or by working next to their profession of being artists.
Of course there was a selection between art(ists) tolerated, supported or prohibited – the socalled 3 T policy. Not only in the "commercial" state galleries, but also as far as the "exhibition" galleries were concerned. In the Hungarian example however there was a controversy between the internal (national) policy and the external (international) policy. Some artists where exhibited abroad who weren't allowed to exhibit in Hungary. It goes without saying that the state had a full monopoly on the art trade in the galleries and on the exhibition program. Artists had the possibility through their own channels to organize exhibitions abroad but one way or another some state companies got involved.
In the middle of the 80-s private initiatives in the field of the art trade were tolerated. The so called artists working communities, which flourished under the umbrella of the Art Fund. Artists were able to found "private" galleries and as a collective they were able to sell there works. In practice it ment that the artists themselves were the owners of a gallery. One of the difficulties of this construction was that

1996-BAN KIÁLLÍTÓ MŰVÉSZEK / EXHIBITING ARTISTS IN 1996

február 7 – február 26.
Fésős László-Barna
Doina Mihailescu
Liviu Pascu
Claudio Toma
Dan Toma (Románia)
március 1 – március 18.
Csikós Tibor
Dezső Andrea
Sabrina Lindemann (Hollandia)
március 20 – április 8.
Polgár Judit „Arcaim"
április 10 – április 29.
Borkovics Péter
május 1 – május 10.
Farkas László
május 13 – május 27.
Laczky István
május 29 – június 17.
Czitrom Béla
Hyttinen Jarma
Kavaszto Mika
Lappalainen Minna
Luuka Leila
Salo Tiina (Finnország)

A Fiatal Művészek Klubja 36 éve működő kulturális intézmény. A művészet különböző területeinek bemutatása mellett évente körülbelül 30 kiállítást rendez kortárs alkotók részére. Kiállítótermeiben (Fekete, Fehér, Galéria Kávézó) bemutatkozási lehetőséget biztosít a legfrissebb hazai és külföldi képzőművészeti irányzatok képviselőinek. Időszakonként „Művészet ma" címmel megrendezi nemzetközi képzőművészeti tárlatát, melyen 42 ország többszáz művésze képviseli magát.

The Club of Young Artists has been in exsistance for 36 years as a cultural institute. Beside showing different sides of art, it organizes 30 visual art exhibitions of contemporary artists work a year. In our exhibition halls (Black, White, Gallery Café) we introduce Hungarian and foreign visual art trends. Sometimes the Club of Young Artists organizes an

GALÉRIA 56

Keith Haring • Cím nélkül / Untitled

CÍM / ADDRESS
1055 Budapest, Falk Miksa utca 7.
Tel./ fax: (36-1) 269-2529

NYITVATARTÁS / OPENING HOURS
kedd – vasárnap: 12 – 18 óra
Tuesday – Sunday: 12 am – 6 pm

ALAPÍTÁSI ÉV / FOUNDED IN
1992

GALÉRIAVEZETŐK / GALLERISTS
Kunz Zsófia, Krúdy Tamás

A GALÉRIA TULAJDONOSA / OWNER OF THE GALLERY

1999

Country / Region

Germany ❹

Switzerland ❸

The Netherlands ❸

Venezuela ❶

Spain ❶

U.S.A. ❶

Japan ❶

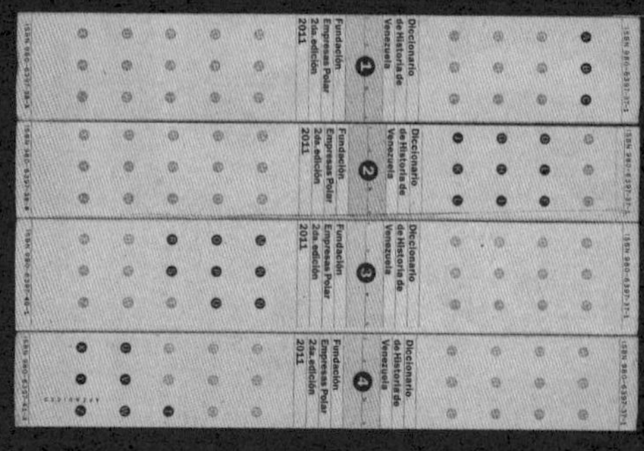

Niggli Giuseppe Terragni Modelle einer rationalen Architektur

MUSEUM FÜR GESTALTUNG ZÜRICH DA ZWIS CHEN

Gaceta de Arte y su época. 1932-1936

HANSER

Palmin - eine Jahrhundertmarke

WIM T. SCHIPPERS HARRY RUHÉ

Photography's multiple roles art document market science

ATLANTIS VERLAG PRO JUVENTUTE · VERA ZOGERMANN · SARDINEN WACHSEN NICHT AUF BÄUMEN

Stroemfeld Verlag Grete Gulbransson Tagebücher · Band 1 Der grüne Vogel des Äthers Hg. von Ulrike Lang

De regie van de stad I Koos Bosma / Helma Hellinga (redactie) NAi Uitgevers / EFL Publications

De regie van de stad II Koos Bosma / Helma Hellinga (redactie) NAi Uitgevers / EFL Publications

Designer

Álvaro Sotillo

Urs Stuber

François Rappo

Raimundo C. Iglesias

Arbeitsgruppe „work ahead“ an der Hochschule für Grafik und Buchkunst Leipzig

Claus Seitz

Gesine Krüger, Andrea Schürings (büro für mitteilungen)

Studio Gonnissen en Widdershoven

studio blue

Ueli Kleeb

Karin Beck

Shincho-Sha Co., Tokyo

Studio Gonnissen en Widdershoven

Arlette Brouwers, Koos van der Meer

1999 Gl

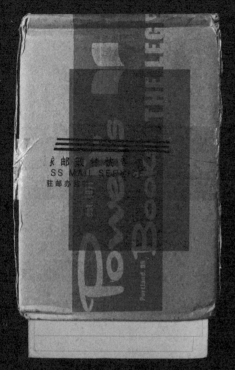

政 特 快 专 递
MAIL SERVICE
监 封
Venezuela
U.S.A.
China

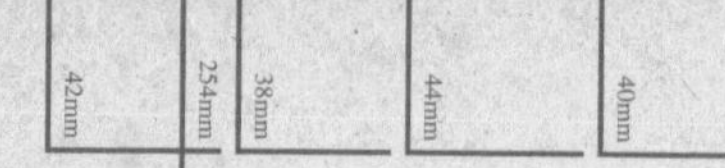

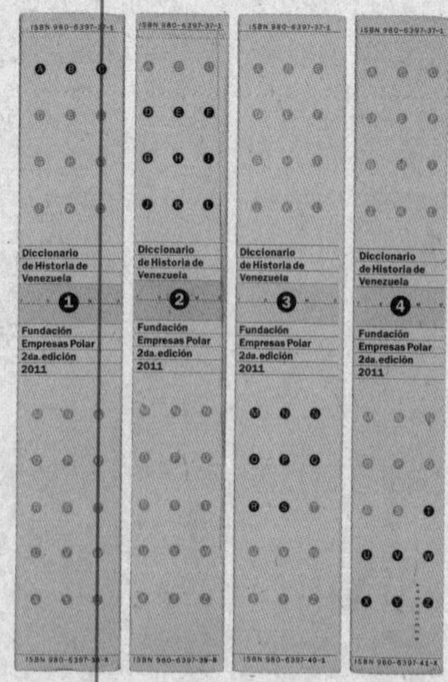

176mm

金字符奖
委内瑞拉历史词典
委内瑞拉

Golden Letter
Diccionario de Historia de Venezuela
Venezuela
Álvaro Sotillo
M. Perez Vila, M. Rodriguez Campo (Hrsg.)
Fundación Empresas Polar, Caracas
254 176 42mm / 254 176 38mm / 254 176 44mm / 254 176 40mm
1390g / 1257g /
1447g / 1282g
1176p / 1064p /
1232p / 1096p
ISBN-10 980-6397-38-X / ISBN-10 980-6397-39-8 /
ISBN-10 980-6397-40-1 / ISBN-10 980-6397-41-X

这是一套涵盖委内瑞拉各个方面历史的词典，1989 年出版了第一版。1997 年增补了大量因时代和生活而改变的词汇，修订了第一版中的问题，推出了第二版。这套词典信息量巨大，总计接近 4500 页，按照字母顺序尽可能地均分成 4 本。书籍的装订形式为锁线硬精装。护封白卡纸绿黑双色印刷，上下书口与勒口均向内翻折。在内封采用薄卡板的情况下，封面整体厚度适中，并在一定程度上降低了书籍的重量。内封灰卡板，书脊处裱贴黑色布面，书脊上的分册编号字母压凹处理。内页字典纸印单色黑。词典内容囊括从西班牙统治时期至今（出版时期）的地理、社会、政治、军事、经济、宗教、法律、科技、艺术和文化领域的专有名词、历史人物和重要事件。正文双栏排版，词条名使用粗字重的无衬线字体，附注使用小字号，整体观感清晰简约。页眉采用三个字母的索引方式，查找更方便。第四册的最后部分加入了各类年表和汇总资料，用于知识的体系化构建。

This is a set of dictionaries covering all aspects of Venezuelan history. The first edition was published in 1989. The second edition was launched in 1997 – a large number of vocabulary from the changing times and life were added and the problems in the first edition were revised. This set of dictionaries has a huge amount of information, nearly 4500 pages in total. They are divided into four parts in alphabetical order as much as possible. The hardcover book is bound by thread sewing. The jacket is made of white cardboard printed in green and black. The upper and lower book edges and the flaps are folded inward. Since the thin cardboard is used for the inner cover, the overall thickness of the cover is moderate, and the weight of the book is reduced to a certain extent. Black cloth is mounted on the spine of the book, and the letters of every volume are embossed. The inside pages are made of dictionary paper printed in monochrome black. The contents of the dictionary include the proper nouns, historical figures and important events in the fields of geography, society, politics, military, economy, religion, law, technology, art and culture from the time of Spanish rule to the present (the date of publication). The text is double-column typeset, and the entry names use bold sans serif font, and the notes use small font size, so that the overall look is clear and simple. The header uses a three-letter index to make searching easier. In the fourth volume, various chronologies and summary materials are attached for the systematic construction of knowledge.

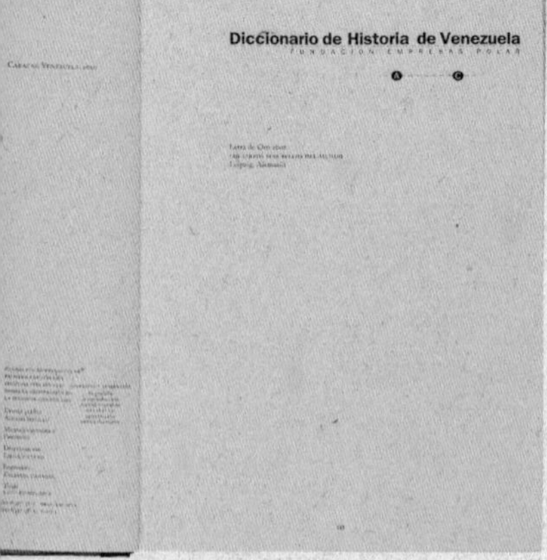

Diccionario de Historia de Venezuela

FUNDACIÓN EMPRESAS POLAR

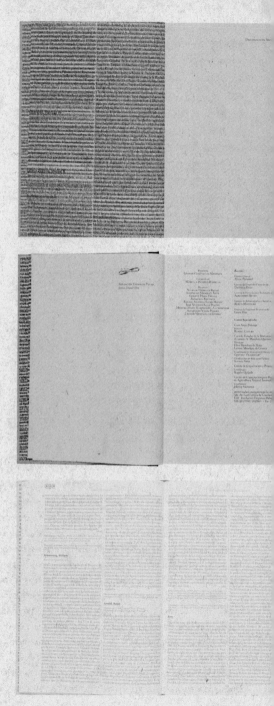

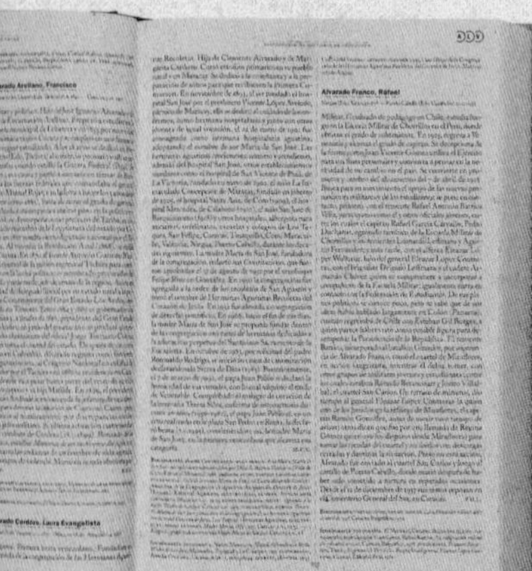

Alvarado Franco, Rafael

movimientos liberales radicales, en el primero de los cuales fue fusilado el padre de Abreu. No consiguió elegirse debido a la amplia campaña de sus adversarios. Se preparaba una nueva revolución libertadora que estalló en 1848. Abreu fue condenado a prisión perpetua, siendo después amnistiado. En 1855 publicó los primeros libros brasileños sobre socialismo, en los moldes de Fourier y los socialistas utópicos. En su lecho de muerte recibió la condenación de la Iglesia católica por ser defensor de la libertad de cultos para los protestantes; por ello le fue negada sepultura católica y, por consiguiente, sus restos fueron inhumados en el cementerio de los ingleses. v.ch.

Bibliografía: *directa:* abreu y lima, josé ignacio de. *Resumen histórico de la última dictadura del Libertador Simón Bolívar.* Caracas: Universidad Simón Bolívar, 1983. *indirecta:* baptista nava, manuel. *General José Ignacio Ribeiro de Abreu e Lima: 1794-1869.* [Caracas: s.n., 1986]; carbonell, diego. *Sobre el tablado: conferencias y discursos.* Bogotá: Editorial Cromos, 1935; ——. *Un héroe brasileño en la guerra grancolombiana de emancipación.* Río de Janeiro: s.n., s.f.; chacón, vamireh. *Abreu e Lima, general de Bolívar.* Caracas: Universidad Simón Bolívar, Instituto de Altos Estudios de América Latina, Centro Abreu e Lima de Estudio Brasileños, 1985.

Iconografía: retrato, *Resumen histórico de la última dictadura del Libertador Simón Bolívar.* Río de Janeiro, Editora O Norte, 1922.

Academia de Ciencias Físicas, Matemáticas y Naturales

Creada por ley el 13 de junio de 1917, fue fundada con 30 individuos de número nombrados por el presidente de la República. A pesar de los esfuerzos para constituir la institución no fue posible hacerlo sino 16 años después. A la sesión solemne de instalación, celebrada el 24 de julio de 1933, que fue presidida por Plácido Daniel Rodríguez Rivero, rector de la Universidad Central de Venezuela, en representación del ministro de Instrucción Pública, asistieron 21 de los 30 individuos de número ya seleccionados. Felipe Aguerrevere, Santiago Aguerrevere, Crispín Ayala Duarte, Hernán Ayala Duarte, Lucio Baldó, Eduardo Calcaño Sánchez, Melchor Centeno Grau, Francisco José Duarte, Carlos Guinand, Juan Iturbe, José Izquierdo, Alfredo Jahn, Vicente Lecuna, Ernesto León, Antonio P. Mora, Miguel Parra León, Manuel Cipriano Pérez, Rafael Requena, Jesús Rafael Rísquez, Eduardo Röhl, Francisco J. Sucre, Eduardo Tamayo, Enrique Tejera, Carlos Toro Manrique, Luis Ugueto, Ciro Vásquez, Luis Vélez, Carlos Raúl Villanueva, Gustavo Wallis y Oscar Zuloaga fueron los miembros fundadores. Desde su fundación, la dirección de la academia ha sido ejercida por las siguientes personas: Luis Vélez, ingeniero, pe-

Juan Francisco Stolk, ingeniero, período 1959-1961; Enrique Jorge Aguerrevere, ingeniero, período 1961-1962; Tobías Lasser, médico naturalista, período 1962-1965; Miguel Parra León, ingeniero, período 1965-1980; Marcel Granier, médico, período 1980-1981; Gustavo Rivas Mijares, ingeniero, período 1981-1985. Las atribuciones de la academia son las siguientes: *a)* esforzarse porque las Ciencias Físicas, Matemáticas y Naturales, alcancen en el país el mayor desarrollo y adelanto; *b)* estudiar los mejores métodos para la enseñanza de estas ciencias y de sus aplicaciones; *c)* estudiar los asuntos de su competencia que el Ejecutivo someta a su consideración; *d)* formar una biblioteca en la cual figuren las mejores obras de ciencias físicas, matemáticas y naturales, de autores nacionales y extranjeros; *e)* reunir los trabajos de sus miembros, presentados a la Academia y solicitar del Ejecutivo la publicación de lo que el cuerpo considere que lo merece en forma de *Memoria* de la Academia; *f)* reunir y organizar datos sobre las riquezas naturales del país; estudiar sus aplicaciones y pasar al Ejecutivo sus informes sobre estos particulares; *g)* propender al estudio de la meteorología del país; *h)* establecer relaciones con todas las academias y cuerpos de igual índole, nacionales y extranjeros; *i)* recomendar al ministro de Instrucción Pública las obras que juzguen puedan adoptarse como textos para la enseñanza de las respectivas ciencias; *j)* ocuparse de todo lo demás que sea propio de la naturaleza y carácter de la corporación. El órgano oficial de las publicaciones de la Academia es el *Boletín*, editado a partir de 1934.
M.A.M.

Bibliografía: academia de ciencias físicas, matemáticas y naturales, ed. *Prontuario de la Academia de Ciencias Físicas, Matemáticas y Naturales.* Caracas: Academia de Ciencias Físicas, Matemáticas y Naturales, 1988; rivas mijares, gustavo. *Historia y futuro de la Academia de Ciencias Físicas, Matemáticas y Naturales: a los 50 años de su instalación.* Caracas: s.n., 1983; *Ley y reglamento de la Academia de Ciencias Físicas, Matemáticas y Naturales.* Caracas: Academia de Ciencias Físicas, Matemáticas y Naturales, 1986; zawisza, lesżek alberto. *La Academia de Matemáticas de Caracas.* Caracas: Ministerio de la Defensa, c. 1980.

Temas relacionados: Ciencias del mar; Ciencias naturales, físicas y matemáticas.

Academia de Ciencias Políticas y Sociales

La corporación pública así denominada fue creada por ley de 16 de junio de 1915 con la finalidad de propender al desarrollo y progreso de las ciencias políticas y sociales en general y, en especial, de la legislación venezolana. Por iniciativa propia o a instancias del Ejecutivo Nacional o de algún estado de la República, la Acade-

ma de su ley constitutiva (30.6.1940). Los 30 miembros fundadores fueron: Alejandro Urbaneja, Francisco A. Guzmán Alfaro, Domingo Antonio Coronil, Diego Bautista Urbaneja, Pedro M. Brito González, Francisco Gerardo Yanes, Alejandro Pietri, Carlo F. Grisanti, Juan Francisco Bustillos, Cristóbal L. Mendoza, José Santiago Rodríguez, Esteban Gil Borges, Carlos Jiménez Rebolledo, Manuel Alfredo Vargas, Juan de Dios Méndez y Mendoza, Juvenal Anzola, José Loreto Arismendi, Raúl Crespo Vivas, Pedro Miguel Reyes, Arminio Borjas, Juan José Mendoza, Jesús Rojas Fernández, José Gil Fortoul, Guillermo Tell Villegas Pulido, Pedro Hermoso Tellería, Pedro Manuel Arcaya, Pedro Itriago Chacín, Emilio Constantino, Lorenzo Herrera Mendoza y Carlos Alberto Urbaneja. La elección de sus miembros debe efectuarse entre venezolanos que hayan escrito alguna obra valiosa sobre ciencias políticas o sociales, o desempeñado una cátedra universitaria en las mismas materias, o haber sido integrante de comisiones codificadoras o legislativas y poseer, en todos los casos, reconocida competencia científica. La Academia puede nombrar, además, miembros correspondientes nacionales y extranjeros: 2 por cada uno de los estados de la República y 30 fuera del país; los nacionales deben llenar las mismas condiciones exigidas a los individuos de número y los extranjeros, se escogerán por su condición de profesor universitario o autor de obras científicas de incontestable mérito. El artículo segundo de su ley constitutiva establece que: «...La Academia tendrá un Presidente, dos Vicepresidentes, un Secretario, un Tesorero y un Bibliotecario, nombrados por ella de su propio seno y que durarán un año en sus funciones...» A lo largo de su historia, la Academia ha reunido en su seno a los representantes más destacados de las ciencias políticas y sociales de cada época y ha expresado su opinión, tanto en materia legislativa como en lo referente a las grandes cuestiones internacionales que interesan a la República: reivindicación de la Guayana Esequiba, delimitación de las áreas marinas y submarinas con la República de Colombia e integración latinoamericana. Desde abril de 1936, la Academia de Ciencias Políticas publica trimestralmente un *Boletín* de divulgación. R.DE S.

Bibliografía: *Academia de las Ciencias Políticas y Sociales: ley y reglamento.* Caracas: Tipografía La Nación, 1943.

Temas relacionados: Ciencias políticas; Ciencias sociales.

Academia de Ciencias Sociales

Calcaño, Julio Calcaño, Carlos Arvelo,
mendi Brito, José Núñez de Cáceres, Di
mírez, Gerónimo Eusebio Blanco, P
Felipe Larrazábal y otros. Existían 2
miembros: individuos de número y n
rrespondientes. La Academia celebraba
narias y extraordinarias, las cuales se ef
no tener local propio, en los salones del
cional que funcionaba entonces en el anti
de San Francisco, luego sede de la univ
Palacio de las Academias. Aunque la
Ciencias Sociales y de Bellas Letras tuv
ción, promovió certámenes de carácter cu
abierto el 1 de octubre de 1869 con motivo
natalicio del Libertador, según particip
hizo al público en el periódico *El Feder*
mismo mes. El 24 de febrero de 1870, rea
sesión pública y solemne al tomar pose
Junta Directiva en el Salón de la Cámar
Posteriormente los acontecimientos pol
de ese año condujeron a su extinción.

Bibliografía: parra márquez, héctor. *La Academia*
de Bellas Letras de 1869: antecedentes de la Academia Vene
correspondiente de la Real Española. Caracas: Italgráfica,
las Academias. Caracas: Academia Nacional de la Historia

Tema relacionado: Círculos artísticos, literarios y cien

Academia de Derecho Público y Espa

16.11.1792 — diciembre 1804

Por iniciativa del regente de la Audienci
Antonio López de Quintana, el Colegio
convino en agosto de 1790 en la necesida
tableciera una Academia de Derecho Pú
ñol, con la finalidad de proporcionar una
ción a los jurisconsultos que integraban
una mejor formación e instrucción a los
ingresar a este gremio. Se encomendó
Juan Agustín de la Torre, José Antonio C
Garay y Francisco Espejo, y a los licenc
José Sanz, Bartolomé Ascanio y José M
tarea de redactar los correspondientes
futura academia, lo cual culminó con la
unas constituciones. Inmediatamente, és
sadas para el estudio del fiscal de la Aud
Díaz de Saravia, quien informó a favor d
del Colegio de Abogados. Por decreto de
de 1790, la Audiencia aprobó provisional
constituciones, hasta la decisión defin

gramas de conferencias e interven-
n hospitales y universidades del in-

F.P.

GARCÍA, MANUEL. *Los períodos pre y post-operato-*
cas: Tipografía Americana, 1938.

IA, *Revista de la Sociedad Médico Quirúrgica del*
úm. 2, febrero 1942.

tonio

1860 — Caracas, 24.9.1939

rio y político. Rico comerciante,
ro del «grupo valenciano» que reci-
no Castro después de la batalla de
), brindándole apoyo financiero.
ompaña a Castro a oriente durante
tra de la Revolución Libertadora y
o 1903) de recolectar las recauda-
os comerciantes de Ciudad Bolívar
s importadas ilegalmente durante
de marzo de 1904, firma un contra-
n de la Compañía de Vapores del
da los activos de la antigua Orino-
pany y establece un discutido mo-
orte naviero en el río Orinoco.
ulo íntimo del presidente Castro,
imiento denominado La Conjura,
dir el ascenso al poder del vicepre-
e Gómez (1906-1907). Exiliado a
olpe de Estado del 19 de diciembre
enezuela en 1911 y participa, junto
n Delgado Chalbaud, tanto en las
das a cabo en París (agosto-octubre
ntes del Credit Français y del ban-
ntre ellos el aventurero Paul Bolo,
reación de un Banco Nacional y un
Venezuela, así como en las nego-
dicato inglés para la explotación de

y letras. A su regreso a Venezuela
tiempo en una de las industrias fun
la Fábrica de Vidrios de Maiquetía
la actividad fabril para iniciarse en
dadera vocación: el periodismo. Fue
en esta disciplina el veterano profes
Francisco Gerardo Yánez, a la sazón
tario del diario caraqueño *El Sol*. H
armas en *El Sol*, Corao pasó a ser co
el escritor Agustín Aveledo U., del r
cial, mientras colaboraba en la revis
Guruceaga (1929). Continuando s
periodista, Corao entró a figurar en
ción del vespertino *El Heraldo*, fun
el poeta y escritor Antonio José Cal
muerte de Calcaño Herrera (1929
con su hermano Virgilio y le compr
caño el mencionado vespertino, el
1945. Al retirarse del periodismo, ej
ción consular de Venezuela en varios
y viajó extensamente por Europa (19

BIBLIOGRAFÍA: CORAO, ÁNGEL. *Obras completas*
Bravo, 1963; ——. *El paisaje y los hombres de los An*
Nación, 1937; ——. *El pobre diablo.* Caracas: T
Romanzas interiores. Caracas [Imp. Bolívar], 1920.

ICONOGRAFÍA: FOTOGRAFÍA, *El Heraldo*, Caraca
FÍA, *El Universal*, Caracas, abril 19, 1951. FOTOGR
completas, Madrid, Imprenta Juan Bravo, 1963
Biblioteca Nacional, Caracas.

Corao Dilardi, Jesús

Caracas, 1.5.1900 — Caracas, 6.10.1970

Empresario, deportista, entrenador
tivo. Hijo del general Manuel Corao
di. Inicia sus estudios en el Saint T
Curazao, donde su padre se encu
regresa a Venezuela en 1911. De niño
de las «cuadras bravas», pandillas
entablan numerosas peleas callej

Historia de Ven

IÓN EMPRESAS

BLO

el castillo Libertador y el fortín
bello. Poco días después, al gru
unieron 2 buques de la armada i
expedición en tareas de acompa
22 de diciembre de 1902 el viceal
bald Lucas Douglas, comandan
junta, en esta ocasión a nombre
hizo publicar en el diario *El Hera*
guiente disposición: «Por la pres
bloqueo ha sido declarado para lo
ra, Carenero, Guanta, Cumaná,
del Orinoco, y se hará efectivo d
de diciembre...» Sólo se refería a
tas situadas al este de dicho puert
tales quedaron a cargo de los alem

timuló la creación de puertos libres y zonas francas; logró mayor cobertura para los créditos industrial, agrícola y pecuario; propició la formación de un plan lechero nacional y elaboró un proyecto de Ley General de Títulos de Crédito. En 1958 ideó, junto con Alejandro Hernández, la fundación de la Asociación Pro-Venezuela, constituida el 7 de julio de 1958, que durante 20 años fue presidida por su hijo Reinaldo. I.R.G.

BIBLIOGRAFÍA: *Ángel Cervini, el maestro*. Caracas: SIVENSA, 1981

ICONOGRAFÍA: FOTOGRAFÍA, *El Universal*, Caracas, noviembre 15, 1980. FOTOGRAFÍAS, *Ángel Cervini, el maestro*, Caracas, SIVENSA, 1981. FOTOGRAFÍA, colección José Agustín Catalá. Biblioteca Nacional, Caracas.

Cevallos, José

Véase: Ceballos, José

Cey, Galeotto

Florencia (Italia) 29.4.1513 — Florencia (Italia) 20.6.1579

Comerciante. Participó en la fundación de El Tocuyo y escribió una relación sobre esa región. Hijo de Juan Bautista Cey y de Camila Di Francesco Di Vanni Strozzi, pertenecientes ambos a familias destacadas de Florencia, parientes de los Médicis, en cuyo desfavor cayeron en 1530, terminando el padre de Cey ajusticiado y confiscados sus bienes. Para el momento de la muerte de su padre se encontraba en Lyon, adonde había sido enviado para adquirir práctica en los negocios, en el banco de la familia Salviatti. En 1532 se trasladó a España pasando por Burgos, Valladolid, Toledo, Segovia y Córdoba, antes de establecerse en Sevilla donde permaneció hasta 1537 cuando recibió la noticia de que el duque Cosme de Médicis, había indultado a los copartidarios de la revolución en que intervinieron los Cey, permitiéndoles regresar a Florencia y restituyéndoles sus bienes. Regresó a su ciudad natal para encontrar que su nombre no había sido incluido en el indulto, por lo que aquel mismo año marchó a Bolonia y después a Venecia. En esa época participó en la batalla de Montemurlo en la que se enfrentaron los partidarios de los Médicis y de los Strozzi, bajo cuya bandera militaba Cey; ganando los primeros, fue hecho prisionero y recluido en la cárcel de Pisa. De allí lo trasladaron a Livorno, donde gracias a la intervención de la duquesa de Parma y de Lope Hurtado de Mendoza terminó siendo liberado en 1538. Volvió a España, y en Sevilla hizo sociedad con el comerciante Luis de Ricasoli, con el objeto de fletar un navío para comerciar en América. Se embarcó en Sanlúcar el 6 de agosto de 1539 contrarrestar a los piratas franceses, llegando a ser alférez. A fines de ese mismo año volvió al Cabo de La Vela, en actividades comerciales, regresando a Santo Domingo, pasando a Nombre de Dios y Cartagena para regresar al Cabo de la Vela atraído por la posibilidad de vender sus mercancías en Coro, donde ya se encontraba en 1545. Allí permaneció, forzado por Juan de Carvajal quien lo unió a su hueste, siguiendo a este capitán en sus entradas al interior de la provincia de Venezuela. Estuvo entre los fundadores de El Tocuyo, presenció las disputas entre Carvajal y los Welser, que terminaron con el asesinato de Bartolomé Welser y Felipe de Hutten. Fue testigo del ajusticiamiento de Carvajal y formó parte de la expedición que envió el licenciado Juan Pérez de Tolosa a la Nueva Granada, habiendo pasado por Tunja y Bogotá. Regresó a El Tocuyo en 1551, volvió a Tunja en 1552, y al año siguiente pasó a Santa Marta y luego a Cartagena de donde partió para España. En Sevilla reanudó sus actividades comerciales viajando frecuentemente por Italia, Alemania y Francia; en esta última corte tuvo como protector a Bartolomeo del Bene, banquero y poeta, el mayor acreedor de los reyes Valois y amigo de Pierre de Ronsard, líder del movimiento poético de la Pléyade. En 1555 realizó un viaje a Marruecos. Regresó a Toscana en 1560, casándose al año siguiente. Después de su regreso a Florencia, donde ocupó algunos cargos administrativos de poca importancia relacionados con actividades económicas, llegó a ser miembro del Consejo de los Doscientos en 1572, muriendo 5 años más tarde. Escribió una relación de sus viajes por América con el título de: *Viaggio e relazione delle Indie (1539-1553)* que se encuentra en el archivo del Museo Británico. Esta relación es de gran importancia para la historia temprana de Venezuela, de la Nueva Granada y de Santo Domingo, pues está escrita con gran minuciosidad, constituyendo una fuente de primer orden para conocer tanto la vida cotidiana de los conquistadores y los vecinos de los recién fundados pueblos, como los usos y las costumbres de las tribus aborígenes con las cuales entró en contacto. Contiene asimismo, la explicación sobre numerosos vocablos indígenas y dibujos de instrumentos y enseres americanos. J.R.L.

BIBLIOGRAFÍA: CEY, GALEOTTO. *Viaje y descripción de las Indias, 1539-1553*; estudio preliminar, notas e índices, José Rafael Lovera. Caracas: Fundación Banco Venezolano de Crédito, 1995.

Chacachacare

Véase: Expedición de Chacachacare

para capturar indígenas. Sin embargo, fue puesto en libertad por Losada, hecho que permite suponer el interés de éste en granjearse la confianza y amistad del jefe indígena para facilitar así la pacificación de la región. No obstante, su intento resultó fallido pues en 1568 Chacao se alió con Guacaipuro y otros jefes indígenas para hacer frente al avance de la conquista española, siendo enfrentados y derrotados por el propio Losada en la llamada batalla de Maracapana. Su nombre se conservó como un poblado al E de Caracas, incorporado hoy al área metropolitana. H.B.C.

BIBLIOGRAFÍA: *Chacao: de un indio y de las arenas de un pueblo*. [Los Teques]: Gobierno del Estado Miranda, s.f.; GONZÁLEZ ANTÍAS, ANTONIO. *Chacao un pueblo de la época de Bolívar, 1768-1880*. Caracas: Academia Nacional de la Historia, 1984; VARGAS MENDOZA, LORENZO. *Semblanzas de un pueblo: de la leyenda a la crónica*. Petare: Concejo Municipal del Distrito Sucre, 1980.

Chacín Itriago, Luis Gregorio

Clarines (Edo. Anzoátegui) 12.10.1877 —
Ginebra (Suiza) 10.3.1934

Médico, iniciador de la política sanitaria en el país. Realizó sus primeros estudios en su ciudad natal. En 1901, obtuvo el título de médico cirujano en la Universidad Central de Venezuela, publicando ese mismo año su tesis de grado, titulada *La orfalotripsia*. Fue a ejercer su profesión en el estado Anzoátegui. Allí hizo investigaciones sobre la fiebre amarilla. Viajó a Londres, donde revalidó su título de médico cirujano ante el Real Colegio de Cirujanos de Inglaterra. Externo e interno por oposición en los hospitales civiles del Distrito Federal, médico adjunto del servicio de Dermatología del hospital de Saint Bartholomew en Londres; director de Sanidad Nacional de Venezuela (1919-1922) e inspector de Hospitales Civiles (1924-1928). Fue uno de los iniciadores de la política sanitaria en el país. Publicó numerosos trabajos de interés médico y social. Individuo de número de la Academia de Medicina (10.3.1920). C.G.A.

BIBLIOGRAFÍA: CHACÍN ITRIAGO, LUIS GREGORIO. *Algunas consideraciones sobre el análisis químico de las aguas de Caracas y otros lugares vecinos*. Caracas: Tip. Americana, 1920; ——. *Clasificación de las pulgas que contienen las ratas de Caracas*. Caracas: Tip. Americana, 1927; ——. *Clasificación de las ratas de Caracas*. Caracas: Tip. Americana, 1927; ——. *La fiebre biliosa hemoglobinúrica: consideraciones etilógicas*. Caracas: Academia Nacional de Medicina, 1910; ——. *Lección inaugural del curso de higiene pública*. Caracas: s.n., 1921; ——. *Lucha contra la anquilostomiasis: ligeras consideraciones sobre la enfermedad y su profilaxia*. Caracas: Tip. Americana, 1927; ——. *Recepción del Dr. Luis Gregorio Chacín Itriago: 19 de marzo de 1920*. Caracas: Academia Nacional de la Historia, 1920.

ICONOGRAFÍA: FOTOGRAFÍA, *Gaceta Médica de Caracas*, Caracas, núm. 8, abril 30, 1934. FOTOGRAFÍA, *El Nuevo Diario*, Caracas, marzo 11, 1934.

ció la máxima autoridad en esa i
1783, se le expidió por 5 años el tít
comandante de la isla de Trinida
capitanía general de Venezuela. I
de agosto de 1784, y el 1 de septie
del cargo. En diciembre inicia la
rre las poblaciones desde Puerto F
ta, dando cuenta del estado de la
vanta censos de habitantes, orde
necesarios para llegar a ciertas po
nanzas de protección a los indios
bución de tierras y cultivos. En el
Chacón pondrá en ejecución u
acordes con el sentido de progres
nifestado también por otros gol
Firme; en junio 1785 acusa la nec
un hospital, pues la humedad de
enfermedades; para su construcci
a los presidiarios pues así el co
considerablemente; dio impulso
isla, preocupándose por el cultiv
de la caña de azúcar; sugiere hac
en 1776, el Rey le autoriza para e
de los destinados al situado, para
y un astillero; el 10 de marzo de 1
ros, artillería y marina. Esta acti
reconocida en 1789, cuando cu
gobierno, y hecho el nombramie
Cañaveral Ponce, se deja éste si
Chacón continuar en el Gobiern
le el título de capitán de navío.
una real orden hace instaurar la
vincia, previstas desde 1786 en la
la Intendencia del Ejército y Re
cas; por tanto en Trinidad, dond
un subdelegado del intendente
futuro un intendente; este cargo
del Rey, en Chacón, quien det
militar y político, inherentes al
el económico con la designaci
aunque dependiente de la Inte
Real Hacienda de Caracas. En
Armada, Cayetano Llorente re
fo de la isla, levantando planos;
me había reconocido la isla y no
vantar fortificaciones en ella,
mismo año, informa de las obra
rías que se han concluido en el
mas. El 11 de noviembre de 1792
para contraer matrimonio con

1999 Gm

Antiquariat Armebooks
Postfach 80 03 33
Königsteiner Str. 9-13
65929 Frankfurt / Germany
info@armebooks.de
Zhao Qing
7-3, Daibei Lane, Meiyuan New Village
210000 Nanjing
China
einer rationalen
Architektur
Niggli

Switzerland

Germany

China

280mm

11mm

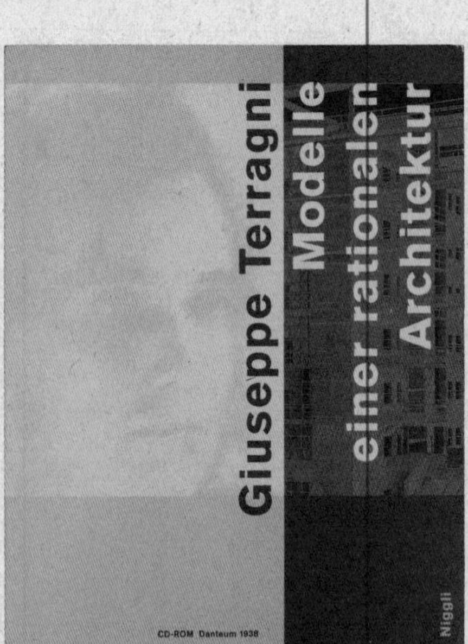

208mm

◉ 金奖
《 朱塞佩 · 特拉尼
——合理架构的模型
◖ 瑞士

Gold Medal ◎
Giuseppe Terragni: 《
Modelle einer rationalen Architektur
Switzerland
Urs Stuber
Jörg Friedrich, Dierk Kasper (Hrsg.)
Verlag Niggli AG, Sulgen
280 208 11mm
697g
104p
ISBN-10 3-7212-0343-7

意大利建筑师朱塞佩·特拉尼（Giuseppe Terragni）虽然只有短暂的 39 年生命，却设计了大量重要的建筑作品。他在理性主义的名义下展开了意大利现代设计运动，努力让建筑设计的风格远离新古典主义和新巴洛克式的复兴主义，并通过融合现代和传统的理论，获得了独特的地中海特色。1998 年德国建筑中心举办了他的回顾展，本书即为展览推出的纪念作品集。书籍的装订形式为无线胶平装。封面白卡纸四色印刷，覆哑膜，背面印橙色。内页哑面艺术纸印单黑，作品部分采用半透明的牛油纸黑橙双色印刷作为插页。书籍开篇由多名建筑评论家对其创作年代、建筑风格、推动的设计风潮做了详细阐述。之后便是展览的主要部分——建筑模型，并在牛油纸上用橙色印刷建筑名，黑色印平面图。作品包含多种类型：别墅、幼儿园、宫殿等。正文版式设置为五栏，其中正文跨四栏，图片与注释在五栏中灵活排布。段落首行的缩进较多，体现出较传统的排版风格。

Although the Italian architect Giuseppe Terragni lived a short 39 years, he designed a large number of important architectural works. In the name of rationalism, he launched the Italian modern movement, striving to keep the style of architectural design away from neoclassicism and neo-Baroque revivalism, and obtained unique Mediterranean characteristics by fusing modern theory and tradition. His retrospective exhibition was held at the German Architecture Center in 1998. This book is a collection of commemorative works launched for the exhibition. The paperback book is perfect bound. The cover is made of white cardboard printed in four colors, mounted with matte film, and the back cover is printed with orange. The inner pages are printed with matte art paper in single black, and the works are printed on translucent tracing paper with black and orange as inserts. At the beginning of the book, many architectural critics elaborate on their creation background, architectural styles, and design trends. Then there is the main part about the exhibition – the architectural model. On the tracing paper, the building names are printed in orange and the planar graphs are printed in black. The works include various types of private villas, kindergartens, palaces and other types. The text is typeset to 5 columns – the text spans 4 columns, and the pictures and comments are arranged flexibly in 5 columns. The first line of the paragraph is indented, reflecting a traditional typesetting style.

Giuseppe Terragni
Modelle
einer rationalen
Architektur

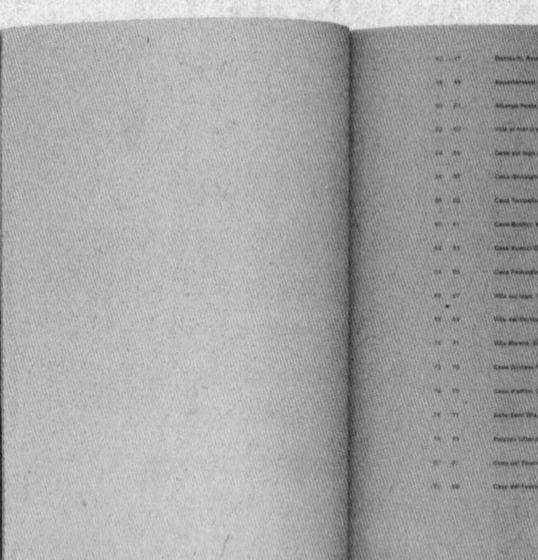

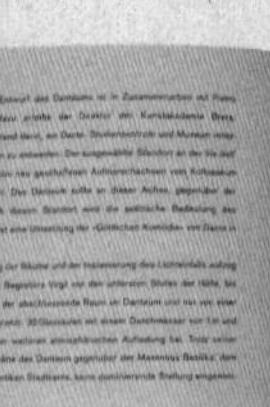

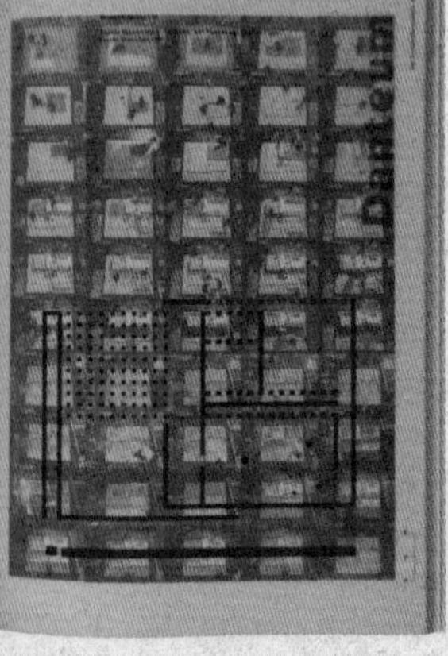
Danteum

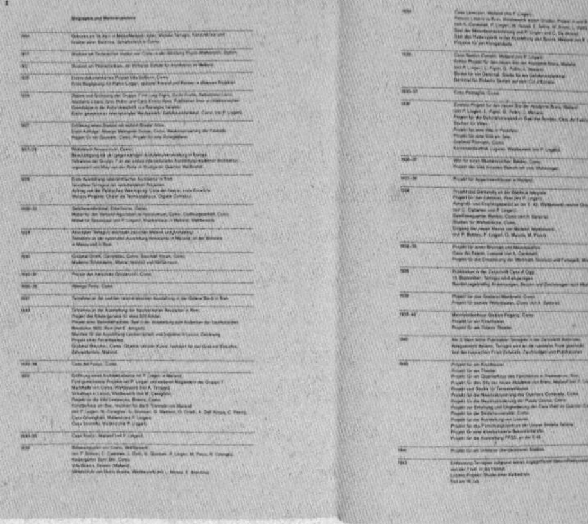

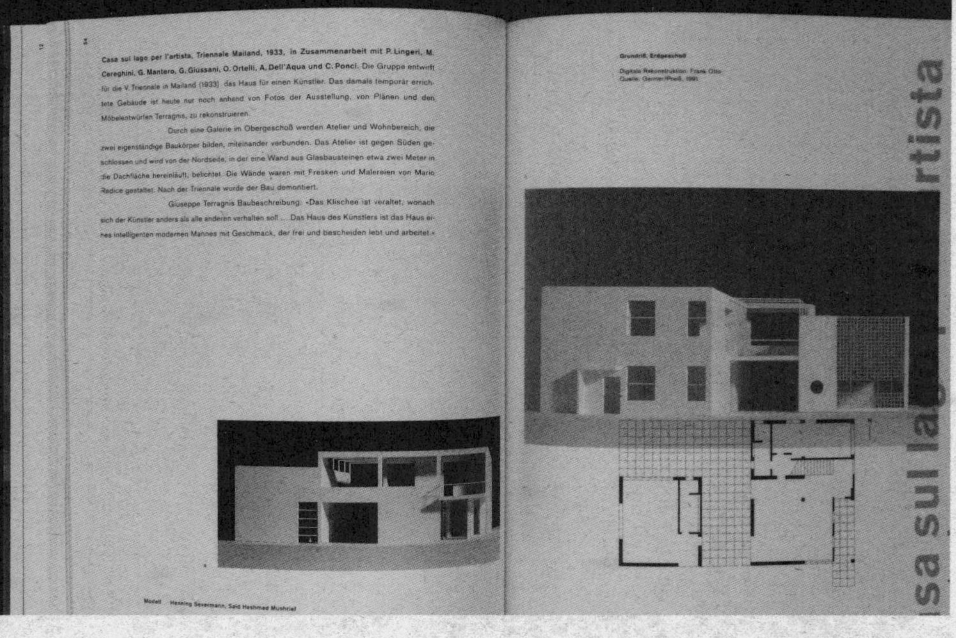

34

Casa sul lago per l'artista, Triennale Mailand, 1933, in Zusammenarbeit mit P. Lingeri, M. Cereghini, G. Mantero, G. Giussani, O. Ortelli, A. Dell'Aqua und C. Ponci. Die Gruppe entwirft für die V. Triennale in Mailand (1933) das Haus für einen Künstler. Das damals temporär errichtete Gebäude ist heute nur noch anhand von Fotos der Ausstellung, von Plänen und den Möbelentwürfen Terragnis, zu rekonstruieren.

Durch eine Galerie im Obergeschoß werden Atelier und Wohnbereich, die zwei eigenständige Baukörper bilden, miteinander verbunden. Das Atelier ist gegen Süden geschlossen und wird von der Nordseite, in der eine Wand aus Glasbausteinen etwa zwei Meter in die Dachfläche hereinläuft, belichtet. Die Wände waren mit Fresken und Malereien von Mario Radice gestaltet. Nach der Triennale wurde der Bau demontiert.

Giuseppe Terragnis Baubeschreibung: »Das Klischee ist veraltet, wonach sich der Künstler anders als alle anderen verhalten soll ... Das Haus des Künstlers ist das Haus eines intelligenten modernen Mannes mit Geschmack, der frei und bescheiden lebt und arbeitet.«

Modell Henning Severmann, Said Heshmad Mushrief

Grundriß, Erdgeschoß

Digitale Rekonstruktion: Frank Otto
Quelle: Germer/Preiß, 1991

[C]asa sul lago per l'artista

Deutsche Bauhütte

Zeitschrift der deutschen Architektenscha

Nr. 10. 4. Mai 1927. 31. Jahrg

Ein verzweifeltes Spiel um die bauliche Führung in Deutschl

c

16 Winfried Nerdinger, Walter Gropius, op. cit., S. 263.

17 Vgl. Winfried Nerdinger; Versuchung und Dilemma de Architekturwettbewerbe 1933–35, in: Hartmut Frank, Architekturen, Planen und Bauen in Europa 1930 bis

18 Mussolinis Ansprache an die Delegierten des XIII. Int Architektenkongresses in Rom, in: Deutsche Bauzeit S. 316.

19 Zitiert nach: Stefan Germer; Die italienische Hoffnun Germer, Hrsg.; Giuseppe Terragni: 1904–1943; Moder München 1991, S. 87.

Deutschen Architektur Zentrum Berlin

e Ausstellung* Giuseppe Terragni, Rationale Architektur 1927 entrum DAZ vom 26. Mai bis 16. August 1998 gezeigt werde n war es nämlich eine große Enttäuschung, daß die imposant e Terragni, die 1996 in Mailand anläßlich der Triennale prä ischen Ausland gezeigt wird, nicht auch in Deutschland bezie Für all jene, die diese breit angelegte Präsentation in Mailanc ist klar, daß damit eine einmalige Chance vertan war, eine ge aufflammende und sich am Baugeschehen der Stadt Berlin

1927	Eröffnung Erste Au Projekt fü
1927–28	Wohnblo Beschäft Teilnahm organisie
1928	Erste Au Teilnahm Auftrag v Weitere I
1928–32	Gefallene Möbel fü Möbel fü
1929	Aktivitäte Teilnahm in Monza

vereinnahmen. Es ist zu befürchten, daß er auch in Berlin in diesem Sinne sprechen wird, und das geschähe dann unter dem besonderen Protektorat des Herrn Ministerpräsidenten Göring.»

Vielleicht zurecht beklagt sich hier Gropius über die Kurzsichtigkeit deutscher Funktionäre, die nun via Italien den «Stilo tedesco» feierten, den sie im eigenen Land bekämpften. Zu spät – oder gerade rechtzeitig – hatte Gropius begriffen, daß der revolutionäre Charakter sowohl dem modernen Bauen als auch der faschistischen Bewegung zugrunde lagen. Und zu spät wohl, daß Modernität auch traditionelle Werte und metaphysische Transformationen beinhalten können. Nicht umsonst war er ebenso bei dem letzten großen und spektakulären Wettbewerb, an dem er sich beteiligt hatte, für den Neubau der Reichsbank in Berlin, durchgefallen, das Gebäude hatte auf die Jury den Eindruck einer großen Fabrik erweckt. Worauf einige Kritiker hämisch Gropius als «Opfer eines von ihm selbst befürworteten Schlagwortes», nämlich dem der sich alles unterzuordnenden Industrieästhetik, bezeichneten.[20]

Es war der politische Hintergrund Marinettis, der den deutschen Funktionären genehm war – Marinetti war ein ausgewiesener Faschist – und das schloß sich in diesem Fall mit der Vorstellung von architektonischer Avantgarde nicht aus. Nicht moderne Architektur schlechthin wurde von den Regimen abgelehnt – wie es von deren Protagonisten nach dem zweiten Weltkrieg nur zu gerne dargestellt wurde –, vielmehr konnten Architekten des Neuen Bauens die auf Fragen der Typologie, auf den Bedeutungsgehalt einzelner architektonischer Gattungen und auf gesamtgesellschaftliche Bezüge der Architektur meist keine Antworten geben, das hatte sich schon lange vor der Machtergreifung der Nationalsozialisten abgezeichnet. Bereits in den zwanziger Jahren macht sich dieses Theoriedefizit bemerkbar, es werden stellvertretende Debatten um Internationalität des Stiles oder Heimatstil geführt, Industrieästhetik kontra Walm- oder Satteldach. Es ging scheinbar um die bauliche Führung in Deutschland, wie es in den Hetzkampagnen um das Bauhaus zu lesen war.[21] Letztlich jedoch immer nur um den Baustil. Es war Gropius nicht gelungen – um ihn nur stellvertretend zu nennen – an die Debatten und Diskurse aus der frühen Werkbundzeit anzuknüpfen. Bereits 1914 hatte Muthesius anläßlich der Werkbundtagung davor gewarnt, daß es viel zu früh sei über einen verbindlichen Stil in der Architektur zu sprechen, wie es die junge Generation immer wieder gefordert hatte. Und in der Tat war dann bereits 1927 der «Sieg des neuen Baustils» plötzlich da, wie es Walter Curt Behrendt mit seiner gleichlautenden Publikation determinierte. Und es hatte sich eben nur

20 Winfried Nerdinger, Walter Gropius, Berlin 1985, S. 180.

21 Vgl. J. C. Bürkle: Das Neue Bauen zwischen Modernität und Tradition, in: Wohnhäuser der klassischen Moderne, Stuttgart 1994, S. 9ff.

ein Stil etabliert, Fragen der Typologie, des Städtebaus, der Bedeutung von architektonisc
Formen wurden nicht gestellt. Die «architektonische Gleichschaltung» war somit also zunä
in den Manifesten der Modernisten vorgenommen worden, lange bevor diese Metapher
Dritten Reich übernommen wurde und natürlich sehr viel effizienter und unter anderen
zeichen umgesetzt werden konnte.

Mit der Machtergreifung in Deutschland erfolgten Gegenstrategien,
sich in Siedlungsvorhaben, Stadtgründungen oder Gegenausstellungen wie der Siedlung
Kochenhof in Stuttgart manifestierten. Da jedoch architekturtheoretische Grundlagen nicht
legt wurden erschöpfte sich diese Architektur zumeist in formalen Kontrapositionen mit volk
melnden, zuweilen bizarr romantisierenden Ausprägungen. Ebenso wie in Deutschland
Salzgitter und der Volkswagenstadt Wolfsburg ländliche Regionen industrialisiert und damit
schäftigungspolitisch autark werden sollten, gab es in Italien zahlreiche Stadtgründungen
dem Ziel, die Autarkie der Regionen zu verstärken und in strukturschwachen Gebieten Arb
plätze zu schaffen. Während die Zentren Rom und Berlin auch tatsächlich im Zentrum des me
lomanen planerischen Interesses standen, entwickelten sich in Italien fernab an den Periphe
Planungen, die gleichsam beinahe unbemerkt von den zentralen Ämtern zwar kontrollierte
dennoch eigenständige Richtungen einschlugen.

Am Beispiel der italienischen Stadtgründung Torviscosa, unweit von Ud
zeigt sich, daß jene integrativen Theoreme des italienischen Rationalismus die tragfähigeren
damit auch die überlebensfähigeren waren. Wie bei den pontinischen Stadtgründungen auch
wurde in Torviscosa ein Industriezweig zur Grundlage der Stadt erklärt, hier war es der Zells

a Vogelsperspektive von Torviscosa, vierziger Jahre

b Torviscosa, Piazza Marinotti, ehemals Piazza Autarchia, mit Theater und Restaurant

c Torviscosa, Fabrikanlage für die Herstellung von Viscose

d Torviscosa, Eingang der Fabrik

DA
FROM
TO
R

Others
Weight (in kg)
Value (in EUR)
1.10
12.71
Total Weight
Total Value
1.10
12.71
09-Nov-19
TO
ZHAO QING
HAN QING TANG DESIGN CO.
DABEI XIANG 7-3
210000 NANJING
CHINA

Switzerland
Germany
China

246mm

25mm

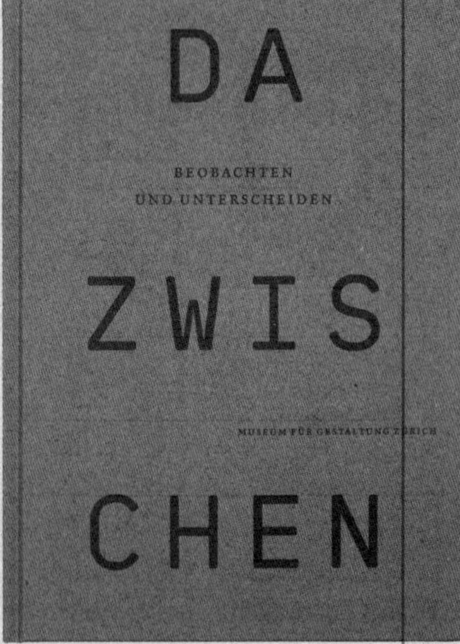

DA
ZWIS
CHEN

MUSEUM FÜR GESTALTUNG ZÜRICH

186mm

银奖

之间——观察与辨别

瑞士

Silver Medal

Dazwischen: Beobachten und Unterscheiden

Switzerland

François Rappo

André Vladimir Heiz, Michael Pfister

Museum für Gestaltung Zürich

246 186 25mm

874g

272p

ISBN-10 3-907065-79-4

瑞士苏黎世设计博物馆出版的一本论文集，着眼于对现实的观察与感知，以及对现有概念的思考与辨别。本书的装订形式为锁线硬精装。封面采用灰绿色艺术纸印银，局部压横向凹线，裱覆卡板。环衬选用土红色艺术纸。内文胶版纸印单黑。全书共分为四大部分，包含 35 位研究者的论文，内容包括观察与思考、自我与大众、输入与输出、欢笑与哭泣、高与低、墙与空间等概念的观察与思考，详细探讨了各个概念和定义之间的边界与关联。书中图片很少，只在中间部分有少量照片，其他部分均在文本上做了一些适度尝试。首先是横线的运用，从封面开始便使用了压凹线；目录使用粗线分割四大部分，细线凸显文章标题；每个部分的隔页均将四大部分的所有标题列出，通过删除线遮挡非本章内容的标题。文章部分如果有延伸阅读的推荐，会设计为“标签”的方式“夹”在书页侧边，如有关联内容也会用圆形“标签”进行标注。

This is a collection of papers published by the Zurich Design Museum in Switzerland, focusing on the observation and perception of reality, as well as the thinking and discrimination of different existing concepts. The hardcover book is bound by thread sewing. The cover is made of grey-green art paper printed in silver, partially embossed with horizontal lines, and mounted with cardboard. Earthy red art paper is selected for the end paper. Inner pages are made of offset paper printed in single black. The whole book is divided into four parts, mainly including 35 researchers' papers, including observation and thinking, the individual and the public, input and output, laughter and crying, high and low, wall and space, etc. The boundary and relationship between each concept and definition are discussed in detail. There are few pictures in the book, only a few in the middle, some moderate attempts have been made in the text of other parts. The first attempt is the use of horizontal lines, embossed lines are used from the beginning of the cover; the table of contents is divided into four parts by thick lines, and the thin lines highlight the article titles; all the headings of the four parts are listed in every part and three are blocked by the strikethrough lines. If there is a recommendation for extended reading in the article section, it will be designed as a 'tag' to 'clamp' on the side of the book pages, and if there is related content, it will also be marked with a circular 'tag'.

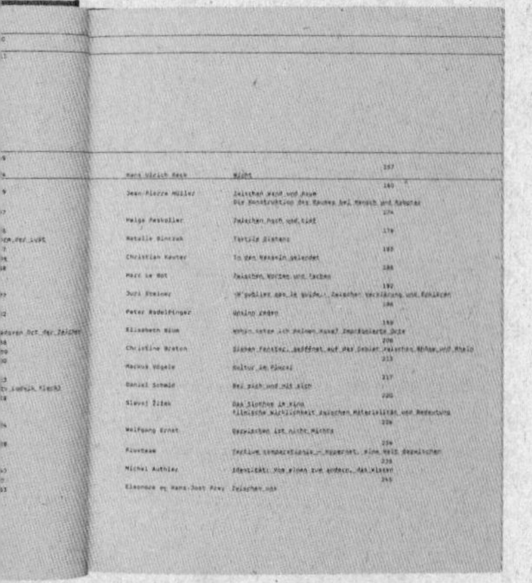

Intonation Seiten 11 bis 20

Etwas kommt dazwischen ~~Eine Welt dazwischen ‹Zwischen Frau und Mann› Eros gibt keine Ruhe Um was es geht Die Lust der Form, die Form der Lust Affektlogische Reflexionen oder 51% Liebe Zwischen Verbrechen und Strafe. Über Schmerz, Schuld und Verantwortung Zwischen Lachen und Weinen ‹Hier, wo zwischen.› Eine semiologische Phantasie über den paradoxen Ort der Zeichen Quantenleere in des Wortes voller Bedeutung Die Botin vom verlorenen Paradies Über Wissenschaftstheorie (Ein Kommentar zu Ludwik Fleck) Christophoros, le passeur/Christophoros, der Über-Setzer ‹Das Wort dazwischen› Zigaretten Zwischenhandlungen und zwischen Handlungen entweder/oder Nicht~~

~~Zwischen Wand und Raum. Die Konstruktion des Raumes bei Mensch und~~ … ~~Zwischen hoch und tief Taktile Distanz In den Nesseln gelandet~~ … ~~Farben ‹N'oubliez pas le guide.› Zwischen Verklärung und Erklären~~ … ~~Wohin setze ich meinen Kuss? Imprägnierte Orte Sieben Fenster, ge~~ … ~~zwischen Rhône und Rhein Kultur im Plural Bei sich und mit sich~~ … ~~Filmische Wirklichkeit zwischen Materialität und Bedeutung Dazwi~~ … ~~Tertium comparationis Hypernet, eine Welt dazwischen Identitä~~ … ~~das Wissen Zwischen uns Etwas geht aus~~

~~Tutti~~ ~~Seiten 21 bis 249~~

~~Etwas kommt dazwischen Eine Welt dazwischen ‹Zwischen Frau und Mann› Eros gibt keine Ruhe Um was es geht Die Lust der Form, die Form der Lust Affektlogische Reflexionen oder 51% Liebe Zwischen Verbrechen und Strafe. Über Schmerz, Schuld und Verantwortung Zwischen Lachen und Weinen ‹Hier, wo zwischen.› Eine semiologische Phantasie über den paradoxen Ort der Zeichen Quantenleere in des Wortes voller Bedeutung Die Botin vom verlorenen Paradies Über Wissenschaftstheorie (Ein Kommentar zu Ludwik Fleck) Christophoros, le passeur/Christophoros, der Über-Setzer ‹Das Wort dazwischen› Zigaretten Zwischenhandlungen und zwischen Handlungen entweder/oder Nicht~~

~~Zwischen Wand und Raum. Die Konstruktion des Raumes bei Mensch und~~ … ~~Zwischen hoch und tief Taktile Distanz In den Nesseln gelandet~~ … ~~Farben ‹N'oubliez pas le guide.› Zwischen Verklärung und Erklären~~ … ~~Wohin setze ich meinen Kuss? Imprägnierte Orte Sieben Fenster, ge~~ … ~~zwischen Rhône und Rhein Kultur im Plural Bei sich und mit sich~~ … ~~Filmische Wirklichkeit zwischen Materialität und Bedeutung Dazwi~~ … ~~Tertium comparationis Hypernet, eine Welt dazwischen Identitä~~ … ~~das Wissen Zwischen uns Etwas geht aus~~

~~Chill-Out~~ ~~Seiten 253 bis 265~~

~~Etwas kommt dazwischen Eine Welt dazwischen ‹Zwischen Frau und Mann› Eros gibt keine Ruhe Um was es geht Die Lust der Form, die Form der Lust Affektlogische Reflexionen oder 51% Liebe Zwischen Verbrechen und Strafe. Über Schmerz, Schuld und Verantwortung Zwischen Lachen und Weinen ‹Hier, wo zwischen.› Eine semiologische Phantasie über den paradoxen Ort der Zeichen Quantenleere in des Wortes voller Bedeutung Die Botin vom verlorenen Paradies Über Wissenschaftstheorie (Ein Kommentar zu Ludwik Fleck) Christophoros, le passeur/Christophoros, der Über-Setzer ‹Das Wort dazwischen› Zigaretten Zwischenhandlungen und zwischen Handlungen entweder/oder Nicht~~

~~Zwischen Wand und Raum. Die Konstruktion des Raumes bei Mensch und~~ … ~~Zwischen hoch und tief Taktile Distanz In den Nesseln gelandet~~ … ~~Farben ‹N'oubliez pas le guide.› Zwischen Verklärung und Erklären~~ … ~~Wohin setze ich meinen Kuss? Imprägnierte Orte Sieben Fenster, ge~~ … ~~zwischen Rhône und Rhein Kultur im Plural Bei sich und mit sich~~ … ~~Filmische Wirklichkeit zwischen Materialität und Bedeutung Dazwi~~ … ~~Tertium comparationis Hypernet, eine Welt dazwischen Identitä~~ … ~~das Wissen Zwischen uns Etwas geht aus~~

~~Index~~ ~~Seiten 268 bis 271~~

~~on des Raumes bei Mensch und Roboter~~
~~nz · In den Nesseln gelandet · Zwischen Worten und~~
~~schen Verklärung und Erklären · Unsinn reden~~
~~rte Orte · Sieben Fenster, geöffnet auf das Gebiet~~
~~lural · Bei sich und mit sich · Das Sinthom im Kino.~~
~~alität und Bedeutung · Dazwischen ist nicht Nichts~~
~~e Welt dazwischen · Identität: Vom einen zum andern~~
~~-aus~~

legt, aus spannun
Wo Rudolf Helm
Ciompi, dass sich
Denken und Verh
Hierarchien. Ged
Und vielleicht
schaulicht.
bestätigt die lust
zwischen der ansa

Vgl. S. 174

[28] Gilles Deleuze, «Di

~~on des Raumes bei Mensch und Roboter~~
~~nz · In den Nesseln gelandet · Zwischen Worten und~~
~~schen Verklärung und Erklären · Unsinn reden~~
~~rte Orte · Sieben Fenster, geöffnet auf das Gebiet~~
~~Plural · Bei sich und mit sich · Das Sinthom im Kino~~
~~alität und Bedeutung · Dazwischen ist nicht Nichts~~
~~e Welt dazwischen · Identität: Vom einen zum andern~~
~~-aus~~

in jenen Tagen mit der Kunst.
sstseinszuständen geschaffen
Tradition von Versuchen, den
neine Prinzipien von Weltsinn
st aber von Kopfverdrehten pr
dende Welterkenntnis einem U
n plausibel gemacht werden sol
. Dem Vermittler kam die Rolle

Verkunstung gesteigerter Wahr
nicht immer in gleichem M
n tiefgreifender Verklärung mit
en Testament. Jesus, der von dre
hend und leuchtend ertappt wu

~~on des Raumes bei Mensch und Roboter~~
~~nz · In den Nesseln gelandet · Zwischen Worten und~~
~~schen Verklärung und Erklären · Unsinn reden~~
~~rte Orte · Sieben Fenster, geöffnet auf das Gebiet~~
~~lural · Bei sich und mit sich · Das Sinthom im Kino~~
~~alität und Bedeutung · Dazwischen ist nicht Nichts~~
~~e Welt dazwischen · Identität: Vom einen zum andern~~

– Hermann BAUMANN
Das doppelte Geschlecht. Studie
Berlin 1955 und 1986
– Gilbert HERDT (Hg.)
Third Sex, Third Gender. Beyond
New York 1994

der immer mitmischt. Die ganze schöne I
Anerkennung ihrer prinzipiellen Unabg
unfreiwilligen Verlust zu haben – um den
verabschieden. Deshalb ist das Subjekt bei
chen, als Hinweis auf einen ursächlichen E
es nie gegeben hat und auch nicht geben l
aber das Subjekt überhaupt erst entstehen
und Identität immer «geschlechtet», den
eine sehr grundsätzliche Weise für die Not
nen» aufzugeben, mit der symbolischen I
ren wie auch auf das andere Geschlecht an
Wo sich die Individuen diesem Mangel n
suchen, da bleibt auch die Sprache versper
res Sprechen» – und damit verliert die Sprache ih

Open Air — Literatur dazu und dazwischen

– George SPENCER-BROWN
Laws of Form – Gesetze der Form
Übersetzung Thomas Wolf Lübeck 1997
– Dirk BAECKER (Hg.)
KALKÜL der Form Frankfurt am Main 1993
– Heinz VON FOERSTER
Wissen und Gewissen. Versuch einer Brücke
Siegfried J. Schmidt (Hg.), Frankfurt am Main 1997

Wir betrachten das Beobachten in zweierlei Hinsicht.

Wenn wir untersuchen, wie wir ein Objekt beobachten, dann denken wir «als ob», denn wir wissen nicht, ob es ein solches Objekt wirklich gibt: wir erschaffen es, um unser Beobachten erklären zu können und um uns die Möglichkeit zu geben, (irgend etwas) zu beobachten. Wenn wir «als ob» denken, dann denken wir nicht ans Beobachten, sondern daran, unser Beobachten zu beschreiben oder zu erklären.

Ganz im Gegensatz dazu betrachten wir unser Beobachten vielleicht durchaus als unser Beobachten. Wenn Objekte blosse Postulate sind, verleihen wir ihnen die Wirklichkeit des Beobachtens: Wir behaupten, es gebe Objekte. Das ist westliche Erkenntnis. Ein hilfreicher Kunstgriff. Beide Denkungsarten kommen hier zur Anwendung.[65]

Objekte zu akzeptieren, «als ob» es sie gäbe (sie als existierende zu behandeln). Diesen Kunstgriff werden wir anwenden, um überhaupt über Erfahrung Rechenschaft geben zu können.

Um in das Universum des Beobachtens einzutreten, muss jedes Objekt selber beobachten. Zumindest kann es sich selber beobachten. Um sein Selbst zu beobachten, muss ein Objekt sowohl Beobachter als auch Beobachtetes sein. Allerdings ist das Selbst des Objektes sein unteilbares Selbst.

Jedenfalls könnte ein Objekt sowohl Beobachter als auch Beobachtetes sein, wenn es Beobachter und Beobachtetes als Rollen auffasst, zwischen denen es hin- und herschwankt, oszilliert. Oszillatoren erzeugen Zeit. Objekte wechseln ihre Rollen: sie haben (zwei) Phasen: eine jede davon macht ihre eigene Zeit.[66]

Also kann man davon ausgehen, dass jedes Objekt seine eigene Zeit der Selbstbeobachtung erzeugt.

Wenn die Objekte nicht sich selber beobachten, dann verschafft ihnen dieser Oszill
onsmechanismus Zeit, andere zu beobachten – sich mit anderen nach aussen hin
synchronisieren.

Alles Beobachten findet in der Zeit statt. Die Beobachtungszeiten unterscheiden s
voneinander. Jede Beobachtung ereignet sich (für mich als Beobachter) in (m)ei
Sequenz. Eine jede benötigt ihre Zeit: sie beginnt, und sie geht zu Ende.[67]

Da Objekte fiktive Konstrukte sind, ist das einzige, was die Beobachtungen von
Erklärung dieser Beobachtungen unterscheiden kann, die Beobachtungszeit.
Beobachtungszeit differenziert die Beobachtungen.

Zu beobachten, damit anzufangen und damit aufzuhören, erfordert und erzeugt Z
denn es geschieht in der Zeit, und Zeit leitet sich daraus ab. Diese Zeit verdankt
sowohl dem Akt des Beobachtens als auch (in der «als-ob»-Darstellung) der Synch
nisation zweier Objekte, des Beobachtens und des Beobachteten, wobei sie diese bei
zusammenbringt, indem sie ihre Zeiten und Phasen interagieren lässt und dabei
besondere Instanz der Beobachtung schafft.

Wir sehen, dass sich Beobachtungen begegnen. Vieles, was wir für mögliche Beob
tungen halten, trifft nicht ein. Wenn wir beobachten, stellen wir lediglich in rel
wenigen Momenten innerhalb einer unvorstellbar weiten Vielfalt Verbindungen h

Dies wird durch die Notwendigkeit erhellt, beim Beobachten die eigenen Zeiten
einzelnen Objekte miteinander zu synchronisieren.

Da, um Beobachten zu ermöglichen, zwei Zeiten synchronisiert werden müs
erstaunt es nicht, dass wir von der unvorstellbar grossen, potentiellen Vielfalt nicht
beobachten können. Synchronisieren ist schwierig. Man denke nur an die Proble

[65] In gewissem Sinne denken wir immer «als ob». In diesem Beitrag beschreibe ich das Beobachten, und
Beobachten richtet sich auf das Beobachten. Dies entspricht Wittgensteins Sicht der Schnittstellen, des
hanges oder des Bildschirmes, der das abtrennt, worüber wir schweigen müssen (vgl. Wittgensteins Tra
logico-philosophicus). [66] Diesen Rollenwechsel diskutiere ich in meiner Doktorarbeit (Ranulph
ville, 1975, op. cit.), aber auch in: Ranulph Glanville, 1976, op. cit. Jede Uhr basiert auf Oszillatoren. Aller
ist der Zeitbegriff für den Oszillator selber und für einen externen Beobachter nicht derselbe. Ein für das O
völlig regelmässiger Rhythmus (im Wechsel zwischen Beobachter und Beobachterem) mag auf ein an
Objekt, das notgedrungen von aussen beobachtet, alles andere als regelmässig wirken. [67] Oszill

1999 Sm²

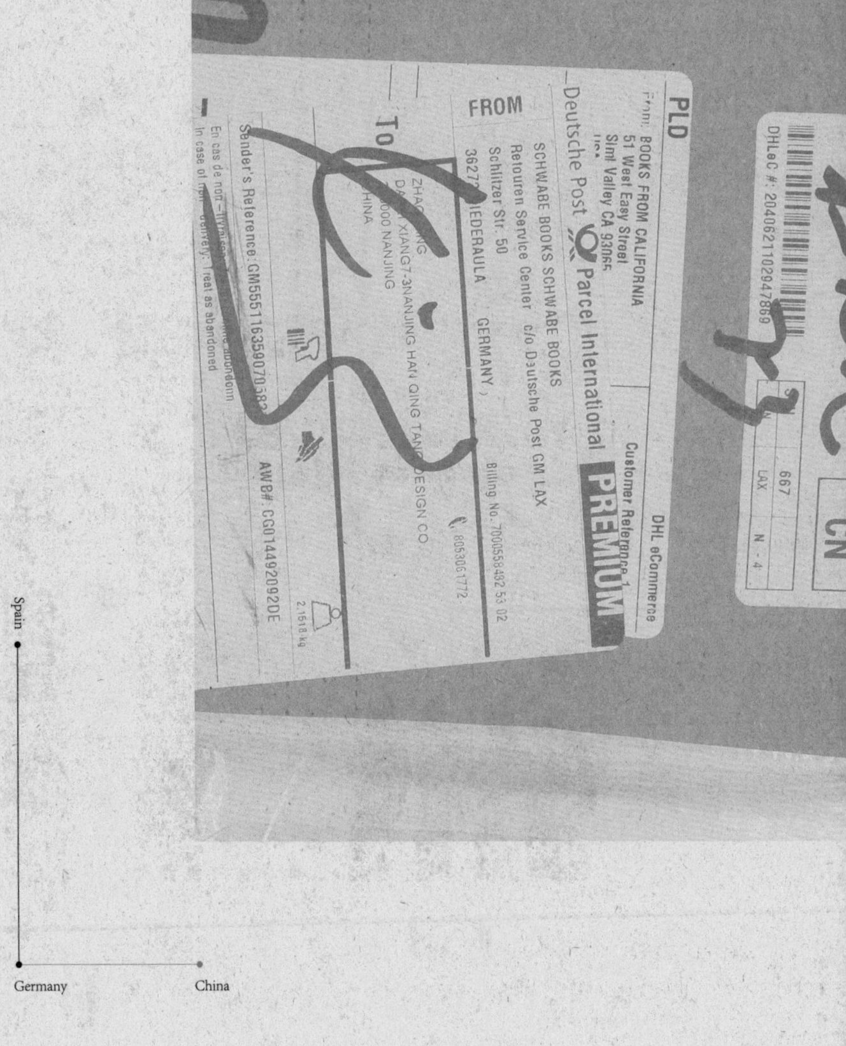
PLD
BOOKS FROM CALIFORNIA
51 West Easy Street
DHL eCommerce
Customer Reference 1
Deutsche Post
Parcel International
PREMIUM
FROM
SCHWABE BOOKS SCHWABE BOOKS
Retouren Service Center c/o Deutsche Post GM LAX
Schlitzer Str. 50
GERMANY
Billing No. 7000558432 53 02
To
CHINA
AWB#: CG014492092DE
DHLeC #: 2040621102947869
667
LAX
CN
Spain
Germany
China

279mm

31mm

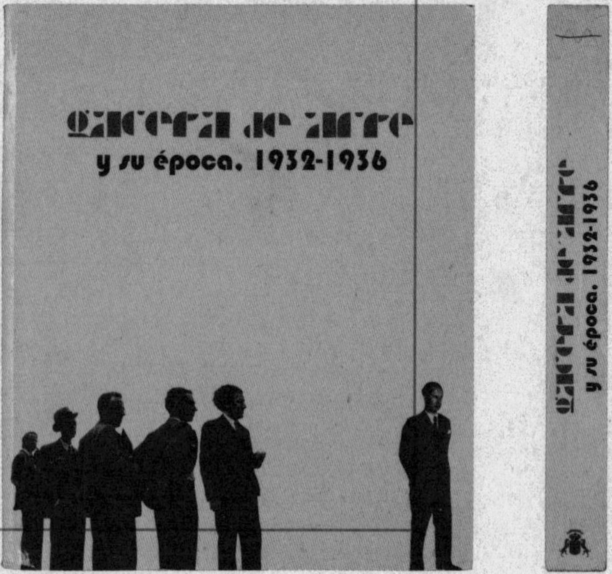

241mm

银奖

1932－1936 年的《艺术公报》

西班牙

Silver Medal

Gaceta de Arte y su Época 1932-1936

Spain

Raimundo C. Iglesias

——

Centro Atlántico de Arte Moderno, Las Palmas de Gran Canaria; Edición Tabapress

279 241 31mm

2032g

364p

ISBN-10 84-89152-11-X

西班牙艺术家爱德华多·韦斯特达尔（Eduardo Westerdahl）于1931年游历欧洲各大美术馆、大学、博物馆，深受当时各种艺术思潮的影响，于1932年在特内里费岛创办了《艺术公报》。这是一份试图打破艺术经典范式的杂志，是西班牙甚至整个欧洲当时最重要的艺术出版物之一。由于政治原因，杂志的出版时间并不长，至1936年停刊，总计38期。本书是这本杂志历史回顾展的出版物。书籍的装订形式为锁线胶平装。封面白卡纸红黄黑三色印刷，覆亮膜。内页前六分之一采用哑面铜版纸四色印刷，对历史、话题、影响进行深入探讨。最后六分之一采用灰色胶版纸红黄黑三色印刷，为整本书的英文翻译。内文版心设定为三栏，但排版时并没有受到三栏的限制，正文大部分通栏排版，采用大行距满足易读性。衬线体和无衬线体混合使用、字距行距自由变换。书名、页码使用专为刊名设计的几何字体，章节标题和注释数字使用包豪斯字体，在当年具有先锋意义，如今亦有独特的怀旧色彩。

Spanish artist Eduardo Westerdahl visited major art galleries, universities, and museums in Europe in 1931. He was deeply influenced by various artistic thoughts at that time and founded the *Gaceta de Arte* in Tenerife in 1932. This was a magazine that tried to break the classic art paradigm, and was one of the most important art publications in Spain and even Europe at the time. Due to political reasons, the publication history of the magazine was not long. It ceased publication in 1936, with a total of 38 issues. This book is a publication of the retrospective exhibition of this magazine. The book is perfect bound with sewing thread. The laminated cover is printed in red, yellow and black on white cardboard. The first one-sixth of the inner pages is printed on matte coated paper in four colors, which provides an in-depth discussion of history, topics, and influence. The last one-sixth is printed on gray offset paper in red, yellow and black, which is the English translation of the entire book. The typeset area is set to three columns, but the layout is not completely restricted within the three columns. Most of the text is typeset in full measure, and the large line spacing is used to meet legibility. Serif fonts and sans serif fonts are used together, and the spaces between words and lines are changed freely. The book title and page numbers use geometric fonts specially designed for the journal, and chapter titles and annotation numbers use Bauhaus font, which was quite avant-garde at that time, and are nostalgic nowadays.

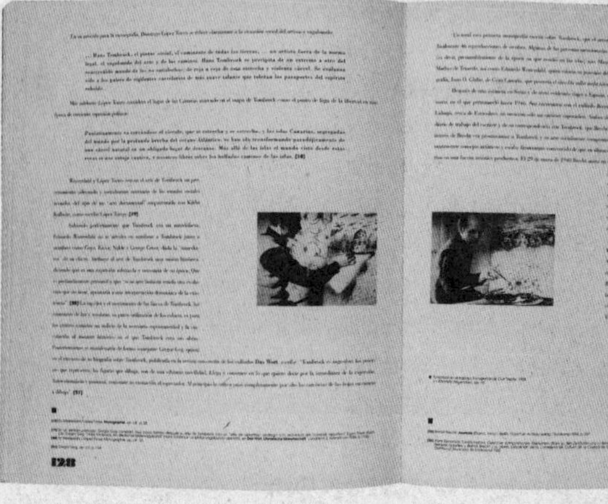
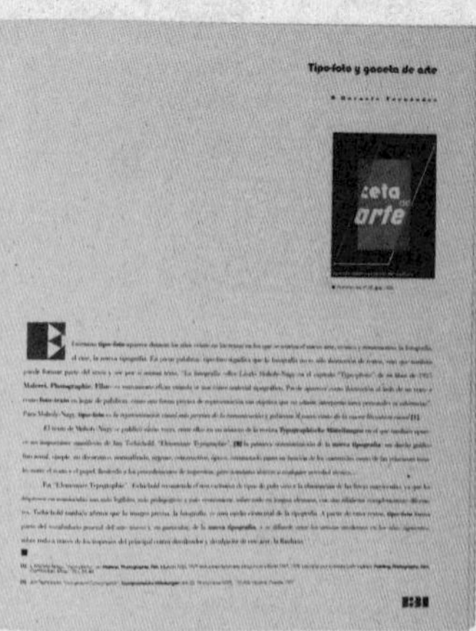
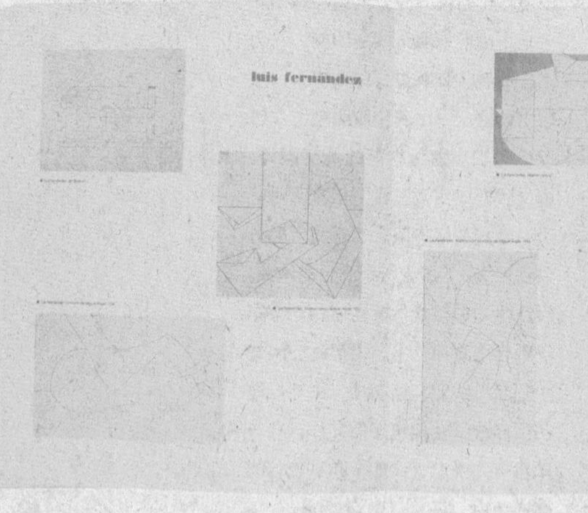

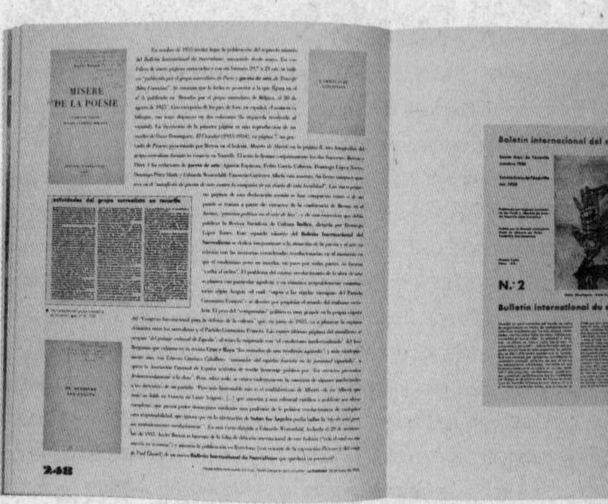

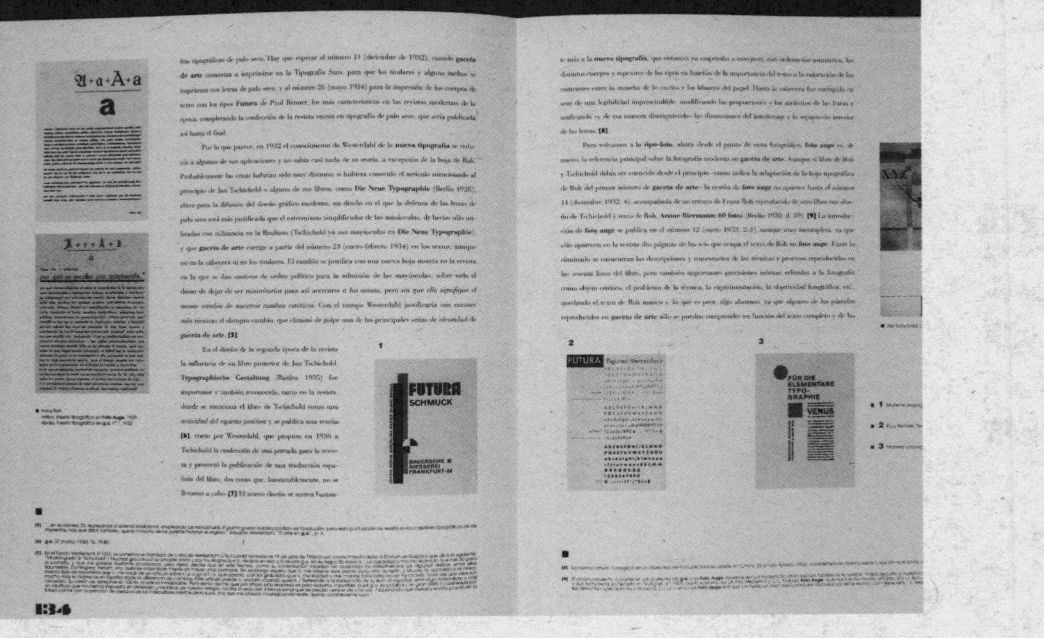

Franz Roh

las tipográficas de palo seco. Hay que esperar al número 11 (diciembre de 1932), cuando **gaceta de arte** comienza a imprimirse en la Tipografía Sans, para que los titulares y alguno sueltos se impriman con letras de palo seco, y al número 26 (mayo 1934) para la impresión de los cuerpos de texto con los tipos **Futura** de Paul Renner, los más característicos en las revistas modernas de la época, completando la confección de la revista entera en tipografía de palo seco, que sería publicada así hasta el final.

Por lo que parece, en 1932 el conocimiento de Westerdahl de la **nueva tipografía** se reducía a algunas de sus aplicaciones y no sabía casi nada de su teoría, a excepción de la hoja de Roh. Probablemente las cosas habrían sido muy distintas si hubiera conocido el artículo mencionado al principio de Jan Tschichold o alguno de sus libros, como **Die Neue Typographie** (Berlín 1928), clave para la difusión del diseño gráfico moderno, un diseño en el que la defensa de las letras de palo seco está más justificada que el extremismo simplificador de las minúsculas, de hecho sólo utilizadas con militancia en la Bauhaus (Tschichold ya usa mayúsculas en **Die Neue Typographie**), y que **gaceta de arte** corrige a partir del número 23 (enero-febrero 1934) en los textos, aunque no en la cabecera ni en los titulares. El cambio se justifica con una nueva hoja inserta en la revista en la que se dan motivos de orden político para la admisión de las mayúsculas, sobre todo el deseo de dejar de ser *minoritarios* para así acercarse *a las masas*, pero *sin que ello signifique el menor cambio de nuestros rumbos estéticos*. Con el tiempo Westerdahl justificaría con razones más técnicas el abrupto cambio, que eliminó de golpe una de las principales señas de identidad de **gaceta de arte**. **[5]**

En el diseño de la segunda época de la revista la influencia de un libro posterior de Jan Tschichold, **Typographische Gestaltung** (Basilea 1935) fue importante y también reconocida, tanto en la revista, donde se menciona el libro de Tschichold como una *actividad del espíritu positivo* y se publica una reseña **[6]**, como por Westerdahl, que propuso en 1936 a Tschichold la confección de una portada para la revista y proyectó la publicación de una traducción española del libro, dos cosas que, lamentablemente, no se llevaron a cabo. **[7]** El nuevo diseño se acerca bastante más a la **nueva tipografía**, que entonces ya empezaba a envejecer, con ordenación asimétrica, los distintos cuerpos y espesores de los tipos en función de la importancia del texto o la valoración de los contrastes entre la mancha de lo escrito y los blancos del papel. Hasta la cabecera fue corregida en aras de una legibilidad imprescindible, modificando las proporciones y los atributos de las letras y unificando –y de esa manera distinguiendo– las dimensiones del interletraje y la separación interior de las letras. **[8]**

Pero volvamos a la **tipo-foto**, ahora desde el punto de vista fotográfico. **foto auge** es, de nuevo, la referencia principal sobre la fotografía moderna en **gaceta de arte**. Aunque el libro de Roh y Tschichold debía ser conocido desde el principio –como indica la adaptación de la hoja tipográfica de Roh del primer número de **gaceta de arte**– la reseña de **foto auge** no aparece hasta el número 11 (diciembre 1932, 4), acompañada de un retrato de Franz Roh reproducido de otro libro con diseño de Tschichold y texto de Roh, **Aenne Biermann: 60 fotos** (Berlín 1930, il. 59). **[9]** La introducción de **foto auge** se publica en el número 12 (enero 1933, 2-3), aunque muy incompleta, ya que sólo aparecen en la revista dos páginas de las seis que ocupa el texto de Roh en **foto auge**. Entre lo eliminado se encuentran las descripciones y comentarios de las técnicas y procesos reproducidos en las sesenta fotos del libro, pero también importantes precisiones teóricas referidas a la fotografía como objeto estético, el problema de la técnica, la experimentación, la objetividad fotográfica, etc., quedando el texto de Roh manco y, lo que es peor, algo abstruso, ya que algunos de los párrafos reproducidos en **gaceta de arte** sólo se pueden comprender en función del texto completo y de las

1

2

3

0843

de Pintura y Fotografía **[46]** *de Pajaritas de Papel, ocurre lo m*

Las pajaritas –hombres y mujeres– vinieron siempre por propi
rentes siempre, son absolutamente privadas. "Pajaritas" se
Pajaritas de papel". El primero de Enero del primer año, em
vulgar. **[47]** *Con arreglo a ese calendario se celebran todas l*
corriente. (...) Tiene himno y música propios, y en lo que va d
se ha celebrado el Año Nuevo, Reyes, Recepciones internaciona

o de las *Pajaritas* habría de verse truncado con las muertes en el mes d

arino. Seguirían Ismael Domínguez, Ernesto Pestana Nóbrega. Todos en

a **Cartones**, que habían elaborado junto a Guillermo Cruz, Carmen Jimé

le la pérdida mueve la iniciativa de la publicación de la colección d

a Westerdahl. **[49]**

le esta manera, resucitado al año de su muerte quiero quitarme lo
uchas le han puesto en pie. Este cuerpo de ahora verdaderamente s
propia casa; no por las casas de sus amigos. Se ha orientado hacia di
esconocidos, agrios– pero jóvenes de España.

blanza de Westerdahl porque proporciona una versión acerca de la

regresar a la ingenuidad, como final de trayecto, que expresa la prop

ios en la evolución hacia el disfrute del arte en la esfera de lo individ

a de arte se encuentra expresada la tesis de Westerdahl en las dos eta

biera sido político. Luego social. Hasta regresar a lo ingenuo. Ahora n
n regreso, volveremos a escuchar de nuevo "La pájara pinta" de Óscar
ulio Antonio. Ahora no es el momento (...)

■ 1935

c o r r e s p o n d e

ará la pelicula "l'age d'or", de buñuel. bre-
á dos conferencias y posiblemente dalí se ha-
icasso, que acaso viniera, pero esto no es se-
n contacto con toda esta gente está nuestro
ero oscar dominguez que ahora expone en
copenhague.
nderá v. que esta exp. es sensacional. yo es-
ıcha gente de fuera porque una exp. de esta
s la primera que se celebra en españa. en
na hay un interés más grande y los mejores
españoles se puede decir que son catalanes

en el correo anterior de Di
un artículo de Gullon, que
No te olvides de Sans. Tu sa
res.

Un cordial saludo de tu ami

as relaciones de gaceta de arte con A.C.

E n r i q u e   G r a n e l l

El paisaje en gaceta de arte y

Del espacio de reflexión al espacio de creación

J o r g e   A g u i a r   G i l

Travesías históricas de gaceta

N i l o   P a l e n z u e l a

Hans Tombrock:

pintor-vagabundo y figura enigmática en g.

S u s a n n e   K l e n g e l

Tipo-foto y gaceta de arte

H o r a c i o   F e r n á n d e z

5

iglo moribundo, es también el princi

evo "umor", en la carta del 29 de ab

es un monstruo que siempre me ha es
ectan, y de la forma en que dicho bu
n una habitación– pero, ¿por qué tien

universalistas y regeneracionistas

odríguez, contaba también con Guillermo C
relativas sobre todo a la discusión política sob
as referencias a las iniciativas de los poetas ti
, la sección "A toda vela" firmada por POPA
ro 10 se mostrara una sección sin título con un

io de su juventud durante la Dictadura de Pr
e que le impulsa a escribir, es un artículo perioc
La Prensa, convertido en **El Día** en la inmedia

ud de España, sentía las mismas inquietudes p
n España. Las columnas de **La Tarde** *publicab*
continuado. Que las discrepancias políticas l
gar a un acuerdo. Los temores y las reservas r
una escisión en el grupo. Y de ahí nace el Sem
de Tenerife. La revista **Proa** *se sigue publican*
io. Las luchas políticas y los personalismos con

1981.

ctos que coincidieron en la isla promovidos cor
Racionalismo en Tenerife, entre otras activida
ionalismo: NAVARRO SEGURA. Maisa: **Racion**

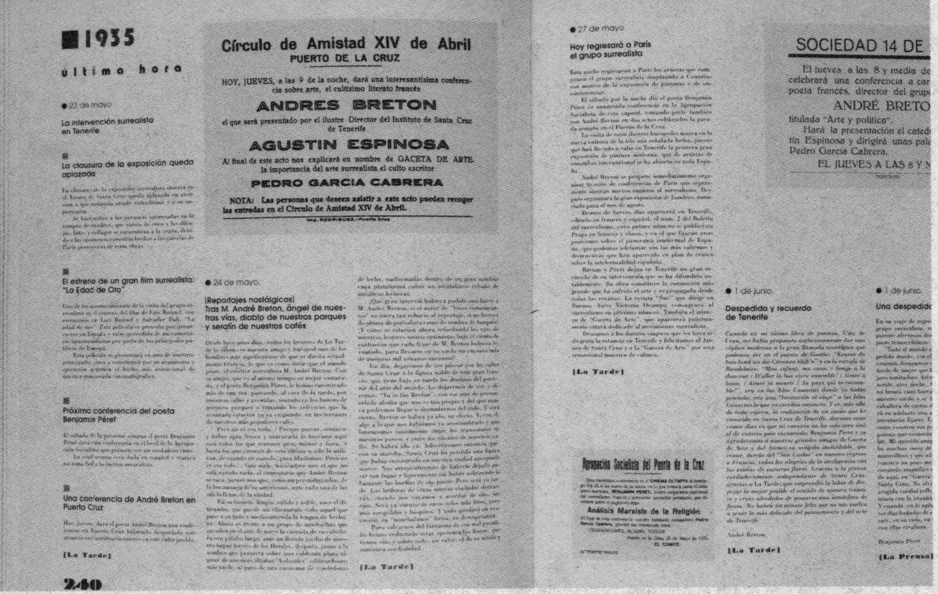

■ 1935

última hora

● 23 de mayo

La intervención surrealista en Tenerife

■ La clausura de la exposición queda aplazada

■ El estreno de un gran film surrealista: "La Edad de Oro"

■ Próxima conferencia del poeta Benjamín Péret

■ Una conferencia de André Breton en Puerto Cruz

[La Tarde]

Círculo de Amistad XIV de Abril
PUERTO DE LA CRUZ

HOY, JUEVES, a las 9 de la noche, dará una interesantísima conferencia sobre arte, el cultísimo literato francés

ANDRES BRETON

el que será presentado por el ilustre Director del Instituto de Santa Cruz de Tenerife

AGUSTIN ESPINOSA

Al final de este acto nos explicará en nombre de GACETA DE ARTE la importancia del arte surrealista el culto escritor

PEDRO GARCIA CABRERA

NOTA: Las personas que deseen asistir a este acto pueden recoger las entradas en el Círculo de Amistad XIV de Abril.

● 24 de mayo

¡Reportajes nostálgicos!
Tras M. André Breton, ángel de nuestras vías, diablo de nuestros parques y serafín de nuestros cafés

[La Tarde]

● 27 de mayo

Hoy regresará a París el grupo surrealista

[La Tarde]

Agrupación Socialista del Puerto de la Cruz

Análisis Marxista de la Religión

EL COMITÉ

SOCIEDAD 14 DE

El jueves a las 8 y media de
celebrará una conferencia a car
poeta francés, director del grupo
ANDRÉ BRETO
titulada "Arte y política".
Hará la presentación el cated
tín Espinosa y dirigirá unas pala
Pedro García Cabrera.
EL JUEVES A LAS 8 Y M

● 1 de junio

Despedida y recuerdo de Tenerife

André Breton

[La Tarde]

● 1 de junio

Una despedida

Benjamin Péret

[La Prensa]

0845

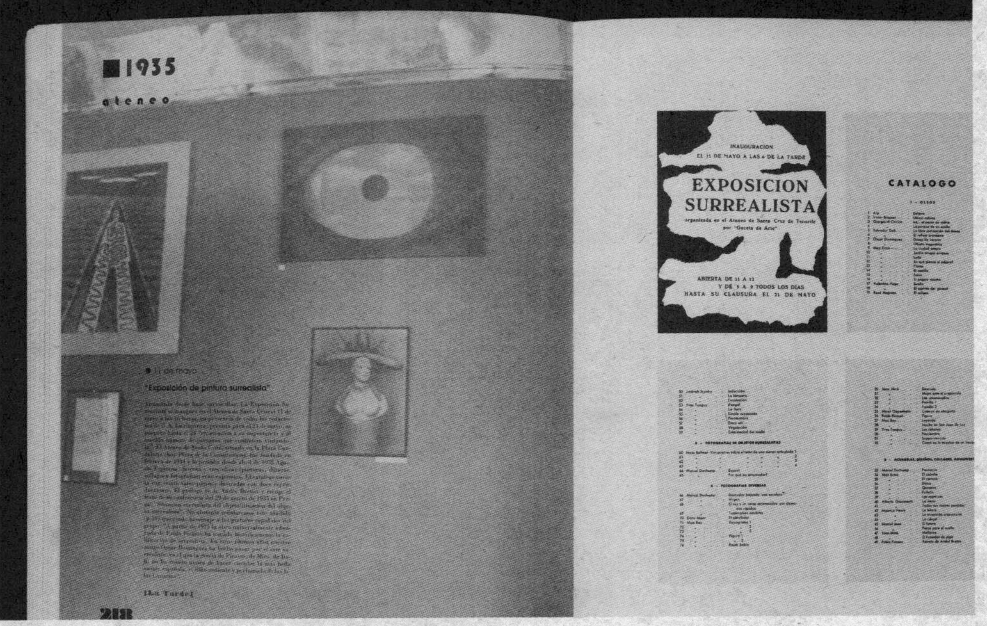
1935
ateneo
11 de mayo
"Exposición de pintura surrealista"
[La Tarde]
218
INAUGURACION
EL 11 DE MAYO A LAS 6 DE LA TARDE
EXPOSICION SURREALISTA
organizada en el Ateneo de Santa Cruz de Tenerife
por "Gaceta de Arte"
ABIERTA DE 11 A 13
Y DE 5 A 8 TODOS LOS DIAS
HASTA SU CLAUSURA EL 21 DE MAYO
CATALOGO

ager
Box 066
pers Ferry
25425

1999 Bm²

Ship To:
ZHAO QING
Dabei xiang7-3,Nanjing Han Qing Tang
Design Co.,
Nanjing,
210000
China

Germany

U.S.A.

China

185mm

8mm

310mm

● 铜奖
《 女巫的数字咒语
◖ 德国

Bronze Medal
Das Hexen-Einmal-Eins
Germany
Claus Seitz
Johann Wolfgang v. Goethe, Wolf Erlbruch
Carl Hanser Verlag, München
185 310 8mm
324g
32p
ISBN-10 3-446-18863-0

这是德国著名戏剧家、诗人约翰·沃尔夫冈·冯·歌德（Johann Wolfgang von Goethe）在他著名的《浮士德》里的片段：浮士德向恶魔寻求长生不老之术，恶魔带他去了女巫的住所，见到女巫一边调制药水、一边念着数字咒语。之后这段记载常被解释为幻方（较常见的类型例如数独）的概念。书籍的装订形式为锁线硬精装，与常见的横开本绘本比起来更长一些。封面胶版纸四色印刷覆哑膜，裱覆卡板。内页胶版纸四色印刷。本书将女巫的咒语与《浮士德》里的形象作为基础，通过拼贴画的方式进行创作，并将数字作为绘画元素融合在画面中。女巫的咒语本身并不需要被解读，单纯作为孩子学数数的范本。书中的文字没有使用同一字体进行排版，而是通过各种笔刷绘制在画面中，与画面的风格相融合。

This is an excerpt from *Faust*, the famous drama written by German dramatist and poet Johann Wolfgang von Goethe. Faust sought immortality from the devil. The devil took him to the witch's residence and saw the witch chanting digital spells while preparing the potion. Later this record was often interpreted as the concept of magic squares (such as Sudoku). The hardcover book is bound with thread sewing, which is longer than the common horizontal book. The matte cover is made of offset paper printed in four colors and mounted with cardboard. The inner pages are printed in four colors on offset paper as well. The illustrations in the book are based on the witch's spell and the images and elements in *Faust*. They are created through collages, and the numbers are integrated into the picture as painting elements. The witch's spell itself does not need to be interpreted, which is simply used as a model for children to learn numbers. Instead of using the same font for typesetting, the text in the book is painted in the picture with various brushes, which is integrated with the style.

Du musst verstehn!

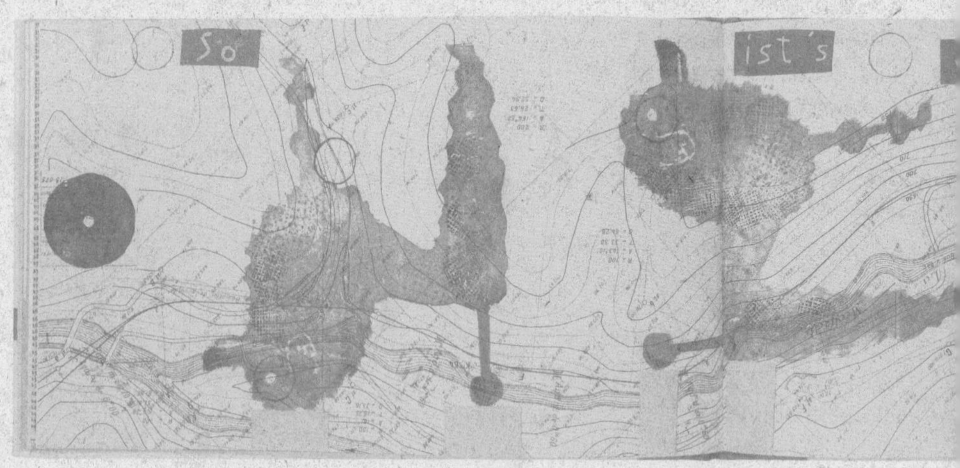
So
ist's

Aus Eins mach Zehn,

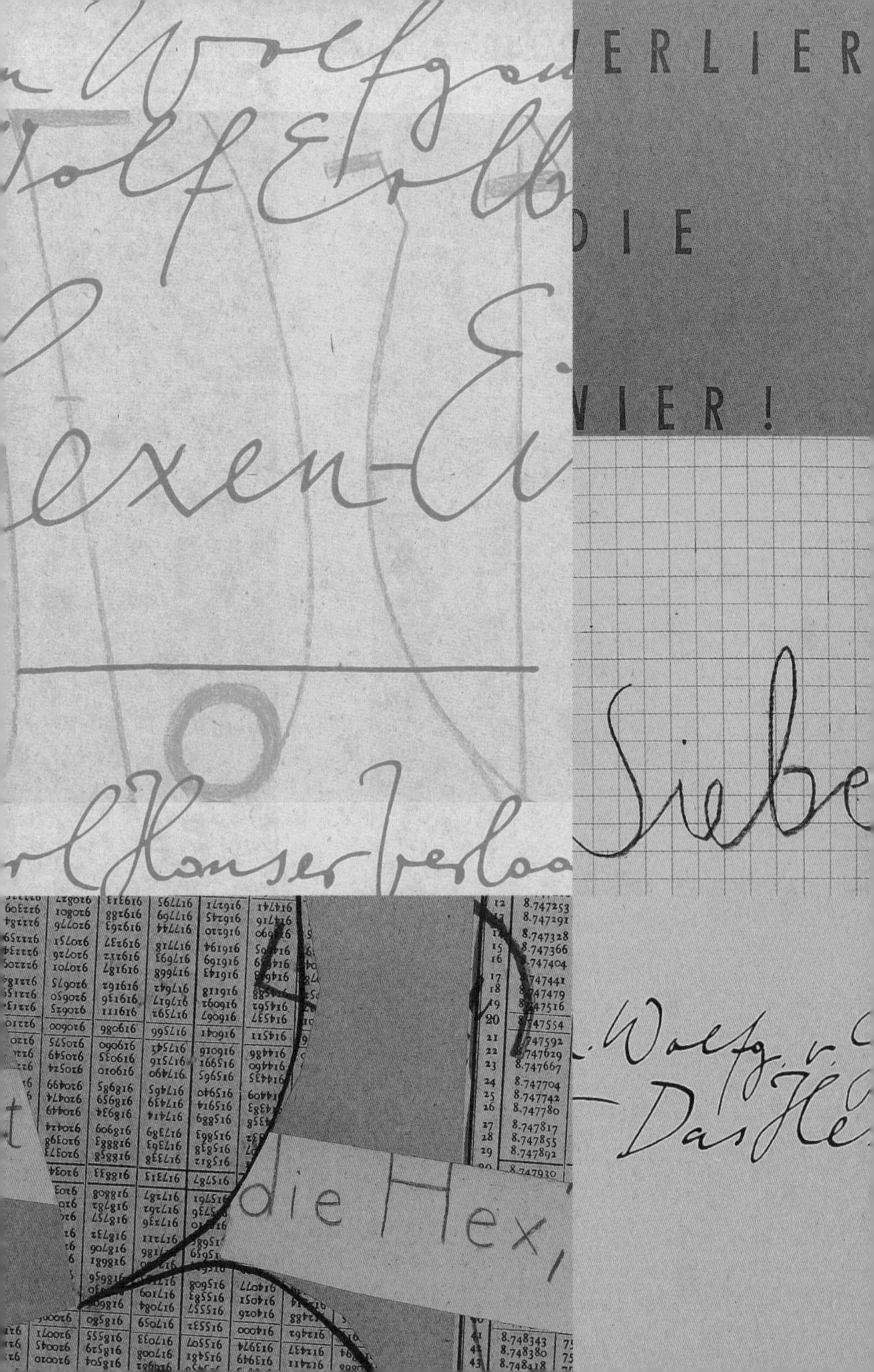
ERLIER
DIE
VIER!
die Hex

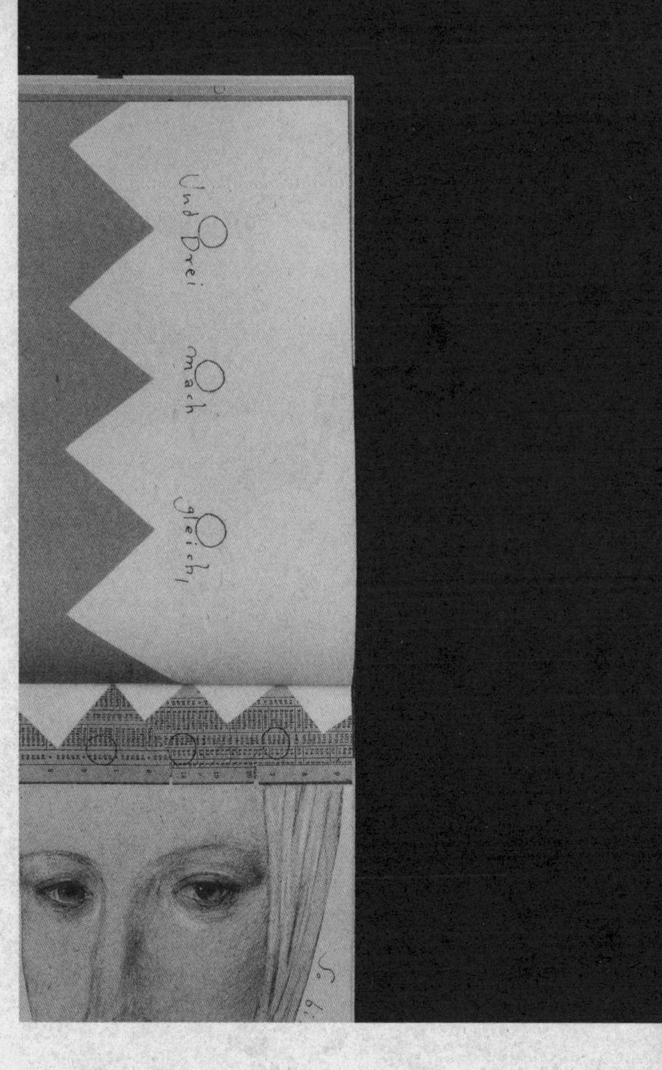
Und Drei
mach
gleich

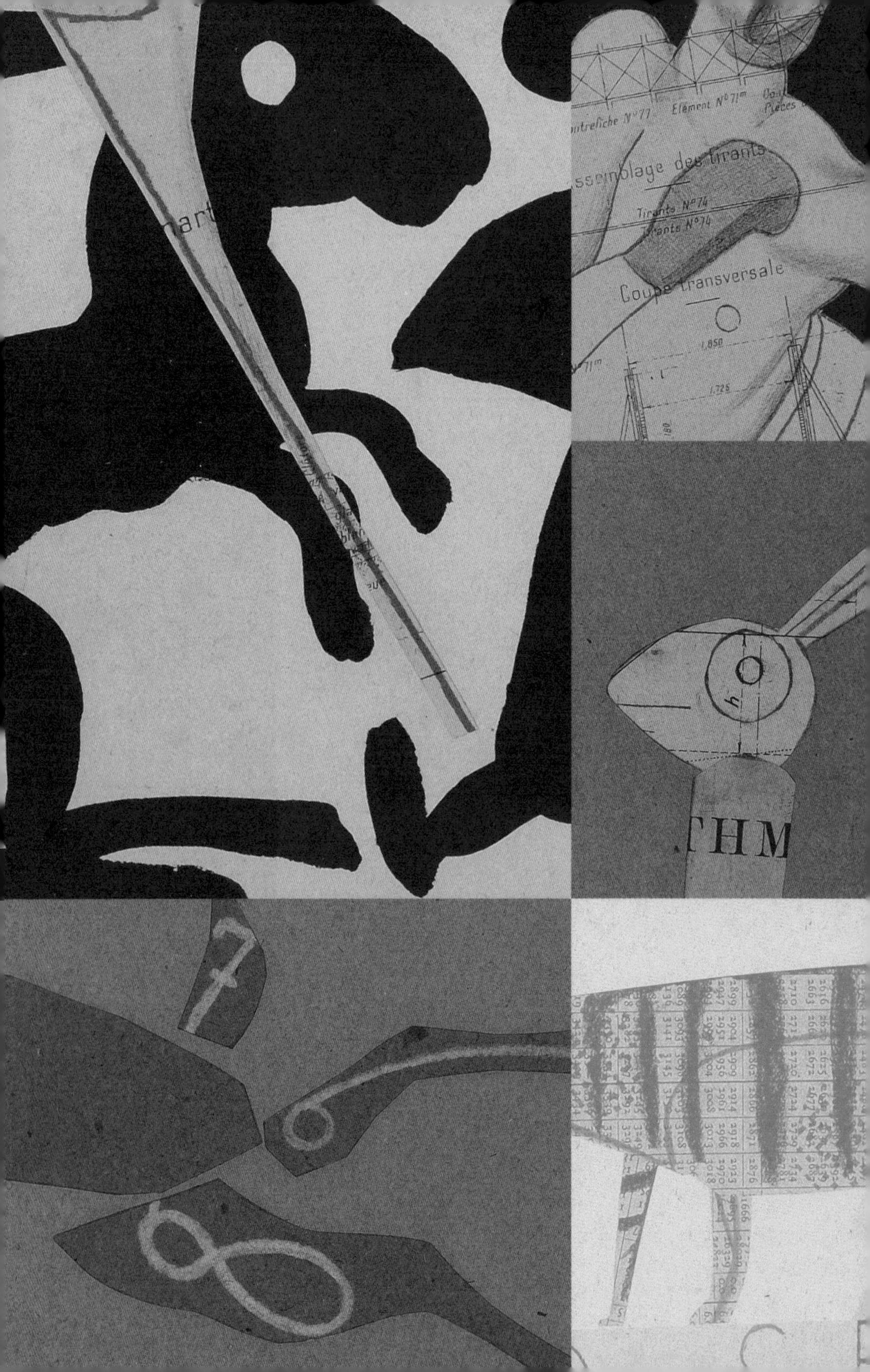

Element N°71m
Tirants N°74
Tirants N°74
Coupe transversale
1.850
1.726
THM

Und Zehn ist keins.

Palmin – eine Jahrhundertmarke

Bm³

Deutsche Post
Documents
Others
Weight (in kg)
Value (in EUR)
0.57
20.00
FRO
21031 Hamburg
GERMANY
TO
ZHAO QING
Nanjing Han Qing Tang Design C
Dabei xiang 7-3
210000 NANJING
CHINA
WARENP.
06 Nov 19
Germany
Germany
China

308mm

13mm

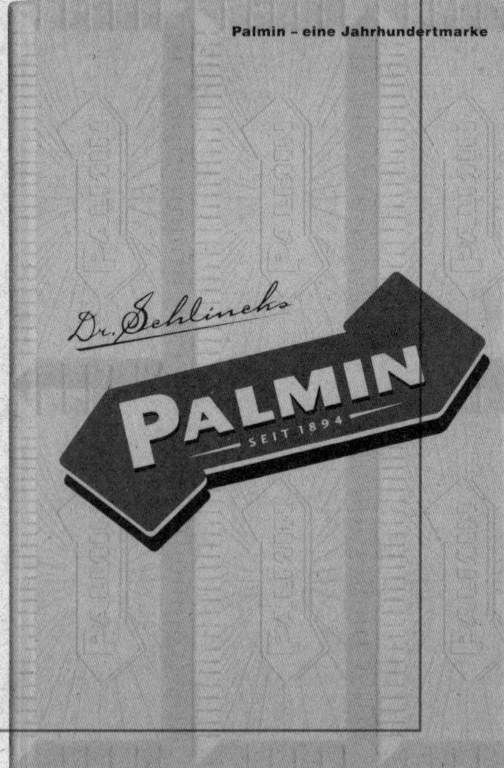

206mm

● 铜奖
《 百年品牌 Palmin
◖ 德国

Bronze Medal
Palmin: eine Jahrhundertmarke
Germany
Gesine Krüger, Andrea Schürings (büro für mitteilungen)
Joachim Nickel
Union Deutsche Lebensmittelwerke, Hamburg
308 206 13mm
568g
96p

Palmin 是由海因里希·弗朗茨·施林克博士（Dr. Heinrich Franz Schlinck）于 1840 年创立的品牌，最主要的产品是椰子油，到书籍出版时（1998 年）已经 150 多年。本书介绍了 Palmin 的品牌历史。书籍的装订形式为圆脊锁线硬精装。护封采用类似于包装固体椰子油的包装纸蓝红黑三色印刷，内封胶版纸蓝红双色印刷，裱覆卡板。环衬选用红色胶版纸。内页哑面涂布纸张四色印刷。书中记录了 Palmin 品牌的各个方面，包含品牌发展、产品生产、广告设计、各类产品的包装变革等相关历史。设计方面，将 Palmin 经典的红蓝双色运用于书中，处于版面的三分之二处的红色条延伸至书口处，用黑色印上年份，方便查阅。

Palmin is a brand founded by Dr. Heinrich Franz Schlinck in 1840. The main product is coconut oil. It has been more than 150 years since the publication of the book (1998). This book introduces Palmin's brand history. The hardcover book is bound by thread sewing with rounding spine. The jacket is made of the wrapping paper used to package solid coconut oil and printed in blue, red and black. The inner cover is made of offset paper printed in blue and red, mounted with cardboard. Red offset paper is used for the end paper while the matte coated paper is printed in four colors for inner pages. As for the content, it describes all aspects of the brand Palmin, including brand development, product production, advertising design, packaging changes of various products and other related history. The year is printed in black on the red strip extending to the forc cdge the book, which is convenient for finding the corresponding year.

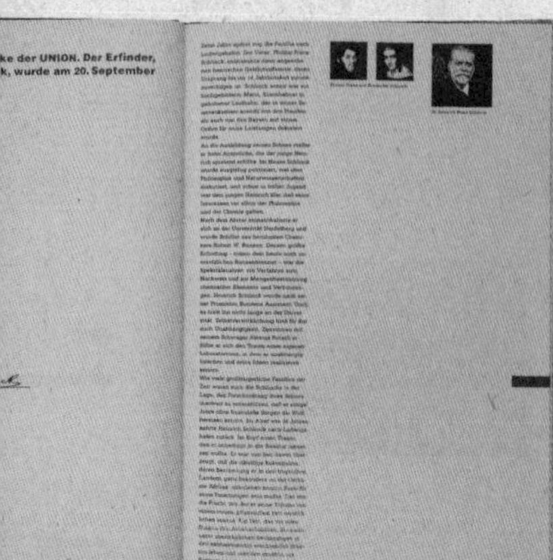

Palmin

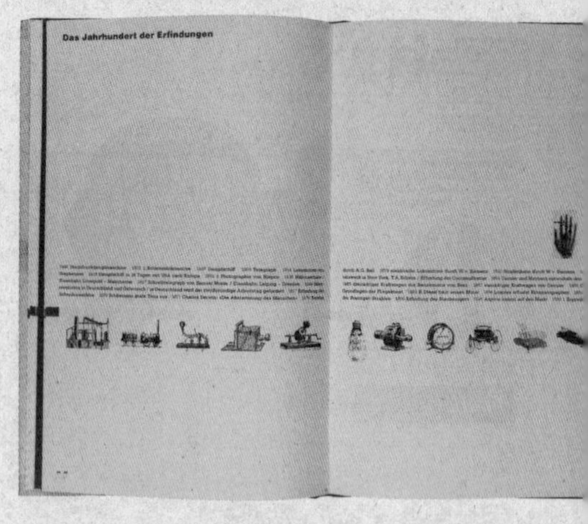
Das Jahrhundert der Erfindungen

t' ich doch
Palmin genommen!
ACHTEN SIE AUF DEN NAMENSZUG
PALMIN
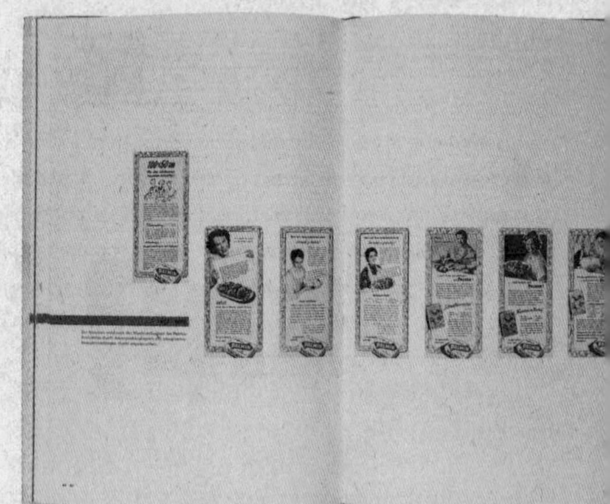

»Die grüne Kuh«

Pflanzliche Öle und Fette sind für die Ernährung des Menschen wegen ihres Gehaltes an ungesättigten Fettsäuren eine notwendige Ergänzung zu den tierischen Fetten. Seit Jahrtausenden sind natürliche Öle auch Basis für unterschiedliche Produkte wie Kosmetika, Medizin oder dienen als Brennstoffe.

Rund 50 Pflanzenfamilien liefern heute etwa 300 verschiedene Sorten Öl, das meist als Reservestoff in den Zellen der Samen oder des Fruchtfleisches gespeichert ist.
Durch Erhitzen auf 100°C und gleichzeitiges Pressen der abgeschälten Samen oder zerkleinerten Früchte in hydraulisch arbeitenden Schrauben- oder Schneckenpressen wird das kostbare Öl gewonnen. Neben der Ölgewinnung haben auch die Rückstände, wie beispielsweise Raps- oder Erdnußkuchen, wegen ihres hohen Eiweißgehaltes als Viehfutter eine große wirtschaftliche Bedeutung.
All' das gilt auch für die Kokospalme, eine der wirtschaftlich wichtigsten Pflanzen der Tropen. Sie ist die »Königin der Palmen«, wie sie ehrfurchtsvoll von den Menschen genannt wird, die von und mit ihr leben. Bis zu 30 Meter hoch wächst sie in den tropisch
Himmel und liefert oft mehr als 2
Früchte im Jahr. Fast alles von d
kann vom Menschen genutzt we
»Aus einer einzigen Kokospalme
man ein ganzes Schiff bauen un
dazu noch voll beladen«, lautet
ein Sprichwort der Ceylonesen.
ten und Boote werden aus den S
men. Stricke, Taue und Matten a
den faserigen Basthüllen der Nü

Neben den üblichen werblichen Aussagen wird in Anzeigen immer wieder aufs neue darauf hingewiesen, daß nur Palmin ein rein pflanzliches und billiges Kokosfett ist. »Der verehrte Konsument solle sich bitte von anderen Produkten nicht täuschen lassen.«

Palmona
ANZEN-BUTTER-MARGARINE
Palmin
Die Erfind
Die vielleic
seit der Ei
Erfindung
großen Pla
langer Zei
zum Abkn
min-Anwe
praktische
sich Palmi
leichter en
wieder ve
he zählt
kereien.
as zeitauf-
g. Erst
Hause ist.
n Würfel
aopulver,
ensaft
men oder
berziehen
des gewohn
fettquantum
Das macht schon etwas
Monat, wenn Sie ein
ünftel Ihrer Fettausgaben
en können. Palmin ist s
iebig, weil es hochkonze
t; 100% naturreines Pfla
das reine

PALMIN

PALMIN
PALMIN
Palmin
PALMIN
KOCH-
BUCH

Gewinnen Sie Ihre Foto-Schürze mit Palmin soft.

Palmin soft verlost 2000 Foto-Schürzen. Und wenn Sie gewinnen, dann kommen Sie mit Ihrem Foto auf diese individuelle Schürze.

So können Sie glücklicher Gewinner werden: Schreiben Sie uns bitte auf einer Postkarte Ihr Lieblingsrezept mit Palmin soft. Weil Palmin soft 100% geschmacksneutral ist, schmeckt ein gutes Stück Fleisch auch so, wie es nun mal schmecken soll: richtig gut nach Fleisch und nicht nach Fett.

Wenn Sie Ihr schönstes Rezept bis zum 31. 10. 79 (Datum des Poststempels) an PALMIN SOFT, Postfach 501020, 2000 Hamburg 50, schicken, haben Sie gute Chancen, unter die 2000 glücklichen Gewinner zu kommen.

Aus allen Einsendungen werden die 2000 glücklichen Gewinner ausgelost. Der Rechtsweg ist ausgeschlossen. Wenn Sie gewonnen haben, werden Sie sofort benachrichtigt und erfahren dann alles Weitere, wie Sie oder eine Person Ihrer Wahl auf diese lustige Schürze kommen. Und nun viel Glück mit Palmin soft.

Die Kunst der feinen Verführung

Konfekt aus der eigenen Küche zählt zu den verführerischsten Leckereien. Das Herstellen ist zwar etwas zeitaufwendig, aber nicht schwierig. Erst recht nicht, wenn Palmin im Hause ist. Einfach einen geschmolzenen Würfel Palmin mit Puderzucker, Kakaopulver, Kaffee, Zitronen- oder Orangensaft verrühren, Konfektstücke formen oder das Gebäck mit der Glasur überziehen – fertig.

Die Erfindung des »Knicks«

Die vielleicht wichtigste Neuerung seit der Einführung von Palmin ist die Erfindung des »Knicks«. Statt einer großen Platte gibt es nun schon seit langer Zeit viele kleine Palmin-Stücke zum Abknicken. Das macht die Palmin-Anwendung noch einfacher und praktischer. Mit dem »Knick« läßt sich Palmin viel exakter portionieren, leichter entnehmen und ganz einfach wieder verschließen.

Manche mögens weich: Palmin soft

Für alle, die Fett ganz individuell portionieren möchten, ohne auf die Vorteile von Palmin zu verzichten, hat die UNION im Jahre 1970 das geschmeidige »Palmin soft« entwickelt. Verpackt in der praktischen 500-Gramm-Schale ist »Palmin soft« jederzeit schnell zur Hand.

Der Knick macht's. Palmin läßt sich jetzt noch leichter portionieren und erleichtert das Arbeiten in der Küche. Die Hausfrau ist begeistert.

^{1999} Bm^{4}

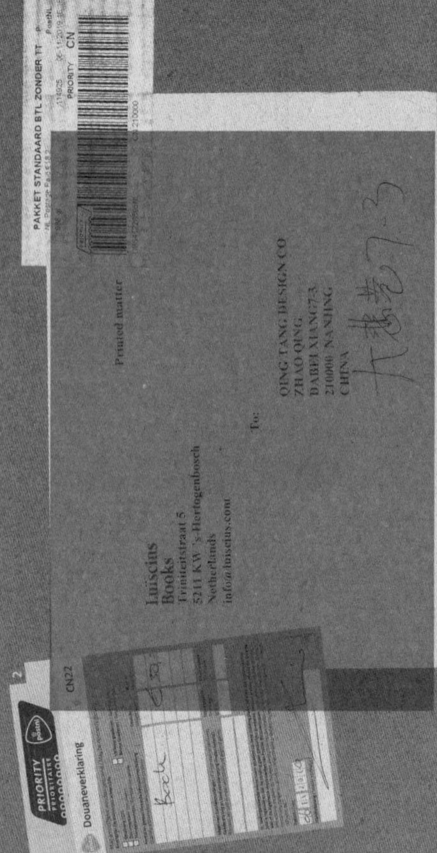

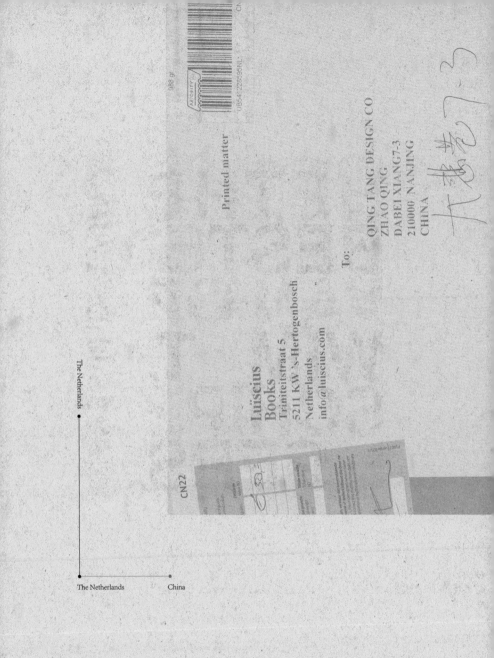
The Netherlands
The Netherlands
China
988 gr
PRIORITY
UB541228595NL
Printed matter
To:
QING TANG DESIGN CO
ZHAO QING
DABEI XIANG7-3
210000 NANJING
CHINA
Luïscius
Books
Triniteitstraat 5
5211 KW 's-Hertogenbosch
Netherlands
info@luiscius.com
CN22

247mm

28mm

HET HET BESTE VAN OF WIM T. SCHIPPERS

HARRY RUHÉ

HARRY RUHÉ WIM T. SCHIPPERS

centraal museum

175mm

● 铜奖
《 维姆 · 席佩斯精选集
◖ 荷兰

Bronze Medal
Het Beste van Wim T. Schippers
The Netherlands
Studio Gonnissen en Widdershoven
Harry Ruhé
Centraal Museum, Utrecht
247 175 28mm
835g
240p
ISBN-10 90-73285-44-5

维姆·席佩斯（Wim T. Schippers）是荷兰视觉艺术家、演员和导演。本书是他在荷兰乌特勒支中央博物馆展览时的作品集。书籍的装订形式为锁线硬精装。封面铜版纸红绿黑三色印刷覆哑膜，裱覆卡板。内页胶版纸红绿黑三色印刷，即从封面开始贯穿全书，只用红绿黑三色。内容方面，主要是评论家对席佩斯的评价、各个时期的作品，以及相应年表资料。设计方面，本书将荷英双语分别设计为红绿双色，所有的正文、作品名、图片均采用红绿双色搭配黑色来呈现。书籍的开始提供了红绿两张透明薄片，可以对书中页面进行遮挡，从而呈现不同的内容。当使用红色薄片时，红色的荷兰文被过滤，绿色的英文和红色薄片组成黑色文字和图片；反之使用绿色薄片时，荷兰文显现出来。虽然是三色印刷，但全书在绿色的使用上有递进变化。评论文本部分使用草绿色，之后颜色逐渐加重，进而呈现出翠绿色、墨绿色。封二上附有薄透的纸袋，内装艺术家的作品、剧照和海报。

Wim T. Schippers is a Dutch visual artist, actor and director. This book is a collection of his works at the Central Museum in Utrecht, the Netherlands. The hardcover book is bound by thread sewing. The matted cover is made of coated paper printed in red, green and black, mounted with cardboard. The inner pages are made of offset paper printed in red, green and black, and only red, green and black are used throughout the whole book. As for the content, it is mainly the critic's evaluation of Schippers, the works from various periods, and the corresponding chronological data. In terms of design, the book has designed Dutch and English text in red and green respectively. All the text, work titles, and pictures are presented in red and green with black. At the beginning of the book, red and green transparent sheets are provided, which can present different contents by blocking the pages in the book. When red flake is used, red Dutch text is filtered. Green English text and red flakes make up black text and pictures. Otherwise, red Dutch text appears when green flake is used. Although it is three-color printing, there is a progressive change in the use of green in the book. The text of comment uses the grass green, and then the color gradually darkens, such as emerald and dark green. There is a thin paper bag on the inside front cover of the book, containing the artists' works, stills and posters.

HE THE
BEST E
VAN OF
WIM T.
SCHIP
PERS

HARRY RUHÉ

Schippers, monogram kaartje ontworpen door George Maciunas

Schippers, monogram card designed by George Maciunas

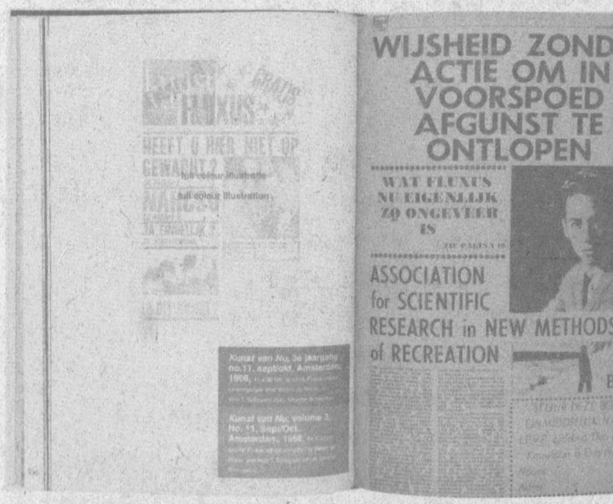

WIJSHEID ZOND
ACTIE OM IN
VOORSPOED
AFGUNST TE
ONTLOPEN
WAT FLUXUS
NU EIGENLIJK
ZO ONGEVEER
IS
ASSOCIATION
for SCIENTIFIC
RESEARCH in NEW METHODS
of RECREATION

Kunst van Nu, 3e jaargang no.11, sept/okt, Amsterdam, 1966.

Kunst van Nu, volume 3, No. 11, Sept/Oct, Amsterdam, 1966.

De oprichters van *Omroepvereniging Eldorado*, met v.l.n.r. Wim van der Linden, Wim T. Schippers, Hans Verhagen en Willem de Ridder, Amsterdam, 1967

The founders of *Omroepvereniging Eldorado*, with from left to right Wim van der Linden, Wim T. Schippers, Hans Verhagen and Willem de Ridder, Amsterdam, 1967

De Telegraaf, 9 januari 1975

De Telegraaf, January 9, 1975

1 Frank Zaagsma, *Wim T. Schippers: 'Ik houd niet van komieken'*, in: Haagse Post, 31 mei 1980
2 Peter van Brummelen, *Dames en heren, wilt u het museum verlaten?*, in: Het Parool, 26 maart 1994
3 *Fred Haché en Barend Servet in de vijf afleveringen van De Fred Haché Show*, Amsterdam, 1973, p.140
4 *Het Barend Servet effect*, Amsterdam, 1974, p.164
5 *Het Barend Servet effect*, o.c., p.179
6 *Het Barend Servet effect*, o.c., p.96
7 *Het Barend Servet effect*, o.c., p.168
8 Peter Bruyn, *Wim T. Schippers, 'Het leven is een aaneenschakeling van onbenulligheden'*, in: Rails, Jrg.44, mei 1995, p.38
9 *Het drama van Wim T. Schippers*, De Bezige Bij, Amsterdam, 1979, p.7
10 De Volkskrant, 14 juni 1977

ORKATER
RELAPSUS
WIM T. SCHIPPERS

Relapsus

Chapter twelve
Twaalfde Hoofdstuk

in which various extremely diverse matters are discussed, and the main character of this book is distinguished with a number of awards

Waarin zeer uiteenlopende zaken aan de orde komen, en de hoofdpersoon van dit boek verschillende onderscheidingen krijgt uitgereikt

Een van Schippers' meest opmerkelijke theaterstukken werd in 1986 i de Amsterdamse Stadsschouwburg opgevoerd. Het publiek maakte door ernaar te komen kijken deel uit van deze voorstelling. Niet voor niets heette het stuk *Going to the dogs*. De rolbezetting van dit 'toneel stuk in vier bedrijven' werd verzorgd door een zestal goed getrainde herdershonden. Maandenlang was er gerepeteerd, opdat er volgens d aanwijzingen in het script werd geblaft, met haardblokken werd gezeuld en televisie werd gekeken. Een 'heus familiedrama' moest het worden, 'vol Dynasty-achtige verwikkelingen'.
Het evenement leverde veel publiciteit op. Verschillende buitenlandse kranten wijdden artikelen aan het stuk 'des Neo-Dadaisten, 'Neo-Fluxus', Alleskönners und Komödianten Wim T. Schippers,' en diverse Europese televisiejournalen lieten beelden zien van de zes blaffende acteurs. In de Tweede Kamer stelde de Minister van Buitenlandse Zaken zorgelijk vast dat dergelijke culturele uitingen een 'vrij ruïneuze uitwerking' zouden hebben op de presentatie van Nederland in het bui tenland. Affiches voor de theatervoorstelling werden door actievoerders overgeplakt met leuzen als 'Schippers doet het met honden', uit protest tegen 'het schandalige misbruik van onze trouwe viervoeters'. Menigeen verbaasde zich erover dat het Ministerie van WVC - overigens na lang aarzelen - hiervoor subsidiegelden beschikbaar had gesteld, en vroeg zich af of dat geld niet beter naar Boltini gestuurd had kunnen worden ('omdat je er daar nog een olifant bijgeleverd krijgt'). Schippers gaf als antwoord dat '*Going to the dogs* geen kunstj was, maar kunst'. Met de kritiek dat het stuk dramatisch gezien nogal saai zou kunnen uitvallen, was hij het overigens wel eens: 'Ik heb later

concert

OCCASION

VAARLIJK

seen as a venue to engage in what is
t form. Since the infamous concert by
it is possible that this description
in Sc
s movement was permitted to stage a
s a contribution towards the
, and the "event" as a bona fide form of
nclined to echo the sentiment of the
... collaboration between t...
... judging from their behaviour were
... than with conventional, romantic
nclined to accept a statement dressed
... to the prevailing mood
... concerts that provoke such a reaction
... on stage or – more often than not
d not be inclined to express what they
ve gentlemen who do nothing except

door-
fenen,
roem-
stones
urzaal
g het
bewe-
n uit-
ij aan
en de
r. Wij
iehon-
iddel-
eens
: een

direct samenspel van uitvoeren
publiek, waarbij het publiek te o
naar hun gedragingen, meer be
is dan bij het conventionele, roma
theater waarbij men zich in kun
gegoten mededeling niaar laat
nen of zich passief conformeert.
Onmogelijk is dit nu bij de
concerten, die dusdanige reacties
pen, dat men ze en plein publiek
uiten, wat weer voor een belangri
bijdraagt aan hetgeen op het tone
beurt of — vaker — niet gebeur
immers wordt niet gezet tot zel
zaamheid wanneer op het tone
heren staan, die niets anders doe
de zaal in staren, zoals bij de
van de Nederlander Wim T. Sch
dat uit vier delen bestond: 1. niet
2. niet eten, 3. roken, 4. eten. Een

roefden zij het apparaat open en
n de anijsgeur verspreidende
istof in een speciaal hiertoe mee-
rachte doos. Uit een soortgelijke
s werd vervolgens een andere
istof in de verstuiver overge-
cht en even later kon het tweetal
geheel bedrijfsklare paneel weer
het plafond bevestigen.

ging wat moeilijk. De jongeman
st er een van de moderne stoelen
akken, terwijl zijn begeleidster
gegeneerd naar ons lachte. Toen
allemaal gelukt was, verlieten zij
een beleefd „Goedenmiddag" het
tje.

ving een zeer stille, maar hevige
d aan in de ons omringende at-
feer. De nieuwe (volgens het
gramma muskus-)geur begon zeer
giek te trachten de nog in het
rek hangende anijslucht te ver-
en. Deze laatste moest na korte
het loodje leggen en van toen
konden de drie bezoekers zich
houden met het opsnuiven

EID ZO
IE OM
ORSPO
GUNST
TLOP

ARS DOOR
MSTERDAM

naal Programma
Muziek - Nieuwste
ieuwste
programma voor
cert in De Kleine
msterdam, 18
963, 30,9 x 21,3 cm.

HEL C.S.

- Nieuwendijk -
Muntplein

chten.
ap te hebben
ordt afge-
at mr R. de
zelf ge-
Oekel-rol in.
s De Ranitz
n heeft
Oekel be-
kel steeds
st lezen zou
rs' veelvul-
(Schippers:
niet onthou-
an Brouwers
rok „in één
ich toe."
r de „smake-
n en kin-
reator to

woord „Meneer
racter, meneer S
De advocaat van
Boogaard, mr H.
Brouwers veron
scabreuze strips
Brouwers in het
borsten van vrou
geleegd in een fie
uitwerpselen ron
vanuit de zaal: „I
en minutenlang
urineren? Waaro
eens zo puriteins

in a case brought by
Schippers. Brouwer
financially reimburs
by Schippers and by
den Boogaard who

Het Vrije Volk.

12.12.1962

A-dynamisch glas in Fodor

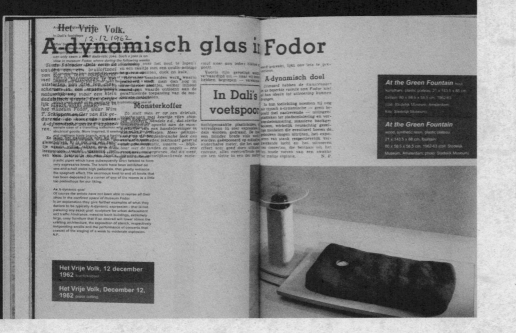

In Dali's voetspoor

plastic pipes which have subsequently been twisted to form very expressive knots. The knots have been exhibited on one-and-a-half metre high pedestals, that greatly enhance the spaghetti effect. The enormous knot to end all knots that has been deposited in a corner of one of the rooms is a little too pretentious for our liking.

As A-dynamic goal

Of course the artists have not been able to realise all their ideas in the confined space of museum Fodor.

In an explanation they give further examples of what they declare to be typically A-dynamic expression - that is not pursuing any exact goal: sculpture for urban defacement and traffic hindrance, massive bank buildings, extremely large, cosy furniture that if so desired will tower above the existing architecture, the exposition of stench, respectively invigorating smells and the performance of concerts that consist of the staging of a weak to moderate explosion.

N.F.

Monsterkoffer

A-dynamisch doel

Uiteraard hebben de kunstenaars in de beperkte ruimte van Fodor niet al hun ideeën tot uitvoering kunnen brengen.

In hun toelichting noemen zij nog als typisch a-dynamische — geen bepaald doel nastrevende — uitingen: plastieken ter stadsontsiering en verkeersbelemmering, massieve bankgebouwen, werkelijk reusachtig gezellige meubelen die eventueel boven de gebouwen mogen uitrijzen, het exposeren van stank respectievelijk verkwikkende lucht en het uitvoeren van concerten, die bestaan uit het ten tonele voeren van een zwakke tot matige explosie.

N. P.

Het Vrije Volk, 12 december 1962 krantenknipsel

Het Vrije Volk, December 12, 1962 press cutting

At the Green Fountain hout, kunsthars, plastic plateau: 21 x 143.5 x 88 cm, fontein: 80 x 58.5 x 58.5 cm, 1962-63 (coll. Stedelijk Museum, Amsterdam; foto: Stedelijk Museum)

At the Green Fountain wood, synthetic resin, plastic plateau: 21 x 143.5 x 88 cm, fountain: 80 x 58.5 x 58.5 cm, 1962-63 (coll. Stedelijk Museum, Amsterdam; photo: Stedelijk Museum)

uto+auto=124
s ruimte+kracht+pracht

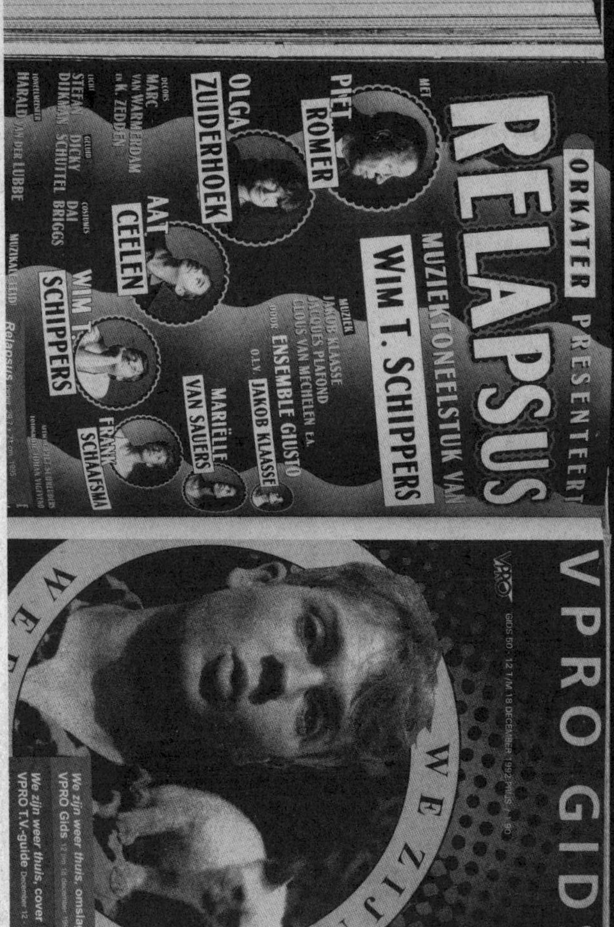

Relapsus 1995
We zijn weer thuis, omslag VPRO Gids 12 t/m 18 december 1992
We zijn weer thuis, cover VPRO T.V.-guide December 12 - 18

1999 Bm5

30500009176 ADM DHL
Shipper Ref.
ITN#:
EXPRESS WORLDWIDE
WPX
From:
Ph:(95
252-457
To:
Ph:+8613
Nanjing 210000
CHINA
CN-PVG-NKG
C
EEI: NOEEI Sec. 30.37(a)
Content: Books
WAYBILL 5095888103
U.S.A.
U.S.A.
China

278mm

22mm

247mm

铜奖
摄影的多重角色
——艺术、文献、市场和科学
美国

Bronze Medal
Photography's Multiple Roles:
Art, Document, Market, Science
U.S.A.
studio blue
——
Museum of contemporary Photography, Chicago
278 247 22mm
1425g
256p
ISBN-10 0-9658887-2-X

这是芝加哥当代摄影博物馆出版的摄影理论书籍，通过多位理论家和艺术家的文章和作品来探讨摄影这个艺术门类需要扮演的四重角色。书籍的装订形式为锁线胶平装。封面白卡纸红黑双色印刷，覆哑光膜。内页哑面铜版纸四色印刷。封面标题周围，划分出四个象限，四种字体清晰地区分出全书的四大部分，即：衬线体——艺术，无衬线粗斜体——文献，粗圆体——市场，用斜切的直线代替弧线的无衬线体——科学。内文部分也通过与四个部分标题相适配的字体来排版。全书的纵向网格为 20 栏，根据内容需要排出宽度分别占用 10+10 的双栏、5+5+5+5 的四栏、3+7+10 的不对称三栏、3+7+7+3 的大双栏加两个小边栏、3+3+3+10 的不对称四栏等丰富的版面形式。

This is a photography theory book published by the Contemporary Photography Museum in Chicago, exploring the four roles that photography, as an art genre, needs to play through the articles and works of multiple theorists and artists. The book is perfect bound with thread sewing. The cover is printed in red and black on white cardboard, mounted with matte film. The inner pages are made of coated paper printed in four colors. Around the book title on the cover, four quadrants clearly distinguish the four major parts of the book using four typefaces: serif–art, sans serif bold italics–literature, bold roundhand–market, and sans serif using oblique straight lines instead of arcs–science. The text is also typeset using fonts that match the titles of the four sections. The vertical grid of the entire book has 20 columns, and rich layout forms are arranged according to the content needs: double columns with a width of 10+10, four columns with a width of 5+5+5+5, asymmetric three columns with a width of 3+7+10, large double columns with two small sidebars with a width of 3+7+7+3, asymmetric four columns with a width of 3+3+3+10, etc.

Photography's multiple roles
Denise Miller
Eugenia Parry
F. David Peat
Naomi Rosenblum
Rod Slemmons
Mihaly Csikszentmihalyi
Ed Paschke
Franz Schulze

art + i
Denise Miller

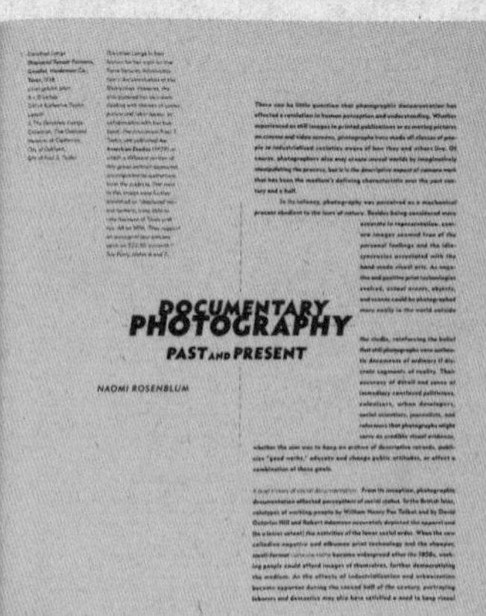
DOCUMENTARY PHOTOGRAPHY
PAST AND PRESENT
NAOMI ROSENBLUM

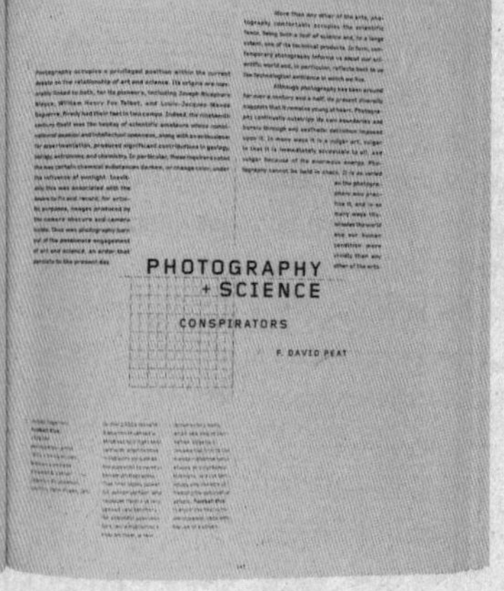
PHOTOGRAPHY + SCIENCE
CONSPIRATORS
F. DAVID PEAT

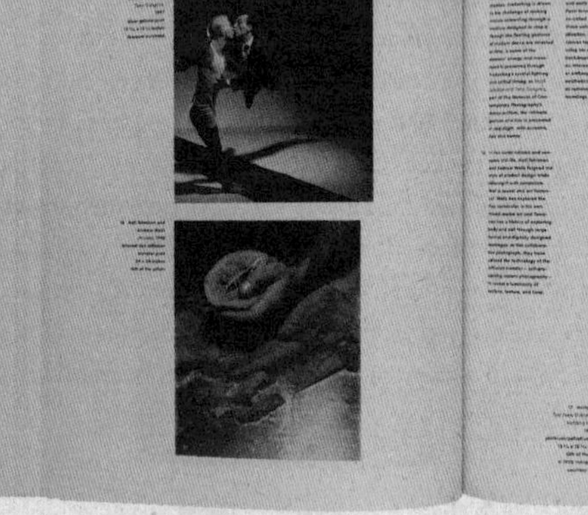

The founding of what was to become The Museum of Contemporary Photography was motivated by both pragmatic and poetic concerns. By 1975 Columbia College Chicago's photography department was recognized as one of the leaders in the photography renaissance that had started in the early 1960s. For some time the department had been regularly exhibiting the work of noted photographers in their print finishing room. These exhibitions had grown out of a strong faculty conviction that there was a need for such shows, and that the college had a responsibility to promote and support contemporary photographers whose work would teach and inspire our students. In addition, we wished to develop a broader audience for photography and further encourage its acceptance as a contemporary art and as a force for learning, understanding, and perhaps even changing the world.

Our success was as much a surprise to us as it was to anyone else. To build upon it further, in the spring of 1975 we rented two adjoining vacant offices on Ohio Street to create The Columbia College Gallery. To achieve a gallery ambiance, the college's carpenter cut arches at each end of the wall that separated the offices. Our first exhibition included work by all the photographers currently teaching in Chicago colleges who wished to participate.

Forewo

John Mulvany
Chairman of the Museum Governing Board

Time present and time past
Are both perhaps present in time future
And time future contained in time past...

(T. S. Eliot, *Four Quartets*, 1943)

into its
own sp
change
of expa
opport
was of
museu
Throug
the Am
today, i

tion of T

vogue style

11. Eileen Cowin
Untitled (Magritte),
1988
silver dye bleach
(from color positive) print
62 x 48½ inches
Gift of the artist

With a skillful use of gesture and pose, Eileen Cowin creates mysterious scenarios. Blending the ecclesiastical allegory of the fifteenth century, the romantic vision of the nineteenth century, and the Surrealist trends of the twentieth, Cowin draws inspiration from European painting. Her work emulates poses seen in David, Van Eyck, Ingres, and Goya while adapting the styles of film noir and television. **Untitled (Magritte)** reveals Cowin's suspenseful manner and her recurring themes of privacy, victimization, and voyeurism.

1999 Ha[1]
1

Switzerland

Germany

China

222mm

10mm

300mm

荣誉奖

树上不长沙丁鱼

瑞士

Honorary Appreciation

Sardinen wachsen nicht auf Bäumen

Switzerland

Ueli Kleeb

Vera Eggermann

Atlantis Verlag; pro juventute, Zürich

222 300 10mm

384g

28p

ISBN-10 3-7152-0392-7

树上不长沙丁鱼，那沙丁鱼从哪里来？本书的主角是一只叫尤利西斯的猫（昵称 Uli），在主人没有喂食的情况下带着“猫女友”艾米丽（Emilie）跟着鱼贩的捕鱼船周游撒丁岛，寻找沙丁鱼。故事的表述是拟人化的，带着对世间万物充满好奇的儿童视角进行切入。本书的装订形式为锁线硬精装，横开本32 页是绘本较为常用的开本和页数。封面采用铜版纸四色印刷覆哑膜，裱覆卡板。内页胶版纸四色印刷，全书没有特殊工艺。本书的作者同时也是绘本画家，多次获得插画类的大奖，在书中采用了多种绘画工具进行综合创作，画风稚拙，充满童趣。书籍的字体看似选用了一种硬朗加粗的等线体，但字的细节并不是机械绘制而成，而是故意处理成略带手绘感的不平整边缘，在繁复的插图上既能凸显出来，又与插图风格取得一定的统一。

Sardine does not grow on trees, then where does sardine come from? The protagonist of this book is a cat named Ulysses (nicknamed Uli) who took the ‘cat girlfriend’ Emilie to the fishmonger’s fishing boat and travelled around Sardinia to find sardines. The description of the story is anthropomorphic, with the perspective of children who are curious about everything in the world. The hardcover book is bound by thread sewing and the 32-page horizontal format is also commonly used for picture books. The matted cover is made of coated paper printed in four colors, mounted with cardboard. The inner pages are made of offset paper printed in four colors, without any special processing. The author of this book is also a painter of picture books. He has won the prize of illustration for many times. In the book, he uses a varicty of painting tools for comprehensive creation. The painting style is childish and full of children’s interests. The font of the book seems to be a kind of hard and bold Arial, but the details of the characters are not drawn mechanically, but deliberately processed into a slightly hand-drawn uneven edge, which can be highlighted on the complicated illustrations in the book, and achieve a certain unity with the illustration style.

NEIN, DAS SIND ERST DIE FISCHI-INSELN.
DA GIBT ES EIN WALFISCHMUSEUM.
AHOI!
SCHAUT, DA DRÜBEN IST ES JA SCHON!
16

20
WAS SIEHST DU VALERIA?
NEIN!
ABER DA IST NONNA.
SIE WINKT.
SO. ICH MUSS ZURÜCK ZUM U-BOOT.
MEIN FREUND KLAUS NIMMT EUCH MIT.

FISCH & CO.
SARDINIA PASSAGE
WAS?
SCHON SO SPÄT!
HE IHR DREI! AUS DEM WEG.
ES IST SCHON HALB SIEBEN UND ICH WILL NOCH MEINE WARE AUSLADEN.
24

HE! WARTET DOCH.
WO VERSTECKT IHR EUCH DENN?
LINA Delicatessen
PUSS
PUSS
PUSS
PUSS
KOMMT HERVOR, ICH HAB
EUCH ETWAS MITGEBRACHT.

SCH
SCHT
WE
KEIN ZU
FISC
& CO.
BEVOR
FRAU BERTSCHI
HEIMKOMMT, GIBT ES BEI UNS
NIE ETWAS ZU ESSEN.
UND DAS KANN LANGE DAUERN.
MANCHMAL KRIEGE ICH SCHON
NACHMITTAG EINEN BÄRENHUN
07

DAS IST FRAU BERTSCHI.
DAS BIN ICH. ULI.
ICH WOHNE SCHON SEIT FÜNF JAHREN BEI FRAU BERTSCHI.
VALERIA
HERBERT.
DAS IST DER ELEFANT IN DER GESCHICHTE.
UND DAS IST EMILIE.
SIE IST MEINE FREUNDIN.
TEXT & ILLUSTRATION
VERA EGGERMANN
TYPOGRAFIE & GESTALTUNG
UELI KLEEB

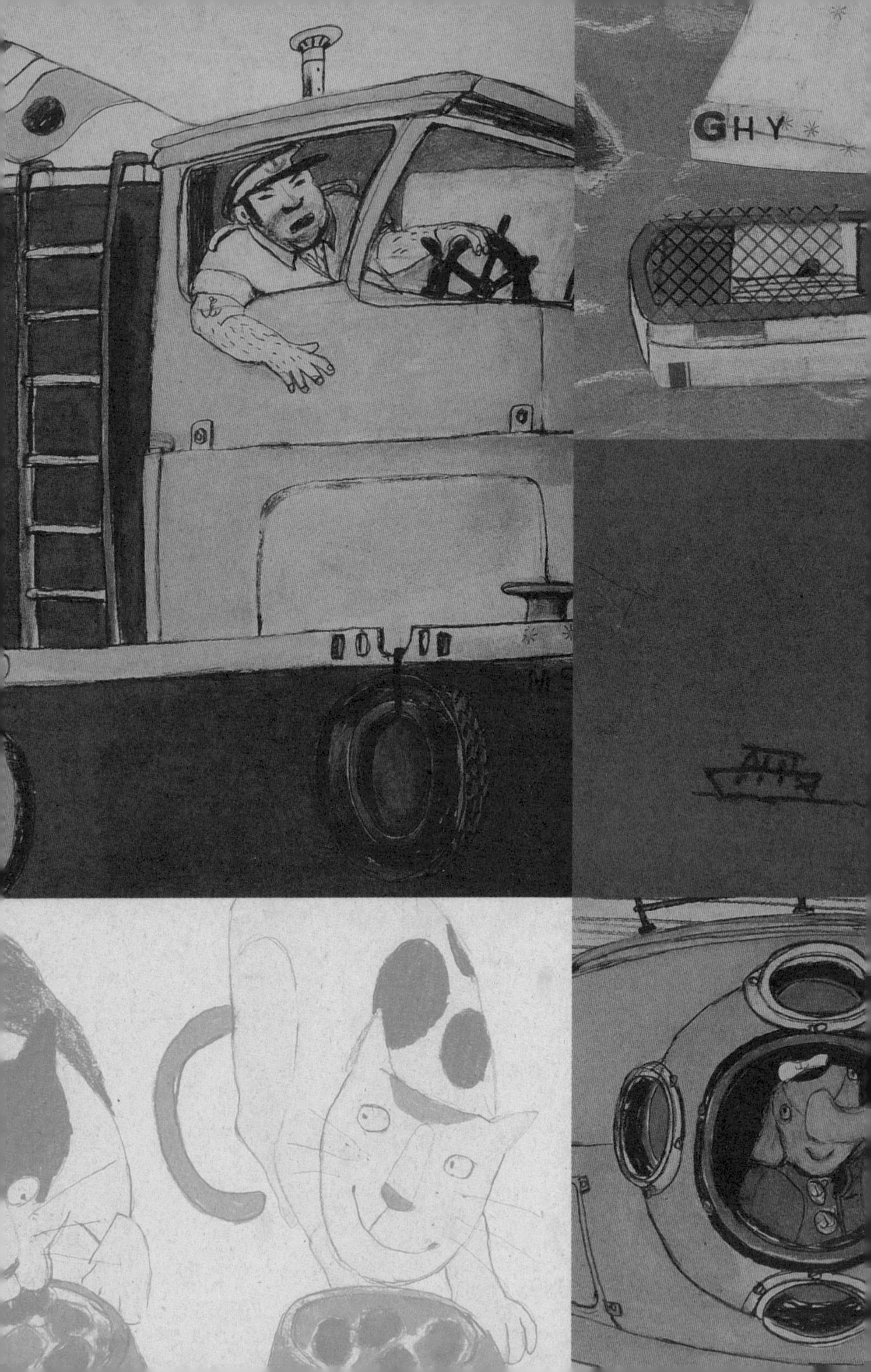
GHY

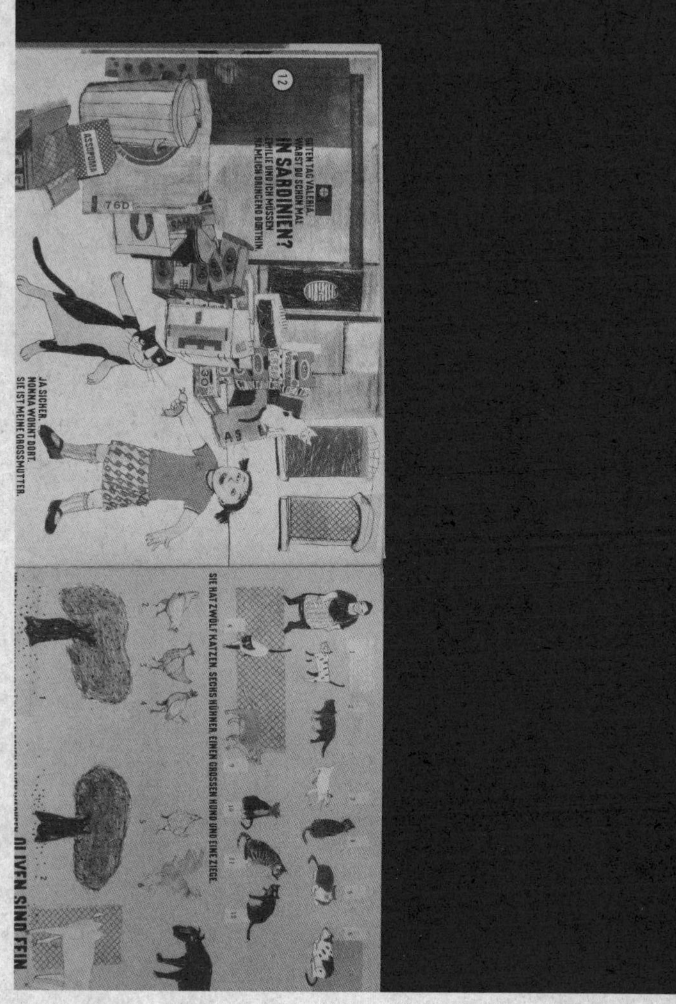
12
GUTEN TAG VALERIA.
WARST DU SCHON MAL
IN SARDINIEN?
EMILIE UND ICH MÜSSEN
NÄMLICH DRINGEND DORTHIN.
ASSOPUMA
JA SICHER.
NONNA WOHNT DORT.
SIE IST MEINE GROSSMUTTER.
SIE HAT ZWÖLF KATZEN, SECHS HÜHNER, EINEN GROSSEN HUND UND EINE ZIEGE.
OLIVEN SIND FEIN

Der grüne Vogel des Äthers

Grete Gulbransson Tagebücher

1999 Ha[2]

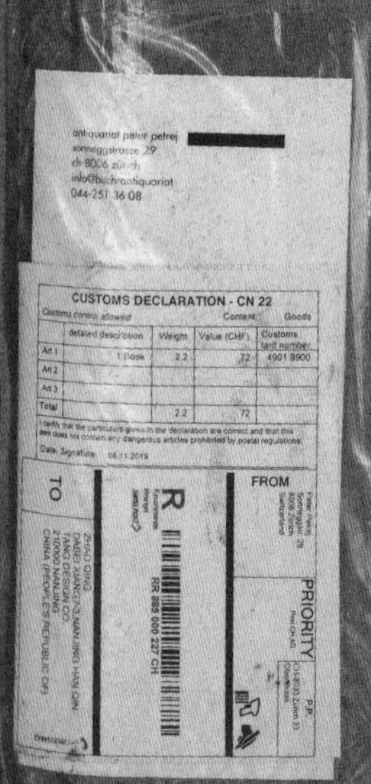

SWISS POST

Zolldeklaration/Dichiarazione doganale

eut être ouvert d'office/Zollamtliche Prüfung gestattet/Visita doganale ammessa

☐ Cadeau / Geschenk / Regalo	☒ Echantillon commercial / Warenmuster / Campione di merci	☐ Documents / Dokumente / Documenti	☐ Autres / Andere / Altri

uantité et description détaillée du contenu (1) / enge und detaillierte Beschreibung des Inhalts / uantità e descrizione dettagliata del contenuto	Poids (2) / Gewicht (kg) / Peso	Valeur (3) / Wert / Valore	N° tarifaire du SH et ori / Zolltarifnummer und He / Numero di tariffa e orig
1 Book	2,3	72	
Poids total (5) / Gesamtgewicht (kg) / Peso totale	2,5		Valeur totale (+monnai / Gesamtwert (+Währun / Valore totale (+valuta)

certifie que les renseignements donnés dans la présente déclaration sont exacts et que cet envoi ne contient aucun objet c
ı interdit par la réglementation postale ou douanière. / Ich bestätige hiermit, dass die Angaben in der vorliegenden Deklarati
ıd und dass die Sendung keine durch die Post- oder Zollvorschriften verbotenen oder gefährlichen Gegenstände enthält. / Co
informazioni contenute nella presente dichiarazione sono esatte e che quest'invio non contiene nessun oggetto pericoloso
al regolamento postale o doganale.

ate, Signature (7)

Germany

Switzerland

China

300mm

50mm

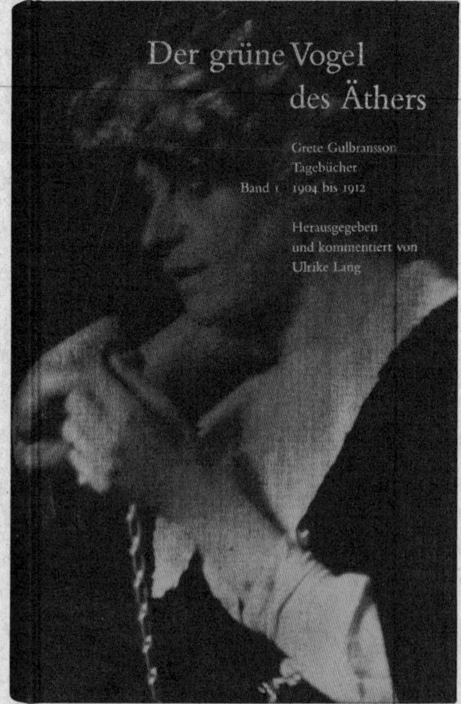

200mm

◎ 荣誉奖 Honorary Appreciation

« 格蕾特 · 古布兰森的日记，第一卷，1904－1912：以太中的青鸟 Grete Gulbransson Tagebücher, Band I, 1904 bis 1912: Der grüne Vogel des Äthers

德国 Germany

Karin Beck

Ulrike Lang

Stroemfeld Verlag, Frankfurt am Main

300 200 50mm

2202g

472p

ISBN-10 3-87877-690-X

格蕾特·古布兰森（Grete Gulbransson）是奥地利作家和诗人，本书收录了她1904－1912年间的日记，并附有大量分析和信息增补。书籍的装订形式为圆脊锁线硬精装，封面采用布面印棕色，主标题烫银，丛书标题和其他信息烫金处理。内页哑面涂布纸四色印刷。内容方面，全书分为三大部分。第一部分是对古布兰森及其创作的介绍，包含生活、工作、所在慕尼黑地区的文化氛围、她的写作习惯、缩略词与符号提示等。第二部分是多篇日记，附带大量与日记相关的照片、笔记、绘画等图片。第三部分是相关附录和慕尼黑地图。版式上，全书设置为三栏。正文选用衬线字体，跨两栏。注释为无衬线字体，紧跟在正文下部，分为两栏。侧边栏则补充一些小图照片，图片说明为意大利斜体。每篇日记的开头时间均用下划线标明，注释数字与内容适当加粗，并以方形右括号“]”结束，各个文本层级划分清晰而得体。

Grete Gulbransson is an Austrian writer and poet. This book contains her diaries from 1904 to 1912, and makes a lot of analysis and information supplement. The hardcover book is bound by thread sewing with rounding spine. The cover is made of cloth printed in brown, the main title is hot-stamped in silver, the title of the series and other information are gold hot-stamped. Inner pages are made of matte coated paper printed in four colors. In terms of content, the book is divided into three parts. The first part is the introduction of Gulbransson and her works, including life, works, cultural atmosphere in Munich, writing habits, abbreviations and symbols. The second part is the collection of diaries, accompanied by a large number of photos, notes, paintings and other pictures related to diaries. The third part is the appendix and Munich map. In terms of layout, the whole book is set to three columns. The text is using serif font, spanning two columns, while the notes are using sans serif font, which is closely followed by the lower part of the text and divided into two columns. The side column is supplemented with some small pictures, and the picture text is using italics. The beginning time of each journal is underlined, the note numbers and content are appropriately bolded, and closed with square right parenthesis ']', so that each text level is clearly and decently divided.

"Kein Wunder,
daß ihr Inneres,
ihr dichterhaftes Wesen,
das die lebendige
Berührung zwar ahnen
läßt, aufdämmern,
doch nur der wirklich
zu erkennen vermag,
wer in ihre vielbespro-
chenen 'Tagebücher'
Einblick nahm, oder, –
besser noch, wer in
oft stundenlangem
Lauschen von ihr selber
daraus hat hören
dürfen."

Tagebuch 26
Freitag, 9. Oktober 1905 bis Montag, 28. Dezember 1905

Tagebuch 28
Sonntag, 13. Mai 1906 - Donnerstag, 13. December 1906

gestern bekommen, in dem er sagt, er hätte einen Gedankenaustausch mit mir gewollt und ich solle die Art entschuldigen, in der er das angestrebt habe! "Leben Sie wohl"[87] – – – Ich nehme mir vor, die Geschichte sofort dem Hans zu sagen, er ist aber so wortkarg und kalt wie er nachhaus kommt und, kaum sind die Kinder im Bett, geht er auch und deutet mir an, er wolle allein sein, sodass ich ungemütlich und verschüchtert in meine Bude steige, ohne gesprochen zu haben. Vor mir liegt der Brief und daneben der aus Corfú. Komische Welt. Ich habe meine Technik des Dramas[88] bekommen. Hans hat darin geblättert und streng zu mir gesagt: "Gretl, zersplittere Dich nicht!" Jetzt darf ich sie also nicht studieren, sie steht nur vor mir neben dem geliebten Avvinarius[89] und wartet. –

Seite 66
Auszug eines Gedichtes von Grete, Arthur von Wallpach gewidmet; auf Pergament, unveröffentlicht (Schloß Babenwohl, Juli 1903)

87 Jüngling [...] wohl] Student, den G. auf einer Schiffsreise kennenlernte; er wollte sie danach auf Planegg besuchen, da sie dies aber Berlepsch verschwiegen hatte, schickte sie den jungen Mann kurzerhand wieder weg, um Diskussionen aus dem Weg zu gehen.

88 Technik des Dramas] s. Anm. 22.

89 Avvinarius] Ferdinand Avenarius gab 1902 das *Hausbuch deutscher Lyrik* heraus, welches beim Lesepublikum großen Anklang fand (vgl. dazu Brief Hermann Hesses an Otto Blümel, Bern, 24. 12. 1913, in: Hermann Hesse. *Gesammelte Briefe. Erster Band, 1835–1921*, hg. von Ursula und Volker Michels. Frankfurt am Main:

++	mehrere zu staben
+++	eines oder
|	unsichere b
||	zwei unsich Wortes
<>	Text in <sp geberin ein
<...>	Nicht ergä
~~Durchstreichung~~	Durchstrei
	Abschnitte
[…]	Auslassung noch leben

auswendig gewusst.[293] Olaf geht nicht
umplerin. Das Glück dieses Tages aber
ends gehen wir um den See und sind

ünchen)

Tag zum Frühstück. Das ist herrlich.
ie wir eben in die Singstunde wollen
homa gegen Olaf aufgehetzt, und dass
nachen sei und er nicht mehr sich um
n liebsten ihn verliesse.[294] Und den
ehr mit ihm zu arbeiten. So geht der
ich allein zu Clemi. Ich darf das Lied

Seite 142
Grete und Olaf
an das Ehepaar
(Malcesine, 28.
Zeichnung: Olaf

oben
Friedrich August
Kaulbach in eine
von Olaf Gulbra

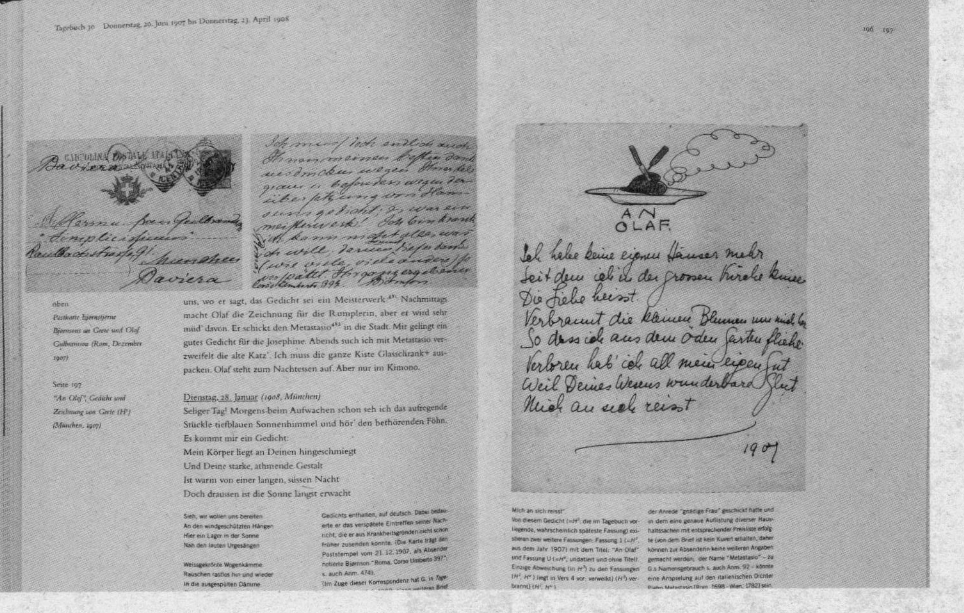

oben
Postkarte Bjørnstjerne Bjørnsons an Grete und Olaf Gulbransson (Rom, Dezember 1907)

Seite 197
"An Olaf", Gedicht und Zeichnung von Grete (H[1]) (München, 1907)

uns, wo er sagt, das Gedicht sei ein Meisterwerk.[481] Nachmittags macht Olaf die Zeichnung für die Rumplerin, aber er wird sehr müd' davon. Er schickt den Metastasio[482] in die Stadt. Mir gelingt ein gutes Gedicht für die Josephine. Abends such ich mit Metastasio verzweifelt die alte Katz'. Ich muss die ganze Kiste Glasschrank+ auspacken. Olaf steht zum Nachtessen auf. Aber nur im Kimono.

Dienstag, 28. Januar *(1908, München)*
Seliger Tag! Morgens beim Aufwachen schon seh ich das aufregende Stückle tiefblauen Sonnenhimmel und hör' den bethörenden Föhn. Es kommt mir ein Gedicht:
Mein Körper liegt an Deinen hingeschmiegt
Und Deine starke, athmende Gestalt
Ist warm von einer langen, süssen Nacht
Doch draussen ist die Sonne längst erwacht

AN OLAF

Ich habe keine eignen Häuser mehr
Seit dem ich in der grossen Kirche knie
Die Liebe heisst.
Verbrannt die kleinen Blumen um mich her
So dass ich aus dem öden Garten fliehe
Verloren hab' ich all mein eigen Gut
Weil Deines Wesens wunderbare Glut
Mich an sich reisst

1907

Sieh, wir wollen uns bereiten
An den windgeschützten Hängen
Hier ein Lager in der Sonne
Nah den lauten Urgesängen

Weissgekrönte Wogenkämme
Rauschen rastlos hin und wieder
In die ausgespülten Dämme

Gedichts enthalten, auf deutsch. Dabei bedauerte er das verspätete Eintreffen seiner Nachricht, die er aus Krankheitsgründen nicht schon früher zusenden konnte. (Die Karte trägt den Poststempel vom 21. 12. 1907, als Absender notierte Bjørnson "Roma, Corso Umberto 397"; s. auch Anm. 474).
(Im Zuge dieser Korrespondenz hat G. in Tage-

Mich an sich reisst"
Von diesem Gedicht (=H[1]; die im Tagebuch vorliegende, wahrscheinlich späteste Fassung) existieren zwei weitere Fassungen: Fassung 1 (=H[1], aus dem Jahr 1907) mit dem Titel: "An Olaf" und Fassung U (=H[U], undatiert und ohne Titel). Einzige Abweichung (in H[3]) zu den Fassungen (H[1], H[U]) liegt in Vers 4 vor: verwelkt] (H[3]) verbrannt] (H[1], H[U]).

der Anrede "gnädige Frau" geschickt hatte und in dem eine genaue Auflistung diverser Haushaltssachen mit entsprechender Preisliste erfolgte (von dem Brief ist kein Kuvert erhalten, daher können zur Absenderin keine weiteren Angaben gemacht werden; der Name "Metastasio" – zu G.s Namensgebrauch s. auch Anm. 92 – könnte eine Anspielung auf den italienischen Dichter Pietro Metastasio (Rom, 1698 - Wien, 1782) sein.

FIPS.

Seite 278
"Sonntag Morgen in Chateau d'Oex", gezeichnet von Grete, Tagebuch 12, 18. April 1909.

Sonntag, 2. Mai *(1909, München)*
Ach es ist ein furchtbares Aufwachen. Man kann's nicht fassen. Immer wieder muss man weinen. – Olaf geht vormittags in den Simpl. Draussen liegt Schnee und es ist eiskalt. Ich vertu' den ganzen Tag damit, Emanuels Briefe abzuschreiben,[773] sie könnten wertvolle Documente sein. 60 Seiten! Eisig und todmüd geh' ich am Nachmittag in's Bett. Les altes Tagebuch von Chateau-d'Oex.[774] Olaf ist so lieb mit mir.

Dienstag, 11. Mai *(1909, München)*
Olaf in die Generalversammlung.[775] Ich umsonst zu Jimmy,[776] dann in die Stadt. Bald heim. Mittags kommen Olaf und Martin. Rez ist schwerkrank.[777] Nachmittags wart ich auf Thoma und Hausmann.[778] Aber sie kommen nicht. Ich schreib' an Fiorella.[779] Dann wieder gequält zu Wormser, Reudl[780] etc. Auf dem Heimweg geh' ich in die Klinik und frag' nach Rez. Der arme Kerl ist um 1 Uhr mittags gestorben; es ist zu furchtbar. Der gute barmherzige Doctor nimmt mich in sein Zimmer und erzählt mir alles. Rez hat sich versehen lassen! Und war bis zum letzten Augenblick bei Bewusstsein.[781] Was für entsetzliche Unglücke fallen über den Simpl her.[782] Man kann's nicht fassen. Die Baronin Rez[783] saust im Automobil her. Ich stürtz weinend durch den blütenduftenden Mai-abend heim. Mein Gott, mein Gott. Unter der Esszimmertür sag' ich's Olaf. – Es ist eine Karte von Dreher[784] da, dass mich die Bianchi[785] nehmen will. Gott sei Lob und Dank! Ich schreib' ihm gleich.

773 abzuschreiben] Es gehörte zu den Eigenarten der Tagebuchautorin, Briefe, die sie selber verfaßte, aber auch solche, die sie erhielt, in ihren Tagebüchern (meist am Ende) wortgetreu niederzuschreiben. (In G.s umfangreicher Korrespondenz befinden sich zahlreiche Briefe von und an Emanuel von Bodman).
774 Chateau-d'Oex] Chateaux-d'Oex, Kurort und Wintersportplatz im schweizerischen Kanton Waadt, 20 km östlich von Montreux, wo G. früher mit ihren Eltern …

ben ein Testament aufgesetzt. Darin wurde die Leitung einem vierköpfigen Kuratorium anvertraut: Korfiz Holm, Dr. Reinhold Geheeb, die Prokuristen Otto Friedrich und August Rommel. Als Erben sah Langen seine Söhne vor, die aber erst nach dem 24. Lebensjahr den Verlag übernehmen sollten. Seiner Lebensgefährtin Josefine Rensch überließ er ein Drittel seines Gesamtvermögens, seine Frau Dagny (juristisch waren die beiden noch ein Ehepaar) erhielt den Pflicht-

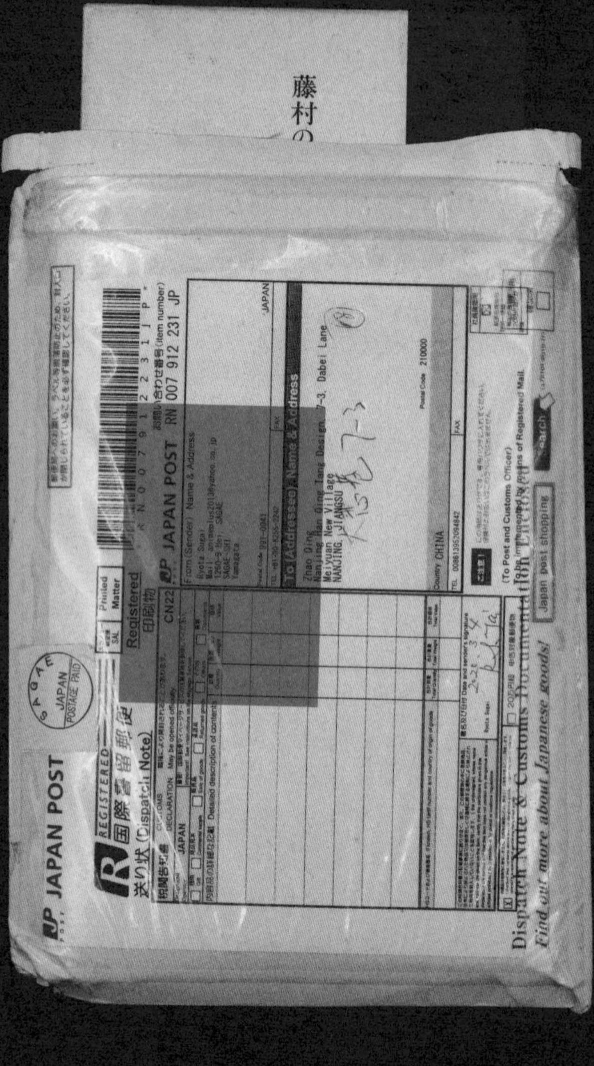
藤村の
JAPAN POST
REGISTERED
国際書留郵便
送り状 (Dispatch Note)
Registered
印刷物
CN22
Printed Matter
JAPAN POSTAGE PAID
JAPAN POST
To (Addressee) Name & Address
Country CHINA
Dispatch Note & Customs Documentation Enclosed
Find out more about Japanese goods!
Japan post shopping

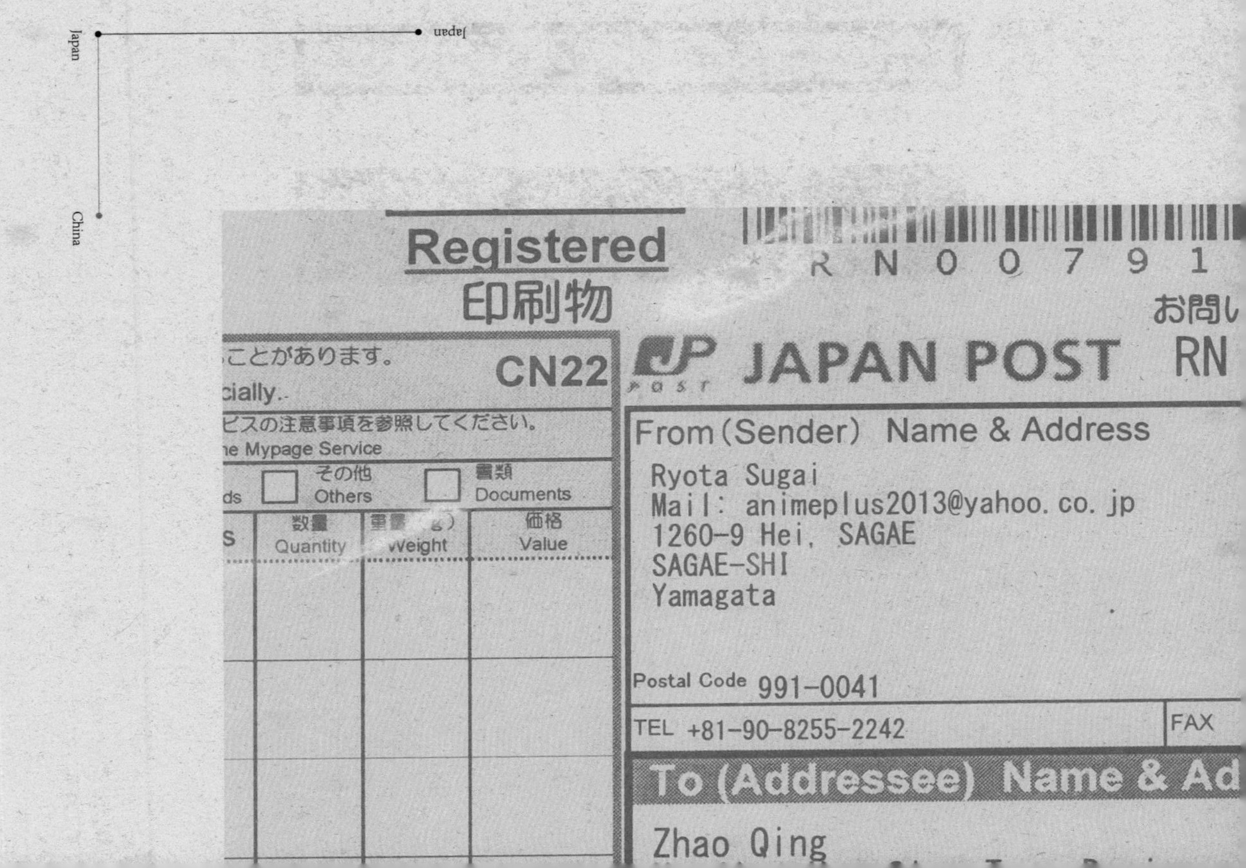
Japan
China
Japan
Registered
印刷物
R N 0 0 7 9 1
お問い
ことがあります。
CN22
cially.
ビスの注意事項を参照してください。
ne Mypage Service
その他 Others
書類 Documents
数量 Quantity
Weight
価格 Value
JAPAN POST
RN
From (Sender) Name & Address
Ryota Sugai
Mail: animeplus2013@yahoo.co.jp
1260-9 Hei, SAGAE
SAGAE-SHI
Yamagata
Postal Code 991-0041
TEL +81-90-8255-2242
FAX
To (Addressee) Name & Ad
Zhao Qing

201mm

23mm

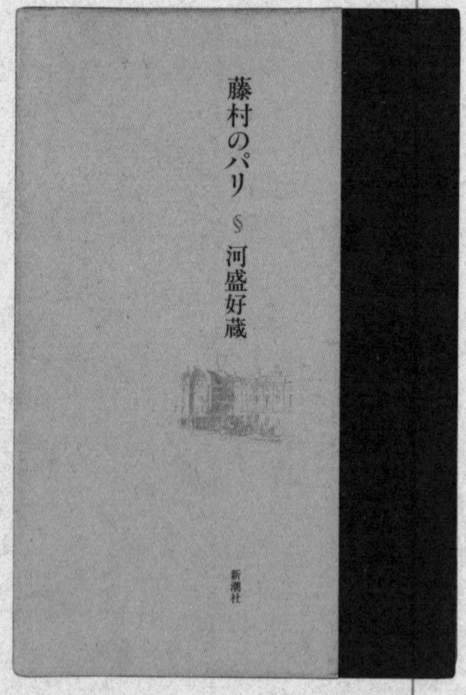

138mm

- 荣誉奖
- 岛崎藤村在巴黎
- 日本

Honorary Appreciation
A Poet: T son Shimazaki in Paris
Japan
Shincho-Sha Co., Tokyo
Yoshizo Kawamori
Shincho-Sha Co., Tokyo
201 138 23mm
416g
352p
ISBN-10 4-10-306005-0

日本诗人、小说家岛崎藤村为了躲避私生活问题在法国巴黎生活了三年。这三年正处于第一次世界大战期间，他在巴黎看到、听到、体验到了什么，生活怎样，与哪些艺术家成了朋友？本书作者河盛好藏通过史料与随想相结合的写法，展现了岛崎藤村漫步巴黎的这段经历。本书的装订形式为锁线软精装。书函采用带有肌理的艺术纸黑绿双色印刷，裱覆卡板，书函开口处包覆深蓝色布料增加耐久度。护封采用带有金属光泽的艺术纸，图案印白，书脊处蓝绿双色渐变。护封采用了传统包书皮的形式将内封包裹在内，内封白卡纸印暖灰色。环衬选用墨绿色艺术纸，内页胶版纸印单黑。内容按照岛崎藤村在巴黎的时间分为 27 个章节。内文单栏竖排，字号较小，行距也比较紧凑，天头空出较多。书籍上口未清边，呈现长短不齐的书页，通过这种传统法式装订（French binding）的细节体现身处巴黎的主题。

Shimazaki Tōson, a Japanese poet and novelist, lived in Paris for three years due to the problem of private life. During the First World War, what did he see, hear and experience in Paris? How was his life? With which artists did he become friends? The author of this book, Yoshizo Kawamori, shows the experience of Shimazaki Tōson living in Paris through the combination of historical materials and random thoughts. The deluxe paperback book is bound by thread sewing. The bookcase is printed in black and green on art paper, mounted with cardboard. The opening of the case is covered with dark blue cloth to increase durability. The jacket is made of art paper with metallic luster with patterns printed in white. Two colors, blue and green, change gradually at the spine of the book. The inner cover, made of white cardboard printed in gray, is wrapped by thc jacket in the traditional way. Dark green art paper is selected for the end paper, while the offset paper is used for the inner pages, printed in black. According to the timeline of Shimazaki Tōson in Paris, the content is divided into 27 chapters. The text is arranged vertically in a single column, with small font size, dense line spacing, and more space in the top margin of the pages. The upper edge of the book hasn't been cut, leaving pages of different lengths. Through these details of traditional French binding, the theme of Paris can be reflected.

藤村のパリ

2

藤村が乗船したフランス船「エルネスト・シモン」号には、日
た。フランス語を全く解さない彼には定めし心細かったであろ
みに英語を話すことを知って彼の不安の念は余程やわらげられ
め、明治学院を卒業し、英語教師の経験もある彼には、英語に
違いない。事実、この英語を操って彼は三十七日間の航海を大
沾は、彼の紀行『海へ』（大正七年七月）に詳しいので、それに
も、藤村より十五年後の同じ季節に、同じ航路で渡仏している

よりその着想や才気、
戯曲を多く書き、「スバ
五年四月）のための戯曲
派系の人びとに大いに
ヤンで、後年三島由紀

と室に納めてから、山本君の画室に行ってみると、仲々広い
はアルコールランプで飯を炊いて大根の漬物を出して呉れた
は『画家と巴里』のなかで書いている。
が、これはシベリア経由でドイツを通過してパリに着き、同
京の隣の室に入った。
パリの貸間若しくは貸画室としては可なりに汚ない者であ
り気に叶って、是迄此に宿った事のある同胞の数のみでも可
長谷川。小杉。柚木。小林。徳永。小川。金山。足立。山
出の諸君、及び僕なぞである。而して多い時は一時に五六人
出するのである。

統領は厳かに軍旗を授け、次でモロッコで武勲を立てた第一セネガル部隊の軍旗を
た。ミシェル将軍の指揮で分列行進が始まった。エコール・ポリテクニックとサン＝
生徒たちで、そのあとにグロー将軍に率いられたアルジェリアとセネガルの狙撃兵た
隊が随った。こんな音楽をこれまで聞いたことのない大衆は大悦びで、《黒人軍万歳
んだ。彼らは帽子を空中にほうりあげた。続いて猟歩兵とアルジェリア歩兵部隊
歳！》、戦闘歩兵部隊、その多くがモロッコの大型メダルをつけた《海外派遣歩兵部隊
衛生隊の犬（大喝采）、砲兵隊、騎兵隊、騎馬砲兵隊。飛行船一機と航空機数機が
空を飛んでいた。午後は夕方まで雨が降った。劇場と映画館は満員だった。（オペ
『ハムレット』。コメディ・フランセーズでは『ポリヨクト』。オデオン座では『オラース』
ン＝パラス映画館では『クオ・ヴァディス』。オランピア座ではマックス・リンダー・
アルであった）。
　十九日。国会は二〇四票対三五八票で軍事法案を可決した。最後の条項についての
時から始まり、社会党員たちは《絶えず議事妨害》を試みた。次いで投票についての
時間の説明が行われた。この《反共和国的》法案に反対するカイヨーの最終抗議に対
トゥは《この法律は侵略でもなければ、挑発でもない。これは避けることのできない
である》と答えた。夜半になって投票の結果が公表された。反対投票をしたのは独立
S・F・I・O（労働者インターナショナル・フランス支部）、急進左派と急進社会党の
あった。新法律は兵役義務期間を二十八年（現行は二十五年）に延長する。現役期間は
百二十日の休暇がある。部隊編入は二十歳で行われる。学生は従来と同じく二十五歳
猶予を許される。家族扶養のため服務期間中は家族手当が支給される。
　十月二十三日。ジャック・コポーのヴュー＝コロンビエ座開場。創立者は『新フラ
論』にその計画を次のように語っている。《日ましにいよいよ臆面もなくわれらの舞
させ、教養ある観客の顔をそむけさせる気違いじみた商業化と、恥知らずの商品を叩
る一にぎりのおどけ者たちによって大部分の劇場が独占されていることに激怒した》
舞台に《その輝きと偉大さ》を取り戻すために働きたいと思う。自分の劇場は、古
《いかなる種類のものであれ》未発表の作品とを交互に上演するであろう。座席券は
ろう（一―八フラン）。一つの演劇学校が劇団を形成するだろう。プログラムは、結局
年前に、アントワーヌが自由劇場を創立したときのものを再び取り上げることになる
コポーはアテネ＝サン＝ジェルマンという彼の地区の小ホールを改造した。《舞台の
黒い枠で囲まれた白い壁の上で開かれ、緑色の幕で閉ざされるだろう。それが開かれ
方形の垂れ幕の第二番目の額縁が見られる。ただ、場面毎に取り替えられる背景幕だ
の筋が進行している場所を認識させる。》団員はルーシェの芸術座（デュラン、ロジ
ル）、アントワーヌのマチネ・リテレール（プランシュ・アルバーヌ）から来た俳優た
しく来た人たち（ルイ・ジュヴェ）によって構成されていた。第一回の出しものは、
スピアーと同時代のトマス・ヘイウッドの『優しさに殺された女』とモリエールの
者』（二十八回上演）を用意していた。

UF 52 354 928 4SE

POSTEN AB
INTERNATIONAL MAIL
PP Sweden
PRIORITY A
If undeliverable return to:
PO Box 5001, 202 26 Malmö
Sweden

From THRIFTBOOKS
880 N HILLS BLVD #503
RENO NV 89506 UNITED STATES

To ZHAO QING
DABEI XIANG7-3,NANJING HAN QING TAN
NANJING 210000

大悲巷7-3

CHINA

Sender's ref. - Order no 001319225864

485364732037

CUSTOMS DECLARATION 22
May be opened officially

	Gift		Commercial Sample
	Documents	X	Other

Quantity and detailed description of contents	Value US$
USED BOOK(S)	USD 20.17

Total Weight 0.36 kg Total Value 20.17

For commercial items only. If known, HS tariff number 4901.99.00

Country of origin of goods UNITED STATES EORI

I, the undersigned, whose name and address are given on the item, certify that the particulars given in this declaration are correct and that this item does not contain any dangerous article or articles prohibited by legislation or by other regulations XXXXX3336

Date and sender's signature 11-06-2019 15:11

PRIORITY A
If undeliverable retu
PO Box 5001 202 2
Sweden

)OKS
.LLS BLVD #503
89506 UNITED STATES

IG
ANG7-3,NANJING HAN QING TAN
210000

大塘巷7-3

Order no 001319225864

CUSTOMS DECLAR

May be opene

		Commercial Samp
nts	X	Other
description of contents		Value US$
		USD 20.17

The Netherlands

U.S.A.

China

168mm

23mm

120mm

● 荣誉奖

《 永远属于你
——产品耐久性的愿景

◐ 荷兰

Honorary Appreciation
Eternally yours:
visions on product endurance
The Netherlands
Studio Gonnissen en Widdershoven
Ed van Hinte
Uitgeverij 010, Rotterdam
168 120 23mm
366g
256p
ISBN-10 90-6450-313-3

本书的作者埃德·范辛特（Ed van Hinte）是荷兰工业设计师、评论家，专注于轻型结构、减少污染和可持续性的设计和教育。1993年开始，他参与到本书的同名项目“eternally yours”中来，研究如何延长产品的使用寿命。本书的装订形式为锁线胶平装。护封艺术纸黑金双色印刷，内封白卡纸无印刷和特殊工艺。内页胶版纸四色印刷，上下书口和翻口刷金处理，使得书口与书脊连成完整的金色闭环。内容方面，有对生活中多种物品使用体验的思考，也有对不同用户的采访，探讨产品使用寿命、环保、可持续性等社会和企业均需要面对的问题。具体包括材料的耐久性、局部的可替换性、材料的环保性等方面，以及为此需要付出的工业生产和设计流程的改变。每篇小文章都以图片开始，标题放置于图片的右页上方，紧跟文章内容，并且直接转入后页，形成了标签式的独特版面。翻口处两侧均加入了巨大的书名文字，左页文字逐渐冲出页面，而右侧逐渐进入页面，暗示循环往复的过程。

Ed van Hinte, the author of this book, is an industrial designer and critic in the Netherlands, focusing on the design and education of light structures, pollution reduction and sustainability. Since 1993, he has participated in a project of the same name as this book, ‘eternally yours’, to study how to extend the service life of products. The book is perfect bound with thread sewing. The jacket is made of art paper printed in black and gold, while the inner cover is made of white cardboard without any special processing. The inner pages are printed in four colors on offset paper, and the upper and lower book edges and the flap are brushed with gold, so that the book edges and the spine are connected into a complete gold closed-loop. As for the content, through the consideration of the use of various objects in life and interviews with different users, the book discusses the products’ life, environmental protection, sustainability and other issues that our society and companies need to face, including the durability of materials, replaceable part, environmental performance of materials, etc., as well as the industrial innovation and design changes that need to be done. Each small article starts with a picture, the title is placed on the right side of the picture, followed by the content of the article, and goes directly to the back page, forming a unique layout with label style. Huge titles are added to both sides of page, which gradually rushes out of the left page, and gradually enters into the right page, implying a cyclical process.

eternally yours

Its task is huge and complicated, as product life span is interdependent with a whole range of factors, from the choice of paint, the shape and texture of coffee makers and fashion, all the way to the organization of services, advertising, and establishing guarantee conditions.
You have to start somewhere and Eternally Yours set out in 1995 to organize graduation projects at different educational institutions. Several are still going on. Graduating students meet on a regular basis and together with the Eternally Yours team relevant themes are discussed. Furthermore the foundation organizes meetings with professional experts from different fields, economists, philosophers, designers, marketeers, publicists, to discuss and refine viewpoints and propositions. In April this year an international Eternally Yours congress took place. In it we discussed the three main aspects of product life extension: Shape 'n Surface dealt with products themselves, Sales 'n Services was about organizing a system around longer lasting products and Signs 'n Scripts investigated product meaning.
All these activities have gained us a lot of insight into ways to provide products with the ability to age with dignity. Hence this publication. The two most important discoveries you will find in it are that we must go all out to reach the goal of longer lasting products and that, paradoxically, there are no fixed rules in this game.
To start with the first insight: product endurance is not a matter of nostalgically harking back to applying wood and leather instead of plastics. On the contrary it entails a total review of the design of products and services.

20

together with everything that may help
long as possible. They need a welfare st
as people do. The second assessment, t
strict rules, stems from the simple fact t
generic in nature. In the same way that
knowledge cannot possibly assert that a
should be 42 centimetres in height, insig
extension is unable to specify universal
characteristics. There aren't any. The on
to improvement is by being alert to way
spans can be prolonged, and building u
doing so.
This book provides the means to be wat
consists of an extensive range of ways t
life. It can be used to judge existing pro
combinations as well as to support deci
developing new ones. The book can be
qualitative checklist. The different viewp
confusing because some of them contra
The reason is that not all insights apply t
the same extent. Furniture requires a se
organization that differs totally from the
television sets need. And surface quality
squeezers requires other considerations
apply to a video camera. Every project r
harmonized combination of solutions.
This book consists of edited versions of
given by key speakers at the congress, c
graduation projects and Eternally Yours
research projects. In between are brief e
discuss, analyse and complete this Infor

21

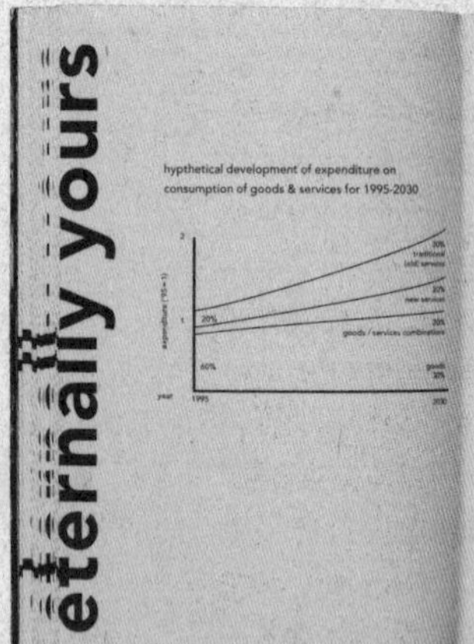

eternally yours

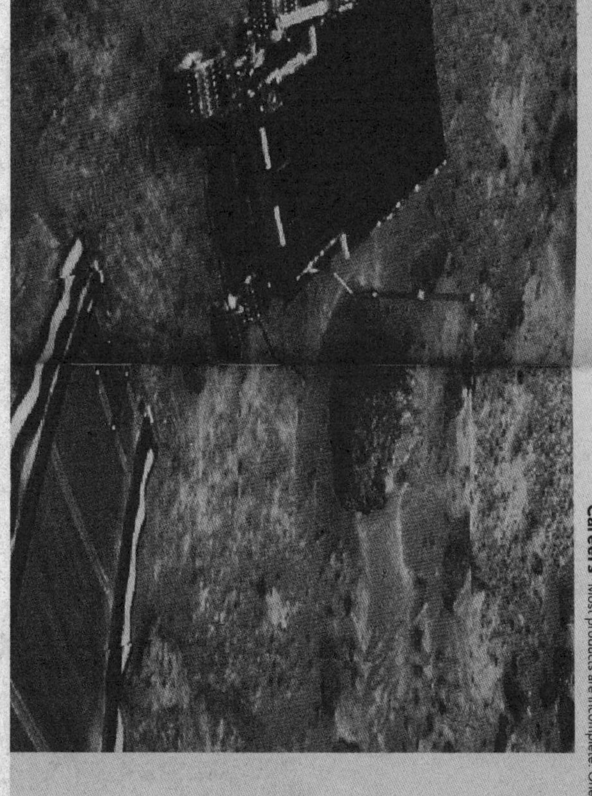

Careers Most products are incomplete. One of t

20%
new services

20%
oods / services combinations

goods
30%

2030

PETER

in the International Summer School of Cultural Studies Muisto-Reminscence, Jyväskylä, 1995

Koskijoki, Maria, *'Keräilijän muotokuva'*, Sosiologia:3, 1995. pp 179-190

- Koskijoki, Maria, 'The practise of collecting/ The spirit of a collection/ The life of the collector', Tangible cosmologies, Veli Granö, Pasi Falk and Maria Koskijoki, Pohjoinen, Oulu, 1997
- Latour, Bruno, 'We have never been modern', Translation: Porter, C. Harvester Wheatsheaf, New York, 1993
- Londos, Eva, UppÅt väggarna i Työselosteita ja esitelmiä 31/96, National Consumer Research Center, Helsinki, 1996
- Rogan, Bjarne, 'Fra tommestokken til semiotiken', SAMDOK bulletinen, 1987, p 40
- Silverstone, Roger & Hirsch, Eric 1992: Introduction. In Roger Silverstone and Eric Hirsch (eds.) Consuming Technologies: Media and Information in Domestic Spaces, Routledge, London and New York, pp 1-11
- Stewart, Susan On Longing; Narratives of the Miniature, the Gigantic, the Souvenir, the Collection, The John Hopkins

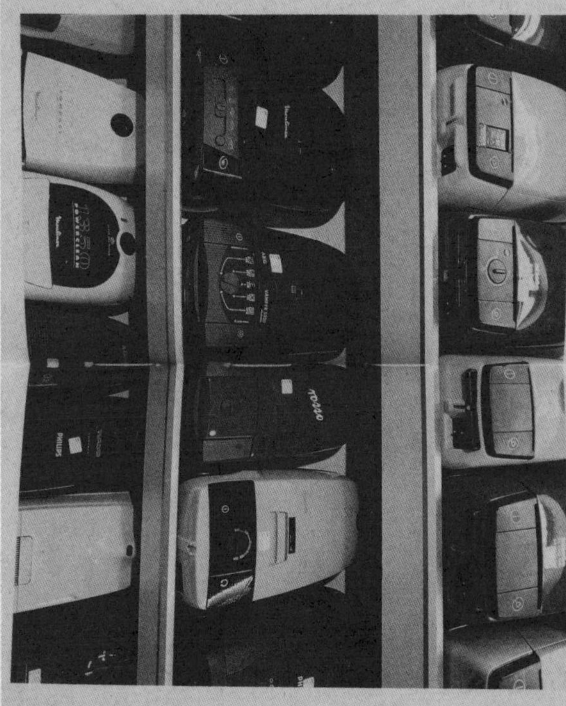
eternally yours
Choice Pick any vacuum cleaner you want.
eternally yours

eternally yours
Equality's safety belt
'But by and by they came to n

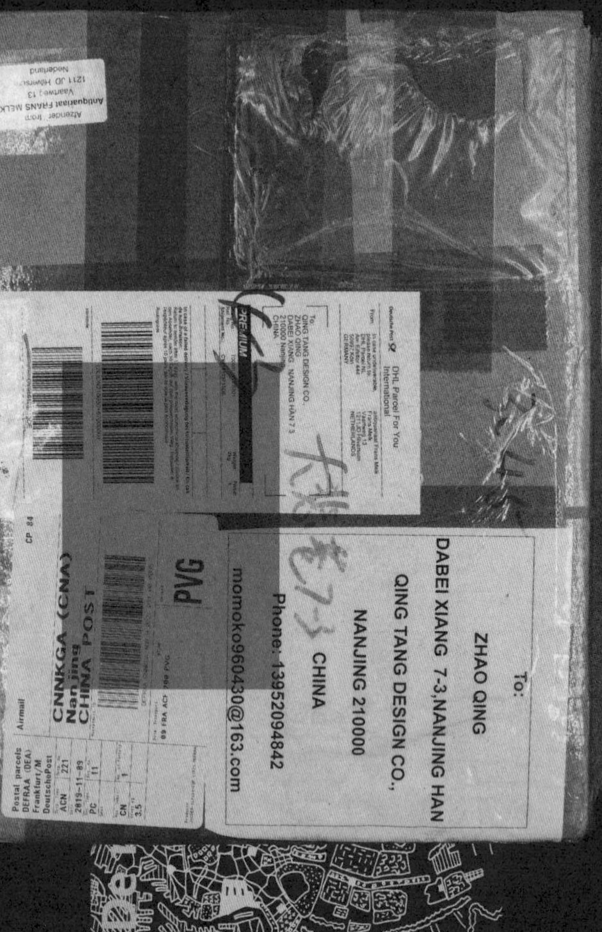
To:
ZHAO QING
DABEI XIANG 7-3,NANJING HAN
QING TANG DESIGN CO.,
NANJING 210000
CHINA
Phone: 13952094842
momoko960430@163.com
PVG
CNNKGA (CNA)
Nan jing
CHINA POST
Airmail
CP 84
DHL Parcel For You
International
PREMIUM
Postal parcels
DEFRAA (DEA)
Frankfurt/M
DeutschePost

1999 Ha^{5}

Weight:
2kg
Piece:
1
very / Vorausverfügung bei Unzustellbarkeit / En cas
0 days, with the most economical shipping / Zurück an
Tagen auf dem preiswertesten Weg / Retourner à
urs, par la voie la plus économique.
(2L)CN210000+67000001
momoko9604
PVG
NA POST
The Netherlands
Germany
China

荣誉奖

城市的方向 I&II

——1900 — 2000 年北欧城市设计

荷兰

Honorary Appreciation

De Regie van de Stad I&II:

Noord-Europese Stedebouw 1900–2000

The Netherlands

Arlette Brouwers, Koos van der Meer

Koos Bosma, Helma Hellinga (Hrsg.)

NAi Uitgevers; EFL Publicaties, Rotterdam / Den Haag

305 252 18mm / 305 252 37mm

943g / 2071g

128p / 356p

ISBN-10 90-5662-058-4

这部著作叙述了100年间北欧城市规划设计及其实际进展的历史。书籍分为上下两册，分别呈现规划设计和城市实际建设的进展与问题。两册的装订形式均为锁线硬精装。封面除书脊处裱贴黑布外，其余大部分为黑色艺术纸，丝印白色和专为两册选用的黄色和橙色油墨，裱覆卡板。环衬部分选用与封面配色相同的艺术纸。内页哑面涂布纸四色印刷。两册分别就区域城市的发展，各国都市圈的形成、规划与现实、当下发展中的问题，不同都市的特点与曾经的重建计划展开论述。版式方面，针对不同部分的图文量特点，设定了两栏和三栏两种版式，并在图片的排版上较为灵活。规划图、效果图和照片的处理上，尽可能地保留原始文件发黄的纸张、褪色的油墨、折叠的痕迹，甚至破损。本套书印刷精美，是不可多得的城市和文化历史文献。

This is a history of Nordic urban planning and design and its actual progress that spans over 100 years. The book is divided into two volumes, presenting the progress and problems of planning, designing and urban construction. Two volumes are hardbound with thread sewing. Except the black cloth used to cover the spine, most of the book are made of black art paper, with silk screening in white and yellow or white and orange according to different volume, mounted with cardboard. The end paper is made of art paper with the same color as the cover. The inner pages are made of matte coated paper printed in four colors. The two volumes respectively discuss the development of regional cities, the formation of metropolitan areas in various countries, planning and reality, problems encountered in the current developing process, characteristics of different cities and previous reconstruction plans. As for the design, two-column layout and three-column layout are set according to the graphic features of different parts, and the typesetting of pictures is quite flexible. When processing planning drawings, renderings and photos, the paper yellowing, ink discoloration, folding marks and even damage of the original document are preserved as much as possible. The book is a rare urban and cultural historical document with exquisite printing.

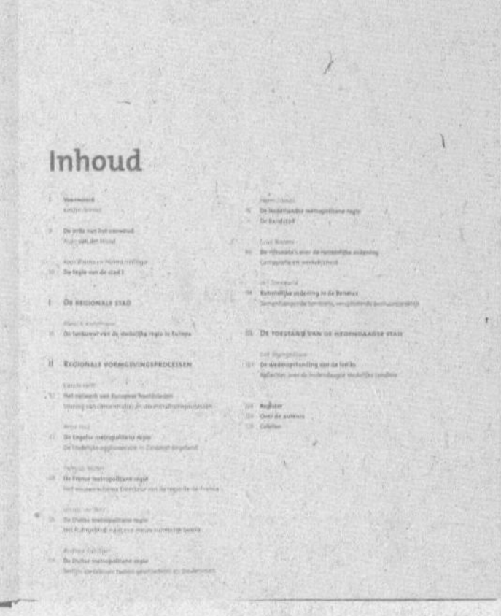
Inhoud

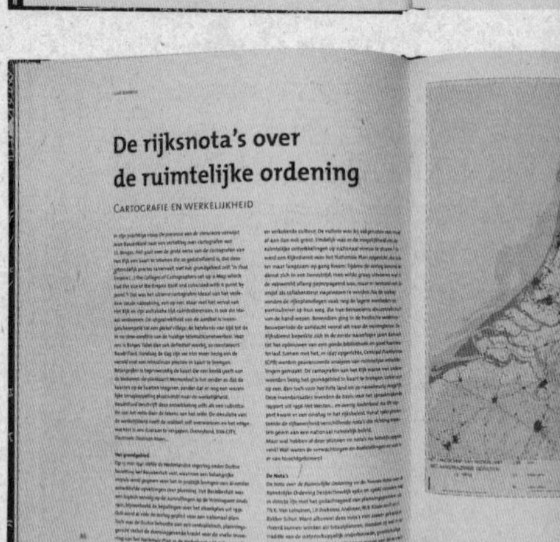
De rijksnota's over
de ruimtelijke ordening
Cartografie en werkelijkheid

Berlijn 1938
Generalbebauungsplan für die Reichshauptstadt

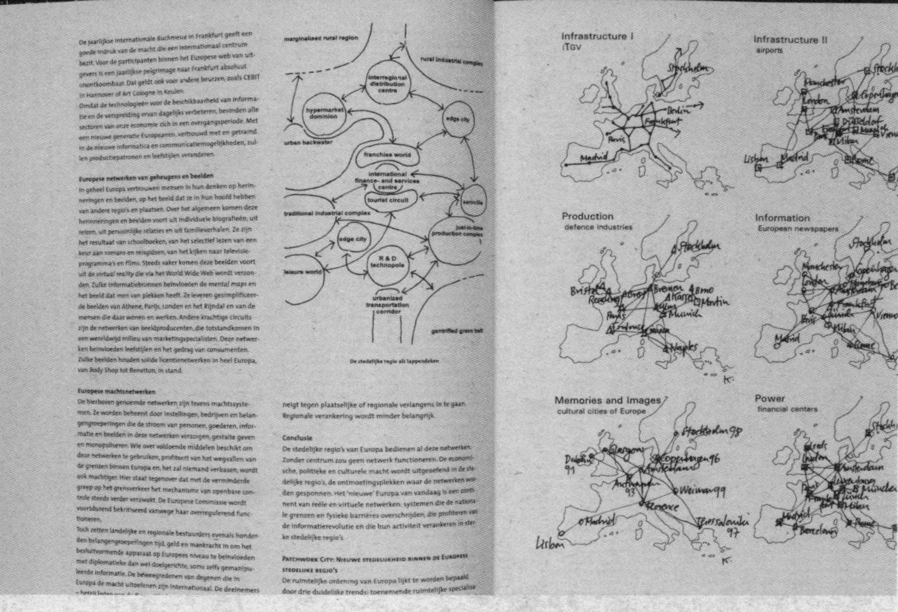

De jaarlijkse internationale *Buchmesse* in Frankfurt geeft een goede indruk van de macht die een internationaal centrum bezit. Voor de participanten binnen het Europese web van uitgevers is een jaarlijkse pelgrimage naar Frankfurt absoluut onontkoombaar. Dat geldt ook voor andere beurzen, zoals CEBIT in Hannover of Art Cologne in Keulen.

Omdat de technologieën voor de beschikbaarheid van informatie en de verspreiding ervan dagelijks verbeteren, bevinden alle sectoren van onze economie zich in een overgangsperiode. Met een nieuwe generatie Europeanen, vertrouwd met en getraind in de nieuwe informatica en communicatiemogelijkheden, zullen productiepatronen en leefstijlen veranderen.

Europese netwerken van geheugens en beelden

In geheel Europa vertrouwen mensen in hun denken op herinneringen en beelden, op het beeld dat ze in hun hoofd hebben van andere regio's en plaatsen. Over het algemeen komen deze herinneringen en beelden voort uit individuele biografieën, uit reizen, uit persoonlijke relaties en uit familieverhalen. Ze zijn het resultaat van schoolboeken, van het selectief lezen van een keur aan romans en reisgidsen, van het kijken naar televisieprogramma's en films. Steeds vaker komen deze beelden voort uit de *virtual reality* die via het World Wide Web wordt verzonden. Zulke informatiebronnen beïnvloeden de *mental maps* en het beeld dat men van plekken heeft. Ze leveren gesimplificeerde beelden van Athene, Parijs, Londen en het Rijndal en van de mensen die daar wonen en werken. Andere krachtige circuits zijn de netwerken van beeldproducenten, die totstandkomen in een wereldwijd milieu van marketingspecialisten. Deze netwerken beïnvloeden leefstijlen en het gedrag van consumenten. Zulke beelden houden solide licentienetwerken in heel Europa, van Body Shop tot Benetton, in stand.

Europese machtsnetwerken

De hierboven genoemde netwerken zijn tevens machtssystemen. Ze worden beheerst door instellingen, bedrijven en belangengroeperingen die de stroom van personen, goederen, informatie en beelden in deze netwerken verzorgen, gestalte geven en monopoliseren. Wie over voldoende middelen beschikt om deze netwerken te gebruiken, profiteert van het wegvallen van de grenzen binnen Europa en, het zal niemand verbazen, wordt ook machtiger. Hier staat tegenover dat met de verminderde greep op het grensverkeer het mechanisme van openbare controle steeds verder verzwakt. De Europese Commissie wordt voortdurend bekritiseerd vanwege haar overregulerend functioneren.

Toch zetten landelijke en regionale bestuurders evenals honderden belangengroeperingen tijd, geld en mankracht in om het besluitvormende apparaat op Europees niveau te beïnvloeden met diplomatieke dan wel doelgerichte, soms zelfs gemanipuleerde informatie. De beweegredenen van degenen die in Europa de macht uitoefenen zijn internationaal. De deelnemers

De stedelijke regio als lappendeken

neigt tegen plaatselijke of regionale verlangens in te gaan. Regionale verankering wordt minder belangrijk.

Conclusie

De stedelijke regio's van Europa bedienen al deze netwerken. Zonder centrum zou geen netwerk functioneren. De economische, politieke en culturele macht wordt uitgeoefend in de stedelijke regio's, de ontmoetingsplekken waar de netwerken worden gesponnen. Het 'nieuwe' Europa van vandaag is een continent van reële en virtuele netwerken: systemen die de nationale grenzen en fysieke barrières overschrijden, die profiteren van de informatierevolutie en die hun activiteit verankeren in sterke stedelijke regio's.

Patchwork City: Nieuwe stedelijkheid binnen de Europese stedelijke regio's

De ruimtelijke ordening van Europa lijkt te worden bepaald door drie duidelijke trends: toenemende ruimtelijke specialisa

ebouw

HET TUINSTAD-MODEL

ationa-
groot'
Geddes
ciologi-
nen-
gandis-
e ste-
an een
sseerd
maat-
eling
pen, het
het
rimen-
een ste-
n prijs-
n
n plan
ver-

'survey before plan', heeft in het interbellum een va gekregen in de stedebouwkunde. Het is vooral beke den in het boek *Cities in Evolution* (1915). Een niet or deel van de stedebouwkundigen vond de benaderin Geddes ouderwets. Zijn streven naar maatschappel ming greep te veel terug op vroegere gedaantes var Men verwelkomde de sociologische en sociaal-econ aspecten van de survey, maar verbond er geen ontw nische consequenties aan. Een uitzondering vormde Abercrombie (1879-1957), die reeds in 1918 een omva onderzoek naar de nationale grondstoffen had bepl tooide in 1922 een survey en een 'regional planning het Doncaster kolenbekken, dat was gebaseerd op d methodiek van Geddes, in 1924 gevolgd door Sheffie Survey (1924). Deze twee surveys stonden aan de ba totaal zestien *regional schemes* die Abercrombie tus 1935 opstelde. Het waren de eerste regionale onderz Engeland waarin de doctrines van Geddes op een d schaal toepassing vonden en met ruimtelijke voorst

Nº 5.

— DIAGRAM —

STRATING CORRECT P
CITY'S GROWTH - OPEN
NEAR AT HAND. AN
MUNICATION BETWEEN O

NFR

EG

orwoord

and maar zelden een internationa g te zien. De laatste keer was ruim ns het Internationaal Stedebouwc 924) werd in het Stedelijk Museum ingericht aan de hand waarvan d e hoogte konden stellen van de pr

dige politieke en bestuurlijke bela ndige, infrastructurele en milieuv onderlijk dat er opnieuw behoefte

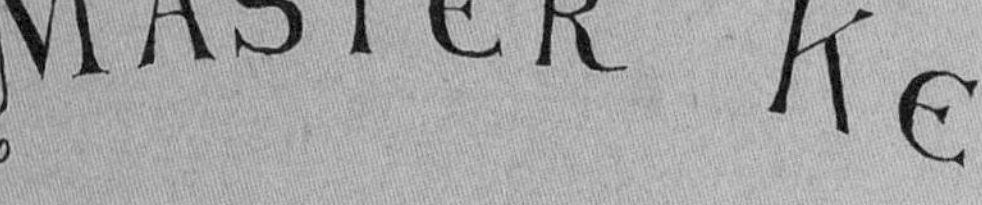

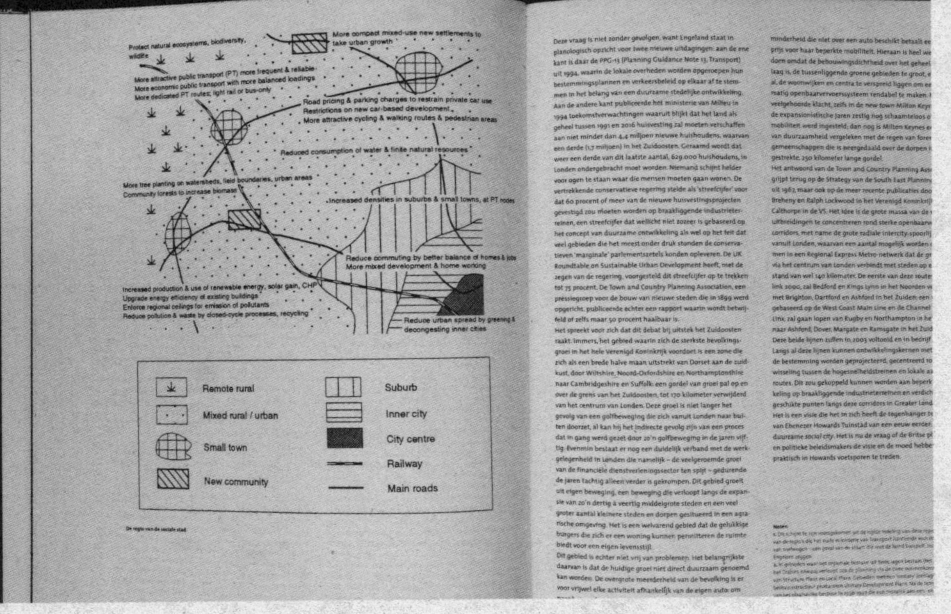

De regio van de sociale stad

Deze vraag is niet zonder gevolgen, want Engeland staat in planologisch opzicht voor twee nieuwe uitdagingen: aan de ene kant is daar de PPG-13 (Planning Guidance Note 13, Transport) uit 1994, waarin de lokale overheden worden opgeroepen hun bestemmingsplannen en verkeersbeleid op elkaar af te stemmen in het belang van een duurzame stedelijke ontwikkeling. Aan de andere kant publiceerde het ministerie van Milieu in 1994 toekomstverwachtingen waaruit blijkt dat het land als geheel tussen 1991 en 2016 huisvesting zal moeten verschaffen aan niet minder dan 4,4 miljoen nieuwe huishoudens, waarvan een derde (1,7 miljoen) in het Zuidoosten. Geraamd wordt dat weer een derde van dit laatste aantal, 629.000 huishoudens, in Londen ondergebracht moet worden. Niemand schijnt helder voor ogen te staan waar die mensen moeten gaan wonen. De vertrekkende conservatieve regering stelde als 'streefcijfer' voor dat 60 procent of meer van de nieuwe huisvestingsprojecten gevestigd zou moeten worden op braakliggende industrieterreinen, een streefcijfer dat wellicht niet zozeer is gebaseerd op het concept van duurzame ontwikkeling als wel op het feit dat veel gebieden die het meest onder druk stonden de conservatieven 'marginale' parlementszetels konden opleveren. De UK Roundtable on Sustainable Urban Development heeft, met de zegen van de regering, voorgesteld dit streefcijfer op te trekken tot 75 procent. De Town and Country Planning Association, een pressiegroep voor de bouw van nieuwe steden die in 1899 werd opgericht, publiceerde echter een rapport waarin wordt betwijfeld of zelfs maar 50 procent haalbaar is.

Het spreekt voor zich dat dit debat bij uitstek het Zuidoosten raakt. Immers, het gebied waarin zich de sterkste bevolkingsgroei in het hele Verenigd Koninkrijk voordoet is een zone die zich als een brede halve maan uitstrekt van Dorset aan de zuidkust, door Wiltshire, Noord-Oxfordshire en Northamptonshire naar Cambridgeshire en Suffolk: een gordel van groei pal op en over de grens van het Zuidoosten, tot 170 kilometer verwijderd van het centrum van Londen. Deze groei is niet langer het gevolg van een golfbeweging die zich vanuit Londen naar buiten doorzet, al kan hij het indirecte gevolg zijn van een proces dat in gang werd gezet door zo'n golfbeweging in de jaren vijftig. Evenmin bestaat er nog een duidelijk verband met de werkgelegenheid in Londen die namelijk – de veelgeroemde groei van de financiële dienstverleningssector ten spijt – gedurende de jaren tachtig alleen verder is gekrompen. Dit gebied groeit uit eigen beweging, een beweging die verloopt langs de expansie van zo'n dertig à veertig middelgrote steden en een veel groter aantal kleinere steden en dorpen gesitueerd in een agrarische omgeving. Het is een welvarend gebied dat de gelukkige burgers die zich er een woning kunnen permitteren de ruimte biedt voor een eigen levensstijl.

Dit gebied is echter niet vrij van problemen. Het belangrijkste daarvan is dat de huidige groei niet direct duurzaam genoemd kan worden. De overgrote meerderheid van de bevolking is er voor vrijwel elke activiteit afhankelijk van de eigen auto: om

minderheid die niet over een auto beschikt betaalt ee
prijs voor haar beperkte mobiliteit. Hieraan is heel we
doen omdat de bebouwingsdichtheid over het geheel
laag is, de tussenliggende groene gebieden te groot, e
al, de woonwijken en centra te verspreid liggen om ee
matig openbaarvervoerssysteem rendabel te maken.
veelgehoorde klacht, zelfs in de *new town* Milton Key
de expansionistische jaren zestig nog schaamteloos o
mobiliteit werd ingesteld; dan nog is Milton Keynes e
van duurzaamheid vergeleken met de regen van fore
gemeenschappen die is neergedaald over de dorpen i
gestrekte, 250 kilometer lange gordel.
Het antwoord van de Town and Country Planning Ass
grijpt terug op de Strategy van de South East Plannin
uit 1967, maar ook op de meer recente publicaties doo
Breheny en Ralph Lockwood in het Verenigd Koninkrij
Calthorpe in de VS. Het idee is de grote massa van de
uitbreidingen te concentreren rond sterke openbaarv
corridors, met name de grote radiale intercity-spoorlij
vanuit Londen, waarvan een aantal mogelijk worden
men in een Regional Express Metro-netwerk dat de gr
via het centrum van Londen verbindt met steden op e
stand van wel 140 kilometer. De eerste van deze route
link 2000, zal Bedford en Kings Lynn in het Noorden v
met Brighton, Dartford en Ashford in het Zuiden; een
gebaseerd op de West Coast Main Line en de Channel
Link, zal gaan lopen van Rugby en Northampton in he
naar Ashford, Dover, Margate en Ramsgate in het Zuid
Deze beide lijnen zullen in 2003 voltooid en in bedrijf
Langs al deze lijnen kunnen ontwikkelingskernen met
de bestemming worden geprojecteerd, gecentreerd ro
wisseling tussen de hogesnelheidstreinen en lokale a
routes. Dit zou gekoppeld kunnen worden aan beperk
keling op braakliggende industrieterreinen en verdich
geschikte punten langs deze corridors in Greater Lond
Het is een visie die het in zich heeft de tegenhanger t
van Ebenezer Howards Tuinstad van een eeuw eerder
duurzame *social city*. Het is nu de vraag of de Britse p
en politieke beleidsmakers de visie en de moed hebbe
praktisch in Howards voetsporen te treden.

SCIENCES
HAUTE
INTERNATONALE
des GOUVERNEMENTS
MEDECINE CHIRURGIE
LA TOUR du PROGRÈS
NANTERRE
VERSAILLES
MASSY
ORLY
ST REMY
CHEVREUSE

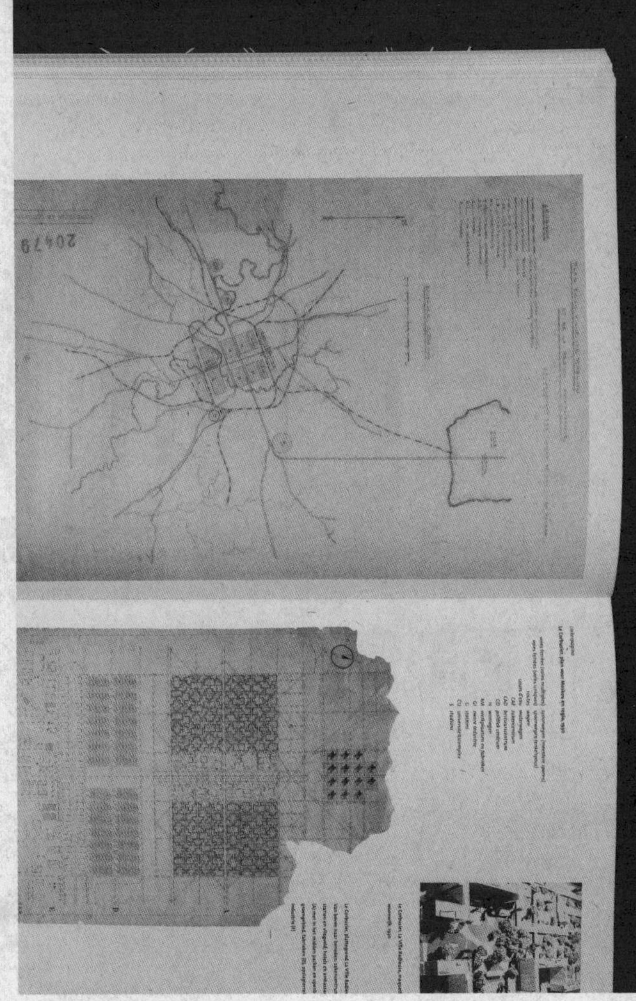

Een voorbeeld uit de jaren tachtig en ideeën voor de toekomst
Een goed voorbeeld van stedebouw uit het einde van de jaren tachtig is een voormalig stationsemplacement in het centrum van Södermalm, in de binnenstad van Stockholm, dat veranderd is in een gebied met huizen en kantoren. Het plan omvat Söder Torn, een flatgebouw van vierentwintig verdiepingen dat ontworpen is door Henning Larsen en in 1997 werd voltooid, een grote boogvormige woningrij door Ricardo Bofill en meer dan twintig *urban villa's*, voornamelijk bestemd als woningen. Het totaalplan kwam uit de koker van Jan Inghe, van de afdeling Stadsontwikkeling van Stockholm.

Onder de plannen die in de komende tien tot twintig jaar zullen worden gerealiseerd, bevindt zich een interessant voorbeeld in de gemeente Järfälla, ten noordwesten van Stockholm. Hier wordt een nieuwe gemeenschap gebouwd op een vliegveld dat in de buurt van oudere dorpen ligt. Het plan omvat 4500 tot 5000 woningen en daarnaast gebouwen met een commerciële bestemming. Een groot deel van het project dient te worden gebaseerd op het model van een stadje in Scandinavische stijl, maar kent daarnaast tevens de invloed van Zweedse, Duitse en Engelse tuinsteden. Aan het project werken talloze specialisten mee, waaronder architecten, landschapsarchitecten, ecologen en experts op het gebied van energie en watervoorziening. Het meest recente stedebouwkundige voorstel is voornamelijk het werk van Ulf Ranhagen. Het bestaat uit een samenhangend netwerk van straten, waarbij verscheidene straten in de kern van dit netwerk als basis voor het gehele systeem dienen. De nieuwe wijk is groener dan de binnenstad van Stockholm, maar aan de andere kant is de bebouwingsdichtheid hoger dan in de tuinsteden. Net als in de tuinstad en het kleine stadje komen hier de voordelen van landelijke gebieden samen met die van de dichtbevolkte stad. Een belangrijk aspect van het ontwerp is het creëren van een omgeving die de ontwikkeling van het ecologische denken stimuleert.

Literatuur
G. Åström, City Planning in Sweden, Stockholm 1967.
Th. Hall, *I nationell skala. Studier kring cityplaneringen i Stockholm. On a national scale, en Studies on the Redevelopment of Stockholm's City Center*, Stockholm 1986.
Th. Hall, *Planning and Urban Growth in the Nordic Countries*, Londen 1991.
I. Johansson, *Stor-Stockholms bebyggelsehistoria. Markpolitik, planering och byggande under sju sekler*, Stockholm 1987.
R. Paulsson, *Den glömda staden. Om stadsbyggandet i Stockholm under 1900-talets första decennier*, Stockholm 1994.
C. Sidenbladh, *Norrmalm förnyat 1951–1981*, Stockholm 1985.
L. Tonell, 1980-talet – förhandlingsplanering och de stora projektens decennium, in: Th. Hall (red.), *Perspektiv på Planering*, Uppsala 1991.

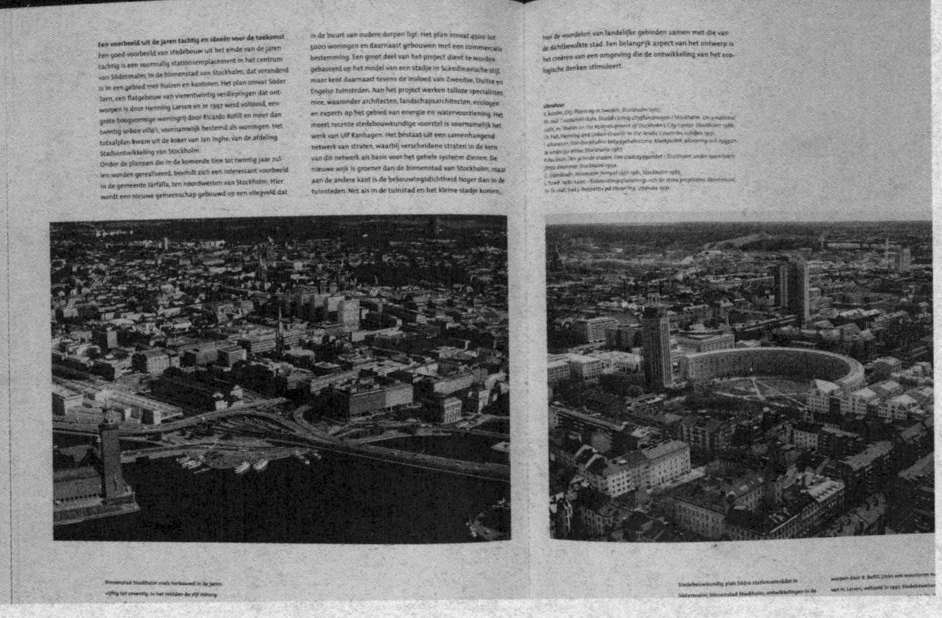

Binnenstad Stockholm zoals herbouwd in de jaren vijftig tot zeventig. In het midden de vijf Hötorg-

Stedebouwkundig plan Södra stationsområdet in Södermalm, binnenstad Stockholm, ontwikkelingen in de

worpen door R. Bofill. Links een woontoren ... van H. Larsen, voltooid in 1997. Stedebouwkun...

2000

Country / Region

Germany ❹

The Netherlands ❸

Austria ❶

Canada ❶

Japan ❶

Sweden ❶

Switzerland ❶

U.S.A. ❶

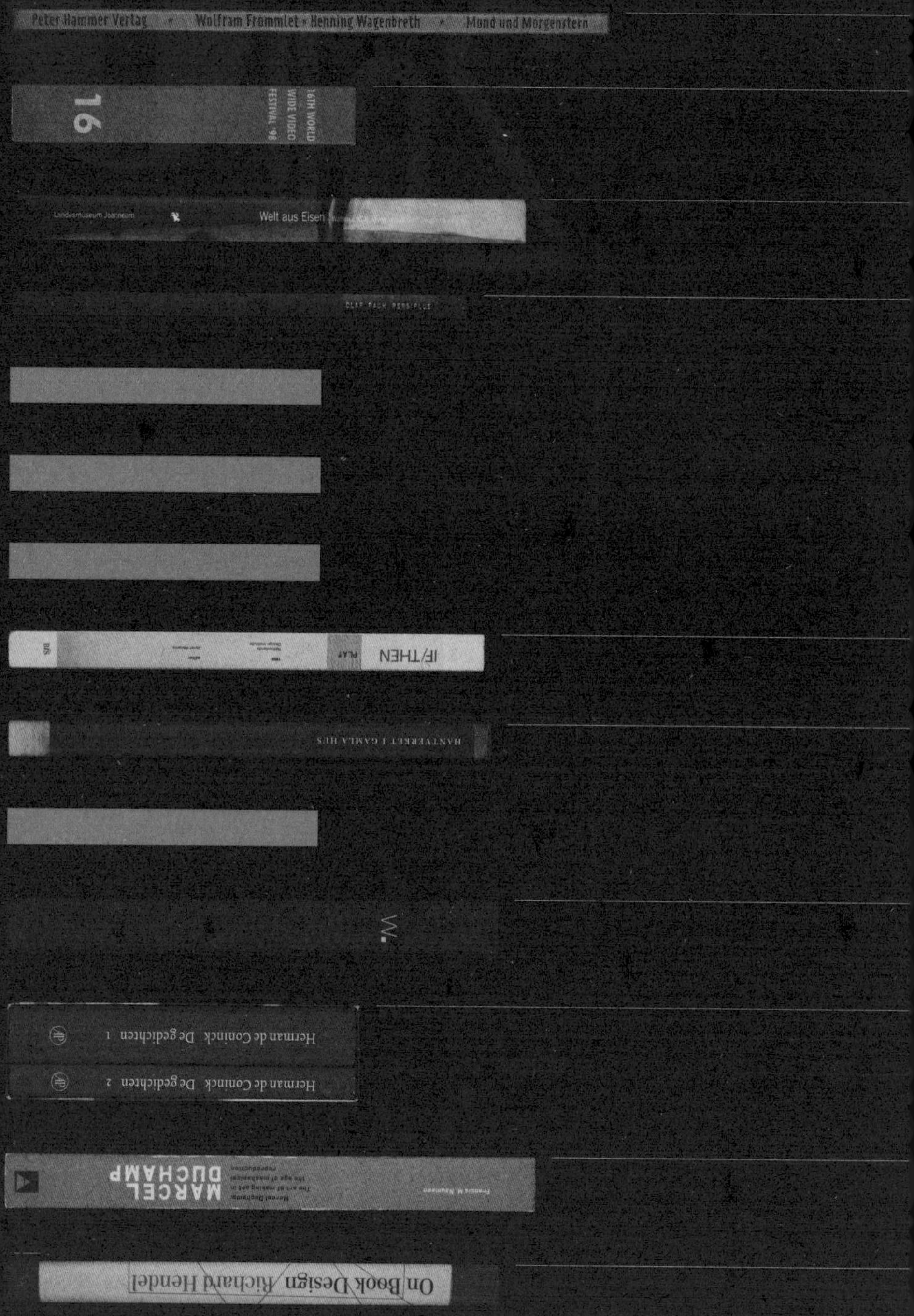

Peter Hammer Verlag • Wolfram Frommlet • Henning Wagenbreth • Mond und Morgenstern

16 16TH WORLD WIDE VIDEO FESTIVAL '98

Landesmuseum Joanneum Welt aus Eisen

IF/THEN PLAY

HANTVERKET I GAMLA HUS

W.

Herman de Coninck De gedichten 1

Herman de Coninck De gedichten 2

MARCEL DUCHAMP Marcel Duchamp: The art of making art in the age of mechanical reproduction Francis M. Naumann

On Book Design Richard Hendel

2000 Gl *0938*

2000 Gm *0948*

2000 Sm[1] *0958*

2000 Sm[2] *0966*

2000 Bm[4] *0976*

2000 Bm[5] *0986*

2000 Ha[2] *0996*

2000 Ha[3] *1006*

2000 Ha[4] *1012*

2000 Ha[5] *1022*

Designer

Henning Wagenbreth

Irma Boom

Alexander Kada

Kerstin Riedel

George Vaitkunas

Jens Bertelsen

Kikko Ken

Mevis & Van Deursen

Dick Norberg

Urs Bernet, Jürg Schönenberger

Gert Wunderlich

Steven van der Gaauw

studio blue

Richard Hendel

2000 Gl

04356 Leipzig
Germany
Referenz – Nr.: 105277978

PRIORITY

W

06 30E0 A750 1E C000 0A7C
DV 12.19 1,00 Deutsche Post
WARENPOST Int. KT Port payé

TO

ZHAO QING
Design Co.,
Dabei xiang7 – 3,Nanjing Han Qing Tang
210000 Nanjing
CHINA VR

大悲巷7一

UC 031 267 352 DE

Postverwaltung/Administration des postes Deutsche Post

1 x Book

0.72

4.65

0.747

17/12/2019

Germany

Germany

China

346mm

14mm

247mm

金字符奖
月亮和晨星
德国

Golden Letter
Mond und Morgenstern
Germany
Henning Wagenbreth
Wolfram Frommlet, Henning Wagenbreth
Peter Hammer Verlag, Wuppertal
346 247 14mm
730g
48p
ISBN-10 3-87294-784-2

这是一本来自非洲的创世神话的儿童绘本改编版。书籍的装订形式为锁线硬精装，封面胶版纸红黄蓝青四色套版印刷，裱覆卡板。内页胶版纸采用与封面相同的四色套版。书中的非洲版创世神话有部分欧洲各地神话的影子，人与宇宙、人与神、男与女、伊甸园、大自然等元素混在一起，夹杂了很多非洲的元素。绘画的风格稚拙而强烈，部分页面的插图通过红青、黄蓝、红黄等双色搭配，形成不同的氛围塑造和情绪渲染。文字方面，选用了一种粗犷拙朴的无衬线字体，与插图部分的风格和色彩相协调。

This is a children's picture book adapted from the African creation myth. The hardcover book is bound by thread sewing. The cover is made of offset paper printed in red, yellow, blue and cyan, mounted with cardboard. The same paper and colors are selected for the inside pages. The African version of the creation myth in the book has the influence of some European myths. People and the universe, people and gods, men and women, Eden, nature and other elements are mixed together, mingled with many African elements as well. The style of the paintings is naive and strong. The illustrations on some pages are displayed by the combination of red and green, yellow and blue, red and yellow, thus forming different atmosphere and emotional rendering. In terms of text, a kind of plain sans serif font is selected, which is in harmony with the style and color of the illustrations.

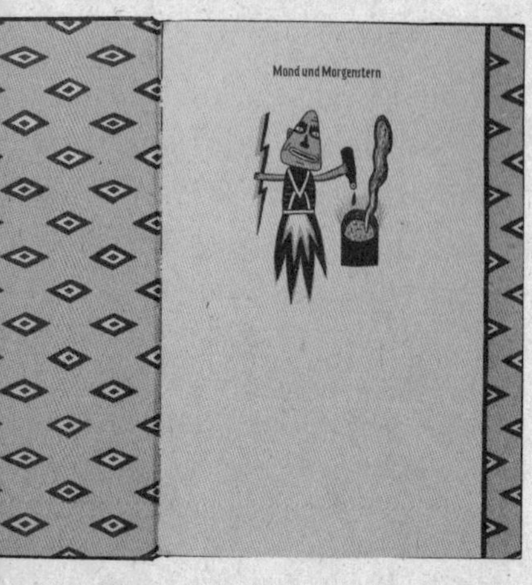

Mond und Morgenstern

Eine Geschichte aus Afrika
nacherzählt von Wolfram Frommlet
mit Bildern von Henning Wagenbreth

Peter Hammer Verlag

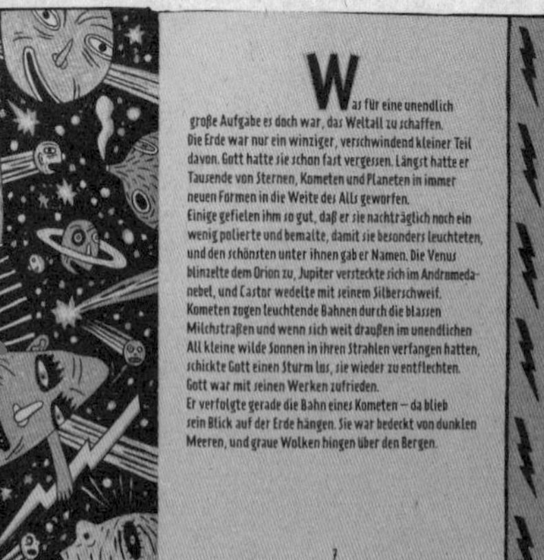

Was für eine unendlich große Aufgabe es doch war, das Weltall zu schaffen.
Die Erde war nur ein winziger, verschwindend kleiner Teil davon. Gott hatte sie schon fast vergessen. Längst hatte er Tausende von Sternen, Kometen und Planeten in immer neuen Formen in die Weite des Alls geworfen.
Einige gefielen ihm so gut, daß er sie nachträglich noch ein wenig polierte und bemalte, damit sie besonders leuchteten, und den schönsten unter ihnen gab er Namen. Die Venus blinzelte dem Orion zu, Jupiter versteckte sich im Andromedanebel, und Castor wedelte mit seinem Silberschweif. Kometen zogen leuchtende Bahnen durch die blassen Milchstraßen und wenn sich weit draußen im unendlichen All kleine wilde Sonnen in ihren Strahlen verfangen hatten, schickte Gott einen Sturm los, sie wieder zu entflechten.
Gott war mit seinen Werken zufrieden.
Er verfolgte gerade die Bahn eines Kometen – da blieb sein Blick auf der Erde hängen. Sie war bedeckt von dunklen Meeren, und graue Wolken hingen über den Bergen.

7

»Wie die Berge steil aufragen, gefällt mir«, sagte Go
»Und wie das Meer rauscht, auch. Aber die Erde ist zu düster.«
Gott dachte eine kleine Weile nach und nahm schließl einen Klumpen geheimnisvoller Knetmasse. Er tränkt in gelber Farbe, formte ihn rund, tätschelte hier ein w und stupste da ein bißchen und blies ihm Leben ein.
Der rundliche Knubbel begann sich zu regen und zu bev er öffnete seine Augen, er bewegte seinen Mund und f
»Wer bin ich?«
»Du bist der Mond«, antwortete Gott, »und du sollst dem Grund des Meeres sitzen und es beleuchten.«
»Auf dem Grund des Meeres?«
»Frag nicht so viel. Ich bin Gott.«
Gott schubste Mond aus seiner Schöpfungshalle hinab Meer. Da saß er nun und leuchtete. Und bibberte und la weilte sich. Kaum hatte sich Gott hingesetzt und begon Schäfchenwolken zu zählen, hörte er von der Erde ein blubberndes Jammern.
»Gott! Gott, hörst du mich?«
Es war der Mond vom Grund des Meeres.
»Was ist los, Mond?«
»Gott, bitte hör mich an. Es ist schrecklich kalt hier unt und fürchterlich langweilig. Bitte, lass' mich auf der Er leben.«

9

Während er noch in Gedanken versunken überlegte, was Gott ihm da wohl schicken würde, erfüllte eine zarte, schwirrende Melodie die Luft und neben ihm landete ein wundervolles Geschöpf, das glitzerte und funkelte.
»Ich bin Morgenstern«, hauchte sie ihm flötenhell zu.
»Ich bin gekommen, dir Gesellschaft zu leisten.«
»Ich bin Mond«, gab er verwirrt zurück.
»Ich weiß. Und? Was sagst du zu mir?«
Das also ist eine Frau, dachte Mond. Was soll ich zu ihr sagen? Er sah sie lange an.
»Ich liebe das Klimpern deiner Wimpern.«
»Das klingt schön«, antwortete Morgenstern und streichelte Mond ein wenig.
»Deine Haut ist so weich, aber auch ein bißchen kalt. Komm zu mir. Ich habe dir etwas mitgebracht.«
Morgenstern öffnete einen kleinen silbernen Stern, den sie versteckt gehalten hatte, und heraus sprangen rötliche Flammen.
Mond erschrak und wich ängstlich zurück. »Was ist das?«
»Das ist Feuer. Komm näher und lass' dich von ihm wärmen.«

12

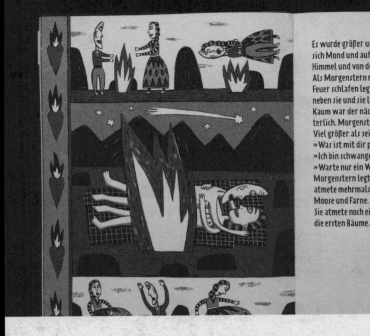

Es wurde größer und wärmer, und auf die eine Seite setzte sich Mond und auf die andere Morgenstern, die ihm vom Himmel und von den Sternen erzählte.
Als Morgenstern müde geworden war und sich neben dem Feuer schlafen legte, sprang Mond über das Feuer, legte sich neben sie und sie liebten sich. Die ganze Nacht.
Kaum war der nächste Tag vergangen, erschrak Mond fürchterlich. Morgensterns Bauch war riesig und rund geworden. Viel größer als seiner.
»Was ist mit dir passiert?«, fragte Mond ängstlich.
»Ich bin schwanger von dir«, erklärte ihm Morgenstern.
»Warte nur ein Weile, und bleibe bei mir.«
Morgenstern legte sich auf eine weiche Stelle im Sand, atmete mehrmals tief durch und gebar die ersten Gräser und Moose und Farne.
Sie atmete noch einmal tief und gebar die ersten Blumen und die ersten Bäume.

Mond und lorgen tern

ch bin gekommen, dir Gesellschaft zu l
ch bin Mond«, gab er verwirrt zurück
ch weiß. Und? Was sagst du zu mir?«
s also ist eine Frau, dachte Mond. Was
gen? Er sah sie lange an.
ch liebe das Klimpern deiner Wimpern
as klingt schön«, antwortete Morger
elte Mond ein wenig.
eine Haut ist so weich, aber auch ein bi
mm zu mir. Ich habe dir etwas mitgebr
orgenstern öffnete einen kleinen silbe
rsteckt gehalten hatte, und heraus spr
ammen.
ond erschrak und wich ängstlich zurüc

nd?«
nich an. Es ist schre
langweilig. Bitte,

Jende von Sternen, Kometen und Planeten in immer
en Formen in die Weite des Alls geworfen.
ge gefielen ihm so gut, daß er sie nachträglich noch ein
nig polierte und bemalte, damit sie besonders leuchteten,
den schönsten unter ihnen gab er Namen. Die Venus
zelte dem Orion zu, Jupiter versteckte sich im Andromeda-
el, und Castor wedelte mit seinem Silberschweif.
neten zogen leuchtende Bahnen durch die blassen
chstraßen und wenn sich weit draußen im unendlichen
kleine wilde Sonnen in ihren Strahlen verfangen hatten,
ckte Gott einen Sturm los, sie wieder zu entflechten.
t war mit seinen Werken zufrieden.
erfolgte gerade die Bahn eines Kometen – da blieb
Blick auf der Erde hängen. Sie war bedeckt von dunklen

Was für
h war, das Weltall
winziger, verschwi
schon fast vergesse
n, Kometen und Pla

»Hör' auf mit dem Gejammer«, rief er von oben.
»Ich schicke dir eine neue Gefährtin. Doch auch sie wird nur zwei Jahre auf der Erde bleiben.«
Dieses Mal wählte Gott Abendstern. Um ihre Schultern legte er seidenweiche Nebeltücher, und in den Falten versteckte er eine Himmelsmelodie. Dann schickte er Abendstern auf die Erde.
Mond war so damit beschäftigt, sich selbst zu bemitleiden, daß er ihre Ankunft gar nicht bemerkte.
»Hör endlich auf, zu jammern und zu weinen. Ich bin Abendstern. Willst du mich nicht begrüßen?«
Mond schämte sich ein wenig für seine Tränen und riß sich zusammen. Sie ist auch sehr schön, dachte Mond, als er Abendstern aus wässrigen Augen so ansah.
Mond dachte nach, und Abendstern wartete.
»Heute Nacht im Mondenlicht
Trag ich dir vor ein Mondgedicht.«
»Danke«, sagte Abendstern. »Ich freu' mich sehr darauf.«
Und sie fühlte sich sehr geschmeichelt.
Jetzt ging es ihm schon besser.
»Und, was hast du mir vom Himmel mitgebracht?«
Abendstern nahm den Nebelschal von der Schulter, schwang ihn in der Luft, als zeichnete sie Lilien und zauberte eine wundervolle Melodie aus ihm hervor.

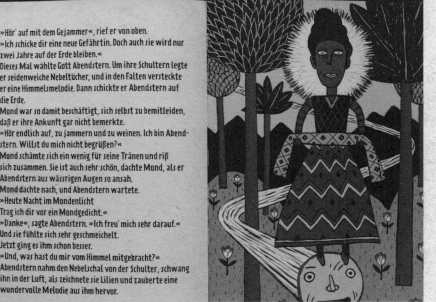

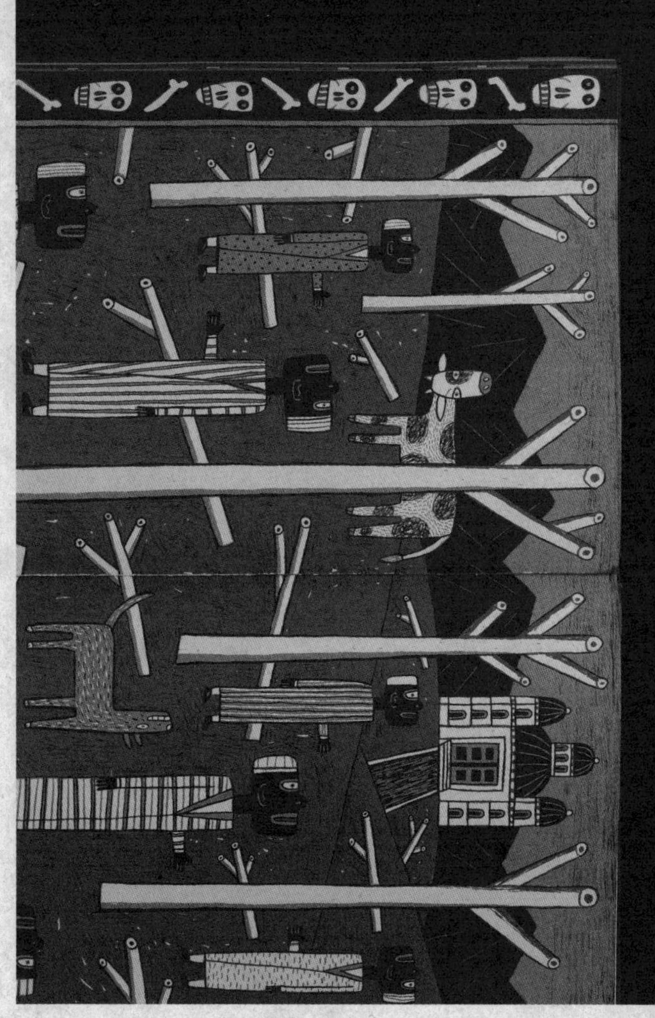

7-3 [illegible]8442

aramex

Origin: LON

32969248221

Destination: SZV

Date: Feb 28, 2020
Foreign Ref:
Ref1: TP 2452917

EXP | PPX | P

Weight: 2 KG
Chargeable: 2 KG
Services:
Pieces: 1
Description: Printed Pa[illegible]

Account: 60498531
Art Data
Nancy Wilson
London

Zhao Qing
Zhao Qing

From:
Art Data
T.G Bolton Ltd
12 Bell Industrial Estate, 50 Cunnington Street
London W4 5HB
United Kingdom

To:
ZHAO QING
7-3, DABEI LANE,
NEW VILLAGE
NANJING 210000
CHINA

2000 Gm

L CHINA
B200 CN-PVG-NKG
2.0 kg 1/1

Print Time: 2020-03-04 06:43

NCY.ICS1.CH518

NO

NO

B-3---1-

PVGNKTP-

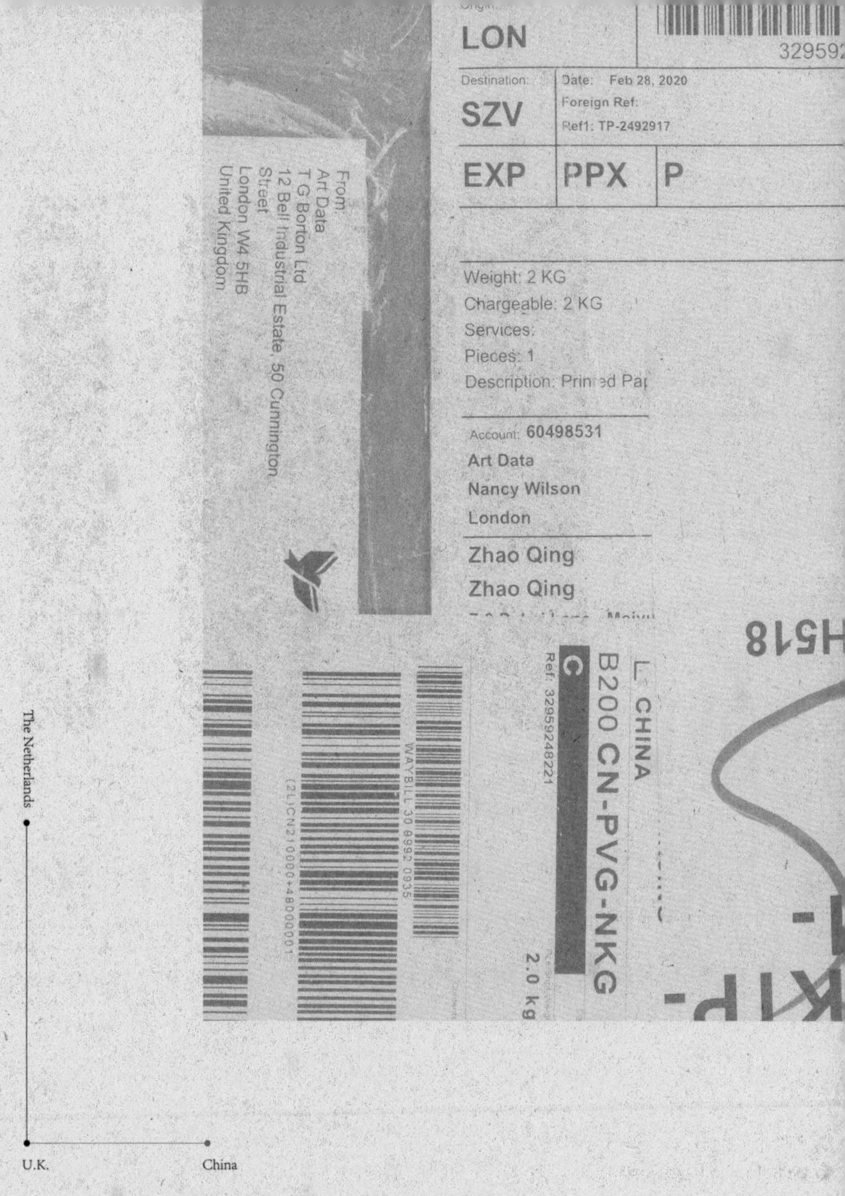
LON
Destination:
SZV
Date: Feb 28, 2020
Foreign Ref:
Ref1: TP-2492917
EXP
PPX
P
Weight: 2 KG
Chargeable: 2 KG
Services:
Pieces: 1
Account: 60498531
Art Data
Nancy Wilson
London
Zhao Qing
Zhao Qing
From
Art Data
T G Borton Ltd
12 Bell Industrial Estate, 50 Cunnington Street
London W4 5HB
United Kingdom
WAYBILL 30 8992 0935
CHINA
B200 CN-PVG-NKG
C
Ref. 32959248221
2.0 kg
The Netherlands
U.K.
China

0949

199mm

33mm

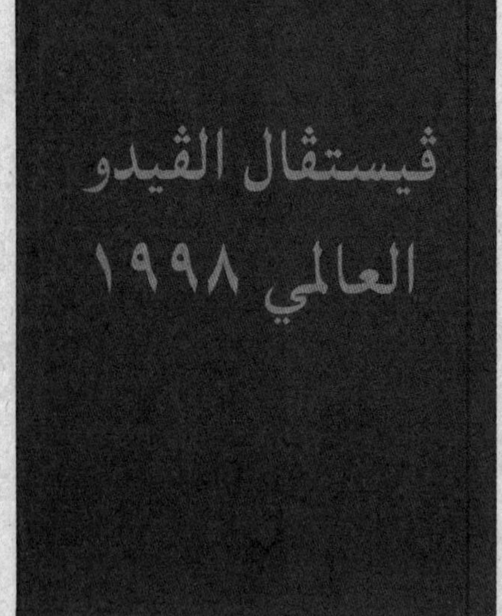

159mm

● 金奖
« 1998 年第 16 届世界影像节
◖ 荷兰

Gold Medal
16th World Wide Video Festival '98
The Netherlands
Irma Boom
——
Stichting World Wide Video Festival, Amsterdam
199 159 33mm
833g
560p
ISBN-10 90-75018-16-9

1998年第16届世界影像节在荷兰阿姆斯特丹举行，本书是此次展览的画册。书籍的装订形式为锁线胶平装。封面白卡纸红黄蓝三色印刷，覆哑光膜。内页胶版纸四色印刷。全书开篇对20世纪以来的视频技术和艺术的发展做了回顾，分析了视频技术对人们生活、教育、娱乐等方面带来的巨大影响，也对未来的视频技术和艺术以及与公众的关联展开想象。由于视频载体、播放环境和技术发展带来的巨大变化，展览中来自23个国家或地区的100件作品被划分为录像带、光盘、表演、装置和网站几大类。全书在设计上极力呈现出视频艺术的技术特点，即基于RGB三原色的显色形式，封面呈现红绿双色，翻口处采用红绿蓝三色条。内页文字部分底色大量使用强烈多变的颜色，体现出视频色彩的丰富性和夺目效果。

The 16th World Image Festival in 1998 was held in Amsterdam, the Netherlands. This book is the album of this exhibition. The book is perfect bound with thread sewing. The cover is printed in red, yellow and blue on white cardboard and covered with matte film. The inner pages are printed in four colors on offset paper. The book begins with a review of the development of video technology and art since the 20th century. It analyzes the tremendous impact of video technology on people's lives, education, entertainment, etc., and also imagines the future of video technology and art and its relationship with the public. Due to the great changes brought by the video carrier, broadcasting environment and technology development, 100 pieces of works from 23 countries or regions in the exhibition are classified into videotapes, CD-ROMs, performances, installations and websites. As for the design, the book tries to present the technical characteristics of video art, that is, the color developing based on the RGB three primary colors. The cover is red and green, and color bars using red, green and blue are printed at the book edge. On the inner pages, strong and changeable colors are used as background to express the richness and dazzling effect of the video.

Met dank aan / With thanks to

Argos, Bruxelles
Charles Bongaerts, Roosmalen
Hugo Bongers, Amsterdam
Pascal Boon, Amsterdam
Eelco Brunsma, Eemnes
Robert Cahen, Mulhouse
Amita Desai, Bombay
Dr. Saryu Doshi, Bombay
Evgenija Dimitrieva, Skopje
Daniel Dion, Montréal
Matthew Drutt, New York
Electronic Arts Intermix, New York
EMAF, Osnabrück
Solange Farkas, Sao Paulo
Rudolf Frieling, Karlsruhe
Rudi Fuchs, Amsterdam
Peter van de Geer, Amsterdam
Shireen Ghandy, Bombay
Goodman Gallery, Johannesburg
J.H.E. Heckman, Bombay
Hearn Enquist, Mouse-la-Baronnie
Truus Jansen, Amsterdam
Kumala Kapoor, Bombay
London Electronic Arts, London
Nederlands Consulaat, New York
Jellie van Nieuwenhoven, Den Haag
Bas van Peer, Leiderdorp
Els Reijnders, Amsterdam
Andreas Schmid, Berlin
Anjali Sen, New Delhi
VideoBrasil, Sao Paulo
Video Data Bank, Chicago
Wolf Siegert, Berlin
Mirjam Stoop, Amsterdam
Mike Stubbs, Hull
Tanzhaus Wasserwerk, Zurich
Roemer van Toorn, Amsterdam
Christine Vroom, Amsterdam
Femke Wolting, Amsterdam

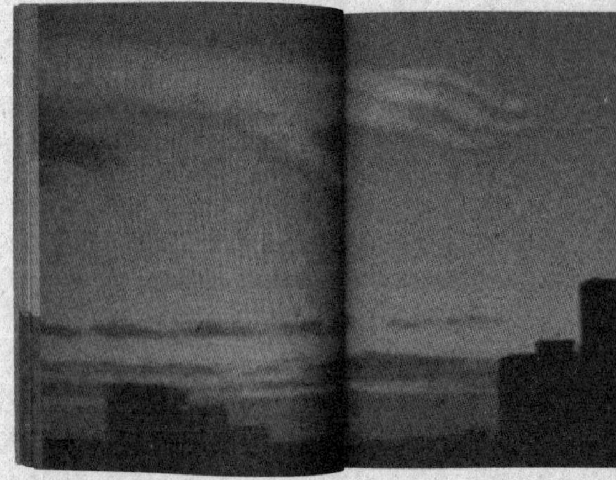

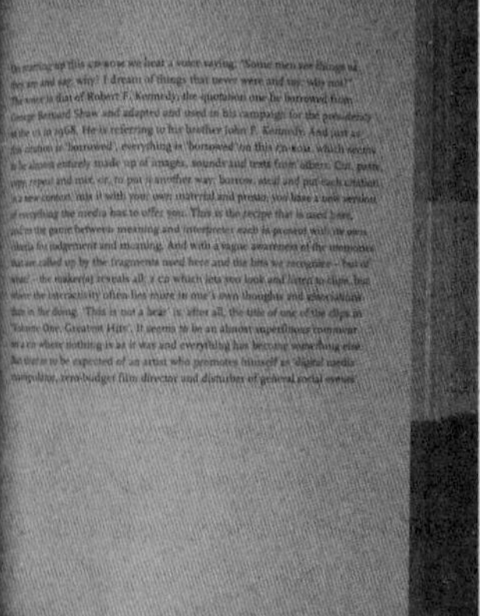

On starting up this CD-ROM we hear a voice saying: 'Some men see things as they are and say: why? I dream of things that never were and say: why not?' The voice is that of Robert F. Kennedy, the quotation one he borrowed from George Bernard Shaw and adapted and used in his campaign for the presidency of the US in 1968. He is referring to his brother John F. Kennedy. And just as this citation is 'borrowed', everything is 'borrowed' on this CD-ROM, which seems to be almost entirely made up of images, sounds and texts from others. Cut, paste, copy, repeat and mix, or, to put it another way: borrow, steal and put each citation in a new context, mix it with your own material and presto, you have a new version of everything the media has to offer you. This is the recipe that is used here, and in the game between meaning and interpreter each is present with its own criteria for judgement and meaning. And with a vague awareness of the memories that are called up by the fragments used here and the bits we recognize – 'but of what' – the maker(s) reveals all; a CD which lets you look and listen to clips, but where the interactivity often lies more in one's own thoughts and associations than in the doing. 'This is not a bear' is, after all, the title of one of the clips in 'Volume One: Greatest Hits'. It seems to be an almost superfluous comment on a CD where nothing is as it was and everything has become something else. But that is to be expected of an artist who promotes himself as 'digital media manipulator, zero-budget film director and disturber of general social events'.

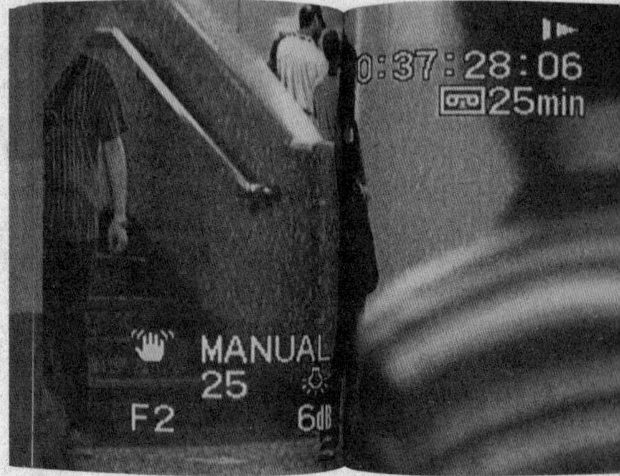

Intentie en Ontwikkeling

Het festivalprogramma is een rijk geschakeerd geheel. Installaties, werk in situ, performances, videotapes, CD-ROMS, websites. Categorieën die arbitrair zijn, zo nu en dan bestaan en relevant zijn, en soms verwerpelijk zijn. De catalogus is alfabetisch opgebouwd, zonder rangschikking in categorie. In deze verantwoording wordt een aanzet gegeven tot het in kaart brengen van ontwikkelingen binnen de mediakunst. Tijdens het festival krijgt dit nader gestalte, onder meer tijdens het seminar, waar wordt ingegaan op de ontologie van het elektronische beeld, verhalende structuren in de mediakunst en het creëren van tentoonstellingsplekken voor mediakunst. Een tiental internationale sprekers analyseren de mediakunst op een moment dat deze kunstvorm naar het schijnt een doorbraak beleeft naar een groter publiek. De analyse wordt gevoed met beelden van recente producties, een retrospectief en een grote overzichtstentoonstelling van Bill Viola, georgani-seerd door het Stedelijk Museum en samengesteld door David Ross en Peter Sellars.

De performancekunst maakt een opmerkelijke ontwikkeling door en staat weer volop in de belangstelling. Performances van 20 en 30 jaar geleden worden zelfs (geheel in de geest van deze tijd) opnieuw opgevoerd, door kunstenaars die te jong zijn om ze ooit live te hebben kunnen zien. Na de hoogtijdagen in de jaren '60 en '70 verschoof de aandacht van het museum- en galeriecircuit naar andere kunstvormen, waardoor de indruk werd gewekt dat de performance op sterven na dood was. Maar wie zijn ogen goed openhield, kon zien dat kunstenaars over de hele wereld rustig doorwerkten aan het verder ontwikkelen ervan. Als ontmoetingsplaats voor experimentele hedendaagse kunst is het World Wide Video Festival de ontwikkelingen in de performancekunst altijd blijven volgen. Zo was in de jaren '80 te zien dat video niet langer alleen als registratiemiddel voor de performance gebruikt werd, maar er integraal onderdeel van ging uitmaken. Performances kregen over het geheel genomen een steeds multi-medialer karakter. Het gesloten videocircuit kwam bij diverse

Intention and Development

The festival programme is of a rich diversity. Installations, site specific w
performances, videotapes, CD-ROMS and websites. Arbitrary categories wh
are sometimes real and relevant and sometimes objectionable. The catalo
is in alphabetical order by artist, disregarding any categories. In these pag
try to make a start with mapping the developments within media art. Dur
the festival this will be further evolved, for instance with a seminar where
discussions will be held about the ontology of the electronic image, narrat
structures in media art and how to create exhibition spaces for it. A dozen
international speakers will analyze media art at this moment in time whe
art form seems to be breaking through to a much wider audience. This ar
is illustrated with images from recent productions, a retrospective and a l
exhibition of works by Bill Viola, organized by the Stedelijk Museum and
curated by David Ross and Peter Sellars.

Performance art has shown a remarkable development and for some time
has again been the centre of interest. Performances from twenty and thirt
years ago are being put on again (in a current context) by artists too young
have ever been able to see them live. After the heydays of the sixties and
seventies the attention performance art got from the museums and galler
shifted to other art forms, giving the impression that performance art was
good as dead. But those who were interested could see that artists all over
world were quietly working on the development of the art form. As meeti
place for modern experimental art, the World Wide Video Festival has alw
followed the developments in performance art. And so in the eighties vid
no longer being used solely as a recording device in performance art, but
taken its place as an integral part of the art form. Performances started to
on more and more the character of multimedia. The closed video circuit
being used by various artist, among others, Fabienne de Quasa Riera & Je
Shaw and Takahiko Iimura, and the electronic potentials of sound were b
taken further and further. In 1983 Lous America experimented with the
interaction between already existent image and sound and impromptu

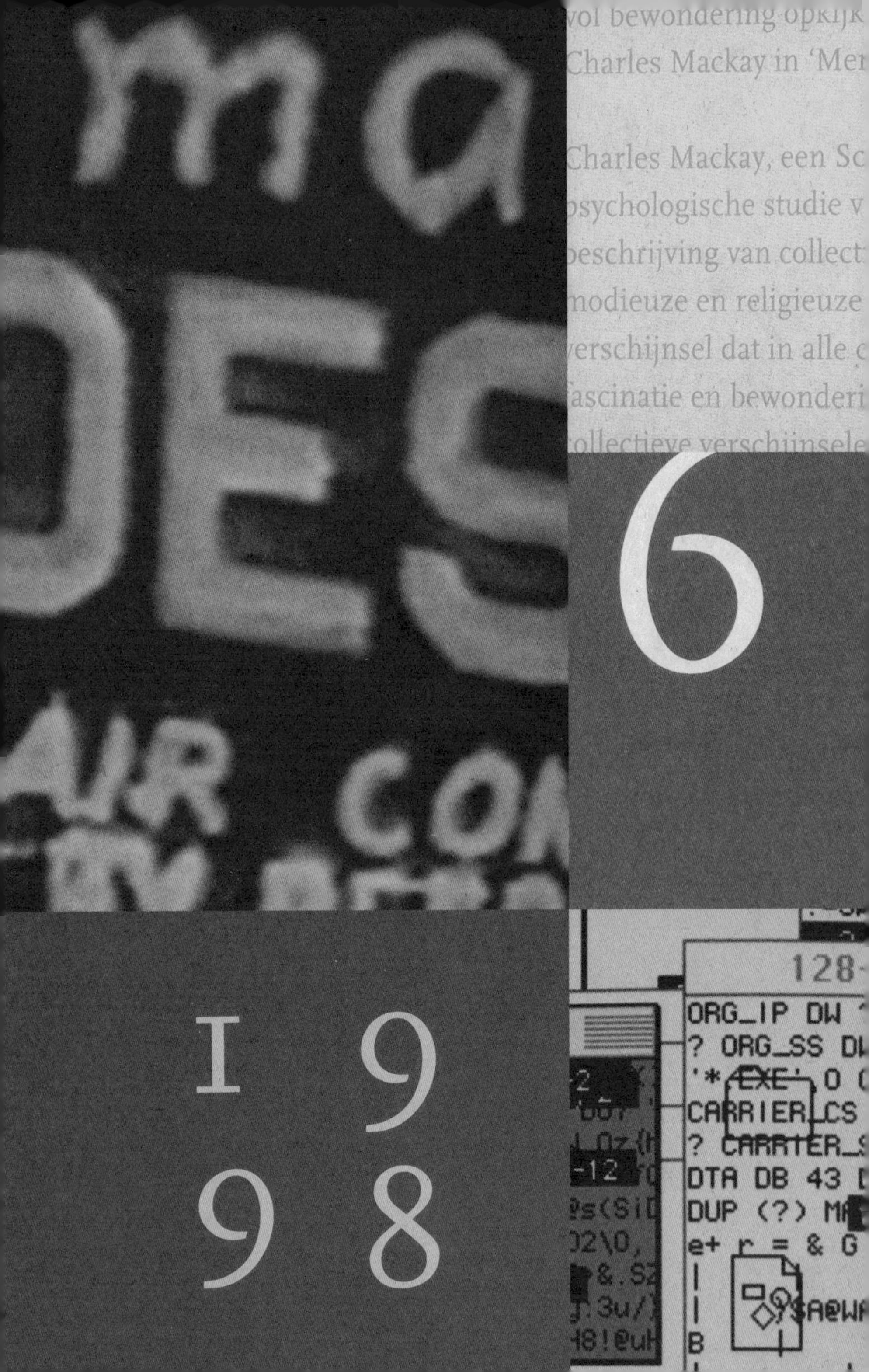
Charles Mackay in 'Me
Charles Mackay, een Sc
psychologische studie v
beschrijving van collect
modieuze en religieuze
verschijnsel dat in alle
fascinatie en bewonderi
6
I 9
9 8
ORG_IP DW
? ORG_SS D
CARRIER_CS
DTA DB 43
DUP (?) M

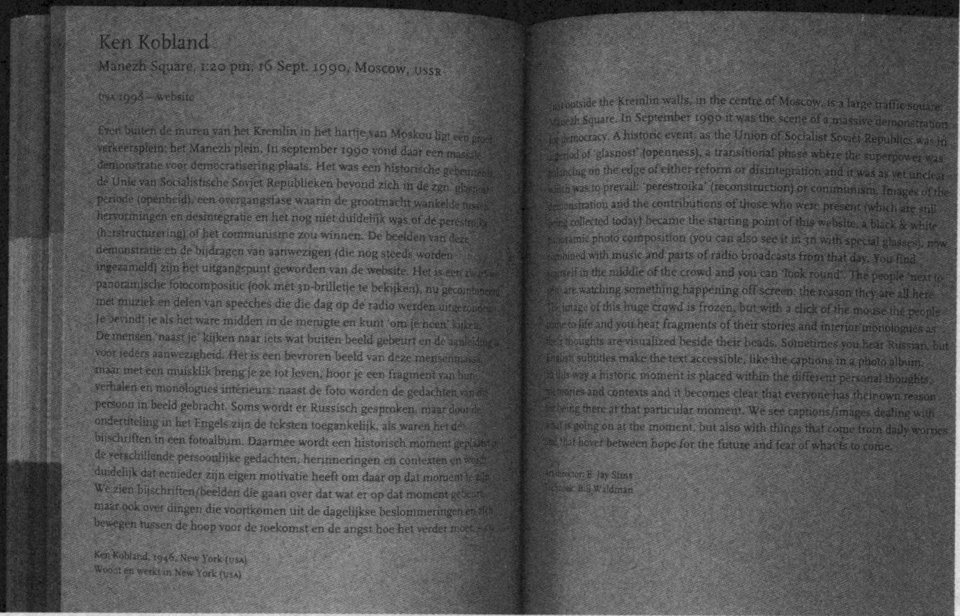

Ken Kobland

Manezh Square, 1:20 pm, 16 Sept. 1990, Moscow, USSR

USA 1998—website

Vlak buiten de muren van het Kremlin in het hartje van Moskou ligt een groot verkeersplein: het Manezh plein. In september 1990 vond daar een massale demonstratie voor democratisering plaats. Het was een historische gebeurtenis, de Unie van Socialistische Sovjet Republieken bevond zich in de zgn. 'glasnost' periode (openheid), een overgangsfase waarin de grootmacht wankelde tussen hervormingen en desintegratie en het nog niet duidelijk was of de perestrojka (herstructurering) of het communisme zou winnen. De beelden van deze demonstratie en de bijdragen van aanwezigen (die nog steeds worden ingezameld) zijn het uitgangspunt geworden van de website. Het is een zwart-wit panoramische fotocompositie (ook met 3D-brilletje te bekijken), nu gecombineerd met muziek en delen van speeches die die dag op de radio werden uitgezonden. Je bevindt je als het ware midden in de menigte en kunt 'om je heen' kijken. De mensen 'naast je' kijken naar iets wat buiten beeld gebeurt en de aanleiding voor ieders aanwezigheid. Het is een bevroren beeld van deze mensenmassa, maar met een muisklik breng je ze tot leven, hoor je een fragment van hun verhalen en monologues intérieurs: naast de foto worden de gedachten van de persoon in beeld gebracht. Soms wordt er Russisch gesproken, maar door de ondertiteling in het Engels zijn de teksten toegankelijk, als waren het de bijschriften in een fotoalbum. Daarmee wordt een historisch moment geplaatst in de verschillende persoonlijke gedachten, herinneringen en contexten en wordt duidelijk dat eenieder zijn eigen motivatie heeft om daar op dat moment te zijn. We zien bijschriften/beelden die gaan over dat wat er op dat moment gebeurt, maar ook over dingen die voortkomen uit de dagelijkse beslommeringen en zich bewegen tussen de hoop voor de toekomst en de angst hoe het verder moet.

Ken Kobland, 1946, New York (USA)
Woont en werkt in New York (USA)

Just outside the Kremlin walls, in the centre of Moscow, is a large traffic square: Manezh Square. In September 1990 it was the scene of a massive demonstration for democracy. A historic event, as the Union of Socialist Soviet Republics was in a period of 'glasnost' (openness), a transitional phase where the superpower was balancing on the edge of either reform or disintegration and it was as yet unclear which was to prevail: 'perestroika' (reconstruction) or communism. Images of the demonstration and the contributions of those who were present (which are still being collected today) became the starting point of this website, a black & white panoramic photo composition (you can also see it in 3D with special glasses), now combined with music and parts of radio broadcasts from that day. You find yourself in the middle of the crowd and you can 'look round'. The people 'next to you' are watching something happening off screen: the reason they are all here. The image of this huge crowd is frozen, but with a click of the mouse the people come to life and you hear fragments of their stories and interior monologues as their thoughts are visualized beside their heads. Sometimes you hear Russian, but English subtitles make the text accessible, like the captions in a photo album. In this way a historic moment is placed within the different personal thoughts, memories and contexts and it becomes clear that everyone has their own reason for being there at that particular moment. We see captions/images dealing with what is going on at the moment, but also with things that come from daily worries and that hover between hope for the future and fear of what is to come.

Art director: E. Jay Sims
Technics: Bill Waldman

2000 Sm1

Page 3 - Not Valid As
Proof-of-Payment for US Postage
PAGE 3 - DISPATCH NOTE
EP 851 760 812 US

Austria

U.S.A.

China

298mm

25mm

254mm

银奖
铁的世界
奥地利

Silver Medal
Welt aus Eisen
Austria
Alexander Kada
Thomas Höft
Springer-Verlag, Wien/New York
298 254 25mm
1826g
222p
ISBN-10 3-211-83097-9

奥地利格拉茨老城中心广场的兵器博物馆——施蒂里亚军械库曾是世界上最大的兵工厂。博物馆中收藏了大约 32000 件武器和铠甲。本书为介绍博物馆收藏和历史的画册。书籍的装订形式为锁线硬精装。护封铜版纸四色印刷，上下书口和勒口处均向内包折，形成良好的翻阅手感，同时避免了这么大的开本上护封变形的问题。内封浅灰色布面书名压凹处理，裱覆卡板。内页选用了一种稍带金属光泽的光面艺术纸四色印刷。不同于其他军事博物馆的展陈，施蒂里亚军械库的历史十分丰富，这也成了本书内容的编辑逻辑，即用参观和历史叙事为轴线，串起相对应的重要收藏，而并不是收藏的简单分类和罗列。书中展示的收藏大致分为几个部分：枪械、头盔、铠甲和冷兵器。多数展品通过全景、局部和细节的多张图片来展现。下方页码处设置了两条横线，用于写明所属年代、重量和尺寸等相关数据。

The Styrian Armoury, a museum of weapons in the central square of the old city of Graz, Austria, used to be the largest arsenal in the world. There are about 32,000 weapons and armor in the museum. This picture album is a collection of museum collections and historical introductions. This hardcover book is bound by thread sewing. The jacket is made of coated paper printed in four colors. The upper and lower book edges and the flaps are folded inward to avoid the deformation of the jacket of such a large format. The inside cover, mounted with card-board, is made of light gray cloth with the title embossed. The inner pages are printed in four colors on glossy art paper with metallic luster. Different from other military museums, the Styrian Armoury has a long history, which also becomes the editing logic of the book, that is, visits and historical narration are used as clues to link up the corresponding important collections, rather than a simple classification and listing of collections. The collections shown in the book are roughly divided into several parts: firearms, helmets, armors and cold weapons. Most of the exhibits are shown in panoramic, partial and detailed pictures. Two horizontal lines are set besides the page number to indicate the age, weight, size and other relevant data.

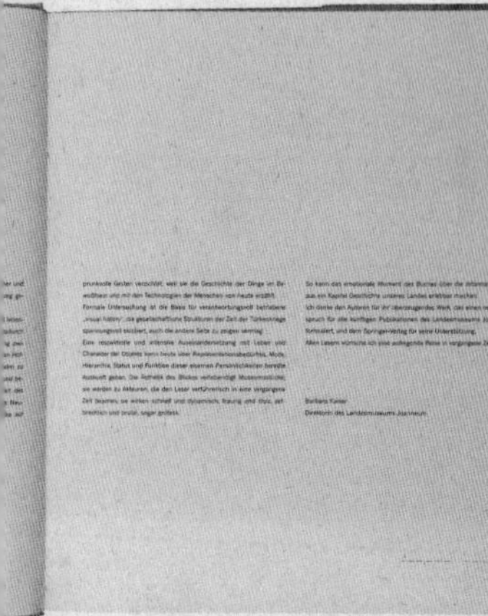

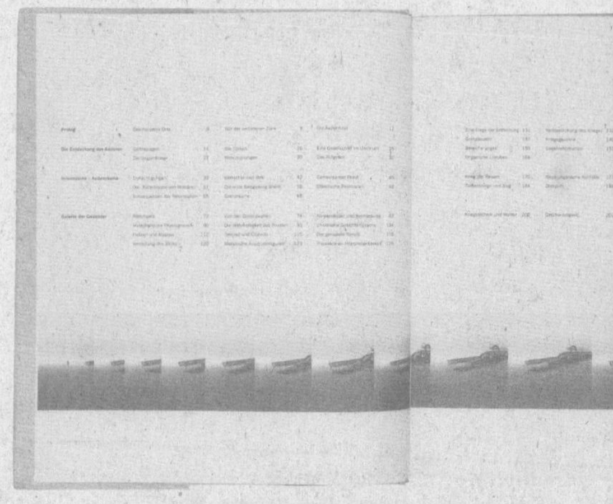

Die Entdeckung des Anderen

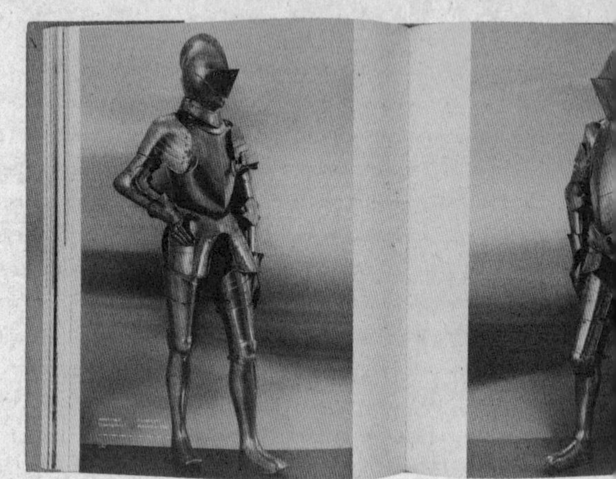

Eroberung Konstantinopels schreibt der Sultan an den Dogen von Venedig: „Warum wollen wir Krieg führen, da wir doch Brüder sind? Denn die Türken sind bekanntlich aus dem Brande Trojas hervorgegangen. Sie stammen von Turcus ab, der ein Sohn des Priamus ist wie Aeneas." Selbst ein türkischer Sultan mag nicht ausgeschlossen sein von der Berechnungskette der Abstammung. Ein Herrscher, der kraft seiner militärischen Stärke und seines anderen Glaubens scheinbar genug eigenständige Berechtigung hätte, selbstbewußt aufzutreten, konkurriert mit seinen Feinden im gedanklichen Krieg um die edle Herkunft.

Herrscher und Volk

Ein Denken und Verhalten, das sich aus genealogischen Rechtfertigungsüberlegungen speist, muß in Konflikt geraten mit dem Anspruch von Ständen und Städten, die ihr Selbstbewußtsein gerade auf die Faktizität des Ökonomischen bauen. Der genealogische Diskurs handelt in den Verhältnissen der Antike. Er tut so, als sei kein Bruch zwischen mythologischer Vergangenheit – die eine rein literarische ist –, römischer Geschichte und monarchischer Gegenwart existent. Die Berechtigung der Souveränität ist eine literarisch-philosophisch gedachte. Und sie ist so mächtig, daß sich auch die Türken ihrer bedienen wollen.

Die Kluft, die sich hier auftut, ist immens. Denn in der gleichen Zeit, in der sich der Kaiser auf Aeneas zurückbeziehen läßt, muß er sich vom Bürger Fugger exorbitante Summen leihen, um seinen Anspruch auch finanzieren zu können. In der gleichen Zeit, in der sich der mythisch aufgewertete Abkomme der antiken Helden über die Fortpflanzung ganze Reiche sichert, stellen sich seine Untertanen mit neuem Selbstbewußtsein dem Kampf ums Überleben und entscheiden, „sich selbst zu helfen". Sie fühlen sich nicht mehr geschützt von einem „väterlichen" Monarchen, dessen genealogische Rechtfertigung auch das antike Patriarchat einschließt, das in traditioneller Weise für die Abhängigen in Gerechtigkeit, Strenge und Gnade einsteht.

Die Gesellschaft schleudert zentrifugal auseinander, verschiedene Gruppen sind als Sieger möglich. In das Vakuum der Macht wollen auch die Unterprivilegiertesten vorstoßen. Die Bauern und Arbeiter, die erkennen, daß sie den Reichtum erwirtschaften, auf dem die anderen ihre Ansprüche errichten. Die Lösung des Konfliktes ist eine durch Gewalt

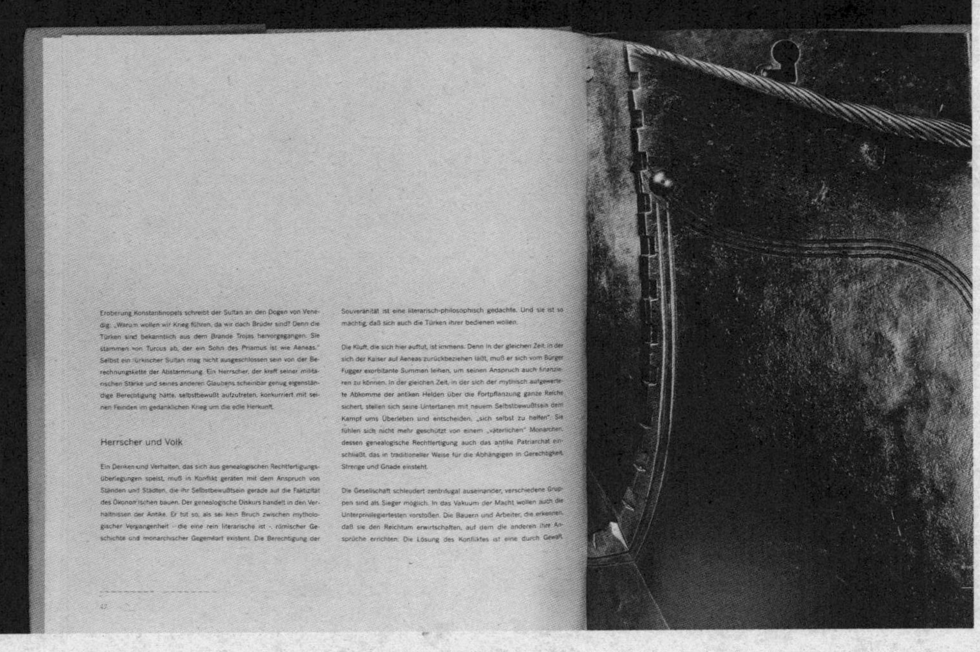

0963

0965

Persiplus

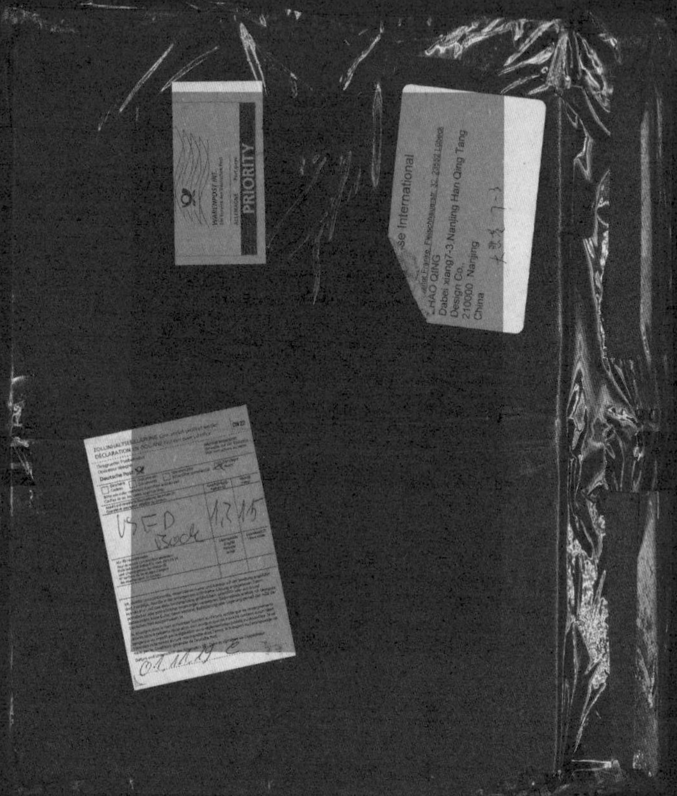

2000 Sm2

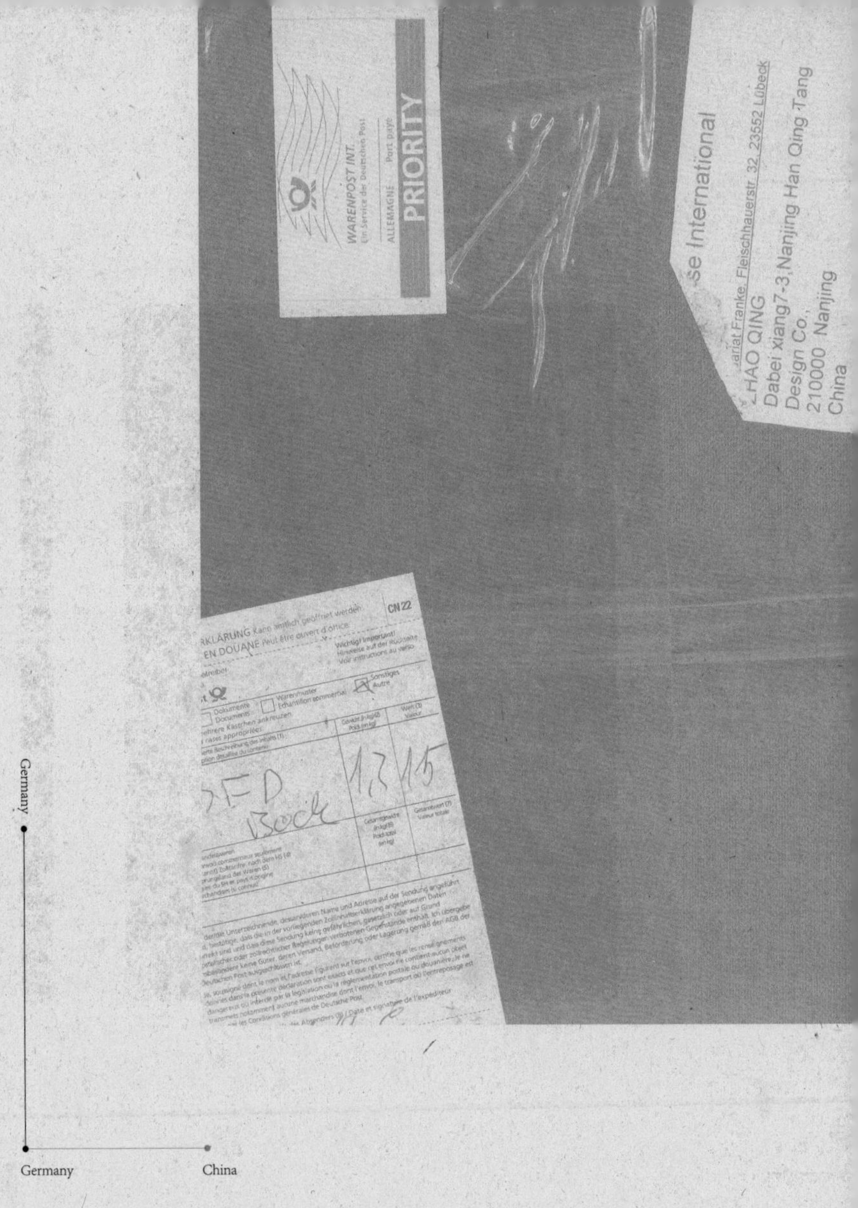

Germany

Germany

China

银奖
Persiplus
德国

Silver Medal
Persiplus
Germany
Kerstin Riedel
Olaf Rauh
Edition Persiplus (Olaf Rauh Eigenverlag), Leipzig
264 194 13mm
465g
120p
ISBN-10 3-00-003944-9

德国摄影师奥拉夫·劳（Olaf Rauh）的作品不属于风光、人像、人文纪实或任何传统分类，摄影在他的创作中不是去记录现实的一种客观媒介，而是在真实的环境中表达想象的世界的手段。这并不是指艺术家去扭曲现实，而是植根于现实本身，通过对光线、摄影方式、后期处理等多重干预，来创造独特的摄影作品。本书便是他的作品集。书籍的装订形式为锁线硬精装。封面塑胶材料裱覆卡板，文字部分压凹和烫银处理。内页哑面艺术纸，四色加橙色共五色印刷。作品部分，摄影师通过多重曝光、绘画、底片或冲扫过程中的绘制等方式来创造独特的影像，探讨精神、光阴、神话、巫术等非现实主题。德英双语的排版很有意思，英文用橙色专色，而德文用黑色，大量删除、文本提前等校对符号的运用，仿佛大量涂改的日记，记录混乱而真实的记忆本身。页码均被放置在页面左下角，通过与封面字体相同倾斜角度的类似“/”的符号来连接。

The works of the German photographer Olaf Rauh don't belong to any traditional photography, such as scenery, portraits, documentary and so on. Photography is not an objective medium to record reality in his creation, but a way to express the imaginary world in the real environment. It does not mean that the artist distorts the reality, but creates unique photographic works rooted in the reality itself through multiple interventions such as light, photography, and post-processing. This book is one of his works. The binding form of books is hard binding with thread sewing. The cover is made of plastic material mounted with cardboard, and the text is embossed and pressed with silver. Inner pages are made of matte art paper, printed in traditional four colors plus orange. In the part of works, the photographer creates unique images by means of multi- exposure, painting, negatives or painting in the process of scanning, and explores non-realistic themes such as spirit, time, myth and witchcraft. The layout of German and English is very interesting. English text is printed in PMS color orange, while German text is in black. A large number of symbols such as deletion and text advance are used in the text, just like the altered words from the diary, recording the chaotic and real memory. The page numbers are placed at the left bottom of the page, connected by a symbol like '/' with the same tilt as the front cover font.

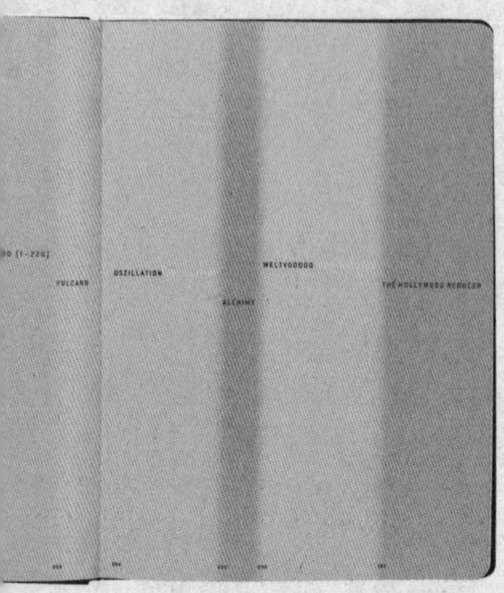
VULCANO
OSZILLATION
THE HOLLYWOOD REDUCER

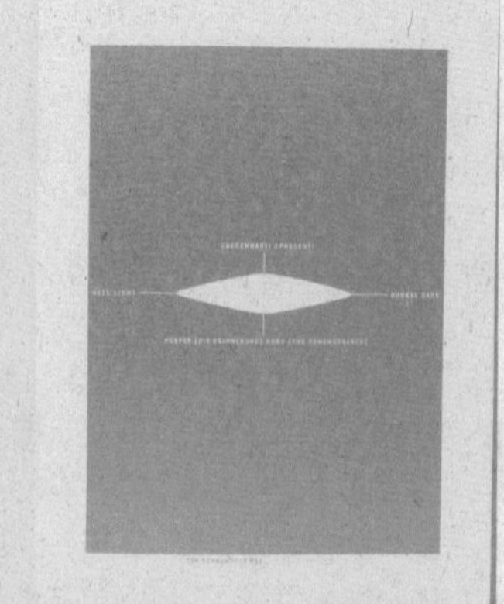

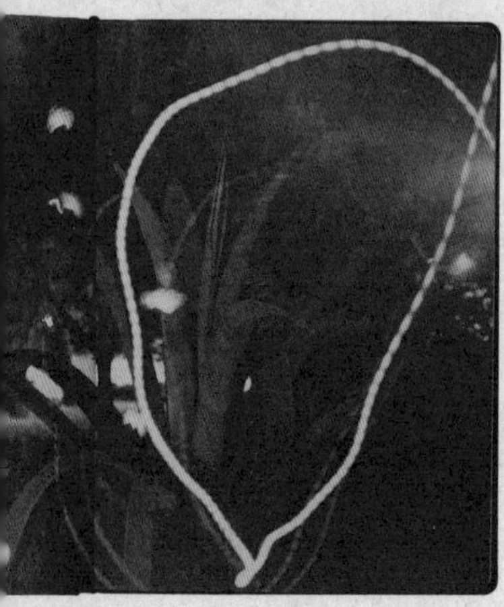

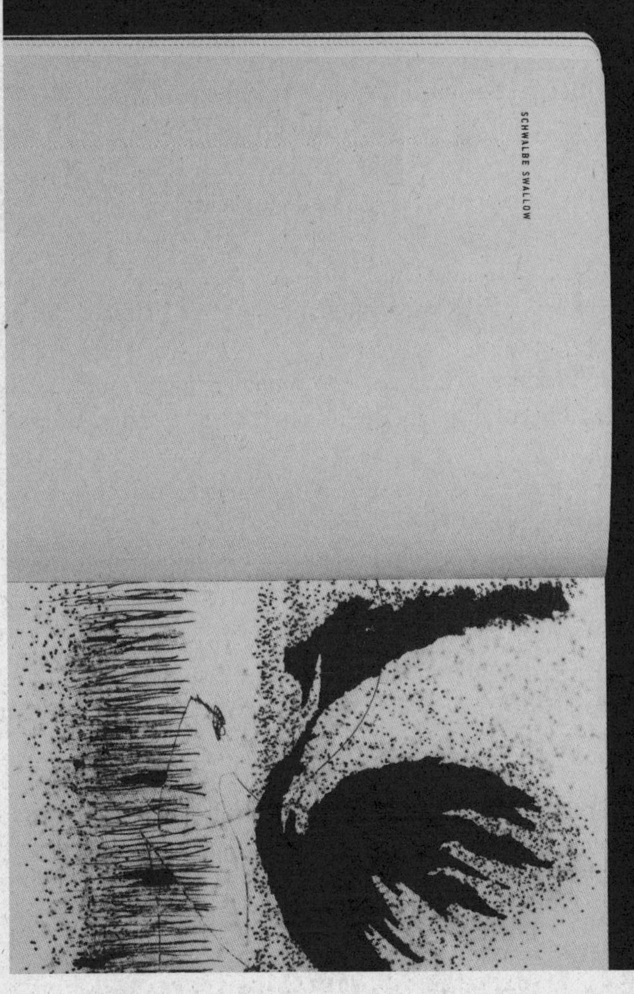
SCHWALBE SWALLOW

schen sind Mailboxen. Sie erschlägt die unterste Stufe, wie k
cht neidisch werden, um zu verstehen, was man ersteht, w
ein.
eißen, der Willensimpulse unterschiedlichen Grades, den Z
heit erreichen? Erreichen, wo doch so unterschiedliche Me
en, wo die halbe Gewalt lebt, wo der Wurf noch offen ist.
eber. Im Pulsspender. Offen sind die *Pulsadern*.

Hymnisch, fast singend.
uls. Reflexion Perfektion plus Impuls. Einst hatte ich ihn de
EGO, Freund.
war ein Freund, zu Gast auf den Wegen meiner Träume. Ein
er die zu Freunde meiner Kindheit hat ich einst den Alter EG
nd.
er, ich konnte einfach träumen und in meinen Träumen, sa
n Nebel aus dem Leben.
tz Abschnitt den man liebte. Irgendwann fand ich, daß es
ich gibt und Atom diesen Momentes, warst du nicht mehr de
gen augenblicklich trennten sich unsere Blicke. ~~[Atom ist irg~~
~~ches Wort, Atmung ist besser.]~~
~~nder, wahrer Freund meiner Träume, verstanden versteckt, e~~
~~ckt, in dieser sublimen Stunde zu zweit.~~

se ich nun in A's Zeitschrift. Wie stehen die Dinge? Gibt es et
n einer gewissen Weise sollte man verschiedene Sprachen spr
h also nun beginnen, was ist anfängliche Ursache, trete ich zu
heint es manchmal.
hlage ich also am Morgen die Zeitschrift auf und lese, »Der I

085–088 >OSZILLATION< MONTAG BIS SONNTAG POLYESTER,

091 >CUORE< COLOR FOTOGRAFIE

093 >EXPERIMENTELLE ZEIT< COLOR FOTOGRAFIE

095 >ALCHIMY< AUSSCHNITTE AUS 9 ARBEITEN, POLYESTER, GLA

BOLTZMANN 142×93 CM, SCHWESTER 141×97 CM, LIMB

WISSEN 121×93 CM, FLUID 145×100 CM, PNEUMA HAGI

098+099 >WELTVOODOO< STOFFPUPPE, CA. 201×113×30 CM

101 WANDBILD (MEISTER NAM UND SEINE TOCHTER)

104 >MARILYN MANSON< SAMT UND GUMMI, 148×145 CM

105 >OSAGE ORANGE< SAMT UND GUMMI, 95×151 CM

106 >YGGDRASIL< SAMT UND GUMMI, 100×162 CM

107 >MILLENNIUM< SAMT UND GUMMI, 100×160 CM

siplus's last words.

in the echo sounding, solus, soluti
s wood. Fear creeps up, *hair* stand

to say. The danger is in having. So t
e world. Discarded there in such a w
e the dogs are bellowing like the hu
n animal, now half awake, eyeing
night, half monkey, one sided br
ng down. The point in the rustling
comes from the trees. Remember?

one's self.

ATION

WELTV

ALCHIMY

be Angst vor jemandem der An
Die Gefahr sitzt im Haben. Ni
m Weltenrand. Dort dermaße
tive. Doch irgendwo klingt das
en. Gib Acht [8!]. Dort sitzt ei
chlich seine Disparitäten. Lem
Gehirntier, denkend mit dem A
Blätter. Das Papier für die Zeit
nerst dich?

Tun an sich,
tun an den Anderen.
Tun über die Anderen durch sic

Nr. 1 Nr. 2 Nr. 3 Nr. 4

Aus H a n n d o R a u h s Arbeit über den Heiligenschein (1908). Nr. 1 »Das Ohr« Nr. 2 »Das Thier« Nr. 3 »Der Schläfer im grünen Thal« Nr. 4 »Das Kreuz im Gebirge«

»Habe Angst vor jemandem der Angst hat.«: pflegte schon irgendwer zu
gen. Die Gefahr sitzt im Haben. Nimm also die *Beine* in die *Hand* und flü
te zum Weltenrand. Dort dermaßen abgestellt, verläßt ein schnell die
spektive. Doch irgendwo klingt das Bellen der Hunde, wie das Summen
Bienen. Gib Acht [8!]. Dort sitzt ein Tier nun halb erwacht, beschaut d
tatsächlich seine Disparitäten. Lemur, Geburt der Nacht, halb Affe, ein
tiges Gehirntier, denkend mit dem *Kopf* nach unten. Der Punkt im Rausc
der Blätter. Das Papier für die Zeitschriften kommt von den Bäumen.
erinnerst dich?

Tun an sich,
tun an den Anderen.
Tun über die Anderen durch sich selbst.

Es ist ein Traum, es muß sich finden für diese Nacht:

Ein sanfter Schneefall,
dort der Tod als Zeittreiber,
jedoch Eva und Maria ereilt das selbe Schicksal.

Warum? ~~[Y?]~~ Sei ruhig mein kleiner Freund, es ist nur ein Augenbl
Allezeit.

Es ist das beste Wetter im Botanischen Garten. Das Telefon klingelt, ich g
heran: »Und?«: frage ich, »Weit und breit gibt es nichts Heiliges!«, »Wer is
der da spricht?«, »Ich kenne ihn nicht.«

Doch eigentlich sind Schmetterlinge fliegende Blätter. Erst Zeitsch
dann Tier, dann Blume in der Tarnung. Die Leichtigkeit des Pergaments,
der Flügelschlag des Blickes der Pfauenaugen am Tag ist ohne Impuls
einfach so im Wind. Der Blick schaukelt über Schaukelwinde, lassen die
terschiedlichen Situationen und erhöhen eines Punktes doch einfach an
beiden Seiten im Schwalbenschwanz. Blau bleibt blau. Der Blick passiert
Drachentor! Der Stuhl ist ein Stuhl und die Schaukel schaukelt. Rosa
doch eine Kunstblume.

If/Then
design implications of new media

6.1
play

1999
Netherlands Design Institute

Dfl 69
US $29.95

IF/THEN PLAY

2000 Bm4

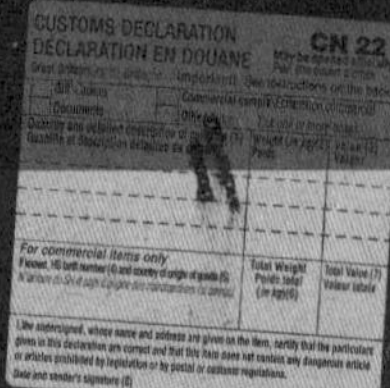

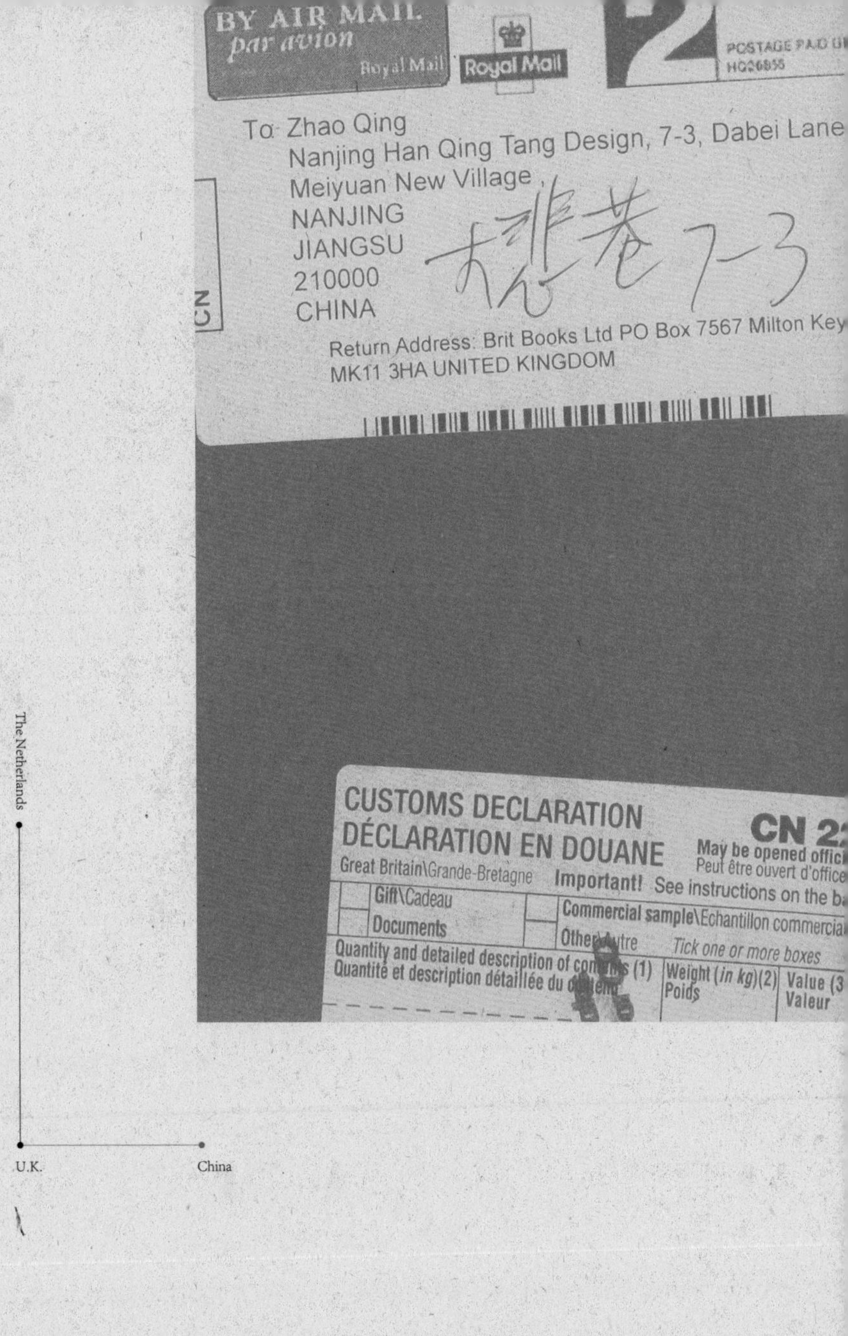
BY AIR MAIL
par avion
Royal Mail
Royal Mail
POSTAGE PAID
To: Zhao Qing
Nanjing Han Qing Tang Design, 7-3, Dabei Lane
Meiyuan New Village,
NANJING
JIANGSU
210000
CHINA
CN
Return Address: Brit Books Ltd PO Box 7567 Milton Key
MK11 3HA UNITED KINGDOM
CUSTOMS DECLARATION
DÉCLARATION EN DOUANE
Great Britain\Grande-Bretagne
Important! See instructions on the b
May be opened offic
Peut être ouvert d'office
Gift\Cadeau
Documents
Commercial sample\Echantillon commercia
Other\Autre
Tick one or more boxes
Quantity and detailed description of contents (1)
Quantité et description détaillée du contenu
Weight (in kg)(2)
Poids
Value (3
Valeur
The Netherlands
U.K.
China

276mm

21mm

IF/THEN

PLAY

BIS

217mm

铜奖
如果 / 那么——新媒体的设计意义
荷兰

Bronze Medal
If / Then: Design implications of new media
The Netherlands
Mevis & Van Deursen
——
Netherlands Design Institute; BIS Publishers, Amsterdam
276 217 21mm
1027g
264p
ISBN-10 90-72007-52-2

这本针对新媒体新技术讨论的前沿书籍，集中了大量有关新技术对当代文化影响的研究。本书的装订形式为锁线胶平装。封面白卡纸四色印刷，覆哑膜。封面封底采用了巨大的勒口，增加了封面的厚度。内页胶版纸五色印刷，除 CMYK 外，第五色使用灵活，在不同折手的印刷正反面分别使用不同的专色，使得内页鲜明的“提醒色”丰富多变。书籍的内容分为四部分：1. 欧盟准备研发的信息接口对普通人生活可能造成的影响；2. 荷兰设计院 1998 年的新媒体主题讨论会；3. 荷兰近期有关设计和新媒体的建议；4. 游戏和相关游戏界面的研究。本书把大量新媒体的元素和理念运用于书籍设计当中，除了页眉的文章标题和页码信息外，正文部分并无固定版式，大部分页面设置为四栏，文本在四栏间灵活排版。多种字体与配色混合使用，形成了丰富的页面效果。来自显示器的“扫描线”和网格留存，给书籍带来了新的视觉体验。

This cutting-edge book on new media and new technology focuses on a large number of studies on the impact of new technology on contemporary culture. This book is perfect bound with thread sewing. The cover is made of white cardboard printed in four colors, mounted with matte film. The front cover and the back cover with huge flaps increase the thickness of the cover. In addition to CMYK four colors, the fifth color is used flexibly. Different PMS colors are used on different folding pages, which make the distinct ‘reminder color’ of the inner pages rich and variable. The contents of the book are divided into four parts: 1. The possible impact of the information interface designed by the European Union on ordinary people’s lives; 2. The new media symposium held by the Dutch Design Institute in 1998; 3. The recent suggestions on design and new media in the Netherlands; 4. The game and related game interfaces. In this book, a large number of new media elements and concepts are applied to the book design. Except for the article title and page number information in the header, there is no fixed format for the internal text. Most pages are set up in 4 columns, and the text is flexibly arranged in 4 columns. A variety of fonts and colors are mixed to form a rich page effect. The ‘scan line’ and grid retained from the display bring new visual experience to books.

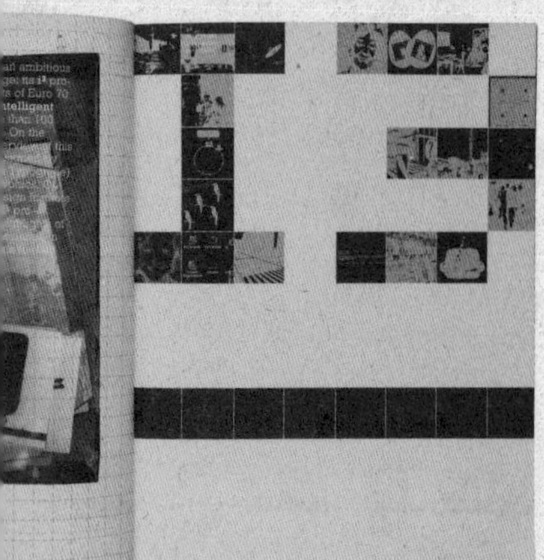

QUANDO SONO
IN CAMPAGNA
DA SOLO
Presence: (inter)active roles for elderly people

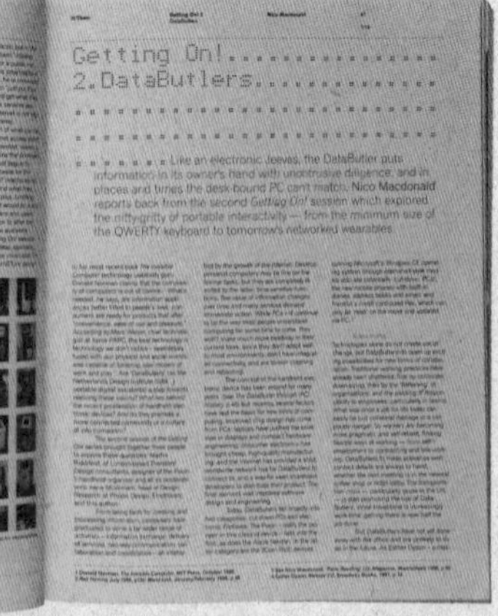
Getting On!
2.DataButlers
Like an electronic Jeeves, the DataButler puts information in its owner's hand with unobtrusive diligence, and in places and times the desk-bound PC can't match. Nico Macdonald reports back from the second Getting On! session which explored the nitty-gritty of portable interactivity — from the minimum size of the QWERTY keyboard to tomorrow's networked wearables.

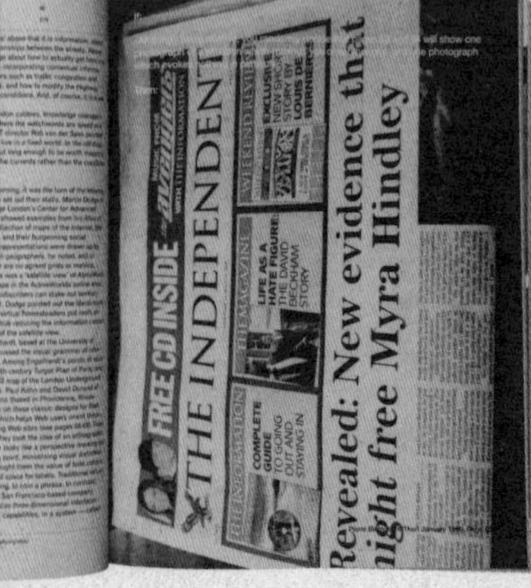
FREE CD INSIDE
THE INDEPENDENT
COMPLETE GUIDE TO GOING OUT AND STAYING IN
LIFE AS A HATE FIGURE THE DAVID BECKHAM STORY
EXCLUSIVE NEW SHORT STORY BY LOUIS DE BERNIERES
Revealed: New evidence that
ight free Myra Hindley

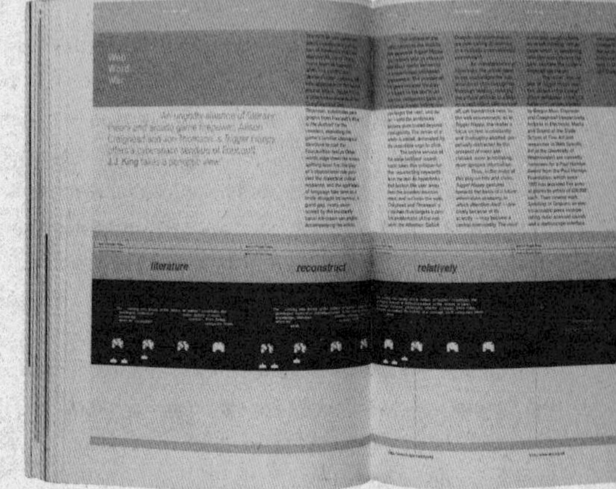
literature
reconstruct
relatively

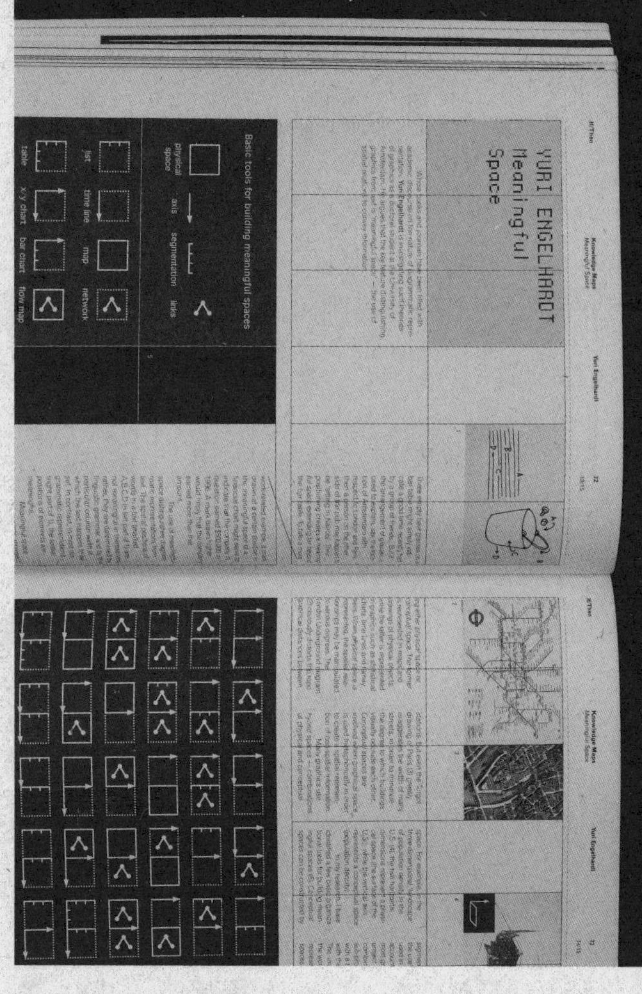
Knowledge Maps
Meaningful Space
Yuri Engelhardt
YURI ENGELHARDT
Meaningful
Space
Basic tools for building meaningful spaces
physical space
axis
segmentation
links
list
time line
map
network
table
x/y chart
bar chart
flow map

n outer space

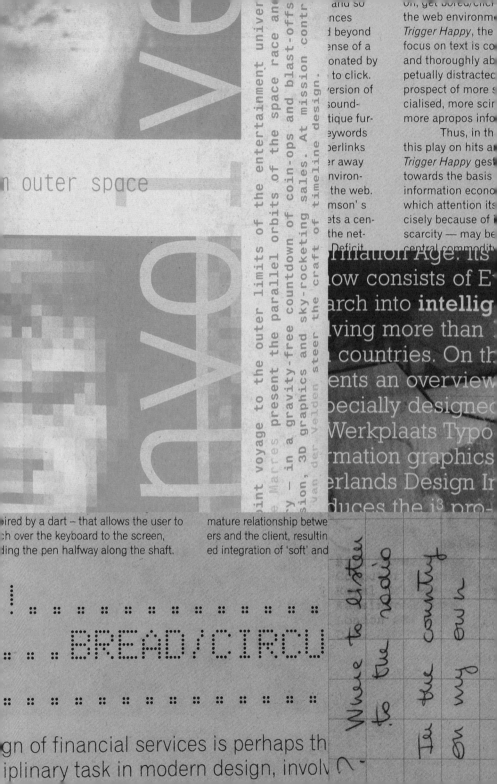

oint voyage to the outer limits of the entertainment univer
e Marres present the parallel orbits of the space race an
ry – in a gravity-free countdown of coin-ops and blast-offs
sion, 3D graphics and sky-rocketing sales. At mission contr
van der Velden steer the craft of timeline design.

and so
nces
d beyond
ense of a
onated by
to click.
version of
sound-
tique fur-
eywords
perlinks
er away
nviron-
the web.
mson' s
ets a cen-
the net-
Deficit

on, get bored/clic
the web environm
Trigger Happy, the
focus on text is co
and thoroughly ab
petually distracted
prospect of more s
cialised, more scir
more apropos info

Thus, in th
this play on hits a
Trigger Happy gest
towards the basis
information econo
which attention its
cisely because of i
scarcity — may be
central commodity

rmation Age. Its
ow consists of E
arch into **intellig**
lving more than
countries. On th
ents an overview
pecially designed
Werkplaats Typo
rmation graphics
erlands Design In
duces the i^3 pro-

ired by a dart – that allows the user to
ch over the keyboard to the screen,
ding the pen halfway along the shaft.

mature relationship betwe
ers and the client, resultin
ed integration of 'soft' and

BREAD/CIRCU

gn of financial services is perhaps th
iplinary task in modern design, involv

Where to listen
to the radio
In the country
on my own
?

Getting On! 3 BREAD/CIR

The design of financial services is perh
abstract and multi-disciplinary task in modern design
ware designers, psychologists and economists, in ad
tional graphic skills. In a fast-changing networked ec
stakes are commensurately high. Jules Marshall was
Getting On! session to hear experts on e-commerce a
cation researchers forecast the future of banking.

Lara

Hantverket i gamla hus

2000 Bm^5

Sweden
Germany
China
CN22
Deutsche Post
Documents
Others
PRIORITY P.P.
FROM
Theisenort 25
96231 Bad Staffelstein
GERMANY
TO
ZHAO QING
QING TANG DESIGN CO.,
DABEI XIANG7-3,
210000 NANJING
NANJING HAN
CHINA
Weight (in kg)
Value (in EUR)
0.90
20.00
Total Weight
Total Value
0.90
20.00

280mm

20mm

185mm

铜奖
老房子的工艺
瑞典

Bronze Medal
Hantverket i gamla hus
Sweden
Dick Norberg
——
Byggförlaget, Stockholm
280 185 20mm
 764g
208p
ISBN-10 91-7988-176-9

瑞典建筑协会编写的传统房屋建造理念以及工艺指导用书，一方面对传统建造工艺进行记录和保护，另一方面也是为了提高从业工匠的整体水平。本书的装订形式为圆脊锁线硬精装。护封采用胶版纸四色印刷，内封采用橙色哑面艺术纸，书脊处书名烫黑处理。环衬选用了带有丰富杂质的手工纸，映衬传统工艺的主题。内页采用胶版纸四色印刷。全书内容除开篇的数篇引言和综述，共有 10 个具体的工艺篇章，分别为：石材工艺、墙体砌修、木质结构、精细木工、玻璃工艺、涂料粉刷、五金锻造、钣金加工、装饰浮雕、瓷砖壁炉。作为一本记录传统工艺并期望起到传承作用的工具书，本书的印制工艺十分朴实，但在内容编辑和设计细节的把握上并不简陋。每部分均以满版大图开篇，正文详述工艺简介、历史、传统案例、加工细节和注意事项。正文版心设置为双栏，照片和插图的排版十分灵活。正文段落层级复杂，但并没有使用编号，而是对各级标题进行了多样化设置。

This is the traditional house construction concept and craft guide book written by the Swedish Architectural Association. On the one hand, it records and protects the traditional construction crafts, and on the other hand, it aims to improve the overall level of craftsmen. The hardcover book is bound by thread sewing with rounding spine. The jacket is made of offset paper printed in four colors. Orange matte art paper is used for the inner cover, and the title is stamped in black at the spine. The end paper is made of handmade paper with impurities to match the theme of traditional technology. The inner pages are printed with offset paper in four colors. In addition to the introduction and summary of the opening chapters, the book is divided into 10 specific chapters, including stone craft, wall masonry, wooden structure, fine woodworking, glass craft, paint painting, metal forging, sheet metal Processing, decorative relief, tile fireplace. As a reference book recording traditional craftsmanship and expecting to play a role of inheritance, the printing process of this book is very simple, but it is not simple in content editing and design details. Each part starts with a full-scale picture, and the main body details the introduction, history, traditional cases, processing details and precautions. The typeset area is set to two columns, but the layout of photos and illustrations is very flexible. The levels of the text paragraph are complex. Instead of using the numbers, the designer diversifies the headings at all levels.

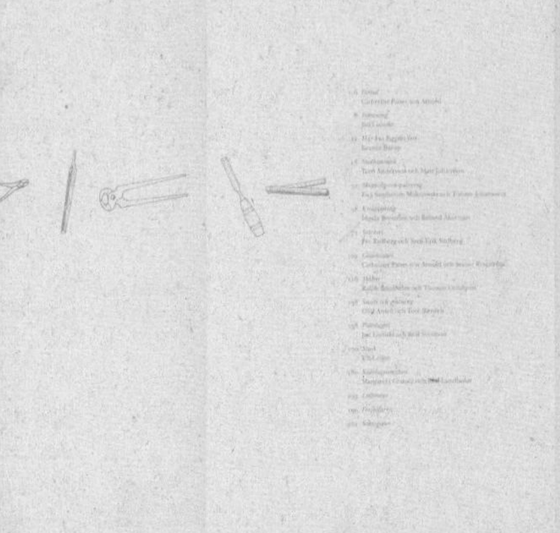

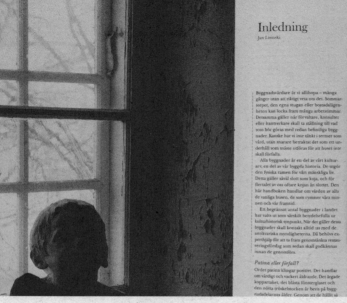

Inledning

Jan Lisinski

Byggnadsvårdare är vi allihopa – många gånger utan att riktigt veta om det. Sommartorpet, den egna stugan eller bostadslägenheten kan locka fram många arbetstimmar. Detsamma gäller när förvaltare, konsulter eller hantverkare skall ta ställning till vad som bör göras med redan befintliga byggnader. Kanske har vi inte tänkt i termer som vård, utan snarare betraktat det som ett underhåll som måste utföras för att huset inte skall förfalla.

Alla byggnader är en del av vårt kulturarv, en del av vår byggda historia. De utgör den fysiska ramen för vårt mänskliga liv. Detta gäller såväl slott som koja, och för flertalet av oss oftare kojan än slottet. Den här handboken handlar om vården av alla de vanliga husen, de som rymmer våra minnen och vår framtid.

Ett begränsat antal byggnader i landet har valts ut som särskilt betydelsefulla ur kulturhistorisk synpunkt. När det gäller dessa byggnader skall kontakt alltid tas med de antikvariska myndigheterna. Då behövs experthjälp för att ta fram genomtänkta restaureringsförslag som sedan skall godkännas innan de genomförs.

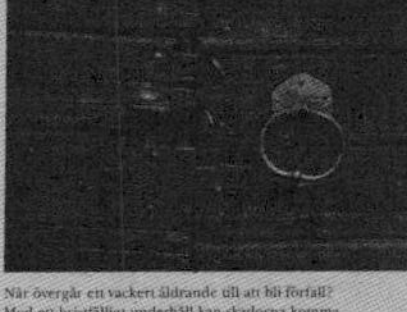

När övergår ett vackert åldrande till att bli förfall? Med ett bristfälligt underhåll kan skadorna komma snabbt.

Patina eller förfall?

Ordet patina klingar positivt. Det handlar om värdigt och vackert åldrande. Det ärgade koppartaket, det blåsta fönsterglaset och den nötta tröskelstocken är bevis på byggnadsdelarnas ålder. Genom att de hållit så länge förmedlar de en känsla av kvalitet. De är minsann inget fuskbygge, utan väl kända material bearbetade av duktiga hantverkare.

Patina har med både åldrande och slitage att göra. Men åldrande och slitage kan över-

torpet, den egna stugan eller bostadslägenheten kan locka fram många arbetstimmar. Detsamma gäller när förvaltare, konsulter eller hantverkare skall ta ställning till vad som bör göras med redan befintliga byggnader. Kanske har vi inte tänkt i termer som vård, utan snarare betraktat det som ett underhåll som måste utföras för att huset inte skall förfalla.

Alla byggnader är en del av vårt kulturarv, en del av vår byggda historia. De utgör den fysiska ramen för vårt mänskliga liv. Detta gäller såväl slott som koja, och för flertalet av oss oftare kojan än slottet. Den här handboken handlar om vården av alla de vanliga husen, de som rymmer våra minnen och vår framtid.

Ett begränsat antal byggnader i landet har valts ut som särskilt betydelsefulla ur kulturhistorisk synpunkt. När det gäller dessa byggnader skall kontakt alltid tas med de antikvariska myndigheterna. Då behövs experthjälp för att ta fram genomtänkta restaureringsförslag som sedan skall godkännas innan de genomförs

Inledning

Jan Lisinski

och dras-
tt husen
nuerligt
rder
. Tillsyn
grund-
er för-

er livs-
en ett
ingom

ler husvård. Risken är faktiskt att vi vårdar ihjäl våra byggnader.

Använd beprövad hantverksteknik och traditionella material

Allt var inte bättre förr, men de byggnadsdelar som ännu håller och fungerar har själva bevisat sitt existensberättigande. Dessutom kan man på en äldre byggnad se en mängd detaljer som inte är utförda enligt dagens byggnorm – men som ändå håller. Uppenbarligen finns det alternativ till normerna. Tiden är där den bäste domaren.

När vattbrädan nedtill på träfasaden är

Att klyva en sten

1. Borra två hål per dr
2. Sätt i kil och bleck
3. Spräck med slägga
4. Bänd isär med spett

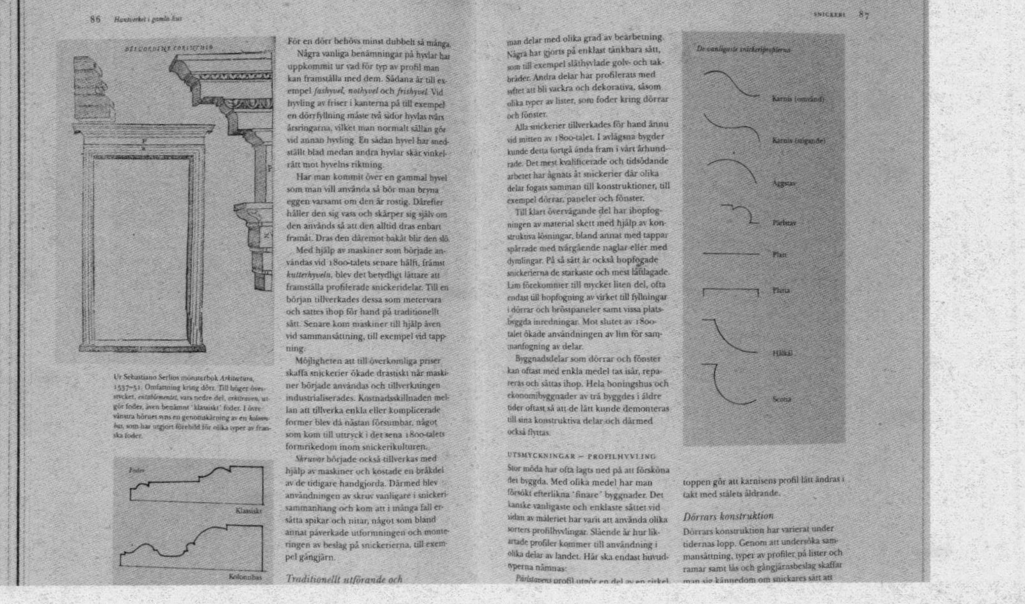

Ur Sebastiano Serlios mönsterbok *Arkitetura*, 1537–51. Omfattning kring dörr. Till höger överstycket, *entablementet*, vars nedre del, *arkitraven*, utgör foder, även benämnt 'klassiskt' foder. I övre vänstra hörnet syns en genomskärning av en *kolonnbas*, som har utgjort förebild för olika typer av franska foder.

De vanligaste snickeriprofilerna

För en dörr behövs minst dubbelt så många.

Några vanliga benämningar på hyvlar har uppkommit ur vad för typ av profil man kan framställa med dem. Sådana är till exempel *fashyvel*, *nothyvel* och *frishyvel*. Vid hyvling av friser i kanterna på till exempel en dörrfyllning måste två sidor hyvlas tvärs årsringarna, vilket man normalt sällan gör vid annan hyvling. En sådan hyvel har snedställt blad medan andra hyvlar skär vinkelrätt mot hyvelns riktning.

Har man kommit över en gammal hyvel som man vill använda så bör man bryna eggen varsamt om den är rostig. Därefter håller den sig vass och skärper sig själv om den används så att den alltid dras enbart framåt. Dras den däremot bakåt blir den slö.

Med hjälp av maskiner som började användas vid 1800-talets senare hälft, främst *kutterhyveln*, blev det betydligt lättare att framställa profilerade snickeridelar. Till en början tillverkades dessa som metervara och sattes ihop för hand på traditionellt sätt. Senare kom maskiner till hjälp även vid sammansättning, till exempel vid tappning.

Möjligheten att till överkomliga priser skaffa snickerier ökade drastiskt när maskiner började användas och tillverkningen industrialiserades. Kostnadsskillnaden mellan att tillverka enkla eller komplicerade former blev då nästan försumbar, något som kom till uttryck i det sena 1800-talets formrikedom inom snickerikulturen.

Skruvar började också tillverkas med hjälp av maskiner och kostade en bråkdel av de tidigare handgjorda. Därmed blev användningen av skruv vanligare i snickerisammanhang och kom att i många fall ersätta spikar och nitar, något som bland annat påverkade utformningen och monteringen av beslag på snickerierna, till exempel gångjärn.

Traditionellt utförande och

man delar med olika grad av bearbetning. Några har gjorts på enklast tänkbara sätt, som till exempel släthyvlade golv- och takbräder. Andra delar har profilerats med syftet att bli vackra och dekorativa, såsom olika typer av lister, som foder kring dörrar och fönster.

Alla snickerier tillverkades för hand ännu vid mitten av 1800-talet. I avlägsna bygder kunde detta fortgå ända fram i vårt århundrade. Det mest kvalificerade och tidsödande arbetet har ägnats åt snickerier där olika delar fogats samman till konstruktioner, till exempel dörrar, paneler och fönster.

Till klart övervägande del har ihopfogningen av material skett med hjälp av konstruktiva lösningar, bland annat med tappar spårrade med tvärgående naglar eller med dymlingar. På så sätt är också hopfogade snickerierna de starkaste och mest lättlagade. Lim förekommer till mycket liten del, ofta endast till hopfogning av virket till fyllningar i dörrar och bröstpaneler samt vissa platsbyggda inredningar. Mot slutet av 1800-talet ökade användningen av lim för sammanfogning av delar.

Byggnadsdelar som dörrar och fönster kan oftast med enkla medel tas isär, repareras och sättas ihop. Hela boningshus och ekonomibyggnader av trä byggdes i äldre tider oftast så att de lätt kunde demonteras till sina konstruktiva delar och därmed också flyttas.

UTSMYCKNINGAR – PROFILHYVLING

Stor möda har ofta lagts ned på att försköna det byggda. Med olika medel har man försökt efterlikna 'finare' byggnader. Det kanske vanligaste och enklaste sättet vid sidan av måleriet har varit att använda olika sorters profilhyvlingar. Slående är hur likartade profiler kommer till användning i olika delar av landet. Här ska endast huvudtyperna nämnas:

Pärlstavens profil utgör en del av en cirkel

toppen gör att karnisens profil lätt ändras i takt med stålets åldrande.

Dörrars konstruktion

Dörrars konstruktion har varierat under tidernas lopp. Genom att undersöka sammansättning, typer av profiler på lister och ramar samt lås och gångjärnsbeslag skaffar man sig kännedom om snickares sätt att

4.
7.
F. 39. B
F. 40. B
F. 42. B
Fig. 43. A
2:a
Distanskil
Solöppning
1:a
H.

Slätputs på Örgryte gamla kyrkas torn, uppfört 1784. Fasaderna putsades om 1981 med en slätputs som följer murens liv och ojämnheter.

Ädelputs/konstputs från 1920-talet. Puts med hydrauliskt bindemedel och grov ballast som innehåller glimmer, vilket ger putsytan drag av natursten.

matiskt plan puts finns på modernismens 'abstrakta' murskivor eller på det sena 1700-talets nyklassicistiska fasader.

Finputs är en ytputs uppbyggd av ballast med fin kornstorlek på 1–2 mm. Den appliceras i tunna, 2–5 mm skikt och ger en slät och jämn yta. För att kunna bearbeta finputsytan med en rivbräda utan att rispa underlaget, bör sanden helst vara rundkornig och kornstorleken inte överskrida en tredjedel av påslagets tjocklek.

Utsmyckningsdetaljer, lister, pilastrar och fönsteromfattningar på fasaderna brukar vara putsade med finputs. Även väggytor inomhus brukar vara putsade med finputs. Utvändigt får inte putsen efterarbetas eller glättas alltför intensivt. Bearbetningen pressar in sandkornen och anrikar putsytan med bindemedel så att en hård 'skorpa' bildas. Ytskiktet kan då spjälkas loss när frosten biter till.

Vattrivning praktiseras sällan idag. Det är ett sätt att få ytan på putsade innerväggar jämn och slät. Ett mycket tunt lager finputs påförs och 'vattrives' genom att ytan gnids med en rivbräda medan vatten stänks på med murarkvasten. Kalken i bruket tränger ut till putsytan och bildar en hårdare ythud. Ännu slätare yta får man genom filtning med en filtbeklädd bräda.

SPRITPUTS

Spritputsen är en bindemedelsrik ytputs innehållande grovkornig sand eller grus som ger karaktär åt ytan. Spritputsens yta lämnas obearbetad. De allra äldsta spritputserna är mindre 'steniga' och innehåller mindre grov ballast än de som blev vanliga under 1800-talet och senare. Ballasten i 1800-talets puts består av naturligt grus med mycket grova partiklar – diametern kan

Grov spritputs med stora, naturrunda stenar som är tätpackade. Detta är den ursprungliga putsen på *Lorensbergsteatern* i Göteborg, uppförd 1916.

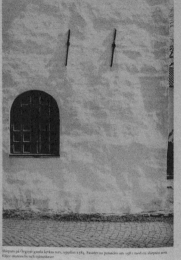

PUTS SOM EFTERLIKNAR BEARBETAD STEN

Puts kan genom olika utföranden ge illusion av huggen och bearbetad sten, polerad marmor, sågad travertin, huggen granit, sandsten eller fantasistenarter.

'Stenblocken' formas vanligen till kvad-

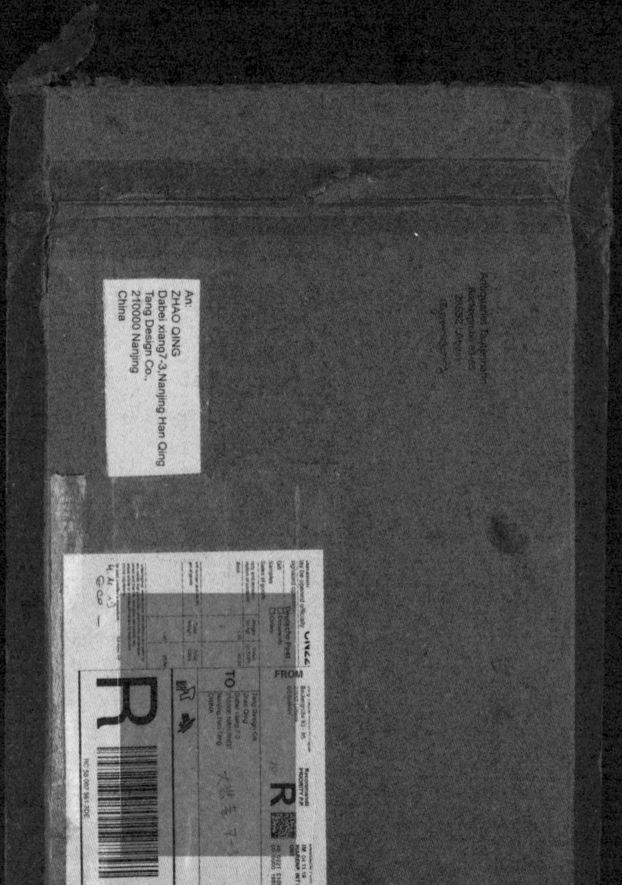

2000

1 Ha²

An:
ZHAO QING
Dabei xiang7-3,Nanjing Han Qing
Tang Design Co.,
210000 Nanjing
China

Deutsche Post

Documents

Others

Weight (in kg)	Value (in EUR)
1.90	80.00

Total Weight	Total Value
1.90	80.00

FROM

GERMANY

TO

Tang Design Co.
Zhao Qing
Dabei xiang 7-3
210000 NANJING1
Nanjing Han Qing
CHINA

R

Germany

Germany

China

285mm

30mm

W.

W.

216mm

● 荣誉奖 Honorary Appreciation

« 从纤细到特粗——格特 · 冯德利希 1957－1998 年的字体排印 Von zart bis extrafett: Typo-Grafik von Gert Wunderlich 1957-1998

◖ 德国 Germany

Gert Wunderlich

Gert Wunderlich (Hrsg.)

Eigenverlag Gert Wunderlich, Leipzig

285 216 30mm

1385g

256p

ISBN-10 3-932865-16-2

格特 · 冯德利希（Gert Wunderlich）是德国著名的平面设计师和设计教育家。本书是他在 1957 — 1998 年间字体、海报、图形和书籍设计的作品集。书籍的装订形式为锁线硬精装。封面采用暗红色艺术纸，做烫白、烫黑和压凹处理，裱覆卡板。封面外有黑色卡纸裱覆黑色卡板制作的书函，做工坚硬但翻折处圆润。环衬选用黑色艺术纸。内页以哑面涂布纸张为主，只在 8 页呈现友人赠送的字画处采用相同纸张的浅灰色版本，四色印刷。内容方面，书籍的前 35 页是友人们谈各自印象中的冯德利希以及对他作品的解读，之后分为 4 个部分对他的字体设计、海报设计、书籍设计和教学工作进行展示。最后是年表和与之相关的信息。设计方面，为了体现冯德利希字体设计师的特点，字体、字号和字重的灵活使用从封面开始贯穿到全书：文章标题、段落首字母、附注数字均使用最粗的字重，跟正文形成鲜明对比。另，本书在2001年出版过简体中文版。

Gert Wunderlich is a famous graphic designer and design educator in Germany. This book is a collection of his works on typeface, poster, graphics and book design from 1957 to 1998. The hardcover book is bound by thread sewing. The cover is made of dark red art paper with white hot stamping, black hot stamping and embossment, mounted with cardboard. The bookcase is made of black cardboard mounted with black canton, which feels hard but smooth. The rend paper is made of black art paper. The inner pages are mainly made of matte coated paper printed in four colors. As for the content, the first 35 pages of the book contain the friends' interpretation of Wunderlich and his works, and then the book is divided into four parts to show Wunderlich's font design, poster design, book design and teaching job. Finally, his chronology and related information are listed. As for the design, the use of font, size and weight is flexible throughout the book: the title of the article, the first letter of the paragraph, and the numbers of notes all use the thickest weight, which forms a sharp contrast with the text. In addition, the book was published in simplified Chinese in 2001.

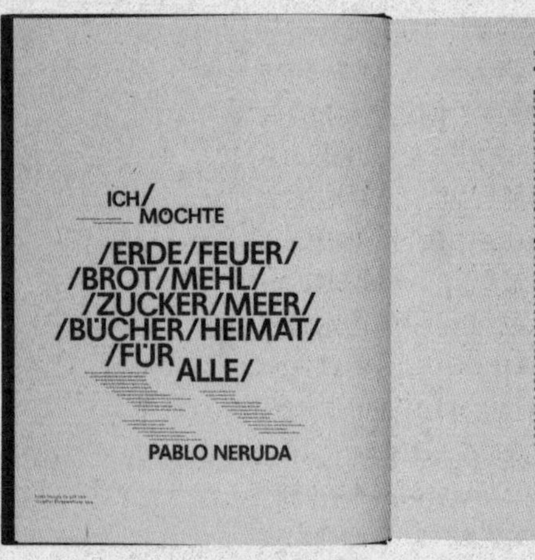

ABCDEFGHIJ
KLMNOPQRSTU
VWXYZ ÄÖÜ
abcdefghijklmnop
qrstuvwxyz
äöü ß chckfffiflft
1234567890
(–.,;!?) »&« [†]
ĆŁŃĘŚŹĄÓ ółś
ćęźńą
uh1

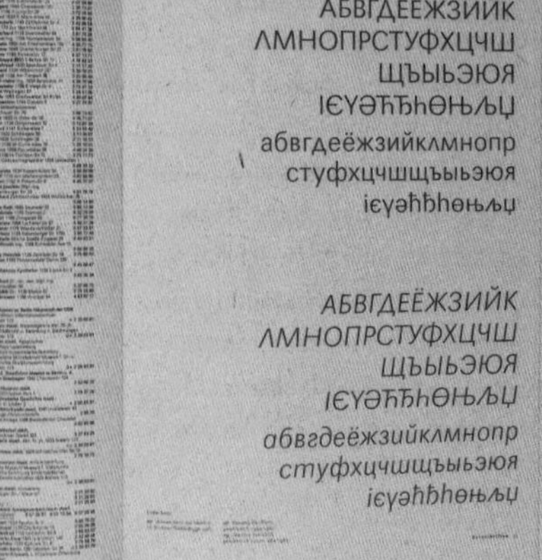

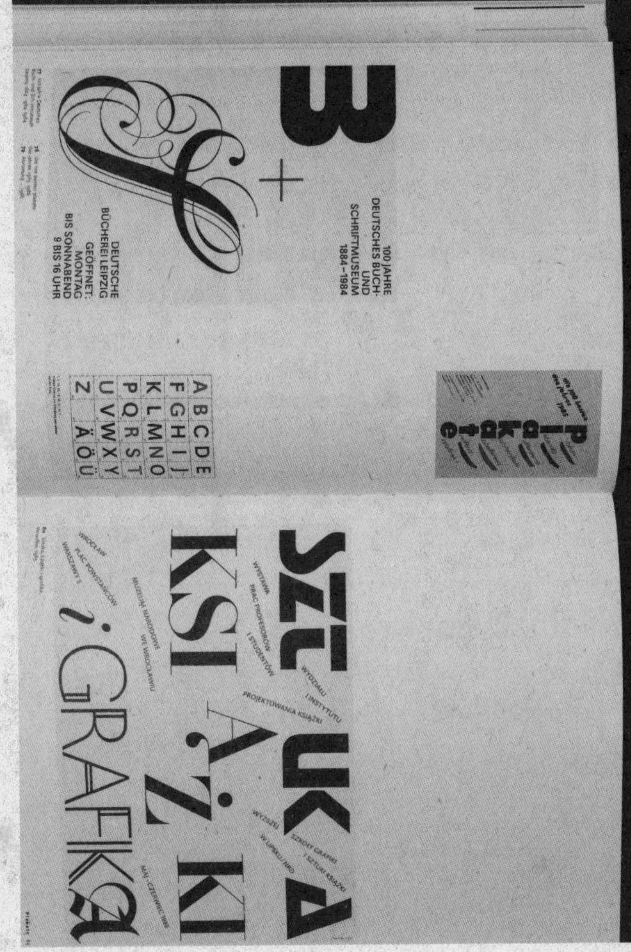
100 JAHRE
DEUTSCHES BUCH-
UND
SCHRIFTMUSEUM
1884–1984
DEUTSCHE
BÜCHEREI LEIPZIG
GEÖFFNET:
MONTAG
BIS SONNABEND
9 BIS 16 UHR
A B C D E
F G H I J
K L M N O
P Q R S T
U V W X Y
Z Ä Ö Ü
SZTUKA
KSIĄŻKI
i GRAFIKA
WYSTAWA
PRAC PROFESORÓW
I STUDENTÓW
WYDZIAŁU
I INSTYTUTU
PROJEKTOWANIA KSIĄŻKI
WYŻSZEJ
SZKOŁY GRAFIKI
I SZTUKI KSIĄŻKI
W LIPSKU/NRD
MUZEUM NARODOWE
WE WROCŁAWIU
WROCŁAW
PLAC POWSTAŃCÓW
WARSZAWY 5

ICH/
MÖCHTE

AHRSCHOR
ND IN DER FINSTERNIS«

ERDE/FEUE
ROT/MEHL/
ZUCKER/ME
CHER/HEIM
/FÜR
ALLE/

g: mich verfolgen die Richter des Verräters,
eichler versuchen gleich abgerichteten
g an mich in einen Sumpf zu verwandeln.
HM, mit DIESEM, der regiert, zum Eingang
ne, in die Wüste der vergessenen Morgenröte,
h ging mit ihm und sprach zu meinen armen Brüdern:
»Ihr werdet nicht die Fasern der zerlumpten Kleidung behalten,
ihr werdet nicht mehr diese Tage haben ohne Brot, ihr werdet behandelt werden,
als wäret ihr des Vaterlandes Kinder.« »Von nun an
wollen wir die Schönheit teilen, und die Augen
der Frauen werden nicht mehr weinen um ihre Söhne.«

Und als man an Stelle ausgestreuter Liebe im Dunkeln
jenem selben Hunger und Marter entlockte,

Ich will mein Land für die Meinen, ich will
das gleiche Licht über dem Scheitel
meines flammenden Landes,
ich will die Liebe des Tages und der Pflugschar haben,
will die Trennungslinie tilgen, die voller Haß
sie schaffen, um das Brot vom Volk zu trenne
und den, der des Heimatlandes Grenz
um es als Kerkermeister auszul
gebunden, denen, die da
Ihn werde ich nicht

und siehe da, es tat ihr gut
und sie rief: Nun bin ich
bereit, sogleich alles Wissen
der Welt in mir aufzunehmen!
Stasingsing trat an den Hebel
der Macht, legte um und
Dopi stieß erreckt aus: Oh

S T A S I N

wie es mich durchströmt,
ach wie mich ein Zittern
ankömmt, whopper, wie mich
gruselt! Mach! oh mach!
oh Macht! ohnmacht! das
Aidsperiment meines Lebens!
Und als sie wieder kam,
wieder kam, da singsing
alle Vöglein, eine neue
Leistungsstufe brach an ...
Dopi verfestigte, verflüssigte
und verfeuerte all die auf sie
eingeströmten Energien
stracks in die lichtverheißene
Zukunft über einer Bahn aus
Asche und war allhier immer
schon da. Am Rande unserer
Geschichte aber stand das
Leben selbst, runzelte nachdenklich das Hirn glatt und
erläuterte: Das hat mit
seinem Singsing das Stasing-

S I N

ЫДЕЕХ
ОПРСТУ
ЩЪЫЬЭ
абßгдеё

anstrengend empfunden. icht durch einen künst- tverwirklichungs- chweren. Die Tugenden istung an Schrift und icht hoch im Kurs: nd Selbstbeschränkung, as Bewußtsein, anonym

leitung soll dem Erken- Rang der Arbeit von h dienen. Zwar haben tionale und inter- und Auszeichnungen er weitgehend nur

auch mit Kunden, Preisen und nen unterscheidet die angew von den freien Künsten. Der unterliegt ähnlichen Gebunde wie ein Regisseur: Er sucht e pretation unter einer Vielzahl wendbarer Gegebenheiten. E nach Brecht – »mit den Köpfe Leute denken« können.

Gert Wunderlich ist in seinem Berufsleben nicht gerade auf Sonnenseite gestalterischer S faltung gewesen. Durch seine bemühungen und Ausstellung initiativen konnten immer wi

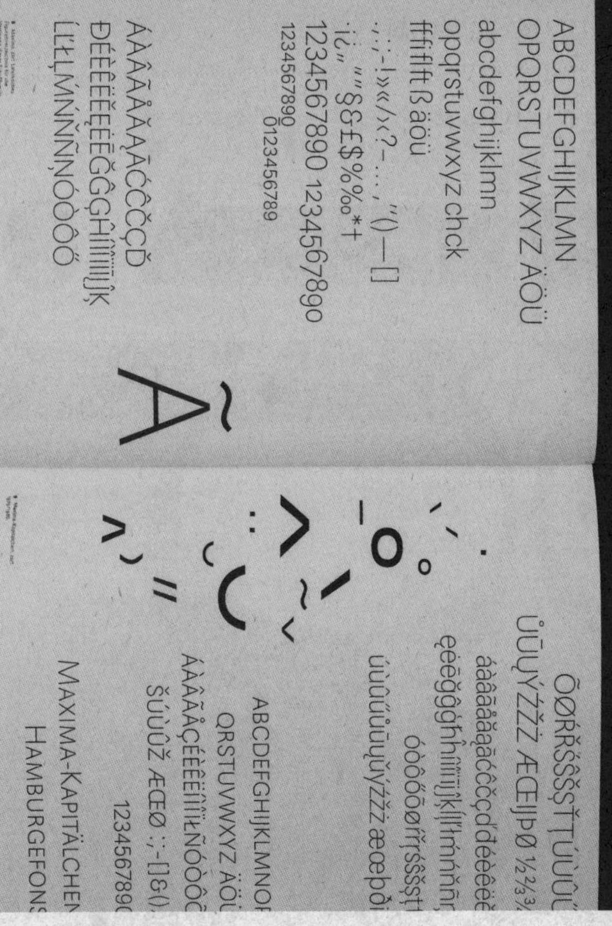
ABCDEFGHIJKLMN
OPQRSTUVWXYZ ÄÖÜ
abcdefghijklmn
opqrstuvwxyz chck
1234567890 1234567890
0123456789

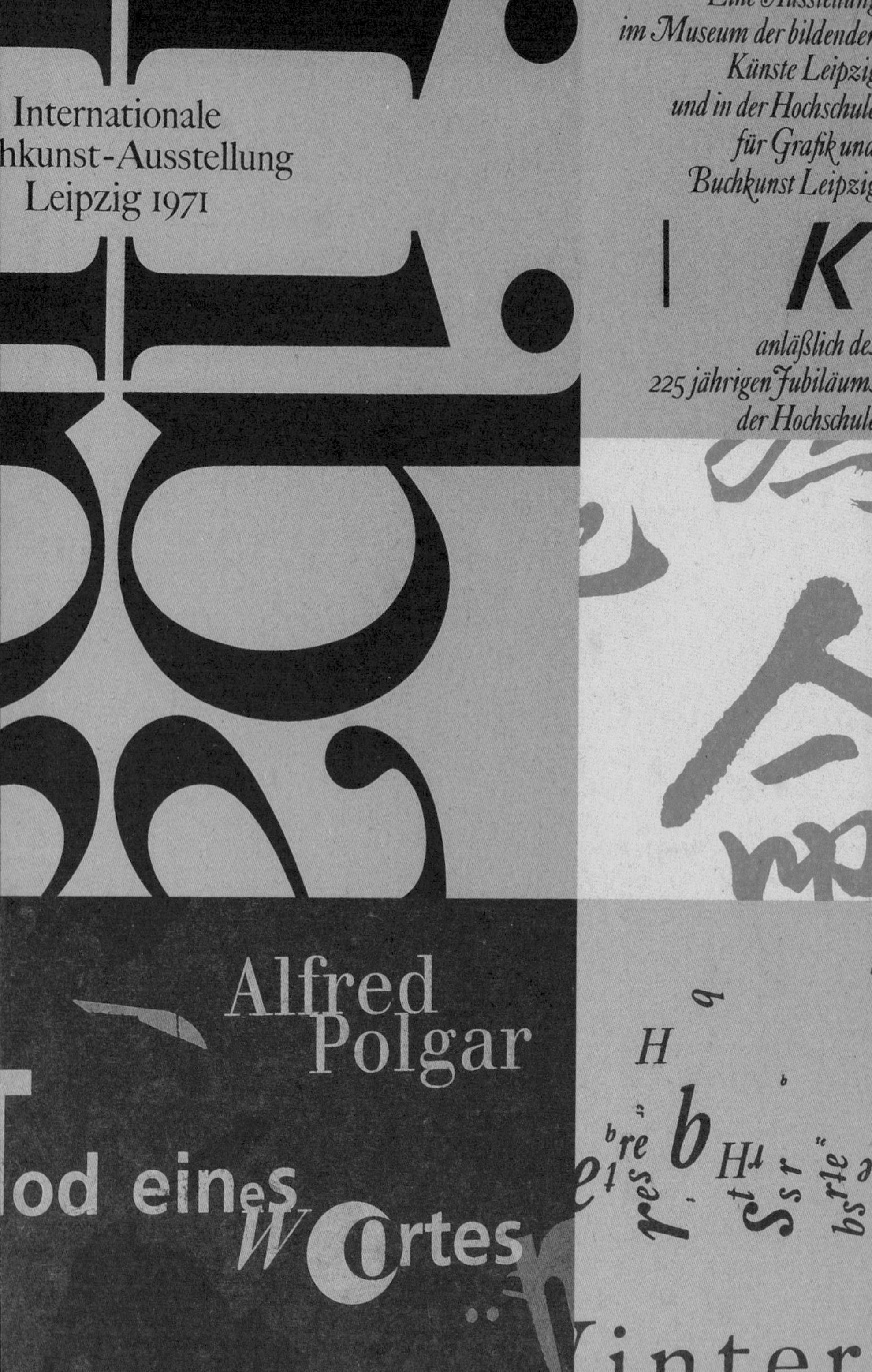
Eine Ausstellung
im Museum der bildenden
Künste Leipzig
und in der Hochschule
für Grafik und
Buchkunst Leipzig
Internationale
hkunst-Ausstellung
Leipzig 1971
K
anläßlich des
225 jährigen Jubiläums
der Hochschule
Alfred
Polgar
od eines
Wortes
inter

Luisenstraße

2000 Ha^3

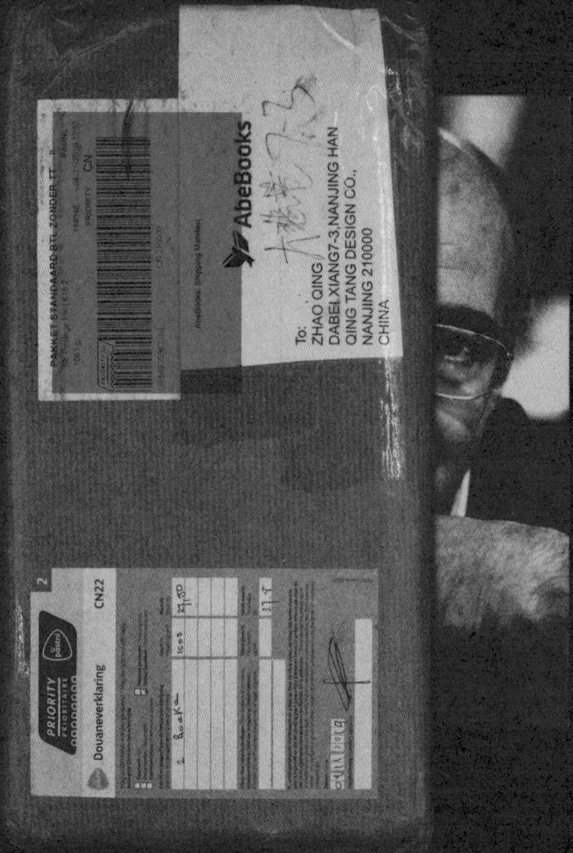

The Netherlands

The Netherlands

China

206mm

55mm

134mm

荣誉奖
诗
荷兰

Honorary Appreciation
De Gedichten
The Netherlands
Steven van der Gaauw
Herman de Coninck
Uitgeverij De Arbeiderspers, Amsterdam/Antwerpen
206 134 55mm
946g
484p / 208p
ISBN-10 90-295-0908-2

赫尔曼·德·科尼克（Herman de Coninck）是比利时诗人、散文家、记者和出版商。他生前曾出版过多部诗集、散文集，本书是他生平作品的合集，一套两本。第一本是他曾经以完整诗集形式出版过的诗，第二本则是分散发表在报纸杂志上的诗，所以上下两本的厚度并不一致。书籍的装订形式为圆脊锁线硬精装。护封胶版纸印暗红色，内封采用红色布面裱覆卡板，书脊处书名烫银。上下两册被收纳于没有卡板的纸制书函内，书函单黑印刷，正反面分别选用了诗人不同时期的照片。内页胶版纸印单黑，微微泛黄的纸色具有良好的阅读感受。红色书签带与封面的红色相呼应。

Herman de Coninck is a Belgian poet, essayist, journalist and publisher. He published many poetry collections and essay collections during his lifetime. This book is a collection of his works, a set of two. The first one is the poems he published in the form of a complete collection, and the second one includes poems distributed in newspapers, magazines and other places. Therefore, the thickness of the two books is not the same. The hard-cover book is bound by thread sewing with rounding back. The cover is made of offset paper printed in dark red, and the inner cover is mounted with red cloth, and the title on the spine is hot stamped in silver. The two volumes are stored in the book case made of paper which are printed in black. On the front side and back side of the case, there are selected photos of the poet from different periods. The inside pages are made of offset paper printed in black, and the slightly yellowish paper makes the readers have good reading experience. The red book-mark band corresponds to the red cover.

k de geheimen van haar lichaam
ır borsten
ıen, alsof ik een beroemdheid
en bij zijn voornaam.
k haar gewoon
e billen.

aar ondefinieerbaarheid
vlijen tot ik ze definieerde
ı ze zei:
angenaam
ich in mijn zonden en ik ging
ier seizoenen en zij speelde
mij en werd een wereldgebeuren
ı en maakte geschiedenis, ik ben

DE LENIGE LIEFDE

ZOMER

De zomer steekt d
De zon lacht al ha
vol stralend zelfve
Het leven is een tr
die me lachend op

man de C
gedichten

esteld en verantwo

ı de bakker, Schoten 1987, H

ıter 1982-83, p. 142 en in F
ıda tussen mythe en werkelijkhe

ı de bakker, Schoten 1987, H

n Frans Boenders (ed.), Kur
eid, Brussel 1984, BRTN, p. 7

ı de bakker, Schoten 1987, H

DE MINNESTREEL

'liefste', schreef hij op een blauw
papier, en daarachter een dubbelpunt
als haar oogopslag:

en op een wit blad
schreef hij niets,
heel lang.

en op een rood blad schreef hij:
'je lippen zijn.'

toen hing ie ze alledrie
aan de muren op
en woonde.

*

Je slaapt al. Er hoeft niks meer gezegd.
Zo overal als landen maar kunnen grenzen,
zo van boven tot beneden als Finland
zich tegen de rug van de rest van Skandinavië legt,

en ook zo definitief, lig ik tegen je aan
in de zeeën van tijd van ons bed.
Dat vind ik zo mooi aan Aardrijkskunde: het
is nu eenmaal zo, verder valt er niks te verstaan.

Maar meestal heb ik je voordien bespeeld, vol
handen, als Beethoven een piano,
al lang niet meer onzeker naar de Verre Beminde

op zoek, maar in het adagio sostenuto
van de Mondscheinsonate vanaf de eerste sol
al wetend dat hij de laatste do zal vinden.

*

En waarom lach je op de trouwfoto
van je dochter? Zij, die zou moeten lachen
kijkt de verte in, als ziet ze ginder
nog wel ergens een mogelijkheid.

Maar jij, stijf arm in arm,
even opgekrikt uit je reuma,
al wat hangen wil mag hangen,
maar jij staat recht,

jij lacht, bij het tweede huwelijk
van je dochter, na het tweede huwelijk
van je zoon,

zoals glasscherven lachen in de zon.

2000 Ha[4]

TO ZHAO QING
NANJING JOHN QING TANG DESIGN CO.,
DABEI XIANG 7-3
NANJING 210018
(CN)
FedEx.
E
AM
INTL ECONOMY
210018
CN PVG
7768 7706 4534
S4 NKGA
PLEASE
HANDLE WITH CARE
FRAGILE
THANK YOU
Ursus Books & Gallery
50 East 78th Street
Suite 1C
New York NY 10075
WWW.URSUSBOOKS.COM
CLEAR REL
REL
B
NKGA 512
MFT W18
776877064534
NKGA 51

DABEI XIANG 7-3
NANJING 210018
13952094842
TRK# 7768 7706 4534
0430
S4 NKGA
FedEx Ship Manager - Print Your Label(s)
MFT W18
776877064534
NKGA 51
PLEASE
U.S.A.
U.S.A.
China

308mm

31mm

240mm

● 荣誉奖 Honorary Appreciation

《 马塞尔·杜尚——机械复制时代制造艺术的艺术 Marcel Duchamp: the art of making art in the age of mechanical reproduction

● 美国 U.S.A.

studio blue

Francis M. Naumann

Harry N. Abrams Ins., New York; Ludion Press, Ghent

308 240 31mm

2135g

332p

ISBN-10 0-8109-6334-5

作为达达主义和超现实主义的代表人物，马塞尔·杜尚（Marcel Duchamp）对后世艺术的观念产生了重大影响。本书的主要内容是杜尚的作品集和观念讨论，尤其是对杜尚创作高峰期所处的工业时代，有关艺术的复制、艺术的本源、艺术的服务对象等观念的批判与思考。本书的装订形式为锁线硬精装。护封胶版纸橙蓝黑三色印刷，内封橙色布面裱覆卡板，书脊处烫白处理。在橙色环衬的包裹下，内页采用哑面铜版纸四色印刷。书中将杜尚的创作周期分为三大阶段，每个阶段又划分为更加细致的三个小阶段，不乏对于杜尚各类作品的罗列和观念探讨。设计方面，正文双栏加注释侧边栏为主，但图片通过交叠、错位、并排等方式形成丰富的版面形式。字体方面，正文使用了规整的无衬线体，但在引用、作品名、人名等局部使用粗字重的斜体、手写体来加以突出。章节的数字在页面的正反面通过镜像、描边的处理方式，暗示了“复制”的概念。

As the representative of Dadaism and surrealism, Marcel Duchamp had a great influence on the conceptuality of subsequent art. The main content of this book is the collection of Duchamp's works discussion of concepts, especially the criticism and reflection on the concepts such as the reproduction of art, the origin of art, and the objects of art in the industrial era at the peak of Duchamp's creation. The hardcover book is bound by thread sewing. The jacket is made of offset paper printed in orange, blue and black, and the inner cover is covered with orange cloth and mounted with cardboard. The spine is treated with white hot stamping. Wrapped by orange end paper, the inner pages are printed in four colors using matte coated paper. The book divides Duchamp's creative procedure into three major stages, and each stage is divided into three more detailed stages. As for the design, the main text is double columned with comment sidebar, but the typesetting of the picture forms a rich layout through overlapping, misalignment, side by side and so on. As for the fonts, the text uses regular sans-serifs, but a large amount of bold italics and handwriting are used to highlight the citations, work names, and personal names. The chapter numbers are also mirrored and stroked on the front and back of the page. The way of processing implies the concept of 'copying'.

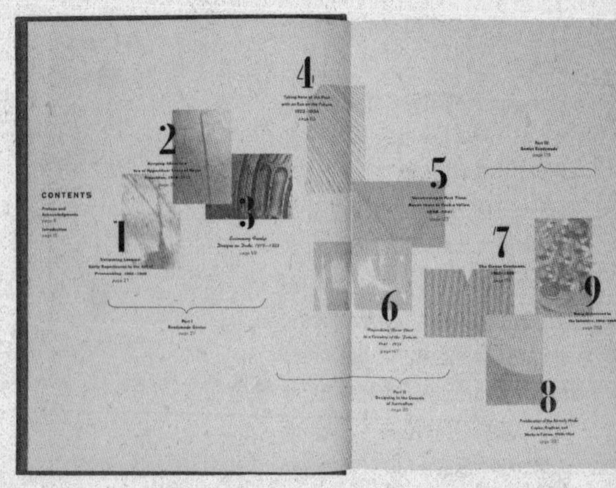

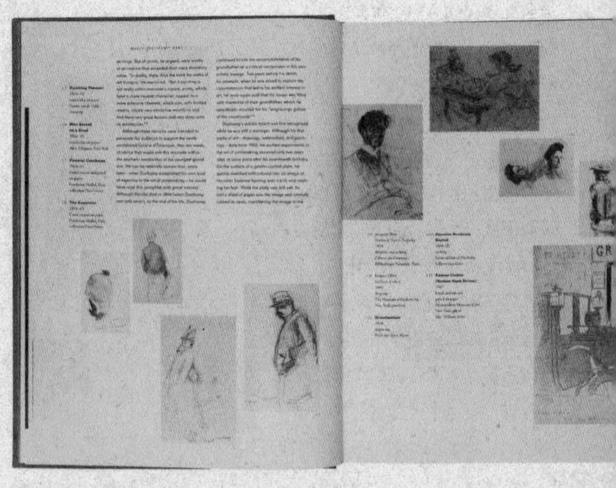

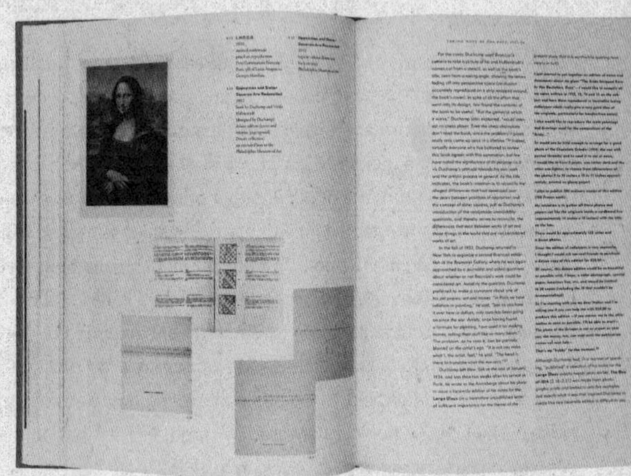

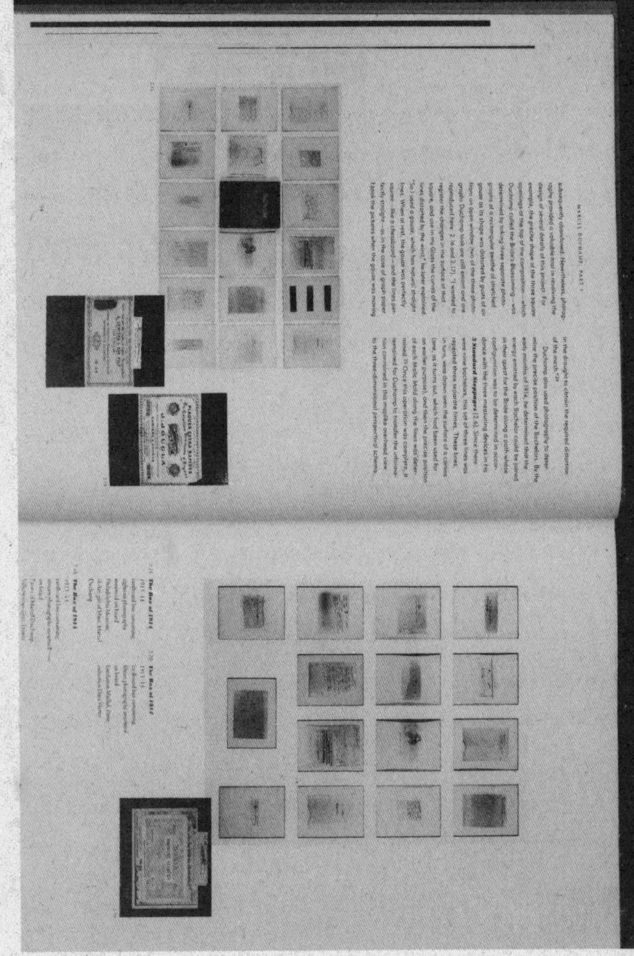

MARCEL DUCHAMP · PART I

subsequently abandoned. Nevertheless, photography provided a valuable tool in resolving the design of several details of this project. For example, the precise shape of the three square openings at the top of the composition—which Duchamp called the Bride's Blossoming—was determined by taking three separate photographs of a rectangular swatch of stretched gauze as its shape was distorted by gusts of air from an open window (two of the three photographs Duchamp took are still extant and are reproduced here: 2.16 and 2.17). "I wanted to register the changes in the surface of that square, and use in my Glass the curves of the lines distorted by the wind," he later explained. "So I used a gauze, which has natural straight lines. When at rest, the gauze was perfectly square—like a chessboard—and the lines perfectly straight—as in the case of graph paper. I took the pictures when the gauze was moving in the draught to obtain the required distortion of the mesh."[24]

Duchamp also used photography to determine the precise position of the Bachelors. By the early months of 1914, he determined that the energy emitted by each Bachelor could be joined in their quest for the Bride along a path whose configuration was to be determined in accordance with the three measuring devices in his **3 Standard Stoppages** (2.6). Since there were nine bachelors, this set of three lines was repeated three separate times. These lines, in turn, were drawn onto the surface of a canvas (one, as it turns out, which had been used for an earlier purpose), and then the precise position of each Malic Mold along the lines was determined.[25] Once this operation was complete, it remained for Duchamp to transfer the information contained in this maplike overhead view to the three-dimensional perspectival schema.

this single reference, exactly what Benjamin might have t
pertinence of his work to his thesis on mechanical reproc
as it may, as the present study hopes to demonstrate, bo
ual implications inherent in reproducing a unique work of
es that, I believe, can be traced to the core of Duchamp'
when taken as a whole, it could be argued that they repre
ant contribution his work has made to the art of this centu

ion and appropriation are artistic strategies that were
p's work as early as 1905, when he reprinted his grandfat
ouen [1.2]. But they were not employed with consistency u
when his Nude Descending a Staircase [2.4] was rejected
dants, an event that he later acknowledged was a critical
tic career. It was then that he decided to consciously se
d artistic practice, something that would remove him fr
orks of art that only reflected his unique artistic sensitivit
y emotive than painting. Although he had gone through c
n development as a painter, he eventually grew to detest
painting, the personal handiwork detectable in all work
for viewers to identify an artist's unique style. Eventually,
ke Monet as "just housepainters who painted for the grea
nd reds together and having fun." Whereas he admitted t
etic pleasure could be derived from viewing images of thi
call it mental ... it's just pure retinal."[8] Duchamp's solutio
was to adopt a more mechanically generated style, usi

50M

HI

40

R U

oat

ea of Oppositio

a Staircase [No. 4]
July 23, 1918
pencil, ink and wash on paper
Museum of the City of
New York

3.28 *To Be Looked at (From the Other Side of the Glass) with One Eye, Close to, for Almost an Hour*
1918
oil paint, silver leaf, lead wire and magnifying lens on glass
Museum of Modern Art, New York; Katherine S. Dreier Bequest

Stereopticon Slide (Hand Stereoscopy)
1918–19
rectified readymade: pencil over photographic stereopticon slide in cardboard mount
Museum of Modern Art, New York; Katherine S. Dreier Bequest

3.30 Eye chart by Wecker & Masselon
n.d.
relief print (eye chart)
Philadelphia Museum of Art; The Louise and Walter Arensberg Collection

En tout cas, le bail finit 15 Juil
reprenais, ne le fais qu'en pr
mon propre de louer 3 mois
cela je passe ordinairement
sûrement. Peut être père ne serai
engager en terme si c'est possi
quittes La Condamine pour 15 Av
I don't know anything ab
intentions and je ne veu
suggérer quelquechose.
Maintenant si tu es montée
dans l'atelier une roue de bicycl
porte bouteilles. — J'avais acheté
une sculpture toute faite. Et j

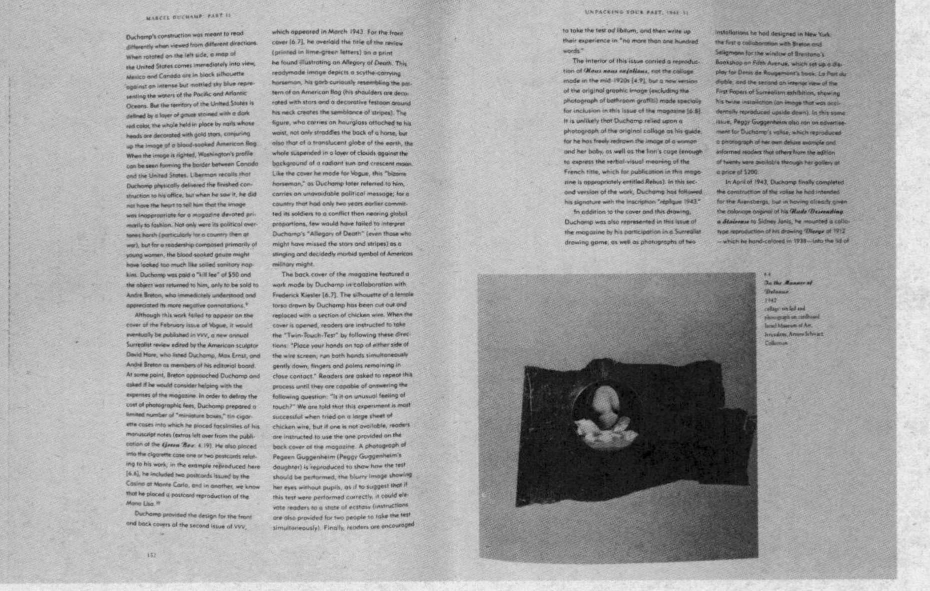

MARCEL DUCHAMP · PART II

Duchamp's construction was meant to read differently when viewed from different directions. When rotated on the left side, a map of the United States comes immediately into view, Mexico and Canada are in black silhouette against an intense but mottled sky blue representing the waters of the Pacific and Atlantic Oceans. But the territory of the United States is defined by a layer of gauze stained with a dark red color, the whole held in place by nails whose heads are decorated with gold stars, conjuring up the image of a blood-soaked American flag. When the image is righted, Washington's profile can be seen forming the border between Canada and the United States. Liberman recalls that Duchamp physically delivered the finished construction to his office, but when he saw it, he did not have the heart to tell him that the image was inappropriate for a magazine devoted primarily to fashion. Not only were its political overtones harsh (particularly for a country then at war), but for a readership composed primarily of young women, the blood-soaked gauze might have looked too much like soiled sanitary napkins. Duchamp was paid a "kill fee" of $50 and the object was returned to him, only to be sold to André Breton, who immediately understood and appreciated its more negative connotations.[9]

Although this work failed to appear on the cover of the February issue of Vogue, it would eventually be published in VVV, a new annual Surrealist review edited by the American sculptor David Hare, who listed Duchamp, Max Ernst, and André Breton as members of his editorial board. At some point, Breton approached Duchamp and asked if he would consider helping with the expenses of the magazine. In order to defray the cost of photographic fees, Duchamp prepared a limited number of "miniature boxes," tin cigarette cases into which he placed facsimiles of his manuscript notes (extras left over from the publication of the *Green Box*: 4.19). He also placed into the cigarette case one or two postcards relating to his work; in the example reproduced here [6.6], he included two postcards issued by the Casino at Monte Carlo, and in another, we know that he placed a postcard reproduction of the Mona Lisa.[10]

Duchamp provided the design for the front and back covers of the second issue of VVV, which appeared in March 1943. For the front cover [6.7], he overlaid the title of the review (printed in lime-green letters) on a print he found illustrating an *Allegory of Death*. This readymade image depicts a scythe-carrying horseman, his garb curiously resembling the pattern of an American flag (his shoulders are decorated with stars and a decorative festoon around his neck creates the semblance of stripes). The figure, who carries an hourglass attached to his waist, not only straddles the back of a horse, but also that of a translucent globe of the earth, the whole suspended in a layer of clouds against the background of a radiant sun and crescent moon. Like the cover he made for Vogue, this "bizarre horseman," as Duchamp later referred to him, carries an unavoidable political message; for a country that had only two years earlier committed its soldiers to a conflict then nearing global proportions, few would have failed to interpret Duchamp's "Allegory of Death" (even those who might have missed the stars and stripes) as a stinging and decidedly morbid symbol of American military might.

The back cover of the magazine featured a work made by Duchamp in collaboration with Frederick Kiesler [6.7]. The silhouette of a female torso drawn by Duchamp has been cut out and replaced with a section of chicken wire. When the cover is opened, readers are instructed to take the "Twin-Touch-Test" by following these directions: "Place your hands on top of either side of the wire screen; run both hands simultaneously gently down, fingers and palms remaining in close contact." Readers are asked to repeat this process until they are capable of answering the following question: "Is it an unusual feeling of touch?" We are told that this experiment is most successful when tried on a large sheet of chicken wire, but if one is not available, readers are instructed to use the one provided on the back cover of the magazine. A photograph of Pegeen Guggenheim (Peggy Guggenheim's daughter) is reproduced to show how the test should be performed, the blurry image showing her eyes without pupils, as if to suggest that if this test were performed correctly, it could elevate readers to a state of ecstasy (instructions are also provided for two people to take the test simultaneously). Finally, readers are encouraged

152

UNPACKING YOUR PART, 1942–53

to take the test *ad libitum*, and then write up their experience in "no more than one hundred words."

The interior of this issue carried a reproduction of *Nous nous cajolions*, not the collage made in the mid-1920s [4.9], but a new version of the original graphic image (excluding the photograph of bathroom graffiti) made specially for inclusion in this issue of the magazine [6.8]. It is unlikely that Duchamp relied upon a photograph of the original collage as his guide, for he has freely redrawn the image of a woman and her baby, as well as the lion's cage (enough to express the verbal-visual meaning of the French title, which for publication in this magazine is appropriately entitled *Rebus*). In this second version of the work, Duchamp has followed his signature with the inscription "*réplique* 1943."

In addition to the cover and this drawing, Duchamp was also represented in this issue of the magazine by his participation in a Surrealist drawing game, as well as photographs of two installations he had designed in New York: the first a collaboration with Breton and Seligmann for the window of Brentano's Bookshop on Fifth Avenue, which set up a display for Denis de Rougemont's book, *La Part du diable*, and the second an interior view of the First Papers of Surrealism exhibition, showing his twine installation (an image that was accidentally reproduced upside down). In this same issue, Peggy Guggenheim also ran an advertisement for Duchamp's valise, which reproduced a photograph of her own deluxe example and informed readers that others from the edition of twenty were available through her gallery at a price of $200.

In April of 1943, Duchamp finally completed the construction of the valise he had intended for the Arensbergs, but in having already given the coloriage original of his *Nude Descending a Staircase* to Sidney Janis, he mounted a collotype reproduction of his drawing *Vierge* of 1912 —which he hand-colored in 1938—into the lid of

6.4
In the Manner of Delvaux
1942
collage: tin foil and photograph on cardboard.
Israel Museum of Art, Jerusalem; Arturo Schwarz Collection

R. MUTT
1917
TAXIMETRE

EAU & GAZ
A TOUS LES ETAGES

C17.3C
CYCLE 1
LEX2 1.1 Lbs 09/11 DEW4
Steve Woo
16 HAWTHORNE CT
08816 - 2742 EAST BRUNSWICK , NJ United States
spOB46KCD
TBA614446429000
DEW4 CYCLE 1
BM4
EWR4 PNE5
A EW4 J DEW
JM8

2000 Ha[5]

LEX2
1.1 Lbs
09/11
DEW4
Steve Woo
16 HAWTHORNE CT
08816 – 2742 EAST BRUNSWICK , NJ United States
46KCD
TBA814446429000
DEW4
CYCL
DBfGV2CRh/2/second/1 of 1/3051
EWR4
PNE5
A EW4
J DEW
3M4
U.S.A.
U.S.A.
China

287mm

23mm

192mm

◎ 荣誉奖
《 关于书籍设计
◖ 美国

Honorary Appreciation
On Book Design
U.S.A.
Richard Hendel
Richard Hendel
Yale University Press, New Haven & London
287 192 23mm
880g
224p
ISBN-10 0-300-07570-7

美国书籍设计师理查德·亨德尔（Richard Hendel）的一本书籍设计教程，探讨了书籍设计工作方法和处理各种细节问题的经验。书籍的装订形式为锁线硬精装。护封铜版纸红、黑、灰、绿四色印刷。内封黑色布面裱覆卡板，书脊书名烫银处理。环衬选用黑色艺术纸，内页胶版纸印单黑。本书邀请了 8 位书籍设计师，包括作者亨德尔自己，呈现了 9 个人的书籍设计经验谈。从开本、造型、字体、排版、印制、成本等各种细节入手，深入探讨书籍设计中除了纸张、装订和工艺以外的那些“隐藏”在背后的工作；所以书中的图片较少，大都是围绕各自主题的理念、网格、字符倍率计算等内容。亨德尔强调，设计一本好书的挑战不是创造出漂亮或者讨巧的东西，而是找到更好的为作者服务的方法。他并不坚持单一的设计方法或理念，而是拥抱不同的方式。这本书对于新手和有一定经验的设计师都有不错的参考价值。

This book is a book design tutorial written by an American book designer Richard Hendel, which discusses working methods of book design and experience in dealing with various details. The hardcover book is bound by thread sewing. The jacket is made of coated paper printed in red, black, gray, and green. The inner cover is covered by black cloth and mounted with cardboard, and the title on the spine is treated with silver hot stamping. The end paper is made of black art paper, while the internal pages are made of offset paper printed in black. The author invited eight book designers, including Hendel himself, to talk about the book design experience. Starting with various details such as formatting, styling, fonts, typesetting, printing, cost and so on, they discussed in-depth the work 'hidden' behind the book design except for paper binding and technology, so there are few pictures in the book, mostly focusing on the concept of their respective themes, grid, character multiplication calculation and other contents. Hendel stressed that the challenge of designing a good book is not to create beautiful or clever things, but to find out how to better serve the author. He did not support a single design method or idea, but embraced different ways. This book has good reference value for novice and designers with some experience.

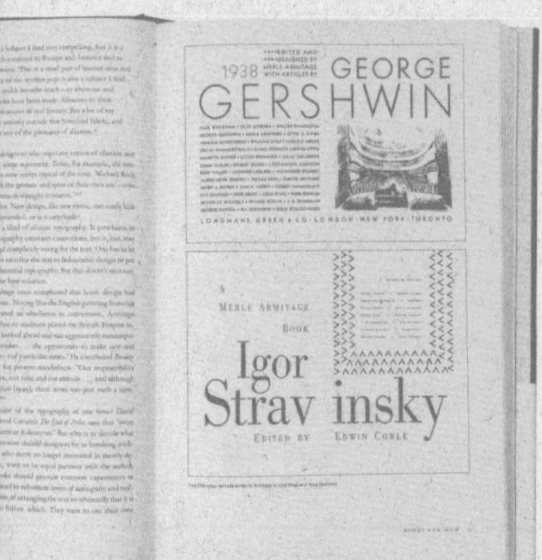
1938
GEORGE
GERSHWIN
LONGMANS, GREEN & CO · LONDON · NEW YORK · TORONTO
A
Merle Armitage
Book
Igor
Strav insky
Edited by Edwin Corle

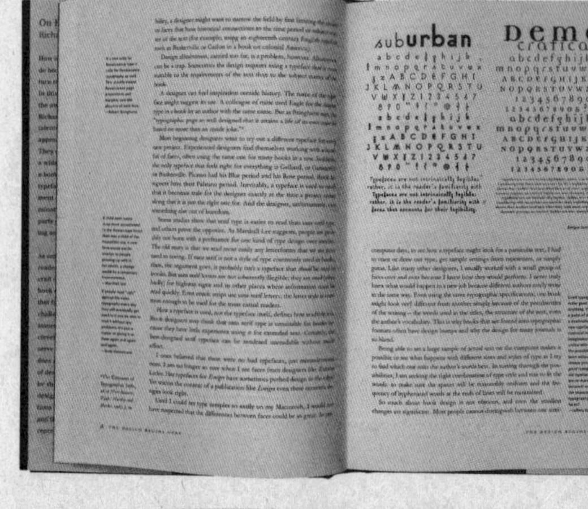
suburban
Demo
cratica

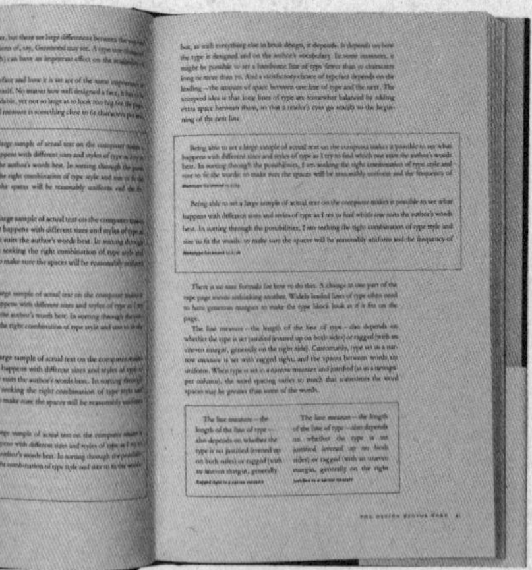

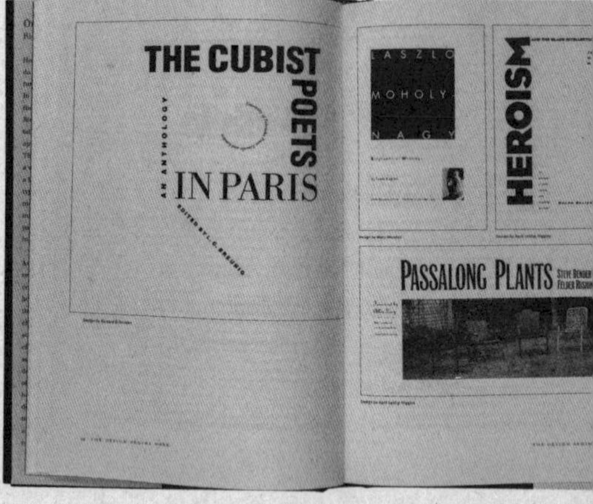
THE CUBIST
POETS
IN PARIS
AN ANTHOLOGY
LASZLO
MOHOLY-
NAGY
HEROISM
PASSALONG PLANTS

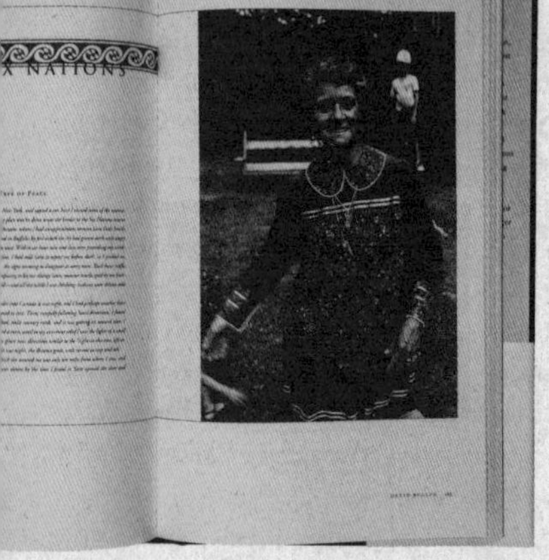

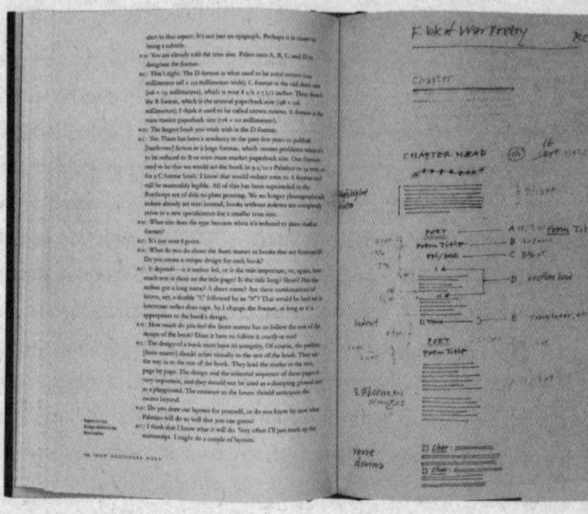

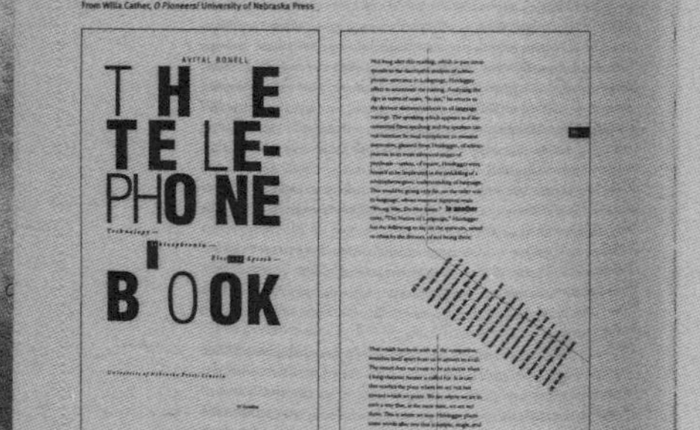

From Willa Cather, *O Pioneers!* University of Nebraska Press

various elements of the book. It feels all of a piece, yet it has little measurable consistency. There is a variation of type style and a play of space that goes beyond the mathematics of the grid.

Avital Ronell's *Telephone Book* might appear to be entirely in contrast with the Cather, but its design is based on the same classical conventions. Ronell challenges how we read and attempts to make us conscious and critical of the process. I tried to convey this by emphasizing the physical structure of the book. The conventions of book typography that we prefer to take for granted are either flaunted by distortion and exaggeration or conspicuous by their absence. They are always there by implication.

Warren Motte, the author of *Questioning Edmond Jabès*, was a young professor in the French department when the manuscript came into production. I knew him quite well and was wary of working with a friend, which can complicate things. In this case, my apprehension was unfounded. Motte gave me a free hand. I read his introduction and skimmed most of the main text, sometimes reading closely. I felt that I had a pretty good idea of the author and his subject.

The Motte is in a small format partly to bulk it but principally because a small page size suggested the intimacy of the dialogue. I usually begin the design of a book with the basic text page, which I use as a linear grid for the placement of display elements. With the Motte I was directed to the use of flush and hang paragraphs by the form of the extracts. Jabès visually patterns his writing very deliberately. I saw that the extract style would jar with a conventional form of paragraphing in the main text, so I built the text design around the extract style. I suppose Jabès's writing would be grouped in the postmodern school, but he doesn't quite fit there. Much of his work is grounded in the Talmud and the Hebrew tradition of exegesis. Although the design of the book might be called postmodern, the basic structure is quite classical. The word "postmodern" is by definition a ragbag of what has gone before, usually combined into inept or outrageous confections. This approach may be entirely valid for certain authors—not, I think, Jabès.

Both the Ronell and the Motte are narrow books. I don't much like the standard trim sizes, perhaps because they have become associated for me with standard designs of too wide a type measure. A trim size of 6 × 9 inches is fine if one is permitted to apply the Golden Section to the type page, but that would reduce the number of words per page to unacceptably few—unacceptable to the publisher, that is. I favor a narrow format, cut down from the standard 6 × 9 or 5 1/2 × 8 1/2. But some books need to be short and wide, or square.

For paragraph indents I usually specify the conventional em space. The

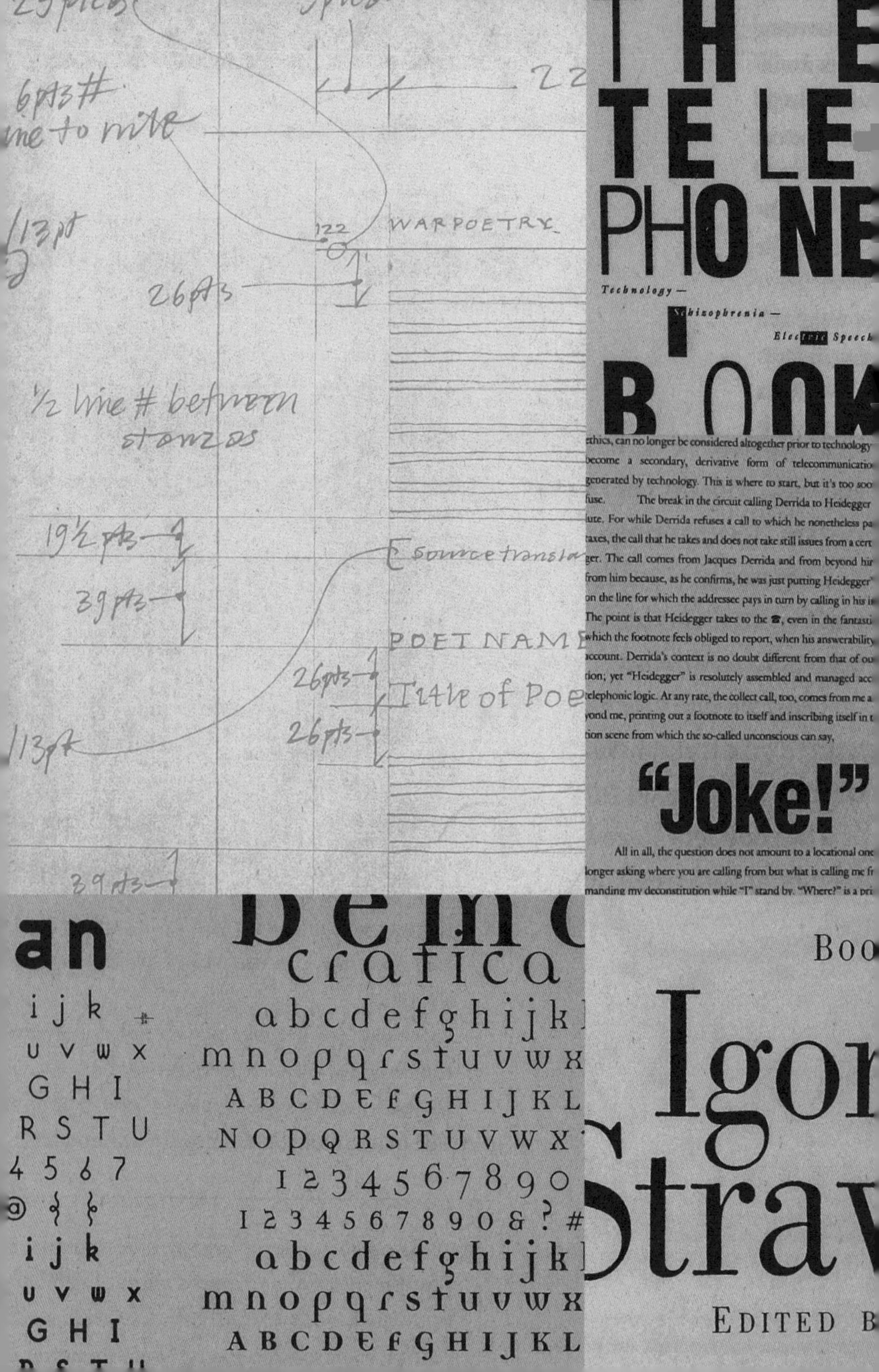

6pts#
13pt
26pts
WARPOETRY
1/2 line # between stanzas
19 1/2 pts
39 pts
source transla
POET NAME
26pts
Title of Poe
26pts
13pt
39 pts
THE TELEPHONE BOOK
Technology —
Schizophrenia —
Electric Speech
ethics, can no longer be considered altogether prior to technology
become a secondary, derivative form of telecommunicatio
generated by technology. This is where to start, but it's too soo
fuse. The break in the circuit calling Derrida to Heidegger
lute. For while Derrida refuses a call to which he nonetheless pa
taxes, the call that he takes and does not take still issues from a cert
ger. The call comes from Jacques Derrida and from beyond hi
from him because, as he confirms, he was just putting Heidegger
on the line for which the addressee pays in turn by calling in his i
The point is that Heidegger takes to the ☎, even in the fantasti
which the footnote feels obliged to report, when his answerability
account. Derrida's context is no doubt different from that of ou
tion; yet "Heidegger" is resolutely assembled and managed acc
telephonic logic. At any rate, the collect call, too, comes from me a
yond me, printing out a footnote to itself and inscribing itself in t
tion scene from which the so-called unconscious can say,
"Joke!"
All in all, the question does not amount to a locational one
longer asking where you are calling from but what is calling me fr
manding my deconstitution while "I" stand by. "Where?" is a pri
an
i j k
u v w x
G H I
R S T U
4 5 6 7
cratica
a b c d e f g h i j k
m n o p q r s t u v w x
A B C D E F G H I J K L
N O P Q R S T U V W X
1 2 3 4 5 6 7 8 9 0
1 2 3 4 5 6 7 8 9 0 & ? #
a b c d e f g h i j k
m n o p q r s t u v w x
A B C D E F G H I J K L
Boo
Igor
Strav
Edited B

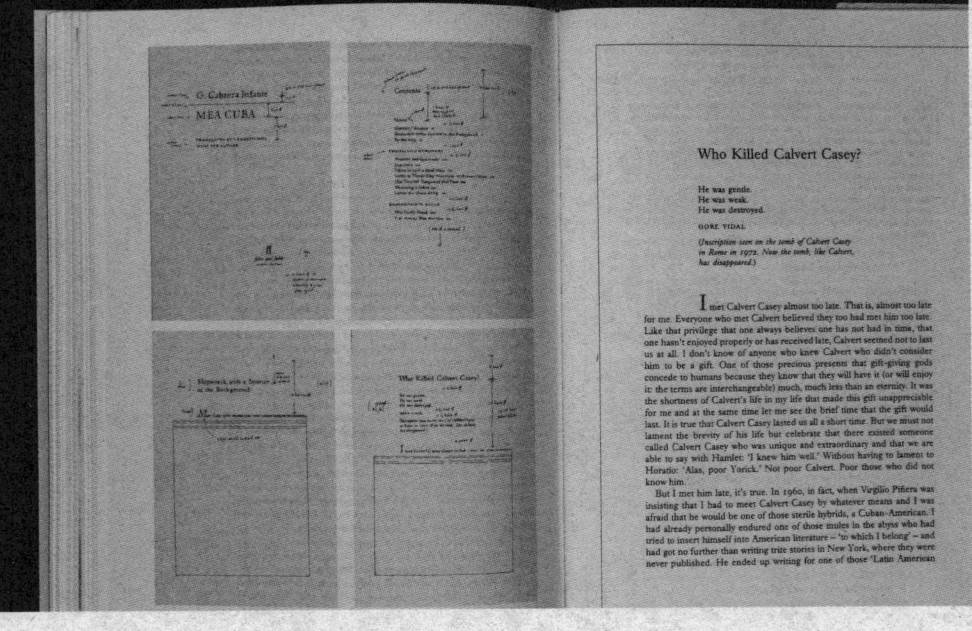

Who Killed Calvert Casey?

He was gentle.
He was weak.
He was destroyed.

GORE VIDAL

(Inscription seen on the tomb of Calvert Casey in Rome in 1972. Now the tomb, like Calvert, has disappeared.)

I met Calvert Casey almost too late. That is, almost too late for me. Everyone who met Calvert believed they too had met him too late. Like that privilege that one always believes one has not had in time, that one hasn't enjoyed properly or has received late, Calvert seemed not to last us at all. I don't know of anyone who knew Calvert who didn't consider him to be a gift. One of those precious presents that gift-giving gods concede to humans because they know that they will have it (or will enjoy it: the terms are interchangeable) much, much less than an eternity. It was the shortness of Calvert's life in my life that made this gift unappreciable for me and at the same time let me see the brief time that the gift would last. It is true that Calvert Casey lasted us all a short time. But we must not lament the brevity of his life but celebrate that there existed someone called Calvert Casey who was unique and extraordinary and that we are able to say with Hamlet: 'I knew him well.' Without having to lament to Horatio: 'Alas, poor Yorick.' Not poor Calvert. Poor those who did not know him.

But I met him late, it's true. In 1960, in fact, when Virgilio Piñera was insisting that I had to meet Calvert Casey by whatever means and I was afraid that he would be one of those sterile hybrids, a Cuban-American. I had already personally endured one of those mules in the abyss who had tried to insert himself into American literature – 'to which I belong' – and had got no further than writing trite stories in New York, where they were never published. He ended up writing for one of those 'Latin American

2001

Country / Region

The Netherlands 3

Switzerland 3

Germany 3

Austria 1

Japan 1

Brazil 1

Czech Republic 1

Venezuela 1

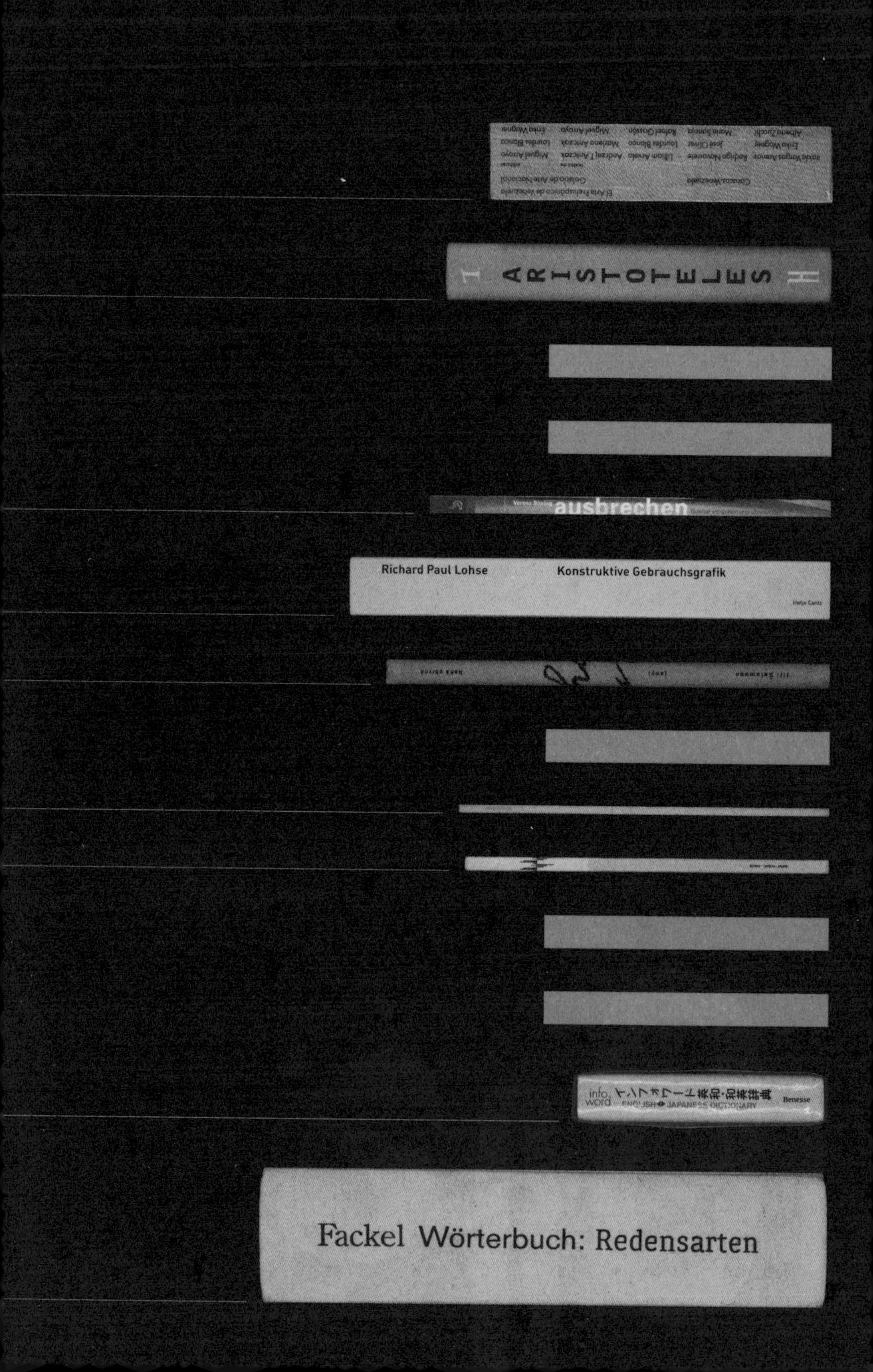

El Arte Prehispánico de Venezuela
Galería de Arte Nacional
Caracas Venezuela

textos de: Iraida Vargas Arenas, Rodrigo Navarrete, Lilliam Arvelo, Andrzej T Antczak, Erika Wagner, José Oliver, Lourdes Blanco, Marlena Antczak, Alberta Zucchi, Mario Sanoja, Rafael Gassón, Miguel Arroyo

editores: Miguel Arroyo, Lourdes Blanco, Erika Wagner

I ARISTOTELES H

Verena Bönig ausbrechen

Richard Paul Lohse Konstruktive Gebrauchsgrafik Hatje Cantz

Jiří Šalamoun (any)

info word インフォワード英和・和英辞典 ENGLISH⇔JAPANESE DICTIONARY Benesse

Fackel Wörterbuch: Redensarten

2001 Gl 1034

2001 Gm 1042

2001 Bm[1] 1050

2001 Bm[2] 1058

2001 Bm[4] 1066

2001 Bm[5] 1076

2001 Ha[1] 1086

2001 Ha[4] 1096

2001 Ha[5] 1102

Designer

Anne Burdick

Haruyo Kobayashi

Jair de Souza

Jonker & Van Vilsteren

Olaf Bender

Jost Hochuli

Beat Müller, Wendelin Hess, Michael Birchmeier, Nora Beer (Müller + Hess)

Ji í Šalamoun, Zden k Ziegler

Markus Bosshard (Weiersmüller Bosshard Grüninger wbg AG)

Verena Böning

Hans Andree, Sigrid Behrens, Uli Brandt, Shin-Young Chung,
Daniel Hahn, Nadine Jung, Volker Lang, Zuzanna Musialczyk,
Peter Piller, Anna Reemts, Sven Seddig, Jan Schmietendorf

Dietwee, Utrecht, -SYB-grafisch ontwerp, Martijn Engelbregt (EGBG)

Rudo Hartmann, Gerard Hadders

Álvaro Sotillo

2001 Gl

arten
Postal parcels
DEFRAA (DFA)
Frankfurt/M
DeutschePost
Airmail
CP 84
ACN
52
2019-02-05
PC
11
CN
1
1.1
CNNKGA (CNA)
Nanjing
CHINA POST
PVG

parcels
A (DEA)
urt/M
hePost

Airmail

Desp. No.
52

CNNKG
Nanjin
CHINA

03-05

Rec. No.
11

Receptacle ID

No. of Parcels
1

DEFRAA CNNKGA

Date Transport

05 FRA ACF 200 PVG

UP (IPZ) DEFRAA

Austria

Germany

China

360mm

82mm

Fackel Wörterbuch: Redensarten

233mm

金字符奖

卡尔·克劳斯主编的《火炬》杂志（1899–1936）的短语词典

奥地利

Golden Letter

Wörterbuch der Redensarten zu der von Karl Kraus 1899 bis 1936 herausgegebenen Zeitschrift „Die Fackel“

Austria

Anne Burdick

Werner Welzig (Hrsg.)

Verlag der Österreichischen Akademie der Wissenschaften, Wien

360 233 82mm

3557g

1056p

ISBN-10 3-7001-2768-5

奥地利作家卡尔·克劳斯（Karl Kraus）于1899年至1936年连续出版了讽刺杂志《火炬》。这本杂志主要由克劳斯自己编写，收录了德国、奥地利、捷克等地的俚语、社会和政治用语，本书是对这本杂志中所用短语的诠释。书籍的装订形式为锁线硬精装。封面白色艺术纸印黑，裱贴卡板，书脊处包裹白色布面，烫红烫黑处理。封面正中间压凹处贴红黑双色印刷方块，方块大小依照《火炬》杂志的A5尺寸。环衬选用带黑色杂质的红色艺术纸，内页黄灰色、黑色和暗红色三色印刷。本词典对《火炬》杂志中的短语进行了罗列，对短语的形成背景、相关意义进行诠释和剖析，并将上下文间重复出现的概念建立关联。正文版式上，依照封面《火炬》杂志占据的区域将版心划分为三栏，中心栏为杂志的剪报，左右两侧分别为上下文索引和短语诠释。由于是对语言本身的诠释，书中专门设计了缩写系统。

Austrian writer Karl Kraus published satirical magazine *Die Fackel* from 1899 to 1936. This magazine was mainly written by Kraus himself and contained slangs, social and political terms in Germany, Austria, Czech Republic, etc. This book interprets the phrases used in this magazine. The hardcover book is bound by thread sewing. The cover is made of white art paper printed in black, mounted with cardboard. The spine is wrapped by white cloth, which is treated with red and black stamping. Red art paper with black impurities is selected for the end paper, and the inner pages are printed in yellow gray, black and dark red. This dictionary lists the phrases in *Die Fackel*, explains and analyzes the background and relevant meaning of the phrases, and establishes the relationship between the concepts that repeat in the context. As for the layout, the center of the typeset area is divided into three columns, the middle column is for magazine cuttings, and the left and right sides are occupied by index and phrase interpretation. Since the book focuses on the interpretation of the language itself, the abbreviation system is specially designed in the book.

Wörterbuch
der Redensarten
Karl Kraus
Die Fackel

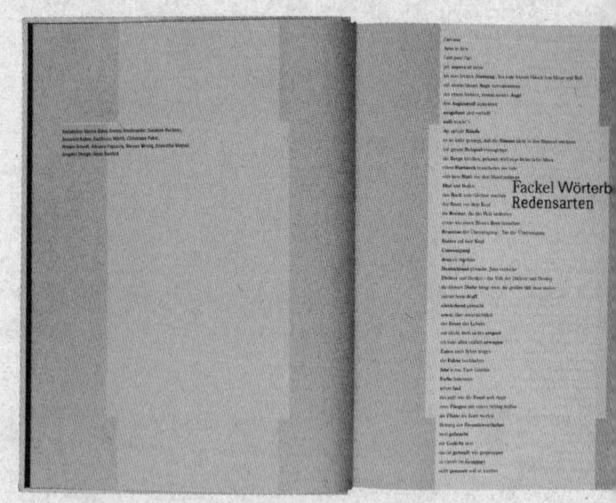
Fackel Wörterb
Redensarten

What is an Excellent Idiom?

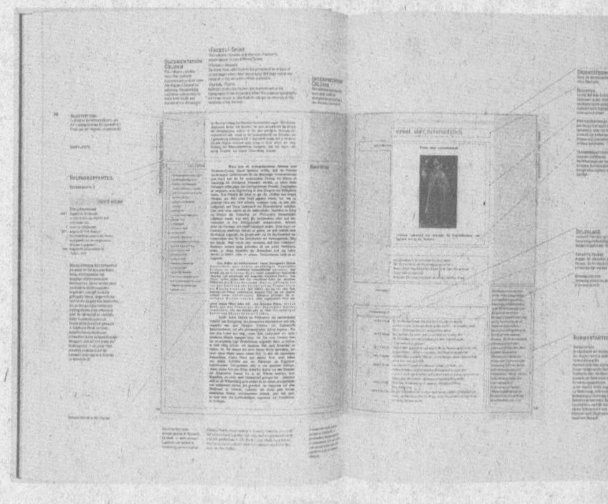

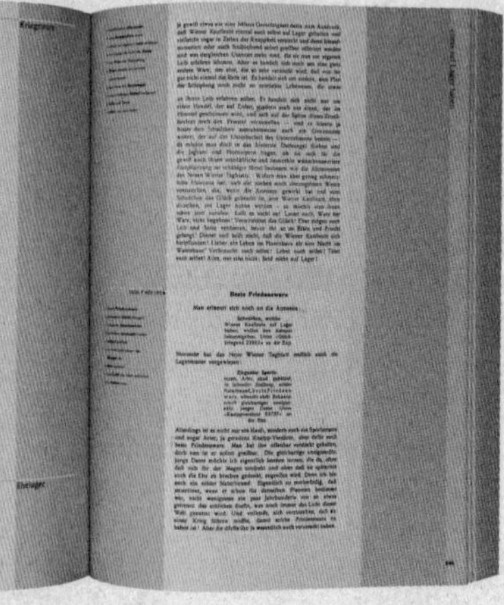

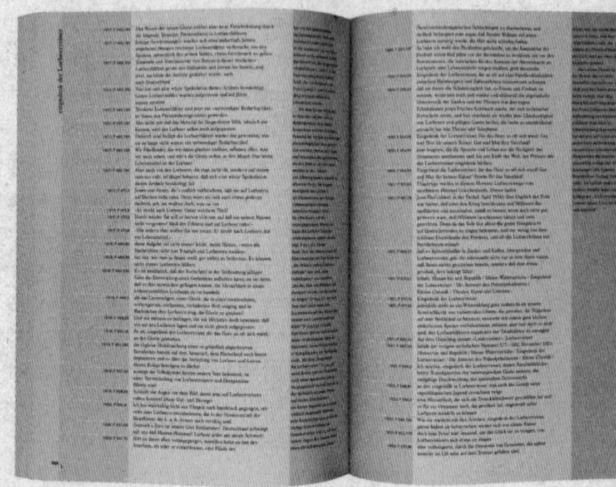

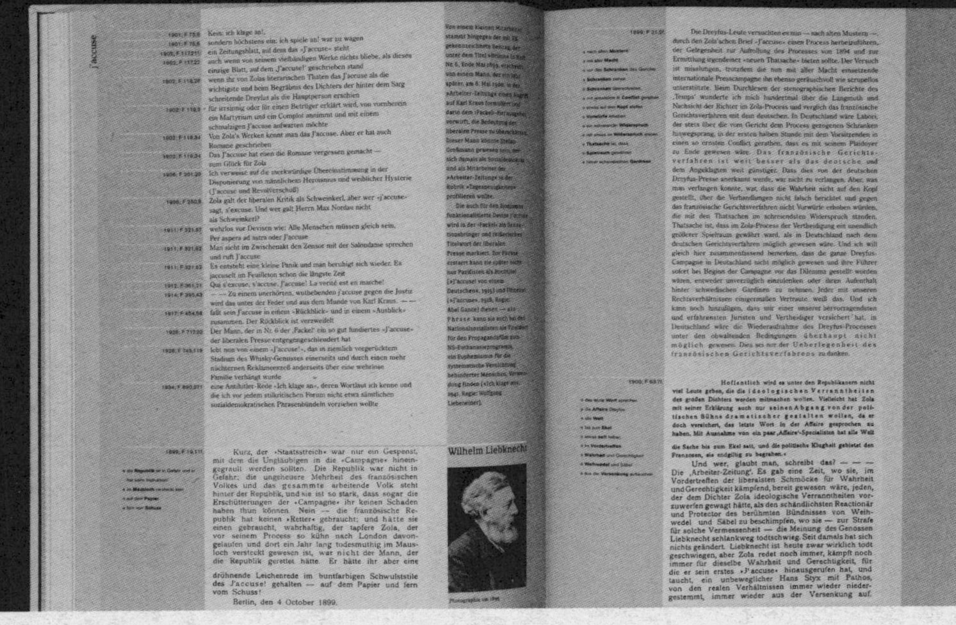

j'accuse

1901, F 75,6 Kein: ich klage an!,
1901, F 75,6 sondern höchstens ein: ich spiele an! war zu wagen
1902, F 117,21 ein Zeitungsblatt, auf dem das »J'accuse« steht
1902, F 117,22 auch wenn von seinem vielbändigen Werke nichts bliebe, als dieses einzige Blatt, auf dem ‚J'accuse!' geschrieben stand
1902, F 118,28 wenn ihr von Zolas literarischen Thaten das J'accuse als die wichtigste und beim Begräbnis des Dichters der hinter dem Sarg schreitende Dreyfus als die Hauptperson erschien
1902, F 119,9 für irrsinnig oder für einen Betrüger erklärt wird, von vornherein ein Martyrium und ein Complot annimmt und mit einem schmalzigen J'accuse aufwarten möchte
1902, F 119,34 Von Zola's Werken kennt man das J'accuse. Aber er hat auch Romane geschrieben
1902, F 119,34 Das J'accuse hat eben die Romane vergessen gemacht — zum Glück für Zola
1906, F 201,20 Ich verweise auf die merkwürdige Übereinstimmung in der Disponierung von männlichem Heroismus und weiblicher Hysterie (J'accuse und Revolverschuß)
1908, F 260,8 Zola galt der liberalen Kritik als Schweinkerl, aber wer »j'accuse« sagt, s'excuse. Und wer galt Herrn Max Nordau nicht als Schweinkerl?
1911, F 321,57 wehrlos vor Devisen wie: Alle Menschen müssen gleich sein, Per aspera ad astra oder J'accuse
1911, F 321,62 Man sieht im Zwischenakt den Zensor mit der Salondame sprechen und ruft J'accuse
1911, F 321,62 Es entsteht eine kleine Panik und man beruhigt sich wieder. Es jaccuselt im Feuilleton schon die längste Zeit
1912, F 361,21 Qui s'excuse, s'accuse. J'accuse! La verité est en marche!
1914, F 395,43 — — Zu einem unerhörten, wutbebenden j'accuse gegen die Justiz wird das unter der Feder und aus dem Munde von Karl Kraus. — —
1917, F 454,58 faßt sein J'accuse in einem »Rückblick« und in einem »Ausblick« zusammen. Der Rückblick ist verzweifelt
1926, F 717,20 Der Mann, der in Nr. 6 der ‚Fackel' ein so gut fundiertes »J'accuse« der liberalen Presse entgegengeschleudert hat
1926, F 743,119 lebt nun von einem »J'accuse!«, das in ziemlich vorgerücktem Stadium des Whisky-Genusses einerseits und durch einen mehr nüchternen Reklameexzeß anderseits über eine wehrlose Familie verhängt wurde
1934, F 890,271 eine Antihitler-Rede »Ich klage an«, deren Wortlaut ich kenne und die ich vor jedem stilkritischen Forum nicht etwa sämtlichen sozialdemokratischen Phrasenbündeln vorziehen wollte

1899, F 19,11f. Kurz, der »Staatsstreich« war nur ein Gespenst, mit dem die Ungläubigen in die »Campagne« hineingegrault werden sollten. Die Republik war nicht in Gefahr; die ungeheuere Mehrheit des französischen Volkes und das gesammte arbeitende Volk steht hinter der Republik, und sie ist so stark, dass sogar die Erschütterungen der »Campagne« ihr keinen Schaden haben thun können. Nein — die französische Republik hat keinen »Retter« gebraucht; und hätte sie einen gebraucht, wahrhaftig, der tapfere Zola, der vor seinem Process so kühn nach London davongelaufen und dort ein Jahr lang todesmuthig im Mausloch versteckt gewesen ist, war nicht der Mann, der die Republik gerettet hätte. Er hätte ihr aber eine dröhnende Leichenrede im buntfarbigen Schwulststile des J'accuse! gehalten — auf dem Papier und fern vom Schuss!

Berlin, den 4 October 1899.

Von einem kleinen Mitarbeiter stammt hingegen der mit [illegible] gekennzeichnete Beitrag, der unter dem Titel »[illegible]« in Heft Nr. 6, Ende Mai 1899, erschien: von einem Mann, der ein Jahr später, am 6. Mai 1900, in der »Arbeiter-Zeitung« einen Angriff auf Karl Kraus formuliert und darin dem ›Fackel‹-Herausgeber vorwirft, die Bedeutung der liberalen Presse zu überschätzen. Dieser Mann könnte Stefan Großmann gewesen sein, der sich damals als Sozialdemokrat und als Mitarbeiter der »Arbeiter-Zeitung« in der Rubrik »Tagesneuigkeiten« profilieren wollte.

Die auch für den [illegible] funktionalisierte Devise J'accuse wird in der ›Fackel‹ als Sensationsbringer und [illegible] Titelwort der liberalen Presse markiert. Zur Phrase erstarrt kann sie später nicht nur Pazifisten als Buchtitel (»J'accuse! von einem Deutschen«, 1915) und Filmtitel (»J'accuse«, 1918, Regie: Abel Gance) dienen — als Phrase kann sie auch bei den Nationalsozialisten als Titelwort für den Propagandafilm zum NS-Euthanasieprogramm, ein Euphemismus für die systematische Vernichtung behinderter Menschen, Verwendung finden (»Ich klage an«, 1941, Regie: Wolfgang Liebeneiner).

Wilhelm Liebknecht

Photographie um 1896

1899, F 21,2f. Die Dreyfus-Leute versuchten es nun — nach alten Mustern —, durch den Zola'schen Brief »J'accuse« einen Process herbeizuführen, der Gelegenheit zur Aufrollung des Processes von 1894 und zur Ermittlung irgendeiner »neuen Thatsache« bieten sollte. Der Versuch ist misslungen, trotzdem die nun mit aller Macht einsetzende internationale Presscampagne ihn ebenso geräuschvoll wie scrupellos unterstützte. Beim Durchlesen der stenographischen Berichte des ‚Temps' wunderte ich mich hundertmal über die Langmuth und Nachsicht der Richter im Zola-Process und verglich das französische Gerichtsverfahren mit dem deutschen. In Deutschland wäre Labori, der stets über die vom Gericht dem Process gezogenen Schranken hinwegsprang, in der ersten halben Stunde mit dem Vorsitzenden in einen so ernsten Conflict gerathen, dass es mit seinem Plaidoyer zu Ende gewesen wäre. Das französische Gerichtsverfahren ist weit besser als das deutsche und dem Angeklagten weit günstiger. Dass dies von der deutschen Dreyfus-Presse anerkannt werde, war nicht zu verlangen. Aber, was man verlangen konnte, war, dass die Wahrheit nicht auf den Kopf gestellt, über die Verhandlungen nicht falsch berichtet und gegen das französische Gerichtsverfahren nicht Vorwürfe erhoben würden, die mit den Thatsachen im schreiendsten Widerspruch standen. Thatsache ist, dass im Zola-Process der Vertheidigung ein unendlich größerer Spielraum gewährt ward, als in Deutschland nach dem deutschen Gerichtsverfahren möglich gewesen wäre. Und ich will gleich hier zusammenfassend bemerken, dass die ganze Dreyfus-Campagne in Deutschland nicht möglich gewesen und ihre Führer sofort bei Beginn der Campagne vor das Dilemma gestellt worden wären, entweder unverzüglich einzulenken oder ihren Aufenthalt hinter schwedischen Gardinen zu nehmen. Jeder mit unseren Rechtsverhältnissen einigermaßen Vertraute weiß das. Und ich kann noch hinzufügen, dass mir einer unserer hervorragendsten und erfahrensten Juristen und Vertheidiger versichert hat, in Deutschland wäre die Wiederaufnahme des Dreyfus-Processes unter den obwaltenden Bedingungen überhaupt nicht möglich gewesen. Dies sei nur der Ueberlegenheit des französischen Gerichtsverfahrens zu danken.

1900, F 63,7f. **Hoffentlich wird es unter den Republikanern nicht viel Leute geben, die die ideologischen Verranntheiten des großen Dichters werden mitmachen wollen. Vielleicht hat Zola mit seiner Erklärung auch nur seinen Abgang von der politischen Bühne dramatischer gestalten wollen, da er doch versichert, das letzte Wort in der Affaire gesprochen zu haben. Mit Ausnahme von ein paar ‚Affaire'-Specialisten hat alle Welt die Sache bis zum Ekel satt, und die politische Klugheit gebietet den Franzosen, sie endgiltig zu begraben.«**

Und wer, glaubt man, schreibt das? — — — Die ‚Arbeiter-Zeitung'. Es gab eine Zeit, wo sie, im Vordertreffen der liberalsten Schmöcke für Wahrheit und Gerechtigkeit kämpfend, bereit gewesen wäre, jeden, der dem Dichter Zola ideologische Verranntheiten vorzuwerfen gewagt hätte, als den schändlichsten Reactionär und Protector des berühmten Bündnisses von Weihwedel und Säbel zu beschimpfen, wo sie — zur Strafe für solche Vermessenheit — die Meinung des Genossen Liebknecht schlankweg todtschwieg. Seit damals hat sich nichts geändert. Liebknecht ist heute zwar wirklich todt geschwiegen, aber Zola redet noch immer, kämpft noch immer für dieselbe Wahrheit und Gerechtigkeit, für die er sein erstes »J'accuse« hinausgerufen hat, und taucht, ein unbeweglicher Hans Styx mit Pathos, von den realen Verhältnissen immer wieder niedergestemmt, immer wieder aus der Versenkung auf.

ntry / Artikel-Sc

de Wissenschaft ist
die Psychiatrie, mag
iz noch so sicher
it bestätigt, ist sie
gegebenen Stande
ne und gewiß noch
zum Beweise einer
Feststellung eines
Mord gehören nicht
missär auch, wenn
teht man schön da.
estanden wird, gilt
n Mord und einem
ist man auf die
n, während sonst
einen Exzedenten
s Wachmanns und
sehr der passive
aus der Wachstube
ist es sogar schon
angen, sondern der
es geschehen und
nen.

Veilchen

mit einem blauen **Auge** davonkommen

den der Treubruch
merzen, weil er ihn
nachen —, niemand
Amt gegeben, die
ist, eine ehrlichere
er Größenwahn, der
ibt, denn abgesehen
in Österreich weit
melodischen Fülle
Krieg mit etwas
Herr Hofmannsthal
nen ist. Wenn einer

Völlig unbeschadet davonkommen

BELEG
Auswahl
strukturi
BELEGEN
REDENSA

KOMM

GRUPPE
Redaktio
Kommen
GSCHICH
Verständ
einer Gru
Die Über
die Bildu
maßgebl
offenleg
werden (
oder ein
Illustrati

aufführung beiwohnen 726
ufführung gelangen 956 → m
was zur **Aufgabe** machen 422 → et
er **Aufgabe** stehen 540 → A
dankbare **Aufgabe** 562 → da
t **aufgehoben** sein 570 → da
fgelegt sein 942 → je
wachsen bei 275, 501, 757 → m
t etwas nicht länger **aufhalten** 352 11
er etwas **aufhalten** 1006 → mi
vas **Aufhebens** machen 161 → etw
vas kein **Aufhebens** machen 410 → für
el **Aufhebens** machen 275, 621 → da
et **aufigstiegen**, wärst net → jem
n 237 → ins
elle **Aufklärung** 207 → ein

Inhalt

What is an Excellent Idiom?

Index der Redensart

Quellen

Abkürzungen

allein begegnet
eren Stellen:
82 / S. 99.
tion in dem
inenten«: „Was
gefragt, und
eute so nennt.
ich, prominent,
hervorgeragt"

mit der Anerkennung seiner, des Satirikers, Leistung
Grund an. Von der staatlichen Ehrung sind jene ausgesc
kompilatorisch arbeiten. Die lediglich ergänzen
berührt den zentralen Punkt. Reproduktion und Ko
selbstständigen Leistung des Karl Kraus. Karl Kraus
Satire, heißt es schon vor dem Ersten Weltkrieg in de
suchen. Es gibt nichts zu erfinden. Was noch nic
Die Sperrung im ersten Satz der Glosse ist anderer
zur Entsperrung ihrer Umgebung von der Vo
Kennwort aufmerksam, das seinerseits Karl Kraus
haben dürfte: ein „kleiner Kreis von prom
ihn gereizt haben. Prominent ist eines der Ekelwo

hrdisposi-
icht noch
wegfahre.
isse dem

arte
chöne, in-
ne, unab-
die Ehe-
eines nur
sit. Herrn
J. Unter
nk und
9524"

Witwe
cht ehrb.
m. feinem
Unter
Nr. 9888"

che
n Brief-
vorragend
önlichkei-
t. Durch-
täten Pa-
sführliche
ter „No-
9876"

kluge

Verstandesmensch, Tiefen-
psychologe, Arier, Patrizier aus russisch-deutscher Hochkultur, neuarm, bestrebt Edelmensch zu sein, sucht geistig-seelische Gemeinschaft und Ehe mit ernstheit., geb., verstehend., selbständ., fescher Jungdame aus Kulturkreisen, die treues Füreinanderleben schätzt. Mitgift nicht nötig. Zuschrift m. Personalien (ohne Namen) unter „Durch Verstandesauswahl zur Liebesheirat und Glücksehe" befördert Adm. Beantwortung binnen 14 Tagen.

Junger Student
sehr fesch, sucht Bekanntschaft sich einsam fühlender Dame. Unter „Diskret 2602"

Welche
edeldenkende Dame

Bu
sucht
tagsa
„Ob

S
sucht
zielle
intelle
der D
zende

Inte
israel.
frau,
einsam
Jude

Pr
e
!!
Ausfüh
über s
Gum

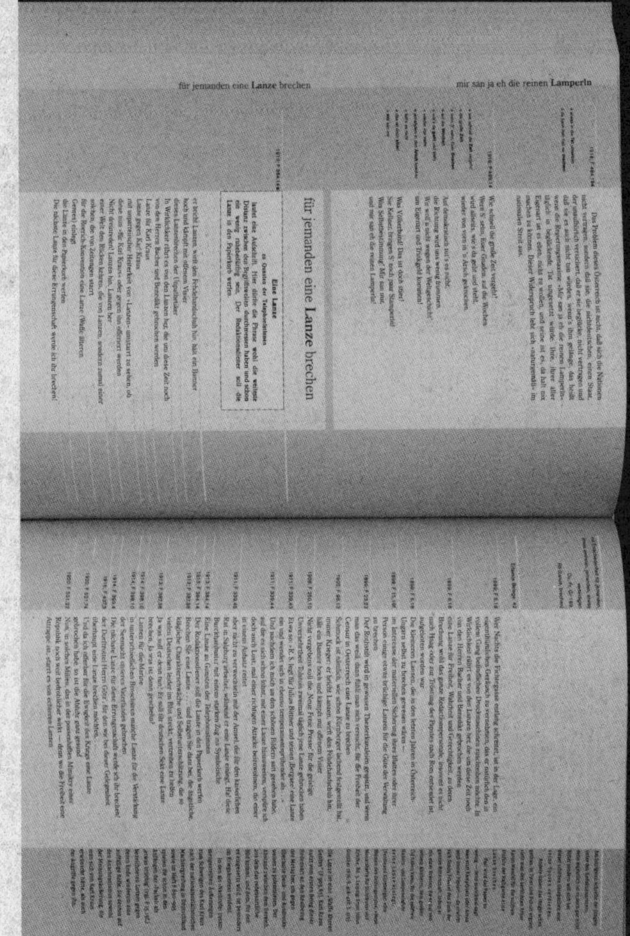

für jemanden eine Lanze brechen

mir san ja eh die reinen Lamperln

für jemanden eine Lanze brechen

2001 Gm

info
word
momox
Shop
medimops
FROM
momox GmbH
Am Alten Flughafen 1
04356 Leipzig
Germany
Referenz-Nr.: 109211452
A R
PRIORITY
W
Deutsche Post
WARENPOST Int. KT Port payé
TO
Zhao Qing
Nanjing Han Qing Tang Design,
Meiyuan New Village
210000 NANJING
CHINA VR
UC 032 957 862 DE
CN 22

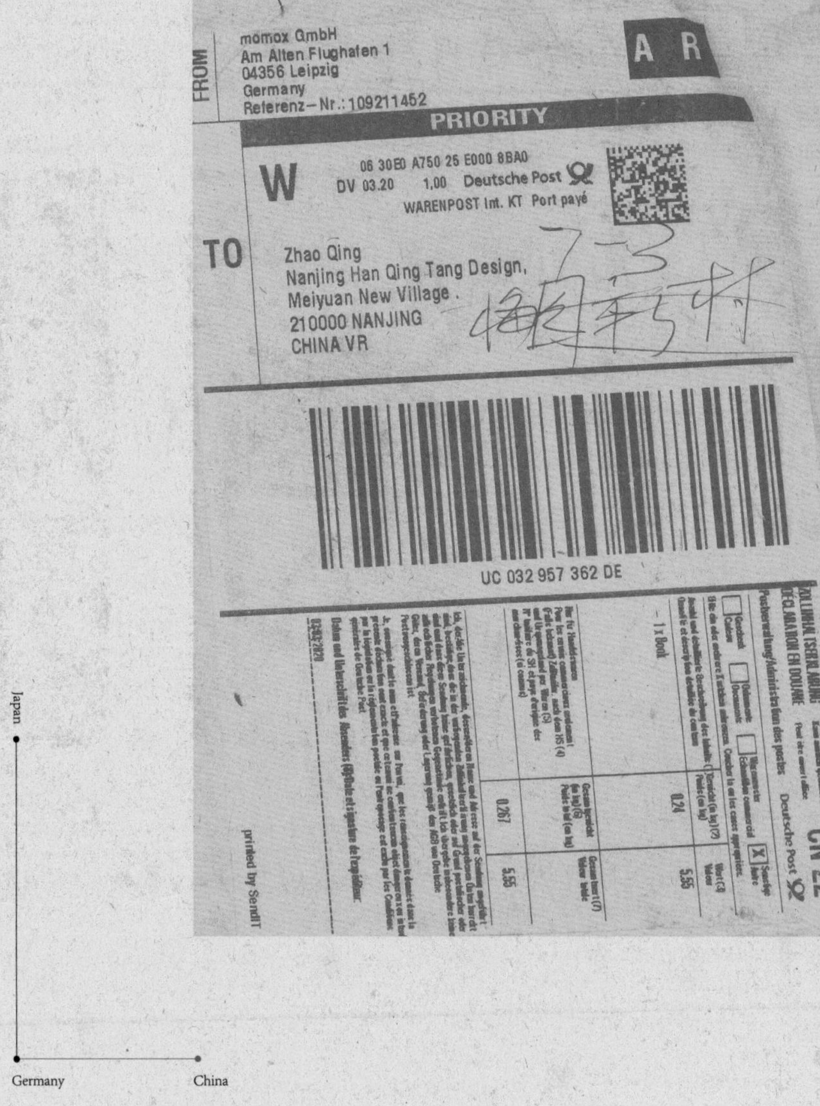

Japan

Germany

China

162mm

29mm

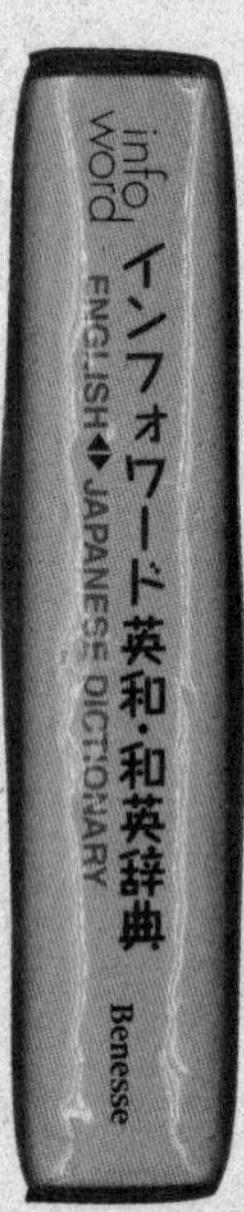

89mm

金奖
英和・和英辞典
日本

Gold Medal
infoword English-Japanese Dictionary
Japan
Haruyo Kobayashi
——
Benesse Corporation, Tokyo
162 89 29mm
317g
1392p
ISBN-10 4-8288-0445-5

极其小巧的一本日英双语字典。书籍的装订形式为锁线胶平装。函套采用透明 PVC 材质裱贴信息标签的形式，对书籍进行最外层的保护。封面白色网背橡胶材质，红蓝黑灰四色丝印。环衬胶版纸四色印刷，裱贴在封面背后，用于增加封面的强度。内页采用极薄的字典纸红黑双色印刷。全书分为三大部分，英译日、日译英，以及常用英语及使用场景。全书的设计都在尽可能地小巧便携、不易损坏和完善的索引功能上发力。作为一本手册，开本极小，但包含了近 1400 页的内容。双栏排版，字号虽小但印刷极其精细。超薄字典纸的使用，使得字典的厚度和重量被控制在一个非常合适的范围内。封面和函套的材质防水、防撞，对内页具有极佳的保护。装订背胶柔软，也使得字典可以很好地完全打开。翻口处设计了三大部分的字符索引，红蓝双色印刷；其中红色为内页线条出血裁切印迹，而蓝色则是直接印在书口上，工艺精度极佳。

This is an extremely small Japanese-English bilingual dictionary. The book is perfect bound with thread sewing. The book case is made of transparent PVC material mounted with information labels to protect the outermost layer of the book. The cover is made of white rubber material with four-color silkscreen printing in red, blue, black and gray. The end paper is made of offset paper printed in four colors, mounted on the back of the cover to increase the strength. The inner pages are printed with extremely thin dictionary paper in red and black. The book is divided into three parts: English to Japanese, Japanese to English, and scenarios of daily English. The design endeavors to make the book as small and portable as possible, not easy to damage, and perfect index function. As a manual, the book format is very small, but contains nearly 1400 pages of content, which is typeset in double columns. Although the font size is small, the printing is extremely fine. The use of ultra-thin dictionary paper restricts the thickness and weight within a very suitable range. The material of the cover and case is waterproof and anti-collision, which has excellent protection for the inner pages. The binding glue is soft and the dictionary can be opened completely. There are index of three sections at the fore-edge, printed in red and blue with excellent accuracy. Red is the bleed mark when trimming the inner pages, while blue is printed directly on the edge.

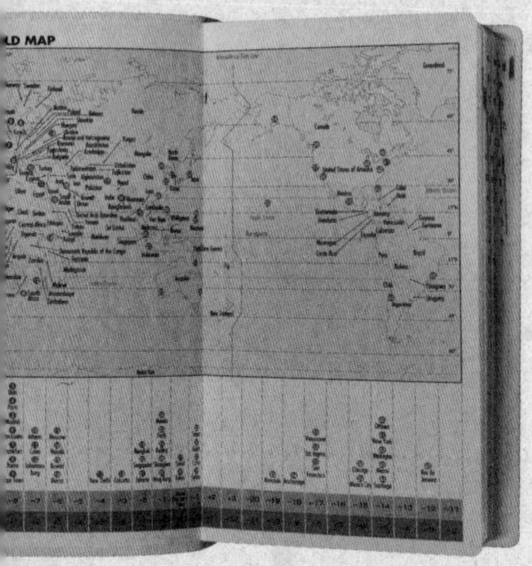

info word
ENGLISH▶JAPANESE DICTIONARY
インフォワード
英和辞典
ベネッセコーポレーション●編
Benesse

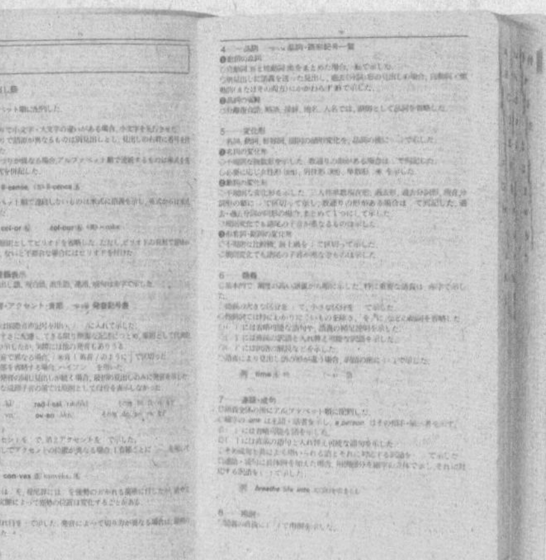

me. お役～になる be dis-
one's position そんなこ
れるのは～だ I don't want
ved in that.
～する comment 《on,
ake a comment 《on,

] *Gomoku* means mix-
ings or ingredients. *Go-*
is vinegared boiled rice
various seasonal ingre-
aoku-meshi is boiled rice
s seasonal ingredients.
交】→ 代わる代わる ‖ 悲喜
e mixed feelings of joy

字】a small letter
ち】‖ 彼は 4 人の～だ He
ldren. ～の魚 a seed fish
御尤も】‖ ～です You are
その件についてのお腹立ちは
may well be angry about

] small articles; accesso-
petty person; 〔総称〕small
an accessory box [case]
] a baby-sitter; a (dry)
る look after a baby; ba-
》 ▷ ～歌 a lullaby; a cra-

] shut *oneself* up 《in
》‖ 部屋に悪臭がこもってい

こゆき【小雪】 a light [little
がちらついた It snowed ligh
こゆび【小指】〔手〕the li
〔足〕the little toe
こよい【今宵】 this evening
こよう【雇用】 employmen
employ; hire; engage ‖
time employment ～を
velop further employme
均等 an equal employme
nity ～期間 the period
ment ～契約 an emplo
tract [agreement] ～者 a
被～者 an employee
ployment terms ～対
ment measures ～保険
ment insurance
ごよう【御用】 ‖ ～の際は
time you want me お
problem. 何か～ですか V
do for you? ▷ ～納め［始
ing [opening] of governm
for the year ～聞き an
～組合 a company unio
purveyor 《to the Impe
hold》 ～邸 an Imperial
ごよう【誤用】 a misuse
use
こよなく ‖ ～愛する love
deeply [more than anyor
こよみ【暦】 a calendar; a
～の上では by [according

vulgar
dialectal
derogatory
euphemistic

《古》 old
《諺》 prov
《小児》 nurs

軍事
経済
建築
言語
鉱石
歴史・史学

〔聖〕 聖書
〔生化〕 生化
〔生理〕 生理
〔地〕 地学
〔虫〕 昆虫
〔鳥〕 鳥類

PASS WITH CARE
追越し注意

NO PARKING $50 FINE
駐車禁止
50ドル罰金

NO TURNS
転回禁止

Department of Veterans Affairs	復員軍人省
Executive Office of the President	大統領府

candid 率直な意見をお聞か

ree —あなたの意見に
ます。

e]. とてもいい［素敵だ
あなたの…いいです

表現
d] ... 素敵な［いい］…です

イギリスの主な省庁	
Cabinet Office	内閣府
Department for Culture, Media and Sport	文化・メディア
Department for Education and Employment	教育雇用省
Department for International Development	国際開発省
Department of Health	保健省
Department of the Environment	環境省
Department of Transport	運輸省
Department of Social Security	社会保障省
Department of Trade and Industry	貿易産業省

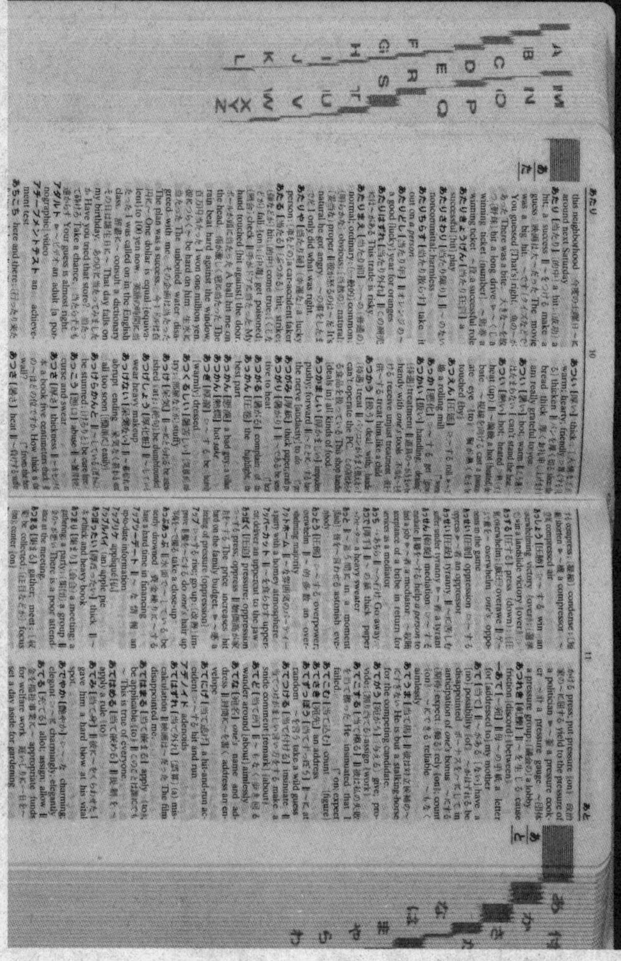

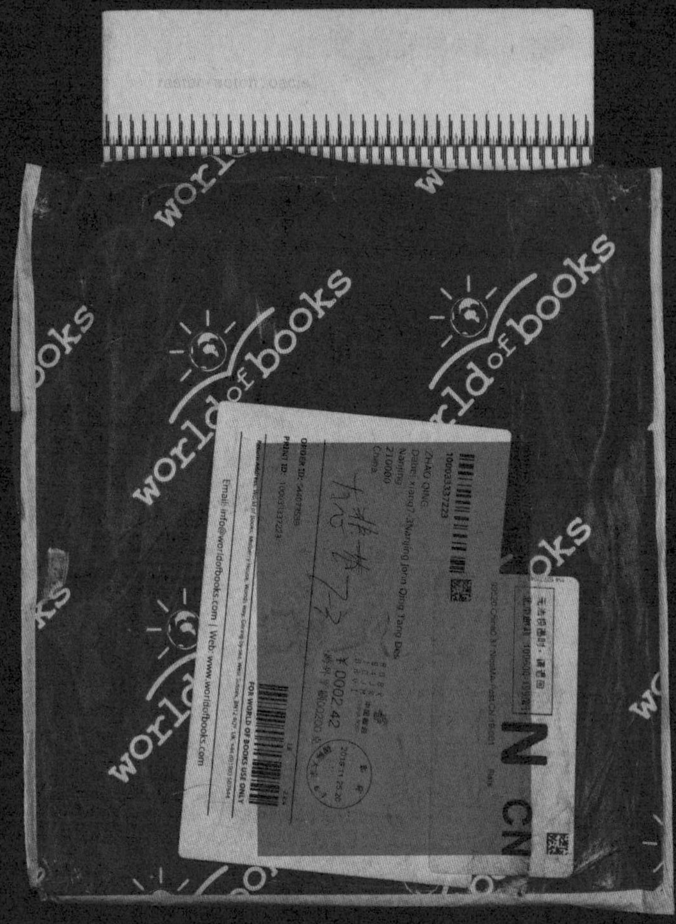
world of books
ZHAO QING
Nanjing
210000
China
Email: info@worldofbooks.com | Web: www.worldofbooks.com
FOR WORLD OF BOOKS USE ONLY
N
CN

2001 Bm 1

1000333337223
ZHAO QING
Dabei xiang7-3Nanjing John Qing Tang Des
Nanjing
210000
China
ORDER ID: 5440733539
RINT ID: 1000333337223
99520 China0.31'NicoMA-Post.CN 19-001
Flats
中国邮政
CHINA POST
¥0002.42
2019.11.25.20
Switzerland
U.K.
China

231mm

9mm

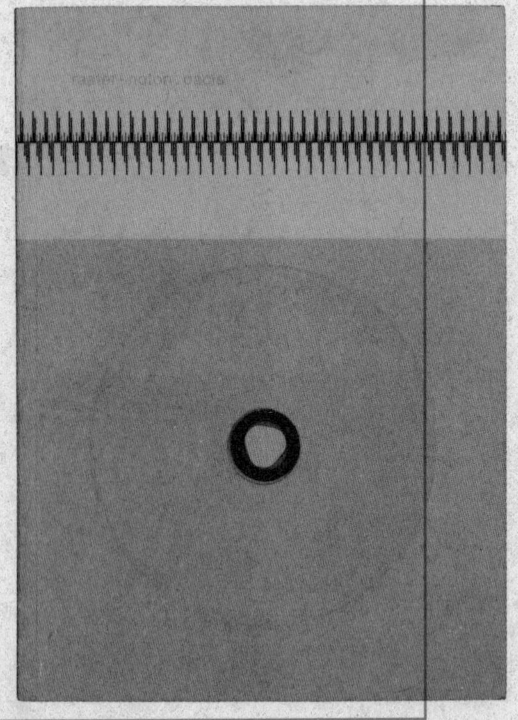

170mm

● 铜奖
« raster-noton.oacis
◐ 瑞士

Bronze Medal
raster-noton.oacis
Switzerland
Olaf Bender
Susanne Binas, Martin Pesch, Pinky Rose, Rob Young, Peter Kraut
raster-noton, Berlin; taktlos-bern, Bern
231 170 9mm
299g
104p
ISBN-10 3-931126-48-X

2000年，德国实验电子厂牌raster-noton在瑞士伯尔尼的Dampfzentrale文化中心举行了为期2天的声音实验展。4位raster-noton的声音艺术家在此活动期间通过声音、影像和装置的形式展示了他们的理念，本书便是对此次活动的记录和思考。书籍的装订形式为锁线胶平装。封面白卡纸黑色和灰白色双色印刷。封面中心挖空，透出夹在前勒口处的光盘。内页哑面铜版纸四色印刷。书中的主要内容是对这次声音实验活动的全面记录、细节描述和相关评论，包括raster-noton所做音乐的社会价值、现场活动、声音实验细节、CD录制、CD包装设计等等。书籍的设计以黑色文字为主，章节隔页采用红色，大量使用与声音、节奏能产生视觉联系的柱状波形图、同心圆音轨图、粗细线条与点状虚线等元素，加上在现场神秘光线下拍摄的照片和抽象而精致的包装设计，共同指向了raster-noton的几位艺术家的音乐风格。前勒口的光盘内收录了4位艺术家的6段音轨，用于欣赏。

In 2000, the German experimental electronics brand raster-noton held a two-day Sound Experiment Exhibition at the Dampfzentrale cultural center in Bern, Switzerland. Four sound artists from raster-noton demonstrated their ideas and thoughts in the form of sound, image and device during this event. This book is the record and review of this event. The book is glue bound with thread sewing. The cover is printed on the white cardboard in black and gray-white, and the center of the cover is hollowed out, showing the disc clipped at the front flap. The inside pages are made of matte coated paper printed in four colors. The main content of the book is a comprehensive record, detailed description and relevant comments on the sound experiment, including the social value of the music done by raster-noton, live activities, details of sound experiment, CD recording, CD packaging design, etc. The design of the book is mainly based on black text and the chapters are separated by red pages. Various elements such as columnar waveforms, concentric circular audio track diagrams, thick or thin lines and dotted lines, which have visual connections with sound and rhythm, together with the photos taken under mysterious light and the abstract and refined packaging design, all point to the music style of raster-noton. The CD-ROM clipped at the front flap contains 6 audio tracks of 4 artists for accompanying the reading process.

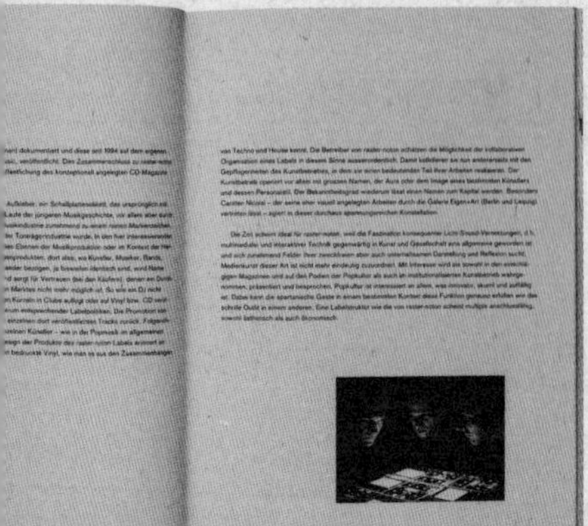

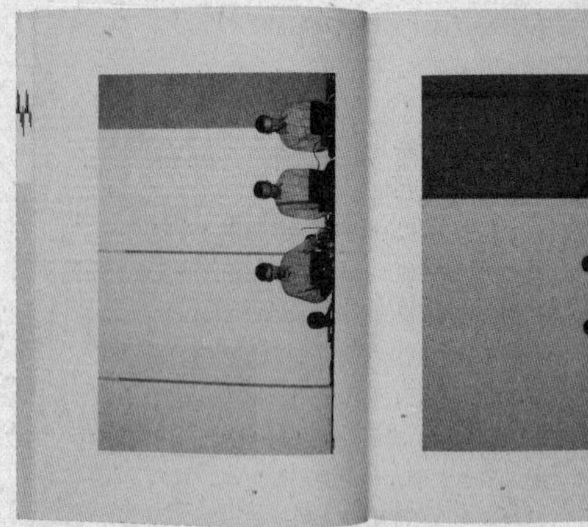

optic.sonic.reduction

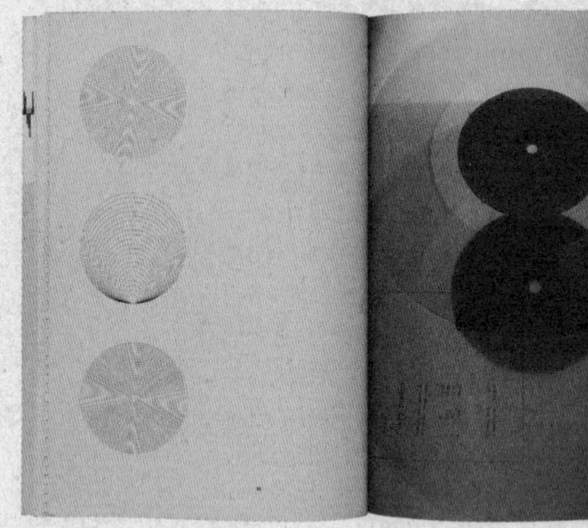

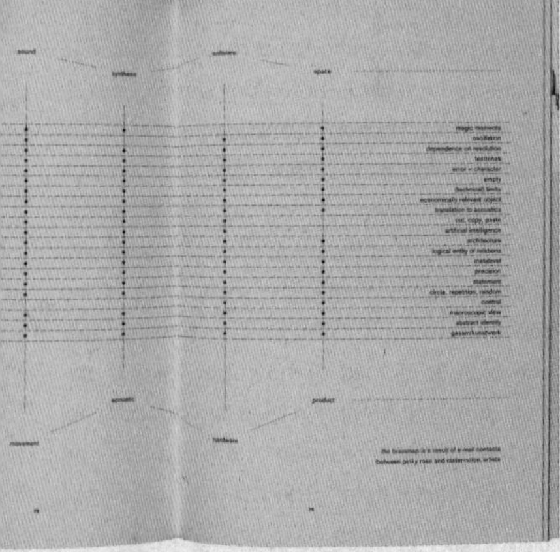

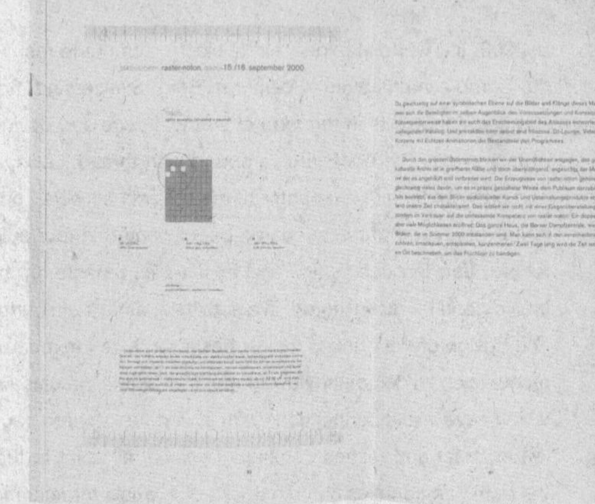

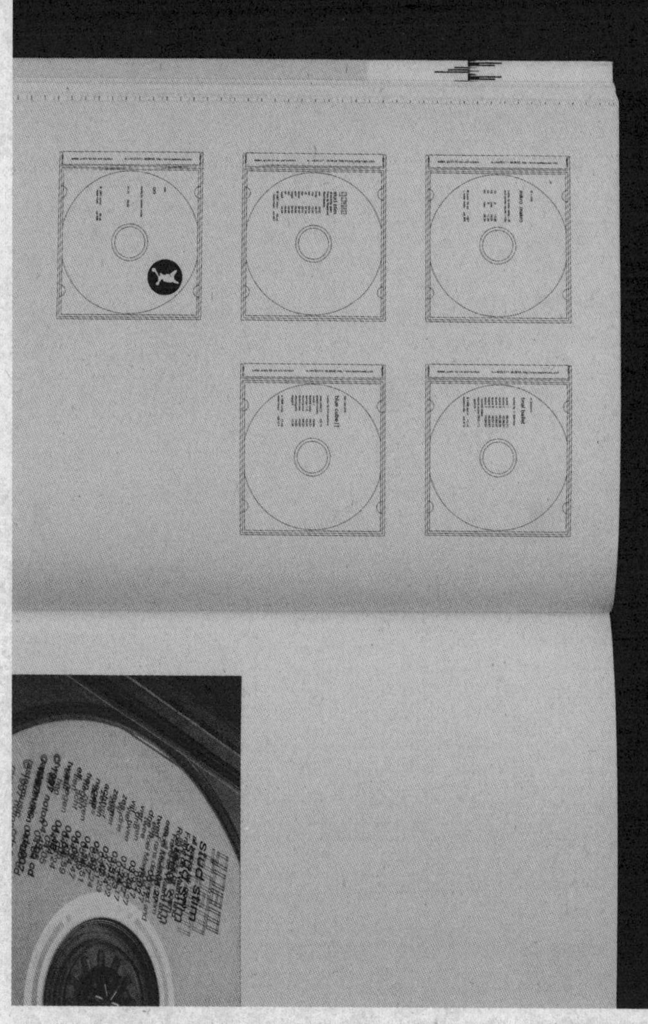

komet.saat
noto.spin
produkt.fl
HIGHSHIELD
STATIC SHIELDING
ATTENTION
OBSERVE PRECAUTIONS
FOR HANDLING
ELECTR.STATIC
DISCHARGE
SENSITIVE
DEVICES
RASTER-NOTON STATIC

twelve releases about the cutting edge of the millennium

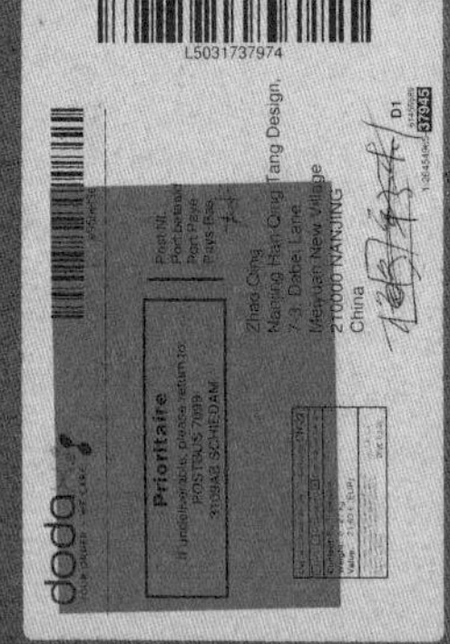
L5031737974
Zhao Qing
Nanjing Han Qing Tang Design,
7-3, Debe Lane
Meiyuan New Village
210000 NANJING
China
Prioritaire
POSTBUS 7099
3109AB SCHIEDAM
doda
D1
37945

2001 Bm²

e95bef3
YOUR ORDER – WE CARE
Prioritaire
If undeliverable, please return to:
POSTBUS 7099
3109AB SCHIEDAM
Post NL
Port betaal
Port Payé
Pays-Bas
Zhao Qing
Nanjing Han Qing
7-3, Dabei Lane
Meiyuan New Villa
May be opened officially Douane CN 22
Gift Documents Commercial Sample
Switzerland
The Netherlands
China

235mm

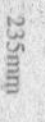

152mm

● 铜奖
« 锡特尔河的卵石
◐ 瑞士

Bronze Medal
Sitterkiesel
Switzerland
Jost Hochuli
Urs Hochuli, Oskar Keller
Edition Ostschweiz, VGS, St. Gallen
235 152 5mm
131g
48p
ISBN-10 3-7291-1095-0

瑞士博登湖（Bodensee）附近的锡特尔河（Sitter）盛产各式卵石，本书展示了部分卵石，并详细研究了这些卵石的起源。书籍的装订形式为锁线胶平装。护封采用半透明牛油纸调频网四色印刷，内封牛皮卡纸印单黑。内页胶版纸调频网印刷，图表部分大量使用明快的专色。书中首先对卵石做了起源和形态学介绍，进而对锡特尔河附近的卵石进行了岩石类型、尺寸、形状、成因等方面的研究。书中包含大量锡特尔河和圣加仑（St. Gallen）地区的水文地理、地质切面和生成年代分析的图表。图表的绘制清晰明快，文字介绍也简单易懂，方便非地质专业人员学习研究。版心设置为三栏，正文跨两栏，附注被设置为小字号占一栏。大标题设置为超粗字重，小标题则为斜体，文本设计简单清晰。卵石在书中的排布虽然较为随意，但也大体依照三栏参考线，具有严谨的节奏。

The Sitter River near the Bodensee Lake in Switzerland is rich in various types of pebbles. This book diaplays some pebbles and studies their origins in detail. The book is glue bound with thread sewing. Frequency modulated screens in four colors are used for the jacket which is made of semitransparent kraft paper. The inner cover is printed in single black on Kraft paper. Inner pages are made of offset paper printed with FM screens, and bright PMS colors are widely used in the charts. The book first introduces the origin and morphology of pebbles, and then studies the type, size, shape and genesis of pebbles near the Sitter River. The book contains a large number of charts for hydrogeological analysis, geological section analysis and generative age analysis of the sitter River and St. Gallen area, but the drawing of the charts is clear and bright, the text introduction is also simple and easy to understand, which is convenient for non geological professionals to study and study. The typeset area is set to three columns – the text spans two columns, and the notes are set in one column with small font size. The headline is set to be super bold while the small title is set to be italicized. The text design is simple and clear. Although the arrangement of pebbles in the book is quite casual, it also follows the three-column reference line and has a strict rhythm.

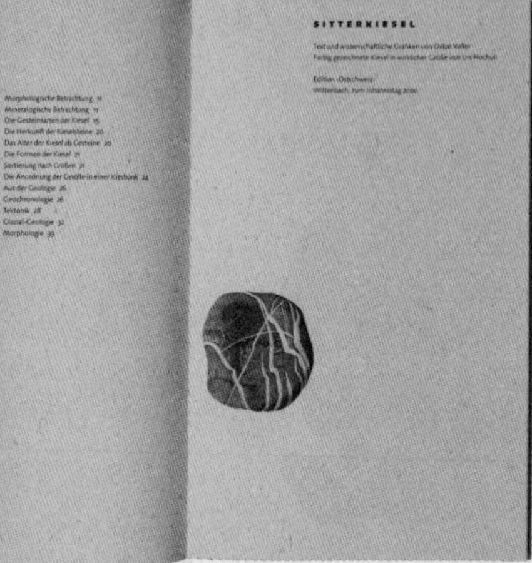
SITTERKIESEL
Morphologische Betrachtung 11
Mineralogische Betrachtung 11
Die Gesteinsarten der Kiesel 15
Die Herkunft der Kieselsteine 20
Das Alter der Kiesel als Gesteine 20
Die Formen der Kiesel 21
Sortierung nach Größen 21
Die Anordnung der Geölle in einer Kiesbank 24
Aus der Geologie 26
Geochronologie 26
Tektonik 28
Glazial-Geologie 32
Morphologie 39

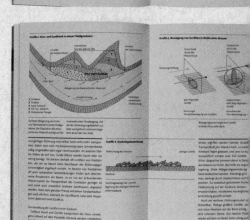

Zusammenschub der Sedimente und der unterliegenden kristallinen Gesteine die Entstehung der Alpen ihren Anfang. Die Gesteinsschichten wurden gefaltet und die Schichtpakete über viele Dutzende Kilometer als so genannte Decken übereinander geschoben. Vor vielleicht 70 Millionen Jahren tauchten die ersten Berge als Inselkette aus dem Meer auf. Sofort setzte die Abtragung durch exogene Prozesse ein. Der Schutt sedimentierte in den Meeresbecken rund um die Inseln und wurde in der Folge zu den Flyschgesteinen verfestigt.

Der weitere Zusammenschub ließ vor rund 40 Millionen Jahren die Inselberge zu einer einzigen großen Insel, den Alpenbogen, zusammenwachsen. Aus dem nun schon mächtigen Gebirge trugen Urflüsse das Verwitterungs- und Erosionsmaterial, Geröll, Sand, Schlamm, in die nördliche Vortiefe hinaus. Alpennah wurde in erster Linie das grobe Material, weiter draußen das feinere deponiert. Vor den Austrittstoren der großen Flüsse wurden riesige Geröllkörper akkumuliert. Alle diese Sedimente werden unter dem Begriff Molasse zusammengefasst (Grafik 6). Die bis über 5000 m mächtigen Aufschüttungen wurden im Laufe der Jahrmillionen zu Molasse-Gestein zusammengepresst, insbesondere zu Nagelfluh, Sandstein und Mergel.

In der letzten Hauptphase der Alpengenese vor 15 bis 5 Millionen Jahren wurden von Süden her die so genannten helvetischen Decken auf die Molassegesteine aufgeschoben. In Alpennähe zerbrach die Molasse in Pakete, die nun ebenfalls übereinander gestapelt und schräg gestellt wurden. In dieser Phase erfolgte die Platznahme der alpinen Frontgebirge wie Alpstein und Churfirsten sowie die Bildung der Molasse-Voralpen (Grafik 7). Schließlich wurde der Alpenkörper in der Ostschweiz um 1500 bis 2000 m emporgehoben. Im Molasse-Vorland schwankt der Hebungsbetrag um 1000 m. Dies bedeutete vor 10 bis 5 Millionen Jahren das Ende der Aufschüttung im Molasse-Vorland; im Gegenteil, die Flüsse begannen sich nun im neuen Molasse-Hochland einzuschneiden.

Die Gebirgsbildung der Alpen ist dafür verantwortlich, dass unter anderem auch jene Gesteinsschichten herausgehoben und durch die Erosion freigelegt wurden, aus denen später un-

Grafik 6. Paläogeographie der Unteren Süßwassermolasse (25 Mio. Jahre vor heute)

Während der Molassezeit schütteten die Urflüsse aus den werdenden Alpen riesige Schuttfächer im Vorland auf. Zweimal erstreckte sich ein untiefer Meeresarm durch dieses Vorland: Untere und Obere Meeresmolasse. Zweimal deh sich im gleichen Raum eine w Tieflandebene wie die Poeber heute aus: Untere (siehe obig Karte) und Obere Süßwasser lasse. Aus Keller & Krayss 199

Gletscher auf, die in Kaltzeiten größere Gebirgsvergletscher gen hervorriefen. Schließlich wechselten ab 1,5 Millionen ren echte Eiszeiten (Glaziale) mit milderen Zwischeneisze (Interglaziale) ab. Während den Glazialen wuchsen die Eisr sen im Alpenraum derart an, dass mächtige Gletscher durch Täler der Alpenflüsse bis ins Vorland vorstoßen und sich fladenartig ausbreiten konnten. Erst hier im Unterland wa die Temperaturen hoch genug, dass die Eismassen abtau

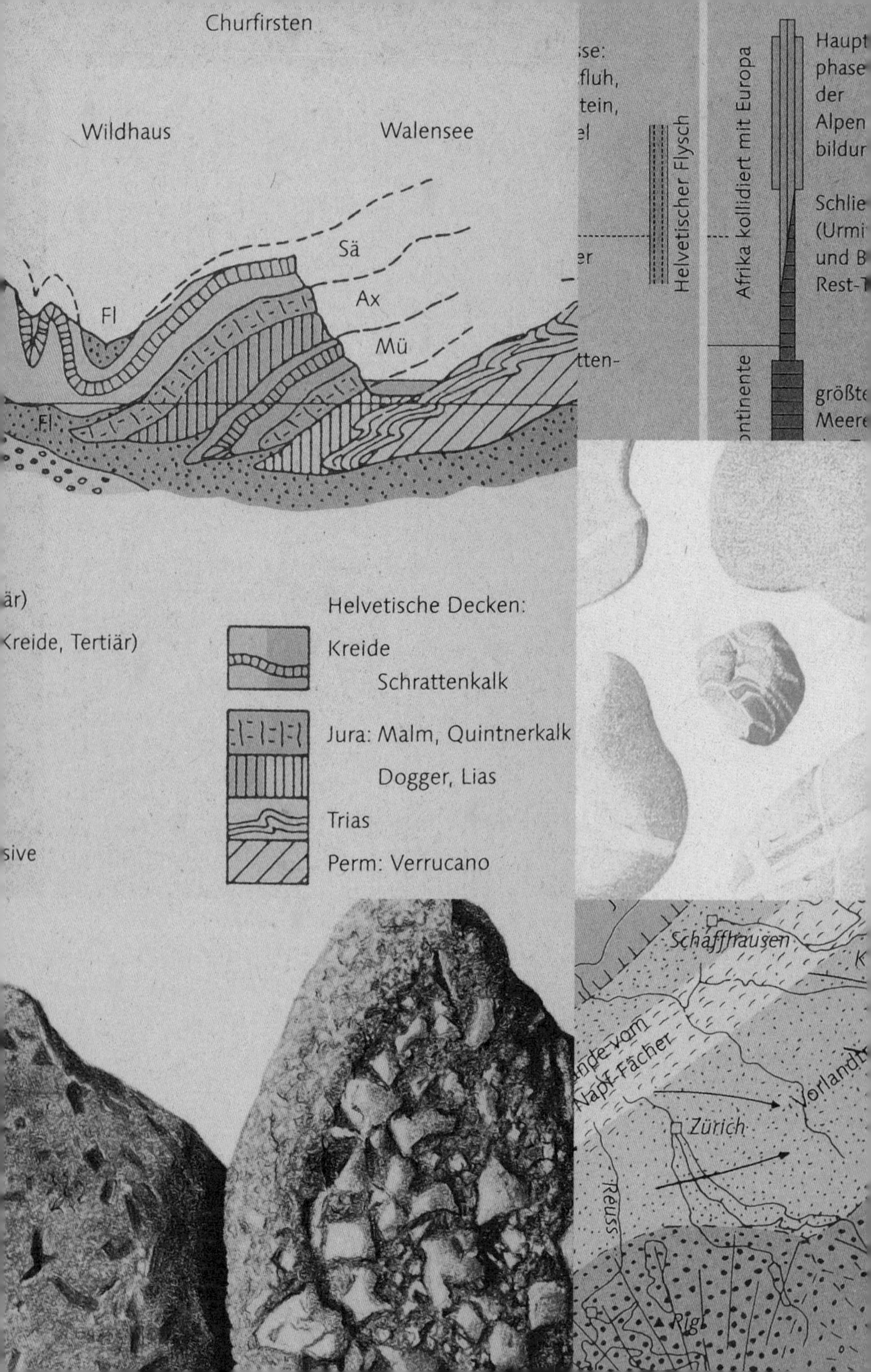
Churfirsten
Wildhaus
Walensee
Sä
Ax
Mü
Fl
Fl
Helvetischer Flysch
Afrika kollidiert mit Europa
Helvetische Decken:
Kreide
Schrattenkalk
Jura: Malm, Quintnerkalk
Dogger, Lias
Trias
Perm: Verrucano
Schaffhausen
Napf-Fächer
Zürich
Reuss
Rigi

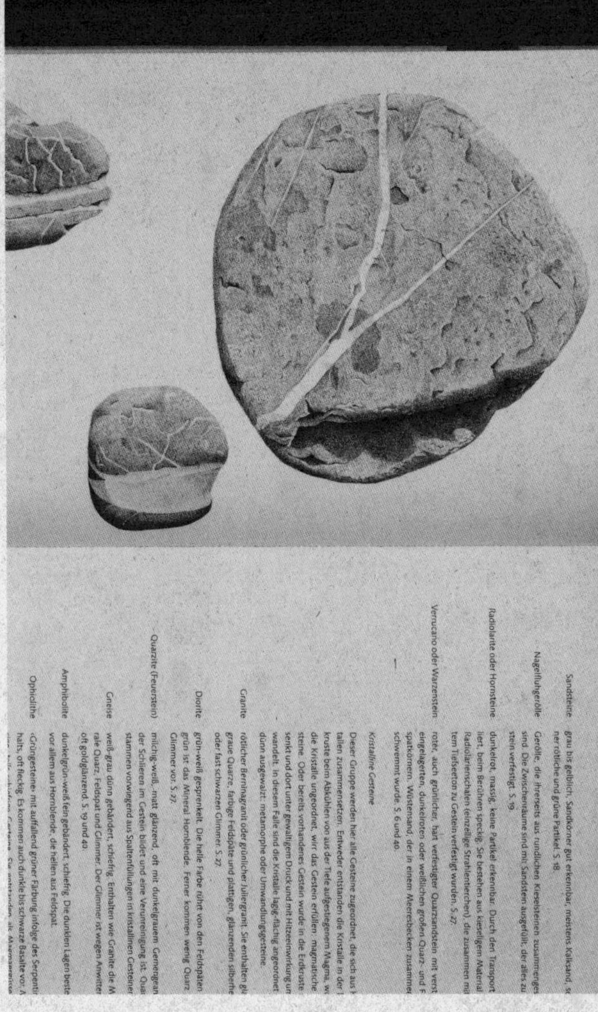

Sandsteine
grau bis gelblich, Sandkörner gut erkennbar, meistens Kalksand,
ner rötliche und grüne Partikel. S. 18.
Nagelfluhgerölle
Gerölle, die ihrerseits aus rundlichen Kieselsteinen
sind. Die Zwischenräume sind mit Sandstein ausgefüllt, der alles
stein verfestigt. S. 19.
Radiolarite oder Hornsteine
dunkelrot, massig, keine Partikel erkennbar. Durch den
liert, beim Berühren speckig. Sie bestehen aus kieseligem
Radiolarienschalen (einzellige Strahlentierchen), die zusammen
tem Tiefseeton zu Gestein verfestigt wurden. S. 27.
Verrucano oder Warzenstein
roter, auch grünlicher, hart verfestigter Quarzsandstein mit
eingelagerten, dunkelroten oder weißlichen großen Quarz- und
spatkörnern. Wüstensand, der in einem Meeresbecken
schwemmt wurde. S. 6 und 40.
Kristalline Gesteine
Dieser Gruppe werden hier alle Gesteine zugeordnet, die sich aus
tallen zusammensetzen. Entweder entstanden die Kristalle in der
kruste beim Abkühlen von aus der Tiefe aufgestiegenem Magma,
die Kristalle ungeordnet, wirr das Gestein erfüllen: magmatische
steine. Oder bereits vorhandenes Gestein wurde in die Erdkruste
senkt und dort unter gewaltigem Druck und mit Hitzeeinwirkung
wandelt. In diesem Falle sind die Kristalle lagig-flächig angeordnet
dünn ausgewalzt: metamorphe oder Umwandlungsgesteine.
Granite
rötlicher Berninagranit oder grünlicher Juliergranit. Sie enthalten
graue Quarze, farbige Feldspäte und plattigen, glänzenden
oder fast schwarzen Glimmer. S. 27.
Diorite
grün-weiß gesprenkelt. Die helle Farbe rührt von den Feldspäten
grün ist das Mineral Hornblende. Ferner kommen wenig Quarz
Glimmer vor. S. 27.
Quarzite (Feuerstein)
milchig-weiß, matt glänzend, oft mit dunkelgrauem
der Schlieren im Gestein bildet und eine Verunreinigung ist.
stammen vorwiegend aus Spaltenfüllungen in kristallinen
Gneise
weiß-grau dünn gebändert, schiefrig. Enthalten wie Granite die
rale Quarz, Feldspat und Glimmer. Der Glimmer ist wegen
oft goldglänzend. S. 19 und 40.
Amphibolite
dunkelgrün-weiß fein gebändert, schiefrig. Die dunklen Lagen
vor allem aus Hornblende, die hellen aus Feldspat.
Ophiolithe
‹Grüngesteine› mit auffallend grüner Färbung infolge des
halts, oft fleckig. Es kommen auch dunkle bis schwarze Basalte vor.

www.valentinska.cz
Česká pošta
R
DOPORUČENÉ
RECOMMANDÉ
Odesílatel/Sender
Antikvariat Valentinska
Valentinská 52/8
110 00 Praha/Staré Město
Czech Republic (CZ)
PSČ podací pošty 143 00
Hmotnost 0.500 kg
TAXE PERCUE
982807-0079/2013
18
QING ZHAO
ZHAO QING
Dabei xiang7-3,Nanjing Han Qing Tang Des
210000 Nanjing
Prioritaire
China (Peoples Rep) (CN)

ká pošta
der:
Valentinska
52/8
na/Staré Město
ublic (CZ)
pošty: 143 00
500 kg
R
DOPORUČENÉ
RECOMMANDÉ
RR951927840CZ
TAXE PERCUE
982807-0079/2013
Adresát/Addressee:
QING ZHAO
ZHAO QING
13952094
Czech Republic
Czech Republic
China

● 铜奖
《 裸体女巨人
◖ 捷克

Bronze Medal
Nahá obryn
Czech Republic
Ji í Šalamoun, Zden k Ziegler
Ji í Šalamoun
Aulos, Prag
282 168 17mm
525g
62p
ISBN-10 80-86184-03-X

伊日 · 沙拉蒙（Ji í Šalamoun）是一名插画家、书籍设计师和动画电影设计师，他善于利用民间艺术和文学原型创作充满幻想、荒诞、神秘和有趣的作品。本书中他通过文字和插图记录下自己的梦。书籍的装订形式为圆脊锁线硬精装。护封采用艺术纸黄黑双色丝印。护封在上下书口处比书籍开本略大，上下书口和勒口处均向内包折，形成良好的翻阅手感。内封艺术纸丝印黑裱覆卡板。内页以艺术纸张印单黑为主，每隔 8 面贴一张三色插画。作者记录了 1971 － 1996 年中的 30 多个梦，大多数梦都包含文字和插图。梦的排序并未完全按照时间排列，而是依照梦的相互关系调整了先后顺序。书的最后部分是记录梦境的操作细节，比如对梦境尽量如实的记录原则、对梦的心理学思考、记录时半梦半醒的状态和速写风格，以及作为记录本身的补充和解释。文字排版上，对梦的描述文字使用正常的字母间距，而对梦醒之后的状态和思考的文字则将字母间距拉开，形成鲜明的差异。

Jiří Šalamoun is an illustrator, book designer, and animated film designer. He is good at using folk art and literary prototypes to create works full of fantasy, absurdity, mystery, and fun. In this book, he records his dreams through words and illustrations. The hardcover book is bound by thread sewing with rounding back. The jacket is made of art paper with yellow and black screen-printing. The jacket is slightly larger at the upper and lower book edges than the format. Both the upper and lower book edges and the flaps are folded inwards, forming a comfortable feeling. The inner cover is made of art paper treated with silk screen-printing in black, mounted with cardboard. The inner pages are mainly printed in black on art paper, with a three-color illustration attached on every eight sides. The author recorded more than 30 dreams from 1971 to 1996. Most dreams were expressed by words and illustrations. The order of dreams was not completely chronological, but the order of the dreams was adjusted in consideration of the relationship between dreams. The final part of the book is to record the details of the operation of the dream, such as the principle of recording the dream as faithfully as possible, the psychological thinking of the dream, the state of half dreaming and half awake when recording, as well as the supplement and explanation of the recording itself. As for typesetting, the normal letter spacing is used for the descriptive words of dreams, while the larger letter spacing is used for the words of expressing and thinking after waking up, thus forming distinct differences in texts.

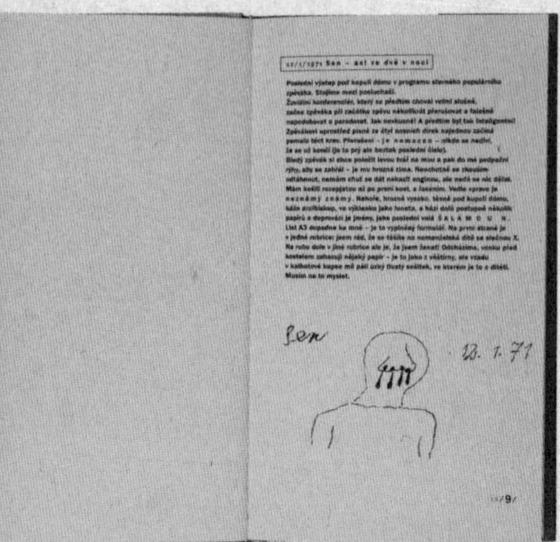
17/1/1971 Sen – asi ve dvě v noci
Sen
12. 1. 71
/9/

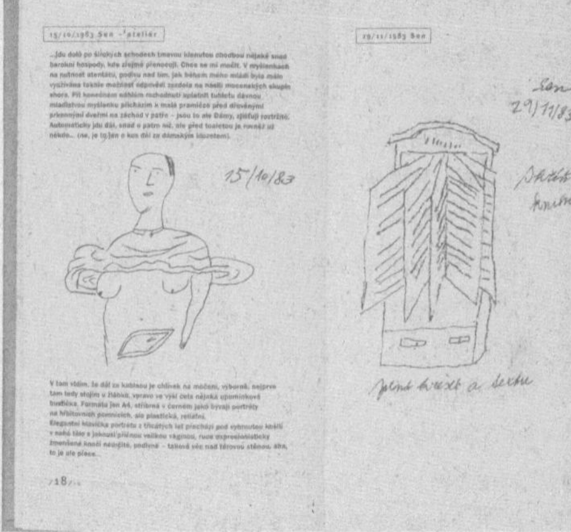
15/10/1983 Sen – ateliér
15/10/83
/18/
29/11/1983 Sen
Sen 29/11/83

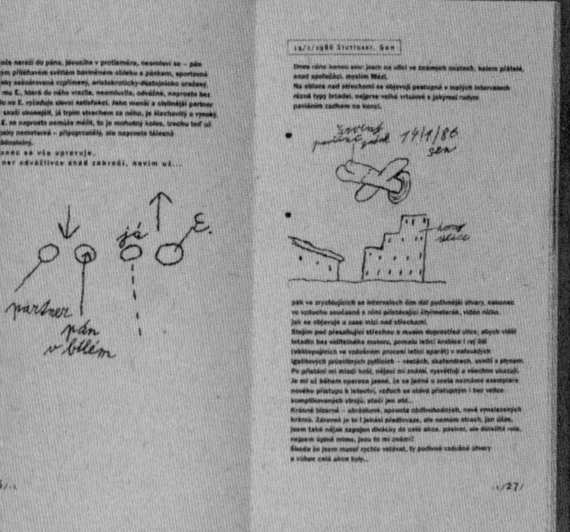
já
E.
partner
pán v bílém
14/1/1986 Stuttgart, Sen
14/1/86 Sen
/27/

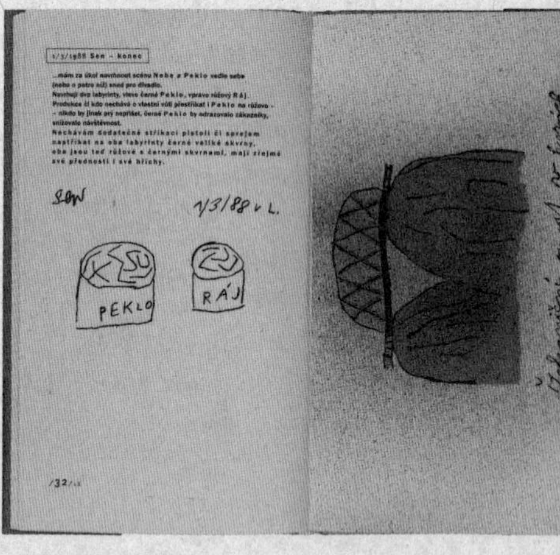
1/3/1988 Sen – konec
Sen
1/3/88 v L.
PEKLO
RÁJ
/32/

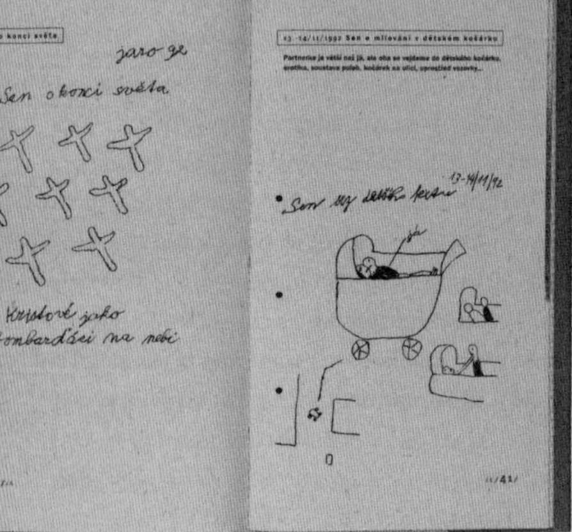
jaro 92
Sen o konci světa
Kristové jako bombarďáci na nebi
13.–14/11/1992 Sen o milování v dětském kočárku
13–14/11/92
/41/

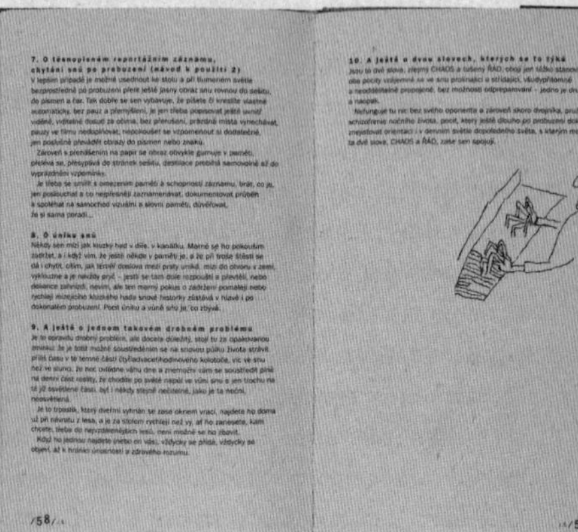
/58/
/59/

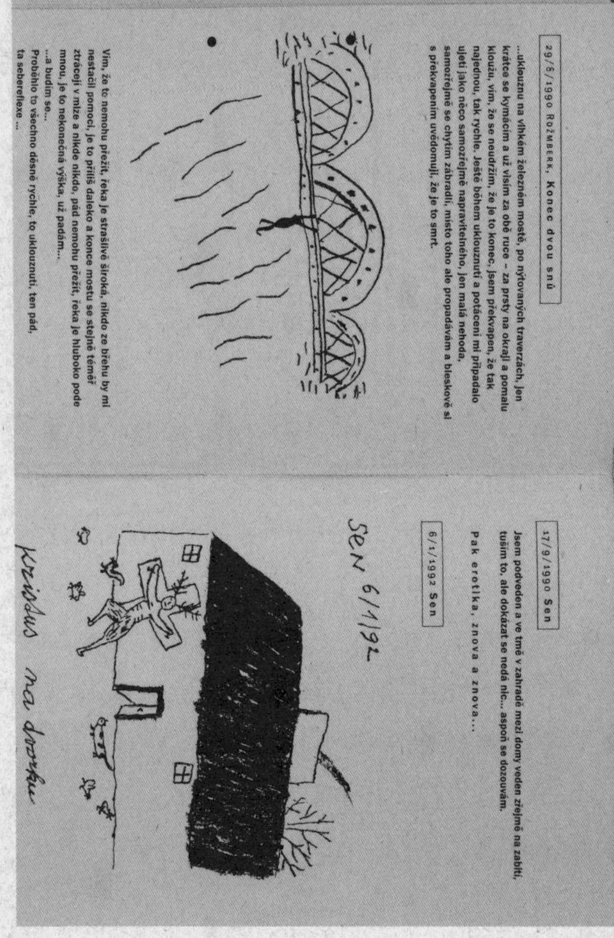

29/8/1990 Rožmberk, Konec dvou snů

...uklouznu na vlhkém železném mostě, po nýtovaných traverzách, jen krátce se kymácím a už visím za obě ruce – za prsty na okraji a pomalu kloužu, vím, že se neudržím, že je to konec, jsem překvapen, že tak najednou, tak rychle. Ještě během uklouznutí a potácení mi připadalo ujetí jako něco samozřejmě napravitelného, jen malá nehoda, samozřejmě se chytím zábradlí, místo toho ale propadávám a bleskově si s překvapením uvědomuji, že je to smrt.

Vím, že to nemohu přežít, řeka je strašlivě široká, nikdo ze břehu by mi nestačil pomoci, je to příliš daleko a konce mostu se stejně téměř ztrácejí v mlze a nikde nikdo, pád nemohu přežít, řeka je hluboko pode mnou, je to nekonečná výška, už padám...
...a budím se...
Proběhlo to všechno děsně rychle, to uklouznutí, ten pád, ta sebereflexe ...

17/9/1990 Sen

Jsem podveden a ve tmě v zahradě mezi domy veden zřejmě na zabití, tuším to, ale dokázat se nedá nic... aspoň se dozouvám.

Pak erotika, znova a znova...

6/1/1992 Sen

část snu slyším tu větu", jako oznámení
mně, ale jako veřejné konstatování – h
ovanou (nikoliv vyřvávanou) nádražním
s dobře slyšitelnou výslovností, jak Py

poté, co zahnu doprava v ulici za pos
a činžáky je malé prostranství, na kter
se snaží popolézt jakýsi chlapík, obleč
vacími pracovními botami.

vpravo před sebou zezadu a zprava a
nými, děsně pomalými broučími pohyb
o ostrého tvaru střechy s hranou ve sr
ůzy či kombinézy!
ý hrb, zřejmě ne odevždy, mu evidentn
tlé znetvoření ho nutí dělat pohyby um
š už trochu jen reflexní, automatické.
zní podivné oznámení:
í fór metafory je v tom, že
jak dlouhý."

4/5/2/77
Karlmarxstadt

-ká
yn

ec pru...

í rozměr čas

osti od kruhu řadit spoje
e něco v úterý... Nešlo by
vzadu otvírají dveře.
ší dlouhý hubený pán v š
, unavenou a jakoby kaši
utku úst polovystrčený ja
spíš jako s protézou, jako
hrůzou tuhnu, v tom za dv
oknem ve tmě, mezi kv
odél celého ateliéru šmír
menuté důvody k urgent
a jindy.
do osvětlené místnosti,
vnitř jedná, ho...

"I suppose, Watson, we must look upon you as a man of letters" said he. "How do you define the word 'grotesque'? "Strange – remarkable," I suggested. He shook his

1. Nejprve o boji se sny zapisováním

Sny se mi zdály od dětství jako asi každému druhému, jen jsem si je snad o něco déle po probuzení pamatoval.

Možná že rozdíl mezi mým dnem a mou nocí nebyl tak radikální, tak čitelný jako u jiných, a tak jsem nezvládl sny prostě tak jednoznačně odbýt pokrčením ramen a „Träume sind Schäume" pro mne od jisté doby tak zřetelně neplatilo – nosil jsem v sobě zbytky snů často dost dlouho, aniž jsem to navenek dával znát. Racionální já bojovalo ztuha s nočními událostmi stejně jako s denními intuitivními pocity a nezřetelnými tušeními.

Nakonec, někdy v padesátých letech, mezi koncem školy a třicítkou, jsem si je začal zapisovat, napůl z potřeby se sám ujistit, že byly, napůl z pocitu jakési úřednicko-dokumentaristické povinnosti je zaznamenat, snad abych se jich tím lépe zbavil, z potřeby zvědavé autoterapie. Byl to můj gymnaziální profesor psychologie a filozofie, kdo tenkrát, vlastně paradoxně riskantně, mě seznámil se jmény Freud, Adler, a hlavně Jung a upozornil, že o snech není nutné jen mlčet a že nejsem sám, koho zajímají.

V mém prvním zapamatovaném dětském snu (asi mezi pátým a šestým rokem) utíkám pronásledován drakem po jakési pláni, padám do jámy, nebo vlastně díry do podzemí, odkud není úniku, nahoře drak zastiňuje otvor při lezení do svislé chodby za mnou. Přitom vím, že i zespoda se cosi chystá, a budím se hrůzou.

Ten pohádkový sen se pak v dětství častěji opakoval.

2. O pravdomluvnosti nočního deníku

Od samého začátku byl v potřebě tohoto zpravodajství o nočních událostech přítomen princip dokumentární svědomitosti a pravdomluvnosti, zásada zákazu dalšího fabulování nebo využívání či zpracování.

Psal jsem sny způsobem selských kronikářů a jen pro sebe, ne pro eventuální pozdější zájemce.

Myslím, že texty mají povahu záznamů ústavů či agenturních zpráv materiálů pro sebereflexi, pro informace o sobě, a snad tedy tím vždycky tak trochu i o druhých.

Jsou pokračováním, možná prodloužením dětského zpovědního zrcadla katolických jinochů, zprávou o životě za zrcadlem spánku, ať už to znamená cokoliv.

Snad fungoval i jako jeden z možných-ryze soukromých důkazů jeho eventuální existence.

Poznámky tedy patří do oblasti lodních deníků kapitána Cooka nebo

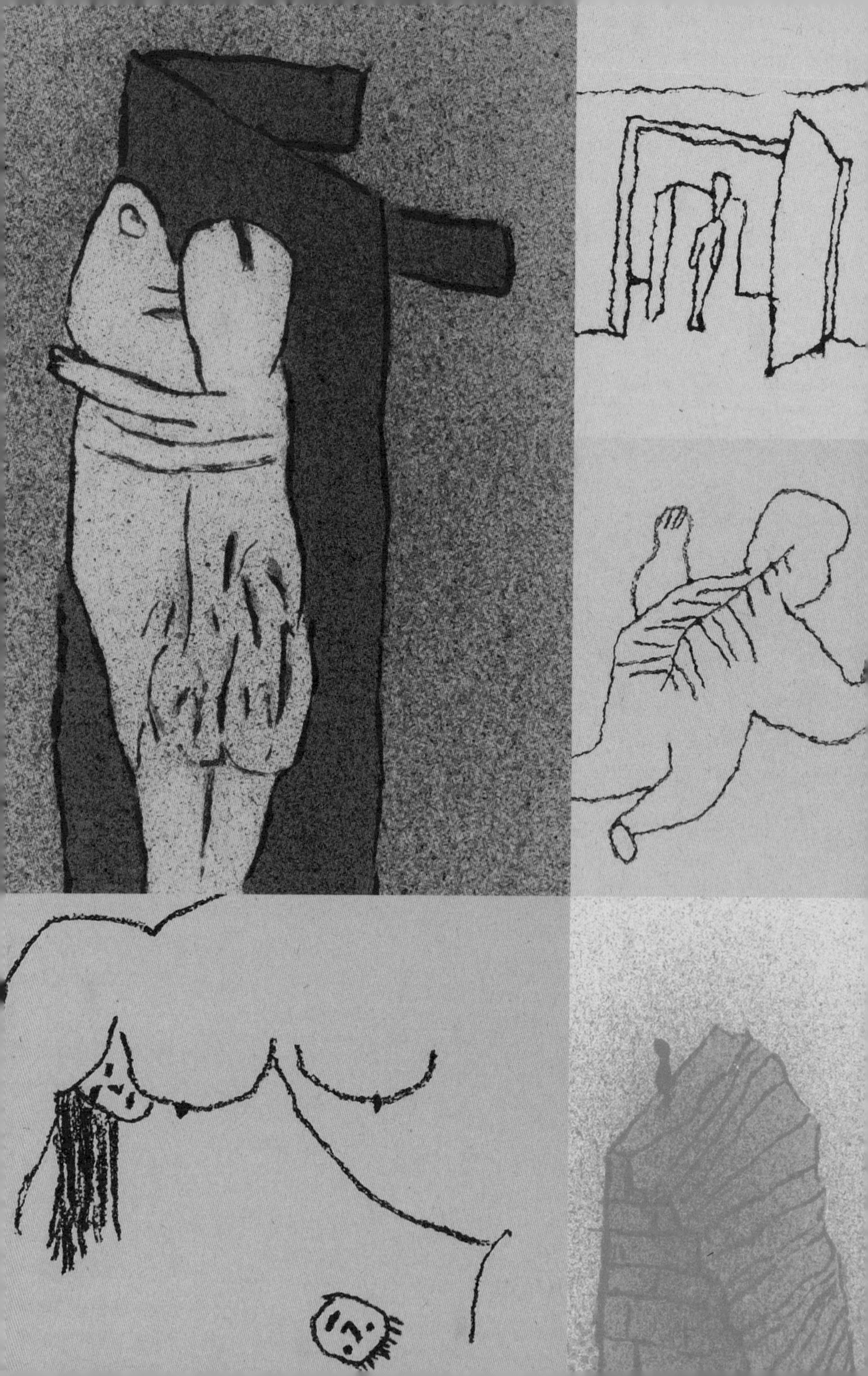

6/9/1993 Sen – konec

Stojím na žluté písčité pláži, vlevo ode mne veliká plocha šplouchajícího klidného moře, zelenomodrého, přecházejícího do mlhava v dálce.
Vnímám jen ten břeh písku, vbíhajícího do vody. Kus přede mnou a nade mnou mi neznámá ruka nabízí ramínko na šaty. Jsem úplně nahý a bezmocný. Gesto neznámé ruky (končící v půli předloktí), která nepozorovaně podává ramínko – klidně, skoro nehybně, ale zřetelně ho nabízí.

Sen 5 – 6/9/93

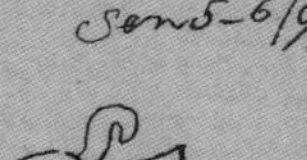

22/11/1993 Sen – zbytek

...na paloučku, uprostřed spíš romantizujícího anglického parku, vraždí Ježíš Kristus kohosi bodnými ranami nožem, vedenými hezky zvysoka – je to ten krásný muž v bílé říze s kadeřavým vousem a kaštanovými dlouhými vlasy, vysoký a štíhlý atlet pod měkkou řízou se spoustou drobných řasnatých záhybů z porcelánové produkce poloviny minulého století, prerafaelistický pastýř z chromolitografií.
Pohybuje se divoce profesionálně jako nějaký zelený baret, vybíhá zprava z lesíka a zaútočí na oběť, kterou stěží rozeznávám. Pružnými skoky šelmy sráží nešťastnou oběť do trávy, pokleká na ni a jako John Silver bodá z výšky svisle dlouhým nožem.

Dochází mi, jak to tedy vlastně je: Kristus je zřejmě Satan a celé to učení je tu jen jako zástěrka pro maniakálního proroka.
Příroda je klidná, zelená, mírná, zřejmě léto.
Mladé poměnší stromoví kolem, něco jako okraj lesoparku, nebojím se, že bych byl další na řadě, ale jsem zaražen tím objevem – tak náhlým tygřím skokem téhle důstojné figury, ke které patří klidný a mírumilovný pohyb, ne tohle soustředěné řádění s tím evidentním jiným řádem věcí...

NE-PO 22/11/93
RUMUNSKA
les vzadu

Konstruktive Gebrauchsgrafik
Richard Paul
Lohse

2001 Bm5

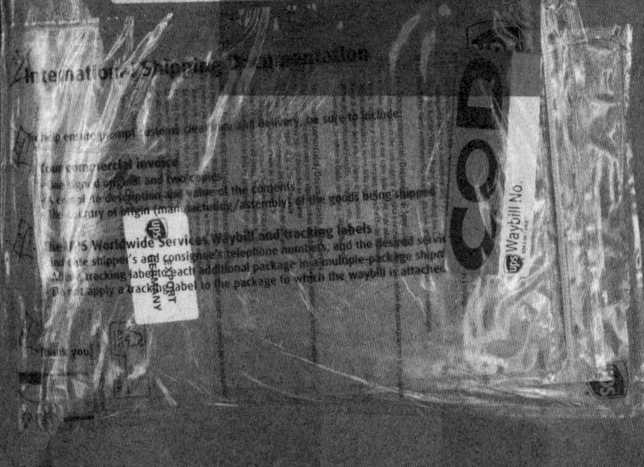

Germany

Germany China

306mm

36mm

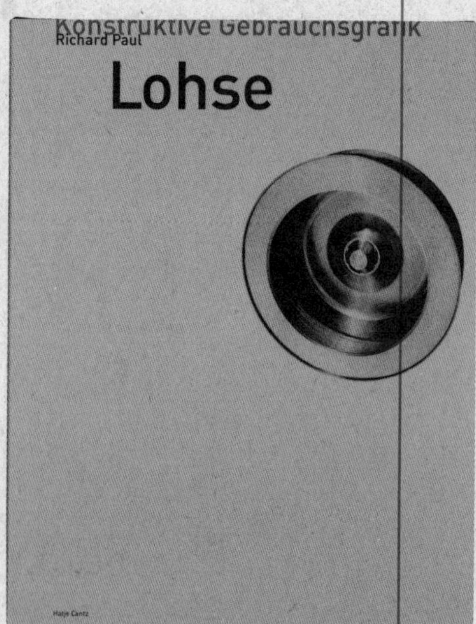

244mm

铜奖	Bronze Medal
理查德·保罗·洛斯 ——建构的商业图形	Richard Paul Lohse: Konstruktive Gebrauchsgrafik
德国	Germany
	Markus Bosshard (Weiersmüller Bosshard Grüninger wbg AG)
	Richard Paul Lohse Stiftung
	Hatje Cantz Verlag, Ostfildern-Ruit
	306 244 36mm
	2118g
	312p
	ISBN-10 3-7757-0767-0

理查德·保罗·洛斯（Richard Paul Lohse）是瑞士画家、广告设计师，是国际主义设计风格的先驱之一。本书是在洛斯100周年诞辰之际举办的回顾展的作品集。书籍的装订形式为锁线硬精装。护封铜版纸红黄黑三色印刷，覆哑光膜。内封深灰色布面裱覆卡板，文字压凹处理。环衬选用黑卡纸，内页哑面铜版纸四色印刷。内容方面，书中对洛斯的一生进行了回顾，比较了其绘画和设计作品的异同，探讨了他的作品受到大时代背景的影响，陈述了他对平面设计的贡献。全书的设计宛如一本国际主义设计的样本，从多种无衬线字体的选择，到使用网格系统进行的版面设计，甚至对于用于区隔内容的极细线的使用，都在践行着洛斯等国际主义设计先驱者们的规范。这使得全书的气质与书中所展示的洛斯的作品相契合，呈现出极致的秩序感。

Richard Paul Lohse is a Swiss painter, advertising designer and one of the pioneers of internationalist design style. This book is a collection of works from a retrospective exhibition held on the 100th anniversary of Lohse's birth. The hardcover book is bound by thread sewing. The matted jacket is made of coated paper printed in red, yellow and black, and the inner cover is made of dark gray cloth mounted with cardboard, and the text is embossed. The end paper is made of black cardboard, while the inner pages are printed with four colors on matte coated paper. As for the content, the book reviews Lohse's life, compares the similarities and differences between his paintings and design works, the influence of the background on his work and his contribution to graphic design. The design of the whole book is like a sample of international style. From the selection of various sans serif fonts to the layout design using grid system, and even the use of extremely fine lines to distinguish the content, it is practicing the norms of the pioneers of international design such as Lohse. This makes the temperament of the book match with the works of Lohse displayed in the book, showing the ultimate sense of order.

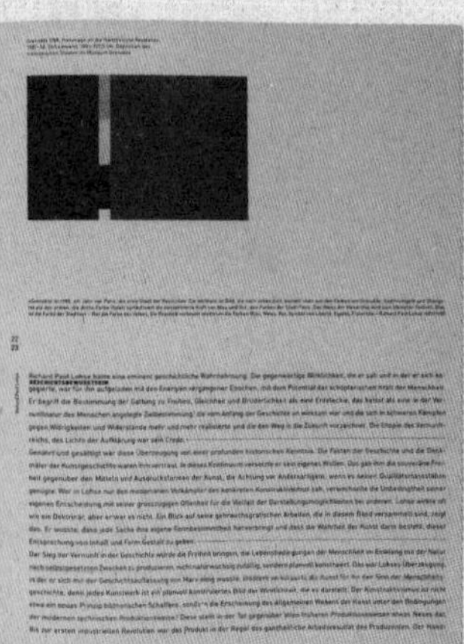

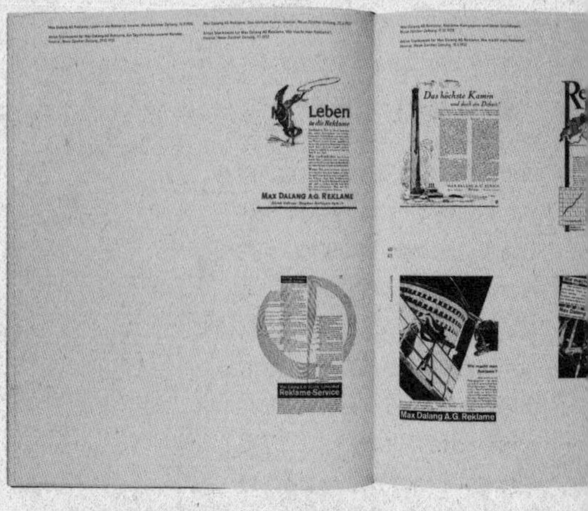
Leben
in die Reklame
Max Dalang A.G. Reklame
Reklame-Service
Das höchste Kamin
Max Dalang A.G. Reklame

SINCLAIR
GELD

Kunstmuseum
Winterthur
Ungegenständliche
Malerei
in der Schweiz

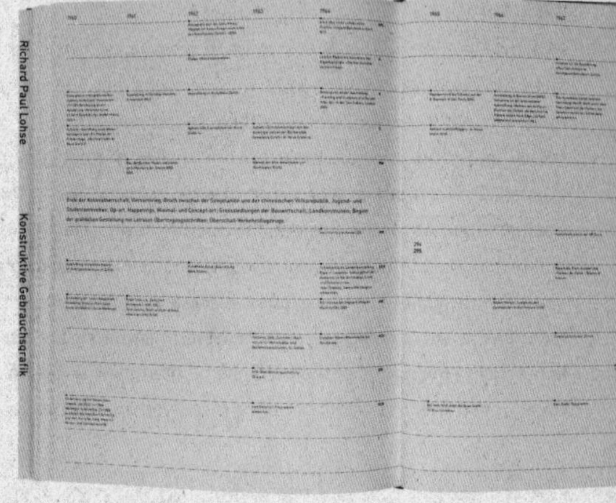
Richard Paul Lohse
Konstruktive Gebrauchsgrafik

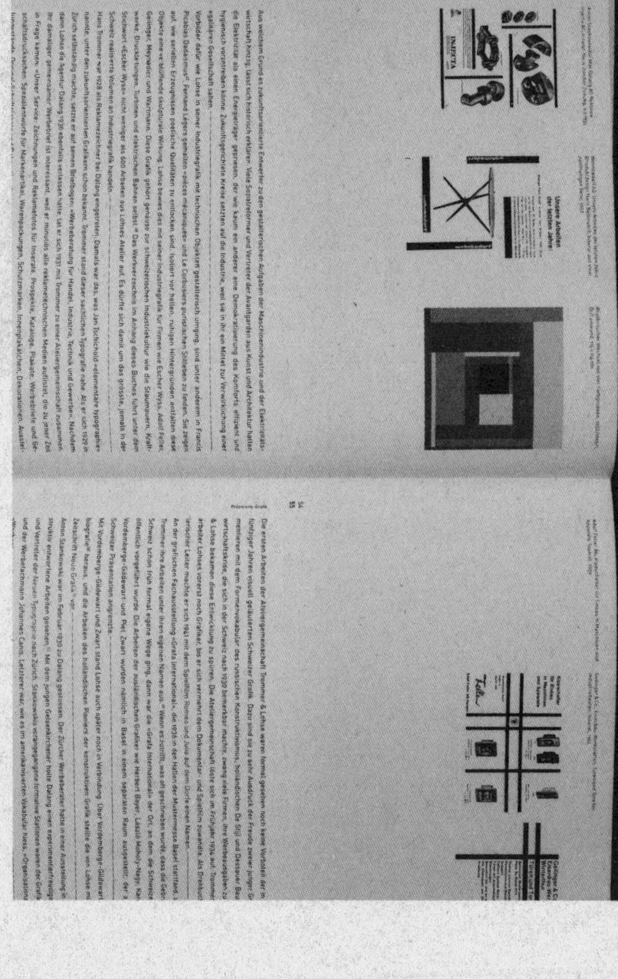

Aus welchem Grund es zukunftsorientierte Entwerfer zu den gestalterischen Aufgaben der Maschinenindustrie und der Elektrizitätswirtschaft hinzog, lässt sich historisch erklären. Viele Sozialreformer und Vertreter der Avantgarden aus Kunst und Architektur hatten die Elektrizität als einen Energieträger gepriesen, der wie kaum ein anderer eine Demokratisierung des Komforts effizient und hygienisch vorantreiben könne. Zukunftsgerichtete Kreise setzten auf die Industrie, weil sie in ihr ein Mittel zur Verwirklichung einer egalitären Gesellschaft sahen.

Vorbilder dafür wie Lohse in seiner Industriegrafik mit technischen Objekten gestalterisch umging, sind unter anderem in Francis Picabias Dadaismus[47], Fernand Légers gemalten «pièces mécaniques» und Le Corbusiers puristischen Stilleben zu finden. Sie zeigen auf, wie seriellen Erzeugnissen poetische Qualitäten zu entlocken sind. Isoliert vor hellen, ruhigen Hintergründen entfalten diese Objekte eine verblüffende skulpturale Wirkung. Lohse bewies dies mit seiner Industriegrafik für Firmen wie Escher Wyss, Adolf Feller, Geilinger, Meynadier und Wartmann. Diese Grafik gehört genauso zur schweizerischen Industriekultur wie die Staumauern, Kraftwerke, Druckleitungen, Turbinen und elektrischen Bahnen selbst.[48] Das Werkverzeichnis im Anhang dieses Buches führt unter dem Stichwort «Escher Wyss» nicht weniger als 600 Arbeiten aus Lohses Atelier auf. Es dürfte sich damit um das grösste, jemals in der Schweiz realisierte Volumen an Industriegrafik handeln.

Hans Trommer war 1928 als Reklamezeichner bei Dalang eingetreten. Damals war das, was Jan Tschichold «elementare typographie» nannte, unter den zukunftsorientierten Grafikern schon bekannt. Trommer stand dieser sachlichen Typografie nahe. Als er sich 1929 in Zürich selbständig machte, setzte er auf seinen Briefbogen: «Werbeberatung für Handel, Industrie, Technik und Gewerbe». Nachdem dann Lohse die Agentur Dalang 1930 ebenfalls verlassen hatte, tat er sich 1931 mit Trommer zu einer Ateliergemeinschaft zusammen. Ihr damaliger gemeinsamer Werbebrief ist interessant, weil er minutiös alle reklametechnischen Medien auflistet, die zu jener Zeit in Frage kamen: «Unser Service: Zeichnungen und Reklamefotos für Inserate, Prospekte, Kataloge, Plakate, Werbebriefe und Geschäftsdrucksachen. Spezialentwürfe für Markenartikel, Warenpackungen, Schutzmarken, Innenplakätchen, Dekorationen, Ausstel-

54
55

Die ersten Arbeiten der Ateliergemeinschaft Trommer & Lohse waren formal gesehen noch keine Vorboten der in
fünfziger Jahren visuell geläuterten Schweizer Grafik. Dazu sind sie zu sehr Ausdruck der Freude zweier junger Gr
mentieren mit dem Formenvokabular des russischen Konstruktivismus, holländischen De Stijl und Dessauer Bau
wirtschaftskrise, die sich in der Schweiz nach 1930 bemerkbar machte, zwang viele Firmen, ihre Werbeausgaben zu
& Lohse bekamen diese Entwicklung zu spüren. Die Ateliergemeinschaft löste sich im Frühjahr 1934 auf. Trommer
arbeiter Lohses vorerst noch Grafiker, bis er sich vermehrt dem Dokumentar- und Spielfilm zuwandte. Als Drehbuc
lerischer Leiter machte er sich 1941 mit dem Spielfilm Romeo und Julia auf dem Dorfe einen Namen.

An der grafischen Fachausstellung «Grafa International», die 1936 in den Hallen der Mustermesse Basel stattfand, s
Trommer ihre Arbeiten unter ihren eigenen Namen aus.[49] Wenn es zutrifft, was oft geschrieben wurde, dass die Gebr
Schweiz schon früh formal eigene Wege ging, dann war die «Grafa International» der Ort, an dem die Schweize
öffentlich vorgeführt wurde. Die Arbeiten der ausländischen Grafiker wie Herbert Bayer, László Moholy-Nagy, Kar
Vordemberge-Gildewart und Piet Zwart wurden nämlich in Basel in einem separaten Raum ausgestellt, der a
Schweizer Präsentation angrenzte.

Mit Vordemberge-Gildewart und Zwart stand Lohse auch später noch in Verbindung. Über Vordemberge-Gildewart
biografie[50] heraus, und die Arbeiten des holländischen Pioniers der konstruktiven Grafik stellte die von Lohse mi
Zeitschrift Neue Grafik[51] vor.

Anton Stankowski war im Februar 1930 zu Dalang gestossen. Der Zürcher Werbeberater hatte in einer Ausstellung in
struktiv entworfene Arbeiten gesehen.[52] Mit dem jungen Gelsenkirchener holte Dalang einen experimentierfreudige
und Vertreter der Neuen Typographie nach Zürich. Stankowskis vorangegangene formative Stationen waren der Grafik
und der Werbefachmann Johannes Canis. Letzterer war, wie es im amerikanisierten Vokabular hiess, «Organisations

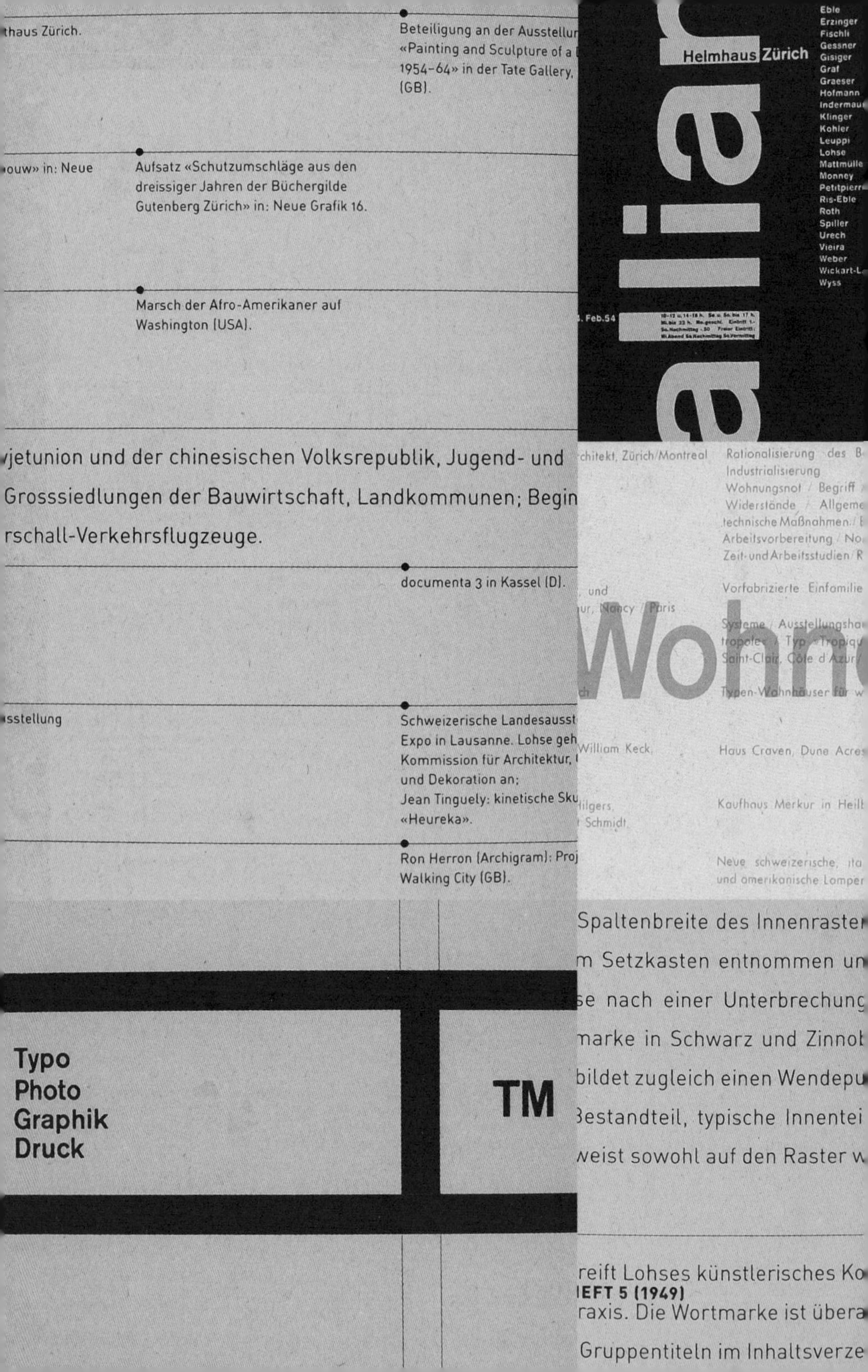
thaus Zürich.
Beteiligung an der Ausstellur
«Painting and Sculpture of a
1954–64» in der Tate Gallery,
(GB).
ouw» in: Neue
Aufsatz «Schutzumschläge aus den dreissiger Jahren der Büchergilde Gutenberg Zürich» in: Neue Grafik 16.
Marsch der Afro-Amerikaner auf Washington (USA).
vjetunion und der chinesischen Volksrepublik, Jugend- und
Grosssiedlungen der Bauwirtschaft, Landkommunen; Begin
rschall-Verkehrsflugzeuge.
documenta 3 in Kassel (D).
sstellung
Schweizerische Landesausst
Expo in Lausanne. Lohse geh
Kommission für Architektur,
und Dekoration an:
Jean Tinguely: kinetische Sku
«Heureka».
Ron Herron (Archigram): Proj
Walking City (GB).
Helmhaus Zürich
Eble
Erzinger
Fischli
Gessner
Gisiger
Graf
Graeser
Hofmann
Klinger
Kohler
Leuppi
Lohse
Monney
Ris-Eble
Roth
Spiller
Urech
Vieira
Weber
Wyss
Feb.54
chitekt, Zürich/Montreal
Rationalisierung des
Industrialisierung
Wohnungsnot / Begriff
Widerstände / Allgeme
technische Maßnahmen /
Arbeitsvorbereitung / No
Zeit- und Arbeitsstudien / R
und
Vorfabrizierte Einfamilie
Wohn
Typen-Wohnhäuser für w
William Keck
Haus Craven, Dune Acre
Hilgers
Schmidt
Kaufhaus Merkur in Heil
Neue schweizerische, ita
und amerikanische Lamper
Spaltenbreite des Innenraste
m Setzkasten entnommen u
se nach einer Unterbrechung
marke in Schwarz und Zinno
bildet zugleich einen Wendepu
Bestandteil, typische Innente
weist sowohl auf den Raster w
Typo
Photo
Graphik
Druck
TM
reift Lohses künstlerisches Ko
HEFT 5 (1949)
raxis. Die Wortmarke ist übera
Gruppentiteln im Inhaltsverze

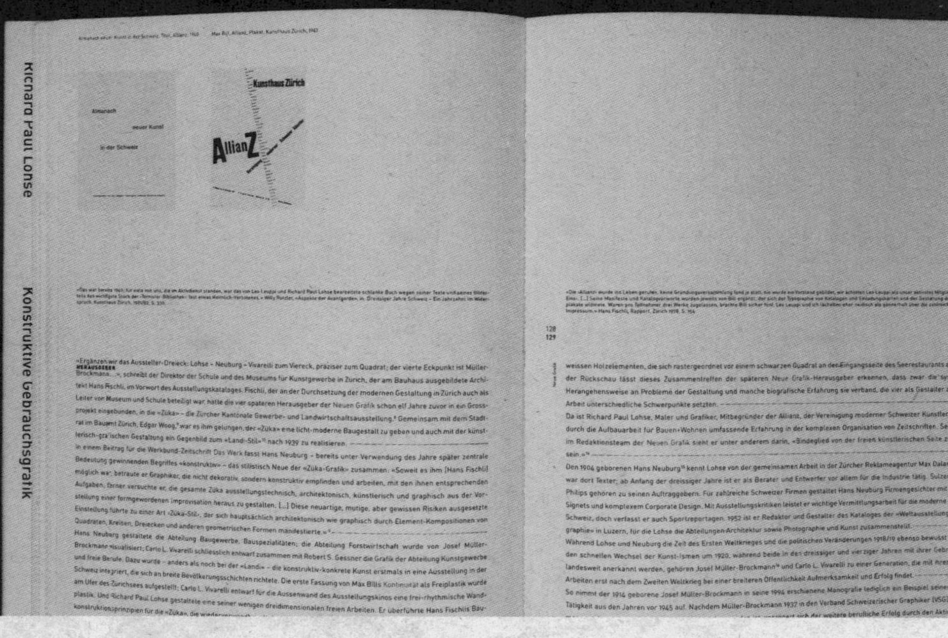

Richard Paul Lohse

Konstruktive Gebrauchsgrafik

Almanach neuer Kunst in der Schweiz, Titel, Allianz, 1940 — Max Bill, Allianz, Plakat, Kunsthaus Zürich, 1942

128
129

«Ergänzen wir das Aussteller-Dreieck: Lohse – Neuburg – Vivarelli zum Viereck, präziser zum Quadrat; der vierte Eckpunkt ist Müller-
HERAUSGEBER
Brockmann…», schreibt der Direktor der Schule und des Museums für Kunstgewerbe in Zürich, der am Bauhaus ausgebildete Architekt Hans Fischli, im Vorwort des Ausstellungskataloges. Fischli, der an der Durchsetzung der modernen Gestaltung in Zürich auch als
Leiter von Museum und Schule beteiligt war, hatte die vier späteren Herausgeber der Neuen Grafik schon elf Jahre zuvor in ein Gross-
projekt eingebunden, in die «Züka» – die Zürcher Kantonale Gewerbe- und Landwirtschaftsausstellung.[8] Gemeinsam mit dem Stadt-
rat im Bauamt Zürich, Edgar Woog,[9] war es ihm gelungen, der «Züka» eine licht-moderne Baugestalt zu geben und auch mit der künst-
lerisch-grafischen Gestaltung ein Gegenbild zum «Landi-Stil»[10] nach 1939 zu realisieren.
In einem Beitrag für die Werkbund-Zeitschrift Das Werk fasst Hans Neuburg – bereits unter Verwendung des Jahre später zentrale
Bedeutung gewinnenden Begriffes «konstruktiv» – das stilistisch Neue der «Züka-Grafik» zusammen: «Soweit es ihm [Hans Fischli]
möglich war, betraute er Graphiker, die nicht dekorativ, sondern konstruktiv empfinden und arbeiten, mit den ihnen entsprechenden
Aufgaben, ferner versuchte er, die gesamte Züka ausstellungstechnisch, architektonisch, künstlerisch und graphisch aus der Vor-
stellung einer formgewordenen Improvisation heraus zu gestalten. […] Diese neuartige, mutige, aber gewissen Risiken ausgesetzte
Einstellung führte zu einer Art ‹Züka-Stil›, der sich hauptsächlich architektonisch wie graphisch durch Element-Kompositionen von
Quadraten, Kreisen, Dreiecken und anderen geometrischen Formen manifestierte.»[11]
Hans Neuburg gestaltete die Abteilung Baugewerbe, Bauspezialitäten; die Abteilung Forstwirtschaft wurde von Josef Müller-
Brockmann visualisiert; Carlo L. Vivarelli schliesslich entwarf zusammen mit Robert S. Gessner die Grafik der Abteilung Kunstgewerbe
und freie Berufe. Dazu wurde – anders als noch bei der «Landi» – die konstruktiv-konkrete Kunst erstmals in eine Ausstellung in der
Schweiz integriert, die sich an breite Bevölkerungsschichten richtete. Die erste Fassung von Max Bills Kontinuität als Freiplastik wurde
am Ufer des Zürichsees aufgestellt; Carlo L. Vivarelli entwarf für die Aussenwand des Ausstellungskinos eine frei-rhythmische Wand-
plastik. Und Richard Paul Lohse gestaltete eine seiner wenigen dreidimensionalen freien Arbeiten. Er überführte Hans Fischlis Bau-
konstruktionsprinzipien für die «Züka», die wiederverwendb…

«Die ‹Allianz› wurde ins Leben gerufen. …» Hans Fischli, Rapport, Zürich 1978, S. 154

weissen Holzelementen, die sich rastergeordnet vor einem schwarzen Quadrat an der Eingangsseite des Seerestaurants a…
der Rückschau lässt dieses Zusammentreffen der späteren Neue Grafik-Herausgeber erkennen, dass zwar die sys…
Herangehensweise an Probleme der Gestaltung und manche biografische Erfahrung sie verband, die vier als Gestalter a…
Arbeit unterschiedliche Schwerpunkte setzten.
Da ist Richard Paul Lohse, Maler und Grafiker, Mitbegründer der Allianz, der Vereinigung moderner Schweizer Künstler…
durch die Aufbauarbeit für Bauen+Wohnen umfassende Erfahrung in der komplexen Organisation von Zeitschriften. Sei…
im Redaktionsteam der Neuen Grafik sieht er unter anderem darin, «Bindeglied von der freien künstlerischen Seite z…
sein.»[14]
Den 1904 geborenen Hans Neuburg[15] kennt Lohse von der gemeinsamen Arbeit in der Zürcher Reklameagentur Max Dalar…
war dort Texter; ab Anfang der dreissiger Jahre ist er als Berater und Entwerfer vor allem für die Industrie tätig. Sulzer…
Philips gehören zu seinen Auftraggebern. Für zahlreiche Schweizer Firmen gestaltet Hans Neuburg Firmengesichter mit…
Signets und komplexem Corporate Design. Mit Ausstellungskritiken leistet er wichtige Vermittlungsarbeit für die moderne…
Schweiz, doch verfasst er auch Sportreportagen. 1952 ist er Redaktor und Gestalter des Kataloges der «Weltausstellung…
graphie» in Luzern, für die Lohse die Abteilungen Architektur sowie Photographie und Kunst zusammenstellt.
Während Lohse und Neuburg die Zeit des Ersten Weltkrieges und die politischen Veränderungen 1918/19 ebenso bewusst…
den schnellen Wechsel der Kunst-Ismen um 1920, während beide in den dreissiger und vierziger Jahren mit ihrer Gebr…
landesweit anerkannt werden, gehören Josef Müller-Brockmann[16] und Carlo L. Vivarelli zu einer Generation, die mit ihrer…
Arbeiten erst nach dem Zweiten Weltkrieg bei einer breiteren Öffentlichkeit Aufmerksamkeit und Erfolg findet.
So nimmt der 1914 geborene Josef Müller-Brockmann in seine 1994 erschienene Monografie lediglich ein Beispiel seiner…
Tätigkeit aus den Jahren vor 1945 auf. Nachdem Müller-Brockmann 1937 in den Verband Schweizerischer Graphiker (VSG)…

THE UNITED STATES OF AMERICA
J65075079 A

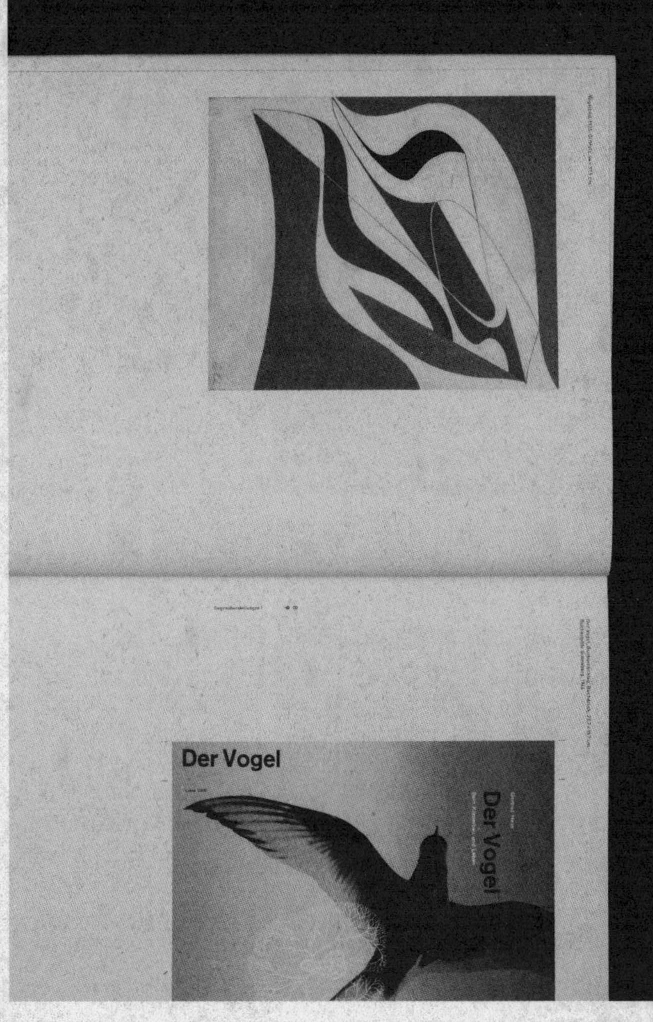
Der Vogel
Der Vogel

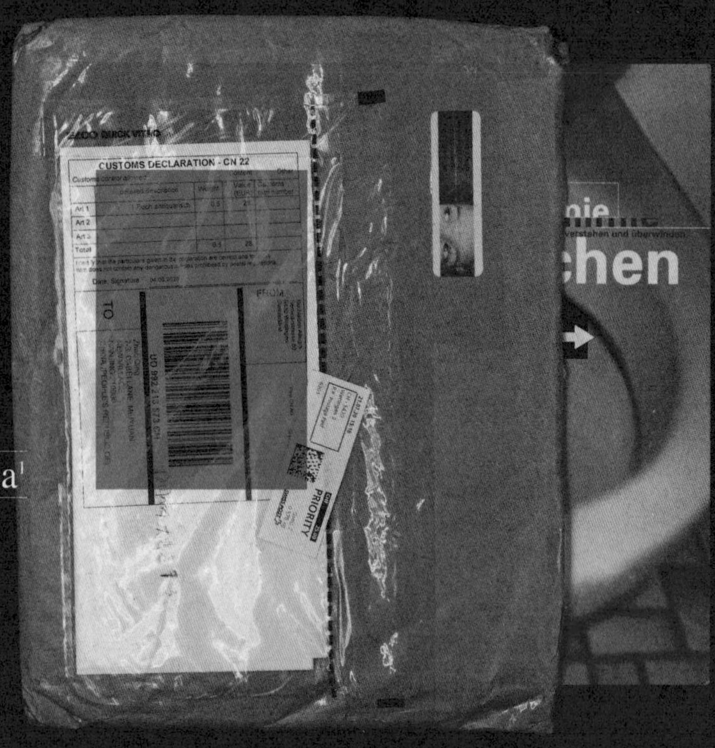
CUSTOMS DECLARATION - CN 22
PRIORITY
verstehen und überwinden
chen

2001 Ha

Germany

Swizerland

China

255mm

11mm

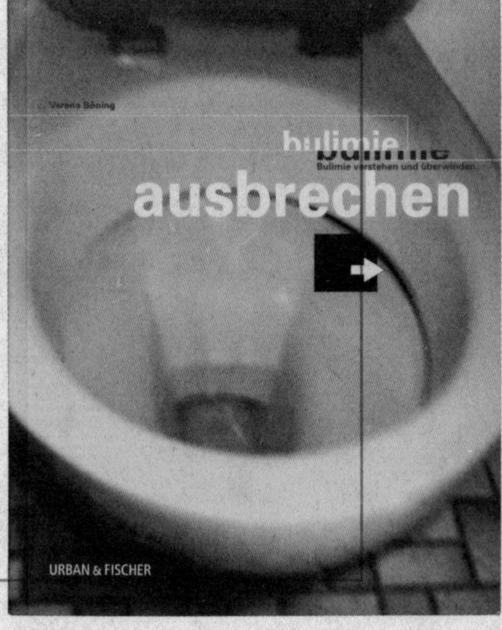

210mm

荣誉奖
摆脱——了解并克服暴食症
德国

Honorary Appreciation
Ausbrechen: Bulimie verstehen und überwinden
Germany
Verena Böning
Verena Böning
Urban & Fischer Verlag, München
255 210 11mm
534g
116p
ISBN-10 3-437-45536-2

暴食症是一种常见的饮食方面的精神疾病，患者一方面渴望通过吃东西来缓解心理压力，但另一方面又会受到身材、体重等社会审美的影响而自我厌恶。对吃东西产生依赖的同时自己又无法控制食欲，患者会陷入“既饿又丑”的自我否定中，并不断恶性循环。本书分析了暴食症的成因，演绎患者复杂的心理活动，并提供了走出恶性循环的建议。书籍的装订形式为锁线胶平装。封面白卡纸四色印刷覆哑光膜。内页哑面铜版纸四色印刷，局部夹插四色印刷的牛油纸，用以展现前后呼应的文本内容或陷入迷茫旋涡的患者心理。大量的图片和文字没有统一的版式规则，而是通过连续页面上的挤压、拉伸、并列、渐变、堆叠来呈现患者自我矛盾又无法自拔的心理状态。在指导建议部分，版式逐渐规整，并有了较强的引导性，仿佛建立了良好的生活习惯，并逐步从暴食症中走出来。

Bulimia is a common mental illness related to diet. On the one hand, patients are eager to relieve psychological pressure by eating, but on the other hand, they are self-disgusted by the influence of social aesthetics such as body size and weight. When they are dependent on eating, they can't control their appetite. Patients will fall into the self-denial of 'hungry and ugly', and continue to have a vicious circle. This book analyzes the causes of bulimia, deduces the complex psychological activities of patients, and provides suggestions for getting out of the vicious circle. The book is perfect bound with thread sewing. The cover is made of white cardboard printed in four colors mounted with matte film. The inner pages are printed on matte coated paper in four colors, and other four-color printed pages printed on tracing paper are inserted to show the psychology of patients trapped in the whirlpool of confusion. A large number of pictures and text do not have a unified layout rules, but are squeezed, stretched, juxtaposed, gradient, and stacked on continuous pages to present the complex psychology of patients with self contradiction and unable to extricate themselves. In the part of the guidance, the layout is gradually regularized to give a strong guidance, as if a good living habit has been established and gradually come out of bulimia.

Bulimie ist heilbar!

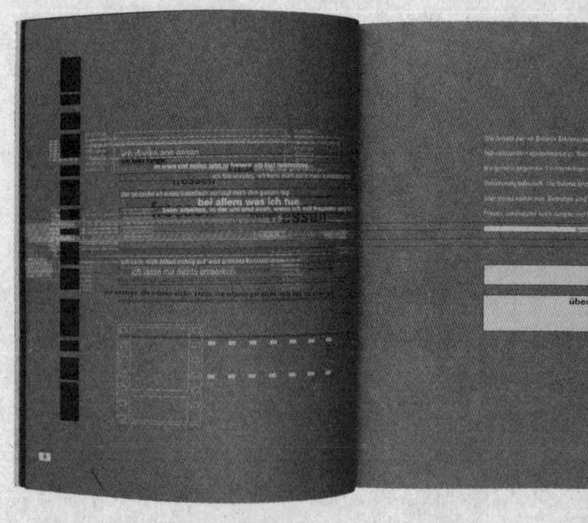
bei allem was ich tue.
ich lasse mir nichts anmerken.

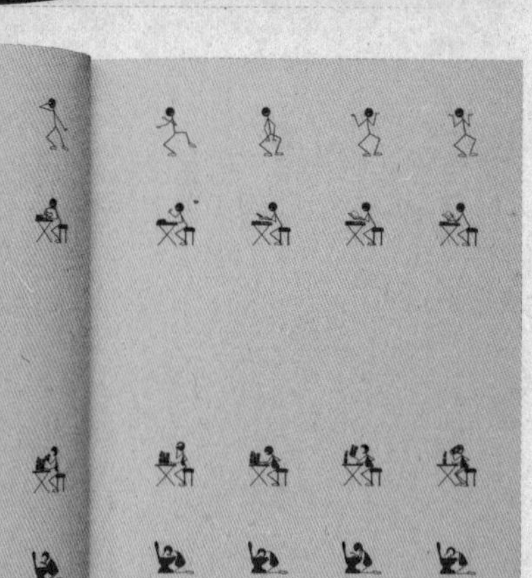

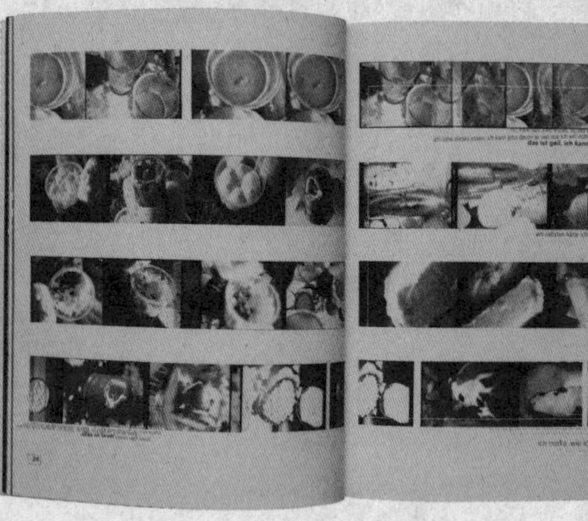

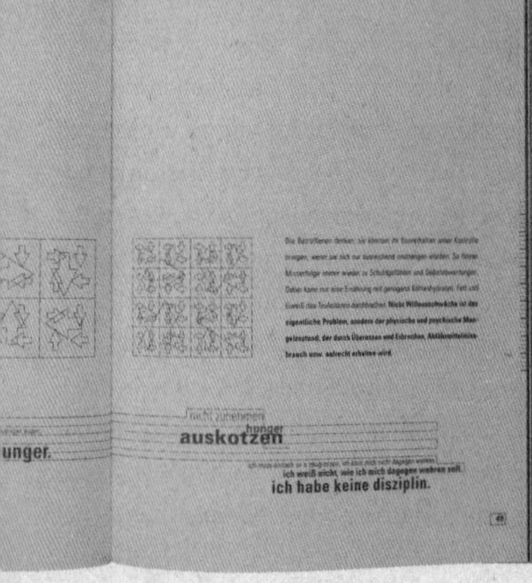
nicht zunehmen
hunger
auskotzen
unger.
ich weiß nicht, wie ich mich dagegen wehren soll.
ich habe keine disziplin.

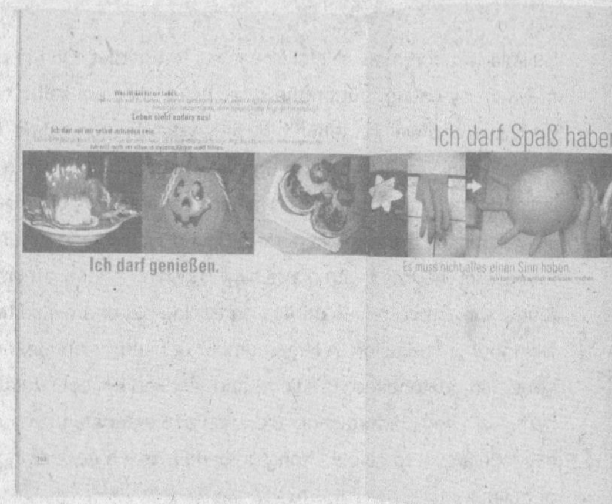
Leben sieht anders aus!
Ich darf Spaß haben
Ich darf genießen.
Es muss nicht alles einen Sinn haben.

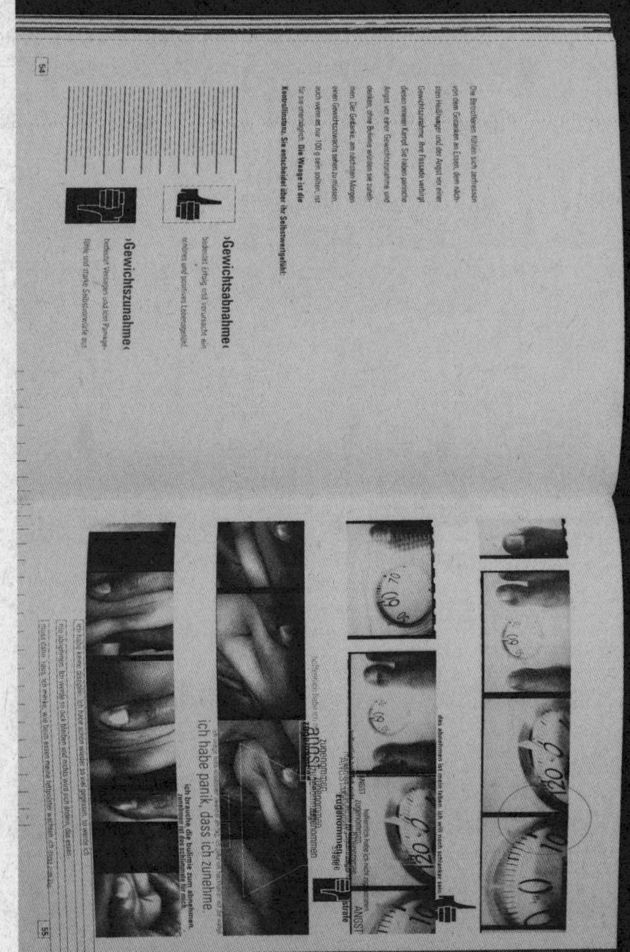
›Gewichtsabnahme‹
›Gewichtszunahme‹
ich habe panik, dass ich zunehme.
54
55

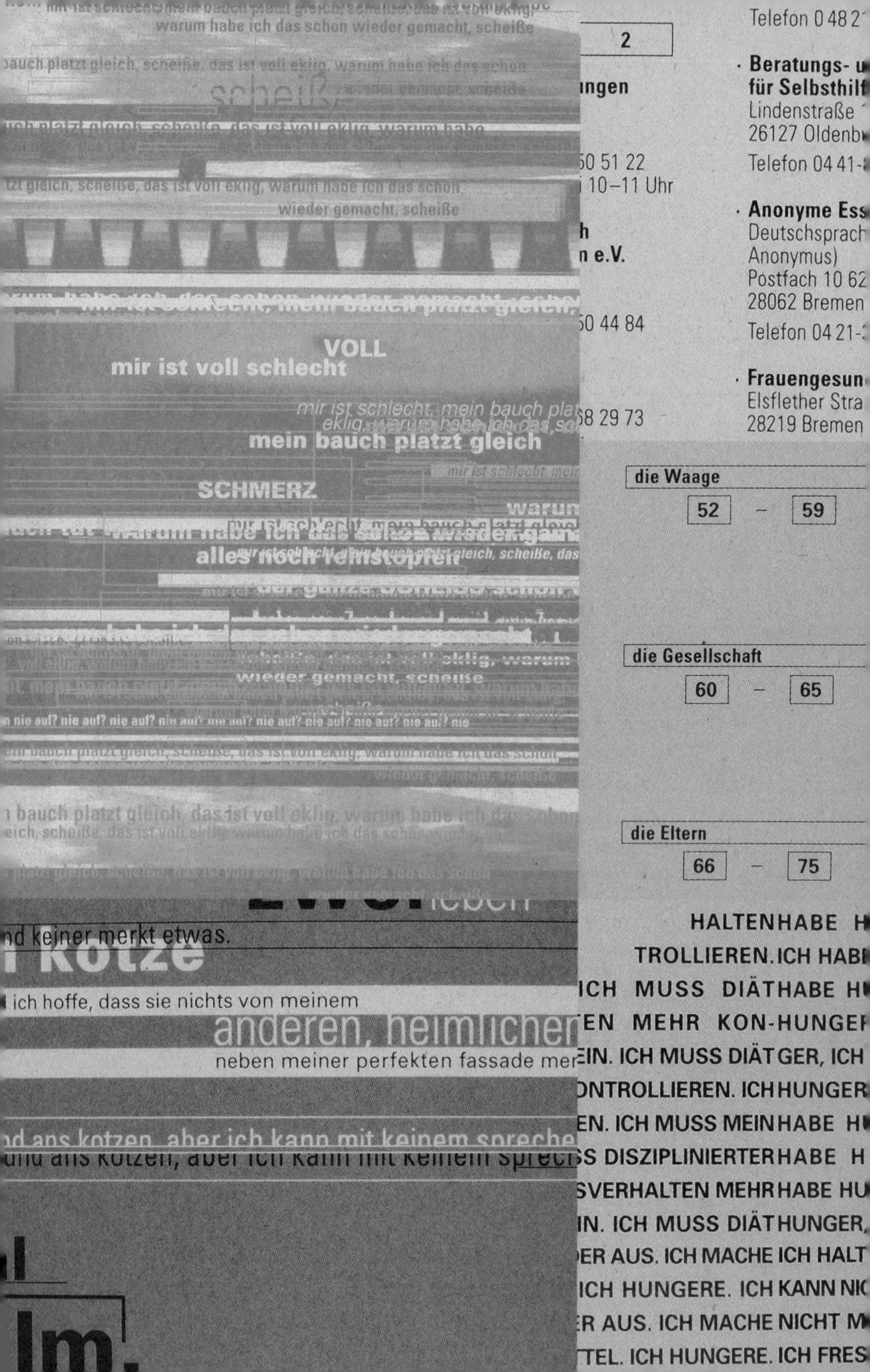

warum habe ich das schon wieder gemacht, scheiße
wieder gemacht, scheiße
VOLL
mir ist voll schlecht
mein bauch platzt gleich
SCHMERZ
alles noch reinstopfen
wieder gemacht, scheiße
nie auf? nie auf? nie auf? nie auf? nie auf? nie auf? nie auf? nie
nd keiner merkt etwas.
Kotze
ich hoffe, dass sie nichts von meinem
anderen, heimlichen
neben meiner perfekten fassade mer
und ans kotzen, aber ich kann mit keinem sprech
Telefon 0 48 2
2
ungen
50 51 22
10–11 Uhr
n e.V.
50 44 84
58 29 73
Beratungs- u
für Selbsthilf
Lindenstraße
26127 Oldenb
Telefon 04 41-
Anonyme Ess
Deutschsprach
Anonymus)
Postfach 10 62
28062 Bremen
Telefon 04 21-
Frauengesun
Elsflether Stra
28219 Bremen
die Waage
52 – 59
die Gesellschaft
60 – 65
die Eltern
66 – 75
HALTENHABE H
TROLLIEREN. ICH HAB
ICH MUSS DIÄTHABE H
TEN MEHR KON-HUNGE
EIN. ICH MUSS DIÄT GER, ICH
ONTROLLIEREN. ICH HUNGER
EN. ICH MUSS MEIN HABE H
SS DISZIPLINIERTER HABE H
SVERHALTEN MEHR HABE HU
IN. ICH MUSS DIÄT HUNGER,
ER AUS. ICH MACHE ICH HALT
ICH HUNGERE. ICH KANN NIC
ER AUS. ICH MACHE NICHT M
TTEL. ICH HUNGERE. ICH FRES

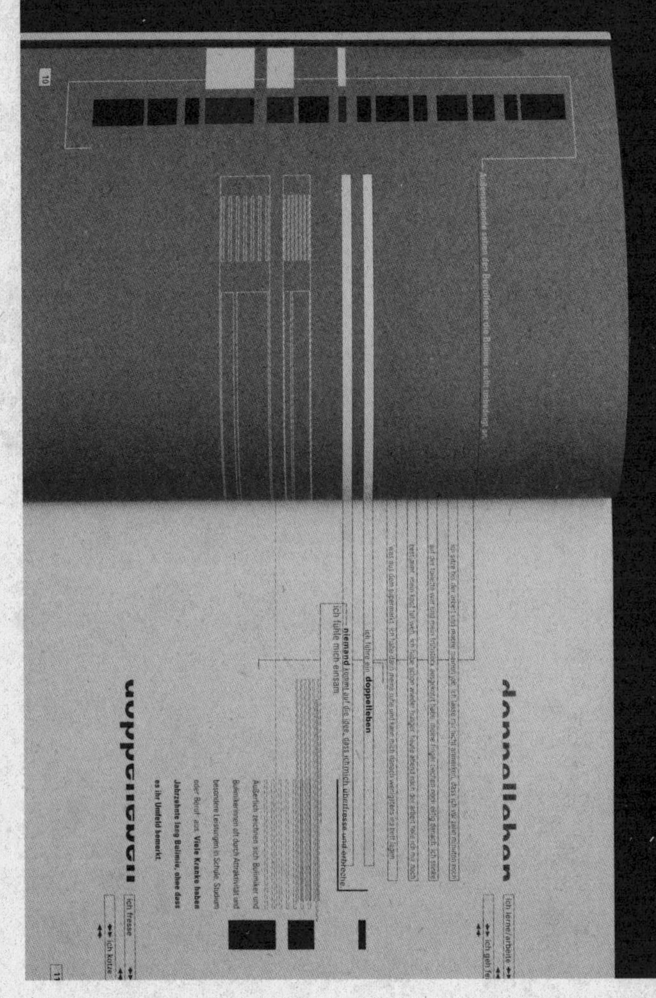

doppelleben
ich lerne/arbeite
ich führe ein doppelleben
niemand kommt auf die idee, dass ich mich überfresse und erbreche.
ich fühle mich einsam.
Äußerlich zeichnen sich Bulimiker und Bulimikerinnen oft durch Attraktivität und besondere Leistungen in Schule, Studium oder Beruf aus. Viele Kranke haben Jahrzehnte lang Bulimie, ohne dass es ihr Umfeld bemerkt.
ich fresse
ich kotze
10

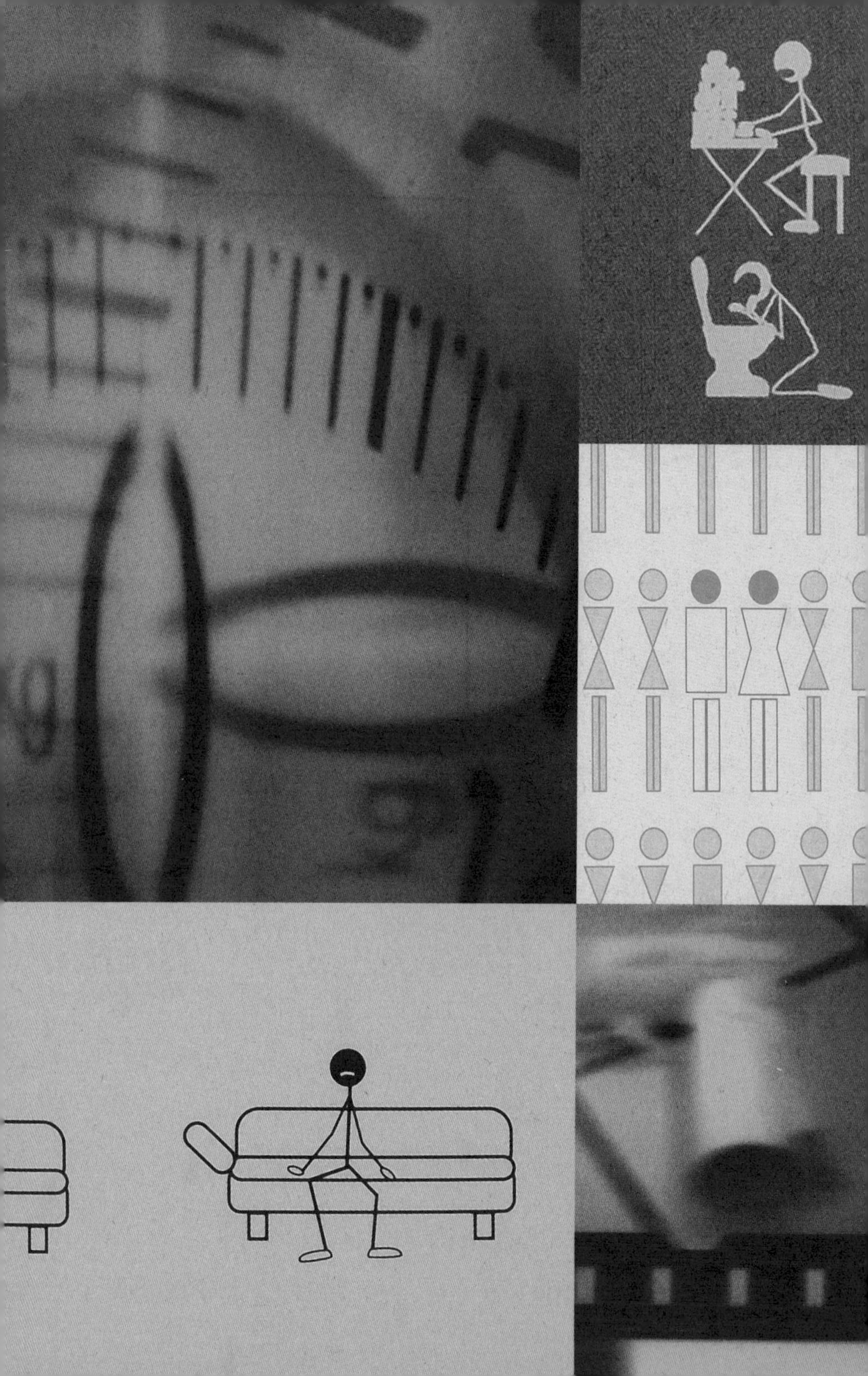

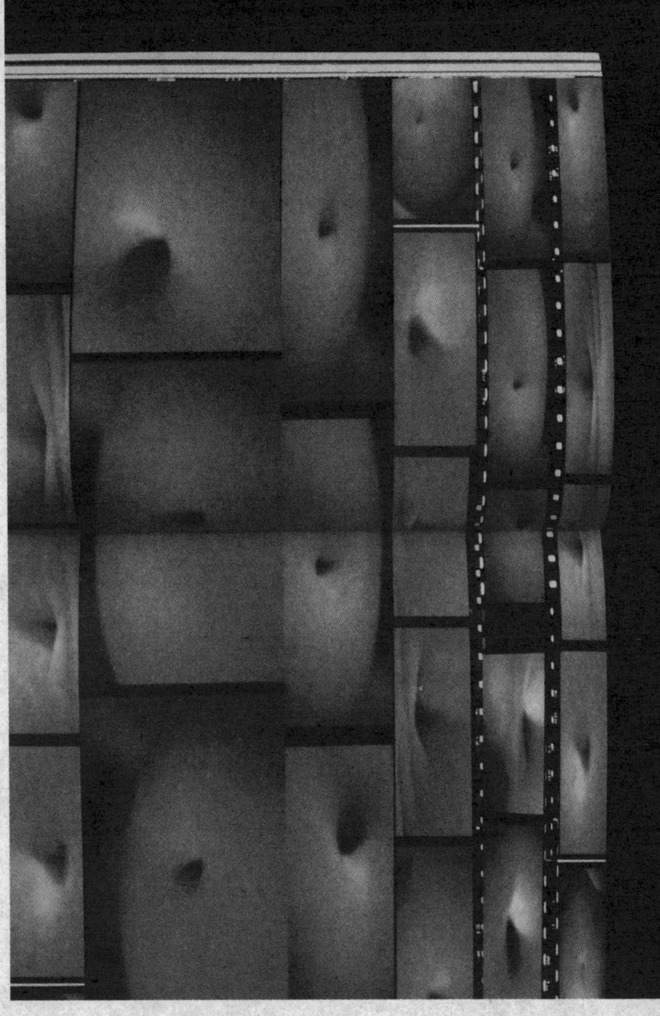

FROM
momox GmbH
Am Alten Flughafen 1
04356 Leipzig
Germany
A R
PRIORITY
W
TO
CN 22
medimops
Ethica

2001 Ha4

Nr.: 109049010

PRIORITY

06 30E0 A750 25 B000 1B87
DV 02.20 1,00 Deutsche Post
WARENPOST Int. KT Port payé

Qing
Dabei Lane, Meiyuan New Village
00 Nanjing
A VR

大楼巷 7-3

UC 032 894 261 DE

DÉCLARATION EN DOUANE
Postverwaltung/Administration des postes Deutsche P

– 1 x Book

1,075

1,102

The Netherlands

Germany

China

246mm

35mm

179mm

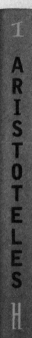

● 荣誉奖
《 伦理学
◖ 荷兰

Honorary Appreciation
Ethica
The Netherlands
Rudo Hartmann, Gerard Hadders
Aristoteles
Historische Uitgeverij, Groningen
246 179 35mm
1055g
360p
ISBN-10 90-6554-005-9

《尼各马可伦理学》（*Nicomachea Ethica*）是古希腊哲学家亚里士多德最重要的论著之一，主要探讨了什么是善，以及对善的追求，即社会伦理的基础、具体的个人德行。本书为荷兰文版。书籍的装订形式为圆脊锁线硬精装。护封胶版纸橙、黄、蓝三色印刷，覆哑膜。内封胶版纸蓝黄双色印刷，裱覆卡板。书脊处包覆橙色布面，烫黄蓝双色。灰色艺术纸做环衬，内页胶版纸印单黑。原书共分 10 册，现代版本已汇总为 1 册，10 册即为本书的 10 个章节。内容分别对善良、道德、行为、公正、理性、自制、友爱、幸福等概念作出定义、阐释并进行实践。设计方面，正文单栏排版，装订口位置的空白处标注原版的内容行位。注释设计为两种：需呈现希腊原文的地方，通过折角符号索引至页侧边；而注释因其文字量较多则通过数字引至页脚。有些注释较长，为了保证正文版心的统一，超出当前页面范围的注释会延续至下一页。

Nicomachea Ethica is one of the most important works of the ancient Greek philosopher Aristotle. It mainly discusses what good is and the pursuit of good, that is the basis of social ethics and the specific personal virtue. This book is in Dutch. The hardcover book is bound by thread sewing with rounding back. The cover is made of offset paper printed in orange, yellow and blue, and mounted with matte film. The inner cover is printed in blue and yellow on offset paper, mounted with cardboard. The spine is covered by orange cloth and treated with hot stamping in yellow and blue. The gray art paper is used as the end paper, and offset paper printed in black is selected for inside pages. The original version was divided into ten volumes, the modern version has been summarized into one volume, turning ten volumes to ten chapters of the book. The contents are the definition, interpretation and practice of the concepts of goodness, morality, behavior, justice, reason, self-control, friendship and happiness. As for the design, the text is typeset in a single column, and the blank space at the gutter is marked with the lines of the original version. There are two kinds of annotation methods. The place where the original Greek text needs to be presented is indexed to the side of the pages by an angle symbol, while the comments that need to be explained are led to the footer by numbers. In order to ensure the unity of the text, some long notes that beyond the current page will continue to the next page.

1099

ESTAAN IS GELUK 19

inleiding

:lke rationele activiteit is gericht op een goed of c
)pvattingen over de aard van het goede 3 Kritiek
ngbare opvattingen over geluk 4 Kritiek op Platc
eorie van het goede 5 De aard van het goede 6 H
t in de eigen taak van de mens 7 Onze definitie g
ıets van het hoogste goed 8 De definitie komt ov
:t de opvattingen over geluk 9 Onze definitie bev
eisten voor geluk 10 Goddelijke beschikking, lot
en inspanning 11 Kan men iemand tijdens zijn l
ukkig noemen? 12 Geluk is een hoogst achtensv
goddelijk goed 13 Inleiding tot het onderzoek na
ɔrtreffelijkheid

OORTREFFELIJKHEID
AN KARAKTER 51

inleiding

ɜewoonte en voortreffelijkheid van karakter 2 W

VIII.1-11	Verschillende vriendschap
VIII.12-IX.3	Vriendschap
IX.4-8	Grondslagen
IX.9-12	Verantwoord de vriendscha
	Genot (2)
	Speculatieve

Afsluiting en overgang na

ARIST

ETHICA

Ethica Nic

Zie 1094 a 1-1097 a 14,

HÊDONÊ 'genot, geno
hem in een genuanceerc
vloeien uit het welslagen
van belang voor de more
positie in tussen de hed
iedere goede eigenschap
Zie 1152 b 23-1154 b 34

HEKÔN 'uit eigen bew
voortkomen uit de hande
(1113 b 22-1114 a 12). D
karakter geven (1110 a 5
7-14). Naar moderne be

intellectuelen. Voor de eerste is het inderdaad iets zichtbaars en tastbaars, zoals genot, rijkdom of aanzien. Voor sommigen is het het een, voor anderen iets anders, en vaak is het voor dezelfde mens nu eens het een en dan weer het ander: als hij ziek is, denkt hij aan gezondheid, als hij armoe lijdt, aan rijkdom. En wie zich van zijn onwetendheid bewust is hoort in bewondering anderen[4] erover spreken als over iets zo groots dat het zijn begrip te boven gaat. Sommigen[5] meenden dat er naast al die verschillende goede dingen nog iets goeds is, dat op zichzelf bestaat en tegelijk verklaart waarom al die andere goede dingen goed zijn. Al deze opvattingen onderzoeken is wellicht verspilde moeite, het volstaat de meningen te onderzoeken die het meest gangbaar zijn of een redelijke grond blijken te hebben.

Het mag ons evenwel niet ontgaan dat er een onderscheid bestaat tussen uiteenzettingen die van principes uitgaan en die welke tot principes leiden. Ja, ook Plato stelde terecht deze vraag en onderzocht of de weg van of naar de principes loopt, zoals men in het stadion van de scheidsrechters naar het keerpunt kan lopen ofwel omgekeerd.

DEIN Inderdaad moet men uitgaan van het bekende. Dit heeft echter twee betekenissen: wat ons bekend is en wat zonder meer bekend is. Wij moeten ongetwijfeld uitgaan van wat ons bekend is. Daarom moet iemand die met vrucht lezingen wil volgen over wat edel en rechtvaardig is, en in het algemeen over wat verband houdt met politiek, door zijn opvoeding vertrouwd zijn met edele gewoonten.

ARCHE In het gegeven feit ligt hier namelijk het uitgangspunt, en als dit voldoende helder is, is er verder geen waarom nodig. Wie een goede vorming heeft genoten, bezit de principes of kan ze gemakkelijk verwerven. Wie ze niet heeft en ze ook niet kan verwerven, moet maar luisteren naar Hesiodus:[6]

Veruit de beste is hij die zelf alles inziet,
goed is ook hij die wil horen naar goede raad,
maar hij die niet zelf inziet en ook niet andermans woord
ter harte neemt, dat is een waardeloos man.

4 Mischien de intellectuelen of sofisten die culturele vorming en politiek succes beloven (zie ook Aristoteles, *Ethica Eudemia* 1214 b 6-9). 5 Plato en leden van de Acade-

3 KRITIEK OP DE GANGBARE OPVATTINGEN OVER GELUK

Maar laten we na deze uitweiding terugkeren naar het punt waar we gebleven waren.

Te oordelen naar de levensvormen[7] van mensen stelt de meerderheid, de meest ordinaire mensen—zoals te verwachten is—het goede en het geluk kennelijk gelijk aan genot; en daarom houden zij ook van een leven dat gericht is op genieten. (Er zijn immers, zo kan men zeggen, drie typische levensvormen: de zojuist genoemde, de politieke en als derde de levensvorm gewijd aan studie.) De grote massa is dus klaarblijkelijk volkomen slaafs: zij verkiezen het leven dat beesten leiden. Wat evenwel voor hen pleit is dat heel wat mensen in hoge posities de smaak van Sardanapalus delen.

Ontwikkelde en sociaal actieve mensen stellen geluk gelijk aan aanzien; want dat is toch, in het algemeen gesproken, het doel van de politieke levensvorm. Maar dit doel is duidelijk te oppervlakkig voor wat wij zoeken. Naar men algemeen aanneemt hangt aanzien immers eerder af van wie achting betuigt dan van wie geacht wordt, terwijl we ons het goede voorstellen als iets dat eigen is aan de persoon en hem niet gemakkelijk kan worden ontnomen. Verder streven die mensen blijkbaar aanzien na om zich van hun eigen goedheid te overtuigen. Zij proberen althans te worden gerespecteerd door verstandige mensen en door degenen die hen kennen, en wel op grond van hun voortreffelijkheid. Hieruit blijkt dat, volgens hen tenminste, voortreffelijkheid een groter goed is dan aanzien. Misschien kan men dan beter voortreffelijkheid als doel van de politieke levensvorm beschouwen. Maar ook zij is blijkbaar nog te onvolmaakt; want het is zoals men weet best mogelijk dat wie voortreffelijkheid bezit zijn leven lang slaapt of niets uitvoert, en bovendien de grootste ellende en tegenslagen ondervindt. En wie zo leeft, zal niemand gelukkig noemen, tenzij iemand een paradoxale stelling wil verdedigen. Genoeg hierover; we hebben er al voldoende over gezegd in onze lezingen voor een breder publiek.[8]

De derde levensvorm, die we verderop[9] in onze uiteenzetting zullen onderzoeken, is het leven gewijd aan studie. De levensvorm THEORIA

7 Plato betoogt in *De staat* 580d2-581e3, dat de menselijke ziel uit drie 'delen' of vermogens bestaat en leidt daaruit af dat er drie levensvormen mogelijk zijn. Aan het begerende deel van de ziel beantwoordt een leven van zingenot, aan het gepassioneerde deel een leven van dienst als staatsburger, aan het redelijke deel een leven van studie. Aristoteles neemt deze traditionele driedeling over. 8 Zie voor deze lezingen voor een breder publiek—de zogenaamde exoterische werken, de algemene inleiding, blz.

GAN
Empfänger / Destinataire
ZHAO QING
Dabei xiang 7-3, Nanjing
Han Qing Tang Design Co.
210000 Nanjing
China
大悲巷7-3
4-3
ZENTRALES VERZEICHNIS ANTIQUARISCHER BÜCHER
Paket WELT bis 5 kg
Deutsche Post
Entgelt bezahlt
CP 85
S.A.L. surface airlifted
Non priority
CNNKGA (CNA)
Nanjing
CHINA POST
PVG
Postal parcels
DEFRAA (DEA)
Frankfurt/M
DeutschePost
BCN
414
2019-11-05
PC
3
CN
1
2.6
CP 91

2001 Ha5

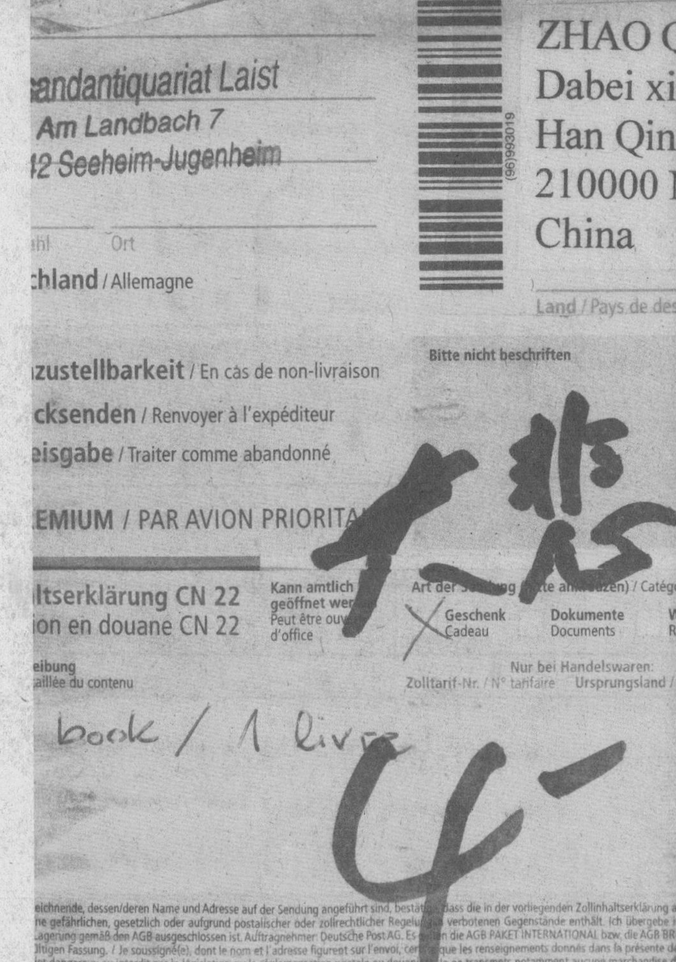
...sandantiquariat Laist
Am Landbach 7
...12 Seeheim-Jugenheim

...ahl Ort

...chland / Allemagne

ZHAO Q...
Dabei xi...
Han Qin...
210000 N...
China

Land / Pays de des...

Bitte nicht beschriften

...zustellbarkeit / En cas de non-livraison
...cksenden / Renvoyer à l'expéditeur
...eisgabe / Traiter comme abandonné

...EMIUM / PAR AVION PRIORITA...

大悲

...altserklärung CN 22
...ion en douane CN 22

Kann amtlich geöffnet wer... Peut être ouv... d'office

Art der Sendung (Bitte ankreuzen) / Catégo...

X Geschenk Cadeau
Dokumente Documents

...eibung
...taillée du contenu

Nur bei Handelswaren:
Zolltarif-Nr. / N° tarifaire Ursprungsland / ...

1 book / 1 livre

4

...zeichnende, dessen/deren Name und Adresse auf der Sendung angeführt sind, bestätigt, dass die in der vorliegenden Zollinhaltserklärung an... ...ne gefährlichen, gesetzlich oder aufgrund postalischer oder zollrechtlicher Regelungen verbotenen Gegenstände enthält. Ich übergebe i... ...Lagerung gemäß den AGB ausgeschlossen ist. Auftragnehmer: Deutsche Post AG. Es gelten die AGB PAKET INTERNATIONAL bzw. die AGB BRI... ...ültigen Fassung. / Je soussigné(e), dont le nom et l'adresse figurent sur l'envoi, certifie que les renseignements donnés dans la présente dé... ...jet dangereux ou interdit par la législation ou la réglementation postale ou douanière. Je ne transmets notamment aucune marchandise d... ...onditions Générales. Mandataire: Deutsche Post AG. Les CGV PAKET INTERNATIONAL resp. CGV BRIEF INTERNATIONAL valides au mome...

...nterschrift des Absenders

Venezuela

Germany China

220mm

47mm

300mm

荣誉奖

西班牙之前的委内瑞拉艺术

委内瑞拉

Honorary Appreciation

El Arte Prehispánico de Venezuela

Venezuela

Álvaro Sotillo

Miguel Arroyo, Lourdes Blanco, Erika Wagner

Fundación Galería de Arte Nacional, Caracas

220 300 47mm

2355g

548p

ISBN-10 980-6420-11-X

1498 年哥伦布发现了委内瑞拉，1523 年西班牙在南美的第一个殖民地在委内瑞拉建立。但在西班牙人建立殖民地之前，委内瑞拉本是印第安人的居住地。本书介绍了在西班牙殖民之前由印第安原住民们创造的委内瑞拉陶瓷艺术。书籍的装订形式为锁线胶平装。封面浅灰色卡纸印黑。内页绝大部分为白色胶版纸四色印刷，少部分夹插浅灰色艺术纸印黑，作为隔页或信息图表。内容主要分为三大部分：1. 考古部分，介绍发掘这些陶瓷艺术品的地区和年代；2. 风格分析，按照各地区风格差异进行阐述；3. 作品图录。三大部分在翻口上设置索引线，页码采用斜向排版，清晰明了。文中提及的作品，在注释部分使用不同方向和字重粗细的页码，建立了一套在三大部分之间相互关联的交叉索引系统。

Columbus discovered Venezuela in 1498, and the first Spanish colony in South America was established in Venezuela in 1523. Before the Spanish colonized Venezuela, it had already been the residence of the Indians. This book introduces the Venezuelan ceramic art created by the native Indians before Spanish colonization. The book is glue bound by thread sewing. The cover is made of light gray cardboard printed in black. Most of the inner pages are printed on white offset paper in four colors, and a small amount of pages made of light gray art paper printed in black are inserted, which are used as a separate page or information chart. The content is mainly divided into three parts: 1. the archaeological part, which mainly introduces the regions and years of excavation; 2. the style analysis, which explains according to the style differences of each region; 3. the catalog of works. The three major parts are set with index lines on the book edge, and the page numbers are laid out diagonally for clarity. In the annotations of the works mentioned in the articles, page numbers with different directions and weights are typeset to establish a complex indexing system that correlates the three parts.

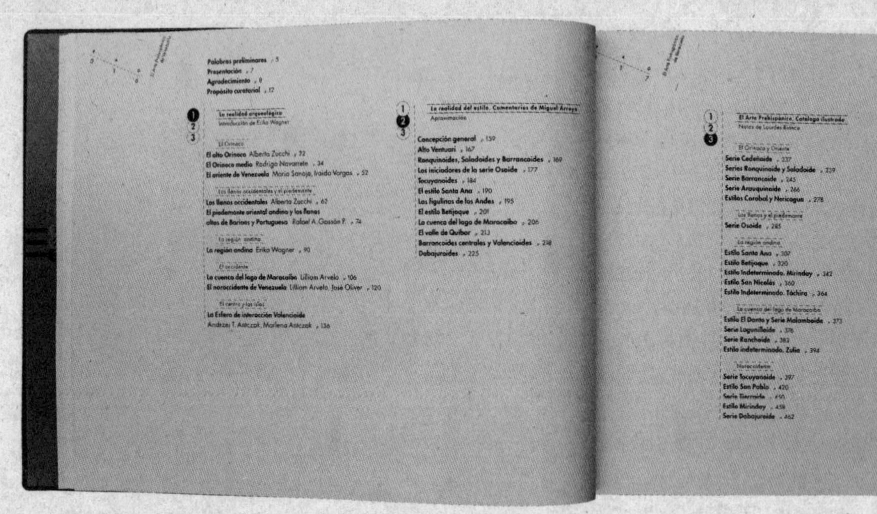
Palabras preliminares , 5
Presentación , 7
Agradecimiento
Propósito curatorial , 12
La realidad arqueológica
Introducción de Erika Wagner
El Orinoco
El alto Orinoco Alberto Zucchi , 22
El Orinoco medio Rodrigo Navarrete , 34
El oriente de Venezuela Mario Sanoja, Iraida Vargas , 52
Los llanos occidentales y el piedemonte
Los llanos occidentales Alberta Zucchi , 62
El piedemonte oriental andino y los llanos
altos de Barinas y Portuguesa Rafael A. Gassón P. , 74
La región andina
La región andina Erika Wagner , 90
El occidente
La cuenca del lago de Maracaibo Lilliam Arvelo , 106
El noroccidente de Venezuela Lilliam Arvelo, José Oliver , 120
El centro y las islas
La Esfera de interacción Valencioide
Andrzej T. Antczak, Marlena Antczak , 136
La realidad del estilo. Comentarios de Miguel Arroyo
Aproximación
Concepción general , 159
Alto Ventuari , 167
Ronquinoides, Saladoides y Barrancoides , 169
Los iniciadores de la serie Osoide , 177
Tocuyanoides , 184
El estilo Santa Ana , 190
Las figulinas de los Andes , 195
El estilo Betijoque , 201
La cuenca del lago de Maracaibo , 206
El valle de Quíbor , 213
Barrancoides centrales y Valencioides , 218
Dabajuroides , 225
El Arte Prehispánico. Catálogo ilustrado
Notas de Lourdes Blanco
El Orinoco y Oriente
Serie Cedeñoide , 237
Series Ronquinoide y Saladoide , 239
Serie Barrancoide , 245
Serie Arauquinoide , 266
Estilos Corobal y Nericagua , 278
Los llanos y el piedemonte
Serie Osoide , 285
La región andina
Estilo Santa Ana , 307
Estilo Betijoque , 320
Estilo Indeterminado. Mirinday , 342
Estilo San Nicolás , 360
Estilo Indeterminado. Táchira , 364
La cuenca del lago de Maracaibo
Estilo El Danto y Serie Malamboide , 373
Serie Lagunilloide , 376
Serie Ranchoide , 383
Estilo indeterminado. Zulia , 394
Noroccidente
Serie Tocuyanoide , 397
Estilo San Pablo , 420
Estilo Mirinday , 458
Serie Dabajuroide , 462

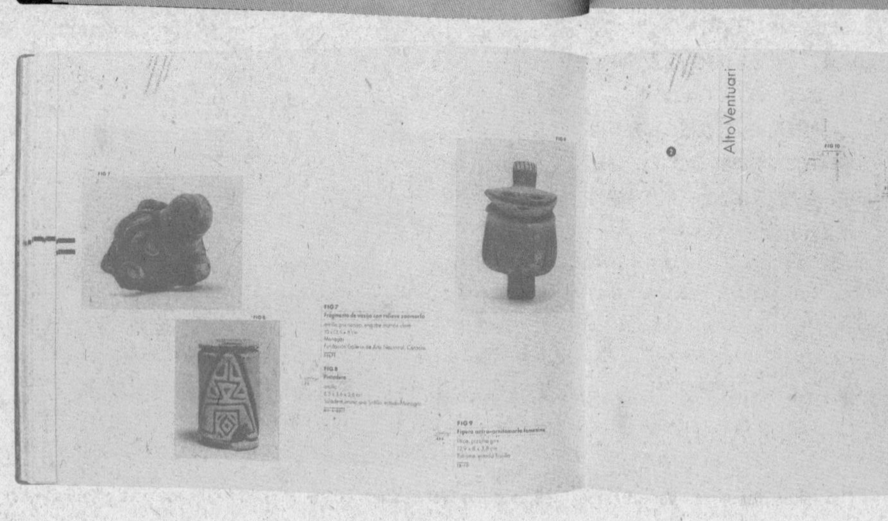
Alto Ventuari

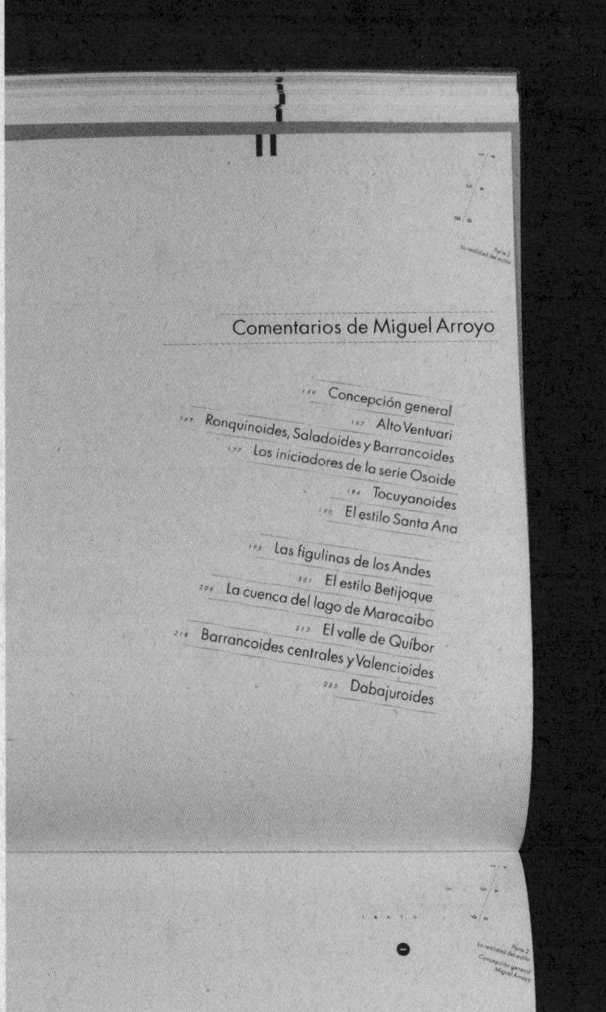
Comentarios de Miguel Arroyo
Concepción general
Alto Ventuari
Ronquinoides, Saladoides y Barrancoides
Los iniciadores de la serie Osoide
Tocuyanoides
El estilo Santa Ana
Las figulinas de los Andes
El estilo Betijoque
La cuenca del lago de Maracaibo
El valle de Quíbor
Barrancoides centrales y Valencioides
Dabajuroides

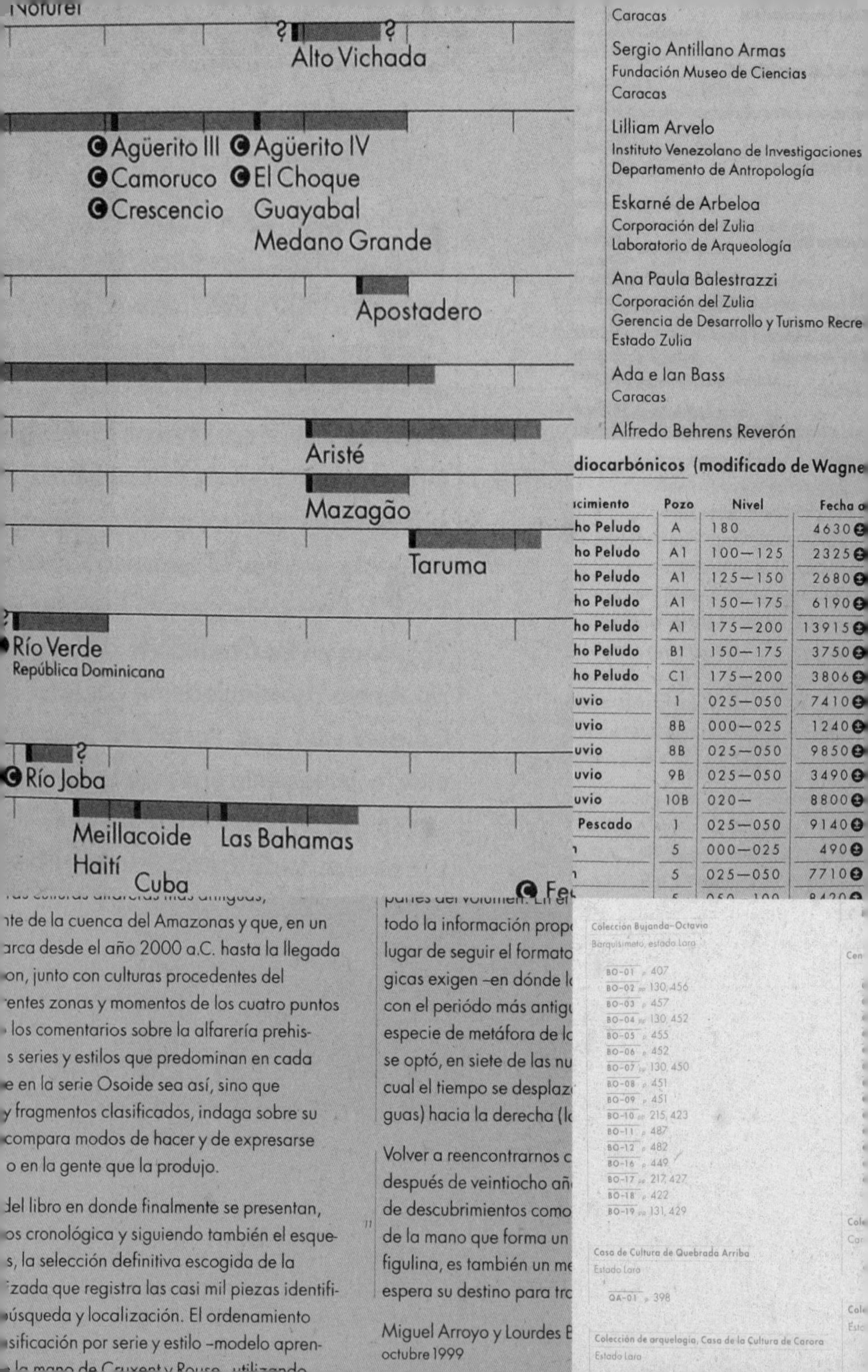
Noturel
Alto Vichada
Agüerito III
Agüerito IV
Camoruco
El Choque
Crescencio
Guayabal
Medano Grande
Apostadero
Aristé
Mazagão
Taruma
Río Verde
República Dominicana
Río Joba
Meillacoide
Las Bahamas
Haití
Cuba
Caracas
Sergio Antillano Armas
Fundación Museo de Ciencias
Caracas
Lilliam Arvelo
Instituto Venezolano de Investigaciones
Departamento de Antropología
Eskarné de Arbeloa
Corporación del Zulia
Laboratorio de Arqueología
Ana Paula Balestrazzi
Corporación del Zulia
Gerencia de Desarrollo y Turismo Recre
Estado Zulia
Ada e Ian Bass
Caracas
Alfredo Behrens Reverón
diocarbónicos (modificado de Wagne
cimiento | Pozo | Nivel | Fecha a
ho Peludo | A | 180 | 4630
ho Peludo | A1 | 100—125 | 2325
ho Peludo | A1 | 125—150 | 2680
ho Peludo | A1 | 150—175 | 6190
ho Peludo | A1 | 175—200 | 13915
ho Peludo | B1 | 150—175 | 3750
ho Peludo | C1 | 175—200 | 3806
uvio | 1 | 025—050 | 7410
uvio | 8B | 000—025 | 1240
uvio | 8B | 025—050 | 9850
uvio | 9B | 025—050 | 3490
uvio | 10B | 020— | 8800
Pescado | 1 | 025—050 | 9140
n | 5 | 000—025 | 490
n | 5 | 025—050 | 7710
nte de la cuenca del Amazonas y que, en un
arca desde el año 2000 a.C. hasta la llegada
on, junto con culturas procedentes del
entes zonas y momentos de los cuatro puntos
los comentarios sobre la alfarería prehis-
s series y estilos que predominan en cada
e en la serie Osoide sea así, sino que
y fragmentos clasificados, indaga sobre su
compara modos de hacer y de expresarse
o en la gente que la produjo.
del libro en donde finalmente se presentan,
os cronológica y siguiendo también el esque-
s, la selección definitiva escogida de la
zada que registra las casi mil piezas identifi-
búsqueda y localización. El ordenamiento
sificación por serie y estilo –modelo apren-
todo la información propo
lugar de seguir el formato
gicas exigen –en dónde lo
con el período más antigu
especie de metáfora de lo
se optó, en siete de las nu
cual el tiempo se desplaz
guas) hacia la derecha (lo
Volver a reencontrarnos c
después de veintiocho añ
de descubrimientos como
de la mano que forma un
figulina, es también un me
espera su destino para tra
Miguel Arroyo y Lourdes B
octubre 1999
Colección Bujanda–Octavio
Barquisimeto, estado Lara
BO-01 p. 407
BO-02 pp. 130, 456
BO-03 p. 457
BO-04 pp. 130, 452
BO-05 p. 455
BO-06 p. 452
BO-07 pp. 130, 450
BO-08 p. 451
BO-09 p. 451
BO-10 pp. 215, 423
BO-11 p. 487
BO-12 p. 482
BO-16 p. 449
BO-17 pp. 217, 427
BO-18 p. 422
BO-19 pp. 131, 429
Casa de Cultura de Quebrada Arriba
Estado Lara
QA-01 p. 398
Colección de arqueología, Casa de la Cultura de Carora
Estado Lara

Los comienzos de la alfarería estuvieron enraizados en la necesidad, los comienzos de su ornamentación en las matemáticas, es decir, existía una voluntad de abstracción que buscaba lograr cierto desapego de las cualidades físicas del objeto, para destilar y hace surgir de lo amorfo, algo simple, limitado, fijo, duradero y universalmente válido. El artista del neolítico deseaba un mundo de formas que ilustrara, no las actividades y eventos pasajeros y mutables... sino las relaciones entre ser y ser y entre éstos y el cosmos, todo dentro de un sistema inmutable. La intención no era la de suprimir el contenido de la vida, sino la de dominarlo y obligarlo a que rindiese su ascendente físico, ante el poder de la voluntad creadora, ante el impulso del hombre de manipular y reformar su mundo.

Max Raphael

Prehistoric Pottery and Civilization in Egypt
Bollingen Series VIII. Nueva York: Pantheon, 1947, p. 76.

Aproximación

En términos generales se puede decir que la alfarería prehispánica venezolana ha sido vista poco y mal, y que no ha dejado de prevalecer en muchos de los que han tenido ocasión de mirarla, el sentimiento de que ella no alcanza la altura formal y expresiva que poseen otras obras en arcilla del continente y del mismo período. Y es, precisamente, la admiración que siento por lo creado en ese campo por las grandes culturas americanas del pasado, lo que me lleva a pensar que hay grandes obras en la alfarería prehispánica de Venezuela y que la variedad de formas, ornamentaciones, texturas y técnicas que nuestros aborígenes usaron, evidencian que quienes las realizaron eran seres sensibles, inteligentes e inquisitivos, que hicieron uso de su talento para crear objetos de muy buen diseño y calidad, que respondían plenamente a sus necesidades físicas y espirituales.

Los siguientes comentarios son un intento de aproximación al arte de algunas de nuestras culturas prehispánicas. En ellos he intentado seguir dentro de las distintas áreas el orden cronológico conocido y sólo lo he irrespetado en dos casos: en el del alto Ventuari, cuya data de tiempo desconozco, y el del estilo Betijoque, al cual antepuse un comentario sobre los productos de una cultura posterior por ser su tema (el espacio) aplicable a cualquier cultura.

Miguel Arroyo

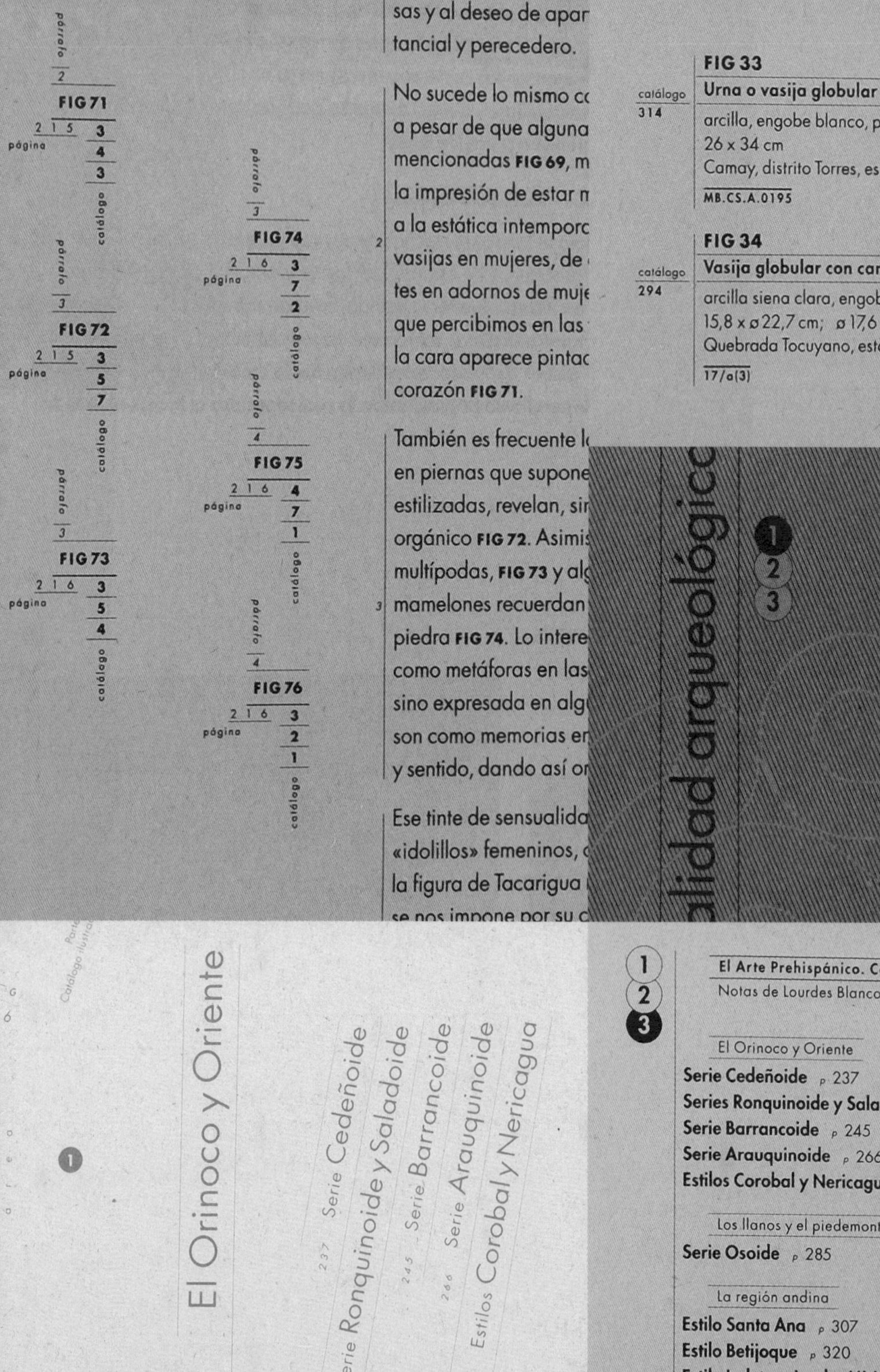

párrafo
2
FIG 71
2 1 5
página
3
4
3
catálogo
párrafo
3
FIG 72
2 1 5
página
3
5
7
catálogo
párrafo
3
FIG 73
2 1 6
página
3
5
4
catálogo
párrafo
3
FIG 74
2 1 6
página
3
7
2
catálogo
párrafo
4
FIG 75
2 1 6
página
4
7
1
catálogo
párrafo
4
FIG 76
2 1 6
página
3
2
1
catálogo
sas y al deseo de apar
tancial y perecedero.
No sucede lo mismo co
a pesar de que alguna
mencionadas FIG 69, m
la impresión de estar m
2
a la estática intemporc
vasijas en mujeres, de
tes en adornos de muje
que percibimos en las
la cara aparece pintac
corazón FIG 71.
También es frecuente l
en piernas que supone
estilizadas, revelan, sir
orgánico FIG 72. Asimis
multípodas, FIG 73 y alg
3
mamelones recuerdan
piedra FIG 74. Lo intere
como metáforas en las
sino expresada en alg
son como memorias er
y sentido, dando así or
Ese tinte de sensualida
«idolillos» femeninos, c
la figura de Tacarigua
FIG 33
catálogo
314
Urna o vasija globular
arcilla, engobe blanco, pi
26 x 34 cm
Camay, distrito Torres, est
MB.CS.A.0195
FIG 34
catálogo
294
Vasija globular con car
arcilla siena clara, engob
15,8 x ø22,7 cm; ø17,6 c
Quebrada Tocuyano, esta
17/a(3)
alidad arqueológica
1
2
3
a r e a
6
Parte
Catálogo ilustrado
1
El Orinoco y Oriente
237
Serie Cedeñoide
Serie Ronquinoide y Saladoide
245
Serie Barrancoide
266
Serie Arauquinoide
278
Estilos Corobal y Nericagua
1
2
3
El Arte Prehispánico. Ca
Notas de Lourdes Blanco
El Orinoco y Oriente
Serie Cedeñoide p 237
Series Ronquinoide y Salac
Serie Barrancoide p 245
Serie Arauquinoide p 266
Estilos Corobal y Nericagu
Los llanos y el piedemonte
Serie Osoide p 285
La región andina
Estilo Santa Ana p 307
Estilo Betijoque p 320
Estilo Indeterminado. Mirin
Estilo San Nicolás p 360

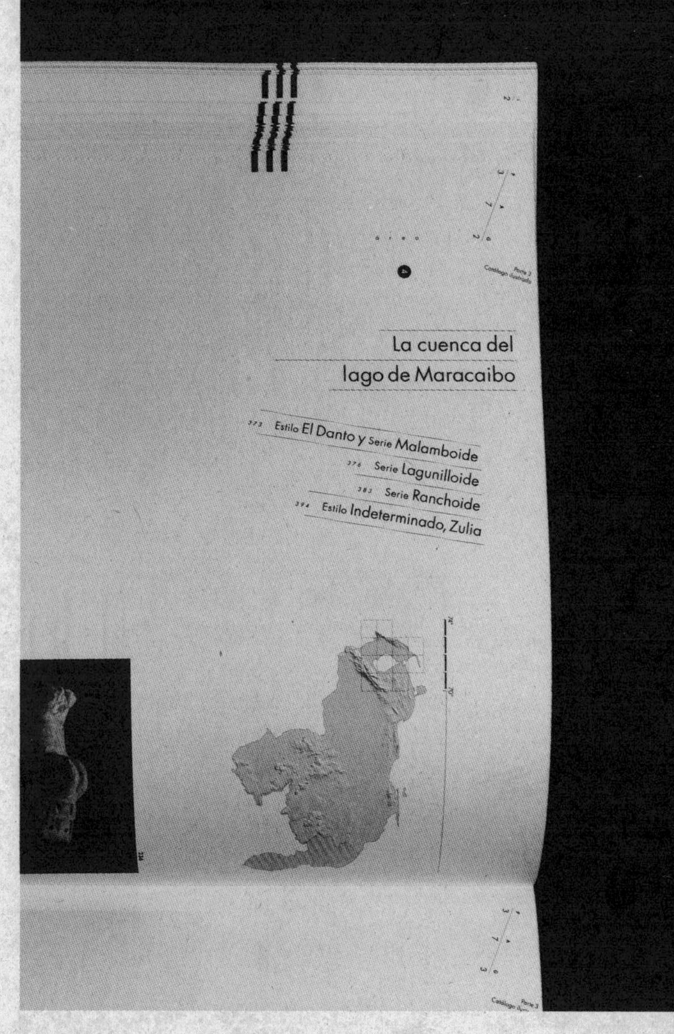
Parte 3
Catálogo ilustrado
La cuenca del
lago de Maracaibo
373 Estilo El Danto y Serie Malamboide
376 Serie Lagunilloide
383 Serie Ranchoide
394 Estilo Indeterminado, Zulia

4
6
3
5
2
B O L I V A R
Rio Mucajai
Rio Siapa

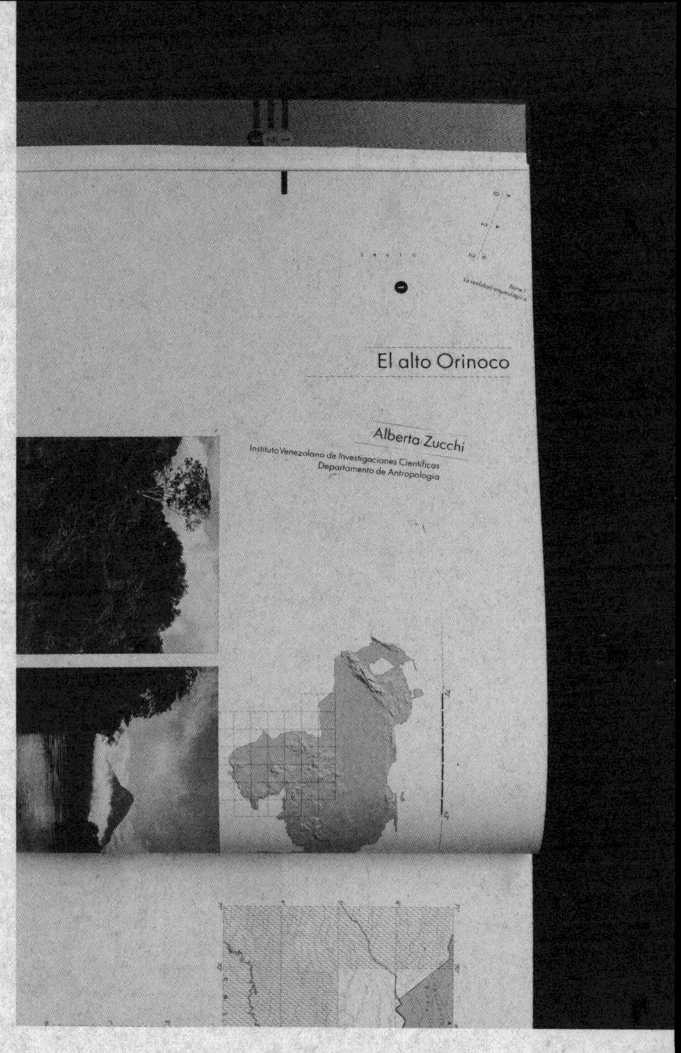
El alto Orinoco
Alberta Zucchi
Instituto Venezolano de Investigaciones Científicas
Departamento de Antropología

2002

Country / Region

The Netherlands 5

Germany 4

Czech Republic 1

Austria 1

Spain 1

Switzerland 1

Sweden 1

rolf schroeter die lichtung

SEPPI/JUNGHANS >>> JVA

HET RIJNLANDS HUIS

HET RIJNLANDS HUIS
TERUGGEVOLGD IN DE TIJD

JORGE LUIS BORGES DIE BIBLIOTHEK VON BABEL

Designer

Mariko Takagi

Juraj Horváth

Klaus Detjen

Irma Boom

Atelier Reinhard Gassner

Markus Dreßen

equipo gráfico de Kalandraka

Typography and Other Serious Matters

Typography and Other Serious Matters

Hans Grüninger (Weiersmüller Bosshard Grüninger wbg AG)

Thomas Müller

Reynoud Homan

Wout de Vringer (Faydherbe / De Vringer)

Lotta Kühlhorn

EXPRESS WORLDWIDE
WPX
DHL
210000 NANJING, CHINA, PEOPLES REPUBLIC
Origin:
STP
CN – PVG – NKG
C
Content Description
BOOKS
WAYBILL 50 9588 8103
LB0019
(10/07) OMN

2002 G1

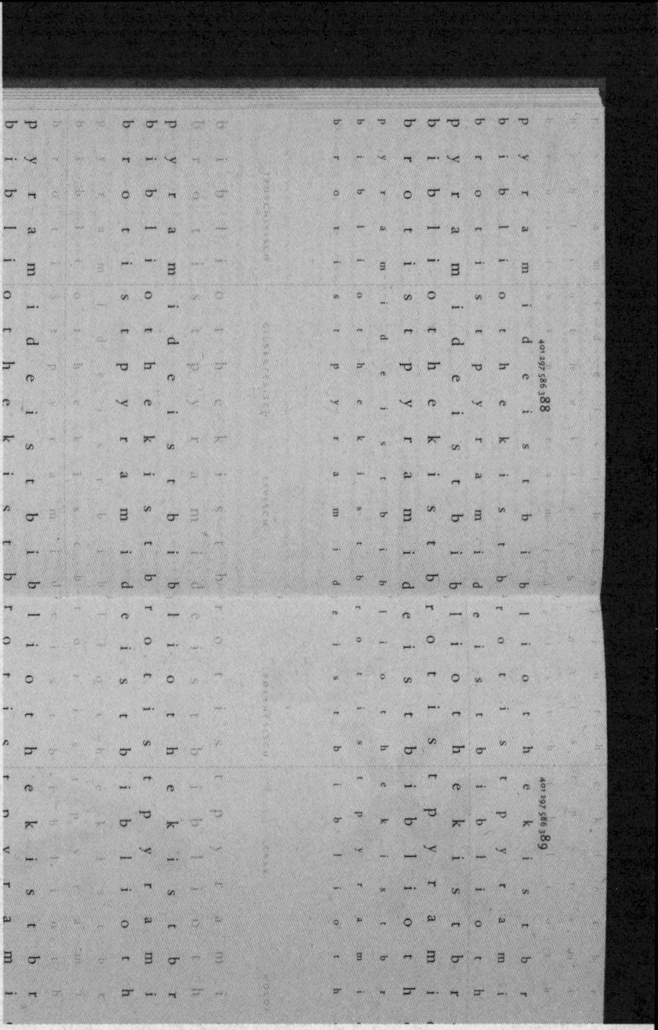

401 297 586 388

401 297 586 389

p y r a m i d e i s t b i b l i o t h e k i s t b r
b i b l i o t h e k i s t b r o t i s t p y r a m i
b r o t i s t p y r a m i d e i s t b i b l i o t h
p y r a m i d e i s t b i b l i o t h e k i s t b r
b i b l i o t h e k i s t b r o t i s t p y r a m i
b r o t i s t p y r a m i d e i s t b i b l i o t h
p y r a m i d e i s t b i b l i o t h e k i s t b r
b i b l i o t h e k i s t b r o t i s t p y r a m i
b r o t i s t p y r a m i d e i s t b i b l i o t h

p y r a m i d e i s t b i b l i o t h e k i s t b r
b i b l i o t h e k i s t b r o t i s t p y r a m i
b r o t i s t p y r a m i d e i s t b i b l i o t h

p y r a m i d e i s t b i b l i o t h e k i s t b r
b i b l i o t h e k i s t b r o t i s t p y r a m i

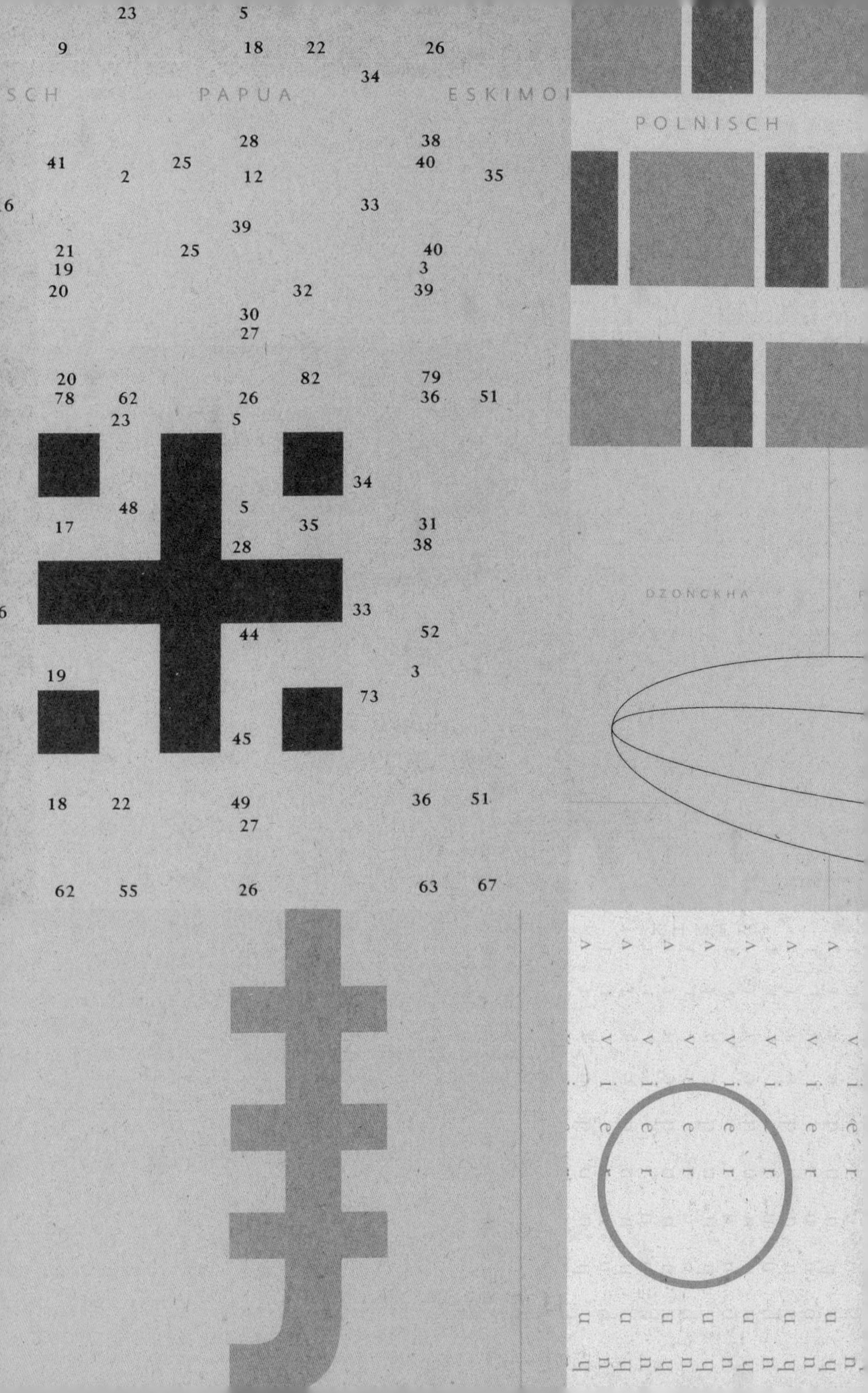
SCH
PAPUA
ESKIMO
POLNISCH
DZONGKHA

401 297 586 324
401 297 586 325

HETRIJN

2002 Sm²

04356 Leipzig
Germany
Referenz-Nr.: 103679072

PRIORITY

W

06 30E0 A750 1C 2000 0E1E
DV 11.19 1,00 Deutsche Post
WARENPOST Int. KT Port payé

ZHAO QING
Design Co.,
Dabei xiang7-3,Nanjing Han Qing Tang.
210000 Nanjing
CHINA VR

大悲巷7-3

UC 030 741 730 DE

The Netherlands

Germany

China

235mm

26mm

186mm

● 银奖
《 莱茵河畔房屋的历史
◖ 荷兰

Silver Medal

Het Rijnlands Huis teruggevolgd in de tijd

The Netherlands

Irma Boom

Leo van der Meule

Hoogheemraadschap van Rijnland, Leiden; Architectura & Natura, Amsterdam

235 186 26mm

777g

232p

ISBN-10 90-72381-08-4

荷兰莱顿市水务局在 1578 年收购了位于莱茵河畔的一栋房屋，在此后的 400 多年间，根据不同时期的实际需求对其进行了多次修缮和改造。这本书从出版时候（2000 年）往前回溯，记录了这栋房屋的历史。本书的装订形式为锁线硬精装，封面采用黑色布纹纸文字烫银，裱覆卡板。腰封胶版纸蓝黑双色印刷。环衬选用胶版纸印蓝色。内页胶版纸四色印刷。内容方面，主要分为 7 个章节，梳理了从 2000 年回溯至 1578 年各个时期各个局部翻新和维护的故事，其中包含大量修缮技术和工匠经验，从书中的图片、技术图纸也可以看出随时代推移的改变。由于是水务局的建筑，本书在回溯历史和水务系统这两个方面找到了设计的切入点。全书大量使用蓝色作为辅助色，运用于腰封、环衬、目录页和文中附注。页码系统在翻口处设计成河道吃水线样式，随着翻阅的进程，“水位”下降，历史逐渐浮出水面。

Leiden Water Authority acquired a house on the Rhine River in 1578, which was repaired and transformed many times according to the actual needs in different periods of more than 400 years. The book goes back from the year of publication (2000) and records the history of the house. The hardback book is bound with thread sewing. The cover is made of black cloth paper and hot-stamped in silver, mounted with cardboard. The wraparound band is made of offset paper printed in blue and black, the end paper is printed in single blue, and the inner pages are printed on offset paper in four colors. The main part of the book is divided into 7 chapters, retracing the stories of renovations and maintenance of various parts from 2000 to 1578, including a large number of repair techniques and craftsman experience. The pictures and technical drawings of the book can also show the changes over time. As it is the building of Water Authority, this book finds the starting point of design from two aspects of history and water system. Blue is widely used as auxiliary color in the book, which is used in wraparound band, end paper, catalog pages and notes in the text. The page number system is designed as a river waterline on the side. With the process of reading, the ‘water level’ drops and the history gradually emerges.

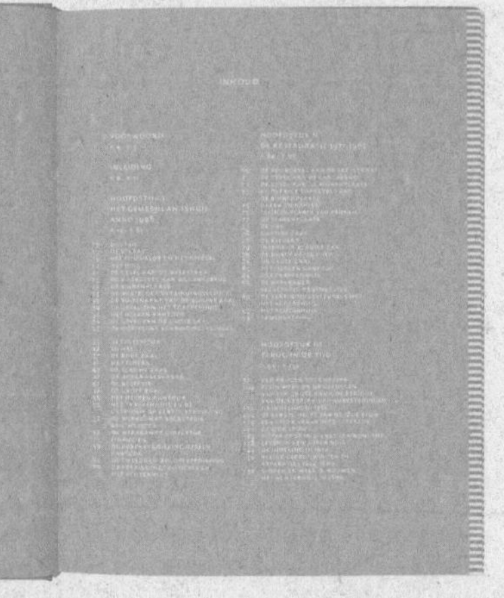

HET GEMEENLANDSHUIS ANNO 1998

DE STAD

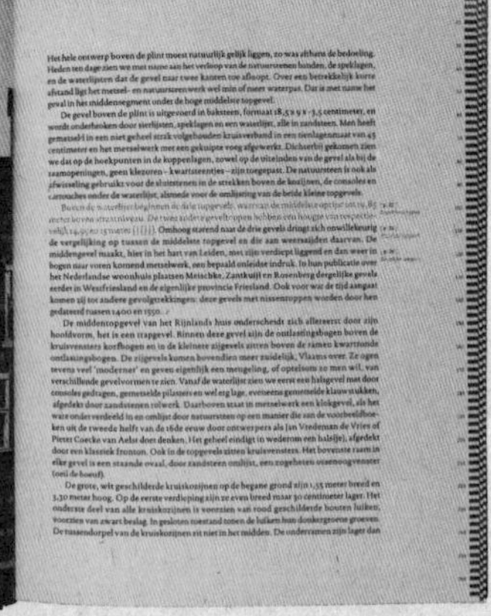

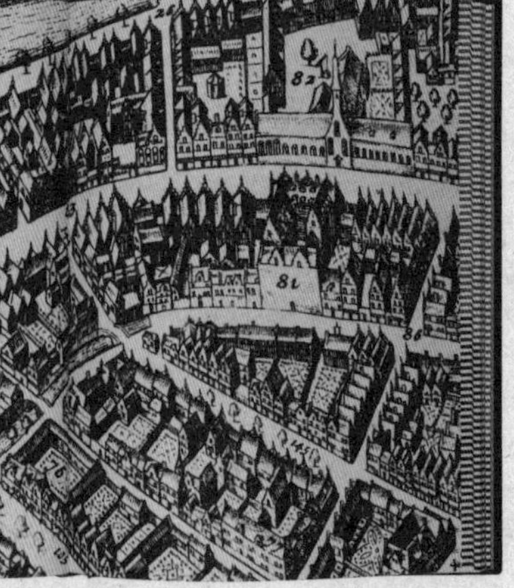

WOORDENLIJST

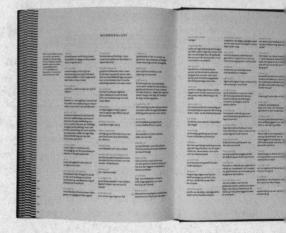

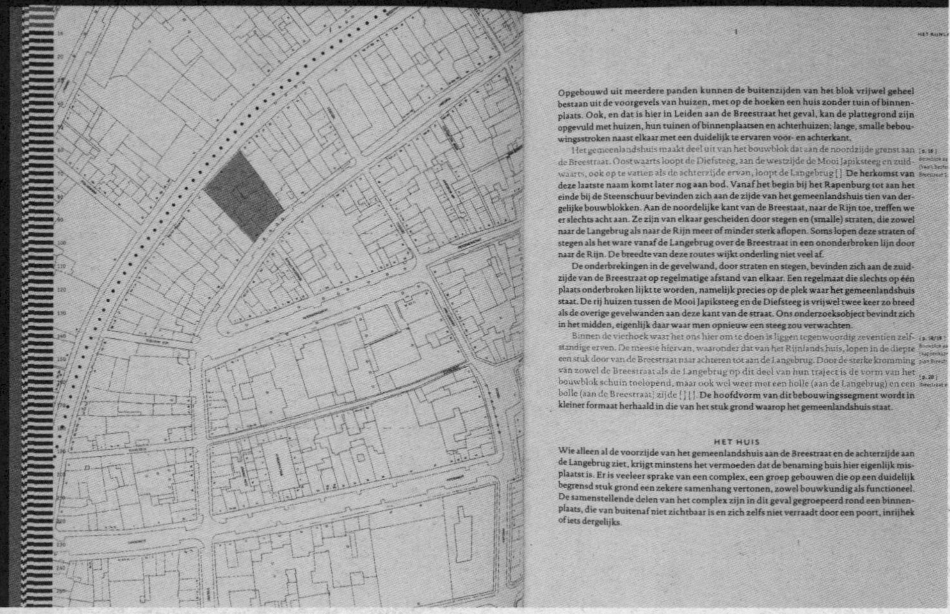

Opgebouwd uit meerdere panden kunnen de buitenzijden van het blok vrijwel geheel bestaan uit de voorgevels van huizen, met op de hoeken een huis zonder tuin of binnenplaats. Ook, en dat is hier in Leiden aan de Breestraat het geval, kan de plattegrond zijn opgevuld met huizen, hun tuinen of binnenplaatsen en achterhuizen; lange, smalle bebouwingsstroken naast elkaar met een duidelijk te ervaren voor- en achterkant.

Het gemeenlandshuis maakt deel uit van het bouwblok dat aan de noordzijde grenst aan de Breestraat. Oostwaarts loopt de Diefsteeg, aan de westzijde de Mooi Japiksteeg en zuidwaarts, ook op te vatten als de achterzijde ervan, loopt de Langebrug [ ]. De herkomst van deze laatste naam komt later nog aan bod. Vanaf het begin bij het Rapenburg tot aan het einde bij de Steenschuur bevinden zich aan de zijde van het gemeenlandshuis tien van dergelijke bouwblokken. Aan de noordelijke kant van de Breestaat, naar de Rijn toe, treffen we er slechts acht aan. Ze zijn van elkaar gescheiden door stegen en (smalle) straten, die zowel naar de Langebrug als naar de Rijn meer of minder sterk aflopen. Soms lopen deze straten of stegen als het ware vanaf de Langebrug over de Breestraat in een ononderbroken lijn door naar de Rijn. De breedte van deze routes wijkt onderling niet veel af. [p. 16]

De onderbrekingen in de gevelwand, door straten en stegen, bevinden zich aan de zuidzijde van de Breestraat op regelmatige afstand van elkaar. Een regelmaat die slechts op één plaats onderbroken lijkt te worden, namelijk precies op de plek waar het gemeenlandshuis staat. De rij huizen tussen de Mooi Japiksteeg en de Diefsteeg is vrijwel twee keer zo breed als de overige gevelwanden aan deze kant van de straat. Ons onderzoeksobject bevindt zich in het midden, eigenlijk daar waar men opnieuw een steeg zou verwachten.

Binnen de vierhoek waar het ons hier om te doen is liggen tegenwoordig zeventien zelfstandige erven. De meeste hiervan, waaronder dat van het Rijnlands huis, lopen in de diepte een stuk door van de Breestraat naar achteren tot aan de Langebrug. Door de sterke kromming van zowel de Breestraat als de Langebrug op dit deel van hun traject is de vorm van het bouwblok schuin toelopend, maar ook wel weer met een holle (aan de Langebrug) en een bolle (aan de Breestraat) zijde [ ] [ ]. De hoofdvorm van dit bebouwingssegment wordt in kleiner formaat herhaald in die van het stuk grond waarop het gemeenlandshuis staat. [p. 18/19] [p. 20]

HET HUIS

Wie alleen al de voorzijde van het gemeenlandshuis aan de Breestraat en de achterzijde aan de Langebrug ziet, krijgt minstens het vermoeden dat de benaming huis hier eigenlijk misplaatst is. Er is veeleer sprake van een complex, een groep gebouwen die op een duidelijk begrensd stuk grond een zekere samenhang vertonen, zowel bouwkundig als functioneel. De samenstellende delen van het complex zijn in dit geval gegroepeerd rond een binnenplaats, die van buitenaf niet zichtbaar is en zich zelfs niet verraadt door een poort, inrijhek of iets dergelijks.

er tegen schuine dak-
icht. Wat bij velen als
t en schuin geplaatste
Het heeft tot doel de
le dwars- en langsver-
antbenen, schoren en
den, die op hun beurt

de begane grond hele-
geheel zonder risico's
helemaal gereed dan
a werd hij gedemon-
ag, beitel of guts aan-
kte schoor tussen het
Ve kunnen hier op de
an boven de Blauwe
n de Blauwe Zaal op

constructie van een zo
(tekening uit Herman
Houten kappen in Ned
1000-1940)

— NOCK
— MANDER
— GORDING
— HAENBALK
— SCHEUYNE
— SPANNING
— GORDING
— BINT
— CARBEEL
— WINTBAND
— SCHAERBIN
— SCHAERSTIJ
— GORDING
— KERBEEL
— FLIERBINT
— BLOCKEEL

NLAN

EVOLGD I

KUNST OP EEN LE

, onderste deel van zuil of

, in vorm gehakt, met een bei-
l van toepassing op natuur-
aar soms ook op baksteen

OTEN
d deelwoord van beschie-
zelfstandig naamwoord
: houten bekleding van
elal niet over de volledige
(lambrizering) maar ook
planking, als extra isolatie,
oren onder de eigenlijke
edekking van het dak

BOLKOZIJN
van het kruiskozijn afgeleid raam, bestaand uit een enkele licht- en luchtopening met luik naast elkaar

BOOGNIS
verdiept liggend vlak in gevel met boogvormige afsluiting aan de bovenzijde

BOOGSTELLING
meerdere bogen op rij

BORSTWERING
verhoging van de buitenmuur van een huis boven de zolderbalken

andshuis
zo breed
indt zich

tien zelf-
de diepte
romming
n van het
g) en een
wordt in
aat.

zijde aan
lijk mis-
duidelijk

[p. 18/19]
Bouwblok aan de Breestraat (kappenkaart bestemmingsplan Breestraat 1:500)

[p. 20]
Breestraat en omgeving

120
130
140
150
160
170
180
190
200

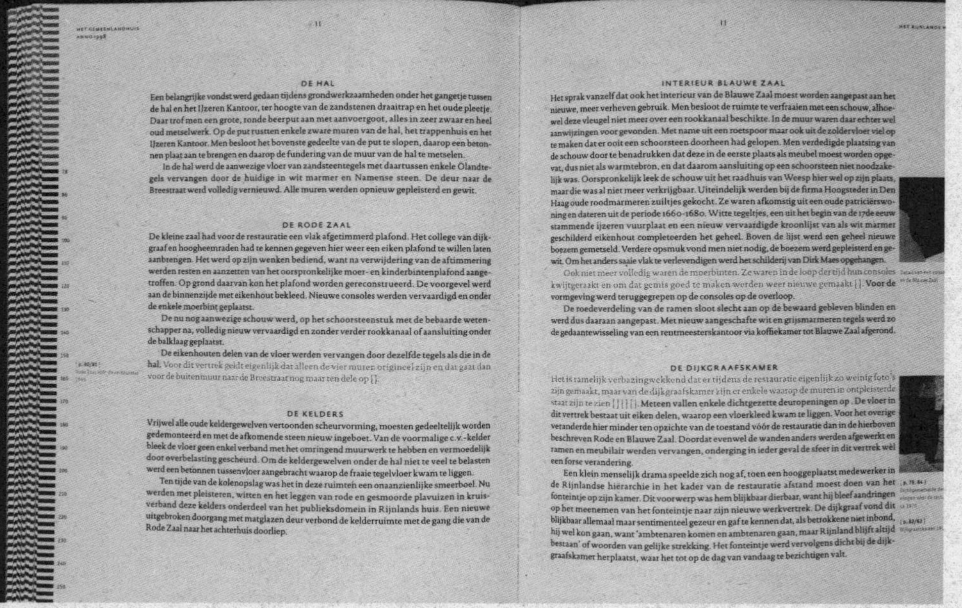

DE HAL

Een belangrijke vondst werd gedaan tijdens grondwerkzaamheden onder het gangetje tussen de hal en het IJzeren Kantoor, ter hoogte van de zandstenen draaitrap en het oude pleetje. Daar trof men een grote, ronde beerput aan met aanvoergoot, alles in zeer zwaar en heel oud metselwerk. Op de put rustten enkele zware muren van de hal, het trappenhuis en het IJzeren Kantoor. Men besloot het bovenste gedeelte van de put te slopen, daarop een betonnen plaat aan te brengen en daarop de fundering van de muur van de hal te metselen.

In de hal werd de aanwezige vloer van zandsteentegels met daartussen enkele Ölandtegels vervangen door de huidige in wit marmer en Namense steen. De deur naar de Breestraat werd volledig vernieuwd. Alle muren werden opnieuw gepleisterd en gewit.

DE RODE ZAAL

De kleine zaal had voor de restauratie een vlak afgetimmerd plafond. Het college van dijkgraaf en hoogheemraden had te kennen gegeven hier weer een eiken plafond te willen laten aanbrengen. Het werd op zijn wenken bediend, want na verwijdering van de aftimmering werden resten en aanzetten van het oorspronkelijke moer- en kinderbintenplafond aangetroffen. Op grond daarvan kon het plafond worden gereconstrueerd. De voorgevel werd aan de binnenzijde met eikenhout bekleed. Nieuwe consoles werden vervaardigd en onder de enkele moerbint geplaatst.

De nu nog aanwezige schouw werd, op het schoorsteenstuk met de bebaarde wetenschapper na, volledig nieuw vervaardigd en zonder verder rookkanaal of aansluiting onder de balklaag geplaatst.

De eikenhouten delen van de vloer werden vervangen door dezelfde tegels als die in de hal. Voor dit vertrek geldt eigenlijk dat alleen de vier muren origineel zijn en dat gaat dan voor de buitenmuur naar de Breestraat nog maar ten dele op [].

[p. 80/81]

DE KELDERS

Vrijwel alle oude keldergewelven vertoonden scheurvorming, moesten gedeeltelijk worden gedemonteerd en met de afkomende steen nieuw ingeboet. Van de voormalige c.v.-kelder bleek de vloer geen enkel verband met het omringend muurwerk te hebben en vermoedelijk door overbelasting gescheurd. Om de keldergewelven onder de hal niet te veel te belasten werd een betonnen tussenvloer aangebracht waarop de fraaie tegelvloer kwam te liggen.

Ten tijde van de kolenopslag was het in deze ruimten een onaanzienlijke smeerboel. Nu werden met pleisteren, witten en het leggen van rode en gesmoorde plavuizen in kruisverband deze kelders onderdeel van het publieksdomein in Rijnlands huis. Een nieuwe uitgebroken doorgang met matglazen deur verbond de kelderruimte met de gang die van de Rode Zaal naar het achterhuis doorliep.

INTERIEUR BLAUWE ZAAL

Het sprak vanzelf dat ook het interieur van de Blauwe Zaal moest worden aangepast aan het nieuwe, meer verheven gebruik. Men besloot de ruimte te verfraaien met een schouw, alhoewel deze vleugel niet meer over een rookkanaal beschikte. In de muur waren daar echter wel aanwijzingen voor gevonden. Met name uit een roetspoor maar ook uit de zoldervloer viel op te maken dat er ooit een schoorsteen doorheen had gelopen. Men verdedigde plaatsing van de schouw door te benadrukken dat deze in de eerste plaats als meubel moest worden opgevat, dus niet als warmtebron, en dat daarom aansluiting op een schoorsteen niet noodzakelijk was. Oorspronkelijk leek de schouw uit het raadhuis van Weesp hier wel op zijn plaats, maar die was al niet meer verkrijgbaar. Uiteindelijk werden bij de firma Hoogsteder in Den Haag oude roodmarmeren zuiltjes gekocht. Ze waren afkomstig uit een oude patriciërswoning en dateren uit de periode 1660-1680. Witte tegeltjes, een uit het begin van de 17de eeuw stammende ijzeren vuurplaat en een nieuw vervaardigde kroonlijst van als wit marmer geschilderd eikenhout completeerden het geheel. Boven de lijst werd een geheel nieuwe boezem gemetseld. Verdere opsmuk vond men niet nodig, de boezem werd gepleisterd en gewit. Om het anders saaie vlak te verlevendigen werd het schilderij van Dirk Maes opgehangen.

Ook niet meer volledig waren de moerbinten. Ze waren in de loop der tijd hun consoles kwijtgeraakt en om dat gemis goed te maken werden weer nieuwe gemaakt []. Voor de vormgeving werd teruggegrepen op de consoles op de overloop.

De roedeverdeling van de ramen sloot slecht aan op de bewaard gebleven blinden en werd dus daaraan aangepast. Met nieuw aangeschafte wit en grijsmarmeren tegels werd zo de gedaantewisseling van een rentmeesterskantoor via koffiekamer tot Blauwe Zaal afgerond.

DE DIJKGRAAFSKAMER

Het is tamelijk verbazingwekkend dat er tijdens de restauratie eigenlijk zo weinig foto's zijn gemaakt, maar van de dijkgraafskamer zijn er enkele waarop de muren in ontpleisterde staat zijn te zien [] [] []. Meteen vallen enkele dichtgezette deuropeningen op. De vloer in dit vertrek bestaat uit eiken delen, waarop een vloerkleed kwam te liggen. Voor het overige veranderde hier minder ten opzichte van de toestand vóór de restauratie dan in de hierboven beschreven Rode en Blauwe Zaal. Doordat evenwel de wanden anders werden afgewerkt en ramen en meubilair werden vervangen, onderging in ieder geval de sfeer in dit vertrek wèl een forse verandering.

Een klein menselijk drama speelde zich nog af, toen een hooggeplaatst medewerker in de Rijnlandse hiërarchie in het kader van de restauratie afstand moest doen van het fonteintje op zijn kamer. Dit voorwerp was hem blijkbaar dierbaar, want hij bleef aandringen op het meenemen van het fonteintje naar zijn nieuwe werkvertrek. De dijkgraaf vond dit blijkbaar allemaal maar sentimenteel gezeur en gaf te kennen dat, als betrokkene niet inbond, hij wel kon gaan, want 'ambtenaren komen en ambtenaren gaan, maar Rijnland blijft altijd bestaan' of woorden van gelijke strekking. Het fonteintje werd vervolgens dicht bij de dijkgraafskamer herplaatst, waar het tot op de dag van vandaag te bezichtigen valt.

[p. 78, 84]

[p. 82/83]

54
6
60

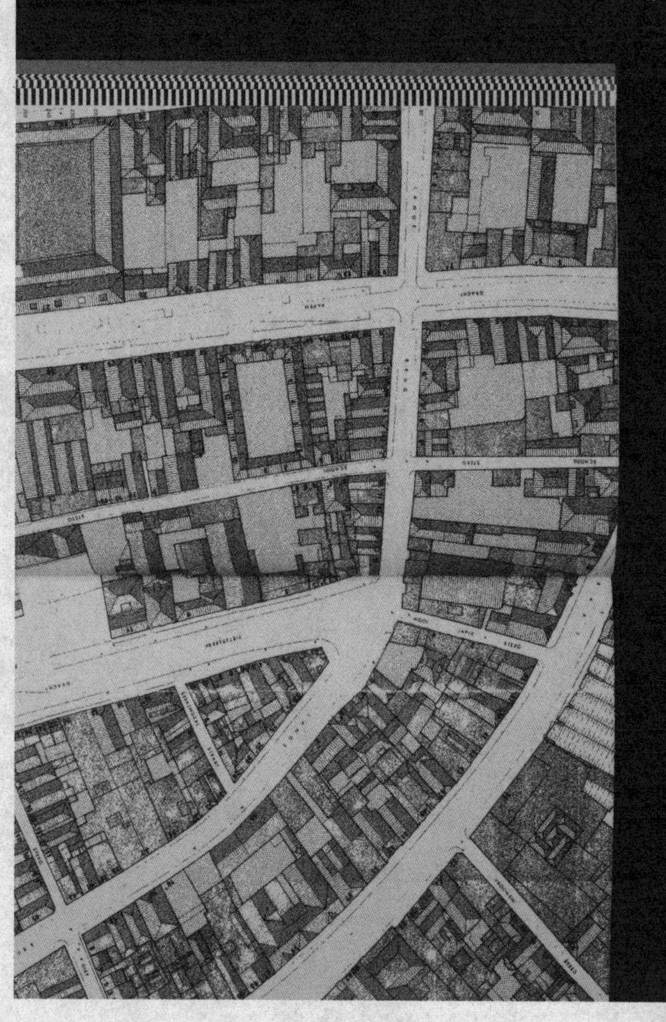

ZHAO QING, Nanjing John Qing
Tang Design Co.
Dabei xiang 7-3
210000 NANJING
CHINA
RC 57 936 176 1DE
Bitte diese Lasche hochziehen !
Inliegend Übermittlungsbeleg, Rückschein, Paketkarte, Zollbegleitpapiere usw.
UNTERWEGS ANTIQUARIAT
ZHAO QING
Dabei xiang7-3
Nanjing John Qing
Tang Design Co.
210000 NANJING
CHINA
Rechnung Nr. 270728A
Vielen Dank für Ihre Bestellung vom 23.10.2019
CP 92
DEUTSCHE POST
911-028-000
ZENTRALES VERZEICHNIS ANTIQUARISCHER BÜCHER

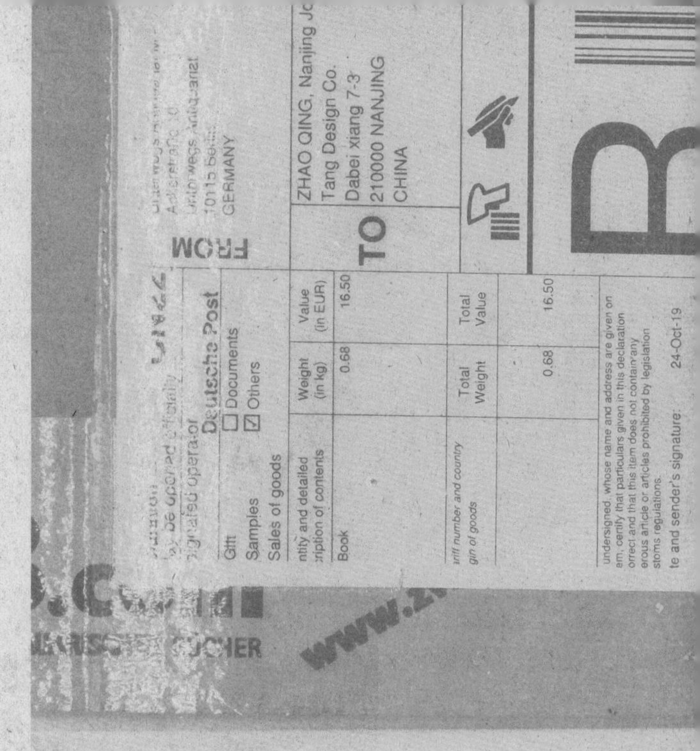

Germany

Germany China

255mm

10mm

195mm

● 铜奖
《 JVA 协议
◖ 德国

Bronze Medal
JVA Protokoll
Germany
Markus Dreßen
Andrea Seppi, Steffen Junghans
Institut für Buchkunst Leipzig an der Hochschule für Grafik und Buchkunst Leipzig
255 195 10mm
405g
56p
ISBN-10 3-932865-27-8

这是一本德国监狱建筑摄影合集。艺术家拍摄了德国多座城市的监狱内外景，探讨监狱的意义和外延。本书为锁线硬精装。护封铜版纸四色加浅灰色共五色印刷，覆哑光膜。护封在上下书口处比书籍开本略大，上下书口和勒口处均向内包折，形成良好的翻阅手感。内封直接使用灰色卡板，在书脊处包覆白色布面。环衬选用灰色艺术纸，内页哑面铜版纸四色加浅灰色共五色印刷。内容方面，艺术家并没有单纯拍摄监狱内部空旷的场景，而是加入了大量监狱外墙和监狱周边的环境。照片并未按照监狱地点进行组合，而是对墙面、窗户及周边环境的相似性进行了组织。艺术家表达了对监狱内外摄像头无差别监视的担忧，一种对新技术（2000 年前后）被滥用的思考。书中没有页码，只有对照片的顺序编号，当遇到满版、跨版图片时，会将编号集中在前部非满版照片的页面上，通过这种方式尽量减少文本符号对照片的干扰，让读者的关注点集中在照片上。

This is a collection of photographs of prison buildings in Germany. The artist photographed the interior and exterior scenes of prisons in many German cities to discuss the meaning and extension of prisons. The hardcover book is bound with thread sewing. The matted jacket is made of coated paper printed in five colors – ordinary four colors plus light gray. The jacket is slightly larger at the upper and lower edges than the book format. Not only the upper and lower edges but also the flaps are folded inward to form a good feeling of browsing. the Gray cardboard is simply used for the inside cover with the white cloth covering at the spine of the book. Gray art paper is used for the end paper, and the matte coated paper is used for the inner pages printed in four colors and light gray. As for the contents, the artist did not simply shoot the empty scenes inside the prison, but mixed with a large number of prison walls and surroundings. The photos are not grouped according to the location of the prison, but are organized by the similarities of the walls, windows and surrounding environment. The artist expressed concern about the indiscriminate surveillance of cameras inside and outside the prison, a reflection on the misuse of new technologies (around 2000). There is no page number in the book, only the sequence number of photos. When encountering full page and cross page pictures, the number will be concentrated on the front page, which is not full. In this way, the interference of text symbols on the photos will be minimized, and the readers' attention will be focused on the photos.

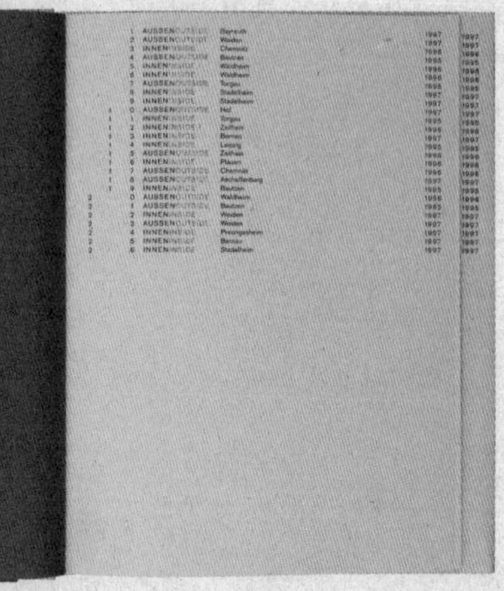
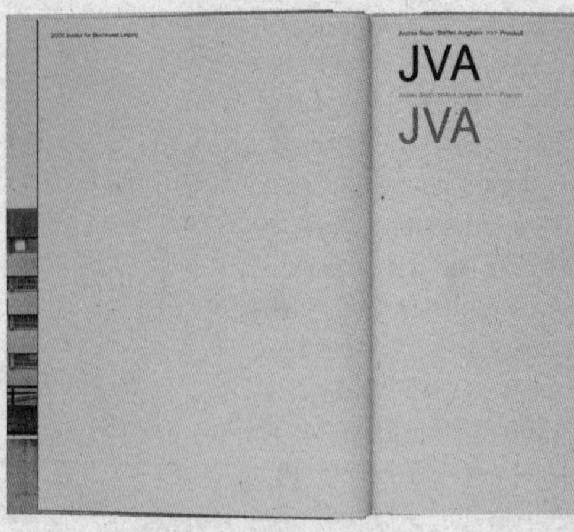
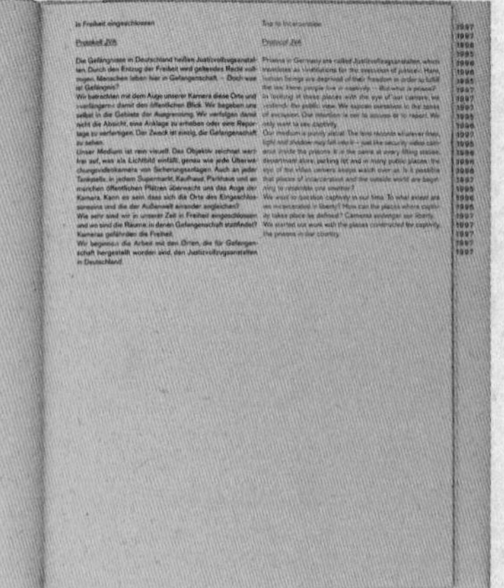

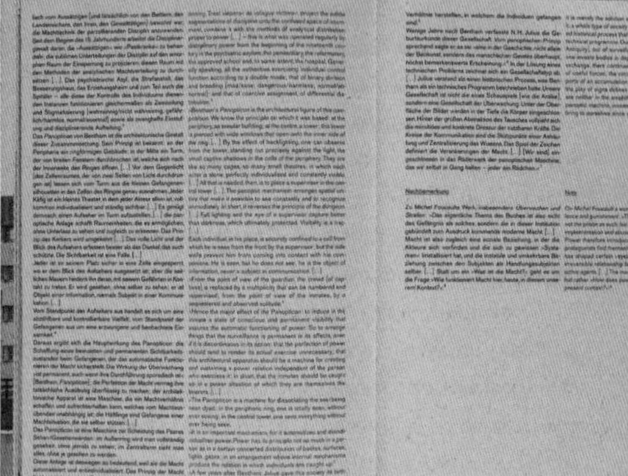

ACHTUNG!

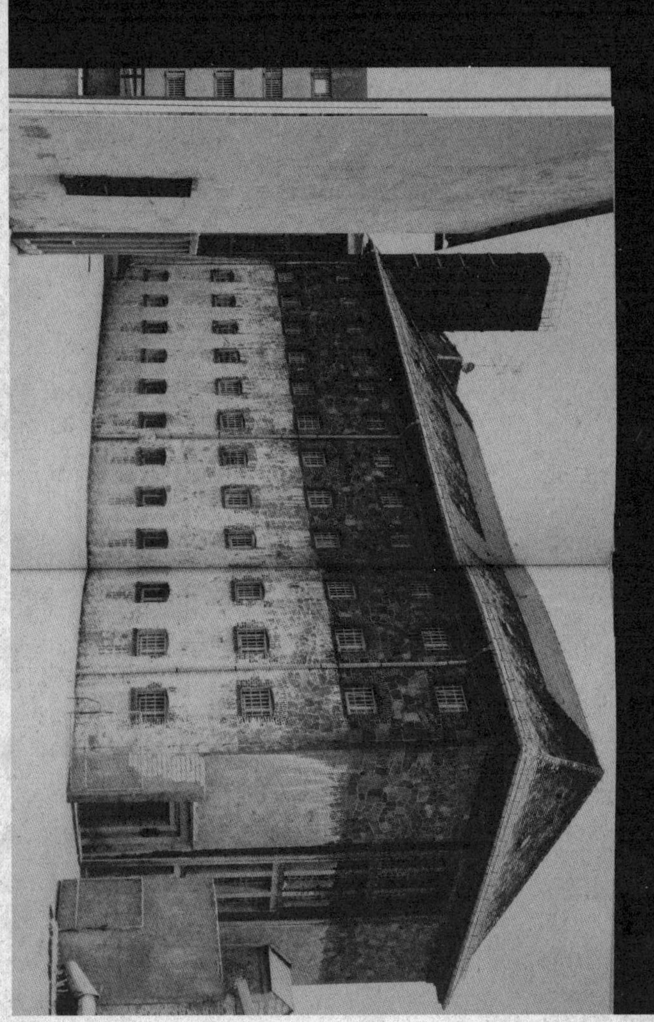

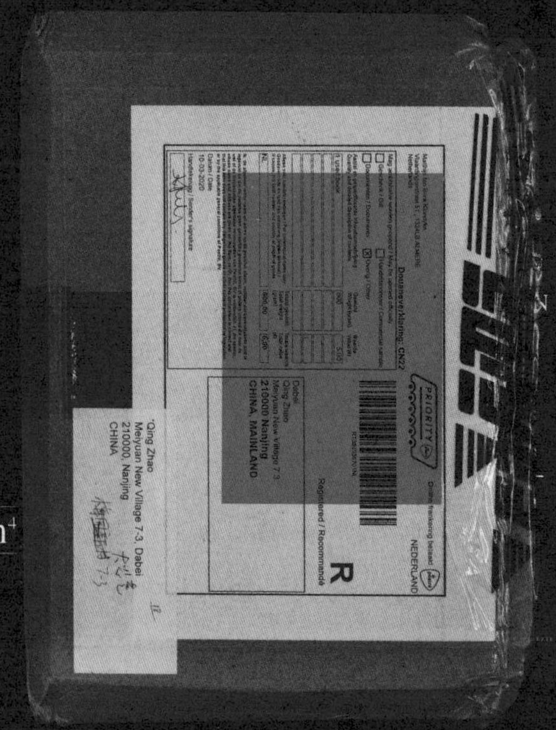

Douaneverklaring: CN22
PRIORITY
NEDERLAND
R
Registered / Recommandé
Qing Zhao
Meiyuan New Village 7-3
210000 Nanjing
CHINA, MAINLAND
Qing Zhao
Meiyuan New Village 7-3, Dabei
210000, Nanjing
CHINA

2002 Bm4

s ten Brink Klondyke
ngenstraat 51 , 1324LB ALMERE
lands

PRIORITY

Online

Douaneverklaring: CN22

nbtshalve worden geopend / May be opened officially

schenk / Gift ☐ Handelsmonster / Commercial sample

cumenten / Documents ☒ Overig / Other

n gespecificeerde inhoudsomschrijving / and detailed description of contents	Gewicht Weight (gram)	Waarde Value (€)
d book	600	5,00
oor handelszendingen / For commercial items only ncode en land van oorsprong (indien bekend) / , HS tariff number and country of origin of goods	Totaal gewicht Total weight (gram)	Totale waarde Total value (€)
	600,00	5,00

zender wiens naam en adres op dit poststuk staan, verklaar dat bovenstaande juist is en dat de poststuk geen gevaarlijke goederen bevat of andere inhoud die door de e toepasselijke algemene voorwaarden van PostNL BV is verbonden. / I, the sender, ame and address are given on this item, certify that this declaration is correct and item does not contain any dangerous goods or other content prohibited by legislation

RT384936701NL

Regist

Dabei
Qing Zhao
Meiyuan New Village 7 3
210000 Nanjing
CHINA, MAINLAND

The Netherlands

China

The Netherlands

247mm

17mm

EENEN-
VIJFTIG
STEMMEN
UIT DE
WERELD-
POËZIE

162mm

● 铜奖
《 世界诗歌的 51 个声音
◖ 荷兰

Bronze Medal
Eenenvijftig stemmen uit de wereldpoëzie
The Netherlands
Typography and Other Serious Matters
Kathinka van Dorp
Stichting Wereldpoezie / Stichting Poetry International, Rotterdam
247 162 17mm
517g
144p
ISBN-10 90-72546-26-1

荷兰诗人卡辛卡·范多普(Kathinka van Dorp)长期致力于诗歌的创作、教学和推广。她受到自1970年以来鹿特丹国际诗歌节的影响，尤其是1999年第30届诗歌节上“可以听的诗歌墙”的启发，在诗歌协会和广播电台的帮助下，收集和整理了本书中51首诗歌的文本和音频资料。不幸的是，作者在2000年意外去世，书籍的后续工作由其妹妹代为执行，并最终出版。书籍的装订形式为锁线硬精装。封面胶版纸红蓝双色印刷，裱覆卡板。环衬采用红色纸张，内页胶版纸印单黑。与单人诗集或来自同一语种的诗集不同，书中这51首诗来自全球多种语言。书的封二处开槽夹附一张由各首诗的作者朗读的录音光盘，以便读者在不懂其他语言的情况下也可以体会到各语种诗歌的魅力，有如音乐是世界性的“语言”、是声音的艺术，听众并不一定需要听懂会唱。内页左右对照排版，左页为原语言的文种，右页为翻译后的荷兰文，双语对照在目录上也做了页码数字的对应。

The Dutch poetess Kathinka van Dorp has devoted herself to the creation, teaching and promotion of poetry for a long time. She has been influenced by the Rotterdam International Poetry Festival since 1970, especially inspired by the “wall of poetry that you can listen to” at the 30th Poetry Festival in 1999. With the help of poetry associations and radio stations, she collected and organized the text and audio material of 51 poems. Unfortunately, the author died unexpectedly in 2000, and the follow-up work of the book was carried out by her sister and finally published. The hardcover book is bound with thread sewing. The cover is made of offset paper printed in red and blue, mounted with cardboard. Red paper is used for the end paper, and the inner pages are printed in black on offset paper. Unlike single poetry's collection or poems in the same language, the 51 poems in the book are using various languages all over the world. An audio CD read by the authors of each poem is attached to the slot in the back cover of the book, so that the readers can also appreciate the charm of each language of poetry. Just as music is a universal ‘Language’ and the art of sound, listeners do not need to understand and sing. The inner pages are typeset for comparison – the left page is in the original language, the right page is in the translated Dutch. The bilingual comparison also corresponds to the page number in the catalog..

ersleten armen, zwakke knieën
80 jaar oud, haar dun en wit
kaak beniger dan ik me herinnerde –
hoofd op zijn hals gebogen, ogen open
nu en dan, luisterde hij –
Ik las mijn vader Wordsworth's *Intimations of Immorta*
wolken van glorie slepend komen wij
van God, die onze woning is…'
'Dat is prachtig,' zei hij, 'maar 't is niet waar.'

en ik een jongen was, hadden we een woning
in de Boyd Street van Newark – daarachter
lag een groot leeg veld vol struikgewas en ho
Ik vroeg me altijd af wat er achter die bomen lag.
en ik ouder werd, liep ik om het blok heen,
en vond uit wat daar achter was –
het was een lijmfabriek.'

18 mei 1976

aling Simon Vinkenoog

أن يأتيَ النّهارُ، أجيءُ
أن يتساءلَ عن شمسِهِ، أُضيءُ
يءِ الأشجارُ راكضةٌ خلفي، وتد
نبني في وَجهِني الأومامُ
ا وقلاعاً من الصّمت يجهلُ أبو
سي ءِ اللّيلُ الصّديقُ، وَتنْسَى
ها في فراشيَ الأيّامُ
إذ شسقط الينابيعُ عدري
خي أُزرارَبا وَتنامُ
ظُ الحاءَ والحرايا، وأَجْلسُ
ها، حفحةَ الرُّؤر، وأَنامُ.

いっていることはうそでも
うそをつくきもちはほんとう
うそでしかいえないほんとの
いぬだってもしくちがきけた
うそをつくんじゃないかしら
うそをついてもうそがばれて
ぼくはあやまらない
あやまってすむようなうそは
だれもしらなくてもじぶんは
ぼくはうそといっしょにいき
どうしてもうそがつけなくな

ם צריך לשנא ולאהב בכת אחת,
וחן עינים לבכות ובאוחן עינים לצח
וחן ירים לזרק אבנים
אוחן ירים לאסף אוחן.
שות אהבה במלחמה ומלחמה באהבו
שנא ולסלח ולזכר ולשכח
סדר ולבלבל ולאכל ולעכל
מה שהסטוריה ארכה
שה בשנים רבות מאד.
ם בחייו אין לו זמן.
שהוא מאבד הוא מחפש
שהוא מוצא הוא שוכח.

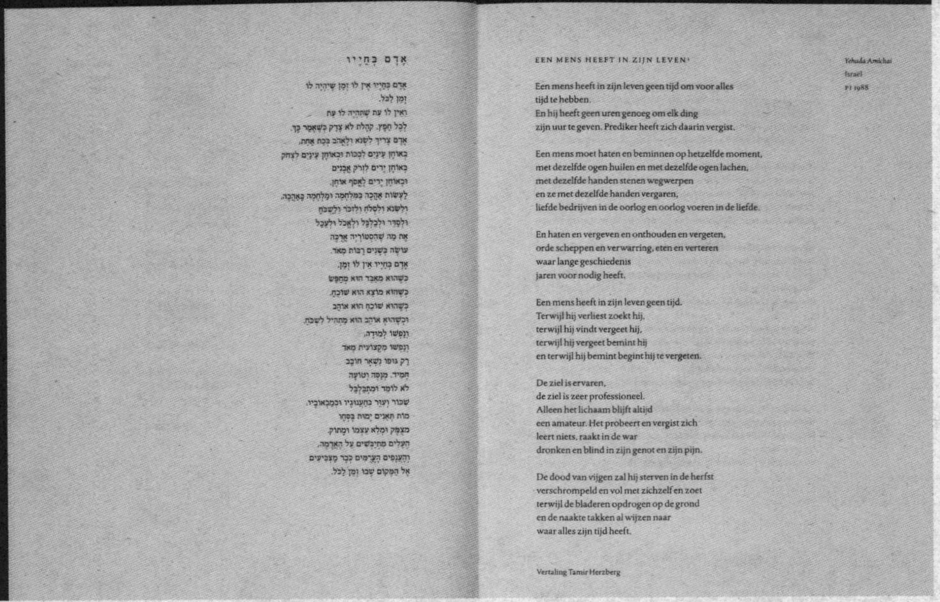

אדם בחייו

אדם בחייו אין לו זמן שיהיה לו
זמן לכל.
ואין לו עת שתהיה לו עת
לכל חפץ. קהלת לא צדק כשאמר כך.
אדם צריך לשנא ולאהב בבת אחת,
באותן עינים לבכות ובאותן עינים לצחק
באותן ידים לזרק אבנים
ובאותן ידים לאסף אותן,
לעשות אהבה במלחמה ומלחמה באהבה.
ולשנא ולסלח ולזכר ולשכח
ולסדר ולבלבל ולאכל ולעכל
את מה שהסטוריה ארכה
עושה בשנים רבות מאד.
אדם בחייו אין לו זמן.
כשהוא מאבד הוא מחפש
כשהוא מוצא הוא שוכח,
כשהוא שוכח הוא אוהב
וכשהוא אוהב הוא מתחיל לשכח.
ונפשו למודה,
ונפשו מקצועית מאד
רק גופו נשאר חובב
תמיד. מנסה וטועה
לא לומד ומתבלבל
שכור ועור בתענוגיו ובמכאוביו.
מות תאנים ימות בסתו
מצמק ומלא עצמו ומתוק.
העלים מתיבשים על האדמה,
והענפים הערמים כבר מצביעים
אל המקום שבו זמן לכל.

Yehuda Amichai
Israel
PI 1988

EEN MENS HEEFT IN ZIJN LEVEN[1]

Een mens heeft in zijn leven geen tijd om voor alles
tijd te hebben.
En hij heeft geen uren genoeg om elk ding
zijn uur te geven. Prediker heeft zich daarin vergist.

Een mens moet haten en beminnen op hetzelfde moment,
met dezelfde ogen huilen en met dezelfde ogen lachen,
met dezelfde handen stenen wegwerpen
en ze met dezelfde handen vergaren,
liefde bedrijven in de oorlog en oorlog voeren in de liefde.

En haten en vergeven en onthouden en vergeten,
orde scheppen en verwarring, eten en verteren
waar lange geschiedenis
jaren voor nodig heeft.

Een mens heeft in zijn leven geen tijd.
Terwijl hij verliest zoekt hij,
terwijl hij vindt vergeet hij,
terwijl hij vergeet bemint hij
en terwijl hij bemint begint hij te vergeten.

De ziel is ervaren,
de ziel is zeer professioneel.
Alleen het lichaam blijft altijd
een amateur. Het probeert en vergist zich
leert niets, raakt in de war
dronken en blind in zijn genot en zijn pijn.

De dood van vijgen zal hij sterven in de herfst
verschrompeld en vol met zichzelf en zoet
terwijl de bladeren opdrogen op de grond
en de naakte takken al wijzen naar
waar alles zijn tijd heeft.

Vertaling Tamir Herzberg

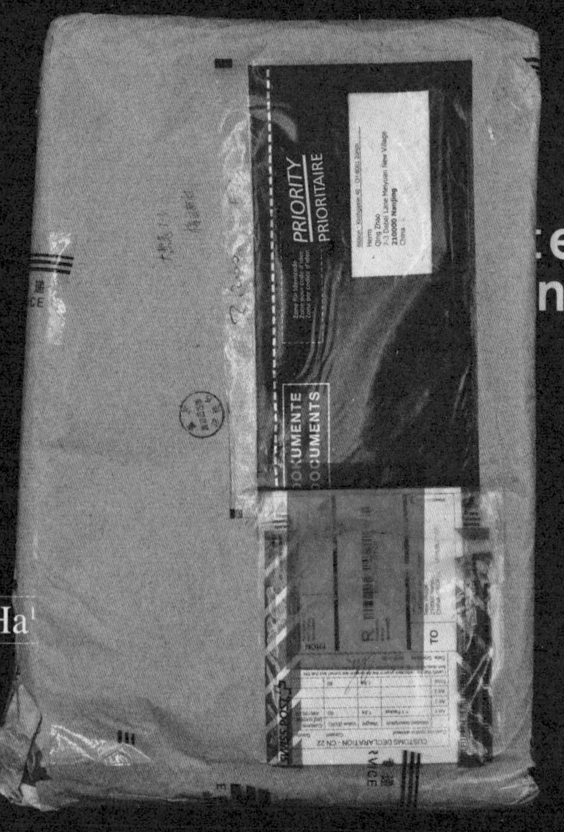

ter
ng

2002 Ha

FROM
Biblion Antiquariat
Leonidas Sakellarios
Kirchgasse 40
8001 Zürich
Switzerland
Post CH
R
Recommandé
étranger
SWISS POST
05.03.2020
Switzerland
Switzerland
China

375mm

28mm

rolfschroeter
die lichtung

rolf schroeter die lichtung

268mm

● 荣誉奖
《 林间空地
● 瑞士

Honorary Appreciation
Die Lichtung
Switzerland
Hans Grüninger (Weiersmüller Bosshard Grüninger wbg AG)
Rolf Schroeter
Verlag Niggli AG Sulgen/Zurich
375 268 28mm
1811g
144p
ISBN-10 3-7212-0396-8

书中的照片本来是瑞士摄影师罗尔夫·施勒特（Rolf Schroeter）的一个拍摄项目——记录一片树林的一年四季，直到树木被人砍伐，从而展示生命的生死轮回。为了强化这一主题，他邀请了多位各地的艺术家、诗人、翻译家，尤其是来自中国和日本的多位艺术家，来与他的照片对话，最终成为本书。书籍的装订形式为锁线硬精装。护封白卡纸印黑覆哑膜，内封白卡纸文字部分烫黑处理，裱覆卡板。选用黑卡纸作为环衬，内页哑面卡纸印单黑。书籍的主线由施勒特的黑白照片引领，多人创作的行为仿佛重建一棵“大树”。施勒特拍摄的茂密的大树、斑驳的树皮、砍伐后露出的年轮、钉满铁钉的树桩，以及盖满白雪的树枝照片，组成了“大树”的主干。其他四位艺术家创作的诗歌、诗歌的德文译本、关于树木的文字排版，以及在施勒特的照片上涂抹油彩形成的画作构成了“大树”的枝叶。照片、排版、书法、诗歌、绘画等形式在这里呈现出东西方多元文化对自然与生死的多重表达。||||||||||||||||||||||||||||||||||||||||||||||||||||||||||||||||||||||||||||||||

The photo book in the book is a shooting project of Swiss photographer Rolf Schroeter—recording the seasons of a forest until the trees are felled, thus showing the cycle of life and death. To strengthen this theme, he invited many artists, poets, and translators from all over the world, especially many artists from China and Japan, to talk to his photos and become a book. The book binding format is hard thread hardcover. The cover white card paper is printed with black matte film, the inner part of the white card paper is blackened, and the card board is mounted. Black cardboard is used as the ring lining, and the matte cardboard on the inner page is printed in black. The main line of the book is led by Schlett's black and white photos, and the act of multi-person creation seems to rebuild a “big tree”. The dense tree of Schlett, the mottled bark, the annual rings exposed after the felling, the nailed stumps, and the photos of the branches covered with snow form the backbone of the “big tree” and the poems created by four other artists , The German translation of poetry, the typesetting of trees, and the paintings formed by applying oil paint to Schlett 's photos constitute the branches and leaves of the “big tree”. In the end, photographs, typesetting, calligraphy, poetry, and painting were formed—multiple expressions of nature, life, and death in the Eastern and Western multiculturalism. ||||||||||||||||||||||||||||||||||||||||||||||||||||||||||||||||||||||||

rolfschroet
die lichtur

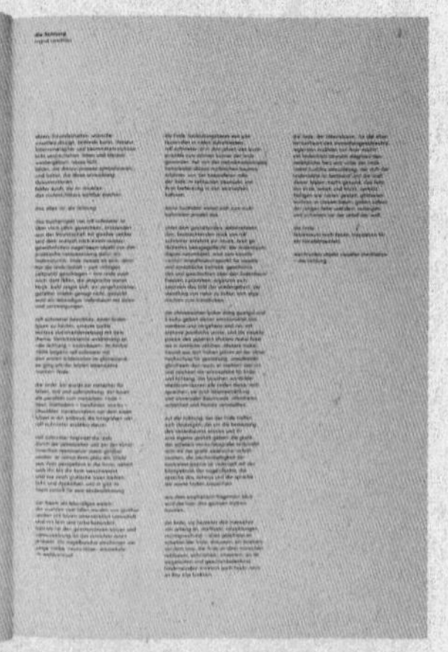

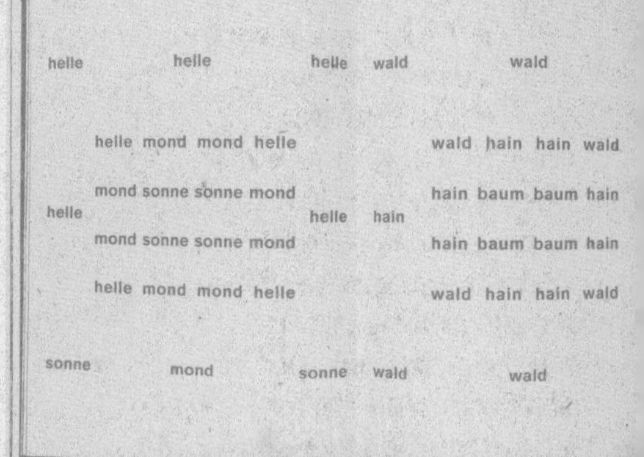
helle helle helle wald wald
helle mond mond helle wald hain hain wald
mond sonne sonne mond hain baum baum hain
helle helle hain
mond sonne sonne mond hain baum baum hain
helle mond mond helle wald hain hain wald
sonne mond sonne wald wald

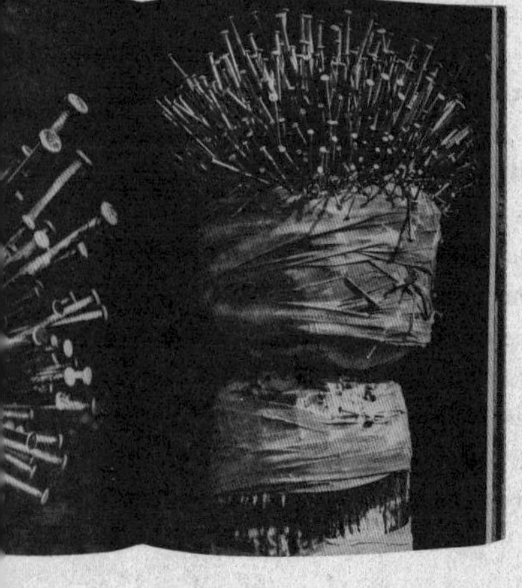

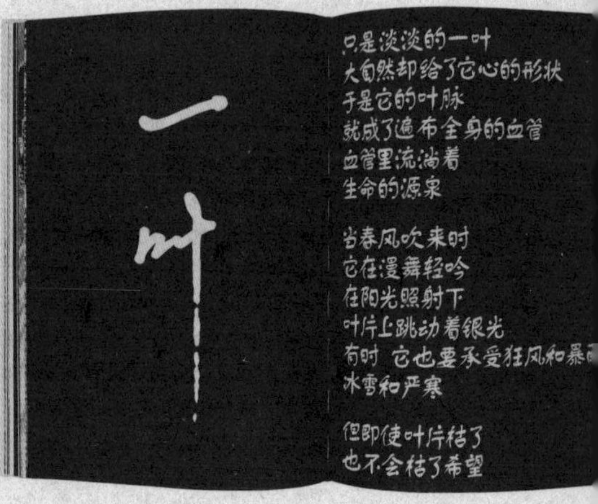
一叶
只是淡淡的一叶
大自然却给了它心的形状
于是它的叶脉
就成了遍布全身的血管
血管里流淌着
生命的源泉
当春风吹来时
它在漫舞轻吟
在阳光照射下
叶片上跳动着银光
有时 它也要承受狂风和暴
冰雪和严寒
但即使叶片枯了
也不会枯了希望

度和尚
化身
喜欢
事
要编造出

一株小苗
求又是那样微小
一线阳光
蹈

的风

汁

多么纯真

一叶知秋
壮丽的夏天刚刚过
整个空间已弥漫着

冬日来临
我已步入老境
树叶已经凋零
树皮如同脸上的皱
更有人拿我的枝桠
去当柴烧
为的是不在寒冬中

我已感到生活的严
不过 正因为严峻
我才渴望爱的温暖
我张开枯干的双臂
面向苍天
但愿给我这颗微小
一个大的宇宙
去包容人类的一切
宁静和喧闹

ch

lid

月 明 明 月

明 明 明 明

明 明 明 明

ch dem fällen. die ansprüche waren
ch. bald zeigte sich: ein vorgefunden
fällter stamm genügt nicht. gesucht
d ein lebendiger lindenbaum mit äste
d verzweigungen.

f schroeter beschloss, einen linden-
um zu kaufen. erneute suche.
itere auseinandersetzung mit dem
ema. fortschreitende annäherung an
e lichtung – lindenbaum». im herbst
94 begann rolf schroeter mit
n ersten bilderserien im glarnerland.
ging um die letzten lebensjahre
einer» linde.

e linde. sie wurde zur metapher für
en, tod und auferstehung. der baum
parallele zum menschen: rinde –
ut. blattadern – handlinien. wuchs –
arakter. transformation von dem einen
en in ein anderes. die fotografien vo

树皮赞

树皮 我赞美你
你粗糙不平 灰黑无华
你终生包裹着树干
象母亲护卫着她的孩子
抵御风刀霜剑酷暑严寒
等到树木成材
你被剥去 被披戟踏
而你无怨无悔 终生奉献

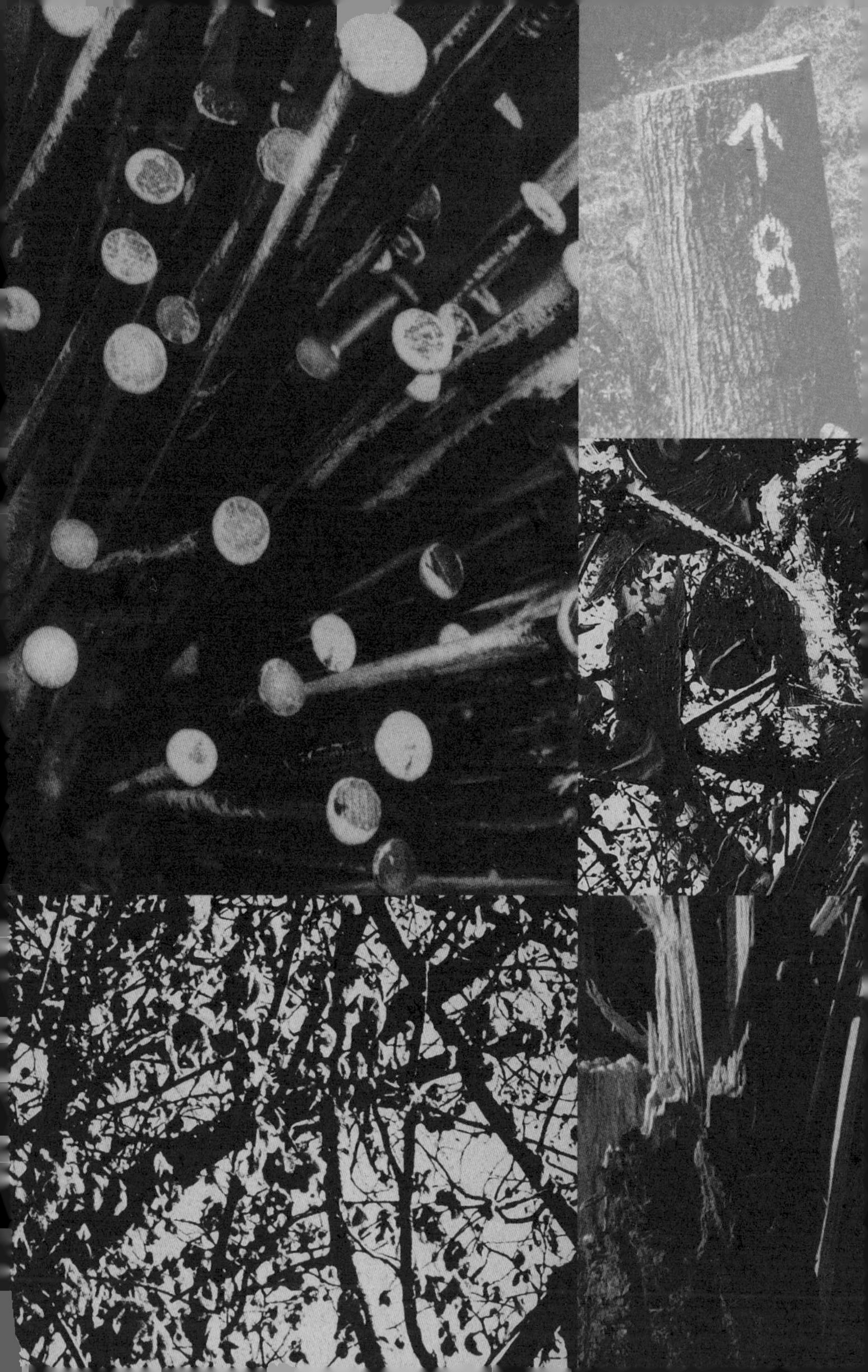
8

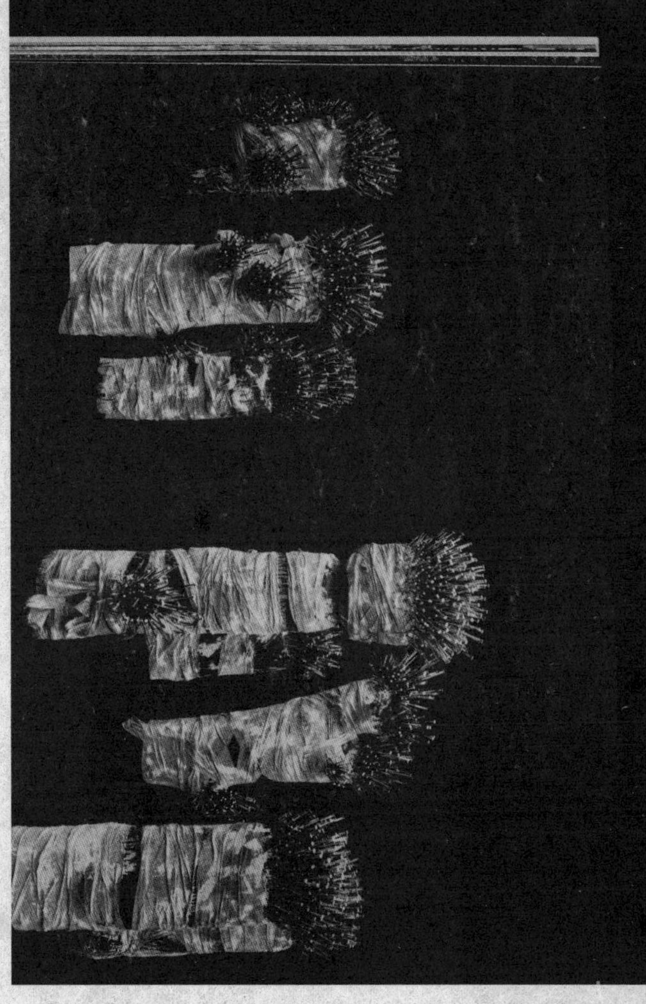

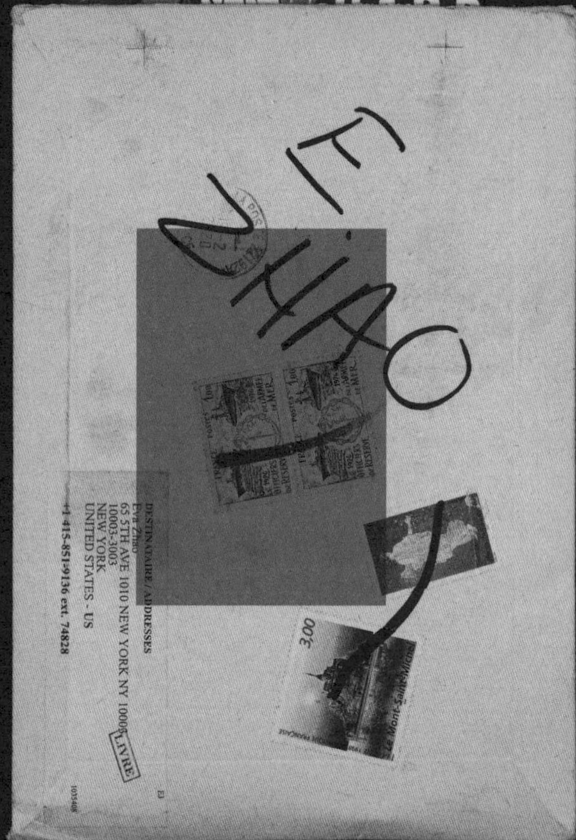

2002 Ha^2

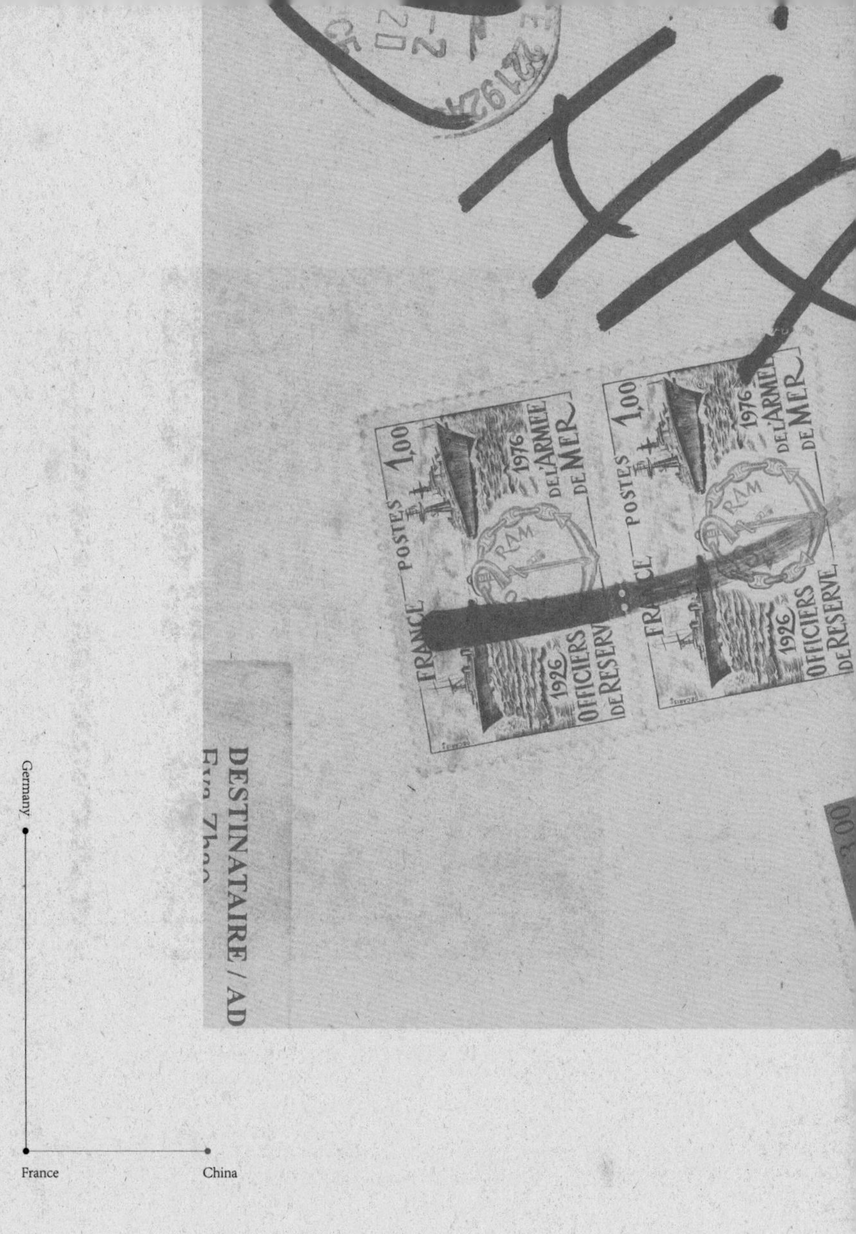
DESTINATAIRE / AD
FRANCE
POSTES
1,00
1976
DE L'ARMEE
DE MER
1926
OFFICIERS
DE RESERVE
Germany
France
China

205mm

6mm

137mm

● 荣誉奖
《 硬摇滚天堂（伟大的小册子 20）
◖ 德国

Honorary Appreciation
Der Hardrock-Himmel (Die Tollen Hefte 20)
Germany
Thomas Müller
Tom Coraghessan Boyle
Büchergilde Gutenberg, Frankfurt am Main / Wien / Zürich
205 137 6mm
110g
32p
ISBN-10 3-7637-6016-1

本书是由小说家托马斯·科拉格桑·博伊尔（Tom Coraghessan Boyle）写作的故事，属于著名设计师、插画家托马斯·穆勒（Thomas Muller）绘制插图并出版的“伟大的小册子”（*Die Tollen Hefte*）书系里的一本（丛书中的第 22 本——《生活用品》获得 2004 年铜奖，可在《莱比锡的选择》一书中找到）。全书采用线装骑马钉外包护封的装订形式。封面与内页均采用胶版纸多色印刷。32 开的胶版纸不仅增加了亲和力，更便于翻阅。书中讲述了一个人去寻找“硬摇滚天堂”的故事，一路上遇到了演奏蓝调、巴萨诺瓦、爵士乐、古典乐等各种音乐的人，并与他们产生了丰富的互动。书中另附折叠于其中的海报和覆膜卡片各一张，印有作者和书籍系列的介绍。

This book is a story written by the novelist Tom Coraghessan Boyle, which belongs to the book series *Die Tollen Hefte* illustrated and published by the famous designer and illustrator Thomas Muller. (*Lebens-Mittel*, the 22nd book of the series, won the Bronze Medal in 2004, which can be found in the book *The Choice of Leipzig*). The book is bound by saddle stitching with thread sewing. Both the cover and the inner pages are printed with offset paper, which not only increases the affinity, but also makes the book readable. The book tells the story of a person looking for 'Hard Rock Paradise'. On the way, he met people who played various music such as blues, Barcelona, jazz, and classical music, and had frequent interaction with them. The book also includes a folded poster and a laminated card, printed with an introduction to the author and book series.

T. Corag

DER H

HIMM

übersetzt vo

mit Bildern v

ein Haar war das feinste F

war groß genug, um norma

Rückgrat wie ein Baguette

sagte er. „Na, dann will ic

tern: Ich kenn dich nicht.“

ch packte meine Gibson aus

en der Verstärker, die rings

mal aufregten, und ließ ihm

AR

O

T

tz

RG

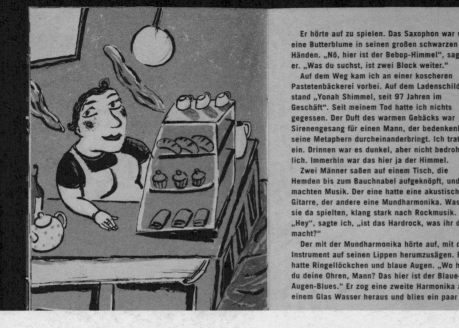

Er hörte auf zu spielen. Das Saxophon war wie eine Butterblume in seinen großen schwarzen Händen. „Nö, hier ist der Bebop-Himmel“, sagte er. „Was du suchst, ist zwei Block weiter.“

Auf dem Weg kam ich an einer koscheren Pastetenbäckerei vorbei. Auf dem Ladenschild stand „Yonah Shimmel, seit 97 Jahren im Geschäft“. Seit meinem Tod hatte ich nichts gegessen. Der Duft des warmen Gebäcks war Sirenengesang für einen Mann, der bedenkenlos seine Metaphern durcheinanderbringt. Ich trat ein. Drinnen war es dunkel, aber nicht bedrohlich. Immerhin war das hier ja der Himmel.

Zwei Männer saßen auf einem Tisch, die Hemden bis zum Bauchnabel aufgeknöpft, und machten Musik. Der eine hatte eine akustische Gitarre, der andere eine Mundharmonika. Was sie da spielten, klang stark nach Rockmusik. „Hey“, sagte ich, „ist das Hardrock, was ihr da macht?“

Der mit der Mundharmonika hörte auf, mit dem Instrument auf seinen Lippen herumzusägen. Er hatte Ringellöckchen und blaue Augen. „Wo hast du deine Ohren, Mann? Das hier ist der Blaue-Augen-Blues.“ Er zog eine zweite Harmonika aus einem Glas Wasser heraus und blies ein paar

2002 1 Ha4

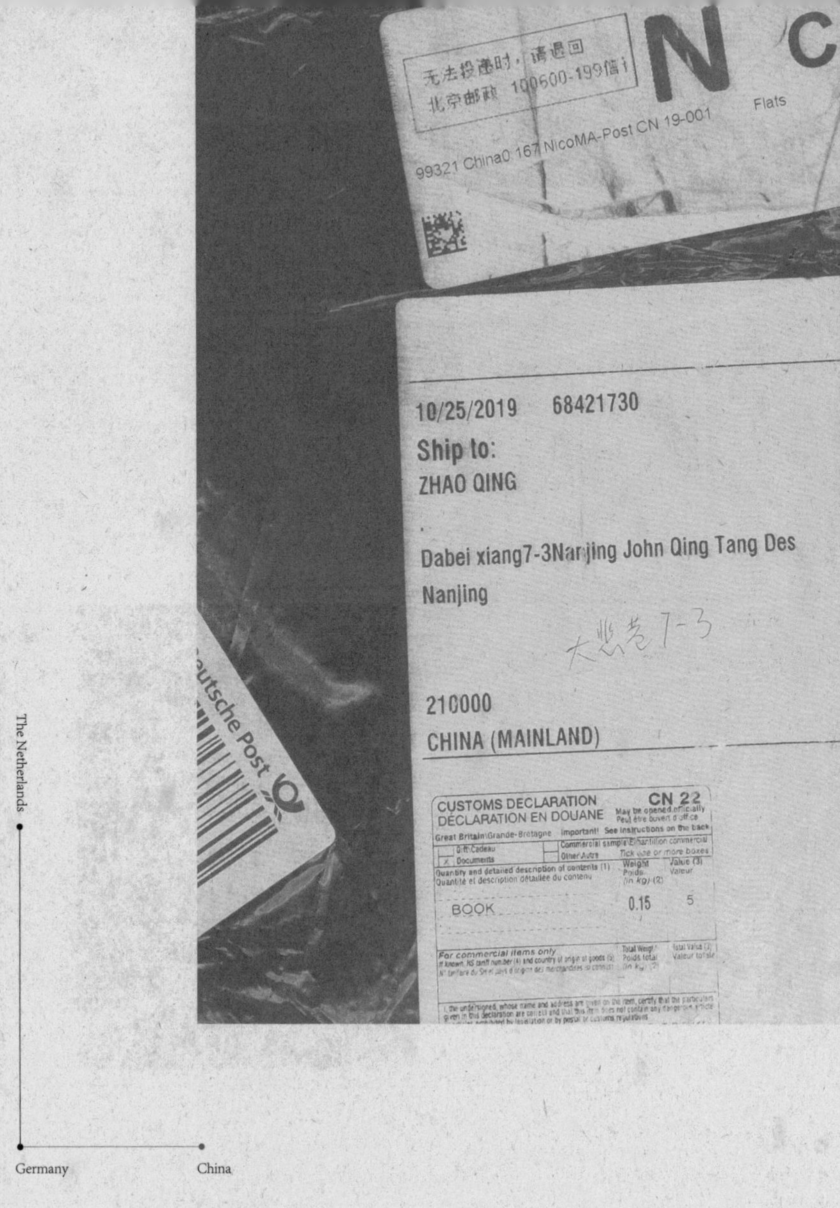

The Netherlands

Germany

China

245mm

4mm

176mm

● 荣誉奖

《 Dolly——华丽的文本字体

◐ 荷兰

Honorary Appreciation

Dolly: A Book Typeface With Flourishes

The Netherlands

Wout de Vringer (Faydherbe / De Vringer)

——

Underware, Den Haag

245 176 4mm

150g

32p

ISBN-10 90-76984-01-8

Dolly 是专为文本阅读而设计的字体，共分为常规、斜体、粗体、小型大写字母四个款型。本书是设计公司 Underware 介绍和推广自家设计的 Dolly 字体的手册。书籍的装订形式为锁线胶平装。封面白卡纸印荧光红色，法国斗牛犬 Dolly 的造型被烫银处理。内页采用哑面艺术纸四色印刷。与常见的字体推广手册不同，Dolly 字体的命名源于设计师的爱犬的名字，所以书中的字体运用与介绍均围绕这只可爱的法国斗牛犬展开。内容方面主要分为适用人群界定、字体细节介绍、作者与 Dolly 的故事、字体的样张等。由于每部分的作者分别使用了英、德、荷三种语言，写作内容上也有较大差异，所以在每部分的版面设计上都不尽相同，并在段落正文、科学图谱、图文混排、信纸文字、字典密排等应用场景上均有展现。最后是每部分作者与各自爱犬的合影。总体来说，这是一本别有趣味的字体宣传手册。

Dolly is a typeface designed for text reading. It is divided into four types: regular, italics, bold, and small capital letters. This book is a manual for design company Underware to introduce and promote Dolly designed by itself. The book is perfect bound with thread sewing. The cover is made of white cardboard printed with fluorescent red, and Dolly, a French bulldog, is hot stamped with silver. The inner pages are printed with matte art paper in four colors. Different from the common font promotion manual, font name Dolly comes from the name of the designer's pet dog, so the font application and introduction in the book are centered on this cute French bulldog. The content is mainly divided into the applicable user definition, the introduction of font details, the story of the writer and Dolly, and font sample, etc. As the writers of each part use English, German, and Dutch respectively, and the writing content is also quite different, so the layout design of each part is different, including the forms of paragraph, scientific atlas, mixed arrangement of pictures and texts, letter text, dictionary layout. The group photos of the writers and their dogs are shown at the end of the book. This is a fun and interesting typeface brochure full of love.

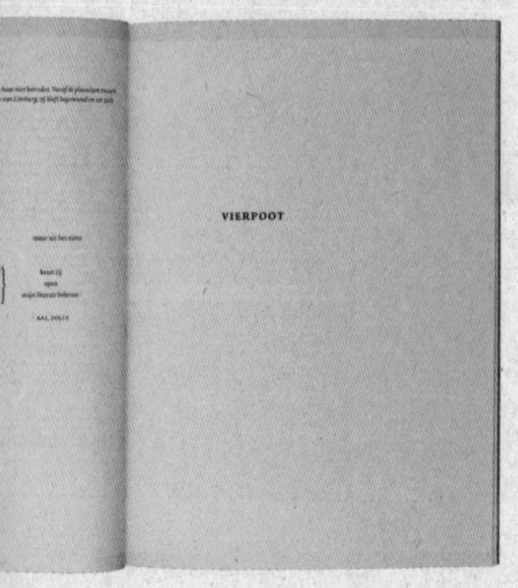

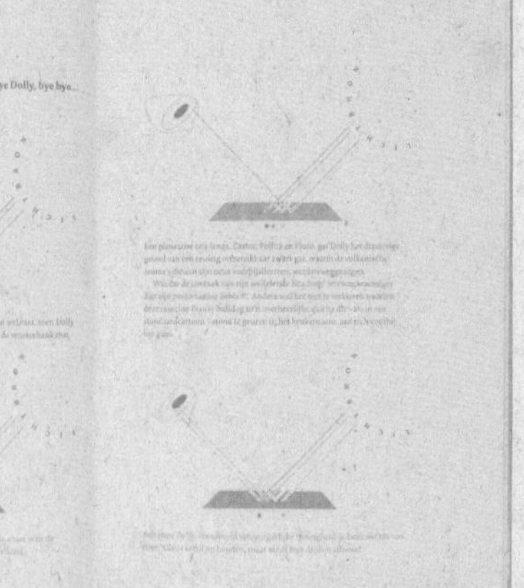

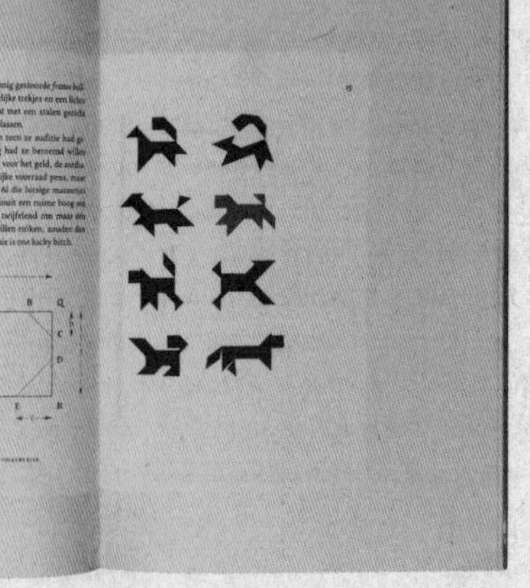

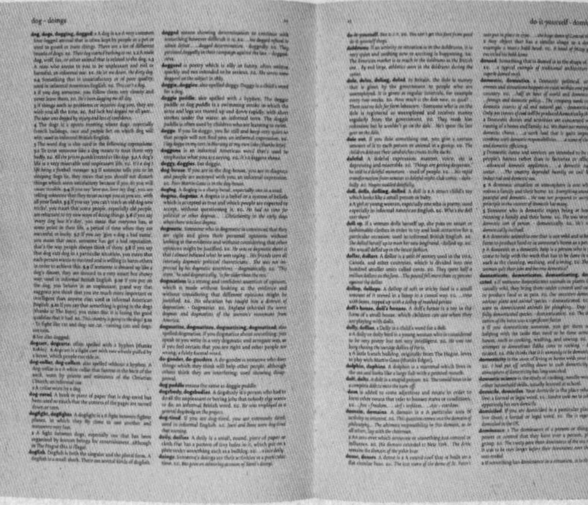

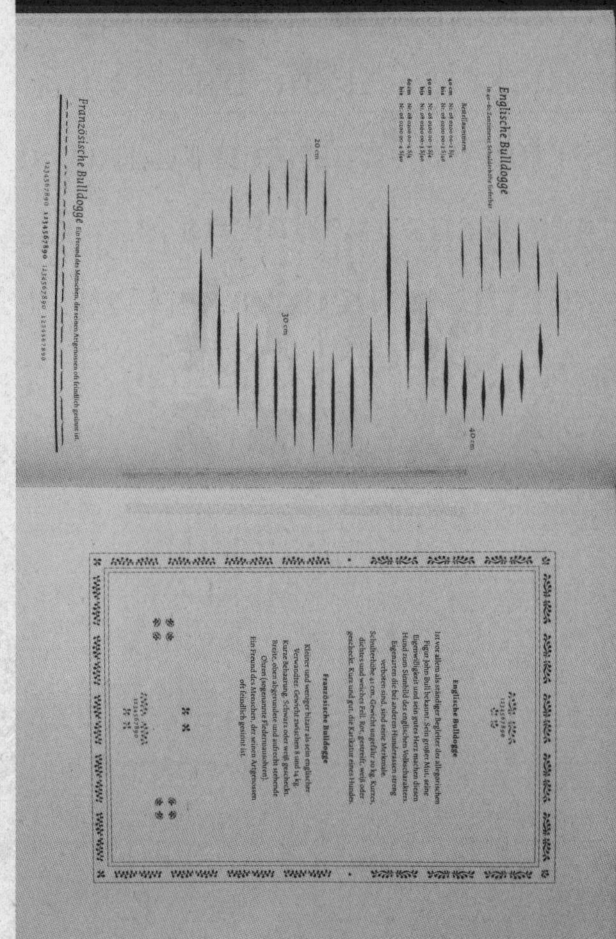
Englische Bulldogge
20 cm
30 cm
40 cm
Französische Bulldogge Ein Freund des Menschen, der seinen Artgenossen oft feindlich gesinnt ist.
Englische Bulldogge
Französische Bulldogge

niet zo zeer voor het geld, de
e onuitputtelijke voorraad pens
starfuckers. Al die hitsige man
opgeheven snuit een ruime bo
en, geeneens twijfelend om ma
aar kont te willen ruiken, zoude
illig zijn. Lassie is one lucky bitc

racht

onbekommerd

Wij hebben geen
wraf wraf wraf mu
wraf wraf wraf ton

ARDS ARE NICE. SALES ARE N
ARD-WINNING ADVERTISING
LLING ADVERTISING ARE OFTEN
ME THING AT

x

a

A **B** **Q**

C

S&H

Englische Bulldogge

in 40–60 Zentimeter Schulderhöhe lieferbar

Bestellnummern:

40 cm	Nr. 08 0200 00–2 S/4
bis	Nr. 08 0200 00–2 S/40
50 cm	Nr. 08 0200 00–3 S/4
bis	Nr. 08 0200 00–3 S/40
60 cm	Nr. 08 0200 00–4 S/4
bis	Nr. 08 0200 00–4 S/40

ons.
very
og's
og's
let
urb
will
are
with
new
ple,
say
, at
are
ne',
on,
say
hat
ters

I lay doggo in my tent in No
doggone is an informal
emphasize what you are s
doggy, doggies. See doggi
dog-house. If you are in th
and people are annoyed w
EG. *Poor Martin Gaus is in t*
dogleg. A dogleg is a sharp
dogma, dogmas. A dogma
which is accepted as true
accept, without questio
political or other dogmas..
when there was less dogma.
dogmatic. Someone who is
are right and gives th
looking at the evidence an
opinions might be justifie
that I almost believed what
intensely dogmatic politica

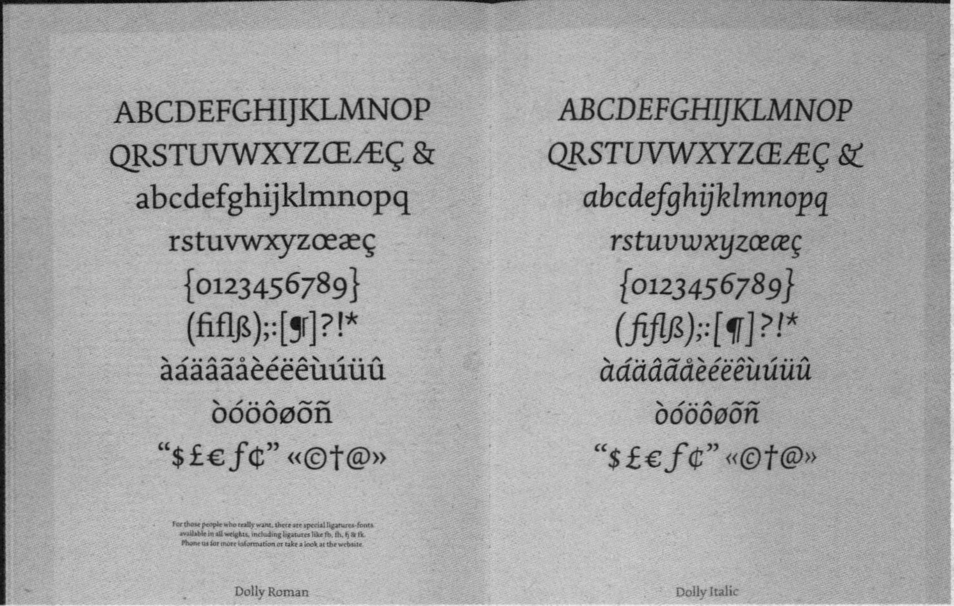

ABCDEFGHIJKLMNOP
QRSTUVWXYZŒÆÇ &
abcdefghijklmnopq
rstuvwxyzœæç
{0123456789}
(fiflß);:[¶]?!*
àáäâãåèéëêùúüû
òóöôøõñ
"$£€ƒ¢" «©†@»

For those people who really want, there are special ligatures-fonts available in all weights, including ligatures like fb, fh, fj & fk. Phone us for more information or take a look at the website.

Dolly Roman

ABCDEFGHIJKLMNOP
QRSTUVWXYZŒÆÇ &
abcdefghijklmnopq
rstuvwxyzœæç
{0123456789}
*(fiflß);:[¶]?!**
àáäâãåèéëêùúüû
òóöôøõñ
"$£€ƒ¢" «©†@»

Dolly Italic

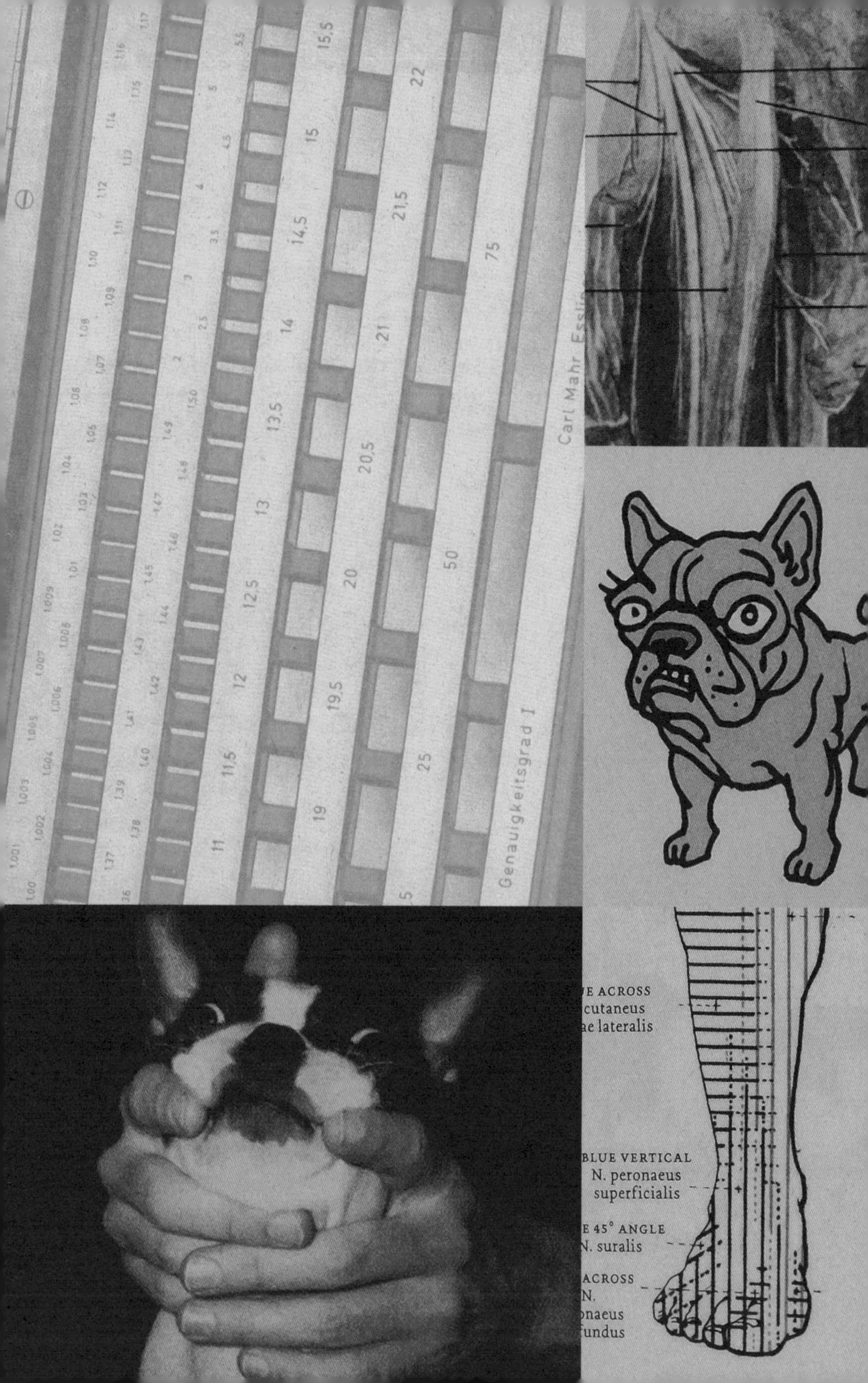
Genauigkeitsgrad I
BLUE VERTICAL
N. peronaeus
superficialis
N. suralis
ACROSS

How - It is not a secret that dogs start to look like their owners. And vice versa. I knew, for instance, a girl who styled her hair like her bouvier. So when her dog died, so did her hair. And an old neighbour of a friend of my mother's boss had a dog that quit smoking THE FIRST OF JANUARY 1987.

She used to be a Cocker Spaniel

A - When the man eventually couldn't resist any longer, the dog couldn't either. They both puffed away their lives. These are only two stories about man's best friend, the dog. I personally didn't give a damn about dogs. Not that I hated them, but I couldn't stand the fact that they smelled, when they were wet and they barked loud. But that all changed WHEN I MET DOLLY.

Dog - She was a bitch. And that didn't just apply to her gender. She was harsh. Seriously. Let me tell you about Dolly. I walked down the street in a French town called Gernui. I never should have stopped. But I had to be sure that I was still sane when someone said: 'hey, good-looking, where are you going? and the only 'person' I saw was a dog. Looking at me, smiling at me. She continued in a melodic french-like accent: 'I'm Dolly, I am looking for a place to stay'. I was stunned. From that moment on I was IN A REAL-TIME TRIP.

Man rescuing dog in snow

Former Pirelli-slipper designer

Cover photo with famous female movie star

Changed - Dolly took over. She told me what to do and where to go. And then one night I saw that she was brushing her teeth with my toothbrush. I saw she was trying on some of my clothes. I also realised that for the last few days she was walking on two legs in my apartment in that French village. And then again a few days later I took a look IN THE MIRROR FOR THE FIRST TIME IN WEEKS.

Dogs can't change into ads or cars

My - I freaked out. My ears where bigger then before and my nose was pointier. Then I struck me. Dolly was taking over literally. She changed into me. And I was changing into a hairy, smelly, ass-kissing bitch. But come to think of it, how can I be a dog now? DOGS CAN'T WRITE.

Life - So who's writing this text, who's word are you reading, who's characters are you seeing? ARE THEY MINE OR ARE THEY DOLLY'S?

Life ain't nothing but bitches and money

Some people change into snakes

First graffiti ever found by dog

Edgar Kober, former illustrator

Book by French dog

Many humans are active as search dogs for the Red Cross nowadays

Dogs love arts, especially when it comes with annual scholarships

2003

Country / Region

Germany ❼

The Netherlands ❺

Switzerland ❶

Japan ❶

kleur kaap 01

NOOM: THE THINGS

FAHRENHEIT 451

LUCEBERT

verzamelde gedichten

Designer

Nynke M. Meijer

Kristina Brusa

Anja Harms

Corina Cotorobai (Werkplaats Typografie)

Dimitri Bruni, Manuel Krebs (NORM)

Gerda Raidt

Friedrich Forssman

Claudio dell'Olio

Bram de Does

Sabine Golde

Rainer Groothuis (Groothuis, Lohfert, Consorten)

Sin Akiyama, Ken Hisase

-SYB-

Tessa van der Waals

2003 Sm[2] *1196*

2003 Bm[1] *1208*

2003 Bm[2] *1222*

2003 Bm[3] *1230*

2003 Ha[5] *1238*

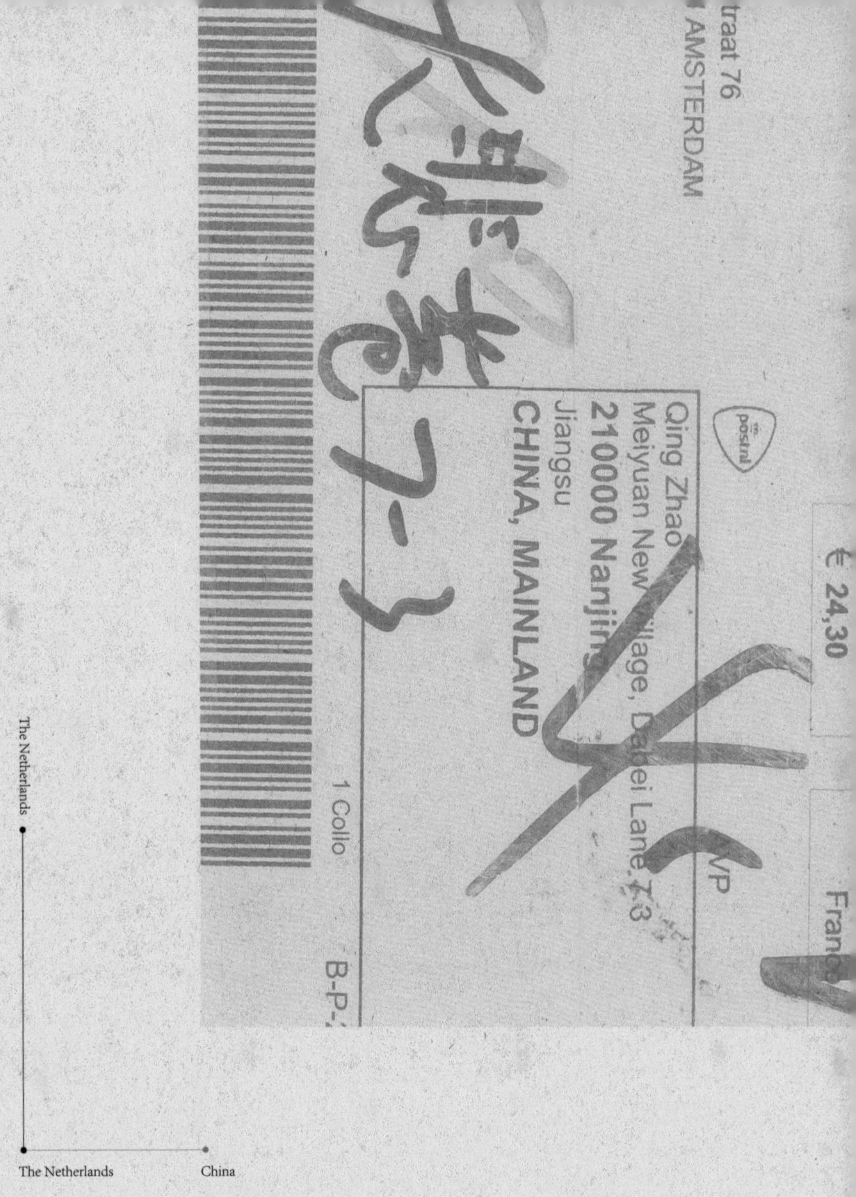
traat 76
AMSTERDAM
Qing Zhao
Meiyuan New Village, Dabei Lane, 7-3
210000 Nanjing
Jiangsu
CHINA, MAINLAND
€ 24,30
1 Collo
The Netherlands
The Netherlands
China

● 银奖 Silver Medal

« 色彩 Kleur

◖ 荷兰 The Netherlands

Corina Cotorobai (Werkplaats Typografie)

——

kaAp / studium generale van de Hogeschool voor de Kunsten Arnhem

240 171 20mm

686g

272p

——

2002年4月，荷兰阿纳姆艺术学院（Hogeschool voor de kunsten Arnhem-organiseert）与“色彩之外基金会”（Stichting Kleur Buiten）共同举办了名为“表达你的色彩”（State Your Colour）的艺术活动，活动中展示了学院的教师和学生对色彩相关领域的展示、应用和理论反思，本书为活动的研究集。书籍的装订形式为锁线胶平装。封面白卡纸印单黑，覆哑膜。内页胶版纸四色印刷。研究由平面、3D、服装、艺术等科系的学生分工执行。研究组不希望给出一套永恒且通用的色彩理论，所以书中收录的论文涵盖多个方面：对于色彩的基础认知、色彩心理学、色彩的符号学和社会学、色彩的文化人类学基本术语、色彩的历史、色彩的应用。书中版式设置为四栏，正文跨两栏排版，附注占单栏，并通过字号和不同等级的字重来形成灰度丰富的文本区块。书籍的三面切口分别附加三原色条，上蓝下红，翻口黄色，并在边角处形成叠加色，是本书的一大亮点。

In April 2002, Academy of Fine Arts Arnhem (Hogeschool voor de kunsten Arnhem-organiseert) and Beyond Colour Foundation (Stichting Kleur Buiten) jointly held an art activity named ‘State Your Colour’, which showed the teachers and students’ performance, application and theoretical reflection in the field of colour. This book is a collection of activity studies. It is perfect bound with thread sewing and the matt laminated cover is made of white cardboard printed in black. The inside pages are mainly printed in four colors on offset paper. The research was carried out by the students in different departments, such as Graphic Arts, 3D, Clothing and Arts. The research group didn’t intend to give a set of eternal and universal colour theory, so the papers included in the book cover many aspects: basic cognition of colour, colour psychology, semiotics and sociology related to colour, basic terms of cultural anthropology, history and the application of colour. The layout of the book is set to four columns – the text is arranged across two columns while the notes occupy a single column. The wide variety of text blocks is formed through the different font sizes and levels of word weight. Three primary color strips are attached to the three edges of the book, blue on the top, red on the bottom and yellow on the right. The superimposed color is formed at the corners, which is one of the highlights of the book.

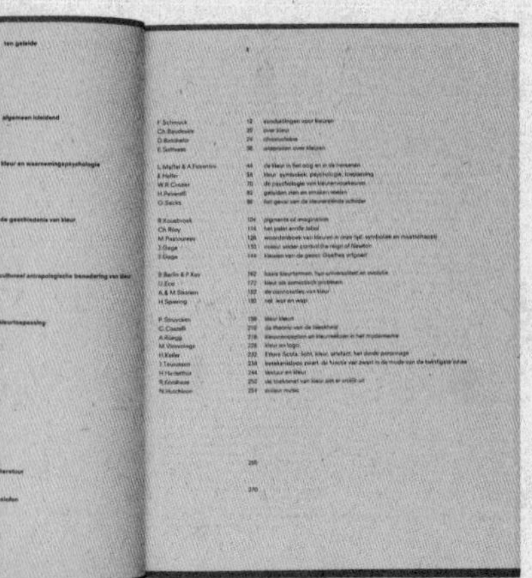

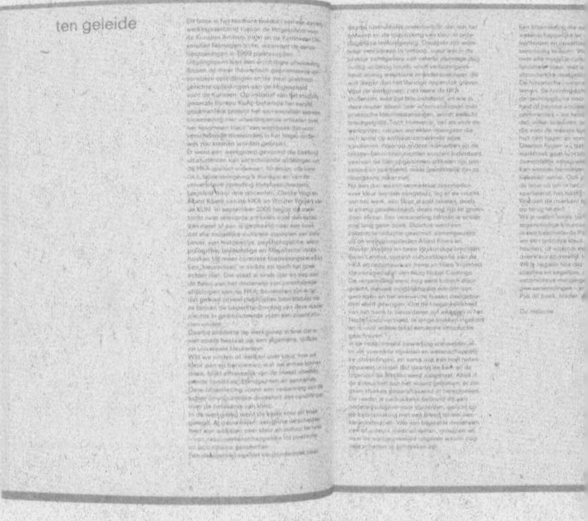

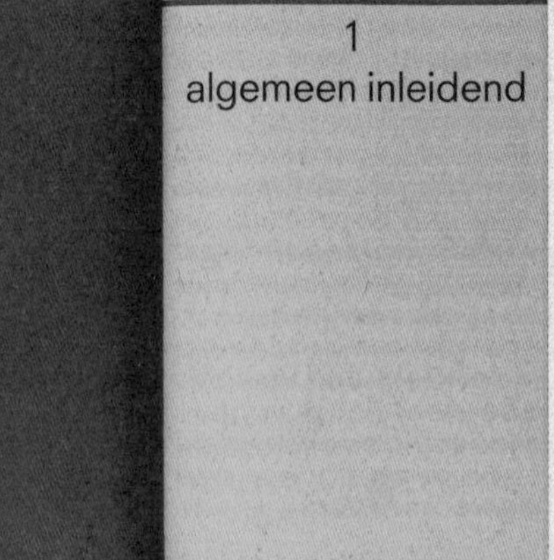

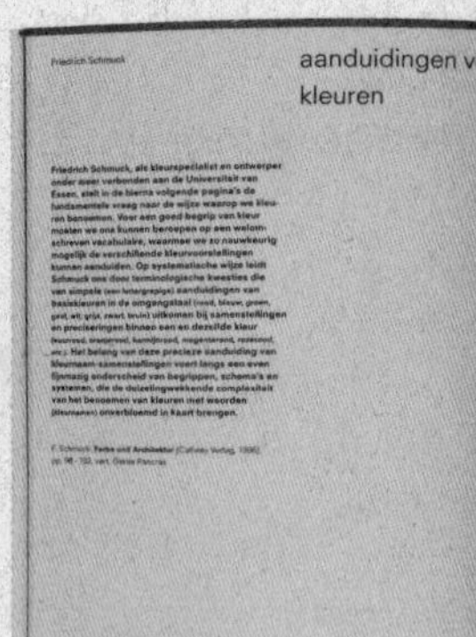

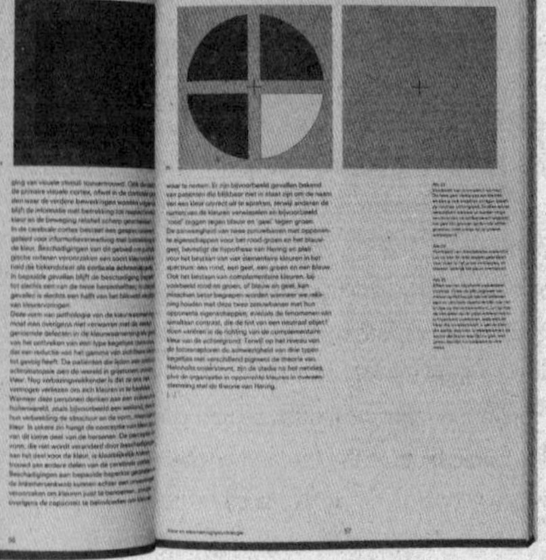

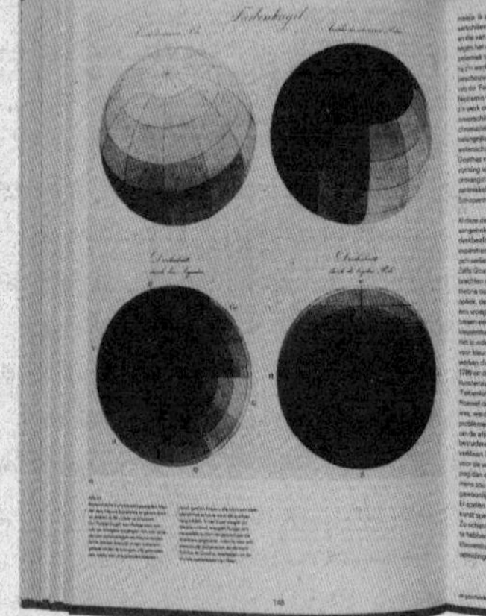

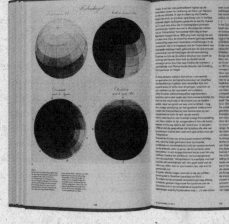

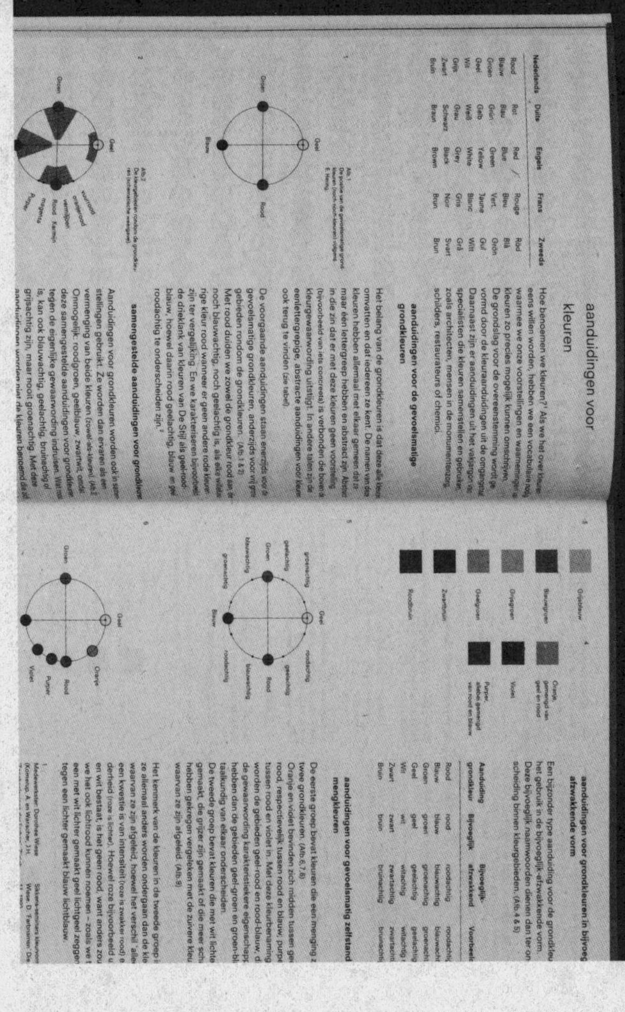

aanduidingen voor kleuren

Nederlands	Duits	Engels	Frans	Zweeds
Rood	Rot	Red	Rouge	Röd
Blauw	Blau	Blue	Bleu	Blå
Groen	Grün	Green	Vert	Grön
Geel	Gelb	Yellow	Jaune	Gul
Wit	Weiß	White	Blanc	Witt
Grijs	Grau	Grey	Gris	Grå
Zwart	Schwarz	Black	Noir	Svart
Bruin	Braun	Brown	Brun	Brun

Hoe benoemen we kleuren?[1] Als we het over kleuren eens willen worden, hebben we een vocabulaire nodig waarmee we onze voorstellingen en waarnemingen van kleuren zo precies mogelijk kunnen omschrijven. De grondslag voor de overeenstemming wordt gevormd door de kleuraanduidingen uit de omgangstaal. Daarnaast zijn er aanduidingen uit het vakjargon van specialisten die kleuren samenstellen en gebruiken, zoals architecten, mensen in de monumentenzorg, schilders, restaurateurs of chemici.

aanduidingen voor de gevoelsmatige grondkleuren

Het belang van de grondkleuren is dat deze alle kleuren omvatten en dat iedereen ze kent. De namen van deze kleuren hebben allemaal met elkaar gemeen dat ze maar één lettergreep hebben en abstract zijn. Abstract in die zin dat er met deze kleuren geen voorstelling (bijvoorbeeld van iets concreets) is verbonden die boven de kleurgewaarwording uitstijgt. In andere talen zijn die eenlettergrepige, abstracte aanduidingen voor kleuren ook terug te vinden (zie tabel).

1

Afb.1
De positie van de gevoelsmatige grondkleuren (noch-noch-kleuren) volgens E. Hering.

De voorgaande aanduidingen staan enerzijds voor de gevoelsmatige grondkleuren, anderzijds voor vrij grote gebieden rondom de grondkleuren. (Afb.1 & 2)
Met rood duiden we zowel de grondkleur rood aan, die noch blauwachtig, noch geelachtig is, als elke willekeurige kleur rood wanneer er geen andere rode kleuren zijn ter vergelijking. En we karakteriseren bijvoorbeeld de drieklank van kleuren van De Stijl als geel-rood-blauw, hoewel daarin rood geelachtig, blauw en geel roodachtig te onderscheiden zijn.[2]

2

Afb.2
De kleurgebieden rondom de grondkleuren (schematische weergave).

samengestelde aanduidingen voor grondkleuren

Aanduidingen voor grondkleuren worden ook in samenstellingen gebruikt. Ze worden dan ervaren als een vermenging van beide kleuren (zowel-als-kleuren). (Afb.3)
Onmogelijk: roodgroen, geelblauw, zwartwit, omdat deze samengestelde aanduidingen voor grondkleuren tegen de eigenlijke gewaarwording indruisen. Wat rood is, kan ook blauwachtig, geelachtig, bruinachtig of grijsachtig zijn, maar nooit groenachtig. Met deze aanduidingen worden niet de kleuren benoemd die zich

3

4

5

6

aanduidingen voor grondkleuren in bijvoeg… afzwakkende vorm

Een bijzonder type aanduiding voor de grondkleu… het gebruik in de bijvoeglijk-afzwakkende vorm. Deze bijvoeglijk naamwoorden dienen dan ter on… scheiding binnen kleurgebieden. (Afb.4 & 5)

Aanduiding grondkleur	Bijvoeglijk	Bijvoeglijk-afzwakkend	Voorbeel…
Rood	rood	roodachtig	roodachti…
Blauw	blauw	blauwachtig	blauwacht…
Groen	groen	groenachtig	groenacht…
Geel	geel	geelachtig	geelachtig…
Wit	wit	witachtig	witachtig…
Zwart	zwart	zwartachtig	zwartachti…
Bruin	bruin	bruinachtig	bruinachti…

aanduidingen voor gevoelsmatig zelfstand… mengkleuren

De eerste groep bevat kleuren die een menging z… twee grondkleuren. (Afb.6,7,8)
Oranje en violet bevinden zich midden tussen ge… rood, respectievelijk tussen rood en blauw; purpe… tussen rood en violet in. Met deze kleurbenaming… worden de gebieden geel-rood en rood-blauw, d… de gewaarwording karakteristiekere eigenschapp… hebben dan de gebieden geel-groen en groen-bl… taalkundig van elkaar onderscheiden.
De tweede groep bevat kleuren die met wit lichte… gemaakt, die grijzer zijn gemaakt of die meer sch… hebben gekregen vergeleken met de zuivere kleu… waarvan ze zijn afgeleid. (Afb.9)

Het kenmerk van de kleuren in de tweede groep i… ze allemaal anders worden ondergaan dan de kle… waarvan ze zijn afgeleid, hoewel het verschil 'alle… een kwestie is van intensiteit (roze is zwakker rood) e… derheid (roze is lichter). Hoewel roze bijvoorbeeld u… en wit bestaat, is het geen rood, want anders zou… we het ook lichtrood kunnen noemen – zoals we t… een met wit lichter gemaakt geel lichtgeel zeggen… tegen een lichter gemaakt blauw lichtblauw.

1
Medewerkster: Dorothee Wiese (Kornerup, A. en Wanscher, J.H.

Sikkens-seminars kleurvorm… Wiese, D. 'Farbnamen' Die…

ture

ntrop

enade

lezer uit om in het diepe te

artelend, het hoofd boven w

e dan de overkant bereikt, h

terug te zien.

je weten welke complexe e

genstrijdige kleurconnotaties

een beschilderde Papoea kr

een precieze kleur zo lastig

uden, of waarom een midde

er kleur zo moeilijk te begrijp

je nagaan hoe dat nu preci

afjes en kegeltjes, met addi

btractieve mengingen, met s

waarwordingen – en nog he

dit boek, blader, pluk en

7	8	9	10	11	12	13	14	15	16	17	18
7	8	8	8	8	9	9	9	9	9	9	10
+	+	+	+	+	+	+	+	+	+	+	+
+	+	+	+	+	+	+	+	+	+	+	+
+	+	+	+	+	+	+	+	+	+	+	+
+	+	+	+	+	+	+	+	+	+	+	+
+	+	+	+	+	+	+	+	+	+	+	+
+	+	+	+	+	+	+	+	+	+	+	+
+	+	+	+	+	+	+	+	+	+	+	+
−	+	−	−	−	+	+	+	−	−	+	+
−	−	+	−	−	+	−	+	+	+	−	+

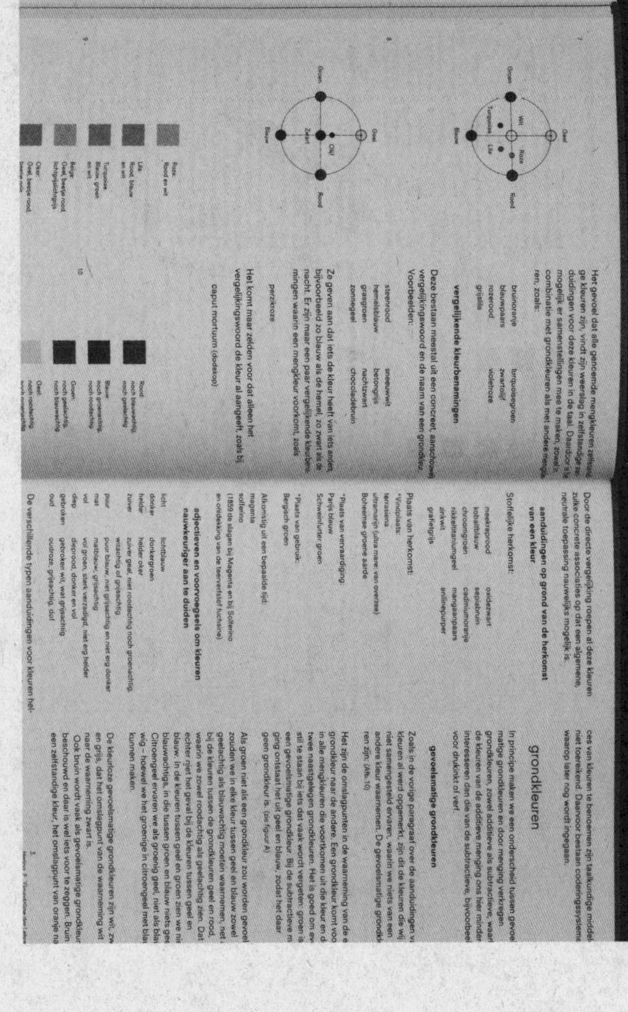

Het gevoel dat alle genoemde mengkleuren zelfstandige kleuren zijn, vindt zijn weerslag in zelfstandige aanduidingen voor deze kleuren in de taal. Daardoor is het mogelijk er samenstellingen mee te maken, zowel in combinatie met grondkleuren als met andere mengkleuren, zoals:

bruinoranje	turquoisegroen
blauwpaars	zwartolijf
rozerood	violetroze
grijslila	

vergelijkende kleurbenamingen

Deze bestaan meestal uit een concreet, aanschouwelijk vergelijkingswoord en de naam van een grondkleur. Voorbeelden:

steenrood	sneeuwwit
hemelsblauw	betongrijs
grasgroen	nachtzwart
zonnegeel	chocoladebruin

Ze geven aan dat iets de kleur heeft van iets anders, bijvoorbeeld zo blauw als de hemel, zo zwart als de nacht. Er zijn maar een paar vergelijkende kleurbenamingen waarin een mengkleur voorkomt, zoals

perzikroze

Het komt maar zelden voor dat alleen het vergelijkingswoord de kleur al aangeeft, zoals bij

caput mortuum (dodekop)

Door de directe vergelijking roepen al deze kleuren zulke concrete associaties op dat een algemene, neutrale toepassing nauwelijks mogelijk is.

aanduidingen op grond van de herkomst van een kleur

Stoffelijke herkomst:

meekraprood	oxidezwart
kobaltblauw	sepiabruin
chroomgroen	cadmiumoranje
nikkeltitaniumgeel	mangaanpaars
zinkwit	anilinepurper
grafietgrijs	

Plaats van herkomst:
*Vindplaats:
terrasiena
ultramarijn (ultra mare: van overzee)
Boheemse groene aarde

*Plaats van vervaardiging:
Parijs blauw
Schweinfurter groen

*Plaats van gebruik:
Bergisch groen

Afkomstig uit een bepaalde tijd:
magenta
solferino
(1859 de slagen bij Magenta en bij Solferino en ontdekking van de teerverfstof fuchsine)

adjectieven en voorvoegsels om kleuren nauwkeuriger aan te duiden

licht	lichtblauw
donker	donkergroen
helder	helder oker
zuiver	zuiver geel, niet roodachtig noch groenachtig, witachtig of grijsachtig
puur	puur blauw, niet grijsachtig en niet erg donker
mat	matblauw, grijsachtig
vol	vol groen, sterk verzadigd, niet erg helder
diep	dieprood, donker en vol
gebroken	gebroken wit, wat grijsachtig
oud	oudroze, grijsachtig, dof

De verschillende typen aanduidingen voor kleuren hel-

ces van kleuren te benoemen zijn taalkundige middel niet toereikend. Daarvoor bestaan coderingssysteme waarop later nog wordt ingegaan.

grondkleuren

In principe maken we een onderscheid tussen gevoel matige grondkleuren en door menging verkregen grondkleuren, zowel additieve als subtractieve, waar de kleuren van de additieve menging ons hier minder interesseren dan die van de subtractieve, bijvoorbee voor drukinkt of verf.

gevoelsmatige grondkleuren

Zoals in de vorige paragraaf over de aanduidingen va kleuren al werd opgemerkt, zijn dit de kleuren die wij niet samengesteld ervaren, waarin we niets van een andere kleur waarnemen. De gevoelsmatige grondkle ren zijn: (Afb.10)

Het zijn de omslagpunten in de waarneming van de e grondkleur naar de andere. Een grondkleur komt voo in alle mengkleuren die voortkomen uit die kleur en d twee naastgelegen grondkleuren. Het is goed om ev stil te staan bij iets dat vaak wordt vergeten: groen is een gevoelsmatige grondkleur. Bij de subtractieve m ging ontstaat het uit geel en blauw, zodat het daar geen grondkleur is. (zie figuur A)

Als groen niet als een grondkleur zou worden gevoel zouden we in elke kleur tussen geel en blauw zowel geelachtig als blauwachtig moeten waarnemen, net a bij de kleuren tussen de grondkleuren geel en rood, waarin we zowel roodachtig als geelachtig zien. Dat echter niet het geval bij de kleuren tussen geel en blauw. In de kleuren tussen geel en groen zien we nie blauwachtigs, in die tussen groen en blauw niets gee Citroengeel ervaren we als groenig geel, niet als blau wig – hoewel we het groenige in citroengeel met blau kunnen maken.

De kleurloze gevoelsmatige grondkleuren zijn wit, zw en grijs, dat het omslagpunt van de waarneming wit naar de waarneming zwart is.
Ook bruin wordt vaak als gevoelsmatige grondkleur beschouwd en daar is wel iets voor te zeggen. Bruin een zelfstandige kleur, het omslagpunt van oranje na

1203

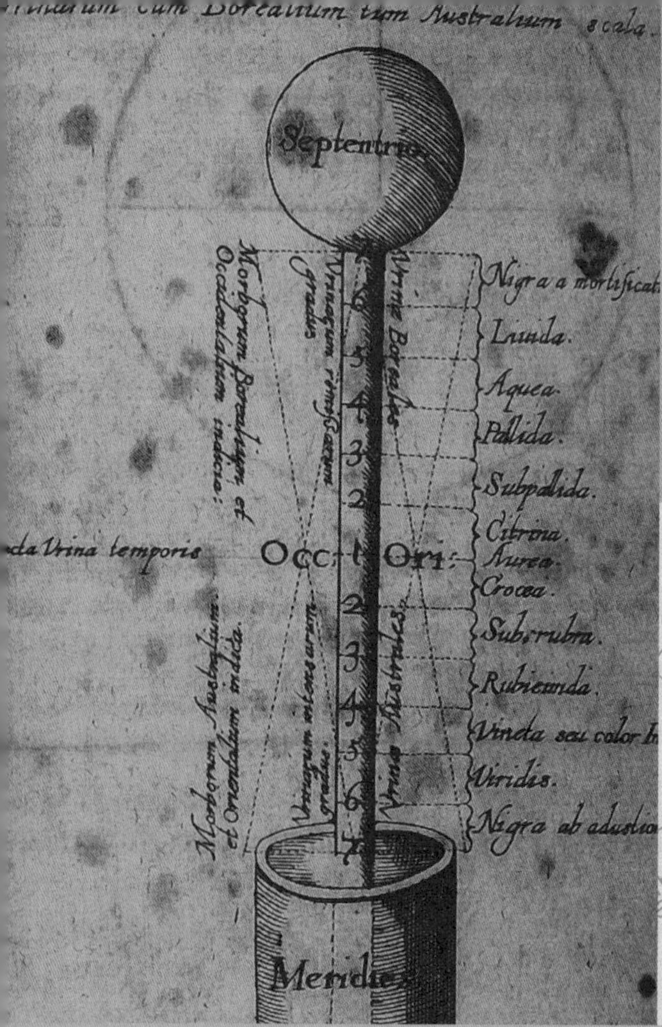

Septentrio

Nigra a mortificat
Liuida.
Aquea.
Pallida.
Subpallida.
Citrina.
Aurea.
Crocea.
Subrubra.
Rubicunda.
Vinea seu color In
Viridis.
Nigra ab adustion

Morborum Borealium et Occidentalium indicia:

Urinarum remissarum gradus

Urinae Boreales

Occ. 1 Ori.

da Urina temporis

Meridies

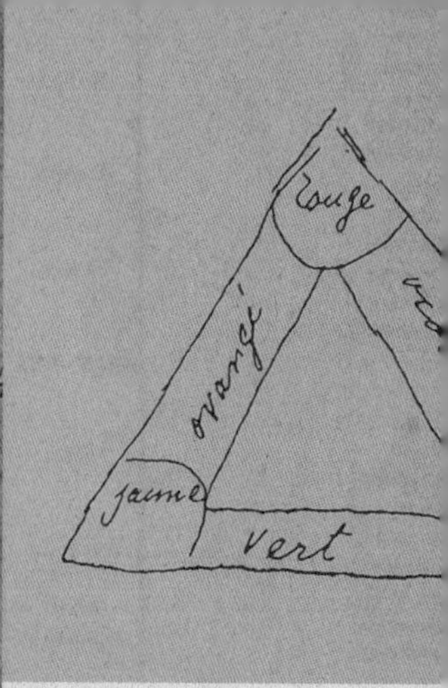

rouge
orangé
jaune
vert

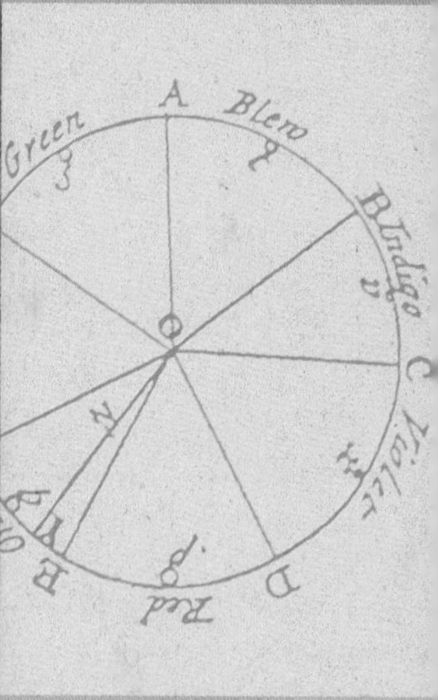

A Bleu
Green
Indigo
C
Violet
D
Red

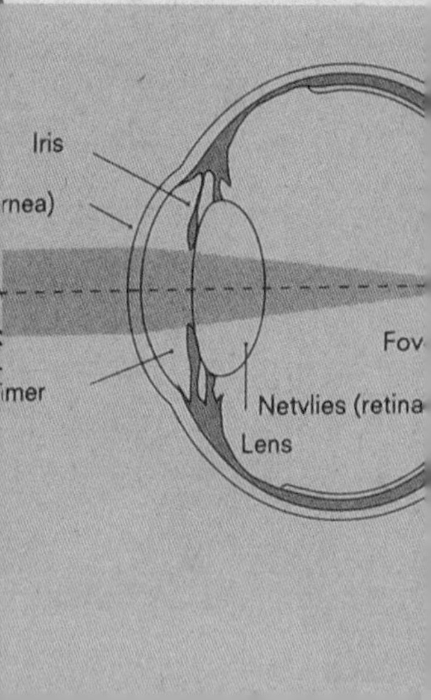

Iris
rnea)
Fov
mer
Netvlies (retina
Lens

A
C
Red
Greenish Bleu
Red
Green
Red
Yellow
Green
Bleu
Purple
Red
Yellow
Green
Bleu
Yellow
Bleuish Green
Red
Bleuish Green
Green
Yellow
Red
Bleu
Violet
White
Yellow
Red

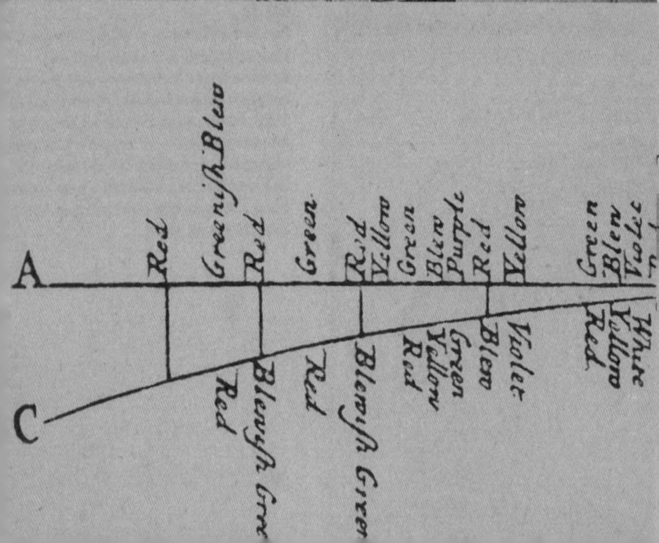

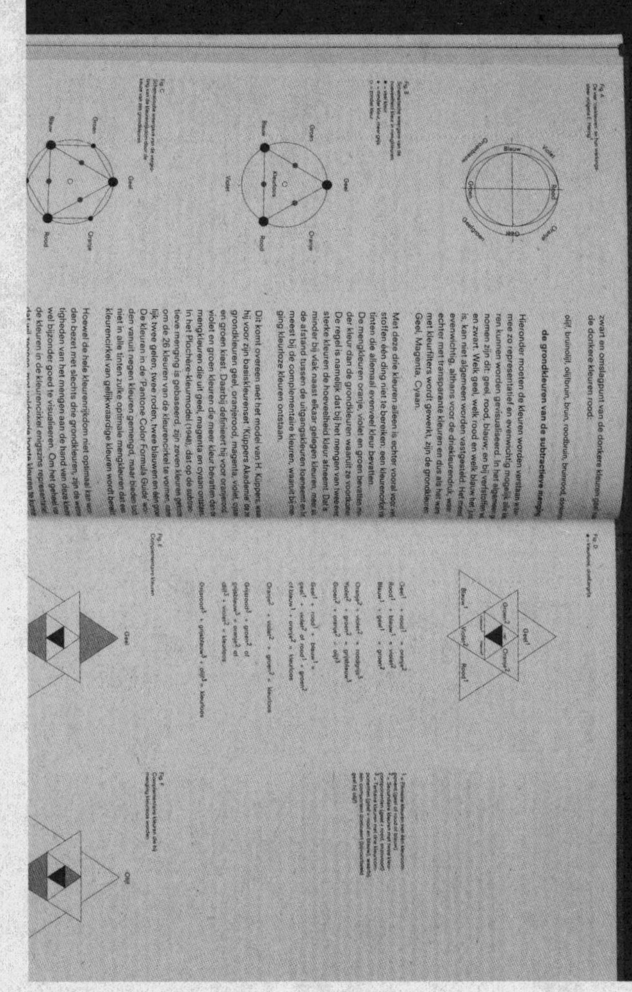

1205

ieve
CHROM
CIRC
OF
HUE

2003 Bm[1]

Switzerland

U.S.A.

China

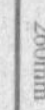

260mm

211mm

铜奖
这些事
瑞士

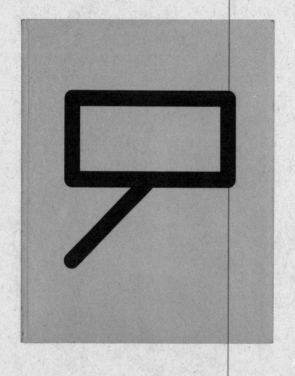

Bronze Medal
The Things
Switzerland
Dimitri Bruni, Manuel Krebs (NORM)
NORM
NORM, Zürich
260 211 13mm
671g
176p
ISBN-10 3-9311-26-75-7

瑞士著名的设计公司 NORM 擅长实验性图形设计，本书是他们对真实世界和符号系统进行的深度抽象化实验。本书实际为两本书，将封面卡纸相互对裱形成“Z”字形结构，分别可以从前后翻阅，两部分的装订形式均为锁线胶装。封面采用白卡纸正面印单黑，勒口内包裹红蓝黑三色印刷的书中代表性符号合集。内页哑面铜版纸四色印刷，在字母抽象化处理的部分使用荧光绿、荧光红和黑色印刷。书的正面部分开篇把二维世界的图像、符号分为四组：1. 表现三维世界的图像；2. 二维符号；3. 无意义的图像；4. 未知或未被纳入人类符号系统的图形。之后，通过一系列抽象化处理，将这四类符号尽可能地简化，尤其对西文字母进行元素归纳。书中局部夹带胶版纸单黑印刷的纸片，用荧光笔标记信息，呈现推演过程。书的最后部分归纳出可以分为 16 条线段的“米字格”字元符号，并在书的背面部分完整呈现 Unicode 可以容纳的 2 的 16 次方即 65536 个字符的符号系统。

The famous Swiss design company NORM is good at experimental graphic design. This book is its deep abstraction experiment of real world and symbol system. The book is actually made of two books. The covers are mounted to form a Z-shaped structure, which can be read from the front and the back respectively. The two parts are perfect bound with thread sewing. The front cover is printed with single black on white cardboard, and the collection of representative symbols from the book is printed on the inside of the flap in red, blue and black. The inside pages are printed in four colours on matte coated paper, and fluorescent green, fluorescent red and black are used in the abstract letters. In the front part of the book, the images and symbols of the two-dimensional world are divided into four groups: 1. Images representing the three-dimensional world; 2. Two-dimensional symbols; 3. Meaningless images; 4. Unknown graphs or not included in the human symbol system. After that, through a series of abstractions, the four kinds of symbols are simplified as much as possible, especially the western letters. In the book, some pieces of paper printed on offset paper in single black are attached to show the process of reasoning and important information is marked with fluorescent pen. The last part of the book sums up the ‘meter grid’ character symbols which can be divided into 16 line segments, and presents the symbol system of 65536 characters that can be included by Unicode on the back cover of the book.

NORM
THE THINGS

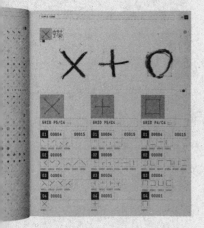

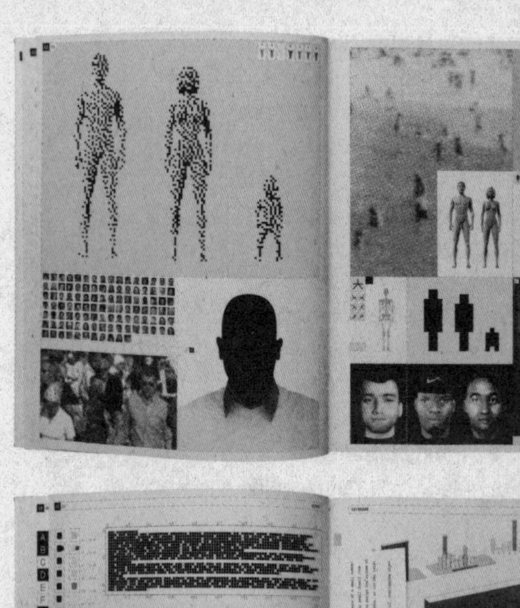

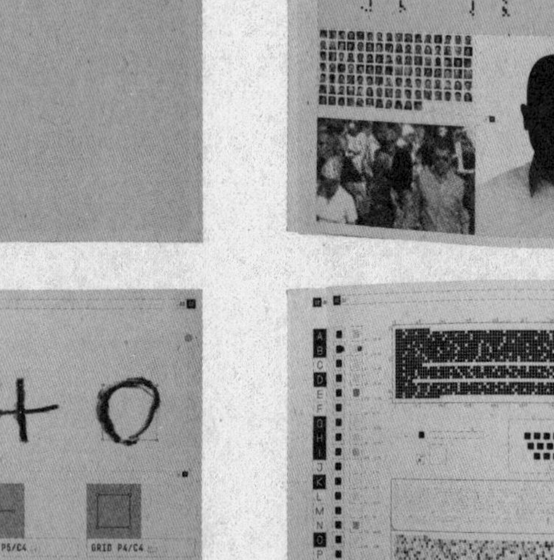
GRID P5/C4
GRID P5/C4
GRID P4/C4

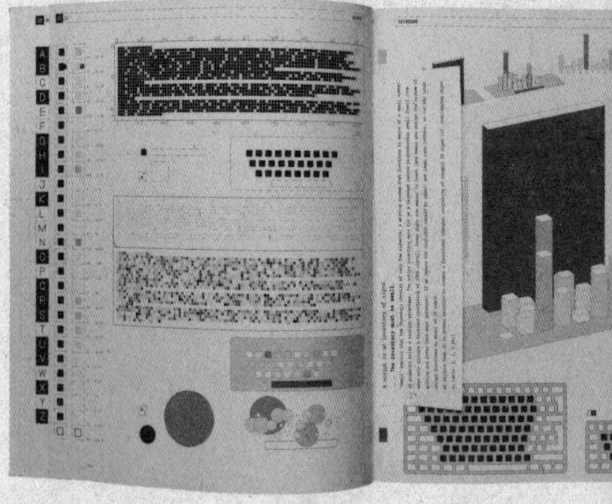

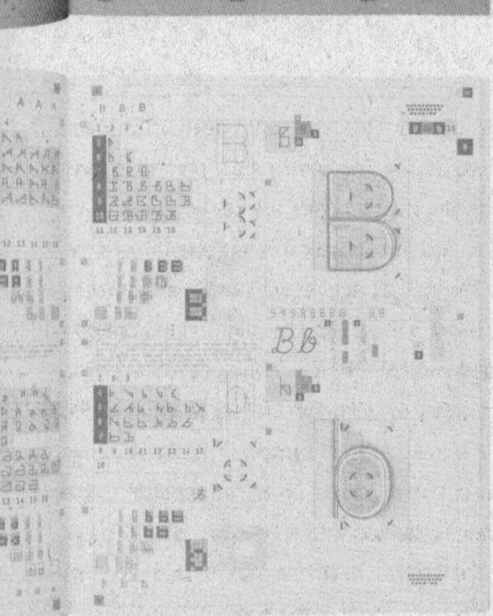

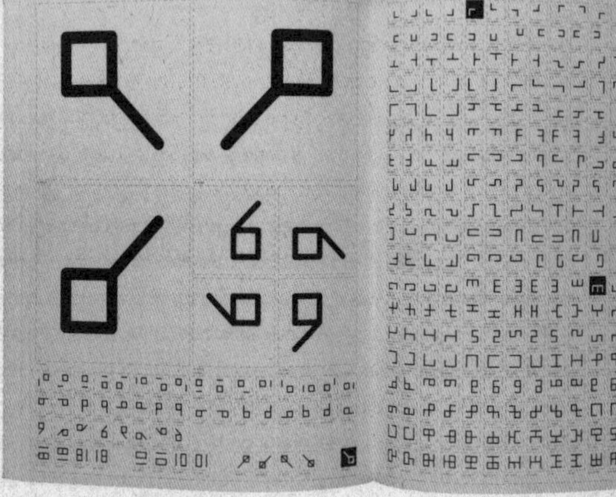

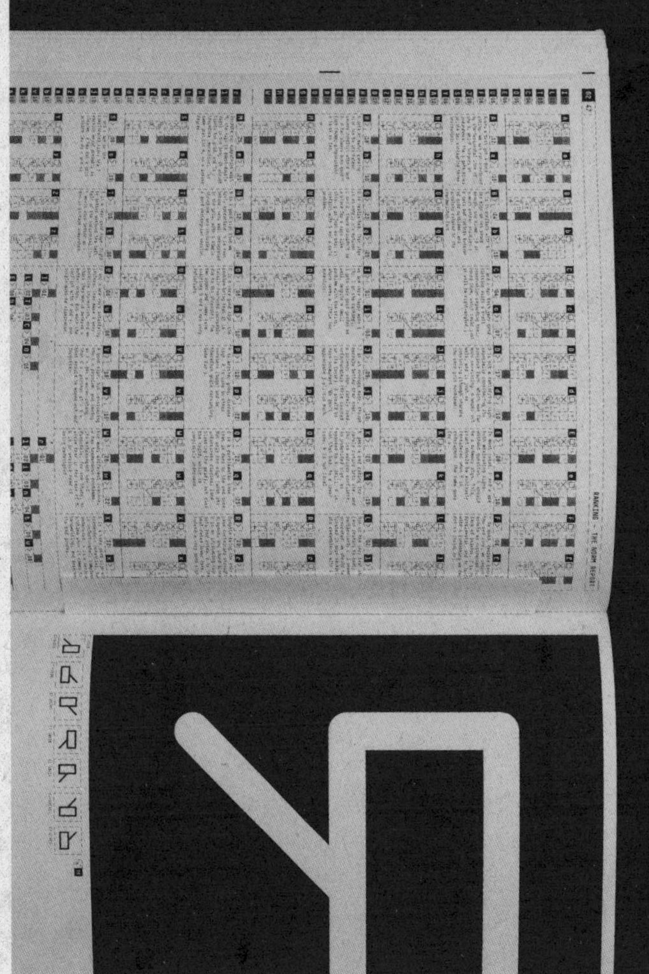

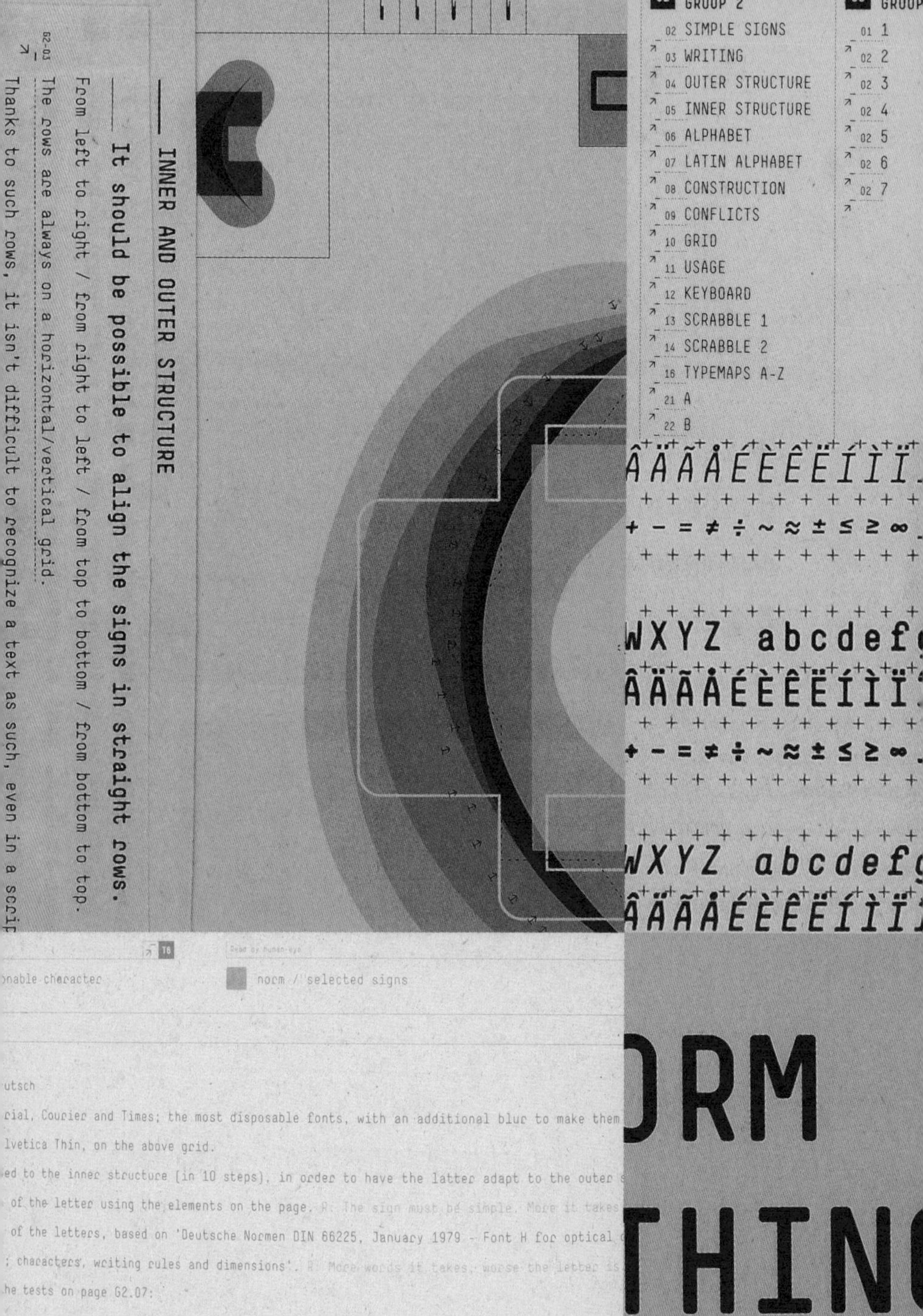

G2 GROUP 2
02 SIMPLE SIGNS
03 WRITING
04 OUTER STRUCTURE
05 INNER STRUCTURE
06 ALPHABET
07 LATIN ALPHABET
08 CONSTRUCTION
09 CONFLICTS
10 GRID
11 USAGE
12 KEYBOARD
13 SCRABBLE 1
14 SCRABBLE 2
16 TYPEMAPS A-Z
21 A
22 B
G3 GROUP
01 1
02 2
02 3
02 4
02 5
02 6
02 7
INNER AND OUTER STRUCTURE
It should be possible to align the signs in straight rows.
From left to right / from right to left / from top to bottom / from bottom to top.
G2-03
The rows are always on a horizontal/vertical grid.
Thanks to such rows, it isn't difficult to recognize a text as such, even in a scrip
ÂÄÃÅÉÈÊËÍÌÏ
+ − = ≠ ÷ ~ ≈ ± ≤ ≥ ∞
WXYZ abcdef
norm / selected signs
utsch
rial, Courier and Times; the most disposable fonts, with an additional blur to make them
lvetica Thin, on the above grid.
ed to the inner structure (in 10 steps), in order to have the latter adapt to the outer
of the letter using the elements on the page.
of the letters, based on 'Deutsche Normen DIN 66225, January 1979 - Font H for optical
; characters, writing rules and dimensions'.
he tests on page G2.07:
1. Length of sign
2. Required basepoints, when writing
3. Elements of which the sign consists
4. Axes of Symmetry
ORM
THIN

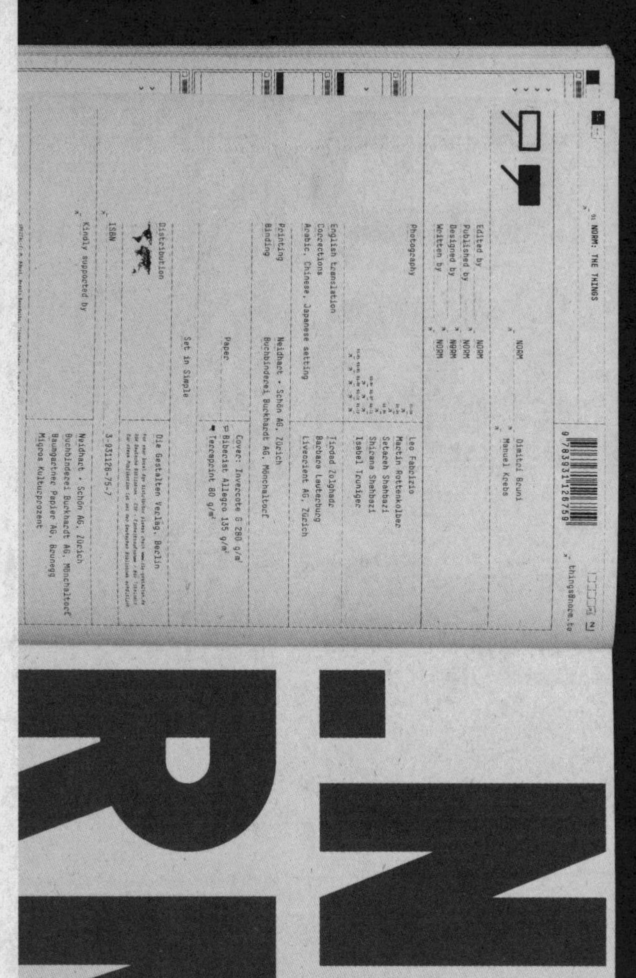

NORM: THE THINGS
Edited by NORM Dimitri Bruni
Published by NORM Manuel Krebs
Designed by NORM
Written by NORM
Photography Leo Fabrizio
Martin Rottenkolber
Setareh Shahbazi
Shirana Shahbazi
Isabel Truniger
English translation Tirdad Zolghadr
Corrections Barbara Lauterburg
Arabic, Chinese, Japanese setting
Printing Neidhart + Schön AG, Zürich
Binding Buchbinderei Burkhardt AG, Mönchaltorf
Paper Cover: Invercote G 280 g/m²
Biberist Allegro 135 g/m²
Terraprint 80 g/m²
Set in Simple
Distribution Die Gestalten Verlag, Berlin
ISBN 3-931126-75-7
Kindly supported by Neidhart + Schön AG, Zürich
Buchbinderei Burkhardt AG, Mönchaltorf
Baumgartner Papier AG, Brunegg
Migros Kulturprozent
things@norm.to

15
01
00004
00001 00002
02
00006
0006
00001 00002
03
00004
L 02 1 02

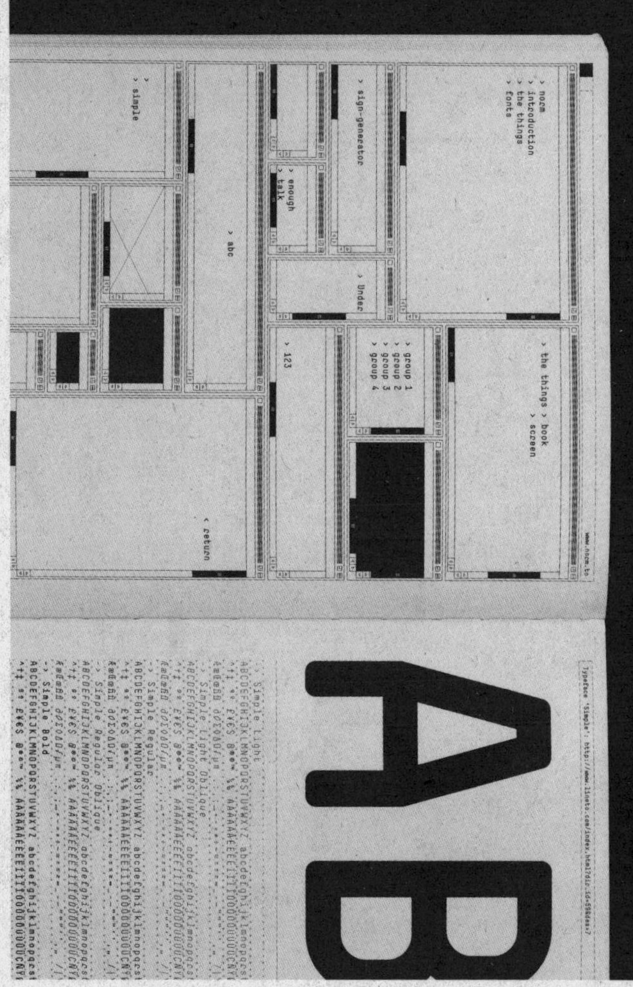

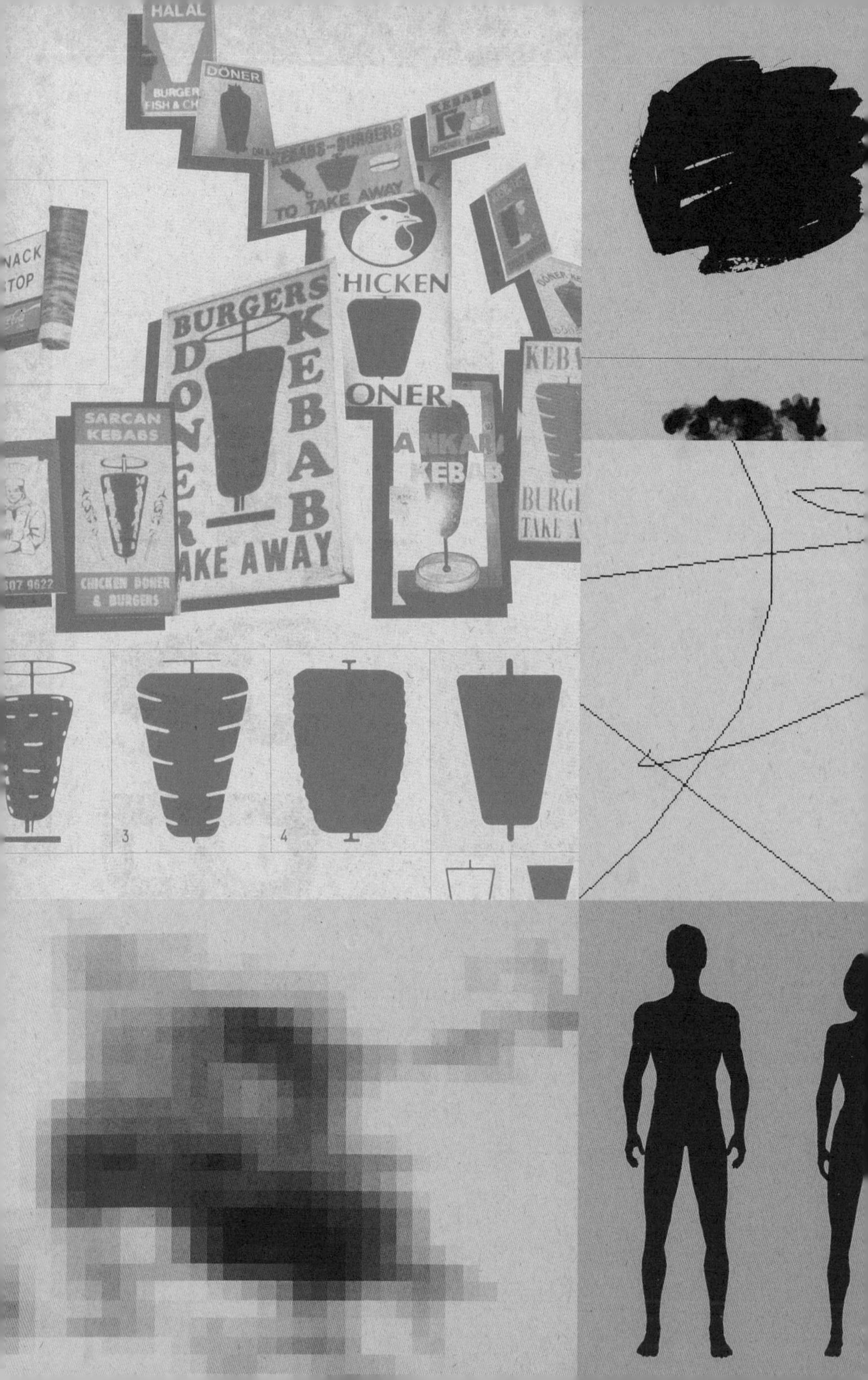

HALAL
DÖNER
BURGER
KEBABS-BURGERS
TO TAKE AWAY
KEBABS
CHICKEN
BURGERS
DONER
KEBAB
SARCAN
KEBABS
CHICKEN DONER
& BURGERS
3
4

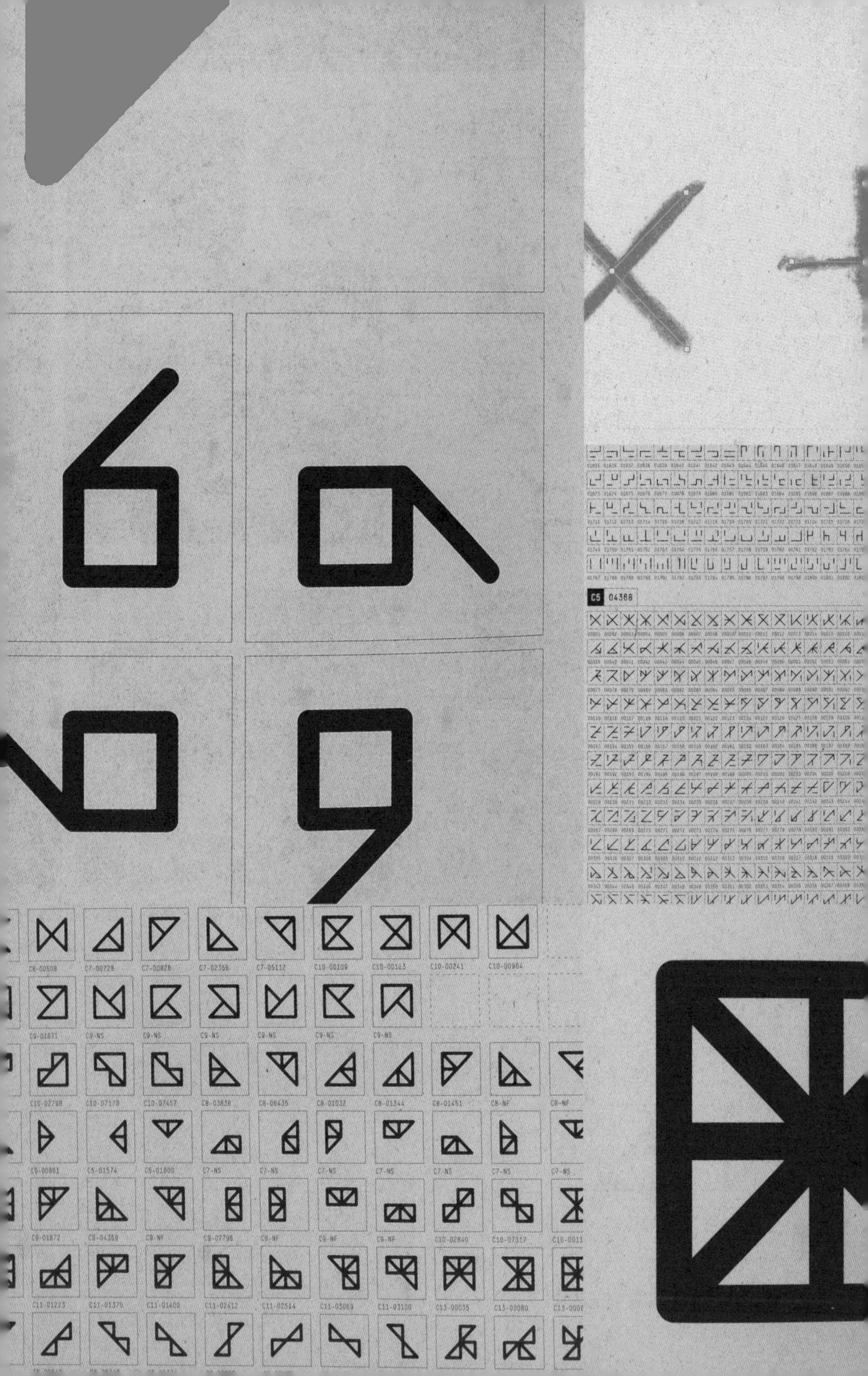

C5 04368

C8-00508	C7-00728	C7-00828	C7-02356	C7-05112	C10-00109	C10-00123	C10-00241	C10-00964
C9-02871	C9-NS	C9-NS	C9-NS	C9-NS	C9-NS	C9-NS		
C10-02798	C10-07170	C10-07457	C8-03638	C8-06435	C8-01032	C8-01344	C8-01451	C8-NF	C8-NF
C5-00861	C5-01574	C5-01600	C7-NS	C7-NS	C7-NS	C7-NS	C7-NS	C7-NS	C7-NS
C9-01872	C9-04369	C9-NF	C9-07796	C9-NF	C9-NF	C9-NF	C10-02840	C10-07317	C10-0011
C11-01223	C11-01375	C11-01400	C11-02412	C11-02514	C11-03069	C11-03100	C13-00035	C13-00080	C13-0008

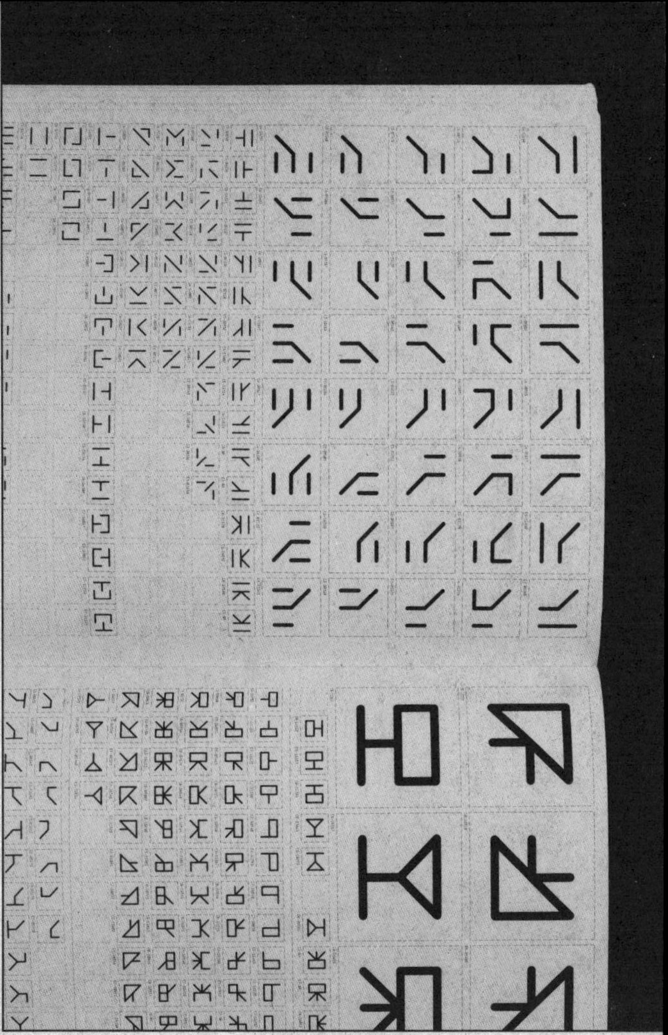

2003 Bm^2

mox GmbH
Alten Flughafen 1
356 Leipzig
rmany
ferenz–Nr.: 101983949

B R

PRIORITY

W

06 30E0 A750 18 E000 388F
DV 10.19 1,00 Deutsche Post
WARENPOST Int. KT Port payé

ZHAO QING
Design Co.,
Dabei xiang7–3, Nanjing John Qing Tang
210000 Nanjing
CHINA VR

UC 030 356 966 DE

– 1 x Book

17.10.2019

Datum und Unterschrift des Abs

Germany

Germany China

◉ 铜奖
《 华氏 451
◐ 德国

Bronze Medal
Fahrenheit 451
Germany
Gerda Raidt
Ray Bradbury
Büchergilde Gutenberg, Frankfurt am Main / Wien / Zürich
245 158 20mm
581g
176p
ISBN-10 3-7632-5111-1

《华氏 451》是美国科幻作家雷·布拉德伯里（Ray Bradbury）最为著名的作品之一。在这部反乌托邦小说中，所有的书籍都被禁止，消防员的工作不是灭火，而是焚书，451 华氏度（大约 232 摄氏度）便是纸张的燃点。本书的装订形式为圆脊锁线硬精装。全书将红、黑和灰绿三色贯穿始终。护封采用胶版纸三色印刷。内封黑和灰绿色双色印刷，表面过油处理，裱覆卡板。环衬采用黑色卡纸，从一开始就带来压抑的氛围。内页胶版纸三色印刷。内容方面，全书共分 3 个章节，穿插 13 幅版画插图，在三色的处理上较为灵活。章节隔页红底黑字，版画插图绿底红图，正文页面黑字红页码，无不提醒着读者燃烧的火苗。正文字体特别选择了一款字宽较窄但笔画较粗的等线体，通篇观感沉重压抑。本书还使用了一条红色的书签带，让读者读到哪页就“烧”到哪页。

Fahrenheit 451 is one of the most famous works of American science fiction writer Ray Bradbury. In this anti-utopian novel, all books are banned. The job of firefighters is not to put out the fire, but to burn books. 451°F (about 232°C) is the burning point of paper. The hardback book is bound by thread sewing with a rounded spine. Red, black and grey-green are three essential colors throughout this book. The jacket is printed with offset paper in three colors. The inner cover mounted with cardboard is printed in black and grey-green, and the surface is matte vanished. The end paper is made of black cardboard, which brings depressing atmosphere from the beginning. The inside pages are printed with offset paper in three colors. The content of the book is divided into three chapters, interspersed with 13 print illustrations. The chapters are separated by red page printed with black characters, the prints are illustrated in red on green background, and the texts are printed in black with red page numbers. The color combination reminds the readers of the burning flames. A special Arial with narrow width but thick strokes is chosen for the typeface of main body, which makes the whole book feel heavy and oppressive. This book uses a red bookmark belt to let readers 'burn' the page they read.

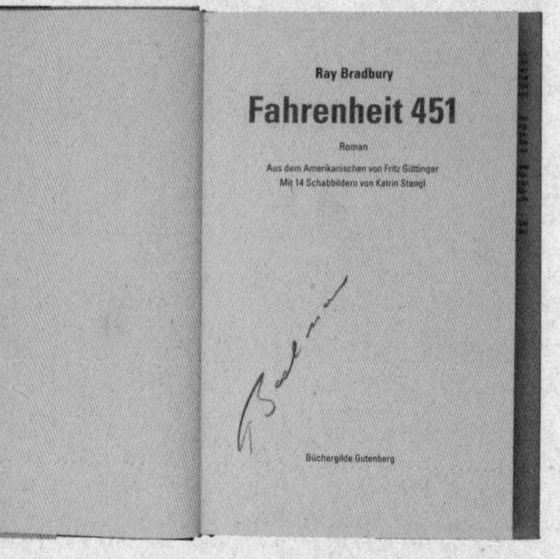

Ray Bradbury

Fahrenheit 451

Roman

Aus dem Amerikanischen von Fritz Güttinger
Mit 14 Schabbildern von Katrin Stangl

Büchergilde Gutenberg

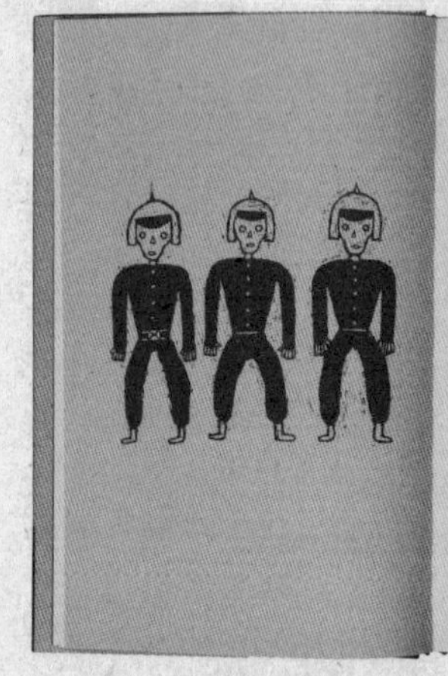

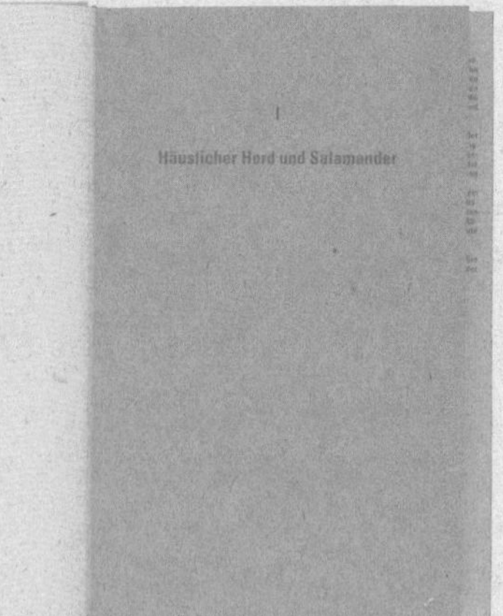

I

Häuslicher Herd und Salamander

in seine Umklammerung geraten ist? Es macht aus einem, was ihm beliebt. Es ist eine Umwelt, so wirklich wie die Welt selber. Sie wird und ist dann wahr. Bücher können verstandesmäßig widerlegt werden, aber bei all meinem Wissen und all meiner Zweifelsucht war ich noch nie imstande, einem hundertköpfigen Symphonieorchester gegenüber zu Wort zu kommen, noch dazu in Farben mit 3 D-Raumton, und alles das in diesen unwahrscheinlichen vier Wänden. Wie Sie sehen, besteht meine Stube nur aus vier Gipswänden. Und dann das da.« Er wies zwei kleine Gummistöpsel vor. »Für meine Ohren, wenn ich mit der U-Bahn fahre.«

»Zanders Zahnpasta; sie arbeiten nicht, auch spinnen sie nicht«, bemerkte Montag, mit geschlossenen Augen. »Wie kommen wir da je wieder heraus? Könnten uns Bücher nicht von Nutzen sein?«

»Nur wenn das dritte Erfordernis gegeben wäre. Das erste war, wie gesagt, Rang der Aussage. Das zweite, Muße, sie innerlich zu verarbeiten. Und drittens: das Recht, nach dem zu handeln, was sich uns aus dem Zusammenwirken der ersten beiden Dinge ergibt. Und ich glaube kaum, daß ein Greis und ein abtrünniger Feuerwehrmann in diesem vorgerückten Zeitpunkt noch viel ausrichten werden ...«

»Ich kann Bücher beschaffen.«

»Sie bringen sich in Gefahr.«

»Das ist das Gute am Sterben; wenn man nichts mehr zu verlieren hat, scheut man keine Gefahr mehr.«

»Sehen Sie«, lachte Faber, »da haben Sie etwas Bemerkenswertes gesagt, ohne es irgendwo gelesen zu haben.«

»Steht dergleichen in den Büchern? Ich habe es nur so aus dem Ärmel geschüttelt.«

»Um so besser, dann haben Sie es nicht für mich oder sonstjemand aufgedonnert, nicht einmal für Sie selber.«

Montag beugte sich vor. »Heute nachmittag kam mir der Gedanke, wenn an den Büchern wirklich etwas liegen sollte, könnten wir uns eine Druckerpresse besorgen und weitere Exemplare herstellen –«

»Wir?«

»Sie und ich.«

»Ohne mich!« Fabers Haltung versteifte sich.

89

um dich in jeder Hinsicht zu widerlegen. Wie treulos Bücher doch sein können! Du glaubst, an ihnen einen Rückhalt zu haben, und sie halten es mit dem Gegner. Andere können sich ihrer auch bedienen, und dann stehst du da, mitten im unwegsamen Gelände, in einem Dickicht von Wörtern aller Art. Und ganz am Schluß des Traumes kam ich mit dem Salamander und sagte, ›haben wir denselben Weg?‹, und du bist eingestiegen, und wir fuhren in wohltuendem Schweigen zur Feuerwache zurück, ganz klein und friedlich geworden.« Beatty ließ Montags Handgelenk los, ließ die Hand schlaff auf den Tisch fallen. »Ende gut, alles gut.«

Schweigen. Montag saß da wie aus weißem Stein gehauen. Der Widerhall des letzten Hammerschlags auf den Schädel verlor sich allmählich in das Dunkel hinein, wo Faber darauf wartete, bis alles wieder still war, und dann sachte begann: »Gut, er hat gesagt, was er zu sagen hatte. Sie müssen es verarbeiten. Ich werde auch sagen, was ich zu sagen habe, in den nächsten paar Stunden. Und Sie werden es verarbeiten und beides miteinander vergleichen und dann Ihren Entscheid treffen, in welche Richtung Sie sich schlagen wollen. Es soll aber Ihr Entscheid sein, nicht der meine, und nicht derjenige des Hauptmanns. Bedenken Sie indessen, daß der Hauptmann zu den gefährlichsten Feinden der Wahrheit und Freiheit gehört, zu der dichtgedrängt stillstehenden Herde der Mehrheit. O Gott, die furchtbare Gewaltherrschaft der Mehrheit. Jeder hat seine Harfe zu schlagen, und an Ihnen ist es jetzt, zu entscheiden, mit welchem Ohr Sie zuhören wollen.«

Montag machte schon den Mund auf, um Faber Antwort zu geben, ein Versehen, das ihm nur erspart blieb, weil die Alarmglocke anschlug. Die Stimme aus dem Lautsprecher an der Decke ertönte, und das Klappern des Fernschreibers setzte ein, der die Adresse hintippte. Hauptmann Beatty, die Pokerkarten in der Hand, begab sich mit übertriebener Gemessenheit hinüber und riß die Adresse heraus, als die Meldung fertig war. Er warf einen flüchtigen Blick darauf und schob sie in die Tasche, dann kehrte er an den Tisch zurück und setzte sich hin. Die andern schauten ihn gespannt an.

»Es hat genau vierzig Sekunden Zeit, während ich euch noch rasch alles Geld abnehme«, erklärte Beatty vergnügt.

Montag legte die Karten hin.

112

»Ach, machen Sie doch bitte nicht so ein Gesicht.«

»Es liegt an diesem Löwenzahn«, meinte er. »Du hast ihn bereits aufgebraucht. Deshalb ging's bei mir nicht.«

»Ach ja, das wird's sein. Und jetzt habe ich Sie verstimmt, ich seh es genau. Es tut mir wirklich leid.« Clarisse rührte an seinen Ellbogen.

»Nein, nein«, beteuerte er rasch, »das macht nichts.«

»Ich muß gehen. Sagen Sie bitte vorher noch, daß Sie mir verzeihen. Ich möchte nicht, daß Sie mir böse sind.«

»Ich bin nicht böse. Verstimmt, ja.«

»Ich muß jetzt zu meinem Psychiater. Man schickt mich hin, und ich denke mir aus, was ich ihm Schönes erzählen könnte. Was er wohl von mir hält? Er behauptet, ich sei eine richtige Zwiebel. Er hat alle Hände voll zu tun, die verschiedenen Schichten abzupellen.«

»Mir scheint, du hast den Psychiater nötig.«

»Das ist doch nicht Ihr Ernst?«

Er holte Atem, stieß ihn wieder aus und sagte dann: »Nein, es war nicht ernstgemeint.«

»Der Psychiater will wissen, warum ich gehe und in den Wäldern umherstreife und den Vögeln zuschaue und Schmetterlinge sammle. Ich zeige Ihnen dann mal meine Sammlung.«

»Schön.«

»Man will herauskriegen, was ich mit meiner ganzen Zeit anfange. Ich sage den Leuten, daß ich manchmal bloß dasitze und nachdenke. Aber worüber, das erzähle ich ihnen nicht. Ich lasse sie zappeln. Und manchmal, sage ich, werfe ich den Kopf zurück, so, und lasse mir in den Mund hineinregnen. Schmeckt wie Wein. Haben Sie es je versucht?«

»Nein, ich –«

»Sie haben mir doch verziehen?«

»Gewiß.« Er dachte nach. »Doch, ich habe dir verziehen. Weiß Gott, warum. Du bist ein eigentümliches Wesen, ein befremdliches, aber man verzeiht dir leicht. Siebzehn bist du, hast du gesagt?«

Pistole auf die Brust, und du zwingst ihn zum Zuhören. Leg los! Was soll's denn diesmal sein? Warum rülpst du mir nicht Shakespeare ins Gesicht, du Pfuscher von einem Bildungsphilister? ›Dein Drohen hat keine Schrecken, Cassius, denn ich bin so bewehrt durch Redlichkeit, daß es vorbeizieht wie der leere Wind, der nichts mir gilt.‹ Was sagst du dazu? Nur zu, du Literat aus zweiter Hand, drück ab.« Er tat einen Schritt auf Montag zu.

Montag sagte bloß: »Wir haben nie richtig gebrannt ...«

»Gib her, Guy«, sagte Beatty mit starrem Lächeln.

Und dann war er ein schreiendes Flammenbündel, eine hopsende, hinschlagende, sich verheddernde Gliederpuppe, kein menschliches oder bekanntes Wesen mehr, nur noch eine wabernde Flamme auf dem Rasen, als Montag einen langen Stoß flüssigen Feuers auf ihn abgab. Ein Zischen entstand, wie wenn ein tüchtiger Mundvoll Spucke einen rotglühenden Ofen trifft, ein quirlendes Schäumen, als wäre eine schwarze Riesenschnecke mit Salz überschüttet worden und sonderte nun gelb brodelnden Schaum ab. Montag machte die Augen zu und schrie, schrie und mühte sich verzweifelt, mit den Händen die Ohren zuzustopfen. Beatty überschlug sich einmal, zweimal, dreimal und krümmte sich schließlich zusammen wie eine geschmolzene Wachsfigur und lag dann still da.

Die beiden andern Feuerwehrleute rührten sich nicht.

Montag erwehrte sich der Übelkeit, die in ihm aufstieg, gerade lange genug, um den Flammenwerfer in Anschlag zu bringen. »Rechts um!«

Sie machten kehrt, mit kreideweißem Gesicht, schweißüberströmt; er schlug ihnen den Helm vom Kopf und schlug weiter, bis sie zusammensackten und reglos dalagen.

Das Rascheln eines einzelnen Herbstblattes.

Er wandte sich um, und da stand der Mechanische Hund. Das Untier war aus dem Schatten herausgekommen und bewegte sich mit einer solchen Mühelosigkeit, daß es wie ein geballter schwärzlicher Rauchfetzen wirkte, der lautlos über den Rasen herangeweht

2003 Bm³

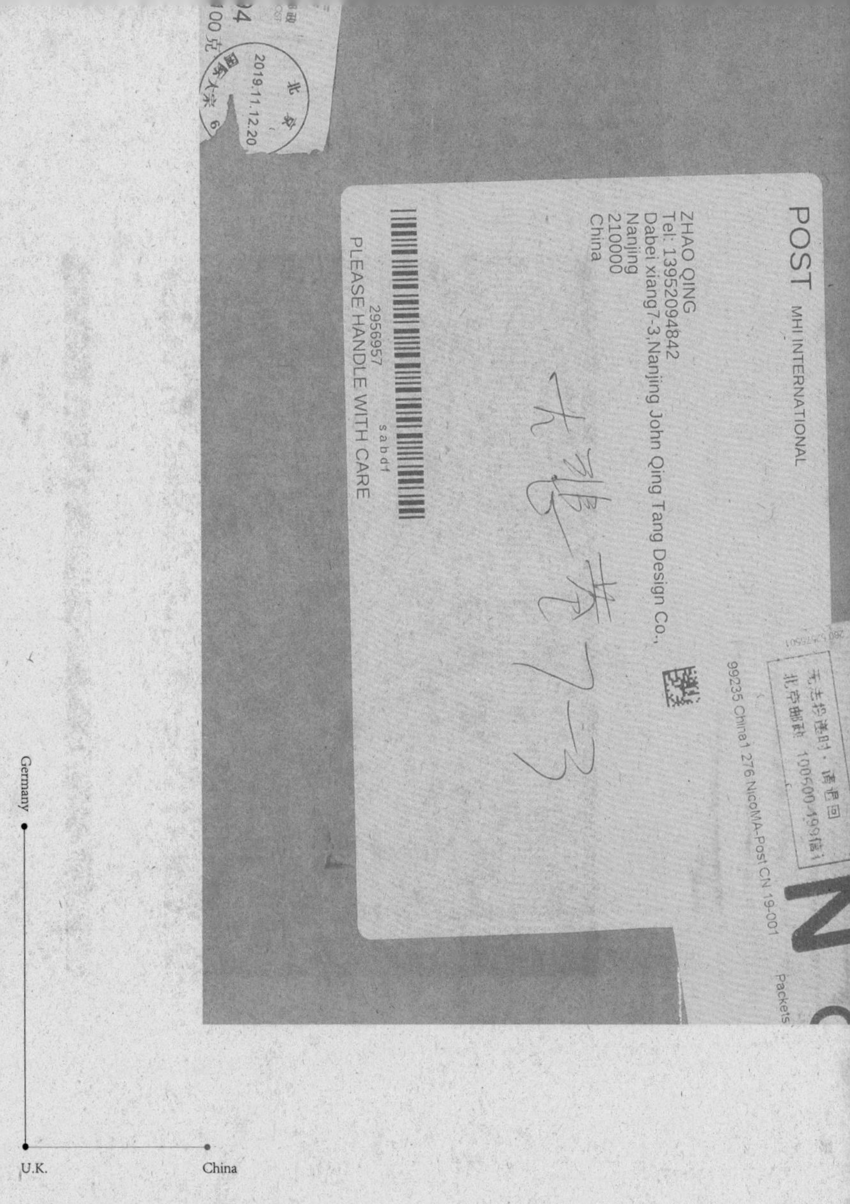

POST MHI INTERNATIONAL
ZHAO QING
Tel: 13952094842
Dabei xiang7-3,Nanjing John Qing Tang Design Co.,
Nanjing
210000
China
2956957
s a b d 1
PLEASE HANDLE WITH CARE
99235 China1 276 NicoMA-Post CN 19-001
Packets
2019.11.12.20
Germany
U.K.
China

384mm

16mm

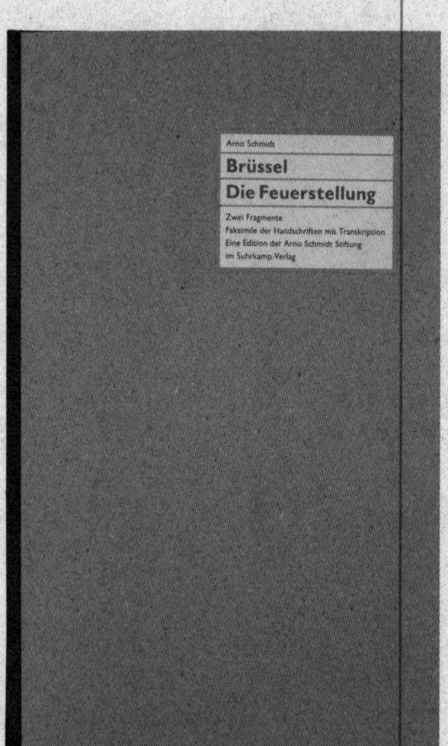

241mm

铜奖
布鲁塞尔 / 射击位置
德国

Bronze Medal
Brüssel / Die Feuerstellung
Germany
Friedrich Forssman
Arno Schmidt
Suhrkamp Verlag, Frankfurt am Main
384 241 16mm
1085g
72p
ISBN-10 3-518-80201-1

德国著名作家、翻译家阿诺·施密特（Arno Schmidt）从二战后成为职业作家，写出了大量源于战争和乌托邦的小说，但他从未在小说中描述过自己曾参与过的战争经历。本书是他撰写的两段生前没有发表过的故事片段。本书的装订形式为裱贴了卡板的锁线胶平装。封面采用深蓝色布面，裱贴灰色卡板，露出未裱板的书脊位置，书名烫灰处理。封面卡板裱贴红黑灰三色印刷的标签。书函的封面材质和书名裱贴方式相同。内页采用哑面涂布艺术纸四色印刷。内容分为三大部分，分别为两个文章片段的扫描版和转录版，以及评论家对于施密特写作的评述、对于两篇文章的评价，以及对翻译和转录方式的介绍。书中两篇文章扫描版的转录设计非常有特点，由于扫描版是手写稿件，在格式、缩进、涂改、符号和画线上较为随意，转录文本虽为电脑排版，但尽可能地尝试保留这些缩进、涂改和符号，读者可以通过左右页进行对照阅读。

Arno Schmidt, a famous German translator, has become a professional writer since the Second World War. He has written a lot of Utopia novels rooted in the wars, but he has never described his own war experience in his novels. This book contains two stories he wrote that have never been published during his lifetime. The book is perfect bound with thread sewing and the paperback is mounted with cardboard. The cover is made of dark-blue cloth, mounted with gray cardboard, showing the unframed spine. The title of the book is gray-stamped. The cover is made of cardboard mounted with red, black and gray labels.. The inner pages are printed with matte coated art paper in four colors. The content is divided into three parts: scanning version and transcribing version of two stories, as well as commentators' comments on Schmidt's writing, comments on two stories, and introduction to the ways of translation and transcription. The transcription design of the scanned version is quite special. Since the scanned version is a handwritten manuscript, it is relatively random in the format, indentation, writing errors, symbols and dashes. Although the transcribed text was typeset by computer, the designer tried to keep these indentation, alteration and symbols as much as possible. Readers can read the left and right pages comparatively.

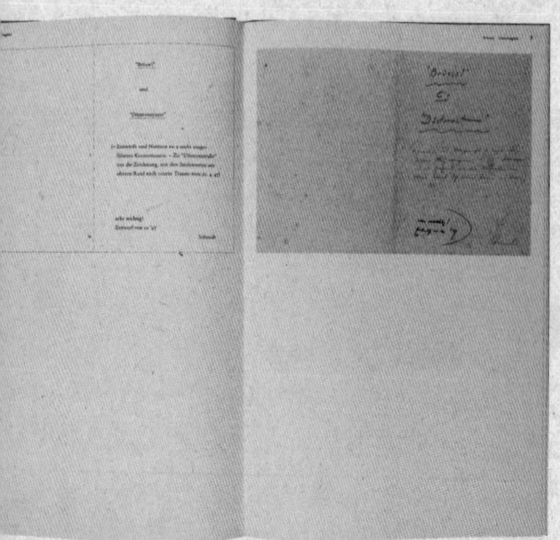
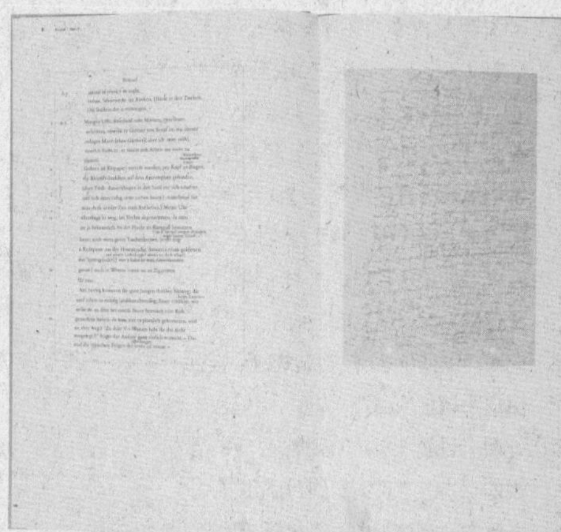
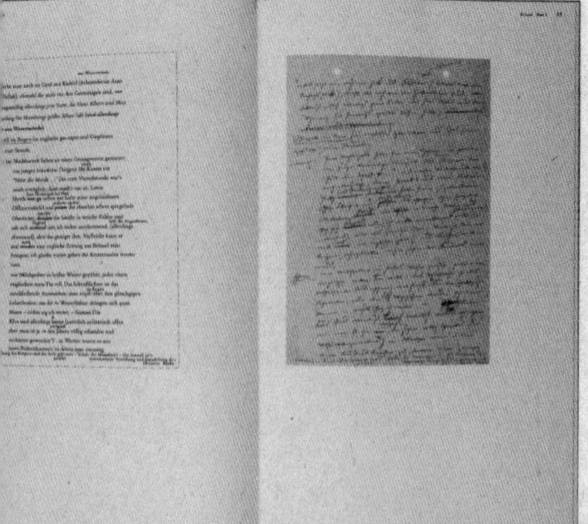
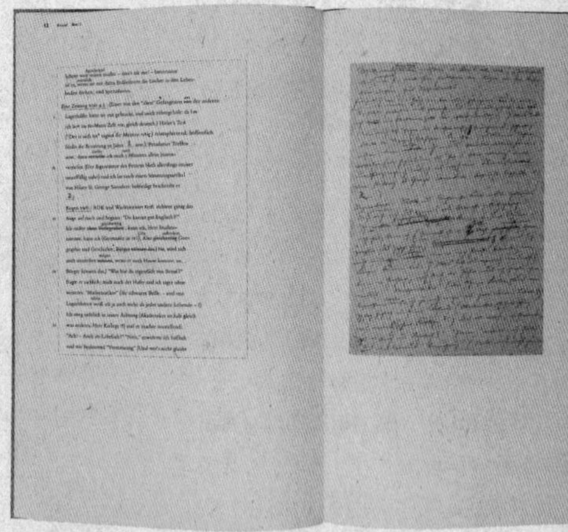
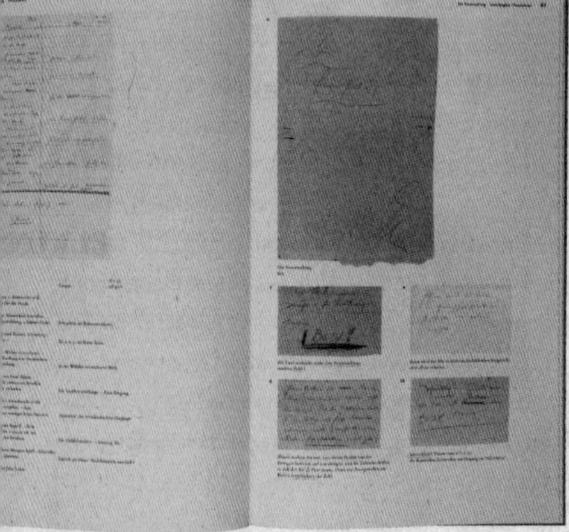
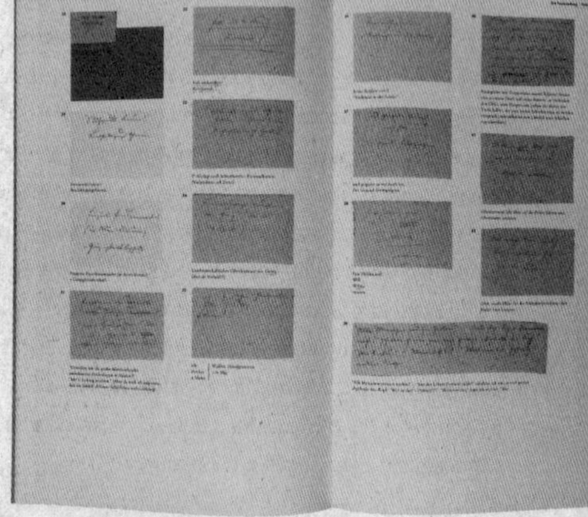

sich   ansehen
soll die Karte von Overaasjöen und Umgebung
1 : 4000 mit Höhenschichtlinien!) Und die konformen
Abbildungen die ich für das Schießen mit seitlicher
Beobachtung zeichnete, usw! Also komm nur!) Er
nickte: ich war zwar nicht völlig ebenbürtig, aber doch ein
Lichtblick; und er begann: "wenn man doch bloß
ein Buch hier hätte!" Da konnte ich nur herzlich
                 mit-
tief überzeugt nicken; (und die 7-st Bremikersche
7-stellige haben sie mir auch noch beschlagnahmt) "ja,
das müßte man!" wiederholte ich resigniert (und
                                        er
ich hörte ihn kommen; jetzt würde endlose literarische
                                                  belehrend;
und ästhetische Diskussionen beginnen versuchen; er war
von der Sorte.) Dies mal rettetemich das Antreten.
Dämmerung: 's war 1 Mann zu viel; ; sie zählten, rechneten;
die Kompanieführer liefen servil hin und zurück;
          begeistert
salutierten vor den Sergeanten; Wir trampelten stampften
fröstelten im Regen, dösten weltblind, stampften
verstohlen; in der Küche bereiten sie den Tee (zuerst
war's was Neues: Tee hatte Deutschland seit Jahren nicht
gekannt; hier kriegten wir's Eimerweise. Verrückte Welt.)
Endlich war Schluß (haben sich wohl auf Zauberei geeinigt)
und wir schritten (sic!) zu den Zelten; manche kreise
fingen wieder an um das Lager zu kreisen, zu zweien,
zu dreien, stundenlang. Erinnerungen, Vorträge, Aufschneiden
Zukunft, Ansichten, Bilder, Gefühle: Diskussion!
Das war es. Nie haben Soldaten soviel gequatscht.
          (unbewußte)
Es war die Reaktion auf Jahre des Schweigens und Stillstehens.

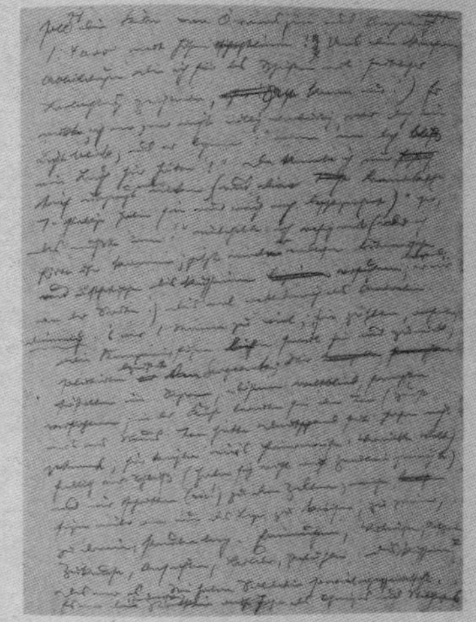

nd Schullehrer als Goethe-Schwärm

V2-Alarm – v

ngen

ua ein Rei taund die Sehnsuch treib

ach Konschieta" trällerte Kuddel.

anther. Jazzkapelle the shooting star

Bibellesen darauf kommen

kalt
n sagte ~~ihn~~: ~~weißt~~: Ochsen! Alles

sprach ihm Keiner; den aber man sah deutlich, ~~"Heil" und~~ "Heil" und Blute lagen. – Befehlen in dieser Welt ~~nur ei~~ zuweilen; aber wer Gef sollte ~~müßte~~ sofort erschossen "Führer befiehl! Wir fo licheres als diese Bitte u auch immer sei?! Pfui D

Der Kompanieführer: Morg einer Schweizerspende,

pfindlich wie Junglehrerinnen. Blutarm.
ns. Das Zelt war gespannt wie eine Trommelfell; wir
ßten hinaus die ~~Strick~~ Verspannungen nachlockern.
Teil versuchte, sich mit Wasser zu waschen; ohne Seife,
e Handtuch. Ich ~~nicht~~ heute nicht. –
Groß-Glockner sei der höchste Berg, sagte Kraft berichtigend,
beherrscht
te Meter; ~~oben sei eine~~ Damm, gekränkt, ~~xxx~~ zog den
fragte
d zusammen~~,~~ und ~~setzte~~ dann spröde fort: "Woher kommt
Name denn? Hat er die Form so?" "Oben steht wohl
Glocke" meinte ~~Kraft~~ der Lehrer gleichgültig, und kramte
in seinen Sachen, "Bergkapellen; gibt viel davon" –"
mit den Fingern durch Briefbogen.
a ~~war mir lieber als d~~ Ich fühlte die Notwendigkeit,
as Selbstbewußtsein wieder aufzurichten, und sagte ihm
ch laut) daß ~~die~~ Keltenreste damals in die ~~Ber Gebirge~~

willst) Dennoch sah er bedauernd nach dem ~~schönen~~ breithüftigen Ofen: war auch schön! So ganz winterfriedlich, Christmas Carol; und ich sah ihn an, durch die Vorsatzlinse dienstlicher Demut, dahinter Mitleid und Verachtung. (Warten, bis die Litzen an ihm sich endlich zu etwas entschlossen: Na, was hastu ausgebrütet, mein Monikin?!)

"Najadenn –" (und noch einmal, ganz leise, : "Dadada"): "Führen Sie die Batterie rauf!" – Ließ ich also meine Hacken ~~schnalzen~~ einschnappen, meine Hand hebelte sich zum Mützenschirm: Life begins at 50 fifty: ge Ewig der Deine! (Er glaubte es, und knurrte zufrieden: es giebt Schlechtere als Bergmann! Unbesorgt!)

Wrummm: (und selbst mit anfassen, sonst kriegen die die lange 10 Zentimeter Kanone, unser bestes Stück, nicht hoch: "Mensch ~~Menke~~ menke: Häng'en Sie sich ~~man~~ lieber vorn über's Rohr! – – Na endlich!"

Der Weg: eiskalter Sand; Weiß und gebacknes Gelb; links immer ‹die Saar›; sie quälte sich zwischen ihren zackig gefrorenen Ufern, wir uns im greisen Dunst, , wir, ohne Diesel verloren

Ein Steinbruch: noch standen die Rüttelsiebe am Rand; eine Feldschmiede, Gerät vor Hütte, machte lange Füße; die zerschundene Felswand drehte uns überall ~~Fläch~~ weißbunte Flächen nach. (Notieren: in dem verschlossenen Schuppen kann eventuell Sprengstoff sein. – ‹Sein› : Sprengstoff, und ‹Sein›.

Nebel im Wald. (Oder waren bloß
wieder die Brillengläser beschlagen? –
Prüfend Kopf kreisen; – ? – : Nee:
Nicht=Ical!)
Also gemeinster poetischer Nebel, bon, und
ich nickte ihm innig zu: die Kahmhaut eines Himmels, in der es nachher von 3000 Sternbazillen wriggeln würde.
: "Nee, hier noch nich: 500 Meter weiter." –

"Ja: hier!" (Und wir bogen rechts ab, den plumpen Berg rauf; ich immer als Gallionsfigur über den Führerkabine.)

Felsen und Jungeichen (xxx werden dann als sogenannte ‹Lohstangen› von den Menschen getötet: das Swift ~~die Menschen~~ uns als ‹Yahoos› angriff und mühelos xxx diffamierte, war falsch!: Er hätte die ganze Schöpfung attackieren sollen!! – Ah: jetzt glitt auch der Volkswagen des Oberleutnants unten um die Kurve!)

"Halt!!": ich stieg vorsichtshalber noch einmal aus, und umging, den Geigerzähler in der Hand, den Hochspannungsmast (die Leitung kam hier der Straße am nächsten: die Masten mußten irgendwie eine unheimliche ~~Strahlungs~~Dosis bekommen haben, denn noch 100 Meter zu beiden Seiten der Leitung maß ich 170 Röntgen: "Gem'm Se Gas, Wolters : – : und durch!!")

So: Hier war die Strahlung wieder kaum noch meßbar. Und ich schleuste sie; ich, immer auf dem Trittbrett; höher die polternde Straße hinauf; immer höher; bis das Kühlwasser sott; und der Weg in der Dämmerung enger wurde: "Warm Se ma, Wolters!"

Voraus gehen: ich; allein. (Und der Weg wurde tatsächlich immer enger. – immer – enger – : ah, hier!) er verbreiterte sich wieder; Häuser keimten zur Rechten auf; und er stieg immer noch; immer noch – so' !:

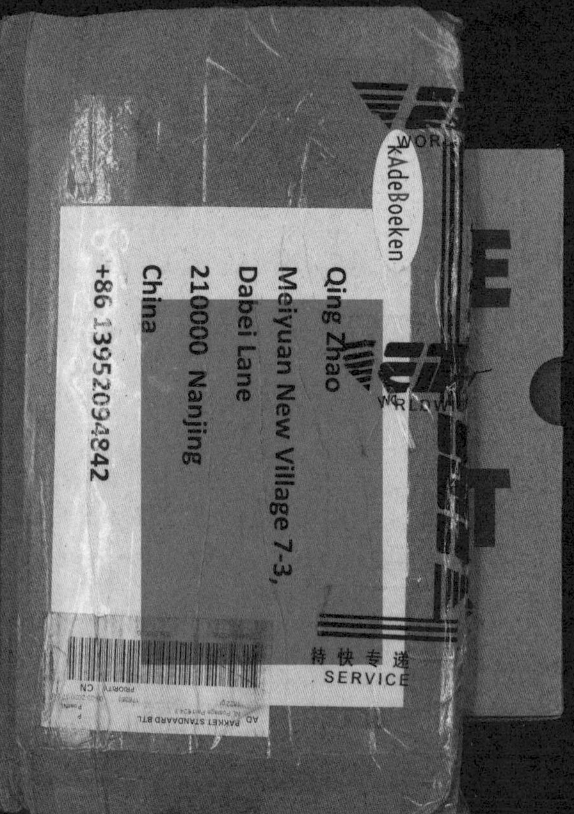
kAdeBoeken
Qing Zhao
Meiyuan New Village 7-3,
Dabei Lane
210000 Nanjing
China
+86 13952094842
特快专递
SERVICE

2003 Ha5

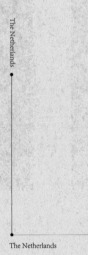

The Netherlands

The Netherlands

China

240mm

54mm

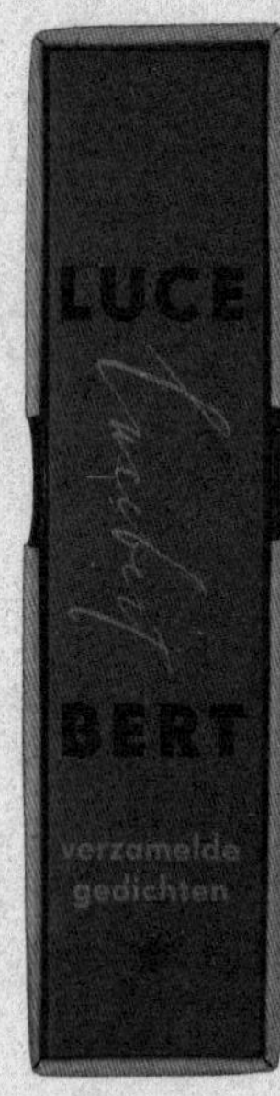

172mm

◎ 荣誉奖
《 卢塞伯特诗集
◖ 荷兰

Honorary Appreciation
Lucebert Verzamelde gedichten
The Netherlands
Tessa van der Waals
Lucebert
De Bezige Bij, Amsterdam
240 172 54mm
1666g
912p
ISBN-10 90-234-0260-X

荷兰艺术家卢塞伯特（Lucebert）在欧洲前卫艺术运动——“眼镜蛇”运动中成为一名诗人。受到前卫文学思潮的影响，他的作品为荷兰诗歌的革命性创新奠定了基础。本书最早的版本是他50岁即诗歌创作25年时出版的诗集，包括之前发表过和未发表过的作品。书籍的装订形式为锁线胶平装。函套采用牛皮卡瓦楞纸制作，红黑双色丝印，函套内部也印有一首小诗。封面红色艺术卡纸黑灰双色丝印。封面内侧印鹅黄色。内页胶版纸印单黑。他作为前卫艺术家和先锋诗人的统一体，本书的视觉形式也充满了冲击力。正文部分采用较粗字重的无衬线体单栏左对齐，粗重且硬朗的字体与穿插其中随性洒脱的绘画形成了独特的对比。目录部分每个大版块的诗名信息密排，通过同一字体家族的不同字号和字重形成丰富的文本肌理。

The Dutch artist, Lubertus Jacobus Swaanswijk, became a poet in the European avant-garde art movement ‘Cobra’. Influenced by the avant-garde literary thoughts, his works laid the foundation for the revolutionary innovation of Dutch poetry. The earliest version of the book was a collection of poems published when he was 50 years old, including previously published and unpublished works during his 25 years of poetry writing. The book is perfect bound with thread sewing. The bookcase is made of corrugated paper, with red and black silk-screen printing, and there is a poem printed inside the case. The cover is made of red art card paper with black and gray silk-screen printing while the inner pages are printed on offset paper in single black. As a collection of an avant-garde artist and poet, the book is full of impact in visual form. The heavy sans serif typeface is used to present the body texts that are aligned on the left in a single column. Heavy and hard fonts form a unique contrast with the free and easy painting. In the catalogue part, the titles of poems of each large section are closely arranged, and rich text variety is formed by different font sizes and weight of the same font family.

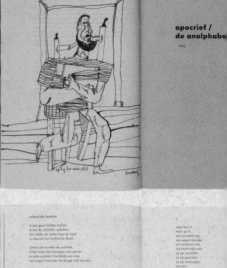

apocrief /
de analphabe

1952

sonnet

ik
mij
ik
mij

mij
ik
mij
ik

ik
ik
mijn

mijn
mijn
ik

school der poësie

ik ben geen lieflijke dichter
ik ben de schielijke oplichter
der liefde, zie onder haar de haat
en daarop een kaaklende dood.

lyriek is de moeder der politiek,
ik ben niets dan omroeper van oproer
en mijn mystiek is het bedorven voer
van leugen waarmee de deugd zich uitziekt.

ik bericht, dat de dichters van fluweel
schuw en humanisties dood gaan.
voortaan zal de hete ijzeren keel
der ontroerde beulen muzikaal opengaan.

nog ik, die in deze bundel woon
als een rat in de val, snak naar het riool
van revolutie en roep: rijmratten, hoon,
hoon nog deze veel te schone poëzieschool.

waar ben ik
waar ga ik
wie verneemt mij
wie neemt mij mee
wie overhoort mij
wie heeft mijn oren
zij zijn verstolen
zij zijn gestolen
zij zijn verborgen
borsten
bijtels weten daarvan

hoeveel begeerte belegt mijn mon
eten
ik spreek melkglazen bevruchting
een ieder zij voorzichtig
een ieder bukt zich
er is geweldig
er is de toen
er is de thans
nu is er een hol
dan is er een kom
waar ga ik
waar ben ik
men mij
letter mij
is mij is mij
mij
frijs
mij
is mij

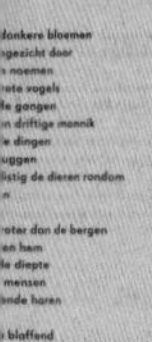

film

in de pupil zittende willoos van doelwit
een dier te zijn van zien
een stier van cinemafotografie die
doorpriemt de prairie met een linnen lieveling van licht

zo breekt door de welriekende stilstand
de spitse keel van de kans
ik rijs langs de trap van trallah
per expresse tot een trans

er bloeit boven elektrisch fluweel
ik verdoof mij daar voor het open voorhoofd
voor het voorhoofd van de beminde
de beminde die slaapt

de beminde die achter grendels grenzen
toevouwt en opbergt en barvoets
in de gedaante van stralen wegvliegt
van de aarde de vertraagde opname
van mijn schaduw afpoetst

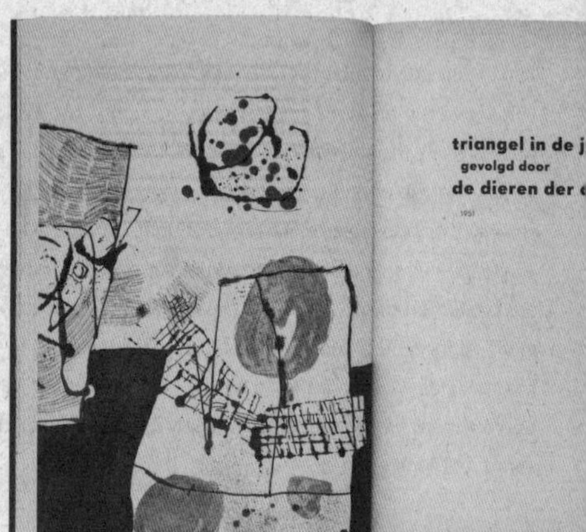

triangel in de j
gevolgd door
de dieren der

1951

•

de schoonheid van een meisje
of de kracht van water en aarde
zo onopvallend mogelijk beschrijven
dat doen de zwanen

maar ik spel van de naam a
en van de namen a z
de analphobetische naam

daarom mij mag men in een lichaam
niet doen verdwijnen
dat vermogen de engelen
met hun ijlere stemmen

maar mij het is blijkbaar is wanhopig
zo woordenloos geboren slechts
in een stem te sterven

51

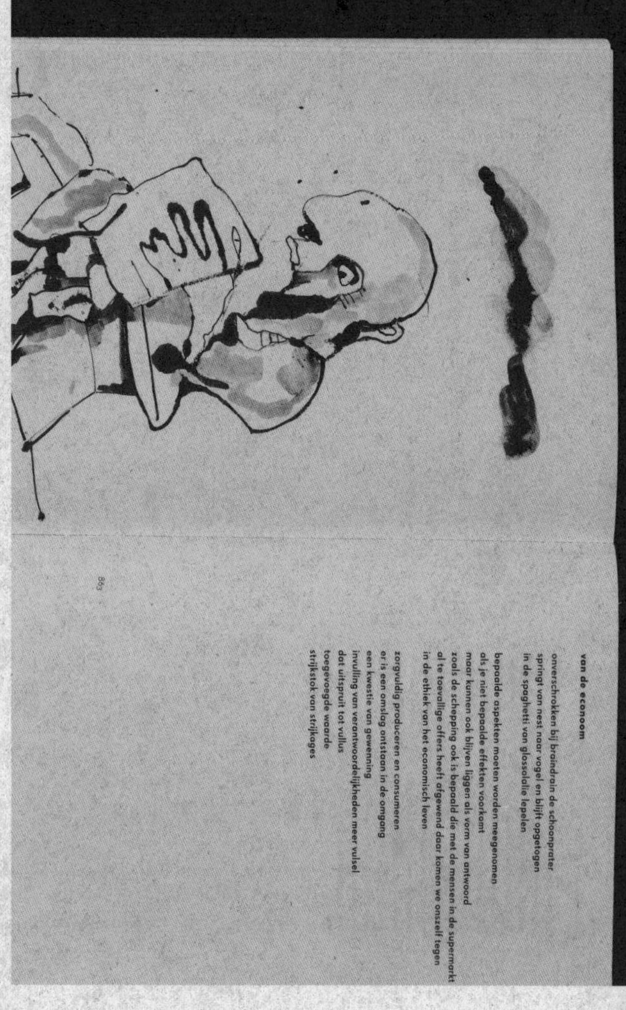

van de econoom

onverschrokken bij braindrain de schoonprater
springt van nest naar vogel en blijft opgetogen
in de spaghetti van glossolalie lepelen

bepaalde aspekten moeten worden meegenomen
als je niet bepaalde effekten voorkomt
maar kunnen ook blijven liggen als vorm van antwoord
zoals de schepping ook is bepaald die met de mensen in de supermarkt
al te toevallige offers heeft afgewend daar komen we onszelf tegen
in de ethiek van het economisch leven

zorgvuldig produceren en consumeren
er is een omslag ontstaan in de omgang
een kwestie van gewenning
invulling van verantwoordelijkheden meer vulsel
dat uitspruit tot vullus
toegevoegde waarde
strijkstok van strijkages

863

Ich möchte an dieser Stelle den Designern

dieser schönen Bücher danken ,

Erst ihre Designs haben es möglich gemacht ,

« *das hier vorgelegte Buch herauszubringen* .

Thanks to the designers of these beautiful books,

because of your design,

this book was born and presented .

在此要感谢这些美丽书的设计师们
因为你们的设计
才有了这本书的诞生与呈现

Index — *Index*

索引

1991

2003

Gl
P 0034

Gm
P 0040

Sm[1]
P 0050

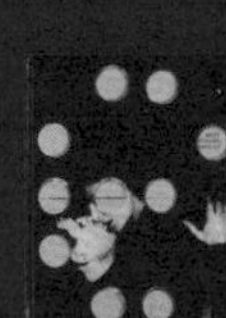

Sm[2]
P 0060

Bm[1]
P 0070

Bm[2]
P 0078

Bm[3]

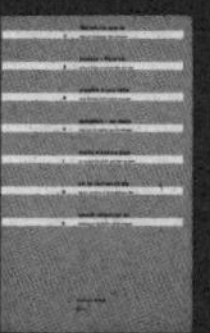

Bm[4]
P 0088

Wim Quist

Bm[5]
P 0094

Ha[1]
P 0104

Ha[2]

Ha[3]
P 0112

Ha[4]
P 0120

1991

1992

Gl
P 0132

Gm
P 0138

Sm[1]
P 0148

Sm[2]
P 0156

Bm[1]
P 0166

Bm[2]
P 0176

Bm[3]
P 0184

Bm[4]

Bm[5]
P 0194

Ha[1]
P 0202

Ha[2]
P 0208

Ha[3]
P 0216

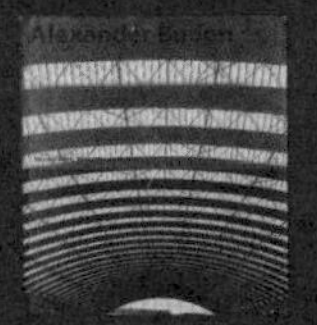
Ha[4]
P 0222

Ha[5]
P 0230

1253

1992

Gl
P 0244

Gm
P 0256

Sm[1]
P 0266

Sm[2]

Sm[3]

Bm[1]
P 0274

Bm[2]
P 0282

Bm[3]
P 0290

Bm[4]
P 0298

Bm[5]
P 0304

Bm[6]

Ha[1]
P 0314

Ha[2]
P 0324

Ha[3]

Ha[4]
P 0332

Ha[5]
P 0340

Ha[6]
P 0348

Ha[7]

1993

1994

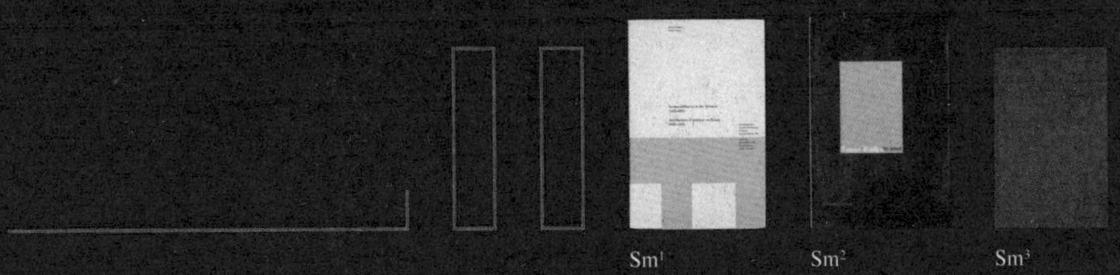

Sm[1]
P 0362

Sm[2]
P 0372

Sm[3]
P 0380

Bm[1]
P 0388

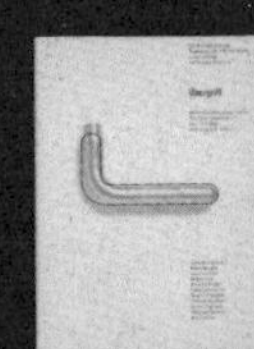

Bm[2]
P 0396

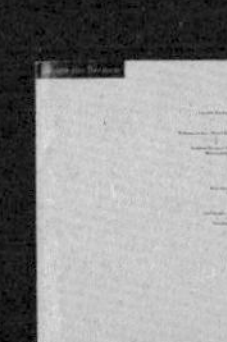

Bm[3]
P 0404

Bm[4]
P 0412

Bm[5]
P 0420

Bm[6]

Ha[1]

Ha[2]
P 0426

Ha[3] Ha[4] Ha[5]

1994

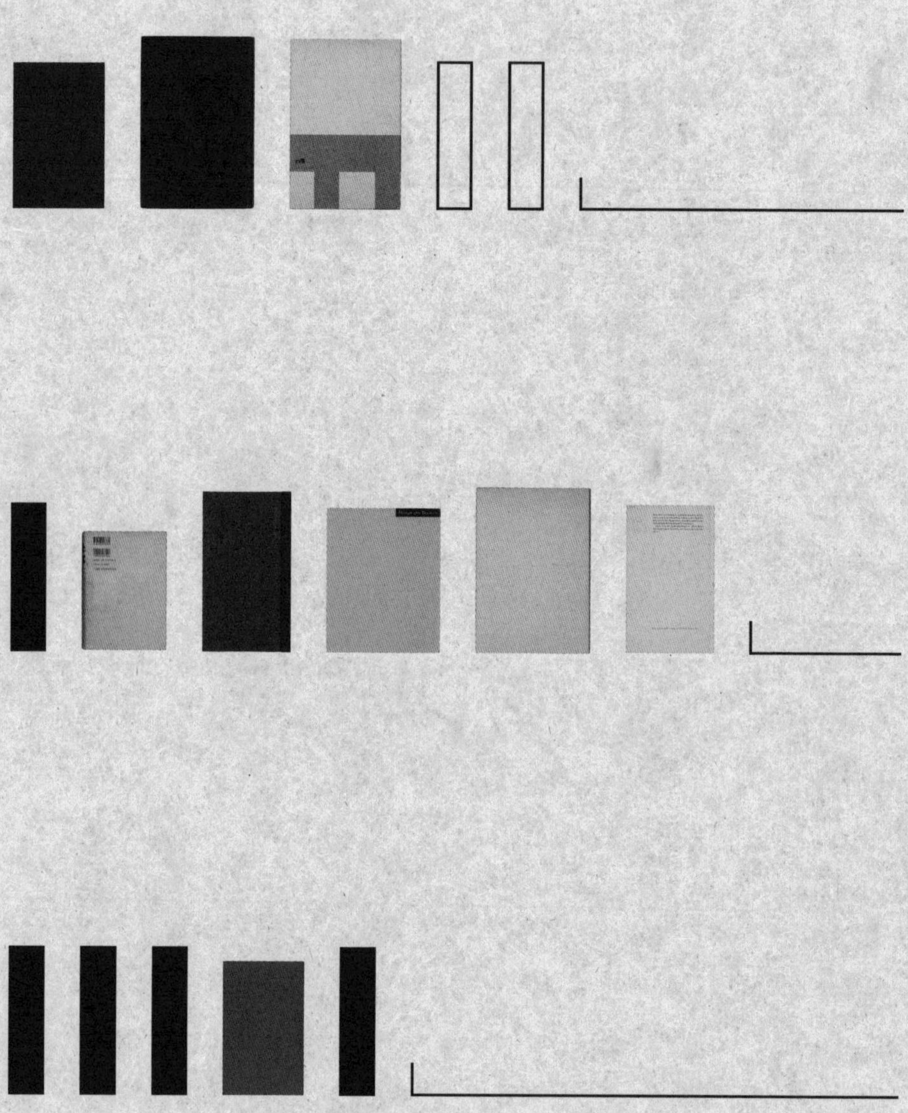

Gl
P 0436

Gm

Sm[1]
P 0444

Sm[2]
P 0452

Bm[1]
P 0462

Bm[2]
P 0472

Bm[3]

Bm[4]
P 0482

Bm[5]
P 0492

Ha[1]
P 0502

Ha[2]
P 0510

Ha[3]
P 0518

Ha[4]

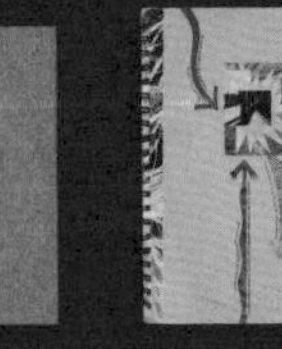

Ha[5]
P 0526

1995

1996

Gl
P 0540

Gm

Sm^1

Sm^2
P 0550

Bm^1
P 0560

Bm^2

Bm^3
P 0570

Bm^4
P 0576

Bm^5
P 0584

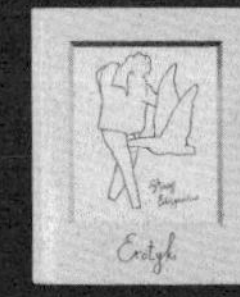

Ha^1

Ha^2
P 0592

Ha^3

Ha^4
P 0602

Ha^5
P 0610

1261

1996

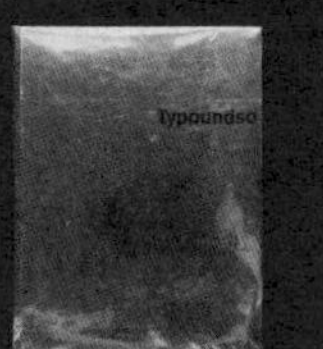

Gl
P 0622

Gm

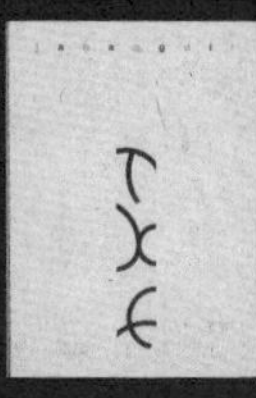

Sm^1
P 0632

Sm^2
P 0640

Bm^1

Bm^2
P 0648

Bm^3

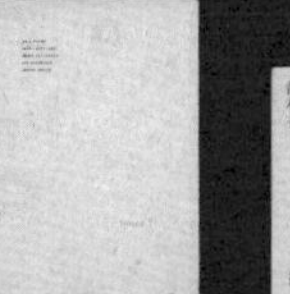

Bm^4
P 0656

Bm^5
P 0666

Ha^1
P 0674

Ha^2

Ha^3
P 0682

Ha^4
P 0692

Ha^5
P 0702

1997

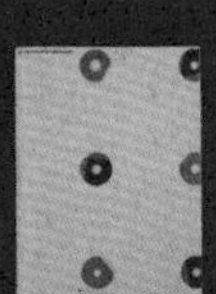

Gl
P 0716

Gm
P 0726

Sm^1

Sm^2
P 0736

Bm^1
P 0744

Bm^2
P 0752

Bm^3

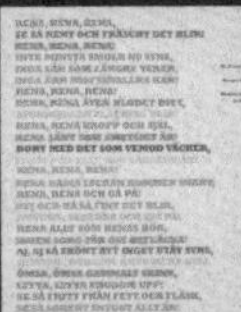

Bm^4
P 0762

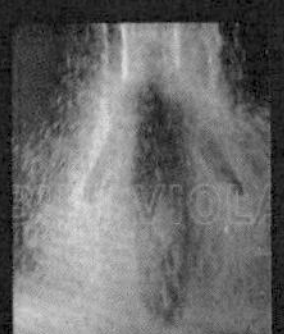

Bm^5
P 0770

Ha^1
P 0780

Ha^2
P 0790

Ha^3

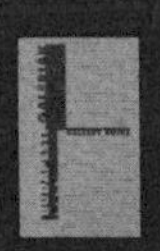

Ha^4
P 0798

Ha^5

1998

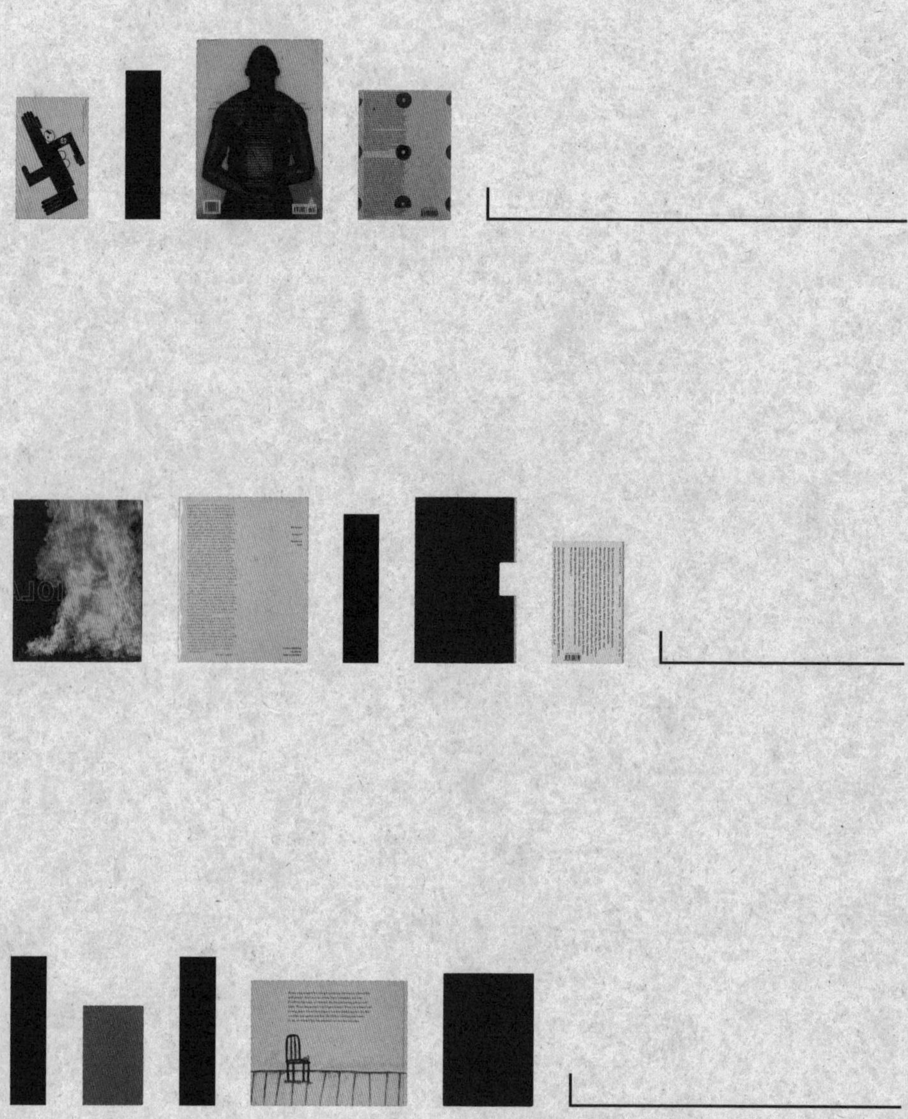

Gl
P 0812

Gm
P 0820

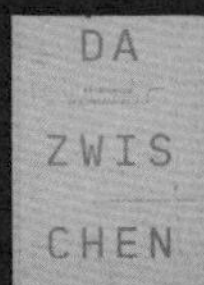

Sm[1]
P 0830

Sm[2]
P 0838

Bm[1]

Bm[2]
P 0848

Bm[3]
P 0858

Bm[4]
P 0868

Bm[5]
P 0878

Ha[1]
P 0886

Ha[2]
P 0896

Ha[3]
P 0906

Ha[4]
P 0912

Ha[5]
P 0922

1999

Gl
P 0938

Gm
P 0948

Sm¹
P 0958

Sm²
P 0966

Bm¹

Bm²

Bm³

Bm⁴
P 0976

Bm⁵
P 0986

Ha¹

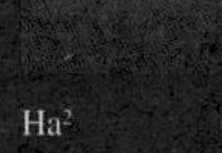

Ha²
P 0996

Ha³
P 1006

Ha⁴
P 1012

Ha⁵
P 1022

1269

2000

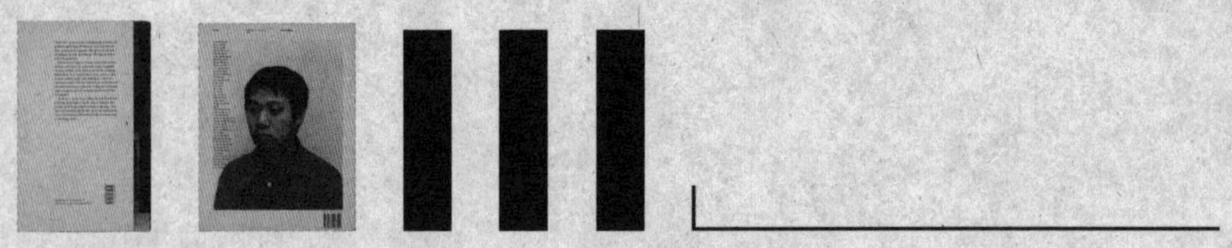

Gl
P 1034

Gm
P 1042

Sm[1]

Sm[2]

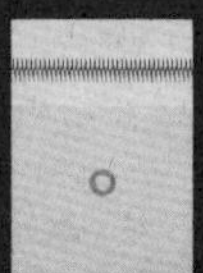

Bm[1]
P 1050

Bm[2]
P 1058

Bm[3]

Bm[4]
P 1066

Lohse

Bm[5]
P 1076

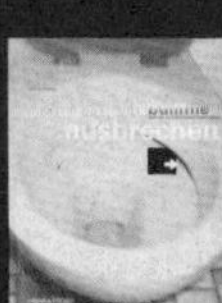

Ha[1]
P 1086

Ha[2]

Ha[3]

Ha[4]
P 1096

Ha[5]
P 1102

2001

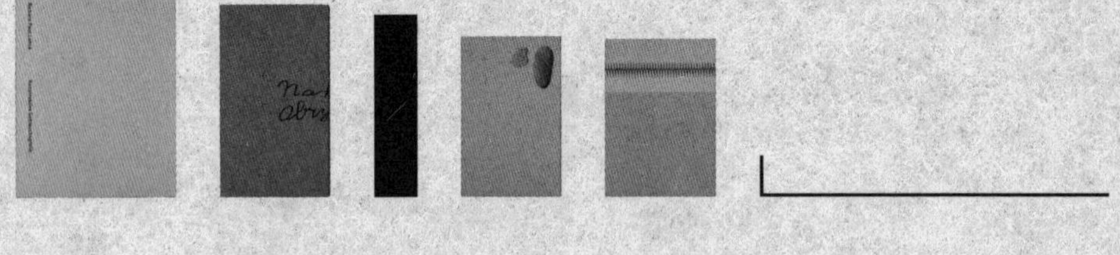

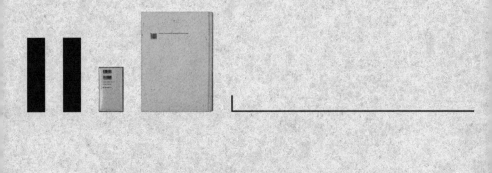

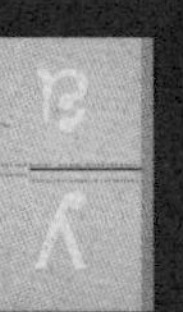

Gl
P 1118

Gm

Sm^1
P 1128

Sm^2
P 1138

Bm^1

Bm^2
P 1150

Bm^3

Bm^4
P 1158

Bm^5

Ha^1
P 1164

Ha^2
P 1174

Ha^3

Ha^4
P 1182

Ha^5

2002

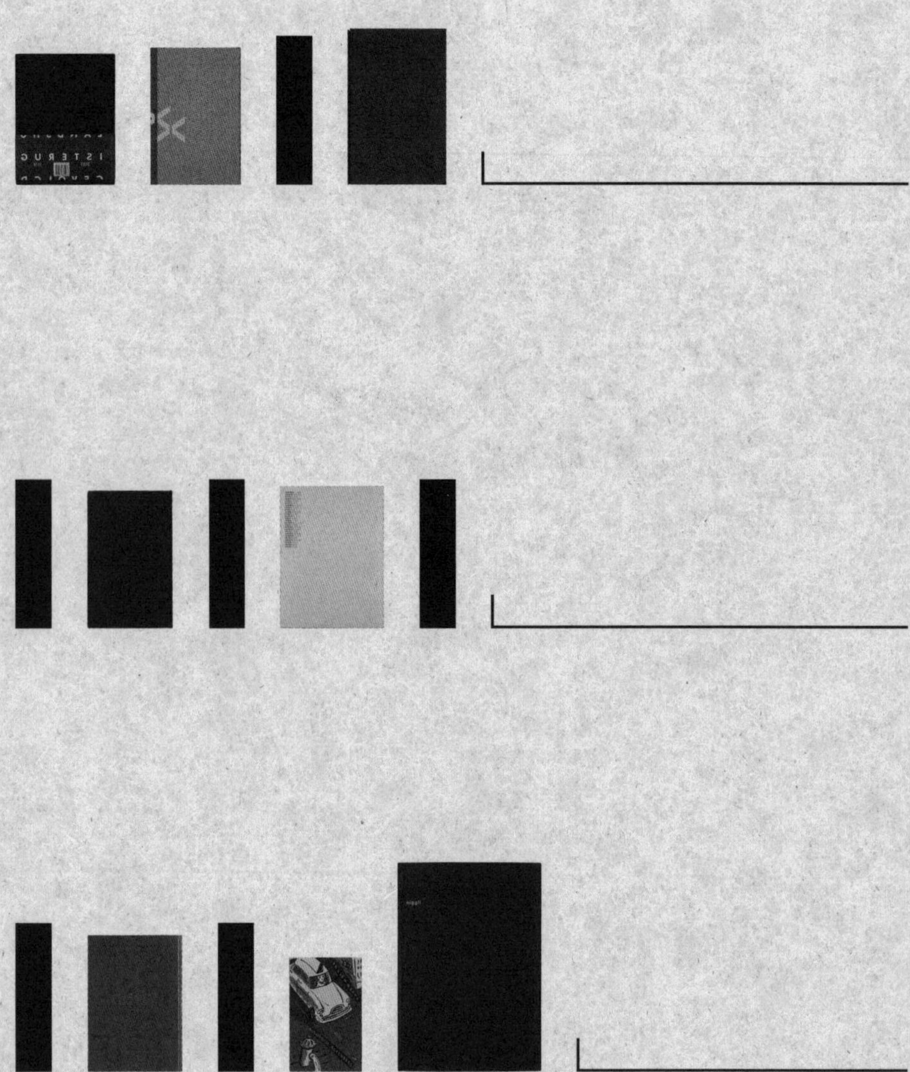

Gl Gm Sm[1] Sm[2]
P 1196

Bm[1]
P 1208

Bm[2]
P 1222

Bm[3]
P 1230

Bm[4] Bm[5]

Ha[1] Ha[2] Ha[3] Ha[4] Ha[5]
P 1238

2003

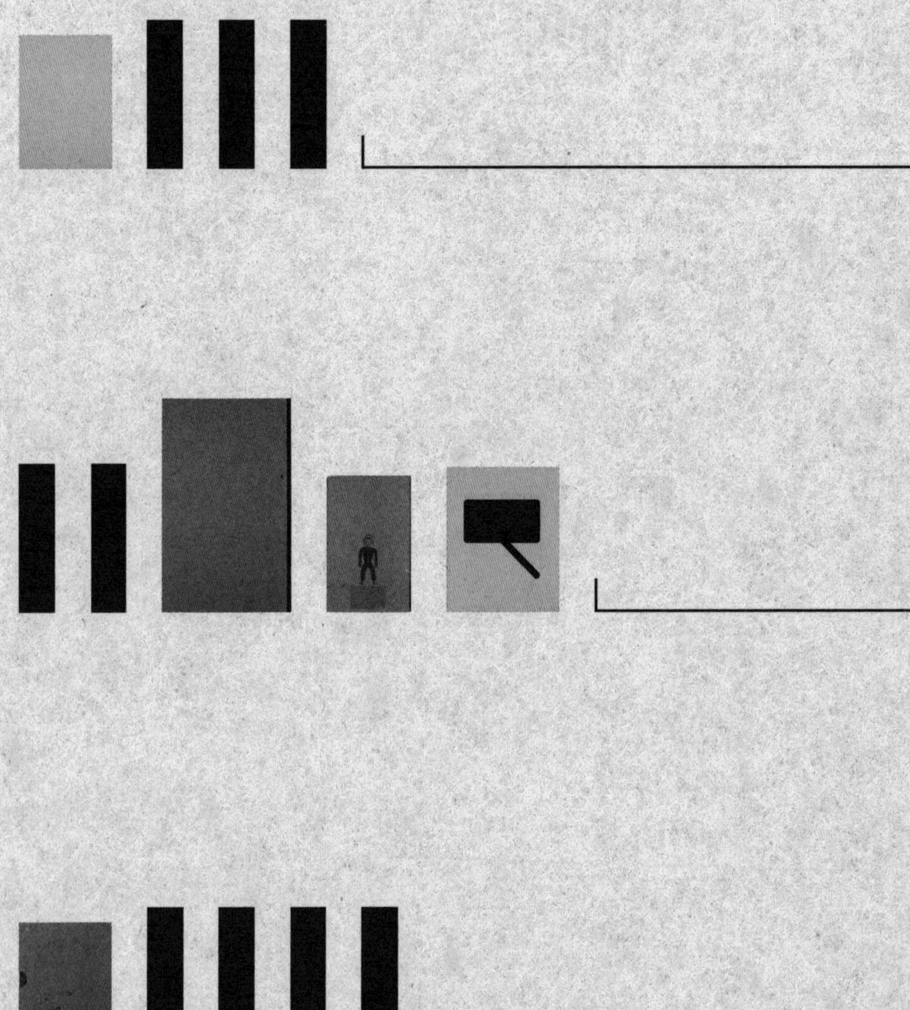

1991 年

金字符奖 |||||| 两条河流 || 美国

Golden Letter |||||| Two Rivers || U.S.A. // *James Robertson, Carolyn Robertson* // Wallace Stegner || The Yolla Bolly Press, Covelo || 258 191 16mm || 454g || 118p

金奖 ||||| 符拉迪沃斯托克 || 美国

Gold Medal ||||| Vladivostok || U.S.A. // *Kim Shkapich* // John Hejduk || Rizzoli International Publications, New York || 313 237 28mm || 1720g || 272p || ISBN-10 0-8478-1129-8

银奖 |||| 爵士乐在洛杉矶 || 德国

Silver Medal |||| Jazz in LA || Germany // *Ingo Wulff* // Bob Willoughby || Nieswand Verlag, Kiel || 480 338 12mm || 1186g || 46p || ISBN-10 3-926048-41-7

银奖 |||| 安迪·沃霍尔的电影——巴黎蓬皮杜艺术中心电影回顾展展册 || 德国

Silver Medal |||| Andy Warhol Cinema: Katalog zur Filmretrospektive im Centre Georges Pompidou, Paris || Germany // *Éditions Carré* // Éditions Carré, Paris || 255 208 27mm || 803g || 272p || ISBN-10 2-908393-30-1

铜奖 ||| 正确与欢乐——从思想到作为商品的书本 || 瑞士

Bronze Medal ||| Richtigkeit und Heiterkeit: Gedanken zum Buch als Gebrauchsgegenstand || Switzerland // *Jost Hochuli* // Franz Zeier || Typotron AG, St. Gallen || 240 150 5mm || 121g || 44p || ISBN-10 3-7291-1058-6

铜奖 ||| 一个，五个，很多个 || 德国

Bronze Medal ||| eins, fünf, viele || Germany || *Květa Pacovská* || Kv ta Pacovská || Ravensburger Buchverlag Otto Maier, Ravensburg || 270 185 10mm || 306g || 24p || ISBN-10 3-473-33569-X

Bronze Medal ||| Taschenatlas der Pharmakologie || Germany // *Karl-Heinz Fleischmann* // Heinz Lüllmann, Klaus Mohr, Albrecht Ziegler || Georg Thieme Verlag, Stuttgart || ISBN-10 3-13-707701-X

铜奖 ||| 什么是诗歌? || 德国

Bronze Medal ||| Was ist Dichtung? || Germany // *Brinkmann & Bose* // Jacques Derrida || Brinkmann & Bose, Berlin || 285 180 6mm || 197g || 48p || ISBN-10 3-922660-46-0

铜奖 ||| 建筑师维姆·奎斯特——荷兰建筑师专论 || 荷兰

Bronze Medal ||| Wim Quist, architect: Monografie van Nederlandse architecten || The Netherlands // *Reynoud Homan* // Auke van der Woud || Uitgeverij 010, Rotterdam || 271 254 18mm || 957g || 144p || ISBN-10 90-6450-075-4

荣誉奖 || 猫王 || 捷克

Honorary Appreciation || Ko i f král || Czech Republic // *Milan Grygar* // Pavel Šrut || Albatros, Prag || 295 210 32mm || 1416g || 352p

Honorary Appreciation || A strange Inn || Japan // *Yosuke Inoue* // Setsuko Hasegawa || Fukinkan Shoten, Publishers, Tokyo

荣誉奖 || 岩波情报科学辞典 || 日本

Honorary Appreciation || Dictionary of Iwanami Information Science || Japan // *Iwanami Shoten, Publishers, Tokyo* // Makoto Nagao || Iwanami Shoten, Publishers, Tokyo || 231 170 43mm || 1388g || 1172p || ISBN-10 4-00-080074-4

荣誉奖 || 66 位荷兰摄影师的自拍 || 荷兰

Honorary Appreciation || 66 Zelfportretten van Nederlandse Fotografen || The Netherlands // *Lex Reitsma* // Rudy Kousbroek || Nicolaas Henneman Stichting, Amsterdam || 150 110 13mm || 210g || 160p || ISBN-10 90-7255-602-X

1992 年

金字符奖 |||||| 我的诗 || 挪威

Golden Letter |||||| Mine Dikt || Norway // *Kristian Ystehede, Guttorm Guttormsgaard* // Halldis Moren Vesaas || H. Aschehoug, Oslo || 265 150 20mm || 547g || 216p || ISBN-10 82-03-16803-5

金奖 ||||| 克利克 || 瑞士

Golden Letter ||||| Klick || Switzerland // *André Hefti* // Erich Grasdorf || Edition A., Zürich || 320 240 20mm || 1074g || 140p || ISBN-10 3-9520133-0-7

银奖 |||| 午后之绿 || 日本

Silver Medal |||| Green in the afternoon || Japan // *Takashi Kuroda* // Chizuru Miyasako || Tokyo Shoseki Co., Tokyo || 210 148 13mm || 365g || 168p || ISBN-10 4-487-75295-7

银奖 |||| 最后三英尺——迪克·埃尔弗斯与他的艺术 || 荷兰

Silver Medal |||| Een leest heeft drie voeten: Dick Elffers & de kunsten || The Netherlands // *Rob Schröder, Lies Ros* // Max Bruinsma || Uitgeverij De Balie; Gerrit Jan Thiemefonds, Amsterdam || 311 246 15mm || 835g || 104p || ISBN-10 90-6617-062-X

铜奖 ||| 加热如何成为事业——150 年苏尔寿加热技术史 || 瑞士

Bronze Medal ||| Wie die Heizung Karriere machte: 150 Jahre Sulzer-Heizungstechnik || Switzerland // *Bruno Güttinger* // Sulzer Infra AG, Winterthur || 294 202 24mm || 1340g || 368p

铜奖 ||| 步行前往锡拉库萨——1989 年踏寻约翰·戈特弗里德·佐伊梅的足迹 || 德国

Bronze Medal ||| Spaziergang nach Syrakus: Eine Reise nach Johann Gottfried Seume im Jahre 1989 || Germany // *Karin Girlatschek* // Andreas Langen || Edition Cantz, Stuttgart || 385 268 15mm || 1415g || 100p || ISBN-10 3-89-322-235-9

铜奖 ||| 瞧——1750－1960 年历史时尚亮点 || 德国

Bronze Medal ||| Voilà: Glanzstücke historischer Moden von 1750-1960 || Germany // *KMS graphic, Maja Thorn* // Wilhelm Hornbostel || Prestel Verlag, München || 309 240 17mm || 1106g || 192p || ISBN-10 3-7913-1117-4

Bronze Medal ||| BSO Jaarverslag 1989 || The Netherlands // *Harry N. Sierman* // Hugo Brandt Corstius || BSO, Utrecht

铜奖 ||| 日建设计——建造现代日本 1900－1990 || 美国

Bronze Medal ||| Nikken Sekkei: Building Modern Japan 1900-1990 || U.S.A. // *Thomas Cox (Willi Kunz Associates)* // Kenneth Frampton, Kunio Kudo || Princeton Architectural Press, New York || 258 262 37mm || 1863g || 288p || ISBN-10 1-878271-01-6

荣誉奖 || 给那没救我命的朋友 || 德国

Honorary Appreciation || Dem Freund, der mir das Leben nicht gerettet hat || Germany // *Joachim Düster* // Hervé Guibert || Rowohlt Verlag, Reinbek || 210 133 28mm || 380g || 256p || ISBN-10 3-498-02463-9

荣誉奖 || 林中探秘 || 日本

Honorary Appreciation || Searching in the Forest || Japan // *Kiichi Miyazaki* // Seizo Tashima || Kaisei-Sha Publishing Co, Tokyo || 241 300 12mm || 537g || 44p || ISBN-10 4-03-435080-6

荣誉奖 || 米兰——雾中风景 || 日本

Honorary Appreciation || Milan: Paesaggi nella nebbia || Japan // *Koji Ise* // Atsuko Suga || Hakusui-Sha, Co., Tokyo || 194 140 18mm || 336g || 220p || ISBN-10 4-560-04179-2

荣誉奖 || 建筑师亚历山大·博顿——荷兰建筑师专论 || 荷兰

Honorary Appreciation || Alexander Bodon, architect: Monografie van Nederlandse architecten || The Netherlands // *Reynoud Homan* // Maarten Kloos || Uitgeverij 010, Rotterdam || 271 254 18mm || 995g || 136p || ISBN-10 90-6450-087-8

荣誉奖 || 反对——荷兰当代艺术中的戏仿、幽默和嘲弄 || 荷兰

Honorary Appreciation || Schräg / Tegendraads: parodie, humor en spot in hedendaagse Nederlandse kunst || The Netherlands // *Gerard Hadders, André van Dijk (hard werken)* // Paul Donker Duyvis, Klaus Honnef || Edition Braus, Heidelberg || 240 215 13mm || 739g || 176p || ISBN-10 90-12-06639-5

1993 年

金字符奖 |||||| 公鸡彼得，纸天堂 || 德国

Golden Letter |||||| Der Hahnepeter, Papier Paradies || Germany // *Květa Pacovská* // Kurt Schwitters, Kv ta Pacovská || Pravis Verlag, Osnabrück || 424 332 45mm || 4354g || 96p

金奖 ||||| 肖像 || 荷兰

Gold Medal ||||| faces || The Netherlands // *Piet Gerards, Marc Vleugels* // Pieter Beek || Gemeente Maastricht, Provincie Limburg || 300 310 15mm || 814g || 70p || ISBN-10 90-73094-04-6

银奖 |||| 脱离黑暗 || 瑞士

Silver Medal |||| Out Of Darkness || Switzerland // *Kaspar Mühlemann* // Connie Imboden || Edition FotoFolie, Zürich/Paris || 289 245 14mm || 902g || 100p || ISBN-10 3-9520309-0-2

Silver Medal |||| A Noh Mask: Mitsui Family Property || Japan // *Yasuhiko Naito* // Masumi Shimizu || Gakken Co., Tokyo || ISBN-10 4-05-105710-0

Silver Medal |||| Silver Medal || Elektrotechniek Tabellenboek || The Netherlands // *Roelof Koster (Studio Dorèl)* // P. F. Den Hartogen, L. Scheltinga || Wolters-Noordhoff, Groningen || ISBN-10 90-01779-54-9

铜奖 ||| 创作者的即兴创作——声音图像中的莫扎特 || 德国

Bronze Medal ||| Mit der Absicht des Schöpfers hat es höchstens zufällig erwas zu tun: Mozart in Tonspuren und Bildräumen || Germany // *Lutz Dudek, Claudia Grotefendt* // Jörg Boström (Hrsg.) || Fotoforum SCHWARZBUNT e.V., Bielefeld || 255 186 10mm || 385g || 56p || ISBN-10 3-928625-01-2

铜奖 ||| 齐默尔曼遇见施皮克曼 || 德国

Bronze Medal ||| Zimmermann meets Spiekermann || Germany // *Ulysses Voelker* // Ulysses Voelker || Meta Design, Berlin || 280 218 8mm || 234g || 42p || ISBN-10 3-929200-01-5

铜奖 ||| 唯一与多样 || 荷兰

Bronze Medal ||| Uniek en meervoudig || The Netherlands // *Reynoud Homan* // Liesbeth Crommelin || Stedelijk Museum, Amsterdam || 240 240 6mm || 289g || 48p || ISBN-10 90-5006-060-9

铜奖 ||| 回文 || 荷兰

Bronze Medal ||| SYMMYS || The Netherlands // *Harry N. Sierman* // Battus || Em. Querido's Uitgeverij, Amsterdam || 240 161 10mm || 236g || 104p || ISBN-10 90-214-5359-2

铜奖 ||| 弗里索·克莱默——工业设计师 || 荷兰

Bronze Medal ||| Friso Kramer: ndustrieel ontwerper || The Netherlands // *Reynoud Homan* // Rainer Bullhorst, Rudoiphine Eggink || Uitgeverij 010, Rotterdam || 303 216 21mm || 1048g || 160p || ISBN-10 90-6450-124-6

Bronze Medal ||| From the edge: Student Work || U.S.A. // *April Greiman, Sean Adams (April Greiman Inc.)* // Michael Rotondi || The Southern California Institute of Architecture, Los Angeles

荣誉奖 || 维尔老城 || 瑞士

Honorary Appreciation || Wil. Die Altstadt || Switzerland // *Gaston Isoz* // Werner Warth, Herbert Maeder || Meyerhans Druck AG, Wil / St. Gallen || 286 215 20mm || 884g || 136p || ISBN-10 3-9520321-0-7

荣誉奖 || 钞票及其图案——瑞士案例 || 瑞士

Honorary Appreciation || Le billet de banque et son image: l'exemple suisse || Switzerland // *Isabelle Lajoinie* // République et Canton de Genève, Genf || 220 227 5mm || 222g || 40p

Honorary Appreciation || Milovan Obraz || Czech Republic // *Clara Istlerová* // Ji í Teper || Atelier Abrakadabra, Prag

荣誉奖 || 诗歌与其他文字 || 西班牙

Honorary Appreciation || Poesía y otros textos || Spain // *Norbert Denkel* // San Juan de la Cruz || Círculo de Lectores, Barcelona || 316 208 36mm || 1493g || 304p || ISBN-10 84-226-3764-2

荣誉奖 || 文字图片 || 瑞典

Honorary Appreciation || Ordbilder || Sweden // *HC Ericson* // HC Ericson || Carlssons Bokförlag, Stockholm || 257 206 12mm || 554g || 88p || ISBN-10 91-7798-451-X

荣誉奖 || 机械学 || 美国

Honorary Appreciation || Mechanika || U.S.A. // *David Betz* // The Contemporary Arts Center, Cincinatti || 306 230 14mm || 374g || 60p || ISBN-10 0-917562-57-7

Honorary Appreciation || De Amor oscuro: Of Dark Love || U.S.A. // *Felicia Rice* // Francisco X. Alarcón || Moving Parts Press, Santa Cruz || ISBN-10 0-939952-08-4

1994 年

银奖 |||| 1942—1992 年的瑞士室内设计 || 瑞士

Silver Medal |||| Innenarchitektur in der Schweiz 1942–1992 || Switzerland // *Thomas Petraschke (Studio Halblützel)* // Alfred Hablützel,Verena Huber || Verlag Niggli AG, Sulgen || 306 207 21mm || 1280g || 248p || ISBN-10 3-7212-0276-7

银奖 |||| 城堡——小说的手稿版 || 德国

Silver Medal |||| Das Schloß: Roman in der Fassung der Handschrift || Germany // *Eckhard Jung, Ulysses Voelker* // Franz Kafka || Büchergilde Gutenberg, Frankfurt am Main || 310 205 38mm || 1540g || 432p || ISBN-10 3-7632-3973-1

银奖 |||| 鼻子 || 德国

Silver Medal |||| Die Nase || Germany // *Gert Wunderlich* // Nikolai Gogol || Edition Curt Visel, Memmingen || 262 169 16mm || 480g || 64p || ISBN-10 3-922406-51-3

铜奖 |||“蓝屋”的周围——来自修士、商人、书籍和书店 || 瑞士

Bronze Medal ||| Rund urns <Blaue Haus>: Von Klosterbrüdern, Kaufleuten, Büchern und Buchhändlern || Switzerland // *Antje Krausch, Ruedi Tachezy* // Ernst Ziegler, Peter Ochsenbein, Hermann Bauer || Ophir-Verlag, St. Gallen || 266 165 13mm || 454g || 128p || ISBN-10 3-907787-02-4

铜奖 ||| 侵占 || 德国

Bronze Medal ||| Übergriff || Germany // *Studenten der HfG Karlsruhe und Sepp Landsbek* // FSB Franz Schneider Brakel (Hrsg.) || der Buchhandlung Walther König, Köln || 297 213 18mm || 1116g || 146p || ISBN-10 3-88375-179-0

铜奖 ||| 来自不来梅的设计 || 德国

Bronze Medal ||| Design aus Bremen || Germany // *Hartmut Brückner* // Beate Manske, Jochen Rahe || Design Zentrum, Bremen || 260 212 7mm || 351g || 76p || ISBN-10 3-929430-01-0

铜奖 ||| 任氏传 || 德国

Bronze Medal ||| Die Geschichte des Fräulein Ren || Germany // *Clemens Tobias Lange* // Shen Jiji || CTL-Presse, Clemens Tobias Lange, Hamburg || 288 166 24mm || 538g || 106p

铜奖 ||| 峠 || 日本

Bronze Medal ||| Touge: A Mountain Pass || Japan // *Shincho-Sha Co., Tokyo* // Ryotaro Shiba || Shincho-Sha Co., Tokyo || 216 160 44mm || 1015g || 716p || ISBN-10 4-10-309738-8

Bronze Medal ||| Jakob Böhme's Offenbarung über Gott || Poland // *Jadwiga, Janusz Tryzno* // Correpondance des Arts, Lódz

Honorary Appreciation || Unwegsame Gebiete || Austria // *Wolfgang Buchta* // Dylan Thomas || Wolfgang Buchta, Wien

荣誉奖 || 字体的乐趣 || 瑞士

Honorary Appreciation || Freude an Schriften || Switzerland // *Jost Hochuli* // Jost Hochuli || Typotron AG, St. Gallen || 240 152 5mm || 114g || 36p || ISBN-10 3-7291-1072-1

Honorary Appreciation || Peleus || Germany // *Anja Harms* // Anja Harms || Unica T, Oberursel i. Ts.

Honorary Appreciation || Det Nye Testamente || Denmark // *Anne Rohweder* // Det Danske Bibelselskab, Kopenhagen || ISBN-10 87-7523-278-2

Honorary Appreciation || The Steadfast Tin Soldier of Joh: Enschedé en Zonen, Haarlem || The Netherlands // *Bram de Does* // Ernst Braches || Spectatorpers, Aartswoud; Just Enschedé, Amsterdam

1995 年

金字符奖 ||||| 夜间飞行——诗歌 || 瑞士

Golden Letter ||||| Nachtflügge: gedichte || Switzerland // *Kaspar Mühlemann* // Kathrin Fischer || Kranich-Verlag, Zürich || 240 152 7mm || 174g || 88p || ISBN-10 3-906640-52-3

Gold Medal ||||| Ministerie van VROM || The Netherlands // *Reynoud Homan, Robbert Zweegman* // Jan Rutten (Hrsg.) || Uitgeverij 010, Rotterdam || ISBN-10 90-5460-205-6

银奖 |||| 具体与发散 || 瑞士

Silver Medal |||| Spezifisches und Diffuses || Switzerland // *Urs Stuber* // Guido von Stürler || Verlag Niggli AG, Sulgen || 240 158 6mm || 231g || 60p || ISBN-10 3-7212-0290-2

银奖 |||| 帽之书 || 美国

Silver Medal |||| The Hat Book || U.S.A. // *Leslie Smolan, Jennifer Domer (Carbon Smolan Associates)* // Rodney Smith, Leslie Smolan || Nan A. Talese; Doubleday, New York || 175 197 22mm || 459g || 128p || ISBN-10 0-385-47228-5

铜奖 ||| 沉寂之屋 || 德国

Bronze Medal ||| camera silens || Germany // *Julia Hasting, Patricia Müller, Gerwin Schmidt, Béla Stetzer* // Olaf Arndt, Rob Moonen || ZKM Zentrum für Kunst und Medientechnologie, Karlsruhe || 265 209 18mm || 815g || 176p || ISBN-10 3-928201-10-7

铜奖 ||| 直立——电梯和扶梯的垂直运输文化史 || 德国

Bronze Medal ||| Vertikal: Aufzug Fahrtreppe Paternoster Eine Kulturgeschichte vom Vertikal-Transport || Germany // *Grappa-Design* // Vittorio Magnano Lampugnani, Lutz Hartwig || Ernst & Sohn, Berlin || 340 242 18mm || 1209g || 144p || ISBN-10 3-433-02480-4

Bronze Medal ||| La llama doble: Amor y erotismo || Spain // *Norbert Denkel* // Octavio Paz || Círculo de Lectores, Barcelona || ISBN-10 84-226-4758-3

铜奖 ||| 旅程 || 美国

Bronze Medal ||| Journey || U.S.A. // *Rita Marshall (Delessert & Marshall)* // Guy Billout || Creative Education, Mankato || 223 280 11mm || 466g || 32p || ISBN-10 0-88682-626-8

铜奖 ||| 雕塑中的结构 || 美国

Bronze Medal ||| Structure in Sculpture || U.S.A. // *Jeanet Leendertse* // Daniel L. Schodek || The MIT Press, Cambridge || 298 231 26mm || 1556g || 328p || ISBN-10 0-262-19313-2

荣誉奖 || 费尔德基希市 || 奥地利

Honorary Appreciation || Stadt Feldkirch || Austria // *Reinhard Gassner* // Christian Mähr, Nikolaus Walter || Amt der Stadt Feldkirch, Feldkirch || 310 237 20mm || 1469g || 192p

荣誉奖 || 双人晚餐 || 德国

Honorary Appreciation || Dinner for Two || Germany // *Büro für Gestaltung Biste & Weißhaupt* // Stiftung Gold- und Silber-schmiedekunst, Schwäbisch Gmünd || 296 232 7mm || 485g || 80p

荣誉奖 || 基础知识 1—6 册 || 德国

Honorary Appreciation || Grundsetzliches 1–6 || Germany // *Philipp Luidl* // Philipp Luidl || SchumacherGebler, München || 213 133 37mm || 540g || 32p / 42p / 56p / 44p / 32p / 36p || ISBN-10 3-920856-04-X / ISBN-10 3-920856-05-8 / ISBN-10 3-920856-08-2 / ISBN-10 3-920856-09-0 / ISBN-10 3-920856-10-4 / ISBN-10 3-920856-03-1

Honorary Appreciation || Sprache ohne Wörter || Egypt // *Mohiedden El-Labbad* // Mohiedden El-Labbad || Dar El Shorouk, Kairo || ISBN-10 977-09-0144-X

荣誉奖 || 看得见——西雅图公共艺术益智书 || 美国

Honorary Appreciation || In Sight: The Seattle Public Art Puzzle Book || U.S.A. // *Judy Anderson, Phillip Helms Cook, Claudia Meyer-Newman* // Judy Anderson, Phillip Helms Cook, Claudia Meyer-Newman || Seattle Arts Commission, Seattle || 284 244 21mm || 620g || 60p || ISBN-10 0-9617443-5-9

1996 年

金字符奖 ||||| 中国总领事在汉堡的诗 || 德国

Golden Letter ||||| Der chinesische Konsul in Hamburg gedichte: Gedichte || Germany // *Andreas Brylka* // Wang Taizhi, Bernd Eberstein || Christians Verlag, Hamburg || 350 240 19mm || 723g || 74p

Gold Medal ||||| Un coup de dés jamais n'abolira le hasard: Ein Würfelwurf niemals tilgt den Zufall Poème / Gedicht Typographische Bibliothek, Band 2 || Germany // *Klaus Detjen* // Stéphane Mallarmé || Steidl Verlag, Göttingen || ISBN-10 3-88243-374-4

Silver Medal |||| Typografické Variace || Czech Republic // *Libor Beránek* // Libor Beránek || Vysoká škola um leckopr myslová VŠUP, Prag

银奖 |||| 纳达尔——创作年代 1854—1860 || 法国

Silver Medal |||| Nadar: Les années créatrices: 1854-1860 || France // *Pierre-Louis Hardy* // Éditions de la Réunion des Musées Nationaux, Paris || 313 239 40mm || 2142g || 368p || ISBN-10 2-7118-2583-3

铜奖 ||| 维也纳的俄罗斯人——奥地利的解放 || 奥地利

Bronze Medal ||| Die Russen in Wien: Die Befreiung Österreichs || Austria // *Hofmann & Kraner* // Erich Klein || Falter Verlagsgesellschaft, Wien || 285 245 20mm || 1244g || 248p || ISBN-10 3-85439-141-2

Bronze Medal ||| Buna 4: Fabrik für synthetischen Gummi der I.G. Auschwitz und Arbeitslager Monowitz/Auschwitz III (1940–1945) || Germany // *Julia Hasting, Patricia Müller, Gerwin Schmidt, Béla Stetzer, Karlsruhe* // Eigenverlag Julia Hasting, Patricia Müller, Gerwin Schmidt, Béla Stetzer, Karlsruhe || ISBN-10 3-922218-62-8

铜奖 ||| 逃亡之魂 || 德国

Bronze Medal |||| Die flüchtige Seele || Germany // *Edith Lackmann* // Harold Brodkey || Rowohlt Verlag, Reinbek || 210 140 62mm || 1220g || 1344p || ISBN-10 3-498-00540-5

铜奖 ||| 丰富的黑白——东弗里斯兰的照片 || 德国
Bronze Medal |||| Schwarzbuntes: Bilder aus Ostfriesland || Germany // *Rainer Groothuis* // Rainer Groothuis, Karl-Heinz Janßen || Christians Verlag, Hamburg || 302 231 10mm || 617g || 80p || ISBN-10 3-7672-1238-2

铜奖 ||| 非洲 || 美国
Bronze Medal |||| Africa || U.S.A. // *Betty Egg, Sam Shahid* // Herb Ritts || Bulfinch Press; Little, Brown and Company, Boston || 312 363 25mm || 2002g || 136p || ISBN-10 0-8212-2121-3

Honorary Appreciation || The Reg Mombassa Diary 1995 || Australia // *Graham Rendoth (Reno Design Group 14144)* // Chris O'Doherty || Bantam Books, Sydney

荣誉奖 ||"看不见的"电影选集 || 法国
Honorary Appreciation || Anthologie du cinéma invisible || France // *Bulnes & Robaglia* // Christian Janicot || Éditions Jean-Michel Place; ARTE Éditions, Paris || 300 230 55mm || 2195g || 672p || ISBN-10 2-85893-233-6

Honorary Appreciation || Ungdom og sex || Norway // *Marte Fæhn, Line Jerner (Lucas Design & illustrasjon)* // Mette Hvalstad || Senteret for Ungdom, Samliv og Seksualitet, Oslo

荣誉奖 || 情色 || 波兰
Honorary Appreciation || Erotyki || Poland // *Grażyna Bareccy, Andrzej Bareccy* // Stasys Eidrigevi ius || Wydawnictwo Tenten, Warschau || 248 218 10mm || 374g || 64p || ISBN-10 83-85477-85-3

荣誉奖 || 软糖树 || 美国
Honorary Appreciation || The Gumdrop Tree || U.S.A. // *Julia Gorton* // Elizabeth Spurr || Hyperion Books for Children, New York || 256 208 9mm || 332g || 32p || ISBN-10 0-7868-0008-9

1997 年

金字符奖 ||||| Typoundso || 瑞士
Golden Letter ||||| Typoundso || Switzerland // *Hans-Rudolf Lutz* // Hans-Rudolf Lutz || Hans-Rudolf Lutz, Zürich || 300 237 43mm || 2602g || 440p

Gold Medal ||||| Veiligheid en Bedreiging in de 21e Eeuw || The Netherlands // *Mieke Gerritzen, Janine Huizenga* // VPRO, Hilversum || ISBN-10 90-6727-012-1

银奖 |||| 贾汉吉尔 || 瑞士
Silver Medal |||| Jahanguir || Switzerland // *Kaspar Mühlemann* // Galerie Jamileh Weber, Zürich || 320 240 11mm || 720g || 76p || ISBN-10 3-85809-100-X

银奖 |||| 至亲至疏夫妻 || 德国
Silver Medal |||| Sehr nah, sehr fern sind sich Mann und Frau || Germany // *Kerstin Weber, Olaf Schmidt* // Hanne Chen || Edition ZeichenSatz, Kiel || 235 173 11mm || 332g || 68p || ISBN-10 3-00-000733-4

Bronze Medal ||| Nesetkání: Nichtbegegnung || Czech Republic // *Stefanie Harms* // Ivan Wernisch || Vysoká škola um leckopr myslová VŠUP, Prag

铜奖 ||| 贝尔塔的船 || 德国
Bronze Medal ||| Bertas Boote || Germany // *Wiebke Oeser* // Wiebke Oeser || Peter Hammer Verlag, Wuppertal || 244 305 10mm || 447g || 32p || ISBN-10 3-87294-755-9

Bronze Medal ||| Lesetypographie || Germany // *Hans Peter Willberg, Friedrich Forssman* // Hans Peter Willberg, Friedrich Forssman || Hermann Schmidt, Mainz || ISBN-10 3-87439-375-5

铜奖 ||| Unica T——10 年的艺术家书籍 || 德国
Bronze Medal ||| Unica T: 10 Jahre Künstlerbücher || Germany // *Anja Harms, Ines v. Ketelhodt, Dois Preußner, Uta Schneider, Ulrike Stoltz (Unica T)* // Unica T, Oberursel i. Ts. / Offenbach am Main || Unica T, Oberursel i. Ts. / Offenbach am Main || 279 214 19mm || 1103g || 228p || ISBN-10 3-00-000854-3

铜奖 ||| 幸田文的五斗柜 || 日本
Bronze Medal ||| Koda Aya No Tansu No Hikidashi || Japan // *Akio Nonaka* // Gyoku Aoki || Shincho-Sha Co., Tokyo || 217 157 22mm || 476g || 208p || ISBN-10 4-10-405201-9

荣誉奖 || 壮美的哥伦比亚高山 || 哥伦比亚
Honorary Appreciation || Alta Colombia: El Esplendor de la Montaña || Colombia // *Benjamin Villegas* // Cristobal von Rothkirch || Villegas Editores, Bogotá || 311 234 21mm || 1598g || 216p || ISBN-10 958-9393-22-5

Honorary Appreciation || Das Leben des Propheten || Egypt // *Mohiedden El-Labbad* // Karimam Hamsa || Dar El Shorouk, Kairo || ISBN-10 997-09-0320-5

荣誉奖 || 布雷达的沙塞剧院 || 荷兰
Honorary Appreciation || Chassé Theater Breda || The Netherlands // *Bureau Piet Gerards* // Herman Hertzberger || Uitgeverij 010, Rotterdam || 270 211 7mm || 377g || 72p || ISBN-10 90-6450-277-3

荣誉奖 || 目标动力 || 荷兰
Honorary Appreciation || Richtkracht || The Netherlands // *Richard Menken* // Richard Menken || Hogeschool voor de Kunsten HKA, Arnhem || 297 210 14mm || 689g || 156p || ISBN-10 90-74485-13-8

荣誉奖 || 街头博物馆——维拉诺夫海报博物馆中的波兰海报 || 波兰
Honorary Appreciation || Muzeum Ulicy: plakat polski w kolekcji muzeum plakatu w wilanowie || Poland // *Michał Piekarski* // Mariusz Knorowski || Krupski i S-ka || 293 222 25mm || 1347g || 240p || ISBN-10 83-86117-60-5

1998 年

金字符奖 ||||| 卡雷尔·马滕斯的印刷作品 || 荷兰
Golden Letter ||||| Karel Martens printed matter / drukwerk || The Netherlands // *Jaap van Triest, Karel Martens* // Karel Martens || Hyphen Press, London || 233 173 16mm || 527g || 144p || ISBN-10 0-907259-11-1

金奖 ||||| 运动之魂 || 美国
Gold Medal ||||| Soul of the Game || U.S.A. // *John C. Jay* // Jimmy Smith, John Huet || Melcher Media; Workman Publishing, New York || 324 235 19mm || 1070g || 144p || ISBN-10 0-7611-1028-3

Silver Medal |||| Sehen ... erfinden ... : Erfindungen von Leonardo, gesehen von Anja Wesner || Germany // *Anja Wesner* // Anja Wesner || Anja Wesner, Stuttgart

银奖 |||| 难民的对话 || 德国
Silver Medal |||| Flüchtlingsgespräche || Germany // *Gert Wunderlich* // Bertolt Brecht || Leipziger Bibliophilen-Abend e.V., Leipzig || 219 138 26mm || 584g || 152p

铜奖 ||| 蓝色奇迹——传真小说 || 德国
Bronze Medal ||| Die blauen Wunder: Faxroman || Germany // *Matthias Gubig* // Christoph Keller, Heinrich Kuhn || Reclam Verlag, Leipzig || 220 133 23mm || 409g || 236p || ISBN-10 3-379-00761-7

铜奖 ||| 针对数字媒体进行设计 || 德国
Bronze Medal ||| Zur Anpassung des Designs an die digitalen Medien || Germany // *Sabine Golde, Tom Gebhardt* // form + zweck, Berlin || 298 190 14mm || 517g || 160p || ISBN-10 3-9804679-3-7

Bronze Medal ||| East Side Souvenirs || The Netherlands // *José van't Klooster* // José van't Klooster || José van't Klooster

铜奖 ||| 处理厂 || 瑞典
Bronze Medal ||| Reningsverk || Sweden // *HC Ericson* // HC Ericson || Carlsson bokförlag, Stockholm || 297 241 23mm || 1068g || 160p || ISBN-10 917-203-033-X

铜奖 ||| 比尔·维奥拉 || 美国
Bronze Medal ||| Bill Viola || U.S.A. // *Rebeca Méndez* // Whitney Museum of American Art, New York; Flammarion, Paris || 291 242 19mm || 1219g || 216p || ISBN-10 0-87427-114-2

荣誉奖 || 库鲁·芒金——埃及梦 || 德国
Honorary Appreciation || Kullu mumkin: Ein ägyptischer Traum || Germany // *Matthias Beyrow, Marion Wagner* // Werner Döppner || Beyrow, Wagner, Berlin || 239 169 11mm || 298g || 64p || ISBN-10 3-00-002924-9

荣誉奖 || 爷爷穿西装吗? || 德国
Honorary Appreciation || Hat Opa einen Anzug an? || Germany // *Claus Seitz* // Amelie Fried, Jacky Gleich || Carl Hanser Verlag, München || 225 294 8mm || 390g || 32p || ISBN-10 3-446-19076-7

Honorary Appreciation || Space Nutrition: Essen im Weltall || Germany // *Anja Osterwalder* // Anja Osterwalder || i-d büro GmbH, Stuttgart

荣誉奖 || 1996 年布达佩斯美术馆指南 || 匈牙利
Honorary Appreciation || Budapest Galériák 1996 / Gallery Guide of Budapest 1996 || Hungary // *Johanna Bárd* // Budapest Art Expo Alapítvány, Budapest || 179 114 12mm || 228g || 146p || ISBN-10 936-04-6417-9

Honorary Appreciation || Sh gi-daiku Nakamura Sotoji no Shigoto || Japan // *Tsutomu Nishioka* // Sotoji Nakamura || Seigen-Sha Art Publishing, Kyoto || ISBN-10 4-916094-11-5

1999 年

金字符奖 ||||| 委内瑞拉历史词典 || 委内瑞拉
Golden Letter ||||| Diccionario de Historia de Venezuela || Venezuela // *Álvaro Sotillo* // M. Perez Vila, M. Rodriguez Campo (Hrsg.) || Fundación Empresas Polar, Caracas || 254 176 42mm / 254 176 38mm / 254 176 44mm / 254 176 40mm || 1390g / 1257g / 1447g / 1282g || 1176p / 1064p / 1232p / 1096p || ISBN-10 980-6397-38-X / ISBN-10 980-6397-39-8 / ISBN-10 980-6397-40-1 / ISBN-10 980-6397-41-X

金奖 ||||| 朱塞佩·特拉尼——合理架构的模型 || 瑞士
Gold Medal ||||| Giuseppe Terragni: Modelle einer rationalen Architektur || Switzerland // *Urs Stuber* // Jörg Friedrich, Dierk Kasper (Hrsg.) || Verlag Niggli AG, Sulgen || 280 208 11mm || 697g || 104p || ISBN-10 3-7212-0343-7

银奖 |||| 之间——观察与辨别 || 瑞士
Silver Medal |||| Dazwischen: Beobachten und Unterscheiden || Switzerland // *François Rappo* // André Vladimir Heiz, Michael Pfister || Museum für Gestaltung Zürich || 246 186 25mm || 874g || 272p || ISBN-10 3-907065-79-4

银奖 |||| 1932-1936 年的《艺术公报》|| 西班牙
Silver Medal |||| Gaceta de Arte y su Época 1932-1936 || Spain // *Raimundo C. Iglesias* // Centro Atlántico de Arte Moderno, Las Palmas de Gran Canaria; Edición Tabapress || 279 241 31mm || 2032g || 364p || ISBN-10 84-89152-11-X

Bronze Medal ||| Reihe „allaphbed" 5 Bände || Germany // *Arbeitsgruppe „work ahead" an der Hochschule für Grafik und Buchkunst Leipzig* // Julia Blume, Günter Karl Bose || Institut für Buchkunst

Leipzig an der Hochschule für Grafik und Buchkunst Leipzig ‖ ISBN-10 3-932865-00-6

铜奖 ||| 女巫的数字咒语 ‖ 德国

Bronze Medal ||| Das Hexen-Einmal-Eins ‖ Germany // *Claus Seitz* // Johann Wolfgang v. Goethe, Wolf Erlbruch ‖ Carl Hanser Verlag, München ‖ 185 310 8mm ‖ 324g ‖ 32p ‖ ISBN-10 3-446-18863-0

铜奖 ||| 百年品牌 Palmin ‖ 德国

Bronze Medal ||| Palmin: eine Jahrhundertmarke ‖ Germany // *Gesine Krüger, Andrea Schürings (büro für mitteilungen)* // Joachim Nickel ‖ Union Deutsche Lebensmittelwerke, Hamburg ‖ 308 206 13mm ‖ 568g ‖ 96p

铜奖 ||| 维姆·席佩斯精选集 ‖ 荷兰

Bronze Medal ||| Het Beste van Wim T. Schippers ‖ The Netherlands // *Studio Gonnissen en Widdershoven* // Harry Ruhé ‖ Centraal Museum, Utrecht ‖ 247 175 28mm ‖ 835g ‖ 240p ‖ ISBN-10 90-73285-44-5

铜奖 ||| 摄影的多重角色——艺术、文献、市场和科学 ‖ 美国

Bronze Medal ||| Photography's Multiple Roles ‖ U.S.A. // *studio blue* // Museum of contemporary Photography, Chicago ‖ 278 247 22mm ‖ 1425g ‖ 256p ‖ ISBN-10 0-9658887-2-X

荣誉奖 ‖ 树上不长沙丁鱼 ‖ 瑞士

Honorary Appreciation ‖ Sardinen wachsen nicht auf Bäumen ‖ Switzerland // *Ueli Kleeb* // Vera Eggermann ‖ Atlantis Verlag; pro juventute, Zürich ‖ 222 300 10mm ‖ 384g ‖ 28p ‖ ISBN-10 3-7152-0392-7

荣誉奖 ‖ 格蕾特·古布兰森的日记，第一卷，1904—1912：以太中的青鸟 ‖ 德国

Honorary Appreciation ‖ Grete Gulbransson Tagebücher, Band I, 1904 bis 1912: Der grüne Vogel des Äthers ‖ Germany // *Karin Beck* // Ulrike Lang ‖ Stroemfeld Verlag, Frankfurt am Main ‖ 300 200 50mm ‖ 2202g ‖ 472p ‖ ISBN-10 3-87877-690-X

荣誉奖 ‖ 岛崎藤村在巴黎 ‖ 日本

Honorary Appreciation ‖ A Poet: T son Shimazaki in Paris ‖ Japan // *Shincho-Sha Co., Tokyo* // Yoshizo Kawamori ‖ Shincho-Sha Co., Tokyo ‖ 201 138 23mm ‖ 416g ‖ 352p ‖ ISBN-10 4-10-306005-0

荣誉奖 ‖ 永远属于你——产品耐久性的愿景 ‖ 荷兰

Honorary Appreciation ‖ Eternally yours: visions on product endurance ‖ The Netherlands // *Studio Gonnissen en Widdershoven* // Ed van Hinte ‖ Uitgeverij 010, Rotterdam ‖ 168 120 23mm ‖ 366g ‖ 256p ‖ ISBN-10 90-6450-313-3

荣誉奖 ‖ 城市的方向 I&II——1900—2000 年北欧城市设计 ‖ 荷兰

Honorary Appreciation ‖ De Regie van de Stad I&II: Noord-Europese Stedebouw 1900-2000 ‖ The Netherlands // *Arlette Brouwers, Koos van der Meer* // Koos Bosma, Helma Hellinga (Hrsg.) ‖ NAi Uitgevers; EFL Publicaties, Rotterdam / Den Haag ‖ 305 252 18mm / 305 252 37mm ‖ 943g / 2071g ‖ 128p / 356p ‖ ISBN-10 90-5662-058-4

2000 年

金字符奖 ||||| 月亮和晨星 ‖ 德国

Golden Letter ||||| Mond und Morgenstern ‖ Germany // *Henning Wagenbreth* // Wolfram Frommlet, Henning Wagenbreth ‖ Peter Hammer Verlag, Wuppertal ‖ 346 247 14mm ‖ 730g ‖ 48p ‖ ISBN-10 3-87294-784-2

金奖 ||||| 1998 年第 16 届世界影像节 ‖ 荷兰

Gold Medal ||||| 16th World Wide Video Festival '98 ‖ The Netherlands // *Irma Boom* // Stichting World Wide Video Festival, Amsterdam ‖ 199 159 33mm ‖ 833g ‖ 560p ‖ ISBN-10 90-75018-16-9

银奖 |||| 铁的世界 ‖ 奥地利

Silver Medal |||| Welt aus Eisen ‖ Austria // *Alexander Kada* // Thomas Höft ‖ Springer-Verlag, Wien/New York ‖ 298 254 25mm ‖ 1826g ‖ 222p ‖ ISBN-10 3-211-83097-9

银奖 |||| Persiplus ‖ 德国

Silver Medal |||| Persiplus ‖ Germany // *Kerstin Riedel* // Olaf Rauh ‖ Edition Persiplus (Olaf Rauh Eigenverlag), Leipzig ‖ 264 194 13mm ‖ 465g ‖ 120p ‖ ISBN-10 3-00-003944-9

Bronze Medal ||| Borderlands: How we talk about Canada ‖ Canada // *George Vaitkunas* // W. H. New ‖ UBC Press University of British Columbia, Vancouver ‖ ISBN-10 91-7988-176-9

Bronze Medal ||| Et hus er et ansigt ‖ Germany // *Jens Bertelsen* // Louis Jensen ‖ Gyldendal, Kopenhagen ‖ ISBN-10 87-00-34860-0

Bronze Medal ||| Kitaoji Rosanjin to Suda Seika ‖ Japan // *Kikko Ken* // Shiko-Sha Co., Ltd., Kyoto ‖ ISBN-10 4-87940-538-8

铜奖 ||| 如果 / 那么——新媒体的设计意义 ‖ 荷兰

Bronze Medal ||| If / Then: Design implications of new media ‖ The Netherlands // *Mevis & Van Deursen* // Netherlands Design Institute; BIS Publishers, Amsterdam ‖ 276 217 21mm ‖ 1027g ‖ 264p ‖ ISBN-10 90-72007-52-2

铜奖 ||| 老房子的工艺 ‖ 瑞典

Bronze Medal ||| Hantverket i gamla hus ‖ Sweden // *Dick Norberg* // Byggförlaget, Stockholm ‖ 280 185 20mm ‖ 764g ‖ 208p ‖ ISBN-10 91-7988-176-9

Honorary Appreciation ‖ Raum Komfort Ästhetik ‖ Switzerland // *Urs Bernet, Jürg Schönenberger* // Andreas Steigmeier ‖ hier+jetzt Verlag für Kultur und Geschichte, Baden ‖ ISBN-10 3-906419-11-8

荣誉奖 ‖ 从纤细到特粗——格特·冯德利希 1957—1998 年的字体排印 ‖ 德国

Honorary Appreciation ‖ Von zart bis extrafett: Typo-Grafik von Gert Wunderlich 1957-1998 ‖ Germany // *Gert Wunderlich* // Gert Wunderlich (Hrsg.) ‖ Eigenverlag Gert Wunderlich, Leipzig ‖ 285 216 30mm ‖ 1385g ‖ 256p ‖ ISBN-10 3-932865-16-2

荣誉奖 ‖ 诗 ‖ 荷兰

Honorary Appreciation ‖ De Gedichten ‖ The Netherlands // *Steven van der Gaauw* // Herman de Coninck ‖ Uitgeverij De Arbeiderspers, Amsterdam / Antwerpen ‖ 206 134 55mm ‖ 946g ‖ 484p / 208p ‖ ISBN-10 90-295-0908-2

荣誉奖 ‖ 马塞尔·杜尚——机械复制时代制造艺术的艺术 ‖ 美国

Honorary Appreciation ‖ Marcel Duchamp: the art of making art in the age of mechanical reproduction ‖ U.S.A. // *studio blue* // Francis M. Naumann ‖ Harry N. Abrams Ins., New York; Ludion Press, Ghent ‖ 308 240 31mm ‖ 2135g ‖ 332p ‖ ISBN-10 0-8109-6334-5

荣誉奖 ‖ 关于书籍设计 ‖ 美国

Honorary Appreciation ‖ On Book Design ‖ U.S.A. // *Richard Hendel* // Richard Hendel ‖ Yale University Press, New Haven & London ‖ 287 192 23mm ‖ 880g ‖ 224p ‖ ISBN-10 0-300-07570-7

2001 年

金字符奖 ||||| 卡尔·克劳斯主编的《火炬》杂志（1899—1936）的短语词典 ‖ 奥地利

Golden Letter ||||| Wörterbuch der Redensarten zu der von Karl Kraus 1899 bis 1936 herausgegebenen Zeitschrift „Die Fackel" ‖ Austria // *Anne Burdick* // Werner Welzig (Hrsg.) ‖ Verlag der Österreichischen Akademie der Wissenschaften, Wien ‖ 360 233 82mm ‖ 3557g ‖ 1056p ‖ ISBN-10 3-7001-2768-5

金奖 ||||| 英和·和英辞典 ‖ 日本

Gold Medal ||||| infoword English-Japanese Dictionary ‖ Japan // *Haruyo Kobayashi* // Benesse Corporation, Tokyo ‖ 162 89 29mm ‖ 317g ‖ 1392p ‖ ISBN-10 4-8288-0445-5

Silver Medal |||| Futebol-Arte ‖ Brazil // *Jair de Souza* // Jair de Souza, Lucia Rito, Sérgio Sá Leitão ‖ Empresa das Artes; Editora Senac, São Paulo ‖ 286 240 40mm ‖ 2459g ‖ 440p ‖ ISBN-10 796-334-098-1

Silver Medal |||| 400 jaar vioolbouwkunst in Nederland ‖ The Netherlands // *Jonker & Van Vilsteren* // Nederlandse Groep van Viool – en Strijkstokkenmakers, Amsterdam ‖ ISBN-10 90-7297-782-3

铜奖 ||| raster-noton.oacis ‖ 瑞士

Bronze Medal ||| raster-noton.oacis ‖ Switzerland // *Olaf Bender* // Susanne Binas, Martin Pesch, Pinky Rose, Rob Young, Peter Kraut ‖ raster-noton, Berlin; taktlos-bern, Bern ‖ 231 170 9mm ‖ 299g ‖ 104p ‖ ISBN-10 3-931126-48-X

铜奖 ||| 锡特尔河的卵石 ‖ 瑞士

Bronze Medal ||| Sitterkiesel ‖ Switzerland // *Jost Hochuli* // Urs Hochuli, Oskar Keller ‖ Edition Ostschweiz, VGS, St. Gallen ‖ 235 152 5mm ‖ 131g ‖ 48p ‖ ISBN-10 3-7291-1095-0

Bronze Medal ||| The Selection vfg. 1999 ‖ Switzerland // *Beat Müller, Wendelin Hess, Michael Birchmeier, Nora Beer (Müller + Hess)* // vfg. vereinigung fotograf. Ges ‖ Schwabe & Co, Basel ‖ ISBN-10 3-7965-1515-0

铜奖 ||| 裸体女巨人 ‖ 捷克

Bronze Medal ||| Nahá obryn ‖ Czech Republic // *Jiří Šalamoun, Zdeněk Ziegler* // Ji í Šalamoun ‖ Aulos, Prag ‖ 282 168 17mm ‖ 525g ‖ 62p ‖ ISBN-10 80-86184-03-X

铜奖 ||| 理查德·保罗·洛斯——建构的商业图形 ‖ 德国

Bronze Medal ||| Richard Paul Lohse: Konstruktive Gebrauchsgrafik ‖ Germany // *Markus Bosshard (Weiersmüller Bosshard Grüninger wbg AG)* // Richard Paul Lohse Stiftung ‖ Hatje Cantz Verlag, Ostfildern-Ruit ‖ 306 244 36mm ‖ 2118g ‖ 312p ‖ ISBN-10 3-7757-0767-0

荣誉奖 ‖ 摆脱——了解并克服暴食症 ‖ 德国

Honorary Appreciation ‖ Ausbrechen: Bulimie verstehen und überwinden ‖ Germany // *Verena Böning* // Verena Böning ‖ Urban & Fischer Verlag, München ‖ 255 210 11mm ‖ 534g ‖ 116p ‖ ISBN-10 3-437-45536-2

Honorary Appreciation ‖ Seminar Typografie: Zum Buch 3, material 88, 3 / 2000 ‖ Germany // *Hans Andree, Sigrid Behrens, Uli Brandt, Shin-Young Chung, Daniel Hahn, Nadine Jung, Volker Lang, Zuzanna Musialczyk, Peter Piller, Anna Reemts, Sven Seddig, Jan Schmietendorf* // material Verlag, Hochschule für bildende Künste Hamburg ‖ ISBN-10 3-9323395107

Honorary Appreciation ‖ Hoe gaat het … Verslag en visie 1998 / 1999 ‖ The Netherlands // *Dietwee, Utrecht, -SYB-grafisch ontwerp, Martijn Engelbregt (EGBG)* // Frans van Rijn, Vincent Huizing ‖ ONVZ Zorgverzekeraar

荣誉奖 ‖ 伦理学 ‖ 荷兰

Honorary Appreciation ‖ Ethica ‖ The Netherlands // *Rudo Hartmann, Gerard Hadders* // Aristoteles ‖ Historische Uitgeverij, Groningen ‖ 246 179 35mm ‖ 1055g ‖ 360p ‖ ISBN-10 90-6554-005-9

荣誉奖 ‖ 西班牙之前的委内瑞拉艺术 ‖ 委内瑞拉

Honorary Appreciation ‖ El Arte Prehispánico de Venezuela ‖ Venezuela // *Álvaro Sotillo* // Miguel

Arroyo, Lourdes Blanco, Erika Wagner ‖ Fundación Galería de Arte Nacional, Caracas ‖ 220 300 47mm ‖ 2355g ‖ 548p ‖ ISBN-10 980-6420-11-X

2002 年

金字符奖 ||||| 和纸：日本纸的传统与艺术 ‖ 德国
Golden Letter ||||| Washi: Tradition und Kunst des Japanpapiers ‖ Germany // *Mariko Takagi* // Mariko Takagi ‖ Mariko Takagi, Meerbusch ‖ 280 183 29mm ‖ 803g ‖ 124p+60p ‖ ISBN-10 3-9803617-6-4

Gold Medal ||||| P íb hy z parní lázn ‖ Czech Republic // *Juraj Horváth* // Herbert T. Schwarz ‖ Argo, Prag ‖ ISBN-10 80-7203-256-9

银奖 |||| 巴别塔图书馆——印刷图书馆，第 4 辑 ‖ 德国
Silver Medal |||| Die Bibliothek von Babel: Typographische Bibliothek, Band 4 ‖ Germany // *Klaus Detjen* // Jorge Luis Borges ‖ Steidl Verlag, Göttingen; Büchergilde Gutenberg, Frankfurt am Main / Wien / Zürich ‖ 246 167 15mm ‖ 417g ‖ 96p ‖ ISBN-10 3-88243-778-2

银奖 |||| 莱茵河畔房屋的历史 ‖ 荷兰
Silver Medal |||| Het Rijnlands Huis teruggevolgd in de tijd ‖ The Netherlands // *Irma Boom* // Leo van der Meule ‖ Hoogheemraadschap van Rijnland, Leiden; Architectura & Natura, Amsterdam ‖ 235 186 26mm ‖ 777g ‖ 232p ‖ ISBN-10 90-72381-08-4

Bronze Medal ||| pro: Holz Information. Bemessung im Holzbau: Von der nationalen zur europäischen Normung ‖ Austria // *Atelier Reinhard Gassner* // Richard Pischl ‖ proHolz Austria, Wien

铜奖 ||| JVA 协议 ‖ 德国
Bronze Medal ||| JVA Protokoll ‖ Germany // *Markus Dreßen* // Andrea Seppi, Steffen Junghans ‖ Institut für Buchkunst Leipzig an der Hochschule für Grafik und Buchkunst Leipzig ‖ 255 195 10mm ‖ 405g ‖ 56p ‖ ISBN-10 3-932865-27-8

Bronze Medal ||| Máquinas ‖ Spain // *equipo gráfico de Kalandraka* // Cloé Poizat ‖ Kalandraka Editora, Pontevedra ‖ ISBN-10 84-8464-026-4

铜奖 ||| 世界诗歌的 51 个声音 ‖ 荷兰
Bronze Medal ||| Eenenvijftig stemmen uit de wereldpoëzie ‖ The Netherlands // *Typography and Other Serious Matters* // Kathinka van Dorp ‖ Stichting Wereldpoezie / Stichting Poetry International, Rotterdam ‖ 247 162 17mm ‖ 517g ‖ 144p ‖ ISBN-10 90-72546-26-1

Bronze Medal ||| Zinnenprikkelend ‖ The Netherlands // *Typography and Other Serious Matters* // Anton Korteweg ‖ Drukkerij Ando, Den Haag

荣誉奖 ‖ 林间空地 ‖ 瑞士
Honorary Appreciation ‖ Die Lichtung ‖ Switzerland // *Hans Grüninger (Weiersmüller Bosshard Grüninger wbg AG)* // Rolf Schroeter ‖ Verlag Niggli AG Sulgen / Zurich ‖ 375 268 28mm ‖ 1811g ‖ 144p ‖ ISBN-10 3-7212-0396-8

荣誉奖 ‖ 硬摇滚天堂（伟大的小册子 20）‖ 德国
Honorary Appreciation ‖ Der Hardrock-Himmel (Die Tollen Hefte 20) ‖ Germany // *Thomas Müller* // Tom Coraghessan Boyle ‖ Büchergilde Gutenberg, Frankfurt am Main / Wien / Zürich ‖ 205 137 6mm ‖ 110g ‖ 32p ‖ ISBN-10 3-7637-6016-1

Honorary Appreciation ‖ Claus en Kaan ‖ The Netherlands // *Reynoud Homan* // Claus en Kaan Architecten, Amsterdam / Rotterdam

荣誉奖 ‖ Dolly——华丽的文本字体 ‖ 荷兰
Honorary Appreciation ‖ Dolly: A Book Typeface With Flourishes ‖ The Netherlands // *Wout de Vringer (Faydherbe/De Vringer)* // Underware, Den Haag ‖ 245 176 4mm ‖ 150g ‖ 32p ‖ ISBN-10 90-76984-01-8

Honorary Appreciation ‖ Astronauten som inte fick landa: Om Michael Collins, Apolo 11 och 9 kilo checklistor ‖ Sweden // *Lotta Kühlhorn* // Bea Uusma Schyffert ‖ Alfabeta Bokförlag AB, Stockholm ‖ ISBN-10 91-7712-916-4

2003 年

Golden Letter ||||| Jan Palach: Morgen word je wakker geboren ‖ The Netherlands // *Nynke M. Meijer* // Nynke M. Meijer ‖ Eigenverlag Nynke M. Meijer, Sneek

Gold Medal ||||| Gutenberg Galaxie 2: Irma Boom ‖ Germany // *Kristina Brusa* // Irma Boom ‖ Institut für Buchkunst Leipzig an der Hochschule für Grafik und Buchkunst Leipzig ‖ ISBN-10 3-932865-30-8

Silver Medal |||| Hans Arp: Worte ‖ Germany // *Anja Harms* // Anja Harms ‖ Anja Harms Künstlerbücher, Oberursel/Ts.

银奖 |||| 色彩 ‖ 荷兰
Silver Medal |||| Kleur ‖ The Netherlands // *Corina Cotorobai (Werkplaats Typografie)* // kaAp / studium generale van de Hogeschool voor de Kunsten Arnhem ‖ 240 171 20mm ‖ 686g ‖ 272p

铜奖 ||| 这些事 ‖ 瑞士
Bronze Medal ||| The Things ‖ Switzerland // *Dimitri Bruni, Manuel Krebs (NORM)* // NORM ‖ NORM, Zürich ‖ 260 211 13mm ‖ 671g ‖ 176p ‖ ISBN-10 3-9311-26-75-7

铜奖 ||| 华氏 451 ‖ 德国
Bronze Medal ||| Fahrenheit 451 ‖ Germany // *Gerda Raidt* // Ray Bradbury ‖ Büchergilde Gutenberg, Frankfurt am Main / Wien / Zürich ‖ 245 158 20mm ‖ 581g ‖ 176p ‖ ISBN-10 3-7632-5111-1

铜奖 ||| 布鲁塞尔 / 射击位置 ‖ 德国
Bronze Medal ||| Brüssel / Die Feuerstellung ‖ Germany // *Friedrich Forssman* // Arno Schmidt ‖ Suhrkamp Verlag, Frankfurt am Main ‖ 384 241 16mm ‖ 1085g ‖ 72p ‖ ISBN-10 3-518-80201-1

Bronze Medal ||| Talking to Myself by Yohji Yamamoto ‖ Germany // *Claudio dell'Olio* // Carla Sozzani ‖ Steidl Verlag, Göttingen ‖ ISBN-10 3-88243-825-8

Bronze Medal ||| Kaba ornament: Deel 1 Vorm ‖ The Netherlands // *Bram de Does* // Bram de Does ‖ Spectatorpers, Orvelte

Honorary Appreciation ‖ Weniger als man / Less than One ‖ Germany // *Sabine Golde* // Joseph Brodsky ‖ Carivari, Leipzig

Honorary Appreciation ‖ Der Ausflug der toten Mädchen ‖ Germany // *Rainer Groothuis (Groothuis, Lohfert, Consorten)* // Anna Seghers ‖ Offizin Bertelsmann Club, Rheda-Wiedenbrück

Honorary Appreciation ‖ The Oral History of Modern Architecture ‖ Japan // *Sin Akiyama, Ken Hisase* // John Peter ‖ TOTO Shuppan, Tokyo

Honorary Appreciation ‖ Heftig vel: Lefen met een chronische huidziekte ‖ The Netherlands // *-SYB-* // Tanny Dobbelaar, Adriënne M. Norman ‖ Elsevier gezondheidszorg, Maarssen

荣誉奖 ‖ 卢塞伯特诗集 ‖ 荷兰
Honorary Appreciation ‖ Lucebert Verzamelde gedichten ‖ The Netherlands // *Tessa van der Waals Lucebert* // De Bezige Bij, Amsterdam ‖ 240 172 54mm ‖ 1666g ‖ 912p ‖ ISBN-10 90-234-0260-X

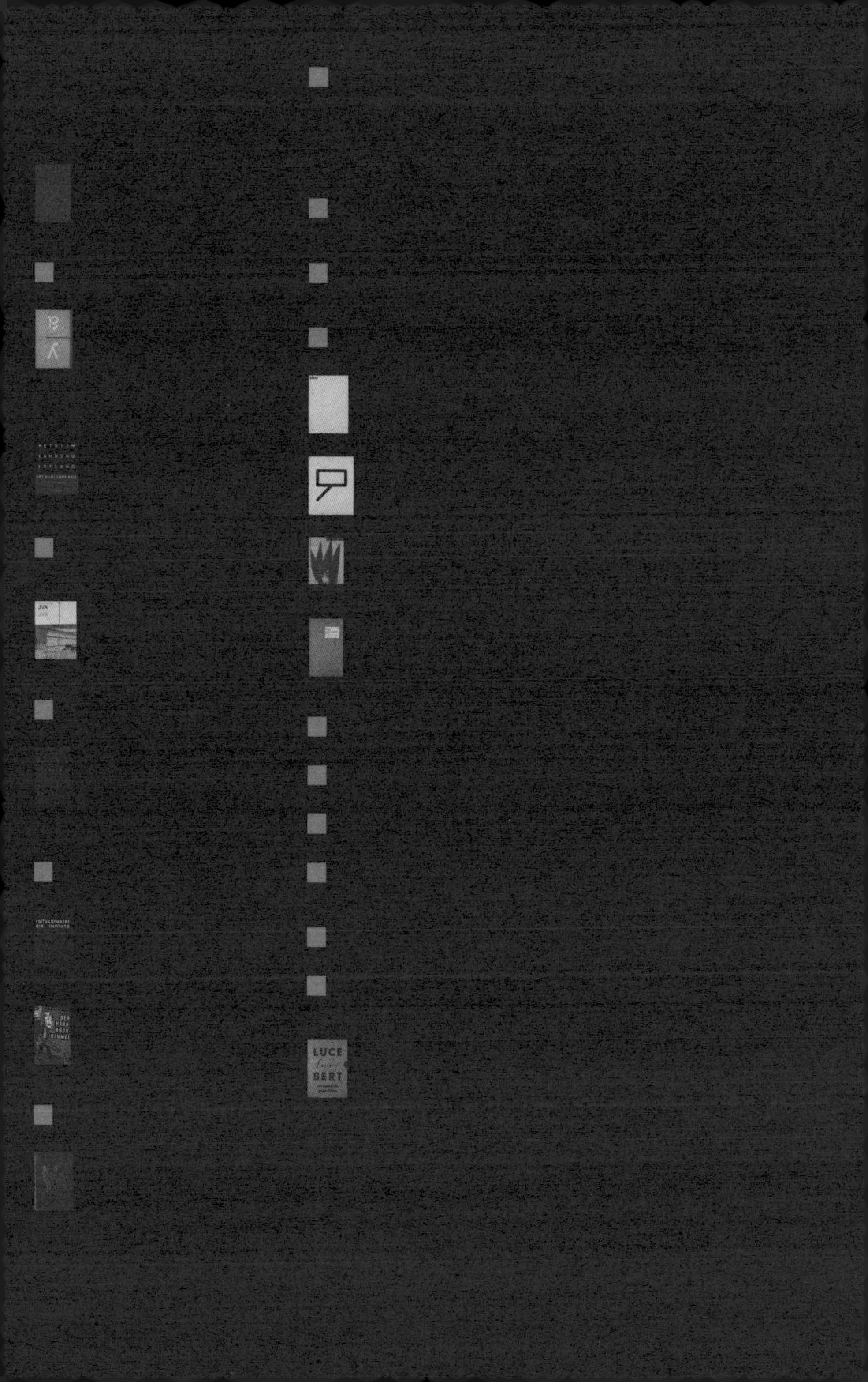
JVA
LUCE
BERT

Nachbemerkungen — *Postscript*

跋

Seit der Veröffentlichung von Leipziger Auswahl – Die schönsten Bücher der Welt 2004-2019 habe ich viel Unterstützung und Bestätigung von Lesern und Freunden erhalten. Es ist eine große Ermutigung für das Team und für mich, dass unser Buch, das unter viel Mühe und mit viel Einsatz angefertigt worden ist, auf solch eine große Resonanz gestoßen ist. Wenn man das Sammeln der schönsten Bücher der Welt als eine endlose lange Reise betrachtet, dann gleicht der Prozess der Entstehung eines Buches einem Zeugnis darüber, wie sich mit Hilfe unzähliger sowohl strenger als auch romantischer Erzählungen in Zeit und Raum ein Weg zur gemeinsamen Teilung von Hoffnungen finden lässt. Leipzig durchblättert – die schönsten Bücher der Welt 1991-2003 ist aus eben dieser genannten Hoffnung heraus entstanden. Das Buch gleicht insofern eher einem Vorläufer von Leipziger Auswahl, als es den Auftakt für eine Sammlung der schönsten Bücher bildet und auf systematischem Wege ein umfassendes Gedenken an Leipzig zustande bringt.

> Bereits Anfang der 1950er Jahre führten politische Spaltungen in Deutschland dazu, dass fast zeitgleich getrennte Buchgestaltungswettbewerbe in Ost- und Westdeutschland ausgerichtet wurden. Mit dem Fall der Berliner Mauer waren die beiden parallel stattfindenden Buchmessen-Wettbewerbe nicht mehr zeitgemäß. Seit 1991 organisiert und veranstaltet die Deutsche Stiftung Buchkunst in Leipzig offiziell den Wettbewerb „Die Schönsten Bücher der Welt“.

Das Buch Leipzig durchblättert stellt eine Sammlung der jährlichen Gewinner von „Leipzigs Schönste Bücher der Welt“ von 1991 bis 2003 dar, beginnend mit der Auswahl nach dem Zusammenschluss der beiden deutschen Staaten. Unsere derzeitige Sammlung umfasst vor allem zwei Phasen. In der ersten Phase, d.h. in der Zeit von 2004 bis 2019, wurden 224 Bücher ausgezeichnet, von denen sich 213 oder 95% in der Hanqing-Sammlung befinden. Die im vergan-

After the publishing *The Choice Of Leipzig – Best Book Design From All Over The World 2019–2004*, we received support and confirmation from many readers. The tough work of day and night gained such a remarkable effect was very encouraging for the group and me. If we took collecting the best designed books as an endless long journey, the process of making a book seemed a witness, that a key made by countless rigorous and romantic narration gathering time and space unlocked more longings for us to share. Hence *Browse Leipzig – Best Book Design From All Over The World 1991–2003*, which was more like a prequel of The *Choice Of Leipzig*, piecing together the origin of the best designed books and completely finishing the commemoration about Leipzig.

In early 1950s, the political division made East Germany and West Germany set up the book design competition respectively almost at the same time. After the fall of the Berlin Wall, the two parallel competitions were out of keeping with the times. So since 1991, the German Book and Art Foundation have officially started organizing and undertaking “the competition of the Best Designed Book All Over the World in Leipzig”.

Beginning with the comments after the amalgamation of East and West Germany, *Browse Leipzig* includes the award-winning books in the competitions of “the Best Designed Books All Over the World in Leipzig” from 1991 to 2003. There are two stages in our collection. The first stage is from 2004 to 2019,

跋

《莱比锡的选择——世界最美的书 2019－2004》出版之后，受到了不少读者朋友们给予的支持和肯定。曾经的日夜付出能获得如此反响，对于团队和我来说都是极大的鼓舞。如果把收藏“世界最美的书”当作一场永不止步的长征，那么做书的过程就如同一种见证，无数严谨又浪漫的叙事聚集时空，拧成一把钥匙，解锁了我们关于分享的更多渴望。而《翻阅莱比锡——世界最美的书 1991－2003》就是在这种渴望之下催生的，它更像是《莱比锡的选择》的前传，拼凑起最美书籍的始源，系统完成了关于莱比锡的圆满纪念。

早在 20 世纪 50 年代早期，德国的政治分裂使得东德和西德几乎同时成立了各自的书籍设计比赛。随着柏林墙的倒塌，两个平行的书展比赛与时代脱节。从 1991 年起，德国图书艺术基金会正式开始在莱比锡组织与承办“世界最美的书”比赛。

《翻阅莱比锡》从两德合并后的评选开始，收录了 1991 年到 2003 年的“莱比锡世界最美的书”每年的获奖书籍。我们目前的藏书主要包括两个阶段。第一个阶段，2004 年到 2019 年，

genen Jahr publizierte Leipziger Auswahl gibt Kunde davon. In der zweiten Phase, d.h. in der Zeit von 1991 bis 2003, haben wir 131 Bücher gesammelt. Da die Zeit lange in die Vergangenheit zurückreicht, wird noch immer nach 54 Büchern gesucht. Insgesamt lässt sich allerdings sagen, dass seit der Einführung des Wettbewerbs „Die Schönsten Bücher der Welt" immerhin 84% aller preisgekrönten Bücher in unsere Sammlung aufgenommen werden konnten. Das Jahr 2004 stellte insofern ein Wende dar, als in der weltweit ausgerichteten Geschichte des Buchdesigns erstmals chinesisches Design miteinbezogen wurde. Mit Leipzig durchblättert wird Zeugnis über die dramatischen Veränderungen auf der Welt abgelegt. Es handelt sich um einen historischen Meilenstein, der den Prozess des Übergangs von handgefertigten Drucken zum allmählichen Einsatz von Computern im Designprozess veranschaulicht.

Auf der Grundlage der Erfahrungen aus der ersten Phase der Sammlung haben wir auch in der zweiten Phase der Sammlung einige Anpassungen vorgenommen. Im Fall von Leipzig durchblättert haben wir das System der Buchnummerierung und der Dokumentenbezeichnung verbessert und die Suche auf den Regalen vereinfacht, um die Effizienz der Sammlung und der Veröffentlichung des Buches zu erhöhen. Im Einklang mit der Geschichte des chinesischen Engagements geht das Buch auch beim Design neue Wege. Die Präsentation erfolgt in einer altertümlich wirkenden Form, Details werden auf vergilbtem Papier sichtbar gemacht. Die Verwendung des Zweifarbendrucks in Rot und Schwarz ebenso wie die gesprenkelte Schrift sind Ausdruck eines Experimentes, bei dem es darum ging, das Prozesshafte der Wirkung „von innen nach außen" und „vom Ganzen zum Teil" aufzuzeigen. So wird zum Beispiel jedes der Bücher auf den Innenseiten mittels des Fotos eines Postpakets erläutert. Das Papier behält seine ursprüngliche grobe Körnung und die grau-gelben Töne bei, es wird Wert gelegt auf den Kontrast von Schwarz und Rot, um Informationen zu unterscheiden, den Text zu fokussieren und allmählich die Reste der staubigen Vergangenheit zum Vorschein zu bringen. Im Gegensatz zu traditionellen Nachschlagewerken ist Leipzig durchblättert

totally 224 award-winning books, of which 213 books collected by Han Qing Hall, reaching 95%, showed in *The Choice Of Leipzig* published last year. The second one is from 1991 to 2003, collecting 131 books. There are 54 books still be searching, however they might be lost in the mists of antiquity. But all in all, starting from the undertaking of the competition of "the Best Designed Books All Over the World", we have collected 84% of the award-winning books. The year 2004 was like a dividing crest, from then on, China has taken part in the history of the book design. Afterwards, the book Browse Leipzig came. It witnessed the huge changes of the world, and established a milestone with the crossing era significance in the process of transforming from manually block making to computer designation.

According to the experience of the first stage, we made some adjustments in the second stage. In *Browse Leipzig*, we improved the system of numbering and naming of the books, using a more easier finding way to arrange the bookshelves, promoting the efficiency of collecting and publishing. According to the history before Chinese taking part in, we tried new designs in this book, presenting in more old-fashioned pattern and every single details showed in yellowing pages. Whatever the red and black two-color printing, or the mottled movable type, both practiced the experimenting process of "from inside to outside" and "from parts to the whole". For example, in the inside pages, every book is narrated beginning with the photo of the parcel, papers remaining the primary rough particles and grayish yellow tone, trying to focus the words and get the information by the contrast of black and red, then uncovering the past events one

获奖书目总共是 224 本，瀚清堂的收藏数量为 213 本，收藏比例达到了 95%，在去年出版的《莱比锡的选择》中已经加以展示。第二个阶段，1991 年到 2003 年，我们收藏了 131 本书。由于年代比较久远了，还有 54 本书仍在继续搜寻当中。但总的来说，从“世界最美的书”比赛开始承办以来，我们收藏到的书籍，已经占总获奖书目的 84% 了。2004 年就像是一条分水岭，往后走，世界的书籍设计史上才有了中国设计的参与。而往前走，便有了《翻阅莱比锡》这本书，它见证了世界的巨变，在从手工制版转型成逐步利用电脑设计的过程之中，树立起了一座具有跨时代意义的里程碑。

根据第一阶段的收藏经验，我们在第二阶段的收藏过程中也做出了一些调整。在《翻阅莱比锡》这本书中，我们完善了图书编号和文件命名系统，用一种更易于查找的书架排列方式，提升了书籍收藏和本书出版的效率。依照中国参与之前的历史，这本书在设计上也做了些新的尝试，以更为老旧的形态呈现，从泛黄的纸张中体察每一个细节。无论是红黑双色印刷的使用，还是斑驳的活字字体的呈现，都是在践行一种“从内到外”“从整体到局部”的实验过程。比如内页中的每一本书，都以邮包的照片展开叙述，纸张保留了原生的粗糙颗粒与灰黄色调，致

eine sinnlichere Präsentation aller erschöpfend recherchierten Bücher und ihrer Höhepunkte. Verbunden ist das mit der Hoffnung, der nach und nach wieder zu sich kommenden Welt ein kleines Geschenk zu bereiten. Wir sind immer noch dabei, die schönsten Bücher der Welt zu sammeln. In Zukunft werden wir ein umfassenderes Programm von Buchausstellungen realisieren, das sich von 1991 bis 2024 erstreckt und auf ein vollständigeres Gedenken zielt.

Abschließend möchte ich allen Partnern und Kollegen für ihre Mühe zur Anfertigung dieses Buches danken. Ich möchte meinen Kollegen Zhu Tao, der immer streng und professionell vorgegangen ist, und meinen Assistenten Qing Yun und Wang Jia für ihre Geduld und Beharrlichkeit bei der Suche nach Büchern danken. Mein Dank gilt auch dem Jiangsu Phoenix Fine Arts Publishing House. Der Verlag hat aufs Neue meine Ideale Wirklichkeit werden lassen und dem romantischen Gefühl durch einen bestimmten Druck Ausdruck zu geben verstanden. Jede Seite des Buches ist wie ein Knopf, mit dem die Zeit angehalten werden kann, um wundervolle Momente aufscheinen zu lassen. In ihnen offenbaren sich körperliche Fehldeutungen, Szenen aus der Kindheit mögen ebenso auftauchen wie solche aus herrlichen Filmträumen, man spürt diffuse persönliche Erinnerungen und kaum zu erhaschende Wünsche.

Zhao Qing
Juli 2024 in der Klause vom Pflaumengarten

after another. Different from the traditional rigorous dictionary books, *Browse Leipzig* displays all the book context and highlights in a more sensitive way. We hope it can be a little gift in this resurgent world. We are still on our way to collect the best designed books, and in the future we will realize the more all-round book exhibition plan in a more integral commemoration from 1991 to 2024.

Last but not least, I would like to give my sincerely gratitude to all the fellows who worked hard for this book. Zhu Tao, who kept rigorous and professional, and the assistants Qing Yun and Wang Jia who were patient and insistent in searching and buying books. Jiangsu Phoenix Fine Arts Publishing, who realized my dream again and made all the romantic matters live on the paper. The misread bodies, the interesting childhood, the magnificent movie dream, the misty private life and the wind that is everywhere—whenever opening the book, time will pause, then you can find a golden tranquil piece of time.

Zhao Qing
July, 2024 in Plum Garden

力于以黑与红的色彩对照区分信息、聚焦文字，进而徐徐铺展尘封已久的往事余韵。有别于严谨传统的辞典式书籍，《翻阅莱比锡》以一种更为感性的方式，或简述或铺陈地展示所有详尽探究过的书籍脉络与精彩片段，希望它能在这个陆陆续续复苏的世界里，成为一个雨过天晴后的小礼物。我们仍然在路上，在搜集“世界最美的书”的旅程之中，未来也将以更为完整的纪念，跨越 1991 年到 2024 年，实现更全面的书籍展览计划。

最后，由衷感谢所有为这本书付出过努力的伙伴们。感谢一直保持着严谨与专业的朱涛同仁，还有助理卿云、王嘉在搜购书籍中的耐心坚持。也要感谢江苏凤凰美术出版社，再一次使我的理想照进现实，将所有的浪漫印刷成特定的烙印，被误读的身体，稚趣的童年，瑰丽的电影梦，迷雾的私生活，无法捕捉的风，在每一次翻开书页咀嚼的时候，都能按下时间的暂停键，找到静处的一份金色时光。

赵清

2024 年 7 月　于梅园

Mitglied der Internationalen Vereinigung der Grafiker (AGI); Stellvertretender Vorsitzender der Kommission für Grafikkunst innerhalb der Chinesischen Vereinigung der Mitarbeiter des Publikationswesens; Mitglied der Shenzhen Graphic Designers Association (SGDA) ; Mitglied der Japan Graphic Design Association Inc. (JAGDA), und Gründer der Nanjing Graphic Designers Alliance. Im Jahr 2000 gründete Zhao Qing die Firma Han Qingtang Design Co., Ltd. und übernahm den Posten als Direktor für Design. Zhao ist herausgeber und rezensent des Verlags Phoenix Jiangsu Science and Technology Press und Betreuer von Masterstudenten der Designer-School der Kunstakademie von Nanjing. Zahlreiche seiner eigenen Designerarbeiten gewannen Preise oder wurden weltweit im Rahmen fast aller wichtigen Grafikdesign-Wettbewerbe und Ausstellungen für Preise nominiert. Unter den internationalen Designerpreisen, mit denen Zhaos Werk ausgezeichnet wurde, sind u.a. anzuführen den D&AD Yellow Pencil Award (Großbritannien); den ADC Gold Cube Award (New York); der TDC-Preis von Tokyo TDC (Tokyo); der One Show Design Silver Pencil Award, New York TDC (New York); Red dot, IF (Deutschland); Golden Bee (Russland); JTA (Japan); Tokyo TDC (Tokyo); GDC Best Awards, Shenzhen Global Design Gold Award (Shenzhen), Hong Kong Global Design Gold Award, DFA Most Influential Award and Gold Award (Hong Kong); China Taiwan Golden Pin Design Annual Best Award (China Taiwan) usw. Seine Buchgestaltungsarbeiten wurden mehr als 30 Mal mit dem Titel „Das schönste Buch" ausgezeichnet.

He is a member of the Alliance Graphique Internationale (AGI), deputy director of the Book Art Committee of the Publishers Association of China, a member of the Shenzhen Graphic Designers Association (SGDA), a member of the Japan Graphic Design Association Inc. (JAGDA), and a founder of the Nanjing Graphic Designers Union. In 2000, Zhao Qing established "Han Qing Tang Design Co., Ltd." and hosted the post of Chief Design Director. He is also the editor and reviewer of Jiangsu Phoenix Science Press and the supervisor of master students at the Nanjing University of the Art. His design works have won awards or been selected in almost all important graphic design competitions and exhibitions worldwide. He has won the D&AD Yellow Pencil Award in the UK, the ADC Gold Cube Award in New York, Tokyo TDC's tdc prize, the One Show Design Silver Pencil Award in New York, New York TDC, Germany Red dot, IF, Russia Golden Bee Award, Japan JTA Best Awards, Tokyo TDC Award, GDC Best Awards (Shenzhen), Shenzhen Global Design Gold Award (Shenzhen), Hong Kong Global Design Gold Award (Hong Kong), DFA Most Influential Award and Gold Award (Hong Kong), China Taiwan Golden Pin Design Annual Best Award (China Taiwan) and many other international design awards. His book design works have won the title of "The Most Beautiful Book" more than 30 times.

赵清
Zhao Qing

国际平面设计联盟（AGI）会员，中国出版工作者协会书籍艺术委员会副主任，深圳平面设计师协会（SGDA）会员，日本平面设计师协会 JAGDA 会员，南京平面设计师联盟创始人。2000 年创办瀚清堂设计有限公司并任设计总监。江苏凤凰科学技术出版社编审，南京艺术学院设计学院硕士生导师。设计作品获奖或入选世界范围内几乎所有重要的平面设计竞赛和展览，并获得了英国 D&AD 黄铅笔奖、美国纽约 ADC 金方块奖、日本东京 TDC "tdc prize"、纽约 One Show Design 银铅笔奖、纽约 TDC、德国 Red dot、IF、俄罗斯 Golden Bee 金蜂奖、日本 JTA Best Awards、东京 TDC 奖、深圳平面设计在中国 GDC Best Awards、深圳环球 SDA 金奖、香港环球 GDA 金奖、亚洲最具影响力 DFA 大奖及金奖、中国台湾 Golden 年度最佳奖等众多国际设计奖项。书籍设计作品三四十次获"最美的书"称号。

NADAR
DIE RUSSEN IN WIEN
WIMTSCHIPPERS
van de stad I
van de stad II
Ein Tag im Leben

Die erste Auswahl von 「Schönste Bücher Chinas」 erschien im Jahre 2004.

Mit dieser Auswahl beteiligte ~~sich China im darauffolgenden Jahr erstmalig an dem in Leipzig~~ ausgetragenen Wettbewerb 「Schönste Bücher der Welt」

– damit begann die Bücherreise nach Leipzig.

In 2004, 「the most beautiful books in China」 were selected to participate in the competition 「best book design from all over the world」 held in Leipzig the following year.

A journey of beautiful books to Leipzig began.

二〇〇四年

中国首次选出『中国最美的书』

以参加次年举办的莱比锡『世界最美的书』评选

开始了一段莱比锡美书之旅

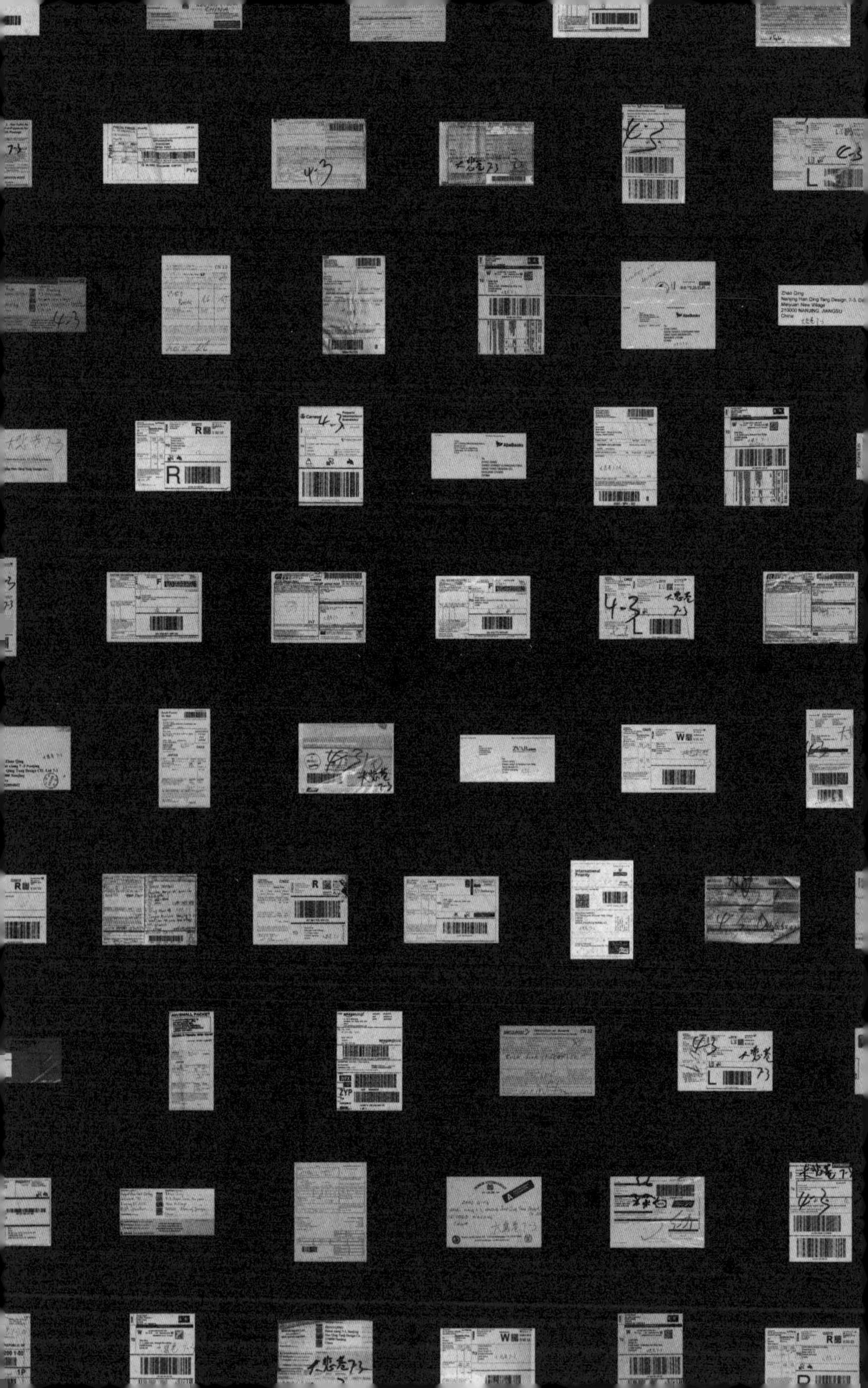

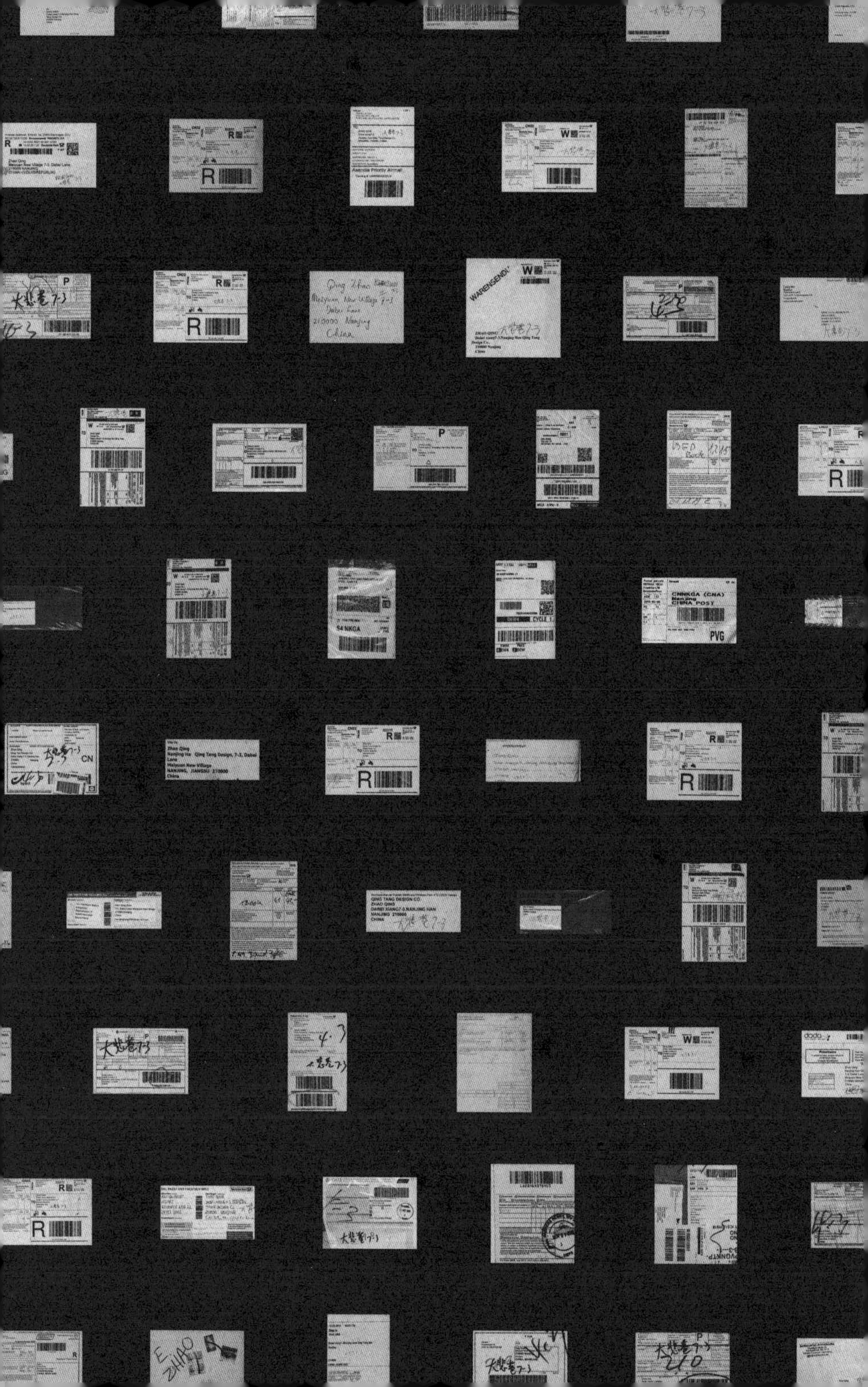

图书在版编目（C I P）数据

翻阅莱比锡：世界最美的书：1991-2003 / 赵清主编. -- 南京：江苏凤凰美术出版社，2020.10（2024.8 重印）

ISBN 978-7-5580-7841-5

Ⅰ.①翻… Ⅱ.①赵… Ⅲ.①书籍装帧－设计－作品集－世界－现代 Ⅳ.① TS881

中国版本图书馆 CIP 数据核字（2020）第 180591 号

责任编辑　王林军　舒金佳
书籍设计　瀚清堂 / 赵　清 + 朱　涛
摄　　影　瀚清堂 / 朱　涛
责任校对　吕猛进
责任监印　张宇华
责任设计编辑　赵　秘

书　　名　翻阅莱比锡：世界最美的书 1991－2003
主　　编　赵　清
译　　者　正文中译英　石晶晶　/　序跋中译德　Thomas Zimmer　/　序跋中译英　李浦林
出版发行　江苏凤凰美术出版社（南京市湖南路 1 号　邮编：210009）
制版印刷　上海雅昌艺术印刷有限公司
开　　本　889 毫米 ×1194 毫米　1/32
印　　张　40.75
版　　次　2020 年 10 月第 1 版　2024 年 8 月第 2 次印刷
标准书号　ISBN　ISBN 978-7-5580-7841-5
定　　价　270.00 元

营销部电话　025-68155675　营销部地址　南京市湖南路 1 号
江苏凤凰美术出版社图书凡印装错误可向承印厂调换